Drop

* mu… m

Jester.

Applied Calculus With Linear Programming

FOR BUSINESS, ECONOMICS,
LIFE SCIENCES, AND SOCIAL SCIENCES

College Mathematics Series

This book is part of a comprehensive series designed to accommodate a wide variety of courses. Difficulty level and topic coverage are determined by the choice of a particular book or a combination of books from the series. Many topics in individual books are independent and may be selected in any order or omitted. All books include reviews of relevant algebraic topics.

Essentials of College Mathematics: *One or two semesters*
A brief introduction to finite mathematics and calculus that is suitable for a one- or two-semester course or a one- or two-quarter course

College Mathematics: *Two semesters*
A more comprehensive introduction to finite mathematics and calculus that is suitable for a two-semester course or a two- or three-quarter course

Applied Mathematics: *Two or three semesters*
An even more comprehensive introduction to finite mathematics and calculus for schools requiring this level and degree of coverage; additional topics are included, and the definite integral is given a more formal treatment

Finite Mathematics: *One semester*
The finite mathematics portion of *College Mathematics*, with an added chapter on games and decisions; may be used in combination with either *Calculus* or *Applied Calculus* to create a two- or three-semester course

Calculus: *One semester*
The calculus portion of *College Mathematics*, with an added chapter on trigonometric functions

Applied Calculus: *One or two semesters*
A more extensive treatment of calculus for schools requiring this level and degree of coverage; the book can be completed in two semesters, however, with appropriate topic selection—since many of the topics are independent—the book also can be used for a strong one-semester course

Applied Calculus With Linear Programming

FOR BUSINESS, ECONOMICS, LIFE SCIENCES, AND SOCIAL SCIENCES

RAYMOND A. BARNETT
Merritt College

MICHAEL R. ZIEGLER
Marquette University

PRENTICE HALL
Englewood Cliffs, New Jersey 07632

On the cover: The detail on the cover is from a metal collage by Los Angeles artist Tony Berlant. His work can be described as part painting, part sculpture. It is resplendent with color and texture. The collage is formed by nailing fragments of printed and colored tin onto a wood backing. Berlant's work is represented by the L. A. Louver Gallery in Venice, California and the Louver Gallery in New York City. His work may also be seen in the permanent collections at the Whitney Museum of American Art in New York City, the Academy of Fine Art in Philadelphia, the Hirshorn Museum in Washington, D.C., the Art Institute of Chicago, and the Los Angeles County Museum.

A Simon & Schuster Company
Englewood Cliffs, New Jersey 07632

Printed in the United States of America

10 9 8 7 6 5 4 3 2

ISBN 0-02-306508-7

Prentice-Hall International (UK) Limited, *London*
Prentice-Hall of Australia Pty. Limited, *Sydney*
Prentice-Hall Canada Inc., *Toronto*
Prentice-Hall Hispanoamericana, S.A., *Mexico*
Prentice-Hall of India Private Limited, *New Delhi*
Prentice-Hall of Japan, Inc., *Tokyo*
Simon & Schuster Asia Pte. Ltd., *Singapore*
Editora Prentice-Hall do Brasil, Ltda., *Rio de Janeiro*

SPECIAL EDITION

Applied Calculus with Linear Programming
for Business, Economics, Life Sciences and Social Sciences

is a special edition of Barnett and Ziegler's

Applied Calculus, Fifth Edition
for Business, Economics, Life Sciences and Social Sciences

This special edition contains all the material in Chapters 1–12 of *Applied Calculus*, fifth edition, along with two chapters from the authors' *College Mathematics for Business, Economics, Life Sciences, and Social Sciences*, sixth edition. The two new chapters from *College Mathematics* are Chapter 5 on Systems of Linear Equations and Matrices and Chapter 6 on Linear Inequalities and Linear Programming. These two new chapters replace Appendices A and B of *Applied Calculus*, fifth edition. The answers for the two new chapters follow the answers for Chapters 1–12. An index for the two new chapters also follows the index for Chapters 1–12. The pages for the two new chapters and their answers are marked with color tabs that are visible when the book is closed, making it easy to find this material.

The preface in this book is the same as the preface for *Applied Calculus*, fifth edition. For information pertaining to the two new chapters, consult the preface of *College Mathematics*, sixth edition. A full set of supplementary material can be obtained by combining the supplements listed in the preface in this text with those listed in the *College Mathematics* preface.

For more information concerning this special edition and its supplementary material, contact

Michael R. Ziegler
509 West Dean Court
Fox Point, WI 53217
(414) 352-4197

CONTENTS

◆ APPENDIX A

CHAPTER 5 Systems of Linear Equations; Matrices

◆ APPENDIX B

CHAPTER 6 Linear Inequalities and Linear Programming

◆ APPENDIX C Tables

PREFACE

The fifth edition of *Applied Calculus for Business, Economics, Life Sciences, and Social Sciences* is designed for a one- or two-term course in calculus and for students who have had $1\frac{1}{2}-2$ years of high school algebra or the equivalent. The choice and independence of topics make the text readily adaptable to a variety of courses (see the chapter dependency chart on page ix). It is one of six books in the authors' College Mathematics series (see page ii for a brief comparison of all six books).

Improvements in this edition evolved out of the generous response from a large number of users of the last and previous editions as well as survey results from instructors, mathematics departments, course outlines, and college catalogs. Fundamental to a book's growth and effectiveness is classroom use and feedback. Now in its fifth edition, *Applied Calculus for Business, Economics, Life Sciences, and Social Sciences* has had the benefit of a substantial amount of both.

◆ PRINCIPAL CHANGES FROM THE FIFTH EDITION

Many reviewers indicated a strong interest in using graphic calculators or computers to increase the effectiveness of their instruction and to stimulate student interest in applications of mathematics. Our response to this desire to incorporate technology into the classroom is threefold.

First, problems requiring the use of a **graphic calculator or computer** have been added to many exercise sets. These problems have been carefully selected to illustrate methods for using technology to increase student understanding of mathematical concepts and to provide an opportunity for the instructor to discuss interesting applications that are not easily solved by hand. Graphic calculator or computer problems are clearly identified with the icon shown in the margin and may be omitted without loss of continuity.

Second, a **graphics calculator manual** by Carolyn L. Meitler is now available for those who want to make more extensive use of these calculators. This manual provides a great deal of support for both the instructor and the student, including a flexible organization that permits instructors to select the portions of the course where they wish to emphasize calculator use. Detailed examples of the specific keystrokes required (on three of the most popular calculator models) to solve problems in the text are included. While well-suited for a class where all students purchase the same calculator, this manual is especially effective where students are using a variety of different calculators—an important consideration as more and more students arrive at college having already purchased one of these calculators. See Student Aids and Instructor Aids, later in this preface, for more information about this manual.

Third, *Visual Calculus*, by David Schneider is new **computer software** available to those who want to incorporate computers into the instructional process. This extensive collection of interactive and very user-friendly programs can be used to provide additional insight into many of the mathematical concepts discussed in the text and to stimulate discussion of applications that are suited to computer solutions. See Student Aids and Instructor Aids, later in this preface, for more information about this software.

All **exercise sets** were carefully reviewed, and additional **application problems** were added.

Specific improvements are as follows:

1. At the request of many reviewers, the material on **equations and inequalities** was moved to Appendix A. This permits the course to start at a higher level with **graphs and functions,** yet still provides a **complete review of pertinent algebra topics** in Appendix A for those students who need it.
2. Several of the sections in Appendix A were revised. In the section dealing with basic operations on polynomials, the FOIL method was deleted, by popular demand, and more emphasis is now placed on the use of distributive properties to accomplish multiplication. The section on **factoring polynomials** was completely rewritten. It is now less formal and has **a systematic approach to factoring** that does not depend on trial-and-error. In the section on equations and inequalities, formal set notation was generally replaced with less formal verbal statements. A new application of the quadratic formula for **factoring second-degree polynomials** was added to the section on quadratic equations.
3. In Chapter 1, the set definition of function was deleted.
4. In Chapter 2, more material on solving exponential equations and the **change-of-base formula** for logarithmic functions were added.
5. In Chapter 4, exercise sets were expanded to include more problems dealing with **fundamental graph properties** and **sign charts.**
6. In Chapter 12, the discussion of the **normal probability distribution** of a continuous random variable, as well as its use in **approximating a binomial distribution** of a discrete random variable, were expanded.

◆ IMPORTANT FEATURES

Emphasis and Style

The text is **written for student comprehension.** Great care has been taken to write a book that is mathematically correct and accessible to students. Emphasis is on computational skills, ideas, and problem solving rather than mathematical theory. Most derivations and proofs are omitted except where their inclusion adds significant insight into a particular concept. General concepts and results are usually presented only after particular cases have been discussed.

Examples and Matched Problems

Over 400 completely worked examples are included. Each example is followed by a similar problem for the student to work while reading the material. This actively involves the student in the learning process. The answers to these matched problems are included at the end of each section for easy reference.

Exercise Sets

The book contains over 4,500 problems. Each exercise set is designed so that an average or below-average student will experience success and a very capable student will be challenged. Exercise sets are mostly divided into A (routine, easy mechanics), B (more difficult mechanics), and C (difficult mechanics and some theory) levels.

Applications

Enough applications are included to convince even the most skeptical student that mathematics is really useful. The majority of the applications are included at the end of exercise sets and are generally divided into business and economics, life science, and social science groupings. An instructor with students from all three disciplines can let them choose applications from their own field of interest; but if most students are from one of the three areas, then special emphasis can be placed there. Most of the applications are simplified versions of actual real-world problems taken from professional journals and books. No specialized experience is required to solve any of the applications.

◆ STUDENT AIDS

1. **Think boxes** (dashed boxes) are used to enclose steps that are usually performed mentally (see Sections 1-1 and 3-4).
2. **Annotation** of examples and developments, in color type, is found throughout the text to help students through critical stages (see Sections 2-2 and 3-2).
3. **Functional use of color** improves the clarity of many illustrations, graphs, and developments, and guides students through certain critical steps (see Sections 2-2 and 3-2).
4. **Boldface type** is used to introduce new terms and highlight important comments.
5. **Shaded boxes** are used to highlight important definitions, theorems, results, and step-by-step processes.
6. **Answers** to odd-numbered problems are included in the back of the book.
7. **Chapter review** sections include a review of all important terms and symbols and a comprehensive review exercise. Answers to all review exercises are included in the back of the book.
8. A **student's solution manual** is available at a nominal cost through a book store. The manual includes detailed solutions to all odd-numbered problems and all review exercises.

9. **Computer software** for IBM-compatible computers is available at a nominal cost through a book store. *Visual Calculus* by David Schneider contains over twenty routines that provide additional insight into the topics discussed in the text. Although this software has much of the computing power of standard calculus software packages, it is primarily a teaching tool that focuses on understanding mathematical concepts, rather than on computing. These routines incorporate graphics whenever possible to illustrate topics such as secant lines, tangent lines, velocity, optimization, the relationship between the graphs of f, f', and f'', and the various approaches to approximating definite integrals. All the routines in this software package are menu-driven and very easy to use. The software is accompanied by a manual with instructions and additional exercises for the student. Hardware requirements are an IBM-compatible computer with at least 640 K of memory and a graphics adapter: CGA, EGA, VGA, or Hercules.
10. A **graphics calculator manual** by Carolyn L. Meitler is available at a nominal cost through a book store. This manual contains examples illustrating the use of a graphics calculator to solve problems similar to those discussed in the text. The manual is organized in terms of the topics in the text, making it easy to find examples in the manual illustrating appropriate calculator solution methods for corresponding problems in the text.
11. A **Supplemental Applications and Topics** manual by Jon E. Baum is available at a nominal cost through a book store. Part I of the manual expands the application exercises in the text and reinforces the important role of the mathematics presented. These exercises provide the student with a richer and more varied experience in solving real-world problems. Part II of the manual presents some applications that are not covered in the text, including transportation problems, assignment problems, sensitivity analysis, and a variety of finance topics. After completing the prerequisite material in the text, students interested in these more specialized topics will realize substantial benefits by studying this portion of the manual.

◆ INSTRUCTOR AIDS

For additional information concerning the instructor aids described below, contact:

Dellen Publishing Company
113 Sylvan Avenue, Route 9W
Englewood Cliffs, NJ 07632

For a summary of all available supplementary materials and detailed information regarding examination copy requests and orders, see page xix.

1. A unique **computer-generated random test system** for either IBM-compatible or Apple Macintosh® computers is available to instructors without cost.

The test system has been greatly expanded and now contains over 400 different problem algorithms directly related to material in the text. These carefully constructed algorithms use random number generators to produce different, yet equivalent, versions of each of these problems. The test system is available now in both **free-response and multiple-choice editions.** An almost unlimited number of quizzes, review exercises, chapter tests, midterms, and final examinations, each different from the other, can be generated quickly and easily. At the same time, the system will produce answer keys and student work sheets, if desired. In addition, the Macintosh version incorporates a unique **editing function** that allows the instructor to alter any of the existing problems in the test and to create new problems. The editor includes a complete set of mathematical notation and also supports importation of text and graphics from other Macintosh applications. Upon request, the publisher will supply institutions using this text with **Dellen Test 4.0 (MS-DOS Free-Response or Multiple-Choice Edition)** or **Dellen Test MAC 2.0 (Apple Macintosh® Free-Response or Multiple-Choice Edition)** on 3.5 inch floppy disks. **User notes** and **annotated problem printouts** are included with the disks. The notes provide step-by-step instructions for using the testing system and a complete description of the options in this menu-driven program. The annotated printouts identify by chapter and number each question the system is capable of generating, and also correlate each question with the prerequisite section from the text. When used in conjunction with the user notes, the annotated printouts enable instructors to select any combination of questions for an examination. The MS-DOS editions will produce high-quality output on IBM-compatible dot-matrix printers and on Hewlett-Packard Laserjet II®-compatible laser printers or Deskjet® printers. The Macintosh® editions require a minimum of two megabytes of RAM and System 6.0.5 or higher.

2. An **instructor's test battery** is also available to instructors without cost. The battery, organized by chapter, contains three equivalent versions (with answers) of over 400 different problems.
3. An **instructor's resource manual** provides over 175 transparency masters, a detailed discussion of chapter and topic dependencies, a comparison of this edition with the previous edition, and a detailed topic chart for comparing this book with other books in the author's College Mathematics series.
4. An **instructor's answer manual** containing all the answers not included in the text is available to instructors without charge.
5. A **student's solution manual** (see Student Aids) is available to instructors without charge from the publisher.
6. **Computer software and accompanying manual** for *Visual Calculus* by David Schneider (see Student Aids) are available to instructors without charge. The manual contains complete instructions for using the software (eliminating the need to spend class time discussing these details) and examples and exercises for the student. In addition to providing students with the opportunity to use the computer as an effective tool in the learning process, instructors will find the software very useful for preparing examples for

class, constructing test questions, classroom demonstrations, and similar activities.

7. A **graphics calculator manual** by Carolyn L. Meitler (see Student Aids) is available to instructors without charge from the publisher. The flexible organization of this manual allows the instructor to select the portions of the course where graphics calculator use will be emphasized. The manual contains all the necessary information for a student with no previous experience with these calculators, eliminating the need for the instructor to prepare materials related to calculator use. In particular, separate appendixes for the TI-81, TI-85, and Casio fx-7700G graphics calculators contain detailed instructions, including calculator-specific keystrokes, for performing the various operations required to effectively use each of these calculators to solve problems in the text. While well-suited for a class where all students purchase the same calculator, this manual is especially effective where students are using a variety of different calculators—an important consideration as more and more students arrive at college having already purchased one of these calculators.
8. A **Supplemental Applications and Topics** manual by Jon E. Baum (see Student Aids) is available to instructors without charge from the publisher. Instructors can use Part I of this manual to supplement the exercise sets in the text, providing students with additional experience in solving applications utilizing the mathematics presented. Part II of the manual can be used to provide coverage of applications not covered in the text, such as transportation problems, assignment problems, sensitivity analysis, and a variety of finance topics, either as part of the syllabus for a course or as subjects for independent study.
9. **Z-graph**, a HyperCard® graphing stack for the Apple Macintosh® computer, allows a user to graph most of the mathematical functions likely to be encountered quickly, accurately, and with considerable control over axes, scales, graph size, and labeling. In addition to graphing functions, this program will perform a variety of mathematical operations related to numerical integration, root approximation, interpolating polynomials, least-square polynomials, and approximate solutions of differential equations. Instructors will find this program useful for preparing examination material, transparency masters, and handouts. The publisher will supply this program free of charge to instructors using this book, and the program may be freely distributed to students.

◆ ERROR CHECK

Because of the careful checking and proofing by a number of mathematics instructors (acting independently), the authors and publisher believe this book to be substantially error-free. For any errors remaining, the authors would be grateful if they were sent to: Dellen Publishing Company, 113 Sylvan Avenue, Route 9W, Englewood Cliffs, NJ 07632.

◆ ACKNOWLEDGMENTS

In addition to the authors, many others are involved in the successful publication of a book. We wish to thank personally:

Carol Adjemian, Pepperdine University
Charles K. Atherton, University of the District of Columbia
Jay Belanger, Wilkes University
Chris Boldt, Eastfield College
Sister Mary Maurice Boyle, Salve Regina College
Bob Bradshaw, Ohlone College
Jeffrey R. Butz, Bridgewater State College
Lynette Cardenas, Texas A & M
Bruce Chaffee, Long Beach City College
Robert Chaney, Sinclair Community College
Chand Chauhan, Indiana University/Purdue University—Fort Wayne
Dianne Clark, Ball State University
Charles E. Cleaver, The Citadel
Barbara Cohen, West Los Angeles College
Richard L. Conlon, University of Wisconsin—Stevens Point
Madhu Deshpande, Marquette University
Kenneth A. Dodaro, Florida State University
Michael W. Ecker, Pennsylvania State University-Wilkes-Barre
Jerry R. Ehman, Franklin University
Lucinda Gallagher, Florida State University
Martha M. Harvey, Midwestern State University
John L. Heath, East Texas State University
Sue Henderson, Dekalb College
Lloyd R. Hicks, Edison Community College
Louis F. Hoelzle, Bucks County Community College
Paul Hutchens, Florissant Valley Community College
John Inglis, Kwantlen College
K. Wayne James, University of South Dakota
Terry L. Jenkins, University of Wyoming
Robert H. Johnston, Virginia Commonwealth University
Ellen King, Anderson College
Robert Krystock, Mississippi State University
Suzanne Lenhart, University of Tennessee
James T. Loats, Metropolitan State College of Denver
Frank Lopez, Eastfield College
Roy H. Luke, Los Angeles Pierce College
Ann Megaw, University of Texas—Austin
Mel Mitchell, Clarion University of Pennsylvania
John E. Olson, Pennsylvania State University
Kenneth A. Peters, Jr., University of Louisville
Tom Plavchak, Wilkes University

Bob Prielipp, University of Wisconsin — Oshkosh
Stephen Rodi, Austin Community College
Sharon Roosevelt, Texas A & M
Sheldon Rothman, Long Island University
Elaine Russell, Angelina College
Daniel E. Scanlon, Orange Coast College
George R. Schriro, Long Island University
Arnold L. Schroeder, Long Beach City College
Sherry F. Senn, Anderson College
Hari Shanker, Ohio University
Joan Smith, Vincennes University
Steven Terry, Ricks College
Raymond E. F. Weaver, Community College of Allegheny County
Delores A. Williams, Pepperdine University
Richard K. Williams, Southern Methodist University
Caroline Woods, Marquette University
Charles W. Zimmerman, Robert Morris College
Pat Zrolka, Dekalb College

We also wish to thank:

John Williams for a strong and effective cover design

John Drooyan and Mark McKenna for the many sensitive and beautiful photographs throughout the book

Stephen Merrill, Robert Mullins, and Susan Pustejovsky for providing a careful and thorough check of all the mathematical calculations in the book, the student solution manual, and the answer manual (a tedious but extremely important job)

Jon Baum, Carolyn Meitler, and David Schneider for developing the supplemental manuals that are so important to the success of a text

Jeanne Wallace for accurately and efficiently producing most of the manuals that supplement the text

All the people at IPS Publishing who contributed their efforts to the production of the computerized testing system

Janet Bollow for another outstanding book design

Phyllis Niklas for guiding the book smoothly through all publication details

Don Dellen, the publisher, who continues to provide all the support services and encouragement an author could hope for

Producing this new edition with the help of all these extremely competent people has been a most satisfying experience.

R. A. Barnett
M. R. Ziegler

Ordering Information

When requesting examination copies or placing orders for this textbook or any of the related supplementary material listed below, please refer to the corresponding ISBN numbers.

Title	ISBN Number
Essentials of College Mathematics for Business, Economics, Life Sciences, and Social Sciences, Third Edition	0-02-305931-1
Computer-Generated Random Test System for Essentials of College Mathematics, Third Edition	
Dellen Test 4.0 Disks:	
MS-DOS Free-Response Edition	0-02-305934-6
MS-DOS Multiple-Choice Edition	0-02-305935-4
DellenTest MAC 2.0 Disks:	
Apple Macintosh® Free-Response Edition	0-02-305936-2
Apple Macintosh® Multiple-Choice Edition	0-02-305937-0
MS-DOS or Apple Macintosh® Edition user notes and annotated problem printouts are included with each edition of the disks.	
Instructor's Answer Manual to accompany Essentials of College Mathematics, Third Edition	0-02-305938-9
Instructor's Resource Manual to accompany Essentials of College Mathematics, Third Edition	0-02-305933-8
Instructor's Test Battery to accompany Essential of College Mathematics, Third Edition	0-02-305932-X
Explorations in Finite Mathematics (3.5 inch disk and manual)	0-02-407732-1
Visual Calculus (3.5 inch disk and manual)	0-02-408004-7
Student's Solution Manual to accompany Essentials of College Mathematics, Third Edition	0-02-334387-7
Graphics Calculator Manual to accompany the Barnett and Ziegler College Mathematics Series	0-02-380158-1
Supplemental Applications and Topics to accompany the Barnett and Ziegler College Mathematics Series	0-02-306770-5
Z-graph Macintosh Disk	0-02-306255-X

PART ONE

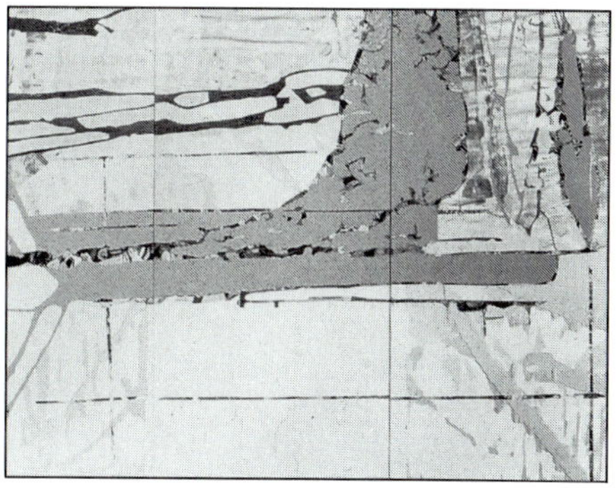

Preliminaries

CHAPTER 1

Graphs and Functions

CHAPTER 1

Contents

The function concept is one of the most important ideas in mathematics. The study of applications of mathematics beyond the most elementary level requires a firm understanding of functions and their graphs. This chapter provides a thorough introduction to the general concept of a function and discusses techniques for graphing certain specific types of functions. The ideas presented in this chapter will be used extensively throughout the remainder of this book. Effort made to understand and use this concept correctly from the beginning will be rewarded many times.

SECTION 1-1 Cartesian Coordinate System and Straight Lines

- CARTESIAN COORDINATE SYSTEM
- GRAPHING LINEAR EQUATIONS IN TWO VARIABLES
- SLOPE
- EQUATIONS OF LINES — SPECIAL FORMS
- APPLICATION

◆ CARTESIAN COORDINATE SYSTEM

Recall that a **Cartesian (rectangular) coordinate system** in a plane is formed by taking two mutually perpendicular real number lines **(coordinate axes)** — one horizontal and one vertical — intersecting at their origins, and then assigning

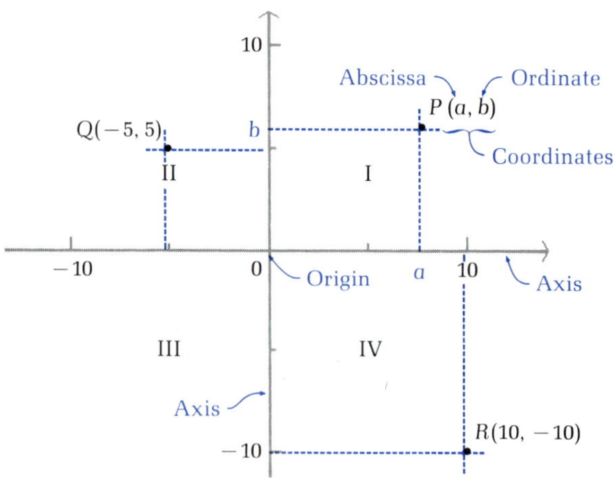

FIGURE 1
The Cartesian coordinate system

unique **ordered pairs** of numbers **(coordinates)** to each point P in the plane (Fig. 1). The first coordinate **(abscissa)** is the distance of P from the vertical axis, and the second coordinate **(ordinate)** is the distance of P from the horizontal axis. In Figure 1, the coordinates of point P are (a, b). By reversing the process, each ordered pair of real numbers can be associated with a unique point in the plane. The coordinate axes divide the plane into four parts **(quadrants)**, numbered I to IV in a counterclockwise direction.

◆ GRAPHING LINEAR EQUATIONS IN TWO VARIABLES

A linear equation in two variables is an equation that can be written in the form

STANDARD FORM $\quad Ax + By = C$

with A and B not both 0. For example,

$$2x - 3y = 5 \qquad x = 7 \qquad y = \tfrac{1}{2}x - 3 \qquad y = -3$$

all can be considered linear equations in two variables. The first is in standard form, while the other three can be written in standard form as follows:

	Standard Form
$x = 7$	$x + 0y = 7$
$y = \frac{1}{2}x - 3$	$-\frac{1}{2}x + y = -3$ or $x - 2y = 6$
$y = -3$	$0x + y = -3$

A **solution** of an equation in two variables is an ordered pair of real numbers that satisfy the equation. For example, $(0, -3)$ is a solution of $3x - 4y = 12$. The **solution set** of an equation in two variables is the set of all solutions of the equation. When we say that we **graph an equation** in two variables, we mean that we graph its solution set on a rectangular coordinate system.

We state the following important theorem without proof:

THEOREM 1

Graph of a Linear Equation in Two Variables

The graph of any equation of the form

STANDARD FORM $\quad Ax + By = C \qquad (1)$

where A, B, and C are constants (A and B not both 0), is a straight line. Every straight line in a Cartesian coordinate system is the graph of an equation of this type.

Also, the graph of any equation of the form

$$y = mx + b \qquad (2)$$

where m and b are constants, is a straight line. Form (2) is simply a special case of (1) for $B \neq 0$. To graph either (1) or (2), we plot any two points of their solution set and use a straightedge to draw the line through these two points. The points where the line crosses the axes—called the **intercepts**—are often the easiest to find when dealing with form (1). To find the ***y* intercept,** we let $x = 0$ and solve for y; to find the ***x* intercept,** we let $y = 0$ and solve for x. It is sometimes wise to find a third point as a check.

◆ EXAMPLE 1

x	y
0	?
?	0
Check	point

Check point optional.

x	y
0	−3
4	0
8	3

(A) The graph of $3x - 4y = 12$ is

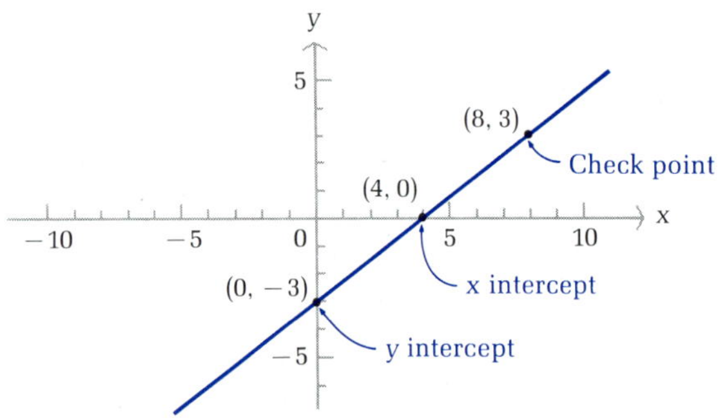

(B) The graph of $y = 2x - 1$ is

x	y
0	−1
4	7
−2	−5

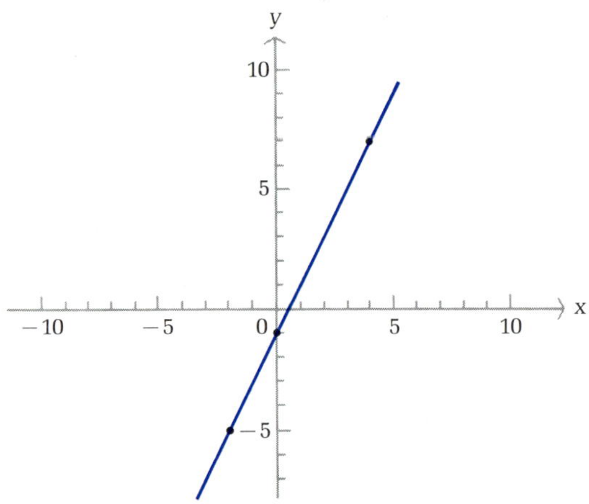

◆

PROBLEM 1* Graph: (A) $4x - 3y = 12$ (B) $y = \frac{x}{2} + 2$ ◆

◆ SLOPE

It is very useful to have a numerical measure of the "steepness" of a line. The concept of *slope* is widely used for this purpose. The **slope** of a line through the two points (x_1, y_1) and (x_2, y_2) is given by the following formula:

* Answers to matched problems are found near the end of each section, before the exercise set.

Slope

$$m = \frac{y_2 - y_1}{x_2 - x_1} \qquad x_1 \neq x_2$$

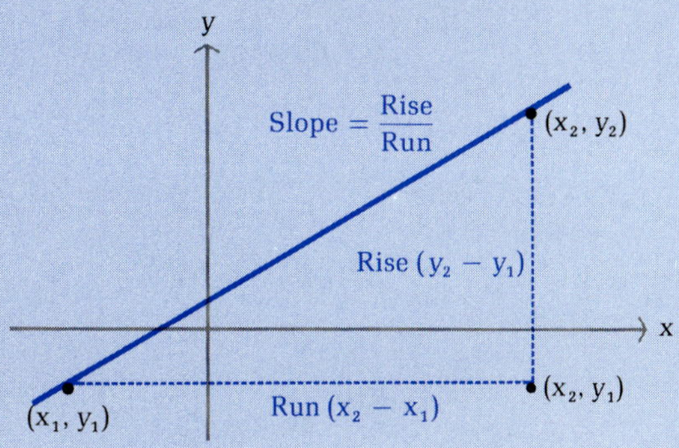

Slope indicates how much a line rises or falls for each unit moved to the right. The slope of a vertical line is not defined. (Why? See Example 2B.)

◆ EXAMPLE 2 Find the slope of the line through each pair of points:

(A) $(-2, 5), (4, -7)$ (B) $(-3, -1), (-3, 5)$

Solutions (A) Let $(x_1, y_1) = (-2, 5)$ and $(x_2, y_2) = (4, -7)$. Then

$$m = \frac{y_2 - y_1}{x_2 - x_1} = \frac{-7 - 5}{4 - (-2)} = \frac{-12}{6} = -2$$

Note that we also could have let $(x_1, y_1) = (4, -7)$ and $(x_2, y_2) = (-2, 5)$, since this simply reverses the sign in both the numerator and denominator and the slope does not change:

$$m = \frac{5 - (-7)}{-2 - 4} = \frac{12}{-6} = -2$$

(B) Let $(x_1, y_1) = (-3, -1)$ and $(x_2, y_2) = (-3, 5)$. Then

$$m = \frac{y_2 - y_1}{x_2 - x_1} = \frac{5 - (-1)}{-3 - (-3)} = \frac{6}{0} \qquad \text{Not defined!}$$

Notice that $x_1 = x_2$. This is always true for a vertical line, since the abscissa (first coordinate) of every point on a vertical line is the same. Thus, the slope of a vertical line is not defined (that is, the slope does not exist). ◆

PROBLEM 2 Find the slope of the line through each pair of points:

(A) $(3, -6), (-2, 4)$ (B) $(-7, 5), (3, 5)$ ◆

TABLE 1

Going from Left to Right

LINE	SLOPE	EXAMPLE
Rising	Positive	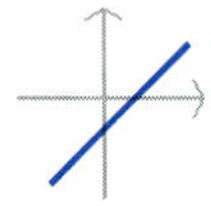
Falling	Negative	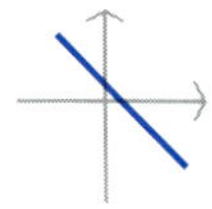
Horizontal	Zero	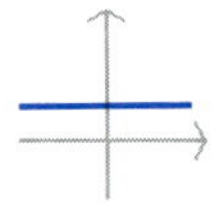
Vertical	Not defined	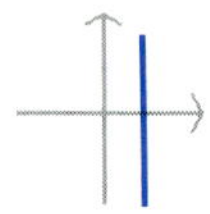

In general, the slope of a line may be positive, negative, zero, or not defined. Each of these cases is interpreted geometrically in Table 1.

◆ EQUATIONS OF LINES—SPECIAL FORMS

The constants m and b in the equation

$$y = mx + b \tag{3}$$

have special geometric significance.

If we let $x = 0$, then $y = b$, and we observe that the graph of (3) crosses the y axis at $(0, b)$. The constant b is the *y intercept*. For example, the y intercept of the graph of $y = -4x - 1$ is -1.

To determine the geometric significance of m, we proceed as follows: If $y = mx + b$, then by setting $x = 0$ and $x = 1$, we conclude that $(0, b)$ and $(1, m + b)$ lie on its graph (a line). Hence, the slope of this graph (line) is given by:

$$\text{Slope} = \frac{y_2 - y_1}{x_2 - x_1} = \frac{(m + b) - b}{1 - 0} = m$$

Thus, m is the slope of the line given by $y = mx + b$.

Slope–Intercept Form

The equation

$$y = mx + b \qquad m = \text{Slope},\ b = y \text{ intercept} \tag{4}$$

is called the **slope–intercept form** of an equation of a line.

◆ EXAMPLE 3 (A) Find the slope and y intercept, and graph $y = -\frac{2}{3}x - 3$.
(B) Write the equation of the line with slope $\frac{2}{3}$ and y intercept -2.

Solutions (A) Slope $= m = -\frac{2}{3}$
y intercept $= b = -3$

(B) $m = \frac{2}{3}$ and $b = -2$; thus, $y = \frac{2}{3}x - 2$

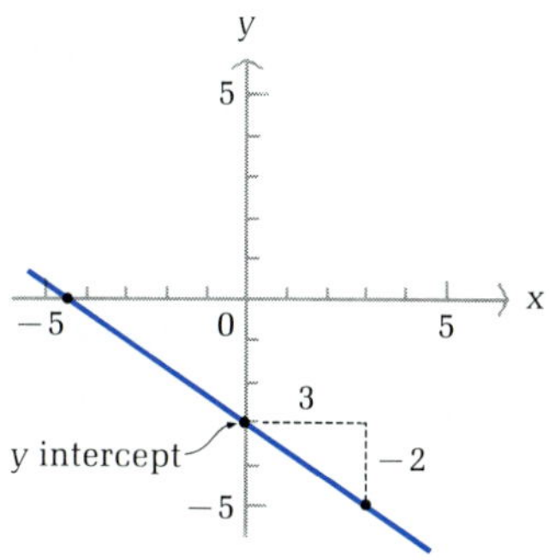

◆

PROBLEM 3 Write the equation of the line with slope $\frac{1}{2}$ and y intercept -1. Graph. ◆

Suppose a line has slope m and passes through a fixed point (x_1, y_1). If the point (x, y) is any other point on the line (Fig. 2), then

$$\frac{y - y_1}{x - x_1} = m$$

that is,

$$y - y_1 = m(x - x_1) \tag{5}$$

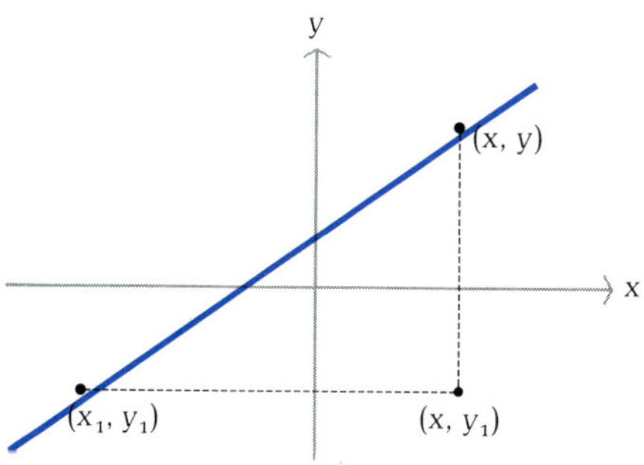

FIGURE 2

We now observe that (x_1, y_1) also satisfies equation (5) and conclude that equation (5) is an equation of a line with slope m that passes through (x_1, y_1).

Point–Slope Form

An equation of a line with slope m that passes through (x_1, y_1) is

$$y - y_1 = m(x - x_1) \tag{5}$$

which is called the **point–slope form** of an equation of a line.

The point–slope form is extremely useful, since it enables us to find an equation for a line if we know its slope and the coordinates of a point on the line or if we know the coordinates of two points on the line.

◆ EXAMPLE 4

(A) Find an equation for the line that has slope $\frac{1}{2}$ and passes through $(-4, 3)$. Write the final answer in the form $Ax + By = C$.

(B) Find an equation for the line that passes through the two points $(-3, 2)$ and $(-4, 5)$. Write the resulting equation in the form $y = mx + b$.

Solutions

(A) Use $y - y_1 = m(x - x_1)$. Let $m = \frac{1}{2}$ and $(x_1, y_1) = (-4, 3)$. Then

$$\begin{aligned} y - 3 &= \tfrac{1}{2}[x - (-4)] \\ y - 3 &= \tfrac{1}{2}(x + 4) \qquad \text{Multiply by 2.} \\ 2y - 6 &= x + 4 \\ -x + 2y &= 10 \quad \text{or} \quad x - 2y = -10 \end{aligned}$$

(B) First, find the slope of the line by using the slope formula:

$$m = \frac{y_2 - y_1}{x_2 - x_1} = \frac{5 - 2}{-4 - (-3)} = \frac{3}{-1} = -3$$

Now use $y - y_1 = m(x - x_1)$ with $m = -3$ and $(x_1, y_1) = (-3, 2)$:

$$\begin{aligned} y - 2 &= -3[x - (-3)] \\ y - 2 &= -3(x + 3) \\ y - 2 &= -3x - 9 \\ y &= -3x - 7 \end{aligned}$$

◆

PROBLEM 4

(A) Find an equation for the line that has slope $\frac{2}{3}$ and passes through $(6, -2)$. Write the resulting equation in the form $Ax + By = C$, $A > 0$.

(B) Find an equation for the line that passes through $(2, -3)$ and $(4, 3)$. Write the resulting equation in the form $y = mx + b$. ◆

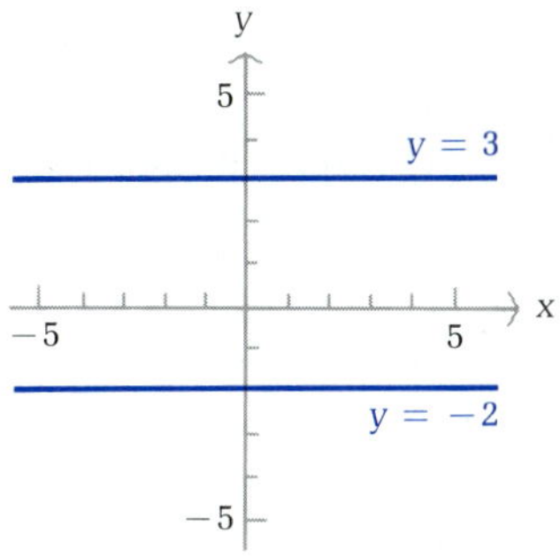

FIGURE 3

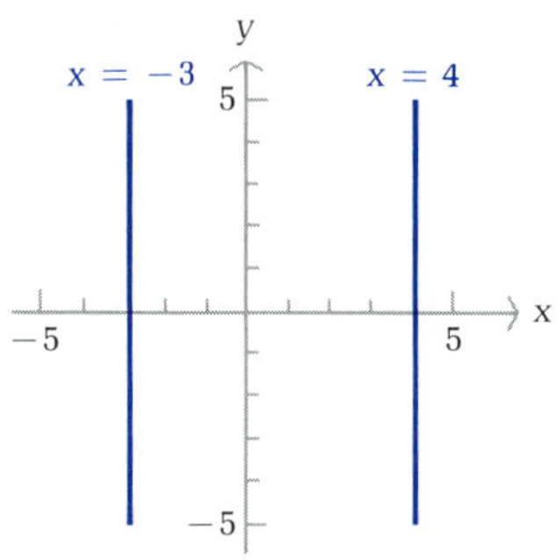

FIGURE 4

The simplest equations of a line are those for horizontal and vertical lines. A **horizontal line** has slope 0; thus, its equation is of the form

$$y = 0x + c \qquad \text{Slope} = 0,\ y \text{ intercept} = c$$

or, simply,

$$y = c$$

Figure 3 illustrates the graphs of $y = 3$ and $y = -2$.

If a line is vertical, then its slope is not defined. All x values (abscissas) of points on a vertical line are equal, while y can take on any value (Fig. 4). Thus, a **vertical line** has an equation of the form

$$x + 0y = c \qquad x \text{ intercept} = c$$

or, simply,

$$x = c$$

Figure 4 illustrates the graphs of $x = -3$ and $x = 4$.

Equations of Horizontal and Vertical Lines

HORIZONTAL LINE WITH y INTERCEPT c:	$y = c$
VERTICAL LINE WITH x INTERCEPT c:	$x = c$

◆ EXAMPLE 5 The equation of a horizontal line through $(-2, 3)$ is $y = 3$, and the equation of a vertical line through the same point is $x = -2$. ◆

PROBLEM 5 Find the equations of the horizontal and vertical lines through $(4, -5)$. ◆

It can be shown that if two nonvertical lines are parallel, then they have the same slope. And if two lines have the same slope, they are parallel. It also can be shown that if two nonvertical lines are perpendicular, then their slopes are the negative reciprocals of each other (that is, $m_2 = -1/m_1$, or, equivalently, $m_1m_2 = -1$). And if the slopes of two lines are the negative reciprocals of each other, the lines are perpendicular. Symbolically:

Parallel and Perpendicular Lines

Given nonvertical lines L_1 and L_2 with slopes m_1 and m_2, respectively, then:

$L_1 \| L_2$ if and only if $m_1 = m_2$

$L_1 \perp L_2$ if and only if $m_1 m_2 = -1$ or $m_2 = -\dfrac{1}{m_1}$

[*Note:* $\|$ means "is parallel to" and $\perp$ means "is perpendicular to."]

◆ EXAMPLE 6 Given the line $x - 2y = 4$, find the equation of a line that passes through $(2, -3)$ and is:

(A) Parallel to the given line (B) Perpendicular to the given line

Write final equations in the form $y = mx + b$.

Solutions First find the slope of the given line by writing $x - 2y = 4$ in the form $y = mx + b$:

$$\begin{aligned} x - 2y &= 4 \\ y &= \tfrac{1}{2}x - 2 \end{aligned}$$

The slope of the given line is $\frac{1}{2}$.

(A) The slope of a line parallel to the given line is also $\frac{1}{2}$. We have only to find the equation of a line through $(2, -3)$ with slope $\frac{1}{2}$ to solve part A:

$$\begin{aligned} y - y_1 &= m(x - x_1) \qquad m = \tfrac{1}{2} \text{ and } (x_1, y_1) = (2, -3) \\ y - (-3) &= \tfrac{1}{2}(x - 2) \\ y + 3 &= \tfrac{1}{2}x - 1 \\ y &= \tfrac{1}{2}x - 4 \end{aligned}$$

(B) The slope of the line perpendicular to the given line is the negative reciprocal of $\frac{1}{2}$; that is, -2. We have only to find the equation of a line through $(2, -3)$ with slope -2 to solve part B:

$$\begin{aligned} y - y_1 &= m(x - x_1) \qquad m = -2 \text{ and } (x_1, y_1) = (2, -3) \\ y - (-3) &= -2(x - 2) \\ y + 3 &= -2x + 4 \\ y &= -2x + 1 \end{aligned}$$

◆

PROBLEM 6 Given the line $2x = 6 - 3y$, find the equation of a line that passes through $(-3, 9)$ and is:

(A) Parallel to the given line (B) Perpendicular to the given line

Write final equations in the form $y = mx + b$. ◆

◆ APPLICATION

We will now see how equations of lines occur in certain applications.

◆ EXAMPLE 7

Cost Equation

The management of a company that manufactures roller skates has fixed costs (costs at zero output) of \$300 per day and total costs of \$4,300 per day at an output of 100 pairs of skates per day. Assume that cost C is linearly related to output x.

(A) Find the slope of the line joining the points associated with outputs of 0 and 100; that is, the line passing through (0, 300) and (100, 4,300).

(B) Find an equation of the line relating output to cost. Write the final answer in the form $C = mx + b$.

(C) Graph the cost equation from part B for $0 \leq x \leq 200$.

Solutions

(A) $m = \dfrac{y_2 - y_1}{x_2 - x_1} = \dfrac{4{,}300 - 300}{100 - 0} = \dfrac{4{,}000}{100} = 40$

(B) We must find an equation of the line that passes through (0, 300) with slope 40. We use the slope–intercept form:

$$C = mx + b$$

$$C = 40x + 300$$

(C)

x	C
0	300
100	4,300
200	8,300

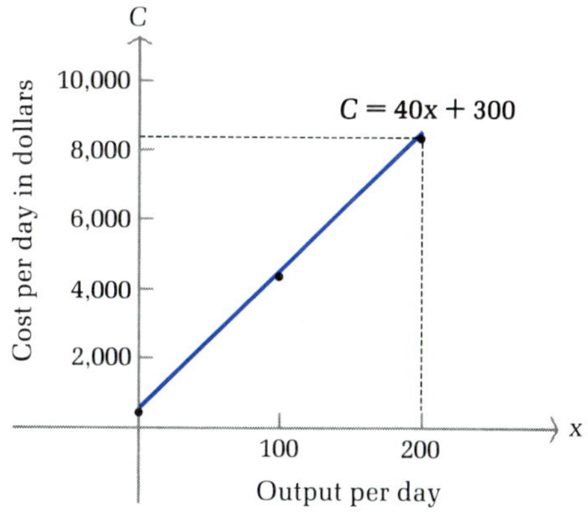

◆

In Example 7, the **fixed cost** of \$300 per day covers plant cost, insurance, and so on. This cost is incurred whether or not there is any production. The **variable cost** is $40x$, which depends on the day's output. Note that the slope 40 is the cost of producing one pair of skates. The total daily cost of operating the plant is given by

(Total cost) = (Variable cost) + (Fixed cost)

PROBLEM 7 Answer parts A and B in Example 7 for fixed costs of \$250 per day and total costs of \$3,450 per day at an output of 80 pairs of skates per day. ◆

Answers to Matched Problems

1. (A)

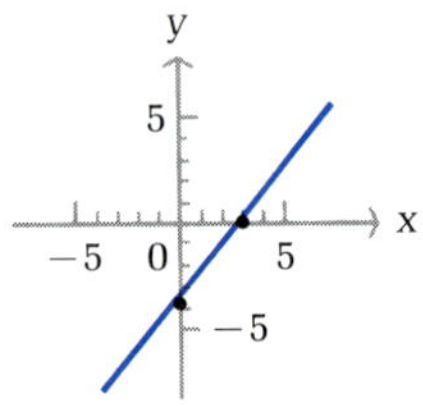

(B)

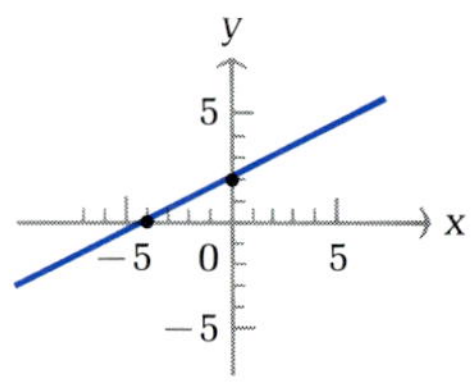

2. (A) −2
   (B) 0 (Zero is a number—it exists! It is the slope of a horizontal line.)
3. $y = \frac{1}{2}x - 1$

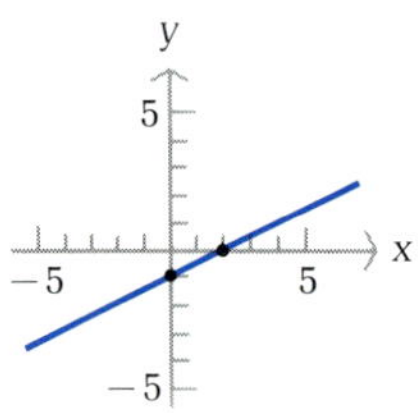

4. (A) $2x - 3y = 18$
   (B) $y = 3x - 9$
5. $y = -5$; $x = 4$
6. (A) $y = -\frac{2}{3}x + 7$
   (B) $y = \frac{3}{2}x + \frac{27}{2}$
7. (A) $m = 40$
   (B) $C = 40x + 250$

EXERCISE 1-1

A

Graph in a rectangular coordinate system.

1. $y = 2x - 3$ **2.** $y = \frac{x}{2} + 1$ **3.** $2x + 3y = 12$ **4.** $8x - 3y = 24$

Find the slope and y intercept of the graph of each equation.

5. $y = 2x - 3$ **6.** $y = \frac{x}{2} + 1$ **7.** $y = -\frac{2}{3}x + 2$ **8.** $y = \frac{3}{4}x - 2$

Write an equation of the line with the indicated slope and y intercept.

9. Slope $= -2$
y intercept $= 4$

10. Slope $= -\frac{2}{3}$
y intercept $= -2$

11. Slope $= -\frac{3}{5}$
y intercept $= 3$

12. Slope $= 1$
y intercept $= -2$

B *Graph in a rectangular coordinate system.*

13. $y = -\frac{2}{3}x - 2$
14. $y = -\frac{3}{2}x + 1$
15. $3x - 2y = 10$
16. $5x - 6y = 15$
17. $x = 3$ and $y = -2$
18. $x = -3$ and $y = 2$

Find the slope of the graph of each equation. (First write the equation in the form $y = mx + b$.)

19. $3x + y = 5$
20. $2x - y = -3$
21. $2x + 3y = 12$
22. $3x - 2y = 10$

Write an equation of the line through each indicated point with the indicated slope. Transform the equation into the form $y = mx + b$.

23. $m = -3$; $(4, -1)$
24. $m = -2$; $(-3, 2)$
25. $m = \frac{2}{3}$; $(-6, -5)$
26. $m = \frac{1}{2}$; $(-4, 3)$

Find the slope of the line that passes through the given points.

27. $(1, 3)$ and $(7, 5)$
28. $(2, 1)$ and $(10, 5)$
29. $(-5, -2)$ and $(5, -4)$
30. $(3\ 7)$ and $(-6, 4)$

Write an equation of the line through each indicated pair of points. Write the final answer in the form $Ax + By = C$, $A > 0$.

31. $(1, 3)$ and $(7, 5)$
32. $(2, 1)$ and $(10, 5)$
33. $(-5, -2)$ and $(5, -4)$
34. $(3, 7)$ and $(-6, 4)$

Write equations of the vertical and horizontal lines through each point.

35. $(3, -5)$
36. $(-2, 7)$
37. $(-1, -3)$
38. $(6, -4)$

Find an equation of the line, given the information in each problem. Write the final answer in the form $y = mx + b$.

39. The line passes through $(-2, 5)$ with slope $-\frac{1}{2}$.
40. The line passes through $(3, -1)$ with slope $-\frac{2}{3}$.
41. The line passes through $(-2, 2)$ and is:
    (A) Parallel to $y = -\frac{1}{2}x + 5$ (B) Perpendicular to $y = -\frac{1}{2}x + 5$
42. The line passes through $(-4, -3)$ and is:
    (A) Parallel to $y = 2x - 3$ (B) Perpendicular to $y = 2x - 3$
43. The line passes through $(-2, -1)$ and is:
    (A) Parallel to $x - 2y = 4$ (B) Perpendicular to $x - 2y = 4$
44. The line passes through $(-3, 2)$ and is:
    (A) Parallel to $2x + 3y = -6$ (B) Perpendicular to $2x + 3y = -6$

C

45. Graph $y = mx - 2$ for $m = 2$, $m = \frac{1}{2}$, $m = 0$, $m = -\frac{1}{2}$, and $m = -2$, all on the same coordinate system.
46. Graph $y = -\frac{1}{2}x + b$ for $b = -4$, $b = 0$, and $b = 4$, all on the same coordinate system.

Write an equation of the line through the indicated points. Be careful!

47. (2, 7) and (2, −3)
48. (−2, 3) and (−2, −1)
49. (2, 3) and (−5, 3)
50. (−3, −3) and (0, −3)

Problems 51–56 require the use of a graphic calculator or a computer.

51. (A) Graph all three equations in the same viewing rectangle:

$$y = \tfrac{1}{2}x - 3 \qquad y = \tfrac{1}{2}x \qquad y = \tfrac{1}{2}x + 3$$

Use [−10, 10] for the range of x and the range of y.

(B) What is the slope of each line?

(C) Are the lines parallel? Explain.

(D) What is the y intercept of each line?

52. Repeat Problem 51 for the following three equations:

$$y = -x - 5 \qquad y = -x \qquad y = -x + 5$$

53. Repeat Problem 51 for the following three equations:

$$y = -\tfrac{1}{2}x + 3 \qquad y = 3 \qquad y = \tfrac{1}{2}x + 3$$

54. Repeat Problem 51 for the following three equations:

$$y = -\tfrac{1}{3}x - 4 \qquad y = -4 \qquad y = \tfrac{1}{3}x - 4$$

55. (A) Graph both equations in the same viewing rectangle:

$$y = 2x - 4 \qquad y = -\tfrac{1}{2}x + 4$$

(B) Are the lines perpendicular? Explain.

56. Repeat Problem 55 for the following two equations:

$$y = x - 2 \qquad y = -x + 2$$

APPLICATIONS

Business & Economics

57. *Simple interest.* If \$$P$ (the principal) is invested at an interest rate of r, then the amount A that is due after t years is given by

$$A = Prt + P$$

If \$100 is invested at 6% ($r = 0.06$), then $A = 6t + 100$, $t \geq 0$.

(A) What will \$100 amount to after 5 years? After 20 years?

(B) Graph the equation for $0 \leq t \leq 20$.

(C) What is the slope of the graph? (The slope indicates the increase in the amount A for each additional year of investment.)

58. *Simple interest.* Repeat Problem 57 for \$1,000 invested at 7.5%.

59. *Cost equation.* The management of a company manufacturing surfboards has fixed costs (zero output) of $200 per day and total costs of $3,800 per day at a daily output of 20 boards.

(A) Assuming the total cost per day (C) is linearly related to the total output per day (x), write an equation relating these two quantities.
(B) What are the total costs for an output of 12 boards per day?
(C) Graph the equation for $0 \leq x \leq 20$.

[*Note:* The slope of the line found in part A is the increase in total cost for each additional unit produced and is called the *marginal cost.* More will be said about this concept later.]

60. *Cost equation.* Repeat Problem 59 if the fixed cost is $300 per day and the total cost per day at an output of 20 boards is $5,100.

61. *Demand equation.* A manufacturing company is interested in introducing a new power mower. Its market research department gave the management the demand–price forecast listed in the table.

PRICE	ESTIMATED DEMAND
$ 70	7,800
$120	4,800
$160	2,400
$200	0

(A) Plot these points, letting d represent the number of mowers people are willing to buy (demand) at a price of $\$p$ each.
(B) Note that the points in part A lie along a straight line. Find an equation of that line.

[*Note:* The slope of the line found in part B indicates the decrease in demand for each $1 increase in price.]

62. *Depreciation.* Office equipment was purchased for $20,000 and is assumed to have a scrap value of $2,000 after 10 years. If its value is depreciated linearly (for tax purposes) from $20,000 to $2,000:

(A) Find the linear equation that relates value (V) in dollars to time (t) in years.
(B) What would be the value of the equipment after 6 years?
(C) Graph the equation for $0 \leq t \leq 10$.

[*Note:* The slope found in part A indicates the decrease in value per year.]

Life Sciences

63. *Nutrition.* In a nutrition experiment, a biologist wants to prepare a special diet for the experimental animals. Two food mixes, A and B, are available. If mix A contains 20% protein and mix B contains 10% protein, what combination of each mix will provide exactly 20 grams of protein? Let x be the amount of A used and let y be the amount of B used. Then write a linear equation relating x, y, and 20. Graph this equation for $x \geq 0$ and $y \geq 0$.

64. Ecology. As one descends into the ocean, pressure increases linearly. The pressure is 15 pounds per square inch on the surface and 30 pounds per square inch 33 feet below the surface.

(A) If p is the pressure in pounds and d is the depth below the surface in feet, write an equation that expresses p in terms of d. [*Hint:* Find an equation of the line that passes through (0, 15) and (33, 30).]
(B) What is the pressure at 12,540 feet (the average depth of the ocean)?
(C) Graph the equation for $0 \leq d \leq 12{,}540$.

[*Note:* The slope found in part A indicates the change in pressure for each additional foot of depth.]

Social Sciences

65. Psychology. In an experiment on motivation, J. S. Brown trained a group of rats to run down a narrow passage in a cage to obtain food in a goal box. Using a harness, he then connected the rats to an overhead wire that was attached to a spring scale. A rat was placed at different distances d (in centimeters) from the goal box, and the pull p (in grams) of the rat toward the food was measured. Brown found that the relationship between these two variables was very close to being linear and could be approximated by the equation

$$p = -\tfrac{1}{5}d + 70 \qquad 30 \leq d \leq 175$$

(See J. S. Brown, *Journal of Comparative and Physiological Psychology*, 1948, 41:450–465.)

(A) What was the pull when $d = 30$? When $d = 175$?
(B) Graph the equation.
(C) What is the slope of the line?

SECTION 1-2 Functions

- DEFINITION OF A FUNCTION
- FUNCTIONS SPECIFIED BY EQUATIONS
- FUNCTION NOTATION
- APPLICATIONS

The function concept is one of the most important concepts in mathematics. The idea of correspondence plays a central role in its formulation. You have already had experiences with correspondences in everyday life. For example:

To each person there corresponds an annual income.

To each item in a supermarket there corresponds a price.

To each day there corresponds a maximum temperature.

For the manufacture of x items there corresponds a cost.

For the sale of x items there corresponds a revenue.

To each square there corresponds an area.

To each number there corresponds its cube.

One of the most important aspects of any science is the establishment of correspondences among various types of phenomena. Once a correspondence is known, predictions can be made. A cost analyst would like to predict costs for various levels of output in a manufacturing process; a medical researcher would like to know the correspondence between heart disease and obesity; a psychologist would like to predict the level of performance after a subject has repeated a task a given number of times; and so on.

◆ DEFINITION OF A FUNCTION

What do all the above examples have in common? Each describes the matching of elements from one set with the elements in a second set. Consider the tables of the cube, square, and square root given in Tables 2–4.

TABLE 2

DOMAIN Number	RANGE Cube
−2 →	−8
−1 →	−1
0 →	0
1 →	1
2 →	8

TABLE 3

DOMAIN Number	RANGE Square
−2 →	4
−1 →	1
0 →	0
1 →	1
2 →	4

TABLE 4

DOMAIN Number	RANGE Square Root
0 →	0
1 →	1
1 →	−1
4 →	2
4 →	−2
9 →	3
9 →	−3

Tables 2 and 3 specify functions, but Table 4 does not. Why not? The definition of the very important term *function* will explain.

Definition of a Function

A **function** is a rule (process or method) that produces a correspondence between two sets of elements such that to each element in the first set there corresponds one and only one element in the second set. The first set is called the **domain,** and the set of all corresponding elements in the second set is called the **range.**

Tables 2 and 3 specify functions, since to each domain value there corresponds exactly one range value (for example, the cube of -2 is -8 and no other number). On the other hand, Table 4 does not specify a function, since to at least one domain value there corresponds more than one range value (for example, to the domain value 9 there corresponds -3 and 3, both square roots of 9).

◆ FUNCTIONS SPECIFIED BY EQUATIONS

Most of the domains and ranges included in this text will be (infinite) sets of real numbers, and the rules associating range values with domain values will be equations in two variables. Consider, for example, the equation for the area of a rectangle with width 1 inch less than its length. If x is the length, then the area y is given by

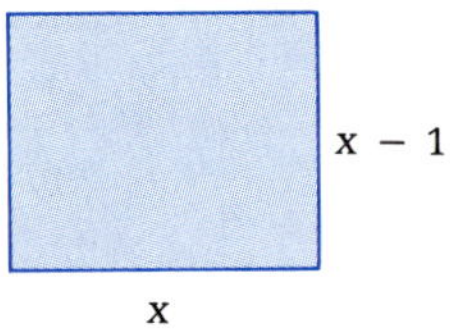

$$y = x(x - 1) \qquad x \geq 1$$

For each **input** x (length), we obtain an **output** y (area). For example:

If $x = 5$, then $y = 5(5 - 1) = 5 \cdot 4 = 20$.

If $x = 1$, then $y = 1(1 - 1) = 1 \cdot 0 = 0$.

If $x = 4.32$, then $y = 4.32(4.32 - 1) = (4.32)(3.32) = 14.3424$.

The input values are domain values, and the output values are range values. The equation (a rule) assigns each domain value x a range value y. The variable x is called an *independent variable* (since values can be "independently" assigned to x from the domain), and y is called a *dependent variable* (since the value of y "depends" on the value assigned to x). In general, any variable used as a placeholder for domain values is called an **independent variable;** any variable that is used as a placeholder for range values is called a **dependent variable.**

When does an equation specify a function?

Equations and Functions

In an equation in two variables, if there corresponds exactly one value of the dependent variable (output) to each value of the independent variable (input), then the equation specifies a function. If there is any value of the independent variable to which there corresponds more than one value of the dependent variable, then the equation does not specify a function.

◆ EXAMPLE 8 Determine which of the following equations specify functions with independent variable x.

(A) $4y - 3x = 8$, x a real number (B) $y^2 - x^2 = 9$, x a real number

Solutions (A) Solving for the dependent variable y, we have

$$4y - 3x = 8 \tag{1}$$
$$4y = 8 + 3x$$
$$y = 2 + \tfrac{3}{4}x$$

Since each input value x corresponds to exactly one output value ($y = 2 + \frac{3}{4}x$), we see that equation (1) specifies a function.

(B) Solving for the dependent variable y, we have

$$y^2 - x^2 = 9 \tag{2}$$
$$y^2 = 9 + x^2$$
$$y = \pm\sqrt{9 + x^2}$$

Since $9 + x^2$ is always a positive real number for any real number x and since each positive real number has two square roots, to each input value x there corresponds two output values ($y = -\sqrt{9 + x^2}$ and $y = \sqrt{9 + x^2}$). For example, if $x = 4$, then equation (2) is satisfied for $y = 5$ and for $y = -5$. Thus, equation (2) does not specify a function. ◆

PROBLEM 8 Determine which of the following equations specify functions with independent variable x.

(A) $y^2 - x^4 = 9$, x a real number (B) $3y - 2x = 3$, x a real number ◆

Since the graph of an equation is the graph of all the ordered pairs that satisfy the equation, it is very easy to determine whether an equation specifies a function by examining its graph. The graphs of the two equations we considered in Example 8 are shown in Figure 5. (The graph in Figure 5B was obtained using point-by-point plotting, a technique we will discuss later in this chapter.)

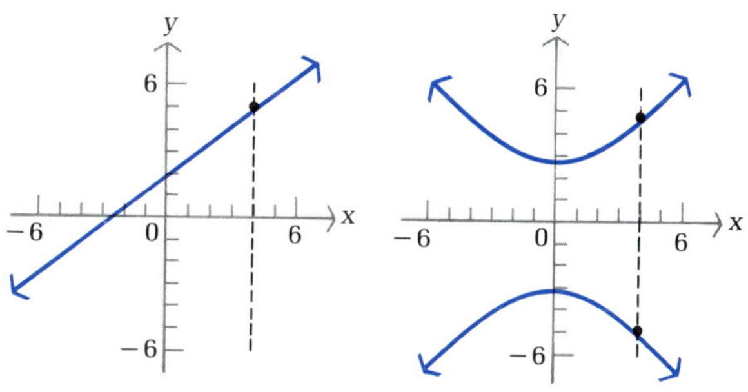

(A) $4y - 3x = 8$ (B) $y^2 - x^2 = 9$

FIGURE 5

In Figure 5A notice that any vertical line will intersect the graph of the equation $4y - 3x = 8$ in exactly one point. This shows that to each x value there corresponds exactly one y value and confirms our conclusion that this equation specifies a function. On the other hand, Figure 5B shows that there exist vertical lines that intersect the graph of $y^2 - x^2 = 9$ in two points. This indicates that there exist x values to which there correspond two different y values and verifies our conclusion that this equation does not specify a function. These observations are generalized in Theorem 2.

THEOREM 2

Vertical Line Test for a Function

An equation specifies a function if each vertical line in the coordinate system passes through at most one point on the graph of the equation. If any vertical line passes through two or more points on the graph of an equation, then the equation does not specify a function.

In Example 8, the domains were explicitly stated along with the given equations. In many cases, this will not be done. Unless stated to the contrary, we shall adhere to the following convention regarding domains and ranges for functions specified by equations.

Agreement on Domains and Ranges

If a function is specified by an equation and the domain is not indicated, then we assume that the domain is the set of all real number replacements of the independent variable (inputs) that produce *real values* for the dependent variable (outputs). The range is the set of all outputs corresponding to input values.

◆ EXAMPLE 9 Find the domain of the function specified by the equation $y = \sqrt{x + 4}$, assuming x is the independent variable.

Solution For y to be real, $x + 4$ must be greater than or equal to 0; that is,

$$\begin{aligned} x + 4 &\geqslant 0 \\ x &\geqslant -4 \end{aligned}$$

Thus,

Domain: $x \geqslant -4$ or $[-4, \infty)$ ◆

PROBLEM 9 Find the domain of the function specified by the equation $y = \sqrt{x - 2}$, assuming x is the independent variable. ◆

◆ FUNCTION NOTATION

We have just seen that a function involves two sets, a domain and a range, and a rule of correspondence that enables us to assign to each element in the domain exactly one element in the range. We use different letters to denote names for numbers; in essentially the same way, we will now use different letters to denote names for functions. For example, f and g may be used to name the functions specified by the equations $y = 2x + 1$ and $y = x^2 + 2x - 3$:

$$\begin{aligned} f&: \quad y = 2x + 1 \\ g&: \quad y = x^2 + 2x - 3 \end{aligned} \tag{3}$$

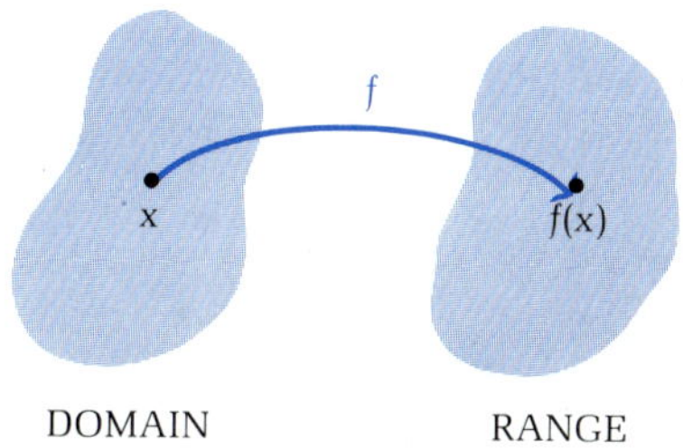

FIGURE 6

If x represents an element in the domain of a function f, then we frequently use the symbol

$$f(x)$$

in place of y to designate the number in the range of the function f to which x is paired (Fig. 6). This symbol does not represent the product of f and x. The symbol $f(x)$ is read as "f of x," "f at x," or "the value of f at x." Whenever we write $y = f(x)$, we assume that the variable x is an independent variable and that both y and $f(x)$ are dependent variables.

Using function notation, we can now write functions f and g in (3) in the form

$$f(x) = 2x + 1 \qquad \text{and} \qquad g(x) = x^2 + 2x - 3$$

Let us find $f(3)$ and $g(-5)$. To find $f(3)$, we replace x with 3 wherever x occurs in $f(x) = 2x + 1$ and evaluate the right side:

$$\begin{aligned} f(x) &= 2x + 1 \\ f(3) &= 2 \cdot 3 + 1 \\ &= 6 + 1 = 7 \end{aligned}$$

Thus,

$f(3) = 7$ — The function f assigns the range value 7 to the domain value 3.

To find $g(-5)$, we replace x by -5 wherever x occurs in $g(x) = x^2 + 2x - 3$ and evaluate the right side:

$$\begin{aligned} g(x) &= x^2 + 2x - 3 \\ g(-5) &= (-5)^2 + 2(-5) - 3 \\ &= 25 - 10 - 3 = 12 \end{aligned}$$

Thus,

$g(-5) = 12$ — The function g assigns the range value 12 to the domain value -5.

It is very important to understand and remember the definition of $f(x)$:

The Symbol $f(x)$

For any element x in the domain of the function f, the symbol $\boldsymbol{f(x)}$ represents the element in the range of f corresponding to x in the domain of f. If x is an input value, then $f(x)$ is the corresponding output value. If x is an element that is not in the domain of f, then f *is not defined at* x and $f(x)$ *does not exist.*

◆ EXAMPLE 10 If

$$f(x) = \frac{12}{x - 2} \qquad g(x) = 1 - x^2 \qquad h(x) = \sqrt{x - 1}$$

then:

(A) $f(6) \; \boxed{= \frac{12}{6 - 2}}\,^{*} = \frac{12}{4} = 3$ (B) $g(-2) \; \boxed{= 1 - (-2)^2} = 1 - 4 = -3$

(C) $h(-2) \; \boxed{= \sqrt{-2 - 1}} = \sqrt{-3}$

Since $\sqrt{-3}$ is not a real number, -2 is not in the domain of h and $h(-2)$ is not defined.

(D) $f(0) + g(1) - h(10) \; \boxed{= \frac{12}{0 - 2} + (1 - 1^2) - \sqrt{10 - 1}}$

$$= \frac{12}{-2} + 0 - \sqrt{9} = -6 - 3 = -9$$ ◆

PROBLEM 10 Use the functions in Example 10 to find:

(A) $f(-2)$ (B) $g(-1)$ (C) $h(-8)$ (D) $\dfrac{f(3)}{h(5)}$ ◆

◆ EXAMPLE 11 Find the domains of functions f, g, and h:

$$f(x) = \frac{12}{x - 2} \qquad g(x) = 1 - x^2 \qquad h(x) = \sqrt{x - 1}$$

Domain of f $12/(x - 2)$ represents a real number for all replacements of x by real numbers except for $x = 2$ (division by 0 is not defined). Thus, $f(2)$ does not exist, and the domain of f is the set of all real numbers except 2. We often indicate this by writing

$$f(x) = \frac{12}{x - 2} \qquad x \neq 2$$

---

* Dashed boxes are used throughout the book to represent steps that are usually performed mentally.

Domain of g The domain is R, the set of all real numbers, since $1 - x^2$ represents a real number for all replacements of x by real numbers.

Domain of h The domain is the set of all real numbers x such that $\sqrt{x-1}$ is a real number—that is, such that

$$x - 1 \geqslant 0$$
$$x \geqslant 1 \quad \text{or} \quad [1, \infty)$$

PROBLEM 11 Find the domains of functions F, G, and H:

$$F(x) = x^2 - 3x + 1 \qquad G(x) = \frac{5}{x+3} \qquad H(x) = \sqrt{2-x}$$

EXAMPLE 12 For $f(x) = x^2 - 2x + 7$, find:

(A) $f(a)$ (B) $f(a+h)$ (C) $\dfrac{f(a+h) - f(a)}{h}$

Solutions

(A) $f(a) = a^2 - 2a + 7$

(B) $f(a+h) = (a+h)^2 - 2(a+h) + 7$
$= a^2 + 2ah + h^2 - 2a - 2h + 7$

(C) $$\frac{f(a+h) - f(a)}{h} = \frac{(a^2 + 2ah + h^2 - 2a - 2h + 7) - (a^2 - 2a + 7)}{h}$$
$$= \frac{2ah + h^2 - 2h}{h} = \frac{h(2a + h - 2)}{h} = 2a + h - 2$$

PROBLEM 12 Repeat Example 12 for $f(x) = x^2 - 4x + 9$.

APPLICATIONS

EXAMPLE 13 A rectangular dog pen is to be made with 120 feet of wire fencing.

Construction

(A) If x represents the width of the pen, express its area $A(x)$ in terms of x.

(B) What is the domain of the function A? (The domain is determined by the physical constraints.)

(C) Find the area for each of the following widths: 10, 20, 30, 40, 50. Notice how the area changes as the width changes.

Solutions

(A) Draw a figure and label the sides:

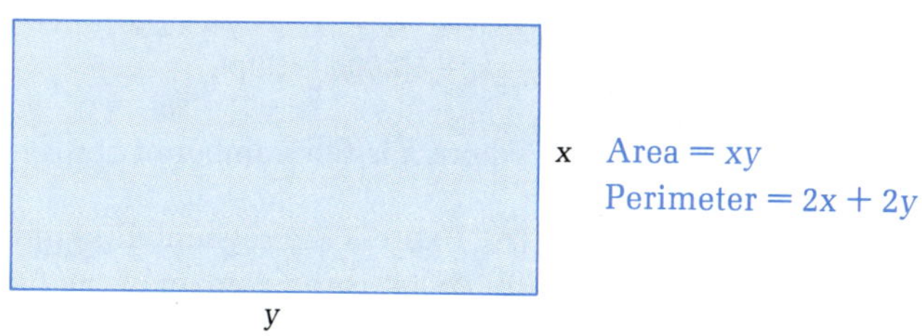

Write y in terms of x:

$$2x + 2y = 120 \qquad \text{Perimeter}$$
$$x + y = 60$$
$$y = 60 - x$$

Area $= xy = x(60 - x)$:

$$A(x) = x(60 - x) \qquad \text{Area depends on width } x.$$

(B) The area cannot be negative. Thus, x cannot be negative nor greater than 60.

Domain: $0 \le x \le 60$ Inequality notation
$[0, 60]$ Interval notation

Note: We use $0 \le x \le 60$ to mean "the set of all real numbers x such that $0 \le x \le 60$," or, more formally, $\{x \in R | 0 \le x \le 60\}$. We will continue to use the less formal representation for domain descriptions, however, with this understanding.

(C) $A(10) = 10(60 - 10) = 10 \cdot 50 = 500 \text{ ft}^2$
$A(20) = 20(60 - 20) = 20 \cdot 40 = 800 \text{ ft}^2$
$A(30) = 30(60 - 30) = 30 \cdot 30 = 900 \text{ ft}^2$
$A(40) = 40(60 - 40) = 40 \cdot 20 = 800 \text{ ft}^2$
$A(50) = 50(60 - 50) = 50 \cdot 10 = 500 \text{ ft}^2$

(An important related problem, which we will not consider now, is to find the width that gives the maximum area.) ◆

PROBLEM 13 Work Example 13 with the added assumption that an existing wooden fence is to be used as one side of the pen, as shown in the figure.

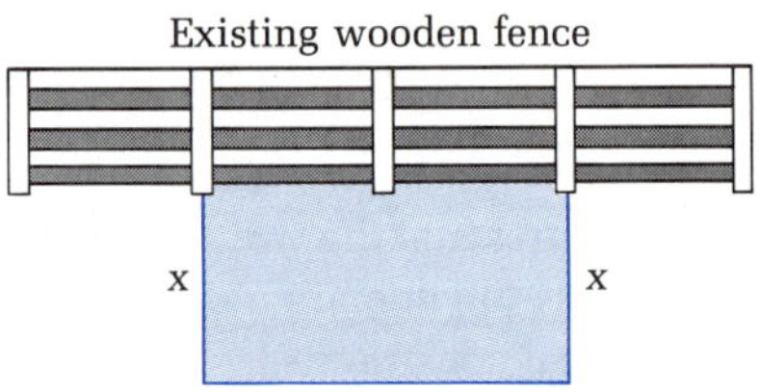

◆

◆ EXAMPLE 14 The market research department of a company that manufactures memory chips for microcomputers has determined that the demand equation for 256k chips is

Revenue

$$x = 10{,}000 - 50p$$

where x is the number of chips that can be sold at a price of \$p per chip.

(A) Express the revenue $R(x)$ in terms of the demand x.
(B) What is the domain of the function R?

Solutions (A) The revenue from the sale of x units at \$p per unit is $R = xp$. To express R in terms of x, we first solve the demand equation for p in terms of x:

$$x = 10{,}000 - 50p$$
$$50p = 10{,}000 - x$$
$$p = 200 - \tfrac{1}{50}x$$

Thus,

$$R(x) = xp = x(200 - \tfrac{1}{50}x) = 200x - \tfrac{1}{50}x^2$$

(B) Since neither price nor demand can be negative, x must satisfy

$$x \geqslant 0 \qquad \text{and} \qquad 200 - \tfrac{1}{50}x \geqslant 0$$

Solving the second inequality and combining it with the first, the domain of the revenue function is

$$0 \leqslant x \leqslant 10{,}000 \quad \text{or} \quad [0, 10{,}000]$$ ◆

PROBLEM 14 Repeat Example 14 for the demand equation $x = 15{,}000 - 30p$. ◆

Answers to Matched Problems

8. (A) Does not specify a function (B) Specifies a function
9. $x \geqslant 2$ (inequality notation) or $[2, \infty)$ (interval notation)
10. (A) -3 (B) 0 (C) Does not exist (D) 6
11. Domain of F: R; Domain of G: All real numbers except -3; Domain of H: $x \leqslant 2$ (inequality notation) or $(-\infty, 2]$ (interval notation)
12. (A) $a^2 - 4a + 9$ (B) $a^2 + 2ah + h^2 - 4a - 4h + 9$ (C) $2a + h - 4$
13. (A) $A(x) = x(120 - 2x)$ (B) Domain: $0 \leqslant x \leqslant 60$
    (C) $A(10) = 1{,}000$ ft²; $A(20) = 1{,}600$ ft²; $A(30) = 1{,}800$ ft²; $A(40) = 1{,}600$ ft²; $A(50) = 1{,}000$ ft²
14. (A) $R(x) = x(500 - \tfrac{1}{30}x) = 500x - \tfrac{1}{30}x^2$
    (B) Domain: $0 \leqslant x \leqslant 15{,}000$ (inequality notation) or $[0, 15{,}000]$ (interval notation)

EXERCISE 1-2

A *Indicate whether each table specifies a function.*

1.

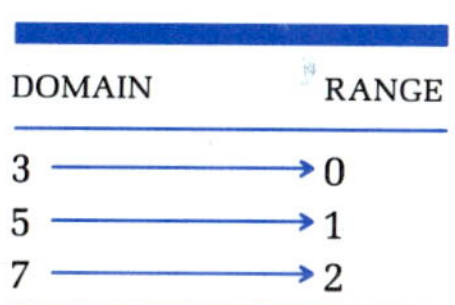

2.

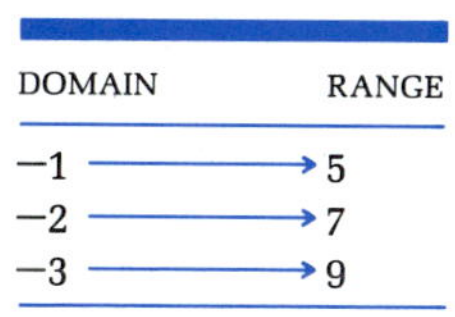

3.

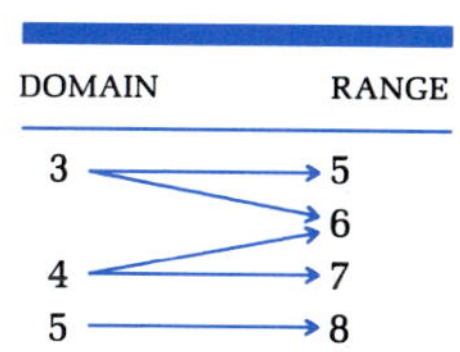

4. 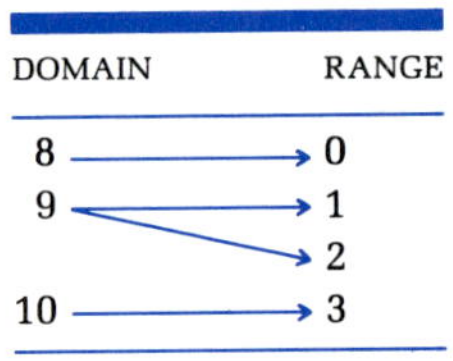

5. 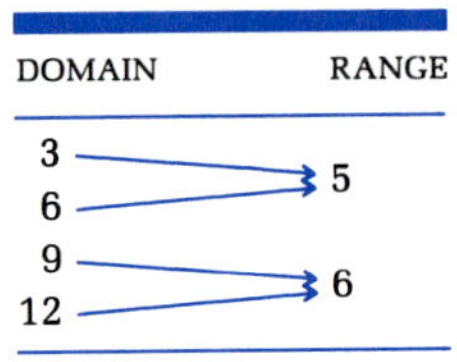

6. 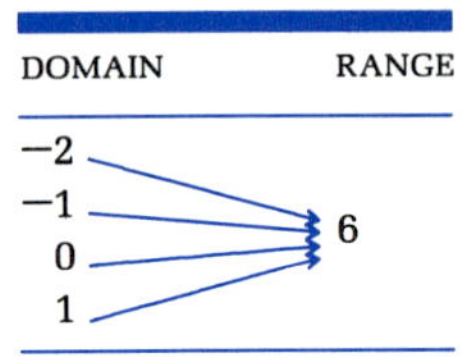

Indicate whether each graph specifies a function.

7. 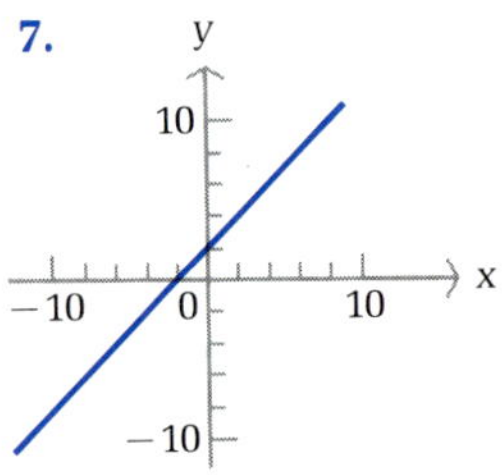

8. 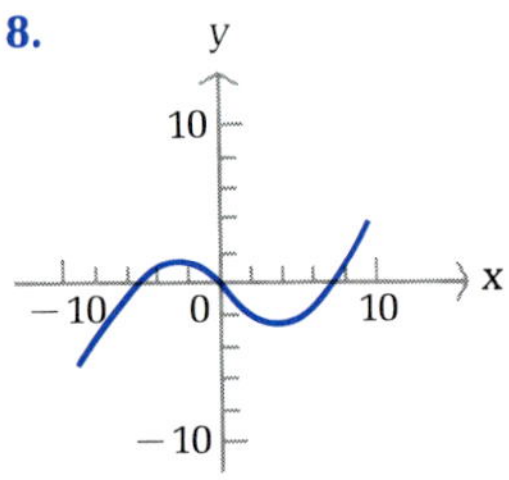

9. 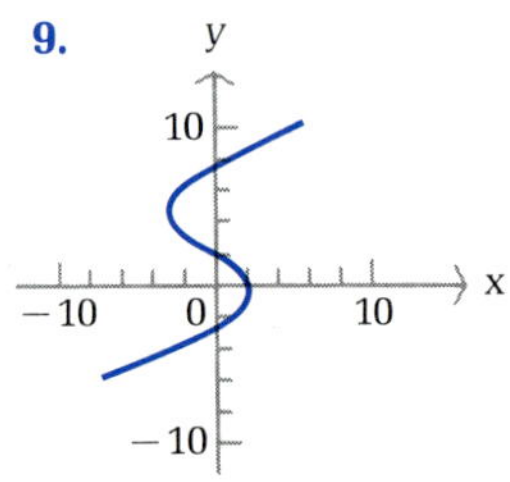

10. 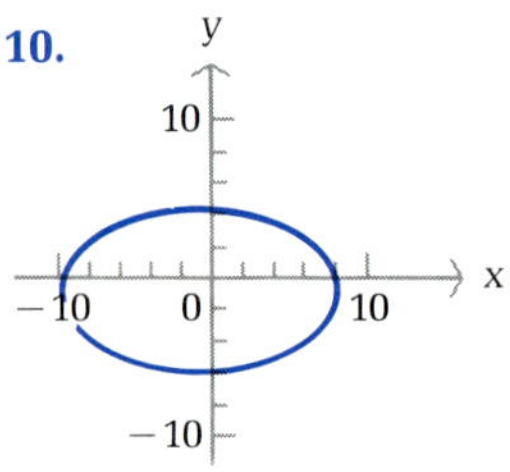

11. 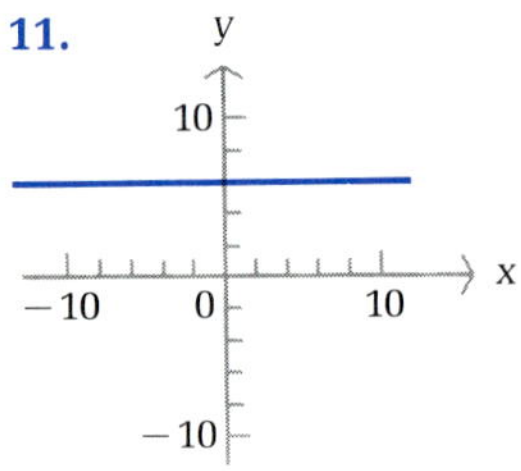

12. 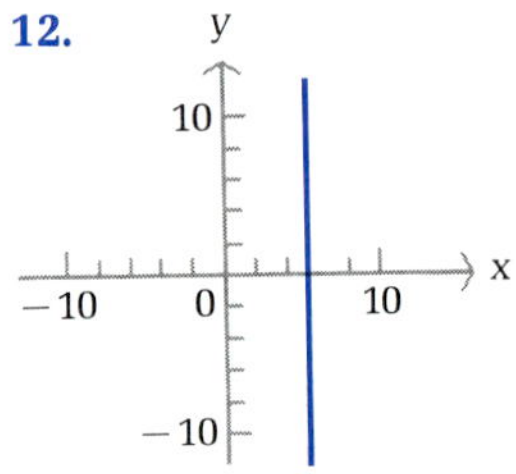

If $f(x) = 3x - 2$ and $g(x) = x - x^2$, find each of the following:

13. $f(2)$ **14.** $f(1)$ **15.** $f(-1)$ **16.** $f(-2)$

17. $g(3)$ **18.** $g(1)$ **19.** $f(0)$ **20.** $f(\frac{1}{3})$

21. $g(-3)$ **22.** $g(-2)$ **23.** $f(1) + g(2)$ **24.** $g(1) + f(2)$

25. $g(2) - f(2)$ **26.** $f(3) - g(3)$ **27.** $g(3) \cdot f(0)$ **28.** $g(0) \cdot f(-2)$

29. $\dfrac{g(-2)}{f(-2)}$ **30.** $\dfrac{g(-3)}{f(2)}$

B *Find the domain of each function in Problems 31–44.*

31. $f(x) = \sqrt{x}$ **32.** $f(x) = \dfrac{1}{\sqrt{x}}$ **33.** $f(x) = \dfrac{x - 3}{(x - 5)(x + 3)}$

34. $f(x) = \dfrac{x + 1}{x - 2}$ **35.** $f(x) = \sqrt{x + 5}$ **36.** $f(x) = \sqrt{7 - x}$

37. $f(x) = \dfrac{x^2 + 1}{x^2 - 1}$ **38.** $f(x) = \dfrac{x^2 + 5}{x^2 - 9}$ **39.** $f(x) = \dfrac{x}{x^2 + 3x - 4}$

40. $f(x) = \dfrac{x}{x^2 + x - 6}$ 41. $f(x) = \dfrac{x + 4}{x^2 - 4x + 5}$ 42. $f(x) = \dfrac{x - 7}{x^2 + 6x + 10}$

43. $f(x) = \dfrac{3}{\sqrt{x + 2}}$ 44. $f(x) = \dfrac{-2}{\sqrt{1 - x}}$

Determine which of the equations in Problems 45–54 specify functions with independent variable x. For those that do, find the domain. For those that do not, find a value of x to which there corresponds more than one value of y.

45. $4x - 5y = 20$ 46. $3y - 7x = 15$ 47. $x^2 - y = 1$
48. $x - y^2 = 1$ 49. $x + y^2 = 10$ 50. $x^2 + y = 10$
51. $xy - 4y = 1$ 52. $xy + y - x = 5$ 53. $x^2 + y^2 = 25$
54. $x^2 - y^2 = 16$

55. If $F(t) = 4t + 7$, find: $\dfrac{F(3 + h) - F(3)}{h}$

56. If $G(r) = 3 - 5r$, find: $\dfrac{G(2 + h) - G(2)}{h}$

57. If $g(w) = w^2 - 4$, find: $\dfrac{g(1 + h) - g(1)}{h}$

58. If $f(m) = 2m^2 + 5$, find: $\dfrac{f(4 + h) - f(4)}{h}$

59. If $Q(x) = x^2 - 5x + 1$, find: $\dfrac{Q(2 + h) - Q(2)}{h}$

60. If $P(x) = 2x^2 - 3x - 7$, find: $\dfrac{P(3 + h) - P(3)}{h}$

C *In Problems 61–68, find and simplify:* $\dfrac{f(a + h) - f(a)}{h}$

61. $f(x) = 4x - 3$ 62. $f(x) = -3x + 9$ 63. $f(x) = 4x^2 - 7x + 6$
64. $f(x) = 3x^2 + 5x - 8$ 65. $f(x) = x^3$ 66. $f(x) = x^3 - x$

67. $f(x) = \sqrt{x}$ 68. $f(x) = \dfrac{1}{x}$

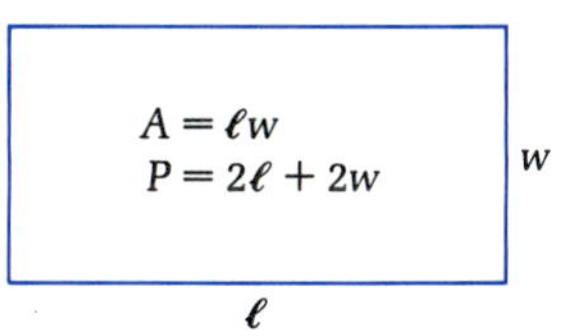

Problems 69–72 refer to the area A and perimeter P of a rectangle with length ℓ and width w (see the figure).

69. The area of a rectangle is 25 square inches. Express the perimeter $P(w)$ as a function of the width w, and state the domain of this function.
70. The area of a rectangle is 81 square inches. Express the perimeter $P(\ell)$ as a function of the length ℓ, and state the domain of this function.
71. The perimeter of a rectangle is 100 meters. Express the area $A(\ell)$ as a function of the length ℓ, and state the domain of this function.
72. The perimeter of a rectangle is 160 meters. Express the area $A(w)$ as a function of the width w, and state the domain of this function.

Business & Economics

73. *Cost function.* The fixed costs (tooling and overhead) for manufacturing a particular stereo system are \$96,000, and the variable costs per unit (labor, material, etc.) are \$80. If x units are manufactured, express the cost $C(x)$ as a function of x. Find the cost of producing 500 stereos.

74. *Cost function.* A company that specializes in manufacturing reproductions of classic automobiles has fixed costs of \$100,000 and variable costs of \$15,000 per automobile produced. If x cars are manufactured, express the cost $C(x)$ as a function of x. Find the cost of producing 48 automobiles.

75. *Revenue function.* After extensive surveys, the research department of a stereo manufacturing company produced the demand equation

$$x = 8{,}000 - 40p$$

where x is the number of units that retailers are likely to purchase at a price of \$p per unit. Express the revenue $R(x)$ in terms of the demand x. Find the domain of R.

76. *Revenue function.* Repeat Problem 75 for the demand equation

$$x = 9{,}000 - 60p$$

77. *Profit function.* Generally speaking, **Profit = Revenue − Cost.** Suppose the cost equation for manufacturing a calculator is $C(x) = 100{,}000 + 20x$ and the demand equation is $x = 48{,}000 - 400p$, where x is the number of units sold per month at a price of \$p per unit.

(A) Find the revenue, $R(x)$, in terms of x.
(B) Find the profit, $P(x)$, in terms of x.
(C) What is the monthly profit or loss for a production level of 10,000 units? 20,000 units? 40,000 units?

78. *Profit function.* Repeat Problem 77 if the demand equation is $x = 50{,}000 - 500p$.

79. *Packaging.* A candy box is to be made out of a piece of cardboard that measures 8 by 12 inches. Equal-sized squares x inches on a side will be cut out of each corner, and then the ends and sides will be folded up to form a rectangular box.

(A) Express the volume of the box $V(x)$ in terms of x.
(B) What is the domain of the function V (determined by the physical restrictions)?
(C) Complete the table:

x	V(x)
1	
2	
3	

Notice how the volume changes with different choices of x.

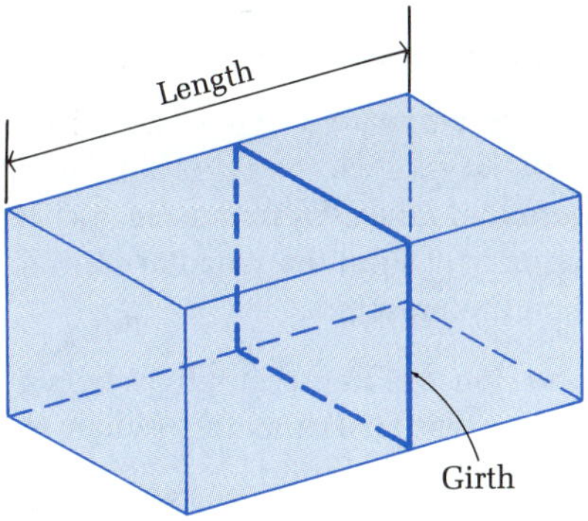

80. *Packaging.* A parcel delivery service will only deliver packages with length plus girth (distance around) not exceeding 108 inches. A rectangular shipping box with square ends x inches on a side is to be used.

(A) If the full 108 inches is to be used, express the volume of the box $V(x)$ in terms of x.

(B) What is the domain of the function V (determined by the physical restrictions)?

(C) Complete the table:

x	V(x)
5	
10	
15	
20	
25	

Notice how the volume changes with different choices of x.

81. *Construction.* A veterinarian wants to construct a kennel with five individual pens, as indicated in the figure. Local ordinances require that each pen have a gate that is 3 feet wide and an area of 45 square feet. If x is the width of one pen, express the total amount of fencing $P(x)$ (excluding the gates) required for construction of the kennel as a function of x, and complete the table in the margin.

x	P(x)
3	
4	
5	
6	

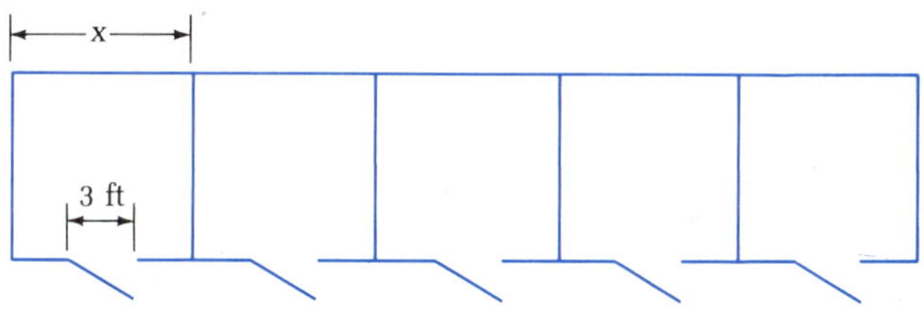

82. *Construction.* A horse breeder wants to construct three adjacent corrals, as indicated in the figure. If each corral must have an area of 600 square feet and x is the width of one corral, express the total amount of fencing $P(x)$ as a function of x, and complete the following table:

x	P(x)
10	
15	
20	
25	

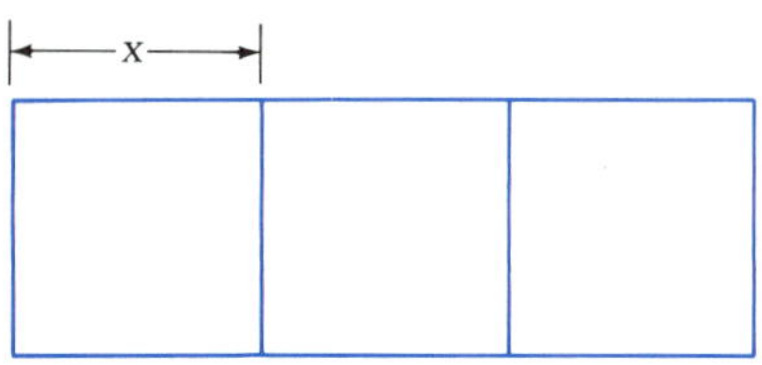

Life Sciences

83. *Muscle contraction.* In a study of the speed of muscle contraction in frogs under various loads, noted British biophysicist and Nobel prize winner A. W. Hill determined that the weight w (in grams) placed on the muscle and the speed of contraction v (in centimeters per second) are approximately related by an equation of the form

$$(w + a)(v + b) = c$$

where a, b, and c are constants. Suppose that for a certain muscle, $a = 15$, $b = 1$, and $c = 90$. Express v as a function of w. Find the speed of contraction if a weight of 16 grams is placed on the muscle.

Social Sciences

84. *Politics.* The percentage s of seats in the House of Representatives won by Democrats and the percentage v of votes cast for Democrats (when expressed as decimal fractions) are related by the equation

$$5v - 2s = 1.4 \qquad 0 < s < 1, \quad 0.28 < v < 0.68$$

(A) Express v as a function of s, and find the percentage of votes required for the Democrats to win 51% of the seats.

(B) Express s as a function of v, and find the percentage of seats won if Democrats receive 51% of the votes.

SECTION 1-3 Linear and Quadratic Functions

- GRAPHS OF FUNCTIONS
- LINEAR FUNCTIONS AND THEIR GRAPHS
- QUADRATIC FUNCTIONS AND THEIR GRAPHS
- PIECEWISE-DEFINED FUNCTIONS
- APPLICATION: MARKET RESEARCH

◆ GRAPHS OF FUNCTIONS

The **graph of a function f** is the graph of the set of ordered pairs of numbers (x, y), where

$$y = f(x)$$

and x is in the domain of f. When functions are graphed, domain values are usually associated with the horizontal axis and range values with the vertical axis (see Fig. 7 at the top of the next page).

The first coordinate (abscissa) of a point where the graph of a function crosses the x axis is called an **x intercept** of the function. The x intercepts are determined by finding the real solutions of the equation $f(x) = 0$, if any exist. The second coordinate (ordinate) of a point where the graph of a function crosses the y axis is called the **y intercept** of the function. The y intercept is given by $f(0)$,

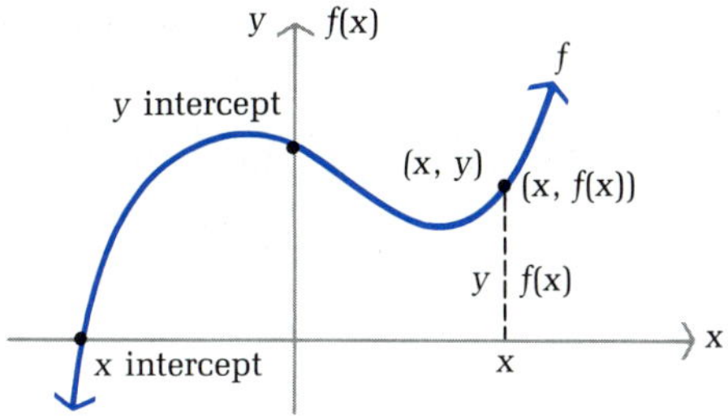

FIGURE 7

provided 0 is in the domain of f. Note that a function can have more than one x intercept, but can never have more than one y intercept (see the vertical line test in the preceding section).

◆ LINEAR FUNCTIONS AND THEIR GRAPHS

A function specified by an equation of the form

$$f(x) = mx + b$$

is called a **linear function.** In the special case $m = 0$, the function has the form

$$f(x) = b$$

and is also called a **constant function.** Graphing linear functions is equivalent to graphing the equation

$$y = mx + b \qquad \text{Slope} = m,\ y \text{ intercept} = b$$

which we discussed in detail in Section 1-1. Since the expression $mx + b$ represents a real number for all real number replacements of x, the domain of any linear function is R, the set of all real numbers. The range of a nonconstant linear function is also R, while the range of a constant function is the single real number b. See the graphs in the box.

Graph of $f(x) = mx + b$

The graph of a linear function f is a nonvertical straight line with slope m and y intercept b.

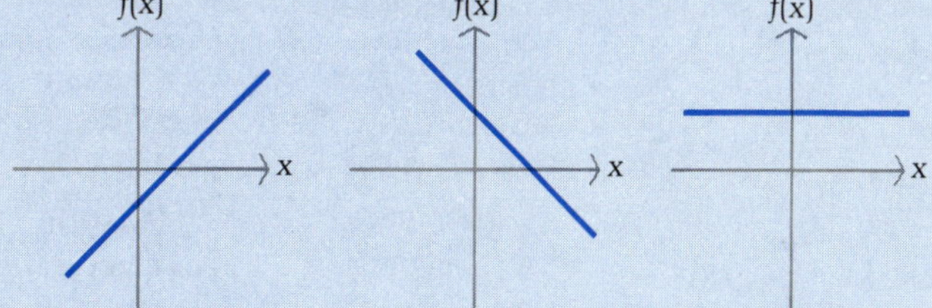

$m > 0$ Positive slope | $m < 0$ Negative slope | $m = 0$ Zero slope

◆ EXAMPLE 15 Graph the linear function defined by

$$f(x) = -\frac{x}{2} + 3 \quad \boxed{= -\frac{1}{2}x + 3}$$

and indicate its slope and intercepts.

Solution

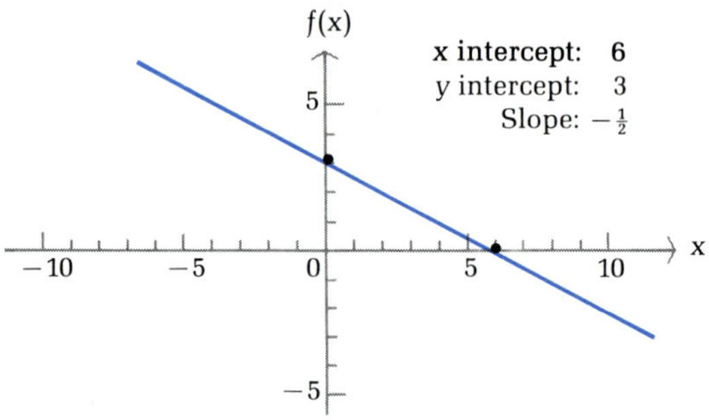

◆

PROBLEM 15 Graph the linear function defined by

$$f(x) = \frac{x}{3} + 1$$

and indicate its slope and intercepts. ◆

◆ QUADRATIC FUNCTIONS AND THEIR GRAPHS

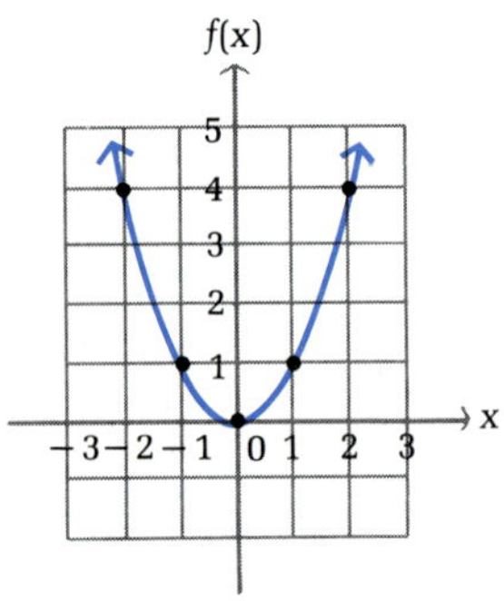

FIGURE 8

Any function defined by an equation of the form

$$f(x) = ax^2 + bx + c \qquad a \neq 0$$

where a, b, and c are constants and x is a variable, is called a **quadratic function.**

Let us start by graphing the simple quadratic function

$$f(x) = x^2$$

We evaluate this function for integer values from its domain, find corresponding range values, then plot the resulting ordered pairs listed in Table 5, and join these points with a smooth curve, as shown in Figure 8. The first two steps are usually done mentally or on scratch paper.

TABLE 5

Graphing $f(x) = x^2$

DOMAIN VALUES x	RANGE VALUES $y = f(x)$	ELEMENTS OF f $(x, f(x))$
-2	$y = f(-2) = (-2)^2 = 4$	$(-2, 4)$
-1	$y = f(-1) = (-1)^2 = 1$	$(-1, 1)$
0	$y = f(0) = 0^2 = 0$	$(0, 0)$
1	$y = f(1) = 1^2 = 1$	$(1, 1)$
2	$y = f(2) = 2^2 = 4$	$(2, 4)$

The curve shown in Figure 8 is called a **parabola.** It can be shown (in a course in analytic geometry) that the graph of any quadratic function is also a parabola. In general:

Graph of $f(x) = ax^2 + bx + c$, $a \neq 0$

The graph of a quadratic function f is a parabola that has its **axis** (line of symmetry) parallel to the vertical axis. It opens upward if $a > 0$ and downward if $a < 0$. The intersection point of the axis and parabola is called the **vertex.**

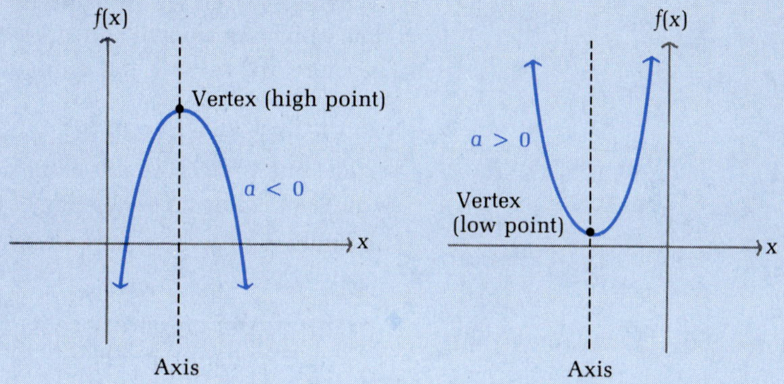

In addition to the point-by-point method of graphing quadratic functions described above, let us consider another approach that will give us added insight into these functions. (A brief review of completing the square, which is discussed in Appendix A-9, may prove useful first.) We illustrate the method through an example, and then generalize the results.

Consider the quadratic function given by

$$f(x) = 2x^2 - 8x + 5$$

If we can find the vertex of the graph, then the rest of the graph can be sketched with relatively few points. In addition, we will then have found the maximum or minimum value of the function. We start by transforming the equation into the form

$$f(x) = a(x - h)^2 + k \qquad a, h, k \text{ constants}$$

by completing the square:

$$\begin{aligned} f(x) &= 2x^2 - 8x + 5 \\ &= 2(x^2 - 4x) + 5 \end{aligned}$$

Factor the coefficient of x^2 out of the first two terms.

$f(x) = 2(x^2 - 4x + ?) + 5$	Complete the square within parentheses.
$\quad = 2(x^2 - 4x + 4) + 5 - 8$	We added 4 to complete the square inside the parentheses; but because of the 2 on the outside, we have actually added 8, so we must subtract 8.
$\quad = 2(x - 2)^2 - 3$	The transformation is complete.

Thus,

$$f(x) = \underbrace{2(x-2)^2}_{\text{Never negative (Why?)}} - 3$$

When $x = 2$, the first term on the right vanishes, and we add 0 to -3. For *any* other value of x we will add a positive number to -3, thus making $f(x)$ larger. Therefore, $f(2) = -3$ is the minimum value of $f(x)$ for all x. A very important result!

The point $(2, -3)$ is the lowest point on the parabola and is also the vertex. The vertical line $x = 2$ is the axis of the parabola. We plot the vertex and the axis and a couple of points on either side of the axis to complete the graph (Fig. 9).

x	$f(x)$
2	-3
1	-1
3	-1
0	5
4	5

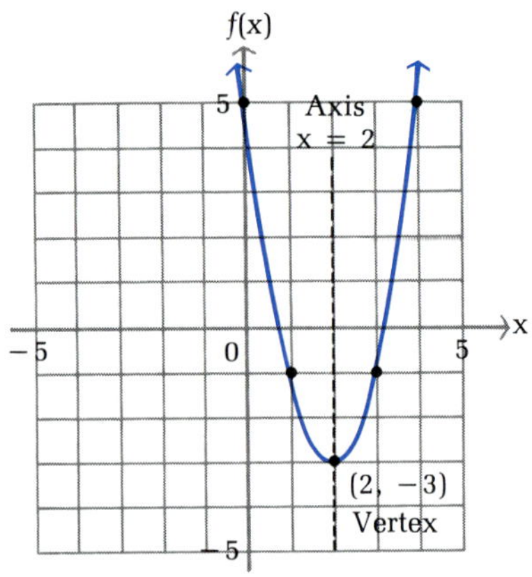

FIGURE 9

Examining the graph, we see that $f(x)$ can assume any value greater than or equal to -3, but no values less than -3. Thus, the range of f is the set of all y such that

$$y \geqslant -3 \qquad \text{or} \qquad [-3, \infty)$$

We also note that f has two x intercepts, which are solutions of the equation $f(x) = 0$. Since we have already completed the square, we use this method to find the x intercepts:

$$f(x) = 0$$
$$2(x - 2)^2 - 3 = 0$$

$$2(x-2)^2 = 3$$
$$(x-2)^2 = \tfrac{3}{2}$$
$$x - 2 = \pm\sqrt{\tfrac{3}{2}}$$
$$= \pm\tfrac{1}{2}\sqrt{6}$$
$$x = 2 \pm \tfrac{1}{2}\sqrt{6}$$

Thus, the x intercepts are

$$x = 2 + \tfrac{1}{2}\sqrt{6} \approx 3.22 \qquad \text{and} \qquad x = 2 - \tfrac{1}{2}\sqrt{6} \approx 0.78$$

For completeness, we observe that the y intercept is $f(0) = 5$, obtained as part of the table of values of $f(x)$.

Note all the important results we have with this approach:

1. Axis of the parabola
2. Vertex of the parabola
3. Minimum value of $f(x)$
4. Graph of $y = f(x)$
5. Range of f
6. x intercepts

If we start with the general quadratic function defined by

$$f(x) = ax^2 + bx + c \qquad a \neq 0$$

and complete the square (see Appendix A-9), we obtain

$$f(x) = a\left(x + \frac{b}{2a}\right)^2 + \frac{4ac - b^2}{4a}$$

Using the same reasoning as above, we obtain the following general results:

Properties of $f(x) = ax^2 + bx + c$, $a \neq 0$

1. Axis of symmetry: $x = -\dfrac{b}{2a}$
2. Vertex: $\left(-\dfrac{b}{2a}, f\left(-\dfrac{b}{2a}\right)\right)$
3. Maximum or minimum value of $f(x)$:

$$f\left(-\frac{b}{2a}\right) = \begin{cases} \text{Minimum} & \text{if } a > 0 \\ \text{Maximum} & \text{if } a < 0 \end{cases}$$

4. Domain: All real numbers
   Range: Determine from graph
5. y intercept: $f(0) = c$
   x intercepts: Solutions of $f(x) = 0$, if any exist

To graph a quadratic function using this method, we can actually complete the square as in the preceding example or use the properties in the box. Some of you can probably more readily remember a formula, others a process. We use the properties in the box in the next example.

◆ EXAMPLE 16 Graph, finding the axis, vertex, maximum or minimum of $f(x)$, range, and intercepts:

$$f(x) = -x^2 - 4x - 5$$

Solution Note that $a = -1$, $b = -4$, and $c = -5$.

$$\text{Axis of symmetry:} \quad x = -\frac{b}{2a} = -\frac{-4}{2(-1)} = -2$$

$$\text{Vertex:} \quad \left(-\frac{b}{2a}, f\left(-\frac{b}{2a}\right)\right) = (-2, f(-2)) = (-2, -1)$$

Maximum value of $f(x)$ (since $a = -1 < 0$): Max $f(x) = f(-2) = -1$

To graph f, locate the axis and vertex; then plot several points on either side of the axis:

x	$f(x)$
−4	−5
−3	−2
−2	−1
−1	−2
0	−5

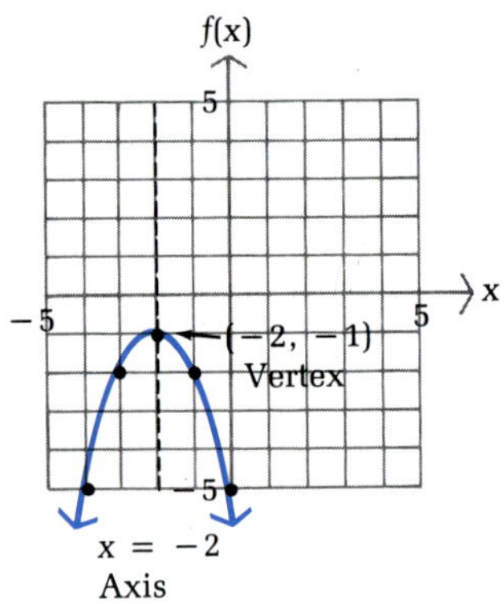

From the graph, we see that the range of f is

$$(-\infty, f(-2)] = (-\infty, -1]$$

The y intercept is $f(0) = -5$. Since the graph of f does not cross the x axis, there are no x intercepts. The equation $f(x) = 0$ has no real solutions (verify this). ◆

PROBLEM 16 Graph, finding the axis, vertex, maximum or minimum of $f(x)$, range, and intercepts:

$$f(x) = x^2 - 4x + 4$$ ◆

◆ PIECEWISE-DEFINED FUNCTIONS

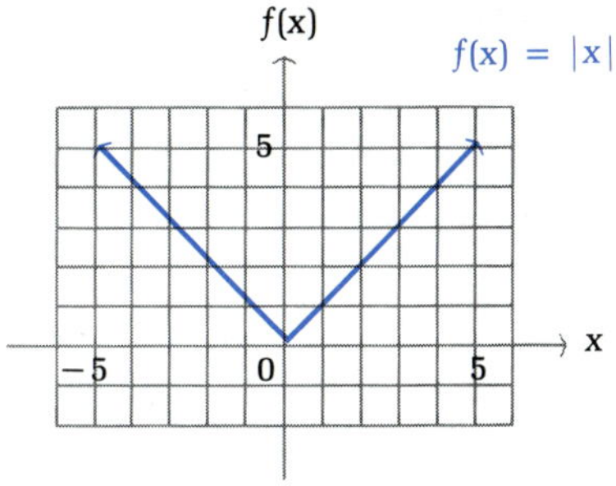

FIGURE 10

The **absolute value** of a real number a, denoted $|a|$, is the (positive) distance on a number line from a to the origin. Thus, $|4| = 4$ and $|-5| = 5$. More formally, the **absolute value function** is defined by

$$f(x) = |x| = \begin{cases} -x & \text{if } x < 0 \\ x & \text{if } x \geq 0 \end{cases}$$

The graph of f is shown in Figure 10. Notice that this function is defined by different formulas for different parts of its domain. A function whose definition involves more than one formula is called a **piecewise-defined function.** As the next example illustrates, piecewise-defined functions occur naturally in many applications.

◆ EXAMPLE 17

Service Charges

The time charges for a service call by a telephone company are \$16.20 for the first 6 minute period and \$5.40 for each additional 6 minute period (or fraction thereof). Let $C(x)$ be the total time charges for a service call that lasts x minutes. Graph C for $0 < x \leq 30$.

Solution

The total time charges are given by the following piecewise definition:

$$C(x) = \begin{cases} \$16.20 & \text{if } \ 0 < x \leq 6 \\ \$21.60 & \text{if } \ 6 < x \leq 12 \\ \$27.00 & \text{if } 12 < x \leq 18 \\ \text{And so on} \end{cases}$$

To graph C, we graph each rule in this definition for the indicated values of x:

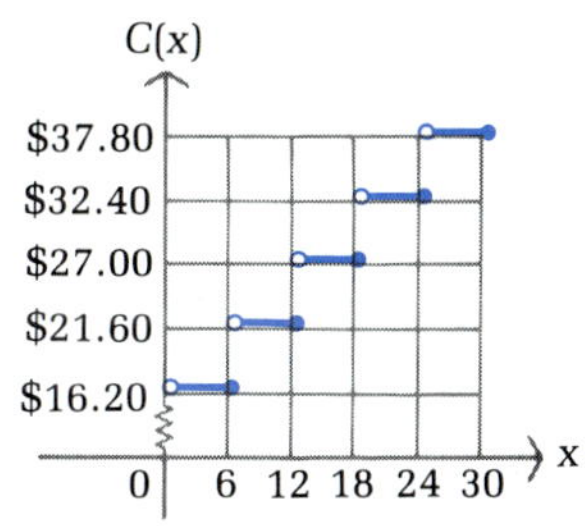

Note: A solid dot on the graph indicates that the point is part of the graph; an open dot indicates that the point is not part of the graph. ◆

PROBLEM 17

The time charges for a company that services home appliances are \$24 for the first 15 minute period and \$12 for each additional 15 minute period (or fraction thereof). Let $C(x)$ be the total time charges for a service call that lasts x minutes. Graph C for $0 < x \leq 60$. ◆

◆ APPLICATION: MARKET RESEARCH

The market research department of a company recommended to management that the company manufacture and market a promising new product. After extensive surveys, the research department backed up the recommendation with the **demand equation**

$$x = f(p) = 6{,}000 - 30p \tag{1}$$

where x is the number of units that retailers are likely to buy per month at \$p per unit. Notice that as the price goes up, the number of units goes down. From the financial department, the following **cost equation** was obtained:

$$C = g(x) = 72{,}000 + 60x \tag{2}$$

where \$72,000 is the fixed cost (tooling and overhead) and \$60 is the variable cost per unit (materials, labor, marketing, transportation, storage, etc.). The **revenue equation** (the amount of money, R, received by the company for selling x units at \$p per unit) is

$$R = xp \qquad \text{Revenue} = (\text{Number of units})(\text{Price per unit}) \tag{3}$$

And, finally, the **profit equation** is

$$P = R - C \qquad \text{Profit} = \text{Revenue} - \text{Cost} \tag{4}$$

where P is profit, R is revenue, and C is cost.

Notice that the cost equation (2) expresses C as a function of x and the demand equation (1) expresses x as a function of p. Substituting (1) into (2), we obtain cost C as a linear function of price p:

$$\begin{aligned} C &= 72{,}000 + 60(6{,}000 - 30p) \\ &= 432{,}000 - 1{,}800p \end{aligned} \qquad \text{Linear function} \tag{5}$$

Similarly, substituting (1) into (3), we obtain revenue R as a quadratic function of price p:

$$\begin{aligned} R &= (6{,}000 - 30p)p \\ &= 6{,}000p - 30p^2 \end{aligned} \qquad \text{Quadratic function} \tag{6}$$

Now let us graph equations (5) and (6) in the same coordinate system. We obtain Figure 11. Notice how much information is contained in this graph.

Let us compute the **break-even points;** that is, the prices at which cost equals revenue (the points of intersection of the two graphs in Figure 11). Find p so that

$$\begin{aligned} C &= R \\ 432{,}000 - 1{,}800p &= 6{,}000p - 30p^2 \\ 30p^2 - 7{,}800p + 432{,}000 &= 0 \\ p^2 - 260p + 14{,}400 &= 0 \\ p &= \frac{260 \pm \sqrt{260^2 - 4(14{,}400)}}{2} \\ &= \frac{260 \pm 100}{2} \\ &= \$80, \quad \$180 \end{aligned}$$

Solve using the quadratic formula (Appendix A-9).

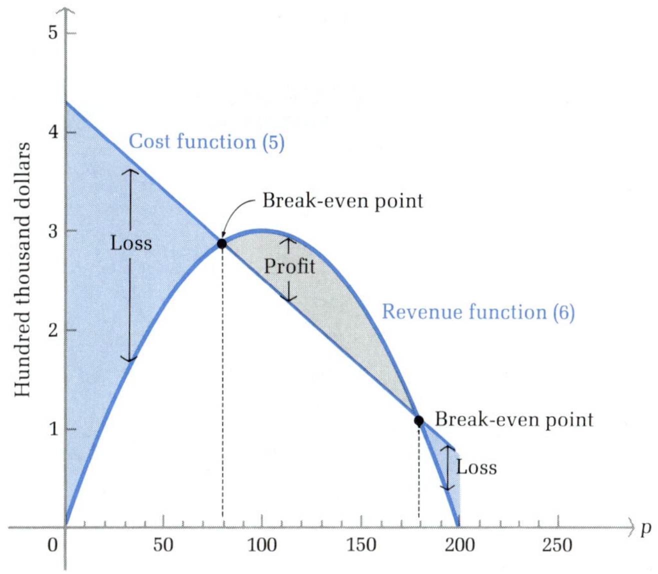

FIGURE 11

Thus, at a price of \$80 or \$180 per unit the company will break even. Between these two prices it is predicted that the company will make a profit.

At what price will a maximum profit occur? To find out, we write

$$\begin{aligned} P &= R - C \\ &= (6{,}000p - 30p^2) - (432{,}000 - 1{,}800p) \\ &= -30p^2 + 7{,}800p - 432{,}000 \end{aligned}$$

Since this is a quadratic function, the maximum profit occurs at

$$p = -\frac{b}{2a} = -\frac{7{,}800}{2(-30)} = \$130$$

Note that this is not the price at which the maximum revenue occurs. The latter occurs at $p = \$100$, as shown in Figure 11.

Answers to Matched Problems

15. Slope: $\frac{1}{3}$
y intercept: 1
x intercept: -3

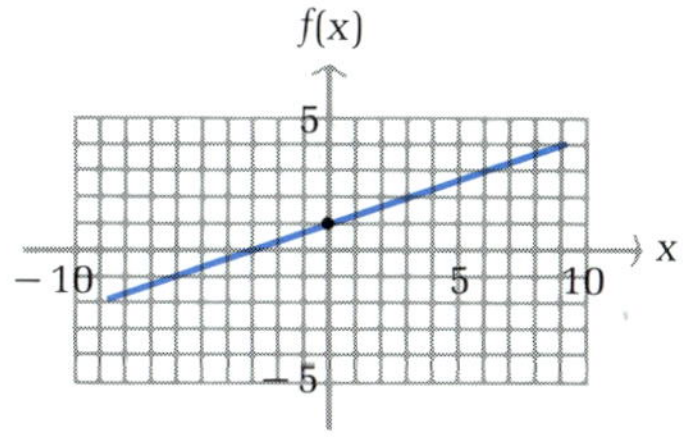

16. Minimum: $f(2) = 0$
Range: $[0, \infty)$
y intercept: 4
x intercept: 2

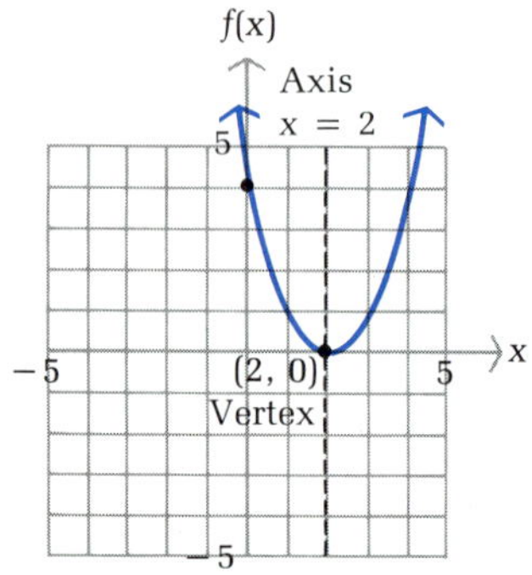

17.

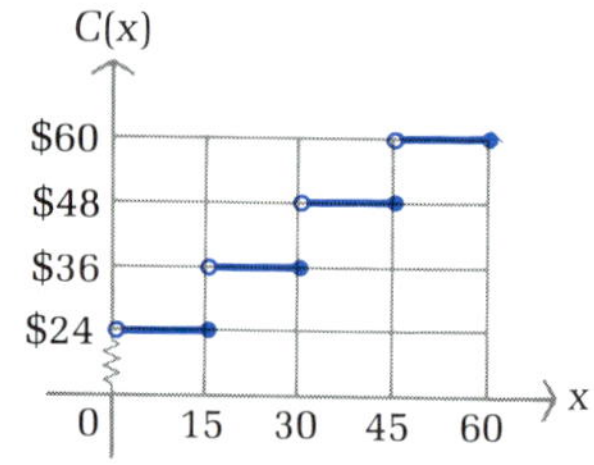

EXERCISE 1-3

A *Graph each linear function, and indicate its slope and intercepts.*

1. $f(x) = 2x - 4$
2. $g(x) = \dfrac{x}{2}$
3. $h(x) = 4 - 2x$
4. $f(x) = -\dfrac{x}{2} + 3$
5. $g(x) = -\frac{2}{3}x + 4$
6. $f(x) = 3$

Find a linear function with the indicated slope and y intercept.

7. Slope -2, y intercept 6
8. Slope 3, y intercept -5

Find a linear function whose graph passes through the indicated points.

9. $(-1, 5)$, $(5, 2)$
10. $(1, -2)$, $(7, 6)$

B *Graph, finding the axis, vertex, maximum or minimum, intercepts, and range.*

11. $f(x) = (x - 3)^2 - 1$
12. $g(x) = -(x + 2)^2 + 4$
13. $h(x) = -(x + 1)^2 + 9$
14. $k(x) = (x - 2)^2 - 16$
15. $f(x) = x^2 + 8x + 16$
16. $h(x) = x^2 - 2x - 3$
17. $f(u) = u^2 - 2u + 4$
18. $f(x) = x^2 - 10x + 25$
19. $h(x) = 2 + 4x - x^2$
20. $g(x) = -x^2 - 6x - 4$
21. $f(x) = 6x - x^2$
22. $G(x) = 16x - 2x^2$
23. $F(s) = s^2 - 4$
24. $g(t) = t^2 + 4$
25. $F(x) = 4 - x^2$
26. $G(x) = 9 - x^2$

Graph each function, and state its domain and range.

27. $f(x) = \begin{cases} 1 & \text{if } 0 \le x \le 2 \\ 3 & \text{if } 2 < x \le 3 \\ 5 & \text{if } 3 < x \le 5 \end{cases}$

28. $g(x) = \begin{cases} -2 & \text{if } -4 \le x \le -2 \\ 1 & \text{if } -2 < x < 2 \\ 2 & \text{if } 2 \le x \le 4 \end{cases}$

29. $f(x) = \begin{cases} x & \text{if } -2 \leq x < 1 \\ -x + 2 & \text{if } 1 \leq x \leq 2 \end{cases}$

30. $f(x) = \begin{cases} x + 1 & \text{if } -1 \leq x < 0 \\ -x + 1 & \text{if } 0 \leq x \leq 1 \end{cases}$

31. $h(x) = \begin{cases} -x^2 - 2 & \text{if } x < 0 \\ x^2 + 2 & \text{if } x > 0 \end{cases}$

32. $g(x) = \begin{cases} x^2 + 1 & \text{if } x < 0 \\ -x^2 - 1 & \text{if } x > 0 \end{cases}$

33. $G(x) = \begin{cases} -3 & \text{if } x < -3 \\ x & \text{if } -3 \leq x \leq 3 \\ 3 & \text{if } x > 3 \end{cases}$

34. $F(x) = \begin{cases} 1 & \text{if } x < -2 \\ 5 - x^2 & \text{if } -2 \leq x \leq 2 \\ 1 & \text{if } x > 2 \end{cases}$

C *Graph, finding the axis, vertex, maximum or minimum, intercepts, and range.*

35. $f(x) = x^2 - 7x + 10$

36. $g(t) = t^2 - 5t + 2$

37. $h(x) = 2 - 5x - x^2$

38. $g(t) = 4 + 3t - t^2$

Graph f and g on the same set of axes and find any points of intersection.

39. $f(x) = 2 + \frac{1}{2}x; \quad g(x) = 8 - x$

40. $f(x) = -2x - 1; \quad g(x) = x - 4$

41. $f(x) = x^2 - 6x + 8; \quad g(x) = x - 2$

42. $f(x) = 8 + 2x - x^2; \quad g(x) = x + 6$

43. $f(x) = 11 + 2x - x^2; \quad g(x) = x^2 - 1$

44. $f(x) = x^2 - 4x - 10; \quad g(x) = 14 - 2x - x^2$

APPLICATIONS

Business & Economics

45. **Service charges.** On weekends and holidays, an emergency plumbing repair service charges \$60 for the first 30 minute period (or fraction thereof) of a service call and \$20 for each additional 15 minute period (or fraction thereof). Let $C(x)$ be the cost of a service call that lasts x minutes. Graph $C(x)$ for $0 < x \leq 90$.

46. **Delivery charges.** A nationwide package delivery service charges \$20.00 for overnight delivery of packages weighing 1 pound or less. Each additional pound (or fraction thereof) costs an additional \$2.50. Let $D(x)$ be the charge for overnight delivery of a package weighing x pounds. Graph $D(x)$ for $0 < x \leq 10$.

47. **Market research.** Suppose that in the market research example in this section the demand equation (1) is changed to $x = 9{,}000 - 30p$ and the cost equation (2) is changed to $C = 90{,}000 + 30x$.

    (A) Express cost C as a linear function of price p.
    (B) Express revenue R as a quadratic function of price p.
    (C) Graph the cost and revenue functions found in parts A and B in the same coordinate system, and identify the regions of profit and loss on your graph.
    (D) Find the break-even points; that is, find the prices to the nearest dollar at which $R = C$. (A calculator might prove useful here.)
    (E) Find the price that produces the maximum revenue.
    (F) Find the price that produces the maximum profit.

48. *Market research.* Repeat Problem 47 if the demand equation (1) is changed to $x = 5{,}000 - 50p$ and the cost equation (2) is changed to $C = 40{,}000 + 12x$.

49. *Market research.* Repeat Problem 47 if the demand equation (1) is changed to $x = 8{,}000 - 40p$ and the cost equation (2) is changed to $C = 100{,}000 + 20x$.

50. *Market research.* Repeat Problem 47 if the demand equation (1) is changed to $x = 12{,}000 - 60p$ and the cost equation (2) is changed to $C = 120{,}000 + 30x$.

Problems 51–54 require the use of a graphic calculator or a computer.

51. *Market research.* Use approximation techniques to answer part D in Problem 47.

52. *Market research.* Use approximation techniques to answer part D in Problem 48.

53. *Market research.* Use approximation techniques to answer part D in Problem 49.

54. *Market research.* Use approximation techniques to answer part D in Problem 50.

55. *Break-even analysis.* A publisher is planning to produce a new textbook. The fixed costs (reviewing, editing, typesetting, etc.) are \$240,000, and the variable costs (printing, sales commissions, etc.) are \$20 per book. The wholesale price (the amount received by the publisher) will be \$35 per book. Let x be the number of books.

(A) Express the cost C as a linear function of x.
(B) Express the revenue R as a linear function of x.
(C) Graph the cost and revenue functions found in parts A and B on the same set of axes.
(D) Find the number of books the publisher has to sell in order to break even.

56. *Break-even analysis.* A computer software company is planning to market a new word processor for a microcomputer. The fixed costs (programming, debugging, etc.) are \$300,000, and the variable costs (disk duplication, manual production, etc.) are \$25 per unit. The wholesale price of the product will be \$100 per unit. Let x be the number of units.

(A) Express the cost C as a linear function of x.
(B) Express the revenue R as a linear function of x.
(C) Graph the cost and revenue functions found in parts A and B on the same set of axes.
(D) Find the number of units the company has to sell in order to break even.

Life Sciences

57. Medicine. The French physician Poiseuille was the first to discover that blood flows faster near the center of an artery than near the edge. Experimental evidence has shown that the rate of flow v (in centimeters per second) at a point x centimeters from the center of an artery (see the figure) is given by

$$v = f(x) = 1{,}000(0.04 - x^2) \qquad 0 \leq x \leq 0.2$$

Graph this quadratic function for the indicated values of x.

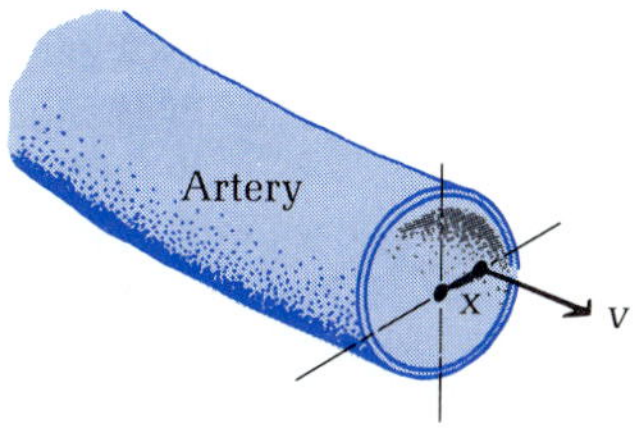

58. *Air pollution.* On an average summer day in a large city, the pollution index at 8:00 AM is 20 parts per million, and it increases linearly by 15 parts per million each hour until 3:00 PM. Let $P(x)$ be the amount of pollutants in the air x hours after 8:00 AM.

(A) Express $P(x)$ as a linear function of x.
(B) What is the air pollution index at 1:00 PM?
(C) Graph the function P for $0 \leq x \leq 7$.
(D) What is the slope of the graph? (The slope is the amount of increase in pollution for each additional hour of time.)

Social Sciences

59. *Psychology—sensory perception.* One of the oldest studies in psychology concerns the following question: Given a certain level of stimulation (light, sound, weight lifting, electric shock, and so on), how much should the stimulation be increased for a person to notice the difference? In the middle of the nineteenth century, E. H. Weber (a German physiologist) formulated a law that still carries his name: If Δs is the change in stimulus that will just be noticeable at a stimulus level s, then the ratio of Δs to s is a constant:

$$\frac{\Delta s}{s} = k$$

Hence, the amount of change that will be noticed is a linear function of the stimulus level, and we note that the greater the stimulus, the more it takes to notice a difference. In an experiment on weight lifting, the constant k for a given individual was found to be $\frac{1}{30}$.

(A) Find Δs (the difference that is just noticeable) at the 30 pound level; at the 90 pound level.
(B) Graph $\Delta s = s/30$ for $0 \leq s \leq 120$.
(C) What is the slope of the graph?

SECTION 1-4 Aids to Graphing Functions

- GRAPHS OF BASIC FUNCTIONS
- GRAPHING AIDS

In the preceding section, we used point-by-point plotting to obtain the basic shape of the graph of a parabola, and then we used this basic shape and properties of quadratic functions to graph other parabolas. In this section, we use point-by-point plotting to establish the basic shapes of the graphs of some additional functions. Then we develop methods that will enable us to use our knowledge of the graph of a given function as an aid in graphing related functions.

◆ GRAPHS OF BASIC FUNCTIONS

In order to apply the graphing aids discussed in this section, we need to be familiar with the graphs of some basic functions. The graphs of these basic functions can then be used to form the graphs of more complicated functions. To begin, let us review the graphs of some functions we have already considered. The graphs of the **identity function** (a linear function), the **squaring function** (a quadratic function), and the **absolute value function** (a piecewise-defined function) are shown in Figure 12.

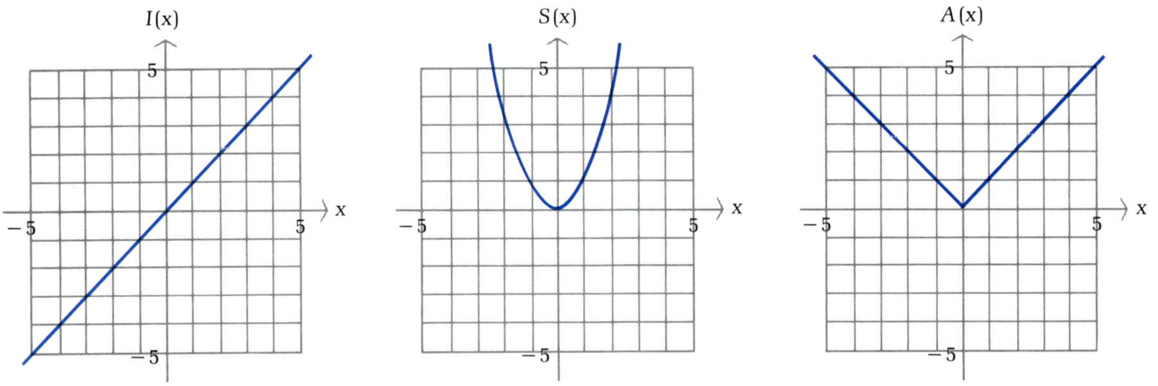

(A) Identity function: $I(x) = x$ Domain: R Range: R

(B) Squaring function: $S(x) = x^2$ Domain: R Range: $[0, \infty)$

(C) Absolute value function: $A(x) = |x|$ Domain: R Range: $[0, \infty)$

FIGURE 12

Now we want to consider the graphs of three new functions: the **square root function,** the **cubing function,** and the **cube root function:**

SQUARE ROOT FUNCTION:	$R(x) = \sqrt{x}$
CUBING FUNCTION:	$C(x) = x^3$
CUBE ROOT FUNCTION:	$T(x) = \sqrt[3]{x}$

The graphs of these basic functions are illustrated in Figure 13.

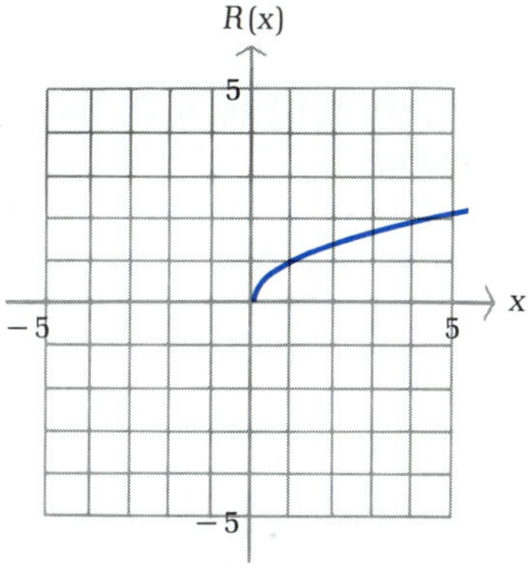

(A) Square root function:
$R(x) = \sqrt{x}$
Domain: $[0, \infty)$
Range: $[0, \infty)$

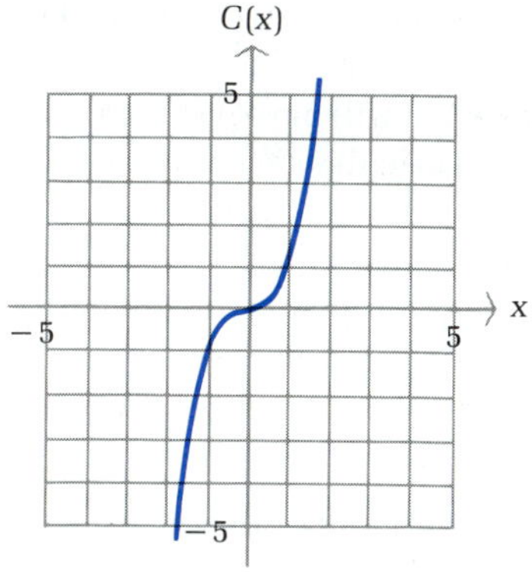

(B) Cubing function:
$C(x) = x^3$
Domain: R
Range: R

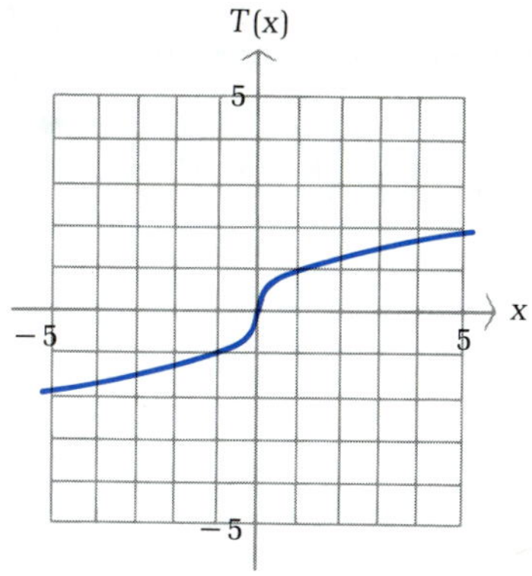

(C) Cube root function:
$T(x) = \sqrt[3]{x}$
Domain: R
Range: R

FIGURE 13

Each of the graphs in Figure 13 was obtained using point-by-point plotting, as illustrated in the following example.

◆ EXAMPLE 18 Use point-by-point plotting to graph $R(x) = \sqrt{x}$ for $0 \le x \le 5$.

Solution First, we use a calculator to construct a table of values of $R(x)$ (values are rounded to two decimal places for graphing purposes). The more points we include in this table, the more accurate our graph will be. Then we plot these points and connect them with a smooth curve.

x	R(x)
0	0
0.2	0.45
0.4	0.63
0.6	0.77
0.8	0.89
1	1
2	1.41
3	1.73
4	2
5	2.24

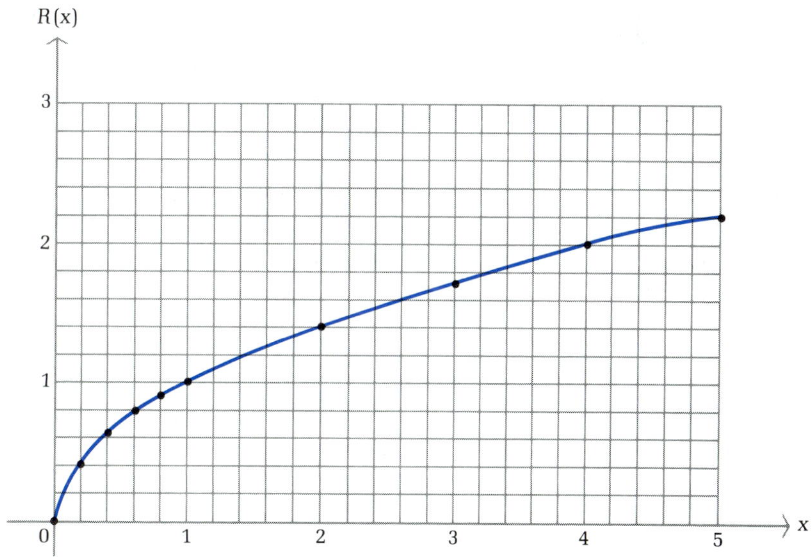

Notice that including additional points between $x = 0$ and $x = 1$ provides important information about the shape of the graph near $x = 0$, where the graph is very steep. We will have more to say about this type of graph later in the book, when we use calculus techniques as an aid to graphing. ◆

PROBLEM 18 Use point-by-point plotting to graph the functions $C(x) = x^3$ and $T(x) = \sqrt[3]{x}$. Include sufficient points between $x = -1$ and $x = 1$ to obtain graphs that have the same basic shapes as the graphs in Figures 13B and 13C. ◆

Carefully study the graphs of the six basic functions illustrated in Figures 12 and 13 until you can draw each graph without reference to the figure and without extensive point-by-point plotting.

◆ GRAPHING AIDS

Now that you are familiar with the graphs of some basic functions, we want to use this knowledge as an aid in graphing related functions. We will consider relationships of the forms

$$p(x) = f(x) + k \qquad q(x) = f(x + k) \qquad r(x) = kf(x)$$

where f is a function whose graph is known and k is a constant. We illustrate these graphing aids through examples.

◆ EXAMPLE 19 Graph $R(x) = \sqrt{x}$, $p(x) = \sqrt{x} + 1$, and $q(x) = \sqrt{x} - 4$.

Solution We already know the basic shape of the graph of the square root function, as shown in part (A) of the figure on page 49. Since R and p are related by

$$p(x) = \sqrt{x} + 1 = R(x) + 1$$

in order to graph $p(x) = \sqrt{x} + 1$, we simply add 1 to each ordinate value (second coordinate) of the graph of $R(x) = \sqrt{x}$. The result is shown in part (B) of the figure.

Since q and R are related by

$$q(x) = \sqrt{x} - 4 = R(x) - 4$$

in order to graph $q(x) = \sqrt{x} - 4$, we simply subtract 4 from each ordinate value. The graph of $q(x)$ is shown in part (C) of the figure.

Thus, the graph of $p(x) = \sqrt{x} + 1$ is just the graph of $R(x) = \sqrt{x}$ shifted upward one unit, and the graph of $q(x) = \sqrt{x} - 4$ is just the graph of $R(x) = \sqrt{x}$ shifted downward four units.

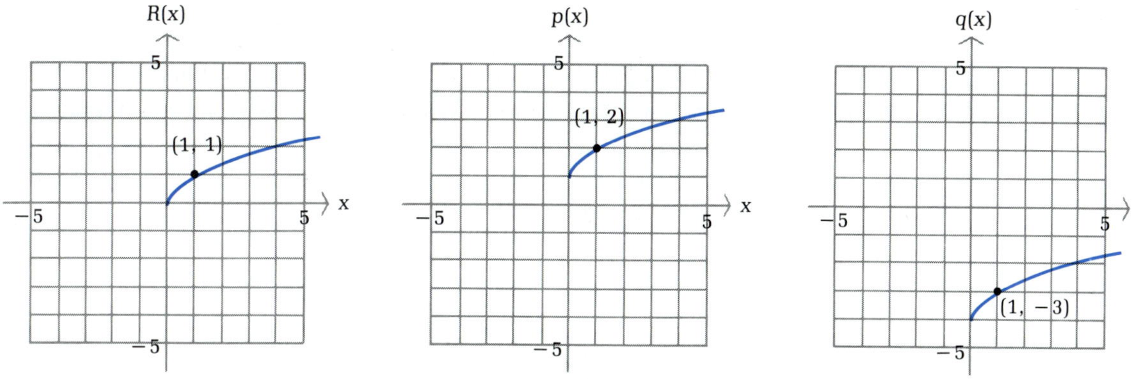

(A) $R(x) = \sqrt{x}$

(B) $p(x) = \sqrt{x} + 1$
The graph of $p(x)$ is the same as the graph of $R(x) = \sqrt{x}$ shifted up one unit.

(C) $q(x) = \sqrt{x} - 4$
The graph of $q(x)$ is the same as the graph of $R(x) = \sqrt{x}$ shifted down four units. ◆

In general, given a function f and a positive constant k:

The graph of $y = f(x) + k$ is the graph of $y = f(x)$ shifted upward k units, and the graph of $y = f(x) - k$ is the graph of $y = f(x)$ shifted downward k units.

Shifting a graph upward or downward is called a **vertical shift** or a **vertical translation.**

PROBLEM 19 Graph $T(x) = \sqrt[3]{x}$, $u(x) = \sqrt[3]{x} + 3$, and $v(x) = \sqrt[3]{x} - 2$. ◆

We now turn to graphs formed by shifting a graph horizontally.

◆ EXAMPLE 20 Graph $C(x) = x^3$, $m(x) = (x - 3)^3$, and $n(x) = (x + 2)^3$.

Solution The functions C and m are related by

$$m(x) = (x - 3)^3 = C(x - 3)$$

If (a, a^3) is any point on the graph of $C(x) = x^3$ [see part (A) of the figure on the next page], then

$$m(a + 3) = (a + 3 - 3)^3 = a^3$$

implies that $(a + 3, a^3)$ is a point on the graph of $m(x) = (x - 3)^3$ with the same ordinate as (a, a^3), as shown in part (B) of the figure. Thus, the graph of $m(x) = (x - 3)^3$ is the same as the graph of $C(x) = x^3$ shifted three units to the right.

Using the same reasoning,

$$n(a-2) = (a-2+2)^3 = a^3$$

implies that $(a-2, a^3)$ is a point on the graph of $n(x) = (x+2)^3$ with the same ordinate as the point (a, a^3) on the graph of $C(x) = x^3$. Thus, the graph of $n(x) = (x+2)^3$ is the same as the graph of $C(x) = x^3$ shifted two units to the left, as shown in part (C) of the figure.

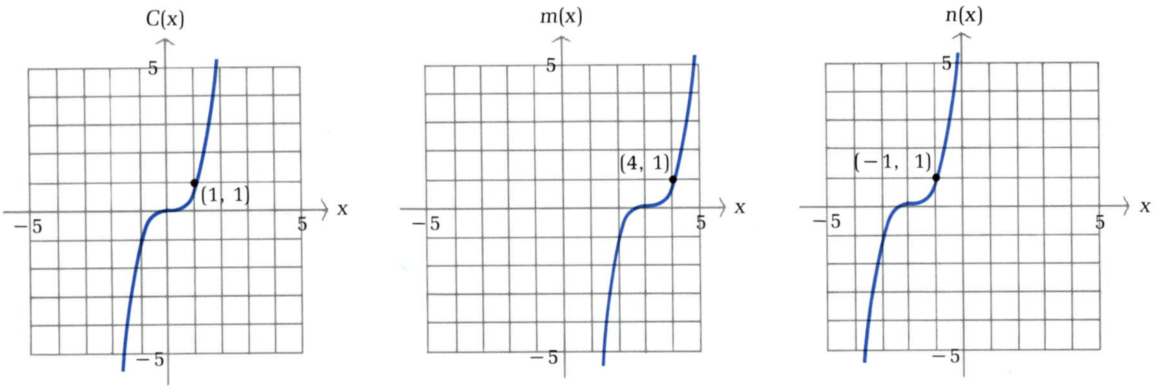

(A) $C(x) = x^3$

(B) $m(x) = (x-3)^3$
The graph of $m(x)$ is the same as the graph of $C(x) = x^3$ shifted right three units.

(C) $n(x) = (x+2)^3$
The graph of $n(x)$ is the same as the graph of $C(x) = x^3$ shifted left two units. ◆

In general, given a function f and a positive constant h:

The graph of $y = f(x-h)$ is the graph of $y = f(x)$ shifted h units to the right, and the graph of $y = f(x+h)$ is the graph of $y = f(x)$ shifted h units to the left.

Shifting a graph left or right is called a **horizontal shift** or a **horizontal translation.**

PROBLEM 20 Graph $I(x) = x$, $p(x) = x + 1$, and $q(x) = x - 2$. ◆

Horizontal and vertical translations shift a graph, but they do not change its shape. Now we will consider some operations that change the shape of a graph.

◆ EXAMPLE 21 Graph $A(x) = |x|$, $u(x) = -|x|$, $v(x) = 2|x|$, and $w(x) = \frac{1}{2}|x|$.

Solution The functions A and u are related by

$$u(x) = -|x| = -A(x)$$

If we take the negative of each ordinate value on the graph of $A(x) = |x|$ [part (A) of the figure below], we obtain the graph of $u(x) = -|x|$ [part (B) of the figure]. The graph of u is the **reflection** of the graph of A in the x axis.

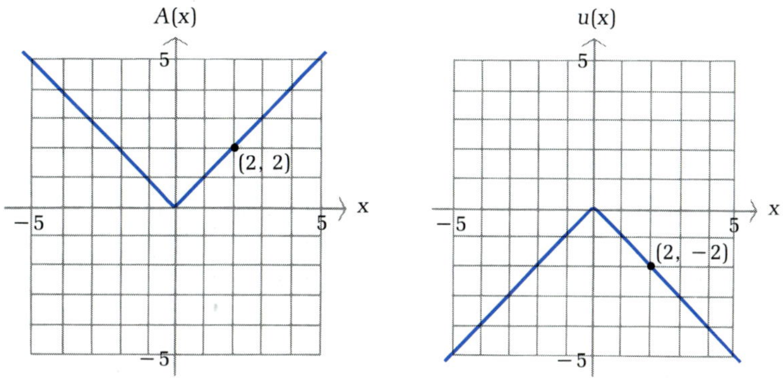

(A) $A(x) = |x|$

(B) $u(x) = -|x|$
Reflection

The functions v and w are related to A by

$$v(x) = 2|x| = 2A(x) \qquad \text{and} \qquad w(x) = \tfrac{1}{2}|x| = \tfrac{1}{2}A(x)$$

If we **expand** the graph of $A(x) = |x|$ by multiplying each ordinate value by 2, we obtain the graph of $v(x) = 2|x|$ [part (C) of the figure below]. If we **contract** the graph of $A(x) = |x|$ by multiplying each ordinate value by $\frac{1}{2}$, we obtain the graph of $w(x) = \frac{1}{2}|x|$ [part (D) of the figure].

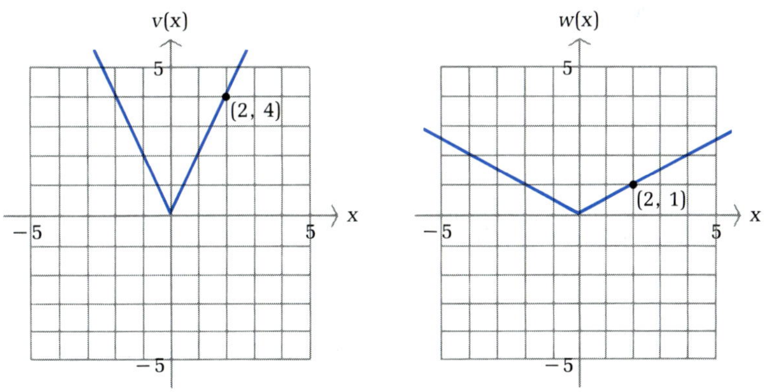

(C) $v(x) = 2|x|$
Expansion

(D) $w(x) = \frac{1}{2}|x|$
Contraction

◆

In general:

The graph of $y = -f(x)$ is the reflection of the graph of $y = f(x)$ in the x axis. The graph of $y = Cf(x)$ is the graph of $y = f(x)$ expanded by a factor of C if $C > 1$, and contracted by a factor of C if $0 < C < 1$.

PROBLEM 21 Graph $S(x) = x^2$, $n(x) = -x^2$, $p(x) = 2x^2$, and $q(x) = \frac{1}{2}x^2$. ◆

The results illustrated in the preceding examples are summarized in the box for convenient reference.

Graphing Aids

VERTICAL TRANSLATION	
$y = f(x) + k,\quad k > 0$	Shift the graph of $y = f(x)$ upward k units.
$y = f(x) - k,\quad k > 0$	Shift the graph of $y = f(x)$ downward k units.
HORIZONTAL TRANSLATION	
$y = f(x - h),\quad h > 0$	Shift the graph of $y = f(x)$ to the right h units.
$y = f(x + h),\quad h > 0$	Shift the graph of $y = f(x)$ to the left h units.
REFLECTION	
$y = -f(x)$	Reflect the graph of $y = f(x)$ in the x axis.
EXPANSION AND CONTRACTION	
$y = Cf(x),\quad C > 1$	Expand the graph of $y = f(x)$ by multiplying each ordinate value by C.
$y = Cf(x),\quad 0 < C < 1$	Contract the graph of $y = f(x)$ by multiplying each ordinate value by C.

More complicated functions often can be graphed by applying several of these graphing aids. The next two examples illustrate this process.

◆ EXAMPLE 22 Graph $f(x) = 3 - \sqrt[3]{x}$.

Solution To graph $f(x) = 3 - \sqrt[3]{x}$, we first reflect the graph of $T(x) = \sqrt[3]{x}$ [part (A) of the figure at the top of the next page] across the x axis, as shown in part (B) of the figure. Then we shift the graph of $g(x) = -\sqrt[3]{x}$ up three units, as shown in part (C).

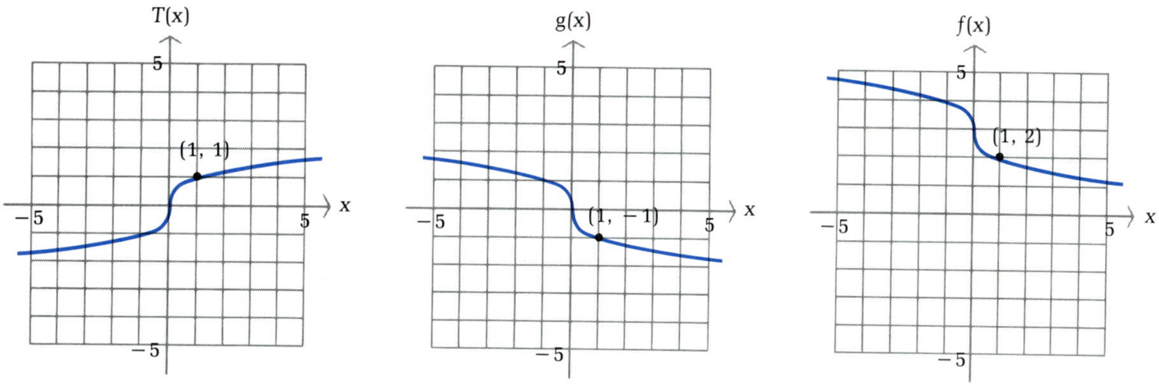

(A) $T(x) = \sqrt[3]{x}$ (B) $g(x) = -\sqrt[3]{x}$ (C) $f(x) = 3 - \sqrt[3]{x}$

PROBLEM 22 Graph $f(x) = 2 - |x|$.

EXAMPLE 23 Graph $f(x) = 2\sqrt{x+1}$.

Solution To graph $f(x) = 2\sqrt{x+1}$, we first shift the graph of $R(x) = \sqrt{x}$ [part (A) of the figure below] one unit to the left, as shown in part (B) of the figure. Then we multiply the ordinate values by 2, as shown in part (C).

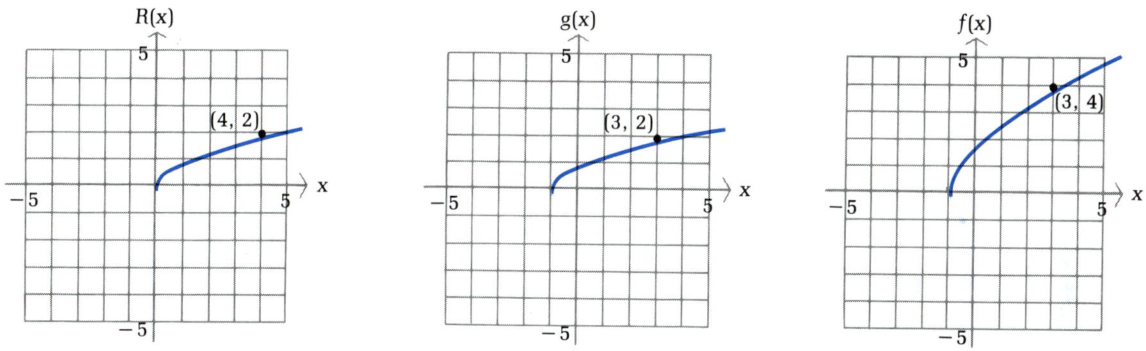

(A) $R(x) = \sqrt{x}$ (B) $g(x) = \sqrt{x+1}$ (C) $f(x) = 2\sqrt{x+1}$

PROBLEM 23 Graph $f(x) = \frac{1}{2}(x - 2)^3$.

Answers to Matched Problems

18. See parts (B) and (C) of Figure 13.

19.

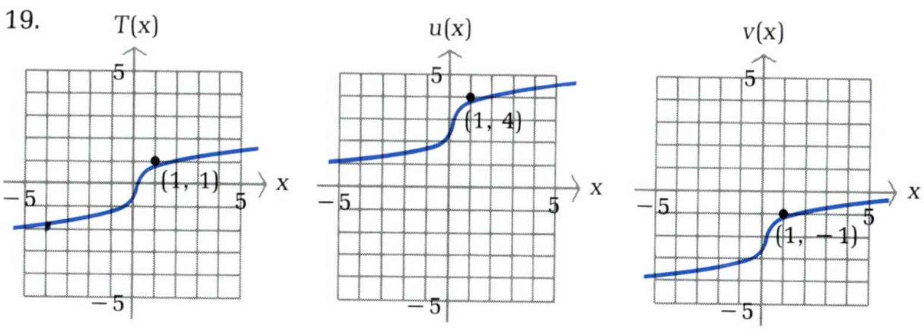

20.

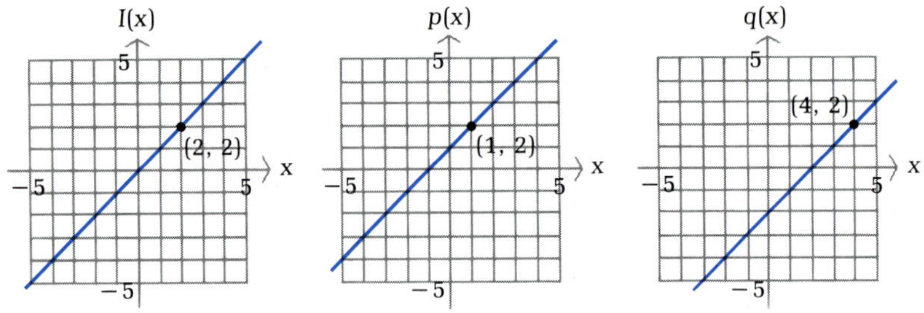

21.

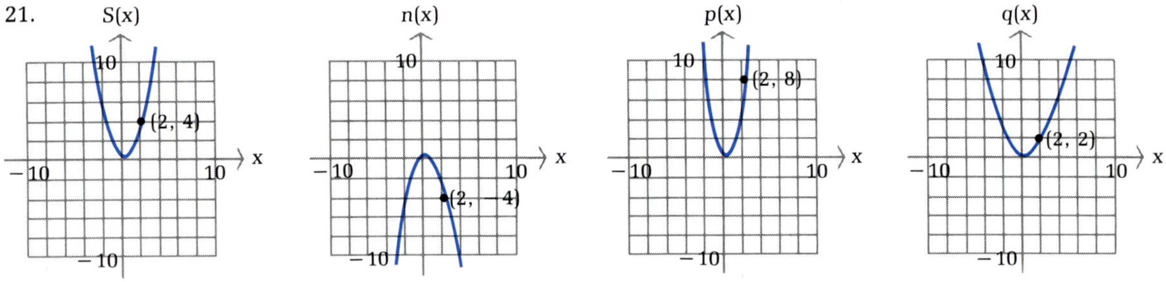

22.

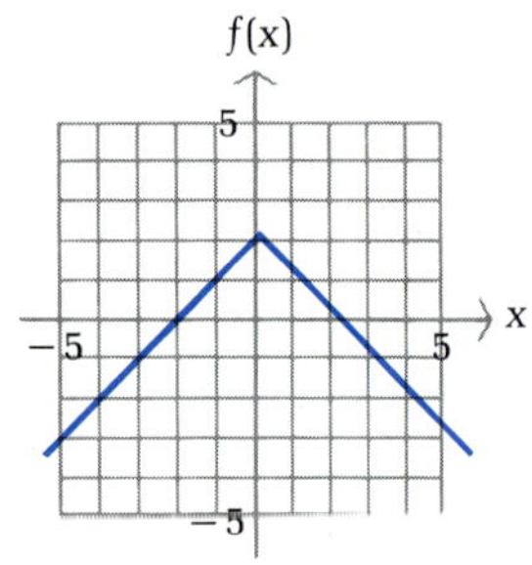

23.

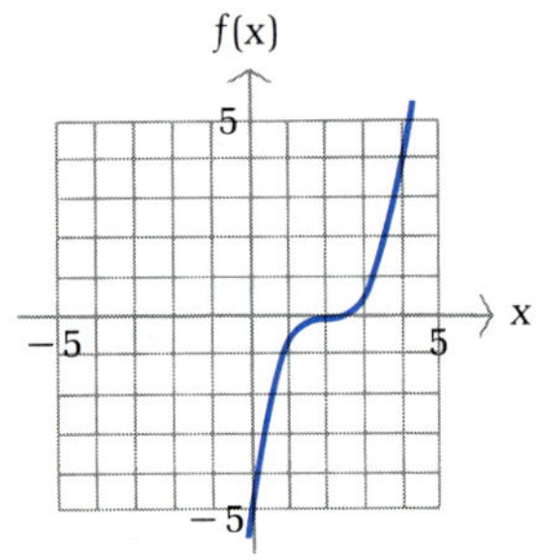

EXERCISE 1-4

A *Problems 1–12 refer to the functions f and g given by the following graphs (the domain of each function is $[-4, 4]$):*

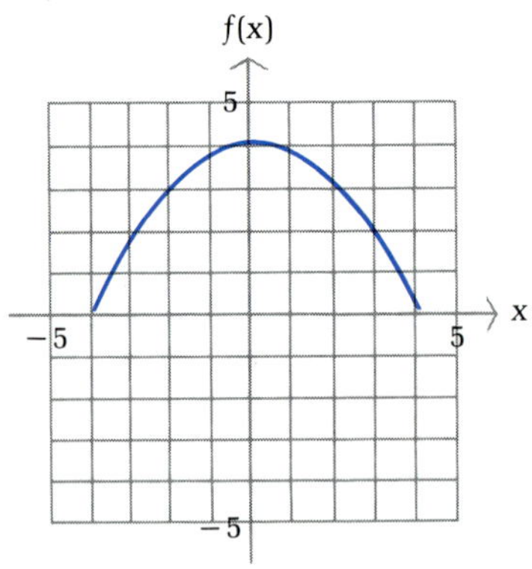

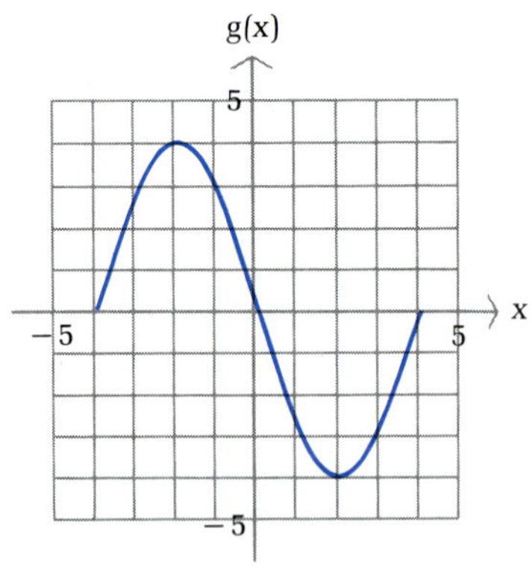

Use the graph of f or g, as appropriate, to graph each of the following:

1. $f(x) + 4$ **2.** $f(x) - 3$ **3.** $g(x) - 4$ **4.** $g(x) + 5$
5. $f(x - 2)$ **6.** $f(x + 5)$ **7.** $g(x + 4)$ **8.** $g(x - 3)$
9. $-f(x)$ **10.** $-g(x)$ **11.** $\frac{3}{2}f(x)$ **12.** $\frac{5}{4}g(x)$

B

Graph as indicated.

13. $f(x) = x^2, \quad p(x) = x^2 + 2, \quad q(x) = x^2 - 4$
14. $f(x) = |x|, \quad p(x) = |x| + 3, \quad q(x) = |x| - 1$
15. $f(x) = \sqrt{x}, \quad p(x) = \sqrt{x + 3}, \quad q(x) = \sqrt{x - 2}$
16. $f(x) = \sqrt[3]{x}, \quad p(x) = \sqrt[3]{x + 1}, \quad q(x) = \sqrt[3]{x - 3}$
17. $f(x) = x^3, \quad u(x) = -x^3, \quad v(x) = 2x^3, \quad w(x) = \frac{1}{2}x^3$
18. $f(x) = \sqrt{x}, \quad u(x) = -\sqrt{x}, \quad v(x) = 3\sqrt{x}, \quad w(x) = \frac{3}{4}\sqrt{x}$

Indicate how the graph of each function is related to the graph of one of the six basic functions in Figures 12 and 13. Graph each function.

19. $f(x) = -(x - 2)^2$ **20.** $f(x) = -(x + 3)^2$
21. $f(x) = |x + 1| + 2$ **22.** $f(x) = |x - 4| + 1$
23. $f(x) = 4\sqrt{x} - 4$ **24.** $f(x) = 3\sqrt{x} - 3$

Graph each function using the aids to graphing discussed in this section.

25. $f(x) = -\frac{1}{2}(x + 2)^2$ **26.** $g(x) = -3|x + 2|$
27. $p(x) = -2|x - 1| + 2$ **28.** $q(x) = \frac{1}{2}(x - 1)^2 - 2$
29. $u(x) = -\frac{1}{3}\sqrt[3]{x}$ **30.** $v(x) = -\frac{1}{4}x^3$
31. $w(x) = \frac{1}{2}x^3 - 2$ **32.** $m(x) = 2\sqrt[3]{x} + 1$
33. $f(x) = 4 - 2\sqrt{x - 1}$ **34.** $g(x) = 3 - 2\sqrt{x + 3}$

C

In Problems 35–38, use the method of completing the square to transform each quadratic function f into the form

$$f(x) = C(x + k)^2 + h$$

where C, k, and h are constants. Indicate how the graph of f is related to the graph of the squaring function $S(x) = x^2$. Graph $y = f(x)$.

35. $f(x) = 2x^2 - 12x + 9$ **36.** $f(x) = 2x^2 + 4x + 3$
37. $f(x) = -\frac{1}{2}x^2 + 4x - 3$ **38.** $f(x) = -\frac{1}{3}x^2 - 2x + 1$

In Problems 39–42, use point-by-point plotting to graph $y = f(x)$ for the indicated values of x. Then use the graphing properties discussed in this section to graph $y = m(x)$ and $y = n(x)$.

39. $f(x) = x^4, \quad -2 \leq x \leq 2; \quad m(x) = 15 - x^4, \quad n(x) = \frac{1}{2}(x - 2)^4$
40. $f(x) = x^{3/2}, \quad 0 \leq x \leq 4; \quad m(x) = 4 - x^{3/2}, \quad n(x) = \frac{3}{4}(x + 1)^{3/2}$
41. $f(x) = x^{5/3}, \quad -3 \leq x \leq 3; \quad m(x) = 2 - x^{5/3}, \quad n(x) = \frac{1}{2}(x - 1)^{5/3}$
42. $f(x) = x^{4/3}, \quad -3 \leq x \leq 3; \quad m(x) = 1 - x^{4/3}, \quad n(x) = \frac{3}{2}(x + 2)^{4/3}$

Problems 43–54 require the use of a graphic calculator. Use [−10, 10] for both the x range and the y range for all graphs.

In Problems 43–46, graph $f(x)$, $f(|x|)$, and $f(-|x|)$ in separate viewing rectangles.

43. $f(x) = 5x - x^2$ **44.** $f(x) = 4x + x^2$

45. $f(x) = |x - 3| - 5$ **46.** $f(x) = 9 - 2|x + 3|$

47. Describe the relationship between the graphs of $f(x)$ and $f(|x|)$ in Problems 43–46.

48. Describe the relationship between the graphs of $f(x)$ and $f(-|x|)$ in Problems 43–46.

In Problems 49–52, graph $f(x)$, $|f(x)|$, and $-|f(x)|$ in separate viewing rectangles.

49. $f(x) = 0.2x^2 - 9$ **50.** $f(x) = 8 - 0.3x^2$

51. $f(x) = 2|x + 2| - 6$ **52.** $f(x) = 9 - 2|x - 3|$

53. Describe the relationship between the graphs of $f(x)$ and $|f(x)|$ in Problems 49–52.

54. Describe the relationship between the graphs of $f(x)$ and $-|f(x)|$ in Problems 49–52.

APPLICATIONS

Business & Economics

55. *Price–demand.* The price p (in dollars) and the weekly demand x for a certain product are related by

$$p = d(x) = 100 - 5\sqrt{x} \qquad 0 \le x \le 400$$

Graph $p = d(x)$.

56. *Price–supply.* The price p (in dollars) and the number of units x that manufacturers are willing to supply weekly for a certain product are related by

$$p = s(x) = 5\sqrt{x} \qquad 0 \le x \le 400$$

Graph $p = s(x)$.

57. *Cost function.* If the total cost of producing x units of a product is expressed in the form

$$C(x) = K + f(x)$$

where K is a constant and f is a function satisfying $f(0) = 0$, then $K = C(0)$ represents the fixed costs (costs that do not vary with changes in output), and $f(x)$ represents the variable costs (costs that vary with changes in output). For example, payments for interest on loans, property taxes, and insurance premiums are fixed costs, while payments for material, labor, and transportation are variable costs. Use the graph of the variable cost function $f(x)$ shown in the margin to graph the total cost function if the fixed costs are $2,000.

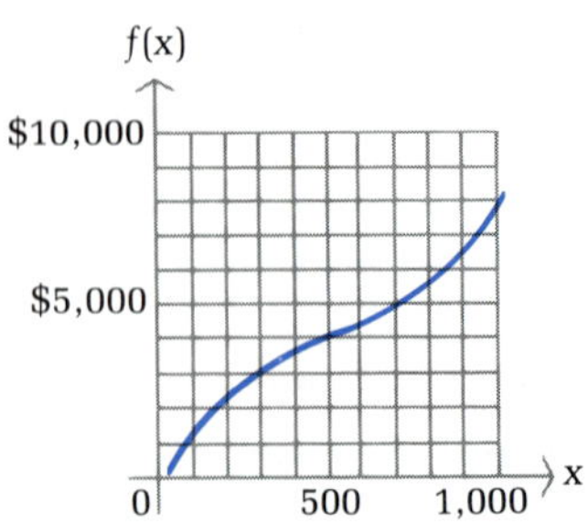

58. *Cost function.* Refer to the variable cost function f in Problem 57. Suppose a new labor contract and a sharp rise in the cost of materials result in a 25% increase in the variable cost at all levels of output. If F is the new variable cost function, then $F(x) = 1.25f(x)$. Use the graph of f to graph $y = F(x)$.

Life Sciences

59. *Timber harvesting.* In order to determine when a forest should be harvested, forest managers must have a convenient method for estimating the number of board feet a tree will produce (1 board foot = 1 square foot of wood, 1 inch thick). If d is the diameter of a tree (in inches) at chest height, then a typical formula for estimating the number of board feet b is given by

$$b = f(d) = 20 + 0.005(d - 10)^3$$

Graph $b = f(d)$ for $10 \leq d \leq 30$.

Social Sciences

60. *Safety research.* Under ideal conditions, if a person driving a vehicle slams on the brakes and skids to a stop, the speed of the vehicle v (in miles per hour) is given approximately by

$$v = f(x) = C\sqrt{x}$$

where x is the length of the skid marks (in feet) and C is a constant that depends on the road conditions and the weight of the vehicle. On the same set of axes, graph $v = f(x)$, $0 \leq x \leq 100$, for $C = 4$, 5, and 6.

SECTION 1-5 Chapter Review

Important Terms and Symbols

1-1 *Cartesian Coordinate System and Straight Lines.* Cartesian coordinate system; rectangular coordinate system; coordinate axes; ordered pair; coordinates; abscissa; ordinate; quadrants; solution of an equation in two variables; solution set; graph of an equation; standard form; x intercept; y intercept; slope; slope–intercept form; point–slope form; horizontal line; vertical line; parallel lines; perpendicular lines

$$Ax + By = C; \quad m = \frac{y_2 - y_1}{x_2 - x_1}; \quad y = mx + b;$$
$$y - y_1 = m(x - x_1); \quad y = c; \quad x = c$$

1-2 *Functions.* Function; domain; range; input; output; independent variable; dependent variable; vertical line test; function notation; revenue function

$$f(x); \quad f: x \to f(x); \quad (x, f(x))$$

1-3 *Linear and Quadratic Functions.* Graph of a function; x intercept; y intercept; linear function; constant function; quadratic function; parabola; axis of a parabola; vertex of a parabola; maximum; minimum; absolute value;

absolute value function; piecewise-defined function; demand equation; cost equation; revenue equation; profit equation; break-even point

$$f(x) = ax + b; \quad f(x) = ax^2 + bx + c,\ a \neq 0; \quad |x|$$

1-4 *Aids to Graphing Functions.* Identity function; squaring function; absolute value function; square root function; cubing function; cube root function; vertical shift or translation; horizontal shift or translation; reflection; expansion; contraction

$$f(x) \pm k; \quad f(x \pm h); \quad -f(x); \quad Cf(x)$$

EXERCISE 1-5

Chapter Review

Work through all the problems in this chapter review and check your answers in the back of the book. (Answers to all review problems are there.) Where weaknesses show up, review appropriate sections in the text.

A

1. Graph the equation below in a rectangular coordinate system. Indicate the slope and the y intercept.

$$y = \frac{x}{2} - 2$$

2. Write the equation of a line that passes through (4, 3) with slope $\frac{1}{2}$. Write the final answer in the form $y = mx + b$.

3. Graph $x - y = 2$ in a rectangular coordinate system. Indicate the slope.

4. For $f(x) = 2x - 1$ and $g(x) = x^2 - 2x$, find $f(-2) + g(-1)$.

5. Graph the linear function f given by the equation

$$f(x) = \tfrac{2}{3}x - 1$$

Indicate the slope and intercepts.

B

6. Find the maximum or minimum value of $f(x) = x^2 - 8x + 7$ without graphing. What are the coordinates of the vertex of the parabola?

7. How are the graphs of the following related to the graph of $S(x) = x^2$?

(A) $f(x) = -x^2$ (B) $g(x) = x^2 + 4$ (C) $h(x) = (x - 5)^2$

8. Graph $3x + 6y = 18$ in a rectangular coordinate system. Indicate the slope and intercepts.

9. Find an equation of the line that passes through (−2, 3) and (6, −1). Write the answer in the form $Ax + By = C$, $A > 0$. What is the slope of the line?

10. Write the equations of the vertical line and the horizontal line that pass through (−5, 2). Graph both equations on the same coordinate system.

11. Find an equation of the line that passes through $(-2, 5)$ and $(2, -1)$. Write the answer in the form $y = mx + b$.

12. For $f(x) = 10x - 7$, $g(t) = 6 - 2t$, $F(u) = 3u^2$, and $G(v) = v - v^2$, find:

    (A) $2g(-1) - 3G(-1)$ (B) $4G(-2) - g(-3)$

    (C) $\dfrac{f(2) \cdot g(-4)}{G(-1)}$ (D) $\dfrac{F(-1) \cdot G(2)}{g(-1)}$

13. Find the domains of the functions f and g if

$$f(x) = 2x - x^2 \qquad g(x) = \frac{1}{x - 2}$$

14. For $f(x) = 2x - 1$, find: $\dfrac{f(3 + h) - f(3)}{h}$

15. Determine which of the following equations specify functions with independent variable x. For those that do, find the domain. For those that do not, find a value of x that corresponds to more than one value of y.

    (A) $4x - 3y = 11$ (B) $y^2 - 4x = 1$ (C) $xy + 3y + 5x = 4$

16. Graph $g(x) = 8x - 2x^2$, finding the axis, vertex, maximum or minimum, range, and intercepts.

17. Sketch the graph and find the domain and range of

$$f(x) = \begin{cases} 1 - x^2 & \text{if } -1 \leq x < 0 \\ 1 + x^2 & \text{if } \quad 0 \leq x \leq 1 \end{cases}$$

Graph each function in Problems 18–23.

18. $f(x) = |x| - 1$ 19. $g(x) = |x - 1|$ 20. $p(x) = 2\sqrt[3]{x}$
21. $q(x) = \frac{1}{4}x^3$ 22. $n(x) = 2\sqrt{x} - 3$ 23. $m(x) = 4 - (x - 1)^2$

24. Write an equation of the line that passes through the points $(4, -3)$ and $(4, 5)$.

25. Write an equation of the line that passes through $(2, -3)$ and is:

    (A) Parallel to $2x - 4y = 5$ (B) Perpendicular to $2x - 4y = 5$

    Write the final answers in the form $Ax + By = C$, $A > 0$.

26. Find the domain of the function f specified by each equation.

    (A) $f(x) = \dfrac{5}{x - 3}$ (B) $f(x) = \sqrt{x - 1}$

C

27. For $f(x) = x^2 + 7x - 9$, find and simplify: $\dfrac{f(a + h) - f(a)}{h}$

28. Graph $f(x) = 2x - 7$ and $g(x) = x^2 - 6x + 5$ on the same set of axes, and find any points of intersection.

29. How is the graph of $4 - 3(x + 2)^3$ related to the graph of $C(x) = x^3$?

30. Graph: $f(x) = 3 - \frac{1}{2}|x + 2|$

Business & Economics

31. *Linear depreciation.* A word-processing system was purchased by a company for \$12,000 and is assumed to have a salvage value of \$2,000 after 8 years (for tax purposes). If its value is depreciated linearly from \$12,000 to \$2,000:

(A) Find the linear equation that relates value V in dollars to time t in years.

(B) What would be the value of the system after 5 years?

32. *Pricing.* A sporting goods store sells a tennis racket that cost \$30 for \$48 and a pair of jogging shoes that cost \$20 for \$32.

(A) If the markup policy of the store for items that cost over \$10 is assumed to be linear and is reflected in the pricing of these two items, write an equation that relates retail price R to cost C.

(B) What should be the retail price of a pair of skis that cost \$105?

33. *Break-even analysis.* A street-corner hot dog vendor has fixed costs of \$30 a day and total costs of \$60 a day when 75 hot dogs are sold daily. The selling price of a hot dog is \$1.40.

(A) Assuming the total daily cost C is linearly related to the total number x of hot dogs sold daily, express C as a function of x.

(B) Express the total daily revenue R as a function of x.

(C) Graph C and R on the same set of axes.

(D) How many hot dogs must the vendor sell each day to break even?

34. *Parking fees.* Short-term parking fees at an airport are \$1.50 per hour with a maximum daily fee of \$7. Let $f(x)$ be the total fee for parking a car for x hours. Graph f for $0 < x \leq 8$.

35. *Market research.* The market research department of an electronics company has determined that the demand and cost equations for the production of an AM/FM clock radio are

$$x = 500 - 10p \qquad \text{and} \qquad C = 3{,}000 + 10x$$

(A) Express the revenue as a function of the price p.

(B) Express the cost as a function of the price p.

(C) Graph R and C on the same set of axes, and identify the regions of profit and loss.

(D) Find the break-even points.

(E) Find the price that will produce a maximum profit.

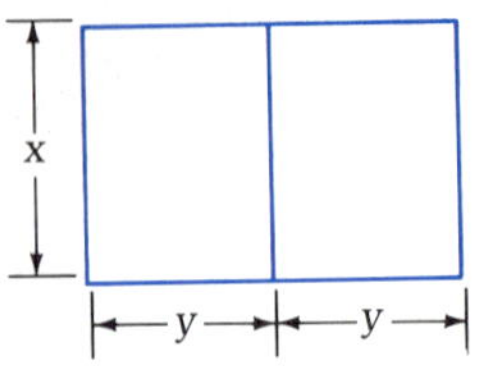

36. *Construction.* A farmer has 180 feet of fencing to be used in the construction of two identical rectangular pens sharing a common side (see the figure in the margin).

(A) Express the total area $A(x)$ enclosed by both pens as a function of the width of x.

(B) From physical considerations, what is the domain of the function A?

(C) Find the dimensions of the pens that will make the total enclosed area maximum.

CHAPTER 2

Exponential and Logarithmic Functions

CHAPTER 2

Contents

In this chapter we will define and investigate the properties of two new and important classes of functions called *exponential functions* and *logarithmic functions*. These functions are used in describing and solving a wide variety of real-world problems, including growth of money at compound interest; growth of populations of people, animals, and bacteria; radioactive decay (negative growth); and learning curves associated with the mastery of a new computer or an assembly process in a manufacturing plant. We will consider applications in these areas plus many more in sections that follow.

SECTION 2-1

Exponential Functions

- EXPONENTIAL FUNCTIONS
- BASIC EXPONENTIAL GRAPHS
- ADDITIONAL PROPERTIES
- APPLICATIONS

In this section we will define exponential functions, look at some of their important properties—including graphs—and consider several significant applications.

EXPONENTIAL FUNCTIONS

Let us start by noting that the functions f and g given by

$$f(x) = 2^x \qquad \text{and} \qquad g(x) = x^2$$

are not the same function. Whether a variable appears as an exponent with a constant base or as a base with a constant exponent, makes a big difference. The function g is a quadratic function, which we have already discussed. The function f is a new type of function called an *exponential function*.

Exponential Function

The equation

$$f(x) = b^x \qquad b > 0, b \neq 1$$

defines an **exponential function** for each different constant b, called the **base.** The independent variable x may assume any real value.

Thus, the **domain of *f*** is the set of all real numbers, and it can be shown that the **range of *f*** is the set of all positive real numbers. We require the base b to be positive to avoid nonreal numbers such as $(-2)^{1/2}$. We exclude the case $b = 1$ since $f(x) = 1^x = 1$ is a constant function, not an exponential function.

◆ BASIC EXPONENTIAL GRAPHS

Many students, if asked to graph equations such as $y = 2^x$ or $y = 2^{-x}$, would not hesitate at all. [*Note:* $2^{-x} = 1/2^x = (\frac{1}{2})^x$.] They would likely make up tables by assigning integers to x, plot the resulting points, and then join these points with a smooth curve as in Figure 1. The only catch is that 2^x has not been defined at this point for all real numbers. From Appendix A-7 we know what 2^5, 2^{-3}, $2^{2/3}$, $2^{-3/5}$, $2^{1.4}$, and $2^{-3.15}$ mean (that is, 2^p, where p is a rational number), but what does

$$2^{\sqrt{2}}$$

mean? The question is not easy to answer at this time. In fact, a precise definition of $2^{\sqrt{2}}$ must wait for more advanced courses, where we can show that

$$2^x$$

names a real number for x any real number, and that the graph of $y = 2^x$ is as indicated in Figure 1A. We also can show that for x irrational, 2^x can be approximated as closely as we like by using rational number approximations for x. Since $\sqrt{2} = 1.414\ 213...$, for example, the sequence

$$2^{1.4}, 2^{1.41}, 2^{1.414}, \ldots$$

approximates $2^{\sqrt{2}}$, and as we use more decimal places, the approximation improves.

It is useful to note that the graph of

$$f(x) = b^x \qquad b > 1$$

will look very much like Figure 1A, and the graph of

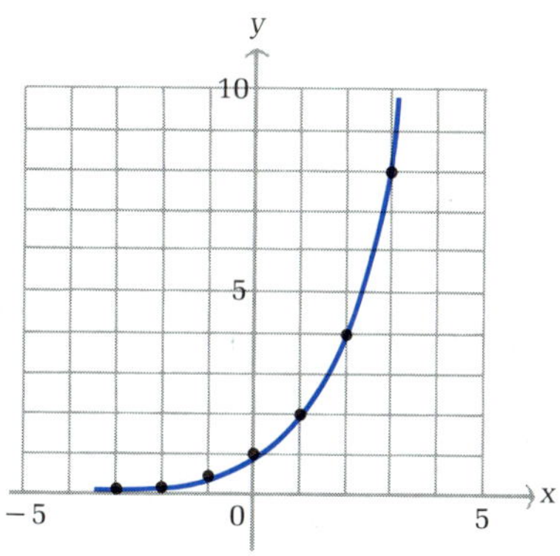

(A) $y = 2^x$

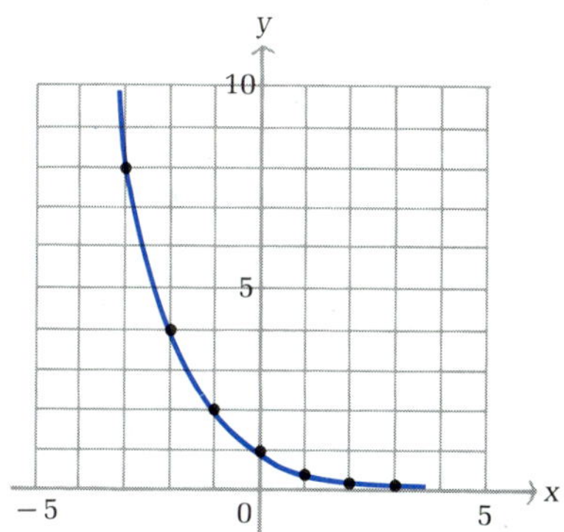

(B) $y = 2^{-x} = \left(\frac{1}{2}\right)^x$

FIGURE 1

$$f(x) = b^x \qquad 0 < b < 1$$

will look very much like Figure 1B. Observe that in both cases the graphs approach the x axis, but never touch it. The x axis is called a *horizontal asymptote* for each graph. In general, a line $y = c$ is a **horizontal asymptote** for the graph of the equation $y = f(x)$ if $f(x)$ approaches c as x increases without bound or as x decreases without bound. Asymptotes, if they exist, provide a useful aid in graphing some functions.

The graphs in Figure 1 suggest the following important general properties of exponential functions, which we state without proof:

Basic Properties of the Graph of $f(x) = b^x$, $b > 0$, $b \neq 1$

1. All graphs will pass through the point (0, 1). $b^0 = 1$ for any permissible base b.
2. All graphs are continuous curves, with no holes or jumps.
3. The x axis is a horizontal asymptote.
4. If $b > 1$, then b^x increases as x increases.
5. If $0 < b < 1$, then b^x decreases as x increases.

The use of the $\boxed{y^x}$ key on a scientific calculator makes the plotting of accurate graphs of exponential functions almost routine. Example 1 illustrates the process.

◆ EXAMPLE 1 Graph $y = (\frac{1}{2})4^x$ for $-3 \leq x \leq 3$.

Solution

x	y
−3	0.01
−2	0.03
−1	0.13
0	0.50
1	2.00
2	8.00
3	32.00

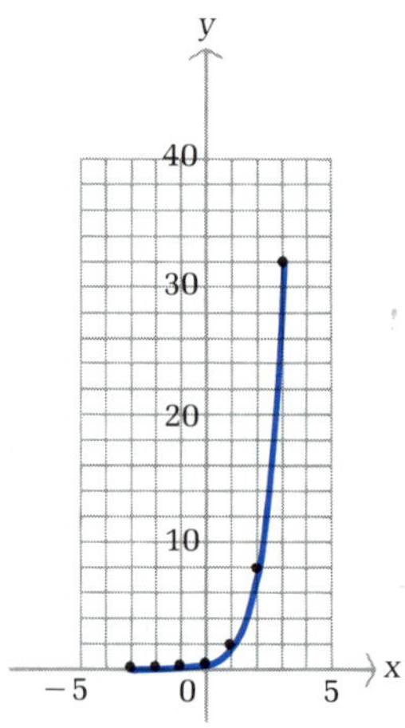

◆

PROBLEM 1 Graph $y = (\frac{1}{2})4^{-x}$ for $-3 \leq x \leq 3$. ◆

◆ ADDITIONAL PROPERTIES

Exponential functions, which include irrational exponents, obey the familiar laws of exponents we discussed earlier for rational exponents. We summarize these exponent laws and add two other important and useful properties in the following box:

Exponential Function Properties

For a and b positive, $a \neq 1$, $b \neq 1$, and x and y real:

1. Exponent laws:

$$a^x a^y = a^{x+y} \qquad \frac{a^x}{a^y} = a^{x-y} \qquad (a^x)^y = a^{xy}$$

$$(ab)^x = a^x b^x \qquad \left(\frac{a}{b}\right)^x = \frac{a^x}{b^x} \qquad \frac{4^{2y}}{4^{5y}} = 4^{2y-5y} = 4^{-3y}$$

2. $a^x = a^y$ if and only if $x = y$ If $7^{5t+1} = 7^{3t-3}$, then $5t + 1 = 3t - 3$, and $t = -2$.

3. For $x \neq 0$,

   $a^x = b^x$ if and only if $a = b$ If $a^5 = 2^5$, then $a = 2$.

◆ APPLICATIONS

We will now consider three applications that utilize exponential functions in their analysis: population growth (an example of exponential growth), radioactive decay (an example of negative exponential growth), and compound interest (another example of exponential growth).

Our first example involves the growth of populations, such as people, animals, insects, and bacteria. Populations tend to grow exponentially and at different rates. A convenient and easily understood measure of growth rate is the **doubling time**—that is, the time it takes for a population to double. Over short periods of time the **doubling-time growth model** is often used to model population growth:

$$P = P_0 2^{t/d}$$

where

P = Population at time t

P_0 = Population at time $t = 0$

d = Doubling time

Note that when $t = d$,

$$P = P_0 2^{d/d} = P_0 2$$

and the population is double the original, as it should be. We use this model to solve a population growth problem in Example 2.

◆ EXAMPLE 2

Population Growth

Ethiopia has a population of around 42 million people, and it is estimated that the population will double in 22 years. If population growth continues at the same rate, what will be the population (to the nearest million):

(A) 10 years from now? (B) 35 years from now?

Solutions

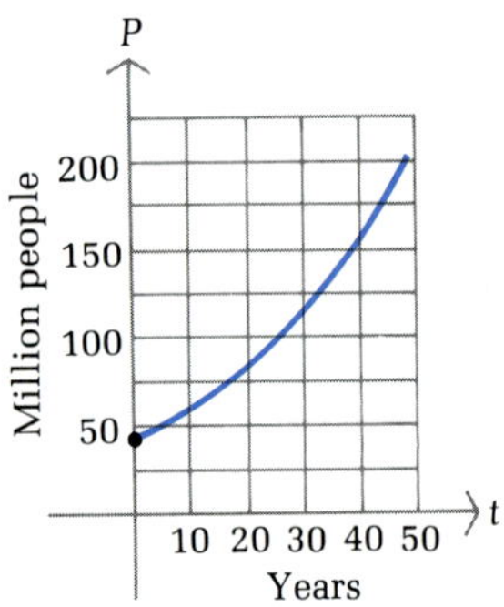

We use the doubling-time growth model:

$$P = P_0 2^{t/d}$$

Substituting $P_0 = 42$ and $d = 22$, we obtain

$$P = 42(2^{t/22})$$ See the graph in the margin.

(A) Find P when $t = 10$ years:

$$P = 42(2^{10/22})$$ Use a calculator.

$$P \approx 58 \text{ million people}$$

(B) Find P when $t = 35$ years:

$$P = 42(2^{35/22})$$ Use a calculator.

$$P \approx 127 \text{ million people}$$ ◆

PROBLEM 2

The bacterium *Escherichia coli* (*E. coli*) is found naturally in the intestines of many mammals. In a particular laboratory experiment, the doubling time for *E. coli* is found to be 25 minutes. If the experiment starts with a population of 1,000 *E. coli* and there is no change in growth rate, how many bacteria will be present:

(A) In 10 minutes? (B) In 5 hours? ◆

Our second application involves radioactive decay (negative growth). Radioactive materials are used extensively in medical diagnosis and therapy, as power sources in satellites, and as power sources (although controversial) in many countries. If we start with an amount A_0 of a particular radioactive isotope, the amount declines exponentially in time; the rate of decay varies from isotope to isotope. A convenient and easily understood measure of the rate of decay is the **half-life** of the isotope—that is, the time it takes for half of a particular material to decay. In this section we use the following **half-life decay model:**

$$A = A_0\left(\tfrac{1}{2}\right)^{t/h} = A_0 2^{-t/h}$$

where

$$A = \text{Amount at time } t$$
$$A_0 = \text{Amount at time } t = 0$$
$$h = \text{Half-life}$$

Note that when $t = h$,

$$A = A_0 2^{-h/h} = A_0 2^{-1} = \frac{A_0}{2}$$

and the amount of isotope is half the original amount, as it should be.

◆ EXAMPLE 3

Radioactive Decay

Radioactive gold-198 (^{198}Au), used in imaging the structure of the liver, has a half-life of 2.67 days. If we start with 50 milligrams of the isotope, how many milligrams will be left after:

(A) $\frac{1}{2}$ day? (B) 1 week?

Compute answers to two decimal places.

Solutions

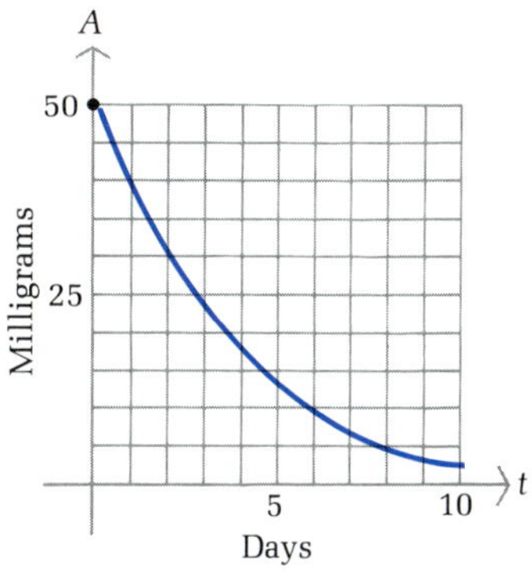

We use the half-life decay model:

$$A = A_0\left(\tfrac{1}{2}\right)^{t/h} = A_0 2^{-t/h}$$

Using $A_0 = 50$ and $h = 2.67$, we obtain

$$A = 50(2^{-t/2.67}) \quad \text{See the graph in the margin.}$$

(A) Find A when $t = 0.5$ day:

$$A = 50(2^{-0.5/2.67}) \quad \text{Use a calculator.}$$
$$\approx 43.91 \text{ milligrams}$$

(B) Find A when $t = 7$ days:

$$A = 50(2^{-7/2.67}) \quad \text{Use a calculator.}$$
$$\approx 8.12 \text{ milligrams}$$

◆

PROBLEM 3

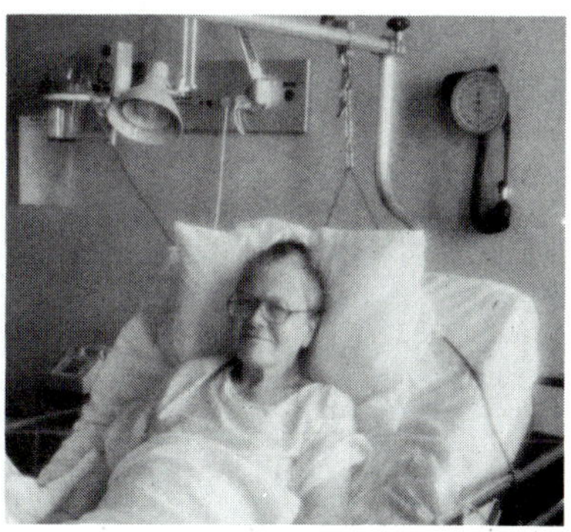

The radioactive isotope gallium-67 (^{67}Ga), used in the diagnosis of malignant tumors, has a biological half-life of 46.5 hours. If we start with 100 milligrams of the isotope, how many milligrams will be left after:

(A) 24 hours? (B) 1 week?

Compute answers to two decimal places. ◆

Our third application deals with the growth of money at compound interest. The fee paid to use another's money is called **interest**. It is usually computed as a percent (called **interest rate**) of the principal over a given period of time. If, at the end of a payment period, the interest due is reinvested at the same rate, then the interest earned as well as the principal will earn interest during the next payment period. Interest paid on interest reinvested is called **compound interest**, and may be calculated using the following compound interest formula:

Compound Interest

If a **principal** P is invested at an annual **rate** r (expressed as a decimal) compounded m times a year, then the **amount** A **(future value)** in the account at the end of t years is given by

$$A = P\left(1 + \frac{r}{m}\right)^{mt}$$

◆ EXAMPLE 4 *Compound Growth*

If $1,000 is invested in an account paying 10% compounded monthly, how much will be in the account at the end of 10 years? Compute the answer to the nearest cent.

Solution We use the compound interest formula as follows:

$$\begin{aligned} A &= P\left(1 + \frac{r}{m}\right)^{mt} \\ &= 1{,}000\left(1 + \frac{0.10}{12}\right)^{(12)(10)} \qquad \text{Use a calculator.} \\ &= \$2{,}707.04 \end{aligned}$$

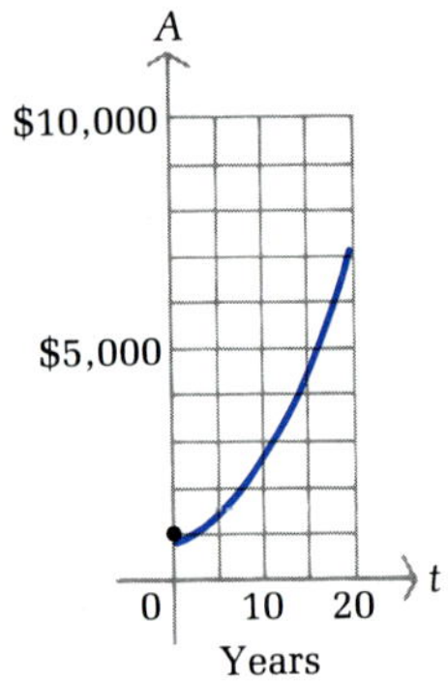

The graph of

$$A = 1{,}000\left(1 + \frac{0.10}{12}\right)^{12t}$$

for $0 \le t \le 20$ is shown in the margin. ◆

PROBLEM 4 If you deposit $5,000 in an account paying 9% compounded daily, how much will you have in the account in 5 years? Compute the answer to the nearest cent. ◆

Answers to Matched Problems

1. $y = (\frac{1}{2})4^{-x}$

x	y
−3	32.00
−2	8.00
−1	2.00
0	0.50
1	0.13
2	0.03
3	0.01

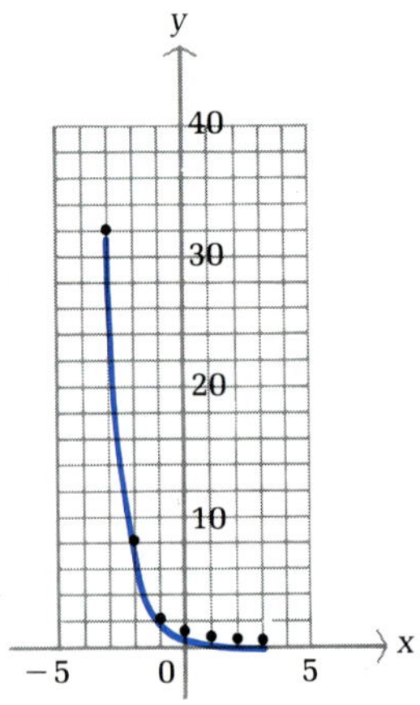

2. (A) $P \approx 1{,}320$ bacteria
   (B) $P \approx 4{,}096{,}000$ bacteria
3. (A) $A \approx 69.92$ mg
   (B) $A \approx 8.17$ mg
4. $7,841.13

EXERCISE 2-1

A *Graph each function over the indicated interval.*

1. $y = 5^x$; $[-2, 2]$
2. $y = 3^x$; $[-3, 3]$
3. $y = (\frac{1}{5})^x = 5^{-x}$; $[-2, 2]$
4. $y = (\frac{1}{3})^x = 3^{-x}$; $[-3, 3]$
5. $f(x) = -5^x$; $[-2, 2]$
6. $g(x) = -3^{-x}$; $[-3, 3]$
7. $f(x) = 4(5^x)$; $[-2, 2]$
8. $h(x) = 5(3^x)$; $[-3, 3]$
9. $y = 5^{x+2} + 4$; $[-4, 0]$
10. $y = 3^{x+3} - 5$; $[-6, 0]$

Simplify.

11. $(4^{3x})^{2y}$
12. $10^{3x-1}10^{4-x}$
13. $\dfrac{5^{x-3}}{5^{x-4}}$
14. $\dfrac{3^x}{3^{1-x}}$
15. $(2^x 3^y)^z$
16. $\left(\dfrac{4^x}{5^y}\right)^{3z}$

B *Solve for x.*

17. $10^{2-3x} = 10^{5x-6}$
18. $5^{3x} = 5^{4x-2}$
19. $4^{5x-x^2} = 4^{-6}$
20. $7^{x^2} = 7^{2x+3}$
21. $5^3 = (x+2)^3$
22. $(1-x)^5 = (2x-1)^5$
23. $9^{x-1} = 3^x$
24. $2^x = 4^{x+1}$

Graph each function over the indicated interval.

25. $f(t) = 2^{t/10}$; $[-30, 30]$
26. $G(t) = 3^{t/100}$; $[-200,200]$
27. $y = 7(2^{-2x})$; $[-2, 2]$
28. $y = 11(3^{-x/2})$; $[-9, 9]$
29. $f(x) = 2^{|x|}$; $[-3, 3]$
30. $g(x) = 2^{-|x|}$; $[-3, 3]$
31. $y = 100(1.03)^x$; $[0, 20]$
32. $y = 1{,}000(1.08)^x$; $[0, 10]$
33. $y = 3^{-x^2}$; $[-2, 2]$
34. $y = 2^{-x^2}$; $[-2, 2]$

C *Simplify.*

35. $(3^x - 3^{-x})(3^x + 3^{-x})$
36. $(6^x + 6^{-x})(6^x - 6^{-x})$
37. $(3^x - 3^{-x})^2 + (3^x + 3^{-x})^2$
38. $(6^x + 6^{-x})^2 - (6^x - 6^{-x})^2$

Graph each function over the indicated interval.

39. $h(x) = x(2^x)$; $[-5, 0]$
40. $m(x) = x(3^{-x})$; $[0, 3]$
41. $g(x) = \dfrac{3^x + 3^{-x}}{2}$; $[-3, 3]$
42. $f(x) = \dfrac{2^x + 2^{-x}}{2}$; $[-3, 3]$

CALCULATOR PROBLEMS

Calculate each using a scientific calculator. Compute answers to four decimal places.

43. $3^{-\sqrt{2}}$
44. $5^{\sqrt{3}}$
45. $\pi^{-\sqrt{3}}$
46. $\pi^{\sqrt{2}}$
47. $\dfrac{3^\pi - 3^{-\pi}}{2}$
48. $\dfrac{2^\pi + 2^{-\pi}}{2}$

Business & Economics

49. *Finance.* Suppose \$2,500 is invested at 7% compounded quarterly. How much money will be in the account in:

(A) $\frac{3}{4}$ year? (B) 15 years?

Compute answers to the nearest cent.

50. *Finance.* Suppose \$4,000 is invested at 11% compounded weekly. How much money will be in the account in:

(A) $\frac{1}{2}$ year? (B) 10 years?

Compute answers to the nearest cent.

51. *Finance.* A person wishes to have \$15,000 cash for a new car 5 years from now. How much should be placed in an account now, if the account pays 9.75% compounded weekly? Compute the answer to the nearest dollar.

52. *Finance.* A couple just had a new child. How much should they invest now at 8.25% compounded daily in order to have \$40,000 for the child's education 17 years from now? Compute the answer to the nearest dollar.

Life Sciences

53. *Population growth.* If Kenya has a population of 23,000,000 people and a doubling time of 19 years, and if the growth continues at the same rate, find the population in:

(A) 10 years (B) 30 years

Compute answers to the nearest million.

54. *Bacterial growth.* If bacteria in a certain culture double every $\frac{1}{2}$ hour, write an equation that gives the number of bacteria N in the culture after t hours, assuming the culture has 100 bacteria at the start. Graph the equation for $0 \leq t \leq 5$.

55. *Radioactive tracers.* The radioactive isotope technetium-99m (^{99m}Tc) is used in imaging the brain. This isotope has a half-life of 6 hours. If 12 milligrams are used, how much will be present after:

(A) 3 hours? (B) 24 hours?

Compute answers to two decimal places.

56. *Insecticides.* The use of the insecticide DDT is no longer allowed in many countries because of its long-term adverse effects. If a farmer uses 25 pounds of active DDT, assuming its half-life is 12 years, how much will still be active after:

(A) 5 years? (B) 20 years?

Compute answers to the nearest pound.

Use a graphic calculator or a computer for the graphs in Problems 57–62.

Business & Economics

57. *Finance.* In Problem 49 write the formula for the amount of money in the account at the end of t years and graph it for $0 \leqslant t \leqslant 30$.
58. *Finance.* In Problem 50 write the formula for the amount of money in the account at the end of t years and graph it for $0 \leqslant t \leqslant 15$.

Life Sciences

59. *Population growth.* In Problem 53 write the formula for the population at the end of t years and graph it for $0 \leqslant t \leqslant 40$.
60. *Bacterial growth.* In Problem 54 write the formula for the size of the culture at the end of t hours and graph it for $0 \leqslant t \leqslant 4$.
61. *Radioactive tracers.* In Problem 55 write the formula for the amount of radioactive isotope left after t hours and graph it for $0 \leqslant t \leqslant 36$.
62. *Insecticides.* In Problem 56 write the formula for the amount of DDT left at the end of t years and graph it for $0 \leqslant t \leqslant 50$.

SECTION 2-2 The Exponential Function with Base e

- BASE e EXPONENTIAL FUNCTION
- GROWTH AND DECAY APPLICATIONS REVISITED
- CONTINUOUS COMPOUND INTEREST

The number π is probably the most important irrational number you have encountered until now. In this section we will introduce another irrational number, e, that is just as important in mathematics and its applications.

BASE e EXPONENTIAL FUNCTION

Consider the following expression:

$$\left(1 + \frac{1}{m}\right)^m$$

TABLE 1

m	$\left(1 + \frac{1}{m}\right)^m$
1	2
10	2.593 74...
100	2.704 81...
1,000	2.716 92...
10,000	2.718 14...
100,000	2.718 27...
1,000,000	2.718 28...
.	.
.	.
.	.

What happens to the value of the expression as m increases without bound? (Think about this for a moment before proceeding.) Maybe you guessed that the value approaches 1 using the following reasoning: As m gets large, $1 + (1/m)$ approaches 1 (since $1/m$ approaches 0), and 1 raised to any power is 1. Let us see if this reasoning is correct by actually calculating the value of the expression for larger and larger values of m. Table 1 summarizes the results.

Interestingly, the value of $[1 + (1/m)]^m$ is never close to 1 but seems to be approaching a number close to 2.7183. As m increases without bound, the value of $[1 + (1/m)]^m$ approaches an irrational number that we call **e**. The irrational number e to twelve decimal places is

$$e = 2.718\ 281\ 828\ 459$$

Exactly who discovered the constant e is still being debated. It is named after the great Swiss mathematician Leonhard Euler (1707–1783). This constant turns out to be an ideal base for an exponential function, because in calculus and higher mathematics many operations take on their simplest form using this base. This is why you will see e used extensively in expressions and formulas that model real-world phenomena.

Exponential Function with Base *e*

For x a real number, the equation

$$f(x) = e^x$$

defines the **exponential function with base e.**

The exponential function with base e is used so frequently that it is often referred to as *the* exponential function. Because of its importance, all scientific calculators have an $\boxed{e^x}$ key or its equivalent—consult your user's manual.

The important constant e, along with the two other important constants, $\sqrt{2}$ and π, are shown on the number line in Figure 2A. Using the properties of the graph of $f(x) = b^x$ discussed in Section 2-1, we obtain the graphs of $y = e^x$ and $y = (1/e)^x = e^{-x}$ shown in Figure 2B. Notice that neither graph crosses the x axis. Thus, we conclude that $e^x > 0$ and $e^{-x} > 0$ for all real numbers x, an important observation when solving certain equations involving the exponential function.

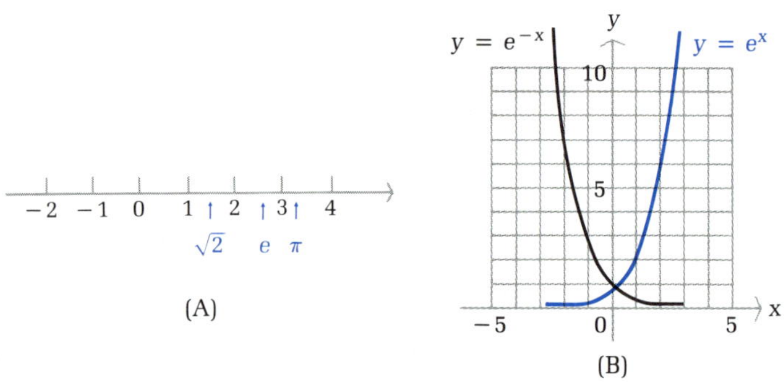

FIGURE 2

◆ GROWTH AND DECAY APPLICATIONS REVISITED

Most exponential growth and decay problems are modeled using base e exponential functions. We present two applications here and many more in Exercise 2-2.

◆ EXAMPLE 5

Exponential Growth

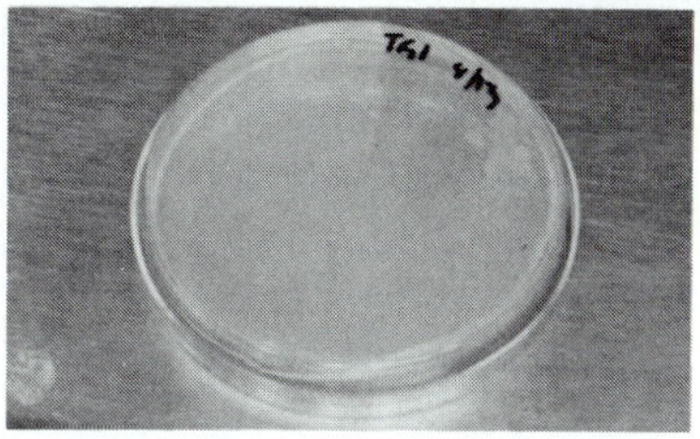

Cholera, an intestinal disease, is caused by a cholera bacterium that multiplies exponentially by cell division as given approximately by

$$N = N_0e^{1.386t}$$

where N is the number of bacteria present after t hours and N_0 is the number of bacteria present at $t = 0$. If we start with 10 bacteria, how many bacteria will be present:

(A) In 0.6 hour? (B) In 3.5 hours?

Solutions

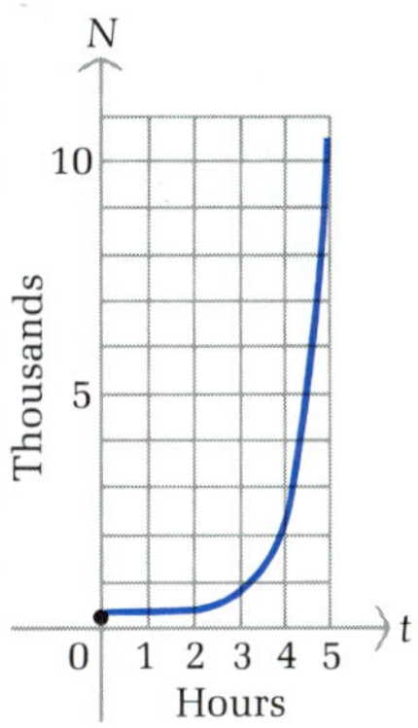

Substituting $N_0 = 10$ into the above equation, we obtain

$N = 10e^{1.386t}$ The graph is shown in the margin.

(A) Solve for N when $t = 0.6$:

$$N = 10e^{1.386(0.6)} \quad \text{Use a calculator.}$$
$$= 23 \text{ bacteria}$$

(B) Solve for N when $t = 3.5$:

$$N = 10e^{1.386(3.5)} \quad \text{Use a calculator.}$$
$$= 1{,}279 \text{ bacteria}$$

◆

PROBLEM 5

Refer to the exponential growth model for cholera in Example 5. If we start with 50 bacteria, how many bacteria will be present:

(A) In 0.85 hour? (B) In 7.25 hours? ◆

◆ EXAMPLE 6

Exponential Decay

Cosmic-ray bombardment of the atmosphere produces neutrons, which in turn react with nitrogen to produce radioactive carbon-14 (^{14}C). Radioactive ^{14}C enters all living tissues through carbon dioxide, which is first absorbed by plants. As long as a plant or animal is alive, ^{14}C is maintained in the living organism at a constant level. Once the organism dies, however, ^{14}C decays according to the equation

$$A = A_0e^{-0.000124t}$$

where A is the amount present after t years and A_0 is the amount present at time $t = 0$. If 500 milligrams of ^{14}C are present in a sample from a skull at the time of death, how many milligrams will be present in the sample in:

(A) 15,000 years? (B) 45,000 years?

Compute answers to two decimal places.

Solutions Substituting $A_0 = 500$ in the decay equation, we have

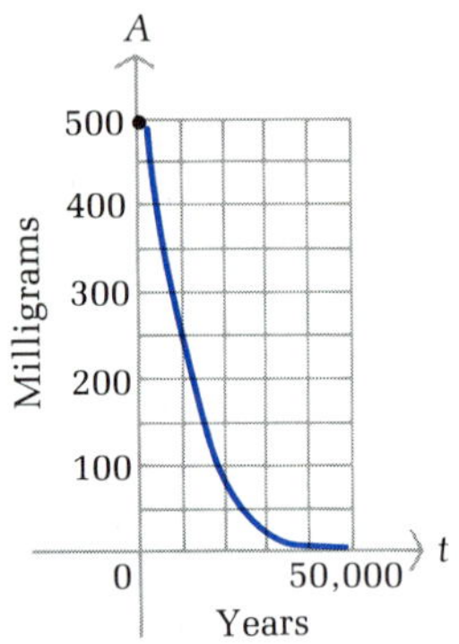

$$A = 500e^{-0.000124t} \qquad \text{See the graph in the margin.}$$

(A) Solve for A when $t = 15{,}000$:

$$\begin{aligned} A &= 500e^{-0.000124(15{,}000)} \qquad \text{Use a calculator.} \\ &= 77.84 \text{ milligrams} \end{aligned}$$

(B) Solve for A when $t = 45{,}000$:

$$\begin{aligned} A &= 500e^{-0.000124(45{,}000)} \qquad \text{Use a calculator.} \\ &= 1.89 \text{ milligrams} \end{aligned}$$

◆

PROBLEM 6 Refer to the exponential decay model in Example 6. How many milligrams of ^{14}C would have to be present at the beginning in order to have 25 milligrams present after 18,000 years? Compute the answer to the nearest milligram. ◆

◆ CONTINUOUS COMPOUND INTEREST

The constant *e* occurs naturally in the study of compound interest. Returning to the compound interest formula discussed in Section 2-1,

COMPOUND INTEREST $$A = P\left(1 + \frac{r}{m}\right)^{mt}$$

recall that *P* is the principal invested at an annual rate *r* compounded *m* times a year and *A* is the amount in the account after *t* years. Suppose *P*, *r*, and *t* are held fixed and *m* is increased without bound. Will the amount *A* increase without bound or will it tend to some limiting value?

Starting with $P = \$100$, $r = 0.08$, and $t = 2$ years, we construct Table 2 for several values of *m* with the aid of a calculator. Notice that the largest gain appears in going from annual to semiannual compounding. Then, the gains slow down as *m* increases. It appears that *A* gets closer and closer to \$117.35 as *m* gets larger and larger.

TABLE 2

COMPOUNDING FREQUENCY	m	$A = 100\left(1 + \frac{0.08}{m}\right)^{2m}$
Annually	1	\$116.6400
Semiannually	2	116.9859
Quarterly	4	117.1659
Weekly	52	117.3367
Daily	365	117.3490
Hourly	8,760	117.3510

It can be shown that

$$P\left(1+\frac{r}{m}\right)^{mt}$$

gets closer and closer to Pe^{rt} as the number of compounding periods m gets larger and larger. The latter is referred to as the **continuous compound interest formula,** a formula that is widely used in business, banking, and economics.

Continuous Compound Interest Formula

If a principal P is invested at an annual rate r (expressed as a decimal) compounded continuously, then the amount A in the account at the end of t years is given by

$$A = Pe^{rt}$$

◆ EXAMPLE 7 What amount will an account have after 2 years if $5,000 is invested at an annual rate of 8%:

(A) Compounded daily? (B) Compounded continuously?

Compute answers to the nearest cent.

Solutions (A) Use the compound interest formula

$$A = P\left(1+\frac{r}{m}\right)^{mt}$$

with $P = 5{,}000$, $r = 0.08$, $m = 365$, and $t = 2$:

$$A = 5{,}000\left(1+\frac{0.08}{365}\right)^{(365)(2)} \quad \text{Use a calculator.}$$
$$= \$5{,}867.45$$

(B) Use the continuous compound interest formula

$$A = Pe^{rt}$$

with $P = 5{,}000$, $r = 0.08$, and $t = 2$:

$$A = 5{,}000e^{(0.08)(2)} \quad \text{Use a calculator.}$$
$$= \$5{,}867.55$$

◆

PROBLEM 7 What amount will an account have after 1.5 years if $8,000 is invested at an annual rate of 9%:

(A) Compounded weekly? (B) Compounded continuously?

Compute answers to the nearest cent. ◆

The formulas for simple interest, compound interest, and continuous compound interest are summarized in the box for convenient reference.

Interest Formulas

$A = P(1 + rt)$	Simple interest
$A = P\left(1 + \frac{r}{m}\right)^{mt}$	Compound interest
$A = Pe^{rt}$	Continuous compound interest

Answers to Matched Problems

5. (A) 162 bacteria (B) 1,156,054 bacteria
6. 233 mg
7. (A) \$9,155.23 (B) \$9,156.29

EXERCISE 2-2

A

Graph over the indicated interval.

1. $y = -e^{-x}$; $[-3, 3]$
2. $y = -e^{x}$; $[-3, 3]$
3. $y = 100e^{0.1x}$; $[-5, 5]$
4. $y = 10e^{0.2x}$; $[-10, 10]$
5. $g(t) = 10e^{-0.2t}$; $[-5, 5]$
6. $f(t) = 100e^{-0.1t}$; $[-5, 5]$

B

7. $y = -3 + e^{1+x}$; $[-4, 2]$
8. $y = 2 + e^{x-2}$; $[-1, 5]$
9. $y = e^{|x|}$; $[-3, 3]$
10. $y = e^{-|x|}$; $[-3, 3]$
11. $C(x) = \dfrac{e^x + e^{-x}}{2}$; $[-5, 5]$
12. $M(x) = e^{x/2} + e^{-x/2}$; $[-5, 5]$

Simplify.

13. $e^x(e^{-x} + 1) - e^{-x}(e^x + 1)$
14. $(e^x + e^{-x})^2 + (e^x - e^{-x})^2$
15. $\dfrac{e^x(e^x + e^{-x}) - (e^x - e^{-x})e^x}{e^{2x}}$
16. $\dfrac{e^{-x}(e^x - e^{-x}) + e^{-x}(e^x + e^{-x})}{e^{-2x}}$

C

Solve each equation (remember, $e^x \neq 0$ and $e^{-x} \neq 0$).

17. $(x - 3)e^x = 0$
18. $2xe^{-x} = 0$
19. $3xe^{-x} + x^2e^{-x} = 0$
20. $x^2e^x - 5xe^x = 0$

Graph over the indicated interval.

21. $N = \dfrac{100}{1 + e^{-t}}$; $[0, 5]$
22. $N = \dfrac{200}{1 + 3e^{-t}}$; $[0, 5]$

Business & Economics

23. *Money growth.* If you invest \$7,500 in an account paying 8.35% compounded continuously, how much money will be in the account at the end of:

(A) 5.5 years? (B) 12 years?

24. *Money growth.* If you invest \$5,250 in an account paying 11.38% compounded continuously, how much money will be in the account at the end of:

(A) 6.25 years? (B) 17 years?

25. *Money growth.* *Barron's* (a national business and financial weekly) published the following "Top Savings Deposit Yields" for 1 year certificate of deposit accounts:

(A) Alamo Savings, 8.25% compounded quarterly
(B) Lamar Savings, 8.05% compounded continuously

Compute the value of \$10,000 invested in each account at the end of 1 year.

26. *Money growth.* Refer to Problem 25. In another issue of *Barron's*, $2\frac{1}{2}$ year certificate of deposit accounts included the following:

(A) Gill Saving, 8.30% compounded continuously
(B) Richardson Savings and Loan, 8.40% compounded quarterly
(C) USA Savings, 8.25% compounded daily

Compute the value of \$1,000 invested in each account at the end of $2\frac{1}{2}$ years.

27. *Present value.* A promissory note will pay \$50,000 at maturity $5\frac{1}{2}$ years from now. How much should you be willing to pay for the note now if money is worth 10% compounded continuously?

28. *Present value.* A promissory note will pay \$30,000 at maturity 10 years from now. How much should you be willing to pay for the note now if money is worth 9% compounded continuously?

29. *Advertising.* A company is trying to introduce a new product to as many people as possible through television advertising in a large metropolitan area with 2 million possible viewers. A model for the number of people N (in millions) who are aware of the product after t days of advertising was found to be

$$N = 2(1 - e^{-0.037t})$$

Graph this function for $0 \leq t \leq 50$. What value does N tend to as t increases without bound?

30. *Learning curve.* People assigned to assemble circuit boards for a computer manufacturing company undergo on-the-job training. From past experience it was found that the learning curve for the average employee is given by

$$N = 40(1 - e^{-0.12t})$$

where N is the number of boards assembled per day after t days of training. Graph this function for $0 \leq t \leq 30$. What is the maximum number of boards an average employee can be expected to produce in 1 day?

Life Sciences

31. *Marine biology.* Marine life is dependent upon the microscopic plant life that exists in the *photic zone*, a zone that goes to a depth where about 1% of the surface light still remains. In some waters with a great deal of sediment, the photic zone may go down only 15–20 feet. In some murky harbors, the intensity of light d feet below the surface is given approximately by

$$I = I_0 e^{-0.23d}$$

What percentage of the surface light will reach a depth of:

(A) 10 feet? (B) 20 feet?

32. *Marine biology.* Refer to Problem 31. Light intensity I relative to depth d (in feet) for one of the clearest bodies of water in the world, the Sargasso Sea in the West Indies, can be approximated by

$$I = I_0 e^{-0.00942d}$$

where I_0 is the intensity of light at the surface. What percent of the surface light will reach a depth of:

(A) 50 feet? (B) 100 feet?

33. *AIDS epidemic.* In 1989 AIDS among intravenous drug users in the United States was spreading at the rate of about 21% compounded continuously. Assuming this rate does not change and there were 21,000 cases of AIDS among intravenous drug users at the end of 1989, how many cases should we expect by the end of:

(A) 1996? (B) 2000?

34. *AIDS epidemic.* At the end of 1989, there were approximately 100,000 diagnosed cases of AIDS among the general population in the United States. As of this writing, the disease is estimated to be spreading at the rate of 9% compounded continuously. Assuming this rate does not change, how many cases of AIDS should be expected by the end of:

(A) 1996? (B) 2000?

Social Sciences

35. *Population growth.* If the population in Mexico was around 100 million people in 1987 and if the population continues to grow at 2.3% compounded continuously, what will the population be in 1995? Compute the answer to the nearest million.

36. *Population growth.* If the world population was 5 billion people in 1988 and if the population continues to grow at 1.7% compounded continuously, what will the population be in 1999? Compute the answer to the nearest billion.

SECTION 2-3 Logarithmic Functions

- DEFINITION OF LOGARITHMIC FUNCTIONS
- FROM LOGARITHMIC TO EXPONENTIAL FORM AND VICE VERSA
- PROPERTIES OF LOGARITHMIC FUNCTIONS
- CALCULATOR EVALUATION OF COMMON AND NATURAL LOGARITHMS
- CALCULATOR EVALUATION OF ARBITRARY LOGARITHMS
- APPLICATION

Now we are ready to consider logarithmic functions, which are closely related to exponential functions.

DEFINITION OF LOGARITHMIC FUNCTIONS

If we start with an exponential function *f* defined by

$$y = 2^x \tag{1}$$

and interchange the variables, we obtain an equation that defines a new relationship g between x and y:

$$x = 2^y \tag{2}$$

Any ordered pair of numbers on the graph of *f* will be on the graph of g if we interchange the order of the components. For example, (3, 8) satisfies equation (1) and (8, 3) satisfies equation (2). Thus, the domain of *f* becomes the range of g and the range of *f* becomes the domain of g. Graphing *f* and g on the same coordinate system (Fig. 3), we see that g is also a function. We call this new function the **logarithmic function with base 2,** and write

$$y = \log_2 x \qquad \text{if and only if} \qquad x = 2^y$$

Note that if we fold the paper along the dashed line $y = x$ in Figure 3, the two graphs match exactly. The line $y = x$ is a line of symmetry for the two graphs.

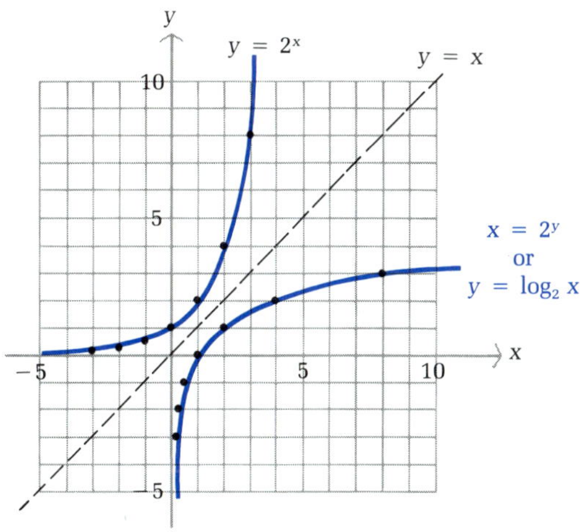

EXPONENTIAL FUNCTION x	$y = 2^x$	LOGARITHMIC FUNCTION $x = 2^y$	y
−3	$\frac{1}{8}$	$\frac{1}{8}$	−3
−2	$\frac{1}{4}$	$\frac{1}{4}$	−2
−1	$\frac{1}{2}$	$\frac{1}{2}$	−1
0	1	1	0
1	2	2	1
2	4	4	2
3	8	8	3

Ordered pairs reversed

FIGURE 3

In general, we define the **logarithmic functions with base *b*** as follows:

Definition of Logarithmic Function

For $b > 0$ and $b \neq 1$,

$y = \log_b x$	is equivalent to	$x = b^y$
$y = \log_{10} x$	is equivalent to	$x = 10^y$
$y = \log_e x$	is equivalent to	$x = e^y$

(The log to the base b of x is the exponent to which b must be raised to obtain x.)

Remember: A logarithm is an exponent.

Since the domain of an exponential function includes all real numbers and its range is the set of positive real numbers, the **domain** of a logarithmic function is the set of all positive real numbers and its **range** is the set of all real numbers. Thus, $\log_{10} 3$ is defined, but $\log_{10} 0$ and $\log_{10}(-5)$ are not defined (3 is a logarithmic domain value, but 0 and -5 are not). Typical logarithmic curves are shown in Figure 4.

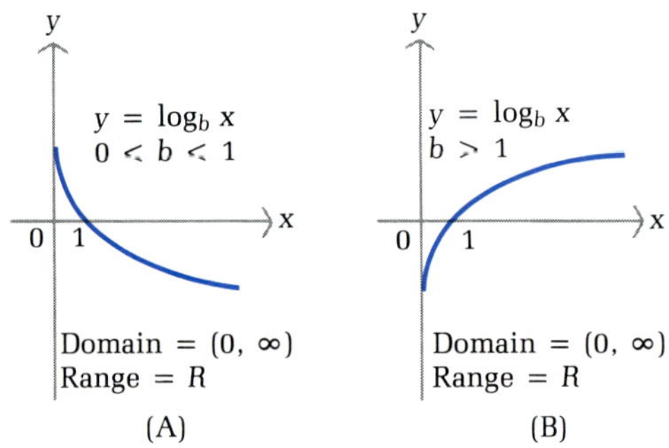

FIGURE 4
Typical logarithmic graphs

◆ FROM LOGARITHMIC TO EXPONENTIAL FORM AND VICE VERSA

We now consider the matter of converting logarithmic forms to equivalent exponential forms and vice versa.

◆ EXAMPLE 8 Change from logarithmic form to exponential form:

(A) $\log_5 25 = 2$ is equivalent to $25 = 5^2$
(B) $\log_9 3 = \frac{1}{2}$ is equivalent to $3 = 9^{1/2}$
(C) $\log_2(\frac{1}{4}) = -2$ is equivalent to $\frac{1}{4} = 2^{-2}$ ◆

PROBLEM 8 Change to an equivalent exponential form:

(A) $\log_3 9 = 2$ (B) $\log_4 2 = \frac{1}{2}$ (C) $\log_3(\frac{1}{9}) = -2$ ◆

◆ **EXAMPLE 9** Change from exponential form to logarithmic form:

(A) $64 = 4^3$ is equivalent to $\log_4 64 = 3$
(B) $6 = \sqrt{36}$ is equivalent to $\log_{36} 6 = \frac{1}{2}$
(C) $\frac{1}{8} = 2^{-3}$ is equivalent to $\log_2(\frac{1}{8}) = -3$ ◆

PROBLEM 9 Change to an equivalent logarithmic form:

(A) $49 = 7^2$ (B) $3 = \sqrt{9}$ (C) $\frac{1}{3} = 3^{-1}$ ◆

◆ **EXAMPLE 10** Find y, b, or x, as indicated.

(A) Find y: $y = \log_4 16$ (B) Find x: $\log_2 x = -3$
(C) Find y: $y = \log_8 4$ (D) Find b: $\log_b 100 = 2$

Solutions (A) $y = \log_4 16$ is equivalent to $16 = 4^y$. Thus,

$$y = 2$$

(B) $\log_2 x = -3$ is equivalent to $x = 2^{-3}$. Thus,

$$x = \frac{1}{2^3} = \frac{1}{8}$$

(C) $y = \log_8 4$ is equivalent to

$$4 = 8^y \quad \text{or} \quad 2^2 = 2^{3y}$$

Thus,

$$3y = 2$$
$$y = \tfrac{2}{3}$$

(D) $\log_b 100 = 2$ is equivalent to $100 = b^2$. Thus,

$b = 10$ Recall that b cannot be negative. ◆

PROBLEM 10 Find y, b, or x, as indicated.

(A) Find y: $y = \log_9 27$ (B) Find x: $\log_3 x = -1$
(C) Find b: $\log_b 1{,}000 = 3$ ◆

◆ **EXAMPLE 11** Graph $y = \log_2(x + 1)$ by converting to an equivalent exponential form first. Do not use a calculator or table.

Solution Changing $y = \log_2(x + 1)$ to an equivalent exponential form, we have

$$x + 1 = 2^y \quad \text{or} \quad x = 2^y - 1$$

Even though x is the independent variable and y is the dependent variable, it is easier to assign y values and solve for x.

x	y
$-\frac{7}{8}$	-3
$-\frac{3}{4}$	-2
$-\frac{1}{2}$	-1
0	0
1	1
3	2
7	3

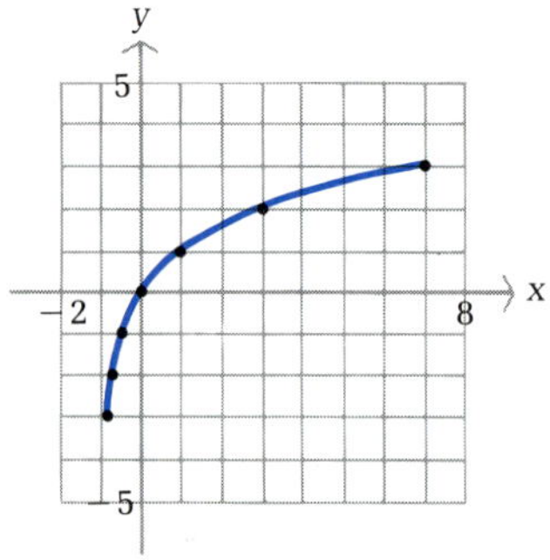

◆

PROBLEM 11 Graph $y = \log_3(x - 1)$ by converting to an equivalent exponential form first. ◆

◆ PROPERTIES OF LOGARITHMIC FUNCTIONS

Logarithmic functions have several very useful properties that follow directly from their definitions. These properties will enable us to convert multiplication problems into addition problems, division problems into subtraction problems, and power and root problems into multiplication problems. We will also be able to solve exponential equations such as $2 = 1.06^n$.

Logarithmic Properties ($b > 0$, $b \neq 1$, $M > 0$, $N > 0$)

1. $\log_b b^x = x$
2. $\log_b MN = \log_b M + \log_b N$
3. $\log_b \frac{M}{N} = \log_b M - \log_b N$
4. $\log_b M^p = p \log_b M$
5. $\log_b M = \log_b N$ if and only if $M = N$
6. $\log_b 1 = 0$

The first property follows directly from the definition of a logarithmic function. Here, we will sketch a proof for property 2. The other properties are established in a similar way. Let

$$u = \log_b M \qquad \text{and} \qquad v = \log_b N$$

Or, in equivalent exponential form,

$$M = b^u \qquad \text{and} \qquad N = b^v$$

Now, see if you can provide reasons for each of the following steps:

$$\log_b MN = \log_b b^u b^v = \log_b b^{u+v} = u + v = \log_b M + \log_b N$$

◆ EXAMPLE 12

(A) $\log_b \frac{wx}{yz} = \log_b wx - \log_b yz$

$= \log_b w + \log_b x - (\log_b y + \log_b z)$

$= \log_b w + \log_b x - \log_b y - \log_b z$

(B) $\log_b(wx)^{3/5} = \frac{3}{5}\log_b wx = \frac{3}{5}(\log_b w + \log_b x)$ ◆

PROBLEM 12

Write in simpler logarithmic forms, as in Example 12.

(A) $\log_b \frac{R}{ST}$ (B) $\log_b \left(\frac{R}{S}\right)^{2/3}$ ◆

The following examples and problems, though somewhat artificial, will give you additional practice in using basic logarithmic properties.

◆ EXAMPLE 13

Find x so that: $\frac{3}{2}\log_b 4 - \frac{2}{3}\log_b 8 + \log_b 2 = \log_b x$

Solution

$\frac{3}{2}\log_b 4 - \frac{2}{3}\log_b 8 + \log_b 2 = \log_b x$

$\log_b 4^{3/2} - \log_b 8^{2/3} + \log_b 2 = \log_b x$ Property 4

$\log_b 8 - \log_b 4 + \log_b 2 = \log_b x$

$\log_b \frac{8 \cdot 2}{4} = \log_b x$ Properties 2 and 3

$\log_b 4 = \log_b x$

$x = 4$ Property 5 ◆

PROBLEM 13

Find x so that: $3\log_b 2 + \frac{1}{2}\log_b 25 - \log_b 20 = \log_b x$ ◆

◆ EXAMPLE 14

Solve: $\log_{10} x + \log_{10}(x + 1) = \log_{10} 6$

Solution

$\log_{10} x + \log_{10}(x + 1) = \log_{10} 6$

$\log_{10} x(x + 1) = \log_{10} 6$ Property 2

$x(x + 1) = 6$ Property 5

$x^2 + x - 6 = 0$ Solve by factoring.

$(x + 3)(x - 2) = 0$

$x = -3, 2$

We must exclude $x = -3$, since the domain of the function $\log_{10}(x + 1)$ is $x > -1$ or $(-1, \infty)$; hence, $x = 2$ is the only solution. ◆

PROBLEM 14

Solve: $\log_3 x + \log_3(x - 3) = \log_3 10$ ◆

◆ CALCULATOR EVALUATION OF COMMON AND NATURAL LOGARITHMS

Of all possible logarithmic bases, the base e and the base 10 are used almost exclusively. Before we can use logarithms in certain practical problems, we need to be able to approximate the logarithm of any positive number either to base 10 or to base e. And conversely, if we are given the logarithm of a number to base 10 or base e, we need to be able to approximate the number. Historically, tables were used for this purpose, but now calculators make computations faster and far more accurate.

Common logarithms (also called **Briggsian logarithms**) are logarithms with base 10. **Natural logarithms** (also called **Napierian logarithms**) are logarithms with base e. Most scientific calculators have a key labeled "log" (or "LOG") and a key labeled "ln" (or "LN"). The former represents a common (base 10) logarithm and the latter a natural (base e) logarithm. In fact, "log" and "ln" are both used extensively in mathematical literature, and whenever you see either used in this book without a base indicated they will be interpreted as follows:

Logarithmic Notation

COMMON LOGARITHM: $\log x = \log_{10} x$
NATURAL LOGARITHM: $\ln x = \log_e x$

Finding the common or natural logarithm using a scientific calculator is very easy: You simply enter a number from the domain of the function and push [log] or [ln].

◆ EXAMPLE 15 Use a scientific calculator to evaluate each to six decimal places:

(A) log 3,184 (B) ln 0.000 349 (C) log(−3.24)

Solutions

	Enter	Press	Display
(A)	3184	[log]	3.502973
(B)	0.000 349	[ln]	−7.960439
(C)	−3.24	[log]	Error

An error is indicated in part C because −3.24 is not in the domain of the log function. [*Note:* The manner in which error messages are displayed varies from one brand of calculator to the next.] ◆

PROBLEM 15 Use a scientific calculator to evaluate each to six decimal places:

(A) log 0.013 529 (B) ln 28.693 28 (C) ln(−0.438) ◆

We now turn to the second problem mentioned above: Given the logarithm of a number, find the number. We make direct use of the logarithmic–exponential relationships, which follow from the definition of logarithmic function given at the beginning of this section.

Logarithmic–Exponential Relationships

$\log x = y$	is equivalent to	$x = 10^y$
$\ln x = y$	is equivalent to	$x = e^y$

◆ EXAMPLE 16 Find x to four decimal places, given the indicated logarithm:

(A) $\log x = -2.315$ (B) $\ln x = 2.386$

Solutions

(A) $\log x = -2.315$ Change to equivalent exponential form.

$x = 10^{-2.315}$ Evaluate with a calculator.

$x = 0.0048$

(B) $\ln x = 2.386$ Change to equivalent exponential form.

$x = e^{2.386}$ Evaluate with a calculator.

$x = 10.8699$ ◆

PROBLEM 16 Find x to four decimal places, given the indicated logarithm:

(A) $\ln x = -5.062$ (B) $\log x = 2.0821$ ◆

◆ EXAMPLE 17 Solve for x to four decimal places:

(A) $10^x = 2$ (B) $e^x = 3$ (C) $3^x = 4$

Solutions

(A) $10^x = 2$ Take common logarithms of both sides.

$\log 10^x = \log 2$ Property 1

$x = \log 2$ Use a calculator.

$x = 0.3010$

(B) $e^x = 3$ Take natural logarithms of both sides.

$\ln e^x = \ln 3$ Property 1

$x = \ln 3$ Use a calculator.

$x = 1.0986$

(C) $3^x = 4$ Take either natural or common logarithms of both sides. (We choose common logarithms.)

$\log 3^x = \log 4$ Property 4

$x \log 3 = \log 4$ Solve for x.

$x = \dfrac{\log 4}{\log 3}$ Use a calculator.

$x = 1.2619$ ◆

PROBLEM 17 Solve for x to four decimal places:

(A) $10^x = 7$ (B) $e^x = 6$ (C) $4^x = 5$ ◆

Common Error

In Example 17C do not mistake

$$\frac{\log A}{\log B} \quad \text{for} \quad \log \frac{A}{B}$$

For the latter we can write

$$\log \frac{A}{B} = \log A - \log B$$

but

$$\frac{\log A}{\log B} \neq \log A - \log B$$

◆ CALCULATOR EVALUATION OF ARBITRARY LOGARITHMS

Since most calculators have only [log] and [ln] keys, how can we find the logarithm of a positive number to a base other than 10 or e? The *change-of-base formula* provides the answer:

Change-of-Base Formula

$$\log_b N = \frac{\log_a N}{\log_a b}$$

In words, this formula states that the logarithm of a number to a given base is the logarithm of that number to a new base divided by the logarithm of the old base to the new base. It is instructive to see how the change-of-base formula is derived.

$\log_b N = y$	Write in exponential form.
$N = b^y$	Take the log of each side to another base a, $a \neq 1$.
$\log_a N = \log_a b^y$	Solve for y.
$\log_a N = y \log_a b$	

$$y = \frac{\log_a N}{\log_a b} \qquad \text{Replace } y \text{ with } \log_b N \text{ (see first step).}$$

$$\log_b N = \frac{\log_a N}{\log_a b} \qquad \text{Change-of-base formula}$$

◆ EXAMPLE 18 Evaluate $\log_{16} 5.3227$ to four decimal places.

Solution We can use either the common logarithm or the natural logarithm for the right side of the change-of-base formula. We do it both ways:

Method 1. Use base 10 as the new base:

$$\log_{16} 5.3227 = \frac{\log 5.3227}{\log 16} = 0.6030$$

Method 2. Use base e as the new base:

$$\log_{16} 5.3227 = \frac{\ln 5.3227}{\ln 16} = 0.6030$$

◆

PROBLEM 18 Evaluate: $\log_{0.8} 22.3409$ ◆

◆ APPLICATION

A convenient and easily understood way of comparing different investments is to use their **doubling times**—the length of time it takes the value of an investment to double. Logarithm properties, as you will see in Example 19, provide us with just the right tool for solving some doubling-time problems.

◆ EXAMPLE 19 How long (to the next whole year) will it take money to double if it is invested at 20% compounded annually?

Solution We use the compound interest formula discussed in Section 2-1:

$$A = P\left(1 + \frac{r}{m}\right)^{mt} \qquad \text{Compound interest}$$

The problem is to find t, given $r = 0.20$, $m = 1$, and $A = 2P$; that is,

$$2P = P(1 + 0.2)^t$$

$$2 = 1.2^t$$

$$1.2^t = 2 \qquad \text{Solve for } t \text{ by taking the natural or common}$$

$$\ln 1.2^t = \ln 2 \qquad \text{logarithm of both sides (we choose the natural logarithm).}$$

$$t \ln 1.2 = \ln 2 \qquad \text{Property 4}$$

$$t = \frac{\ln 2}{\ln 1.2}$$ Use a calculator. [*Note:* $(\ln 2)/(\ln 1.2) \neq \ln 2 - \ln 1.2$]

$= 3.8$ years

≈ 4 years To the next whole year

When interest is paid at the end of 3 years, the money will not be doubled; when paid at the end of 4 years, the money will be slightly more than doubled. ◆

PROBLEM 19 How long (to the next whole year) will it take money to double if it is invested at 13% compounded annually? ◆

It is interesting and instructive to graph the doubling times for various rates compounded annually. We proceed as follows:

$$A = P(1 + r)^t$$
$$2P = P(1 + r)^t$$
$$2 = (1 + r)^t$$
$$(1 + r)^t = 2$$
$$\ln(1 + r)^t = \ln 2$$
$$t \ln(1 + r) = \ln 2$$
$$t = \frac{\ln 2}{\ln(1 + r)}$$

Figure 5 shows the graph of this equation (doubling time in years) for interest rates compounded annually from 1% to 70% (expressed as decimals). Note the dramatic change in doubling time as rates change from 1% to 20% (from 0.01 to 0.20).

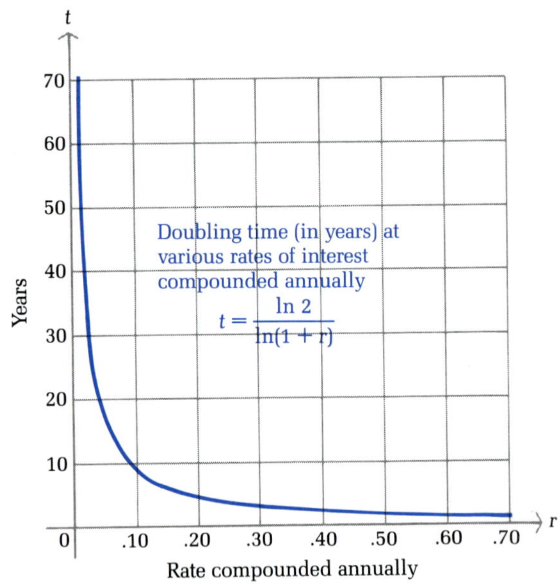

FIGURE 5

Answers to Matched Problems

8. (A) $9 = 3^2$ (B) $2 = 4^{1/2}$ (C) $\frac{1}{9} = 3^{-2}$
9. (A) $\log_7 49 = 2$ (B) $\log_9 3 = \frac{1}{2}$ (C) $\log_3(\frac{1}{3}) = -1$
10. (A) $y = \frac{3}{2}$ (B) $x = \frac{1}{3}$ (C) $b = 10$
11. $y = \log_3(x - 1)$ is equivalent to $x = 3^y + 1$

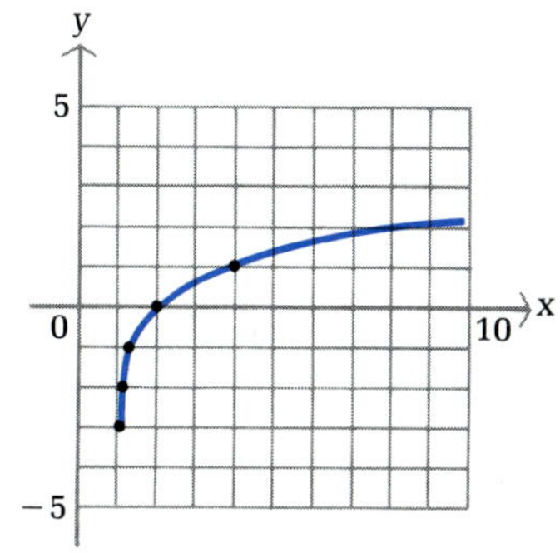

12. (A) $\log_b R - \log_b S - \log_b T$ (B) $\frac{2}{3}(\log_b R - \log_b S)$ 13. $x = 2$
14. $x = 5$ 15. (A) $-1.868\ 734$ (B) $3.356\ 663$ (C) Not defined
16. (A) 0.0063 (B) 120.8092
17. (A) 0.8451 (B) 1.7918 (C) 1.1610
18. -13.9212 19. 6 yr

EXERCISE 2-3

A *Rewrite in exponential form.*

1. $\log_3 27 = 3$
2. $\log_2 32 = 5$
3. $\log_{10} 1 = 0$
4. $\log_e 1 = 0$
5. $\log_4 8 = \frac{3}{2}$
6. $\log_9 27 = \frac{3}{2}$

Rewrite in logarithmic form.

7. $49 = 7^2$
8. $36 = 6^2$
9. $8 = 4^{3/2}$
10. $9 = 27^{2/3}$
11. $A = b^u$
12. $M = b^x$

Evaluate each of the following:

13. $\log_{10} 10^3$
14. $\log_{10} 10^{-5}$
15. $\log_2 2^{-3}$
16. $\log_3 3^5$
17. $\log_{10} 1{,}000$
18. $\log_6 36$

Write in terms of simpler logarithmic forms, as in Example 12.

19. $\log_b \dfrac{P}{Q}$
20. $\log_b FG$
21. $\log_b L^5$
22. $\log_b w^{15}$
23. $\log_b \dfrac{p}{qrs}$
24. $\log_b PQR$

B *Find x, y, or b without a table or calculator.*

25. $\log_3 x = 2$
26. $\log_2 x = 2$
27. $\log_7 49 = y$
28. $\log_3 27 = y$
29. $\log_b 10^{-4} = -4$
30. $\log_b e^{-2} = -2$

31. $\log_4 x = \frac{1}{2}$

32. $\log_{25} x = \frac{1}{2}$

33. $\log_{1/3} 9 = y$

34. $\log_{49}(\frac{1}{7}) = y$

35. $\log_b 1{,}000 = \frac{3}{2}$

36. $\log_b 4 = \frac{2}{3}$

Write in terms of simpler logarithmic forms, going as far as you can with logarithmic properties (see Example 12).

37. $\log_b \dfrac{x^5}{y^3}$

38. $\log_b x^2y^3$

39. $\log_b \sqrt[3]{N}$

40. $\log_b \sqrt[5]{Q}$

41. $\log_b(x^2 \sqrt[3]{y})$

42. $\log_b \sqrt[3]{\dfrac{x^2}{y}}$

43. $\log_b(50 \cdot 2^{-0.2t})$

44. $\log_b(100 \cdot 1.06^t)$

45. $\log_b P(1 + r)^t$

46. $\log_e Ae^{-0.3t}$

47. $\log_e 100e^{-0.01t}$

48. $\log_{10}(67 \cdot 10^{-0.12x})$

Find x.

49. $\log_b x = \frac{2}{3} \log_b 8 + \frac{1}{2} \log_b 9 - \log_b 6$

50. $\log_b x = \frac{2}{3} \log_b 27 + 2 \log_b 2 - \log_b 3$

51. $\log_b x = \frac{3}{2} \log_b 4 - \frac{2}{3} \log_b 8 + 2 \log_b 2$

52. $\log_b x = 3 \log_b 2 + \frac{1}{2} \log_b 25 - \log_b 20$

53. $\log_b x + \log_b(x - 4) = \log_b 21$

54. $\log_b(x + 2) + \log_b x = \log_b 24$

55. $\log_{10}(x - 1) - \log_{10}(x + 1) = 1$

56. $\log_{10}(x + 6) - \log_{10}(x - 3) = 1$

Graph by converting to exponential form first.

57. $y = \log_2(x - 2)$

58. $y = \log_3(x + 2)$

In Problems 59 and 60, evaluate to five decimal places using a scientific calculator.

59. (A) log 3,527.2 (B) log 0.006 913 2
(C) ln 277.63 (D) ln 0.040 883

60. (A) log 72.604 (B) log 0.033 041
(C) ln 40,257 (D) ln 0.005 926 3

In Problems 61 and 62, find x to four decimal places.

61. (A) $\log x = 1.1285$ (B) $\log x = -2.0497$
(C) $\ln x = 2.7763$ (D) $\ln x = -1.8879$

62. (A) $\log x = 2.0832$ (B) $\log x = -1.1577$
(C) $\ln x = 3.1336$ (D) $\ln x = -4.3281$

Evaluate each of the following to three decimal places using a calculator.

63. $n = \dfrac{\log 2}{\log 1.15}$

64. $n = \dfrac{\log 2}{\log 1.12}$

65. $n = \dfrac{\ln 3}{\ln 1.15}$

66. $n = \dfrac{\ln 4}{\ln 1.2}$

67. $x = \dfrac{\ln 0.5}{-0.21}$

68. $x = \dfrac{\ln 0.1}{-0.0025}$

In Problems 69–74, solve each equation to four decimal places.

69. $10^x = 12$ **70.** $10^x = 153$ **71.** $e^x = 4.304$

72. $e^x = 0.3059$ **73.** $1.03^x = 2.475$ **74.** $1.075^x = 1.837$

In Problems 75–80, find the indicated logarithms to three decimal places using the change-of-base formula and a calculator.

75. $\log_8 25$ **76.** $\log_9 33$ **77.** $\log_{32} 4.017$

78. $\log_{64} 404$ **79.** $\log_{0.5} 0.377$ **80.** $\log_{0.8} 5.29$

Graph, using a calculator.

81. $y = \ln x$ **82.** $y = -\ln x$ **83.** $y = |\ln x|$

84. $y = \ln|x|$ **85.** $y = 2 \ln(x + 2)$ **86.** $y = 2 \ln x + 2$

87. $y = 4 \ln x - 3$ **88.** $y = 4 \ln(x - 3)$

C

89. Find the logarithm of 1 for any permissible base.

90. Why is 1 not a suitable logarithmic base? [*Hint:* Try to find $\log_1 8$.]

91. Write $\log_{10} y - \log_{10} c = 0.8x$ in an exponential form that is free of logarithms.

92. Write $\log_e x - \log_e 25 = 0.2t$ in an exponential form that is free of logarithms.

APPLICATIONS

Business & Economics

93. *Doubling time.* How long (to the next whole year) will it take money to double if it is invested at 6% interest compounded annually?

94. *Doubling time.* How long (to the next whole year) will it take money to double if it is invested at 3% interest compounded annually?

95. *Tripling time.* Write a formula similar to the doubling-time formula in Figure 5 for the tripling time of money invested at 100r% interest compounded annually.

96. *Tripling time.* How long (to the next whole year) will it take money to triple if it is invested at 15% interest compounded annually?

97. *Investing.* How many years, to two decimal places, will it take \$1,000 to grow to \$1,800 if it is invested at 6% compounded quarterly? Compounded continuously?

98. *Investing.* How many years, to two decimal places, will it take \$5,000 to grow to \$7,500 if it is invested at 8% compounded semiannually? Compounded continuously?

Life Sciences

99. *Sound intensity—decibels.* Because of the extraordinary range of sensitivity of the human ear (a range of over 1,000 million millions to 1), it is helpful to use a logarithmic scale, rather than an absolute scale, to measure sound intensity over this range. The unit of measure is called the *decibel*, after the inventor of the telephone, Alexander Graham Bell. If we let N be the number of decibels, I the power of the sound in question (in watts per square centimeter), and I_0 the power of sound just below the threshold of hearing (approximately 10^{-16} watt per square centimeter), then

$$I = I_0 10^{N/10}$$

Show that this formula can be written in the form

$$N = 10 \log \frac{I}{I_0}$$

100. *Sound intensity—decibels.* Use the formula in Problem 99 (with $I_0 = 10^{-16}$ watt/cm^2) to find the decibel ratings of the following sounds:

(A) Whisper: 10^{-13} watt/cm^2
(B) Normal conversation: 3.16×10^{-10} watt/cm^2
(C) Heavy traffic: 10^{-8} watt/cm^2
(D) Jet plane with afterburner: 10^{-1} watt/cm^2

Social Sciences

101. *World population.* If the world population is now 4 billion (4×10^9) people and if it continues to grow at 2% per year compounded annually, how long will it be before there is only 1 square yard of land per person? (The earth contains approximately 1.68×10^{11} square yards of land.)

102. *Archaeology—carbon-14 dating.* The radioactive carbon-14 (^{14}C) in an organism at the time of its death decays according to the equation

$$A = A_0 e^{-0.000124t}$$

where t is time in years and A_0 is the amount of ^{14}C present at time $t = 0$. (See Example 6 in Section 2-2.) Estimate the age of a skull uncovered in an archaeological site if 10% of the original amount of ^{14}C is still present. [*Hint:* Find t such that $A = 0.1A_0$.]

SECTION 2-4 Chapter Review

Important Terms and Symbols

2-1 *Exponential Functions.* Exponential function; base; domain; range; basic graphs; horizontal asymptote; basic properties; exponential growth; exponential decay; doubling-time growth model; half-life decay model; compound interest

$$b^x; \quad P = P_0 2^{t/d}; \quad A = A_0(\tfrac{1}{2})^{t/h}; \quad A = P\left(1 + \frac{r}{m}\right)^{mt}$$

2-2 *The Exponential Function with Base e.* Irrational number e; exponential function with base e; exponential growth; exponential decay; continuous compound interest

$$N = N_0e^{kt}; \quad A = A_0e^{-kt}; \quad A = Pe^{rt}$$

2-3 *Logarithmic Functions.* Logarithmic function; base; domain; range; exponential form; properties; common logarithm; natural logarithm; calculator evaluation; change-of-base formula; doubling time

$$\log_b x; \quad \log x; \quad \ln x; \quad \log_b N = \frac{\log_a N}{\log_a b}$$

EXERCISE 2-4

Chapter Review

Work through all the problems in this chapter review and check your answers in the back of the book. (Answers to all review problems are there.) Where weaknesses show up, review appropriate sections in the text.

A

1. Write in logarithmic form using base e: $u = e^v$
2. Write in logarithmic form using base 10: $x = 10^y$
3. Write in exponential form using base e: $\ln M = N$
4. Write in exponential form using base 10: $\log u = v$

Simplify.

5. $\dfrac{5^{x+4}}{5^{4-x}}$

6. $\left(\dfrac{e^u}{e^{-u}}\right)^u$

Solve for x exactly without the use of a calculator.

7. $\log_3 x = 2$
8. $\log_x 36 = 2$
9. $\log_2 16 = x$

Solve for x to three decimal places.

10. $10^x = 143.7$
11. $e^x = 503{,}000$
12. $\log x = 3.105$
13. $\ln x = -1.147$

B *Solve for x exactly without the use of a calculator.*

14. $\log(x + 5) = \log(2x - 3)$
15. $2 \ln(x - 1) = \ln(x^2 - 5)$
16. $9^{x-1} = 3^{1+x}$
17. $e^{2x} = e^{x^2-3}$
18. $2x^2e^x = 3xe^x$
19. $\log_{1/3} 9 = x$
20. $\log_x 8 = -3$
21. $\log_9 x = \frac{3}{2}$

Solve for x to four decimal places.

22. $x = 3(e^{1.49})$
23. $x = 230(10^{-0.161})$
24. $\log x = -2.0144$
25. $\ln x = 0.3618$

26. $35 = 7(3^x)$

27. $0.01 = e^{-0.05x}$

28. $8{,}000 = 4{,}000(1.08^x)$

29. $5^{2x-3} = 7.08$

30. $x = \log_2 7$

31. $x = \log_{0.2} 5.321$

Simplify.

32. $e^x(e^{-x} + 1) - (e^x + 1)(e^{-x} - 1)$

33. $(e^x - e^{-x})^2 - (e^x + e^{-x})(e^x - e^{-x})$

Graph over the indicated interval.

34. $y = 2^{x-1}$; $[-2, 4]$

35. $f(t) = 10e^{-0.08t}$; $t \geq 0$

36. $y = \ln(x + 1)$; $(-1, 10]$

C *Solve Problems 37–40 exactly without the use of a calculator.*

37. $\log x - \log 3 = \log 4 - \log(x + 4)$

38. $\ln(2x - 2) - \ln(x - 1) = \ln x$

39. $\ln(x + 3) - \ln x = 2 \ln 2$

40. $\log 3x^2 = 2 + \log 9x$

41. Write $\ln y = -5t + \ln c$ in an exponential form free of logarithms. Then solve for y in terms of the remaining variables.

42. Explain why 1 cannot be used as a logarithmic base.

APPLICATIONS

The two formulas below will be of use in some of the problems that follow:

$$A = P\left(1 + \frac{r}{m}\right)^{mt} \qquad \text{Compound interest}$$

$$A = Pe^{rt} \qquad \text{Continuous compound interest}$$

Business & Economics

43. Money growth. If \$5,000 is invested at 12% compounded weekly, how much (to the nearest cent) will be in the account 6 years from now?

44. Money growth. If \$5,000 is invested at 12% compounded continuously, how much (to the nearest cent) will be in the account 6 years from now?

45. Finance. Find the tripling time (to the next whole year) for money invested at 15% compounded annually.

46. Finance. Find the doubling time (to two decimal places) for money invested at 10% compounded continuously.

Life Sciences

47. Medicine. One leukemic cell injected into a healthy mouse will divide into 2 cells in about $\frac{1}{2}$ day. At the end of the day these 2 cells will divide into 4. This doubling continues until 1 billion cells are formed; then the animal dies with leukemic cells in every part of the body.

(A) Write an equation that will give the number N of leukemic cells at the end of t days.

(B) When, to the nearest day, will the mouse die?

48. *Marine biology.* The intensity of light entering water is reduced according to the exponential equation

$$I = I_0 e^{-kd}$$

where I is the intensity d feet below the surface, I_0 is the intensity at the surface, and k is the coefficient of extinction. Measurements in the Sargasso Sea in the West Indies have indicated that half of the surface light reaches a depth of 73.6 feet. Find k (to five decimal places), and find the depth (to the nearest foot) at which 1% of the surface light remains.

Social Sciences

49. *Population growth.* Many countries have a population growth rate of 3% (or more) per year. At this rate, how many years (to the nearest tenth of a year) will it take a population to double? Use the annual compounding growth model $P = P_0(1 + r)^t$.

50. *Population growth.* Repeat Problem 49 using the continuous compounding growth model $P = P_0 e^{rt}$.

PART TWO

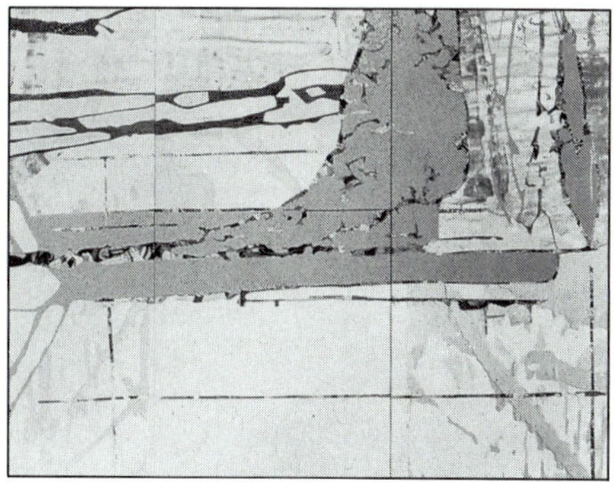

Calculus

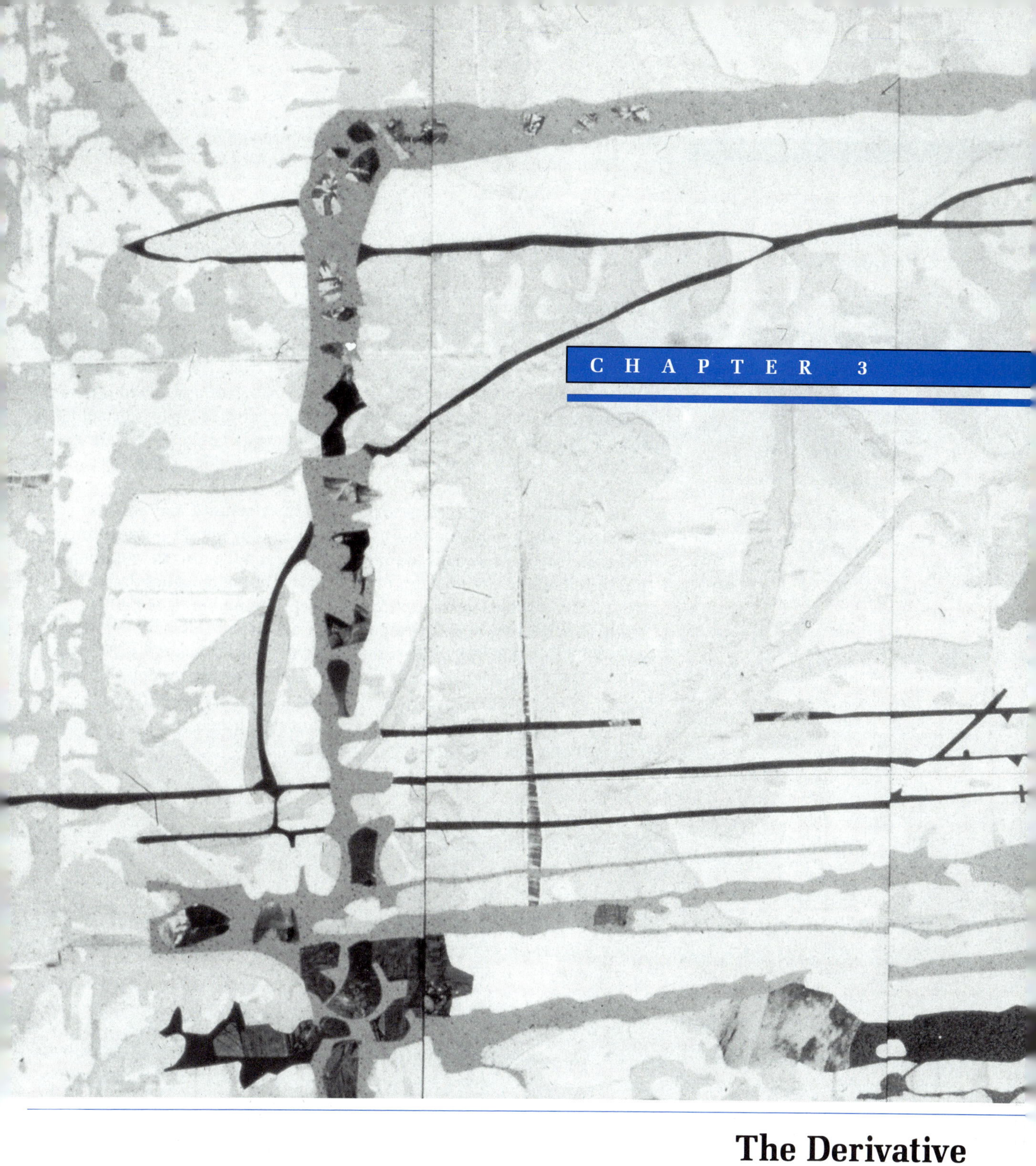

CHAPTER 3

The Derivative

CHAPTER 3

Contents

How do algebra and calculus differ? The two words *static* and *dynamic* probably come as close as any in expressing the difference between the two disciplines. In algebra, we solve equations for a particular value of a variable—a static notion. In calculus, we are interested in how a change in one variable affects another variable —a dynamic notion.

Figure 1 illustrates three basic problems in calculus. It may surprise you to learn that all three problems—as different as they appear—are mathematically related. The solutions to these problems and the discovery of their relationship required the creation of a new kind of mathematics. Isaac Newton (1642–1727) of England and Gottfried Wilhelm von Leibniz (1646–1716) of Germany simultaneously and independently developed this new mathematics, called **the calculus**—it was an idea whose time had come.

In addition to solving the problems described in Figure 1, calculus will enable us to solve many other important problems. Until fairly recently, calculus was used primarily in the physical sciences, but now people in many other disciplines are finding it a useful tool.

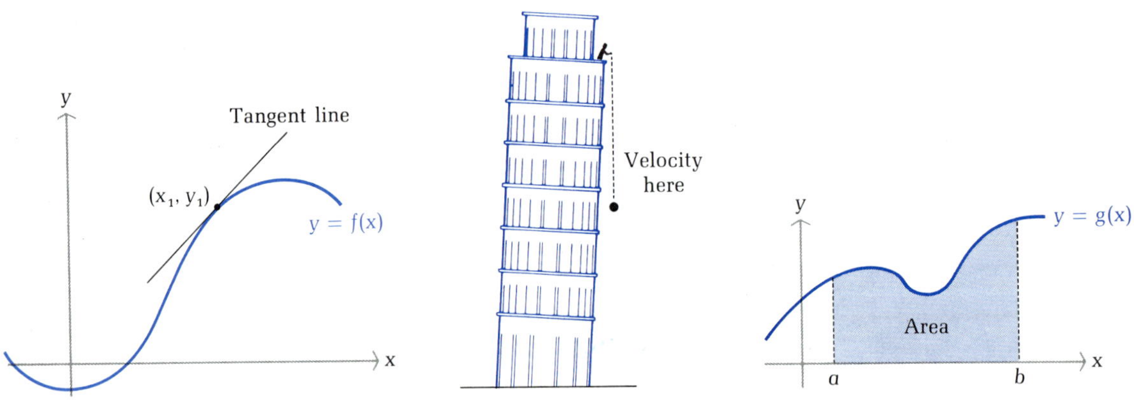

(A) Find the equation of the tangent line at (x_1, y_1) given $y = f(x)$

(B) Find the instantaneous velocity of a falling object

(C) Find the indicated area bounded by $y = g(x)$, $x = a$, $x = b$, and the x axis

FIGURE 1

SECTION 3-1 Limits and Continuity— A Geometric Introduction

- FUNCTIONS AND GRAPHS—A BRIEF REVIEW
- LIMITS
- CONTINUITY
- CONTINUITY PROPERTIES
- APPLICATION

Basic to the study of calculus is the concept of *limit*. This concept helps us to describe in a precise way the behavior of $f(x)$ when x is close to, but not equal to, a particular number c, or when x increases or decreases without bound. Our discussion here will concentrate on concept development and understanding rather than on formal mathematical detail.

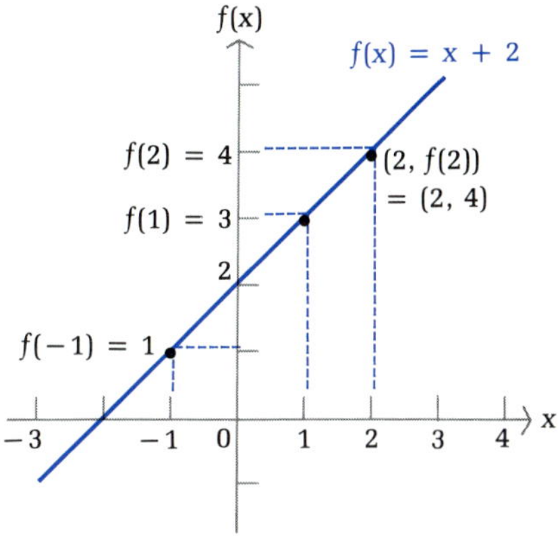

FIGURE 2

FUNCTIONS AND GRAPHS—A BRIEF REVIEW

The graph of the function $y = f(x) = x + 2$ is the graph of the set of all ordered pairs $(x, f(x))$. For example, if $x = 2$, then $f(2) = 4$ and $(2, f(2)) = (2, 4)$ is a point on the graph of f. Figure 2 shows $(-1, f(-1))$, $(1, f(1))$, and $(2, f(2))$ plotted on the graph of f. Notice that the domain values -1, 1, and 2 are associated with the x axis, and the range values $f(-1) = 1$, $f(1) = 3$, and $f(2) = 4$ are associated with the y axis.

Given x, it is sometimes useful to be able to read $f(x)$ directly from the graph of f. Example 1 reviews this process.

EXAMPLE 1 Complete the table below using the given graph of the function g.

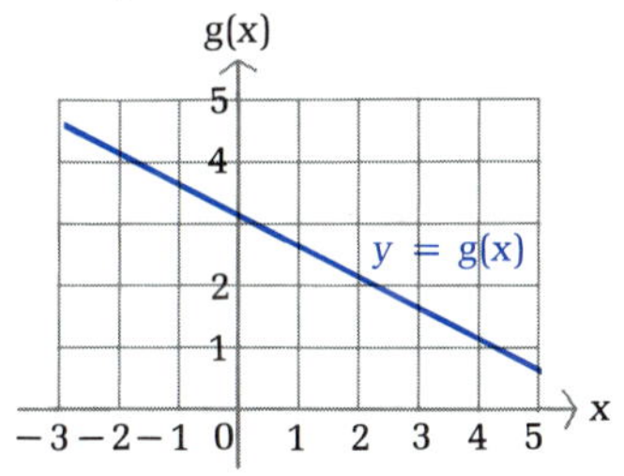

x	g(x)
−2	
1	
3	
4	

Solution To determine $g(x)$, proceed vertically from the x value on the x axis to the graph of g, then horizontally to the corresponding y value, $g(x)$, on the y axis (as indicated by the dashed lines):

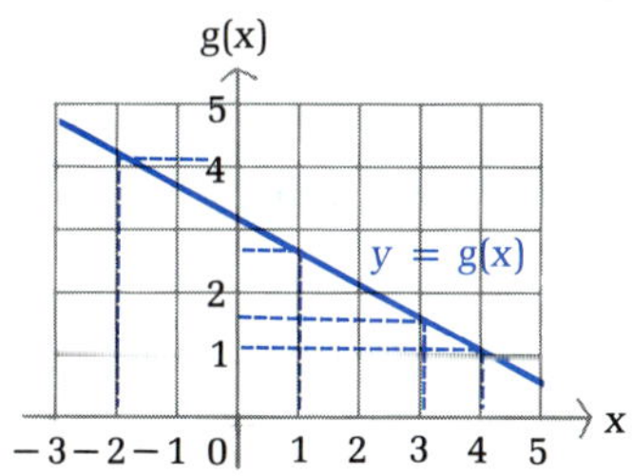

x	g(x)
−2	4.0
1	2.5
3	1.5
4	1.0

◆

PROBLEM 1 Complete the table below using the given graph of the function h.

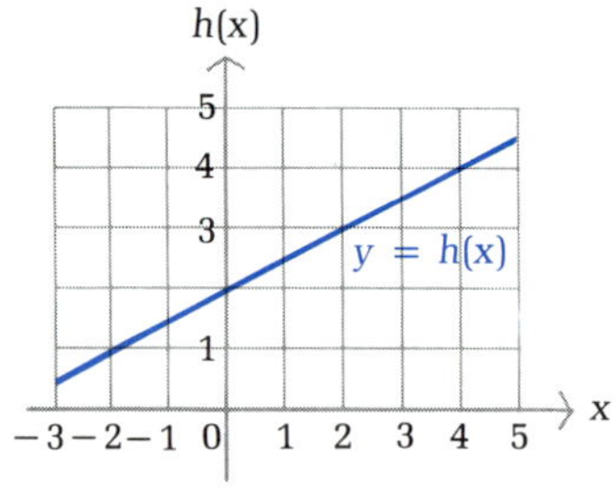

x	$h(x)$
−2	
−1	
0	
1	
2	
3	
4	

◆

◆ LIMITS

We introduce the important notion of *limit* through two examples, after which the limit concept will be defined.

◆ EXAMPLE 2 Let $f(x) = x + 2$. What happens to $f(x)$ when x is chosen closer and closer to 2, but not equal to 2?

Solution We construct a table of values of $f(x)$ for some values of x close to 2 and on either side of 2:

x approaches 2 from the left → 2 ← x approaches 2 from the right

x	1.5	1.8	1.9	1.99	1.999 → 2 ← 2.001	2.01	2.1	2.2	2.5
$f(x)$	3.5	3.8	3.9	3.99	3.999 → ? ← 4.001	4.01	4.1	4.2	4.5

f(x) approaches 4 → 4 ← f(x) approaches 4

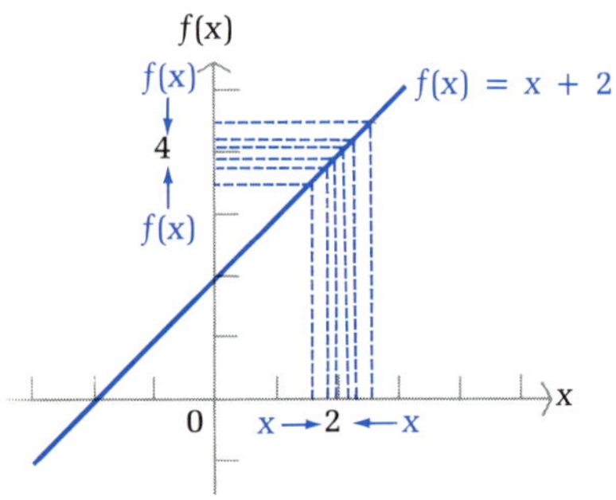

We also draw a graph of f for x near 2, as shown in the margin. Referring to the table and the graph, we see that $f(x)$ approaches 4 as x approaches 2 from either side of 2. We say that "the limit of $x + 2$ as x approaches 2 is 4" and write

$$\lim_{x \to 2} (x + 2) = 4 \qquad \text{or} \qquad x + 2 \to 4 \quad \text{as} \quad x \to 2$$

This means that we can make $f(x) = x + 2$ as close to 4 as we like by restricting x to a sufficiently small interval about 2 but excluding 2.

Also note that $f(2) = 4$. Thus, the value of the function at 2 and the limit of the function at 2 are the same. That is,

$$\lim_{x \to 2} (x + 2) = f(2)$$

Geometrically, this means there is no break, or hole, in the graph of f at $x = 2$.

◆

PROBLEM 2 Let $f(x) = x + 1$.

(A) Complete the following table:

x	0.9	0.99	$0.999 \rightarrow 1 \leftarrow 1.001$	1.01	1.1
$f(x)$	?	?	$? \quad \rightarrow ? \leftarrow ?$	?	?

(B) Referring to the table in part A, find

$$\lim_{x \to 1} (x + 1)$$

(That is, what value does $x + 1$ approach as x approaches 1 from either side of 1, but is not equal to 1?)

(C) Graph $f(x) = x + 1$ for $-1 \leq x \leq 4$.

(D) Referring to the graph in part C, find

$$\lim_{x \to 0} (x + 1) \qquad \text{and} \qquad \lim_{x \to 3} (x + 1)$$ ◆

Example 2 and Problem 2 were fairly obvious. The next example is a little less obvious.

◆ EXAMPLE 3 Let

$$g(x) = \frac{x^2 - 4}{x - 2} \qquad x \neq 2$$

Even though the function is not defined when $x = 2$ (both the numerator and denominator are 0), we can still ask how $g(x)$ behaves when x is near 2, but not equal to 2. Can you guess what happens to $g(x)$ as x approaches 2? The numerator tending to 0 is a force pushing the fraction toward 0. The denominator tending to 0 is another force pushing the fraction toward larger values. How do these two forces balance out?

Solution We could proceed by constructing a table of values to the left and to the right of 2, as in Example 2. Instead, we take advantage of some algebraic simplification using factoring:

$$g(x) = \frac{x^2 - 4}{x - 2} = \frac{(x - 2)(x + 2)}{x - 2} = x + 2 \qquad x \neq 2$$

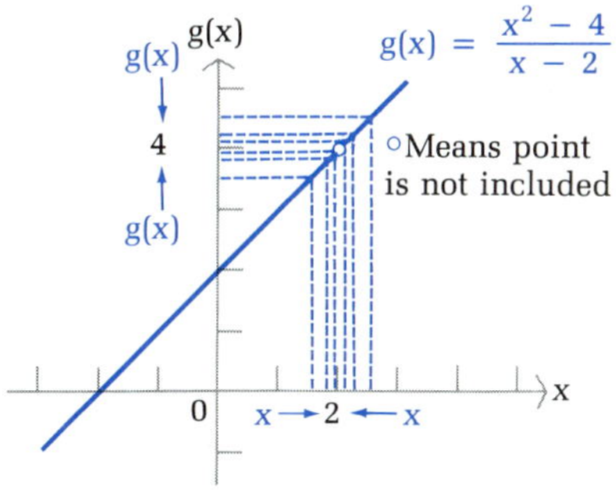

Thus, we see that the graph of function g is the same as the graph of function f in Example 2, except that the graph of g has a hole at the point with coordinates (2, 4), as shown in the margin.

Since the behavior of $(x^2 - 4)/(x - 2)$ for x near 2, but not equal to 2, is the same as the behavior of $x + 2$ for x near 2, but not equal to 2, we have

$$\lim_{x \to 2} \frac{x^2 - 4}{x - 2} = \lim_{x \to 2} (x + 2) = 4$$

And we see that the limit of the function g at 2 exists even though the function is not defined there (the graph has a hole at $x = 2$). ◆

PROBLEM 3 Let

$$g(x) = \frac{x^2 - 1}{x - 1} \qquad x \neq 1$$

[Before proceeding with parts A and B, try to guess what g(x) approaches as x approaches 1, but does not equal 1. Experiment a little with a calculator.]

(A) Graph g for $-1 \leq x \leq 4$. [*Hint:* Proceed as in Example 3.]
(B) Using the graph in part A, find

$$\lim_{x \to 1} \frac{x^2 - 1}{x - 1}$$

◆

We now present an informal definition of the important concept of limit. A precise definition is not needed for our discussion, but one is given in the footnote.*

Limit

We write

$$\lim_{x \to c} f(x) = L \qquad \text{or} \qquad f(x) \to L \quad \text{as} \quad x \to c$$

if the functional value $f(x)$ is close to the single real number L whenever x is close to, but not equal to, c (on either side of c).

[*Note:* The existence of a limit at c has nothing to do with the value of the function at c. In fact, c may not even be in the domain of f (see Examples 2 and 3). However, the function must be defined on both sides of c.]

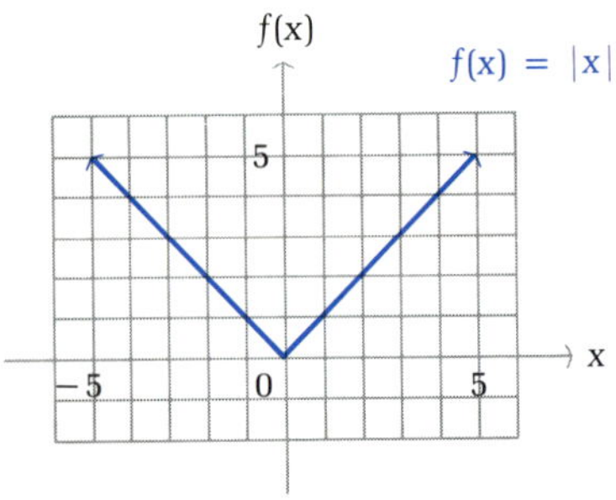

The next example involves the **absolute value function:**

$$f(x) = |x| = \begin{cases} -x & \text{if } x < 0 \\ x & \text{if } x \geq 0 \end{cases} \qquad \begin{aligned} &f(3) = |3| = 3 \\ &f(-2) = |-2| = -(-2) = 2 \end{aligned}$$

The graph of f is shown in the margin.

◆ EXAMPLE 4 Let $h(x) = |x|/x$. Explore the behavior of $h(x)$ for x near 0, but not equal to 0, using a table and a graph. Find $\lim_{x \to 0} h(x)$, if it exists.

* To make the informal definition of limit precise, the use of the word *close* must be made more precise. This is done as follows: We write $\lim_{x \to c} f(x) = L$ if for each $e > 0$, there exists a $d > 0$ such that $|f(x) - L| < e$ whenever $0 < |x - c| < d$. This definition is used to establish particular limits and to prove many useful properties of limits that will be helpful to us in finding particular limits. [Even though intuitive notions of limit existed for a long time, it was not until the nineteenth century that a precise definition was given by the German mathematician, Karl Weierstrass (1815–1897).]

Solution The function h is defined for all real numbers except 0. For example,

$$h(-2) = \frac{|-2|}{-2} = \frac{2}{-2} = -1$$

$$h(0) = \frac{|0|}{0} = \frac{0}{0} \qquad \text{Not defined}$$

$$h(2) = \frac{|2|}{2} = \frac{2}{2} = 1$$

In general, $h(x)$ is -1 for all negative x and 1 for all positive x. The following table and graph illustrate the behavior of $h(x)$ for x near 0:

	x approaches 0 from the left →					0	← x approaches 0 from the right				
x	−2	−1	−0.1	−0.01	−0.001 →	0	← 0.001	0.01	0.1	1	2
h(x)	−1	−1	−1	−1	−1 → −1		1 ← 1	1	1	1	1
					h(x) → −1		1 ← h(x)				

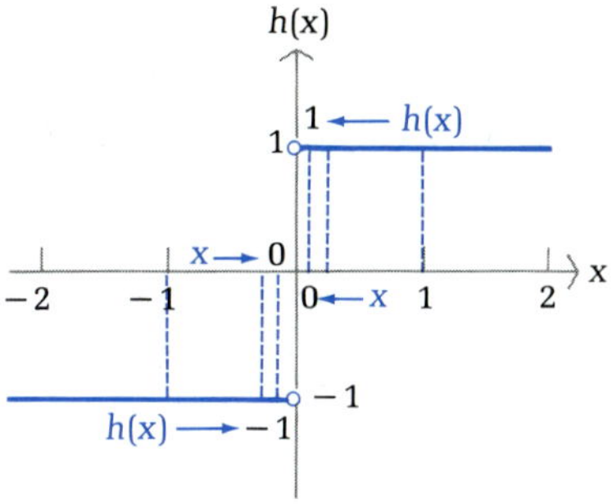

When x is near 0 (on either side of 0), is $h(x)$ near one specific number? The answer is "No," because $h(x)$ is -1 for $x < 0$ and 1 for $x > 0$. Consequently, we say that

$$\lim_{x \to 0} \frac{|x|}{x} \quad \text{does not exist}$$

Thus, neither $h(x)$ nor the limit of $h(x)$ exist at $x = 0$. However, the limit from the left and the limit from the right both exist at 0, but they are not equal. (We will discuss this further below.) ◆

PROBLEM 4 Graph

$$h(x) = \frac{x - 2}{|x - 2|} \qquad -1 \leq x \leq 5$$

and find $\lim_{x \to 2} h(x)$, if it exists. ◆

In Example 2, we found it helpful to examine the values of the function $f(x)$ as x approached 2 from the left and then from the right. In Example 4, we saw that the values of the function $h(x)$ approached two different numbers, depending on the direction of approach, and it was natural to refer to these values as "the limit from the left" and "the limit from the right." These experiences suggest that the notion of **one-sided limits** will be very useful when discussing basic limit concepts.

We write

$$\lim_{x \to c^-} f(c) = K \qquad x \to c^- \text{ is read "x approaches c from the left" and means } x \to c \text{ and } x < c.$$

and call K the **limit from the left** (or **left-hand limit**) if $f(x)$ is close to K whenever x is close to c, but to the left of c on the real number line. We write

$$\lim_{x \to c^+} f(c) = L \qquad x \to c^+ \text{ is read "x approaches c from the right" and means } x \to c \text{ and } x > c.$$

and call L the **limit from the right** (or **right-hand limit**) if $f(x)$ is close to L whenever x is close to c, but to the right of c on the real number line.

We now make the following important observation:

On the Existence of a Limit

In order for a limit to exist, the limit from the left and the limit from the right must exist and be equal.

In Example 4,

$$\lim_{x \to 0^-} \frac{|x|}{x} = -1 \qquad \text{and} \qquad \lim_{x \to 0^+} \frac{|x|}{x} = 1$$

Since the left- and right-hand limits are not the same,

$$\lim_{x \to 0} \frac{|x|}{x} \quad \text{does not exist}$$

◆ EXAMPLE 5

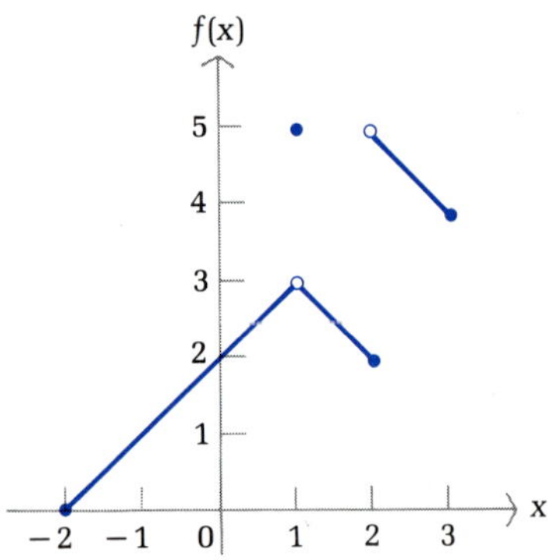

Given the graph of the function f shown in the margin, we discuss the behavior of $f(x)$ for x near -1, 1, 2, and 3:

(A) Behavior of $f(x)$ for x near -1:

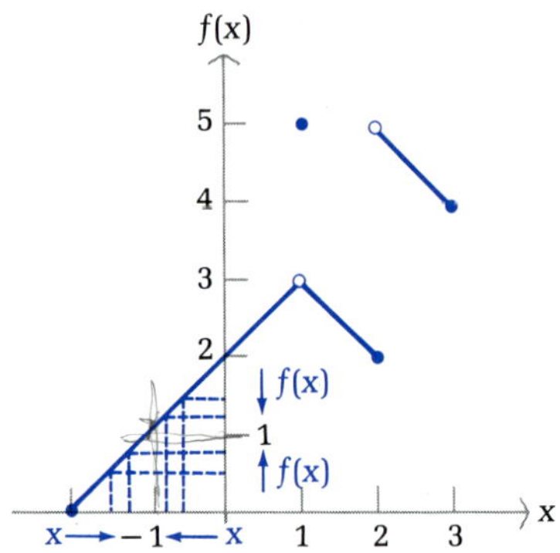

$$\lim_{x \to -1^-} f(x) = 1$$
$$\lim_{x \to -1^+} f(x) = 1$$
$$\lim_{x \to -1} f(x) = 1$$
$$f(-1) = 1$$

(B) Behavior of $f(x)$ for x near 1:

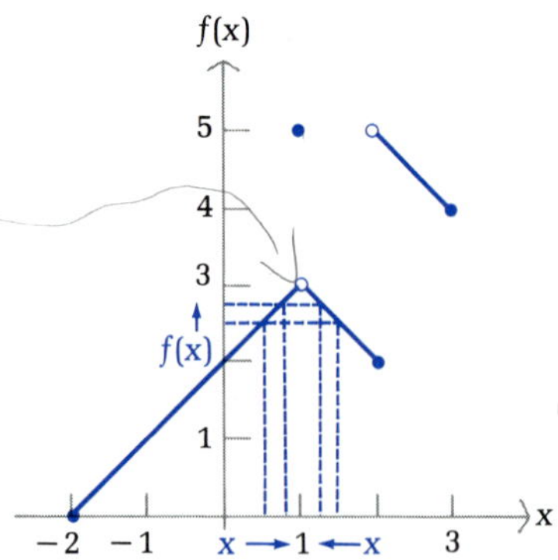

$$\lim_{x \to 1^-} f(x) = 3$$
$$\lim_{x \to 1^+} f(x) = 3$$
$$\lim_{x \to 1} f(x) = 3$$
$$f(1) = 5$$

(C) Behavior of $f(x)$ for x near 2:

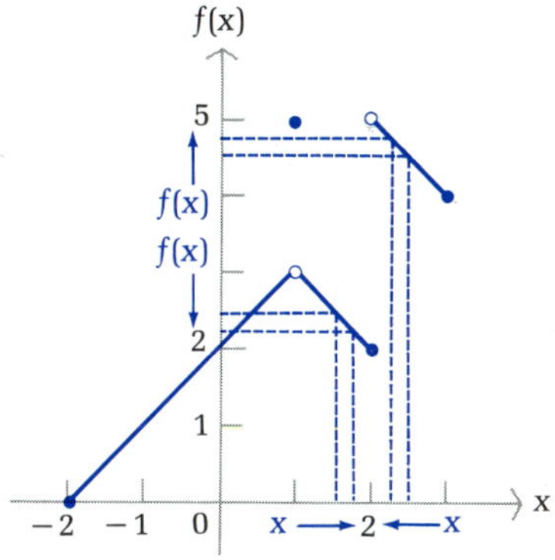

$$\lim_{x\to 2^-} f(x) = 2$$

$$\lim_{x\to 2^+} f(x) = 5$$

$\lim_{x\to 2} f(x)$ does not exist

$$f(2) = 2$$

(D) Behavior of $f(x)$ for x near 3:

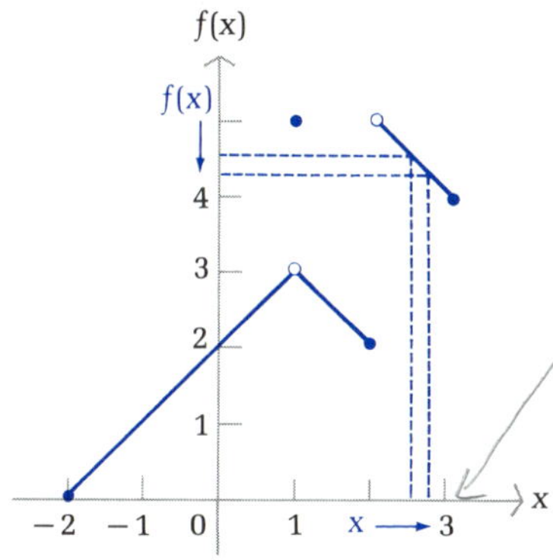

$$\lim_{x\to 3^-} f(x) = 4$$

$\lim_{x\to 3^+} f(x)$ does not exist

f is not defined for $x > 3$

$\lim_{x\to 3} f(x)$ does not exist

$$f(3) = 4$$

◆

PROBLEM 5

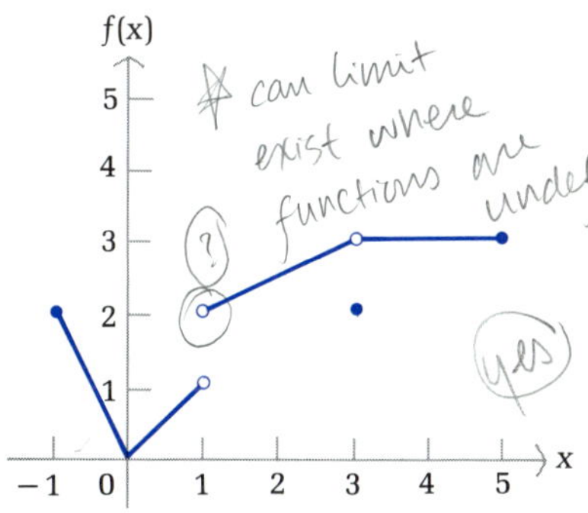

Given the graph of the function f shown in the margin, discuss the following, as we did in Example 5:

(A) Behavior of $f(x)$ for x near −1
(B) Behavior of $f(x)$ for x near 0
(C) Behavior of $f(x)$ for x near 1
(D) Behavior of $f(x)$ for x near 3 ◆

◆ CONTINUITY

Compare the graphs from Examples 2–4, which are repeated in Figure 3 (at the top of the next page). Notice that two of the graphs are broken; that is, they cannot be drawn without lifting a pen off the paper. Informally, a function is *continuous over an interval* if its graph over the interval can be drawn without removing a pen from the paper. A function whose graph is broken (disconnected) at $x = c$ is said to be *discontinuous* at $x = c$. Function f (Figure 3A) is continuous for all x. Function g (Figure 3B) is discontinuous at $x = 2$, but is continuous over any interval that does not include 2. Function h (Figure 3C) is discontinuous at $x = 0$, but is continuous over any interval that does not include 0.

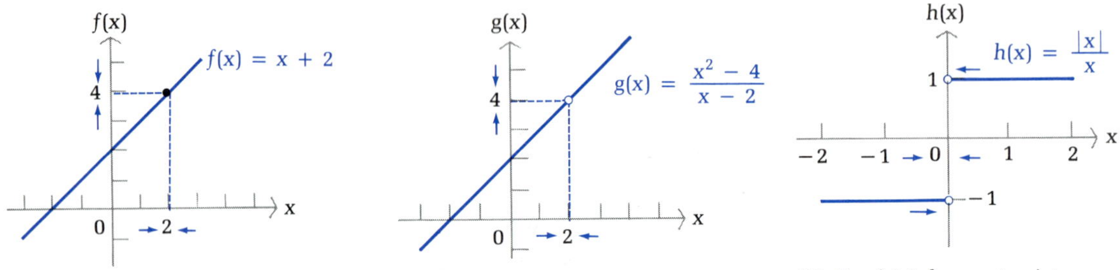

(A) $\lim_{x \to 2} f(x) = 4$
$f(2) = 4$

(B) $\lim_{x \to 2} g(x) = 4$
$g(2)$ is not defined

(C) $\lim_{x \to 0} h(x)$ does not exist
$h(0)$ is not defined

FIGURE 3

Most graphs of natural phenomena are continuous, whereas many graphs in business and economics applications have discontinuities. Figure 4A illustrates temperature variation over a 24 hour period—a continuous phenomenon. Figure 4B illustrates warehouse inventory over a 1 week period—a discontinuous phenomenon.

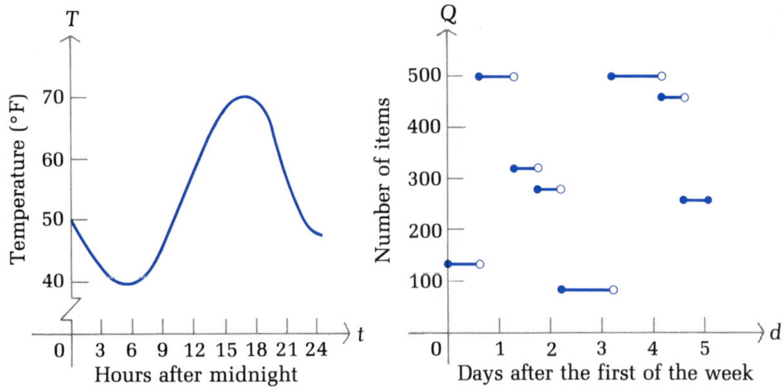

(A) Temperature for a 24 hour period

(B) Inventory in a warehouse during 1 week

FIGURE 4

The preceding discussion, examples, and figures lead to the following formal definition of continuity:

Continuity

A function f is **continuous at the point $x = c$** if

1. $\lim_{x \to c} f(x)$ exists 2. $f(c)$ exists 3. $\lim_{x \to c} f(x) = f(c)$

A function is **continuous on the open interval (a, b)** if it is continuous at each point on the interval.

If one or more of the three conditions in the definition fails, then the function is **discontinuous** at $x = c$.

◆ EXAMPLE 6 Using the definition of continuity, discuss the continuity of each function at the indicated point(s).

(A) $f(x) = x + 2$ at $x = 2$ (B) $g(x) = \dfrac{x^2 - 4}{x - 2}$ at $x = 2$

(C) $h(x) = \dfrac{|x|}{x}$ at $x = 0$ and at $x = 1$ limit DNE.

Solutions (A) f is continuous at $x = 2$, since

$$\lim_{x \to 2} f(x) = 4 = f(2)$$

(See Example 2 and Figure 3A.)

(B) g is not continuous at $x = 2$, since $g(2) = 0/0$ is not defined. (See Example 3 and Figure 3B.)

(C) h is not continuous at $x = 0$, since $h(0) = |0|/0$ is not defined; also, $\lim_{x \to 0} h(x)$ does not exist.

h is continuous at $x = 1$, since

$$\lim_{x \to 1} \frac{|x|}{x} = 1 = h(1)$$

(See Example 4 and Figure 3C.) ◆

PROBLEM 6 Using the definition of continuity, discuss the continuity of each function at the indicated point(s). (Compare with Problems 2–4.)

(A) $f(x) = x + 1$ at $x = 1$ (B) $g(x) = \dfrac{x^2 - 1}{x - 1}$ at $x = 1$

(C) $h(x) = \dfrac{x - 2}{|x - 2|}$ at $x = 2$ and at $x = 0$ ◆

We can also talk about one-sided continuity, just as we talked about one-sided limits. For example, a function is said to be **continuous on the right** at $x = c$ if $\lim_{x \to c^+} f(x) = f(c)$ and **continuous on the left** at $x = c$ if $\lim_{x \to c^-} f(x) = f(c)$. A function is **continuous on the closed interval** $[a, b]$ if it is continuous on the open interval (a, b) and is continuous on the right at a and continuous on the left at b.

Figure 5A (at the top of the next page) illustrates a function that is continuous on the closed interval $[-1, 1]$. Figure 5B illustrates a function that is continuous on a half-closed interval $[0, \infty)$.

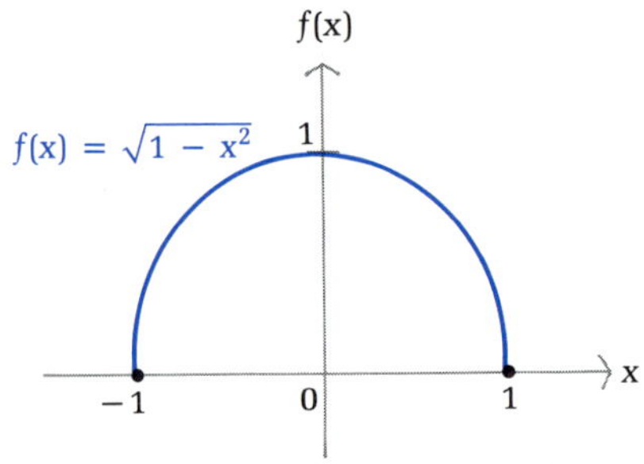

(A) f is continuous on the closed interval $[-1, 1]$

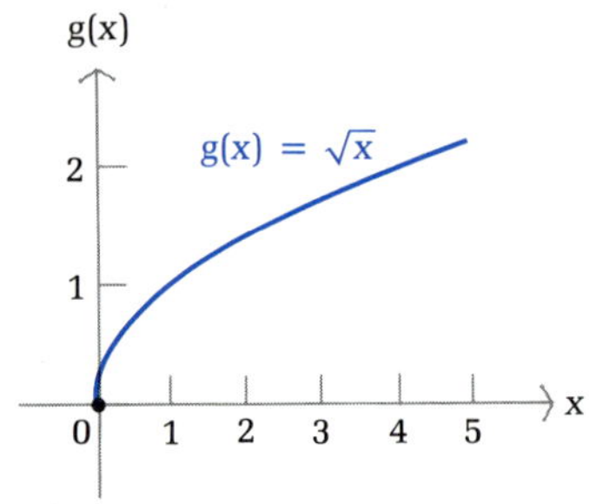

(B) g is continuous on the half-closed interval $[0, \infty)$

FIGURE 5
Continuity on closed and half-closed intervals

◆ CONTINUITY PROPERTIES

Functions have some useful **general continuity properties:**

If two functions are continuous on the same interval, then their sum, difference, product, and quotient are continuous on the same interval, except for values of x that make a denominator 0.

These properties, along with Theorem 1 below, enable us to determine intervals of continuity for some important classes of functions without having to look at their graphs or use the three conditions in the definition.

THEOREM 1

Continuity Properties of Some Specific Functions

(A) A constant function $f(x) = k$, where k is a constant, is continuous for all x.
$f(x) = 7$ is continuous for all x.

(B) For n a positive integer, $f(x) = x^n$ is continuous for all x.
$f(x) = x^5$ is continuous for all x.

(C) A polynomial function is continuous for all x.
$2x^3 - 3x^2 + x - 5$ is continuous for all x.

(D) A rational function is continuous for all x except those values that make a denominator 0.
$\frac{x^2 + 1}{x - 1}$ is continuous for all x except $x = 1$, a value that makes the denominator 0.

(E) For n an odd positive integer greater than 1, $\sqrt[n]{f(x)}$ is continuous wherever $f(x)$ is continuous.
$\sqrt[3]{x^2}$ is continuous for all x.

(F) For n an even positive integer, $\sqrt[n]{f(x)}$ is continuous wherever $f(x)$ is continuous and nonnegative.
$\sqrt[4]{x}$ is continuous on the interval $[0, \infty)$.

Notice that Theorem 1C follows from parts A, B, and the general continuity properties stated above. Also, note that part D follows from part C and the general continuity properties, since a rational function is a function that can be expressed as the quotient of two polynomials.

◆ EXAMPLE 7

Using Theorem 1 and the general properties of continuity, determine where each function is continuous.

(A) $f(x) = x^2 - 2x + 1$

(B) $f(x) = \dfrac{x}{(x + 2)(x - 3)}$

(C) $f(x) = \sqrt[3]{x^2 - 4}$

(D) $f(x) = \sqrt{x - 2}$

Solutions

(A) Since f is a polynomial function, f is continuous for all x.

(B) Since f is a rational function, f is continuous for all x except -2 and 3 (values that make the denominator 0).

(C) The polynomial function $x^2 - 4$ is continuous for all x. Since $n = 3$ is odd, f is continuous for all x.

(D) The polynomial function $x - 2$ is continuous for all x and nonnegative for $x \geq 2$. Since $n = 2$ is even, f is continuous for $x \geq 2$, or on the interval $[2, \infty)$. ◆

PROBLEM 7

Using Theorem 1 and the general properties of continuity, determine where each function is continuous.

(A) $f(x) = x^4 + 2x^2 + 1$

(B) $f(x) = \dfrac{x^2}{(x + 1)(x - 4)}$

(C) $f(x) = \sqrt{x - 4}$

(D) $f(x) = \sqrt[3]{x^3 + 1}$ ◆

◆ APPLICATION

A bicycle messenger service in a downtown financial district uses the weight of a package to determine the charge for a delivery. The charge is \$12 for the first pound (or any fraction thereof) and \$1 for each additional pound (or fraction thereof) up to 10 pounds. If C(x) is the charge for delivering a package weighing x pounds, then

$$C(x) = \begin{cases} \$12 & \text{if } 0 < x \leq 1 \\ \$13 & \text{if } 1 < x \leq 2 \\ \$14 & \text{if } 2 < x \leq 3 \\ \text{and so on} & \end{cases}$$

The function C is graphed for $0 < x \leq 3$ in Figure 6 (at the top of the next page).

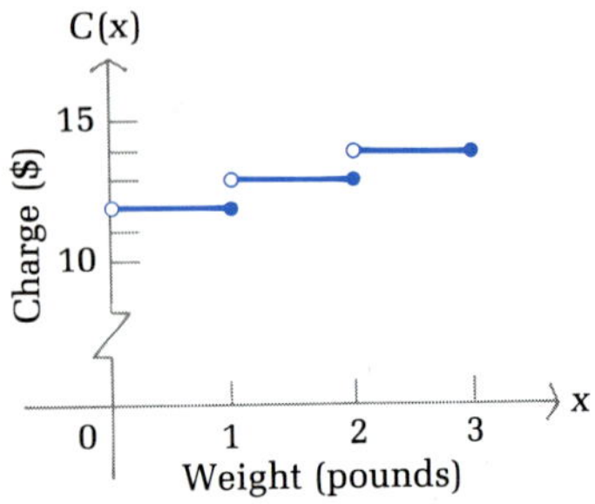

FIGURE 6
Delivery cost function

The following are a few observations about limits and continuity relative to the cost function C over the interval $0 < x \leq 3$:

1. $\lim_{x\to 1.5^-} C(x) = 13$

   $\lim_{x\to 1.5^+} C(x) = 13$

   $\lim_{x\to 1.5} C(x) = 13$

   $C(1.5) = 13$

2. $\lim_{x\to 1^-} C(x) = 12$

   $\lim_{x\to 1^+} C(x) = 13$

   $\lim_{x\to 1} C(x)$ does not exist

   $C(1) = 12$

3. $\lim_{x\to 2^-} C(x) = 13$

   $\lim_{x\to 2^+} C(x) = 14$

   $\lim_{x\to 2} C(x)$ does not exist

   $C(2) = 13$

4. C is continuous at $x = 1.5$, since

$$\lim_{x\to 1.5} C(x) = C(1.5) = 13$$

5. C is discontinuous at 1 and 2, since

$$\lim_{x\to 1} C(x) \quad \text{does not exist}$$

$$\lim_{x\to 2} C(x) \quad \text{does not exist}$$

6. C is continuous from the left, but not from the right, at 1:

$$\lim_{x\to 1^-} C(x) = 12 = C(1)$$

$$\lim_{x\to 1^+} C(x) = 13 \neq C(1) = 12$$

7. C is continuous on the intervals (0, 1], (1, 2], and (2, 3].

Answers to Matched Problems

1.

x	−2	−1	0	1	2	3	4
h(x)	1.0	1.5	2.0	2.5	3.0	3.5	4.0

2. (A)

x	0.9	0.99	0.999 → 1 ← 1.001	1.01	1.1
f(x)	1.9	1.99	1.999 → 2 ← 2.001	2.01	2.1

(B) $\lim_{x\to 1} (x + 1) = 2$

(C)

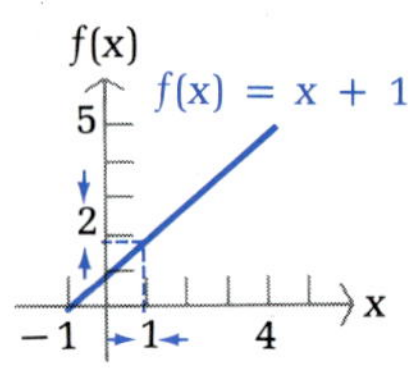

(D) $\lim_{x\to 0} (x + 1) = 1;$

$\lim_{x\to 3} (x + 1) = 4$

3. (A)

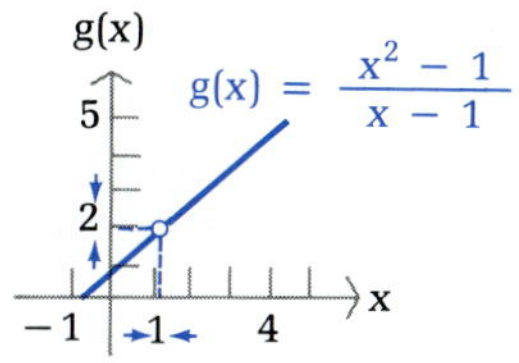

(Note that 1 is not in the domain of g.)

(B) $\lim_{x \to 1} \dfrac{x^2 - 1}{x - 1} = \lim_{x \to 1} (x + 1) = 2$

4.

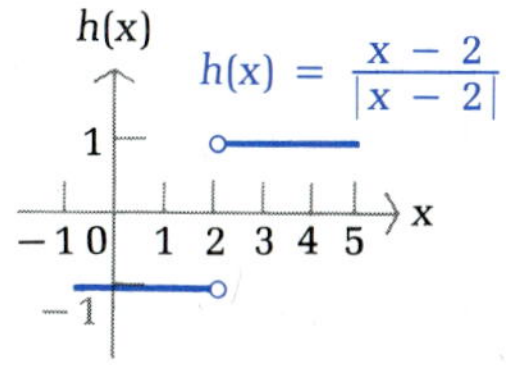

$\lim_{x \to 2} \dfrac{x - 2}{|x - 2|}$ does not exist

5. (A) $\lim_{x \to -1^-} f(x)$ does not exist

$\lim_{x \to -1^+} f(x) = 2$

$\lim_{x \to -1} f(x)$ does not exist

$f(-1) = 2$

(B) $\lim_{x \to 0^-} f(x) = 0$

$\lim_{x \to 0^+} f(x) = 0$

$\lim_{x \to 0} f(x) = 0$

$f(0) = 0$

(C) $\lim_{x \to 1^-} f(x) = 1$

$\lim_{x \to 1^+} f(x) = 2$

$\lim_{x \to 1} f(x)$ does not exist

$f(1)$ not defined

(D) $\lim_{x \to 3^-} f(x) = 3$

$\lim_{x \to 3^+} f(x) = 3$

$\lim_{x \to 3} f(x) = 3$

$f(3) = 2$

6. (A) f is continuous at $x = 1$, since $\lim_{x \to 1} f(x) = 2 = f(1)$. (See answer to Problem 2 above.)
   (B) g is not continuous at $x = 1$, since $g(1) = 0/0$ is not defined. (See answer to Problem 3 above.)
   (C) h is not continuous at $x = 2$, since $h(2) = 0/|0| = 0/0$ is not defined; also, $\lim_{x \to 2} h(x)$ does not exist. h is continuous at $x = 0$, since $\lim_{x \to 0} h(x) = -1 = h(0)$. (See answer to Problem 4 above.)
7. (A) Since f is a polynomial function, it is continuous for all x.
   (B) Since f is a rational function, f is continuous for all x except -1 and 4 (values that make the denominator 0).
   (C) The polynomial function $x - 4$ is continuous for all x and nonnegative for $x \geqslant 4$. Since $n = 2$ is even, f is continuous for $x \geqslant 4$, or on the interval $[4, \infty)$.
   (D) The polynomial function $x^3 + 1$ is continuous for all x. Since $n = 3$ is odd, f is continuous for all x.

EXERCISE 3-1

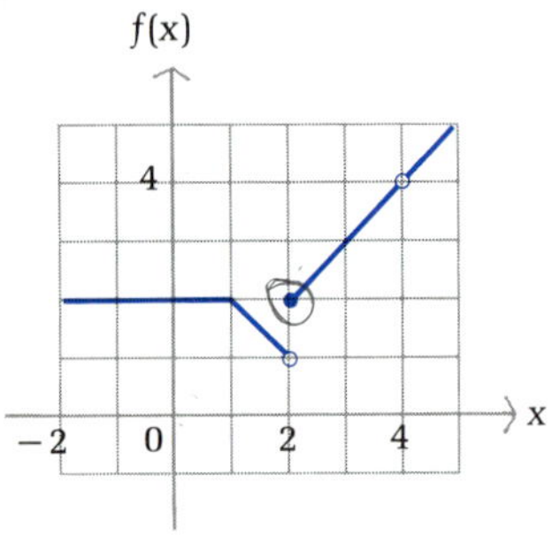

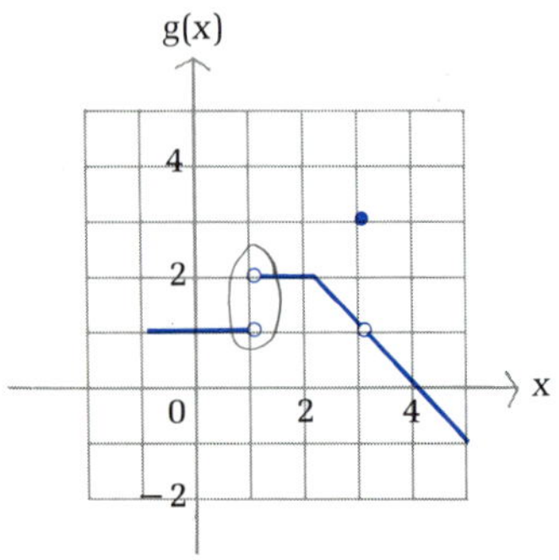

A

1. Use the graph of the function f shown in the margin to estimate each limit, if it exists.
   (A) $\lim_{x \to 0} f(x)$
   (B) $\lim_{x \to 1} f(x)$
   (C) $\lim_{x \to 2} f(x)$
   (D) $\lim_{x \to 4} f(x)$

2. Use the graph of the function g shown in the margin to estimate each limit, if it exists.
   (A) $\lim_{x \to 0} g(x)$
   (B) $\lim_{x \to 1} g(x)$
   (C) $\lim_{x \to 2} g(x)$
   (D) $\lim_{x \to 3} g(x)$

3. Use the graph of the function f in Problem 1 to estimate each value, if it is defined:
   (A) $f(0)$ (B) $f(1)$ (C) $f(2)$ (D) $f(4)$

4. Use the graph of the function g in Problem 2 to estimate each value, if it is defined:
   (A) $g(0)$ (B) $g(1)$ (C) $g(2)$ (D) $g(3)$

5. Referring to the graph of function f in Problem 1, for which of the values $c = 0, 1, 2,$ or 4 does $\lim_{x \to c} f(x) = f(c)$?

6. Referring to the graph of function g in Problem 2, for which of the values $c = 0, 1, 2,$ or 3 does $\lim_{x \to c} g(x) = g(c)$?

7. For which of the following values of x is the function f in Problem 1 discontinuous?

   0, 1, 2, 4

8. For which of the following values of x is the function g in Problem 2 discontinuous?

   0, 1, 2, 3

Use Theorem 1 to determine where each function in Problems 9–14 is continuous.

9. $f(x) = 2x - 3$
10. $g(x) = 3 - 5x$
11. $h(x) = \dfrac{2}{x - 5}$
12. $k(x) = \dfrac{x}{x + 3}$
13. $g(x) = \dfrac{x - 5}{(x - 3)(x + 2)}$
14. $F(x) = \dfrac{1}{x(x + 7)}$

B *Problems 15–20 refer to the function f shown in the following graph. Use the graph to estimate limits.*

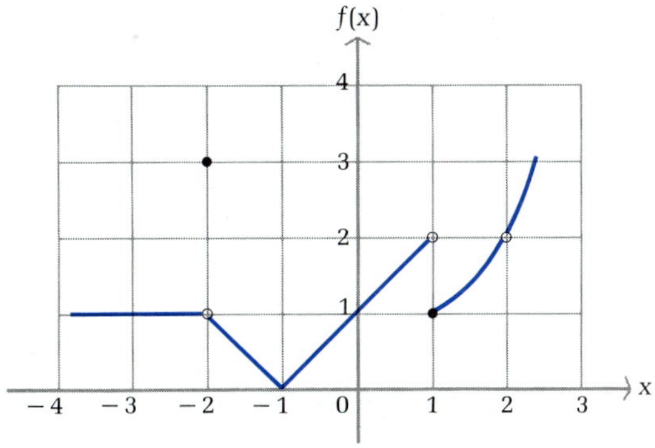

For each value of c:

(A) Find $\lim_{x\to c^-} f(x)$, $\lim_{x\to c^+} f(x)$, $\lim_{x\to c} f(x)$, *and* $f(c)$.

(B) Is f continuous at $x = c$*? Explain.*

15. $c = 0$ **16.** $c = -1$ **17.** $c = 1$

18. $c = 2$ **19.** $c = -2$ **20.** $c = 0.5$

21. Given the following function *f*:

$$f(x) = \begin{cases} 2 & \text{if } x \text{ is an integer} \\ 1 & \text{if } x \text{ is not an integer} \end{cases}$$

(A) Graph *f*. (B) $\lim_{x\to 2} f(x) = ?$ (C) $f(2) = ?$

(D) Is *f* continuous at $x = 2$? (E) Where is *f* discontinuous?

22. Given the following function g:

$$g(x) = \begin{cases} -1 & \text{if } x \text{ is an even integer} \\ 1 & \text{if } x \text{ is not an even integer} \end{cases}$$

(A) Graph g. (B) $\lim_{x\to 1} g(x) = ?$ (C) $g(1) = ?$

(D) Is g continuous at $x = 1$?

(E) Where is g discontinuous?

In Problems 23–30, find all limits that exist using intuition, graphing, or a little algebra, as needed.

23. $\lim_{x\to 2} (2x + 1)$ **24.** $\lim_{x\to 1} (3x - 2)$ **25.** $\lim_{x\to 2} 7$

26. $\lim_{x\to 5} 9$ **27.** $\lim_{x\to -3} \frac{x^2 - 9}{x + 3}$ **28.** $\lim_{x\to -5} \frac{x^2 - 25}{x + 5}$

29. $\lim_{x\to 1^+} \frac{|x - 1|}{x - 1}$ **30.** $\lim_{x\to 3^-} \frac{x - 3}{|x - 3|}$

Use Theorem 1 to determine where each function in Problems 31–38, is continuous. Express the answer in interval notation.

31. $F(x) = 2x^8 - 3x^4 + 5$

32. $h(x) = \dfrac{x^4 - 3x + 5}{x^2 + 2x}$

33. $g(x) = \sqrt{x - 5}$

34. $f(x) = \sqrt{3 - x}$

35. $K(x) = \sqrt[3]{x - 5}$

36. $H(x) = \sqrt[3]{3 - x}$

37. $f(x) = \dfrac{x^2 - 1}{x^2 - 3x + 2}$

38. $k(x) = \dfrac{x^2 - 4}{x^2 + x - 2}$

Complete the following table for each function in Problems 39–42:

x	0.9	0.99	0.999 → 1 ← 1.001	1.01	1.1
f(x)			→ ? ←		

From the completed table, guess the following (a calculator will be helpful for some):

(A) $\lim_{x \to 1^-} f(x)$ (B) $\lim_{x \to 1^+} f(x)$ (C) $\lim_{x \to 1} f(x)$

39. $f(x) = \dfrac{|x - 1|}{x - 1}$

40. $f(x) = \dfrac{x - 1}{|x| - 1}$

41. $f(x) = \dfrac{x^3 - 1}{x - 1}$

42. $f(x) = \dfrac{x^4 - 1}{x - 1}$

In Problems 43–48, graph f and locate all points of discontinuity.

43. $f(x) = \begin{cases} 1 + x & \text{if } x < 1 \\ 5 - x & \text{if } x \geqslant 1 \end{cases}$

44. $f(x) = \begin{cases} x^2 & \text{if } x \leqslant 1 \\ 2x & \text{if } x > 1 \end{cases}$

45. $f(x) = \begin{cases} 1 + x & \text{if } x \leqslant 2 \\ 5 - x & \text{if } x > 2 \end{cases}$

46. $f(x) = \begin{cases} x^2 & \text{if } x \leqslant 2 \\ 2x & \text{if } x > 2 \end{cases}$

47. $f(x) = \begin{cases} -x & \text{if } x < 0 \\ 1 & \text{if } x = 0 \\ x & \text{if } x > 0 \end{cases}$

48. $f(x) = \begin{cases} 1 & \text{if } x < 0 \\ 0 & \text{if } x = 0 \\ 1 + x & \text{if } x > 0 \end{cases}$

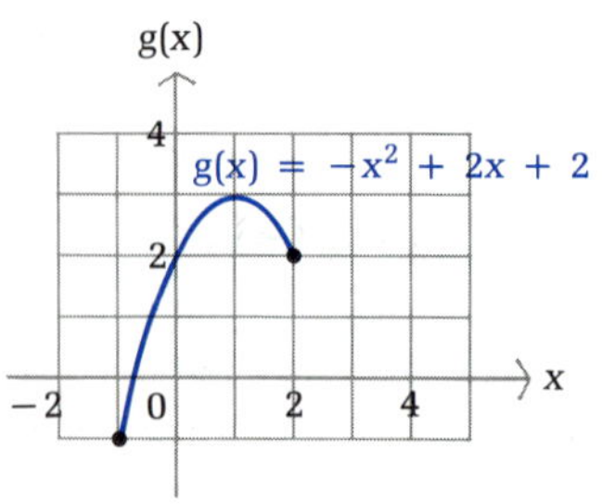

C

49. Use the graph of the function g shown in the margin to answer the following questions:

(A) Is g continuous on the open interval $(-1, 2)$?

(B) Is g continuous from the right at $x = -1$? That is, does $\lim_{x \to -1^+} g(x) = g(-1)$?

(C) Is g continuous from the left at $x = 2$? That is, does $\lim_{x \to 2^-} g(x) = g(2)$?

(D) Is g continuous on the closed interval $[-1, 2]$?

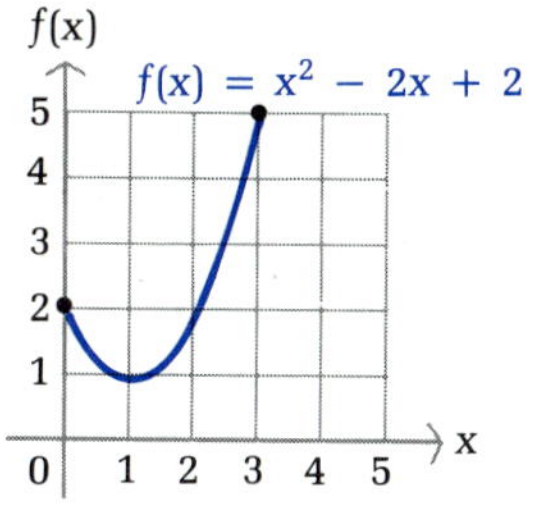

50. Use the graph of the function f shown in the margin to answer the following questions:

(A) Is f continuous on the open interval $(0, 3)$?
(B) Is f continuous from the right at $x = 0$? That is, does $\lim_{x\to 0^+} f(x) = f(0)$?
(C) Is f continuous from the left at $x = 3$? That is, does $\lim_{x\to 3^-} f(x) = f(3)$?
(D) Is f continuous on the closed interval $[0, 3]$?

Problems 51 and 52 refer to the ***greatest integer function,*** *which is denoted by* $[\![x]\!]$ *and is defined as follows:*

$$[\![x]\!] = \textit{Greatest integer} \leq x$$

For example,

$$[\![-3.6]\!] = \textit{Greatest integer} \leq -3.6 = -4$$
$$[\![2]\!] = \textit{Greatest integer} \leq 2 = 2$$
$$[\![2.5]\!] = \textit{Greatest integer} \leq 2.5 = 2$$

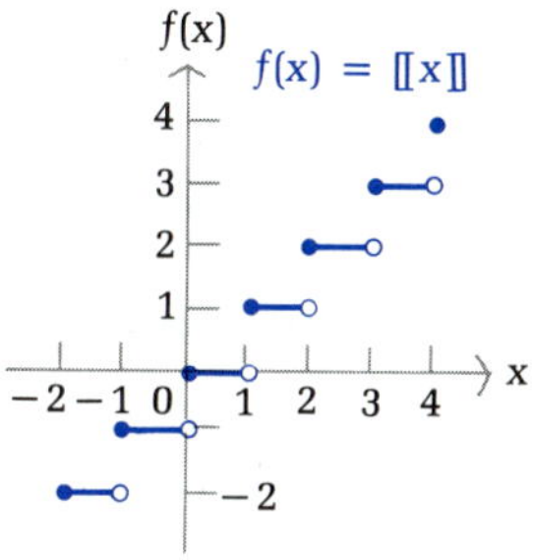

The graph of $f(x) = [\![x]\!]$ *is shown in the margin. There, we can see that*

$$
\begin{aligned}
[\![x]\!] &= -2 && \textit{for} && -2 \leq x < -1\\
[\![x]\!] &= -1 && \textit{for} && -1 \leq x < 0\\
[\![x]\!] &= 0 && \textit{for} && 0 \leq x < 1\\
[\![x]\!] &= 1 && \textit{for} && 1 \leq x < 2\\
[\![x]\!] &= 2 && \textit{for} && 2 \leq x < 3
\end{aligned}
$$

and so on.

51. (A) Is f continuous from the right at $x = 0$?
(B) Is f continuous from the left at $x = 0$?
(C) Is f continuous on the open interval $(0, 1)$?
(D) Is f continuous on the closed interval $[0, 1]$?
(E) Is f continuous on the half-closed interval $[0, 1)$?

52. (A) Is f continuous from the right at $x = 2$?
(B) Is f continuous from the left at $x = 2$?
(C) Is f continuous on the open interval $(1, 2)$?
(D) Is f continuous on the closed interval $[1, 2]$?
(E) Is f continuous on the half-closed interval $[1, 2)$?

Use your intuition and current knowledge about continuous functions to answer Problems 53 and 54. (The ideas in these two problems will be generalized in later sections.)

53. If a function f is continuous and never equal to 0 on the interval $(1, 5)$, and if $f(2) = 3$, can $f(4)$ be negative? Can $f(x)$ be negative for any x on the interval $(1, 5)$?

54. If a function g is continuous and never equal to 0 on the interval $(-1, 8)$, and if $g(0) = -3$, can $g(1)$ be positive? Is $g(x)$ negative for all x on the interval $(-1, 8)$?

Problems 55–62 require the use of a graphic calculator or a computer. Graph each function and use the graph to estimate the left- and right-hand limits at the indicated value(s) of c.

55. $f(x) = \dfrac{x^2 + 3|x|}{x}$, $c = 0$

56. $f(x) = \dfrac{x^2 - 5|x|}{x}$, $c = 0$

57. $f(x) = \dfrac{x^3 + 4|x|}{x}$, $c = 0$

58. $f(x) = \dfrac{|x|^3 + 2|x|}{x}$, $c = 0$

59. $f(x) = \dfrac{x^2 - 4}{|x - 2|}$, $c = 2$

60. $f(x) = \dfrac{1 - x^2}{|x + 1|}$, $c = -1$

61. $f(x) = \dfrac{x^3 - 9x}{|x^2 - 9|}$, $c = -3$ and $c = 3$

62. $f(x) = \dfrac{4x - x^3}{|x^2 - 4|}$, $c = -2$ and $c = 2$

APPLICATIONS

Business & Economics

63. *Postal rates.* First-class postage in 1989 was \$0.25 for the first ounce (or any fraction thereof) and \$0.20 for each additional ounce (or fraction thereof) up to 11 ounces. If P(x) is the amount of postage for a letter weighing x ounces, then we can write

$$P(x) = \begin{cases} \$0.25 & \text{if } 0 < x \le 1 \\ \$0.45 & \text{if } 1 < x \le 2 \\ \$0.65 & \text{if } 2 < x \le 3 \\ \text{and so on} \end{cases}$$

(A) Graph P for $0 < x \le 5$.
(B) Find $\lim_{x \to 4.5} P(x)$ and $P(4.5)$.
(C) Find $\lim_{x \to 4} P(x)$ and $P(4)$.
(D) Is P continuous at $x = 4.5$? At $x = 4$?

64. *Telephone rates.* A person placing a station-to-station call on Saturday from San Francisco to New York is charged \$0.30 for the first minute (or any fraction thereof) and \$0.20 for each additional minute (or fraction thereof). If the length of a call is x minutes, then the long-distance charge R(x) is

$$R(x) = \begin{cases} \$0.30 & \text{if } 0 < x \le 1 \\ \$0.50 & \text{if } 1 < x \le 2 \\ \$0.70 & \text{if } 2 < x \le 3 \\ \text{and so on} \end{cases}$$

(A) Graph R for $0 < x \leq 6$.
(B) Find $\lim_{x \to 2.5} R(x)$ and $R(2.5)$.
(C) Find $\lim_{x \to 2} R(x)$ and $R(2)$.
(D) Is R continuous at $x = 2.5$? At $x = 2$?

65. *Income.* A personal computer salesperson receives a base salary of \$1,000 per month and a commission of 5% of all sales over \$10,000 during the month. If the monthly sales are \$20,000 or more, the salesperson is given an additional \$500 bonus. Let $E(s)$ represent the person's earnings during the month as a function of the monthly sales s.

(A) Graph $E(s)$ for $0 \leq s \leq 30{,}000$.
(B) Find $\lim_{s \to 10{,}000} E(s)$ and $E(10{,}000)$.
(C) Find $\lim_{s \to 20{,}000} E(s)$ and $E(20{,}000)$.
(D) Is E continuous at $s = 10{,}000$? At $s = 20{,}000$?

66. *Equipment rental.* An office equipment rental and leasing company rents electric typewriters for \$10 per day (and any fraction thereof) or for \$50 per 7 day week. Let $C(x)$ be the cost of renting a typewriter for x days.

(A) Graph $C(x)$ for $0 \leq x \leq 10$.
(B) Find $\lim_{x \to 4.5} C(x)$ and $C(4.5)$.
(C) Find $\lim_{x \to 8} C(x)$ and $C(8)$.
(D) Is C continuous at $x = 4.5$? At $x = 8$?

Life Sciences

67. *Animal supply.* A medical laboratory raises its own rabbits. The number of rabbits $N(t)$ available at any time t depends on the number of births and deaths. When a birth or death occurs, the function N generally has a discontinuity, as shown in the figure.

(A) Where is the function N discontinuous?
(B) $\lim_{t \to t_5} N(t) = ?$; $N(t_5) = ?$ (C) $\lim_{t \to t_3} N(t) = ?$; $N(t_3) = ?$

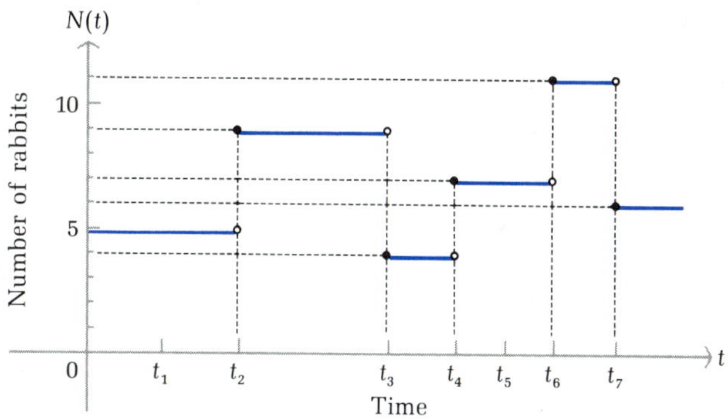

Social Sciences

68. *Learning.* The graph shown on the next page might represent the history of a particular person learning the material on limits and continuity in this book. At time t_2, the student's mind goes blank during a quiz. At time t_4, the instructor explains a concept particularly well, and suddenly, a big jump in understanding takes place.

(A) Where is the function p discontinuous?

(B) $\lim_{t \to t_1} p(t) = ?$; $p(t_1) = ?$

(C) $\lim_{t \to t_2} p(t) = ?$; $p(t_2) = ?$

(D) $\lim_{t \to t_4} p(t) = ?$; $p(t_4) = ?$

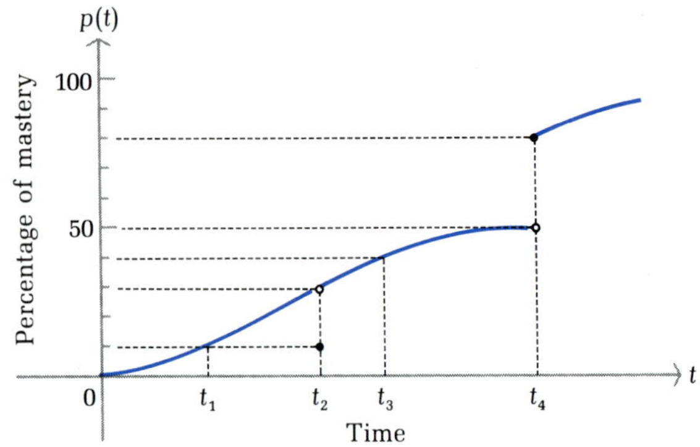

SECTION 3-2 Computation of Limits

- LIMITS AT POINTS OF CONTINUITY
- LIMITS AND INFINITY
- LIMIT PROPERTIES
- LIMITS AT INFINITY
- SUMMARY OF LIMIT FORMS

To introduce the concept of limit in the preceding section, we relied heavily on tables, graphs, and intuition. But we will find it very useful to be able to compute limits of functions without having to resort to graphs or tables. In this section, we will introduce and use various properties of limits to find limits of a wide variety of functions, in the same way we used continuity properties earlier. We will also discuss the use of the infinity symbol, ∞, relative to limits.

LIMITS AT POINTS OF CONTINUITY

If we know that a function is continuous at a point, then, using the definition of continuity at a point from the preceding section, it is easy to find the limit of the function at that point—we simply evaluate the function at that point.

Limit at a Point of Continuity

If a function f is continuous at $x = c$, then

$$\lim_{x \to c} f(x) = f(c)$$

For example,

$$\lim_{x \to 2} (x^2 - x + 1) = 2^2 - 2 + 1 = 3$$

(since a polynomial function is continuous for all x).

◆ EXAMPLE 8 Find the following limits:

(A) $\lim_{x \to 3} \dfrac{x^2 - 3x}{x + 7}$ (B) $\lim_{x \to 5} \sqrt{x - 2}$ (C) $\lim_{x \to 3} 5$

Solutions (A) Since $(x^2 - 3x)/(x + 7)$ is continuous for all x except $x = -7$, it is continuous at $x = 3$. Thus,

$$\lim_{x \to 3} \frac{x^2 - 3x}{x + 7} = \frac{3^2 - 3(3)}{3 + 7} = \frac{9 - 9}{10} = \frac{0}{10} = 0$$

(B) Since $x - 2$ is continuous for all x and nonnegative for $x \geq 2$, $\sqrt{x - 2}$ is continuous for all $x \geq 2$. In particular, $\sqrt{x - 2}$ is continuous for $x = 5$. Thus,

$$\lim_{x \to 5} \sqrt{x - 2} = \sqrt{5 - 2} = \sqrt{3}$$

(C) Since $f(x) = 5$ is a constant function, it is continuous for all x. In particular, it is continuous when $x = 3$. Thus,

$$\lim_{x \to 3} f(x) = f(3) = 5$$

That is, $\lim_{x \to 3} 5 = 5$. ◆

PROBLEM 8 Find the following limits:

(A) $\lim_{x \to 4} (x + \sqrt{x})$ (B) $\lim_{x \to -7} \sqrt[3]{1 - x}$ (C) $\lim_{x \to -2} 10$ ◆

◆ LIMITS AND INFINITY

How does $f(x) = 1/x$ behave for x near 0? When a number gets small (close to 0), the reciprocal of its absolute value gets large. For example, when $x = 0.001$, $1/x = 1{,}000$; when $x = -0.001$, $1/x = -1{,}000$. Thus, as x approaches 0 from the right, $f(x)$ is positive and increases without bound. We indicate this behavior by writing

$$\lim_{x \to 0^+} \frac{1}{x} = \infty$$

Similarly, as x approaches 0 from the left, $f(x)$ is negative and decreases without bound. We indicate this behavior by writing

$$\lim_{x \to 0^-} \frac{1}{x} = -\infty$$

In neither case does the limit exist, since neither ∞ nor $-\infty$ are numbers. Nevertheless, the notation tells us something about the behavior of $f(x)$ for x near 0. The graph of f is illustrated in Figure 7A. Figure 7B illustrates the behavior of $g(x) = 1/x^2$ for x near 0. Note that for $g(x)$ we can write

$$\lim_{x \to 0} \frac{1}{x^2} = \infty$$

since $g(x)$ remains positive and increases without bound as x approaches 0 from either side.

For both functions f and g, the line $x = 0$ (the vertical axis) is called a *vertical asymptote*. In general, a line $x = a$ is a **vertical asymptote** for the graph of $y = f(x)$ if $f(x)$ either increases or decreases without bound as x approaches a

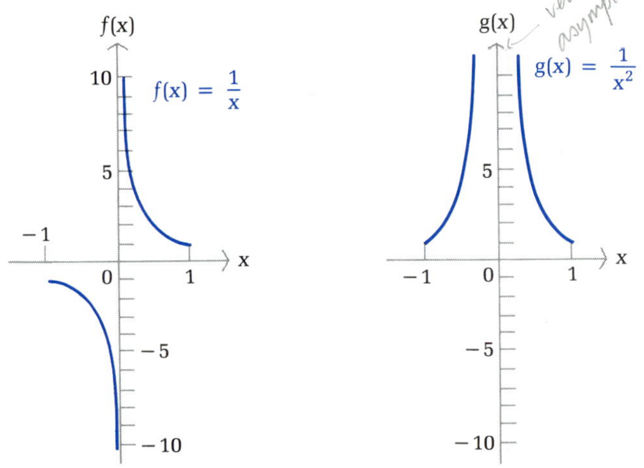

(A) $\lim_{x \to 0^-} \frac{1}{x} = -\infty$

$\lim_{x \to 0^+} \frac{1}{x} = \infty$

$\lim_{x \to 0} \frac{1}{x}$ does not exist

(B) $\lim_{x \to 0^-} \frac{1}{x^2} = \infty$

$\lim_{x \to 0^+} \frac{1}{x^2} = \infty$

$\lim_{x \to 0} \frac{1}{x^2} = \infty$

FIGURE 7

from either the left or the right. Symbolically, $x = a$ is a vertical asymptote if any of the following hold:

$$\lim_{x \to a^-} f(x) = -\infty \text{ (or } \infty)$$

$$\lim_{x \to a^+} f(x) = -\infty \text{ (or } \infty)$$

$$\lim_{x \to a} f(x) = -\infty \text{ (or } \infty)$$

◆ LIMIT PROPERTIES

We now turn to some basic properties of limits that will enable us to evaluate many limits when either the function involved is not continuous at the point or it is not certain whether the function is continuous at the point. [Remember, if a function f is continuous at $x = c$, then we can find $\lim_{x \to c} f(x)$ by evaluating the function f at $x = c$.]

THEOREM 2

Properties of Limits

Let f and g be two functions, and assume that

$$\lim_{x \to c} f(x) = L \qquad \lim_{x \to c} g(x) = M$$

where L and M are real numbers (both limits exist). Then,

1. $\lim_{x \to c} [f(x) + g(x)] = \lim_{x \to c} f(x) + \lim_{x \to c} g(x) = L + M$
2. $\lim_{x \to c} [f(x) - g(x)] = \lim_{x \to c} f(x) - \lim_{x \to c} g(x) = L - M$
3. $\lim_{x \to c} kf(x) = k \lim_{x \to c} f(x) = kL$ for any constant k
4. $\lim_{x \to c} [f(x) \cdot g(x)] = \left[\lim_{x \to c} f(x)\right]\left[\lim_{x \to c} g(x)\right] = LM$
5. $\lim_{x \to c} \frac{f(x)}{g(x)} = \frac{\lim_{x \to c} f(x)}{\lim_{x \to c} g(x)} = \frac{L}{M}$ if $M \neq 0$
6. $\lim_{x \to c} \sqrt[n]{f(x)} = \sqrt[n]{\lim_{x \to c} f(x)} = \sqrt[n]{L}$ $L > 0$ for n even

Suppose

$$\lim_{x \to 2} f(x) = 3 \qquad \text{and} \qquad \lim_{x \to 2} g(x) = 7$$

then no other information is needed about the functions f and g to determine that

$$\lim_{x \to 2} [f(x) - g(x)] = \lim_{x \to 2} f(x) - \lim_{x \to 2} g(x) = 3 - 7 = -4$$

$$\lim_{x \to 2} 9f(x) = 9 \lim_{x \to 2} f(x) = 9 \cdot 3 = 27$$

$$\lim_{x\to 2}\frac{f(x)}{g(x)}=\frac{\lim_{x\to 2}f(x)}{\lim_{x\to 2}g(x)}=\frac{3}{7}$$

$$\lim_{x\to 2}\sqrt{f(x)}=\sqrt{\lim_{x\to 2}f(x)}=\sqrt{3}$$

Now, suppose

$$\lim_{x\to c}f(x)=0 \qquad \text{and} \qquad \lim_{x\to c}g(x)=0$$

then finding

$$\lim_{x\to c}\frac{f(x)}{g(x)} \tag{1}$$

may present some difficulties, since limit property 5 (the limit of a quotient) does not apply when $\lim_{x\to c} g(x)=0$. We often have to use algebraic manipulation or other devices to determine the outcome. Recall from Examples 3 and 4 (in Section 3-1) that

$$\lim_{x\to 2}\frac{x^2-4}{x-2}=\lim_{x\to 2}\frac{(x-2)(x+2)}{x-2}=\lim_{x\to 2}(x+2)=4$$

and

$$\lim_{x\to 0}\frac{|x|}{x} \quad \text{does not exist}$$

From these two examples, it is clear that knowing only that $\lim_{x\to c} f(x)=0$ and $\lim_{x\to c} g(x)=0$ is not enough to determine limit (1). Depending on the choice of functions f and g, the limit (1) may or may not exist. Consequently, if we are given (1) and $\lim_{x\to c} f(x)=0$ and $\lim_{x\to c} g(x)=0$, then (1) is said to be **indeterminate,** or, more specifically, a **0/0 indeterminate form.**

The 0/0 indeterminate form may seem unusual at first—a form one might not encounter too often—but it turns out that this is not the case. This form turns up in a number of very important developments in calculus. Example 9 provides a preview of a form that will be considered in detail in the following sections.

◆ EXAMPLE 9 Find $\lim_{h\to 0}\frac{f(2+h)-f(2)}{h}$ for:

(A) $f(x)=3x-4$ (B) $f(x)=\sqrt{x}$ (C) $f(x)=|x-2|$

Solutions

(A) $\lim_{h\to 0}\frac{f(2+h)-f(2)}{h}$ $\qquad f(x)=3x-4$

$$=\lim_{h\to 0}\frac{[3(\mathbf{2+h})-4]-[3(\mathbf{2})-4]}{h}$$

Since this is a 0/0 indeterminate form and property 5 in Theorem 2 does not apply, we proceed with algebraic simplification.

$$= \lim_{h \to 0} \frac{6 + 3h - 4 - 6 + 4}{h}$$

$$= \lim_{h \to 0} \frac{3h}{h} = \lim_{h \to 0} 3 = 3 \qquad h \neq 0$$

(B) $\lim_{h \to 0} \dfrac{f(2 + h) - f(2)}{h}$ $\qquad f(x) = \sqrt{x}$

$$= \lim_{h \to 0} \frac{\sqrt{2 + h} - \sqrt{2}}{h}$$

This is a 0/0 indeterminate form, so property 5 in Theorem 2 does not apply. Rationalizing the numerator will be of help.

$$= \lim_{h \to 0} \frac{\sqrt{2 + h} - \sqrt{2}}{h} \cdot \frac{\sqrt{2 + h} + \sqrt{2}}{\sqrt{2 + h} + \sqrt{2}} \qquad (A - B)(A + B) = A^2 - B^2$$

$$= \lim_{h \to 0} \frac{2 + h - 2}{h(\sqrt{2 + h} + \sqrt{2})}$$

$$= \lim_{h \to 0} \frac{1}{\sqrt{2 + h} + \sqrt{2}}$$

Use properties 1, 5, and 6 ($h \neq 0$).

$$= \frac{1}{\sqrt{2} + \sqrt{2}} = \frac{1}{2\sqrt{2}}$$

(C) $\lim_{h \to 0} \dfrac{f(2 + h) - f(2)}{h}$ $\qquad f(x) = |x - 2|$

$$= \lim_{h \to 0} \frac{|(2 + h) - 2| - |2 - 2|}{h}$$

Since this is a 0/0 indeterminate form and property 5 in Theorem 2 does not apply, we proceed with algebraic simplification.

$$= \lim_{h \to 0} \frac{|h|}{h} \quad \text{does not exist}$$

See Example 4. ◆

PROBLEM 9 Find $\lim_{h \to 0} \dfrac{g(1 + h) - g(1)}{h}$ for:

(A) $g(x) = 1 - x$ (B) $g(x) = \sqrt{x}$ (C) $g(x) = |x - 1|$ ◆

◆ LIMITS AT INFINITY

Earlier in this section, we discussed the behavior of $f(x) = 1/x$ for x near 0. We now investigate the behavior of $f(x)$ as x increases or decreases without bound. When the absolute value of a number increases, its reciprocal gets smaller. For example, if $x = 10{,}000$, then $1/x = 0.0001$; if $x = -10{,}000$, then $1/x = -0.0001$. Thus, as x increases without bound, indicated by $x \to \infty$, $f(x)$ approaches 0. We indicate this behavior by writing

$$\lim_{x \to \infty} \frac{1}{x} = 0$$

Similarly, as x decreases without bound, indicated by $x \to -\infty$, $f(x)$ approaches 0. We indicate this behavior by writing

$$\lim_{x \to -\infty} \frac{1}{x} = 0$$

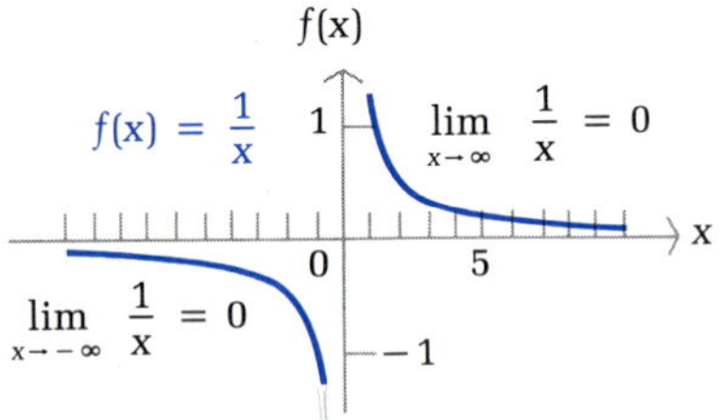

FIGURE 8

Figure 8 illustrates the behavior of $f(x) = 1/x$ as x increases and decreases without bound.

It is important to understand that the symbol ∞ does not represent an actual number that x is approaching, but is used to indicate only that the value of x is increasing with no upper limit on its size. In particular, the statement "$x = \infty$" is meaningless, since ∞ is not a symbol for a real number.

For the function f in Figure 8, the line $y = 0$ (the x axis) is called a *horizontal asymptote*. In general, a line $y = b$ is a **horizontal asymptote** for the graph of $y = f(x)$ if $f(x)$ approaches b as x either increases or decreases without bound. Symbolically, $y = b$ is a horizontal asymptote if either

$$\lim_{x \to -\infty} f(x) = b \qquad \text{or} \qquad \lim_{x \to \infty} f(x) = b$$

In the first case, the graph of f will be close to the horizontal line $y = b$ for large (in absolute value) negative x. In the second case, the graph will be close to the horizontal line $y = b$ for large positive x.

How can we evaluate limits of the form $\lim_{x \to -\infty} f(x)$ and $\lim_{x \to \infty} f(x)$ without drawing graphs or performing complicated calculations? Fortunately, **limit properties 1–6, listed earlier in this section in Theorem 2, are valid if we replace the statement $x \to c$ with $x \to -\infty$ or $x \to \infty$.** These properties, together with Theorem 3, enable us to evaluate limits at infinity for many functions.

THEOREM 3

Two Special Limits at Infinity

If p is a positive real number, and k is any real constant, then

$$\lim_{x \to -\infty} \frac{k}{x^p} = 0 \qquad \text{and} \qquad \lim_{x \to \infty} \frac{k}{x^p} = 0$$

provided that x^p names a real number for negative values of x.

Figure 9 illustrates Theorem 3 for several values of p. (Note that the x axis is a horizontal asymptote in all three cases.)

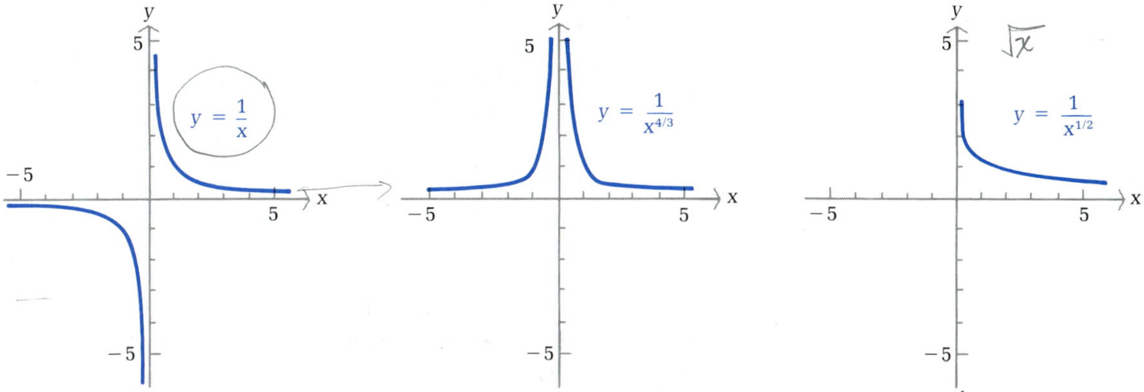

(A) $\lim_{x\to\infty} \frac{1}{x} = 0, \ \lim_{x\to-\infty} \frac{1}{x} = 0$

(B) $\lim_{x\to\infty} \frac{1}{x^{4/3}} = 0, \ \lim_{x\to-\infty} \frac{1}{x^{4/3}} = 0$

(C) $\lim_{x\to\infty} \frac{1}{x^{1/2}} = 0$

$x^{1/2}$ is not defined for negative values of x

FIGURE 9

Now, consider the following limit of a polynomial:

$$\lim_{x\to\infty} (2x^3 - 4x^2 - 9x - 100)$$

At first glance, it is not clear which part of the polynomial, $-4x^2 - 9x - 100$ or $2x^3$, dominates as x increases without bound. To help resolve the conflict, we factor out the highest power of x (x^3):

$$\lim_{x\to\infty} x^3 \left(2 - \frac{4}{x} - \frac{9}{x^2} - \frac{100}{x^3}\right)$$

Looking at the factors separately, we have

$$\lim_{x\to\infty} x^3 = \infty$$

and (using Theorems 2 and 3)

$$\lim_{x\to\infty} \left(2 - \frac{4}{x} - \frac{9}{x^2} - \frac{100}{x^3}\right) = 2 - 0 - 0 - 0 = 2$$

Now, intuitively, we would like to write

$$\lim_{x\to\infty} x^3 \left(2 - \frac{4}{x} - \frac{9}{x^2} - \frac{100}{x^3}\right) = \infty$$

because if one factor is increasing without bound and at the same time the other factor is approaching a positive constant, it seems reasonable to think that their product must increase without bound. This reasoning turns out to be correct; that is:

If $\lim_{x\to\infty} f(x) = \infty$ **and** $\lim_{x\to\infty} g(x) = L > 0$

then $\lim_{x\to\infty} [f(x) \cdot g(x)] = \infty$

This statement also holds if $x \to \infty$ is replaced with $x \to -\infty$ or $x \to c$, for c any real number. (Other variations of this statement are also possible.)

We are now ready to turn our attention to the limits of rational functions as x increases without bound.

EXAMPLE 10 Find the following limits, if they exist.

(A) $\lim_{x\to\infty} \frac{5x^2+3}{3x^2-2}$ (B) $\lim_{x\to\infty} \frac{3x^4-x^2+1}{8x^6-10}$ (C) $\lim_{x\to\infty} \frac{2x^5-x^3-1}{6x^3+2x^2-7}$

Solutions In all three cases, we begin (as in the polynomial example above) by factoring out the highest power of x from the numerator and the highest power of x from the denominator.

(A) $$\lim_{x\to\infty} \frac{5x^2+3}{3x^2-2} = \lim_{x\to\infty} \frac{x^2\left(5+\frac{3}{x^2}\right)}{x^2\left(3-\frac{2}{x^2}\right)} = \lim_{x\to\infty} \frac{5+\frac{3}{x^2}}{3-\frac{2}{x^2}}$$ Use Theorems 2 and 3.

$$= \frac{5+0}{3-0} = \frac{5}{3}$$

(B) $$\lim_{x\to\infty} \frac{3x^4-x^2+1}{8x^6-10} = \lim_{x\to\infty} \frac{x^4\left(3-\frac{1}{x^2}+\frac{1}{x^4}\right)}{x^6\left(8-\frac{10}{x^6}\right)}$$ Use Theorems 2 and 3.

$$= \lim_{x\to\infty} \frac{1}{x^2} \cdot \lim_{x\to\infty} \frac{3-\frac{1}{x^2}+\frac{1}{x^4}}{8-\frac{10}{x^6}} = 0 \cdot \frac{3-0+0}{8-0} = 0 \cdot \frac{3}{8} = 0$$

(C) $$\lim_{x\to\infty} \frac{2x^5-x^3-1}{6x^3+2x^2-7} = \lim_{x\to\infty} \frac{x^5\left(2-\frac{1}{x^2}-\frac{1}{x^5}\right)}{x^3\left(6+\frac{2}{x}-\frac{7}{x^3}\right)}$$

$$= \lim_{x\to\infty} x^2 \left(\frac{2-\frac{1}{x^2}-\frac{1}{x^5}}{6+\frac{2}{x}-\frac{7}{x^3}}\right)$$

$$= \infty$$ Since $\lim_{x\to\infty} x^2 = \infty$ and $\lim_{x\to\infty} \frac{2-\frac{1}{x^2}-\frac{1}{x^5}}{6+\frac{2}{x}-\frac{7}{x^3}} = \frac{2}{6}$ ◆

PROBLEM 10 Find the following limits, if they exist:

(A) $\lim_{x\to\infty} \frac{4x^3-5x+8}{2x^4-7}$ (B) $\lim_{x\to\infty} \frac{5x^6+3x}{2x^5-x-5}$ (C) $\lim_{x\to\infty} \frac{2x^3-x+7}{4x^3+3x^2-100}$ ◆

◆ SUMMARY OF LIMIT FORMS

Table 1 provides a summary of some of the limit forms we have considered.

TABLE 1

Summary of Some Important Limit Forms

NOTATION	DESCRIPTION	EXAMPLE
$\lim_{x\to c} f(x) = L$	$f(x)$ approaches L as x approaches c from either side of c, but is not equal to c	$\lim_{x\to 2} \frac{x^2-4}{x-2} = 4$
$\lim_{x\to c^-} f(x) = L$	$f(x)$ approaches L as x approaches c from the left	$\lim_{x\to 0^-} \frac{\lvert x\rvert}{x} = -1$
$\lim_{x\to c^+} f(x) = L$	$f(x)$ approaches L as x approaches c from the right	$\lim_{x\to 0^+} \frac{\lvert x\rvert}{x} = 1$
$\lim_{x\to c^-} f(x) = -\infty$	$f(x)$ decreases without bound as x approaches c from the left	$\lim_{x\to 0^-} \frac{1}{x} = -\infty$
$\lim_{x\to c^+} f(x) = \infty$	$f(x)$ increases without bound as x approaches c from the right	$\lim_{x\to 0^+} \frac{1}{x} = \infty$
$\lim_{x\to c} f(x) = \infty$	$f(x)$ increases without bound as x approaches c from either side of c, but is not equal to c	$\lim_{x\to 0} \frac{1}{x^2} = \infty$
$\lim_{x\to \infty} f(x) = L$	$f(x)$ approaches L as x increases without bound	$\lim_{x\to\infty} \frac{2x^2+1}{3x^2-1} = \frac{2}{3}$
$\lim_{x\to \infty} f(x) = \infty$	$f(x)$ increases without bound as x increases without bound	$\lim_{x\to\infty} \frac{2x^3}{x^2-7} = \infty$

Answers to Matched Problems

8. (A) $\lim_{x\to 4} (x + \sqrt{x}) = 4 + \sqrt{4} = 6$ (B) $\lim_{x\to -7} \sqrt[3]{1-x} = \sqrt[3]{1-(-7)} = 2$
   (C) $\lim_{x\to -2} 10 = 10$
9. (A) -1 (B) $\frac{1}{2}$ (C) Does not exist
10. (A) 0 (B) ∞ (does not exist) (C) $\frac{1}{2}$

EXERCISE 3-2

A *Given $\lim_{x\to 3} f(x) = 5$ and $\lim_{x\to 3} g(x) = 9$, find the indicated limits in Problems 1–10.*

1. $\lim_{x\to 3} [f(x) - g(x)]$
2. $\lim_{x\to 3} [f(x) + g(x)]$
3. $\lim_{x\to 3} 4g(x)$
4. $\lim_{x\to 3} (-2)f(x)$
5. $\lim_{x\to 3} \frac{f(x)}{g(x)}$
6. $\lim_{x\to 3} [f(x) \cdot g(x)]$
7. $\lim_{x\to 3} \sqrt{f(x)}$
8. $\lim_{x\to 3} \sqrt{g(x)}$
9. $\lim_{x\to 3} \frac{f(x) + g(x)}{2f(x)}$
10. $\lim_{x\to 3} \frac{g(x) \quad f(x)}{3g(x)}$

Find each limit in Problems 11–26, if it exists. (Algebraic manipulation is useful in some cases.)

11. $\lim_{x\to 5} (2x^2 - 3)$

12. $\lim_{x\to 2} (x^2 - 8x + 2)$

13. $\lim_{x\to 4} (x^2 - 5x)$

14. $\lim_{x\to -2} (3x^3 - 9)$

15. $\lim_{x\to 2} \dfrac{5x}{2 + x^2}$

16. $\lim_{x\to 10} \dfrac{2x + 5}{3x - 5}$

17. $\lim_{x\to 2} (x + 1)^3(2x - 1)^2$

18. $\lim_{x\to 3} (x + 2)^2(2x - 4)$

19. $\lim_{x\to 0} \dfrac{x^2 - 3x}{x}$

20. $\lim_{x\to 0} \dfrac{2x^2 + 5x}{x}$

21. $\lim_{x\to \infty} \dfrac{3}{x^2}$

22. $\lim_{x\to \infty} \dfrac{6}{x^4}$

23. $\lim_{x\to \infty} \left(5 - \dfrac{3}{x} + \dfrac{2}{x^2}\right)$

24. $\lim_{x\to \infty} \left(4 + \dfrac{1}{x^2} - \dfrac{3}{x^4}\right)$

25. $\lim_{h\to 0} \dfrac{2(3 + h) - 2(3)}{h}$

26. $\lim_{h\to 0} \dfrac{[(4 + h) - 2] - (4 - 2)}{h}$

B *Find each limit in Problems 27–54, if it exists. (Use algebraic manipulation where necessary. Also, use $-\infty$ or ∞ where appropriate.)*

27. $\lim_{x\to 1} (3x^4 - 2x^2 + x - 2)$

28. $\lim_{x\to -1} (5x^3 - 3x^2 - 5x + 3)$

29. $\lim_{x\to 1} \dfrac{x - 2}{x^2 - 2x}$

30. $\lim_{x\to 1} \dfrac{x + 3}{x^2 + 3x}$

31. $\lim_{x\to 2} \dfrac{x - 2}{x^2 - 2x}$

32. $\lim_{x\to -3} \dfrac{x + 3}{x^2 + 3x}$

33. $\lim_{x\to \infty} \dfrac{x - 2}{x^2 - 2x}$

34. $\lim_{x\to \infty} \dfrac{x + 3}{x^2 + 3x}$

35. $\lim_{x\to 2} \dfrac{x^2 - x - 6}{x + 2}$

36. $\lim_{x\to 3} \dfrac{x^2 + x - 6}{x + 3}$

37. $\lim_{x\to -2} \dfrac{x^2 - x - 6}{x + 2}$

38. $\lim_{x\to -3} \dfrac{x^2 + x - 6}{x + 3}$

39. $\lim_{x\to \infty} \dfrac{x^2 - x - 6}{x + 2}$

40. $\lim_{x\to \infty} \dfrac{x^2 + x - 6}{x + 3}$

41. $\lim_{x\to \infty} \dfrac{2x + 4}{x}$

42. $\lim_{x\to \infty} \dfrac{3x^2 + 5}{x^2}$

43. $\lim_{x\to \infty} \dfrac{3x^3 - x + 1}{5x^3 - 7}$

44. $\lim_{x\to \infty} \dfrac{3x^4 - x^3 + 5}{2x^4 - 10}$

45. $\lim_{h\to 0} \dfrac{(2 + h)^2 - 2^2}{h}$

46. $\lim_{h\to 0} \dfrac{(3 + h)^2 - 3^2}{h}$

47. $\lim_{x\to 3} \left(\dfrac{x}{x + 3} + \dfrac{x - 3}{x^2 - 9}\right)$

48. $\lim_{x\to 2} \left(\dfrac{1}{x + 2} + \dfrac{x - 2}{x^2 - 4}\right)$

49. $\lim\limits_{x \to 0^-} \dfrac{x - 2}{x^2 - 2x}$

50. $\lim\limits_{x \to 0^-} \dfrac{x + 3}{x^2 + 3x}$

51. $\lim\limits_{x \to 0^+} \dfrac{x - 2}{x^2 - 2x}$

52. $\lim\limits_{x \to 0^+} \dfrac{x + 3}{x^2 + 3x}$

53. $\lim\limits_{x \to 0} \dfrac{x - 2}{x^2 - 2x}$

54. $\lim\limits_{x \to 0} \dfrac{x + 3}{x^2 + 3x}$

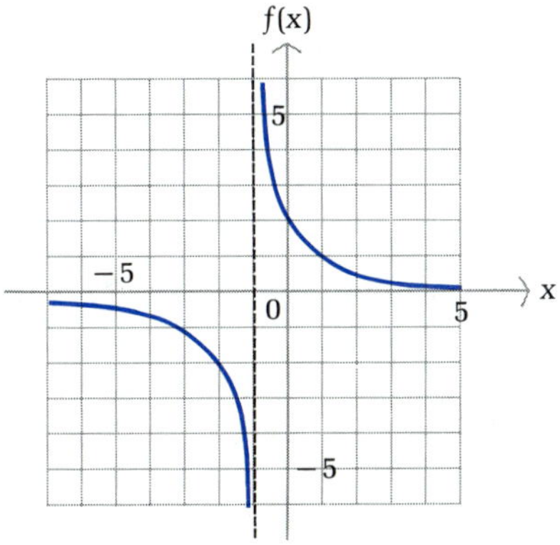

Use the graph of the function f shown in the margin to answer Problems 55 and 56.

55. (A) $\lim\limits_{x \to -\infty} f(x) = ?$ (B) $\lim\limits_{x \to \infty} f(x) = ?$

(C) Write the equation of any horizontal asymptote.

56. (A) $\lim\limits_{x \to -1^-} f(x) = ?$ (B) $\lim\limits_{x \to -1^+} f(x) = ?$ (C) $\lim\limits_{x \to -1} f(x) = ?$

(D) Write the equation of any vertical asymptote.

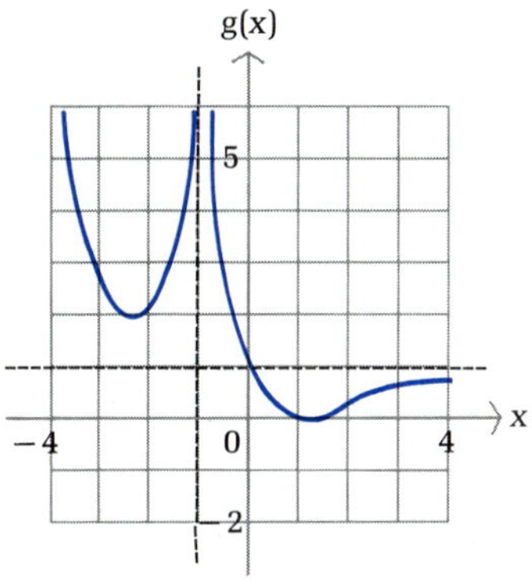

Use the graph of the function g shown in the margin to answer Problems 57 and 58.

57. (A) $\lim\limits_{x \to -1^-} g(x) = ?$ (B) $\lim\limits_{x \to -1^+} g(x) = ?$ (C) $\lim\limits_{x \to -1} g(x) = ?$

(D) Write the equation of any vertical asymptote.

58. (A) $\lim\limits_{x \to \infty} g(x) = ?$ (B) $\lim\limits_{x \to -\infty} g(x) = ?$

(C) Write the equation of any horizontal asymptote.

Compute the following limit for each function in Problems 59–68:

$$\lim_{h \to 0} \frac{f(2 + h) - f(2)}{h}$$

59. $f(x) = 3x + 1$

60. $f(x) = 5x - 1$

61. $f(x) = x^2 + 1$

62. $f(x) = x^2 - 2$

63. $f(x) = 5$

64. $f(x) = -2$

65. $f(x) = \sqrt{x} - 2$

66. $f(x) = 1 + \sqrt{x}$

67. $f(x) = |x - 2| - 3$

68. $f(x) = 2 + |x - 2|$

C *Find each limit in Problems 69–82, if it exists. (Use algebraic manipulation where necessary and $-\infty$ or ∞ where appropriate.) [Recall: $a^3 - b^3 = (a - b)(a^2 + ab + b^2)$ and $a^3 + b^3 = (a + b)(a^2 - ab + b^2)$.]*

69. $\lim\limits_{x \to 1} \sqrt{x^2 + 2x}$

70. $\lim\limits_{x \to 1} \sqrt{2x^4 + 1}$

71. $\lim\limits_{x \to 4} \sqrt[3]{x^2 - 3x}$

72. $\lim\limits_{x \to 2} \sqrt[3]{x^3 - 2x}$

73. $\lim\limits_{x \to 2} \dfrac{4}{(x - 2)^2}$

74. $\lim\limits_{x \to 0} \dfrac{3}{|x|}$

75. $\lim\limits_{x \to \infty} \left(\dfrac{1}{x^2} + \dfrac{1}{\sqrt{x}}\right)$

76. $\lim\limits_{x \to \infty} \left(\dfrac{1}{\sqrt[3]{x}} - \dfrac{2}{x^3}\right)$

77. $\lim\limits_{x \to 0} \left(\sqrt{x^2 + 9} - \dfrac{x^2 + 3x}{x}\right)$

78. $\lim\limits_{x \to 1} \left(\dfrac{x^2 - 1}{x - 1} + \sqrt{x^2 + 3}\right)$

79. $\lim_{x \to 4} \frac{\sqrt{x} - 2}{x - 4}$

80. $\lim_{x \to 0} \frac{\sqrt{x + 4} - 2}{x}$

81. $\lim_{x \to 2} \frac{x^3 - 8}{x - 2}$

82. $\lim_{x \to -1} \frac{x^2 - 1}{x^3 + 1}$

83. Find each limit, if it exists. (Use $-\infty$ or ∞, as appropriate.)

(A) $\lim_{x \to -2^-} \frac{2}{x + 2}$ (B) $\lim_{x \to -2^+} \frac{2}{x + 2}$ (C) $\lim_{x \to -2} \frac{2}{x + 2}$

(D) Is $x = -2$ a vertical asymptote?

84. Find each limit, if it exists. (Use $-\infty$ or ∞, as appropriate.)

(A) $\lim_{x \to 1^-} \frac{1}{x - 1}$ (B) $\lim_{x \to 1^+} \frac{1}{x - 1}$ (C) $\lim_{x \to 1} \frac{1}{x - 1}$

(D) Is $x = 1$ a vertical asymptote?

Find each limit in Problems 85–88, where a is a real constant.

85. $\lim_{h \to 0} \frac{(a + h)^2 - a^2}{h}$

86. $\lim_{h \to 0} \frac{[3(a + h) - 2] - (3a - 2)}{h}$

87. $\lim_{h \to 0} \frac{\sqrt{a + h} - \sqrt{a}}{h}, \quad a > 0$

88. $\lim_{h \to 0} \frac{\frac{1}{a + h} - \frac{1}{a}}{h}, \quad a \neq 0$

APPLICATIONS

Business & Economics

89. *Average cost.* The cost equation for manufacturing a particular compact disk album is

$$C(x) = 20{,}000 + 3x$$

where x is the number of disks produced. The average cost per disk, denoted by $\overline{C}(x)$, is found by dividing C(x) by x:

$$\overline{C}(x) = \frac{C(x)}{x} = \frac{20{,}000 + 3x}{x}$$

If only 10 disks were manufactured, for example, the average cost per disk would be \$2,003. Find:

(A) $\overline{C}(1{,}000)$ (B) $\overline{C}(100{,}000)$ (C) $\lim_{x \to 10{,}000} \overline{C}(x)$ (D) $\lim_{x \to \infty} \overline{C}(x)$

90. *Employee training.* A company producing computer components has established that on the average, a new employee can assemble N(t) components per day after t days of on-the-job training, as given by

$$N(t) = \frac{100t}{t + 9}$$

(See the figure below.) Find:

(A) $N(1)$ (B) $N(11)$ (C) $\lim_{t \to 11} N(t)$ (D) $\lim_{t \to \infty} N(t)$

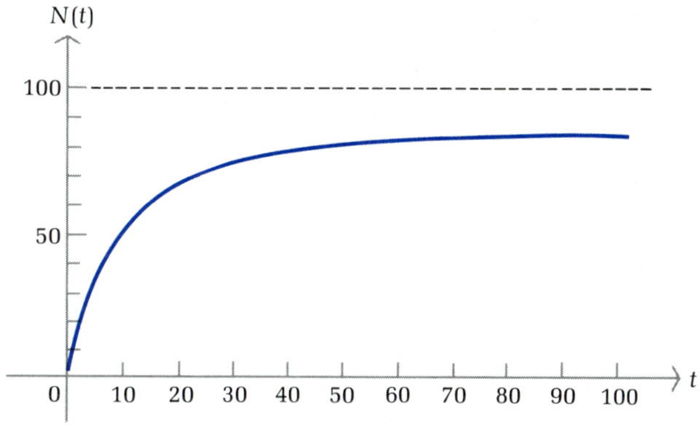

91. *Compound interest.* If \$100 is invested at 8% compounded n times per year, then the amount in the account $A(n)$ at the end of 1 year is given by

$$A(n) = 100\left(1 + \frac{0.08}{n}\right)^n$$

(A) Use a calculator with a y^x key to complete the table in the margin. (Give each entry to the nearest cent.)

(B) Using the results of part A, guess the following limit:

$$\lim_{n \to \infty} A(n) = ?$$

(This problem leads to the important concept of compounding continuously, which will be discussed in detail in Section 5-1.)

COMPOUNDED	n	$A(n)$
Annually	1	\$108.00
Semiannually	2	\$108.16
Quarterly	4	\$108.24
Monthly	12	
Weekly	52	
Daily	365	
Hourly	8,760	

92. *Pollution.* In Silicon Valley (in California), a number of computer-related manufacturing firms were found to be contaminating underground water supplies with toxic chemicals stored in leaking underground containers. A water quality control agency ordered the companies to take immediate corrective action and to contribute to a monetary pool for testing and cleanup of the underground contamination. Suppose the required monetary pool (in millions of dollars) for the testing and cleanup is estimated by

$$P(x) = \frac{2x}{1 - x}$$

where x is the percentage (expressed as a decimal fraction) of the total contaminant removed.

(A) Complete the table in the margin.

(B) Find $\lim_{x \to 0.80} P(x)$.

(C) What happens to the required monetary pool as the desired percentage of contaminant removed approaches 100% (x approaches 1 from the left)?

PERCENTAGE REMOVED	POOL REQUIRED
0.50 (50%)	\$2 million
0.60 (60%)	\$3 million
0.70 (70%)	
0.80 (80%)	
0.90 (90%)	
0.95 (95%)	
0.99 (99%)	

Life Sciences

93. *Medicine.* A drug is injected into the bloodstream of a patient through her right arm. The concentration of the drug in the bloodstream of the left arm t hours after the injection is given by

$$C(t) = \frac{0.14t}{t^2 + 1}$$

(See the figure below.) Find:

(A) $C(0.5)$ (B) $C(1)$ (C) $\lim_{t\to 1} C(t)$ (D) $\lim_{t\to\infty} C(t)$

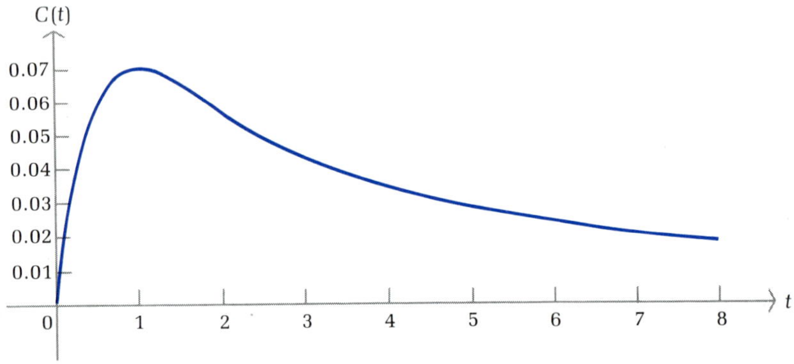

94. *Physiology.* In a study on the speed of muscle contraction in frogs under various loads, researchers W. O. Fems and J. Marsh found that the speed of contraction decreases with increasing loads. More precisely, they found that the relationship between speed of contraction S (in centimeters per second) and load w (in grams) is given approximately by

$$S(w) = \frac{26 + 0.06w}{w} \qquad w > 5$$

Find:

(A) $S(10)$ (B) $S(50)$ (C) $\lim_{w\to 50} S(w)$ (D) $\lim_{w\to\infty} S(w)$

Social Sciences

95. *Psychology—learning theory.* In 1917, L. L. Thurstone, a pioneer in quantitative learning theory, proposed the function

$$f(x) = \frac{a(x + c)}{(x + c) + b}$$

to describe the number of successful acts per unit time that a person could accomplish after x practice sessions. Suppose that for a particular person enrolling in a typing school,

$$f(x) = \frac{60(x + 1)}{x + 5}$$

where $f(x)$ is the number of words per minute the person is able to type after x weeks of lessons. (See the figure below.) Find:

(A) $f(3)$ (B) $f(10)$ (C) $\lim_{x \to 10} f(x)$ (D) $\lim_{x \to \infty} f(x)$

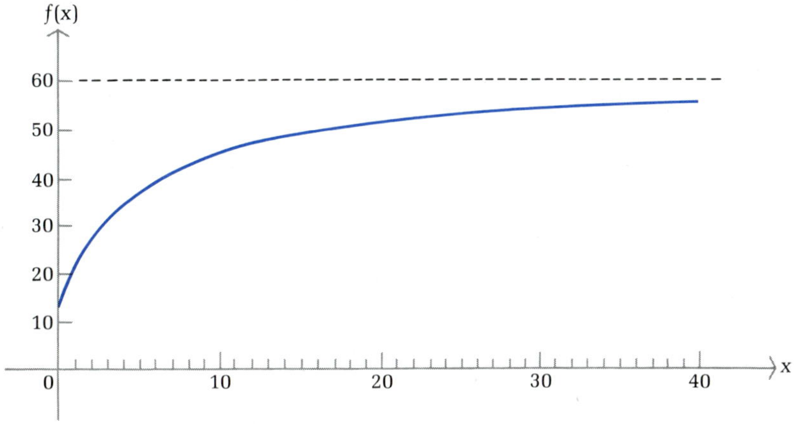

96. *Psychology—retention.* An experiment on retention is conducted in a psychology class. Each student in the class is given 1 day to memorize the same list of 30 special characters. The lists are turned in at the end of the day, and for each succeeding day for 30 days each student is asked to turn in a list of as many of the symbols as can be recalled. Averages are taken, and it is found that

$$N(t) = \frac{5t + 20}{t} \qquad t \geq 1$$

provides a good approximation of the average number of symbols, $N(t)$, retained after t days. (See the figure below.) Find:

(A) $N(2)$ (B) $N(10)$ (C) $\lim_{t \to 10} N(t)$ (D) $\lim_{t \to \infty} N(t)$

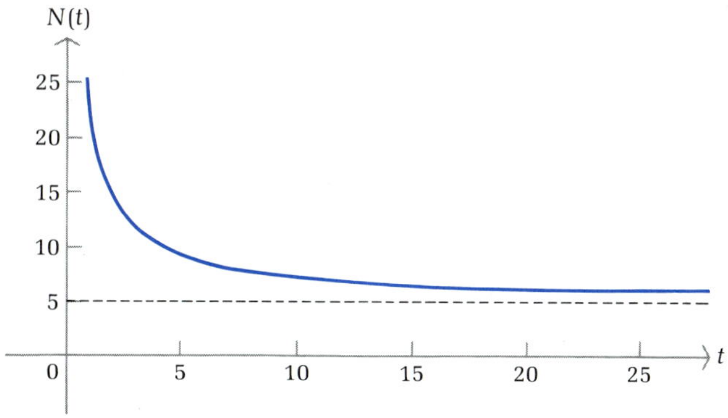

SECTION 3-3

The Derivative

- TANGENT LINES
- AVERAGE AND INSTANTANEOUS RATES OF CHANGE
- THE DERIVATIVE
- NONEXISTENCE OF THE DERIVATIVE
- SUMMARY

We now use the concept of limit to solve two of the important problems illustrated in Figure 1 at the beginning of this chapter. These two problems are repeated in Figure 10. The solution of these two apparently unrelated problems involves a common concept called *the derivative*, which we will be able to define after we have solved the problems illustrated in Figure 10.

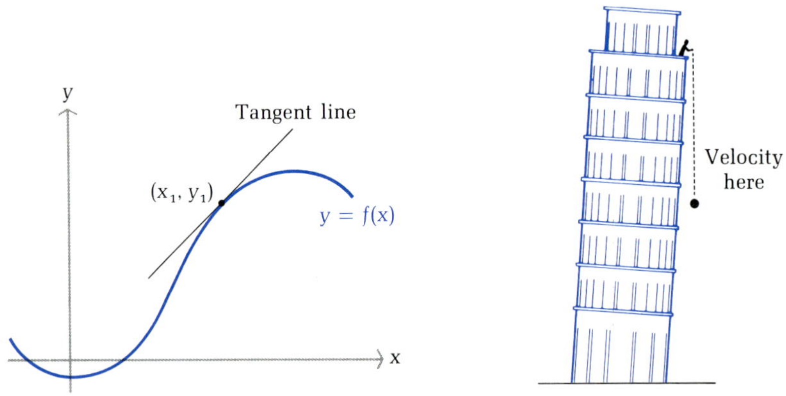

(A) Find the equation of the tangent line at (x_1, y_1) given $y = f(x)$

(B) Find the instantaneous velocity of a falling object

FIGURE 10

TANGENT LINES

From plane geometry, we know that a tangent to a circle is a line that passes through one and only one point on the circle. This definition is not satisfactory for graphs of functions in general, as shown in Figure 11.

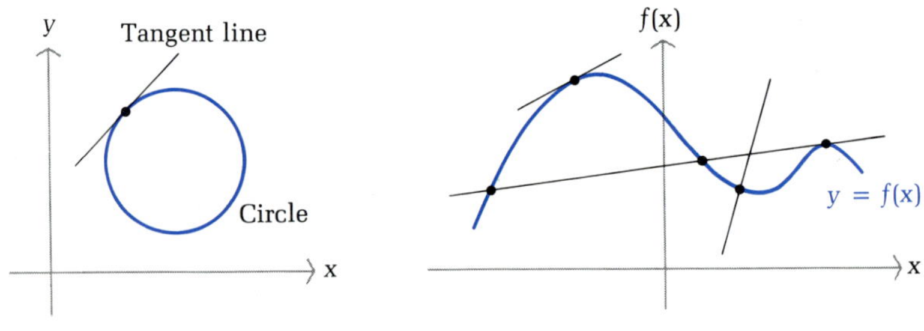

(A) (B)

FIGURE 11

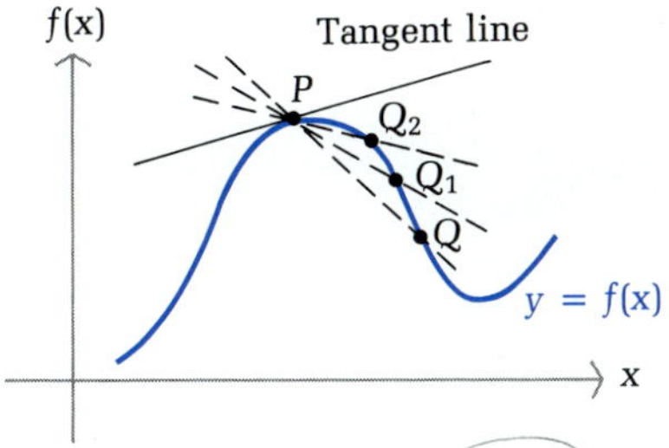

FIGURE 12

To define a tangent line for the graph of a function f at a point P on its graph, we proceed as follows. Select another point Q on the graph of f, and draw a line through P and Q. This line is called a **secant line.** Let Q approach P along the graph, as indicated by Q_1 and Q_2 in Figure 12. If the slopes of the corresponding secant lines have a limit m as Q approaches P, then m is defined to be the *slope of the tangent line* at P and the *slope of the graph of* f at P. (These definitions will be made more precise below.) If we know the slope of the tangent line and the coordinates of P, we can write the equation of the tangent line at P using the point–slope formula for a line (from Section 1-1).

To find the slope of the tangent line at a particular point $P(x_1, f(x_1))$ on the graph of f, we locate any other point Q on the graph, as indicated in Figure 13. Since h can be positive or negative (but not equal to 0), Q can be on the right or on the left of P (but cannot coincide with P). Both cases are illustrated in Figure 13.

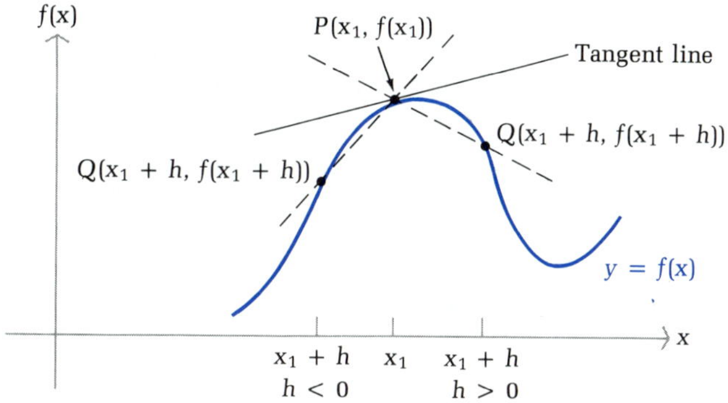

FIGURE 13

We now write the slope of the secant line through P and Q, using the slope formula, $m = (y_2 - y_1)/(x_2 - x_1)$, from Section 1-1:

$$\text{Slope of secant line} = \frac{f(x_1 + h) - f(x_1)}{(x_1 + h) - x_1}$$

$$= \frac{f(x_1 + h) - f(x_1)}{h} \qquad (1)$$

We refer to (1) as a **difference quotient.** Now, as h approaches 0 from either side of 0, but not equal to 0, Q will approach P along the graph of f (assuming f is continuous at $x = x_1$). If the limit of the difference quotient (1) exists as $h \to 0$, we define this limit to be the slope of the tangent line to the graph of f at the point P.

It is useful to note that the limit used to define the slope of a tangent line (and the slope of a graph) is usually a 0/0 indeterminate form—one of the reasons we placed special emphasis on this form earlier.

Tangent Line

Given the graph of $y = f(x)$, the **tangent line** at $x = x_1$ is the line that passes through the point $(x_1, f(x_1))$ with slope

$$\textbf{Slope of tangent line} = \lim_{h \to 0} \frac{f(x_1 + h) - f(x_1)}{h}$$

if the limit exists. The slope of the tangent line is also referred to as the **slope of the graph** at $(x_1, f(x_1))$.

◆ EXAMPLE 11 Given $f(x) = x^2$:

(A) Find the slope of the tangent line at $x = 1$.

(B) Find the equation of the tangent line at $x = 1$; that is, through $(1, f(1))$.

(C) Sketch the graph of f, the tangent line at $(1, f(1))$, and the secant line passing through $(1, f(1))$ and $(2, f(2))$.

Solutions (A) Slope of the tangent line:

Step 1. Write the difference quotient and simplify:

$$\frac{f(1+h) - f(1)}{h} = \frac{(1+h)^2 - 1^2}{h}$$

This is the slope of a secant line passing through $(1, f(1))$ and $(1 + h, f(1 + h))$.

$$= \frac{1 + 2h + h^2 - 1}{h}$$

$$= \frac{2h + h^2}{h} = \frac{h(2 + h)}{h} = 2 + h \qquad h \neq 0$$

Step 2. Find the limit of the difference quotient:

$$\text{Slope of tangent line} = \lim_{h \to 0} \frac{f(1+h) - f(1)}{h}$$

$$= \lim_{h \to 0} (2 + h) = 2$$

This is also the slope of the graph of $f(x) = x^2$ at $(1, f(1))$.

(B) Equation of the tangent line: The tangent line passes tlhrough $(1, f(1)) = (1, 1)$ with slope $m = 2$ (from part A). The point–slope formula from Section 1-1 gives its equation:

$$y - y_1 = m(x - x_1) \qquad \text{Point–slope formula}$$

$$y - 1 = 2(x - 1)$$

$$y = 2x - 1 \qquad \text{Tangent line equation}$$

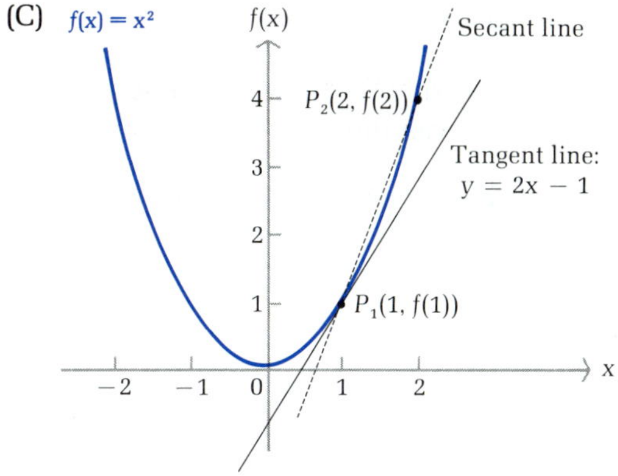

PROBLEM 11 Find the slope of the tangent line for the graph of $f(x) = x^2$ at $x = 2$, and write the equation of the tangent line in the form $y = mx + b$.

AVERAGE AND INSTANTANEOUS RATES OF CHANGE

We now turn to the second problem stated at the beginning of this section.

EXAMPLE 12 A small steel ball dropped from a tower will fall a distance of y feet in x seconds, as given approximately by the formula (from physics)

$$y = f(x) = 16x^2$$

Figure 14 shows the position of the ball on a coordinate line (positive direction down) at the end of 0, 1, 2, and 3 seconds. Note that the distances fallen between these times increases with time; thus, the velocity of the ball is increasing as it falls. Our main objective in this example is to find the velocity of the ball at a given instant, say, at the end of 2 seconds. (*Velocity* is the rate of change of distance with respect to time.)

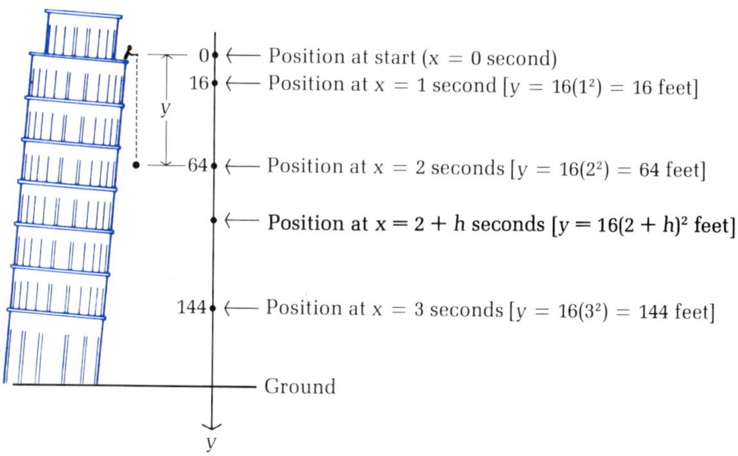

FIGURE 14
Note: Positive y direction is down.

(A) Find the average velocity from $x = 2$ seconds to $x = 3$ seconds.
(B) Find the average velocity from $x = 2$ seconds to $x = 2 + h$ seconds, $h \neq 0$.
(C) Find the limit of the expression from part B as $h \to 0$, if it exists. (What do you think the limit represents?)

Solutions (A) Average velocity from $x = 2$ seconds to $x = 3$ seconds: Recall the formula $d = rt$, which can be written in the form

$$r = \frac{d}{t} = \frac{\text{Distance covered}}{\text{Elapsed time}} = \text{Average rate}$$

For example, if a person drives from San Francisco to Los Angeles (a distance of about 420 miles) in 7 hours, then the average velocity is

$$r = \frac{d}{t} = \frac{420}{7} = 60 \text{ miles per hour}$$

Sometimes the person will be traveling faster and sometimes slower, but the average velocity is 60 miles per hour. In our present problem, the average velocity of the steel ball is given by

$$\begin{aligned}\text{Average velocity} &= \frac{\text{Distance covered}}{\text{Elapsed time}}\\ &= \frac{f(3) - f(2)}{3 - 2}\\ &= \frac{16 \cdot 3^2 - 16 \cdot 2^2}{1} = 80 \text{ feet per second}\end{aligned}$$

(B) Average velocity from $x = 2$ seconds to $x = 2 + h$ seconds, $h \neq 0$: Proceeding as in part A,

$$\begin{aligned}\text{Average velocity} &= \frac{\text{Distance covered}}{\text{Elapsed time}}\\ &= \frac{f(2 + h) - f(2)}{h} && \text{Difference quotient}\\ &= \frac{16(2 + h)^2 - 16(2)^2}{h} && \text{0/0 indeterminate form requires simplification (for part C).}\\ &= \frac{64 + 64h + 16h^2 - 64}{h}\\ &= \frac{h(64 + 16h)}{h} = 64 + 16h && h \neq 0\end{aligned}$$

Note that if $h = 1$, the average velocity is 80 feet per second (our result in part A); if $h = 0.5$, the average velocity is 72 feet per second; if $h = -0.01$, the average velocity is 63.84 feet per second; and so on. The closer to 0 that h gets (on either side of 0), the closer the average velocity is to 64 feet per second.

(C) Limit of the expression from part B as $h \to 0$:

$$\lim_{h\to 0} \frac{f(2+h)-f(2)}{h} = \lim_{h\to 0} (64+16h)$$
$$= 64 \text{ feet per second}$$ ◆

We call the answer to Example 12C, 64 feet per second, the *instantaneous velocity* (or the *instantaneous rate of change*) at $x = 2$ seconds, and we have solved the second basic problem stated at the beginning of this chapter!

Average and Instantaneous Rates of Change

For $y = f(x)$, the **average rate of change from $x = x_1$ to $x = x_1 + h$** is

$$\frac{f(x_1+h)-f(x_1)}{h} \qquad h \neq 0$$

and the **instantaneous rate of change at $x = x_1$** is

$$\lim_{h\to 0} \frac{f(x_1+h)-f(x_1)}{h}$$

if the limit exists.

Note that the limit used to define instantaneous rate of change is usually a 0/0 indeterminate form—another reason we placed special emphasis on this form earlier.

PROBLEM 12 For the falling steel ball in Example 12, find:

(A) The average velocity from $x = 1$ second to $x = 2$ seconds.
(B) The average velocity from $x = 1$ second to $x = 1 + h$ seconds, $h \neq 0$.
(C) The instantaneous velocity (instantaneous rate of change) at $x = 1$. ◆

◆ THE DERIVATIVE

In the last two examples, we found that the special limit

$$\lim_{h\to 0} \frac{f(x_1+h)-f(x_1)}{h} \tag{2}$$

if it exists, gives us the slope of the tangent line to the graph of $y = f(x)$ at $(x_1, f(x_1))$ and also the instantaneous rate of change of y with respect to x at $x = x_1$. Many other applications give rise to this form. In fact, the limit (2) is of such basic importance to calculus and to the applications of calculus that we give it a special name and study it in detail. To keep form (2) simple and general, we drop the subscript on x_1 and think of the difference quotient

$$\frac{f(x+h)-f(x)}{h}$$

as a function of h, with x held fixed as we let h tend to 0. We are now ready to define one of the basic concepts in calculus, *the derivative*:

The Derivative

For $y = f(x)$, we define the **derivative of f at x**, denoted by $f'(x)$, to be

$$f'(x) = \lim_{h \to 0} \frac{f(x+h)-f(x)}{h} \qquad \text{if the limit exists}$$

If $f'(x)$ exists for each x in the open interval (a, b), then f is said to be **differentiable** over (a, b).

(Differentiability from the left or from the right is defined using $h \to 0^-$ or $h \to 0^+$, respectively, in place of $h \to 0$ in the above definition.)

The process of finding the derivative of a function is called **differentiation.** That is, the derivative of a function is obtained by **differentiating** the function.

The Function $f'(x)$

The derivative of a function f is a new function f' that gives us, among other things, the slope of the tangent line to the graph of $y = f(x)$ for each x and the instantaneous rate of change of $y = f(x)$ with respect to x. The domain of f' is a subset of the domain of f.

◆ EXAMPLE 13 Find $f'(x)$, the derivative of f at x, for $f(x) = 4x - x^2$.

Solution To find $f'(x)$, we use a two-step process:

Step 1. Form the difference quotient and simplify:

$$\begin{aligned}\frac{f(x+h)-f(x)}{h} &= \frac{[4(x+h)-(x+h)^2]-(4x-x^2)}{h}\\ &= \frac{4x+4h-x^2-2xh-h^2-4x+x^2}{h}\\ &= \frac{4h-2xh-h^2}{h}\\ &= \frac{h(4-2x-h)}{h}\\ &= 4-2x-h \qquad h \neq 0\end{aligned}$$

Step 2. Find the limit of the difference quotient:

$$f'(x) = \lim_{h \to 0} \frac{f(x+h) - f(x)}{h}$$
$$= \lim_{h \to 0} (4 - 2x - h) = 4 - 2x$$

Thus, if $f(x) = 4x - x^2$, then $f'(x) = 4 - 2x$. The derivative f' is a new function derived from the function f. ◆

PROBLEM 13 Find $f'(x)$, the derivative of f at x, for $f(x) = 8x - 2x^2$. ◆

◆ EXAMPLE 14 In Example 13, we started with the function specified by $f(x) = 4x - x^2$ and found the derivative of f at x to be $f'(x) = 4 - 2x$. Thus, the slope of a tangent line to the graph of f at any point $(x, f(x))$ on the graph of f is

$$m = f'(x) = 4 - 2x$$

(A) Find the slope of the graph of f at $x = 0$, 2, and 3.
(B) Graph $y = f(x) = 4x - x^2$, and use the slopes found in part A to make a rough sketch of the tangent lines to the graph at $x = 0$, 2, and 3.

Solutions (A) Using $f'(x) = 4 - 2x$, we have

$f'(0) = 4 - 2(0) = 4$ Slope at $x = 0$
$f'(2) = 4 - 2(2) = 0$ Slope at $x = 2$
$f'(3) = 4 - 2(3) = -2$ Slope at $x = 3$

(B)

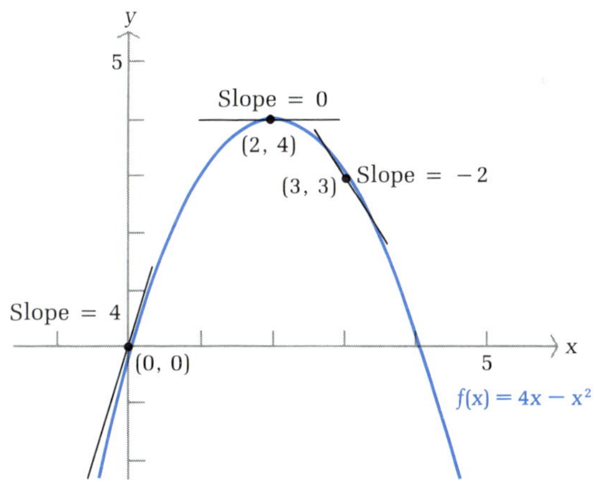

◆

PROBLEM 14 In Problem 13, we started with the function specified by $f(x) = 8x - 2x^2$. Using the derivative found there:

(A) Find the slope of the graph of f at $x = 1$, 2, and 4.
(B) Graph $y = f(x) = 8x - 2x^2$, and use the slopes from part A to make a rough sketch of the tangent lines to the graph at $x = 1$, 2, and 4. ◆

◆ EXAMPLE 15 Find $f'(x)$, the derivative of f at x, for $f(x) = \sqrt{x} + 2$.

Solution To find $f'(x)$, we find

$$\lim_{h\to 0} \frac{f(x+h) - f(x)}{h}$$

using the two-step process.

Step 1. Form the difference quotient and simplify:

$$\frac{f(x+h) - f(x)}{h} = \frac{(\sqrt{x+h} + 2) - (\sqrt{x} + 2)}{h}$$
$$= \frac{\sqrt{x+h} - \sqrt{x}}{h}$$

Since this is a 0/0 indeterminate form, we change the form by rationalizing the numerator:

$$\frac{\sqrt{x+h} - \sqrt{x}}{h} \cdot \frac{\sqrt{x+h} + \sqrt{x}}{\sqrt{x+h} + \sqrt{x}} = \frac{x + h - x}{h(\sqrt{x+h} + \sqrt{x})}$$
$$= \frac{h}{h(\sqrt{x+h} + \sqrt{x})}$$
$$= \frac{1}{\sqrt{x+h} + \sqrt{x}} \qquad h \neq 0$$

Step 2. Find the limit of the difference quotient:

$$f'(x) = \lim_{h\to 0} \frac{f(x+h) - f(x)}{h}$$
$$= \lim_{h\to 0} \frac{1}{\sqrt{x+h} + \sqrt{x}}$$
$$= \frac{1}{\sqrt{x} + \sqrt{x}} = \frac{1}{2\sqrt{x}} \qquad x > 0$$

Thus, the derivative of $f(x) = \sqrt{x} + 2$ is $f'(x) = 1/(2\sqrt{x})$, a new function. The domain of f is $[0, \infty)$. Since $f'(0)$ is not defined, the domain of f' is $(0, \infty)$, a subset of the domain of f. ◆

PROBLEM 15 Find $f'(x)$ for $f(x) = x^{-1}$. ◆

◆ NONEXISTENCE OF THE DERIVATIVE

The existence of a derivative at $x = a$ depends on the existence of a limit at $x = a$; that is, on the existence of

$$f'(a) = \lim_{h\to 0} \frac{f(a+h) - f(a)}{h} \tag{3}$$

If the limit does not exist at $x = a$, we say that the function f is **nondifferentiable at $x = a$**, or **$f'(a)$ does not exist.**

How can we recognize the points on the graph of f where $f'(a)$ does not exist? It is impossible to describe all the ways that the limit in (3) can fail to exist. However, we can illustrate some common situations where $f'(a)$ does fail to exist:

1. If f is not continuous at $x = a$, then $f'(a)$ does not exist (Figure 15A). Or, equivalently, **if f is differentiable at $x = a$, then f must be continuous at $x = a$.**
2. If the graph of f has a sharp corner at $x = a$, then $f'(a)$ does not exist and the graph has no tangent line at $x = a$ (Figure 15B). (In Figure 15B, the left- and right-hand derivatives exist but are not equal.)
3. If the graph of f has a vertical tangent line at $x = a$, then $f'(a)$ does not exist (Figure 15C and D).

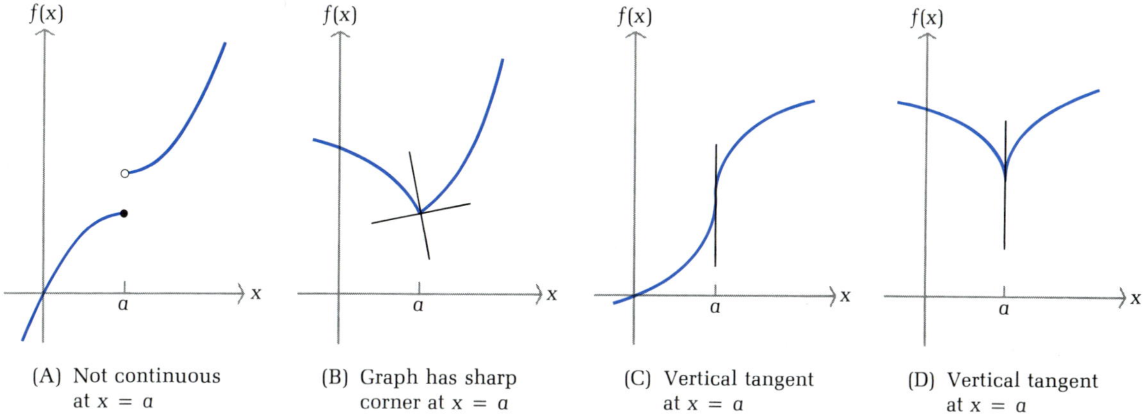

FIGURE 15
The function f is nondifferentiable at $x = a$.

◆ SUMMARY

The concept of the derivative is a very powerful mathematical idea, and its applications are many and varied. In the next three sections we will develop formulas and general properties of derivatives that will enable us to find the derivatives of many functions without having to go through the two-step limiting process each time.

Answers to Matched Problems

11. $y = 4x - 4$
12. (A) 48 ft/sec (B) $32 + 16h$ (C) 32 ft/sec
13. $f'(x) = 8 - 4x$
14. (A) $f'(1) = 4, f'(2) = 0, f'(4) = -8$ (B)

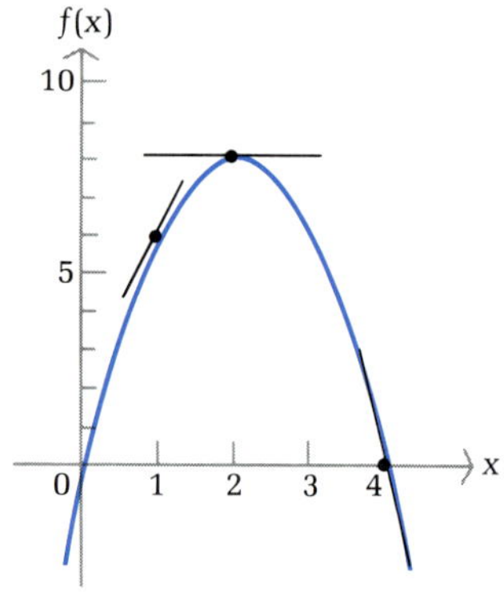

15. $f'(x) = -1/x^2$, or $-x^{-2}$

EXERCISE 3-3

A

Use $f(x) = 3x - 2$ in Problems 1 and 2 to find the given difference quotients and limits.

1. (A) $\dfrac{f(1+h) - f(1)}{h}$ (B) $\lim\limits_{h\to 0} \dfrac{f(1+h) - f(1)}{h}$

2. (A) $\dfrac{f(2+h) - f(2)}{h}$ (B) $\lim\limits_{h\to 0} \dfrac{f(2+h) - f(2)}{h}$

Use $f(x) = 2x^2$ in Problems 3 and 4 to find the given difference quotients and limits.

3. (A) $\dfrac{f(2+h) - f(2)}{h}$ (B) $\lim\limits_{h\to 0} \dfrac{f(2+h) - f(2)}{h}$

4. (A) $\dfrac{f(3+h) - f(3)}{h}$ (B) $\lim\limits_{h\to 0} \dfrac{f(3+h) - f(3)}{h}$

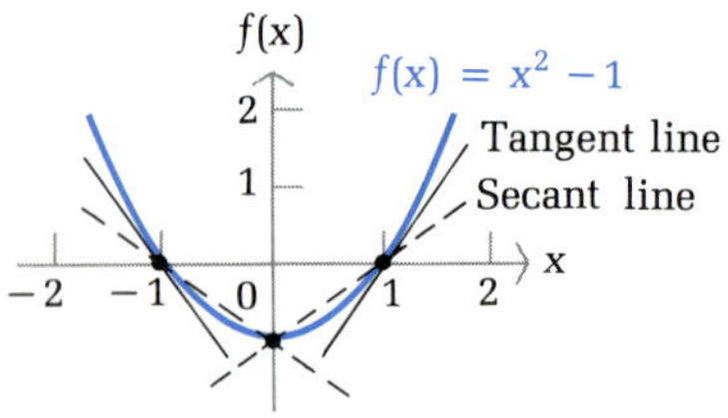

Solve Problems 5 and 6 for the graph of $y = f(x) = x^2 - 1$ (see the figure in the margin).

5. (A) Find the slope of the secant line joining $(0, f(0))$ and $(1, f(1))$.
(B) Find the slope of the secant line joining $(1, f(1))$ and $(1 + h, f(1 + h))$.
(C) Find the slope of the tangent line at $x = 1$.

6. (A) Find the slope of the secant line joining $(-1, f(-1))$ and $(0, f(0))$.
(B) Find the slope of the secant line joining $(-1, f(-1))$ and $(-1 + h, f(-1 + h))$.
(C) Find the slope of the tangent line at $x = -1$.

In Problems 7–10, find f′(x) using the two-step process:

Step 1. *Simplify:* $\frac{f(x+h)-f(x)}{h}$

Step 2. *Evaluate:* $\lim_{h\to 0}\frac{f(x+h)-f(x)}{h}$

Then find f′(1), f′(2), and f′(3).

7. $f(x) = 2x - 3$
8. $f(x) = 4x + 3$
9. $f(x) = 2 - x^2$
10. $f(x) = 2x^2 + 5$

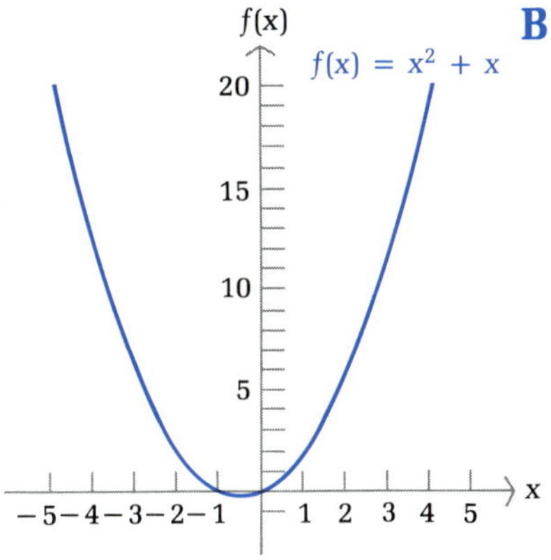

B

Problems 11 and 12 refer to the graph of $y = f(x) = x^2 + x$ shown in the margin.

11. (A) Find the slope of the secant line joining $(1, f(1))$ and $(3, f(3))$.
    (B) Find the slope of the secant line joining $(1, f(1))$ and $(1 + h, f(1 + h))$.
    (C) Find the slope of the tangent line at $(1, f(1))$.
    (D) Find the equation of the tangent line at $(1, f(1))$.
12. (A) Find the slope of the secant line joining $(2, f(2))$ and $(4, f(4))$.
    (B) Find the slope of the secant line joining $(2, f(2))$ and $(2 + h, f(2 + h))$.
    (C) Find the slope of the tangent line at $(2, f(2))$.
    (D) Find the equation of the tangent line at $(2, f(2))$.

In Problems 13 and 14, suppose an object moves along the y axis so that its location is $y = f(x) = x^2 + x$ at time x (y is in meters and x is in seconds). Find:

13. (A) The average velocity (the average rate of change of y with respect to x) for x changing from 1 to 3 seconds
    (B) The average velocity for x changing from 1 to $1 + h$ seconds
    (C) The instantaneous velocity at $x = 1$ second
14. (A) The average velocity (the average rate of change of y with respect to x) for x changing from 2 to 4 seconds
    (B) The average velocity for x changing from 2 to $2 + h$ seconds
    (C) The instantaneous velocity at $x = 2$ seconds

In Problems 15–20, find f′(x) using the two-step limiting process. Then find f′(1), f′(2), and f′(3).

15. $f(x) = 6x - x^2$
16. $f(x) = 2x - 3x^2$
17. $f(x) = \sqrt{x} - 3$
18. $f(x) = 2 - \sqrt{x}$
19. $f(x) = \frac{-1}{x}$
20. $f(x) = \frac{1}{x + 1}$

Problems 21–28 refer to the function F in the graph below. Use the graph to determine whether F′(x) exists at the indicated value of x.

21. $x = a$ **22.** $x = b$ **23.** $x = c$ **24.** $x = d$
25. $x = e$ **26.** $x = f$ **27.** $x = g$ **28.** $x = h$

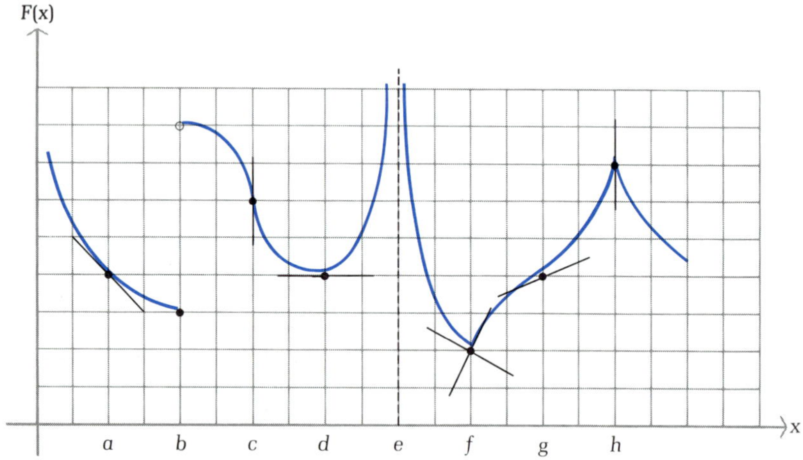

29. Given $f(x) = x^2 - 4x$:
(A) Find $f'(x)$.
(B) Find the slopes of the tangent lines to the graph of f at $x = 0$, 2, and 4.
(C) Graph f, and sketch in the tangent lines at $x = 0$, 2, and 4.

30. Given $f(x) = x^2 + 2x$:
(A) Find $f'(x)$.
(B) Find the slopes of the tangent lines to the graph of f at $x = -2, -1$, and 1.
(C) Graph f, and sketch in the tangent lines at $x = -2, -1$, and 1.

31. If an object moves along a line so that it is at $y = f(x) = 4x^2 - 2x$ at time x (in seconds), find the instantaneous velocity function $v = f'(x)$, and find the velocity at times $x = 1$, 3, and 5 seconds (y is measured in feet).

32. Repeat Problem 31 with $f(x) = 8x^2 - 4x$.

C *In Problems 33 and 34, sketch the graph of f and determine where f is nondifferentiable.*

33. $f(x) = \begin{cases} 2x & \text{if } x < 1 \\ 2 & \text{if } x \geq 1 \end{cases}$

34. $f(x) = \begin{cases} 2x & \text{if } x < 2 \\ 6 - x & \text{if } x \geq 2 \end{cases}$

In Problems 35–38, determine whether f is differentiable at x = 0 by considering

$$\lim_{h \to 0} \frac{f(0 + h) - f(0)}{h}$$

35. $f(x) = |x|$ **36.** $f(x) = 1 - |x|$ **37.** $f(x) = x^{1/3}$ **38.** $f(x) = x^{2/3}$

39. Show that $f(x) = 2x - x^2$ is differentiable over the closed interval [0, 2] by showing that each of the following limits exists:

(A) $\lim_{h \to 0} \frac{f(x + h) - f(x)}{h}, \quad 0 < x < 2$ (B) $\lim_{h \to 0^+} \frac{f(0 + h) - f(0)}{h}, \quad x = 0$

(C) $\lim_{h \to 0^-} \frac{f(2 + h) - f(2)}{h}, \quad x = 2$

40. Show that $f(x) = \sqrt{x}$ is differentiable over the open interval $(0, \infty)$ but not over the half-closed interval $[0, \infty)$ by considering

$$\lim_{h \to 0} \frac{f(x + h) - f(x)}{h} \qquad 0 < x < \infty$$

and

$$\lim_{h \to 0^+} \frac{f(0 + h) - f(0)}{h} \qquad x = 0$$

APPLICATIONS

Business & Economics

Problems 41 and 42 provide a first look at the concept of "marginal analysis," an important topic in business and economics that is developed further in succeeding sections and in detail in Section 3-7.

41. *Cost analysis.* The total cost per day, C(x) (in hundreds of dollars), for manufacturing x windsurfing boards is given by

$$C(x) = 3 + 10x - x^2 \qquad 0 \leq x \leq 5$$

(A) Find: $\frac{C(4) - C(3)}{4 - 3}$

(This is the increase in cost for a 1 unit increase in production at the level of production of 3 boards per day.)

(B) Find: $\frac{C(3 + h) - C(3)}{h}$

(C) Find: $C'(3)$
(This is called the *marginal cost* at the level of production of 3 boards per day; it is the instantaneous rate of change of cost relative to production and approximates the result found in part A.

(D) Find $C'(x)$, the *marginal cost function*.

(E) Using $C'(x)$, find $C'(1)$, $C'(2)$, $C'(3)$, and $C'(4)$, and interpret. [*Note:* As production levels increase, the rate of change of cost relative to production (marginal cost) decreases.]

42. *Cost analysis.* Repeat Problem 41 for

$$C(x) = 5 + 12x - x^2 \qquad 0 \leq x \leq 5$$

SECTION 3-4 Derivatives of Constants, Power Forms, and Sums

- DERIVATIVE OF A CONSTANT
- POWER RULE
- DERIVATIVE OF A CONSTANT TIMES A FUNCTION
- DERIVATIVES OF SUMS AND DIFFERENCES
- APPLICATIONS

In the preceding section, we defined the derivative of f at x as

$$f'(x) = \lim_{h \to 0} \frac{f(x + h) - f(x)}{h}$$

if the limit exists, and we used this definition and a two-step process to find the derivatives of several functions. In this and the next two sections, we will develop some rules based on this definition that will enable us to determine the derivatives of a rather large class of functions without having to go through the two-step process each time.

Before starting on these rules, we list some symbols that are widely used to represent derivatives:

Derivative Notation

Given $y = f(x)$, then

$$f'(x) \qquad y' \qquad \frac{dy}{dx} \qquad D_x f(x)$$

all represent the derivative of f at x.

Each of these symbols for the derivative has its particular advantage in certain situations. All of them will become familiar to you after a little experience.

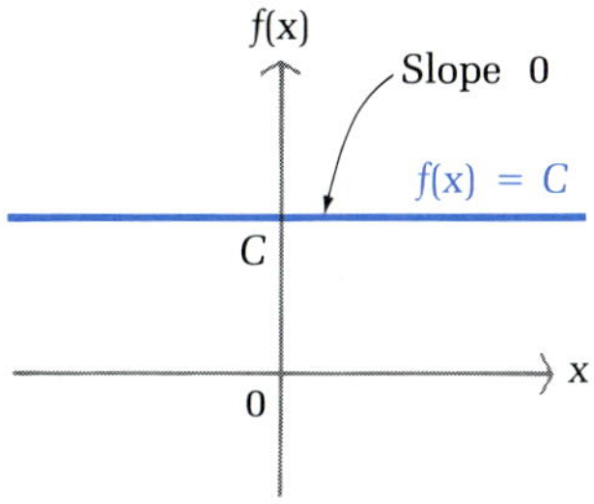

FIGURE 16

◆ DERIVATIVE OF A CONSTANT

Suppose

$f(x) = C$ $\quad$ C a constant $\quad$ A constant function

Geometrically, the graph of $f(x) = C$ is a horizontal straight line with slope 0 (see Figure 16); hence, we would expect $f'(x) = 0$. We will show that this is actually the case using the definition of the derivative and the two-step process introduced earlier. We want to find

$$f'(x) = \lim_{h \to 0} \frac{f(x+h) - f(x)}{h} \qquad \text{Definition of } f'(x)$$

Step 1. $\dfrac{f(x+h) - f(x)}{h} = \dfrac{C - C}{h} = \dfrac{0}{h} = 0 \qquad h \neq 0$

Step 2. $\lim_{h \to 0} 0 = 0$

Thus,

$f'(x) = 0$

We conclude that:

The derivative of any constant is 0.

Derivative of a Constant

If $y = f(x) = C$, then

$f'(x) = 0$

Also, $y' = 0$, $dy/dx = 0$, and $D_x\, C = 0$.

[*Note:* When we write $D_x\, C = 0$, we mean $D_x\, f(x) = 0$, where $f(x) = C$.]

◆ EXAMPLE 16 (A) If $f(x) = 3$, then $f'(x) = 0$. (B) If $y = -1.4$, then $y' = 0$. (C) If $y = \pi$, then $dy/dx = 0$. (D) $D_x\,(23) = 0$ ◆

PROBLEM 16 Find:

(A) $f'(x)$ for $f(x) = -24$ (B) y' for $y = 12$
(C) dy/dx for $y = -\sqrt{7}$ (D) $D_x(-\pi)$ ◆

◆ POWER RULE

Using the definition of derivative and the two-step process introduced in the preceding section, we can show that:

If $f(x) = x$, then $f'(x) = 1$.
If $f(x) = x^2$, then $f'(x) = 2x$.
If $f(x) = x^3$, then $f'(x) = 3x^2$.
If $f(x) = x^4$, then $f'(x) = 4x^3$.

In general, for any positive integer n:

If $f(x) = x^n$, then $f'(x) = nx^{n-1}$. (1)

In fact, more advanced techniques can be used to show that (1) holds for *any* real number n. We will assume this general result for the remainder of this book.

> **Power Rule**
>
> If $y = f(x) = x^n$, where n is a real number, then
>
> $$f'(x) = nx^{n-1}$$

◆ EXAMPLE 17

(A) If $f(x) = x^5$, then $f'(x) = 5x^{5-1} = 5x^4$.
(B) If $y = x^{25}$, then $y' = 25x^{25-1} = 25x^{24}$.
(C) If $y = x^{-3}$, then $dy/dx = -3x^{-3-1} = -3x^{-4}$.
(D) $D_x\, x^{5/3} = \frac{5}{3}x^{(5/3)-1} = \frac{5}{3}x^{2/3}$ ◆

PROBLEM 17

Find:

(A) $f'(x)$ for $f(x) = x^6$ (B) y' for $y = x^{30}$
(C) dy/dx for $y = x^{-2}$ (D) $D_x\, x^{3/2}$ ◆

In some cases, properties of exponents must be used to rewrite an expression before the power rule is applied.

◆ EXAMPLE 18

(A) If $f(x) = 1/x^4$, then we can write $f(x) = x^{-4}$ and

$$f'(x) = -4x^{-4-1} = -4x^{-5} \quad \text{or} \quad \frac{-4}{x^5}$$

(B) If $y = \sqrt{x}$, then we can write $y = x^{1/2}$ and

$$y' = \frac{1}{2}x^{(1/2)-1} = \frac{1}{2}x^{-1/2} \quad \text{or} \quad \frac{1}{2\sqrt{x}}$$

(C) $$D_x \frac{1}{\sqrt[3]{x}} = D_x\, x^{-1/3} = -\frac{1}{3}x^{(-1/3)-1} = -\frac{1}{3}x^{-4/3} \quad \text{or} \quad \frac{-1}{3\sqrt[3]{x^4}}$$ ◆

PROBLEM 18 Find:

(A) $f'(x)$ for $f(x) = \frac{1}{x}$ (B) y' for $y = \sqrt[3]{x^2}$ (C) $D_x \frac{1}{\sqrt{x}}$ ◆

◆ DERIVATIVE OF A CONSTANT TIMES A FUNCTION

Let $f(x) = ku(x)$; where k is a constant and u is differentiable at x. Then, using the two-step process, we have the following:

Step 1. $$\frac{f(x+h) - f(x)}{h} = \frac{ku(x+h) - ku(x)}{h} = k\left[\frac{u(x+h) - u(x)}{h}\right]$$

Step 2. $$\lim_{h\to 0} \frac{f(x+h) - f(x)}{h} = \lim_{h\to 0} k\left[\frac{u(x+h) - u(x)}{h}\right] \qquad \lim_{x\to c} kg(x) = k \lim_{x\to c} g(x)$$

$$= k \lim_{h\to 0}\left[\frac{u(x+h) - u(x)}{h}\right] \qquad \text{Definition of } u'(x)$$

$$= ku'(x)$$

Thus:

The derivative of a constant times a differentiable function is the constant times the derivative of the function.

Constant Times a Function Rule

If $y = f(x) = ku(x)$, then

$$f'(x) = ku'(x)$$

Also, $y' = ku'$, $dy/dx = k\,du/dx$, and $D_x\, ku(x) = k\, D_x\, u(x)$.

◆ EXAMPLE 19 (A) If $f(x) = 3x^2$, then $f'(x) = 3 \cdot 2x^{2-1} = 6x$.

(B) If $y = \frac{x^3}{6} = \frac{1}{6}x^3$, then $\frac{dy}{dx} = \frac{1}{6} \cdot 3x^{3-1} = \frac{1}{2}x^2$.

(C) If $y = \frac{1}{2x^4} = \frac{1}{2}x^{-4}$, then $y' = \frac{1}{2}(-4x^{-4-1}) = -2x^{-5}$ or $\frac{-2}{x^5}$.

(D) $D_x \frac{4}{\sqrt{x^3}} = D_x \frac{4}{x^{3/2}} = D_x\, 4x^{-3/2} = 4\left[-\frac{3}{2}x^{(-3/2)-1}\right]$

$= -6x^{-5/2}$ or $-\frac{6}{\sqrt{x^5}}$ ◆

PROBLEM 19 Find:

(A) $f'(x)$ for $f(x) = 4x^5$ (B) $\frac{dy}{dx}$ for $y = \frac{x^4}{12}$

(C) y' for $y = \frac{1}{3x^3}$ (D) $D_x \frac{9}{\sqrt[3]{x}}$

DERIVATIVES OF SUMS AND DIFFERENCES

Let $f(x) = u(x) + v(x)$, where $u'(x)$ and $v'(x)$ exist. Then, using the two-step process, we have the following:

Step 1.
$$\frac{f(x+h) - f(x)}{h} = \frac{[u(x+h) + v(x+h)] - [u(x) + v(x)]}{h}$$
$$= \frac{u(x+h) + v(x+h) - u(x) - v(x)}{h}$$
$$= \frac{u(x+h) - u(x)}{h} + \frac{v(x+h) - v(x)}{h}$$

Step 2.
$$\lim_{h\to 0} \frac{f(x+h) - f(x)}{h} = \lim_{h\to 0} \left[\frac{u(x+h) - u(x)}{h} + \frac{v(x+h) - v(x)}{h}\right]$$

$\lim_{x\to c} [g(x) + h(x)] = \lim_{x\to c} g(x) + \lim_{x\to c} h(x)$

$$= \lim_{h\to 0} \frac{u(x+h) - u(x)}{h} + \lim_{h\to 0} \frac{v(x+h) - v(x)}{h}$$
$$= u'(x) + v'(x)$$

Thus:

The derivative of the sum of two differentiable functions is the sum of the derivatives.

Similarly, we can show that:

The derivative of the difference of two differentiable functions is the difference of the derivatives.

Together, we then have the **sum and difference rule** for differentiation:

Sum and Difference Rule

If $y = f(x) = u(x) \pm v(x)$, then

$$f'(x) = u'(x) \pm v'(x)$$

[*Note:* This rule generalizes to the sum and difference of any given number of functions.]

With this and the other rules stated previously, we will be able to compute the derivatives of all polynomials and a variety of other functions.

◆ EXAMPLE 20

(A) If $f(x) = 3x^2 + 2x$, then

$$f'(x) \boxed{= (3x^2)' + (2x)' = 3(2x) + 2(1)} = 6x + 2$$

(B) If $y = 4 + 2x^3 - 3x^{-1}$, then

$$y' \boxed{= (4)' + (2x^3)' - (3x^{-1})' = 0 + 2(3x^2) - 3(-1)x^{-2}} = 6x^2 + 3x^{-2}$$

(C) If $y = \sqrt[3]{x} - 3x$, then

$$\frac{dy}{dx} = \frac{d}{dx}x^{1/3} - \frac{d}{dx}3x = \frac{1}{3}x^{-2/3} - 3$$

(D) $$D_x\left(\frac{5}{3x^2} - \frac{2}{x^4} + \frac{x^3}{9}\right) \boxed{= D_x\frac{5}{3}x^{-2} - D_x 2x^{-4} + D_x\frac{1}{9}x^3}$$

$$= \frac{5}{3}(-2)x^{-3} - 2(-4)x^{-5} + \frac{1}{9}\cdot 3x^2 = -\frac{10}{3x^3} + \frac{8}{x^5} + \frac{1}{3}x^2$$ ◆

PROBLEM 20

Find:

(A) $f'(x)$ for $f(x) = 3x^4 - 2x^3 + x^2 - 5x + 7$

(B) y' for $y = 3 - 7x^{-2}$ (C) $\frac{dy}{dx}$ for $y = 5x^3 - \sqrt[4]{x}$

(D) $D_x\left(-\frac{3}{4x} + \frac{4}{x^3} - \frac{x^4}{8}\right)$ ◆

◆ APPLICATIONS

◆ EXAMPLE 21
Instantaneous Velocity

An object moves along the y axis (marked in feet) so that its position at time x (in seconds) is

$$f(x) = x^3 - 6x^2 + 9x$$

(A) Find the instantaneous velocity function v.
(B) Find the velocity at $x = 2$ and $x = 5$ seconds.
(C) Find the time(s) when the velocity is 0.

Solutions

(A) $v = f'(x) \boxed{= (x^3)' - (6x^2)' + (9x)'} = 3x^2 - 12x + 9$

(B) $f'(2) = 3(2)^2 - 12(2) + 9 = -3$ feet per second
$f'(5) = 3(5)^2 - 12(5) + 9 = 24$ feet per second

(C) $$\begin{aligned} v = f'(x) = 3x^2 - 12x + 9 &= 0 \\ 3(x^2 - 4x + 3) &= 0 \\ 3(x-1)(x-3) &= 0 \\ x &= 1, 3 \end{aligned}$$

Thus, $v = 0$ at $x = 1$ and $x = 3$ seconds. ◆

PROBLEM 21 Repeat Example 21 for $f(x) = x^3 - 15x^2 + 72x$. ◆

◆ EXAMPLE 22 Let $f(x) = x^4 - 8x^2 + 10$.

Tangents

(A) Find $f'(x)$.
(B) Find the equation of the tangent line at $x = 1$.
(C) Find the values of x where the tangent line is horizontal.

Solutions

(A) $f'(x) = (x^4)' - (8x^2)' + (10)'$

$= 4x^3 - 16x$

(B)

$$\begin{aligned} y - y_1 &= m(x - x_1) && y_1 = f(x_1) = f(1) = (1)^4 - 8(1)^2 + 10 = 3 \\ y - 3 &= -12(x - 1) && m = f'(x_1) = f'(1) = 4(1)^3 - 16(1) = -12 \\ y &= -12x + 15 && \text{Tangent line at } x = 1 \end{aligned}$$

(C) Since a horizontal line has 0 slope, we must solve $f'(x) = 0$ for x:

$$\begin{aligned} f'(x) = 4x^3 - 16x &= 0 \\ 4x(x^2 - 4) &= 0 \\ 4x(x - 2)(x + 2) &= 0 \\ x &= 0, 2, -2 \end{aligned}$$

Thus, the tangent line to the graph of f will be horizontal at $x = -2$, $x = 0$, and $x = 2$. (In the next chapter, we will see how this information is used to help sketch the graph of f.) ◆

PROBLEM 22 Repeat Example 22 for $f(x) = x^4 - 4x^3 + 7$. ◆

In business and economics, one is often interested in the rate at which something is taking place. A manufacturer, for example, is not only interested in the total cost C(x) at certain production levels x, but is also interested in the rate of change of costs at various production levels.

In economics, the word **marginal** refers to a rate of change; that is, to a derivative. Thus, if

$C(x)$ = Total cost of producing x units during some unit of time

then

$C'(x)$ = Marginal cost
= Rate of change in cost per unit change in production at an output level of x units

Just as with instantaneous velocity, marginal cost $C'(x)$ is an instantaneous rate of change of total costs $C(x)$ with respect to production at a production level of x units. If this rate remains constant as production is increased by 1 unit, then it represents the change in cost for a 1 unit change in production. If the rate does not remain constant, then the instantaneous rate is an approximation of what actually happens during the next unit change in production.

The Marginal Cost Function

If the marginal cost function $C'(x)$ is a constant function, then $C'(x)$ represents the cost of producing 1 more unit at any production level x. If $C'(x)$ is not a constant function, then $C'(x)$ approximates the cost of producing 1 more unit at a production level of x units.

Example 23 should help to clarify these ideas.

◆ EXAMPLE 23 *Marginal Cost*

Suppose the total cost $C(x)$, in thousands of dollars, for manufacturing x sailboats per year is given by the function

$$C(x) = 575 + 25x - \frac{x^2}{4} \qquad 0 \le x \le 50$$

shown in the figure below.

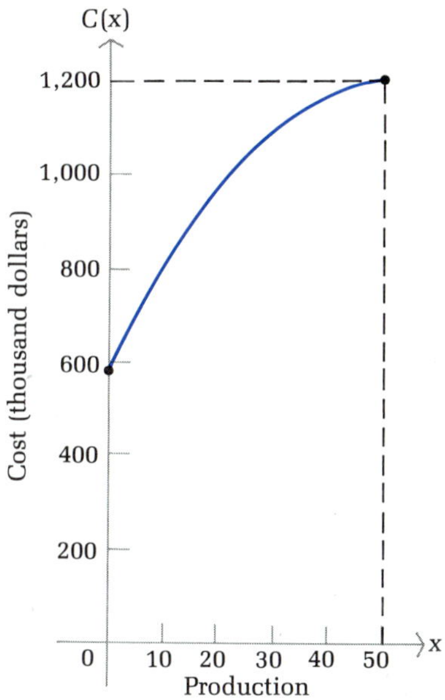

(A) Find the marginal cost at a production level of x boats.

(B) Find the marginal cost at a production level of 40 boats, and interpret the result.

(C) Find the actual cost of producing the 41st boat, and compare this cost with the result found in part B.

(D) Find $C'(30)$, and interpret the result.

Solutions

(A) The marginal cost at a production level of x boats is

$$C'(x) \boxed{= (575)' + (25x)' - \left(\frac{x^2}{4}\right)'} = 25 - \frac{x}{2}$$

(B) The marginal cost at a production level of 40 boats is

$$C'(40) = 25 - \frac{40}{2} = 5 \quad \text{or} \quad \$5{,}000 \text{ per boat}$$

At a production level of 40 boats, the rate of change of total cost relative to production is \$5,000 per boat. Thus, the cost of producing 1 more boat at a production level of 40 boats is approximately \$5,000.

(C) The actual cost of producing the 41st boat is

$$\begin{aligned} &\begin{pmatrix}\text{Total cost of}\\ \text{producing}\\ \text{41 boats}\end{pmatrix} - \begin{pmatrix}\text{Total cost of}\\ \text{producing}\\ \text{40 boats}\end{pmatrix} \\ &= C(41) \qquad - \qquad C(40) \\ &= 1{,}179.75 - 1{,}175.00 = 4.75 \quad \text{or} \quad \$4{,}750\end{aligned}$$

The marginal cost of \$5,000 per boat found in part B is a close approximation to this value.

(D) $$C'(30) = 25 - \frac{30}{2} = 10 \quad \text{or} \quad \$10{,}000 \text{ per boat}$$

At a production level of 30 boats, the rate of change of total cost relative to production is \$10,000 per boat. Thus, the cost of producing 1 more boat at this level of production is approximately \$10,000. ◆

In Example 23, we observe that as production goes up, the marginal cost goes down.

PROBLEM 23 Suppose the total cost $C(x)$, in thousands of dollars, for manufacturing x sailboats per year is given by the function

$$C(x) = 500 + 24x - \frac{x^2}{5} \qquad 0 \le x \le 50$$

(A) Find the marginal cost at a production level of x boats.
(B) Find the marginal cost at a production level of 35 boats, and interpret the result.
(C) Find the actual cost of producing the 36th boat, and compare this cost with the result found in part B.
(D) Find $C'(40)$, and interpret the result. ◆

Answers to Matched Problems

16. All are 0. 17. (A) $6x^5$ (B) $30x^{29}$ (C) $-2x^{-3}$ (D) $\frac{3}{2}x^{1/2}$
18. (A) $-x^{-2}$ or $-1/x^2$ (B) $\frac{2}{3}x^{-1/3}$ or $2/(3\sqrt[3]{x})$
(C) $-\frac{1}{2}x^{-3/2}$ or $-1/(2\sqrt{x^3})$
19. (A) $20x^4$ (B) $x^3/3$ (C) $-x^{-4}$ or $-1/x^4$ (D) $-3x^{-4/3}$ or $-3/\sqrt[3]{x^4}$
20. (A) $12x^3 - 6x^2 + 2x - 5$ (B) $14x^{-3}$
(C) $15x^2 - \frac{1}{4}x^{-3/4}$ (D) $3/(4x^2) - (12/x^4) - (x^3/2)$
21. (A) $v = 3x^2 - 30x + 72$ (B) $f'(2) = 24$ ft/sec; $f'(5) = -3$ ft/sec
(C) $x = 4$ and $x = 6$ sec

22. (A) $f'(x) = 4x^3 - 12x^2$ (B) $y = -8x + 12$ (C) $x = 0$ and $x = 3$

23. (A) $C'(x) = 24 - (2x/5)$

(B) $C'(35) = 10$ or \$10,000 per boat; at a production level of 35 boats, the rate of change of total cost relative to production is \$10,000 per boat; thus, the cost of producing 1 more boat at this level of production is approx. \$10,000

(C) $C(36) - C(35) = 9.8$ or \$9,800; the marginal cost of \$10,000 per boat found in part B is a close approximation to this value

(D) $C'(40) = 8$ or \$8,000 per boat; at a production level of 40 boats, the rate of change of total cost relative to production is \$8,000 per boat; thus, the cost of producing 1 more boat at this level of production is approx. \$8,000

EXERCISE 3-4

Find each of the following:

A

1. $f'(x)$ for $f(x) = 12$
2. $\frac{dy}{dx}$ for $y = -\sqrt{3}$
3. $D_x\, 23$
4. y' for $y = \pi$
5. $\frac{dy}{dx}$ for $y = x^{12}$
6. $D_x\, x^5$
7. $f'(x)$ for $f(x) = x$
8. y' for $y = x^7$
9. y' for $y = x^{-7}$
10. $f'(x)$ for $f(x) = x^{-11}$
11. $\frac{dy}{dx}$ for $y = x^{5/2}$
12. $D_x\, x^{7/3}$
13. $D_x \frac{1}{x^5}$
14. $f'(x)$ for $f(x) = \frac{1}{x^9}$
15. $f'(x)$ for $f(x) = 2x^4$
16. $\frac{dy}{dx}$ for $y = -3x$
17. $D_x(\frac{1}{3}x^6)$
18. y' for $y = \frac{1}{2}x^4$
19. $\frac{dy}{dx}$ for $y = \frac{x^5}{15}$
20. $f'(x)$ for $f(x) = \frac{x^6}{24}$

B

21. $D_x(2x^{-5})$
22. y' for $y = -4x^{-1}$
23. $f'(x)$ for $f(x) = \frac{4}{x^4}$
24. $\frac{dy}{dx}$ for $y = \frac{-3}{x^6}$
25. $D_x \frac{-1}{2x^2}$
26. y' for $y = \frac{1}{6x^3}$
27. $f'(x)$ for $f(x) = -3x^{1/3}$
28. $\frac{dy}{dx}$ for $y = -8x^{1/4}$
29. $D_x(2x^2 - 3x + 4)$
30. y' for $y = 3x^2 + 4x - 7$
31. $\frac{dy}{dx}$ for $y = 3x^5 - 2x^3 + 5$
32. $f'(x)$ for $f(x) = 2x^3 - 6x + 5$
33. $D_x(3x^{-4} + 2x^{-2})$
34. y' for $y = 2x^{-3} - 4x^{-1}$

35. $\frac{dy}{dx}$ for $y = \frac{1}{2x} - \frac{2}{3x^3}$

36. $f'(x)$ for $f(x) = \frac{3}{4x^3} + \frac{1}{2x^5}$

37. $D_x(3x^{2/3} - 5x^{1/3})$

38. $D_x(8x^{3/4} + 4x^{-1/4})$

39. $D_x\left(\frac{3}{x^{3/5}} - \frac{6}{x^{1/2}}\right)$

40. $D_x\left(\frac{5}{x^{1/5}} - \frac{8}{x^{3/2}}\right)$

41. $D_x \frac{1}{\sqrt[3]{x}}$

42. y' for $y = \frac{10}{\sqrt[5]{x}}$

43. $\frac{dy}{dx}$ for $y = \frac{12}{\sqrt{x}} - 3x^{-2} + x$

44. $f'(x)$ for $f(x) = 2x^{-3} - \frac{6}{\sqrt[3]{x^2}} + 7$

For Problems 45–48, find:

(A) $f'(x)$
(B) The slope of the graph of f at $x = 2$ and $x = 4$.
(C) The equations of the tangent lines at $x = 2$ and $x = 4$.
(D) The value(s) of x where the tangent line is horizontal.

45. $f(x) = 6x - x^2$

46. $f(x) = 2x^2 + 8x$

47. $f(x) = 3x^4 - 6x^2 - 7$

48. $f(x) = x^4 - 32x^2 + 10$

If an object moves along the y axis (marked in feet) so that its position at time x (in seconds) is given by the indicated function in Problems 49–52, find:

(A) The instantaneous velocity function $v = f'(x)$.
(B) The velocity when $x = 0$ and $x = 3$ seconds.
(C) The time(s) when $v = 0$.

49. $f(x) = 176x - 16x^2$

50. $f(x) = 80x - 10x^2$

51. $f(x) = x^3 - 9x^2 + 15x$

52. $f(x) = x^3 - 9x^2 + 24x$

C *In Problems 53–56, find each derivative.*

53. $f'(x)$ for $f(x) = \frac{10x + 20}{x}$

54. $\frac{dy}{dx}$ for $y = \frac{x^2 + 25}{x^2}$

55. $D_x \frac{x^4 - 3x^3 + 5}{x^2}$

56. y' for $y = \frac{2x^5 - 4x^3 + 2x}{x^3}$

In Problems 57 and 58, use the definition of derivative and the two-step process to verify each statement.

57. $D_x\, x^3 = 3x^2$

58. $D_x\, x^4 = 4x^3$

Problems 59–66 require the use of a graphic calculator or a computer. For each problem, find $f'(x)$ and approximate to two decimal places the value(s) of x where the graph of f has a horizontal tangent line.

59. $f(x) = x^2 - 3x - 4\sqrt{x}$

60. $f(x) = x^2 + x - 10\sqrt{x}$

61. $f(x) = 3\sqrt[3]{x^4} - 1.5x^2 - 3x$

62. $f(x) = 3\sqrt[3]{x^4} - 2x^2 + 4x$

63. $f(x) = 0.05x^4 + 0.1x^3 - 1.5x^2 - 1.6x + 3$
64. $f(x) = 0.02x^4 - 0.06x^3 - 0.78x^2 + 0.94x + 2.2$
65. $f(x) = 0.2x^4 - 3.12x^3 + 16.25x^2 - 28.25x + 7.5$
66. $f(x) = 0.25x^4 - 2.6x^3 + 8.1x^2 - 10x + 9$

APPLICATIONS

Business & Economics

67. *Marginal cost.* The total cost (in dollars) of producing x tennis rackets per day is

$$C(x) = 800 + 60x - \frac{x^2}{4} \qquad 0 \leq x \leq 120$$

(A) Find the marginal cost at a production level of x rackets.
(B) Find the marginal cost at a production level of 60 rackets, and interpret the result.
(C) Find the actual cost of producing the 61st racket, and compare this cost with the result found in part B.
(D) Find $C'(80)$, and interpret the result.

68. *Marginal cost.* The total cost (in dollars) of producing x portable radios per day is

$$C(x) = 1{,}000 + 100x - \frac{x^2}{2} \qquad 0 \leq x \leq 100$$

(A) Find the marginal cost at a production level of x radios.
(B) Find the marginal cost at a production level of 80 radios, and interpret the result.
(C) Find the actual cost of producing the 81st radio, and compare this cost with the result found in part B.
(D) Find $C'(50)$, and interpret the result.

69. *Advertising.* Using past records, it is estimated that a company will sell $N(x)$ units of a product after spending \$x thousand on advertising, as given by

$$N(x) = 60x - x^2 \qquad 5 \leq x \leq 30$$

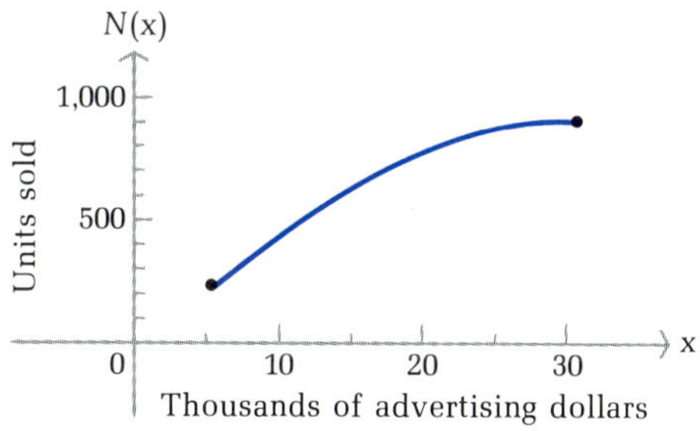

(A) Find $N'(x)$, the instantaneous rate of change of sales with respect to the amount of money spent on advertising at the $\$x$ thousand level.
(B) Find $N'(10)$ and $N'(20)$, and interpret the results.

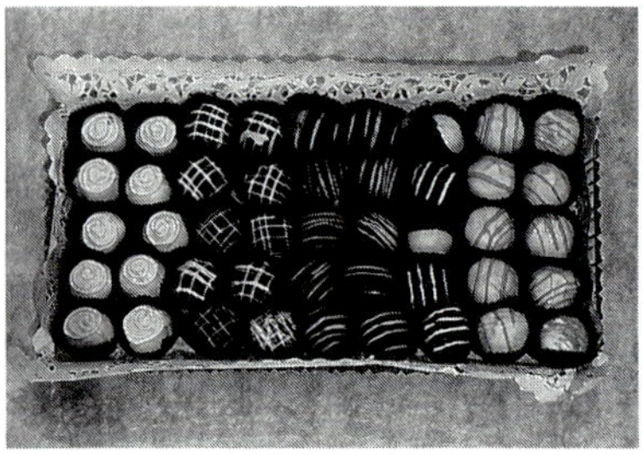

70. *Demand function.* Suppose that in a given gourmet food store, people are willing to buy $D(x)$ pounds of chocolate candy per day at $\$x$ per quarter pound, as given by the demand function

$$D(x) = 100 - x^2 \qquad 1 \leq x \leq 10$$

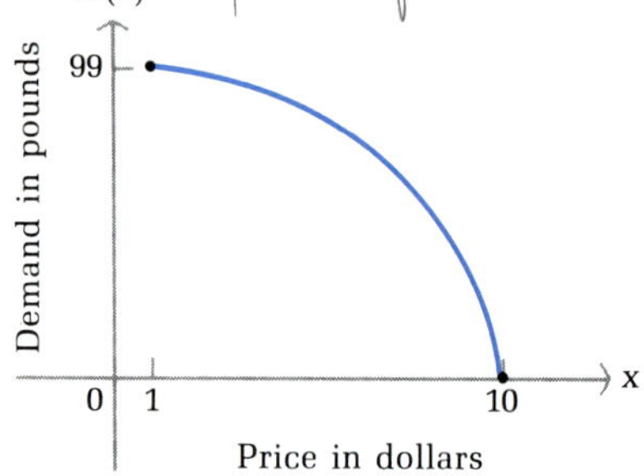

(A) Find $D'(x)$, the instantaneous rate of change of demand with respect to price at the $\$x$ price level.
(B) Find $D'(2)$ and $D'(8)$, and interpret the results.

Life Sciences

71. *Medicine.* A person x inches tall has a pulse rate of y beats per minute, as given approximately by

$$y = 590x^{-1/2} \qquad 30 \leq x \leq 75$$

What is the instantaneous rate of change of pulse rate at the:

(A) 36 inch level? (B) 64 inch level?

72. *Ecology.* A coal-burning electrical generating plant emits sulfur dioxide into the surrounding air. The concentration $C(x)$, in parts per million, is given approximately by

$$C(x) = \frac{0.1}{x^2}$$

where x is the distance from the plant in miles. Find the instantaneous rate of change of concentration at:

(A) $x = 1$ mile (B) $x = 2$ miles

Social Sciences

73. *Learning.* Suppose a person learns y items in x hours, as given by

$$y = 50\sqrt{x} \qquad 0 \leq x \leq 9$$

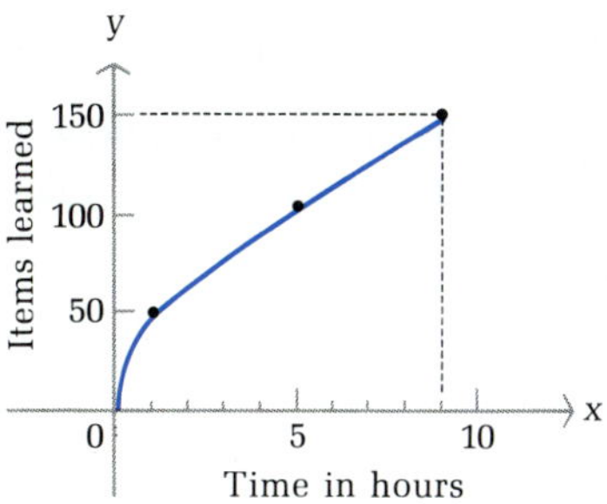

(see the figure in the margin). Find the rate of learning at the end of:

(A) 1 hour (B) 9 hours

74. *Learning.* If a person learns y items in x hours, as given by

$$y = 21\sqrt[3]{x^2} \qquad 0 \le x \le 8$$

find the rate of learning at the end of:

(A) 1 hour (B) 8 hours

SECTION 3-5 Derivatives of Products and Quotients

- DERIVATIVES OF PRODUCTS
- DERIVATIVES OF QUOTIENTS

The derivative rules discussed in the preceding section added substantially to our ability to compute and apply derivatives to many practical problems. In this and the next section, we will add a few more rules that will increase this ability even further.

◆ DERIVATIVES OF PRODUCTS

In Section 3-4, we found that the derivative of a sum is the sum of the derivatives. Is the derivative of a product the product of the derivatives? Let us take a look at a simple example. Consider

$$f(x) = u(x)v(x) = (x^2 - 3x)(2x^3 - 1) \tag{1}$$

where $u(x) = x^2 - 3x$ and $v(x) = 2x^3 - 1$. The product of the derivatives is

$$u'(x)v'(x) = (2x - 3)6x^2 = 12x^3 - 18x^2 \tag{2}$$

To see if this is equal to the derivative of the product, we multiply the right side of (1) and use the derivative formulas we already know:

$$f(x) = (x^2 - 3x)(2x^3 - 1) = 2x^5 - 6x^4 - x^2 + 3x$$

Thus,

$$f'(x) = 10x^4 - 24x^3 - 2x + 3 \tag{3}$$

Since (2) and (3) are not equal, we conclude that the derivative of a product is *not* the product of the derivatives. There is a product rule for derivatives, but it is slightly more complicated than you might expect.

Using the definition of derivative and the two-step process, we can show that:

The derivative of the product of two functions is the first function times the derivative of the second function plus the second function times the derivative of the first function.

That is:

Product Rule

If

$$y = f(x) = F(x)S(x)$$

and if $F'(x)$ and $S'(x)$ exist, then

$$f'(x) = F(x)S'(x) + S(x)F'(x)$$

Also,

$$y' = FS' + SF' \qquad \frac{dy}{dx} = F\frac{dS}{dx} + S\frac{dF}{dx}$$

$$D_x[F(x)S(x)] = F(x)\,D_x\,S(x) + S(x)\,D_x\,F(x)$$

◆ EXAMPLE 24 Use two different methods to find $f'(x)$ for $f(x) = 2x^2(3x^4 - 2)$.

Solution *Method 1.* Use the product rule:

$$\begin{aligned} f'(x) &= 2x^2(3x^4 - 2)' + (3x^4 - 2)(2x^2)' \\ &= 2x^2(12x^3) + (3x^4 - 2)(4x) \\ &= 24x^5 + 12x^5 - 8x \\ &= 36x^5 - 8x \end{aligned}$$

First times derivative of second plus second times derivative of first

Method 2. Multiply first; then take derivatives:

$$f(x) = 2x^2(3x^4 - 2) = 6x^6 - 4x^2$$
$$f'(x) = 36x^5 - 8x$$

◆

PROBLEM 24 Use two different methods to find $f'(x)$ for $f(x) = 3x^3(2x^2 - 3x + 1)$. ◆

At this point, all the products we will encounter can be differentiated by either of the methods illustrated in Example 24. In the next and later sections, we will see that there are situations where the product rule must be used. Unless instructed otherwise, you should use the product rule to differentiate all products in this section to gain experience with the use of this important differentiation rule.

◆ EXAMPLE 25 Let $f(x) = (2x - 9)(x^2 + 6)$.

(A) Find the equation of the line tangent to the graph of $f(x)$ at $x = 3$.
(B) Find the value(s) of x where the tangent line is horizontal.

Solutions (A) First, find $f'(x)$:

$$f'(x) = (2x - 9)(x^2 + 6)' + (x^2 + 6)(2x - 9)'$$
$$= (2x - 9)(2x) + (x^2 + 6)(2)$$

Now, find the equation of the tangent line at $x = 3$:

$$y - y_1 = m(x - x_1) \qquad y_1 = f(x_1) = f(3) = -45$$
$$y - (-45) = 12(x - 3) \qquad m = f'(x_1) = f'(3) = 12$$
$$y = 12x - 81 \qquad \text{Tangent line at } x = 3$$

(B) The tangent line is horizontal at any value of x such that $f'(x) = 0$, so

$$f'(x) = (2x - 9)2x + (x^2 + 6)2 = 0$$
$$6x^2 - 18x + 12 = 0$$
$$x^2 - 3x + 2 = 0$$
$$(x - 1)(x - 2) = 0$$
$$x = 1, 2$$

The tangent line is horizontal at $x = 1$ and at $x = 2$. ◆

PROBLEM 25 Repeat Example 25 for $f(x) = (2x + 9)(x^2 - 12)$. ◆

As Example 25 illustrates, the way we write $f'(x)$ depends on what we want to do with it. If we are interested only in evaluating $f'(x)$ at specified values of x, the form in part A is sufficient. However, if we want to solve $f'(x) = 0$, we must multiply and collect like terms, as we did in part B.

◆ DERIVATIVES OF QUOTIENTS

As in the case with a product, the derivative of a quotient of two functions is *not* the quotient of the derivatives of the two functions.

Let

$$f(x) = \frac{T(x)}{B(x)} \qquad \text{where } T'(x) \text{ and } B'(x) \text{ exist}$$

Starting with the definition of a derivative, it can be shown that

$$f'(x) = \frac{B(x)T'(x) - T(x)B'(x)}{[B(x)]^2}$$

Thus:

The derivative of the quotient of two functions is the bottom function times the derivative of the top function minus the top function times the derivative of the bottom function, all over the bottom function squared.

Quotient Rule

If

$$y = f(x) = \frac{T(x)}{B(x)}$$

and if $T'(x)$ and $B'(x)$ exist, then

$$f'(x) = \frac{B(x)T'(x) - T(x)B'(x)}{[B(x)]^2}$$

Also,

$$y' = \frac{BT' - TB'}{B^2} \qquad \frac{dy}{dx} = \frac{B\dfrac{dT}{dx} - T\dfrac{dB}{dx}}{B^2}$$

$$D_x \frac{T(x)}{B(x)} = \frac{B(x)\, D_x\, T(x) - T(x)\, D_x\, B(x)}{[B(x)]^2}$$

◆ EXAMPLE 26

(A) If $f(x) = \dfrac{x^2}{2x - 1}$, find $f'(x)$.

(B) Find: $D_x \dfrac{x^2 - x}{x^3 + 1}$

(C) Find $D_x \dfrac{x^2 - 3}{x^2}$ by using the quotient rule and also by splitting the fraction into two fractions.

Solutions

(A)

$$\begin{aligned} f'(x) &= \frac{(2x - 1)(x^2)' - x^2(2x - 1)'}{(2x - 1)^2} \\ &= \frac{(2x - 1)(2x) - x^2(2)}{(2x - 1)^2} \\ &= \frac{4x^2 - 2x - 2x^2}{(2x - 1)^2} \\ &= \frac{2x^2 - 2x}{(2x - 1)^2} \end{aligned}$$

The bottom times the derivative of the top minus the top times the derivative of the bottom, all over the square of the bottom

(B)

$$\begin{aligned} D_x \frac{x^2 - x}{x^3 + 1} &= \frac{(x^3 + 1)\, D_x(x^2 - x) - (x^2 - x)\, D_x(x^3 + 1)}{(x^3 + 1)^2} \\ &= \frac{(x^3 + 1)(2x - 1) - (x^2 - x)(3x^2)}{(x^3 + 1)^2} \\ &= \frac{2x^4 - x^3 + 2x - 1 - 3x^4 + 3x^3}{(x^3 + 1)^2} \\ &= \frac{-x^4 + 2x^3 + 2x - 1}{(x^3 + 1)^2} \end{aligned}$$

(C) *Method 1.* Use the quotient rule:

$$D_x \frac{x^2 - 3}{x^2} = \frac{x^2 D_x(x^2 - 3) - (x^2 - 3) D_x x^2}{(x^2)^2}$$

$$= \frac{x^2(2x) - (x^2 - 3)2x}{x^4}$$

$$= \frac{2x^3 - 2x^3 + 6x}{x^4} = \frac{6x}{x^4} = \frac{6}{x^3}$$

Method 2. Split into two fractions:

$$\frac{x^2 - 3}{x^2} = \frac{x^2}{x^2} - \frac{3}{x^2} = 1 - 3x^{-2}$$

$$D_x(1 - 3x^{-2}) = 0 - 3(-2)x^{-3} = \frac{6}{x^3}$$

Comparing methods 1 and 2, we see that it often pays to change an expression algebraically before blindly using a differentiation formula. ◆

PROBLEM 26 Find:

(A) $f'(x)$ for $f(x) = \dfrac{2x}{x^2 + 3}$ (B) y' for $y = \dfrac{x^3 - 3x}{x^2 - 4}$

(C) $D_x \dfrac{2 + x^3}{x^3}$ two ways ◆

◆ EXAMPLE 27

Sales Analysis

When a successful home video game is first introduced, the monthly sales generally increase rapidly for a period of time, and then begin to decrease. Suppose that the monthly sales $S(t)$, in thousands of games, t months after the game is introduced are given by

$$S(t) = \frac{200t}{t^2 + 100}$$

(A) Find $S'(t)$. (B) Find $S(5)$ and $S'(5)$, and interpret the results.
(C) Find $S(30)$ and $S'(30)$, and interpret the results.

Solutions

(A) $$S'(t) = \frac{(t^2 + 100)(200t)' - 200t(t^2 + 100)'}{(t^2 + 100)^2}$$

$$= \frac{(t^2 + 100)200 - 200t(2t)}{(t^2 + 100)^2}$$

$$= \frac{200t^2 + 20{,}000 - 400t^2}{(t^2 + 100)^2}$$

$$= \frac{20{,}000 - 200t^2}{(t^2 + 100)^2}$$

(B) $S(5) = \dfrac{200(5)}{5^2 + 100} = 8$ and $S'(5) = \dfrac{20{,}000 - 200(5)^2}{(5^2 + 100)^2} = 0.96$

The sales for the 5th month are 8,000 units. At this point in time, sales are increasing at the rate of 0.96(1,000) = 960 units per month.

(C) $S(30) = \dfrac{200(30)}{30^2 + 100} = 6$ and $S'(30) = \dfrac{20{,}000 - 200(30)^2}{(30^2 + 100)^2} = -0.16$

The sales for the 30th month are 6,000 units. At this point in time, sales are decreasing at the rate of $0.16(1{,}000) = 160$ units per month. ◆

The function $S(t)$ in Example 27 is graphed in Figure 17. Notice that the maximum monthly sales seem to occur during the 10th month. In the next chapter, we will see how the derivative $S'(t)$ is used to help sketch the graph of $S(t)$ and to find the maximum monthly sales.

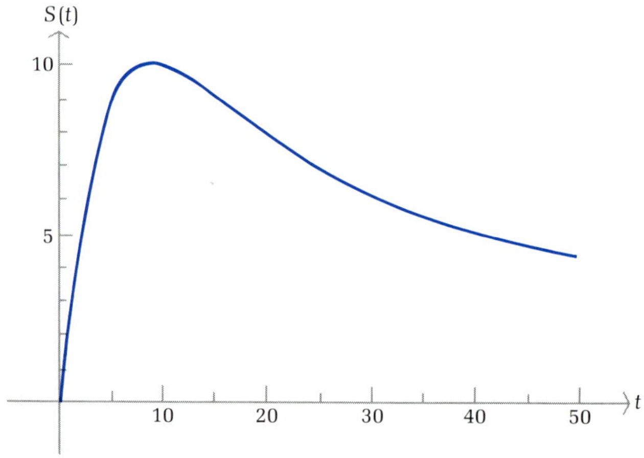

FIGURE 17

PROBLEM 27 Refer to Example 27. Suppose that the monthly sales $S(t)$, in thousands of games, t months after the game is introduced are given by

$$S(t) = \frac{200t}{t^2 + 64}$$

(A) Find $S'(t)$. (B) Find $S(4)$ and $S'(4)$, and interpret the results.
(C) Find $S(24)$ and $S'(24)$, and interpret the results. ◆

Answers to Matched Problems

24. $30x^4 - 36x^3 + 9x^2$ 25. (A) $y = 84x - 297$ (B) $x = -4, x = 1$

26. (A) $\dfrac{(x^2 + 3)2 - (2x)(2x)}{(x^2 + 3)^2} = \dfrac{6 - 2x^2}{(x^2 + 3)^2}$

(B) $\dfrac{(x^2 - 4)(3x^2 - 3) - (x^3 - 3x)(2x)}{(x^2 - 4)^2} = \dfrac{x^4 - 9x^2 + 12}{(x^2 - 4)^2}$ (C) $-\dfrac{6}{x^4}$

27. (A) $S'(t) = \dfrac{12{,}800 - 200t^2}{(t^2 + 64)^2}$

(B) $S(4) = 10$; $S'(4) = 1.5$; at $t = 4$ months, monthly sales are 10,000 and increasing at 1,500 games per month

(C) $S(24) = 7.5$; $S'(24) = -0.25$; at $t = 24$ months, monthly sales are 7,500 and decreasing at 250 games per month

EXERCISE 3-5

A *For f(x) as given, find f′(x) and simplify.*

1. $f(x) = 2x^3(x^2 - 2)$
2. $f(x) = 5x^2(x^3 + 2)$
3. $f(x) = (x - 3)(2x - 1)$
4. $f(x) = (3x + 2)(4x - 5)$
5. $f(x) = \dfrac{x}{x - 3}$
6. $f(x) = \dfrac{3x}{2x + 1}$
7. $f(x) = \dfrac{2x + 3}{x - 2}$
8. $f(x) = \dfrac{3x - 4}{2x + 3}$
9. $f(x) = (x^2 + 1)(2x - 3)$
10. $f(x) = (3x + 5)(x^2 - 3)$
11. $f(x) = \dfrac{x^2 + 1}{2x - 3}$
12. $f(x) = \dfrac{3x + 5}{x^2 - 3}$
13. $f(x) = (x^2 + 2)(x^2 - 3)$
14. $f(x) = (x^2 - 4)(x^2 + 5)$
15. $f(x) = \dfrac{x^2 + 2}{x^2 - 3}$
16. $f(x) = \dfrac{x^2 - 4}{x^2 + 5}$

B *Find each of the following and simplify:*

17. $f'(x)$ for $f(x) = (2x + 1)(x^2 - 3x)$
18. y' for $y = (x^3 + 2x^2)(3x - 1)$
19. $\dfrac{dy}{dx}$ for $y = (2x - x^2)(5x + 2)$
20. $D_x[(3 - x^3)(x^2 - x)]$
21. y' for $y = \dfrac{5x - 3}{x^2 + 2x}$
22. $f'(x)$ for $f(x) = \dfrac{3x^2}{2x - 1}$
23. $D_x \dfrac{x^2 - 3x + 1}{x^2 - 1}$
24. $\dfrac{dy}{dx}$ for $y = \dfrac{x^4 - x^3}{3x - 1}$

In Problems 25–28, find f′(x) and find the equation of the line tangent to the graph of f at x = 2.

25. $f(x) = (1 + 3x)(5 - 2x)$
26. $f(x) = (7 - 3x)(1 + 2x)$
27. $f(x) = \dfrac{x - 8}{3x - 4}$
28. $f(x) = \dfrac{2x - 5}{2x - 3}$

In Problems 29–32, find f′(x) and find the value(s) of x where f′(x) = 0.

29. $f(x) = (2x - 15)(x^2 + 18)$
30. $f(x) = (2x - 3)(x^2 - 6)$
31. $f(x) = \dfrac{x}{x^2 + 1}$
32. $f(x) = \dfrac{x}{x^2 + 9}$

In Problems 33–36, find f′(x) two ways; by using the product or quotient rule and by simplifying first.

33. $f(x) = x^3(x^4 - 1)$
34. $f(x) = x^4(x^3 - 1)$
35. $f(x) = \dfrac{x^3 + 9}{x^3}$
36. $f(x) = \dfrac{x^4 + 4}{x^4}$

C *Find each of the following. Do not simplify.*

37. $f'(x)$ for $f(x) = (2x^4 - 3x^3 + x)(x^2 - x + 5)$

38. $\frac{dy}{dx}$ for $y = (x^2 - 3x + 1)(x^3 + 2x^2 - x)$

39. $D_x \frac{3x^2 - 2x + 3}{4x^2 + 5x - 1}$

40. y' for $y = \frac{x^3 - 3x + 4}{2x^2 + 3x - 2}$

41. $\frac{dy}{dx}$ for $y = 9x^{1/3}(x^3 + 5)$

42. $D_x[(4x^{1/2} - 1)(3x^{1/3} + 2)]$

43. $f'(x)$ for $f(x) = \frac{6\sqrt[3]{x}}{x^2 - 3}$

44. y' for $y = \frac{2\sqrt{x}}{x^2 - 3x + 1}$

45. $D_x \frac{x^3 - 2x^2}{\sqrt[3]{x^2}}$

46. $\frac{dy}{dx}$ for $y = \frac{x^2 - 3x + 1}{\sqrt[4]{x}}$

47. $f'(x)$ for $f(x) = \frac{(2x^2 - 1)(x^2 + 3)}{x^2 + 1}$

48. y' for $y = \frac{2x - 1}{(x^3 + 2)(x^2 - 3)}$

APPLICATIONS

Business & Economics

49. *Sales analysis.* The monthly sales S (in thousands) for a record album are given by

$$S(t) = \frac{200t}{t^2 + 36}$$

where t is the number of months since the release of the album.

(A) Find $S'(t)$, the instantaneous rate of change of monthly sales with respect to time.
(B) Find $S(2)$ and $S'(2)$, and interpret the results.
(C) Find $S(8)$ and $S'(8)$, and interpret the results.

50. *Sales analysis.* A communications company has installed a cable television system in a city. The total number N (in thousands) of subscribers t months after the installation of the system is given by

$$N(t) = \frac{200t}{t + 5}$$

(A) Find $N'(t)$, the instantaneous rate of change of the total number of subscribers with respect to time.
(B) Find $N(5)$ and $N'(5)$, and interpret the results.
(C) Find $N(15)$ and $N'(15)$, and interpret the results.

51. *Price–demand function.* According to classical economic theory, the demand $d(x)$ for a commodity in a free market decreases as the price x increases. Suppose that the number $d(x)$ of transistor radios people are will-

ing to buy per week in a given city at a price \$x is given by

$$d(x) = \frac{50{,}000}{x^2 + 10x + 25} \qquad 2 \le x \le 25$$

(A) Find $d'(x)$, the instantaneous rate of change of demand with respect to price change.

(B) Find $d'(5)$ and $d'(15)$, and interpret the results.

52. *Employee training.* A company producing computer components has established that on the average, a new employee can assemble $N(t)$ components per day after t days of on-the-job training, as given by

$$N(t) = \frac{100t}{t + 9}$$

(A) Find $N'(t)$, the instantaneous rate of change of units assembled with respect to time.

(B) Find $N'(1)$ and $N'(11)$, and interpret the results.

Use a graphic calculator or a computer to draw the graphs in Problems 53–56.

53. *Sales analysis.* Refer to Problem 49. Find the equations of the tangent lines at $t = 2$ and $t = 8$. Graph $S(t)$ and these tangent lines in the same viewing rectangle, with t range [0, 15] and S range [0, 20].

54. *Sales analysis.* Refer to Problem 50. Find the equations of the tangent lines at $t = 5$ and $t = 15$. Graph $N(t)$ and these tangent lines in the same viewing rectangle, with t range [0, 25] and N range [0, 200].

55. *Price–demand function.* Refer to Problem 51. Find the equations of the tangent lines at $x = 5$ and $x = 15$. Graph $d(x)$ and these tangent lines in the same viewing rectangle, with x range [0, 25] and d range [0, 1,000].

56. *Employee training.* Refer to Problem 52. Find the equations of the tangent lines at $t = 1$ and $t = 11$. Graph $N(t)$ and these tangent lines in the same viewing rectangle, with t range [0, 20] and N range [0, 80].

Life Sciences

57. *Medicine.* A drug is injected into the bloodstream of a patient through her right arm. The concentration of the drug in the bloodstream of the left arm t hours after the injection is given by

$$C(t) = \frac{0.14t}{t^2 + 1}$$

(A) Find $C'(t)$, the instantaneous rate of change of drug concentration with respect to time.

(B) Find $C'(0.5)$ and $C'(3)$, and interpret the results.

58. *Drug sensitivity.* One hour after x milligrams of a particular drug are given to a person, the change in body temperature $T(x)$, in degrees Fahrenheit, is given approximately by

$$T(x) = x^2\left(1 - \frac{x}{9}\right) \qquad 0 \le x \le 7$$

The rate at which T changes with respect to the size of the dosage x, $T'(x)$, is called the *sensitivity* of the body to the dosage.

(A) Find $T'(x)$, using the product rule.
(B) Find $T'(1)$, $T'(3)$, and $T'(6)$.

Social Sciences

59. *Learning.* In the early days of quantitative learning theory (around 1917), L. L. Thurstone found that a given person successfully accomplished $N(x)$ acts after x practice acts, as given by

$$N(x) = \frac{100x + 200}{x + 32}$$

(A) Find the instantaneous rate of change of learning, $N'(x)$, with respect to the number of practice acts x.
(B) Find $N'(4)$ and $N'(68)$.

SECTION 3-6 Chain Rule: Power Form

- CHAIN RULE: POWER FORM
- COMBINING RULES OF DIFFERENTIATION

CHAIN RULE: POWER FORM

We have already made extensive use of the power rule,

$$D_x\, x^n = nx^{n-1} \qquad n \text{ any real number} \tag{1}$$

Now we want to generalize this rule so that we can differentiate functions of the form $[u(x)]^n$. Is rule (1) still valid if we replace x with a function $u(x)$? We begin by considering a simple example. Let $u(x) = 2x$ and $n = 4$. Then

$$[u(x)]^n = (2x)^4 = 2^4x^4 = 16x^4$$

and

$$D_x[u(x)]^n = D_x\, 16x^4 = 64x^3 \tag{2}$$

But

$$n[u(x)]^{n-1} = 4(2x)^3 = 32x^3 \tag{3}$$

Comparing (2) and (3), we see that

$$D_x[u(x)]^n \neq n[u(x)]^{n-1}$$

for this particular choice of $u(x)$ and n. (In fact, it can be shown that the only time this last equation is valid is if $u(x) = x + k$, k a constant.) Thus, we cannot generalize the power rule by simply substituting $u(x)$ for x in rule (1).

How can we find a formula for $D_x[u(x)]^n$ where $u(x)$ is an arbitrary differentiable function? Let us first find $D_x[u(x)]^2$ and $D_x[u(x)]^3$ to see if a general pattern emerges. Since $[u(x)]^2 = u(x)u(x)$, we use the product rule with $F(x) = u(x)$ and $S(x) = u(x)$ to write

$$\begin{aligned} D_x[u(x)]^2 &= D_x[u(x)u(x)] = u(x)u'(x) + u(x)u'(x) \\ &= 2u(x)u'(x) \end{aligned} \tag{4}$$

Since $[u(x)]^3 = [u(x)]^2u(x)$, we now use the product rule with $F(x) = [u(x)]^2$ and (4) to write

$$\begin{aligned} D_x[u(x)]^3 &= D_x[u(x)]^2u(x) = [u(x)]^2\, D_x\, u(x) + u(x)\, D_x[u(x)]^2 \\ &= [u(x)]^2u'(x) + u(x)[2u(x)u'(x)] \\ &= 3[u(x)]^2u'(x) \end{aligned}$$

Continuing in this fashion, it can be shown that

$$D_x[u(x)]^n = n[u(x)]^{n-1}u'(x) \qquad n \text{ a positive integer} \tag{5}$$

Using more advanced techniques, formula (5) can be established for all real numbers n. Thus, we have the **general power rule.**

General Power Rule

If n is any real number, then

$$D_x[u(x)]^n = n[u(x)]^{n-1}u'(x)$$

provided $u'(x)$ exists. This rule is often written more compactly as

$$D_x\, u^n = nu^{n-1}\frac{du}{dx} \qquad u = u(x)$$

The general power rule is a special case of a very important and useful differentiation rule called the **chain rule.** In essence, the chain rule will enable us to differentiate a composition form $f[g(x)]$ if we know how to differentiate $f(x)$ and $g(x)$. We defer a complete discussion of the chain rule until Chapter 5.

◆ EXAMPLE 28

Find $f'(x)$:

(A) $f(x) = (3x + 1)^4$ (B) $f(x) = (x^3 + 4)^7$

(C) $f(x) = \dfrac{1}{(x^2 + x + 4)^3}$ (D) $f(x) = \sqrt{3 - x}$

Solutions (A) $f(x) = (3x + 1)^4$ — Let $u = 3x + 1$, $n = 4$.

$$f'(x) = 4(3x + 1)^3 D_x(3x + 1) \qquad nu^{n-1}\frac{du}{dx}$$

$$= 4(3x + 1)^3 3 \qquad \frac{du}{dx} = 3$$

$$= 12(3x + 1)^3$$

(B) $f(x) = (x^3 + 4)^7$ — Let $u = (x^3 + 4)$, $n = 7$.

$$f'(x) = 7(x^3 + 4)^6 D_x(x^3 + 4) \qquad nu^{n-1}\frac{du}{dx}$$

$$= 7(x^3 + 4)^6 3x^2 \qquad \frac{du}{dx} = 3x^2$$

$$= 21x^2(x^3 + 4)^6$$

(C) $f(x) = \dfrac{1}{(x^2 + x + 4)^3} = (x^2 + x + 4)^{-3}$ — Let $u = x^2 + x + 4$, $n = -3$.

$$f'(x) = -3(x^2 + x + 4)^{-4} D_x(x^2 + x + 4) \qquad nu^{n-1}\frac{du}{dx}$$

$$= -3(x^2 + x + 4)^{-4}(2x + 1) \qquad \frac{du}{dx} = 2x + 1$$

$$= \frac{-3(2x + 1)}{(x^2 + x + 4)^4}$$

(D) $f(x) = \sqrt{3 - x} = (3 - x)^{1/2}$ — Let $u = 3 - x$, $n = \frac{1}{2}$.

$$f'(x) = \frac{1}{2}(3 - x)^{-1/2} D_x(3 - x) \qquad nu^{n-1}\frac{du}{dx}$$

$$= \frac{1}{2}(3 - x)^{-1/2}(-1) \qquad \frac{du}{dx} = -1$$

$$= -\frac{1}{2(3 - x)^{1/2}} \quad \text{or} \quad -\frac{1}{2\sqrt{3 - x}}$$ ◆

PROBLEM 28 Find $f'(x)$:

(A) $f(x) = (5x + 2)^3$ (B) $f(x) = (x^4 - 5)^5$

(C) $f(x) = \dfrac{1}{(x^2 + 4)^2}$ (D) $f(x) = \sqrt{4 - x}$ ◆

Notice that we used two steps to differentiate each function in Example 28. First, we applied the general power rule; then we found du/dx. As you gain experience with the general power rule, you may want to combine these two steps. If you do this, be certain to multiply by du/dx. For example,

$D_x(x^5 + 1)^4 = 4(x^5 + 1)^3 5x^4$ Correct

$D_x(x^5 + 1)^4 \neq 4(x^5 + 1)^3$ $du/dx = 5x^4$ is missing

If we let $u(x) = x$, then $du/dx = 1$, and the general power rule reduces to the (ordinary) power rule discussed in Section 3-4. Compare the following:

$D_x\, x^n = nx^{n-1}$ Yes—power rule

$D_x\, u^n = nu^{n-1}\dfrac{du}{dx}$ Yes—general power rule

$D_x\, u^n \neq nu^{n-1}$ Unless $u(x) = x + k$ so that $du/dx = 1$

◆ COMBINING RULES OF DIFFERENTIATION

The following examples illustrate the use of the general power rule in combination with other rules of differentiation.

◆ EXAMPLE 29 Find the equation of the line tangent to the graph of f at $x = 2$ for $f(x) = x^2\sqrt{2x + 12}$.

Solution

$$f(x) = x^2\sqrt{2x+12} = x^2(2x+12)^{1/2}$$

Apply the product rule with $F(x) = x^2$ and $S(x) = (2x + 12)^{1/2}$.

$$\begin{aligned} f'(x) &= x^2\, D_x(2x+12)^{1/2} + (2x+12)^{1/2}\, D_x\, x^2 \\ &= x^2[\tfrac{1}{2}(2x+12)^{-1/2}](2) + (2x+12)^{1/2}(2x) \\ &= \frac{x^2}{\sqrt{2x+12}} + 2x\sqrt{2x+12} \end{aligned}$$

Use the general power rule to differentiate $(2x + 12)^{1/2}$ and the ordinary power rule to differentiate x^2.

$$f'(2) = \frac{4}{\sqrt{16}} + 4\sqrt{16} = 1 + 16 = 17$$

$$f(2) = 4\sqrt{16} = 16$$

$(x_1, y_1) = (2, f(2)) = (2, 16)$ Point

$m = f'(2) = 17$ Slope

$y - 16 = 17(x - 2)$ $y - y_1 = m(x - x_1)$

$y = 17x - 18$ Tangent line ◆

PROBLEM 29 Find the equation of the line tangent to the graph of f at $x = 3$ for $f(x) = x\sqrt{15 - 2x}$. ◆

◆ EXAMPLE 30 Find the value(s) of x where the tangent line is horizontal for

$$f(x) = \frac{x^3}{(2 - 3x)^5}$$

Solution Use the quotient rule with $T(x) = x^3$ and $B(x) = (2 - 3x)^5$:

$$\begin{aligned} f'(x) &= \frac{(2 - 3x)^5 D_x x^3 - x^3 D_x(2 - 3x)^5}{[(2 - 3x)^5]^2} \\ &= \frac{(2 - 3x)^5 3x^2 - x^3 5(2 - 3x)^4(-3)}{(2 - 3x)^{10}} \\ &= \frac{(2 - 3x)^4 3x^2[(2 - 3x) + 5x]}{(2 - 3x)^{10}} \\ &= \frac{3x^2(2 + 2x)}{(2 - 3x)^6} = \frac{6x^2(x + 1)}{(2 - 3x)^6} \end{aligned}$$

Use the ordinary power rule to differentiate x^3 and the general power rule to differentiate $(2 - 3x)^5$.

Since a fraction is 0 when the numerator is 0 and the denominator is not, we see that $f'(x) = 0$ at $x = -1$ and $x = 0$. Thus, the graph of f will have horizontal tangent lines at $x = -1$ and $x = 0$. ◆

PROBLEM 30 Find the value(s) of x where the tangent line is horizontal for

$$f(x) = \frac{x^3}{(3x - 2)^2}$$

◆

◆ EXAMPLE 31 Starting with the function f in Example 30, write f as a product and then differentiate.

Solution

$$f(x) = \frac{x^3}{(2 - 3x)^5} = x^3(2 - 3x)^{-5}$$

$$\begin{aligned} f'(x) &= x^3 D_x(2 - 3x)^{-5} + (2 - 3x)^{-5} D_x x^3 \\ &= x^3(-5)(2 - 3x)^{-6}(-3) + (2 - 3x)^{-5}3x^2 \\ &= 15x^3(2 - 3x)^{-6} + 3x^2(2 - 3x)^{-5} \end{aligned}$$

At this point, we have an unsimplified form for $f'(x)$. This may be satisfactory for some purposes, but not for others. For example, if we need to solve the equation $f'(x) = 0$, we must simplify algebraically:

$$\begin{aligned} f'(x) &= \frac{15x^3}{(2 - 3x)^6} + \frac{3x^2}{(2 - 3x)^5} = \frac{15x^3}{(2 - 3x)^6} + \frac{3x^2(2 - 3x)}{(2 - 3x)^6} \\ &= \frac{15x^3 + 3x^2(2 - 3x)}{(2 - 3x)^6} = \frac{3x^2(5x + 2 - 3x)}{(2 - 3x)^6} \\ &= \frac{3x^2(2 + 2x)}{(2 - 3x)^6} = \frac{6x^2(1 + x)}{(2 - 3x)^6} \end{aligned}$$

◆

PROBLEM 31 Refer to the function f in Problem 30, above. Write f as a product and then differentiate. Do not simplify. ◆

As Example 31 illustrates, any quotient can be converted to a product and differentiated by the product rule. However, if the derivative must be simplified, it is usually easier to use the quotient rule. (Compare the algebraic simplifications in Example 31 with those in Example 30.) There is one special case where using negative exponents is the preferred method—a fraction whose numerator is a constant.

◆ EXAMPLE 32 Find $f'(x)$ two ways for: $f(x) = \dfrac{4}{(x^2+9)^3}$

Solution *Method 1.* Use the quotient rule:

$$f'(x) = \frac{(x^2+9)^3\, D_x\, 4 - 4\, D_x(x^2+9)^3}{[(x^2+9)^3]^2}$$

$$= \frac{(x^2+9)^3(0) - 4[3(x^2+9)^2(2x)]}{(x^2+9)^6}$$

$$= \frac{-24x(x^2+9)^2}{(x^2+9)^6} = \frac{-24x}{(x^2+9)^4}$$

Method 2. Rewrite as a product, and use the general power rule:

$$f(x) = \frac{4}{(x^2+9)^3} = 4(x^2+9)^{-3}$$

$$f'(x) = 4(-3)(x^2+9)^{-4}(2x)$$

$$= \frac{-24x}{(x^2+9)^4}$$

Which method do you prefer? ◆

PROBLEM 32 Find $f'(x)$ two ways for: $f(x) = \dfrac{5}{(x^3+1)^2}$ ◆

Answers to Matched Problems

28. (A) $15(5x+2)^2$ (B) $20x^3(x^4-5)^4$ (C) $-4x/(x^2+4)^3$
(D) $-1/(2\sqrt{4-x})$
29. $y = 2x+3$ 30. $x=0, x=2$
31. $-6x^3(3x-2)^{-3} + 3x^2(3x-2)^{-2}$ 32. $-30x^2/(x^3+1)^3$

EXERCISE 3-6

A *In Problems 1–12, find $f'(x)$ using the general power rule and simplify.*

1. $f(x) = (2x+5)^3$ **2.** $f(x) = (3x-7)^5$ **3.** $f(x) = (5-2x)^4$
4. $f(x) = (9-5x)^2$ **5.** $f(x) = (3x^2+5)^5$ **6.** $f(x) = (5x^2-3)^6$

7. $f(x) = (x^3 - 2x^2 + 2)^8$ 8. $f(x) = (2x^2 + x + 1)^7$ 9. $f(x) = (2x - 5)^{1/2}$
10. $f(x) = (4x + 3)^{1/2}$ 11. $f(x) = (x^4 + 1)^{-2}$ 12. $f(x) = (x^5 + 2)^{-3}$

In Problems 13–16, find $f'(x)$ and the equation of the line tangent to the graph of f at the indicated value of x. Find the value(s) of x where the tangent line is horizontal.

13. $f(x) = (2x - 1)^3; \quad x = 1$
14. $f(x) = (3x - 1)^4; \quad x = 1$
15. $f(x) = (4x - 3)^{1/2}; \quad x = 3$
16. $f(x) = (2x + 8)^{1/2}; \quad x = 4$

B *In Problems 17–34, find dy/dx using the general power rule.*

17. $y = 3(x^2 - 2)^4$
18. $y = 2(x^3 + 6)^5$
19. $y = 2(x^2 + 3x)^{-3}$
20. $y = 3(x^3 + x^2)^{-2}$
21. $y = \sqrt{x^2 + 8}$
22. $y = \sqrt[3]{3x - 7}$
23. $y = \sqrt[3]{3x + 4}$
24. $y = \sqrt{2x - 5}$
25. $y = (x^2 - 4x + 2)^{1/2}$
26. $y = (2x^2 + 2x - 3)^{1/2}$
27. $y = \dfrac{1}{2x + 4}$
28. $y = \dfrac{1}{3x - 7}$
29. $y = \dfrac{1}{(x^3 + 4)^5}$
30. $y = \dfrac{1}{(x^2 - 3)^6}$
31. $y = \dfrac{1}{4x^2 - 4x + 1}$
32. $y = \dfrac{1}{2x^2 - 3x + 1}$
33. $y = \dfrac{4}{\sqrt{x^2 - 3x}}$
34. $y = \dfrac{3}{\sqrt[3]{x - x^2}}$

In Problems 35–40, find $f'(x)$, and find the equation of the line tangent to the graph of f at the indicated value of x.

35. $f(x) = x(4 - x)^3; \quad x = 2$
36. $f(x) = x^2(1 - x)^4; \quad x = 2$
37. $f(x) = \dfrac{x}{(2x - 5)^3}; \quad x = 3$
38. $f(x) = \dfrac{x^4}{(3x - 8)^2}; \quad x = 4$
39. $f(x) = x\sqrt{2x + 2}; \quad x = 1$
40. $f(x) = x\sqrt{x - 6}; \quad x = 7$

In Problems 41–46, find $f'(x)$, and find the value(s) of x where the tangent line is horizontal.

41. $f(x) = x^2(x - 5)^3$
42. $f(x) = x^3(x - 7)^4$
43. $f(x) = \dfrac{x}{(2x + 5)^2}$
44. $f(x) = \dfrac{x - 1}{(x - 3)^3}$
45. $f(x) = \sqrt{x^2 - 8x + 20}$
46. $f(x) = \sqrt{x^2 + 4x + 5}$

C *In Problems 47–58, find each derivative and simplify.*

47. $D_x[3x(x^2 + 1)^3]$
48. $D_x[2x^2(x^3 - 3)^4]$
49. $D_x \dfrac{(x^3 - 7)^4}{2x^3}$
50. $D_x \dfrac{3x^2}{(x^2 + 5)^3}$
51. $D_x[(2x - 3)^2(2x^2 + 1)^3]$
52. $D_x[(x^2 - 1)^3(x^2 - 2)^2]$
53. $D_x(4x^2\sqrt{x^2 - 1})$
54. $D_x(3x\sqrt{2x^2 + 3})$

55. $D_x \dfrac{2x}{\sqrt{x-3}}$

56. $D_x \dfrac{x^2}{\sqrt{x^2+1}}$

57. $D_x \sqrt{(2x-1)^3(x^2+3)^4}$

58. $D_x \sqrt{\dfrac{4x+1}{2x^2+1}}$

APPLICATIONS

Business & Economics

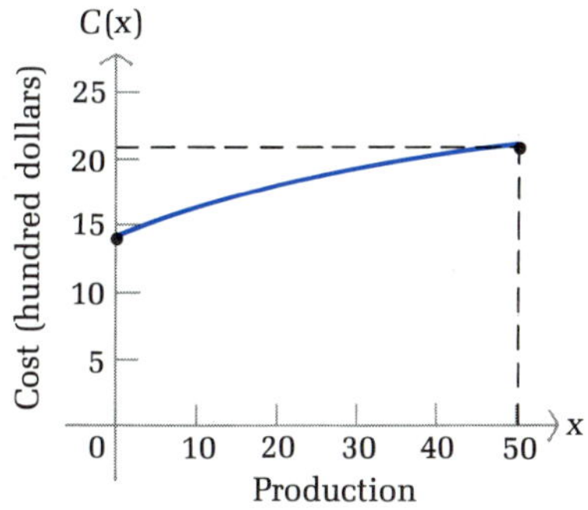

59. *Marginal cost.* The total cost (in hundreds of dollars) of producing x calculators per day is

$$C(x) = 10 + \sqrt{2x+16} \qquad 0 \leq x \leq 50$$

(see the figure in the margin).

(A) Find the marginal cost at a production level of x calculators.
(B) Find $C'(24)$ and $C'(42)$, and interpret the results.

60. *Marginal cost.* The total cost (in thousands of dollars) of producing x cameras per week is

$$C(x) = 6 + \sqrt{4x+4} \qquad 0 \leq x \leq 30$$

(A) Find the marginal cost at a production level of x cameras.
(B) Find $C'(15)$ and $C'(24)$, and interpret the results.

61. *Compound interest.* If $1,000 is invested at an annual interest rate r compounded monthly, the amount in the account at the end of 4 years is given by

$$A = 1{,}000(1 + \tfrac{1}{12}r)^{48}$$

Find the rate of change of the amount A with respect to the interest rate r.

62. *Compound interest.* If $100 is invested at an annual interest rate r compounded semiannually, the amount in the account at the end of 5 years is given by

$$A = 100(1 + \tfrac{1}{2}r)^{10}$$

Find the rate of change of the amount A with respect to the interest rate r.

Life Sciences

63. *Bacteria growth.* The number y of bacteria in a certain colony after x days is given approximately by

$$y = (3 \times 10^6)\left[1 - \frac{1}{\sqrt[3]{(x^2-1)^2}}\right]$$

Find dy/dx.

64. *Pollution.* A small lake in a resort area became contaminated with harmful bacteria because of excessive septic tank seepage. After treating the lake with a bactericide, the Department of Public Health estimated the bacteria concentration (number per cubic centimeter) after t days to be given by

$$C(t) = 500(8-t)^2 \qquad 0 \leq t \leq 7$$

(A) Find $C'(t)$ using the general power rule.
(B) Find $C'(1)$ and $C'(6)$, and interpret the results.

Social Sciences

65. *Learning.* In 1930, L. L. Thurstone developed the following formula to indicate how learning time T depends on the length of a list n:

$$T = f(n) = \frac{c}{k}\, n\sqrt{n - a}$$

where a, c, and k are empirical constants. Suppose that for a particular person, time T (in minutes) for learning a list of length n is

$$T = f(n) = 2n\sqrt{n - 2}$$

(A) Find dT/dn, the instantaneous rate of change in time with respect to n.
(B) Find $f'(11)$ and $f'(27)$, and interpret the results.

SECTION 3-7 Marginal Analysis in Business and Economics

- MARGINAL COST, REVENUE, AND PROFIT
- APPLICATION
- MARGINAL AVERAGE COST, REVENUE, AND PROFIT

MARGINAL COST, REVENUE, AND PROFIT

One important use of calculus in business and economics is in *marginal analysis*. We introduced the concept of *marginal cost* earlier. There is no reason to stop there. Economists also talk about *marginal revenue* and *marginal profit*. Recall that the word "marginal" refers to an instantaneous rate of change—that is, a derivative. Thus, we define the following:

Marginal Cost, Revenue, and Profit

If x is the number of units of a product produced in some time interval, then

Total cost $= C(x)$ Total revenue $= R(x)$
Marginal cost $= C'(x)$ **Marginal revenue** $= R'(x)$

Total profit $= P(x) = R(x) - C(x)$
Marginal profit $= P'(x) = R'(x) - C'(x)$
$=$ (Marginal revenue) $-$ (Marginal cost)

Marginal cost (or revenue or profit) is the instantaneous rate of change of cost (or revenue or profit) relative to production at a given production level.

Marginal functions have several important economic interpretations. Earlier in this chapter, we discussed interpretations of marginal cost, which we summarize here. Similar interpretations can be made for marginal revenue and marginal profit.

Returning to the definition of a derivative, we observe the following (assuming the limit exists):

$$C'(x) = \lim_{h \to 0} \frac{C(x + h) - C(x)}{h} \qquad \text{Marginal cost}$$

$$C'(x) \approx \frac{C(x + h) - C(x)}{h} \qquad h \neq 0$$

$$C'(x) \approx \frac{C(x + 1) - C(x)}{1}$$

$= C(x + 1) - C(x)$

= Exact change in total cost for 1 unit change in production at the x level of production

= Exact cost of producing the $(x + 1)$st item at the x level of production

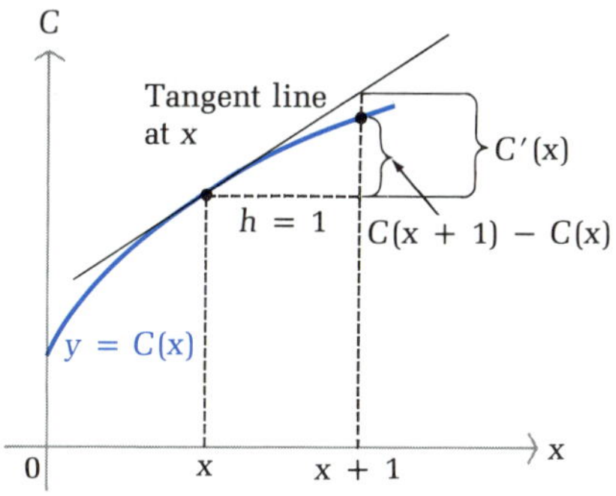

FIGURE 18
$C'(x) \approx C(x + 1) - C(x)$

Thus, at a production level of x units:

Marginal cost $C'(x)$ approximates the change in total cost that results from a 1 unit change in production.

In other words,

Marginal cost $C'(x)$ approximates the cost of producing the $(x + 1)$st item.

These observations are illustrated in Figure 18. Note that if $h = 1$ as shown in the figure, then $C(x + 1) - C(x)$ is the exact change in total cost per unit change in production at a production level of x units. The marginal cost, $C'(x)$, is the slope of the tangent line, and is approximately equal to the change in total cost C per unit change in production at a production level of x units.

◆ APPLICATION

We now present an example in market research to show how marginal cost, revenue, and profit are tied together.

◆ EXAMPLE 33

Production Strategy

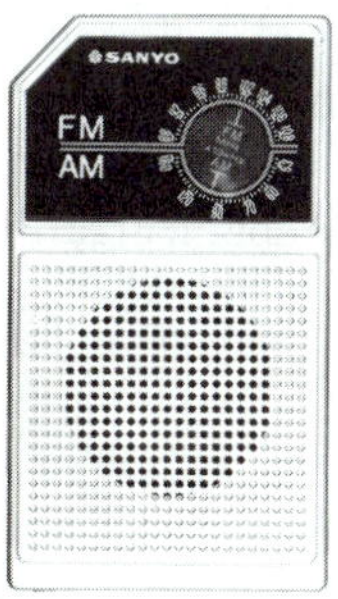

The market research department of a company recommends that the company manufacture and market a new transistor radio. After suitable test marketing, the research department presents the following **demand equation:**

$$x = 10{,}000 - 1{,}000p \qquad x \text{ is demand at \$}p\text{ per radio} \tag{1}$$

or, solving (1) for p,

$$p = 10 - \frac{x}{1{,}000} \tag{2}$$

where x is the number of radios retailers are likely to buy per week at \$p per radio. Equation (2) is simply equation (1) solved for p in terms of x. Notice that as price goes up, demand goes down.

The financial department provides the following **cost equation:**

$$C(x) = 7{,}000 + 2x \qquad (3)$$

where \$7,000 is the estimated fixed costs (tooling and overhead), and \$2 is the estimated variable costs (cost per unit for materials, labor, marketing, transportation, storage, etc.).

The **marginal cost** is

$$C'(x) = 2$$

Since this is a constant, it costs an additional \$2 to produce 1 more radio at all production levels.

The **revenue** (the amount of money R received by the company for manufacturing and selling x units at \$p per unit) is

$$R = (\text{Number of units sold})(\text{Price per unit}) = xp$$

In general, the revenue R can be expressed in terms of p by using equation (1) or in terms of x by using equation (2). In marginal analysis (problems involving marginal cost, marginal revenue, or marginal profit), cost, revenue, and profit must be expressed in terms of the number of units x. Thus, the **revenue equation** in terms of x is

$$R(x) = xp = x\left(10 - \frac{x}{1{,}000}\right) \qquad \text{Using equation (2)} \qquad (4)$$
$$= 10x - \frac{x^2}{1{,}000}$$

The **marginal revenue** is

$$R'(x) = 10 - \frac{x}{500}$$

For production levels of $x = 2{,}000$, 5,000, and 7,000, we have

$$R'(2{,}000) = 6 \qquad R'(5{,}000) = 0 \qquad R'(7{,}000) = -4$$

This means that at production levels of 2,000, 5,000, and 7,000, the respective approximate changes in revenue per unit change in production are \$6, \$0, and −\$4. That is, at the 2,000 output level, revenue increases as production increases; at the 5,000 output level, revenue does not change with a "small" change in production; and at the 7,000 output level, revenue decreases with an increase in production.

When we graph R(x) and C(x) in the same coordinate system, we obtain Figure 19.

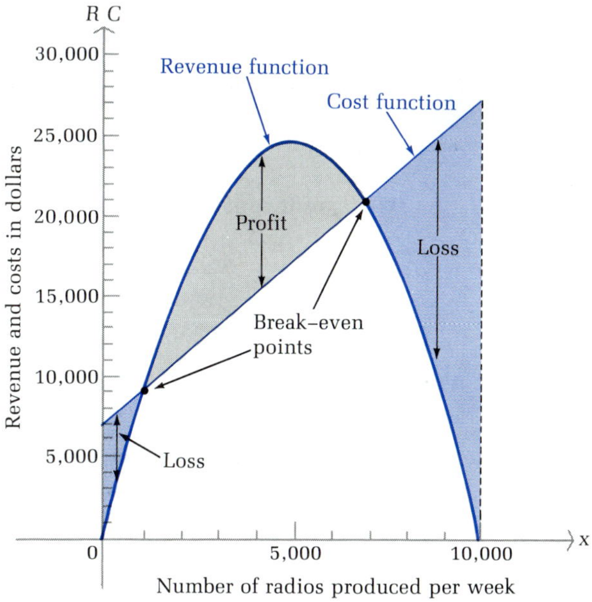

FIGURE 19

The **break-even points** (the points where revenue equals cost) are obtained as follows:

$$C(x) = R(x)$$

$$7{,}000 + 2x = 10x - \frac{x^2}{1{,}000}$$

$$\frac{x^2}{1{,}000} - 8x + 7{,}000 = 0$$

$$x^2 - 8{,}000x + 7{,}000{,}000 = 0$$

Solve using the quadratic formula (see Appendix A-9).

$$x = \frac{8{,}000 \pm \sqrt{8{,}000^2 - 4(7{,}000{,}000)}}{2}$$

$$= \frac{8{,}000 \pm \sqrt{36{,}000{,}000}}{2}$$

$$= \frac{8{,}000 \pm 6{,}000}{2}$$

$$= 1{,}000, \quad 7{,}000$$

$$R(1{,}000) = 10(1{,}000) - \frac{1{,}000^2}{1{,}000} = 9{,}000$$

$$C(1{,}000) = 7{,}000 + 2(1{,}000) = 9{,}000$$

$$R(7{,}000) = 10(7{,}000) - \frac{7{,}000^2}{1{,}000} = 21{,}000$$

$$C(7{,}000) = 7{,}000 + 2(7{,}000) = 21{,}000$$

Thus, the break-even points are (1,000, 9,000) and (7,000, 21,000), as shown in Figure 19.

The **profit equation** is

$$\begin{aligned} P(x) &= R(x) - C(x) \\ &= \left(10x - \frac{x^2}{1{,}000}\right) - (7{,}000 + 2x) \\ &= -\frac{x^2}{1{,}000} + 8x - 7{,}000 \end{aligned}$$

The graph in Figure 19 also provides some useful information concerning the profit equation. At a production level of 1,000 or 7,000, revenue equals cost; hence, profit is 0 and the company will break even. For any production level between 1,000 and 7,000, revenue is greater than cost; hence, $P(x)$ is positive and the company will make a profit. For production levels less than 1,000 or greater than 7,000, revenue is less than cost; hence, $P(x)$ is negative and the company will have a loss.

The **marginal profit** is

$$P'(x) = -\frac{x}{500} + 8$$

For production levels of 1,000, 4,000, and 6,000, we have

$$P'(1{,}000) = 6 \qquad P'(4{,}000) = 0 \qquad P'(6{,}000) = -4$$

This means that at production levels of 1,000, 4,000, and 6,000, the respective approximate changes in profit per unit change in production are \$6, \$0, and −\$4. That is, at the 1,000 output level, profit will be increased if production is increased; at the 4,000 output level, profit does not change for "small" changes in production; and at the 6,000 output level, profits will decrease if production is increased. It seems the best production level to produce a maximum profit is 4,000.

Example 33 warrants careful study, since a number of important ideas in economics and calculus are involved. In the next chapter, we will develop a systematic procedure for finding the production level (and, using the demand equation, the selling price) that will maximize profit. ◆

PROBLEM 33 Refer to the revenue and profit equations in Example 33.

(A) Find $R'(3{,}000)$ and $R'(6{,}000)$, and interpret the results.
(B) Find $P'(2{,}000)$ and $P'(7{,}000)$, and interpret the results. ◆

◆ MARGINAL AVERAGE COST, REVENUE, AND PROFIT

Sometimes, it is desirable to carry out marginal analysis relative to **average cost (cost per unit), average revenue (revenue per unit), and average profit (profit per unit).** The relevant definitions are summarized in the following box:

Marginal Average Cost, Revenue, and Profit

If x is the number of units of a product produced in some time interval, then

COST PER UNIT:

$$\textbf{Average cost} = \overline{C}(x) = \frac{C(x)}{x}$$

$$\textbf{Marginal average cost} = \overline{C}'(x) = D_x\,\overline{C}(x)$$

REVENUE PER UNIT:

$$\textbf{Average revenue} = \overline{R}(x) = \frac{R(x)}{x}$$

$$\textbf{Marginal average revenue} = \overline{R}'(x) = D_x\,\overline{R}(x)$$

PROFIT PER UNIT:

$$\textbf{Average profit} = \overline{P}(x) = \frac{P(x)}{x}$$

$$\textbf{Marginal average profit} = \overline{P}'(x) = D_x\,\overline{P}(x)$$

As was the case with marginal cost:

The marginal average cost approximates the change in average cost that results from a unit increase in production.

Similar statements can be made for marginal average revenue and marginal average profit.

◆ EXAMPLE 34

Cost Analysis

A small machine shop manufactures drill bits used in the petroleum industry. The shop manager estimates that the total daily cost (in dollars) of producing x bits is

$$C(x) = 1{,}000 + 25x - \frac{x^2}{10} \qquad \text{Total cost function}$$

Find the average cost and the marginal average cost at a production level of 10 bits and interpret the results.

Solution

$$\overline{C}(x) = \frac{C(x)}{x} = \frac{1{,}000}{x} + 25 - \frac{x}{10} \qquad \text{Average cost function}$$

$$\overline{C}'(x) = D_x\,\overline{C}(x) = -\frac{1{,}000}{x^2} - \frac{1}{10} \qquad \text{Marginal average cost function}$$

$$\overline{C}(10) = \frac{1{,}000}{10} + 25 - \frac{10}{10} = \$124$$

Average cost per unit if 10 units are produced

$$\overline{C}'(10) = -\frac{1{,}000}{100} - \frac{1}{10} = -\$10.10$$

A unit increase in production will decrease the average cost per unit by approximately \$10.10 at a production level of 10 units. ◆

PROBLEM 34 Consider the cost function $C(x) = 7{,}000 + 2x$ from Example 33.

(A) Find $\overline{C}(x)$ and $\overline{C}'(x)$.
(B) Find $\overline{C}(1{,}000)$ and $\overline{C}'(1{,}000)$, and interpret the results. ◆

Answers to Matched Problems

33. (A) $R'(3{,}000) = 4$ (at a production level of 3,000, a unit increase in production will increase revenue by approx. \$4); $R'(6{,}000) = -2$ (at a production level of 6,000, a unit increase in production will decrease revenue by approx. \$2)
(B) $P'(2{,}000) = 4$ (at a production level of 2,000, a unit increase in production will increase profit by approx. \$4); $P'(7{,}000) = -6$ (at a production level of 7,000, a unit increase in production will decrease profit by approx. \$6)

34. (A) $\overline{C}(x) = \frac{7{,}000}{x} + 2$, $\overline{C}'(x) = -\frac{7{,}000}{x^2}$
(B) $\overline{C}(1{,}000) = 9$ (at a production level of 1,000, the average cost per unit is \$9); $\overline{C}'(1{,}000) = -0.007$ (at a production level of 1,000, a unit increase in production will decrease the average cost per unit by approx. 0.7¢)

EXERCISE 3-7

APPLICATIONS

Business & Economics

1. *Cost analysis.* The total cost (in dollars) of producing x food processors is

$$C(x) = 2{,}000 + 50x - \frac{x^2}{2}$$

(A) Find the exact cost of producing the 21st food processor.
(B) Use the marginal cost to approximate the cost of producing the 21st food processor.

2. *Cost analysis.* The total cost (in dollars) of producing x electric guitars is

$$C(x) = 1{,}000 + 100x - \frac{x^2}{4}$$

(A) Find the exact cost of producing the 51st guitar.
(B) Use the marginal cost to approximate the cost of producing the 51st guitar.

3. *Cost analysis.* The total cost (in dollars) of manufacturing x auto body frames is

$$C(x) = 60{,}000 + 300x$$

(A) Find the average cost per unit if 500 frames are produced.
(B) Find the marginal average cost at a production level of 500 units, and interpret the results.

4. *Cost analysis.* The total cost (in dollars) of printing x dictionaries is

$$C(x) = 20{,}000 + 10x$$

(A) Find the average cost per unit if 1,000 dictionaries are produced.
(B) Find the marginal average cost at a production level of 1,000 units, and interpret the results.

5. *Revenue analysis.* The total revenue (in dollars) from the sale of x clock radios is

$$R(x) = 100x - \frac{x^2}{40}$$

Evaluate the marginal revenue at the given values of x, and interpret the results.

(A) $x = 1{,}600$ (B) $x = 2{,}500$

6. *Revenue analysis.* The total revenue (in dollars) from the sale of x steam irons is

$$R(x) = 50x - \frac{x^2}{20}$$

Evaluate the marginal revenue at the given values of x, and interpret the results.

(A) $x = 400$ (B) $x = 650$

7. *Profit analysis.* The total profit (in dollars) from the sale of x skateboards is

$$P(x) = 30x - \frac{x^2}{2} - 250$$

(A) Find the exact profit from the sale of the 26th skateboard.
(B) Use the marginal profit to approximate the profit from the sale of the 26th skateboard.

8. *Profit analysis.* The total profit (in dollars) from the sale of x portable stereos is

$$P(x) = 22x - \frac{x^2}{10} - 400$$

(A) Find the exact profit from the sale of the 41st stereo.
(B) Use the marginal profit to approximate the profit from the sale of the 41st stereo.

9. *Profit analysis.* The total profit (in dollars) from the sale of x video cassettes is

$$P(x) = 5x - \frac{x^2}{200} - 450$$

Evaluate the marginal profit at the given values of x, and interpret the results.

(A) $x = 450$ (B) $x = 750$

10. *Profit analysis.* The total profit (in dollars) from the sale of x cameras is

$$P(x) = 12x - \frac{x^2}{50} - 1{,}000$$

Evaluate the marginal profit at the given values of x, and interpret the results.

(A) $x = 200$ (B) $x = 350$

11. *Profit analysis.* Refer to the profit equation in Problem 9.
(A) Find the average profit per unit if 150 cassettes are produced.
(B) Find the marginal average profit at a production level of 150 units, and interpret the results.

12. *Profit analysis.* Refer to the profit equation in Problem 10.
(A) Find the average profit per unit if 200 cameras are produced.
(B) Find the marginal average profit at a production level of 200 units, and interpret the results.

13. *Revenue, cost, and profit.* In Example 33, suppose we have the demand equation

$$x = 6{,}000 - 30p \qquad \text{or} \qquad p = 200 - \frac{x}{30}$$

and the cost equation

$$C(x) = 72{,}000 + 60x$$

(A) Find the marginal cost.
(B) Find the revenue equation in terms of x.
(C) Find the marginal revenue.
(D) Find $R'(1{,}500)$ and $R'(4{,}500)$, and interpret the results.

(E) Graph the cost function and the revenue function on the same coordinate system for $0 \leq x \leq 6{,}000$. Find the break-even points, and indicate regions of loss and profit.

(F) Find the profit equation in terms of x.

(G) Find the marginal profit.

(H) Find $P'(1{,}500)$ and $P'(3{,}000)$, and interpret the results.

14. *Revenue, cost, and profit.* In Example 33, suppose we have the demand equation

$$x = 9{,}000 - 30p \qquad \text{or} \qquad p = 300 - \frac{x}{30}$$

and the cost equation

$$C(x) = 150{,}000 + 30x$$

(A) Find the marginal cost.

(B) Find the revenue equation in terms of x.

(C) Find the marginal revenue.

(D) Find $R'(3{,}000)$ and $R'(6{,}000)$, and interpret the results.

(E) Graph the cost function and the revenue function on the same coordinate system for $0 \leq x \leq 9{,}000$. Find the break-even points, and indicate regions of loss and profit.

(F) Find the profit equation in terms of x.

(G) Find the marginal profit.

(H) Find $P'(1{,}500)$ and $P'(4{,}500)$, and interpret the results.

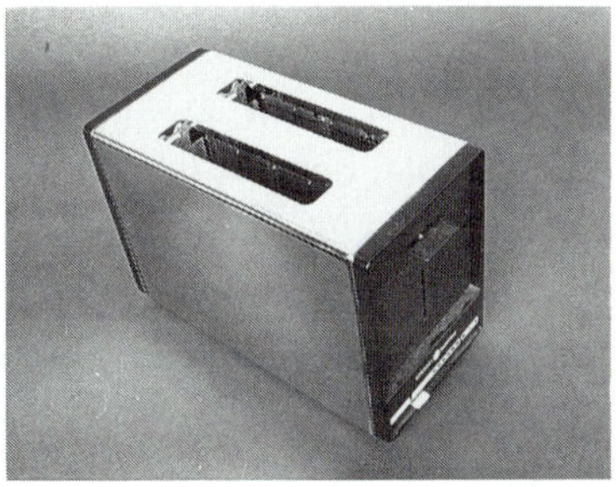

15. *Revenue, cost, and profit.* A company is planning to manufacture and market a new two-slice electric toaster. After conducting extensive market surveys, the research department provides the following estimates: a weekly demand of 200 toasters at a price of \$16 per toaster and a weekly demand of 300 toasters at a price of \$14 per toaster. The financial department estimates that weekly fixed costs will be \$1,400 and variable costs (cost per unit) will be \$4.

(A) Assume that the demand equation is linear. Use the research department's estimates to find the demand equation.

(B) Find the revenue equation in terms of x.

(C) Assume that the cost equation is linear. Use the financial department's estimates to find the cost equation.

(D) Graph the cost function and the revenue function on the same coordinate system for $0 \leq x \leq 1{,}000$. Find the break-even points, and indicate regions of loss and profit.

(E) Find the profit equation in terms of x.

(F) Evaluate the marginal profit at $x = 250$ and $x = 475$, and interpret the results.

16. *Revenue, cost, and profit.* The company in Problem 15 is also planning to manufacture and market a four-slice toaster. For this toaster, the research department's estimates are a weekly demand of 300 toasters at a price of \$25 per toaster and a weekly demand of 400 toasters at a price of \$20. The financial department's estimates are fixed weekly costs of \$5,000 and variable costs of \$5 per toaster. Assume the demand and cost equations are linear (see Problem 15, parts A and C).

(A) Use the research department's estimates to find the demand equation.
(B) Find the revenue equation in terms of x.
(C) Use the financial department's estimates to find the cost equation in terms of x.
(D) Graph the cost function and the revenue function on the same coordinate system for $0 \leq x \leq 800$. Find the break-even points, and indicate regions of loss and profit.
(E) Find the profit equation in terms of x.
(F) Evaluate the marginal profit at $x = 325$ and $x = 425$, and interpret the results.

17. *Revenue, cost, and profit.* The total cost and the total revenue (in dollars) for the production and sale of x ski jackets are given by

$$C(x) = 24x + 21{,}900 \qquad \text{and} \qquad R(x) = 200x - 0.2x^2 \qquad 0 \leq x \leq 1{,}000$$

(A) Find the value of x where the graph of $R(x)$ has a horizontal tangent line.
(B) Find the profit function $P(x)$.
(C) Find the value of x where the graph of $P(x)$ has a horizontal tangent line.
(D) Graph $C(x)$, $R(x)$, and $P(x)$ on the same coordinate system for $0 \leq x \leq 1{,}000$. Find the break-even points. Find the x intercepts for the graph of $P(x)$.

18. *Revenue, cost, and profit.* The total cost and the total revenue (in dollars) for the production and sale of x hair dryers are given by

$$C(x) = 5x + 2{,}340 \qquad \text{and} \qquad R(x) = 40x - 0.1x^2 \qquad 0 \leq x \leq 400$$

(A) Find the value of x where the graph of $R(x)$ has a horizontal tangent line.
(B) Find the profit function $P(x)$.
(C) Find the value of x where the graph of $P(x)$ has a horizontal tangent line.
(D) Graph $C(x)$, $R(x)$, and $P(x)$ on the same coordinate system for $0 \leq x \leq 400$. Find the break-even points. Find the x intercepts for the graph of $P(x)$.

Problems 19 and 20 require the use of a graphic calculator or a computer.

19. *Break-even analysis.* The demand equation and the cost equation for the production of garden hoses are given by

$$p = 20 - \sqrt{x} \qquad \text{and} \qquad C(x) = 500 + 2x$$

where x is the number of garden hoses that can be sold at a price of \$p per unit and C(x) is the total cost (in dollars) of producing x garden hoses.

(A) Express the revenue equation in terms of x.

(B) Graph the cost function and the revenue function in the same viewing rectangle for $0 \leq x \leq 400$. Use approximation techniques to find the break-even points correct to the nearest unit.

20. *Break-even analysis.* The demand equation and the cost equation for the production of hand-woven silk scarfs are given by

$$p = 60 - 2\sqrt{x} \qquad \text{and} \qquad C(x) = 3{,}000 + 5x$$

where x is the number of scarfs that can be sold at a price of \$p per unit and C(x) is the total cost (in dollars) of producing x scarfs.

(A) Express the revenue equation in terms of x.

(B) Graph the cost function and the revenue function in the same viewing rectangle for $0 \leq x \leq 900$. Use approximation techniques to find the break-even points correct to the nearest unit.

SECTION 3-8 Chapter Review

Important Terms and Symbols

3-1 *Limits and Continuity—A Geometric Introduction.* Limit as x approaches c; one-sided limits; limit as x approaches c from the left; left-hand limit; limit as x approaches c from the right; right-hand limit; continuous curve; continuity at a point; discontinuity at a point; continuity on an open interval; continuity on a closed or half-closed interval; continuity properties

$$\lim_{x\to c} f(x); \quad \lim_{x\to c^-} f(x); \quad \lim_{x\to c^+} f(x)$$

3-2 *Computation of Limits.* Limits at points of continuity; limits and infinity; vertical asymptote; limit properties; 0/0 indeterminate form; limits at infinity; horizontal asymptote

$$\lim_{x\to c} f(x) = f(c) \quad \text{if } f \text{ is continuous at } x = c;$$

$$\lim_{x\to c} f(x) = \infty; \quad \lim_{x\to c} f(x) = -\infty; \quad \lim_{x\to\infty} f(x); \quad \lim_{x\to-\infty} f(x)$$

3-3 *The Derivative.* Secant lines; tangent lines; difference quotient; slope of a tangent line; slope of a graph; average rate of change; instantaneous rate of change; the derivative; differentiable at a point; nondifferentiable at a point; differentiable over an open interval; differentiable over a closed or half-closed interval; differentiation

$$f'(x) = \lim_{h\to 0} \frac{f(x+h) - f(x)}{h}$$

3-4 *Derivatives of Constants, Power Forms, and Sums.* Derivative notation; derivative of a constant; power rule; constant times a function rule; sum and difference rule; marginal cost

$$f'(x); \quad y'; \quad \frac{dy}{dx}; \quad D_x f(x)$$

3-5 *Derivatives of Products and Quotients.* Product rule; quotient rule

3-6 *Chain Rule: Power Form.* General power rule; chain rule; combining rules of differentiation

3-7 *Marginal Analysis in Business and Economics.* Marginal cost; marginal revenue; marginal profit; demand equation; cost equation; revenue equation; break-even points; profit equation; average cost; marginal average cost; average revenue; marginal average revenue; average profit; marginal average profit

$$C'(x); \quad \overline{C}(x); \quad \overline{C}'(x); \quad R'(x); \quad \overline{R}(x); \quad \overline{R}'(x); \quad P'(x); \quad \overline{P}(x); \quad \overline{P}'(x)$$

Summary of Rules of Differentiation

$$D_x k = 0 \qquad D_x[F(x)S(x)] = F(x)S'(x) + S(x)F'(x)$$

$$D_x x^n = nx^{n-1} \qquad D_x \frac{T(x)}{B(x)} = \frac{B(x)T'(x) - T(x)B'(x)}{[B(x)]^2}$$

$$D_x kf(x) = kf'(x)$$

$$D_x[u(x) \pm v(x)] = u'(x) \pm v'(x) \qquad D_x[u(x)]^n = n[u(x)]^{n-1}u'(x)$$

EXERCISE 3-8

Chapter Review

Work through all the problems in this chapter review and check your answers in the back of the book. (Answers to all review problems are there.) Where weaknesses show up, review appropriate sections in the text.

A

In Problems 1–10, find $f'(x)$ for $f(x)$ as given.

1. $f(x) = 3x^4 - 2x^2 + 1$

2. $f(x) = 2x^{1/2} - 3x$

3. $f(x) = 5$

4. $f(x) = \frac{1}{2x^2} + \frac{x^2}{2}$

5. $f(x) = (2x - 1)(3x + 2)$

6. $f(x) = (x^2 - 1)(x^3 - 3)$

7. $f(x) = \frac{2x}{x^2 + 2}$

8. $f(x) = \frac{1}{3x + 2}$

9. $f(x) = (2x - 3)^3$

10. $f(x) = (x^2 + 2)^{-2}$

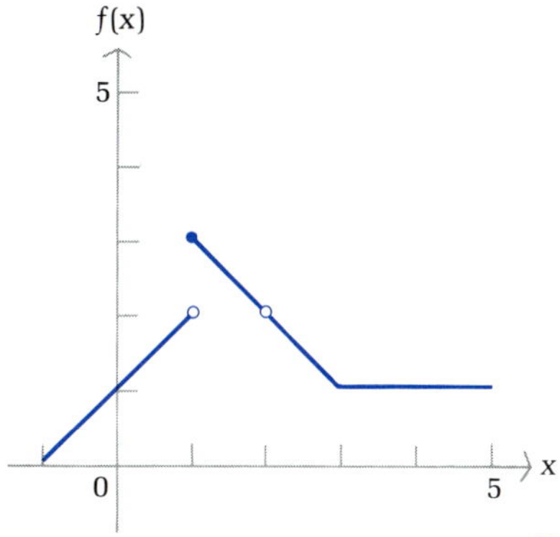

In Problems 11–13, use the graph of the function f shown in the margin to answer each question.

11. (A) $\lim_{x\to 1} f(x) = ?$ (B) $f(1) = ?$
(C) Is f continuous at $x = 1$?

12. (A) $\lim_{x\to 2} f(x) = ?$ (B) $f(2) = ?$
(C) Is f continuous at $x = 2$?

13. (A) $\lim_{x\to 3} f(x) = ?$ (B) $f(3) = ?$
(C) Is f continuous at $x = 3$?

B

In Problems 14–21, find the indicated derivative.

14. $\dfrac{dy}{dx}$ for $y = 3x^4 - 2x^{-3} + 5$

15. y' for $y = (2x^2 - 3x + 2)(x^2 + 2x - 1)$

16. $f'(x)$ for $f(x) = \dfrac{2x - 3}{(x - 1)^2}$

17. y' for $y = 2\sqrt{x} + \dfrac{4}{\sqrt{x}}$

18. $D_x[(x^2 - 1)(2x + 1)^2]$

19. $D_x \sqrt[3]{x^3 - 5}$

20. $\dfrac{dy}{dx}$ for $y = \dfrac{3x^2 + 4}{x^2}$

21. $D_x \dfrac{(x^2 + 2)^4}{2x - 3}$

22. For $y = f(x) = x^2 + 4$, find:
(A) The slope of the graph at $x = 1$
(B) The equation of the tangent line at $x = 1$ in the form $y = mx + b$

23. Repeat Problem 22 for $f(x) = x^3(x + 1)^2$.

In Problems 24–27, find the value(s) of x where the tangent line is horizontal.

24. $f(x) = 10x - x^2$

25. $f(x) = (x + 3)(x^2 - 45)$

26. $f(x) = \dfrac{x}{x^2 + 4}$

27. $f(x) = x^2(2x - 15)^3$

28. If an object moves along the y axis (scale in feet) so that it is at $y = f(x) = 16x^2 - 4x$ at time x (in seconds), find:
(A) The instantaneous velocity function
(B) The velocity at time $x = 3$ seconds

29. An object moves along the y axis (scale in feet) so that at time x (in seconds) it is at $y = f(x) = 96x - 16x^2$. Find:
(A) The instantaneous velocity function
(B) The time(s) when the velocity is 0

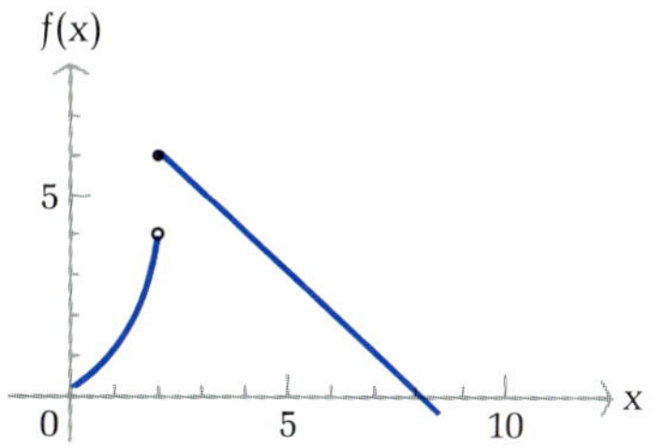

$$f(x) = \begin{cases} x^2 & 0 \le x < 2 \\ 8 - x & x \ge 2 \end{cases}$$

Problems 30 and 31 refer to the function f described in the figure in the margin.

30. (A) $\lim_{x\to 2^-} f(x) = ?$ (B) $\lim_{x\to 2^+} f(x) = ?$ (C) $\lim_{x\to 2} f(x) = ?$
(D) $f(2) = ?$ (E) Is f continuous at $x = 2$?

31. (A) $\lim_{x\to 5^-} f(x) = ?$ (B) $\lim_{x\to 5^+} f(x) = ?$ (C) $\lim_{x\to 5} f(x) = ?$
(D) $f(5) = ?$ (E) Is f continuous at $x = 5$?

32. Find each limit. (Use $-\infty$ or ∞, if appropriate.)

(A) $\lim_{x\to 0^-} \frac{1}{|x|}$ (B) $\lim_{x\to 0^+} \frac{1}{|x|}$ (C) $\lim_{x\to 0} \frac{1}{|x|}$

33. Find each limit. (Use $-\infty$ or ∞, if appropriate.)

(A) $\lim_{x\to 1^-} \frac{1}{x-1}$ (B) $\lim_{x\to 1^+} \frac{1}{x-1}$ (C) $\lim_{x\to 1} \frac{1}{x-1}$

In Problems 34–38, determine where f is continuous. Express the answer in interval notation.

34. $f(x) = 2x^2 - 3x + 1$

35. $f(x) = \frac{1}{x+5}$

36. $f(x) = \frac{x-3}{x^2-x-6}$

37. $f(x) = \sqrt{x-3}$

38. $f(x) = \sqrt[3]{1-x^2}$

In Problems 39–52, find each limit, if it exists. (Use $-\infty$ or ∞, if appropriate.)

39. $\lim_{x\to 3} \frac{2x-3}{x+5}$

40. $\lim_{x\to 3} (2x^2 - x + 1)$

41. $\lim_{x\to 0} \frac{2x}{3x^2-2x}$

42. $\lim_{h\to 0} \frac{[(2+h)^2 - 1] - [2^2 - 1]}{h}$

43. $\lim_{h\to 0} \frac{f(2+h) - f(2)}{h}$ for $f(x) = x^2 + 4$

44. $\lim_{x\to 3} \frac{x-3}{x^2-9}$

45. $\lim_{x\to -3} \frac{x-3}{x^2-9}$

46. $\lim_{x\to 7} \frac{\sqrt{x}-\sqrt{7}}{x-7}$

47. $\lim_{x\to -2} \sqrt{\frac{x^2+4}{2-x}}$

48. $\lim_{x\to\infty} \left(3 + \frac{1}{x^{1/3}} + \frac{2}{x^3}\right)$

49. $\lim_{x\to\infty} (3x^3 - 2x^2 - 10x - 100)$

50. $\lim_{x\to\infty} \frac{2x^2+3}{3x^2+2}$

51. $\lim_{x\to\infty} \frac{2x+3}{3x^2+2}$

52. $\lim_{x\to\infty} \frac{2x^2+3}{3x+2}$

In Problems 53 and 54, use the definition of the derivative to find $f'(x)$.

53. $f(x) = x^2 - x$

54. $f(x) = \sqrt{x} - 3$

C *Problems 55–58 refer to the function f in the figure. Determine whether f is differentiable at the indicated value of x.*

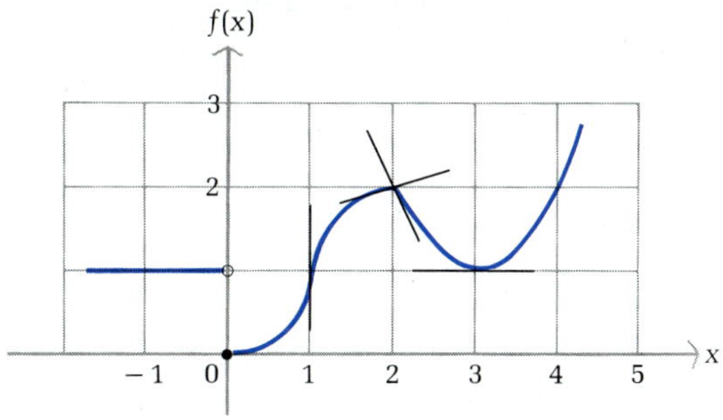

55. $x = 0$ **56.** $x = 1$ **57.** $x = 2$ **58.** $x = 3$

In Problems 59 and 60, graph f and find all discontinuities.

59. $f(x) = \begin{cases} 4 - x^2 & \text{if } x < 0 \\ 2 + x^2 & \text{if } x \geq 0 \end{cases}$

60. $f(x) = \begin{cases} 4 - x^2 & \text{if } x < 1 \\ 3x & \text{if } x \geq 1 \end{cases}$

In Problems 61–64, find f′(x) and simplify.

61. $f(x) = (x - 4)^4(x + 3)^3$

62. $f(x) = \dfrac{x^5}{(2x + 1)^4}$

63. $f(x) = \dfrac{\sqrt{x^2 - 1}}{x}$

64. $f(x) = \dfrac{x}{\sqrt{x^2 + 4}}$

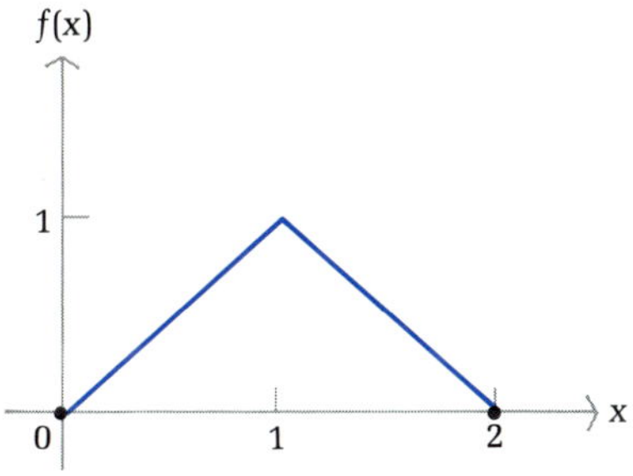

Answer the questions in Problems 65–67 for the function f shown in the figure in the margin and given below:

$$f(x) = 1 - |x - 1| \qquad 0 \leq x \leq 2$$

65. (A) $\lim_{x \to 1^-} f(x) = ?$ (B) $\lim_{x \to 1^+} f(x) = ?$

(C) $\lim_{x \to 1} f(x) = ?$ (D) Is f continuous at $x = 1$?

66. (A) Is f continuous on the open interval $(0, 2)$?

(B) Is f continuous from the right at $x = 0$?

(C) Is f continuous from the left at $x = 2$?

(D) Is f continuous on the closed interval $[0, 2]$?

67. (A) $\lim_{h \to 0^-} \dfrac{f(1 + h) - f(1)}{h} = ?$ (B) $\lim_{h \to 0^+} \dfrac{f(1 + h) - f(1)}{h} = ?$

(C) $\lim_{h \to 0} \dfrac{f(1 + h) - f(1)}{h} = ?$ (D) Does $f'(1)$ exist?

APPLICATIONS

Business & Economics

68. Cost analysis. The total cost (in dollars) of producing x television sets is

$$C(x) = 10{,}000 + 200x - 0.1x^2$$

(A) Find the exact cost of producing the 101st television set.

(B) Use the marginal cost to approximate the cost of producing the 101st television set.

69. *Cost analysis.* The total cost (in dollars) of producing x bicycles is

$$C(x) = 5{,}000 + 40x + 0.05x^2$$

(A) Find the total cost and the marginal cost at a production level of 100 bicycles and interpret the results.

(B) Find the average cost and the marginal average cost at a production level of 100 bicycles and interpret the results.

70. *Marginal analysis.* Let

$$p = 20 - x \qquad \text{and} \qquad C(x) = 2x + 56 \qquad 0 \leq x \leq 20$$

be the demand equation and the cost function, respectively, for a certain commodity.

(A) Find the marginal cost, average cost, and marginal average cost functions.

(B) Express the revenue in terms of x, and find the marginal revenue, average revenue, and marginal average revenue functions.

(C) Find the profit, marginal profit, average profit, and marginal average profit functions.

(D) Find the break-even point(s).

(E) Evaluate the marginal profit at $x = 7$, 9, and 11, and interpret the results.

(F) Graph $R = R(x)$ and $C = C(x)$ on the same coordinate system, and locate regions of profit and loss.

71. *Employee training.* A company producing computer components has established that on the average, a new employee can assemble $N(t)$ components per day after t days of on-the-job training, as given by

$$N(t) = \frac{40t}{t + 2}$$

(A) Find the average rate of change of $N(t)$ from 3 days to 6 days.

(B) Find the instantaneous rate of change of $N(t)$ at 3 days.

(C) Find $\lim_{t\to\infty} N(t)$.

72. *Sales analysis.* Past sales records for a swimming pool manufacturer indicate that the number of pools, N (in thousands), sold during each month of the year is given by

$$N(t) = 5 + t\sqrt{12 - t}$$

where t is the number of months since the beginning of the year. Find $N(3)$ and $N'(3)$ and interpret the results.

Life Sciences

73. Pollution. A sewage treatment plant disposes of its effluent through a pipeline that extends 1 mile toward the center of a large lake. The concentration of effluent $C(x)$, in parts per million, x meters from the end of the pipe is given approximately by

$$C(x) = 500(x + 1)^{-2}$$

What is the instantaneous rate of change of concentration at 9 meters? At 99 meters?

Social Sciences

74. Learning. If a person learns N items in t hours, as given by

$$N(t) = 20\sqrt{t}$$

find the rate of learning after:

(A) 1 hour (B) 4 hours

CHAPTER 4

Graphing and Optimization

Contents

CHAPTER 4

SECTION 4-1

First Derivative and Graphs

- SOLVING INEQUALITIES USING CONTINUITY PROPERTIES
- INCREASING AND DECREASING FUNCTIONS
- CRITICAL VALUES AND LOCAL EXTREMA
- FIRST-DERIVATIVE TEST

Since the derivative is associated with the slope of the graph of a function at a point, we might expect that it is also associated with other properties of a graph. As we will see in this and the next section, the derivative can tell us a great deal about the shape of the graph of a function. In addition, this investigation will lead to methods for finding absolute maximum and minimum values for functions that do not require graphing. Manufacturing companies can use these methods to find production levels that will minimize cost or maximize profit. Pharmacologists can use them to find levels of drug dosages that will produce maximum sensitivity to a drug. And so on.

We digress for a moment to discuss the use of continuity and *sign charts* in solving inequalities, a process that will see frequent use in this and succeeding sections.

◆ SOLVING INEQUALITIES USING CONTINUITY PROPERTIES

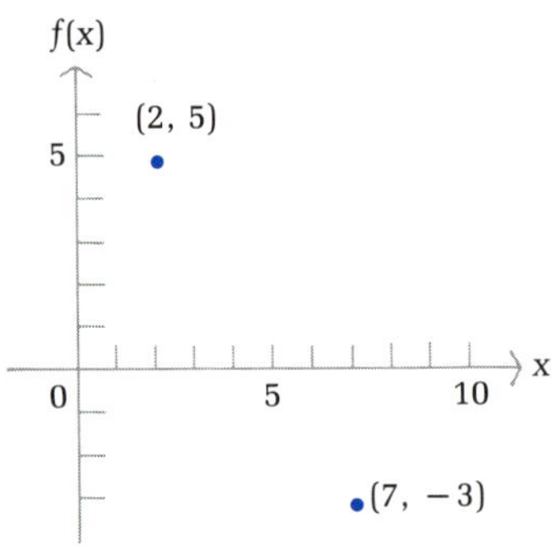

FIGURE 1

In our informal discussion of continuity in Section 3-1, we said that a function is continuous over an interval if we can draw its graph over the interval without lifting a pencil from the paper. Suppose a function f is continuous over the interval $(1, 8)$ and $f(x) \neq 0$ for any x in $(1, 8)$. Also suppose $f(2) = 5$, a positive number. Is it possible for $f(x)$ to be negative for any x in $(1, 8)$? The answer is "no." If $f(7)$ were -3, for example, as shown in Figure 1, how would it be possible to join the points $(2, 5)$ and $(7, -3)$ with the graph of a continuous function without crossing the x axis between 1 and 8 at least once? [Crossing the x axis would violate our assumption that $f(x) \neq 0$ for any x in $(1, 8)$.] Thus, we conclude that $f(x)$ must be positive for all x in $(1, 8)$. If $f(2)$ were negative, then, using the same type of reasoning, $f(x)$ would have to be negative over the whole interval $(1, 8)$.

In general, **if f is continuous and $f(x) \neq 0$ on the interval (a, b), then $f(x)$ cannot change sign on (a, b).** This is the essence of Theorem 1.

THEOREM 1

Sign Properties on an Interval (a, b)

If f is continuous on (a,b) and $f(x) \neq 0$ for all x in (a, b), then either $f(x) > 0$ for all x in (a,b) or $f(x) < 0$ for all x in (a,b).

Theorem 1 provides the basis for an effective method of solving many types of inequalities. Example 1 illustrates the process.

◆ EXAMPLE 1 Solve: $\dfrac{x+1}{x-2} > 0$

Solution We start by using the left side of the inequality to form the function f:

$$f(x) = \frac{x+1}{x-2}$$

The rational function f is discontinuous at $x = 2$, and $f(x) = 0$ for $x = -1$ (a fraction is 0 when the numerator is 0 and the denominator is not 0). We plot $x = 2$ and $x = -1$, which we call *partition numbers*, on a real number line:

(Note that the dot at 2 is open, because the function is not defined at $x = 2$.) The partition numbers 2 and -1 determine three open intervals: $(-\infty, -1)$, $(-1, 2)$, and $(2, \infty)$. The function f is continuous and nonzero on each of these intervals. From Theorem 1 we know that $f(x)$ does not change sign on any of these intervals. Thus, we can find the sign of $f(x)$ on each of these intervals by selecting a **test number** in each interval and evaluating $f(x)$ at that number. Since any number in each subinterval will do, we choose test numbers that are easy to evaluate: -2, 0, and 3. The table in the margin shows the results.

x	-2	0	3
$f(x)$	$\frac{1}{4}$ +	$-\frac{1}{2}$ −	4 +

Test numbers

The sign of $f(x)$ at each test number is the same as the sign of $f(x)$ over the interval containing that test number. Using this information, we construct a **sign chart** for $f(x)$:

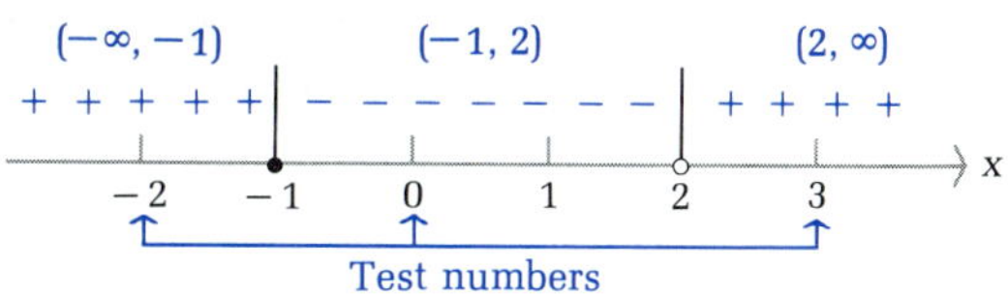

Now using the sign chart, we can easily write the solution for the given nonlinear inequality: $f(x) > 0$ for

$x < -1$ or $x > 2$ Inequality notation
$(-\infty, -1) \cup (2, \infty)$ Interval notation ◆

Most of the inequalities we will encounter will involve strict inequalities ($>$ or $<$). If it is necessary to solve inequalities of the form $\geqslant$ or $\leqslant$, we simply include the end point of any interval if it is a zero of f [that is, if it is a value of x such that $f(x) = 0$]. For example, referring to the sign chart in Example 1, the solution of the inequality

$\frac{x+1}{x-2} \geqslant 0$ is $x \leqslant -1$ or $x > 2$ Inequality notation
$(-\infty, -1] \cup (2, \infty)$ Interval notation

In general, given a function f, we will call all values x such that f is discontinuous at x or $f(x) = 0$ **partition numbers. Partition numbers determine open intervals where $f(x)$ does not change sign.** By using a test number from each interval, we can construct a sign chart for $f(x)$ on the real number line. It is then an easy matter to determine where $f(x) < 0$ or $f(x) > 0$; that is, to solve the inequality $f(x) < 0$ or $f(x) > 0$.

We summarize the procedure for constructing sign charts in the following box:

Constructing Sign Charts

Given a function f:

Step 1. Find all partition numbers. That is:

(A) Find all numbers where f is discontinuous. (Rational functions are discontinuous for values of x that make a denominator 0.)
(B) Find all numbers where $f(x) = 0$. (For a rational function, this occurs where the numerator is 0 and the denominator is not 0.)

Step 2. Plot the numbers found in step 1 on a real number line, dividing the number line into intervals.

Step 3. Select a test number in each open interval determined in step 2, and evaluate $f(x)$ at each test number to determine whether $f(x)$ is positive (+) or negative (−) in each interval.

Step 4. Construct a sign chart using the real number line in step 2. This will show the sign of $f(x)$ on each open interval

[*Note:* From the sign chart, it is easy to find the solution for the inequality $f(x) < 0$ or $f(x) > 0$.]

PROBLEM 1 Solve: $\frac{x^2-1}{x-3} < 0$ ◆

◆ INCREASING AND DECREASING FUNCTIONS

Graphs of functions generally have *rising* or *falling* sections as we scan the graphs from left to right. It would be an aid to graphing if we could determine where these sections occur. Suppose the graph of a function f is as indicated in Figure 2. As we look from left to right, we see that on the interval (a, b) the graph of f is *rising*, $f(x)$ is *increasing*,* and the slope of the graph is positive $[f'(x) > 0]$. On the other hand, on the interval (b, c) the graph of f is *falling*, $f(x)$ is *decreasing*, and the slope of the graph is negative $[f'(x) < 0]$. At $x = b$, the graph of f changes direction (from rising to falling), $f(x)$ changes from increasing to decreasing, the slope of the graph is 0 $[f'(b) = 0]$, and the tangent line is horizontal.

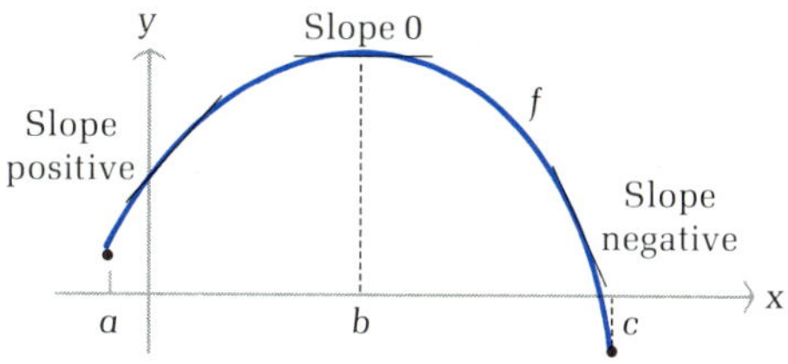

FIGURE 2

In general, if $f'(x) > 0$ (is positive) on the interval (a, b), then $f(x)$ increases (↗) and the graph of f rises as we move from left to right over the interval; if $f'(x) < 0$ (is negative) on an interval (a, b), then $f(x)$ decreases (↘) and the graph of f falls as we move from left to right over the interval. We summarize these important results in the box.

Increasing and Decreasing Functions

For the interval (a, b):

$f'(x)$	$f(x)$	GRAPH OF f	EXAMPLES
+	Increases ↗	Rises ↗	
−	Decreases ↘	Falls ↘	

◆ EXAMPLE 2 Given $f(x) = 8x - x^2$:

(A) Which values of x correspond to horizontal tangent lines?
(B) For which values of x is $f(x)$ increasing? Decreasing?
(C) Sketch a graph of f. Add horizontal tangent lines.

---

* Formally, we say that $f(x)$ is **increasing** on an interval (a, b) if $f(x_2) > f(x_1)$ whenever $a < x_1 < x_2 < b$; f is **decreasing** on (a, b) if $f(x_2) < f(x_1)$ whenever $a < x_1 < x_2 < b$.

Solutions (A) $f'(x) = 8 - 2x = 0$

$$x = 4$$

Thus, a horizontal tangent line exists at $x = 4$ only.

(B) As in Example 1, we will construct a sign chart for $f'(x)$ to determine which values of x make $f'(x) > 0$ and which values make $f'(x) < 0$. To do this, we must first determine the partition numbers (values of x for which $f'(x) = 0$ or for which $f'(x)$ is discontinuous). From part A we know that $f'(x) = 8 - 2x = 0$ at $x = 4$. Since $f'(x) = 8 - 2x$ is a polynomial, it is continuous for all x. Thus, 4 is the only partition number. We construct a sign chart for the intervals $(-\infty, 4)$ and $(4, \infty)$, using test numbers 3 and 5:

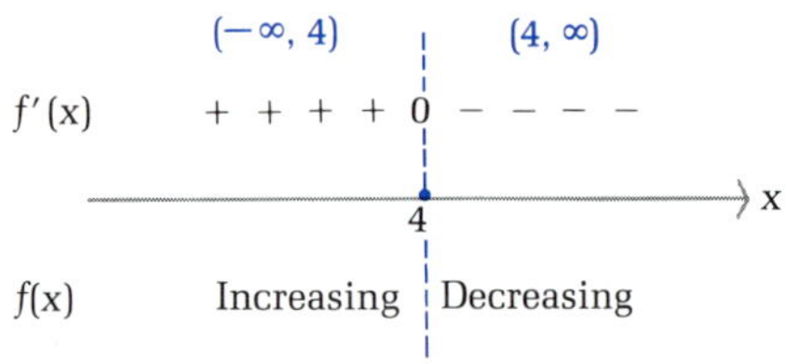

Test Numbers

x	$f'(x)$	
3	2	(+)
5	−2	(−)

Thus, $f(x)$ is increasing on $(-\infty, 4)$ and decreasing on $(4, \infty)$.

x	$f(x)$
0	0
2	12
4	16
6	12
8	0

(C)

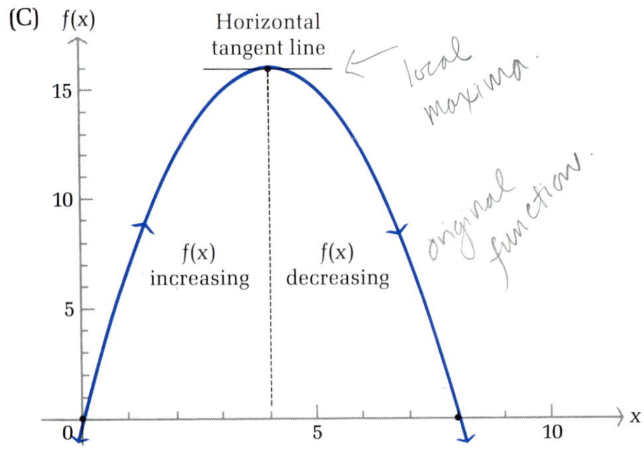

◆

PROBLEM 2 Repeat Example 2 for $f(x) = x^2 - 6x + 10$. ◆

◆ EXAMPLE 3 Determine the intervals where f is increasing and those where f is decreasing for:

(A) $f(x) = 1 + x^3$ (B) $f(x) = (1 - x)^{1/3}$ (C) $f(x) = \dfrac{1}{x - 2}$

Solutions (A) $f(x) = 1 + x^3$ $\quad f'(x) = 3x^2 = 0$

$$x = 0$$

Sign chart for $f'(x) = 3x^2$ (partition number is 0):

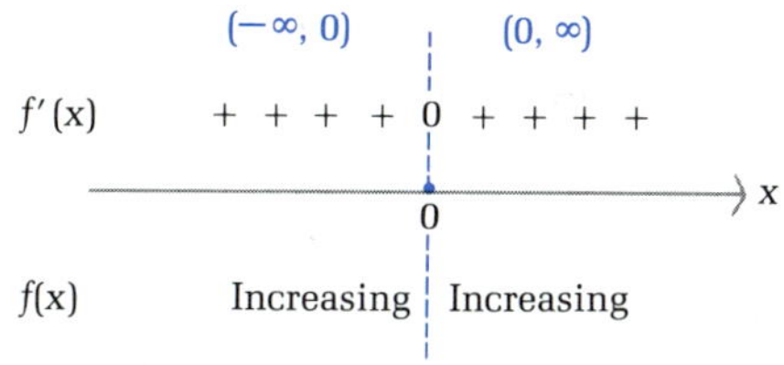

Test Numbers

x	$f'(x)$
-1	3 (+)
1	3 (+)

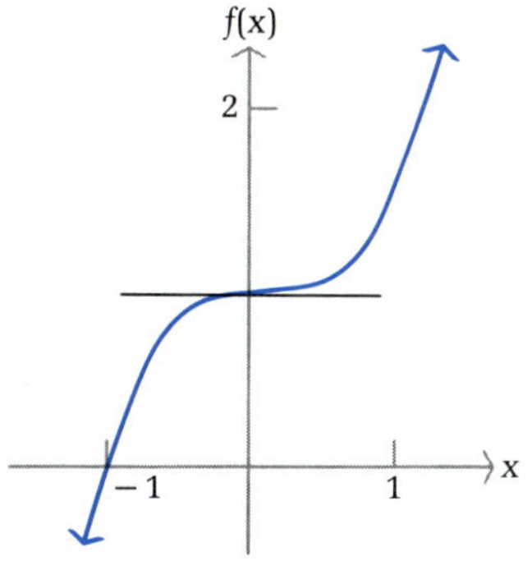

The sign chart indicates that $f(x)$ is increasing on $(-\infty, 0)$ and $(0, \infty)$. Since f is continuous at $x = 0$, it follows that $f(x)$ is increasing for all x. The graph of f is shown in the margin.

(B) $f(x) = (1 - x)^{1/3}$ $\quad f'(x) = -\frac{1}{3}(1 - x)^{-2/3} = \frac{-1}{3(1 - x)^{2/3}}$

To find partition numbers for $f'(x)$, we note that f' is continuous for all x except for values of x for which the denominator is 0; that is, f' is discontinuous at $x = 1$. Since the numerator is the constant -1, $f'(x) \neq 0$ for any value of x. Thus, we plot the partition number $x = 1$ on the real number line and use the abbreviation ND to note the fact that $f'(x)$ is *not defined* at $x = 1$.

Sign chart for $f'(x) = -1/[3(1 - x)^{2/3}]$ (partition number is 1):

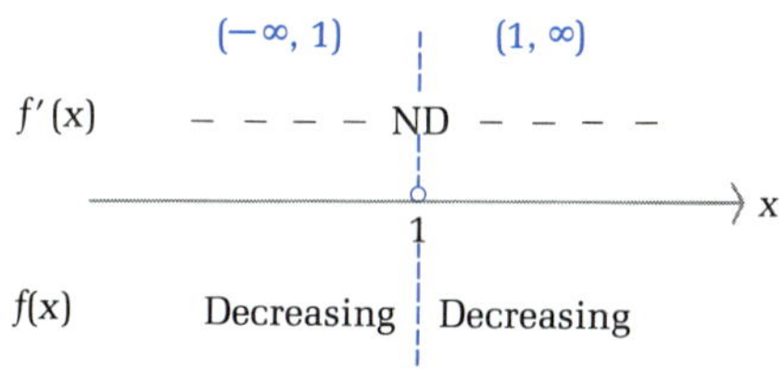

Test Numbers

x	$f'(x)$
0	$-\frac{1}{3}$ (−)
2	$-\frac{1}{3}$ (−)

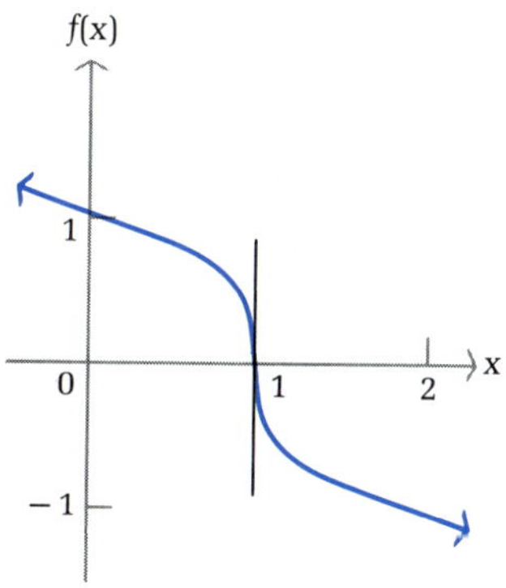

The sign chart indicates that f is decreasing on $(-\infty, 1)$ and $(1, \infty)$. Since f is continuous at $x = 1$, it follows that $f(x)$ is decreasing for all x. Thus, **a continuous function can be decreasing (or increasing) on an interval containing values of x where $f'(x)$ does not exist.** The graph of f is shown in the margin. Notice that the undefined derivative at $x = 1$ results in a vertical tangent line at $x = 1$. In general, **a vertical tangent will occur at $x = c$ if f is continuous at $x = c$ and $|f'(x)|$ becomes larger and larger as x approaches c.**

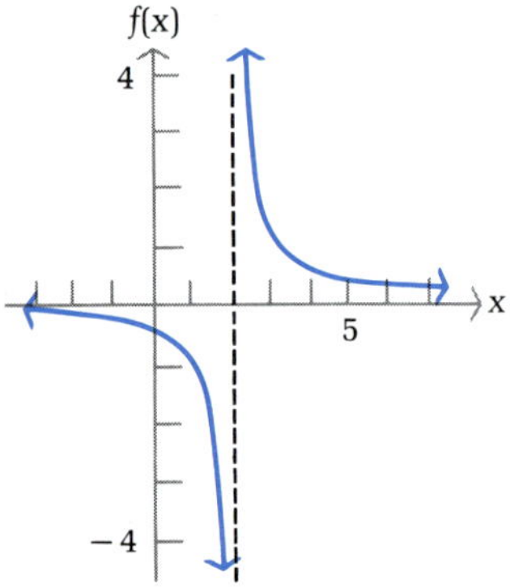

(C) $f(x) = \dfrac{1}{x-2}$ $\qquad f'(x) = \dfrac{-1}{(x-2)^2}$

Sign chart for $f'(x) = -1/(x-2)^2$ (partition number is 2):

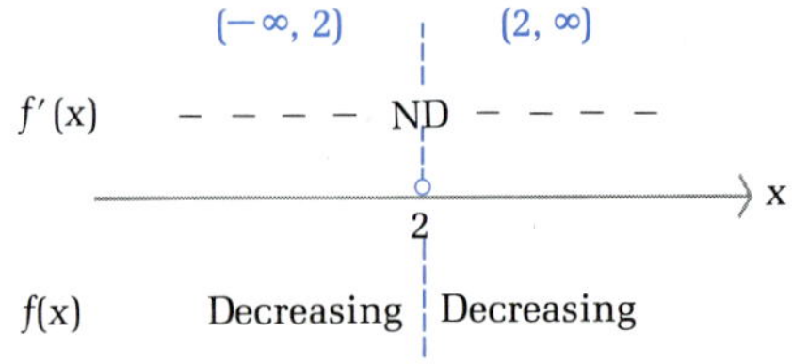

Test Numbers

x	$f'(x)$
1	-1 (−)
3	-1 (−)

Thus, f is decreasing on $(-\infty, 2)$ and $(2, \infty)$. See the graph of f in the margin. ◆

The values where a function is increasing or decreasing must always be expressed in terms of open intervals that are subsets of the domain of the function.

PROBLEM 3 Determine the intervals where f is increasing and those where f is decreasing for:

(A) $f(x) = 1 - x^3$ (B) $f(x) = (1 + x)^{1/3}$ (C) $f(x) = \dfrac{1}{x}$ ◆

◆ CRITICAL VALUES AND LOCAL EXTREMA

When the graph of a continuous function changes from rising to falling, a high point, or *local maximum*, occurs; and when the graph changes from falling to rising, a low point, or *local minimum*, occurs. In Figure 3, high points occur at c_3 and c_6, and low points occur at c_2 and c_4. In general, we call $f(c)$ a **local maximum** if there exists an interval (m, n) containing c such that

$$f(x) \le f(c)$$

for all x in (m, n).

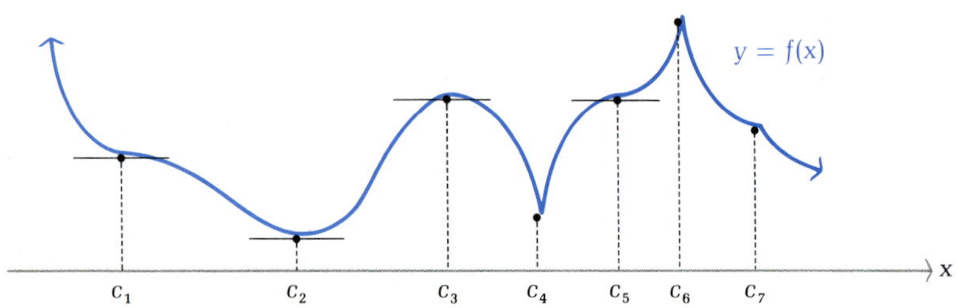

FIGURE 3

The quantity $f(c)$ is called a **local minimum** if there exists an interval (m, n) containing c such that

$$f(x) \geq f(c)$$

for all x in (m, n). The quantity $f(c)$ is called a **local extremum** if it is either a local maximum or a local minimum. Thus, in Figure 3, we see that local maxima occur at c_3 and c_6, local minima occur at c_2 and c_4, and all four of these points are local extrema.

How can we locate local maxima and minima if we are given the equation for a function and not its graph? Figure 3 suggests an approach. It appears that local maxima and minima occur among those values of x such that $f'(x) = 0$ or $f'(x)$ does not exist; that is, among the values $c_1, c_2, c_3, c_4, c_5, c_6$, and c_7. [Recall from Section 3-3 that $f'(x)$ is not defined at points on the graph of f where there is a sharp corner or a vertical tangent line.]

Critical Values of *f*

The values of x in the domain of f where $f'(x) = 0$ or $f'(x)$ does not exist are called the **critical values** of f.

It is possible to prove the following theorem:

THEOREM 2

Existence of Local Extrema

If f is continuous on the interval (a, b) and $f(c)$ is a local extremum, then either $f'(c) = 0$ or $f'(c)$ does not exist (is not defined).

Theorem 2 states that a local extremum can occur only at a critical value, but it does not imply that every critical value produces a local extremum. In Figure 3, c_1 and c_5 are critical values (the slope is 0), but the function does not have a local maximum or local minimum at either of these values.

Our strategy for finding local extrema is now clear. We find all critical values for f and test each one to see if it produces a local maximum, a local minimum, or neither.

◆ FIRST-DERIVATIVE TEST

If $f'(x)$ exists on both sides of a critical value c, then the sign of $f'(x)$ can be used to determine whether the point $(c, f(c))$ is a local maximum, a local minimum, or neither. The various possibilities are summarized in the box and illustrated in Figure 4.

First-Derivative Test for Local Extrema

Let c be a critical value of f [$f(c)$ defined and either $f'(c) = 0$ or $f'(c)$ not defined]. Construct a sign chart for $f'(x)$ close to and on either side of c.

SIGN CHART — $f(c)$

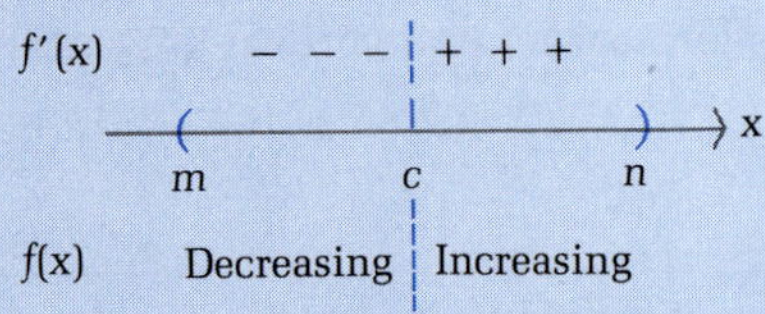

$f(c)$ is a local minimum.
If $f'(x)$ changes from negative to positive at c, then $f(c)$ is a local minimum.

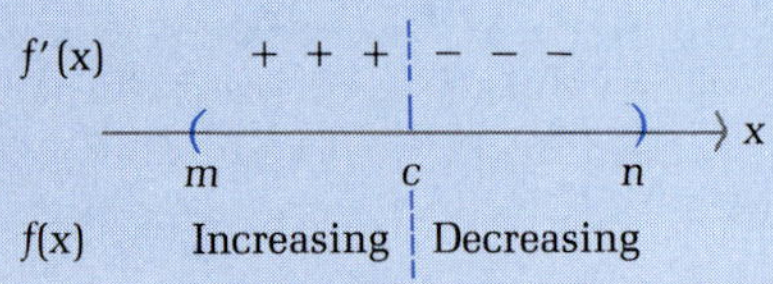

$f(c)$ is a local maximum.
If $f'(x)$ changes from positive to negative at c, then $f(c)$ is a local maximum.

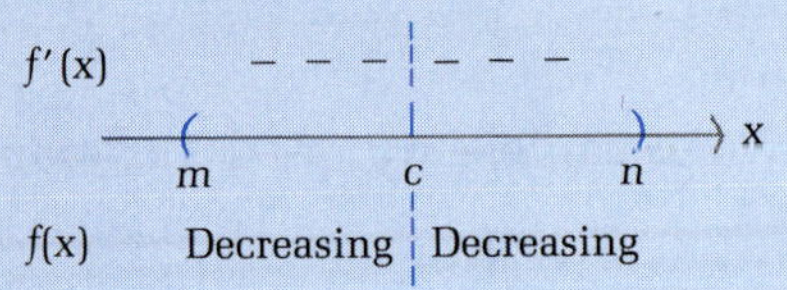

$f(c)$ is not a local extremum.
If $f'(x)$ does not change sign at c, then $f(c)$ is neither a local maximum nor a local minimum.

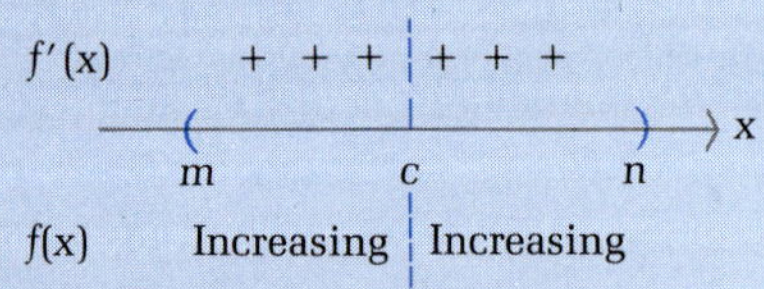

$f(c)$ is not a local extremum.
If $f'(x)$ does not change sign at c, then $f(c)$ is neither a local maximum nor a local minimum.

◆ EXAMPLE 4 Given $f(x) = x^3 - 6x^2 + 9x + 1$:

(A) Find the critical values of f. (B) Find the local maxima and minima.
(C) Sketch the graph of f.

Solutions (A) Find all numbers x in the domain of f where $f'(x) = 0$ or $f'(x)$ does not exist.

$$f'(x) = 3x^2 - 12x + 9 = 0$$
$$3(x^2 - 4x + 3) = 0$$
$$3(x - 1)(x - 3) = 0$$
$$x = 1 \quad \text{or} \quad x = 3$$

$f'(x)$ exists for all x; the critical values are $x = 1$ and $x = 3$.

$f'(c) = 0$
Horizontal tangent

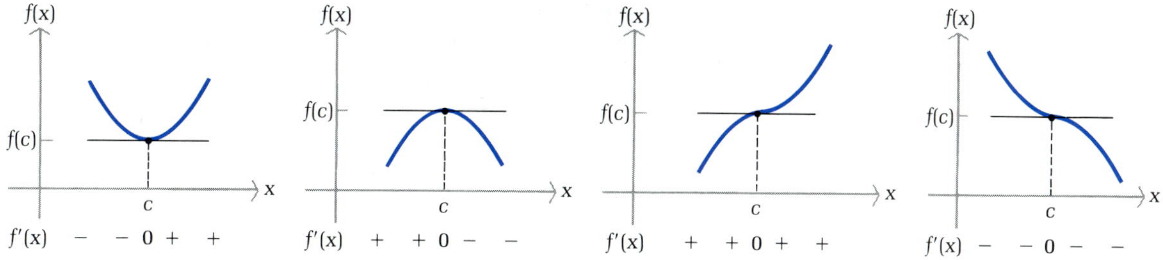

(A) $f(c)$ is a local minimum.

(B) $f(c)$ is a local maximum.

(C) $f(c)$ is neither a local maximum nor a local minimum.

(D) $f(c)$ is neither a local maximum nor a local minimum.

$f'(c)$ is not defined but $f(c)$ is defined

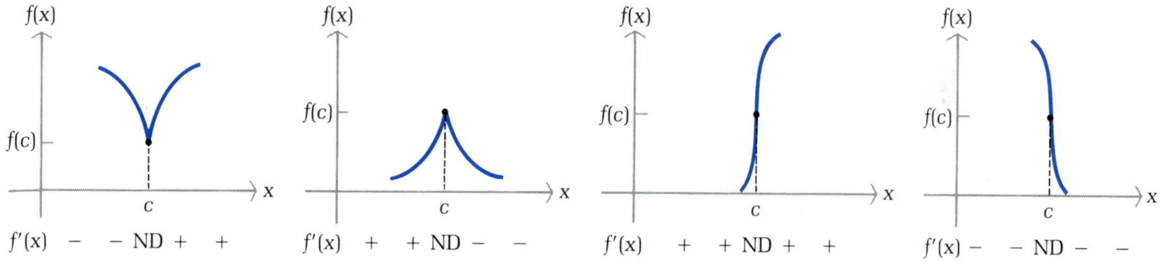

(E) $f(c)$ is a local minimum.

(F) $f(c)$ is a local maximum.

(G) $f(c)$ is neither a local maximum nor a local minimum.

(H) $f(c)$ is neither a local maximum nor a local minimum.

FIGURE 4
Local extrema

(B) The easiest way to apply the first-derivative test for local maxima and minima is to construct a sign chart for $f'(x)$ for all x. Partition numbers for $f'(x)$ are $x = 1$ and $x = 3$ (which also happen to be critical values for f).

Sign chart for $f'(x) = 3(x - 1)(x - 3)$:

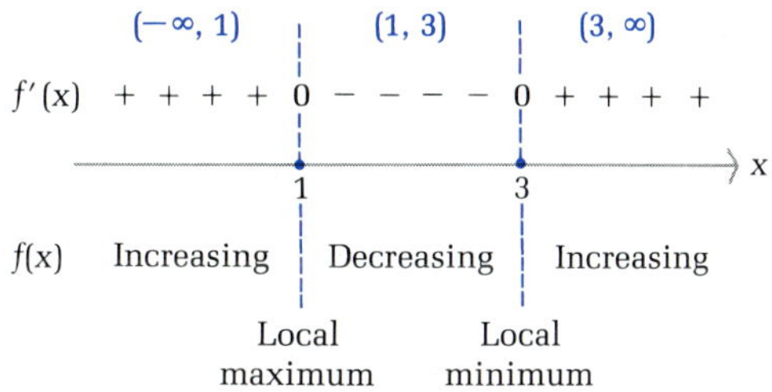

Test Numbers

x	$f'(x)$	
0	9	(+)
2	−3	(−)
4	9	(+)

The sign chart indicates that f increases on $(-\infty, 1)$, has a local maximum at $x = 1$, decreases on $(1, 3)$, has a local minimum at $x = 3$, and increases on $(3, \infty)$. These facts are summarized in the following table:

x	$f'(x)$	$f(x)$	GRAPH OF f
$(-\infty, 1)$	+	Increasing	Rising
$x = 1$	0	Local maximum	Horizontal tangent
$(1, 3)$	−	Decreasing	Falling
$x = 3$	0	Local minimum	Horizontal tangent
$(3, \infty)$	+	Increasing	Rising

(C) We sketch a graph of f using the information from part B and point-by-point plotting.

x	$f(x)$
0	1
1	5
2	3
3	1
4	5

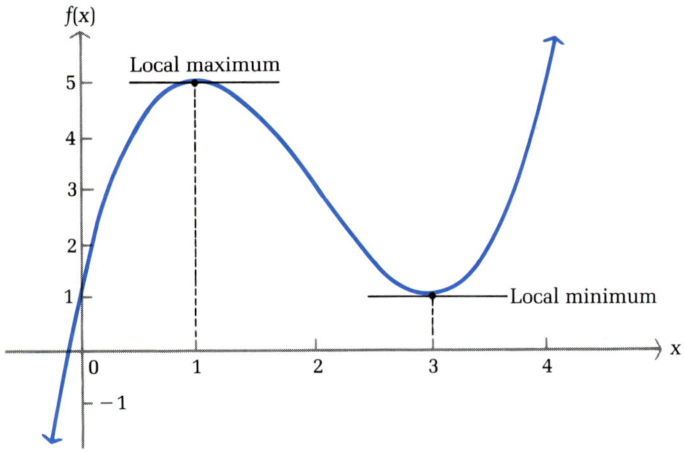

PROBLEM 4 Given $f(x) = x^3 - 9x^2 + 24x - 10$:

(A) Find the critical values of f.
(B) Find the local maxima and minima.
(C) Sketch a graph of f.

In Example 4, the function f had local extrema at both of its critical values. However, as was noted earlier, not every critical value of a function will produce a local extremum. For example, consider the function discussed in Example 3B:

$$f(x) = (1 - x)^{1/3} \qquad \text{and} \qquad f'(x) = \frac{-1}{3(1 - x)^{2/3}}$$

Since $f(1)$ exists and $f'(1)$ does not exist, $x = 1$ is a critical value for this function. However, the sign chart for $f'(x)$ shows that $f'(x)$ does not change sign at $x = 1$:

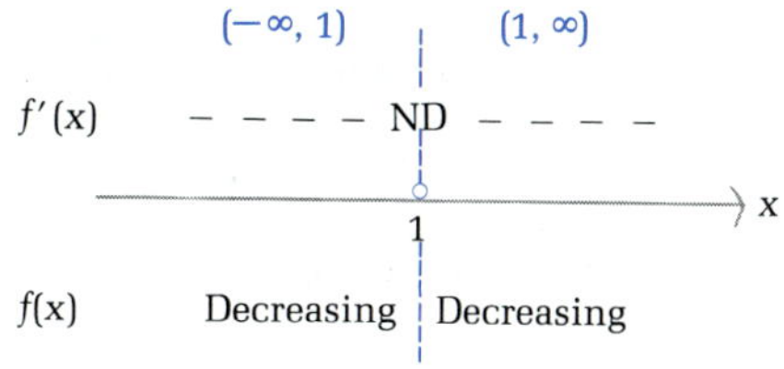

Test Numbers

x	$f'(x)$	
0	$-\frac{1}{3}$	(−)
2	$-\frac{1}{3}$	(−)

Thus, f does not have a local maximum nor a local minimum at $x = 1$.

Finally, it is important to remember that a critical value must be in the domain of the function. Refer to the function discussed in Example 3C:

$$f(x) = \frac{1}{x - 2} \qquad \text{and} \qquad f'(x) = \frac{-1}{(x - 2)^2}$$

The derivative is not defined at $x = 2$, but neither is the function. Thus, $x = 2$ is not a critical value for f (in fact, this function does not have any critical values). Nevertheless, $x = 2$ is a partition number for $f'(x)$ and must be included in the sign chart for $f'(x)$.

In general, every critical value for $f(x)$ is a partition number for $f'(x)$, but some partition numbers for $f'(x)$ may not be critical values for f. [Critical values have the added requirement that $f(x)$ must be defined at that value.]

Answers to Matched Problems

1. $-\infty < x < -1$ or $1 < x < 3$; $(-\infty, -1) \cup (1, 3)$
2. (A) Horizontal tangent line at $x = 3$.
   (B) Decreasing on $(-\infty, 3)$; increasing on $(3, \infty)$
   (C)

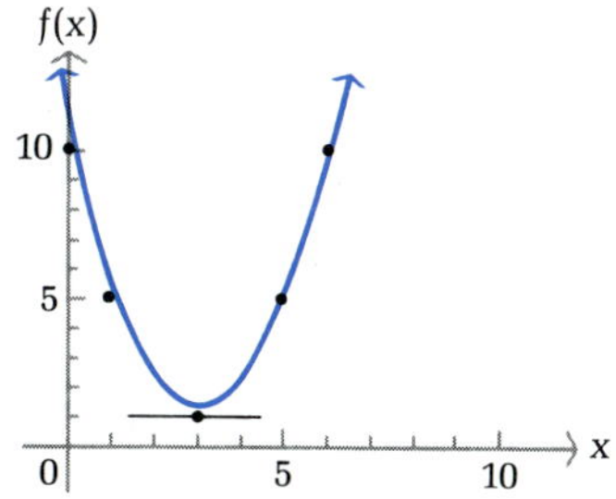

3. (A) Decreasing for all x (B) Increasing for all x
   (C) Decreasing on $(-\infty, 0)$ and $(0, \infty)$

4. (A) Critical values: $x = 2$, $x = 4$
   (B) Local maximum at $x = 2$; local minimum at $x = 4$
   (C)

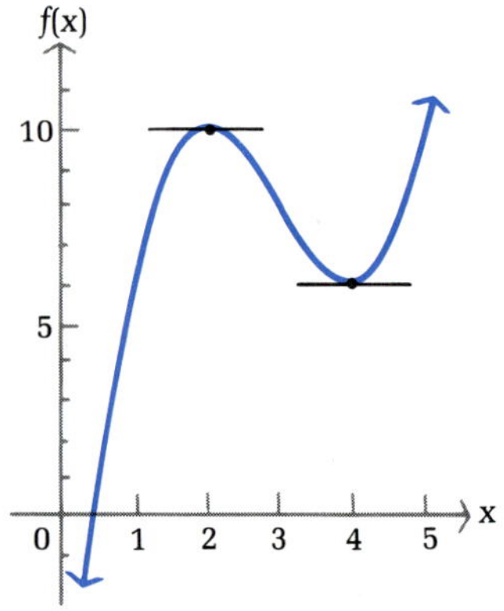

EXERCISE 4-1

A *Problems 1–6 refer to the following graph of $y = f(x)$:*

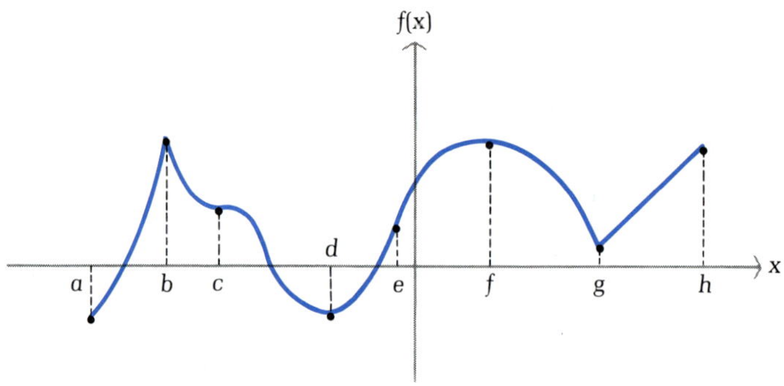

1. Identify the intervals over which $f(x)$ is increasing.
2. Identify the intervals over which $f(x)$ is decreasing.
3. Identify the points where $f'(x) = 0$.
4. Identify the points where $f'(x)$ does not exist.
5. Identify the points where f has a local maximum.
6. Identify the points where f has a local minimum.

In Problems 7 and 8, $f(x)$ is continuous on $(-\infty, \infty)$ and has critical values at $x = a$, b, c, and d. Use the sign chart for $f'(x)$ to determine whether f has a local maximum, a local minimum, or neither at each critical value.

7. 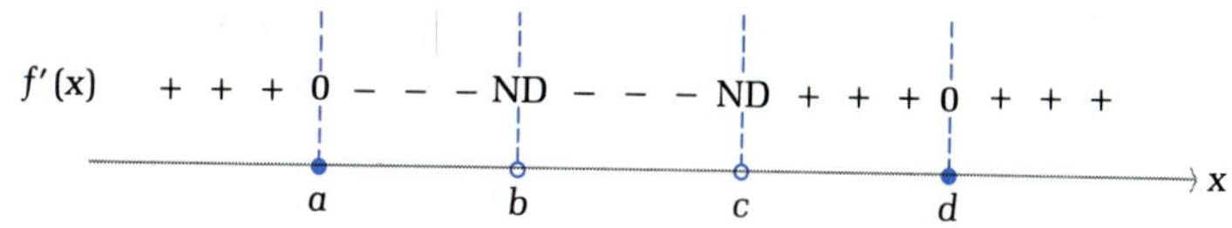

8. 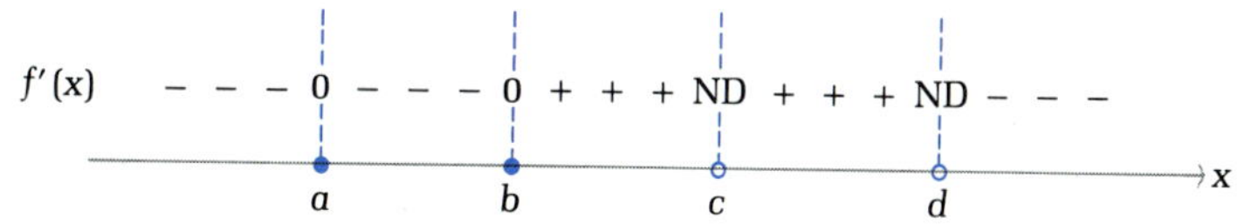

In Problems 9–12, $f(x)$ is continuous on $(-\infty, \infty)$. Use the given information to sketch the graph of f.

9. 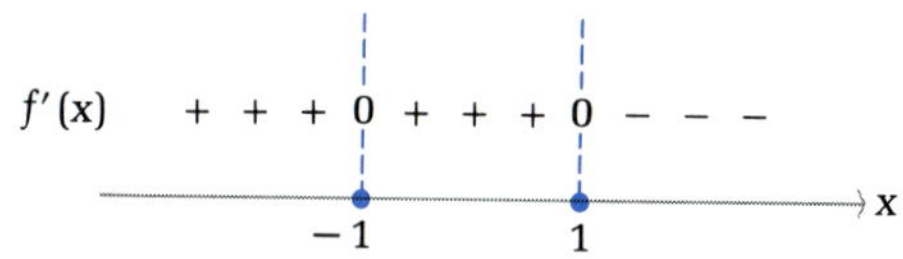

x	−2	−1	0	1	2
f(x)	−1	1	2	3	1

10. 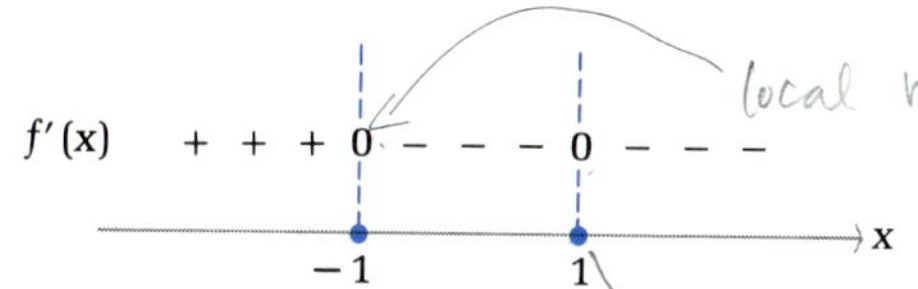

x	−2	−1	0	1	2
f(x)	1	3	2	1	−1

11. 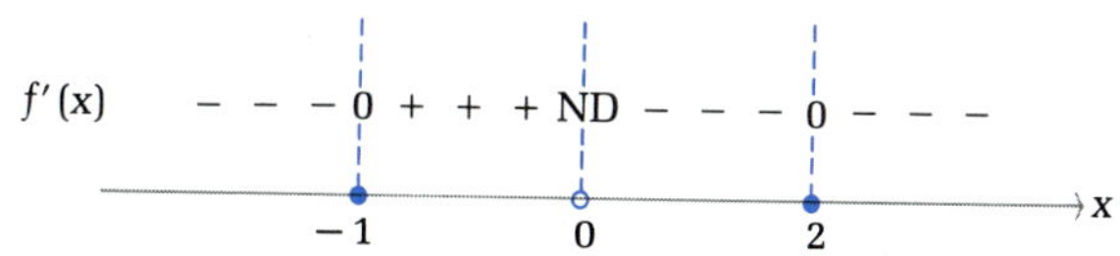

x	−2	−1	0	2	4
f(x)	2	1	2	1	0

12. 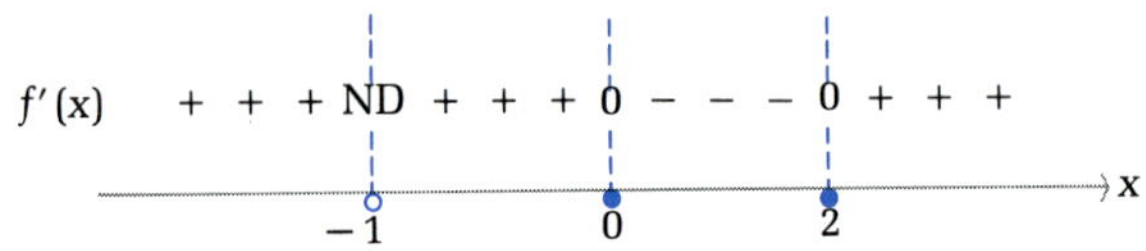

x	−2	−1	0	2	3
f(x)	−3	0	2	−1	0

B *Solve each inequality using a sign chart. Express answers in inequality and interval notation.*

13. $x^2 - x - 12 < 0$
14. $x^2 - 2x - 8 < 0$
15. $x^2 + 21 > 10x$
16. $x^2 + 7x > -10$
17. $\dfrac{x^2 + 5x}{x - 3} > 0$
18. $\dfrac{x - 4}{x^2 + 2x} < 0$

Find the intervals where f(x) is increasing, the intervals where f(x) is decreasing, and the local extrema.

19. $f(x) = x^2 - 16x + 12$
20. $f(x) = x^2 + 6x + 7$
21. $f(x) = 4 + 10x - x^2$
22. $f(x) = 5 + 8x - 2x^2$
23. $f(x) = 2x^3 + 4$
24. $f(x) = 2 - 3x^3$
25. $f(x) = 2 - 6x - 2x^3$
26. $f(x) = x^3 + 9x + 7$
27. $f(x) = x^3 - 12x + 8$
28. $f(x) = 3x - x^3$
29. $f(x) = x^3 - 3x^2 - 24x + 7$
30. $f(x) = x^3 + 3x^2 - 9x + 5$
31. $f(x) = 2x^2 - x^4$
32. $f(x) = x^4 - 8x^2 + 3$

Find the intervals where f(x) is increasing, the intervals where f(x) is decreasing, and sketch the graph. Add horizontal tangent lines.

33. $f(x) = 4 + 8x - x^2$
34. $f(x) = 2x^2 - 8x + 9$
35. $f(x) = x^3 - 3x + 1$
36. $f(x) = x^3 - 12x + 2$
37. $f(x) = 10 - 12x + 6x^2 - x^3$
38. $f(x) = x^3 + 3x^2 + 3x$

C *Find the critical values, the intervals where f(x) is increasing, the intervals where f(x) is decreasing, and the local extrema. Do not graph.*

39. $f(x) = \dfrac{x - 1}{x + 2}$
40. $f(x) = \dfrac{x + 2}{x - 3}$

41. $f(x) = x + \frac{4}{x}$

42. $f(x) = \frac{9}{x} + x$

43. $f(x) = 1 + \frac{1}{x} + \frac{1}{x^2}$

44. $f(x) = 3 - \frac{4}{x} - \frac{2}{x^2}$

45. $f(x) = \frac{x^2}{x - 2}$

46. $f(x) = \frac{x^2}{x + 1}$

47. $f(x) = x^4(x - 6)^2$

48. $f(x) = x^3(x - 5)^2$

49. $f(x) = 3(x - 2)^{2/3} + 4$

50. $f(x) = 6(4 - x)^{2/3} + 4$

51. $f(x) = 2\sqrt{x} - x, \quad x > 0$

52. $f(x) = x - 4\sqrt{x}, \quad x > 0$

Problems 53–56 require the use of a graphic calculator or a computer. Graph $f(x)$ and $f'(x)$ in the same viewing rectangle. (Use $[-5, 5]$ for both the x range and the y range.) Approximate the critical values of $f(x)$ to two decimal places, and find the intervals where $f(x)$ is increasing, the intervals where $f(x)$ is decreasing, and the local extrema.

53. $f(x) = x^4 - 2x^2 + 3x$

54. $f(x) = x^4 - x^2 - 4x$

55. $f(x) = x^4 - 3x^3 + 2x$

56. $f(x) = x^4 + 3x^3 - 3x$

APPLICATIONS

Business & Economics

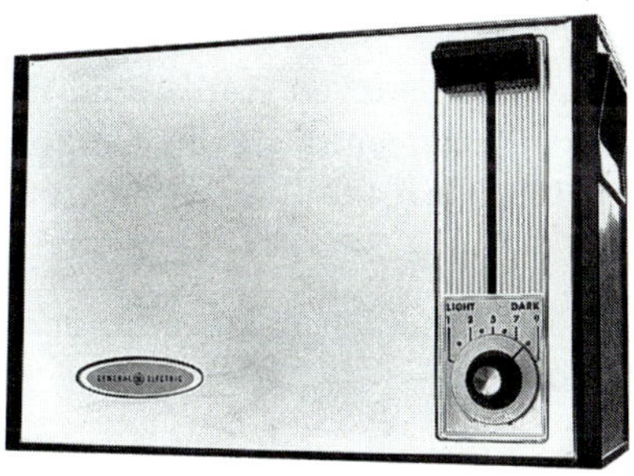

57. *Average cost.* A manufacturer has the following costs in producing x toasters in one day for $0 < x < 150$: fixed costs, \$320; unit production cost, \$20 per toaster; equipment maintenance and repairs, $x^2/20$ dollars. Thus, the cost of manufacturing x toasters in one day is given by

$$C(x) = \frac{x^2}{20} + 20x + 320 \qquad 0 < x < 150$$

and the average cost per toaster is given by

$$\overline{C}(x) = \frac{C(x)}{x} = \frac{x}{20} + 20 + \frac{320}{x} \qquad 0 < x < 150$$

Find the critical values for $\overline{C}(x)$, the intervals where the average cost per toaster is decreasing, the intervals where the average cost per toaster is increasing, and the local extrema. Do not graph.

58. *Average cost.* A manufacturer has the following costs in producing x blenders in one day for $0 < x < 200$: fixed costs, \$450; unit production cost, \$60 per blender; equipment maintenance and repairs, $x^2/18$ dollars.

(A) What is the average cost $\overline{C}(x)$ per blender if x blenders are produced in one day?

(B) Find the critical values for $\overline{C}(x)$, the intervals where the average cost per blender is decreasing, the intervals where the average cost per blender is increasing, and the local extrema. Do not graph.

59. *Marginal analysis.* Show that profit will be increasing over production intervals (a, b) for which marginal revenue is greater than marginal cost. [*Hint:* $P(x) = R(x) - C(x)$]

60. *Marginal analysis.* Show that profit will be decreasing over production intervals (a, b) for which marginal revenue is less than marginal cost.

Life Sciences

61. *Medicine.* A drug is injected into the bloodstream of a patient through the right arm. The concentration of the drug in the bloodstream of the left arm t hours after the injection is approximated by

$$C(t) = \frac{0.14t}{t^2 + 1} \qquad 0 < t < 24$$

Find the critical values for $C(t)$, the intervals where the concentration of the drug is increasing, the intervals where the concentration of the drug is decreasing, and the local extrema. Do not graph.

62. *Medicine.* The concentration $C(t)$, in milligrams per cubic centimeter, of a particular drug in a patient's bloodstream is given by

$$C(t) = \frac{0.16t}{t^2 + 4t + 4} \qquad 0 < t < 12$$

where t is the number of hours after the drug is taken orally. Find the critical values for $C(t)$, the intervals where the concentration of the drug is increasing, the intervals where the concentration of the drug is decreasing, and the local extrema. Do not graph.

Social Sciences

63. *Politics.* Public awareness of a Congressional candidate before and after a successful campaign was approximated by

$$P(t) = \frac{8.4t}{t^2 + 49} + 0.1 \qquad 0 < t < 24$$

where t is time (in months) after the campaign started and $P(t)$ is the fraction of people in the Congressional district who could recall the candidate's (and later, Congressman's) name. Find the critical values for $P(t)$, the time intervals where the fraction is increasing, the time intervals where the fraction is decreasing, and the local extrema. Do not graph.

SECTION 4-2 Second Derivative and Graphs

- CONCAVITY
- INFLECTION POINTS
- SECOND-DERIVATIVE TEST
- APPLICATION

In the preceding section, we saw that the derivative can be used to determine when a graph is rising and falling. Now we want to see what the *second derivative* (the derivative of the derivative) can tell us about the shape of a graph.

CONCAVITY

Consider the functions

$$f(x) = x^2 \qquad \text{and} \qquad g(x) = \sqrt{x}$$

for x in the interval $(0, \infty)$. Since

$$f'(x) = 2x > 0 \qquad \text{for } 0 < x < \infty$$

and

$$g'(x) = \frac{1}{2\sqrt{x}} > 0 \qquad \text{for } 0 < x < \infty$$

both functions are increasing on $(0, \infty)$.

Notice the different shapes of the graphs of f and g, shown in Figure 5. Even though the graph of each function is rising and each graph starts at (0, 0) and goes through (1, 1), the graphs are quite dissimilar. The graph of f opens upward, while the graph of g opens downward. We say that the graph of f is *concave upward*, and the graph of g is *concave downward*. It will help us draw a graph if we can determine the concavity of the graph before we draw it. How can we find a mathematical formulation of concavity?

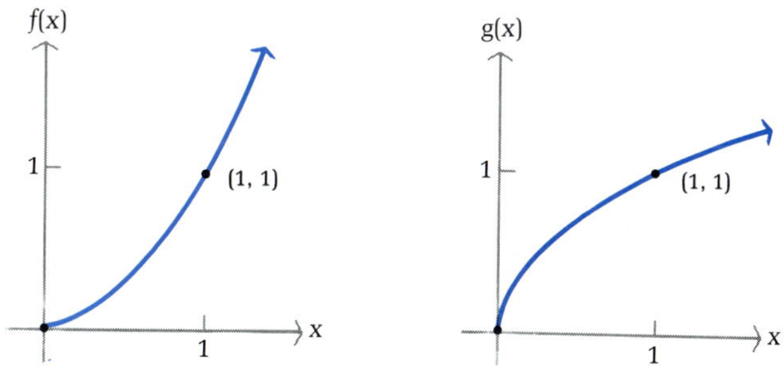

(A) $f(x) = x^2$ (B) $g(x) = \sqrt{x}$

FIGURE 5

It will be instructive to examine the slopes of f and g at various points on their graphs (see Figure 6). We can make two observations about each graph. Looking at the graph of f in Figure 6A, we see that $f'(x)$ (the slope of the tangent line) is *increasing* and that the graph lies *above* each tangent line. Looking at Figure 6B, we see that $g'(x)$ is *decreasing* and that the graph lies *below* each tangent line.

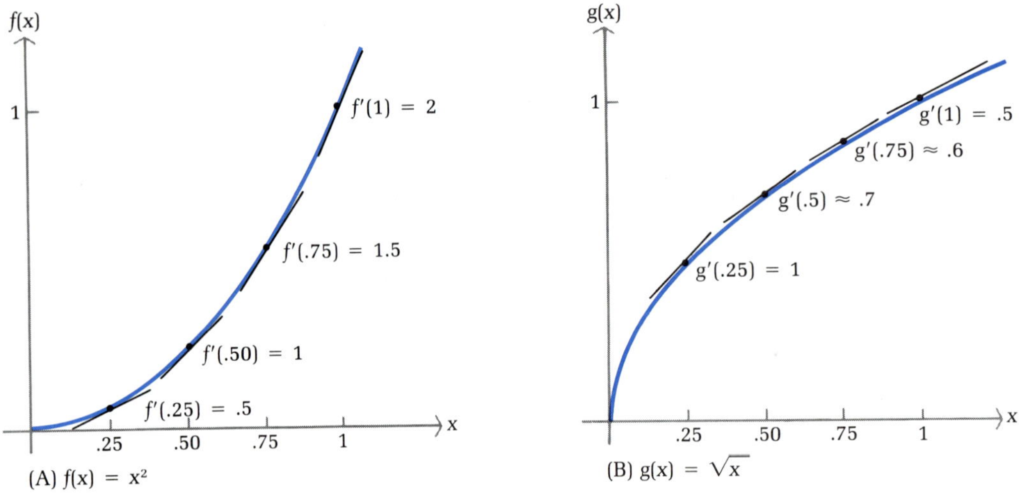

FIGURE 6

With these ideas in mind, we state the general definition of concavity:

The graph of a function f is concave upward on the interval (a, b) if $f'(x)$ is *increasing* on (a, b) and is concave downward on the interval (a, b) if $f'(x)$ is *decreasing* on (a, b).

Geometrically, the graph is concave upward on (a, b) if it lies above its tangent lines in (a, b) and is concave downward on (a, b) if it lies below its tangent lines in (a, b).

How can we determine when $f'(x)$ is increasing or decreasing? In the preceding section, we used the derivative of a function to determine when that function is increasing or decreasing. Thus, to determine when the function $f'(x)$ is increasing or decreasing, we use the derivative of $f'(x)$. The derivative of the derivative of a function is called the *second derivative* of the function. Various notations for the second derivative are given in the following box:

Second Derivative

For $y = f(x)$, the **second derivative** of f, provided it exists, is

$$f''(x) = D_x f'(x)$$

Other notations for $f''(x)$ are

$$\frac{d^2y}{dx^2} \qquad y'' \qquad D_x^2 f(x)$$

Returning to the functions f and g discussed at the beginning of this section, we have

$$f(x) = x^2 \qquad g(x) = \sqrt{x} = x^{1/2}$$

$$f'(x) = 2x \qquad g'(x) = \frac{1}{2} x^{-1/2} = \frac{1}{2\sqrt{x}}$$

$$f''(x) = D_x\, 2x = 2 \qquad g''(x) = D_x \frac{1}{2} x^{-1/2} = -\frac{1}{4} x^{-3/2} = -\frac{1}{4\sqrt{x^3}}$$

For $x > 0$, we see that $f''(x) > 0$; thus, $f'(x)$ is increasing and the graph of f is concave upward (see Fig. 6A). For $x > 0$, we also see that $g''(x) < 0$; thus, $g'(x)$ is decreasing and the graph of g is concave downward (see Fig. 6B). These ideas are summarized in the following box:

Concavity

For the interval (a, b):

$f''(x)$	$f'(x)$	GRAPH OF $y = f(x)$	EXAMPLE
+	Increasing	Concave upward	
−	Decreasing	Concave downward	

Be careful not to confuse concavity with falling and rising. As Figure 7 (on the next page) illustrates, a graph that is concave upward on an interval may be falling, rising, or both falling and rising on that interval. A similar statement holds for a graph that is concave downward.

$f''(x) > 0$ over (a, b)
Concave upward

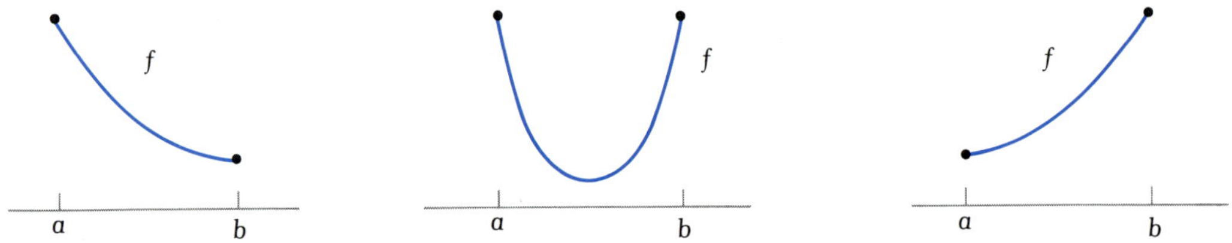

(A) $f'(x)$ is negative and increasing. Graph of f is falling.

(B) $f'(x)$ increases from negative to positive. Graph of f falls, then rises.

(C) $f'(x)$ is positive and increasing. Graph of f is rising.

$f''(x) < 0$ over (a, b)
Concave downward

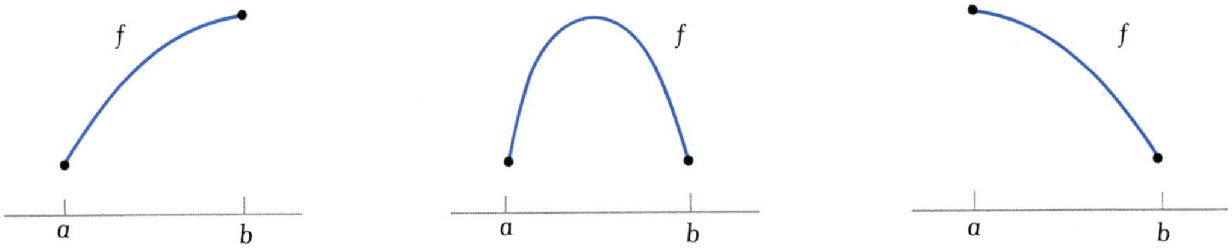

(D) $f'(x)$ is positive and decreasing. Graph of f is rising.

(E) $f'(x)$ decreases from positive to negative. Graph of f rises, then falls.

(F) $f'(x)$ is negative and decreasing. Graph of f is falling.

FIGURE 7
Concavity

◆ EXAMPLE 5 Let $f(x) = x^3$. Find the intervals where the graph of f is concave upward and the intervals where the graph of f is concave downward. Sketch a graph of f.

Solution To determine concavity, we must determine the sign of $f''(x)$.

$$f(x) = x^3 \qquad f'(x) = 3x^2 \qquad f''(x) = 6x$$

Sign chart for $f''(x) = 6x$ (partition number is 0):

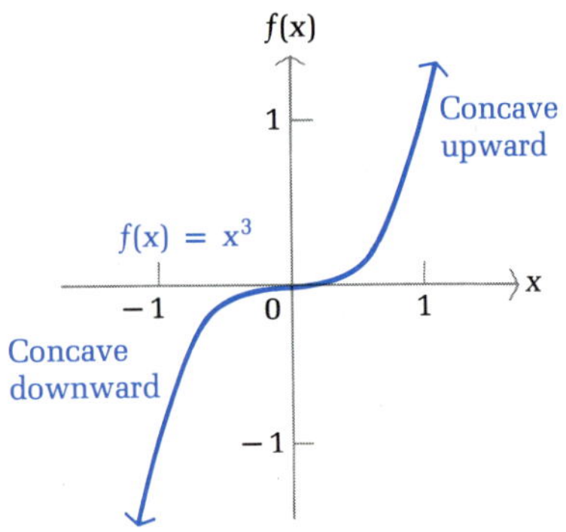

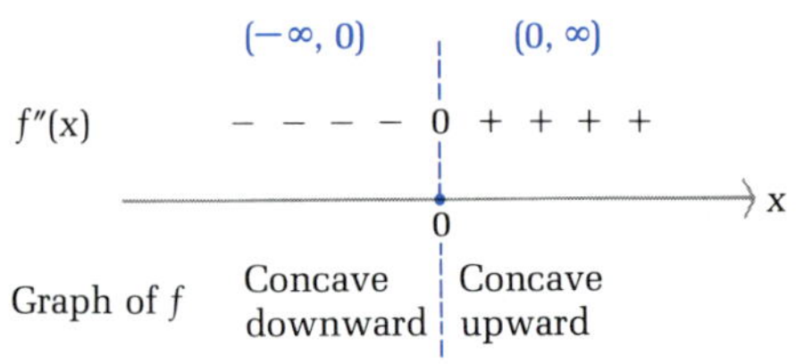

Test Numbers

x	$f''(x)$	
-1	-6	$(-)$
1	6	$(+)$

Thus, the graph of f is concave downward on $(-\infty, 0)$ and concave upward on $(0, \infty)$. The graph of f (without going through other graphing details) is shown in the figure in the margin. ◆

PROBLEM 5 Repeat Example 5 for $f(x) = 1 - x^3$. ◆

The graph in Example 5 changes from concave downward to concave upward at the point (0, 0). This point is called an *inflection point*.

◆ INFLECTION POINTS

In general, an **inflection point** is a point on the graph of a function where the concavity changes (from upward to downward or from downward to upward). In order for the concavity to change at a point, $f''(x)$ must change sign at that point. Reasoning as we did in the previous section, we conclude that the inflection points must occur at points where $f''(x) = 0$ or $f''(x)$ does not exist [but $f(x)$ must exist]. Figure 8 illustrates several typical cases.

If $f'(c)$ exists and $f''(x)$ changes sign at $x = c$, then the tangent line at an inflection point $(c, f(c))$ will always lie below the graph on the side that is concave upward and above the graph on the side that is concave downward (see Figs. 8A, B, and C).

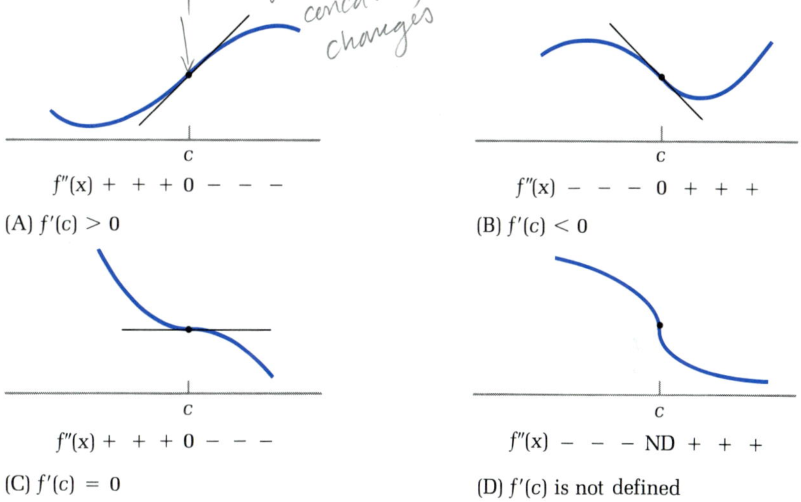

FIGURE 8
Inflection points

◆ EXAMPLE 6 Find the inflection points of $f(x) = x^3 - 6x^2 + 9x + 1$.

Solution Since inflection points occur at values of x where $f''(x)$ changes sign, we construct a sign chart for $f''(x)$.

$$f(x) = x^3 - 6x^2 + 9x + 1$$
$$f'(x) = 3x^2 - 12x + 9$$
$$f''(x) = 6x - 12 = 6(x - 2)$$

Sign chart for $f''(x) = 6(x - 2)$ (partition number is 2):

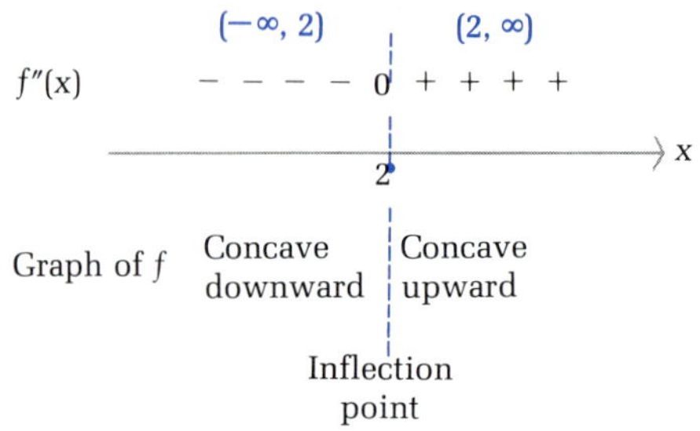

Test Numbers

x	$f''(x)$	
1	−6	(−)
3	6	(+)

From the sign chart, we see that the graph of f has an inflection point at $x = 2$. The graph of f is shown in the figure. (See Example 4 in Section 4-1.)

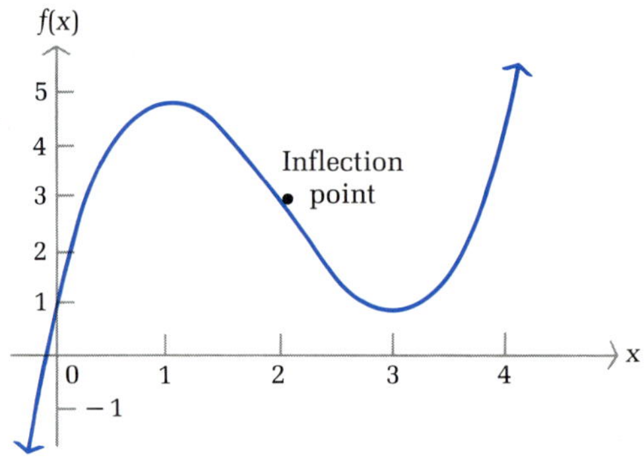

◆

PROBLEM 6 Find the inflection points of $f(x) = x^3 - 9x^2 + 24x - 10$. (See Problem 4 in Section 4-1 for the graph of f.)

◆

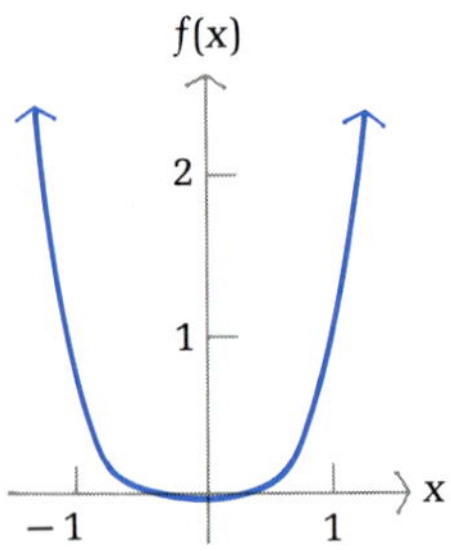

FIGURE 9
$f(x) = x^4$

It is important to remember that the values of x where $f''(x) = 0$ [or $f''(x)$ does not exist] are only candidates for inflection points. The second derivative must change sign at $x = c$ in order for the graph of f to have an inflection point at $x = c$. For example, consider

$$f(x) = x^4 \qquad f'(x) = 4x^3 \qquad f''(x) = 12x^2$$

The second derivative is 0 at $x = 0$, but $f''(x) > 0$ for all other values of x. Since $f''(x)$ does not change sign at $x = 0$, the graph of f does not have an inflection point at $x = 0$, as illustrated in Figure 9.

◆ SECOND-DERIVATIVE TEST

Now we want to see how the second derivative can be used to find local extrema. Suppose f is a function satisfying $f'(c) = 0$ and $f''(c) > 0$. First, note that if $f''(c) > 0$, then it follows from the properties of limits* that $f''(x) > 0$ in some

* Actually, we are assuming that $f''(x)$ is continuous in an interval containing c. It is very unlikely that we will encounter a function for which $f''(c)$ exists, but $f''(x)$ is not continuous in an interval containing c.

interval (m, n) containing c. Thus, the graph of f must be concave upward in this interval. But this implies that $f'(x)$ is increasing in this interval. Since $f'(c) = 0$, $f'(x)$ must change from negative to positive at $x = c$ and $f(c)$ is a local minimum (see Figure 10). Reasoning in the same fashion, we conclude that if $f'(c) = 0$ and $f''(c) < 0$, then $f(c)$ is a local maximum. Of course, it is possible that both $f'(c) = 0$ and $f''(c) = 0$. In this case, the second derivative cannot be used to determine the shape of the graph around $x = c$; $f(c)$ may be a local minimum, a local maximum, or neither.

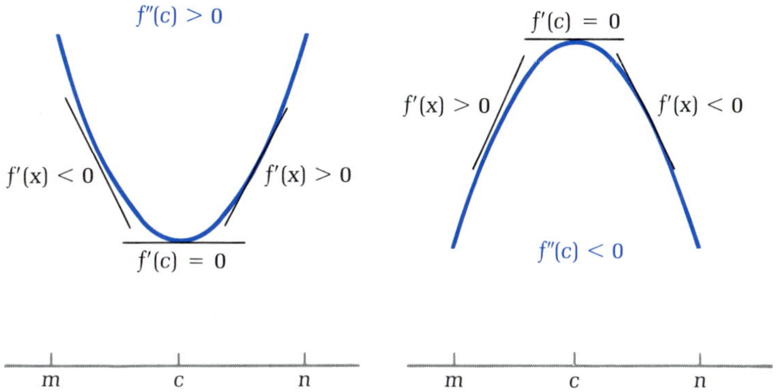

(A) $f'(c) = 0$ and $f''(c) > 0$ implies $f(c)$ is a local minimum.

(B) $f'(c) = 0$ and $f''(c) < 0$ implies $f(c)$ is a local maximum.

FIGURE 10
The second derivative and local extrema

The sign of the second derivative thus provides a simple test for identifying local maxima and minima. This test is most useful when we do not want to draw the graph of the function. If we are interested in drawing the graph and have already constructed the sign chart for $f'(x)$, then the first-derivative test can be used to identify the local extrema.

Second-Derivative Test for Local Maxima and Minima

Let c be a critical value for $f(x)$.

$f'(c)$	$f''(c)$	GRAPH OF f IS	$f(c)$	EXAMPLE
0	+	Concave upward	Local minimum	
0	−	Concave downward	Local maximum	
0	0	?	Test fails	

The first-derivative test must be used whenever $f''(c) = 0$ or $f''(c)$ does not exist.

◆ EXAMPLE 7 Find the local maxima and minima for each function. Use the second-derivative test when it applies.

(A) $f(x) = x^3 - 6x^2 + 9x + 1$ (B) $f(x) = \frac{1}{6}x^6 - 4x^5 + 25x^4$

Solutions (A) Take first and second derivatives and find critical values:

$$f(x) = x^3 - 6x^2 + 9x + 1$$
$$f'(x) = 3x^2 - 12x + 9 = 3(x - 1)(x - 3)$$
$$f''(x) = 6x - 12 = 6(x - 2)$$

Critical values are $x = 1$ and $x = 3$.

$f''(1) = -6 < 0$ f has a local maximum at $x = 1$.
$f''(3) = 6 > 0$ f has a local minimum at $x = 3$.

(B)
$$f(x) = \frac{1}{6}x^6 - 4x^5 + 25x^4$$
$$f'(x) = x^5 - 20x^4 + 100x^3 = x^3(x - 10)^2$$
$$f''(x) = 5x^4 - 80x^3 + 300x^2$$

Critical values are $x = 0$ and $x = 10$.

$f''(0) = 0$ The second-derivative test fails at both critical values, so
$f''(10) = 0$ the first-derivative test must be used.

Sign chart for $f'(x) = x^3(x - 10)^2$ (partition numbers are 0 and 10):

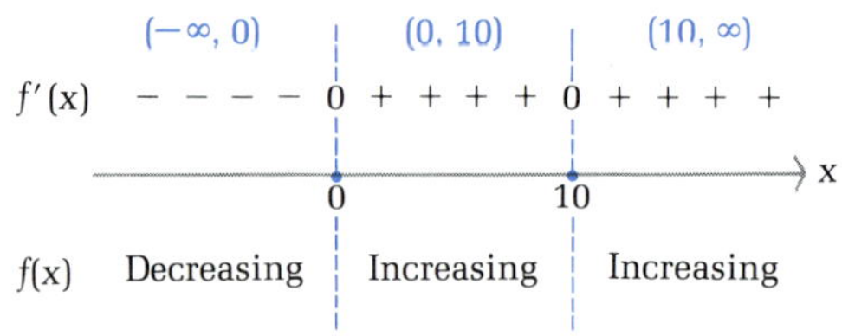

Test Numbers

x	f′(x)	
−1	−121	(−)
1	81	(+)
11	1,331	(+)

From the sign chart, we see that $f(x)$ has a local minimum at $x = 0$ and does not have a local extremum at $x = 10$. ◆

PROBLEM 7 Find the local maxima and minima for each function. Use the second-derivative test when it applies.

(A) $f(x) = x^3 - 9x^2 + 24x - 10$ (B) $f(x) = 10x^6 - 24x^5 + 15x^4$ ◆

A common error is to assume that $f''(c) = 0$ implies that $f(c)$ is not a local extreme point. As Example 7B illustrates, if $f''(c) = 0$, then $f(c)$ may or may not be a local extreme point. **The first-derivative test *must* be used whenever $f''(c) = 0$ or $f''(c)$ does not exist.**

◆ APPLICATION

◆ EXAMPLE 8

Maximum Rate of Change

Using past records, a company estimates that it will sell $N(x)$ units of a product after spending $x thousand on advertising, as given by

$$N(x) = 2{,}000 - 2x^3 + 60x^2 - 450x \qquad 5 \leq x \leq 15$$

When is the rate of change of sales per unit (thousand dollars) change in advertising increasing? Decreasing? What is the maximum rate of change? Graph N and N' on the same coordinate system and interpret.

Solution

The rate of change of sales per unit (thousand dollars) change in advertising expenditure is

$$N'(x) = -6x^2 + 120x - 450 = -6(x - 5)(x - 15)$$

To determine when this rate is increasing and decreasing, we find $N''(x)$, the derivative of $N'(x)$:

$$N''(x) = -12x + 120 = 12(10 - x)$$

The information obtained by analyzing the signs of $N'(x)$ and $N''(x)$ is summarized in the table (sign charts are omitted).

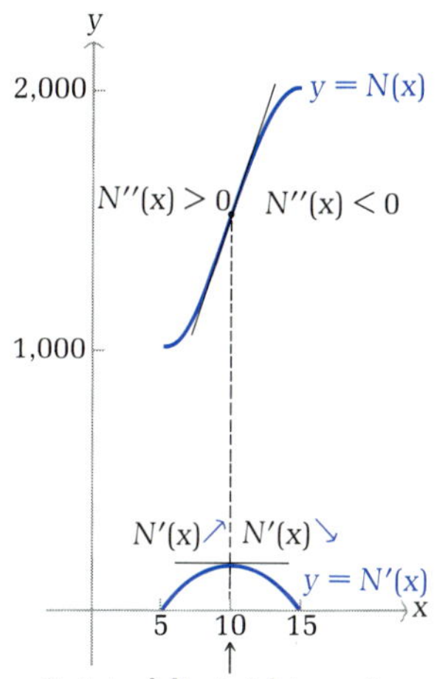

Point of diminishing returns

x	$N''(x)$	$N'(x)$	$N'(x)$	$N(x)$
$5 < x < 10$	+	+	Increasing	Increasing, concave upward
$x = 10$	0	+	Local maximum	Inflection point
$10 < x < 15$	−	+	Decreasing	Increasing, concave downward

Thus, we see that $N'(x)$, the rate of change of sales, is increasing on (5, 10) and decreasing on (10, 15). Both N and N' are graphed in the figure in the margin. An examination of the graph of $N'(x)$ shows that the maximum rate of change is $N'(10) = 150$. Notice that $N'(x)$ has a local maximum and $N(x)$ has an inflection point at $x = 10$. This value of x is referred to as the **point of diminishing returns**, since the rate of change of sales begins to decrease at this point. ◆

PROBLEM 8

Repeat Example 8 for $N(x) = 5{,}000 - x^3 + 60x^2 - 900x$, $10 \leq x \leq 30$. ◆

Answers to Matched Problems

5. Concave upward on $(-\infty, 0)$; concave downward on $(0, \infty)$

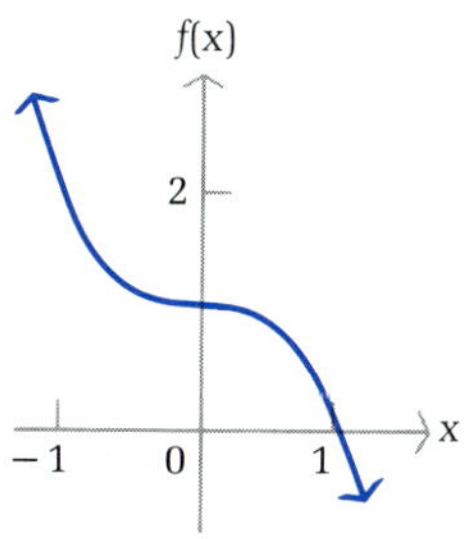

6. Inflection point at $x = 3$
7. (A) $f(2)$ is a local maximum; $f(4)$ is a local minimum
   (B) $f(0)$ is a local minimum; no local extremum at $x = 1$
8. $N'(x)$ is increasing on (10, 20), decreasing on (20, 30); maximum rate of change is $N'(20) = 300$; $x = 20$ is point of diminishing returns

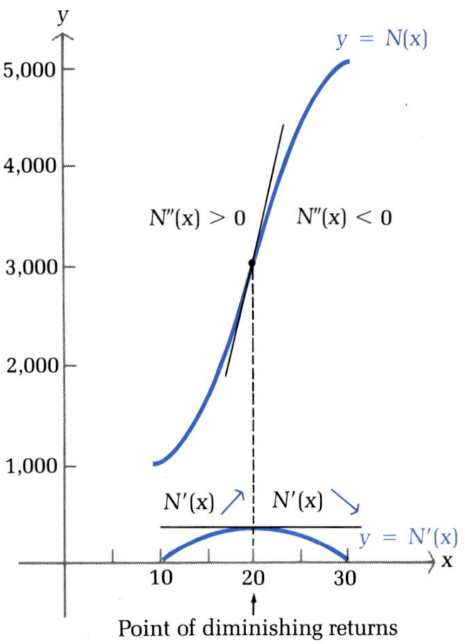

EXERCISE 4-2

A *Problems 1–4 refer to the following graph of $y = f(x)$:*

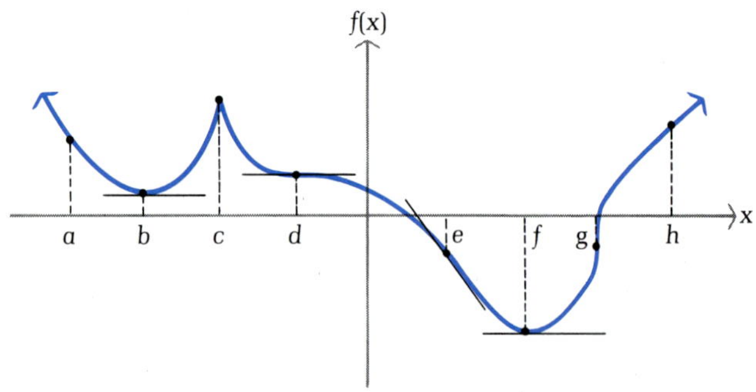

1. Identify intervals over which the graph of f is concave upward.
2. Identify intervals over which the graph of f is concave downward.

3. Identify inflection points.
4. Identify local extrema.

In Problems 5–10, describe the graph of f at the given point relative to the existence of a local maximum or minimum with one of the following phrases: "Local maximum," "Local minimum," "Neither," or "Unable to determine from the given information." Assume that f(x) is continuous on $(-\infty, \infty)$.

5. $(2, f(2))$ if $f'(2) = 0$ and $f''(2) > 0$
6. $(4, f(4))$ if $f'(4) = 1$ and $f''(4) < 0$
7. $(-3, f(-3))$ if $f'(-3) = 0$ and $f''(-3) = 0$
8. $(-1, f(-1))$ if $f'(-1) = 0$ and $f''(-1) < 0$
9. $(6, f(6))$ if $f'(6) = 1$ and $f''(6)$ does not exist
10. $(5, f(5))$ if $f'(5) = 0$ and $f''(5)$ does not exist

In Problems 11–14, f(x) is continuous on $(-\infty, \infty)$. Use the given information to sketch the graph of f.

11.

x	−4	−2	−1	0	2	4
f(x)	0	3	1.5	0	−1	−3

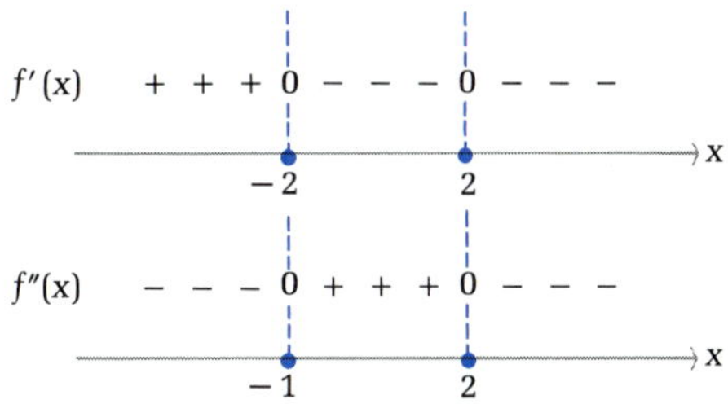

12.

x	−4	−2	−1	0	2	4
f(x)	0	−2	−1	0	1	3

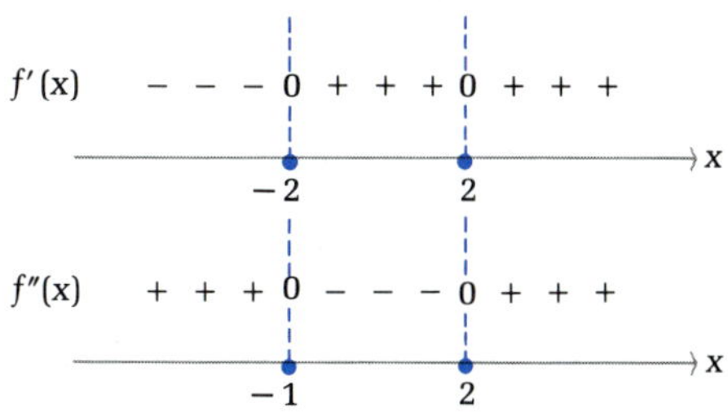

13.

x	-3	0	1	2	4	5
$f(x)$	-4	0	2	1	-1	0

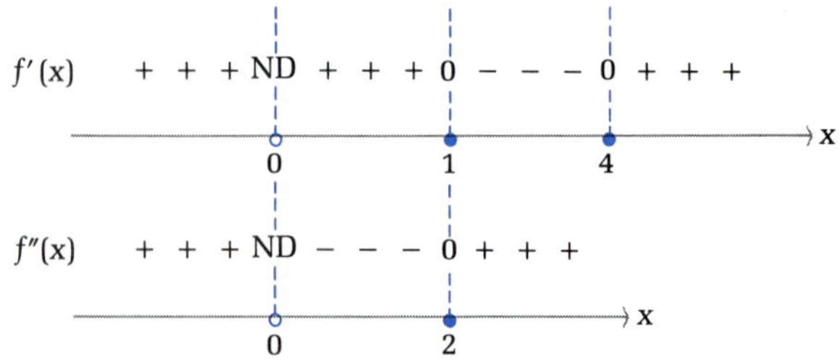

14.

x	-4	-2	0	2	4	6
$f(x)$	0	3	0	-2	0	3

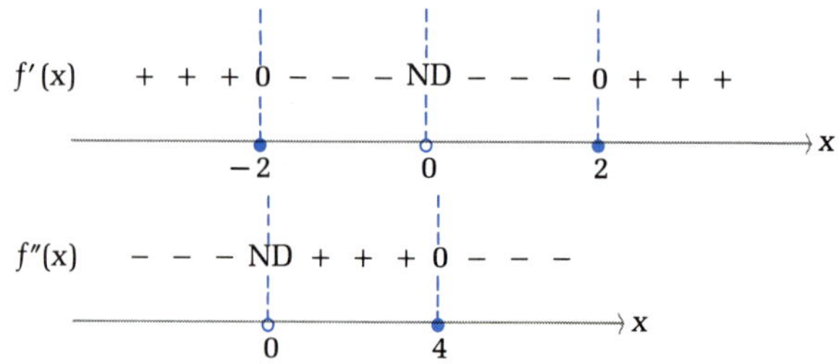

Find the indicated derivative for each function.

15. $f''(x)$ for $f(x) = x^3 - 2x^2 - 1$
16. $g''(x)$ for $g(x) = x^4 - 3x^2 + 5$
17. d^2y/dx^2 for $y = 2x^5 - 3$
18. d^2y/dx^2 for $y = 3x^4 - 7x$
19. $D_x^2(1 - 2x + x^3)$
20. $D_x^2(3x^2 - x^3)$
21. y'' for $y = (x^2 - 1)^3$
22. y'' for $y = (x^2 + 4)^4$
23. $f''(x)$ for $f(x) = 3x^{-1} + 2x^{-2} + 5$
24. $f''(x)$ for $f(x) = x^2 - x^{1/3}$

B *Find all local maxima and minima using the second-derivative test whenever it applies (do not graph). If the second-derivative test fails, use the first-derivative test.*

25. $f(x) = 2x^2 - 8x + 6$
26. $f(x) = 6x - x^2 + 4$
27. $f(x) = 2x^3 - 3x^2 - 12x - 5$
28. $f(x) = 2x^3 + 3x^2 - 12x - 1$
29. $f(x) = 3 - x^3 + 3x^2 - 3x$
30. $f(x) = x^3 + 6x^2 + 12x + 2$

31. $f(x) = x^4 - 8x^2 + 10$
32. $f(x) = x^4 - 18x^2 + 50$
33. $f(x) = x^6 + 3x^4 + 2$
34. $f(x) = 4 - x^6 - 6x^4$
35. $f(x) = x + \dfrac{16}{x}$
36. $f(x) = x + \dfrac{25}{x}$

Find the intervals where the graph of f is concave upward, the intervals where the graph is concave downward, and the inflection points.

37. $f(x) = x^2 - 4x + 5$
38. $f(x) = 9 + 3x - 4x^2$
39. $f(x) = x^3 - 18x^2 + 10x - 11$
40. $f(x) = x^3 + 24x^2 + 15x - 12$
41. $f(x) = x^4 - 24x^2 + 10x - 5$
42. $f(x) = x^4 + 6x^2 + 9x + 11$
43. $f(x) = -x^4 + 4x^3 + 3x + 7$
44. $f(x) = -x^4 - 2x^3 + 12x^2 + 15$

Find local maxima, local minima, and inflection points. Sketch the graph of each function. Include tangent lines at each local extreme point and inflection point.

45. $f(x) = x^3 - 6x^2 + 16$
46. $f(x) = x^3 - 9x^2 + 15x + 10$
47. $f(x) = x^3 + x + 2$
48. $f(x) = 1 - 3x - x^3$
49. $f(x) = (2 - x)^3 + 1$
50. $f(x) = (1 + x)^3 - 1$
51. $f(x) = x^3 - 12x$
52. $f(x) = 27x - x^3$

C *Find the inflection points. Do not graph.*

53. $f(x) = \dfrac{1}{x^2 + 12}$
54. $f(x) = \dfrac{x^2}{x^2 + 12}$
55. $f(x) = \dfrac{x}{x^2 + 12}$
56. $f(x) = \dfrac{x^3}{x^2 + 12}$

Problems 57–60 require the use of a graphic calculator or a computer. Graph f(x) and f″(x) in the same viewing rectangle. (Use [−5, 5] for the x range and [−15, 15] for the y range.) Approximate the inflection points of f to two decimal places. Find the intervals where the graph of f is concave upward and the intervals where the graph is concave downward.

57. $f(x) = x^5 + 2x^4 + 4x^2 - 5$
58. $f(x) = x^5 - 3x^4 + x^3 - x^2 + 10$
59. $f(x) = x^5 - 3x^4 - x^3 + 7x^2 - 2$
60. $f(x) = x^5 - 2x^4 - 3x^3 + 4x^2 + 4x + 5$

APPLICATIONS

Business & Economics

61. Revenue. The marketing research department for a computer company used a large city to test market their new product. They found that the relationship between price p (dollars per unit) and the demand x (units per

week) was given approximately by

$$p = 1{,}296 - 0.12x^2 \qquad 0 < x < 80$$

Thus, the weekly revenue can be approximated by

$$R(x) = xp = 1{,}296x - 0.12x^3 \qquad 0 < x < 80$$

(A) Find the local extrema for the revenue function.
(B) Over which intervals is the graph of the revenue function concave upward? Concave downward?

62. *Profit.* Suppose the cost equation for the company in Problem 61 is

$$C(x) = 830 + 396x$$

(A) Find the local extrema for the profit function.
(B) Over which intervals is the graph of the profit function concave upward? Concave downward?

63. *Advertising.* A company estimates that it will sell $N(x)$ units of a product after spending \$x thousand on advertising, as given by

$$N(x) = -3x^3 + 225x^2 - 3{,}600x + 17{,}000 \qquad 10 \leq x \leq 40$$

(A) When is the rate of change of sales $N'(x)$ increasing? Decreasing?
(B) Find the inflection points for the graph of N.
(C) Graph N and N' on the same coordinate system.
(D) What is the maximum rate of change of sales?

64. *Advertising.* A company estimates that it will sell $N(x)$ units of a product after spending \$x thousand on advertising, as given by

$$N(x) = -2x^3 + 90x^2 - 750x + 2{,}000 \qquad 5 \leq x \leq 25$$

(A) When is the rate of change of sales $N'(x)$ increasing? Decreasing?
(B) Find the inflection points for the graph of N.
(C) Graph N and N' on the same coordinate system.
(D) What is the maximum rate of change of sales?

Life Sciences

65. *Population growth—bacteria.* A drug that stimulates reproduction is introduced into a colony of bacteria. After t minutes, the number of bacteria is given approximately by

$$N(t) = 1{,}000 + 30t^2 - t^3 \qquad 0 \leq t \leq 20$$

(A) When is the rate of growth $N'(t)$ increasing? Decreasing?
(B) Find the inflection points for the graph of N.
(C) Sketch the graphs of N and N' on the same coordinate system.
(D) What is the maximum rate of growth?

66. *Drug sensitivity.* One hour after x milligrams of a particular drug are given to a person, the change in body temperature $T(x)$, in degrees Fahrenheit, is given by

$$T(x) = x^2\left(1 - \frac{x}{9}\right) \qquad 0 \le x \le 6$$

The rate at which $T(x)$ changes with respect to the size of the dosage x, $T'(x)$, is called the *sensitivity* of the body to the dosage.

(A) When is $T'(x)$ increasing? Decreasing?
(B) Where does the graph of T have inflection points?
(C) Sketch the graphs of T and T' on the same coordinate system.
(D) What is the maximum value of $T'(x)$?

Social Sciences

67. *Learning.* The time T (in minutes) it takes a person to learn a list of length n is

$$T(n) = \tfrac{2}{25}n^3 - \tfrac{6}{5}n^2 + 6n \qquad n \ge 0$$

(A) When is the rate of change of T with respect to the length of the list increasing? Decreasing?
(B) Where does the graph of T have inflection points? Graph T and T' on the same coordinate system.
(C) What is the minimum value of $T'(n)$?

SECTION 4-3 Curve Sketching Techniques: Unified and Extended

- ASYMPTOTES
- GRAPHING STRATEGY
- USING THE STRATEGY
- APPLICATION

In this section we will apply, in a systematic way, all the graphing concepts discussed in Sections 4-1 and 4-2. Before outlining a graphing strategy and considering the graphs of specific functions, we need to discuss in more detail the concept of *asymptotes*, which we introduced in Section 3-2.

ASYMPTOTES

To refresh your memory concerning horizontal and vertical asymptotes, refer to Figure 11 on the next page. The lines $y = 2$ and $y = -2$ are horizontal asymptotes, since $\lim_{x\to\infty} f(x) = 2$ and $\lim_{x\to-\infty} f(x) = -2$. The line $x = -4$ is a vertical asymptote, since $\lim_{x\to-4} f(x) = \infty$. The line $x = 4$ is also a vertical asymptote, since $\lim_{x\to4^-} f(x) = -\infty$ and $\lim_{x\to4^+} f(x) = \infty$.

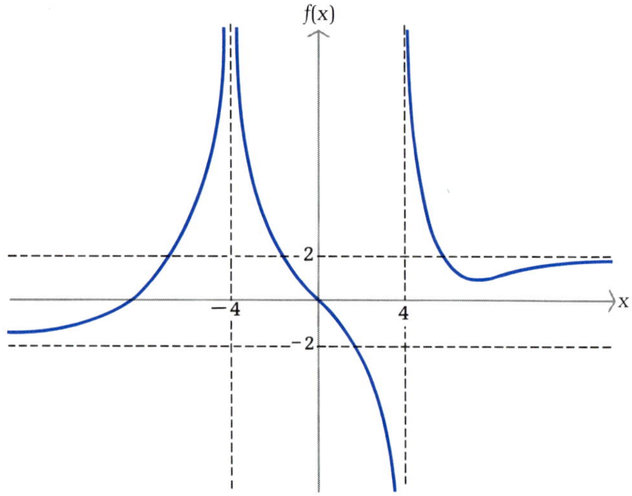

FIGURE 11

We restate the definitions of horizontal and vertical asymptotes in the box for convenient reference.

Horizontal and Vertical Asymptotes

A line $y = b$ is a **horizontal asymptote** for the graph of $y = f(x)$ if

$$\lim_{x \to -\infty} f(x) = b \quad \text{or} \quad \lim_{x \to \infty} f(x) = b$$

A line $x = a$ is a **vertical asymptote** for the graph of $y = f(x)$ if

$$\lim_{x \to a^-} f(x) = \infty \text{ (or } -\infty\text{)}, \lim_{x \to a^+} f(x) = \infty \text{ (or } -\infty\text{)}, \text{ or } \lim_{x \to a} f(x) = \infty \text{ (or } -\infty\text{)}$$

We are interested in locating any horizontal and vertical asymptotes as an aid to graphing a function. Consequently, we are interested in locating horizontal and vertical asymptotes from an equation that defines a function rather than from its graph. Consider the rational function given by

$$f(x) = \frac{2x - 3}{x - 1}$$

To locate horizontal asymptotes, we use Theorem 2 from Section 3-2 to compute $\lim_{x \to -\infty} f(x)$ and $\lim_{x \to \infty} f(x)$ as follows:

$$\lim_{x \to -\infty} \frac{2x - 3}{x - 1} = \lim_{x \to -\infty} \frac{x\left(2 - \frac{3}{x}\right)}{x\left(1 - \frac{1}{x}\right)} = \lim_{x \to -\infty} \frac{2 - \frac{3}{x}}{1 - \frac{1}{x}} = \frac{2 - 0}{1 - 0} = 2$$

$$\lim_{x\to\infty} \frac{2x-3}{x-1} = \lim_{x\to\infty} \frac{x\left(2-\frac{3}{x}\right)}{x\left(1-\frac{1}{x}\right)} = \lim_{x\to\infty} \frac{2-\frac{3}{x}}{1-\frac{1}{x}} = \frac{2-0}{1-0} = 2$$

Thus, $y = 2$ is the only horizontal asymptote, and it could have been obtained using either limit. In general, **a rational function has at most one horizontal asymptote.** (Notice that this implies that the function graphed in Figure 11 is not a rational function, since it has two horizontal asymptotes.)

Since the given function f is continuous for all values of x except those that make the denominator 0, the only candidate for a vertical asymptote is $x = 1$. How can we tell if $x = 1$ is a vertical asymptote? We could try to compute the limits in the definition of a vertical asymptote above—a difficult task in many problems. But Theorem 3, which we state without proof, eliminates the need to compute any of the limits in the definition, and it applies to f, as well as a wide variety of other functions.

THEOREM 3

On Vertical Asymptotes

Let $f(x) = N(x)/D(x)$, where both N and D are continuous at $x = c$. If at $x = c$ the denominator $D(x)$ is 0 and the numerator $N(x)$ is not 0, then the line $x = c$ is a vertical asymptote for the graph of f.

[*Note:* Since a rational function is the ratio of two polynomial functions, and polynomial functions are continuous for all real numbers, this theorem includes rational functions as a special case.]

Returning to the example above, $x = 1$ is a vertical asymptote for the graph of $f(x) = (2x - 3)/(x - 1)$, since the denominator is 0 at $x = 1$ and the numerator is not.

◆ EXAMPLE 9 Find horizontal and vertical asymptotes, if they exist, for:

(A) $f(x) = \dfrac{6x+5}{2x-4}$ (B) $f(x) = \dfrac{x}{x^2+1}$ (C) $f(x) = \dfrac{x^2-4}{x}$

Solutions (A) *Horizontal asymptotes:*

$$\lim_{x\to\infty} \frac{6x+5}{2x-4} = \lim_{x\to\infty} \frac{x\left(6+\frac{5}{x}\right)}{x\left(2-\frac{4}{x}\right)} = \frac{6}{2} = 3$$

The line $y = 3$ is a horizontal asymptote. [Since a rational function cannot have more than one horizontal asymptote, there is no need to compute $\lim_{x\to-\infty} f(x)$.]

Vertical asymptotes:

$$f(x) = \frac{6x + 5}{2x - 4} = \frac{6x + 5}{2(x - 2)}$$

Using Theorem 3, we search for values of x that make the denominator 0 without making the numerator 0 at the same time. The denominator is 0 for $x = 2$, and we see that the line $x = 2$ is a vertical asymptote, since the numerator is not 0 for $x = 2$.

(B) *Horizontal asymptotes:*

$$\lim_{x\to\infty} \frac{x}{x^2 + 1} = \lim_{x\to\infty} \frac{x}{x^2\left(1 + \frac{1}{x^2}\right)} = \lim_{x\to\infty} \left[\frac{1}{x} \cdot \frac{1}{\left(1 + \frac{1}{x^2}\right)}\right] = 0$$

The line $y = 0$ (the x axis) is a horizontal asymptote.

Vertical asymptotes: Since the denominator, $x^2 + 1$, is never 0, there are no vertical asymptotes.

(C) *Horizontal asymptotes:*

$$\lim_{x\to\infty} \frac{x^2 - 4}{x} = \lim_{x\to\infty} \frac{x^2\left(1 - \frac{4}{x^2}\right)}{x} = \lim_{x\to\infty} x\left(1 - \frac{4}{x^2}\right) = \infty$$

There are no horizontal asymptotes.

Vertical asymptotes: The line $x = 0$ (the y axis) is a vertical asymptote, since for $x = 0$, the denominator is 0 and the numerator is not 0. ◆

PROBLEM 9 Find horizontal and vertical asymptotes, if they exist, for:

(A) $f(x) = \frac{3x + 5}{x + 2}$ (B) $f(x) = \frac{x + 1}{x^2}$ (C) $f(x) = \frac{x^3}{x^2 + 4}$ ◆

◆ GRAPHING STRATEGY

We now have powerful tools to determine the shape of a graph of a function, even before we plot any points. We can accurately sketch the graphs of many functions using these tools and point-by-point plotting as needed (often, very little point-by-point plotting is necessary). We organize these tools in the graphing strategy summarized in the box.

A Graphing Strategy for $y = f(x)$

Omit any of the following steps if procedures involved appear to be too difficult or impossible (what may seem too difficult now, will become less so with a little practice).

Step 1. *Use $f(x)$:*

(A) Find the domain of f. [The domain of f is the set of all real numbers x that produce real values for $f(x)$.]

(B) Find intercepts. [The y intercept is $f(0)$, if it exists; the x intercepts are the solutions to $f(x) = 0$, if they exist.]

(C) Find asymptotes. [Find any horizontal asymptotes by calculating $\lim_{x \to \pm\infty} f(x)$. Find any vertical asymptotes by using Theorem 3.]

Step 2. *Use $f'(x)$:* Find any critical values for $f(x)$ and any partition numbers for $f'(x)$. [Remember, every critical value for $f(x)$ is also a partition number for $f'(x)$, but some partition numbers for $f'(x)$ may not be critical values for $f(x)$.] Construct a sign chart for $f'(x)$, determine the intervals where $f(x)$ is increasing and decreasing, and find local maxima and minima.

Step 3. *Use $f''(x)$:* Construct a sign chart for $f''(x)$, determine where the graph of f is concave upward and concave downward, and find any inflection points.

Step 4. *Sketch the graph of f:* Draw asymptotes and locate intercepts, local maxima and minima, and inflection points. Sketch in what you know from steps 1–3. In regions of uncertainty, use point-by-point plotting to complete the graph.

◆ USING THE STRATEGY

Some examples will illustrate the use of the graphing strategy.

◆ EXAMPLE 10 Graph $f(x) = x^4 - 2x^3$ using the graphing strategy.

Solution Step 1. *Use $f(x)$:* $f(x) = x^4 - 2x^3$

(A) Domain: All real x

(B) Intercepts:

y intercept: $f(0) = 0$

x intercepts:
$$\begin{aligned} f(x) &= 0 \\ x^4 - 2x^3 &= 0 \\ x^3(x - 2) &= 0 \\ x &= 0, 2 \end{aligned}$$

(C) Asymptotes: Since f is a polynomial, there are no horizontal or vertical asymptotes.

Step 2. *Use $f'(x)$:* $f'(x) = 4x^3 - 6x^2 = 4x^2(x - \frac{3}{2})$

Critical values: 0 and $\frac{3}{2}$

Partition numbers: 0 and $\frac{3}{2}$

Sign chart for $f'(x)$:

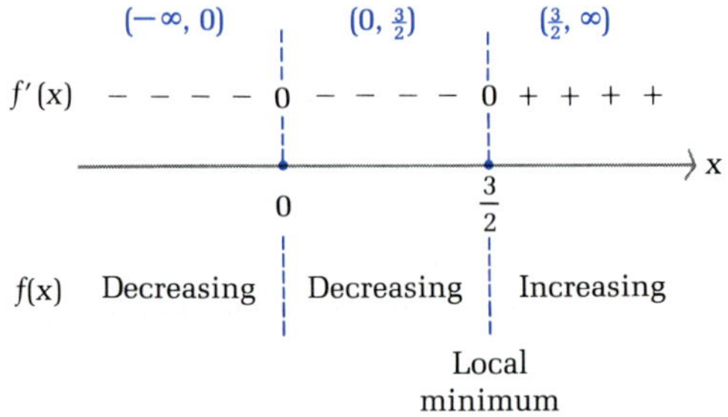

Test Numbers

x	$f'(x)$	
-1	-10	(−)
1	-2	(−)
2	8	(+)

Thus, $f(x)$ is decreasing on $(-\infty, \frac{3}{2})$, increasing on $(\frac{3}{2}, \infty)$, and has a local minimum at $x = \frac{3}{2}$.

Step 3. *Use $f''(x)$:* $f''(x) = 12x^2 - 12x = 12x(x - 1)$

Partition numbers for $f''(x)$: 0 and 1

Sign chart for $f''(x)$:

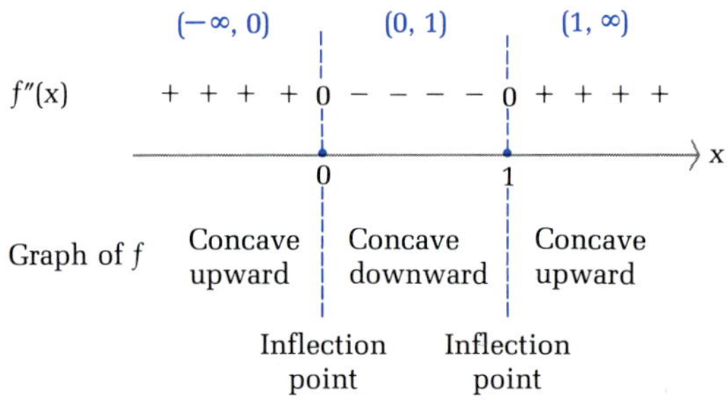

Test Numbers

x	$f''(x)$	
-1	24	(+)
$\frac{1}{2}$	-3	(−)
2	24	(+)

Thus, the graph of f is concave upward on $(-\infty, 0)$ and $(1, \infty)$, concave downward on $(0, 1)$, and has inflection points at $x = 0$ and $x = 1$.

Step 4. *Sketch the graph of f:*

x	$f(x)$
0	0
1	-1
$\frac{3}{2}$	$-\frac{27}{16}$
2	0

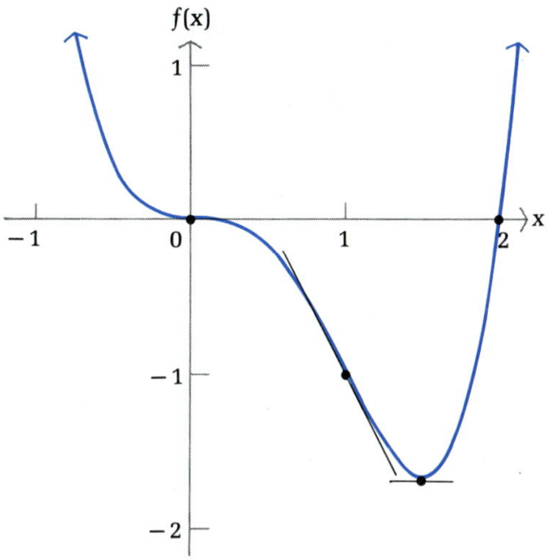

PROBLEM 10 Graph $f(x) = x^4 + 4x^3$ using the graphing strategy.

EXAMPLE 11 Graph $f(x) = \dfrac{x-1}{x-2}$ using the graphing strategy.

Solution Step 1. *Use f(x):* $f(x) = \dfrac{x-1}{x-2}$

(A) Domain: All real x, except $x = 2$

(B) Intercepts:

$$y \text{ intercept: } f(0) = \frac{0-1}{0-2} = \frac{1}{2}$$

x intercepts: Since a fraction is 0 when its numerator is 0 and the denominator is not 0, the x intercept is $x = 1$.

(C) Asymptotes:

$$\text{Horizontal asymptote: } \lim_{x\to\infty} \frac{x-1}{x-2} = \lim_{x\to\infty} \frac{x\left(1-\frac{1}{x}\right)}{x\left(1-\frac{2}{x}\right)} = 1$$

Thus, the line $y = 1$ is a horizontal asymptote.

Vertical asymptote: The denominator is 0 for $x = 2$, and the numerator is not 0 for this value. Therefore, the line $x = 2$ is a vertical asymptote.

Step 2. *Use f′(x):* $f'(x) = \dfrac{(x-2)(1)-(x-1)(1)}{(x-2)^2} = \dfrac{-1}{(x-2)^2}$

Critical values: None

Partition number: $x = 2$

Sign chart for $f'(x)$:

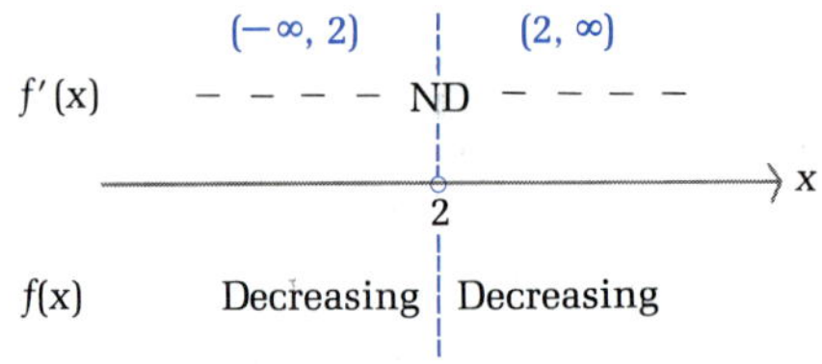

Test Numbers

x	$f'(x)$
1	−1 (−)
3	−1 (−)

Thus, $f(x)$ is decreasing on $(-\infty, 2)$ and $(2, \infty)$. There are no local extrema.

Step 3. *Use f″(x):* $f''(x) = \dfrac{2}{(x-2)^3}$

Partition number for $f''(x)$: 2

Sign chart for $f''(x)$:

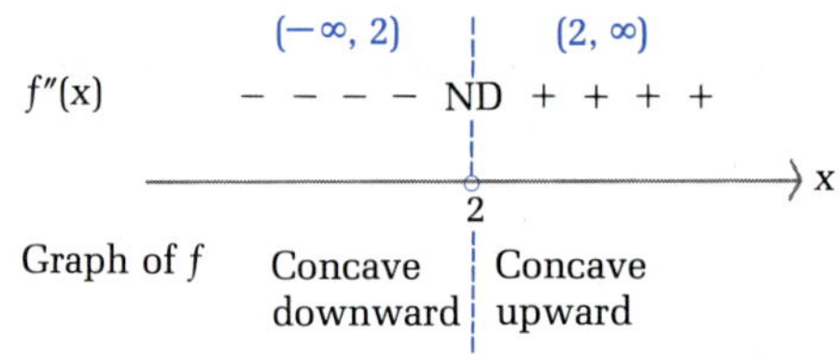

Test Numbers

x	$f''(x)$
1	−2 (−)
3	2 (+)

Thus, the graph of f is concave downward on $(-\infty, 2)$ and concave upward on $(2, \infty)$. Since $f(2)$ is not defined, there is no inflection point at $x = 2$, even though $f''(x)$ changes sign at $x = 2$.

Step 4. *Sketch a graph of f:* Insert intercepts and asymptotes, and plot a few additional points (for functions with asymptotes, plotting additional points is often helpful). Then sketch the graph.

x	$f(x)$
−2	$\frac{3}{4}$
0	$\frac{1}{2}$
1	0
$\frac{3}{2}$	−1
$\frac{5}{2}$	3
3	2
4	$\frac{3}{2}$

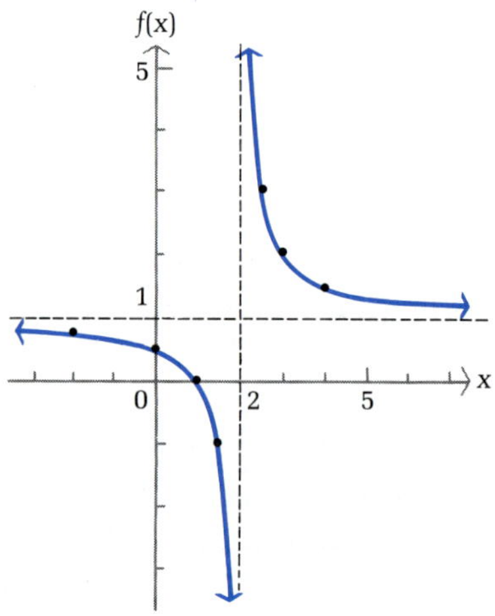

◆

PROBLEM 11 Graph $f(x) = \dfrac{2x}{1 - x}$ using the graphing strategy. ◆

◆ APPLICATION

◆ EXAMPLE 12 Given the cost function $C(x) = 5{,}000 + \frac{1}{2}x^2$, where x is the number of units produced, graph the average cost function and the marginal cost function on the same set of coordinate axes.

Average Cost

Solution The average cost function is

$$\overline{C}(x) = \frac{C(x)}{x} = \frac{5{,}000}{x} + \frac{x}{2} = \frac{10{,}000 + x^2}{2x}$$

Graph $\overline{C}$ using the graphing strategy.

Step 1. *Use $\overline{C}(x)$:*

(A) Domain: Since negative values of x do not make sense and $\overline{C}(0)$ is not defined, the domain is the set of all positive real numbers.

(B) Intercepts: None

(C) Asymptotes:

Horizontal asymptote:

$$\lim_{x\to\infty} \frac{10{,}000 + x^2}{2x} = \lim_{x\to\infty} \frac{x^2\left(\dfrac{10{,}000}{x^2} + 1\right)}{2x} = \lim_{x\to\infty} \frac{x}{2}\left(\frac{10{,}000}{x^2} + 1\right) = \infty$$

Thus, there is no horizontal asymptote.

Vertical asymptote: The line $x = 0$ (the y axis) is a vertical asymptote, since the denominator is 0 and the numerator is not 0 for $x = 0$.

Oblique asymptotes: Some graphs have asymptotes that are neither vertical nor horizontal. These are called **oblique asymptotes.** If we look at

$$\overline{C}(x) = \frac{5{,}000}{x} + \frac{x}{2}$$

we can see that for x near and to the right of 0, $\overline{C}(x)$ is approximated by $5{,}000/x$. On the other hand, as x increases without bound, $\overline{C}(x)$ approaches $x/2$; that is,

$$\lim_{x\to\infty}\left[\overline{C}(x) - \frac{x}{2}\right] = \lim_{x\to\infty} \frac{5{,}000}{x} = 0$$

This implies that the graph of $y = \overline{C}(x)$ approaches the line $y = x/2$ as x approaches ∞. This line is an oblique asymptote for the graph of $y = \overline{C}(x)$.

Step 2. *Use $\overline{C}'(x)$:* $\overline{C}'(x) = -\dfrac{5{,}000}{x^2} + \dfrac{1}{2} = \dfrac{x^2 - 10{,}000}{2x^2}$

Critical value: 100

Partition numbers: 0 and 100

Sign chart for $\overline{C}'(x)$:

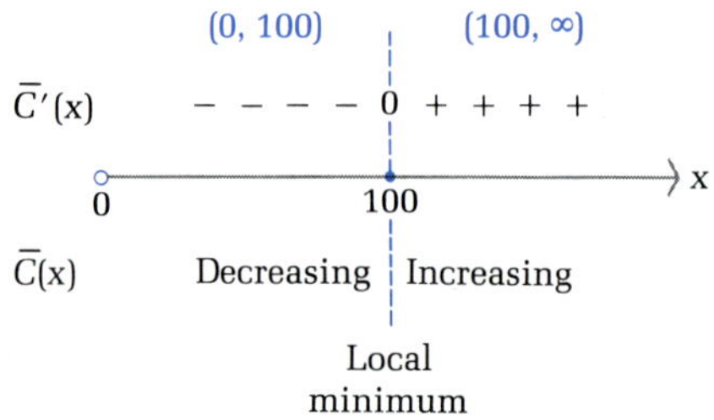

Test Numbers

x	$\overline{C}'(x)$	
50	−1.5	(−)
125	0.18	(+)

Thus, $\overline{C}(x)$ is decreasing on (0, 100), increasing on (100, ∞), and has a local minimum at $x = 100$.

Step 3. *Use $\overline{C}''(x)$:* $\overline{C}''(x) = \dfrac{10{,}000}{x^3}$

$\overline{C}''(x)$ is positive for all positive x; therefore, the graph of $y = \overline{C}(x)$ is concave upward on (0, ∞).

Step 4. *Sketch the graph of $\overline{C}$:* The graph of $\overline{C}$ is shown in the figure. The marginal cost function is $C'(x) = x$. The graph of this linear function is also shown in the figure.

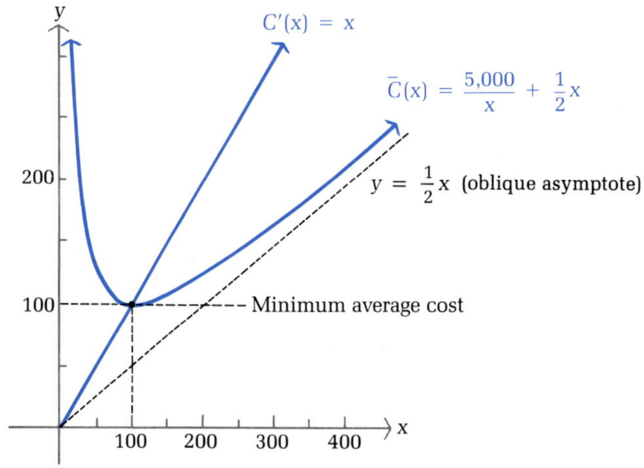

◆

The graph in Example 12 illustrates an important principle in economics.:

The minimum average cost occurs when the average cost is equal to the marginal cost.

PROBLEM 12 Given the cost function $C(x) = 1{,}600 + \frac{1}{4}x^2$, where x is the number of units produced:

(A) Graph the average cost function (using the graphing strategy) and the marginal cost function on the same set of coordinate axes. Include any oblique asymptotes.

(B) Find the minimum average cost. ◆

Answers to Matched Problems

9. (A) Horizontal asymptote: $y = 3$; vertical asymptote: $x = -2$
   (B) Horizontal asymptote: $y = 0$ (x axis); vertical asymptote: $x = 0$ (y axis)
   (C) No horizontal or vertical asymptotes
10. Domain: $(-\infty, \infty)$
    y intercept: $f(0) = 0$; x intercepts: $-4, 0$
    Asymptotes: No horizontal or vertical asymptotes
    Decreasing on $(-\infty, -3)$; increasing on $(-3, \infty)$; local minimum at $x = -3$
    Concave upward on $(-\infty, -2)$ and $(0, \infty)$; concave downward on $(-2, 0)$
    Inflection points at $x = -2$ and $x = 0$

x	f(x)
−4	0
−3	−27
−2	−16
0	0

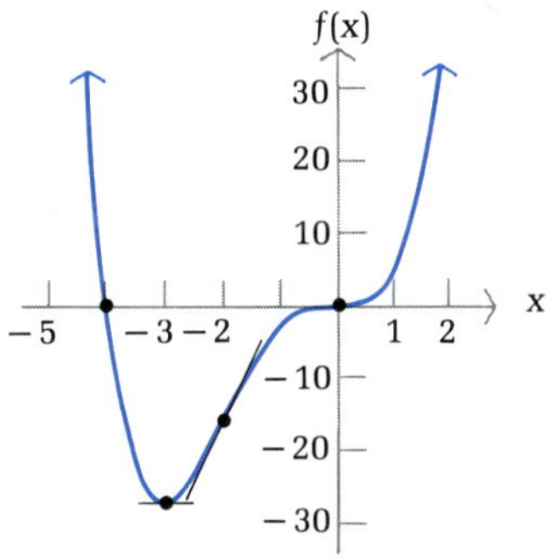

11. Domain: All real x, except $x = 1$
    y intercept: $f(0) = 0$;
    x intercept: 0
    Horizontal asymptote: $y = -2$;
    vertical asymptote: $x = 1$

Increasing on $(-\infty, 1)$ and $(1, \infty)$
Concave upward on $(-\infty, 1)$; concave downward on $(1, \infty)$

x	$f(x)$
-1	-1
0	0
$\frac{1}{2}$	2
$\frac{3}{2}$	-6
2	-4
5	$-\frac{5}{2}$

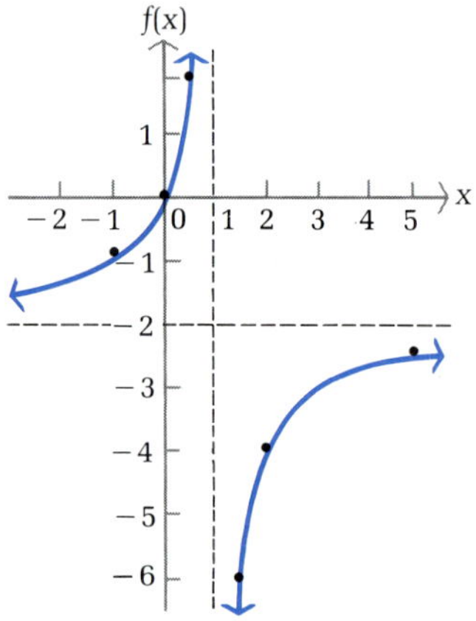

12. (A)

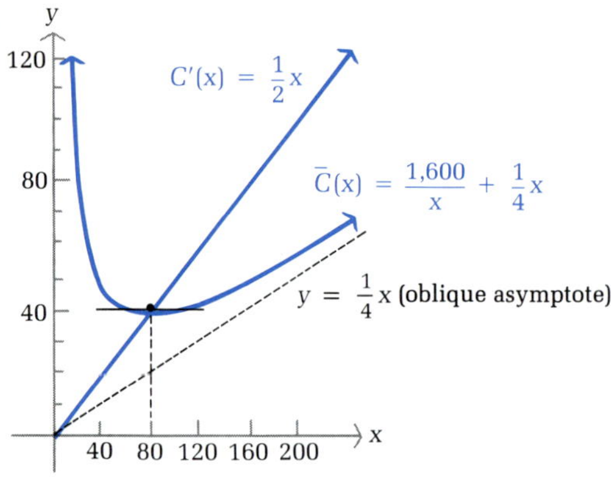

(B) Minimum average cost is 40 at $x = 80$.

EXERCISE 4-3

A *Problems 1–10 refer to the graph of $y = f(x)$ at the top of the next page:*

1. Identify the intervals over which $f(x)$ is increasing.
2. Identify the intervals over which $f(x)$ is decreasing.
3. Identify the points where $f(x)$ has a local maximum.
4. Identify the points where $f(x)$ has a local minimum.
5. Identify the intervals over which the graph of f is concave upward.
6. Identify the intervals over which the graph of f is concave downward.
7. Identify the inflection points.
8. Identify the horizontal asymptotes.
9. Identify the vertical asymptotes.
10. Identify the x and y intercepts.

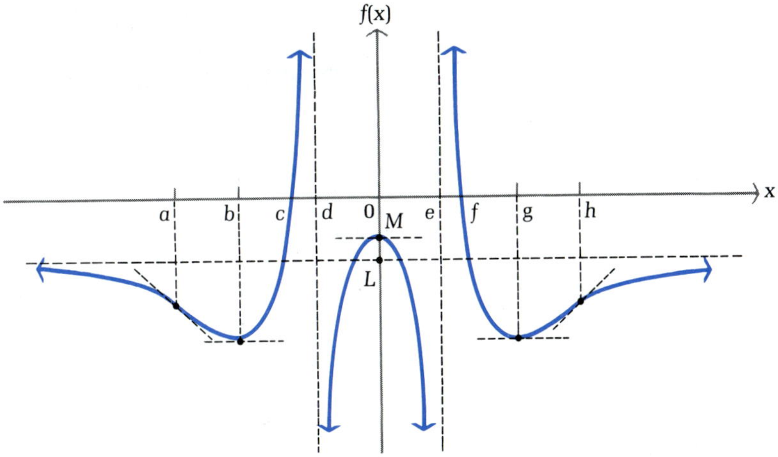

In Problems 11–16, use the given information to sketch the graph of f. Assume that f is continuous on its domain and that all intercepts are included in the table of values.

11. Domain: All real x

x	−4	−2	0	2	4
f(x)	0	1	0	−1	0

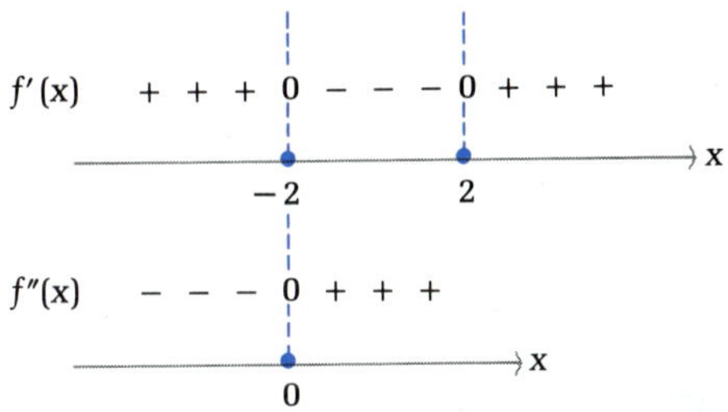

12. Domain: All real x

x	−4	−2	−1	0	1	2	4
f(x)	0	3	2	1	2	3	0

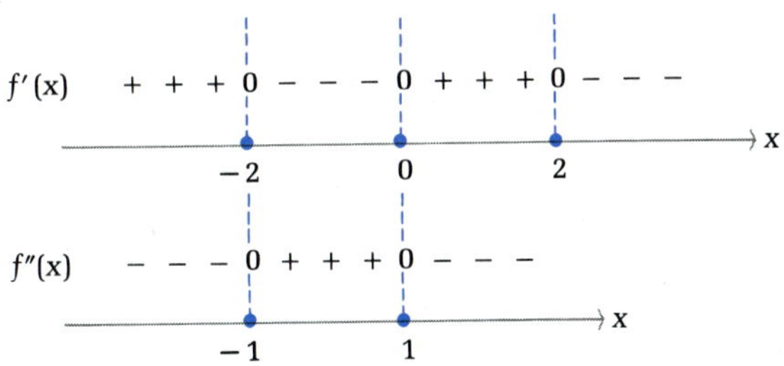

13. Domain: All real x; $\lim_{x\to\pm\infty} f(x) = 2$

x	−4	−2	0	2	4
f(x)	0	−2	0	−2	0

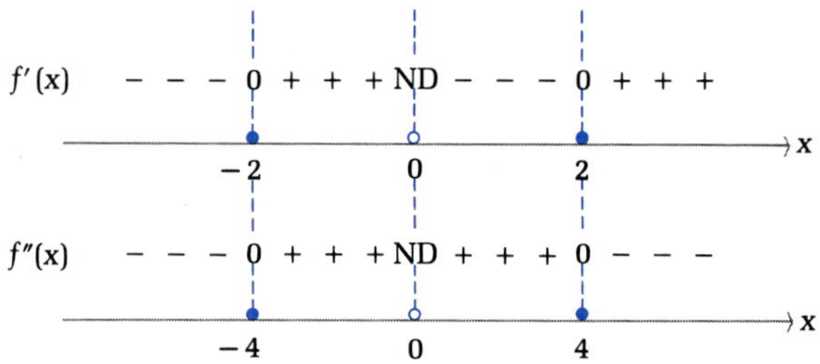

14. Domain: All real x; $\lim_{x\to-\infty} f(x) = -3$; $\lim_{x\to\infty} f(x) = 3$

x	−2	−1	0	1	2
f(x)	0	2	0	−2	0

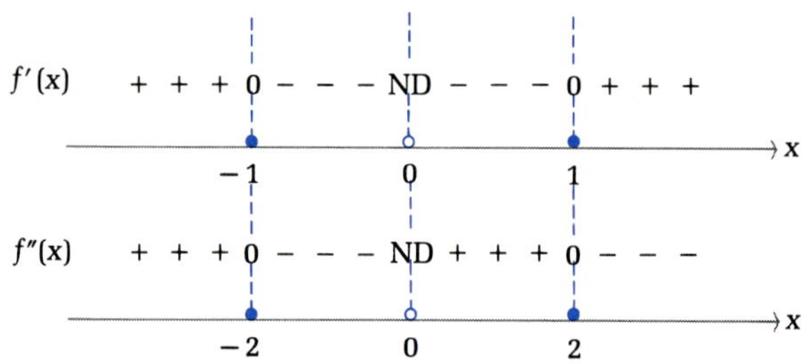

15. Domain: All real x, except x = −2; $\lim_{x\to-2^-} f(x) = \infty$; $\lim_{x\to-2^+} f(x) = -\infty$; $\lim_{x\to\infty} f(x) = 1$

x	−4	0	4	6
f(x)	0	0	3	2

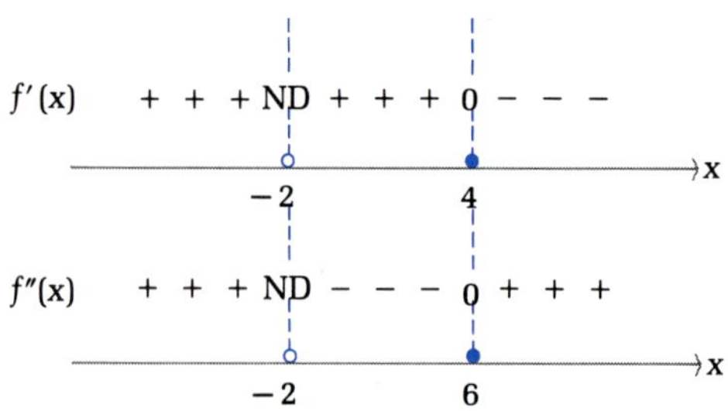

16. Domain: All real x, except $x = 1$; $\lim_{x\to 1^-} f(x) = \infty$; $\lim_{x\to 1^+} f(x) = \infty$; $\lim_{x\to\infty} f(x) = -2$

x	−4	−2	0	2
$f(x)$	0	−2	0	0

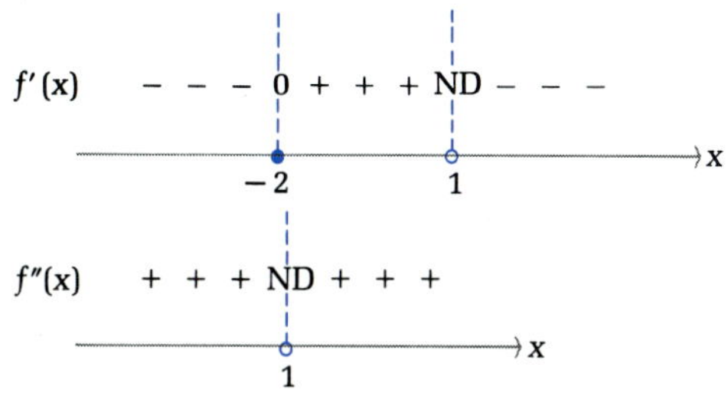

B *Find any horizontal and vertical asymptotes.*

17. $f(x) = \dfrac{2x}{x+2}$
18. $f(x) = \dfrac{3x+2}{x-4}$
19. $f(x) = \dfrac{x^2+1}{x^2-1}$
20. $f(x) = \dfrac{x^2-1}{x^2+2}$
21. $f(x) = \dfrac{x^3}{x^2+6}$
22. $f(x) = \dfrac{x}{x^2-4}$
23. $f(x) = \dfrac{x}{x^2+4}$
24. $f(x) = \dfrac{x^2+9}{x}$
25. $f(x) = \dfrac{x^2}{x-3}$
26. $f(x) = \dfrac{x+5}{x^2}$

Sketch a graph of $y = f(x)$ using the graphing strategy.

27. $f(x) = x^2 - 6x + 5$
28. $f(x) = 3 + 2x - x^2$
29. $f(x) = x^3 - 6x^2$
30. $f(x) = 3x^2 - x^3$
31. $f(x) = (x+4)(x-2)^2$
32. $f(x) = (2-x)(x+1)^2$
33. $f(x) = 8x^3 - 2x^4$
34. $f(x) = x^4 - 4x^3$
35. $f(x) = \dfrac{x+3}{x-3}$
36. $f(x) = \dfrac{2x-4}{x+2}$
37. $f(x) = \dfrac{x}{x-2}$
38. $f(x) = \dfrac{2+x}{3-x}$

C *In Problems 39 and 40, show that the line $y = x$ is an oblique asymptote for the graph of $y = f(x)$, and then use the graphing strategy to sketch a graph of $y = f(x)$.*

39. $f(x) = x + \dfrac{1}{x}$
40. $f(x) = x - \dfrac{1}{x}$

Sketch a graph of $y = f(x)$ using the graphing strategy.

41. $f(x) = x^3 - x$
42. $f(x) = x^3 + x$
43. $f(x) = (x^2+3)(9-x^2)$
44. $f(x) = (x^2+3)(x^2-1)$

45. $f(x) = (x^2 - 4)^2$

46. $f(x) = (x^2 - 1)(x^2 - 5)$

47. $f(x) = 2x^6 - 3x^5$

48. $f(x) = 3x^5 - 5x^4$

49. $f(x) = \dfrac{x}{x^2 - 4}$

50. $f(x) = \dfrac{1}{x^2 - 4}$

51. $f(x) = \dfrac{1}{1 + x^2}$

52. $f(x) = \dfrac{x^2}{1 + x^2}$

In Problems 53–58, use a graphic calculator or a computer to graph f(x). (Use [−5, 5] for both the x range and the y range.) Use the graph to find any horizontal and vertical asymptotes.

53. $f(x) = \dfrac{\sqrt{x^2 + 1}}{x}$

54. $f(x) = \sqrt{\dfrac{x^2 + 1}{x^2}}$

55. $f(x) = \dfrac{\sqrt[3]{x^3 - 1}}{1 - x}$

56. $f(x) = \dfrac{\sqrt[3]{x^3 + 1}}{x + 1}$

57. $f(x) = \dfrac{x\sqrt{x^2 + 1}}{x^2 - 1}$

58. $f(x) = \dfrac{x\sqrt{x^2 + 4}}{4 - x^2}$

APPLICATIONS

Business & Economics

59. *Revenue.* The marketing research department for a computer company used a large city to test market their new product. They found that the relationship between price p (dollars per unit) and the demand x (units per week) was given approximately by

$$p = 1{,}296 - 0.12x^2 \qquad 0 < x < 80$$

Thus, the weekly revenue can be approximated by

$$R(x) = xp = 1{,}296x - 0.12x^3 \qquad 0 < x < 80$$

Graph the revenue function R.

60. *Profit.* Suppose the cost function $C(x)$ (in dollars) for the company in Problem 59 is

$$C(x) = 830 + 396x$$

(A) Write an equation for the profit $P(x)$.

(B) Graph the profit function P.

61. *Pollution.* In Silicon Valley (California), a number of computer-related manufacturing firms were found to be contaminating underground water supplies with toxic chemicals stored in leaking underground containers. A water quality control agency ordered the companies to take immediate corrective action and to contribute to a monetary pool for testing and cleanup of the underground contamination. Suppose the required monetary pool (in millions of dollars) for the testing and cleanup is estimated to be given by

$$P(x) = \frac{2x}{1 - x} \qquad 0 \leqslant x < 1$$

where x is the percentage (expressed as a decimal fraction) of the total contaminant removed.

(A) Where is $P(x)$ increasing? Decreasing?
(B) Where is the graph of P concave upward? Downward?
(C) Find any horizontal and vertical asymptotes.
(D) Find the x and y intercepts.
(E) Sketch a graph of P.

62. *Employee training.* A company producing computer components has established that on the average a new employee can assemble $N(t)$ components per day after t days of on-the-job training, as given by

$$N(t) = \frac{100t}{t + 9} \qquad t \geq 0$$

(A) Where is $N(t)$ increasing? Decreasing?
(B) Where is the graph of N concave upward? Downward?
(C) Find any horizontal and vertical asymptotes.
(D) Find the intercepts.
(E) Sketch a graph of N.

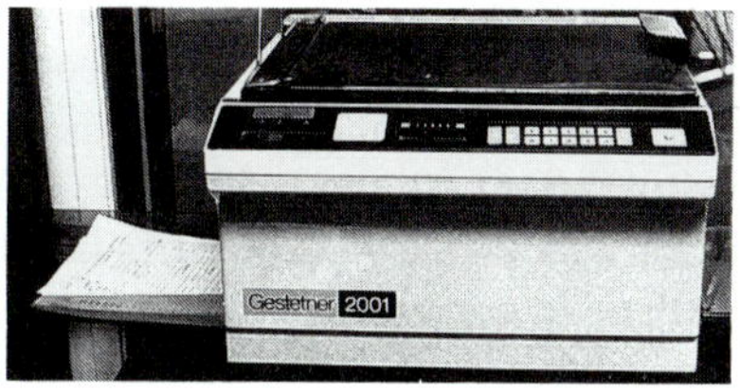

63. *Replacement time.* An office copier has an initial price of $3,200. A maintenance/service contract costs $300 for the first year and increases $100 per year thereafter. It can be shown that the total cost of the copier (in dollars) after n years is given by

$$C(n) = 3{,}200 + 250n + 50n^2$$

(A) Write an expression for the average cost per year, $\overline{C}(n)$, for n years.
(B) Graph the average cost function found in part A.
(C) When is the average cost per year minimum? (This is frequently referred to as the **replacement time** for this piece of equipment.)

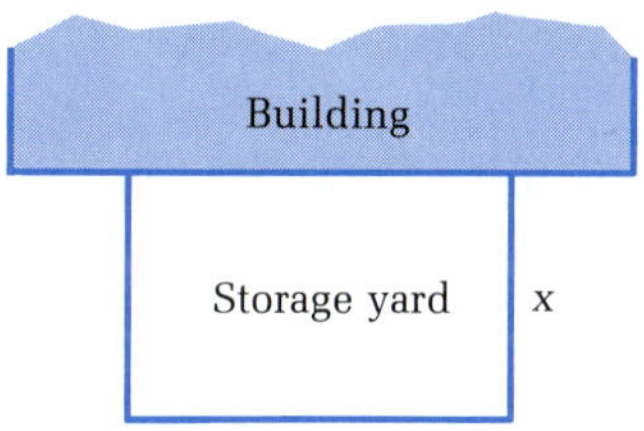

64. *Construction costs.* The management of a manufacturing plant wishes to add a fenced-in rectangular storage yard of 20,000 square feet, using the plant building as one side of the yard (see the figure in the margin). If x is the distance (in feet) from the building to the fence parallel to the building, then show that the length of the fence required for the yard is given by

$$L(x) = 2x + \frac{20{,}000}{x} \qquad x > 0$$

(A) Graph L.
(B) What are the dimensions of the rectangle requiring the least amount of fencing?

65. *Average and marginal costs.* The cost (in dollars) of producing x units of a certain product is given by

$$C(x) = 1{,}000 + 5x + \tfrac{1}{10}x^2$$

(A) Sketch the graphs of the average cost function and the marginal cost function on the same set of coordinate axes. Include any oblique asymptotes.
(B) Find the minimum average cost.

66. *Average and marginal costs.* Repeat Problem 65 for $C(x) = 500 + 2x + \frac{1}{5}x^2$.

Life Sciences

67. *Medicine.* A drug is injected into the bloodstream of a patient through her right arm. The concentration of the drug in the bloodstream of the left arm t hours after the injection is given by

$$C(t) = \frac{0.14t}{t^2 + 1}$$

Graph C.

68. *Physiology.* In a study on the speed of muscle contraction in frogs under various loads, researchers W. O. Fems and J. Marsh found that the speed of contraction decreases with increasing loads. More precisely, they found that the relationship between speed of contraction S (in centimeters per second) and load w (in grams) is given approximately by

$$S(w) = \frac{26 + 0.06w}{w} \qquad w \geqslant 5$$

Graph S.

Social Sciences

69. *Psychology—retention.* An experiment on retention is conducted in a psychology class. Each student in the class is given 1 day to memorize the same list of 30 special characters. The lists are turned in at the end of the day, and for each succeeding day for 30 days each student is asked to turn in a list of as many of the symbols as can be recalled. Averages are taken, and it is found that

$$N(t) = \frac{5t + 20}{t} \qquad t \geqslant 1$$

provides a good approximation of the average number of symbols, $N(t)$, retained after t days. Graph N.

SECTION 4-4 Optimization; Absolute Maxima and Minima

- ABSOLUTE MAXIMA AND MINIMA
- APPLICATIONS

We are now ready to consider one of the most important applications of the derivative, namely, the use of derivatives to find the *absolute maximum* or *minimum* value of a function. As we mentioned earlier, an economist may be interested in the price or production level of a commodity that will bring a maximum profit; a doctor may be interested in the time it takes for a drug to reach its maximum concentration in the bloodstream after an injection; and a city planner might be interested in the location of heavy industry in a city to produce minimum pollution in residential and business areas. Before we launch an attack on problems of this type, which are called *optimization* problems, we have to say a few words about the procedures needed to find absolute maximum and absolute minimum values of functions. We have most of the tools we need from the previous sections.

◆ ABSOLUTE MAXIMA AND MINIMA

First, what do we mean by *absolute maximum* and *absolute minimum*? We say that $f(c)$ is an **absolute maximum** of f if

$$f(c) \geqslant f(x)$$

for all x in the domain of f. Similarly, $f(c)$ is called an **absolute minimum** of f if

$$f(c) \leqslant f(x)$$

for all x in the domain of f. Figure 12 illustrates several typical examples.

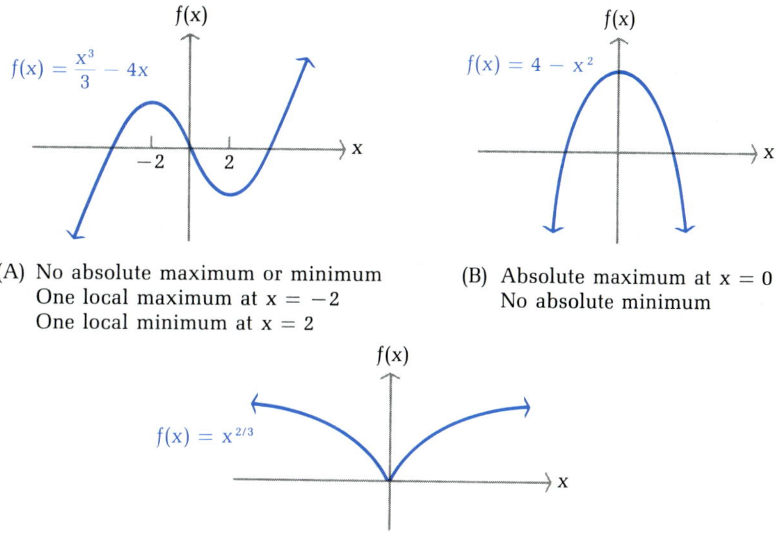

(A) No absolute maximum or minimum
One local maximum at $x = -2$
One local minimum at $x = 2$

(B) Absolute maximum at $x = 0$
No absolute minimum

(C) Absolute minimum at $x = 0$
No absolute maximum

FIGURE 12

In many practical problems, the domain of a function is restricted because of practical or physical considerations. If the domain is restricted to some closed interval, as is often the case, then Theorem 4 can be proved.

THEOREM 4

A function f continuous on a closed interval $[a, b]$ (see Section 3-1) assumes both an absolute maximum and an absolute minimum on that interval.

It is important to understand that the absolute maximum and minimum depend on both the function f and the interval $[a, b]$. Figure 13 (at the top of the next page) illustrates four cases.

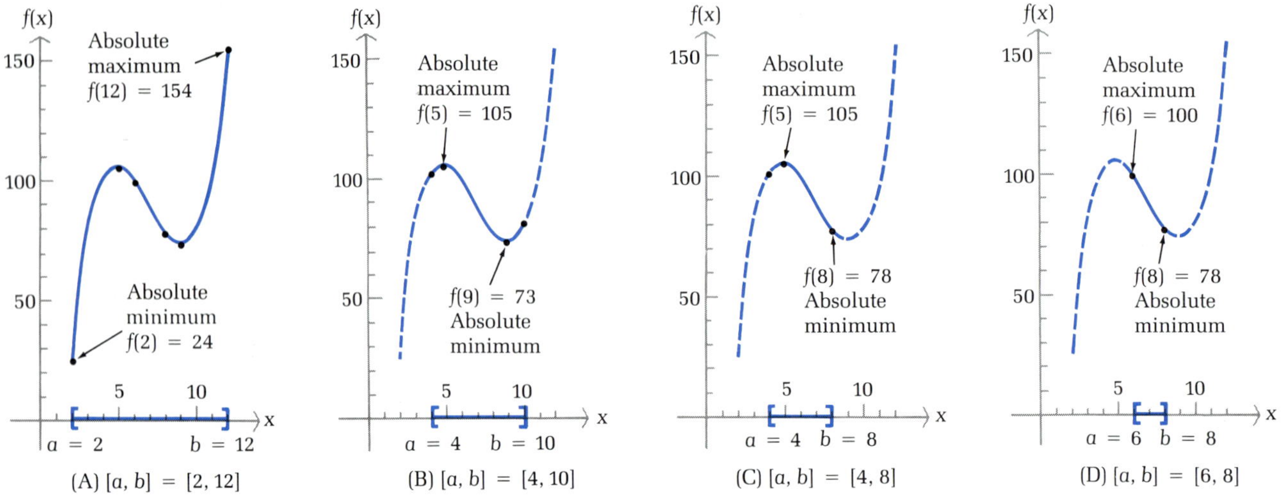

FIGURE 13
Absolute extrema for $f(x) = x^3 - 21x^2 + 135x - 170$ for various closed intervals

In all four cases illustrated in Figure 13, the absolute maximum and absolute minimum both occur at a critical value or an end point. In general:

Absolute extrema (if they exist) must always occur at critical values or at end points.

Thus, to find the absolute maximum or minimum value of a continuous function on a closed interval, we simply identify the end points and the critical values in the interval, evaluate each, and then choose the largest and smallest values out of this group.

Steps in Finding Absolute Maximum and Minimum Values of a Continuous Function f

Step 1. Check to make certain that f is continuous over $[a, b]$.

Step 2. Find the critical values in the interval (a, b).

Step 3. Evaluate f at the end points a and b and at the critical values found in step 2.

Step 4. The absolute maximum $f(x)$ on $[a, b]$ is the largest of the values found in step 3.

Step 5. The absolute minimum $f(x)$ on $[a, b]$ is the smallest of the values found in step 3.

◆ EXAMPLE 13 Find the absolute maximum and absolute minimum values of

$$f(x) = x^3 + 3x^2 - 9x - 7$$

on each of the following intervals:

(A) $[-6, 4]$ (B) $[-4, 2]$ (C) $[-2, 2]$

Solutions (A) The function is continuous for all values of x.

$$f'(x) = 3x^2 + 6x - 9 = 3(x - 1)(x + 3)$$

Thus, $x = -3$ and $x = 1$ are critical values in the interval $(-6, 4)$. Evaluate f at the end points and critical values, -6, -3, 1, and 4, and choose the maximum and minimum from these:

$f(-6) = -61$ Absolute minimum
$f(-3) = 20$
$f(1) = -12$
$f(4) = 69$ Absolute maximum

(B) Interval: $[-4, 2]$

x	f(x)	
−4	13	
−3	20	Absolute maximum
1	−12	Absolute minimum
2	−5	

(C) Interval: $[-2, 2]$

x	f(x)	
−2	15	Absolute maximum
1	−12	Absolute minimum
2	−5	

The critical value $x = -3$ is not included in this table, because it is not in the interval $[-2, 2]$. ◆

PROBLEM 13 Find the absolute maximum and absolute minimum values of

$$f(x) = x^3 - 12x$$

on each of the following intervals:

(A) $[-5, 5]$ (B) $[-3, 3]$ (C) $[-3, 1]$ ◆

Now, suppose we want to find the absolute maximum or minimum value of a function that is continuous on an interval that is not closed. Since Theorem 4 no longer applies, we cannot be certain that the absolute maximum or minimum value exists. Figure 14 (at the top of the next page) illustrates several ways that functions can fail to have absolute extrema.

In general, the best procedure to follow when the interval is not a closed interval (that is, not of the form $[a, b]$) is to sketch the graph of the function. However, one special case that occurs frequently in applications can be ana-

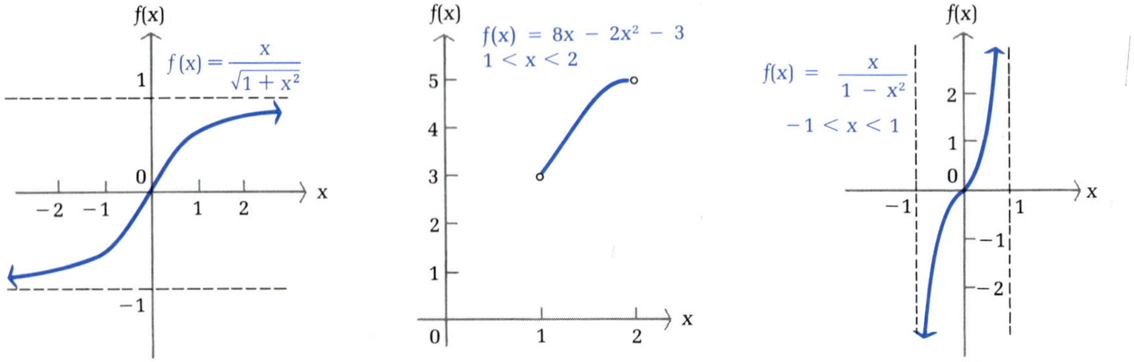

(A) No absolute extrema on $(-\infty, \infty)$:
$-1 < f(x) < 1$ for all x
$[f(x) \neq 1$ or -1 for any x$]$

(B) No absolute extrema on (1, 2):
$3 < f(x) < 5$ for $x \in (1, 2)$
$[f(x) \neq 3$ or 5 for any $x \in (1, 2)]$

(C) No absolute extrema on $(-1, 1)$:
Graph has vertical asymptotes at $x = -1$ and $x = 1$.

FIGURE 14
Functions with no absolute extrema

lyzed without drawing a graph. It often happens that f is continuous on an interval I and has only one critical value c in the interval I (here, I can be any type of interval—open, closed, or half-closed). If this is the case and if $f''(c)$ exists, then we have the second-derivative test for absolute extrema given in the box below.

Second-Derivative Test for Absolute Maximum and Minimum When f Is Continuous on an Interval I and c Is the Only Critical Value in I

$f'(c)$	$f''(c)$	$f(c)$	EXAMPLE
0	+	Absolute minimum	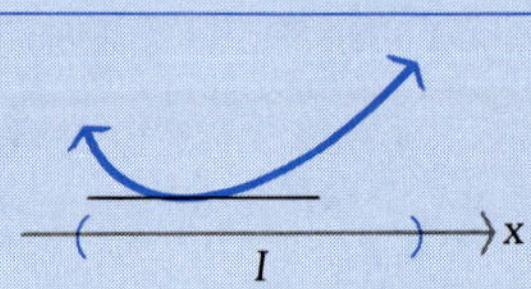
0	−	Absolute maximum	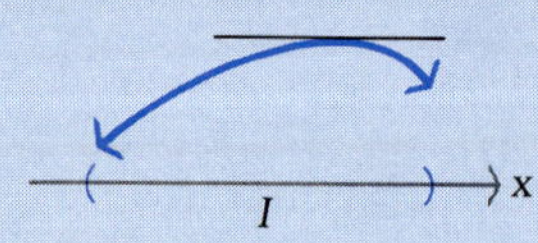
0	0	Test fails	

◆ EXAMPLE 14 Find the absolute minimum value of $f(x) = x + \frac{4}{x}$ on the interval $(0, \infty)$.

Solution

$$f'(x) = 1 - \frac{4}{x^2} = \frac{x^2 - 4}{x^2} = \frac{(x-2)(x+2)}{x^2} \qquad f''(x) = \frac{8}{x^3}$$

The only critical value in the interval $(0, \infty)$ is $x = 2$. Since $f''(2) = 1 > 0$, $f(2) = 4$ is the absolute minimum value of f on $(0, \infty)$. ◆

PROBLEM 14 Find the absolute maximum value of $f(x) = 12 - x - \frac{9}{x}$ on the interval $(0, \infty)$. ◆

◆ APPLICATIONS

Now we want to solve some applied problems that involve absolute extrema. Before beginning, we outline in the next box the steps to follow in solving this type of problem. The first step is the most difficult one. The techniques used to solve optimization problems are best illustrated through a series of examples.

A Strategy for Solving Applied Optimization Problems

Step 1. Introduce variables and a function f, including the domain I of f, and then construct a mathematical model of the form

Maximize (or minimize) $f(x)$ on the interval I

Step 2. Find the absolute maximum (or minimum) value of $f(x)$ on the interval I and the value(s) of x where this occurs.

Step 3. Use the solution to the mathematical model to answer the questions asked in the problem.

◆ EXAMPLE 15 *Cost–Demand*

A company manufactures and sells x transistor radios per week. If the weekly cost and demand equations are

$$C(x) = 5{,}000 + 2x$$

$$p = 10 - \frac{x}{1{,}000} \qquad 0 \le x \le 8{,}000$$

find for each week:

(A) The maximum revenue

(B) The maximum profit, the production level that will realize the maximum profit, and the price that the company should charge for each radio to realize the maximum profit

Solutions (A) The revenue received for selling x radios at \$p per radio is

$$R(x) = xp$$
$$= x\left(10 - \frac{x}{1,000}\right)$$
$$= 10x - \frac{x^2}{1,000}$$

Thus, the mathematical model is

$$\text{Maximize} \quad R(x) = 10x - \frac{x^2}{1,000} \qquad 0 \leq x \leq 8,000$$

$$R'(x) = 10 - \frac{x}{500}$$

$$10 - \frac{x}{500} = 0$$

$$x = 5,000 \qquad \text{Only critical value}$$

Use the second-derivative test for absolute extrema:

$$R''(x) = -\frac{1}{500} < 0 \qquad \text{for all } x$$

Thus, the maximum revenue is

$$\text{Max } R(x) = R(5,000) = \$25,000$$

(B) Profit = Revenue − Cost

$$P(x) = R(x) - C(x)$$
$$= 10x - \frac{x^2}{1,000} - 5,000 - 2x$$
$$= 8x - \frac{x^2}{1,000} - 5,000$$

The mathematical model is

$$\text{Maximize} \quad P(x) = 8x - \frac{x^2}{1,000} - 5,000 \qquad 0 \leq x \leq 8,000$$

$$P'(x) = 8 - \frac{x}{500}$$

$$8 - \frac{x}{500} = 0$$

$$x = 4,000$$

only critical value

$$P''(x) = -\frac{1}{500} < 0 \qquad \text{for all } x$$

Since $x = 4,000$ is the only critical value and $P''(x) < 0$,

$$\text{Max } P(x) = P(4,000) = \$11,000$$

Using the price–demand equation with $x = 4{,}000$, we find

$$p = 10 - \frac{4{,}000}{1{,}000} = \$6$$

Thus, a maximum profit of \$11,000 per week is realized when 4,000 radios are produced weekly and sold for \$6 each. Notice that this is not the same level of production that produces the maximum revenue. ◆

All the results in Example 15 are illustrated in Figure 15. We also note that profit is maximum when

$$P'(x) = R'(x) - C'(x) = 0$$

that is, when the marginal revenue is equal to the marginal cost (the rate of increase in revenue is the same as the rate of increase in cost at the 4,000 output level—notice that the slopes of the two curves are the same at this point).

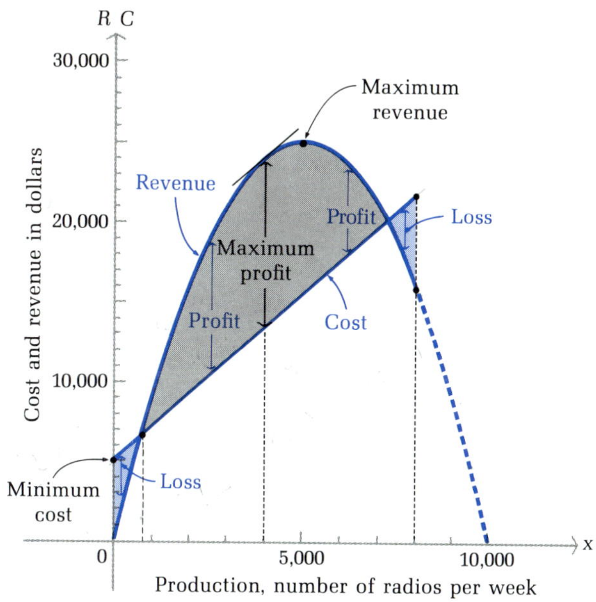

FIGURE 15

PROBLEM 15 Repeat Example 15 for

$$C(x) = 90{,}000 + 30x$$

$$p = 300 - \frac{x}{30} \qquad 0 \le x \le 9{,}000$$ ◆

◆ EXAMPLE 16 *Profit*

In Example 15 the government has decided to tax the company \$2 for each radio produced. Taking into account this additional cost, how many radios should the company manufacture each week in order to maximize its weekly profit? What is the maximum weekly profit? How much should it charge for the radios to realize the maximum weekly profit?

Solution The tax of \$2 per unit changes the company's cost equation:

$$\begin{aligned} C(x) &= \text{Original cost} + \text{Tax} \\ &= 5{,}000 + 2x + 2x \\ &= 5{,}000 + 4x \end{aligned}$$

The new profit function is

$$\begin{aligned} P(x) &= R(x) - C(x) \\ &= 10x - \frac{x^2}{1{,}000} - 5{,}000 - 4x \\ &= 6x - \frac{x^2}{1{,}000} - 5{,}000 \end{aligned}$$

Thus, we must solve the following:

$$\text{Maximize} \quad P(x) = 6x - \frac{x^2}{1{,}000} - 5{,}000 \qquad 0 \le x \le 8{,}000$$

$$P'(x) = 6 - \frac{x}{500}$$

$$6 - \frac{x}{500} = 0$$

$$x = 3{,}000$$

$$P''(x) = -\frac{1}{500} < 0 \qquad \text{for all } x$$

$$\text{Max } P(x) = P(3{,}000) = \$4{,}000$$

Using the price–demand equation with $x = 3{,}000$, we find

$$p = 10 - \frac{3{,}000}{1{,}000} = \$7$$

Thus, the company's maximum profit is \$4,000 when 3,000 radios are produced and sold weekly at a price of \$7.

Even though the tax caused the company's cost to increase by \$2 per radio, the price that the company should charge to maximize its profit increases by only \$1. The company must absorb the other \$1 with a resulting decrease of \$7,000 in maximum profit. ◆

PROBLEM 16 Repeat Example 16 if

$$C(x) = 90{,}000 + 30x$$

$$p = 300 - \frac{x}{30} \qquad 0 \leq x \leq 9{,}000$$

and the government decides to tax the company \$20 for each unit produced. Compare the results with the results in Problem 15B. ◆

◆ EXAMPLE 17 *Maximize Yield*

A walnut grower estimates from past records that if 20 trees are planted per acre, each tree will average 60 pounds of nuts per year. If for each additional tree planted per acre (up to 15) the average yield per tree drops 2 pounds, how many trees should be planted to maximize the yield per acre? What is the maximum yield?

Solution Let x be the number of additional trees planted per acre. Then

$$20 + x = \text{Total number of trees per acre}$$

$$60 - 2x = \text{Yield per tree}$$

$$\text{Yield per acre} = (\text{Total number of trees per acre})(\text{Yield per tree})$$

$$Y(x) = (20 + x)(60 - 2x)$$

$$= 1{,}200 + 20x - 2x^2 \qquad 0 \leq x \leq 15$$

Thus, we must solve the following:

$$\text{Maximize} \quad Y(x) = 1{,}200 + 20x - 2x^2 \qquad 0 \leq x \leq 15$$

$$Y'(x) = 20 - 4x$$

$$20 - 4x = 0$$

$$x = 5$$

$$Y''(x) = -4 < 0 \qquad \text{for all } x$$

Hence,

$$\text{Max } Y(x) = Y(5) = 1{,}250 \text{ pounds per acre}$$

Thus, a maximum yield of 1,250 pounds of nuts per acre is realized if 25 trees are planted per acre. ◆

PROBLEM 17 Repeat Example 17 starting with 30 trees per acre and a reduction of 1 pound per tree for each additional tree planted. ◆

◆ EXAMPLE 18 *Maximize Area*

A farmer wants to construct a rectangular pen next to a barn 60 feet long, using all of the barn as part of one side of the pen. Find the dimensions of the pen with the largest area that the farmer can build if:

(A) 160 feet of fencing material is available
(B) 250 feet of fencing material is available

Solutions

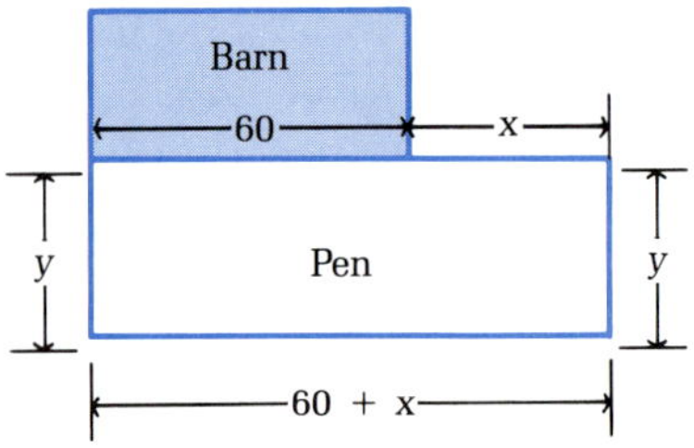

(A) We begin by constructing and labeling the figure in the margin. The area of the pen is

$$A = (x + 60)y$$

Before we can maximize the area, we must determine a relationship between x and y in order to express A as a function of one variable. In this case, x and y are related to the total amount of available fencing material:

$$x + y + 60 + x + y = 160$$
$$2x + 2y = 100$$
$$y = 50 - x$$

Thus,

$$A(x) = (x + 60)(50 - x)$$

Now we need to determine the permissible values of x; that is, the domain of the function A. Since the farmer wants to use all of the barn as part of one side of the pen, x cannot be negative. Since y is the other dimension of the pen, y cannot be negative. Thus,

$$y = 50 - x \geqslant 0$$
$$50 \geqslant x$$

The domain of A is [0, 50]. Thus, we must solve the following:

$$\text{Maximize} \quad A(x) = (x + 60)(50 - x) \qquad 0 \leqslant x \leqslant 50$$

$$A(x) = 3{,}000 - 10x - x^2$$
$$A'(x) = -10 - 2x$$
$$-10 - 2x = 0$$
$$x = -5$$

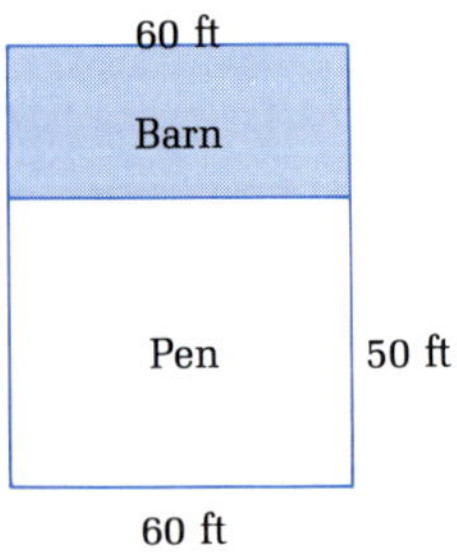

Since $x = -5$ is not in the interval [0, 50], there are no critical values in the interval. A(x) is continuous on [0, 50], so the absolute maximum must occur at one of the end points.

$$A(0) = 3{,}000 \qquad \text{Maximum area}$$
$$A(50) = 0$$

If $x = 0$, then $y = 50$. Thus, the dimensions of the pen with largest area are 60 feet by 50 feet.

(B) If 250 feet of fencing material is available, then

$$x + y + 60 + x + y = 250$$
$$2x + 2y = 190$$
$$y = 95 - x$$

The model becomes

$$\text{Maximize} \quad A(x) = (x + 60)(95 - x) \qquad 0 \leq x \leq 95$$

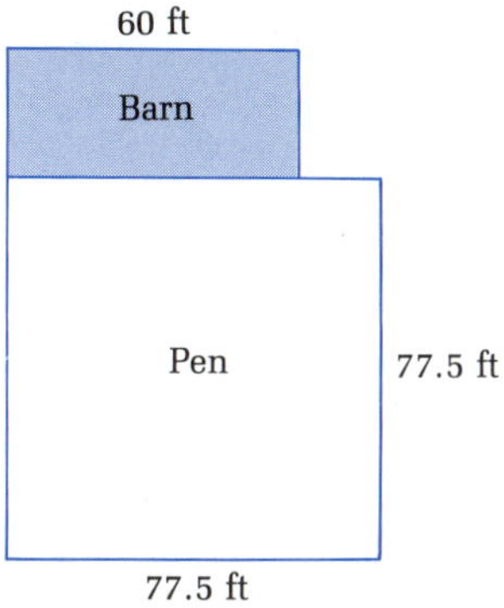

$$A(x) = 5{,}700 + 35x - x^2$$
$$A'(x) = 35 - 2x$$
$$35 - 2x = 0$$
$$x = \tfrac{35}{2} = 17.5 \qquad \text{The only critical value}$$
$$A''(x) = -2 < 0 \qquad \text{for all } x$$
$$\text{Max } A(x) = A(17.5) = 6{,}006.25$$
$$y = 95 - 17.5 = 77.5$$

This time, the dimensions of the pen with the largest area are 77.5 feet by 77.5 feet. ◆

PROBLEM 18 Repeat Example 18 if the barn is 80 feet long. ◆

◆ EXAMPLE 19

Inventory Control

A record company anticipates that there will be a demand for 20,000 copies of a certain album during the following year. It costs the company \$0.50 to store a record for 1 year. Each time it must press additional records, it costs \$200 to set up the equipment. How many records should the company press during each production run in order to minimize its total storage and set-up costs?

Solution

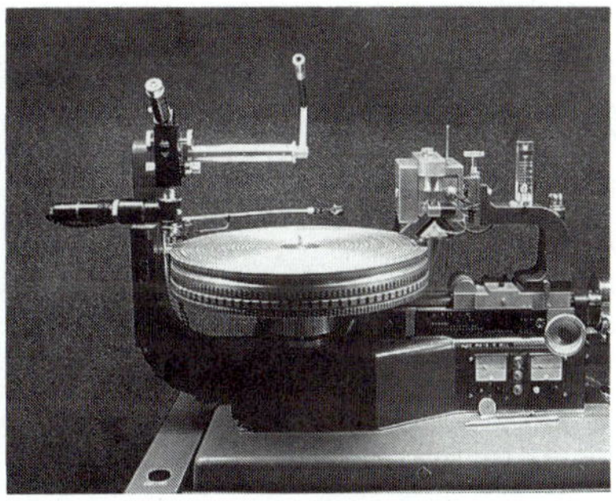

This type of problem is called an **inventory control problem.** One of the basic assumptions made in such problems is that the demand is uniform. For example, if there are 250 working days in a year, then the daily demand would be $20{,}000/250 = 80$ records. The company could decide to produce all 20,000 records at the beginning of the year. This would certainly minimize the set-up costs, but would result in very large storage costs. At the other extreme, it could produce 80 records each day. This would minimize the storage costs, but would result in very large set-up costs. Somewhere between these two extremes is the optimal solution that will minimize the total storage and set-up costs. Let

x = Number of records pressed during each production run

y = Number of production runs

It is easy to see that the total set-up cost for the year is $200y$, but what is the total storage cost? If the demand is uniform, then the number of records in storage between production runs will decrease from x to 0, and the average number in storage each day is $x/2$. This result is illustrated in the figure on page 260.

Since it costs \$0.50 to store a record for 1 year, the total storage cost is $0.5(x/2) = 0.25x$ and the total cost is

$$\text{Total cost} = \text{Set-up cost} + \text{Storage cost}$$
$$C = 200y + 0.25x$$

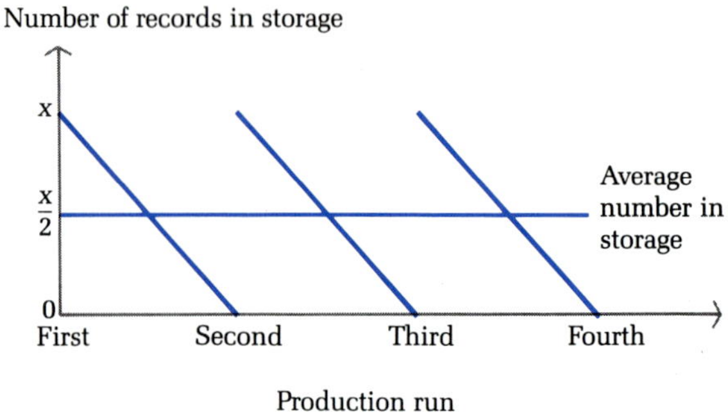

In order to write the total cost C as a function of one variable, we must find a relationship between x and y. If the company produces x records in each of y production runs, then the total number of records produced is xy. Thus,

$$xy = 20{,}000$$

$$y = \frac{20{,}000}{x}$$

Certainly, x must be at least 1 and cannot exceed 20,000. Thus, we must solve the following:

$$\text{Minimize} \quad C(x) = 200\left(\frac{20{,}000}{x}\right) + 0.25x \qquad 1 \leq x \leq 20{,}000$$

$$C(x) = \frac{4{,}000{,}000}{x} + 0.25x$$

$$C'(x) = -\frac{4{,}000{,}000}{x^2} + 0.25$$

$$-\frac{4{,}000{,}000}{x^2} + 0.25 = 0$$

$$x^2 = \frac{4{,}000{,}000}{0.25}$$

$$x^2 = 16{,}000{,}000$$

$$x = 4{,}000$$

$-4{,}000$ is not a critical value since $1 \leq x \leq 20{,}000$

$$C''(x) = \frac{8{,}000{,}000}{x^3} > 0 \qquad \text{for } x \in (1, 20{,}000)$$

Thus,

$$\text{Min } C(x) = C(4{,}000) = 2{,}000$$

$$y = \frac{20{,}000}{4{,}000} = 5$$

The company will minimize its total cost by pressing 4,000 records five times during the year. ◆

PROBLEM 19 Repeat Example 19 if it costs \$250 to set up a production run and \$0.40 to store a record for 1 year. ◆

Answers to Matched Problems

13. (A) Absolute maximum: $f(5) = 65$; absolute minimum: $f(-5) = -65$
    (B) Absolute maximum: $f(-2) = 16$; absolute minimum: $f(2) = -16$
    (C) Absolute maximum: $f(-2) = 16$; absolute minimum: $f(1) = -11$
14. $f(3) = 6$
15. (A) Max $R(x) = R(4{,}500) = \$675{,}000$
    (B) Max $P(x) = P(4{,}050) = \$456{,}750$; $p = \$165$
16. Max $P(x) = P(3{,}750) = \$378{,}750$; $p = \$175$; price increases \$10, profit decreases \$78,000
17. Max $Y(x) = Y(15) = 2{,}025$ lb/acre
18. (A) 80 ft by 40 ft (B) 82.5 ft by 82.5 ft
19. Press 5,000 records four times during the year

EXERCISE 4-4

A

Find the absolute maximum and minimum, if either exists, for each function.

1. $f(x) = x^2 - 4x + 5$
2. $f(x) = x^2 + 6x + 7$
3. $f(x) = 10 + 8x - x^2$
4. $f(x) = 6 - 8x - x^2$
5. $f(x) = 1 - x^3$
6. $f(x) = 1 - x^4$

B

Find the indicated extremum of each function.

7. Absolute maximum value of $f(x) = 24 - 2x - \frac{8}{x}$, $x > 0$
8. Absolute minimum value of $f(x) = 3x + \frac{27}{x}$, $x > 0$
9. Absolute minimum value of $f(x) = 5 + 3x + \frac{12}{x^2}$, $x > 0$
10. Absolute maximum value of $f(x) = 10 - 2x - \frac{27}{x^2}$, $x > 0$

Find the absolute maximum and minimum, if either exists, for each function on the indicated intervals.

11. $f(x) = x^3 - 6x^2 + 9x - 6$
    (A) $[-1, 5]$ (B) $[-1, 3]$ (C) $[2, 5]$
12. $f(x) = 2x^3 - 3x^2 - 12x + 24$
    (A) $[-3, 4]$ (B) $[-2, 3]$ (C) $[-2, 1]$
13. $f(x) = (x - 1)(x - 5)^3 + 1$
    (A) $[0, 3]$ (B) $[1, 7]$ (C) $[3, 6]$

14. $f(x) = x^4 - 8x^2 + 16$

(A) $[-1, 3]$ (B) $[0, 2]$ (C) $[-3, 4]$

Preliminary word problems:

C

15. How would you divide a 10 inch line so that the product of the two lengths is a maximum?
16. What quantity should be added to 5 and subtracted from 5 in order to produce the maximum product of the results?
17. Find two numbers whose difference is 30 and whose product is a minimum.
18. Find two positive numbers whose sum is 60 and whose product is a maximum.
19. Find the dimensions of a rectangle with perimeter 100 centimeters that has maximum area. Find the maximum area.
20. Find the dimensions of a rectangle of area 225 square centimeters that has the least perimeter. What is the perimeter?

APPLICATIONS

Business & Economics

21. *Maximum revenue and profit.* A company manufactures and sells x television sets per month. The monthly cost and demand equations are

$$C(x) = 72{,}000 + 60x$$

$$p = 200 - \frac{x}{30} \qquad 0 \le x \le 6{,}000$$

(A) Find the maximum revenue.

(B) Find the maximum profit, the production level that will realize the maximum profit, and the price the company should charge for each television set.

(C) If the government decides to tax the company \$5 for each set it produces, how many sets should the company manufacture each month in order to maximize its profit? What is the maximum profit? What should the company charge for each set?

22. *Maximum revenue and profit.* Repeat Problem 21 for

$$C(x) = 60{,}000 + 60x$$

$$p = 200 - \frac{x}{50} \qquad 0 \le x \le 10{,}000$$

23. *Car rental.* A car rental agency rents 200 cars per day at a rate of \$30 per day. For each \$1 increase in rate, 5 fewer cars are rented. At what rate

should the cars be rented to produce the maximum income? What is the maximum income?

24. *Rental income.* A 300 room hotel in Las Vegas is filled to capacity every night at \$80 a room. For each \$1 increase in rent, 3 fewer rooms are rented. If each rented room costs \$10 to service per day, how much should the management charge for each room to maximize gross profit? What is the maximum gross profit?

25. *Agriculture.* A commercial cherry grower estimates from past records that if 30 trees are planted per acre, each tree will yield an average of 50 pounds of cherries per season. If for each additional tree planted per acre (up to 20), the average yield per tree is reduced by 1 pound, how many trees should be planted per acre to obtain the maximum yield per acre? What is the maximum yield?

26. *Agriculture.* A commercial pear grower must decide on the optimum time to have fruit picked and sold. If the pears are picked now, they will bring 30¢ per pound, with each tree yielding an average of 60 pounds of salable pears. If the average yield per tree increases 6 pounds per tree per week for the next 4 weeks, but the price drops 2¢ per pound per week, when should the pears be picked to realize the maximum return per tree? What is the maximum return?

27. *Manufacturing.* A candy box is to be made out of a piece of cardboard that measures 8 by 12 inches. Squares of equal size will be cut out of each corner, and then the ends and sides will be folded up to form a rectangular box. What size square should be cut from each corner to obtain a maximum volume?

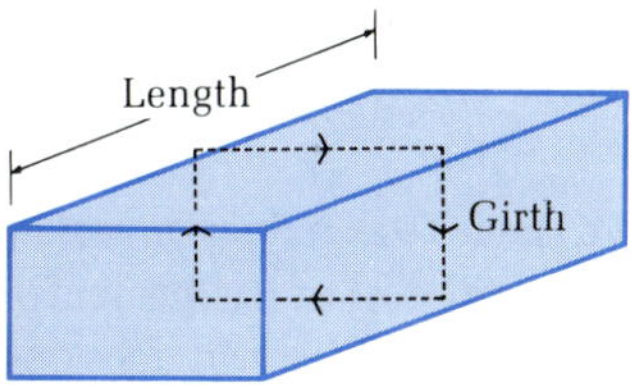

28. *Packaging.* A parcel delivery service will deliver a package only if the length plus girth (distance around) does not exceed 108 inches.

(A) Find the dimensions of a rectangular box with square ends that satisfies the delivery service's restriction and has maximum volume. What is the maximum volume?

(B) Find the dimensions (radius and height) of a cylindrical container that meets the delivery service's requirement and has maximum volume. What is the maximum volume?

29. *Construction costs.* A fence is to be built to enclose a rectangular area of 800 square feet. The fence along three sides is to be made of material that costs \$6 per foot. The material for the fourth side costs \$18 per foot. Find the dimensions of the rectangle that will allow the most economical fence to be built.

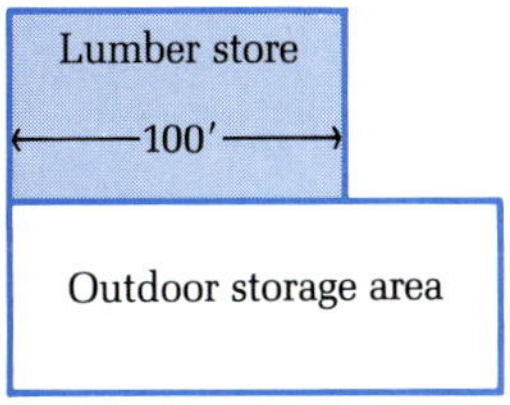

30. *Construction costs.* The owner of a retail lumber store wants to construct a fence to enclose an outdoor storage area adjacent to the store, as indicated in the figure in the margin. Find the dimensions that will enclose the largest area if:

(A) 240 feet of fencing material is used.

(B) 400 feet of fencing material is used.

31. *Inventory control.* A publishing company sells 50,000 copies of a certain book each year. It costs the company \$1.00 to store a book for 1 year. Each time it must print additional copies, it costs the company \$1,000 to set up the presses. How many books should the company produce during each printing in order to minimize its total storage and set-up costs?

32. *Operational costs.* The cost per hour for fuel to run a train is $v^2/4$ dollars, where v is the speed of the train in miles per hour. (Note that the cost goes up as the square of the speed.) Other costs, including labor, are \$300 per hour. How fast should the train travel on a 360 mile trip to minimize the total cost for the trip?

33. *Construction costs.* A freshwater pipeline is to be run from a source on the edge of a lake to a small resort community on an island 5 miles off-shore, as indicated in the figure.

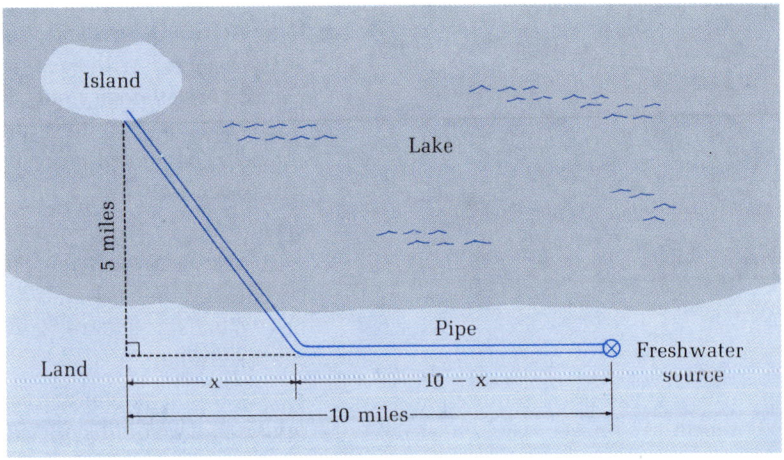

(A) If it costs 1.4 times as much to lay the pipe in the lake as it does on land, what should x be (in miles) to minimize the total cost of the project?

(B) If it costs only 1.1 times as much to lay the pipe in the lake as it does on land, what should x be to minimize the total cost of the project? [*Note:* Compare with Problem 38.]

34. *Manufacturing costs.* A manufacturer wants to produce cans that will hold 12 ounces (approximately 22 cubic inches) in the form of a right circular cylinder. Find the dimensions (radius of an end and height) of the can that will use the smallest amount of material. Assume the circular ends are cut out of squares, with the corner portions wasted, and the sides are made from rectangles, with no waste.

Life Sciences

35. *Bacteria control.* A recreational swimming lake is treated periodically to control harmful bacteria growth. Suppose t days after a treatment, the concentration of bacteria per cubic centimeter is given by

$$C(t) = 30t^2 - 240t + 500 \qquad 0 \leq t \leq 8$$

How many days after a treatment will the concentration be minimal? What is the minimum concentration?

36. *Drug concentration.* The concentration $C(t)$, in milligrams per cubic centimeter, of a particular drug in a patient's bloodstream is given by

$$C(t) = \frac{0.16t}{t^2 + 4t + 4}$$

where t is the number of hours after the drug is taken. How many hours after the drug is given will the concentration be maximum? What is the maximum concentration?

37. *Laboratory management.* A laboratory uses 500 white mice each year for experimental purposes. It costs \$4.00 to feed a mouse for 1 year. Each time mice are ordered from a supplier, there is a service charge of \$10 for processing the order. How many mice should be ordered each time in order to minimize the total cost of feeding the mice and of placing the orders for the mice?

38. *Bird flights.* Some birds tend to avoid flights over large bodies of water during daylight hours. (It is speculated that more energy is required to fly over water than land, because air generally rises over land and falls over water during the day.) Suppose an adult bird with this tendency is taken from its nesting area on the edge of a large lake to an island 5 miles off-shore and is then released (see the accompanying figure).

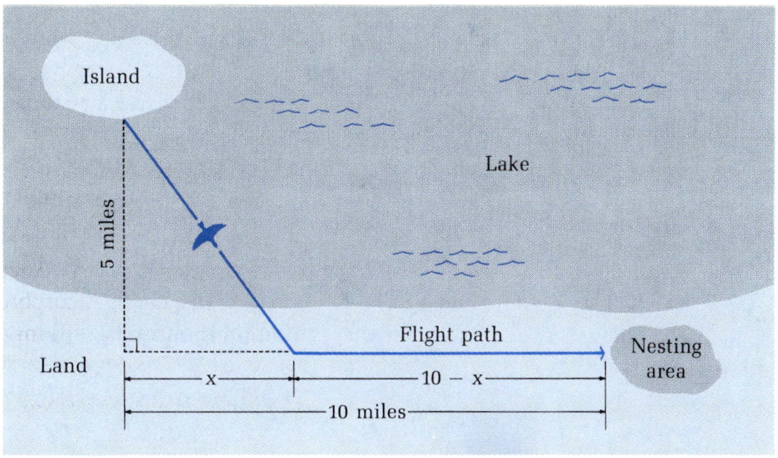

(A) If it takes 1.4 times as much energy to fly over water as land, how far up-shore (x, in miles) should the bird head in order to minimize the total energy expended in returning to the nesting area?

(B) It it takes only 1.1 times as much energy to fly over water as land, how far up-shore should the bird head in order to minimize the total energy expended in returning to the nesting area? [*Note:* Compare with Problem 33.]

39. *Botany.* If it is known from past experiments that the height (in feet) of a given plant after t months is given approximately by

$$H(t) = 4t^{1/2} - 2t \qquad 0 \leq t \leq 2$$

how long, on the average, will it take a plant to reach its maximum height? What is the maximum height?

40. *Pollution.* Two heavy industrial areas are located 10 miles apart, as indicated in the figure. If the concentration of particulate matter (in parts per million) decreases as the reciprocal of the square of the distance from the source, and area A_1 emits eight times the particulate matter as A_2, then the concentration of particulate matter at any point between the two areas is given by

$$C(x) = \frac{8k}{x^2} + \frac{k}{(10 - x)^2} \qquad 0.5 \leq x \leq 9.5, \quad k > 0$$

How far from A_1 will the concentration of particulate matter be at a minimum?

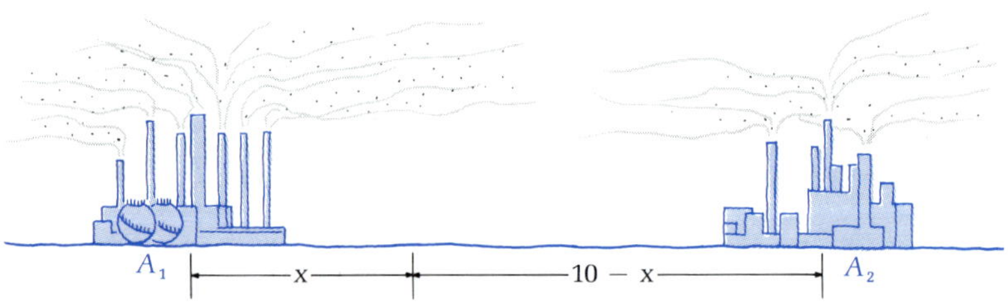

Social Sciences

41. *Politics.* In a newly incorporated city, it is estimated that the voting population (in thousands) will increase according to

$$N(t) = 30 + 12t^2 - t^3 \qquad 0 \leq t \leq 8$$

where t is time in years. When will the rate of increase be most rapid?

42. *Learning.* A large grocery chain found that, on the average, a checker can memorize $P\%$ of a given price list in x continuous hours, as given approximately by

$$P(x) = 96x - 24x^2 \qquad 0 \leq x \leq 3$$

How long should a checker plan to take to memorize the maximum percentage? What is the maximum?

SECTION 4-5 Increments and Differentials

- INCREMENTS
- DIFFERENTIALS
- APPROXIMATIONS USING DIFFERENTIALS

We now introduce the concepts of *increments* and *differentials*. Increments are useful in their own right, and they also provide an alternate notation for defining the derivative. Differentials are often easier to compute than increments, and can be used to approximate increments. Differential notation will be used extensively in Chapters 6 and 7 (where we discuss integral calculus).

◆ INCREMENTS

Given $y = f(x)$, we are sometimes interested in how y is affected by a small change in x. For example, if y represents profit and x represents price per item, we might be interested in the change in profit resulting from a small change in price.

Given

$$y = f(x) = x^3$$

if x changes from 2 to 2.1, then y will change from $y = f(2) = 2^3 = 8$ to $y = f(2.1) = 2.1^3 = 9.261$. Mathematically, the change in x and the corresponding change in y, called *increments in x and y*, respectively, are denoted by Δx and Δy (read "delta x" and "delta y," since Δ is the Greek letter delta). In terms of the given example, we write

$$\Delta x = 2.1 - 2 = 0.1 \qquad \text{Change in } x$$

$$\begin{aligned} \Delta y &= f(2.1) - f(2) \\ &= 2.1^3 - 2^3 \\ &= 9.261 - 8 = 1.261 \qquad \text{Corresponding change in } y \end{aligned}$$

It is important to remember that Δx and Δy do not represent products—each is a variable with replacements from the real numbers. The general definitions are given in the following box:

Increments

For $y = f(x)$,

$$\Delta x = x_2 - x_1 \qquad \Delta y = y_2 - y_1$$
$$x_2 = x_1 + \Delta x \qquad\quad = f(x_2) - f(x_1)$$
$$= f(x_1 + \Delta x) - f(x_1)$$

Δy represents the change in y corresponding to a Δx change in x. Δx can be either positive or negative.

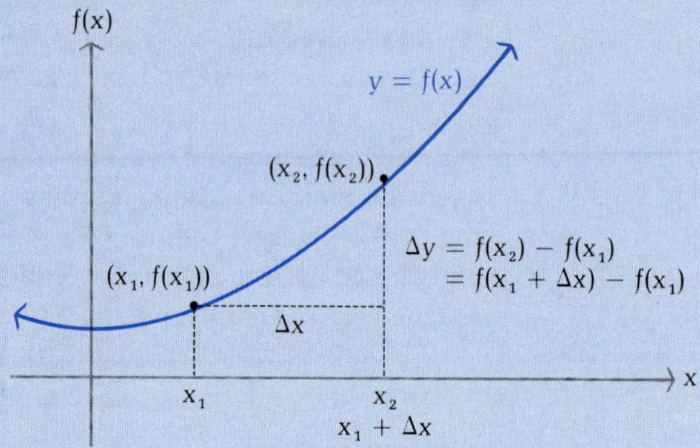

[*Note:* Δy depends on the function f, the input x, and the increment Δx.]

◆ EXAMPLE 20 Given the function: $y = f(x) = \dfrac{x^2}{2}$

(A) Find Δx, Δy, and $\Delta y/\Delta x$ for $x_1 = 1$ and $x_2 = 2$.

(B) Find $\dfrac{f(x_1 + \Delta x) - f(x_1)}{\Delta x}$ for $x_1 = 1$ and $\Delta x = 2$.

Solutions (A) $\Delta x = x_2 - x_1 = 2 - 1 = 1$

$$\Delta y = f(x_2) - f(x_1)$$
$$= f(2) - f(1) = \frac{4}{2} - \frac{1}{2} = \frac{3}{2}$$
$$\frac{\Delta y}{\Delta x} = \frac{f(x_2) - f(x_1)}{x_2 - x_1} = \frac{\frac{3}{2}}{1} = \frac{3}{2}$$

(B) $$\frac{f(x_1 + \Delta x) - f(x_1)}{\Delta x} = \frac{f(1 + 2) - f(1)}{2}$$
$$= \frac{f(3) - f(1)}{2} = \frac{\frac{9}{2} - \frac{1}{2}}{2} = \frac{4}{2} = 2$$ ◆

PROBLEM 20 Given the function: $y = f(x) = x^2 + 1$

(A) Find Δx, Δy, and $\Delta y/\Delta x$ for $x_1 = 2$ and $x_2 = 3$.

(B) Find $\dfrac{f(x_1 + \Delta x) - f(x_1)}{\Delta x}$ for $x_1 = 1$ and $\Delta x = 2$. ◆

In Example 20, we observe another notation for the familiar difference quotient

$$\frac{f(x + h) - f(x)}{h} \tag{1}$$

It is common to refer to h, the change in x, as Δx. Then, the difference quotient (1) takes on the form

$$\frac{f(x + \Delta x) - f(x)}{\Delta x} \quad \text{or} \quad \frac{\Delta y}{\Delta x} \qquad \Delta y = f(x + \Delta x) - f(x)$$

and the derivative is defined by

$$f'(x) = \lim_{\Delta x \to 0} \frac{f(x + \Delta x) - f(x)}{\Delta x}$$

or

$$f'(x) = \lim_{\Delta x \to 0} \frac{\Delta y}{\Delta x} \tag{2}$$

if the limits exists.

◆ DIFFERENTIALS

Assume that the limit in equation (2) exists. Then, for small Δx, the difference quotient $\Delta y/\Delta x$ provides a good approximation for $f'(x)$. Also, $f'(x)$ provides a good approximation for $\Delta y/\Delta x$. We indicate the latter by writing

$$\frac{\Delta y}{\Delta x} \approx f'(x) \qquad \Delta x \text{ is small but } \neq 0 \tag{3}$$

Multiplying both sides of (3) by Δx gives us

$$\Delta y \approx f'(x)\,\Delta x \qquad \Delta x \text{ is small but } \neq 0 \tag{4}$$

From equation (4) we see that $f'(x)\,\Delta x$ provides a good approximation for Δy when Δx is small.

Because of the practical and theoretical importance of $f'(x)\,\Delta x$, we give it the special name **differential** and represent it by special symbols, ***dy*** or ***df***:

$$dy = f'(x)\,\Delta x \qquad \text{or} \qquad df = f'(x)\,\Delta x$$

For example,

$$d(2x^3) = (2x^3)' \, \Delta x = 6x^2 \, \Delta x$$
$$d(x) = (x)' \, \Delta x = 1 \, \Delta x = \Delta x$$

In the second example, we usually drop the parentheses in $d(x)$ and simply write

$$dx = \Delta x$$

In summary, we have the following:

Differentials

If $y = f(x)$ defines a differentiable function, then the **differential dy, or df,** is defined as the product of $f'(x)$ and dx, where $dx = \Delta x$. Symbolically,

$$dy = f'(x) \, dx \qquad \text{or} \qquad df = f'(x) \, dx$$

where

$$dx = \Delta x$$

[*Note:* The differential dy (or df) is actually a function involving two independent variables, x and dx; a change in either one or both will affect dy (or df).]

◆ EXAMPLE 21 Find dy for $f(x) = x^2 + 3x$. Evaluate dy for:

(A) $x = 2$ and $dx = 0.1$ (B) $x = 3$ and $dx = 0.1$ (C) $x = 1$ and $dx = 0.02$

Solution

$$\begin{aligned} dy &= f'(x) \, dx \\ &= (2x + 3) \, dx \end{aligned}$$

(A) When $x = 2$ and $dx = 0.1$,

$$dy = [2(2) + 3]0.1 = 0.7$$

(B) When $x = 3$ and $dx = 0.1$,

$$dy = [2(3) + 3]0.1 = 0.9$$

(C) When $x = 1$ and $dx = 0.02$,

$$dy = [2(1) + 3]0.02 = 0.1$$

◆

PROBLEM 21 Find dy for $f(x) = \sqrt{x} + 3$. Evaluate dy for:

(A) $x = 4$ and $dx = 0.1$ (B) $x = 9$ and $dx = 0.12$ (C) $x = 1$ and $dx = 0.01$

◆

We now have two interpretations of the symbol dy/dx. Referring to the function $y = f(x) = x^2 + 3x$ in Example 21 with $x = 2$ and $dx = 0.1$, we have

$$\frac{dy}{dx} = f'(2) = 7 \qquad \text{Derivative}$$

and

$$\frac{dy}{dx} = \frac{0.7}{0.1} = 7 \qquad \text{Ratio of differentials}$$

◆ APPROXIMATIONS USING DIFFERENTIALS

Earlier, we noted that for small Δx,

$$\frac{\Delta y}{\Delta x} \approx f'(x) \qquad \text{and} \qquad \Delta y \approx f'(x)\,\Delta x$$

Also, since

$$dy = f'(x)\,dx$$

it follows that

$$\boldsymbol{\Delta y \approx dy}$$

and dy can be used to approximate Δy.

To interpret this result geometrically, we need to recall a basic property of slope. The vertical change in a line is equal to the product of the slope and the horizontal change, as shown in Figure 16.

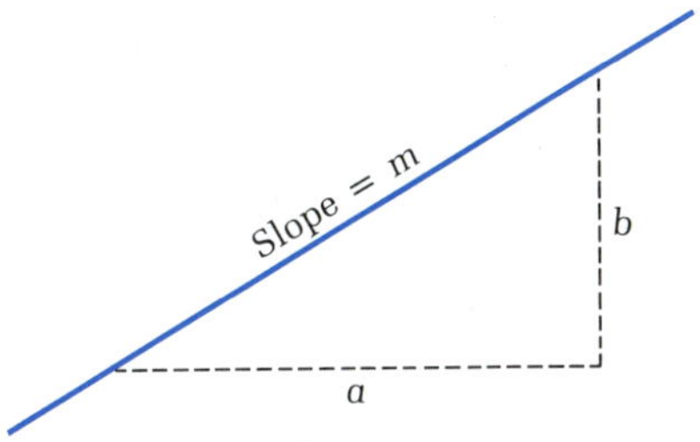

$m = \frac{b}{a}$ or $b = ma$

FIGURE 16

Now consider the line tangent to the graph of $y = f(x)$, as shown in Figure 17. Since $f'(x)$ is the slope of the tangent line and dx is the horizontal change in the tangent line, it follows that the vertical change in the tangent line is given by $dy = f'(x)\,dx$, as indicated in Figure 17.

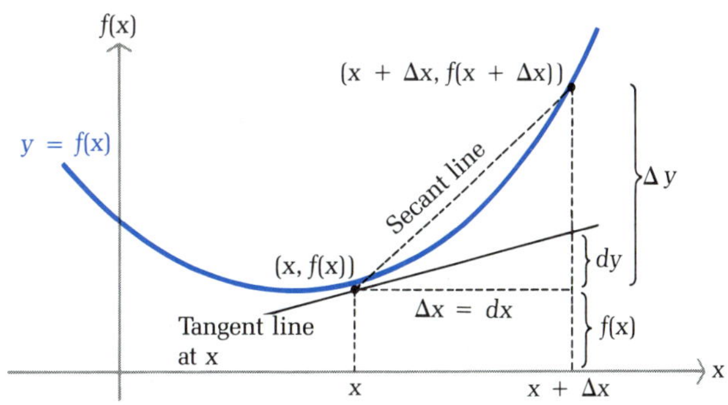

FIGURE 17

Why do we want to use dy to approximate Δy? Because, as we mentioned earlier, dy often is much easier to compute than Δy.

◆ EXAMPLE 22 Find Δy and dy for $f(x) = 6x - x^2$ when $x = 2$ and $\Delta x = dx = 0.1$.

Solution

$$\begin{aligned} \Delta y &= f(x + \Delta x) - f(x) \\ &= f(2.1) - f(2) \\ &= [6(2.1) - (2.1)^2] - [6(2) - 2^2] \\ &= 8.19 - 8 = 0.19 \end{aligned} \qquad \begin{aligned} dy &= f'(x)\,dx \\ &= (6 - 2x)\,dx \\ &= [6 - 2(2)](0.1) \\ &= 0.2 \end{aligned}$$

Notice that dy and Δy differ by only 0.01 in this case. ◆

PROBLEM 22 Repeat Example 22 for $x = 4$ and $\Delta x = dx = -0.2$. ◆

◆ EXAMPLE 23 *Cost–Revenue* A company manufactures and sells x transistor radios per week. If the weekly cost and revenue equations are

$$C(x) = 5{,}000 + 2x \qquad R(x) = 10x - \frac{x^2}{1{,}000} \qquad 0 \le x \le 8{,}000$$

find the approximate changes in revenue and profit if production is increased from 2,000 to 2,010 units per week.

Solution We will approximate ΔR and ΔP with dR and dP, respectively, using $x = 2{,}000$ and $dx = 2{,}010 - 2{,}000 = 10$.

$$R(x) = 10x - \frac{x^2}{1{,}000}$$

$$\begin{aligned} dR &= R'(x)\,dx \\ &= \left(10 - \frac{x}{500}\right)dx \\ &= \left(10 - \frac{2{,}000}{500}\right)10 \\ &= \$60 \text{ per week} \end{aligned}$$

$$\begin{aligned} P(x) = R(x) - C(x) &= 10x - \frac{x^2}{1{,}000} - 5{,}000 - 2x \\ &= 8x - \frac{x^2}{1{,}000} - 5{,}000 \end{aligned}$$

$$\begin{aligned} dP &= P'(x)\,dx \\ &= \left(8 - \frac{x}{500}\right)dx \\ &= \left(8 - \frac{2{,}000}{500}\right)10 \\ &= \$40 \text{ per week} \end{aligned}$$

◆

PROBLEM 23 Repeat Example 23 with production increasing from 6,000 to 6,010. ◆

Comparing the results in Example 23 and Problem 23, we see that an increase in production results in a revenue and profit increase at the 2,000 production level, but a revenue and profit loss at the 6,000 production level.

Before closing, we should mention that even though differentials can be used to approximate certain quantities, the error can be substantial in certain cases. Generally, the smaller Δx is, the better the approximation $\Delta y \approx dy$.

Answers to Matched Problems

20. (A) $\Delta x = 1$, $\Delta y = 5$, $\Delta y/\Delta x = 5$ (B) 4

21. $dy = \frac{1}{2\sqrt{x}}\,dx$ (A) 0.025 (B) 0.02 (C) 0.005

22. $\Delta y = 0.36$, $dy = 0.4$

23. $dR = -\$20/\text{wk}$; $dP = -\$40/\text{wk}$

EXERCISE 4-5

A

In Problems 1–6, find the indicated quantities for $y = f(x) = 3x^2$.

1. Δx, Δy, and $\frac{\Delta y}{\Delta x}$; given $x_1 = 1$ and $x_2 = 4$

2. Δx, Δy, and $\frac{\Delta y}{\Delta x}$; given $x_1 = 2$ and $x_2 = 5$

3. $\frac{f(x_1 + \Delta x) - f(x_1)}{\Delta x}$; given $x_1 = 1$ and $\Delta x = 2$

4. $\frac{f(x_1 + \Delta x) - f(x_1)}{\Delta x}$; given $x_1 = 2$ and $\Delta x = 1$

5. $\frac{\Delta y}{\Delta x}$; given $x_1 = 1$ and $x_2 = 3$

6. $\frac{\Delta y}{\Delta x}$; given $x_1 = 2$ and $x_2 = 3$

In Problems 7–12, find dy for each function

7. $y = 30 + 12x^2 - x^3$

8. $y = 200x - \frac{x^2}{30}$

9. $y = x^2\left(1 - \frac{x}{9}\right)$

10. $y = x^3(60 - x)$

11. $y = \frac{590}{\sqrt{x}}$

12. $y = 52\sqrt{x}$

B

In Problems 13 and 14, find the indicated quantities for $y = f(x) = 3x^2$.

13. (A) $\frac{f(2 + \Delta x) - f(2)}{\Delta x}$ (simplify)

    (B) What does the quantity in part A approach as Δx approaches 0?

14. (A) $\frac{f(3 + \Delta x) - f(3)}{\Delta x}$ (simplify)

    (B) What does the quantity in part A approach as Δx approaches 0?

In Problems 15–18, find dy for each function.

15. $y = (2x + 1)^3$

16. $y = (3x + 5)^5$

17. $y = \frac{x}{x^2 + 9}$

18. $y = \frac{x^2}{(x + 1)^2}$

In Problems 19–22, evaluate dy and Δy for each function for the indicated values.

19. $y = f(x) = x^2 - 3x + 2; \quad x = 5, \quad dx = \Delta x = 0.2$

20. $y = f(x) = 30 + 12x^2 - x^3; \quad x = 2, \quad dx = \Delta x = 0.1$

21. $y = f(x) = 75\left(1 - \frac{2}{x}\right); \quad x = 5, \quad dx = \Delta x = -0.5$

22. $y = f(x) = 100\left(x - \frac{4}{x^2}\right); \quad x = 2, \quad dx = \Delta x = -0.1$

23. A cube with sides 10 inches long is covered with a coat of fiberglass 0.2 inch thick. Use differentials to estimate the volume of the fiberglass shell.

24. A sphere with a radius of 5 centimeters is coated with ice 0.1 centimeter thick. Use differentials to estimate the volume of the ice. [Recall that $V = \frac{4}{3}\pi r^3$, $\pi \approx 3.14$.]

C

25. Find dy if $y = \sqrt[3]{3x^2 - 2x + 1}$.

26. Find dy if $y = (2x^2 - 4)\sqrt{x + 2}$.

27. Find dy and Δy for $y = 52\sqrt{x}$, $x = 4$, and $\Delta x = dx = 0.3$.

28. Find dy and Δy for $y = 590/\sqrt{x}$, $x = 64$, and $\Delta x = dx = 1$.

APPLICATIONS

Use differential approximations in the following problems.

Business & Economics

29. *Advertising.* Using past records, it is estimated that a company will sell N units of a product after spending \$x thousand in advertising, as given by

$$N = 60x - x^2 \qquad 5 \leqslant x \leqslant 30$$

Approximately what increase in sales will result by increasing the advertising budget from \$10,000 to \$11,000? From \$20,000 to \$21,000?

30. *Price–demand.* Suppose, in a grocery chain, the daily demand (in pounds) for chocolate candy at \$x per pound is given by

$$D = 1{,}000 - 40x^2 \qquad 1 \leqslant x \leqslant 5$$

If the price is increased from \$3.00 per pound to \$3.20 per pound, what is the approximate change in demand?

31. *Average cost.* For a company that manufactures tennis rackets, the average cost per racket, $\overline{C}$, is found to be

$$\overline{C} = \frac{400}{x} + 5 + \frac{1}{2}x \qquad x \geqslant 1$$

where x is the number of rackets produced per hour. What will the approximate change in average cost per racket be if production is increased from 20 per hour to 25 per hour? From 40 per hour to 45 per hour?

32. *Revenue and profit.* A company manufactures and sells x televisions per month. If the cost and revenue equations are

$$C(x) = 72{,}000 + 60x \qquad R(x) = 200x - \frac{x^2}{30} \qquad 0 \le x \le 6{,}000$$

what will the approximate changes in revenue and profit be if production is increased from 1,500 to 1,510? From 4,500 to 4,510?

Life Sciences

33. *Pulse rate.* The average pulse rate y (in beats per minute) of a healthy person x inches tall is given approximately by

$$y = \frac{590}{\sqrt{x}} \qquad 30 \le x \le 75$$

Approximately how will the pulse rate change for a height change from 36 to 37 inches? From 64 to 65 inches?

34. *Measurement.* An egg of a particular bird is very nearly spherical. If the radius to the inside of the shell is 5 millimeters and the radius to the outside of the shell is 5.3 millimeters, approximately what is the volume of the shell? [Remember that $V = \frac{4}{3}\pi r^3$, $\pi \approx 3.14$.]

35. *Medicine.* A drug is given to a patient to dilate her arteries. If the radius of an artery is increased from 2 to 2.1 millimeters, approximately how much is the cross-sectional area increased? [Assume the cross-section of the artery is circular; $A = \pi r^2$, $\pi \approx 3.14$.]

36. *Drug sensitivity.* One hour after x milligrams of a particular drug are given to a person, the change in body temperature T (in degrees Fahrenheit) is given by

$$T = x^2\left(1 - \frac{x}{9}\right) \qquad 0 \le x \le 6$$

Approximate the changes in body temperature produced by the following changes in drug dosages:

(A) From 2 to 2.1 milligrams (B) From 3 to 3.1 milligrams
(C) From 4 to 4.1 milligrams

Social Sciences

37. *Learning.* A particular person learning to type has an achievement record given approximately by

$$N = 75\left(1 - \frac{2}{t}\right) \qquad 3 \le t \le 20$$

where N is the number of words per minute typed after t weeks of practice. What is the approximate improvement from 5 to 5.5 weeks of practice?

38. *Learning.* If a person learns y items in x hours, as given approximately by

$$y = 52\sqrt{x} \qquad 0 \le x \le 9$$

what is the approximate increase in the number of items learned when x changes from 1 to 1.1 hours? From 4 to 4.1 hours?

39. *Politics.* In a newly incorporated city, it is estimated that the voting population (in thousands) will increase according to

$$N(t) = 30 + 12t^2 - t^3 \qquad 0 \leq t \leq 8$$

where t is time in years. Find the approximate change in votes for the following time changes:

(A) From 1 to 1.1 years (B) From 4 to 4.1 years
(C) From 7 to 7.1 years

SECTION 4-6 Chapter Review

Important Terms and Symbols

4-1 *First Derivative and Graphs.* Solving inequalities using continuity properties; sign chart; partition numbers; increasing and decreasing functions; rising and falling graphs; local extremum; local maximum; local minimum; critical values; first-derivative test for local extrema

4-2 *Second Derivative and Graphs.* Concave upward; concave downward; second derivative; concavity and the second derivative; inflection point; second-derivative test for local maxima and minima

$$f''(x); \quad \frac{d^2y}{dx^2}; \quad y''; \quad D_x^2 f(x)$$

4-3 *Curve Sketching Techniques: Unified and Extended.* Horizontal asymptote; vertical asymptote; graphing strategy: use $f(x)$ to find the domain, intercepts, horizontal and vertical asymptotes; use $f'(x)$ to find increasing and decreasing regions, and local extrema; use $f''(x)$ to find concave upward and downward regions, and inflection points; oblique asymptote

4-4 *Optimization; Absolute Maxima and Minima.* Absolute maximum; absolute minimum; absolute extrema of a function continuous on a closed interval; second-derivative test for absolute maximum and minimum; optimization problems

4-5 *Increments and Differentials.* Increments; differentials; approximations using differentials

$$\Delta x; \quad \Delta y; \quad \frac{\Delta y}{\Delta x}; \quad \lim_{\Delta x \to 0} \frac{\Delta y}{\Delta x}; \quad dy; \quad df; \quad dx$$

EXERCISE 4-6

Chapter Review

Work through all the problems in this chapter review and check your answers in the back of the book. (Answers to all review problems are there.) Where weaknesses show up, review appropriate sections in the text.

A *Problems 1–8 refer to the following graph of $y = f(x)$:*

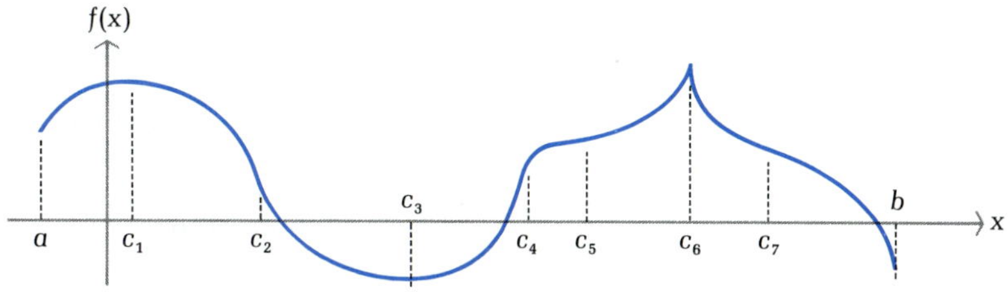

Identify the points or intervals on the x axis that produce the indicated behavior.

1. $f(x)$ is increasing
2. $f'(x) < 0$
3. Graph of f is concave downward
4. Local minima
5. Absolute maxima
6. $f'(x)$ appears to be 0
7. $f'(x)$ does not exist
8. Inflection points
9. Use the given information to sketch the graph of f. Assume that f is continuous on its domain and that all intercepts are included in the table of values.

Domain: All real x

x	−3	−2	−1	0	2	3
f(x)	0	3	2	0	−3	0

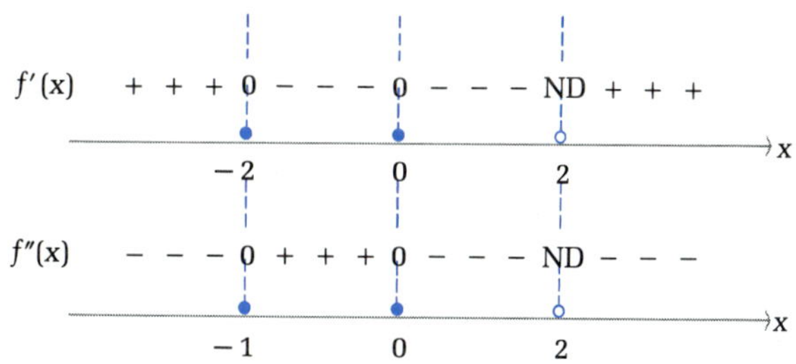

10. Find $f''(x)$ for $f(x) = x^4 + 5x^3$.
11. Find y'' for $y = 3x + 4/x$.
12. Find dy for $y = f(x) = x^3 + 4x$.
13. Find dy for $y = f(x) = (3x^2 - 7)^3$.

B *In Problems 14 and 15, solve each inequality.*

14. $x^2 - x < 12$
15. $\dfrac{x-5}{x^2+3x} > 0$

Problems 16–19 refer to the function: $f(x) = x^3 - 18x^2 + 81x$

16. Using $f(x)$:

(A) Determine the domain of f.
(B) Find any intercepts for the graph of f.
(C) Find any horizontal or vertical asymptotes for the graph of f.

17. Using $f'(x)$:

(A) Find critical values for $f(x)$.
(B) Find partition numbers for $f'(x)$.
(C) Find intervals over which $f(x)$ is increasing; decreasing.
(D) Find any local maxima and minima.

18. Using $f''(x)$:

(A) Find intervals over which the graph of f is concave upward; concave downward.
(B) Find any inflection points.

19. Use the results of Problems 16–18 to graph f.

Problems 20–23 refer to the function: $y = f(x) = \frac{3x}{x+2}$

20. Using $f(x)$:

(A) Determine the domain of f.
(B) Find any intercepts for the graph of f.
(C) Find any horizontal or vertical asymptotes for the graph of f.

21. Using $f'(x)$:

(A) Find critical values for $f(x)$.
(B) Find partition numbers for $f'(x)$.
(C) Find intervals over which $f(x)$ is increasing; decreasing.
(D) Find any local maxima and minima.

22. Using $f''(x)$:

(A) Find intervals over which the graph of f is concave upward; concave downward.
(B) Find any inflection points.

23. Use the results of Problems 20–22 to graph f.

24. Use the second derivative test to find any local extrema for $f(x) = x^3 - 6x^2 - 15x + 12$.

25. Find the absolute maximum and absolute minimum for

$$y = f(x) = x^3 - 12x + 12 \qquad -3 \leq x \leq 5$$

26. Find the absolute minimum for

$$y = f(x) = x^2 + \frac{16}{x^2} \qquad x > 0$$

Find horizontal and vertical asymptotes (if they exist) in Problems 27 and 28.

27. $f(x) = \dfrac{x}{x^2 + 9}$ **28.** $f(x) = \dfrac{x^3}{x^2 - 9}$

29. Find dy and Δy for $f(x) = x^3 - 2x + 1$, $x = 5$, and $\Delta x = dx = 0.1$.

C

30. Find the absolute maximum for $f'(x)$ if

$$f(x) = 6x^2 - x^3 + 8$$

Graph f and f' on the same coordinate system.

31. Find two positive numbers whose product is 400 and whose sum is a minimum. What is the minimum sum?

32. Sketch the graph of $f(x) = (x - 1)^3(x + 3)$ using the graphing strategy discussed in Section 4-3.

33. Find dy and Δy for $y = (2/\sqrt{x}) + 8$, $x = 16$, and $\Delta x = dx = 0.2$.

APPLICATIONS

Business & Economics

34. *Profit.* The profit for a company manufacturing and selling x units per month is given by

$$P(x) = 150x - \frac{x^2}{40} - 50{,}000 \qquad 0 \leq x \leq 5{,}000$$

What production level will produce the maximum profit? What is the maximum profit?

35. *Construction.* A fence is to be built to enclose a rectangular area. The fence along three sides is to be made of material that costs \$5 per foot. The material for the fourth side costs \$15 per foot.

(A) If the area is 5,000 square feet, find the dimensions of the rectangle that will allow the most economical fence to be built.

(B) If \$3,000 is available for the fencing, find the dimensions of the rectangle that will enclose the most area.

36. *Average cost.* The total cost of producing x units per month is given by

$$C(x) = 4{,}000 + 10x + \tfrac{1}{10}x^2$$

Find the minimum average cost. Graph the average cost and the marginal cost functions on the same coordinate system. Include any oblique asymptotes.

37. *Rental income.* A 200 room hotel in Fresno is filled to capacity every night at a rate of \$40 per room. For each \$1 increase in the nightly rate, 4 fewer rooms are rented. If each rented room costs \$8 a day to service, how much should the management charge per room in order to maximize gross profit? What is the maximum gross profit?

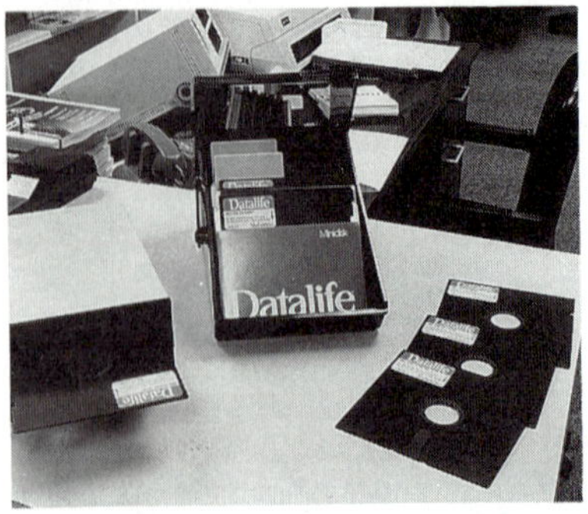

38. *Inventory control.* A computer store sells 7,200 boxes of floppy disks annually. It costs the store \$0.20 to store a box of disks for 1 year. Each time it reorders disks, the store must pay a \$5.00 service charge for processing the order. How many times during the year should the store order disks in order to minimize the total storage and reorder costs?

39. *Rate of change of revenue.* A company is manufacturing a new video game and can sell all it manufactures. The revenue (in dollars) is given by

$$R = 36x - \frac{x^2}{20}$$

where the production output in 1 day is x games. Use dR to approximate the change in revenue if production is increased from 250 to 260 games per day.

Life Sciences

40. *Bacteria control.* If t days after a treatment, the bacteria count per cubic centimeter in a body of water is given by

$$C(t) = 20t^2 - 120t + 800 \qquad 0 \leq t \leq 9$$

in how many days will the count be a minimum?

Social Sciences

41. *Politics.* In a new suburb, it is estimated that the number of registered voters will grow according to

$$N = 10 + 6t^2 - t^3 \qquad 0 \leq t \leq 5$$

where t is time in years and N is in thousands. When will the rate of increase be maximum?

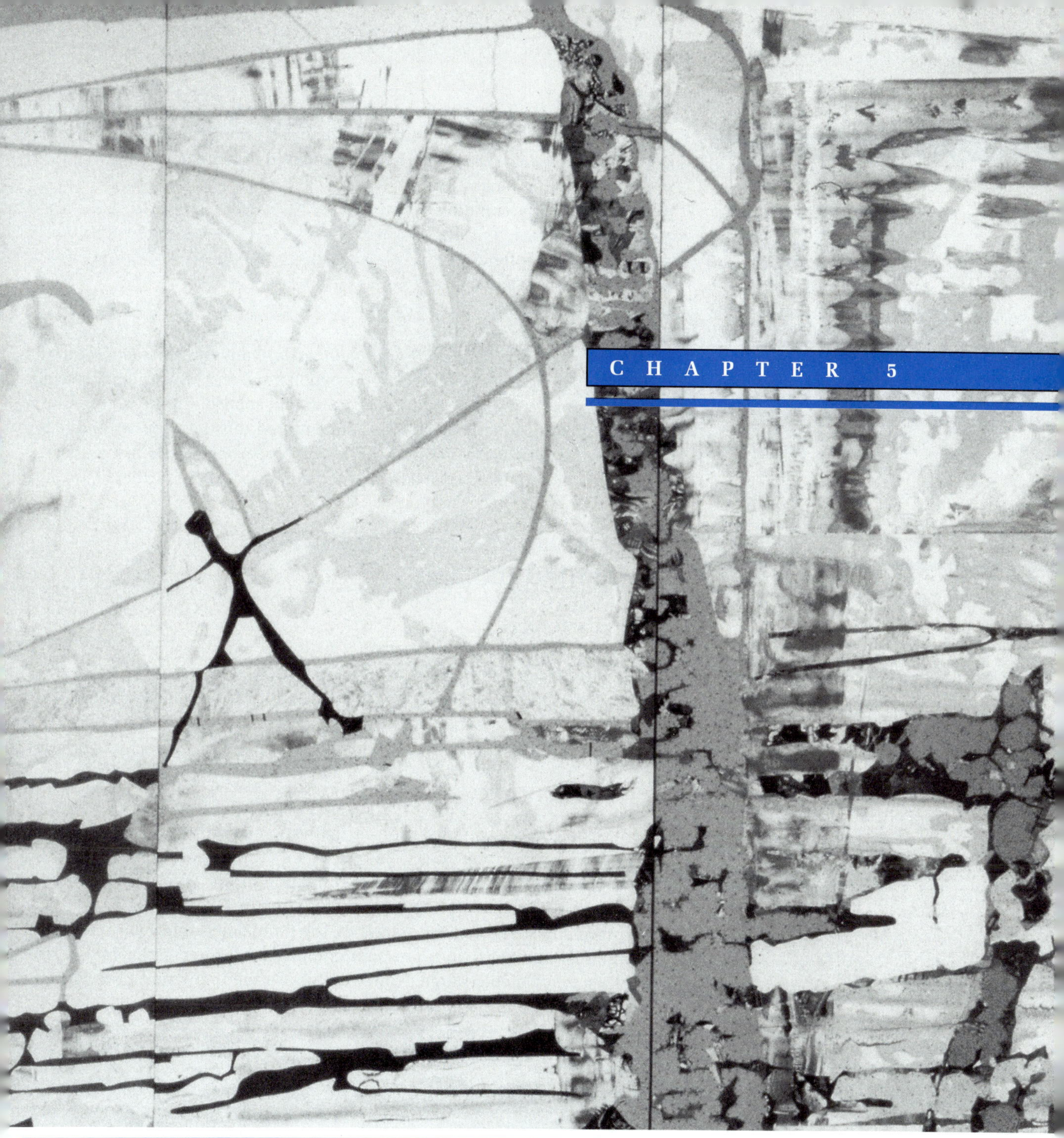

CHAPTER 5

Additional Derivative Topics

CHAPTER 5

Contents

In this chapter we complete our discussion of derivatives by first looking at the differentiation of forms that involve the exponential and logarithmic functions and then considering some additional topics and applications involving all the different types of functions we have encountered thus far. You will probably find it helpful to review some of the important properties of the exponential and logarithmic functions given in Chapter 2 before proceeding further.

SECTION 5-1

The Constant *e* and Continuous Compound Interest

- THE CONSTANT e
- CONTINUOUS COMPOUND INTEREST

In Chapter 2, both the exponential function with base e and continuous compound interest were introduced informally. Now, with limit concepts at our disposal, we can give precise definitions of e and continuous compound interest.

THE CONSTANT e

The special irrational number e is a particularly suitable base for both exponential and logarithmic functions. The reasons for choosing this number as a base will become clear as we develop differentiation formulas for the exponential function e^x and the natural logarithmic function $\ln x$.

In precalculus treatments (Chapter 2), the number e is informally defined as an irrational number that can be approximated by the expression $[1 + (1/n)]^n$ by taking n sufficiently large. Now we will use the limit concept to formally define e as either of the following two limits:

The Number *e*

$$e = \lim_{n\to\infty}\left(1 + \frac{1}{n}\right)^n \quad \text{or, alternately,} \quad e = \lim_{s\to 0}(1 + s)^{1/s}$$

$$e = 2.718\ 281\ 828\ 459\ldots$$

We will use both these limit forms. [*Note:* If $s = 1/n$, then as $n \to \infty$, $s \to 0$.]

The proof that the indicated limits exist and represent an irrational number between 2 and 3 is not easy and is omitted here. Many people reason (incorrectly) that the limits are 1, since "$(1 + s)$ approaches 1 as $s \to 0$, and 1 to any power is 1." A little experimentation with a calculator can convince you otherwise. Consider the table of values for s and $f(s) = (1 + s)^{1/s}$ and the graph shown in Figure 1 for s close to 0.

s approaches 0 from the left $\to 0 \leftarrow$ s approaches 0 from the right

s	-0.5	-0.2	-0.1	$-0.01 \to 0 \leftarrow 0.01$	0.1	0.2	0.5
$(1 + s)^{1/s}$	4.0000	3.0518	2.8680	$2.7320 \to e \leftarrow 2.7048$	2.5937	2.4883	2.2500

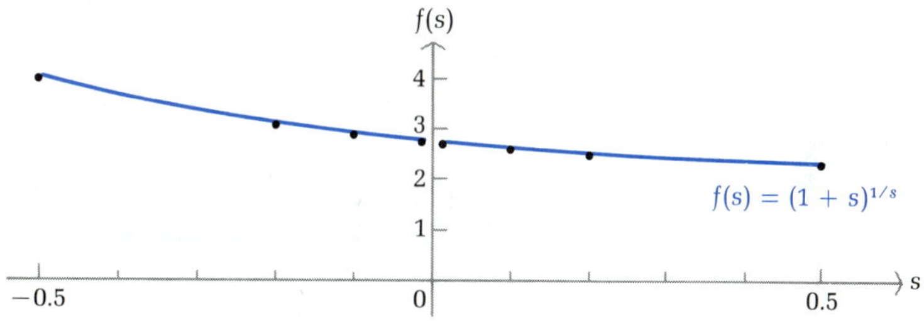

FIGURE 1

Compute some of the table values with a calculator yourself, and also try several values of s even closer to 0. Note that the function is discontinuous at $s = 0$.

Exactly who discovered e is still being debated. It is named after the great mathematician Leonard Euler (1707–1783), who computed e to twenty-three decimal places using $[1 + (1/n)]^n$.

◆ CONTINUOUS COMPOUND INTEREST

Now we will see how e appears quite naturally in the important application of compound interest. Let us start with simple interest, move on to compound interest, and then on to continuous compound interest.

If a principal P is borrowed at an annual rate of r,* then after t years at simple interest the borrower will owe the lender an amount A given by

$$A = P + Prt = P(1 + rt) \qquad \text{Simple interest} \tag{1}$$

* If r is the interest rate written as a decimal, then $100r\%$ is the rate using %. For example, if $r = 0.12$, then we have $100r\% = 100(0.12)\% = 12\%$. The expressions 0.12 and 12% are therefore equivalent. Unless stated otherwise, all formulas in this book use r as a decimal.

On the other hand, if interest is compounded n times a year, then the borrower will owe the lender an amount A given by

$$A = P\left(1 + \frac{r}{n}\right)^{nt} \qquad \text{Compound interest} \tag{2}$$

where r/n is the interest rate per compounding period and nt is the number of compounding periods. Suppose P, r, and t in (2) are held fixed and n is increased. Will the amount A increase without bound, or will it tend to approach some limiting value?

Let us perform a calculator experiment before we attack the general limit problem. If $P = \$100$, $r = 0.06$, and $t = 2$ years, then

$$A = 100\left(1 + \frac{0.06}{n}\right)^{2n}$$

We compute A for several values of n in Table 1. The biggest gain appears in the first step; then the gains slow down as n increases. In fact, it appears that A might be tending to approach \$112.75 as n gets larger and larger.

TABLE 1

COMPOUNDING FREQUENCY	n	$A = 100\left(1 + \frac{0.06}{n}\right)^{2n}$
Annually	1	\$112.3600
Semiannually	2	112.5509
Quarterly	4	112.6493
Weekly	52	112.7419
Daily	365	112.7486
Hourly	8,760	112.7496

Now we turn back to the general problem for a moment. Keeping P, r, and t fixed in equation (2), we compute the following limit and observe an interesting and useful result.

$$\lim_{n\to\infty} P\left(1 + \frac{r}{n}\right)^{nt} = P \lim_{n\to\infty}\left(1 + \frac{r}{n}\right)^{(n/r)rt}$$

Insert r/r in the exponent and let $s = r/n$. Note that $n \to \infty$ implies $s \to 0$.

$$= P \lim_{s\to 0}[(1 + s)^{1/s}]^{rt}$$

Use the limit property given in the footnote below.*

$$= P[\lim_{s\to 0}(1 + s)^{1/s}]^{rt}$$

$\lim_{s\to 0}(1 + s)^{1/s} = e$

$$= Pe^{rt}$$

* The following new limit property is used: If $\lim_{x\to c} f(x)$ exists, then $\lim_{x\to c}[f(x)]^p = [\lim_{x\to c} f(x)]^p$, provided the last expression names a real number.

The resulting formula is called the **continuous compound interest formula,** a very important and widely used formula in business and economics.

Continuous Compound Interest

$$A = Pe^{rt}$$

where

P = Principal

r = Annual nominal interest rate compounded continuously

t = Time in years

A = Amount at time t

◆ EXAMPLE 1 If \$100 is invested at an annual nominal rate of 6% compounded continuously, what amount will be in the account after 2 years?

Solution

$$\begin{aligned} A &= Pe^{rt} \\ &= 100e^{(0.06)(2)} \qquad \text{6\% is equivalent to } r = 0.06. \\ &\approx \$112.7497 \end{aligned}$$

(Compare this result with the values calculated in Table 1.) ◆

PROBLEM 1 What amount (to the nearest cent) will an account have after 5 years if \$100 is invested at an annual nominal rate of 8% compounded annually? Semiannually? Continuously? ◆

◆ EXAMPLE 2 If \$100 is invested at 12% compounded continuously,* graph the amount in the account relative to time for a period of 10 years.

Solution We want to graph

$$A = 100e^{0.12t} \qquad 0 \le t \le 10$$

---

* Following common usage, we will often write the form "at 12% compounded continuously," understanding that this means "at an annual nominal rate of 12% compounded continuously."

We construct a table of values using a calculator, graph the points from the table, and join the points with a smooth curve.

t	A
0	100
1	113
2	127
3	143
4	162
5	182
6	205
7	232
8	261
9	294
10	332

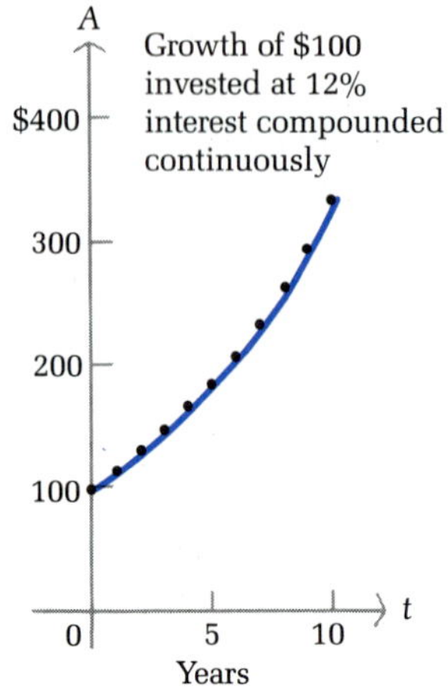

◆

PROBLEM 2 If $5,000 is invested at 20% compounded continuously, graph the amount in the account relative to time for a period of 10 years. ◆

◆ EXAMPLE 3 How long will it take an investment of $5,000 to grow to $8,000 if it is invested at 12% compounded continuously?

Solution Starting with the continuous compound interest formula $A = Pe^{rt}$, we must solve for t:

$$A = Pe^{rt}$$

$$8{,}000 = 5{,}000e^{0.12t}$$ Divide both sides by 5,000 and reverse the equation.

$$e^{0.12t} = 1.6$$ Take the natural logarithm of both sides—recall that $\log_b b^x = x$.

$$\ln e^{0.12t} = \ln 1.6$$

$$0.12t = \ln 1.6$$

$$t = \frac{\ln 1.6}{0.12}$$

$$t \approx 3.92 \text{ years}$$ ◆

PROBLEM 3 How long will it take an investment of $10,000 to grow to $15,000 if it is invested at 9% compounded continuously? ◆

◆ EXAMPLE 4 How long will it take money to double if it is invested at 18% compounded continuously?

Solution Starting with the continuous compound interest formula $A = Pe^{rt}$, we must solve for t given $A = 2P$ and $r = 0.18$:

$$2P = Pe^{0.18t} \quad \text{Divide both sides by } P \text{ and reverse the equation.}$$

$$e^{0.18t} = 2 \quad \text{Take the natural logarithm of both sides.}$$

$$\ln e^{0.18t} = \ln 2$$

$$0.18t = \ln 2$$

$$t = \frac{\ln 2}{0.18}$$

$$t \approx 3.85 \text{ years}$$

◆

PROBLEM 4 How long will it take money to triple if it is invested at 12% compounded continuously? ◆

Answers to Matched Problems

1. \$146.93; \$148.02; \$149.18
2. $A = 5{,}000e^{0.2t}$

t	A
0	5,000
1	6,107
2	7,459
3	9,111
4	11,128
5	13,591
6	16,601
7	20,276
8	24,765
9	30,248
10	36,945

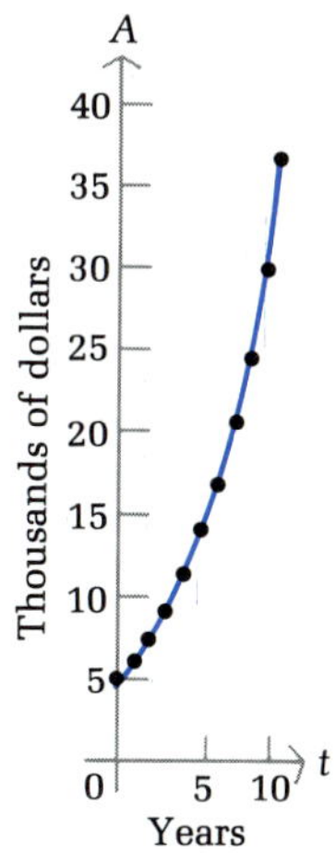

3. 4.51 yr
4. 9.16 yr

EXERCISE 5-1

A *Use a calculator or table to evaluate A to the nearest cent in Problems 1 and 2.*

1. $A = \$1{,}000e^{0.1t}$ for $t = 2, 5,$ and 8
2. $A = \$5{,}000e^{0.08t}$ for $t = 1, 4,$ and 10

B *In Problems 3–8, solve for t or r to two decimal places.*

3. $2 = e^{0.06t}$ **4.** $2 = e^{0.03t}$ **5.** $3 = e^{0.1t}$

6. $3 = e^{0.25t}$ **7.** $2 = e^{5r}$ **8.** $3 = e^{10r}$

C *In Problems 9 and 10, complete each table to five decimal places using a calculator.*

9.

n	$[1 + (1/n)]^n$
10	2.593 74
100	
1,000	
10,000	
100,000	
1,000,000	
10,000,000	
↓	↓
∞	e = 2.718 281 828 459...

10.

s	$(1 + s)^{1/s}$
0.01	2.704 81
−0.01	
0.001	
−0.001	
0.000 1	
−0.000 1	
0.000 01	
−0.000 01	
↓	↓
0	e = 2.718 281 828 459...

Problems 11 and 12 require the use of a graphic calculator or computer.

11. It can be shown that the number e satisfies the inequality

$$\left(1 + \frac{1}{n}\right)^n < e < \left(1 + \frac{1}{n}\right)^{n+1} \qquad n \geq 1$$

Illustrate this graphically by graphing $y = (1 + 1/n)^n$, $y = 2.718\,281\,828 \approx e$, and $y = (1 + 1/n)^{n+1}$ in the same viewing rectangle for $1 \leq n \leq 20$.

12. It can be shown that

$$e^s = \lim_{n\to\infty}\left(1 + \frac{s}{n}\right)^n$$

for any real number s. Illustrate this graphically for $s = 2$ by graphing $y = (1 + 2/n)^n$ and $y = 7.389\ 056\ 099 \approx e^2$ in the same viewing rectangle for $1 \leq n \leq 50$.

Business & Economics

13. *Continuous compound interest.* If \$20,000 is invested at an annual nominal rate of 12% compounded continuously, how much will it be worth in 8.5 years?

14. *Continuous compound interest.* Assume \$1 had been invested at an annual nominal rate of 4% compounded continuously, at the time of the birth of Christ. What would be the value of the account in solid gold Earths in the year 2000? Assume that the Earth weighs approximately 2.11×10^{26} ounces and that gold will be worth \$1,000 an ounce in the year 2000. What would be the value of the account in dollars at simple interest?

15. *Present value.* A note will pay \$20,000 at maturity 10 years from now. How much should you be willing to pay for the note now if money is worth 7% compounded continuously?

16. *Present value.* A note will pay \$50,000 at maturity 5 years from now. How much should you be willing to pay for the note now if money is worth 8% compounded continuously?

17. *Continuous compound interest.* An investor bought stock for \$20,000. Four years later, the stock was sold for \$30,000. If interest is compounded continuously, what annual nominal rate of interest did the original \$20,000 investment earn?

18. *Continuous compound interest.* A family paid \$40,000 cash for a house. Ten years later, they sold the house for \$100,000. If interest is compounded continuously, what annual nominal rate of interest did the original \$40,000 investment earn?

19. *Present value.* Solving $A = Pe^{rt}$ for P, we obtain

$$P = Ae^{-rt}$$

which is the present value of the amount A due in t years if money earns interest at an annual nominal rate r compounded continuously.

(A) Graph $P = 10{,}000e^{-0.08t}$, $0 \leq t \leq 50$.

(B) $\lim_{t\to\infty} 10{,}000e^{-0.08t} = ?$ (Guess, using part A.)

[*Conclusion:* The longer the duration of time until the amount A is due, the smaller its present value, as we would expect.]

20. *Present value.* Referring to Problem 19, in how many years will the \$10,000 have to be due in order for its present value to be \$5,000?

21. *Doubling time.* How long will it take money to double if it is invested at 25% compounded continuously?

22. *Doubling time.* How long will it take money to double if it is invested at 5% compounded continuously?

23. *Doubling rate.* At what nominal rate compounded continuously must money be invested to double in 5 years?

24. *Doubling rate.* At what nominal rate compounded continuously must money be invested to double in 3 years?

25. *Doubling time.* It is instructive to look at doubling times for money invested at various nominal rates of interest compounded continuously. Show that doubling time t at an annual rate r compounded continuously is given by

$$t = \frac{\ln 2}{r}$$

26. *Doubling time.* Graph the doubling-time equation from Problem 25 for $0 < r < 1.00$. Identify vertical and horizontal asymptotes.

Life Sciences

27. *World population.* A mathematical model for world population growth over short periods of time is given by

$$P = P_0 e^{rt}$$

where

$$P_0 = \text{Population at time } t = 0$$
$$r = \text{Continuous compound rate of growth}$$
$$t = \text{Time in years}$$
$$P = \text{Population at time } t$$

How long will it take the world population to double if it continues to grow at its current continuous compound rate of 2% per year?

28. *World population.* Repeat Problem 27 under the assumption that the world population is growing at a continuous compound rate of 1% per year.

29. *Population growth.* Some underdeveloped nations have population doubling times of 20 years. At what continuous compound rate is the population growing? (Use the population growth model in Problem 27.)

30. *Population growth.* Some developed nations have population doubling times of 120 years. At what continuous compound rate is the population growing? (Use the population growth model in Problem 27.)

31. *Radioactive decay.* A mathematical model for the decay of radioactive substances is given by

$$Q = Q_0 e^{rt}$$

where

$$Q_0 = \text{Amount of the substance at time } t = 0$$
$$r = \text{Continuous compound rate of decay}$$
$$t = \text{Time in years}$$
$$Q = \text{Amount of the substance at time } t$$

If the continuous compound rate of decay of radium per year is $r = -0.000\ 433\ 2$, how long will it take an amount of radium to decay to half the original amount? (This period of time is the half-life of the substance.)

32. *Radioactive decay.* The continuous compound rate of decay of carbon-14 per year is $r = -0.000\ 123\ 8$. How long will it take an amount of carbon-14 to decay to half the original amount? (Use the radioactive decay model in Problem 31.)

33. *Radioactive decay.* A cesium isotope has a half-life of 30 years. What is the continuous compound rate of decay? (Use the radioactive decay model in Problem 31.)

34. *Radioactive decay.* A strontium isotope has a half-life of 90 years. What is the continuous compound rate of decay? (Use the radioactive decay model in Problem 31.)

Social Sciences

35. *World population.* If the world population is now 5 billion (5×10^9) people and if it continues to grow at a continuous compound rate of 2% per year, how long will it be before there is only 1 square yard of land per person? (The Earth has approximately 1.68×10^{14} square yards of land.)

SECTION 5-2 Derivatives of Logarithmic and Exponential Functions

- DERIVATIVE FORMULAS FOR $\ln x$ AND e^x
- GRAPHING TECHNIQUES
- APPLICATION

In this section, we discuss derivative formulas for $\ln x$ and e^x. Out of all the possible choices for bases for the logarithmic and exponential functions, $\log_b x$ and b^x, it turns out (as we will see in this and the next section) that the simplest derivative formulas occur when the base b is chosen to be e.

◆ DERIVATIVE FORMULAS FOR $\ln x$ AND e^x

We are now ready to derive a formula for the derivative of

$$f(x) = \ln x = \log_e x \qquad x > 0$$

using the definition of the derivative

$$f'(x) = \lim_{h \to 0} \frac{f(x+h) - f(x)}{h}$$

and the two-step process discussed in Section 3-3.

Step 1. Simplify the difference quotient first:

$$\frac{f(x+h) - f(x)}{h} = \frac{\ln(x+h) - \ln x}{h}$$

$$= \frac{1}{h}[\ln(x+h) - \ln x] \qquad \text{Use } \ln A - \ln B = \ln \frac{A}{B}.$$

$$= \frac{1}{h} \ln \frac{x+h}{x} \qquad \text{Multiply by } 1 = x/x \text{ to change form.}$$

$$= \frac{x}{x} \cdot \frac{1}{h} \ln \frac{x+h}{x}$$

$$= \frac{1}{x}\left[\frac{x}{h} \ln \left(1 + \frac{h}{x}\right)\right] \qquad \text{Use } p \ln A = \ln A^p.$$

$$= \frac{1}{x} \ln \left(1 + \frac{h}{x}\right)^{x/h}$$

Step 2. Find the limit. Let $s = h/x$. For x fixed, if $h \to 0$, then $s \to 0$. Thus,

$$D_x \ln x = \lim_{h \to 0} \frac{f(x+h) - f(x)}{h}$$

$$= \lim_{h \to 0} \left[\frac{1}{x} \ln\left(1 + \frac{h}{x}\right)^{x/h} \right]$$ Let $s = h/x$. Note that $h \to 0$ implies $s \to 0$.

$$= \frac{1}{x} \lim_{s \to 0} [\ln(1+s)^{1/s}]$$ Use the new limit property given in the footnote below.*

$$= \frac{1}{x} \ln[\lim_{s \to 0} (1+s)^{1/s}]$$ Use the definition of e.

$$= \frac{1}{x} \ln e$$ $\ln e = \log_e e = 1$

$$= \frac{1}{x}$$

Thus,

$$D_x \ln x = \frac{1}{x}$$

In the next section, we will show that, in general,

$$D_x \log_b x = \frac{1}{\ln b}\left(\frac{1}{x}\right)$$

which is a somewhat more complicated result than the above—unless $b = e$.

We now apply the two-step process to the exponential function $f(x) = e^x$. In the process, we will use (without proof) the fact that

$$\lim_{h \to 0} \left(\frac{e^h - 1}{h} \right) = 1$$

[Try computing $(e^h - 1)/h$ for values of h closer and closer to 0 and on either side of 0 to convince yourself of the reasonableness of this limit.]

Step 1. Simplify the difference quotient first:

$$\frac{f(x+h) - f(x)}{h} = \frac{e^{x+h} - e^x}{h}$$ Use $e^{a+b} = e^a e^b$.

$$= \frac{e^x e^h - e^x}{h}$$ Factor out e^x.

$$= e^x \left(\frac{e^h - 1}{h} \right)$$

* The following new limit property is used: If $\lim_{x \to c} f(x)$ exists and is positive, then $\lim_{x \to c} [\ln f(x)] = \ln[\lim_{x \to c} f(x)]$.

Step 2. Compute the limit of the result in step 1:

$$D_x e^x = \lim_{h \to 0} \frac{f(x+h) - f(x)}{h}$$

$$= \lim_{h \to 0} e^x\left(\frac{e^h - 1}{h}\right)$$

$$= e^x \lim_{h \to 0} \left(\frac{e^h - 1}{h}\right) \qquad \text{Use the assumed limit given above.}$$

$$= e^x \cdot 1 = e^x$$

Thus,

$$D_x e^x = e^x$$

In the next section, we will show that

$$D_x b^x = b^x \ln b$$

which is, again, a somewhat more complicated result than the above—unless $b = e$.

The two results just obtained explain why e^x is so widely used that it is sometimes referred to as *the* exponential function. These two new and important derivative formulas are restated in the box for reference.

Derivatives of the Natural Logarithmic and Exponential Functions

$$D_x \ln x = \frac{1}{x} \qquad D_x e^x = e^x$$

These new derivative formulas can be combined with the rules of differentiation discussed in Chapter 3 to differentiate a wide variety of functions.

◆ EXAMPLE 5 Find $f'(x)$ for:

(A) $f(x) = 2e^x + 3 \ln x$ (B) $f(x) = \dfrac{e^x}{x^3}$

(C) $f(x) = (\ln x)^4$ (D) $f(x) = \ln x^4$

Solutions

(A) $f'(x) = \boxed{2\, D_x e^x + 3\, D_x \ln x}$

$$= 2e^x + 3\left(\frac{1}{x}\right) = 2e^x + \frac{3}{x}$$

(B) $f'(x) = \boxed{\dfrac{x^3\, D_x e^x - e^x\, D_x x^3}{(x^3)^2}}$ Quotient rule

$$= \frac{x^3e^x - e^x 3x^2}{x^6} = \frac{x^2e^x(x-3)}{x^6} = \frac{e^x(x-3)}{x^4}$$

(C) $D_x(\ln x)^4 = 4(\ln x)^3 D_x \ln x$ Power rule for functions

$$= 4(\ln x)^3 \left(\frac{1}{x}\right) = \frac{4(\ln x)^3}{x}$$

(D) $D_x \ln x^4 = D_x(4 \ln x)$ Property of logarithms

$$= 4\left(\frac{1}{x}\right) = \frac{4}{x}$$

◆

PROBLEM 5 Find $f'(x)$ for:

(A) $f(x) = 4 \ln x - 5e^x$ (B) $f(x) = x^2e^x$
(C) $f(x) = \ln x^3$ (D) $f(x) = (\ln x)^3$ ◆

Common Error

$D_x\, e^x \neq xe^{x-1}$ $D_x\, e^x = e^x$

The power rule cannot be used to differentiate the exponential function. The power rule applies to exponential forms x^n where the exponent is a constant and the base is a variable. In the exponential form e^x, the base is a constant and the exponent is a variable.

◆ GRAPHING TECHNIQUES

Using the techniques discussed in Chapter 4, we can use first and second derivatives to gain useful information about the graphs of $y = \ln x$ and $y = e^x$. Using the derivative formulas given above, we can construct Table 2.

TABLE 2

$\ln x$		e^x	
$y = \ln x$	$x > 0$	$y = e^x$	$-\infty < x < \infty$
$y' = 1/x > 0$	$x > 0$	$y' = e^x > 0$	$-\infty < x < \infty$
$y'' = -1/x^2 < 0$	$x > 0$	$y'' = e^x > 0$	$-\infty < x < \infty$

From the table, we can see that both functions are increasing throughout their respective domains, the graph of $y = \ln x$ is always concave downward, and the graph of $y = e^x$ is always concave upward. It can be shown that the y axis is a vertical asymptote for the graph of $y = \ln x$ ($\lim_{x\to 0^+} \ln x = -\infty$), and the x axis is a horizontal asymptote for the graph of $y = e^x$ ($\lim_{x\to -\infty} e^x = 0$). Both equations are graphed in Figure 2.

Notice that if we fold the page along the dashed line $y = x$, the two graphs match exactly (see Section 2-3). Also notice that both graphs are unbounded as $x \to \infty$. Comparing each graph with the graph of $y = x$ (the dashed line), we

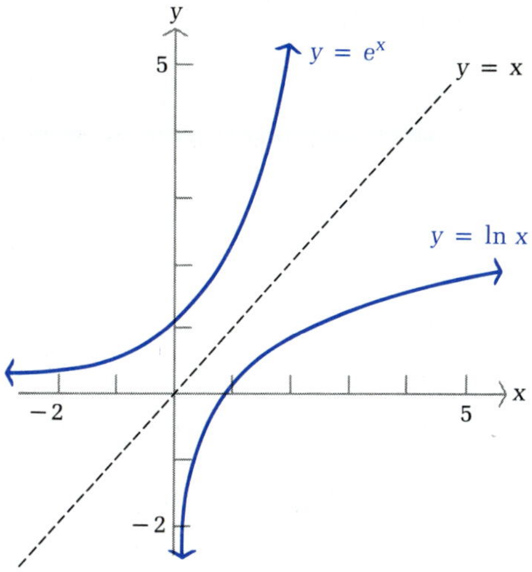

FIGURE 2
e^x is continuous on $(-\infty, \infty)$
$\ln x$ is continuous on $(0, \infty)$

conclude that e^x grows more rapidly than x and ln x grows more slowly than x. In fact, the following limits can be established:

$$\lim_{x \to \infty} \frac{x^p}{e^x} = 0, \quad p > 0 \qquad \text{and} \qquad \lim_{x \to \infty} \frac{\ln x}{x^p} = 0, \quad p > 0$$

These limits indicate that e^x grows more rapidly than any positive power of x, and ln x grows more slowly than any positive power of x.

Now we will apply graphing techniques to a slightly more complicated function.

◆ EXAMPLE 6 Sketch the graph of $f(x) = xe^x$ using the graphing strategy discussed in Section 4-3.

Solution Step 1. Use $f(x)$: $f(x) = xe^x$

(A) Domain: All real numbers
(B) Intercepts:
y intercept: $f(0) = 0$
x intercept: $xe^x = 0$ for $x = 0$ only, since $e^x > 0$ for all x (see Figure 2).
(C) Asymptotes:
Vertical asymptotes: None
Horizontal asymptotes: We have not developed limit techniques for functions of this type to determine the behavior of $f(x)$ as $x \to -\infty$ and $x \to \infty$.

However, the following tables of values suggest the nature of the graph of f as $x \to -\infty$ and $x \to \infty$:

x	1	5	10	$\to \infty$
$f(x)$	2.72	742.07	220,264.66	$\to \infty$

x	-1	-5	-10	$\to -\infty$
$f(x)$	-0.37	-0.03	$-0.000\ 45$	$\to 0$

Step 2. Use $f'(x)$:

$$f'(x) = x\,D_x\,e^x + e^x\,D_x\,x$$
$$= xe^x + e^x = e^x(x + 1)$$

Critical value: -1
Partition number: -1
Sign chart for $f'(x)$:

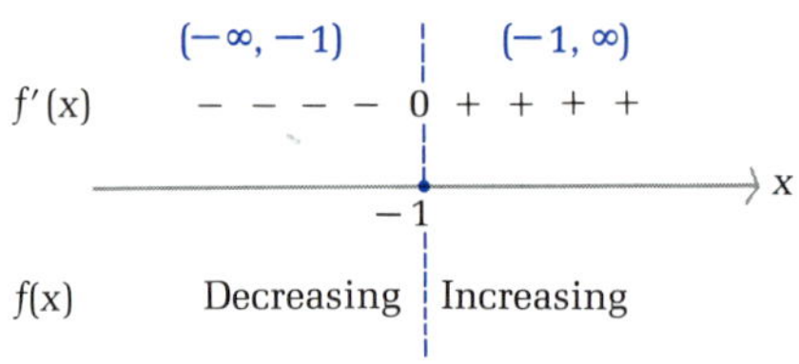

Test Numbers

x	$f'(x)$	
-2	$-e^{-2}$	(−)
0	1	(+)

Thus, $f(x)$ decreases on $(-\infty, -1)$, has a local minimum at $x = -1$, and increases on $(-1, \infty)$. [Since $e^x > 0$ for all x, we do not have to evaluate e^{-2} to conclude that $-e^{-2} < 0$ when using the test number -2.]

Step 3. Use $f''(x)$:

$$f''(x) = e^x\,D_x(x + 1) + (x + 1)\,D_x e^x$$
$$= e^x + (x + 1)e^x = e^x(x + 2)$$

Sign chart for $f''(x)$ (partition number is -2):

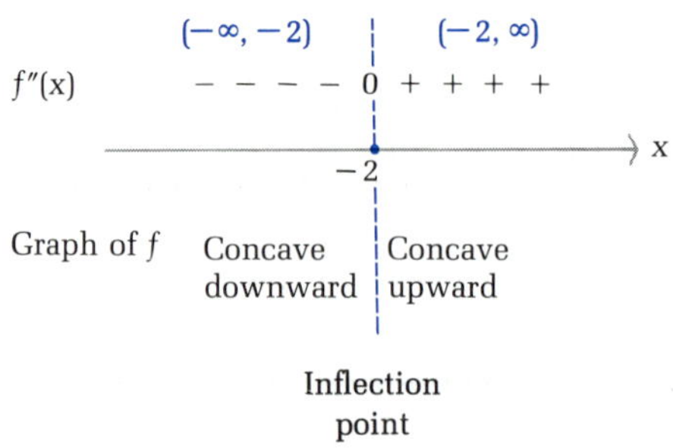

Test Numbers

x	$f''(x)$	
-3	$-e^{-3}$	(−)
-1	e^{-1}	(+)

Thus, the graph of f is concave downward on $(-\infty, -2)$, has an inflection point at $x = -2$, and is concave upward on $(-2, \infty)$.

Step 4. Sketch the graph of f using the information from steps 1–3:

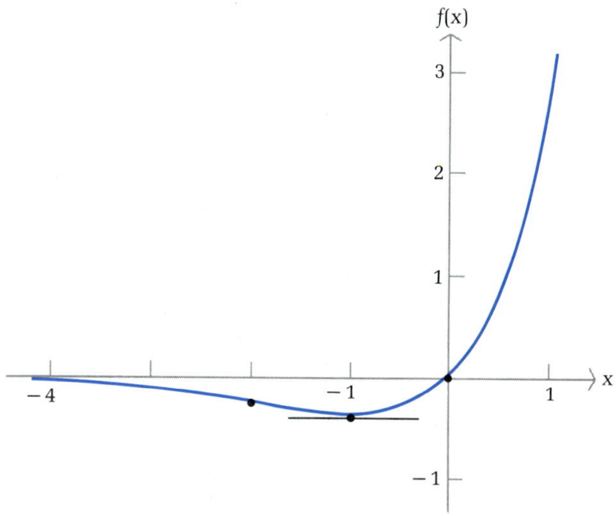

x	$f(x)$
−2	−0.27
−1	−0.37
0	0

PROBLEM 6 Sketch the graph of $f(x) = x \ln x$.

APPLICATION

EXAMPLE 7 *Maximum Profit*

The market research department of a chain of pet stores test marketed their aquarium pumps (as well as other items) in several of their stores in a test city. They found that the weekly demand for aquarium pumps is given approximately by

$$p = 12 - 2 \ln x \qquad 0 < x < 90$$

where x is the number of pumps sold each week and $\$p$ is the price of one pump. If each pump costs the chain \$3, how should it be priced in order to maximize the weekly profit?

Solution

Although we want to find the price that maximizes the weekly profit, it will be easier to first find the number of pumps that will maximize the weekly profit. The revenue equation is

$$R(x) = xp = 12x - 2x \ln x$$

The cost equation is

$$C(x) = 3x$$

and the profit equation is

$$\begin{aligned} P(x) &= R(x) - C(x) \\ &= 12x - 2x \ln x - 3x \\ &= 9x - 2x \ln x \end{aligned}$$

Thus, we must solve the following:

$$\text{Maximize} \quad P(x) = 9x - 2x \ln x \qquad 0 < x < 90$$

$$\begin{aligned} P'(x) &= 9 - 2x\left(\frac{1}{x}\right) - 2 \ln x \\ &= 7 - 2 \ln x = 0 \\ 2 \ln x &= 7 \\ \ln x &= 3.5 \\ x &= e^{3.5} \end{aligned}$$

$$P''(x) = -2\left(\frac{1}{x}\right) = -\frac{2}{x}$$

Since $x = e^{3.5}$ is the only critical value and $P''(e^{3.5}) < 0$, the maximum weekly profit occurs when $x = e^{3.5} \approx 33$ and $p = 12 - 2 \ln e^{3.5} = \$5$. ◆

PROBLEM 7 Repeat Example 7 if each pump costs the chain \$3.50. ◆

Answers to Matched Problems

5. (A) $(4/x) - 5e^x$ (B) $xe^x(x + 2)$ (C) $3/x$ (D) $3(\ln x)^2/x$
6. Domain: $(0, \infty)$
   y intercept: None [$f(0)$ is not defined]
   x intercept: $x = 1$
   Increasing on (e^{-1}, ∞)
   Decreasing on $(0, e^{-1})$
   Local minimum at $x = e^{-1} \approx 0.368$
   Concave upward on $(0, \infty)$

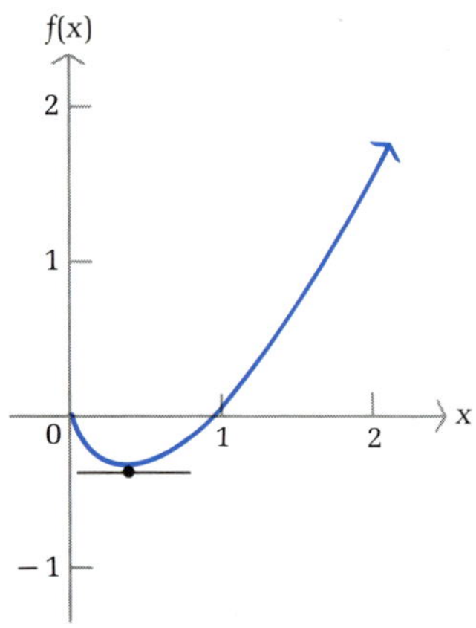

x	5	10	100	$\to\infty$
$f(x)$	8.05	23.03	460.52	$\to\infty$

x	0.1	0.01	0.001	0.000 1	$\to 0$
$f(x)$	−0.23	−0.046	−0.006 9	−0.000 92	$\to 0$

7. Maximum profit occurs for $x = e^{3.25} \approx 26$ and $p = \$5.50$.

EXERCISE 5-2

A *Find $f'(x)$.*

1. $f(x) = 6e^x - 7 \ln x$
2. $f(x) = 4e^x + 5 \ln x$
3. $f(x) = 2x^e + 3e^x$
4. $f(x) = 4e^x - ex^e$
5. $f(x) = \ln x^5$
6. $f(x) = (\ln x)^5$
7. $f(x) = (\ln x)^2$
8. $f(x) = \ln x^2$

B

9. $f(x) = x^4 \ln x$
10. $f(x) = x^3 \ln x$
11. $f(x) = x^3 e^x$
12. $f(x) = x^4 e^x$
13. $f(x) = \dfrac{e^x}{x^2 + 9}$
14. $f(x) = \dfrac{e^x}{x^2 + 4}$
15. $f(x) = \dfrac{\ln x}{x^4}$
16. $f(x) = \dfrac{\ln x}{x^3}$
17. $f(x) = (x + 2)^3 \ln x$
18. $f(x) = (x - 1)^2 \ln x$
19. $f(x) = (x + 1)^3 e^x$
20. $f(x) = (x - 2)^3 e^x$
21. $f(x) = \dfrac{x^2 + 1}{e^x}$
22. $f(x) = \dfrac{x + 1}{e^x}$
23. $f(x) = x(\ln x)^3$
24. $f(x) = x(\ln x)^2$
25. $f(x) = (4 - 5e^x)^3$
26. $f(x) = (5 - \ln x)^4$
27. $f(x) = \sqrt{1 + \ln x}$
28. $f(x) = \sqrt{1 + e^x}$
29. $f(x) = xe^x - e^x$
30. $f(x) = x \ln x - x$
31. $f(x) = 2x^2 \ln x - x^2$
32. $f(x) = x^2 e^x - 2xe^x + 2e^x$

Find the equation of the line tangent to the graph of $y = f(x)$ at the indicated value of x.

33. $f(x) = e^x$; $x = 1$
34. $f(x) = e^x$; $x = 2$
35. $f(x) = \ln x$; $x = e$
36. $f(x) = \ln x$; $x = 1$

C *Find the indicated extremum of each function for $x > 0$.*

37. Absolute maximum value of: $f(x) = 4x - x \ln x$
38. Absolute minimum value of: $f(x) = x \ln x - 3x$
39. Absolute minimum value of: $f(x) = \dfrac{e^x}{x}$
40. Absolute maximum value of: $f(x) = \dfrac{x^2}{e^x}$
41. Absolute maximum value of: $f(x) = \dfrac{1 + 2 \ln x}{x}$
42. Absolute minimum value of: $f(x) = \dfrac{1 - 5 \ln x}{x}$

Sketch the graph of $y = f(x)$.

43. $f(x) = 1 - e^x$
44. $f(x) = 1 - \ln x$
45. $f(x) = x - \ln x$
46. $f(x) = e^x - x$
47. $f(x) = (3 - x)e^x$
48. $f(x) = (x - 2)e^x$
49. $f(x) = x^2 \ln x$
50. $f(x) = \dfrac{\ln x}{x}$

Problems 51–54 require the use of a graphic calculator or a computer. Graph f(x) and f′(x) in the indicated viewing rectangle. Approximate the critical values of f(x) to two decimal places and find the intervals where f(x) is increasing, the intervals where f(x) is decreasing, and the local extrema.

51. $f(x) = e^x - 2x^2$
    x range: $[-2, 4]$
    y range: $[-4, 4]$
52. $f(x) = e^x + x^2$
    x range: $[-3, 3]$
    y range: $[-10, 10]$
53. $f(x) = 20 \ln x - e^x$
    x range: $[0, 4]$
    y range: $[-10, 10]$
54. $f(x) = x^2 - 3x \ln x$
    x range: $[0, 5]$
    y range: $[-3, 3]$

APPLICATIONS

Business & Economics

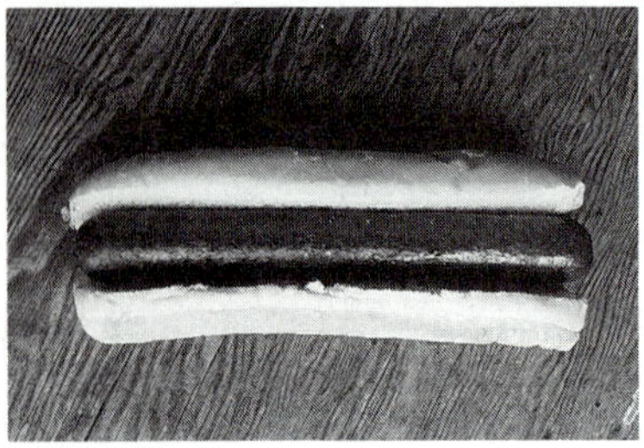

55. *Maximum profit.* A national food service runs food concessions for sporting events throughout the country. Their marketing research department chose a particular football stadium to test market a new jumbo hot dog. It was found that the demand for the new hot dog is given approximately by

$$p = 5 - \ln x \qquad 5 \leq x \leq 50$$

where x is the number of hot dogs (in thousands) that can be sold during one game at a price of \$p. If the concessionaire pays \$1 for each hot dog, how should the hot dogs be priced to maximize the profit per game?

56. *Maximum profit.* On a national tour of a rock band, the demand for T-shirts is given by

$$p = 15 - 4 \ln x \qquad 1 \leq x \leq 40$$

where x is the number of T-shirts (in thousands) that can be sold during a single concert at a price of \$p. If the shirts cost the band \$5 each, how should they be priced in order to maximize the profit per concert?

57. *Minimum average cost.* The cost of producing x units of a product is given by

$$C(x) = 600 + 100x - 100 \ln x \qquad x \geq 1$$

Find the minimum average cost.

58. *Minimum average cost.* The cost of producing x units of a product is given by

$$C(x) = 1{,}000 + 200x - 200 \ln x \qquad x \geq 1$$

Find the minimum average cost.

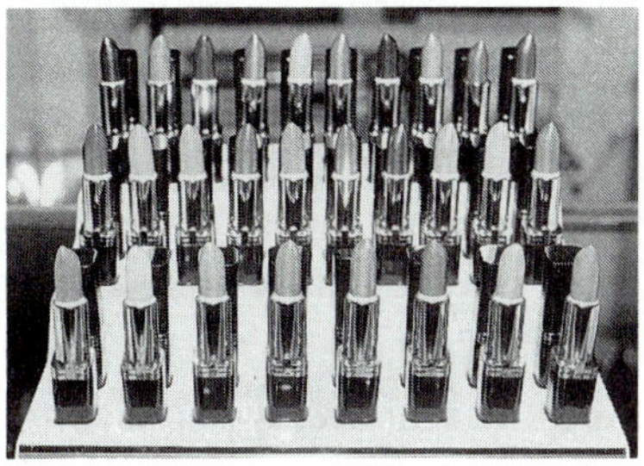

59. *Maximizing revenue.* A cosmetic company is planning the introduction and promotion of a new lipstick line. The marketing research department, after test marketing the new line in a carefully selected large city, found that the demand in that city is given approximately by

$$p = 10e^{-x} \qquad 0 \leq x \leq 2$$

where x thousand lipsticks were sold per week at a price of p dollars each.

(A) At what price will the weekly revenue $R(x) = xp$ be maximum? What is the maximum weekly revenue in the test city?

(B) Graph R for $0 \leq x \leq 2$.

60. *Maximizing revenue.* Repeat Problem 59 using the demand equation $p = 12e^{-x}$, $0 \leq x \leq 2$.

Life Sciences

61. *Blood pressure.* An experiment was set up to find a relationship between weight and systolic blood pressure in normal children. Using hospital records for 5,000 normal children, it was found that the systolic blood pressure was given approximately by

$$P(x) = 17.5(1 + \ln x) \qquad 10 \leq x \leq 100$$

where $P(x)$ is measured in millimeters of mercury and x is measured in pounds. What is the rate of change of blood pressure with respect to weight at the 40 pound weight level? At the 90 pound weight level?

62. *Blood pressure.* Graph the systolic blood pressure equation in Problem 61.

63. *Drug concentration.* The concentration of a drug in the bloodstream t hours after injection is given approximately by

$$C(t) = 4.35e^{-t} \qquad 0 \leq t \leq 5$$

where $C(t)$ is concentration in milligrams per milliliter.

(A) What is the rate of change of concentration after 1 hour? After 4 hours?

(B) Graph C.

64. *Water pollution.* The use of iodine crystals is a popular way of making small quantities of nonpotable water safe to drink. Crystals placed in a 1 ounce bottle of water will dissolve until the solution is saturated. After saturation, half of this solution is poured into a quart container of nonpotable water, and after about an hour, the water is usually safe to drink. The half empty 1 ounce bottle is then refilled to be used again in the same way. Suppose the concentration of iodine in the 1 ounce bottle t minutes after the crystals are introduced can be approximated by

$$C(t) = 250(1 - e^{-t}) \qquad t \geq 0$$

where $C(t)$ is the concentration of iodine in micrograms per milliliter.

(A) What is the rate of change of the concentration after 1 minute? After 4 minutes?

(B) Graph C for $0 \leq t \leq 5$.

Social Sciences

65. *Psychology—stimulus/response.* In psychology, the Weber–Fechner law for stimulus response is

$$R = k \ln\left(\frac{S}{S_0}\right)$$

where R is the response, S is the stimulus, and S_0 is the lowest level of stimulus that can be detected. Find dR/dS.

66. *Psychology—learning.* A mathematical model for the average of a group of people learning to type is given by

$$N(t) = 10 + 6 \ln t \qquad t \geq 1$$

where $N(t)$ is the number of words per minute typed after t hours of instruction and practice (2 hours per day, 5 days per week). What is the rate of learning after 10 hours of instruction and practice? After 100 hours?

SECTION 5-3

Chain Rule: General Form

- COMPOSITE FUNCTIONS
- CHAIN RULE
- GENERALIZED DERIVATIVE RULES
- OTHER LOGARITHMIC AND EXPONENTIAL FUNCTIONS

In Section 3-6, we introduced the power form of the chain rule:

$$D_x[u(x)]^n = n[u(x)]^{n-1} u'(x)$$

For example,

$$D_x(x^2 - 3)^5 = 5(x^2 - 3)^4 D_x(x^2 - 3) = 10x(x^2 - 3)^4$$

This general power rule is a special case of one of the most important derivative rules of all—the *chain rule*—which will enable us to determine the derivatives of some fairly complicated functions in terms of derivatives of more elementary functions.

Suppose you were asked to find the derivative of

$$h(x) = \ln(2x + 1) \qquad \text{or} \qquad m(x) = e^{3x^2-1}$$

We have formulas for computing derivatives of $\ln x$ and e^x, and polynomial functions in general, but not in the indicated combinations. The chain rule is used to compute derivatives of functions that are *compositions* of more elementary functions whose derivatives are known. We start the section with a brief review of *composite functions.*

◆ COMPOSITE FUNCTIONS

Let us look at function h more closely:

$$h(x) = \ln(2x + 1)$$

The function h is a combination of the natural logarithm function and a linear function. To see this more clearly, let

$$y = f(u) = \ln u \qquad \text{and} \qquad u = g(x) = 2x + 1$$

Then we can express y as a function of x as follows:

$$y = f(u) = f[g(x)] = \ln(2x + 1) = h(x)$$

The function h is said to be the *composite* of the two simpler functions f and g. (Loosely speaking, we can think of h as a function of a function.) In general, we have the following:

Composite Functions

A function h is a **composite** of functions f and g if

$$h(x) = f[g(x)]$$

The domain of h is the set of all numbers x such that x is in the domain of g and $g(x)$ is in the domain of f.

◆ EXAMPLE 8 Let $f(u) = e^u$, $g(x) = 3x^2 + 1$, and $m(v) = v^{3/2}$. Find:

(A) $f[g(x)]$ (B) $g[f(u)]$ (C) $m[g(x)]$

Solutions

(A) $f[g(x)] = e^{g(x)} = e^{3x^2+1}$

(B) $g[f(u)] = 3[f(u)]^2 + 1 = 3(e^u)^2 + 1 = 3e^{2u} + 1$

(C) $m[g(x)] = [g(x)]^{3/2} = (3x^2 + 1)^{3/2}$ ◆

PROBLEM 8 Let $f(u) = \ln u$, $g(x) = 2x^3 + 4$, and $m(v) = v^{-5}$. Find:

(A) $f[g(x)]$ (B) $g[f(u)]$ (C) $m[g(x)]$ ◆

◆ EXAMPLE 9 Write each function as a composition of the natural logarithm or exponential function and a polynomial.

(A) $y = \ln(x^3 - 2x^2 + 1)$ (B) $y = e^{x^2+4}$

Solutions

(A) Let

$$y = f(u) = \ln u$$
$$u = g(x) = x^3 - 2x^2 + 1$$

Check: $y = f[g(x)] = \ln[g(x)] = \ln(x^3 - 2x^2 + 1)$

(B) Let

$$y = f(u) = e^u$$
$$u = g(x) = x^2 + 4$$

Check: $y = f[g(x)] = e^{g(x)} = e^{x^2+4}$ ◆

PROBLEM 9 Repeat Example 9 for:

(A) $y = e^{2x^3+7}$ (B) $y = \ln(x^4 + 10)$ ◆

◆ CHAIN RULE

The word "chain" in the name *chain rule* comes from the fact that a function formed by composition (such as those in Example 8) involves a chain of functions—that is, a function of a function. The *chain rule* will enable us to compute the derivative of a composite function in terms of the derivatives of the functions making up the composition.

Suppose

$$y = h(x) = f[g(x)]$$

is a composite of f and g, where

$$y = f(u) \qquad \text{and} \qquad u = g(x)$$

We would like to express the derivative dy/dx in terms of the derivatives of f and g. From the definition of a derivative in increment form (see Section 4-5), we have

$$\frac{dy}{dx} = \lim_{\Delta x \to 0} \frac{h(x + \Delta x) - h(x)}{\Delta x}$$

$$= \lim_{\Delta x \to 0} \frac{\Delta y}{\Delta x} \tag{1}$$

Noting that

$$\frac{\Delta y}{\Delta x} = \frac{\Delta y}{\Delta u}\frac{\Delta u}{\Delta x} \tag{2}$$

we substitute (2) into (1) to obtain

$$\frac{dy}{dx} = \lim_{\Delta x \to 0} \frac{\Delta y}{\Delta u}\frac{\Delta u}{\Delta x}$$

and reason that $\Delta u \to 0$ as $\Delta x \to 0$ so that

$$\frac{dy}{dx} = \left(\lim_{\Delta u \to 0} \frac{\Delta y}{\Delta u}\right)\left(\lim_{\Delta x \to 0} \frac{\Delta u}{\Delta x}\right)$$

$$= \frac{dy}{du}\frac{du}{dx}$$

The result is correct under rather general conditions, and is called the *chain rule*, but our "derivation" is superficial, because it ignores a number of hidden problems. Since a formal proof of the chain rule is beyond the scope we simply state it as follows:

Chain Rule

If $y = f(u)$ and $u = g(x)$, define the composite function

$$y = h(x) = f[g(x)]$$

Then

$$\frac{dy}{dx} = \frac{dy}{du}\frac{du}{dx} \qquad \text{provided } \frac{dy}{du} \text{ and } \frac{du}{dx} \text{ exist}$$

or, equivalently,

$$h'(x) = f'[g(x)]g'(x) \qquad \text{provided } f'[g(x)] \text{ and } g'(x) \text{ exist}$$

◆ EXAMPLE 10 Find dy/dx, given:

(A) $y = \ln(x^2 - 4x + 2)$ (B) $y = e^{2x^3+5}$ (C) $y = (3x^2 + 1)^{3/2}$

Solutions (A) Let $y = \ln u$ and $u = x^2 - 4x + 2$. Then

$$\frac{dy}{dx} = \frac{dy}{du}\frac{du}{dx} \quad *$$

$$= \frac{1}{u}(2x - 4)$$

$$= \frac{1}{x^2 - 4x + 2}(2x - 4) \qquad \text{Since } u = x^2 - 4x + 2$$

$$= \frac{2x - 4}{x^2 - 4x + 2}$$

(B) Let $y = e^u$ and $u = 2x^3 + 5$. Then

$$\frac{dy}{dx} = \frac{dy}{du}\frac{du}{dx}$$

$$= e^u(6x^2)$$

$$= 6x^2e^{2x^3+5} \qquad \text{Since } u = 2x^3 + 5$$

* After some experience with the chain rule, the steps in the dashed boxes are usually done mentally.

(C) We have two methods:

Method 1. Chain rule—general form: Let $y = u^{3/2}$ and $u = 3x^2 + 1$. Then

$$\begin{aligned}\frac{dy}{dx} &= \frac{dy}{du}\frac{du}{dx} \\ &= \tfrac{3}{2}u^{1/2}(6x) \\ &= \tfrac{3}{2}(3x^2+1)^{1/2}(6x) \qquad \text{Since } u = 3x^2 + 1 \\ &= 9x(3x^2+1)^{1/2} \quad \text{or} \quad 9x\sqrt{3x^2+1}\end{aligned}$$

Method 2. Chain rule—power form (general power rule):

$$\begin{aligned}D_x(3x^2+1)^{3/2} &= \tfrac{3}{2}(3x^2+1)^{1/2}\,D_x(3x^2+1) \qquad D_x[u(x)]^n = n[u(x)]^{n-1}\,D_x\,u(x) \\ &= \tfrac{3}{2}(3x^2+1)^{1/2}(6x) \\ &= 9x(3x^2+1)^{1/2} \quad \text{or} \quad 9x\sqrt{3x^2+1}\end{aligned}$$

◆

The general power rule stated in Section 3-6 can be derived using the chain rule as follows: Given $y = [u(x)]^n$, let $y = v^n$ and $v = u(x)$. Then

$$\begin{aligned}\frac{dy}{dx} &= \frac{dy}{dv}\frac{dv}{dx} \\ &= nv^{n-1}\,D_x\,u(x) \\ &= n[u(x)]^{n-1}\,D_x\,u(x) \qquad \text{Since } v = u(x)\end{aligned}$$

PROBLEM 10 Find dy/dx, given:

(A) $y = e^{3x^4+6}$ (B) $y = \ln(x^2 + 9x + 4)$ (C) $y = (2x^3 + 4)^{-5}$ (Use two methods.) ◆

The chain rule can be extended to compositions of three or more functions. For example, if $y = f(w)$, $w = g(u)$, and $u = h(x)$, then

$$\frac{dy}{dx} = \frac{dy}{dw}\frac{dw}{du}\frac{du}{dx}$$

◆ EXAMPLE 11 For $y = h(x) = e^{1+(\ln x)^2}$, find dy/dx.

Solution Note that h is of the form $y = e^w$, where $w = 1 + u^2$ and $u = \ln x$. Thus,

$$\begin{aligned}\frac{dy}{dx} &= \frac{dy}{dw}\frac{dw}{du}\frac{du}{dx} \\ &= e^w(2u)\left(\frac{1}{x}\right) \\ &= e^{1+u^2}(2u)\left(\frac{1}{x}\right) \qquad \text{Since } w = 1 + u^2 \\ &= e^{1+(\ln x)^2}(2\ln x)\left(\frac{1}{x}\right) \qquad \text{Since } u = \ln x \\ &= \frac{2}{x}(\ln x)e^{1+(\ln x)^2}\end{aligned}$$

◆

PROBLEM 11 For $y = h(x) = [\ln(1 + e^x)]^3$, find dy/dx. ◆

◆ GENERALIZED DERIVATIVE RULES

In practice, it is not necessary to introduce additional variables when using the chain rule, as we did in Examples 10 and 11. Instead, the chain rule can be used to extend the derivative rules for specific functions to general derivative rules for compositions. This is what we did above when we showed that the general power rule is a consequence of the chain rule. The same technique can be applied to functions of the form $y = e^{f(x)}$ and $y = \ln[f(x)]$ (see Problems 59 and 60 at the end of this section). The results are summarized in the following box:

General Derivative Rules

$$D_x[f(x)]^n = n[f(x)]^{n-1}f'(x) \quad (3)$$

$$D_x \ln[f(x)] = \frac{1}{f(x)} f'(x) \quad (4)$$

$$D_x e^{f(x)} = e^{f(x)}f'(x) \quad (5)$$

For power, natural logarithm, or exponential forms, we can either use the chain rule discussed earlier or these special differentiation formulas based on the chain rule. Use whichever is easier for you. In Example 12, we will use the general derivative rules.

◆ EXAMPLE 12

(A) $D_x e^{2x} = e^{2x} D_x 2x$ Using (5)
$= e^{2x}(2) = 2e^{2x}$

(B) $D_x \ln(x^2 + 9) = \frac{1}{x^2 + 9} D_x(x^2 + 9)$ Using (4)
$= \frac{1}{x^2 + 9} 2x = \frac{2x}{x^2 + 9}$

(C) $D_x(1 + e^{x^2})^3 = 3(1 + e^{x^2})^2 D_x(1 + e^{x^2})$ Using (3)
$= 3(1 + e^{x^2})^2 e^{x^2} D_x x^2$ Using (5)
$= 3(1 + e^{x^2})^2 e^{x^2}(2x)$
$= 6xe^{x^2}(1 + e^{x^2})^2$ ◆

PROBLEM 12 Find:

(A) $D_x \ln(x^3 + 2x)$ (B) $D_x e^{3x^2+2}$ (C) $D_x(2 + e^{-x^2})^4$ ◆

◆ OTHER LOGARITHMIC AND EXPONENTIAL FUNCTIONS

In most applications involving logarithmic or exponential functions, the number e is the preferred base. However, there are situations where it is convenient to use a base other than e. Derivatives of $y = \log_b x$ and $y = b^x$ can be obtained by expressing these functions in terms of the natural logarithmic and exponential functions.

The change-of-base formula derived in Section 2-3 provides a relationship between $\log_b x$ and $\log_a x$ for any two bases a and b. We repeat the derivation here for $a = e$ and b any base, $b > 0$ and $b \neq 1$. Some of you may prefer to remember the process, and others the formula.

$$\begin{aligned} y &= \log_b x && \text{Change to exponential form.} \\ b^y &= x && \text{Take the natural logarithm of both sides.} \\ \ln b^y &= \ln x && \text{Recall that } \ln b^y = y \ln b. \\ y \ln b &= \ln x && \text{Solve for } y. \\ y &= \frac{1}{\ln b} \ln x \end{aligned}$$

Thus,

$$\log_b x = \frac{1}{\ln b} \ln x \qquad \text{Change-of-base formula with } a = e \tag{6}$$

Differentiating both sides of (6), we have

$$\boldsymbol{D_x \log_b x = \frac{1}{\ln b} D_x \ln x = \frac{1}{\ln b}\left(\frac{1}{x}\right)}$$

◆ EXAMPLE 13 Find $f'(x)$ for:

(A) $f(x) = \log_2 x$ (B) $f(x) = \log(1 + x^3)$

Solutions (A) $f(x) = \log_2 x = \dfrac{1}{\ln 2} \ln x$ Using (6)

$$f'(x) = \frac{1}{\ln 2}\left(\frac{1}{x}\right)$$

(B) $f(x) = \log(1 + x^3)$ Recall that $\log r = \log_{10} r$.

$$= \frac{1}{\ln 10} \ln(1 + x^3) \qquad \text{Using (6)}$$

$$f'(x) = \frac{1}{\ln 10}\left(\frac{1}{1 + x^3} 3x^2\right) = \frac{1}{\ln 10}\left(\frac{3x^2}{1 + x^3}\right)$$ ◆

PROBLEM 13 Find $f'(x)$ for:

(A) $f(x) = \log x$ (B) $f(x) = \log_3(x + x^2)$ ◆

Now we want to find a relationship between b^x and e^x for any base b, $b > 0$ and $b \neq 1$.

$$
\begin{aligned}
y &= b^x && \text{Take the natural logarithm of both sides.} \\
\ln y &= \ln b^x \\
&= x \ln b && \text{If } \ln A = B \text{, then } A = e^B. \\
y &= e^{x \ln b}
\end{aligned}
$$

Thus,

$$b^x = e^{x \ln b} \tag{7}$$

Differentiating both sides of (7), we have

$$\boldsymbol{D_x\, b^x = e^{x \ln b} \ln b = b^x \ln b}$$

◆ EXAMPLE 14 Find $f'(x)$ for:

(A) $f(x) = 2^x$ (B) $f(x) = 10^{x^5+x}$

Solutions

(A) $f(x) = 2^x = e^{x \ln 2}$ Using (7)

$f'(x) = e^{x \ln 2} \ln 2 = 2^x \ln 2$

(B) $f(x) = 10^{x^5+x} = e^{(x^5+x)\ln 10}$ Using (7)

$f'(x) = e^{(x^5+x)\ln 10}(5x^4 + 1) \ln 10$

$= 10^{x^5+x}(5x^4 + 1) \ln 10$ ◆

PROBLEM 14 Find $f'(x)$ for:

(A) $f(x) = 5^x$ (B) $f(x) = 4^{x^2+3x}$ ◆

Answers to Matched Problems

8. (A) $\ln(2x^3 + 4)$ (B) $2(\ln u)^3 + 4$ (C) $(2x^3 + 4)^{-5}$
9. (A) $y = f(u) = e^u$; $u = g(x) = 2x^3 + 7$
   (B) $y = f(u) = \ln u$; $u = g(x) = x^4 + 10$
10. (A) $12x^3e^{3x^4+6}$ (B) $\dfrac{2x + 9}{x^2 + 9x + 4}$ (C) $-30x^2(2x^3 + 4)^{-6}$
11. $\dfrac{3e^x[\ln(1 + e^x)]^2}{1 + e^x}$
12. (A) $\dfrac{3x^2 + 2}{x^3 + 2x}$ (B) $6xe^{3x^2+2}$ (C) $-8xe^{-x^2}(2 + e^{-x^2})^3$
13. (A) $\dfrac{1}{\ln 10}\left(\dfrac{1}{x}\right)$ (B) $\dfrac{1}{\ln 3}\left(\dfrac{1 + 2x}{x + x^2}\right)$
14. (A) $5^x \ln 5$ (B) $4^{x^2+3x}(2x + 3) \ln 4$

EXERCISE 5-3

A *Write each composite function in the form $y = f(u)$ and $u = g(x)$.*

1. $y = (2x + 5)^3$
2. $y = (3x - 7)^5$
3. $y = \ln(2x^2 + 7)$
4. $y = \ln(x^2 - 2x + 5)$
5. $y = e^{x^2-2}$
6. $y = e^{3x^3+5x}$

Express y in terms of x. Use the chain rule to find dy/dx, and then express dy/dx in terms of x.

7. $y = u^2; \quad u = 2 + e^x$
8. $y = u^3; \quad u = 3 - \ln x$
9. $y = e^u; \quad u = 2 - x^4$
10. $y = e^u; \quad u = x^6 + 5x^2$
11. $y = \ln u; \quad u = 4x^5 - 7$
12. $y = \ln u; \quad u = 2 + 3x^4$

Find each derivative.

13. $D_x \ln(x - 3)$
14. $D_w \ln(w + 100)$
15. $D_t \ln(3 - 2t)$
16. $D_y \ln(4 - 5y)$
17. $D_x\, 3e^{2x}$
18. $D_y\, 2e^{3y}$
19. $D_t\, 2e^{-4t}$
20. $D_r\, 6e^{-3r}$

B

21. $D_x\, 100e^{-0.03x}$
22. $D_t\, 1{,}000e^{0.06t}$
23. $D_x \ln(x + 1)^4$
24. $D_x \ln(x + 1)^{-3}$
25. $D_x(2e^{2x} - 3e^x + 5)$
26. $D_t(1 + e^{-t} - e^{-2t})$
27. $D_x\, e^{3x^2-2x}$
28. $D_x\, e^{x^3-3x^2+1}$
29. $D_t \ln(t^2 + 3t)$
30. $D_x \ln(x^3 - 3x^2)$
31. $D_x \ln(x^2 + 1)^{1/2}$
32. $D_x \ln(x^4 + 5)^{3/2}$
33. $D_t[\ln(t^2 + 1)]^4$
34. $D_w\, [\ln(w^3 - 1)]^2$
35. $D_x(e^{2x} - 1)^4$
36. $D_x(e^{x^2} + 3)^5$
37. $D_x \dfrac{e^{2x}}{x^2 + 1}$
38. $D_x \dfrac{e^{x+1}}{x + 1}$
39. $D_x(x^2 + 1)e^{-x}$
40. $D_x(1 - x)e^{2x}$
41. $D_x(e^{-x} \ln x)$
42. $D_x \dfrac{\ln x}{e^x + 1}$
43. $D_x \dfrac{1}{\ln(1 + x^2)}$
44. $D_x \dfrac{1}{\ln(1 - x^3)}$
45. $D_x \sqrt[3]{\ln(1 - x^2)}$
46. $D_t \sqrt[5]{\ln(1 - t^5)}$

C *Sketch the graph of $y = f(x)$.*

47. $f(x) = 1 - e^{-x}$
48. $f(x) = 2 - 3e^{-2x}$
49. $f(x) = \ln(1 - x)$
50. $f(x) = \ln(2x + 4)$
51. $f(x) = e^{-(1/2)x^2}$
52. $f(x) = \ln(x^2 + 4)$

Express y in terms of x. Use the chain rule to find dy/dx, and express dy/dx in terms of x.

53. $y = 1 + w^2; \quad w = \ln u; \quad u = 2 + e^x$
54. $y = \ln w; \quad w = 1 + e^u; \quad u = x^2$

Find each derivative.

55. $D_x \log_2(3x^2 - 1)$
56. $D_x \log(x^3 - 1)$
57. $D_x\, 10^{x^2+x}$
58. $D_x\, 8^{1-2x^2}$
59. Use the chain rule to derive the formula: $D_x \ln[f(x)] = \dfrac{1}{f(x)} f'(x)$
60. Use the chain rule to derive the formula: $D_x\, e^{f(x)} = e^{f(x)}f'(x)$

Business & Economics

61. *Maximum revenue.* Suppose the price–demand equation for x units of a commodity is determined from empirical data to be

$$p = 100e^{-0.05x}$$

where x units are sold per day at a price of \$p each. Find the production level and price that maximize revenue. What is the maximum revenue?

62. *Maximum revenue.* Repeat Problem 61 using the price–demand equation

$$p = 10e^{-0.04x}$$

Problems 63 and 64 require the use of a graphic calculator or a computer.

63. *Maximum profit.* Refer to Problem 61. If the daily fixed cost is \$400 and the cost per unit is \$6, use approximation techniques to find the production level and the price that maximize profit. What is the maximum profit? [*Hint:* Graph $y = P(x)$ and $y = P'(x)$ in the same viewing rectangle.]

64. *Maximum profit.* Refer to Problem 62. If the daily fixed cost is \$30 and the cost per unit is \$0.70, use approximation techniques to find the production level and the price that maximize profit. What is the maximum profit? [*Hint:* Graph $y = P(x)$ and $y = P'(x)$ in the same viewing rectangle.]

65. *Salvage value.* The salvage value S (in dollars) of a company airplane after t years is estimated to be given by

$$S(t) = 300{,}000e^{-0.1t}$$

What is the rate of depreciation (in dollars per year) after 1 year? 5 years? 10 years?

66. *Resale value.* The resale value R (in dollars) of a company car after t years is estimated to be given by

$$R(t) = 20{,}000e^{-0.15t}$$

What is the rate of depreciation (in dollars per year) after 1 year? 2 years? 3 years?

67. *Promotion and maximum profit.* A recording company has produced a new compact disk featuring a very popular recording group. Before launching a national sales campaign, the marketing research department chose to test market the disk in a bellwether city. Their interest is in determining the length of a sales campaign that will maximize total profits. From empirical data, the research department estimates that the proportion of a target group of 50,000 persons buying the disk after t days of television promotion is given by $1 - e^{-0.03t}$. If \$4 is received for each disk sold, then the total revenue after t days of promotion will be approximated by

$$R(t) = (4)(50{,}000)(1 - e^{-0.03t}) \qquad t \geqslant 0$$

Television promotion costs are

$$C(t) = 4{,}000 + 3{,}000t \qquad t \geq 0$$

(A) How many days of television promotion should be used to maximize total profit? What is the maximum total profit? What percentage of the target market will have purchased the disk when the maximum profit is reached?

(B) Graph the profit function.

68. *Promotion and maximum profit.* Repeat Problem 67 using the revenue equation

$$R(t) = (3)(60{,}000)(1 - e^{-0.04t})$$

Life Sciences

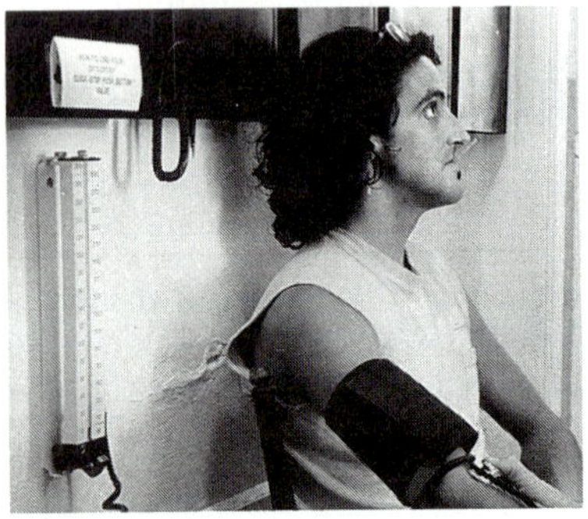

69. *Blood pressure and age.* A research group using hospital records developed the following approximate mathematical model relating systolic blood pressure and age:

$$P(x) = 40 + 25 \ln(x + 1) \qquad 0 \leq x \leq 65$$

where $P(x)$ is pressure measured in millimeters of mercury and x is age in years. What is the rate of change of pressure at the end of 10 years? At the end of 30 years? At the end of 60 years?

70. *Biology.* A yeast culture at room temperature (68°F) is placed in a refrigerator maintaining a constant temperature of 38°F. After t hours, the temperature T of the culture is given approximately by

$$T = 30e^{-0.58t} + 38 \qquad t \geq 0$$

What is the rate of change of temperature of the culture at the end of 1 hour? At the end of 4 hours?

71. *Bacterial growth.* A single cholera bacterium divides every 0.5 hour to produce two complete cholera bacteria. If we start with a colony of 5,000 bacteria, then after t hours there will be

$$A(t) = 5{,}000 \cdot 2^{2t}$$

bacteria. Find $A'(t)$, $A'(1)$, and $A'(5)$, and interpret the results.

72. *Bacterial growth.* Repeat Problem 71 for a starting colony of 1,000 bacteria where a single bacterium divides every 0.25 hour.

Social Sciences

73. *Sociology.* Daniel Lowenthal, a sociologist at Columbia University, made a 5 year study on the sale of popular records relative to their position in the top 20. He found that the average number of sales $N(n)$ of the nth ranking record was given approximately by

$$N(n) = N_1 e^{-0.09(n-1)} \qquad 1 \leq n \leq 20$$

where N_1 was the number of sales of the number one record on the list at a given time. Graph N for $N_1 = 1{,}000{,}000$ records.

74. *Political science.* Thomas W. Casstevens, a political scientist at Oakland University, has studied legislative turnover. He (with others) found that the number $N(t)$ of continuously serving members of an elected legislative body remaining t years after an election is given approximately by a function of the form

$$N(t) = N_0 e^{-ct}$$

In particular, for the 1965 election for the U.S. House of Representatives, it was found that

$$N(t) = 434e^{-0.0866t}$$

What is the rate of change after 2 years? After 10 years?

SECTION 5-4 Implicit Differentiation

- SPECIAL FUNCTION NOTATION
- IMPLICIT DIFFERENTIATION

SPECIAL FUNCTION NOTATION

The equation

$$y = 2 - 3x^2 \tag{1}$$

defines a function f with y as a dependent variable and x as an independent variable. Using function notation, we would write

$$y = f(x) \qquad \text{or} \qquad f(x) = 2 - 3x^2$$

In order to reduce to a minimum the number of symbols involved in a discussion, we will often write equation (1) in the form

$$y = 2 - 3x^2 = y(x)$$

where y is *both* a dependent variable and a function symbol. This is a convenient notation and no harm is done as long as one is aware of the double role of y. Other examples are

$$x = 2t^2 - 3t + 1 = x(t)$$

$$z = \sqrt{u^2 - 3u} = z(u)$$

$$r = \frac{1}{(s^2 - 3s)^{2/3}} = r(s)$$

This type of notation will simplify much of the discussion and work that follows.

Until now we have considered functions involving only one independent variable. There is no reason to stop there. The concept can be generalized to functions involving two or more independent variables, and this will be done in detail in Chapter 8. For now, we will "borrow" the notation for a function involving two independent variables. For example,

$$F(x, y) = x^2 - 2xy + 3y^2 - 5$$

specifies a function F involving two independent variables.

◆ IMPLICIT DIFFERENTIATION

Consider the equation

$$3x^2 + y - 2 = 0 \tag{2}$$

and the equation obtained by solving (2) for y in terms of x,

$$y = 2 - 3x^2 \tag{3}$$

Both equations define the same function using x as the independent variable and y as the dependent variable. For (3), we can write

$$y = f(x)$$

where

$$f(x) = 2 - 3x^2 \tag{4}$$

and we have an **explicit** (clearly stated) rule that enables us to determine y for each value of x. On the other hand, the y in equation (2) is the same y as in equation (3), and equation (2) **implicitly** gives (implies though does not plainly express) y as a function of x. Thus, we say that equations (3) and (4) define the function f explicitly, and equation (2) defines f implicitly.

The direct use of an equation that defines a function implicitly to find the derivative of the dependent variable with respect to the independent variable is called **implicit differentiation.** Let us differentiate (2) implicitly and (3) directly, and compare results.

Starting with

$$3x^2 + y - 2 = 0$$

we think of y as a function of x—that is, $y = y(x)$—and write

$$3x^2 + y(x) - 2 = 0$$

Then we differentiate both sides with respect to x:

$$\begin{aligned} D_x[3x^2 + y(x) - 2] &= D_x\, 0 \\ D_x\, 3x^2 + D_x\, y(x) - D_x\, 2 &= 0 \\ 6x + y' - 0 &= 0 \end{aligned}$$

Since y is a function of x, but is not explicitly given, we simply write $D_x\, y(x) = y'$ to indicate its derivative.

Now, we solve for y':

$$y' = -6x$$

Note that we get the same result if we start with equation (3) and differentiate directly:

$$y = 2 - 3x^2$$
$$y' = -6x$$

Why are we interested in implicit differentiation? In general, why do we not solve for y in terms of x and differentiate directly? The answer is that there are many equations of the form

$$F(x, y) = 0 \tag{5}$$

that are either difficult or impossible to solve for y explicitly in terms of x (try it for $x^2y^5 - 3xy + 5 = 0$ or for $e^y - y = 3x$, for example). But it can be shown that, under fairly general conditions on F, equation (5) will define one or more functions where y is a dependent variable and x is an independent variable. To find y' under these conditions, we differentiate (5) implicitly.

◆ **EXAMPLE 15** Given

$$F(x, y) = x^2 + y^2 - 25 = 0 \tag{6}$$

find y' and the slope of the graph at $x = 3$.

Solution We start with the graph of $x^2 + y^2 - 25 = 0$ (a circle, as shown in the margin) so that we can interpret our results geometrically. From the graph, it is clear that equation (6) does not define a function. But with a suitable restriction on the variables, equation (6) can define two or more functions. For example, the upper half and the lower half of the circle each define a function. A point on each half-circle that corresponds to $x = 3$ is found by substituting $x = 3$ into (6) and solving for y:

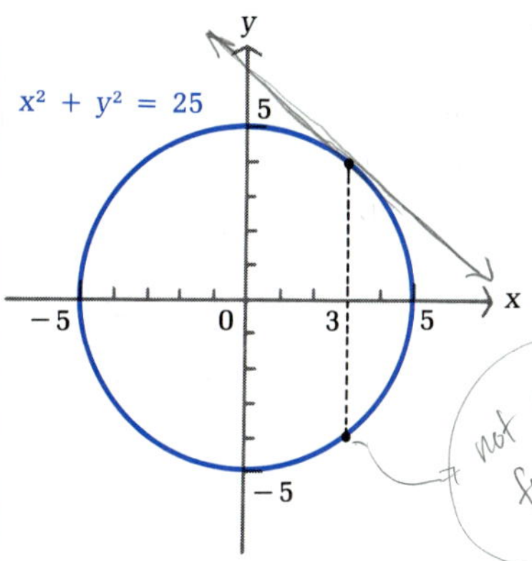

$$\begin{aligned} x^2 + y^2 - 25 &= 0 \\ (3)^2 + y^2 &= 25 \\ y^2 &= 16 \\ y &= \pm 4 \end{aligned}$$

Thus, the point (3, 4) is on the upper half-circle and the point (3, −4) is on the lower half-circle. We will use these results in a moment. We now differentiate (6) implicitly, treating y as a function of x; that is, $y = y(x)$:

$$x^2 + y^2 - 25 = 0$$

$$x^2 + [y(x)]^2 - 25 = 0$$

$$D_x\{x^2 + [y(x)]^2 - 25\} = D_x\, 0$$

$$D_x\, x^2 + D_x[y(x)]^2 - D_x\, 25 = 0 \qquad \text{Use the chain rule.}$$

$$2x + 2[y(x)]^{2-1}y'(x) - 0 = 0$$

$$2x + 2yy' = 0 \qquad \text{Solve for } y' \text{ in terms of } x \text{ and } y.$$

$$y' = -\frac{2x}{2y}$$

$$y' = -\frac{x}{y} \qquad \text{Leave the answer in terms of } x \text{ and } y.$$

We have found y' without first solving $x^2 + y^2 - 25 = 0$ for y in terms of x. And by leaving y' in terms of x and y, we can use $y' = -x/y$ to find y' for *any* point on the graph of $x^2 + y^2 - 25 = 0$ (except where $y = 0$). In particular, for $x = 3$, we found that $(3, 4)$ and $(3, -4)$ are on the graph; thus, the slope of the graph at $(3, 4)$ is

$$y'|_{(3,4)} = -\tfrac{3}{4} \qquad \text{The slope of the graph at } (3, 4)$$

and the slope at $(3, -4)$ is

$$y'|_{(3,-4)} = -\tfrac{3}{-4} = \tfrac{3}{4} \qquad \text{The slope of the graph at } (3, -4)$$

The symbol

$$\boldsymbol{y'|_{(a,b)}}$$

is used to indicate that we are evaluating y' at $x = a$ and $y = b$.

The results are interpreted geometrically on the original graph as follows:

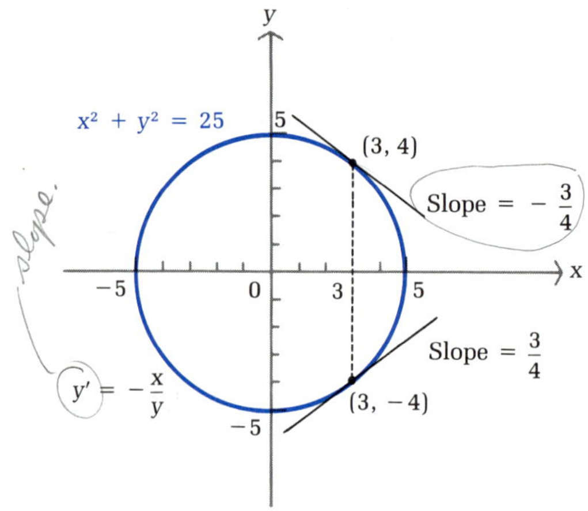

◆

In Example 15, the fact that y' is given in terms of both x and y is not a great disadvantage. We have only to make certain that when we want **to evaluate y' for a particular value of x and y, say, (x_0, y_0), the ordered pair must satisfy the original equation.**

PROBLEM 15 Graph $x^2 + y^2 - 169 = 0$, find y' by implicit differentiation, and find the slope of the graph when $x = 5$. ◆

◆ EXAMPLE 16 Find the equation(s) of the tangent line(s) to the graph of

$$y - xy^2 + x^2 + 1 = 0 \tag{7}$$

at the point(s) where $x = 1$.

Solution We first find y when x = 1:

$$
\begin{aligned}
y - xy^2 + x^2 + 1 &= 0 \\
y - (1)y^2 + (1)^2 + 1 &= 0 \\
y - y^2 + 2 &= 0 \\
y^2 - y - 2 &= 0 \\
(y - 2)(y + 1) &= 0 \\
y &= -1, 2
\end{aligned}
$$

Thus, there are two points on the graph of (7) where $x = 1$; namely, $(1, -1)$ and $(1, 2)$. We next find the slope of the graph at these two points by differentiating (7) implicitly:

$$
\begin{aligned}
y - xy^2 + x^2 + 1 &= 0 \\
D_x\, y - D_x\, xy^2 + D_x\, x^2 + D_x\, 1 &= D_x\, 0 \\
y' - (x \cdot 2yy' + y^2) + 2x &= 0 \\
y' - 2xyy' - y^2 + 2x &= 0 \\
y' - 2xyy' &= y^2 - 2x \\
(1 - 2xy)y' &= y^2 - 2x \\
y' &= \frac{y^2 - 2x}{1 - 2xy}
\end{aligned}
$$

Use the product rule and the chain rule for $D_x\, xy^2$.

Solve for y' by getting all terms involving y' on one side.

Now, find the slope at each point:

$$y'|_{(1,-1)} = \frac{(-1)^2 - 2(1)}{1 - 2(1)(-1)} = \frac{1 - 2}{1 + 2} = \frac{-1}{3} = -\frac{1}{3}$$

$$y'|_{(1,2)} = \frac{(2)^2 - 2(1)}{1 - 2(1)(2)} = \frac{4 - 2}{1 - 4} = \frac{2}{-3} = -\frac{2}{3}$$

Equation of the tangent line at $(1, -1)$:

$$
\begin{aligned}
y - y_1 &= m(x - x_1) \\
y + 1 &= -\tfrac{1}{3}(x - 1) \\
y + 1 &= -\tfrac{1}{3}x + \tfrac{1}{3} \\
y &= -\tfrac{1}{3}x - \tfrac{2}{3}
\end{aligned}
$$

Equation of the tangent line at $(1, 2)$:

$$
\begin{aligned}
y - y_1 &= m(x - x_1) \\
y - 2 &= -\tfrac{2}{3}(x - 1) \\
y - 2 &= -\tfrac{2}{3}x + \tfrac{2}{3} \\
y &= -\tfrac{2}{3}x + \tfrac{8}{3}
\end{aligned}
$$

◆

PROBLEM 16 Repeat Example 16 for $x^2 + y^2 - xy - 7 = 0$ at $x = 1$. ◆

◆ EXAMPLE 17 Find x' for $x = x(t)$ defined implicitly by

$$t \ln x = xe^t - 1$$

and evaluate x' at $(t, x) = (0, 1)$.

Solution It is important to remember that x is the dependent variable and t is the independent variable. Therefore, we differentiate both sides of the equation with respect to t (using product and chain rules where appropriate), and then solve for x':

$$t \ln x = xe^t - 1 \qquad \text{Differentiate implicitly with respect to } t.$$

$$D_t(t \ln x) = D_t(xe^t) - D_t\, 1$$

$$t\frac{x'}{x} + \ln x = xe^t + x'e^t \qquad \text{Clear fractions.}$$

$$\mathbf{x}\left(t\frac{x'}{x}\right) + \mathbf{x} \ln x = \mathbf{x} \cdot xe^t + \mathbf{x}e^t x' \qquad x \neq 0$$

$$tx' + x \ln x = x^2e^t + xe^tx' \qquad \text{Solve for } x'.$$

$$tx' - xe^tx' = x^2e^t - x \ln x \qquad \text{Factor out } x'.$$

$$(t - xe^t)x' = x^2e^t - x \ln x$$

$$x' = \frac{x^2e^t - x \ln x}{t - xe^t}$$

Now, we evaluate x' at (t, x) = (0, 1), as requested:

$$x'|_{(0, 1)} = \frac{(1)^2e^0 - 1 \ln 1}{0 - 1e^0}$$

$$= \frac{1}{-1} = -1$$

◆

PROBLEM 17 Find x' for x = x(t) defined implicitly by

$$1 + x \ln t = te^x$$

and evaluate x' at (t, x) = (1, 0). ◆

Answers to Matched Problems

15. $y' = -x/y$; when $x = 5$, $y = \pm 12$, thus, $y'|_{(5, 12)} = -\frac{5}{12}$ and $y'|_{(5, -12)} = \frac{5}{12}$
16. $y' = \dfrac{y - 2x}{2y - x}$; $y = \frac{4}{5}x - \frac{14}{5}$, $y = \frac{1}{5}x + \frac{14}{5}$
17. $x' = \dfrac{te^x - x}{t \ln t - t^2e^x}$; $x'|_{(1, 0)} = -1$

EXERCISE 5-4

In Problems 1–18, find y' without solving for y in terms of x (use implicit differentiation). Evaluate y' at the indicated point.

A

1. $y - 3x^2 + 5 = 0$; $(1, -2)$
2. $3x^4 + y - 2 = 0$; $(1, -1)$
3. $y^2 - 3x^2 + 8 = 0$; $(2, 2)$
4. $3y^2 + 2x^3 - 14 = 0$; $(1, 2)$
5. $y^2 + y - x = 0$; $(2, 1)$
6. $2y^3 + y^2 - x = 0$; $(3, 1)$

B

7. $xy - 6 = 0$; $(2, 3)$
8. $3xy - 2x - 2 = 0$; $(2, 1)$
9. $2xy + y + 2 = 0$; $(-1, 2)$
10. $2y + xy - 1 = 0$; $(-1, 1)$
11. $x^2y - 3x^2 - 4 = 0$; $(2, 4)$
12. $2x^3y - x^3 + 5 = 0$; $(-1, 3)$
13. $e^y = x^2 + y^2$; $(1, 0)$
14. $x^2 - y = 4e^y$; $(2, 0)$
15. $x^3 - y = \ln y$; $(1, 1)$
16. $\ln y = 2y^2 - x$; $(2, 1)$
17. $x \ln y + 2y = 2x^3$; $(1, 1)$
18. $xe^y - y = x^2 - 2$; $(2, 0)$

In Problems 19 and 20, find x′ for x = x(t) defined implicitly by the given equation. Evaluate x′ at the indicated point.

19. $x^2 - t^2x + t^3 + 11 = 0$; $(-2, 1)$
20. $x^3 - tx^2 - 4 = 0$; $(-3, -2)$

Find the equation(s) of the tangent line(s) to the graphs of the indicated equations at the point(s) with abscissas as indicated.

21. $xy - x - 4 = 0$; $x = 2$
22. $3x + xy + 1 = 0$; $x = -1$
23. $y^2 - xy - 6 = 0$; $x = 1$
24. $xy^2 - y - 2 = 0$; $x = 1$

C

Find y′ and the slope of the tangent line to the graph of each equation at the indicated point.

25. $(1 + y)^3 + y = x + 7$; $(2, 1)$
26. $(y - 3)^4 - x = y$; $(-3, 4)$
27. $(x - 2y)^3 = 2y^2 - 3$; $(1, 1)$
28. $(2x - y)^4 - y^3 = 8$; $(-1, -2)$
29. $\sqrt{7 + y^2} - x^3 + 4 = 0$; $(2, 3)$
30. $6\sqrt{y^3 + 1} - 2x^{3/2} - 2 = 0$; $(4, 2)$
31. $\ln(xy) = y^2 - 1$; $(1, 1)$
32. $e^{xy} - 2x = y + 1$; $(0, 0)$

APPLICATIONS

Business & Economics

For the demand equations in Problems 33–36, find the rate of change of p with respect to x by differentiating implicitly (x is the number of items that can be sold at a price of \$p.)

33. $x = p^2 - 2p + 1{,}000$
34. $x = p^3 - 3p^2 + 200$
35. $x = \sqrt{10{,}000 - p^2}$
36. $x = \sqrt[3]{1{,}500 - p^3}$

Life Sciences

37. **Biophysics.** In biophysics, the equation

$$(L + m)(V + n) = k$$

is called the *fundamental equation of muscle contraction*, where m, n, and k are constants, and V is the velocity of the shortening of muscle fibers for a muscle subjected to a load of L. Find dL/dV using implicit differentiation.

38. **Biophysics.** In Problem 37, find dV/dL using implicit differentiation.

SECTION 5-5 Related Rates

We start this discussion of *related rates* with an example, out of which will come a list of helpful suggestions for solving related rates problems in general.

◆ EXAMPLE 18

A 26 foot ladder is placed against a wall. If the top of the ladder is sliding down the wall at 2 feet per second, at what rate is the bottom of the ladder moving away from the wall when the bottom of the ladder is 10 feet away from the wall?

Solution

Many people reason that since the ladder is of constant length, the bottom of the ladder will move away from the wall at the same rate that the top of the ladder is moving down the wall. This is not the case, as we will see.

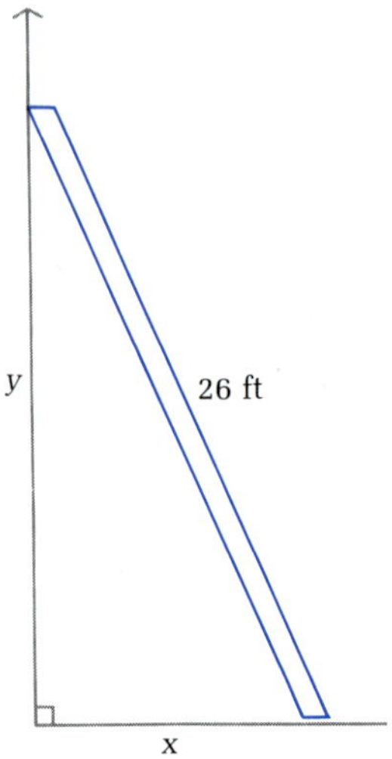

At any moment in time, let x be the distance of the bottom of the ladder from the wall, and let y be the distance of the top of the ladder on the wall (see the figure in the margin). Both x and y are changing with respect to time and can be thought of as functions of time; that is, $x = x(t)$ and $y = y(t)$. Furthermore, x and y are related by the Pythagorean relationship:

$$x^2 + y^2 = 26^2 \tag{1}$$

Differentiating (1) implicitly with respect to time t, and using the chain rule where appropriate, we obtain

$$2x\frac{dx}{dt} + 2y\frac{dy}{dt} = 0 \tag{2}$$

The rates dx/dt and dy/dt are related by equation (2); hence, this type of problem is referred to as a **related rates problem.**

Now our problem is to find dx/dt when $x = 10$ feet, given that $dy/dt = -2$ (y is decreasing at a constant rate of 2 feet per second). We have all the quantities we need in equation (2) to solve for dx/dt, except y. When $x = 10$, y can be found using (1):

$$10^2 + y^2 = 26^2$$

$$y = \sqrt{26^2 - 10^2} = 24 \text{ feet}$$

Substitute $dy/dt = -2$, $x = 10$, and $y = 24$ into (2); then solve for dx/dt:

$$2(10)\frac{dx}{dt} + 2(24)(-2) = 0$$

$$\frac{dx}{dt} = \frac{-2(24)(-2)}{2(10)} = 4.8 \text{ feet per second}$$

Thus, the bottom of the ladder is moving away from the wall at a rate of 4.8 feet per second. ◆

PROBLEM 18

A 26 foot ladder is placed against a wall. If the bottom of the ladder is moving away from the wall at 3 feet per second, at what rate is the top moving down when the top of the ladder is 24 feet up the wall? ◆

Suggestions for Solving Related Rates Problems

Step 1. Sketch a figure if helpful.

Step 2. Identify all relevant variables, including those whose rates are given and those whose rates are to be found.

Step 3. Express all given rates and rates to be found as derivatives.

Step 4. Find an equation connecting the variables in step 2.

Step 5. Implicitly differentiate the equation found in step 4, using the chain rule where appropriate, and substitute in all given values.

Step 6. Solve for the derivative that will give the unknown rate.

◆ EXAMPLE 19 Suppose two motor boats leave from the same point at the same time. If one travels north at 15 miles per hour and the other travels east at 20 miles per hour, how fast will the distance between them be changing after 2 hours?

Solution First, draw a picture:

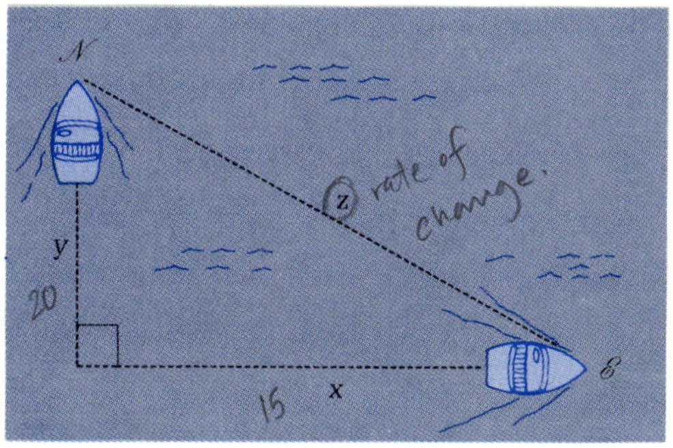

All variables, x, y, and z, are changing with time. Hence, they can be thought of as functions of time; $x = x(t)$, $y = y(t)$, and $z = z(t)$, given implicitly. It now makes sense to take derivatives of each variable with respect to time. From the Pythagorean theorem,

$$z^2 = x^2 + y^2 \tag{3}$$

We also know that

$$\frac{dx}{dt} = 20 \text{ miles per hour} \qquad \text{and} \qquad \frac{dy}{dt} = 15 \text{ miles per hour}$$

We would like to find dz/dt at the end of 2 hours; that is, when $x = 40$ miles and $y = 30$ miles. To do this, we differentiate both sides of (3) with respect to t and solve for dz/dt:

$$2z\frac{dz}{dt} = 2x\frac{dx}{dt} + 2y\frac{dy}{dt} \tag{4}$$

We have everything we need except z. When $x = 40$ and $y = 30$, we find z from (3) to be 50. Substituting the known quantities into (4), we obtain

$$2(50)\frac{dz}{dt} = 2(40)(20) + 2(30)(15)$$

$$\frac{dz}{dt} = 25 \text{ miles per hour}$$

Thus, the boats will be separating at a rate of 25 miles per hour. ◆

PROBLEM 19 Repeat Example 19 for the situation at the end of 3 hours. ◆

◆ EXAMPLE 20 Suppose a point is moving on the graph of $x^2 + y^2 = 25$. When the point is at $(-3, 4)$, its x coordinate is increasing at the rate of 0.4 unit per second. How fast is the y coordinate changing at that moment?

Solution Since both x and y are changing with respect to time, we can think of each as a function of time:

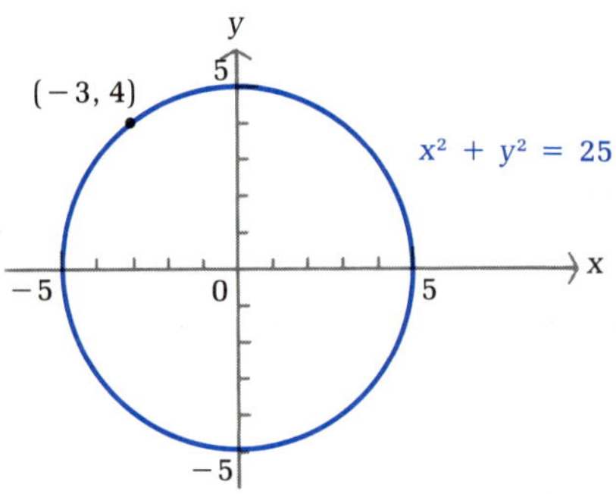

$$x = x(t) \qquad \text{and} \qquad y = y(t)$$

but restricted so that

$$x^2 + y^2 = 25 \tag{5}$$

Our problem is now to find dy/dt, given $x = -3$, $y = 4$, and $dx/dt = 0.4$. Implicitly differentiating both sides of (5) with respect to t, we have

$$x^2 + y^2 = 25$$

$$2x\frac{dx}{dt} + 2y\frac{dy}{dt} = 0 \qquad \text{Divide both sides by 2.}$$

$$x\frac{dx}{dt} + y\frac{dy}{dt} = 0 \qquad \text{Substitute } x = -3,\ y = 4, \text{ and } dx/dt = 0.4, \text{ and solve for } dy/dt.$$

$$(-3)(0.4) + 4\frac{dy}{dt} = 0$$

$$\frac{dy}{dt} = 0.3 \text{ unit per second}$$ ◆

PROBLEM 20 A point is moving on the graph of $y^3 = x^2$. When the point is at $(-8, 4)$, its y coordinate is decreasing at 2 units per second. How fast is the x coordinate changing at that moment? ◆

◆ EXAMPLE 21 Suppose that for a company manufacturing transistor radios, the cost, revenue, and profit equations are given by

$$C = 5{,}000 + 2x \qquad \text{Cost equation}$$

$$R = 10x - \frac{x^2}{1{,}000} \qquad \text{Revenue equation}$$

$$P = R - C \qquad \text{Profit equation}$$

where the production output in 1 week is x radios. If production is increasing at the rate of 500 radios per week when production is 2,000 radios, find the rate of increase in:

(A) Cost (B) Revenue (C) Profit

Solutions If production x is a function of time (it must be, since it is changing with respect to time), then C, R, and P must also be functions of time. These functions are implicitly (rather than explicitly) given. Letting *t* represent time in weeks, we differentiate both sides of each of the three equations above with respect to *t*, and then substitute $x = 2{,}000$ and $dx/dt = 500$ to find the desired rates.

(A)

$$C = 5{,}000 + 2x \qquad \text{Think: } C = C(t) \text{ and } x = x(t).$$

$$\frac{dC}{dt} = \frac{d}{dt}(5{,}000) + \frac{d}{dt}(2x) \qquad \text{Differentiate both sides with respect to } t.$$

$$\frac{dC}{dt} = 0 + 2\frac{dx}{dt} = 2\frac{dx}{dt}$$

Since $dx/dt = 500$ when $x = 2{,}000$,

$$\frac{dC}{dt} = 2(500) = \$1{,}000 \text{ per week}$$

Cost is increasing at a rate of \$1,000 per week.

(B)

$$R = 10x - \frac{x^2}{1{,}000}$$

$$\frac{dR}{dt} = \frac{d}{dt}(10x) - \frac{d}{dt}\frac{x^2}{1{,}000}$$

$$\frac{dR}{dt} = 10\frac{dx}{dt} - \frac{x}{500}\frac{dx}{dt}$$

$$\frac{dR}{dt} = \left(10 - \frac{x}{500}\right)\frac{dx}{dt}$$

Since $dx/dt = 500$ when $x = 2{,}000$,

$$\frac{dR}{dt} = \left(10 - \frac{2{,}000}{500}\right)(500) = \$3{,}000 \text{ per week}$$

Revenue is increasing at a rate of \$3,000 per week.

(C) $P = R - C$

$$\frac{dP}{dt} = \frac{dR}{dt} - \frac{dC}{dt}$$

$= \$3,000 - \$1,000$ Results from parts A and B

$= \$2,000$ per week

Profit is increasing at a rate of \$2,000 per week. ◆

PROBLEM 21 Repeat Example 21 for a production level of 6,000 radios per week. ◆

Answers to Matched Problems

18. $dy/dt = -1.25$ ft/sec 19. $dz/dt = 25$ mi/hr
20. $dx/dt = 6$ units/sec
21. (A) $dC/dt = \$1,000/\text{wk}$ (B) $dR/dt = -\$1,000/\text{wk}$ (C) $dP/dt = -\$2,000/\text{wk}$

EXERCISE 5-5

A

In Problems 1–6, assume $x = x(t)$ and $y = y(t)$. Find the indicated rate, given the other information.

1. $y = 2x^2 - 1$; $dy/dt = ?$, $dx/dt = 2$ when $x = 30$
2. $y = 2x^{1/2} + 3$; $dy/dt = ?$, $dx/dt = 8$ when $x = 4$
3. $x^2 + y^2 = 25$; $dy/dt = ?$, $dx/dt = -3$ when $x = 3$ and $y = 4$
4. $y^2 + x = 11$; $dx/dt = ?$, $dy/dt = -2$ when $x = 2$ and $y = 3$
5. $x^2 + xy + 2 = 0$; $dy/dt = ?$, $dx/dt = -1$ when $x = 2$ and $y = -3$
6. $y^2 + xy - 3x = -3$; $dx/dt = ?$, $dy/dt = -2$ when $x = 1$ and $y = 0$

B

7. A point is moving on the graph of $xy = 36$. When the point is at (4, 9), its x coordinate is increasing at 4 units per second. How fast is the y coordinate changing at that moment?
8. A point is moving on the graph of $4x^2 + 9y^2 = 36$. When the point is at (3, 0), its y coordinate is decreasing at 2 units per second. How fast is its x coordinate changing at that moment?
9. A boat is being pulled toward a dock as indicated in the accompanying figure. If the rope is being pulled in at 3 feet per second, how fast is the distance between the dock and the boat decreasing when it is 30 feet from the dock?

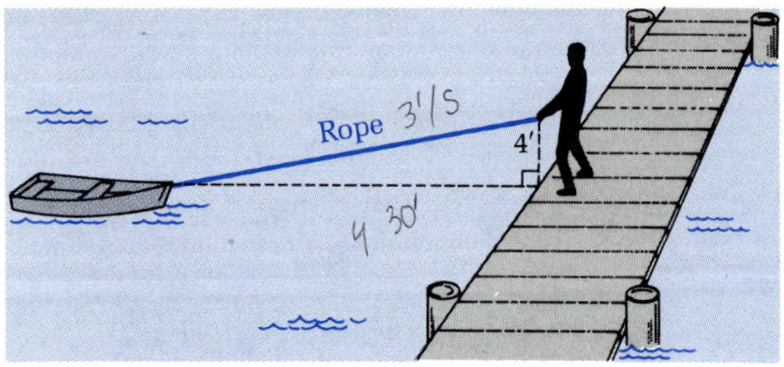

10. Refer to Problem 9. Suppose the distance between the boat and the dock is decreasing at 3.05 feet per second. How fast is the rope being pulled in when the boat is 10 feet from the dock?

11. A rock is thrown into a still pond and causes a circular ripple. If the radius of the ripple is increasing at 2 feet per second, how fast is the area changing when the radius is 10 feet? [Use $A = \pi R^2$, $\pi \approx 3.14$.]

12. Refer to Problem 11. How fast is the circumference of a circular ripple changing when the radius is 10 feet? [Use $C = 2\pi R$, $\pi \approx 3.14$.]

13. The radius of a spherical balloon is increasing at the rate of 3 centimeters per minute. How fast is the volume changing when the radius is 10 centimeters? [Use $V = \frac{4}{3}\pi R^3$, $\pi \approx 3.14$.]

14. Refer to Problem 13. How fast is the surface area of the sphere increasing? [Use $S = 4\pi R^2$, $\pi \approx 3.14$.]

15. Boyle's law for enclosed gases states that if the volume is kept constant, then the pressure P and temperature T are related by the equation

$$\frac{P}{T} = k$$

where k is a constant. If the temperature is increasing at 3 degrees per hour, what is the rate of change of pressure when the temperature is 250° (Kelvin) and the pressure is 500 pounds per square inch?

16. Boyle's law for enclosed gases states that if the temperature is kept constant, then the pressure P and volume V of the gas are related by the equation

$$VP = k$$

where k is a constant. If the volume is decreasing by 5 cubic inches per second, what is the rate of change of the pressure when the volume is 1,000 cubic inches and the pressure is 40 pounds per square inch?

17. A 10 foot ladder is placed against a vertical wall. Suppose the bottom slides away from the wall at a constant rate of 3 feet per second. How fast is the top sliding down the wall (negative rate) when the bottom is 6 feet from the wall? [*Hint:* Use the Pythagorean theorem: $a^2 + b^2 = c^2$, where c is the length of the hypotenuse of a right triangle and a and b are the lengths of the two shorter sides.]

18. A weather balloon is rising vertically at the rate of 5 meters per second. An observer is standing on the ground 300 meters from the point where the balloon was released. At what rate is the distance between the observer and the balloon changing when the balloon is 400 meters high?

C

19. A street light is on top of a 20 foot pole. A person who is 5 feet tall walks away from the pole at the rate of 5 feet per second. At what rate is the tip of the person's shadow moving away from the pole when he is 20 feet from the pole?

20. Refer to Problem 19. At what rate is the person's shadow growing when he is 20 feet from the pole?

Business & Economics

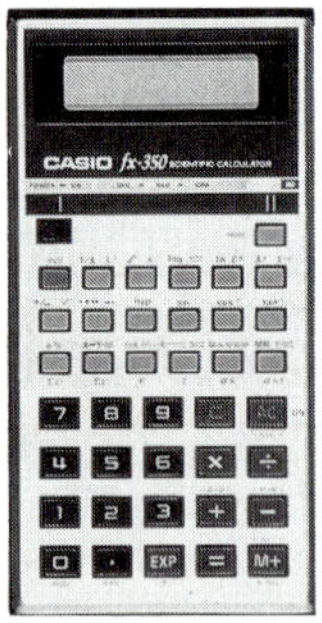

21. *Cost, revenue, and profit rates.* Suppose that for a company manufacturing calculators, the cost, revenue, and profit equations are given by

$$C = 90{,}000 + 30x \qquad R = 300x - \frac{x^2}{30} \qquad P = R - C$$

where the production output in 1 week is x calculators. If production is increasing at a rate of 500 calculators per week when production output is 6,000 calculators, find the rate of increase (decrease) in:

(A) Cost (B) Revenue (C) Profit

22. *Cost, revenue, and profit rates.* Repeat Problem 21 for

$$C = 72{,}000 + 60x \qquad R = 200x - \frac{x^2}{30} \qquad P = R - C$$

where production is increasing at a rate of 500 calculators per week at a production level of 1,500 calculators.

23. *Advertising.* A retail store estimates that weekly sales s and weekly advertising costs x (both in dollars) are related by

$$s = 60{,}000 - 40{,}000e^{-0.0005x}$$

The current weekly advertising costs are \$2,000, and these costs are increasing at the rate of \$300 per week. Find the current rate of change of sales.

24. *Advertising.* Repeat Problem 23 for

$$s = 50{,}000 - 20{,}000e^{-0.0004x}$$

25. *Price–demand.* The price p (in dollars) and demand x for a product are related by

$$2x^2 + 5xp + 50p^2 = 80{,}000$$

(A) If the price is increasing at a rate of \$2 per month when the price is \$30, find the rate of change of the demand.

(B) If the demand is decreasing at a rate of 6 units per month when the demand is 150 units, find the rate of change of the price.

26. *Price–demand.* Repeat Problem 25 for

$$x^2 + 2xp + 25p^2 = 74{,}500$$

Life Sciences

27. *Pollution.* An oil tanker aground on a reef is leaking oil that forms a circular oil slick about 0.1 foot thick (see the figure). To estimate the rate dV/dt (in cubic feet per minute) at which the oil is leaking from the tanker, it was found that the radius of the slick was increasing at 0.32 foot per minute ($dR/dt = 0.32$) when the radius R was 500 feet. Find dV/dt, using $\pi \approx 3.14$

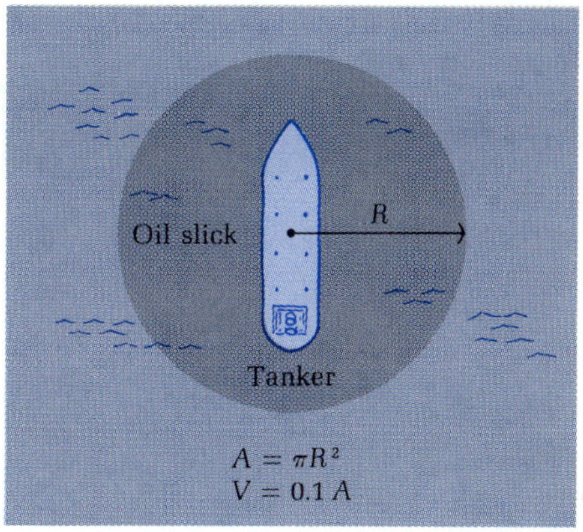

Social Sciences

28. *Learning.* A person who is new on an assembly line performs an operation in T minutes after x performances of the operation, as given by

$$T = 6\left(1 + \frac{1}{\sqrt{x}}\right)$$

If $dx/dt = 6$ operations per hour, where t is time in hours, find dT/dt after 36 performances of the operation.

SECTION 5-6 L'Hôpital's Rule

- L'HÔPITAL'S RULE AND THE INDETERMINATE FORM $0/0$
- ONE-SIDED LIMITS AND LIMITS AT ∞
- L'HÔPITAL'S RULE AND THE INDETERMINATE FORM ∞/∞

The ability to evaluate a wide variety of different types of limits is one of the skills necessary to apply the techniques of calculus successfully. We have already seen that limits play a fundamental role in the development of the derivative and are an important graphing tool. Later in this text (especially in Chapters 7 and 12) we will encounter important applications involving limits. In order to deal effectively with some of these applications, we need to develop some additional methods for evaluating limits. (Before proceeding further, a brief review of Sections 3-1 and 3-2 might prove to be beneficial.)

In this section we discuss a powerful technique for evaluating limits of quotients called *L'Hôpital's rule.* This rule is named after the French mathematician Marquis de L'Hôpital (1661–1704), who is generally credited with writing the first calculus textbook. To use L'Hôpital's rule, it is necessary to be very

TABLE 3

Limits Involving Powers of x

LIMITS AT 0	LIMITS AT ∞	GRAPH
$\lim_{x\to 0} x = 0$	$\lim_{x\to\infty} x = \infty$ $\lim_{x\to -\infty} x = -\infty$	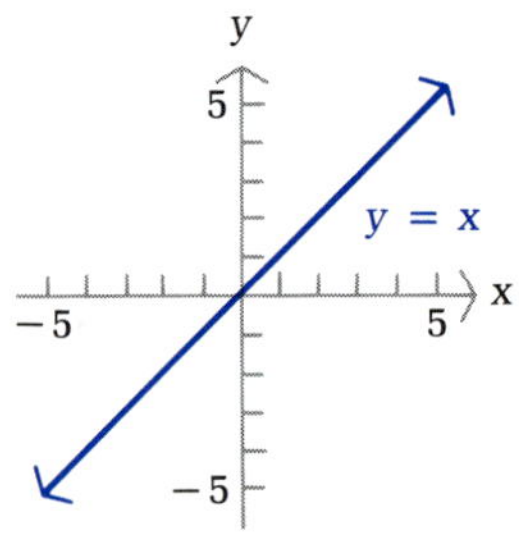
$\lim_{x\to 0} x^2 = 0$	$\lim_{x\to\infty} x^2 = \infty$ $\lim_{x\to -\infty} x^2 = \infty$	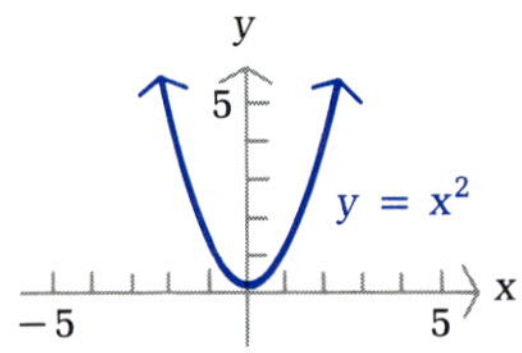
$\lim_{x\to 0^+} \frac{1}{x} = \infty$ $\lim_{x\to 0^-} \frac{1}{x} = -\infty$ $\lim_{x\to 0} \frac{1}{x}$ Does not exist	$\lim_{x\to\infty} \frac{1}{x} = 0$ $\lim_{x\to -\infty} \frac{1}{x} = 0$	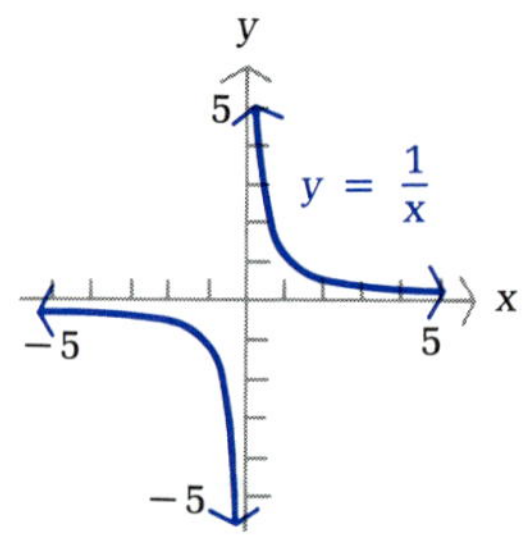
$\lim_{x\to 0^+} \frac{1}{x^2} = \infty$ $\lim_{x\to 0^-} \frac{1}{x^2} = \infty$ $\lim_{x\to 0} \frac{1}{x^2} = \infty$	$\lim_{x\to\infty} \frac{1}{x^2} = 0$ $\lim_{x\to -\infty} \frac{1}{x^2} = 0$	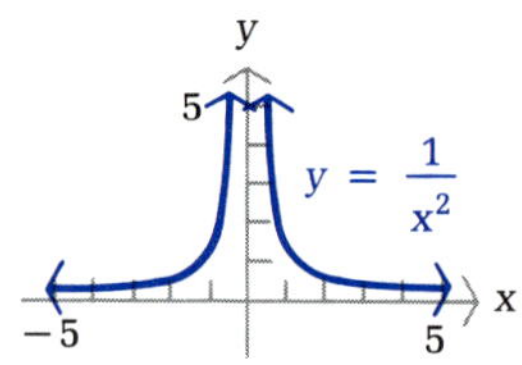

familiar with the limit properties of some basic functions. Tables 3 and 4 review some limits we have discussed in earlier chapters.

The limits in Table 3 are easily extended to functions of the form $f(x) = (x - c)^n$ and $g(x) = 1/(x - c)^n$. In general, if n is an odd integer, limits involving $(x - c)^n$ or $1/(x - c)^n$ as x approaches c (or $\pm\infty$) behave like the limits of x and $1/x$ as x approaches 0 (or $\pm\infty$). If n is an even integer, limits involving these expressions behave like the limits of x^2 and $1/x^2$ as x approaches 0 (or $\pm\infty$).

TABLE 4

Limits Involving Logarithmic and Exponential Functions

LIMITS AT 0	LIMITS AT ∞	GRAPH
$\lim_{x \to 0} e^x = 1$	$\lim_{x \to \infty} e^x = \infty$ $\lim_{x \to -\infty} e^x = 0$	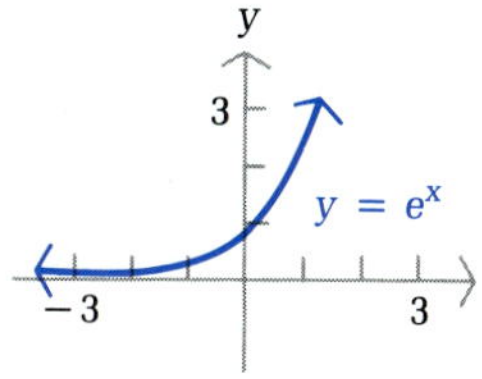
$\lim_{x \to 0} e^{-x} = 1$	$\lim_{x \to \infty} e^{-x} = 0$ $\lim_{x \to -\infty} e^{-x} = \infty$	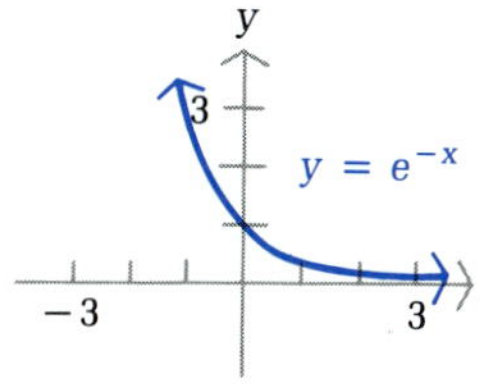
$\lim_{x \to 0^+} \ln x = -\infty$	$\lim_{x \to \infty} \ln x = \infty$	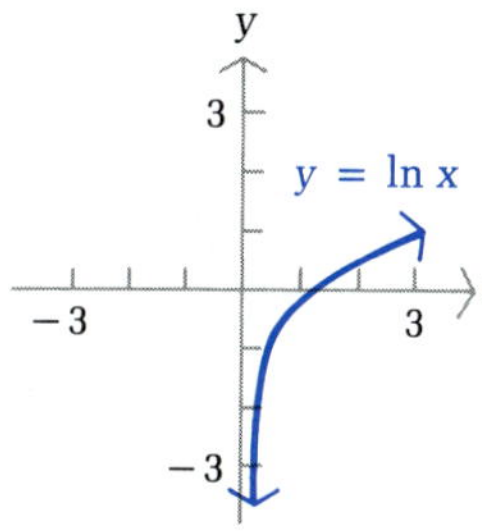

◆ EXAMPLE 22

(A) $\lim_{x \to 2} \frac{5}{(x-2)^4} = \infty$ Compare with $\lim_{x \to 0} \frac{1}{x^2}$ in Table 3.

(B) $\lim_{x \to -1^-} \frac{4}{(x+1)^3} = -\infty$ Compare with $\lim_{x \to 0^-} \frac{1}{x}$ in Table 3.

(C) $\lim_{x \to \infty} \frac{4}{(x-9)^6} = 0$ Compare with $\lim_{x \to \infty} \frac{1}{x^2}$ in Table 3

(D) $\lim_{x \to -\infty} 3x^3 = -\infty$ Compare with $\lim_{x \to -\infty} x$ in Table 3. ◆

PROBLEM 22 Evaluate each limit.

(A) $\lim_{x \to 3^+} \frac{7}{(x-3)^5}$ (B) $\lim_{x \to -4} \frac{6}{(x+4)^6}$

(C) $\lim_{x \to -\infty} \frac{3}{(x+2)^3}$ (D) $\lim_{x \to \infty} 5x^4$ ◆

The limits in Table 4 also generalize to other simple exponential and logarithmic forms.

◆ EXAMPLE 23

(A) $\lim_{x \to \infty} 2e^{3x} = \infty$ Compare with $\lim_{x \to \infty} e^x$

(B) $\lim_{x \to \infty} 4e^{-5x} = 0$ Compare with $\lim_{x \to \infty} e^{-x}$

(C) $\lim_{x \to \infty} \ln(x + 4) = \infty$ Compare with $\lim_{x \to \infty} \ln x$

(D) $\lim_{x \to 2^+} \ln(x - 2) = -\infty$ Compare with $\lim_{x \to 0^+} \ln x$ ◆

PROBLEM 23

Evaluate each limit.

(A) $\lim_{x \to -\infty} 2e^{-6x}$ (B) $\lim_{x \to -\infty} 3e^{2x}$

(C) $\lim_{x \to -4^+} \ln(x + 4)$ (D) $\lim_{x \to \infty} \ln(x - 10)$ ◆

Now that we have reviewed the limit properties of some basic functions, we are ready to consider the main topic of this section, L'Hôpital's rule—a powerful tool for evaluating certain types of limits.

◆ L'HÔPITAL'S RULE AND THE INDETERMINATE FORM 0/0

Recall that the limit

$$\lim_{x \to c} \frac{f(x)}{g(x)}$$

is a 0/0 indeterminate form (Section 3-2) if

$$\lim_{x \to c} f(x) = 0 \qquad \text{and} \qquad \lim_{x \to c} g(x) = 0$$

The quotient property for limits in Section 3-2 does not apply since $\lim_{x \to c} g(x) = 0$.

If we are dealing with a 0/0 indeterminate form, the limit may or may not exist, and we cannot tell which is true without further investigation.

Each of the following is a 0/0 indeterminate form:

$$\lim_{x \to 2} \frac{x^2 - 4}{x - 2} \qquad \text{and} \qquad \lim_{x \to 1} \frac{e^x - e}{x - 1}$$

The first limit can be evaluated by performing some algebraic simplifications, such as

$$\lim_{x \to 2} \frac{x^2 - 4}{x - 2} = \lim_{x \to 2} \frac{(x - 2)(x + 2)}{x - 2} = \lim_{x \to 2} (x + 2) = 4$$

But the second cannot. Instead, we turn to the very powerful **L'Hôpital's rule,** which we now state without proof. This rule can be used whenever a limit is a 0/0 indeterminate form.

THEOREM 1

L'Hôpital's Rule for 0/0 Indeterminate Forms: Version 1

For c a real number:

If $\lim_{x\to c} f(x) = 0$ and $\lim_{x\to c} g(x) = 0$, then

$$\lim_{x\to c} \frac{f(x)}{g(x)} = \lim_{x\to c} \frac{D_x f(x)}{D_x g(x)}$$

provided the second limit exists or is $+\infty$ or $-\infty$.

The use of L'Hôpital's rule is best illustrated through examples.

◆ **EXAMPLE 24** Evaluate: $\lim_{x\to 1} \frac{e^x - e}{x - 1}$

Solution Step 1. Check to see if L'Hôpital's rule applies:

$$\lim_{x\to 1}(e^x - e) = e^1 - e = 0 \quad \text{and} \quad \lim_{x\to 1}(x - 1) = 1 - 1 = 0$$

L'Hôpital's rule does apply.

Step 2. Apply L'Hôpital's rule :

0/0 form

$$\lim_{x\to 1} \frac{e^x - e}{x - 1} = \lim_{x\to 1} \frac{D_x(e^x - e)}{D_x(x - 1)}$$

$$= \lim_{x\to 1} \frac{e^x}{1} \qquad \text{L'Hôpital's rule}$$

$$= \frac{e^1}{1} = e \qquad \text{Limit property}$$

◆

PROBLEM 24 Evaluate: $\lim_{x\to 4} \frac{e^x - e^4}{x - 4}$ ◆

Remark

In L'Hôpital's rule, the symbol $D_x f(x)/D_x g(x)$ represents the derivative of $f(x)$ divided by the derivative of $g(x)$, not the derivative of the quotient $f(x)/g(x)$. **When applying L'Hôpital's rule to a 0/0 indeterminate form, be certain that you differentiate the numerator and denominator separately.**

◆ EXAMPLE 25 Evaluate: $\lim_{x\to 0} \frac{\ln(1+x^2)}{x^4}$

Solution Step 1. Check to see if L'Hôpital's rule applies:

$$\lim_{x\to 0} \ln(1+x^2) = \ln(1) = 0 \qquad \text{and} \qquad \lim_{x\to 0} x^4 = 0$$

L'Hôpital's rule does apply.

Step 2. Apply L'Hôpital's rule:

0/0 form

$$\begin{aligned}
\lim_{x\to 0} \frac{\ln(1+x^2)}{x^4} &= \lim_{x\to 0} \frac{D_x \ln(1+x^2)}{D_x x^4} \\
&= \lim_{x\to 0} \frac{2x/(1+x^2)}{4x^3} && \text{L'Hôpital's rule} \\
&= \lim_{x\to 0} \frac{1}{1+x^2} \cdot \frac{1}{2x^2} && \text{Algebra} \\
&= \infty && \text{Limit property}
\end{aligned}$$

The last step follows, since

$$\lim_{x\to 0} \frac{1}{1+x^2} = 1 \qquad \text{and} \qquad \lim_{x\to 0} \frac{1}{2x^2} = \infty$$

[Notice that we used the property that if $\lim_{x\to c} F(x) = \infty$ and $\lim_{x\to c} G(x) = L > 0$, then $\lim_{x\to c} F(x)G(x) = \infty$. See Section 3-2 for a discussion of limits of this form.] ◆

PROBLEM 25 Evaluate: $\lim_{x\to 1} \frac{\ln x}{(x-1)^3}$ ◆

◆ EXAMPLE 26 Evaluate: $\lim_{x\to 1} \frac{\ln x}{x}$

Solution Step 1. Check to see if L'Hôpital's rule applies:

$$\lim_{x\to 1}(\ln x) = \ln 1 = 0 \qquad \text{but} \qquad \lim_{x\to 1} x = 1 \neq 0$$

L'Hôpital's rule does not apply.

Step 2. Evaluate by another method. The quotient property for limits from Section 3-2 does apply, and we have

$$\lim_{x\to 1}\frac{\ln x}{x}=\frac{\lim_{x\to 1}(\ln x)}{\lim_{x\to 1} x}=\frac{\ln 1}{1}=\frac{0}{1}=0$$

Note that applying L'Hôpital's rule would give us an incorrect result:

$$\lim_{x\to 1}\frac{\ln x}{x}\neq\lim_{x\to 1}\frac{D_x \ln x}{D_x\, x}=\lim_{x\to 1}\frac{1/x}{1}=1$$ ◆

PROBLEM 26 Evaluate: $\lim_{x\to 0}\dfrac{x}{e^x}$ ◆

Remark

As Example 26 illustrates, all limits involving quotients are not 0/0 indeterminate forms.

You must always check to see if L'Hôpital's rule applies before you use it.

◆ EXAMPLE 27 Evaluate: $\lim_{x\to 0}\dfrac{x^2}{e^x-1-x}$

Solution *Step 1.* Check to see if L'Hôpital's rule applies:

$$\lim_{x\to 0} x^2=0 \qquad\text{and}\qquad \lim_{x\to 0}(e^x-1-x)=0$$

L'Hôpital's rule does apply.

Step 2. Apply L'Hôpital's rule:

0/0 form

$$\lim_{x\to 0}\frac{x^2}{e^x-1-x}=\lim_{x\to 0}\frac{D_x\, x^2}{D_x(e^x-1-x)}=\lim_{x\to 0}\frac{2x}{e^x-1}$$

Since $\lim_{x\to 0} 2x=0$ and $\lim_{x\to 0}(e^x-1)=0$, the new limit obtained is also a 0/0 indeterminate form, and L'Hôpital's rule can be applied again.

Step 3. Apply L'Hôpital's rule again:

0/0 form

$$\lim_{x\to 0}\frac{2x}{e^x-1}=\lim_{x\to 0}\frac{D_x\, 2x}{D_x(e^x-1)}=\lim_{x\to 0}\frac{2}{e^x}=\frac{2}{e^0}=2$$

Thus,

$$\lim_{x\to 0}\frac{x^2}{e^x-1-x}=\lim_{x\to 0}\frac{2x}{e^x-1}=\lim_{x\to 0}\frac{2}{e^x}=2$$ ◆

PROBLEM 27 Evaluate: $\lim_{x \to 0} \frac{e^{2x} - 1 - 2x}{x^2}$ ◆

◆ ONE-SIDED LIMITS AND LIMITS AT ∞

In addition to the limit as x approaches c, we have discussed one-sided limits and limits at ∞ in Chapter 3. L'Hôpital's rule is valid in these cases also.

THEOREM 2

L'Hôpital's Rule for 0/0 Indeterminate Forms: Version 2 (For One-Sided Limits and Limits at Infinity)

The first version of L'Hôpital's rule (Theorem 1) remains valid if the symbol $x \to c$ is replaced everywhere it occurs with one of the following symbols:

$$x \to c^+ \qquad x \to c^- \qquad x \to \infty \qquad x \to -\infty$$

For example, if $\lim_{x\to\infty} f(x) = 0$ and $\lim_{x\to\infty} g(x) = 0$, then

$$\lim_{x \to \infty} \frac{f(x)}{g(x)} = \lim_{x \to \infty} \frac{D_x f(x)}{D_x g(x)}$$

provided the second limit exists or is $+\infty$ or $-\infty$. Similar rules can be written for $x \to c^+$, $x \to c^-$, and $x \to -\infty$.

◆ EXAMPLE 28 Evaluate: $\lim_{x \to 1^+} \frac{\ln x}{(x-1)^2}$

Solution Step 1. Check to see if L'Hôpital's rule applies:

$$\lim_{x \to 1^+} \ln x = 0 \qquad \text{and} \qquad \lim_{x \to 1^+} (x-1)^2 = 0$$

L'Hôpital's rule does apply.

Step 2. Apply L'Hôpital's rule :

0/0 form

$$\lim_{x \to 1^+} \frac{\ln x}{(x-1)^2} = \lim_{x \to 1^+} \frac{D_x(\ln x)}{D_x(x-1)^2}$$

$$= \lim_{x \to 1^+} \frac{1/x}{2(x-1)} \qquad \text{L'Hôpital's rule}$$

$$= \lim_{x \to 1^+} \frac{1}{2x} \cdot \frac{1}{x-1} \qquad \text{Algebra}$$

$$= \infty \qquad \text{Limit property}$$

The last step follows, since

$$\lim_{x\to 1^+} \frac{1}{2x} = \frac{1}{2} \qquad \text{and} \qquad \lim_{x\to 1^+} \frac{1}{x-1} = \infty$$

◆

PROBLEM 28 Evaluate: $\lim_{x\to 1^-} \frac{\ln x}{(x-1)^2}$ ◆

◆ EXAMPLE 29 Evaluate: $\lim_{x\to\infty} \frac{\ln(1+e^{-x})}{e^{-x}}$

Solution Step 1. Check to see if L'Hôpital's rule applies:

$$\lim_{x\to\infty} \ln(1+e^{-x}) = \ln(1+0) = \ln 1 = 0 \qquad \text{and} \qquad \lim_{x\to\infty} e^{-x} = 0$$

L'Hôpital's rule does apply.

Step 2. Apply L'Hôpital's rule :

0/0 form

$$\lim_{x\to\infty} \frac{\ln(1+e^{-x})}{e^{-x}} = \lim_{x\to\infty} \frac{D_x[\ln(1+e^{-x})]}{D_x\, e^{-x}}$$

$$= \lim_{x\to\infty} \frac{-e^{-x}/(1+e^{-x})}{-e^{-x}} \qquad \text{L'Hôpital's rule}$$

$$= \lim_{x\to\infty} \frac{1}{1+e^{-x}} \qquad \text{Algebra}$$

$$= \frac{1}{1+0} = 1 \qquad \text{Limit property}$$

◆

PROBLEM 29 Evaluate: $\lim_{x\to -\infty} \frac{\ln(1+2e^{x})}{e^{x}}$ ◆

◆ L'HÔPITAL'S RULE AND THE INDETERMINATE FORM ∞/∞

In Sections 3-2 and 4-3 we discussed techniques for evaluating limits of rational functions such as

$$\lim_{x\to\infty} \frac{2x^2}{x^3+3} \qquad \lim_{x\to\infty} \frac{4x^3}{2x^2+5} \qquad \lim_{x\to\infty} \frac{3x^3}{5x^3+6}$$

Each of these limits is an ∞/∞ *indeterminate form*. In general, if $\lim_{x\to c} f(x) = \pm\infty$ and $\lim_{x\to c} g(x) = \pm\infty$, then

$$\lim_{x\to c} \frac{f(x)}{g(x)}$$

is called an ∞/∞ **indeterminate form.** Furthermore, $x \to c$ can be replaced in all three limits with $x \to c^+$, $x \to c^-$, $x \to \infty$, or $x \to -\infty$. It can be shown that L'Hôpital's rule also applies to these ∞/∞ indeterminate forms.

THEOREM 3 **L'Hôpital's Rule for the Indeterminate Form ∞/∞: Version 3**

Versions 1 and 2 of L'Hôpital's rule for the indeterminate form $0/0$ are also valid if the limit of f and the limit of g are both infinite; that is, both $+\infty$ and $-\infty$ are permissible for either limit.

For example, if $\lim_{x\to c^+} f(x) = \infty$ and $\lim_{x\to c^+} g(x) = -\infty$, then L'Hôpital's rule can be applied to $\lim_{x\to c^+} [f(x)/g(x)]$.

◆ **EXAMPLE 30** Evaluate: $\lim\limits_{x\to\infty} \dfrac{\ln x}{x^2}$

Solution Step 1. Check to see if L'Hôpital's rule applies:

$$\lim_{x\to\infty} \ln x = \infty \qquad \text{and} \qquad \lim_{x\to\infty} x^2 = \infty$$

L'Hôpital's rule does apply.

Step 2. Apply L'Hôpital's rule:

∞/∞ form

$$\begin{aligned}
\lim_{x\to\infty} \frac{\ln x}{x^2} &= \lim_{x\to\infty} \frac{D_x(\ln x)}{D_x\, x^2} \\
&= \lim_{x\to\infty} \frac{1/x}{2x} && \text{L'Hôpital's rule} \\
&= \lim_{x\to\infty} \frac{1}{2x^2} && \text{Algebra} \\
&= 0 && \text{Limit property}
\end{aligned}$$

◆

PROBLEM 30 Evaluate: $\lim\limits_{x\to\infty} \dfrac{\ln x}{x}$ ◆

◆ **EXAMPLE 31** Evaluate: $\lim\limits_{x\to\infty} \dfrac{e^x}{x^2}$

Solution Step 1. Check to see if L'Hôpital's rule applies:

$$\lim_{x\to\infty} e^x = \infty \qquad \text{and} \qquad \lim_{x\to\infty} x^2 = \infty$$

L'Hôpital's rule does apply.

Step 2. Apply L'Hôpital's rule:

∞/∞ form

$$\lim_{x\to\infty} \frac{e^x}{x^2} = \lim_{x\to\infty} \frac{D_x\, e^x}{D_x\, x^2} = \lim_{x\to\infty} \frac{e^x}{2x}$$

Since $\lim_{x\to\infty} e^x = \infty$ and $\lim_{x\to\infty} 2x = \infty$, this limit is an ∞/∞ indeterminate form and L'Hôpital's rule can be applied again.

Step 3. Apply L'Hôpital's rule again:

∞/∞ form

$$\lim_{x\to\infty} \frac{e^x}{2x} = \lim_{x\to\infty} \frac{D_x\, e^x}{D_x\, 2x} = \lim_{x\to\infty} \frac{e^x}{2} = \infty$$

Thus,

$$\lim_{x\to\infty} \frac{e^x}{x^2} = \lim_{x\to\infty} \frac{e^x}{2x} = \lim_{x\to\infty} \frac{e^x}{2} = \infty$$

◆

PROBLEM 31 Evaluate: $\lim_{x\to\infty} \dfrac{e^{2x}}{x^2}$ ◆

Answers to Matched Problems

22. (A) ∞ (B) ∞ (C) 0 (D) ∞
23. (A) ∞ (B) 0 (C) $-\infty$ (D) ∞
24. e^4 25. ∞ 26. 0 27. 2 28. $-\infty$ 29. 2 30. 0 31. ∞

EXERCISE 5-6

A *Find each limit using L'Hôpital's rule.*

1. $\lim_{x\to 2} \dfrac{x^4 - 16}{x^3 - 8}$
2. $\lim_{x\to 1} \dfrac{x^6 - 1}{x^5 - 1}$
3. $\lim_{x\to 2} \dfrac{x^2 + x - 6}{x^2 + 6x - 16}$
4. $\lim_{x\to 4} \dfrac{x^2 - 8x + 16}{x^2 - 5x + 4}$
5. $\lim_{x\to 0} \dfrac{e^x - 1}{x}$
6. $\lim_{x\to 1} \dfrac{\ln x}{x - 1}$

7. $\lim_{x \to 0} \frac{\ln(1 + 4x)}{x}$

8. $\lim_{x \to 0} \frac{e^{2x} - 1}{x}$

9. $\lim_{x \to \infty} \frac{2x^2 + 7}{5x^3 + 9}$

10. $\lim_{x \to \infty} \frac{3x^4 + 6}{2x^2 + 5}$

11. $\lim_{x \to \infty} \frac{e^{3x}}{x}$

12. $\lim_{x \to \infty} \frac{x}{e^{4x}}$

13. $\lim_{x \to \infty} \frac{x^2}{\ln x}$

14. $\lim_{x \to \infty} \frac{\ln x}{x^4}$

B *Find each limit in Problems 15–38. Note that L'Hôpital's rule does not apply to every problem, and some problems will require more than one application of L'Hôpital's rule.*

15. $\lim_{x \to 0} \frac{e^{4x} - 1 - 4x}{x^2}$

16. $\lim_{x \to 0} \frac{3x + 1 - e^{3x}}{x^2}$

17. $\lim_{x \to 2} \frac{\ln(x - 1)}{x - 1}$

18. $\lim_{x \to -1} \frac{\ln(x + 2)}{x + 2}$

19. $\lim_{x \to 0^+} \frac{\ln(1 + x^2)}{x^3}$

20. $\lim_{x \to 0^-} \frac{\ln(1 + 2x)}{x^2}$

21. $\lim_{x \to 0^+} \frac{\ln(1 + \sqrt{x})}{x}$

22. $\lim_{x \to 0^+} \frac{\ln(1 + x)}{\sqrt{x}}$

23. $\lim_{x \to -2} \frac{x^2 + 2x + 1}{x^2 + x + 1}$

24. $\lim_{x \to 1} \frac{2x^3 - 3x^2 + 1}{x^3 - 3x + 2}$

25. $\lim_{x \to -1} \frac{x^3 + x^2 - x - 1}{x^3 + 4x^2 + 5x + 2}$

26. $\lim_{x \to 3} \frac{x^3 + 3x^2 - x - 3}{x^2 + 6x + 9}$

27. $\lim_{x \to 2^-} \frac{x^3 - 12x + 16}{x^3 - 6x^2 + 12x - 8}$

28. $\lim_{x \to 1^+} \frac{x^3 - x^2 - x + 1}{x^3 - 3x^2 + 3x - 1}$

29. $\lim_{x \to \infty} \frac{3x^2 + 5x}{4x^3 + 7}$

30. $\lim_{x \to \infty} \frac{4x^2 + 9x}{5x^2 + 8}$

31. $\lim_{x \to \infty} \frac{x^2}{e^{2x}}$

32. $\lim_{x \to \infty} \frac{e^{3x}}{x^3}$

33. $\lim_{x \to \infty} \frac{1 + e^{-x}}{1 + x^2}$

34. $\lim_{x \to -\infty} \frac{1 + e^{-x}}{1 + x^2}$

35. $\lim_{x \to \infty} \frac{e^{-x}}{\ln(1 + 4e^{-x})}$

36. $\lim_{x \to \infty} \frac{\ln(1 + 2e^{-x})}{\ln(1 + e^{-x})}$

37. $\lim_{x \to 0} \frac{e^{x} - e^{-x} - 2x}{x^3}$

38. $\lim_{x \to 0} \frac{e^{2x} - 1 - 2x - 2x^2}{x^3}$

C

39. Find: $\lim_{x \to 0^+} (x \ln x)$

[*Hint:* Write $x \ln x = (\ln x)/x^{-1}$.]

40. Find: $\lim_{x \to 0^+} (\sqrt{x} \ln x)$

[*Hint:* Write $\sqrt{x} \ln x = (\ln x)/x^{-1/2}$.]

In Problems 41–44, n is a positive integer. Find each limit.

41. $\lim_{x\to\infty} \frac{\ln x}{x^n}$ **42.** $\lim_{x\to\infty} \frac{x^n}{\ln x}$ **43.** $\lim_{x\to\infty} \frac{e^x}{x^n}$ **44.** $\lim_{x\to\infty} \frac{x^n}{e^x}$

In Problems 45–48, show that repeated application of L'Hôpital's rule does not lead to a solution. Use a graphic calculator or a computer to graph each function and estimate the limit from the graph.

45. $\lim_{x\to\infty} \frac{\sqrt{1+x^2}}{x}$

46. $\lim_{x\to-\infty} \frac{x}{\sqrt{4+x^2}}$

47. $\lim_{x\to-\infty} \frac{\sqrt[3]{x^3+1}}{x}$

48. $\lim_{x\to\infty} \frac{x^2}{\sqrt[3]{(x^3+1)^2}}$

SECTION 5-7 Elasticity of Demand

- PRICE AND ELASTICITY OF DEMAND
- REVENUE AND ELASTICITY OF DEMAND

◆ PRICE AND ELASTICITY OF DEMAND

In this section we will study the effects that changes in price have on demand and revenue. Suppose the price \$p and the demand x for a certain product are related by the **price–demand equation:**

$$x + 500p = 10{,}000 \qquad (1)$$

In problems involving revenue, cost, and profit, it is customary to use the demand equation to express price as a function of demand. Since we are now interested in the effects that changes in price have on demand, it will be more convenient to express demand as a function of price. Solving (1) for x, we have

$$x = 10{,}000 - 500p \qquad \text{Demand as a function of price}$$
$$= 500(20 - p)$$

or

$$x = f(p) = 500(20 - p) \qquad 0 \le p \le 20 \qquad (2)$$

Since x and p must be nonnegative quantities, we must restrict p so that $0 \le p \le 20$.

For most products, demand is assumed to be a decreasing function of price. That is, price increases result in lower demand and price decreases result in higher demand (see Figure 3).

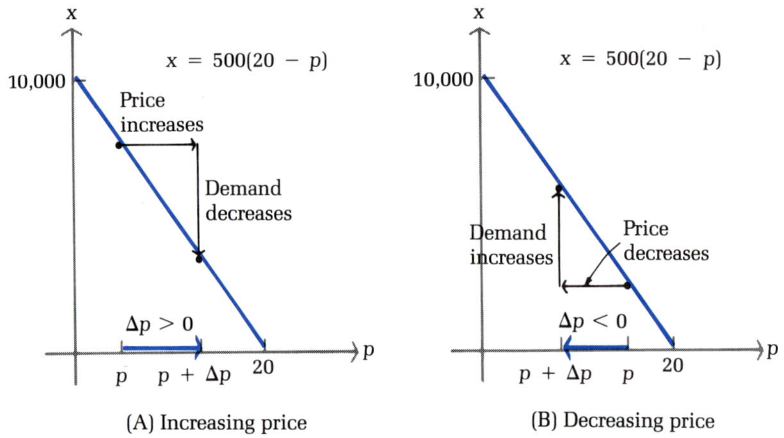

FIGURE 3
Price and demand

Given a price–demand equation $x = f(p)$ and using increment notation (see Section 4-5), suppose the price p is changed by an amount Δp, causing the demand x to change by an amount Δx. Then the **relative change in price** and the **relative change in demand** are, respectively,

$$\frac{\Delta p}{p} \quad \text{and} \quad \frac{\Delta x}{x} = \frac{f(p + \Delta p) - f(p)}{f(p)}$$

Economists use the ratio

$$\frac{\dfrac{\Delta x}{x}}{\dfrac{\Delta p}{p}} = \frac{\text{Relative change in demand}}{\text{Relative change in price}} \tag{3}$$

to study the effect of price changes on demand. Economics texts that do not use calculus call the expression in (3) the **elasticity of demand at price *p*.** However, this expression obviously depends on both p and Δp. Using calculus, we can let $\Delta p \to 0$ and obtain an expression for the point elasticity of demand at price p, denoted $E(p)$:

$$E(p) = \lim_{\Delta p \to 0} \frac{\frac{\Delta x}{x}}{\frac{\Delta p}{p}}$$

$$= \lim_{\Delta p \to 0} \frac{\Delta x}{x} \cdot \frac{p}{\Delta p}$$

$$= \frac{p}{x} \lim_{\Delta p \to 0} \frac{\Delta x}{\Delta p}$$ Since p and $x = f(p)$ are constant with respect to this limit

$$= \frac{p}{x} \cdot \frac{dx}{dp}$$ Definition of derivative

$$= \frac{p}{f(p)} f'(p)$$ Since $x = f(p)$ and $\frac{dx}{dp} = f'(p)$

Thus, we define the **point elasticity of demand** to be

$$E(p) = \frac{pf'(p)}{f(p)}$$

Since p and $f(p)$ are always nonnegative quantities and $f'(p) \leq 0$ (remember, demand is assumed to be a decreasing function of price), $E(p) \leq 0$ for all values of p for which $E(p)$ is defined.

◆ EXAMPLE 32 If $x = f(p) = 500(20 - p)$, find $E(p)$ and calculate $E(p)$ at:
(A) $p = \$4$ (B) $p = \$16$ (C) $p = \$10$

Solutions

$$E(p) = \frac{pf'(p)}{f(p)}$$

$$= \frac{p(-500)}{500(20 - p)}$$

$$= \frac{-p}{20 - p} \qquad 0 \leq p < 20$$

(A) $E(4) = -\frac{4}{16} = -0.25$ (B) $E(16) = -\frac{16}{4} = -4$

(C) $E(10) = -\frac{10}{10} = -1$ ◆

An economist would interpret the results in Example 32 as follows:

Part A. $E(4) = -0.25 > -1$. At this price level ($p = 4$), a percentage change in price will result in a smaller percentage change in demand. For example, if the price is increased by 10%, then the demand will change by approximately

$$-0.25(10\%) = -2.5\%$$

Since this change is negative, a 10% price increase will result in a 2.5% decrease in demand. On the other hand, a 10% price cut will result in a 2.5% increase in demand. Since the demand is not very sensitive to changes in price at this price level, we say that demand is **inelastic** when $E(p) > -1$.

Part B. $E(16) = -4 < -1$. At this price level ($p = 16$), a percentage change in price will result in a larger percentage change in demand. This time a 10% price increase will result in an approximate 40% decrease in demand, while a 10% price cut will result in an approximate 40% increase in demand. Since the demand is very sensitive to changes in price at this price level, we say that demand is **elastic** when $E(p) < -1$.

Part C. $E(10) = -1$. In this case, percentage changes in price will result in approximately equal percentage changes in demand. When $E(p) = -1$, we say that the demand has **unit elasticity.** ◆

PROBLEM 32 If $x = f(p) = 1{,}000(40 - p)$, find $E(p)$ and evaluate $E(p)$ at:

(A) $p = \$8$ (B) $p = \$30$ (C) $p = \$20$ ◆

All the pertinent definitions are summarized in the box.

Point Elasticity of Demand

Let demand x and price p be related by the **price–demand equation**

$$x = f(p)$$

The **point elasticity of demand** is

$$E(p) = \frac{pf'(p)}{f(p)}$$

Demand is **inelastic** if $-1 < E(p) \leq 0$.

Demand is **elastic** if $E(p) < -1$.

Demand has **unit elasticity** if $E(p) = -1$.

◆ EXAMPLE 33 Given $x = f(p) = 9{,}000 - 30p^2$:

Price–Demand

(A) Determine the values of p for which demand is inelastic and the values for which it is elastic.
(B) Discuss the effect of a 10% price cut when $p = \$7$.
(C) Discuss the effect of a 10% price increase when $p = \$15$.

Solutions (A) First, notice that

$$\begin{aligned} f(p) &= 30(300 - p^2) \\ &= 30(10\sqrt{3} - p)(10\sqrt{3} + p) \end{aligned}$$

Since both p and $f(p)$ must be nonnegative, we must restrict p to

$$0 \leq p \leq 10\sqrt{3} \approx 17.3$$

$$E(p) = \frac{pf'(p)}{f(p)}$$

$$= \frac{p(-60p)}{9{,}000 - 30p^2}$$

$$= \frac{-2p^2}{300 - p^2} \qquad 0 \leq p < 10\sqrt{3}$$

The following observations will simplify our calculations:

Demand is inelastic:	$E(p) > -1$	$E(p) + 1 > 0$
Demand is elastic:	$E(p) < -1$	$E(p) + 1 < 0$

Thus, we can determine where demand is inelastic and where it is elastic by constructing a sign chart for $E(p) + 1$:

$$E(p) + 1 = \frac{-2p^2}{300 - p^2} + 1$$

$$= \frac{300 - 3p^2}{300 - p^2}$$

$$= \frac{3(10 - p)(10 + p)}{(10\sqrt{3} - p)(10\sqrt{3} + p)}$$

Test Numbers

p	$E(p) + 1$
5	$\frac{9}{11}$ (+)
15	-5 (−)

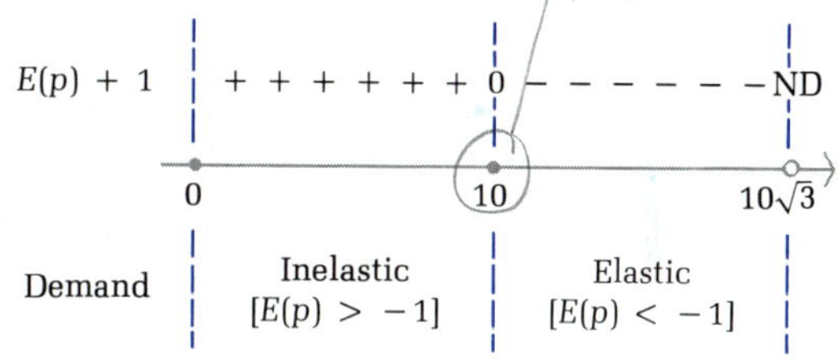

Thus, demand is inelastic for $0 \leq p < 10$ and elastic for $10 < p < 10\sqrt{3}$.

(B) $$E(7) = \frac{-2 \cdot 49}{300 - 49} \approx -0.39$$

Thus, when $p = \$7$, a 10% price cut will result in a change in demand of approximately

$$-0.39(-10\%) = 3.9\%$$

That is, the demand will increase approximately 3.9%.

(C) $$E(15) = \frac{-2 \cdot 225}{300 - 225} = -6$$

Thus, when $p = \$15$, a 10% price increase will result in a change in demand of approximately

$$-6(10\%) = -60\%$$

That is, the demand will decrease approximately 60%. ◆

PROBLEM 33 Given $x = f(p) = 6{,}000 - 5p^2$:

(A) Determine the values of p for which demand is inelastic and those for which it is elastic.
(B) Discuss the effect of a 10% price increase when $p = \$10$.
(C) Discuss the effect of a 10% price decrease when $p = \$25$. ◆

◆ REVENUE AND ELASTICITY OF DEMAND

Now we want to see how revenue and elasticity of demand are related. We begin by considering an example.

◆ EXAMPLE 34 Given the price–demand equation $x = f(p) = 500(20 - p)$, $0 \le p \le 20$:

Price–Demand

(A) Determine the values of p for which revenue is increasing and those for which revenue is decreasing.
(B) Determine the values of p for which demand is inelastic and those for which demand is elastic.

Solutions

(A) Revenue = (Price per unit)(Number of units)

$$\begin{aligned} R(p) &= px \\ &= 500p(20 - p) \\ &= 10{,}000p - 500p^2 \qquad 0 \le p \le 20 \\ R'(p) &= 10{,}000 - 1{,}000p \\ &= 1{,}000(10 - p) \end{aligned}$$

The only critical value is $p = 10$. Thus,

$$R'(p) > 0 \quad \text{for} \quad 0 < p < 10 \qquad \text{Increasing revenue}$$

and

$$R'(p) < 0 \quad \text{for} \quad 10 < p < 20 \qquad \text{Decreasing revenue}$$

(B)

$$E(p) = \frac{pf'(p)}{f(p)} = \frac{-500p}{500(20 - p)} = \frac{-p}{20 - p}$$

$$E(p) + 1 = -\frac{p}{20 - p} + 1 = \frac{20 - 2p}{20 - p} = \frac{2(10 - p)}{20 - p}$$

Since the denominator is positive for $0 < p < 20$, we see that

$$E(p) + 1 > 0 \quad \text{for} \quad 0 < p < 10 \qquad \text{Inelastic demand}$$

and

$$E(p) + 1 < 0 \quad \text{for} \quad 10 < p < 20 \qquad \text{Elastic demand}$$

◆

PROBLEM 34 Repeat Example 34 for $x = f(p) = 1{,}000(40 - p)$, $0 \leq p \leq 40$. ◆

Comparing the answers in parts A and B of Example 34, we see that revenue is increasing precisely when demand is inelastic and revenue is decreasing when demand is elastic. Is this always the case?

In general, let $x = f(p)$ be a demand function and let

$$R(p) = px = pf(p)$$

Then

$$\begin{aligned} R'(p) &= pf'(p) + f(p) \\ &= f(p)\left[\frac{pf'(p)}{f(p)} + 1\right] \\ &= f(p)[E(p) + 1] \end{aligned}$$

Since $x = f(p) > 0$, we conclude:

All Are True or All Are False	All Are True or All Are False
$R'(p) > 0$	$R'(p) < 0$
$E(p) + 1 > 0$	$E(p) + 1 < 0$
Demand is inelastic	Demand is elastic

These facts are summarized in Figure 4 (on the next page) and the following box.

Summary—Revenue and Elasticity of Demand

DEMAND IS INELASTIC:

A price increase will increase revenue.

A price decrease will decrease revenue.

DEMAND IS ELASTIC:

A price increase will decrease revenue.

A price decrease will increase revenue.

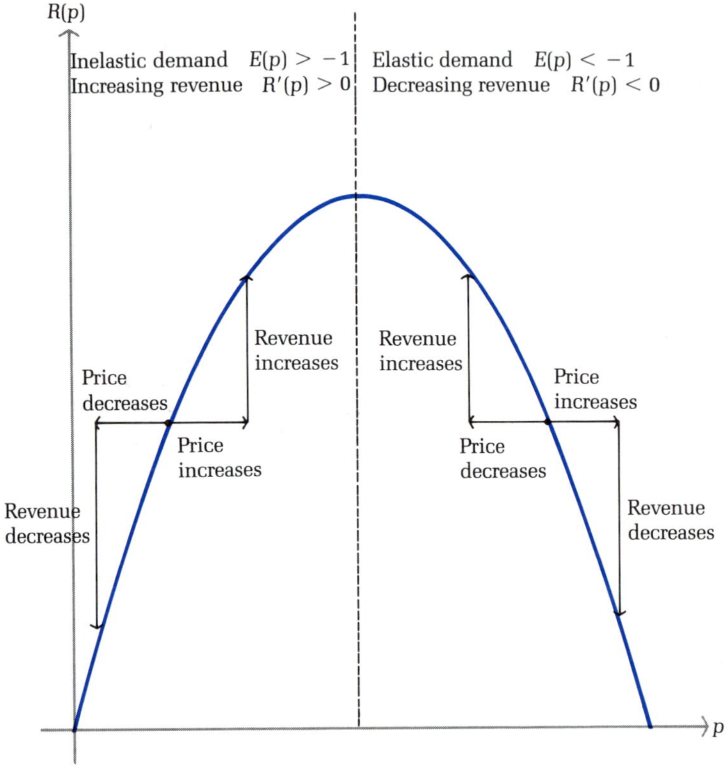

FIGURE 4
Revenue and elasticity of demand

◆ EXAMPLE 35

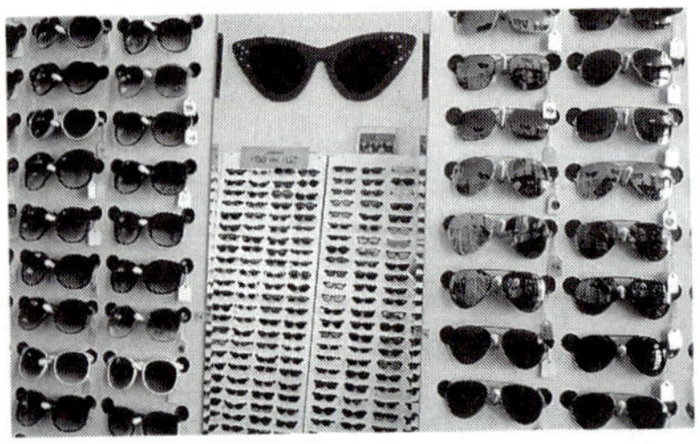

A company can sell 4,500 pairs of sunglasses monthly when the price is \$5.00. When the price of a pair of sunglasses is increased by 10%, the demand drops to 4,250 pairs a month. Assume that the demand equation is linear.

(A) Find the point elasticity of demand at the new price level.

(B) Approximate the change in demand if the price is increased by an additional 10%.

(C) Will a second 10% price increase cause the revenue to increase or decrease?

Solutions

First, we must find the demand equation. We know that the graph of the demand equation is a straight line and that (5, 4,500) is one point on its graph. Furthermore, when the price is increased by 10% from \$5 to $5 + (0.1)5 = \$5.50$, the demand drops to 4,250. Thus, (5.5, 4,250) is a second point on the graph of this line. Remembering that x is the dependent variable and p is the independent variable, the slope of the line is

$$m = \frac{4{,}250 - 4{,}500}{5.5 - 5} = -500 \qquad m = \frac{x_2 - x_1}{p_2 - p_1}$$

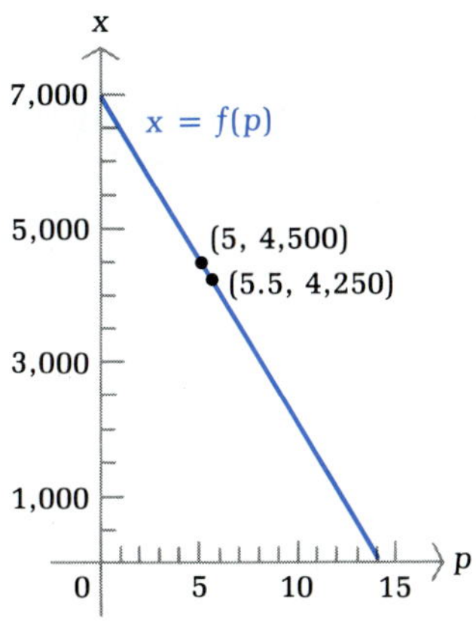

and the equation of the line is

$$x - 4{,}500 = -500(p - 5) \qquad x - x_1 = m(p - p_1)$$
$$x = 7{,}000 - 500p$$

Thus, the demand equation is (see the figure in the margin)

$$x = f(p) = 7{,}000 - 500p \qquad \textit{Check:}\ \ p = 5{:}\ \ x = 7{,}000 - 2{,}500 \stackrel{\checkmark}{=} 4{,}500$$
$$p = 5.5{:}\ \ x = 7{,}000 - 2{,}750 \stackrel{\checkmark}{=} 4{,}250$$

(A) $$E(p) = \frac{pf'(p)}{f(p)} = \frac{-500p}{7{,}000 - 500p} = \frac{-p}{14 - p} \qquad E(5.5) = \frac{-5.5}{14 - 5.5} \approx -0.65$$

Point elasticity of demand at $p = 5.5$

(B) At a price level of \$5.50, a 10% increase in the price will result in a percentage change in demand of approximately

$$E(5.5) \cdot 10\% \approx -0.65(10\%) = -6.5\%$$

Thus, the demand will decrease by

$$0.065(4{,}250) \approx 276$$

(C) Since $E(5.5) \approx -0.65 > -1$, demand is inelastic at this price level and an increase in price will increase revenue. ◆

PROBLEM 35 Repeat Example 35 if the demand drops from 4,500 to 3,250 when the price is increased from \$5.00 to \$5.50. ◆

Answers to Matched Problems

32. $E(p) = -p/(40 - p)$; (A) $E(8) = -0.25$ (B) $E(30) = -3$ (C) $E(20) = -1$
33. (A) Inelastic for $0 < p < 20$; elastic for $20 < p < 20\sqrt{3}$ (B) 1.8% decrease in demand (C) 22% increase in demand
34. (A) Revenue is increasing for $0 < p < 20$, decreasing for $20 < p < 40$ (B) Demand is inelastic for $0 < p < 20$, elastic for $20 < p < 40$
35. (A) $E(5.5) \approx -4.2$ (B) Changes by -42%; decreases by approximately 1,365 pairs (C) Revenue decreases

EXERCISE 5-7

A

1. Given the demand equation

$$p + \tfrac{1}{200}x = 30 \qquad 0 \le p \le 30$$

(A) Express the demand x as a function of the price p.
(B) Find the point elasticity of demand, $E(p)$.

(C) What is the point elasticity of demand when $p = \$10$? If this price is increased by 10%, what is the approximate change in demand?
(D) What is the point elasticity of demand when $p = \$25$? If this price is increased by 10%, what is the approximate change in demand?
(E) What is the point elasticity of demand when $p = \$15$? If this price is increased by 10%, what is the approximate change in demand?

2. Given the demand equation

$$p + \tfrac{1}{100}x = 50 \qquad 0 \leq p \leq 50$$

(A) Express the demand x as a function of the price p.
(B) Find the point elasticity of demand, $E(p)$.
(C) What is the point elasticity of demand when $p = \$10$? If this price is decreased by 5%, what is the approximate change in demand?
(D) What is the point elasticity of demand when $p = \$45$? If this price is decreased by 5%, what is the approximate change in demand?
(E) What is the point elasticity of demand when $p = \$25$? If this price is decreased by 5%, what is the approximate change in demand?

3. Given the demand equation

$$\tfrac{1}{50}x + p = 60 \qquad 0 \leq p \leq 60$$

(A) Express the demand x as a function of the price p.
(B) Express the revenue R as a function of the price p.
(C) Find the point elasticity of demand, $E(p)$.
(D) For which values of p is demand elastic? Inelastic?
(E) For which values of p is revenue increasing? Decreasing?
(F) If $p = \$10$ and the price is cut by 10%, wll revenue increase or decrease?
(G) If $p = \$40$ and the price is cut by 10%, will revenue increase or decrease?

4. Repeat Problem 3 for the demand equation

$$\tfrac{1}{60}x + p = 50 \qquad 0 \leq p \leq 50$$

For each of the following demand equations, determine whether demand is elastic, inelastic, or has unit elasticity at the indicated values of p.

5. $x = f(p) = 12{,}000 - 10p^2$

(A) $p = 10$ (B) $p = 20$ (C) $p = 30$

6. $x = f(p) = 1{,}875 - p^2$

(A) $p = 15$ (B) $p = 25$ (C) $p = 40$

7. $x = f(p) = 950 - 2p - \tfrac{1}{10}p^2$

(A) $p = 30$ (B) $p = 50$ (C) $p = 70$

8. $x = f(p) = 875 - p - \tfrac{1}{20}p^2$

(A) $p = 50$ (B) $p = 70$ (C) $p = 100$

B *For each of the following demand equations, find the values of p for which demand is elastic and the values for which demand is inelastic.*

9. $x = f(p) = 10(p - 30)^2, \quad 0 \le p \le 30$
10. $x = f(p) = 5(p - 60)^2, \quad 0 \le p \le 60$
11. $x = f(p) = \sqrt{144 - 2p}, \quad 0 \le p \le 72$
12. $x = f(p) = \sqrt{324 - 2p}, \quad 0 \le p \le 162$
13. $x = f(p) = \sqrt{2{,}500 - 2p^2}, \quad 0 \le p \le 25\sqrt{2}$
14. $x = f(p) = \sqrt{3{,}600 - 2p^2}, \quad 0 \le p \le 30\sqrt{2}$

For each of the following demand equations, sketch the graph of the revenue function and indicate the regions of inelastic and elastic demand on the graph.

15. $x = f(p) = 20(10 - p), \quad 0 \le p \le 10$
16. $x = f(p) = 10(16 - p), \quad 0 \le p \le 16$
17. $x = f(p) = 40(p - 15)^2, \quad 0 \le p \le 15$
18. $x = f(p) = 10(p - 9)^2, \quad 0 \le p \le 9$
19. $x = f(p) = 30 - 10\sqrt{p}, \quad 0 \le p \le 9$
20. $x = f(p) = 30 - 5\sqrt{p}, \quad 0 \le p \le 36$

C *In economics, it is common to use the demand x as the independent variable. If $p = g(x)$ is the demand equation, then it can be shown that the point elasticity of demand is given by*

$$E(x) = \frac{g(x)}{xg'(x)}$$

Use this formula in Problems 21–24 to find the point elasticity of demand at the indicated value of x.

21. $p = g(x) = 50 - \frac{1}{10}x, \quad x = 200$
22. $p = g(x) = 30 - \frac{1}{20}x, \quad x = 400$
23. $p = g(x) = 50 - 2\sqrt{x}, \quad x = 400$
24. $p = g(x) = 20 - \sqrt{x}, \quad x = 100$

25. Find $E(p)$ for $x = f(p) = Ap^{-k}$, where A and k are positive constants.
26. Find $E(p)$ for $x = f(p) = Ae^{-kp}$, where A and k are positive constants.

In Problems 27–32, use a graphic calculator or a computer to graph $y = E(p)$ and $y = -1$ in the same viewing rectangle and to approximate the point of intersection of these graphs. Use this information to find the values of p for which demand is elastic and the values for which demand is inelastic.

27. From Problem 9:
$f(p) = 10(p - 30)^2$
p range: [0, 30]
y range: [−10, 1]

28. From Problem 10:
$f(p) = 5(p - 60)^2$
p range: [0, 60]
y range: [−10, 1]

29. From Problem 11:
$f(p) = \sqrt{144 - 2p}$
p range: [0, 72]
y range: [−10, 1]

30. From Problem 12:
$f(p) = \sqrt{324 - 2p}$
p range: [0, 162]
y range: [−10, 1]

31. $f(p) = 100 \ln(50 - p)$
p range: [0, 49]
y range: [−10, 1]

32. $f(p) = 200 \ln(100 - p)$
p range: [0, 99]
y range: [−10, 1]

Business & Economics

33. *Revenue and elasticity.* The weekly demand for hamburgers sold by a chain of restaurants is 30,000 when the price of a hamburger is \$2.00. A 10% price increase caused the weekly demand to drop to 28,000 hamburgers. Assume that the demand equation is linear.

    (A) Find the point elasticity of demand at the new price.
    (B) Approximate the change in demand if the price is increased by an additional 10% over the first 10% increase (see Example 35).
    (C) Will the second price increase cause the revenue to increase or decrease?

34. *Revenue and elasticity.* Repeat Problem 33 if the 10% price increase caused the weekly demand to drop to 24,000 hamburgers.

35. *Revenue and elasticity.* The weekly demand for a small personal computer is 2,000 when the price of a computer is \$100. A 10% price cut caused the demand to increase to 2,100 computers per week. Assume that the demand equation is linear.

    (A) Find the point elasticity of demand at the new price.
    (B) Approximate the change in demand if the price is decreased by an additional 10% over the first 10% cut (see Example 35).
    (C) Will the second price decrease cause the revenue to increase or decrease?

36. *Revenue and elasticity.* Repeat Problem 35 if the 10% price cut caused the demand to increase to 2,400 computers per week.

SECTION 5-8 Chapter Review

Important Terms and Symbols

5-1 *The Constant e and Continuous Compound Interest.* Definition of e; continuous compound interest

5-2 *Derivatives of Logarithmic and Exponential Functions.* Derivative formulas for the natural logarithmic and exponential functions; graph properties of $y = \ln x$ and $y = e^x$

5-3 *Chain Rule: General Form.* Composite functions; chain rule; general derivative formulas; derivative formulas for $y = \log_b x$ and $y = b^x$

5-4 *Implicit Differentiation.* Special function notation; function explicitly defined; function implicitly defined; implicit differentiation

$$y = f(x); \quad y = y(x); \quad F(x, y) = 0; \quad y'|_{(a,b)}$$

5-5 *Related Rates.* Related rates

$$x = x(t); \quad y = y(t)$$

5-6 *L'Hôpital's Rule.* Limit properties of basic functions; 0/0 indeterminate forms; L'Hôpital's rule (versions 1, 2, and 3); ∞/∞ indeterminate forms

$$\lim_{x \to c} \frac{f(x)}{g(x)}; \quad \lim_{x \to c^+} \frac{f(x)}{g(x)}; \quad \lim_{x \to c^-} \frac{f(x)}{g(x)}; \quad \lim_{x \to \infty} \frac{f(x)}{g(x)}; \quad \lim_{x \to -\infty} \frac{f(x)}{g(x)}$$

5-7 *Elasticity of Demand.* Price-demand equation; relative change in price; relative change in demand; elasticity of demand at price p; point elasticity of demand; inelastic demand; elastic demand; unit elasticity; relationship between revenue and elasticity of demand

$$E(p) = \frac{pf'(p)}{f(p)}$$

Additional Rules of Differentiation

$$D_x \ln x = \frac{1}{x} \qquad D_x e^x = e^x$$

$$D_x \ln[f(x)] = \frac{1}{f(x)} f'(x) \qquad D_x e^{f(x)} = e^{f(x)} f'(x)$$

$$D_x \log_b x = D_x \frac{1}{\ln b} \ln x = \frac{1}{\ln b}\left(\frac{1}{x}\right)$$

$$D_x b^x = D_x e^{x \ln b} = e^{x \ln b} \ln b = b^x \ln b$$

$$D_x[f(x)]^n = n[f(x)]^{n-1} f'(x)$$

$$\frac{dy}{dx} = \frac{dy}{du}\frac{du}{dx}, \quad \frac{dy}{dx} = \frac{dy}{dw}\frac{dw}{du}\frac{du}{dx}, \quad \text{and so on}$$

EXERCISE 5-8

Chapter Review

Work through all the problems in this chapter review and check your answers in the back of the book. (Answers to all review problems are there.) Where weaknesses show up, review appropriate sections in the text.

A

1. Use a calculator to evaluate $A = 2{,}000e^{0.09t}$ to the nearest cent for $t = 5$, 10, and 20.

Find the indicated derivatives in Problems 2–4.

2. $D_x(2 \ln x + 3e^x)$ 3. $D_x\, e^{2x-3}$ 4. y' for $y = \ln(2x + 7)$

5. Let $y = \ln u$ and $u = 3 + e^x$.
   (A) Express y in terms of x.
   (B) Use the chain rule to find dy/dx, and then express dy/dx in terms of x.
6. Find y' for $y = y(x)$ defined implicitly by $2y^2 - 3x^3 - 5 = 0$, and evaluate at $(x, y) = (1, 2)$.
7. For $y = 3x^2 - 5$, where $x = x(t)$ and $y = y(t)$, find dy/dt if $dx/dt = 3$ when $x = 12$.

8. Given the demand equation

$$p + \tfrac{1}{25}x = 40 \qquad 0 \leq p \leq 40$$

(A) Express the demand x as a function of the price p.
(B) Find the point elasticity of demand, $E(p)$.
(C) What is the point elasticity of demand when $p = \$15$? If this price is increased by 10%, what is the approximate change in demand?
(D) Express the revenue R as a function of the price p.
(E) If $p = \$25$ and the price is cut by 10%, will revenue increase or decrease?

B

9. Graph $y = 100e^{-0.1x}$.

Find the indicated derivatives in Problems 10–15.

10. $D_z[(\ln z)^7 + \ln z^7]$

11. $D_x(x^6 \ln x)$

12. $D_x \dfrac{e^x}{x^6}$

13. y' for $y = \ln(2x^3 - 3x)$

14. $f'(x)$ for $f(x) = e^{x^3 - x^2}$

15. dy/dx for $y = e^{-2x} \ln 5x$

16. Find the equation of the line tangent to the graph of $y = f(x) = 1 + e^{-x}$ at $x = 0$. At $x = -1$.
17. Find y' for $y = y(x)$ defined implicitly by $x^2 - 3xy + 4y^2 = 23$, and find the slope of the graph at $(-1, 2)$.
18. Find x' for $x = x(t)$ defined implicitly by $x^3 - 2t^2x + 8 = 0$, and evaluate at $(t, x) = (-2, 2)$.
19. Find y' for $y = y(x)$ defined implicitly by $x - y^2 = e^y$, and evaluate at $(1, 0)$.
20. Find y' for $y = y(x)$ defined implicitly by $\ln y = x^2 - y^2$, and evaluate at $(1, 1)$.
21. A point is moving on the graph of $y^2 - 4x^2 = 12$ so that its x coordinate is decreasing at 2 units per second when $(x, y) = (1, 4)$. Find the rate of change of the y coordinate.
22. A 17 foot ladder is placed against a wall. If the foot of the ladder is pushed toward the wall at 0.5 foot per second, how fast is the top rising when the foot of the ladder is 8 feet from the wall?
23. Water from a water heater is leaking onto a floor. A circular pool is created whose area is increasing at the rate of 24 square inches per minute. How fast is the radius R of the pool increasing when the radius is 12 inches? $[A = \pi R^2]$
24. Find the values of p for which demand is elastic and the values for which demand is inelastic if the demand equation is

$$x = f(p) = 20(p - 15)^2 \qquad 0 \leq p \leq 15$$

25. Graph the revenue function and indicate the regions of inelastic and elastic demand on the graph if the demand equation is

$$x = f(p) = 5(20 - p) \qquad 0 \leq p \leq 20$$

Find each limit.

26. $\lim_{x \to 0} \frac{e^{3x} - 1}{x}$

27. $\lim_{x \to 2} \frac{x^2 - 5x + 6}{x^2 + x - 6}$

28. $\lim_{x \to 0^-} \frac{\ln(1 + x)}{x^2}$

29. $\lim_{x \to 0} \frac{\ln(1 + x)}{1 + x}$

30. $\lim_{x \to \infty} \frac{e^{4x}}{x^2}$

31. $\lim_{x \to 0} \frac{e^x + e^{-x} - 2}{x^2}$

32. $\lim_{x \to 0^+} \frac{\sqrt{1 + x} - 1}{\sqrt{x}}$

33. $\lim_{x \to \infty} \frac{\ln x}{x^5}$

34. $\lim_{x \to \infty} \frac{\ln(1 + 6x)}{\ln(1 + 3x)}$

35. $\lim_{x \to 0} \frac{\ln(1 + 6x)}{\ln(1 + 3x)}$

C *In Problems 36 and 37, find the absolute maximum value of $f(x)$ for $x > 0$.*

36. $f(x) = 11x - 2x \ln x$

37. $f(x) = 10xe^{-2x}$

Sketch the graph of $y = f(x)$ in Problems 38 and 39.

38. $f(x) = 5 - 5e^{-x}$

39. $f(x) = x^3 \ln x$

40. Let $y = w^3$, $w = \ln u$, and $u = 4 - e^x$.
    (A) Express y in terms of x.
    (B) Use the chain rule to find dy/dx, and then express dy/dx in terms of x.

Find the indicated derivatives in Problems 41–43.

41. y' for $y = 5^{x^2-1}$

42. $D_x \log_5(x^2 - x)$

43. $D_x \sqrt{\ln(x^2 + x)}$

44. Find y' for $y = y(x)$ defined implicitly by $e^{xy} = x^2 + y + 1$, and evaluate at $(0, 0)$.

APPLICATIONS

Business & Economics

45. Doubling time. How long will it take money to double if it is invested at 5% interest compounded:
    (A) Annually? (B) Continuously?

46. Continuous compound interest. If \$100 is invested at 10% interest compounded continuously, the amount (in dollars) at the end of t years is given by

$$A = 100e^{0.1t}$$

Find $A'(t)$, $A'(1)$, and $A'(10)$.

47. *Marginal analysis.* If the price–demand equation for x units of a commodity is

$$p(x) = 1{,}000e^{-0.02x}$$

find the marginal revenue equation.

48. *Maximum revenue.* For the price–demand equation in Problem 47, find the production level and price per unit that produces the maximum revenue. What is the maximum revenue?

49. *Maximum revenue.* Graph the revenue function from Problems 47 and 48 for $0 \leq x \leq 100$.

50. *Minimum average cost.* The cost of producing x units of a product is given by

$$C(x) = 200 + 50x - 50 \ln x \qquad x \geq 1$$

Find the minimum average cost.

51. *Demand equation.* Given the demand equation

$$x = \sqrt{5{,}000 - 2p^3}$$

find the rate of change of p with respect to x by implicit differentiation (x is the number of items that can be sold at a price of $p per item).

52. *Rate of change of revenue.* A company is manufacturing a new video game and can sell all it manufactures. The revenue (in dollars) is given by

$$R = 36x - \frac{x^2}{20}$$

where the production output in 1 day is x games. If production is increasing at 10 games per day when production is 250 games per day, find the rate of increase in revenue.

53. *Revenue and elasticity.* In a metropolitan area, the weekly demand for home delivery of 12 inch pizzas is 4,400 pizzas when the price of a pizza is $8.00. A 10% price increase caused the weekly demand to drop to 4,000 pizzas. Assume that the demand equation is linear.

(A) Find the point elasticity of demand at the new price.

(B) Approximate the change in demand if the price is increased an additional 10% over the first 10% increase.

(C) Will the second price increase cause the revenue to increase or decrease?

Life Sciences

54. *Drug concentration.* The concentration of a drug in the bloodstream t hours after injection is given approximately by

$$C(t) = 5e^{-0.3t}$$

where $C(t)$ is concentration in milligrams per milliliter. What is the rate of change of concentration after 1 hour? After 5 hours?

55. *Wound healing.* A circular wound on an arm is healing at the rate of 45 square millimeters per day (the area of the wound is decreasing at this rate). How fast is the radius R of the wound decreasing when $R = 15$ millimeters? [$A = \pi R^2$]

Social Sciences

56. *Psychology—learning.* In a computer assembly plant, a new employee, on the average, is able to assemble

$$N(t) = 10(1 - e^{-0.4t})$$

units after t days of on-the-job training.

(A) What is the rate of learning after 1 day? After 5 days?
(B) Graph N for $0 \leq t \leq 10$.

57. *Learning.* A new worker on the production line performs an operation in T minutes after x performances of the operation, as given by

$$T = 2\left(1 + \frac{1}{x^{3/2}}\right)$$

If, after performing the operation 9 times, the rate of improvement is $dx/dt = 3$ operations per hour, find the rate of improvement in time dT/dt in performing each operation.

CHAPTER 6

Integration

CHAPTER 6

Contents

The last three chapters dealt with differential calculus. We now begin the development of the second main part of calculus, called *integral calculus*. Two types of integrals will be introduced, the *indefinite integral* and the *definite integral*; each is quite different from the other. But through the remarkable *fundamental theorem of calculus*, we will show that not only are the two integral forms intimately related, but both are intimately related to differentiation.

SECTION 6-1 Antiderivatives and Indefinite Integrals

- ANTIDERIVATIVES
- INDEFINITE INTEGRALS
- INDEFINITE INTEGRALS INVOLVING ALGEBRAIC FUNCTIONS
- INDEFINITE INTEGRALS INVOLVING EXPONENTIAL AND LOGARITHMIC FUNCTIONS
- APPLICATIONS

◆ ANTIDERIVATIVES

Many operations in mathematics have reverses—compare addition and subtraction, multiplication and division, and powers and roots. The function $f(x) = \frac{1}{3}x^3$ has the derivative $f'(x) = x^2$. Reversing this process is referred to as **antidifferentiation.** Thus,

$$\frac{x^3}{3} \quad \text{is an antiderivative of} \quad x^2$$

since

$$D_x\left(\frac{x^3}{3}\right) = x^2$$

In general, we say that $F(x)$ is an **antiderivative** of $f(x)$ if

$$F'(x) = f(x)$$

Note that

$$D_x\left(\frac{x^3}{3} + 2\right) = x^2 \qquad D_x\left(\frac{x^3}{3} - \pi\right) = x^2 \qquad D_x\left(\frac{x^3}{3} + \sqrt{5}\right) = x^2$$

Hence,

$$\frac{x^3}{3} + 2 \qquad \frac{x^3}{3} - \pi \qquad \frac{x^3}{3} + \sqrt{5}$$

are also antiderivatives of x^2, since each has x^2 as a derivative. In fact, it appears that

$$\frac{x^3}{3} + C$$

for any real number C, is an antiderivative of x^2, since

$$D_x\left(\frac{x^3}{3} + C\right) = x^2$$

Thus, antidifferentiation of a given function does not lead to a unique function, but to a whole set of functions.

Does the expression

$$\frac{x^3}{3} + C$$

with C any real number, include all antiderivatives of x^2? Theorem 1 (which we state without proof) indicates that the answer is yes.

THEOREM 1

> If F and G are differentiable functions on the interval (a, b) and $F'(x) = G'(x)$, then $F(x) = G(x) + k$ for some constant k.

◆ INDEFINITE INTEGRALS

In words, Theorem 1 states that **if the derivatives of two functions are equal, then the functions differ by at most a constant.** We use the symbol

$$\int f(x)\, dx$$

called the **indefinite integral,** to represent the set of all antiderivatives of $f(x)$, and we write

$$\int f(x)\, dx = F(x) + C \qquad \text{if} \qquad F'(x) = f(x)$$

The symbol $\int$ is called an **integral sign,** and the function $f(x)$ is called the **integrand.** The symbol dx indicates that the antidifferentiation is performed with respect to the variable x. (We will have more to say about the symbol dx

later.) The arbitrary constant C is called the **constant of integration.** Referring to the preceding discussion, we can write

$$\int x^2\,dx = \frac{x^3}{3} + C \qquad \text{since} \qquad D_x\left(\frac{x^3}{3} + C\right) = x^2$$

Of course, variables other than x can be used in indefinite integrals. Thus, we also can write

$$\int t^2\,dt = \frac{t^3}{3} + C \qquad \text{since} \qquad D_t\left(\frac{t^3}{3} + C\right) = t^2$$

or

$$\int u^2\,du = \frac{u^3}{3} + C \qquad \text{since} \qquad D_u\left(\frac{u^3}{3} + C\right) = u^2$$

The fact that indefinite integration and differentiation are reverse operations, except for the addition of the constant of integration, can be expressed symbolically as

$$D_x\left[\int f(x)\,dx\right] = f(x)$$ The derivative of the indefinite integral of $f(x)$ is $f(x)$.

and

$$\int F'(x)\,dx = F(x) + C$$ The indefinite integral of the derivative of $F(x)$ is $F(x) + C$.

◆ INDEFINITE INTEGRALS INVOLVING ALGEBRAIC FUNCTIONS

Just as with differentiation, we can develop formulas and special properties that will enable us to find indefinite integrals of many frequently encountered functions. To start, we list some formulas that can be established using the definitions of antiderivative and indefinite integral, and the properties of derivatives considered in Chapter 3.

Indefinite Integral Formulas and Properties

For k and C constants:

1. $\int k\,dx = kx + C$
2. $\int x^n\,dx = \dfrac{x^{n+1}}{n+1} + C \qquad n \neq -1$
3. $\int kf(x)\,dx = k\int f(x)\,dx$
4. $\int [f(x) \pm g(x)]\,dx = \int f(x)\,dx \pm \int g(x)\,dx$

We will establish formula 2 and property 3 here (the others may be shown to be true in a similar manner). To establish formula 2, we simply differentiate the right side to obtain the integrand on the left side. Thus,

$$D_x\left(\frac{x^{n+1}}{n+1}+C\right)=\frac{(n+1)x^n}{n+1}+0=x^n \qquad n\neq -1$$

(Notice that formula 2 cannot be used when $n=-1$; that is, when the integrand is x^{-1} or $1/x$. The indefinite integral of $x^{-1}=1/x$ will be considered later in this section.)

To establish property 3, let F be a function such that $F'(x)=f(x)$. Then

$$k\int f(x)\,dx=k\int F'(x)\,dx=k[F(x)+C_1]=kF(x)+kC_1$$

and since $(kF(x))'=kF'(x)=kf(x)$, we have

$$\int kf(x)\,dx=\int kF'(x)\,dx=kF(x)+C_2$$

But $kF(x)+kC_1$ and $kF(x)+C_2$ describe the same set of functions, since C_1 and C_2 are arbitrary real numbers. Thus, property 3 is established.

It is important to remember that property 3 states that **a constant factor can be moved across an integral sign; a variable factor cannot be moved across an integral sign:**

Constant Factor	Variable Factor
$\int 5x^{1/2}\,dx=5\int x^{1/2}\,dx$	$\int xx^{1/2}\,dx\neq x\int x^{1/2}\,dx$

Now let us put the formulas and properties to use.

◆ EXAMPLE 1

(A) $\int 5\,dx=5x+C$

(B) $\int x^4\,dx=\frac{x^{4+1}}{4+1}+C=\frac{x^5}{5}+C$

(C) $\int 5t^7\,dt=5\int t^7\,dt=5\frac{t^8}{8}+C=\frac{5}{8}t^8+C$

(D) $\int (4x^3+2x-1)\,dx=\int 4x^3\,dx+\int 2x\,dx-\int dx$

$=4\int x^3\,dx+2\int x\,dx-\int dx$

$=\frac{4x^4}{4}+\frac{2x^2}{2}-x+C$

$=x^4+x^2-x+C$ only 1 C.

Property 4 can be extended to the sum and difference of an arbitrary number of functions.

(E) $\displaystyle\int \frac{3\,dx}{x^2} = \int 3x^{-2}\,dx = \frac{3x^{-2+1}}{-2+1} + C = -3x^{-1} + C$

(F) $\displaystyle\int 5\sqrt[3]{u^2}\,du = 5\int u^{2/3}\,du = 5\,\frac{u^{(2/3)+1}}{\frac{2}{3}+1} + C$

$$= 5\,\frac{u^{5/3}}{\frac{5}{3}} + C = 3u^{5/3} + C$$ ◆

To check any of the results in Example 1, we differentiate the final result to obtain the integrand in the original indefinite integral. When you evaluate an indefinite integral, do not forget to include the arbitrary constant C.

PROBLEM 1 Find each of the following:

(A) $\displaystyle\int dx$ (B) $\displaystyle\int 3t^4\,dt$ (C) $\displaystyle\int (2x^5 - 3x^2 + 1)\,dx$

(D) $\displaystyle\int 4\sqrt[5]{w^3}\,dw$ (E) $\displaystyle\int \left(2x^{2/3} - \frac{3}{x^4}\right) dx$ ◆

◆ EXAMPLE 2 (A) $\displaystyle\int \frac{x^3 - 3}{x^2}\,dx = \int \left(\frac{x^3}{x^2} - \frac{3}{x^2}\right) dx$

$$= \int (x - 3x^{-2})\,dx$$

$$= \int x\,dx - 3\int x^{-2}\,dx$$

$$= \frac{x^{1+1}}{1+1} - 3\,\frac{x^{-2+1}}{-2+1} + C$$

$$= \tfrac{1}{2}x^2 + 3x^{-1} + C$$

(B) $\displaystyle\int \left(\frac{2}{\sqrt[3]{x}} - 6\sqrt{x}\right) dx = \int (2x^{-1/3} - 6x^{1/2})\,dx$

$$= 2\int x^{-1/3}\,dx - 6\int x^{1/2}\,dx$$

$$= 2\,\frac{x^{(-1/3)+1}}{-\frac{1}{3}+1} - 6\,\frac{x^{(1/2)+1}}{\frac{1}{2}+1} + C$$

$$= 2\,\frac{x^{2/3}}{\frac{2}{3}} - 6\,\frac{x^{3/2}}{\frac{3}{2}} + C$$

$$= 3x^{2/3} - 4x^{3/2} + C$$ ◆

PROBLEM 2 Find each indefinite integral.

(A) $\displaystyle\int \frac{x^4 - 8x^3}{x^2}\,dx$ (B) $\displaystyle\int \left(8\sqrt[3]{x} - \frac{6}{\sqrt{x}}\right) dx$ ◆

◆ INDEFINITE INTEGRALS INVOLVING EXPONENTIAL AND LOGARITHMIC FUNCTIONS

We now give indefinite integral formulas for e^x and $1/x$. (Recall that the form $x^{-1} = 1/x$ is not covered by formula 2, given earlier.)

Indefinite Integral Formulas

5. $\int e^x\,dx = e^x + C$ 6. $\int \frac{1}{x}\,dx = \ln|x| + C \qquad x \neq 0$

Formula 5 follows immediately from the derivative formula for the exponential function discussed in the last chapter. Because of the absolute value, formula 6 does not follow directly from the derivative formula for the natural logarithm function. Let us show that

$$D_x \ln|x| = \frac{1}{x} \qquad x \neq 0$$

We consider two cases, $x > 0$ and $x < 0$:

Case 1. $x > 0$:

$$D_x \ln|x| = D_x \ln x \qquad \text{Since } |x| = x \text{ for } x > 0$$
$$= \frac{1}{x}$$

Case 2. $x < 0$:

$$D_x \ln|x| = D_x \ln(-x) \qquad \text{Since } |x| = -x \text{ for } x < 0$$
$$= \frac{1}{-x} D_x(-x)$$
$$= \frac{-1}{-x} = \frac{1}{x}$$

Thus,

$$D_x \ln|x| = \frac{1}{x} \qquad x \neq 0$$

and hence,

$$\int \frac{1}{x}\,dx = \ln|x| + C \qquad x \neq 0$$

What about the indefinite integral of $\ln x$? We postpone a discussion of $\int \ln x\,dx$ until Section 7-3, where we will be able to find it using a technique called *integration by parts.*

EXAMPLE 3 $\int \left(2e^x + \frac{3}{x}\right) dx = 2\int e^x\,dx + 3\int \frac{1}{x}\,dx$

$$= 2e^x + 3\ln|x| + C$$

PROBLEM 3 Find: $\int \left(\frac{5}{x} - 4e^x\right) dx$

APPLICATIONS

Let us now consider some applications of the indefinite integral to see why we are interested in finding antiderivatives of functions.

EXAMPLE 4 *Curves*

Find the equation of the curve that passes through (2, 5) if its slope is given by $dy/dx = 2x$ at any point x.

Solution We are interested in finding a function $y = f(x)$ such that

$$\frac{dy}{dx} = 2x \tag{1}$$

and

$$y = 5 \qquad \text{when} \qquad x = 2 \tag{2}$$

If $dy/dx = 2x$, then

$$y = \int 2x\,dx$$
$$= x^2 + C \tag{3}$$

Since $y = 5$ when $x = 2$, we determine the *particular value of* C so that

$$5 = 2^2 + C$$

Thus, $C = 1$, and

$$y = x^2 + 1$$

is the *particular antiderivative* out of all those possible from (3) that satisfies both (1) and (2). See Figure 1.

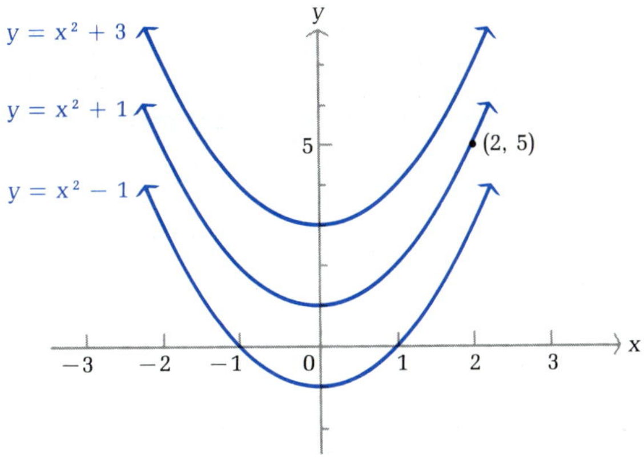

FIGURE 1
$y = x^2 + C$

PROBLEM 4 Find the equation of the curve that passes through (2, 6) if the slope of the curve at any point x is given by $dy/dx = 3x^2$.

In certain situations, it is easier to determine the rate at which something happens than how much of it has happened in a given length of time (for example, population growth rates, business growth rates, rate of healing of a wound, rates of learning or forgetting). If a rate function (derivative) is given and we know the value of the dependent variable for a given value of the independent variable, then—if the rate function is not too complicated—we can often find the original function by integration.

EXAMPLE 5 *Cost Function*

If the marginal cost of producing x units is given by

$$C'(x) = 0.3x^2 + 2x$$

and the fixed cost is \$2,000, find the cost function $C(x)$ and the cost of producing 20 units.

Solution Recall that marginal cost is the derivative of the cost function and that fixed cost is cost at a 0 production level. Thus, the mathematical problem is to find $C(x)$ given

$$C'(x) = 0.3x^2 + 2x \qquad C(0) = 2{,}000$$

We now find the indefinite integral of $0.3x^2 + 2x$ and determine the arbitrary integration constant using $C(0) = 2{,}000$:

$$C'(x) = 0.3x^2 + 2x$$

$$C(x) = \int (0.3x^2 + 2x)\,dx$$

$$= 0.1x^3 + x^2 + K$$ Since C represents the cost, we use K for the constant of integration.

But

$$C(0) = (0.1)0^3 + 0^2 + K = 2{,}000$$

Thus, $K = 2{,}000$, and the particular cost function is

$$C(x) = 0.1x^3 + x^2 + 2{,}000$$

We now find $C(20)$, the cost of producing 20 units:

$$\begin{aligned} C(20) &= (0.1)20^3 + 20^2 + 2{,}000 \\ &= \$3{,}200 \end{aligned}$$

◆

PROBLEM 5

Find the revenue function $R(x)$ when the marginal revenue is

$$R'(x) = 400 - 0.4x$$

and no revenue results at a 0 production level. What is the revenue at a production level of 1,000 units? ◆

◆ EXAMPLE 6

Advertising

An FM radio station is launching an aggressive advertising campaign in order to increase the number of daily listeners. The station currently has 27,000 daily listeners, and management expects the number of daily listeners, $S(t)$, to grow at the rate of

$$S'(t) = 60t^{1/2}$$

listeners per day, where t is the number of days since the campaign began. How long should the campaign last if the station wants the number of daily listeners to grow to 41,000?

Solution

We must solve the equation $S(t) = 41{,}000$ for t, given that

$$S'(t) = 60t^{1/2} \qquad \text{and} \qquad S(0) = 27{,}000$$

First, we use integration to find $S(t)$:

$$\begin{aligned} S(t) &= \int 60t^{1/2}\, dt \\ &= 60\frac{t^{3/2}}{\frac{3}{2}} + C \\ &= 40t^{3/2} + C \end{aligned}$$

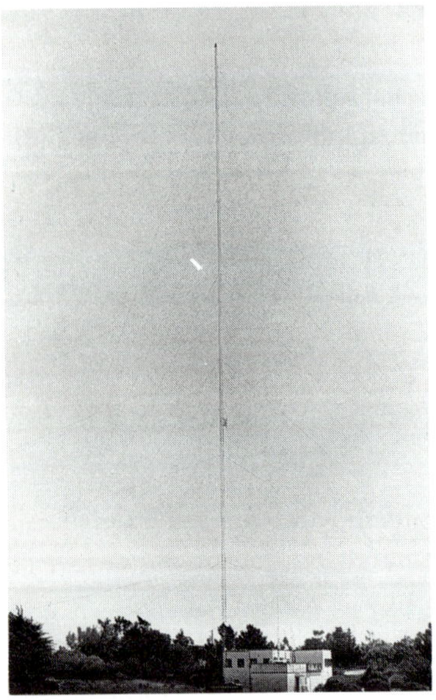

Since

$$S(0) = 40(0)^{3/2} + C = 27{,}000$$

we have $C = 27{,}000$, and

$$S(t) = 40t^{3/2} + 27{,}000$$

Now we solve the equation $S(t) = 41{,}000$ for t:

$$\begin{aligned} 40t^{3/2} + 27{,}000 &= 41{,}000 \\ 40t^{3/2} &= 14{,}000 \\ t^{3/2} &= 350 \\ t &= 350^{2/3} \qquad \text{Use a calculator.} \\ &= 49.664\ 419\ldots \end{aligned}$$

Thus, the advertising campaign should last approximately 50 days. ◆

PROBLEM 6 The current monthly circulation of the magazine *Computing News* is 640,000 copies. Due to competition from a new magazine in the same field, the monthly circulation of *Computing News*, $C(t)$, is expected to decrease at the rate of

$$C'(t) = -6{,}000t^{1/3}$$

copies per month, where t is the time in months since the new magazine began publication. How long will it take for the circulation of *Computing News* to decrease to 460,000 copies per month? ◆

Common Errors

1. $\displaystyle\int e^x\,dx \neq \frac{e^{x+1}}{x+1} + C$

   The power rule only applies to power functions of the form x^n where the exponent n is a real constant not equal to -1 and the base x is the variable. The function e^x is an exponential function with variable exponent x and constant base e. The correct form for this problem is

   $$\int e^x\,dx = e^x + C$$

2. $\displaystyle\int x(x^2 + 2)\,dx \neq \frac{x^2}{2}\left(\frac{x^3}{3} + 2x\right) + C$

   The integral of a product is not equal to the product of the integrals. The correct form for this problem is

   $$\int x(x^2 + 2)\,dx = \int (x^3 + 2x)\,dx = \frac{x^4}{4} + x^2 + C$$

Answers to Matched Problems

1. (A) $x + C$ (B) $\frac{3}{5}t^5 + C$ (C) $(x^6/3) - x^3 + x + C$ (D) $\frac{5}{2}w^{8/5} + C$
   (E) $\frac{6}{5}x^{5/3} + x^{-3} + C$
2. (A) $\frac{1}{3}x^3 - 4x^2 + C$ (B) $6x^{4/3} - 12x^{1/2} + C$ 3. $5\ln|x| - 4e^x + C$
4. $y = x^3 - 2$ 5. $R(x) = 400x - 0.2x^2$; $R(1{,}000) = \$200{,}000$
6. $t = (40)^{3/4} \approx 16$ months

EXERCISE 6-1

A *Find each indefinite integral. (Check by differentiating.)*

1. $\int 7\,dx$ **2.** $\int \pi\,dx$ **3.** $\int x^6\,dx$

4. $\int x^3\,dx$ **5.** $\int 8t^3\,dt$ **6.** $\int 10t^4\,dt$

7. $\int (2u+1)\,du$ **8.** $\int (1-2u)\,du$ **9.** $\int (3x^2+2x-5)\,dx$

10. $\int (2+4x-6x^2)\,dx$ **11.** $\int (s^4-8s^5)\,ds$ **12.** $\int (t^5+6t^3)\,dt$

13. $\int 3e^t\,dt$ **14.** $\int 2e^t\,dt$ **15.** $\int 2z^{-1}\,dz$

16. $\int \frac{3}{s}\,ds$

Find all the antiderivatives for each derivative.

17. $\frac{dy}{dx} = 200x^4$ **18.** $\frac{dx}{dt} = 42t^5$ **19.** $\frac{dP}{dx} = 24 - 6x$

20. $\frac{dy}{dx} = 3x^2 - 4x^3$ **21.** $\frac{dy}{du} = 2u^5 - 3u^2 - 1$ **22.** $\frac{dA}{dt} = 3 - 12t^3 - 9t^5$

23. $\frac{dy}{dx} = e^x + 3$ **24.** $\frac{dy}{dx} = x - e^x$ **25.** $\frac{dx}{dt} = 5t^{-1} + 1$

26. $\frac{du}{dv} = \frac{4}{v} + \frac{v}{4}$

B *Find each indefinite integral. (Check by differentiation.)*

27. $\int 6x^{1/2}\,dx$ **28.** $\int 8t^{1/3}\,dt$

29. $\int 8x^{-3}\,dx$ **30.** $\int 12u^{-4}\,du$

31. $\int \frac{du}{\sqrt{u}}$ **32.** $\int \frac{dt}{\sqrt[3]{t}}$

33. $\int \frac{dx}{4x^3}$ **34.** $\int \frac{6\,dm}{m^2}$

35. $\int \frac{du}{2u^5}$ **36.** $\int \frac{dy}{3y^4}$

37. $\int \left(3x^2 - \frac{2}{x^2}\right) dx$ **38.** $\int \left(4x^3 + \frac{2}{x^3}\right) dx$

39. $\int \left(10x^4 - \frac{8}{x^5} - 2\right) dx$

40. $\int \left(\frac{6}{x^4} - \frac{2}{x^3} + 1\right) dx$

41. $\int \left(3\sqrt{x} + \frac{2}{\sqrt{x}}\right) dx$

42. $\int \left(\frac{2}{\sqrt[3]{x}} - \sqrt[3]{x^2}\right) dx$

43. $\int \left(\sqrt[3]{x^2} - \frac{4}{x^3}\right) dx$

44. $\int \left(\frac{12}{x^5} - \frac{1}{\sqrt[3]{x^2}}\right) dx$

45. $\int \frac{e^x - 3x}{4} dx$

46. $\int \frac{e^x - 3x^2}{2} dx$

47. $\int (2z^{-3} + z^{-2} + z^{-1}) dz$

48. $\int (3x^{-2} - x^{-1}) dx$

In Problems 49–58, find the particular antiderivative of each derivative that satisfies the given condition.

49. $\frac{dy}{dx} = 2x - 3; \quad y(0) = 5$

50. $\frac{dy}{dx} = 5 - 4x; \quad y(0) = 20$

51. $C'(x) = 6x^2 - 4x; \quad C(0) = 3{,}000$

52. $R'(x) = 600 - 0.6x; \quad R(0) = 0$

53. $\frac{dx}{dt} = \frac{20}{\sqrt{t}}; \quad x(1) = 40$

54. $\frac{dR}{dt} = \frac{100}{t^2}; \quad R(1) = 400$

55. $\frac{dy}{dx} = 2x^{-2} + 3x^{-1} - 1; \quad y(1) = 0$

56. $\frac{dy}{dx} = 3x^{-1} + x^{-2}; \quad y(1) = 1$

57. $\frac{dx}{dt} = 4e^t - 2; \quad x(0) = 1$

58. $\frac{dy}{dt} = 5e^t - 4; \quad y(0) = -1$

59. Find the equation of the curve that passes through (2, 3) if its slope is given by

$$\frac{dy}{dx} = 4x - 3$$

for each x.

60. Find the equation of the curve that passes through (1, 3) if its slope is given by

$$\frac{dy}{dx} = 12x^2 - 12x$$

for each x.

C *Find each indefinite integral.*

61. $\int \frac{2x^4 - x}{x^3} dx$

62. $\int \frac{x^{-1} - x^4}{x^2} dx$

63. $\int \frac{x^5 - 2x}{x^4} dx$

64. $\int \frac{1 - 3x^4}{x^2} dx$

65. $\int \frac{x^2e^x - 2x}{x^2} dx$

66. $\int \frac{1 - xe^x}{x} dx$

For each derivative, find an antiderivative that satisfies the given condition.

67. $\frac{dM}{dt} = \frac{t^2 - 1}{t^2}$; $M(4) = 5$

68. $\frac{dR}{dx} = \frac{1 - x^4}{x^3}$; $R(1) = 4$

69. $\frac{dy}{dx} = \frac{5x + 2}{\sqrt[3]{x}}$; $y(1) = 0$

70. $\frac{dx}{dt} = \frac{\sqrt{t^3} - t}{\sqrt{t^3}}$; $x(9) = 4$

71. $p'(x) = -\frac{10}{x^2}$; $p(1) = 20$

72. $p'(x) = \frac{10}{x^3}$; $p(1) = 15$

APPLICATIONS

Business & Economics

73. *Profit function.* If the marginal profit for producing x units is given by

$$P'(x) = 50 - 0.04x \qquad P(0) = 0$$

where $P(x)$ is the profit in dollars, find the profit function P and the profit on 100 units of production.

74. *Natural resources.* The world demand for wood is increasing. In 1975, the demand was 12.6 billion cubic feet, and the rate of increase in demand is given approximately by

$$d'(t) = 0.009t$$

where t is time in years after 1975 (data from the U.S. Department of Agriculture and Forest Service). Noting that $d(0) = 12.6$, find $d(t)$. Also find $d(25)$, the demand in the year 2000.

75. *Revenue function.* The marginal revenue from the sale of x digital sports watches is given by

$$R'(x) = 100 - \tfrac{1}{5}x \qquad R(0) = 0$$

where $R(x)$ is the revenue in dollars. Find the revenue function and the price–demand equation. What is the price when the demand is 700 units?

76. *Cost function.* The marginal average cost for producing x digital sports watches is given by

$$\overline{C}'(x) = -\frac{1{,}000}{x^2} \qquad \overline{C}(100) = 25$$

where $\overline{C}(x)$ is the average cost in dollars. Find the average cost function and the cost function. What are the fixed costs?

77. *Sales analysis.* The monthly sales of a particular personal computer are expected to decline at the rate of

$$S'(t) = -25t^{2/3}$$

computers per month, where t is time in months and $S(t)$ is the number of computers sold each month. The company plans to stop manufacturing this computer when the monthly sales reach 800 computers. If the monthly sales now ($t = 0$) are 2,000 computers, find $S(t)$. How long will the company continue to manufacture this computer?

78. *Sales analysis.* The rate of change of the monthly sales of a new home video game cartridge is given by

$$S'(t) = 500t^{1/4} \qquad S(0) = 0$$

where t is the number of months since the game was released and $S(t)$ is the number of cartridges sold each month. Find $S(t)$. When will the monthly sales reach 20,000 cartridges?

79. *Labor costs and learning.* A defense contractor is starting production on a new missile control system. On the basis of data collected while assembling the first 16 control systems, the production manager obtained the following function describing the rate of labor use:

$$g(x) = 2{,}400x^{-1/2}$$

where $g(x)$ is the number of labor-hours required to assemble the xth unit of the control system. For example, after assembling 16 units, the rate of assembly is 600 labor-hours per unit, and after assembling 25 units, the rate of assembly is 480 labor-hours per unit. The more units assembled, the more efficient the process because of learning. If 19,200 labor-hours are required to assemble the first 16 units, how many labor-hours, $L(x)$, will be required to assemble the first x units? The first 25 units?

80. *Labor costs and learning.* If the rate of labor use in Problem 79 is

$$g(x) = 2{,}000x^{-1/3}$$

and if the first 8 control units require 12,000 labor-hours, how many labor-hours, $L(x)$, will be required for the first x control units? The first 27 control units?

Life Sciences

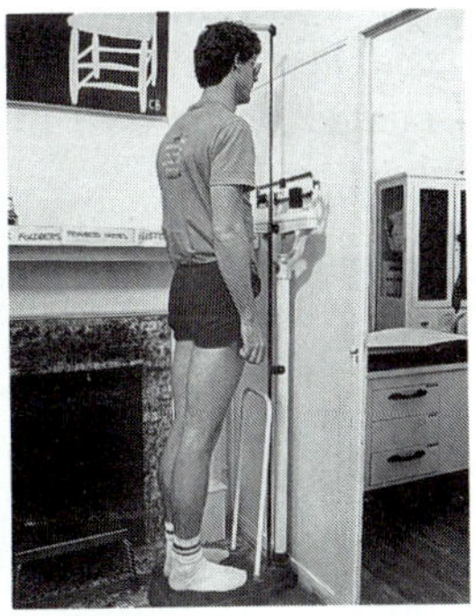

81. *Weight–height.* For an average person, the rate of change of weight W (in pounds) with respect to height h (in inches) is given approximately by

$$\frac{dW}{dh} = 0.0015h^2$$

Find $W(h)$ if $W(60) = 108$ pounds. Also find the weight for a person who is 5 feet 10 inches tall.

82. *Wound healing.* If the area A of a healing wound changes at a rate given approximately by

$$\frac{dA}{dt} = -4t^{-3} \qquad 1 \leq t \leq 10$$

where t is time in days and $A(1) = 2$ square centimeters, what will the area of the wound be in 10 days?

Social Sciences

83. *Urban growth.* The rate of growth of the population, $N(t)$, of a newly incorporated city t years after incorporation is estimated to be

$$\frac{dN}{dt} = 400 + 600\sqrt{t} \qquad 0 \leq t \leq 9$$

If the population was 5,000 at the time of incorporation, find the population 9 years later.

84. *Learning.* A beginning high school language class was chosen for an experiment in learning. Using a list of 50 words, the experiment involved measuring the rate of vocabulary memorization at different times during a continuous 5 hour study session. It was found that the average rate of learning for the whole class was inversely proportional to the time spent studying and was given approximately by

$$V'(t) = \frac{15}{t} \qquad 1 \leq t \leq 5$$

If the average number of words memorized after 1 hour of study was 15 words, what was the average number of words learned after t hours of study for $1 \leq t \leq 5$? After 4 hours of study? (Round answer to the nearest whole number.)

SECTION 6-2 Integration by Substitution

- GENERAL INDEFINITE INTEGRAL FORMULAS
- INTEGRATION BY SUBSTITUTION
- SUBSTITUTION TECHNIQUES
- APPLICATION

GENERAL INDEFINITE INTEGRAL FORMULAS

In Section 5-3, we saw that the chain rule extends the derivative formulas for x^n, e^x, and $\ln x$ to derivative formulas for $[f(x)]^n$, $e^{f(x)}$, and $\ln[f(x)]$. The chain rule can also be used to extend the indefinite integral formulas discussed in Section 6-1. Some general formulas are summarized in the following box:

General Indefinite Integral Formulas

1. $\int [f(x)]^n f'(x)\, dx = \frac{[f(x)]^{n+1}}{n+1} + C \qquad n \neq -1$
2. $\int e^{f(x)} f'(x)\, dx = e^{f(x)} + C$
3. $\int \frac{1}{f(x)} f'(x)\, dx = \ln|f(x)| + C$

Each formula can be verified by using the chain rule to show that the derivative of the function on the right is the integrand on the left. For example,

$$D_x[e^{f(x)} + C] = e^{f(x)}f'(x)$$

verifies formula 2.

◆ EXAMPLE 7 (A) $\int (3x + 4)^{10}3\, dx = \frac{(3x + 4)^{11}}{11} + C$ Formula 1 with $f(x) = 3x + 4$ and $f'(x) = 3$

Check: $D_x \frac{(3x + 4)^{11}}{11} = 11 \frac{(3x + 4)^{10}}{11} D_x(3x + 4) = (3x + 4)^{10}3$

(B) $\int e^{x^2}2x\, dx = e^{x^2} + C$ Formula 2 with $f(x) = x^2$ and $f'(x) = 2x$

Check: $D_x e^{x^2} = e^{x^2}D_x x^2 = e^{x^2}2x$

(C) $\int \frac{1}{1 + x^3} 3x^2\, dx = \ln|1 + x^3| + C$ Formula 3 with $f(x) = 1 + x^3$ and $f'(x) = 3x^2$

Check: $D_x \ln|1 + x^3| = \frac{1}{1 + x^3} D_x(1 + x^3) = \frac{1}{1 + x^3} 3x^2$ ◆

PROBLEM 7 Find each indefinite integral.

(A) $\int (2x^3 - 3)^{20}6x^2\, dx$ (B) $\int e^{5x}5\, dx$ (C) $\int \frac{1}{4 + x^2} 2x\, dx$ ◆

◆ INTEGRATION BY SUBSTITUTION

The key step in using formulas 1, 2, and 3 is recognizing the form of the integrand. Some people find it difficult to identify $f(x)$ and $f'(x)$ in these formulas and prefer to use a *substitution* to simplify the integrand. The **method of substitution,** which we now discuss, becomes increasingly useful as one progresses in studies of integration.

◆ EXAMPLE 8 Find $\int (x^2 + 2x + 5)^5(2x + 2)\, dx$.

Solution If

$$u = x^2 + 2x + 5$$

then the differential of u is (see Section 4-5)

$$du = (2x + 2)\, dx$$

Notice that du is one of the factors in the integrand. Substitute u for $x^2 + 2x + 5$ and du for $(2x + 2)\, dx$ to obtain

$$\int (x^2 + 2x + 5)^5(2x + 2)\, dx = \int u^5\, du$$

$$= \frac{u^6}{6} + C$$

$$= \frac{1}{6}(x^2 + 2x + 5)^6 + C \qquad \text{Since } u = x^2 + 2x + 5$$

Check $D_x \frac{1}{6}(x^2 + 2x + 5)^6 = \frac{1}{6}(6)(x^2 + 2x + 5)^5 D_x(x^2 + 2x + 5)$

$$= (x^2 + 2x + 5)^5(2x + 2)$$ ◆

PROBLEM 8 Find $\int (x^2 - 3x + 7)^4(2x - 3)\, dx$ by substitution. ◆

The substitution method is also called the **change-of-variable method,** since u replaces the variable x in the process. Substituting $u = f(x)$ and $du = f'(x)\, dx$ in formulas 1, 2, and 3 above produces the general indefinite integral formulas in the following box:

General Indefinite Integral Formulas

4. $\int u^n\, du = \frac{u^{n+1}}{n+1} + C \qquad n \neq -1$

5. $\int e^u\, du = e^u + C$

6. $\int \frac{1}{u}\, du = \ln|u| + C$

These formulas are valid if u is an independent variable or if u is a function of another variable and du is its differential with respect to that variable.

The substitution method for evaluating certain indefinite integrals is outlined in the following box:

Integration by Substitution

Step 1. Select a substitution that appears to simplify the integrand. In particular, try to select u so that du is a factor in the integrand.

Step 2. Express the integrand entirely in terms of u and du, completely eliminating the original variable and its differential.

Step 3. Evaluate the new integral, if possible.

Step 4. Express the antiderivative found in step 3 in terms of the original variable.

◆ EXAMPLE 9 Use a substitution to find the following:

(A) $\int (3x + 4)^6 3\, dx$ (B) $\int e^{t^2} 2t\, dt$

Solutions (A) If we let $u = 3x + 4$, then $du = 3\,dx$, and

$$\int (3x+4)^6 3\,dx = \int u^6\,du \qquad \text{Use formula 4.}$$

$$= \frac{u^7}{7} + C$$

$$= \frac{(3x+4)^7}{7} + C \qquad \text{Since } u = 3x + 4$$

Check: $D_x \frac{(3x+4)^7}{7} = \frac{7(3x+4)^6}{7} D_x(3x+4) = (3x+4)^6 3$

(B) If we let $u = t^2$, then $du = 2t\,dt$, and

$$\int e^{t^2} 2t\,dt = \int e^u\,du \qquad \text{Use formula 5.}$$

$$= e^u + C$$

$$= e^{t^2} + C \qquad \text{Since } u = t^2$$

Check: $D_t\, e^{t^2} = e^{t^2} D_t\, t^2 = e^{t^2} 2t$ ◆

PROBLEM 9 Use a substitution to find each indefinite integral.

(A) $\int (2x^3 - 3)^4 6x^2\,dx$ (B) $\int e^{5w} 5\,dw$ ◆

Integration by substitution is an effective procedure for some indefinite integrals, but not all. Substitution is not helpful for $\int e^{x^2}\,dx$ or $\int (\ln x)\,dx$, for example.

◆ SUBSTITUTION TECHNIQUES

In order to use the substitution method, **the integrand must be expressed entirely in terms of *u* and *du*.** In some cases, the integrand will have to be modified before making a substitution and using one of the integration formulas. Example 10 illustrates this process.

◆ EXAMPLE 10 Integrate: (A) $\int \frac{1}{4x+7}\,dx$ (B) $\int t e^{-t^2}\,dt$ (C) $\int 4x^2\sqrt{x^3+5}\,dx$

Solutions (A) If $u = 4x + 7$, then $du = 4\,dx$. We are missing a factor of 4 in the integrand to match formula 6 exactly. Recalling that a constant factor can be moved across an integral sign, we proceed as follows:

$$\int \frac{1}{4x+7}\,dx = \int \frac{1}{4x+7}\frac{4}{4}\,dx$$

$$= \frac{1}{4}\int \frac{1}{4x+7} 4\,dx \qquad \text{Substitute } u = 4x + 7 \text{ and } du = 4\,dx.$$

$$\int \frac{1}{4x+7}\,dx = \frac{1}{4}\int \frac{1}{u}\,du \qquad \text{Use formula 6.}$$

$$= \tfrac{1}{4}\ln|u| + C$$

$$= \tfrac{1}{4}\ln|4x+7| + C \qquad \text{Since } u = 4x+7$$

Check: $D_x \frac{1}{4}\ln|4x+7| = \frac{1}{4}\frac{1}{4x+7}D_x(4x+7) = \frac{1}{4}\frac{1}{4x+7}4 = \frac{1}{4x+7}$

(B) If $u = -t^2$, then $du = -2t\,dt$. Proceed as in part A:

$$\int te^{-t^2}\,dt = \int e^{-t^2}\frac{-2}{-2}t\,dt$$

$$= -\frac{1}{2}\int e^{-t^2}(-2t)\,dt \qquad \text{Substitute } u = -t^2 \text{ and } du = -2t\,dt.$$

$$= -\frac{1}{2}\int e^u\,du \qquad \text{Use formula 5.}$$

$$= -\tfrac{1}{2}e^u + C$$

$$= -\tfrac{1}{2}e^{-t^2} + C \qquad \text{Since } u = -t^2$$

Check: $D_t(-\frac{1}{2}e^{-t^2}) = -\frac{1}{2}e^{-t^2}D_t(-t^2) = -\frac{1}{2}e^{-t^2}(-2t) = te^{-t^2}$

(C) $\int 4x^2\sqrt{x^3+5}\,dx = 4\int \sqrt{x^3+5}(x^2)\,dx$ — Move the 4 across the integral sign and proceed as before.

$$= 4\int \sqrt{x^3+5}\,\frac{3}{3}(x^2)\,dx$$

$$= \frac{4}{3}\int \sqrt{x^3+5}(3x^2)\,dx \qquad \text{Substitute } u = x^3+5 \text{ and } du = 3x^2\,dx.$$

$$= \frac{4}{3}\int \sqrt{u}\,du$$

$$= \frac{4}{3}\int u^{1/2}\,du \qquad \text{Use formula 4.}$$

$$= \frac{4}{3}\frac{u^{3/2}}{\frac{3}{2}} + C$$

$$= \tfrac{8}{9}u^{3/2} + C$$

$$= \tfrac{8}{9}(x^3+5)^{3/2} + C \qquad \text{Since } u = x^3+5$$

Check: $D_x[\frac{8}{9}(x^3+5)^{3/2}] = \frac{4}{3}(x^3+5)^{1/2}D_x(x^3+5)$
$= \frac{4}{3}(x^3+5)^{1/2}3x^2 = 4x^2\sqrt{x^3+5}$ ◆

PROBLEM 10 Integrate:

(A) $\int e^{-3x}\,dx$ (B) $\int \frac{x}{x^2-9}\,dx$ (C) $\int 5t^2(t^3+4)^{-2}\,dt$ ◆

Even if it is not possible to find a substitution that makes an integrand match one of the integration formulas exactly, a substitution may sufficiently simplify the integrand so that other techniques can be used.

◆ EXAMPLE 11 Find: $\int \frac{x}{\sqrt{x+2}}\,dx$

Solution Proceeding as before, if we let $u = x + 2$, then $du = dx$ and

$$\int \frac{x}{\sqrt{x+2}}\,dx = \int \frac{x}{\sqrt{u}}\,du$$

Notice that this substitution is not yet complete, because we have not expressed the integrand entirely in terms of u and du. As we noted earlier, only a constant factor can be moved across an integral sign, so we cannot move x outside the integral sign (as much as we would like to). Instead, we must return to the original substitution, solve for x in terms of u, and use the resulting equation to complete the substitution:

$$u = x + 2 \qquad \text{Solve for } x \text{ in terms of } u.$$
$$u - 2 = x \qquad \text{Substitute this expression for } x.$$

Thus,

$$\begin{aligned}
\int \frac{x}{\sqrt{x+2}}\,dx &= \int \frac{u-2}{\sqrt{u}}\,du && \text{Simplify the integrand.}\\
&= \int \frac{u-2}{u^{1/2}}\,du\\
&= \int (u^{1/2} - 2u^{-1/2})\,du\\
&= \boxed{\int u^{1/2}\,du - 2\int u^{-1/2}\,du}\\
&= \frac{u^{3/2}}{\frac{3}{2}} - 2\frac{u^{1/2}}{\frac{1}{2}} + C\\
&= \tfrac{2}{3}(x+2)^{3/2} - 4(x+2)^{1/2} + C && \text{Since } u = x + 2
\end{aligned}$$

Check:
$$\begin{aligned}
D_x[\tfrac{2}{3}(x+2)^{3/2} - 4(x+2)^{1/2}] &= (x+2)^{1/2} - 2(x+2)^{-1/2}\\
&= \frac{x+2}{(x+2)^{1/2}} - \frac{2}{(x+2)^{1/2}}\\
&= \frac{x}{(x+2)^{1/2}}
\end{aligned}$$
◆

PROBLEM 11 Find: $\int x\sqrt{x+1}\,dx$ ◆

◆ APPLICATION

◆ EXAMPLE 12

Price–Demand

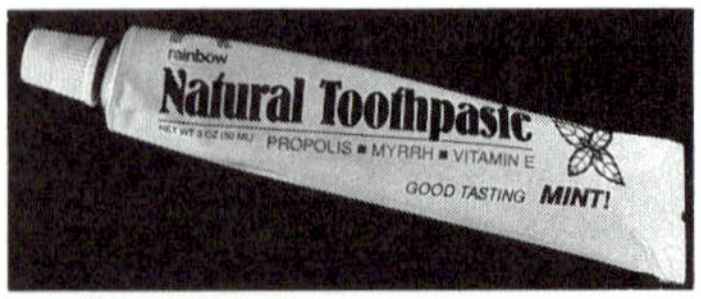

The market research department for a supermarket chain has determined that for one store the marginal price $p'(x)$ at x tubes per week for a certain brand of toothpaste is given by

$$p'(x) = -0.015e^{-0.01x}$$

Find the price–demand equation if the weekly demand is 50 when the price of a tube is \$2.35. Find the weekly demand when the price of a tube is \$1.89.

Solution

$$\begin{aligned}
p(x) &= \int -0.015e^{-0.01x}\,dx \\
&= -0.015 \int e^{-0.01x}\,dx \\
&= -0.015 \int e^{-0.01x} \frac{-0.01}{-0.01}\,dx \\
&= \frac{-0.015}{-0.01} \int e^{-0.01x}(-0.01)\,dx && \text{Substitute } u = -0.01x \text{ and } du = -0.01\,dx. \\
&= 1.5 \int e^{u}\,du \\
&= 1.5e^{u} + C \\
&= 1.5e^{-0.01x} + C && \text{Since } u = -0.01x
\end{aligned}$$

We find C by noting that

$$\begin{aligned}
p(50) = 1.5e^{-0.01(50)} + C &= \$2.35 \\
C &= \$2.35 - 1.5e^{-0.5} && \text{Use a calculator.} \\
C &= \$2.35 - 0.91 \\
C &= \$1.44
\end{aligned}$$

Thus,

$$p(x) = 1.5e^{-0.01x} + 1.44$$

To find the demand when the price is \$1.89, we solve $p(x) = \$1.89$ for x:

$$\begin{aligned}
1.5e^{-0.01x} + 1.44 &= 1.89 \\
1.5e^{-0.01x} &= 0.45 \\
e^{-0.01x} &= 0.3 \\
-0.01x &= \ln 0.3 \\
x &= -100 \ln 0.3 \approx 120 \text{ tubes}
\end{aligned}$$

◆

PROBLEM 12

The marginal price $p'(x)$ at a supply level of x tubes per week for a certain brand of toothpaste is given by

$$p'(x) = 0.001e^{0.01x}$$

Find the price–supply equation if the supplier is willing to supply 100 tubes per week at a price of \$1.65 each. How many tubes would the supplier be willing to supply at a price of \$1.98 each? ◆

Integrals of the form encountered in Example 12 occur so frequently that it is worthwhile to state a general formula for integrals of this type:

$$\int e^{au}\,du = \frac{1}{a}e^{au} + C \qquad \text{where } a \text{ is a constant, } a \neq 0$$

Common Errors

1. $$\int (x^2+3)^2\,dx = \int (x^2+3)^2 \frac{2x}{2x}\,dx$$
$$\neq \frac{1}{2x}\int (x^2+3)^2 2x\,dx$$

Remember: A variable factor cannot be moved across an integral sign! The correct procedure for this problem is

$$\int (x^2+3)^2\,dx = \int (x^4+6x^2+9)\,dx$$
$$= \frac{x^5}{5} + 2x^3 + 9x + C$$

[*Note:* The substitution $u = x^2 + 3$ does not work.]

2. $$\int \frac{1}{10x+3}\,dx = \int \frac{1}{u}\,dx \qquad u = 10x + 3$$
$$\neq \ln|u| + C$$

Remember: An integral must be expressed entirely in terms of *u* and *du* before applying integration formulas 1–6. The correct procedure for this problem is

$$\int \frac{1}{10x+3}\,dx = \frac{1}{10}\int \frac{1}{10x+3}10\,dx \qquad u = 10x+3,\ du = 10\,dx$$
$$= \frac{1}{10}\int \frac{1}{u}\,du$$
$$= \tfrac{1}{10}\ln|u| + C$$
$$= \tfrac{1}{10}\ln|10x+3| + C$$

Answers to Matched Problems

7. (A) $\frac{1}{21}(2x^3-3)^{21} + C$ (B) $e^{5x} + C$
(C) $\ln|4+x^2| + C$ or $\ln(4+x^2) + C$, since $4 + x^2 > 0$
8. $\frac{1}{5}(x^2 - 3x + 7)^5 + C$
9. (A) $\frac{1}{5}(2x^3-3)^5 + C$ (B) $e^{5w} + C$
10. (A) $-\frac{1}{3}e^{-3x} + C$ (B) $\frac{1}{2}\ln|x^2 - 9| + C$ (C) $-\frac{5}{3}(t^3+4)^{-1} + C$
11. $\frac{2}{5}(x+1)^{5/2} - \frac{2}{3}(x+1)^{3/2} + C$ 12. $p(x) = 0.1e^{0.01x} + 1.38$; 179 tubes

EXERCISE 6-2

A *Find each indefinite integral, and check the result by differentiating.*

1. $\int (x^2 - 4)^5 2x\, dx$ **2.** $\int (x^3 + 1)^4 3x^2\, dx$ **3.** $\int e^{4x} 4\, dx$

4. $\int e^{-3x}(-3)\, dx$ **5.** $\int \frac{1}{2t + 3} 2\, dt$ **6.** $\int \frac{1}{5t - 7} 5\, dt$

B **7.** $\int (3x - 2)^7\, dx$ **8.** $\int (5x + 3)^9\, dx$ **9.** $\int (x^2 + 3)^7 x\, dx$

10. $\int (x^3 - 5)^4 x^2\, dx$ **11.** $\int 10e^{-0.5t}\, dt$ **12.** $\int 4e^{0.01t}\, dt$

13. $\int \frac{1}{10x + 7}\, dx$ **14.** $\int \frac{1}{100 - 3x}\, dx$ **15.** $\int xe^{2x^2}\, dx$

16. $\int x^2 e^{4x^3}\, dx$ **17.** $\int \frac{x^2}{x^3 + 4}\, dx$ **18.** $\int \frac{x}{x^2 - 2}\, dx$

19. $\int \frac{t}{(3t^2 + 1)^4}\, dt$ **20.** $\int \frac{t^2}{(t^3 - 2)^5}\, dt$ **21.** $\int \frac{x^2}{(4 - x^3)^2}\, dx$

22. $\int \frac{x}{(5 - 2x^2)^5}\, dx$ **23.** $\int x\sqrt{x + 4}\, dx$ **24.** $\int x\sqrt{x - 9}\, dx$

25. $\int \frac{x}{\sqrt{x - 3}}\, dx$ **26.** $\int \frac{x}{\sqrt{x + 5}}\, dx$ **27.** $\int x(x - 4)^9\, dx$

28. $\int x(x + 6)^8\, dx$ **29.** $\int e^{2x}(1 + e^{2x})^3\, dx$ **30.** $\int e^{-x}(1 - e^{-x})^4\, dx$

31. $\int \frac{1 + x}{4 + 2x + x^2}\, dx$ **32.** $\int \frac{x^2 - 1}{x^3 - 3x + 7}\, dx$ **33.** $\int (2x + 1)e^{x^2 + x + 1}\, dx$

34. $\int (x^2 + 2x)e^{x^3 + 3x^2}\, dx$ **35.** $\int (e^x - 2x)^3(e^x - 2)\, dx$

36. $\int (x^2 - e^x)^4(2x - e^x)\, dx$ **37.** $\int \frac{x^3 + x}{(x^4 + 2x^2 + 1)^4}\, dx$

38. $\int \frac{x^2 - 1}{(x^3 - 3x + 7)^2}\, dx$

C **39.** $\int x\sqrt{3x^2 + 7}\, dx$ **40.** $\int x^2\sqrt{2x^3 + 1}\, dx$ **41.** $\int x(x^3 + 2)^2\, dx$

42. $\int x(x^2 + 2)^2\, dx$ **43.** $\int x^2(x^3 + 2)^2\, dx$ **44.** $\int (x^2 + 2)^2\, dx$

45. $\int \frac{x^3}{\sqrt{2x^4 + 3}}\, dx$ **46.** $\int \frac{x^2}{\sqrt{4x^3 - 1}}\, dx$ **47.** $\int \frac{(\ln x)^3}{x}\, dx$

48. $\int \frac{e^x}{1 + e^x}\, dx$ **49.** $\int \frac{1}{x^2} e^{-1/x}\, dx$ **50.** $\int \frac{1}{x \ln x}\, dx$

Find the antiderivative of each derivative.

51. $\frac{dx}{dt} = 7t^2(t^3 + 5)^6$

52. $\frac{dm}{dn} = 10n(n^2 - 8)^7$

53. $\frac{dy}{dt} = \frac{3t}{\sqrt{t^2 - 4}}$

54. $\frac{dy}{dx} = \frac{5x^2}{(x^3 - 7)^4}$

55. $\frac{dp}{dx} = \frac{e^x + e^{-x}}{(e^x - e^{-x})^2}$

56. $\frac{dm}{dt} = \frac{\ln(t - 5)}{t - 5}$

Use substitution techniques to derive the integration formulas in Problems 57 and 58. Then check your work by differentiation.

57. $\int e^{au}\, du = \frac{1}{a} e^{au} + C, \quad a \neq 0$

58. $\int \frac{1}{au + b}\, du = \frac{1}{a} \ln|au + b| + C, \quad a \neq 0$

APPLICATIONS

Business & Economics

59. *Price–demand equation.* The marginal price for a weekly demand of x bottles of baby shampoo in a drug store is given by

$$p'(x) = \frac{-6,000}{(3x + 50)^2}$$

Find the price–demand equation if the weekly demand is 150 when the price of a bottle of shampoo is \$4. What is the weekly demand when the price is \$2.50?

60. *Price–supply equation.* The marginal price at a supply level of x bottles of baby shampoo per week is given by

$$p'(x) = \frac{300}{(3x + 25)^2}$$

Find the price–supply equation if the distributor of the shampoo is willing to supply 75 bottles a week at a price of \$1.60 per bottle. How many bottles would the supplier be willing to supply at a price of \$1.75 per bottle?

61. *Cost function.* The weekly marginal cost of producing x pairs of tennis shoes is given by

$$C'(x) = 12 + \frac{500}{x + 1}$$

where C(x) is cost in dollars. If the fixed costs are \$2,000 per week, find the cost function. What is the average cost per pair of shoes if 1,000 pairs of shoes are produced each week?

62. *Revenue function.* The weekly marginal revenue from the sale of x pairs of tennis shoes is given by

$$R'(x) = 40 - 0.02x + \frac{200}{x + 1} \qquad R(0) = 0$$

where $R(x)$ is revenue in dollars. Find the revenue function. Find the revenue from the sale of 1,000 pairs of shoes.

63. *Marketing.* An automobile company is ready to introduce a new line of cars with a national sales campaign. After test marketing the line in a carefully selected city, the marketing research department estimates that sales (in millions of dollars) will increase at the monthly rate of

$$S'(t) = 10 - 10e^{-0.1t} \qquad 0 \leq t \leq 24$$

t months after the national campaign has started. What will be the total sales, $S(t)$, t months after the beginning of the national campaign if we assume 0 sales at the beginning of the campaign? What are the estimated total sales for the first 12 months of the campaign?

64. *Marketing.* Repeat Problem 63 if the monthly rate of increase in sales is found to be approximated by

$$S'(t) = 20 - 20e^{-0.05t} \qquad 0 \leq t \leq 24$$

65. *Oil production.* Using data from the first 3 years of production as well as geological studies, the management of an oil company estimates that oil will be pumped from a producing field at a rate given by

$$R(t) = \frac{100}{t+1} + 5 \qquad 0 \leq t \leq 20$$

where $R(t)$ is the rate of production (in thousands of barrels per year) t years after pumping begins. How many barrels of oil, $Q(t)$, will the field produce the first t years if $Q(0) = 0$? How many barrels will be produced the first 9 years?

66. *Oil production.* In Problem 65, if the rate is found to be

$$R(t) = \frac{120t}{t^2+1} + 3 \qquad 0 \leq t \leq 20$$

how many barrels of oil, $Q(t)$, will the field produce the first t years if $Q(0) = 0$? How many barrels will be produced the first 5 years?

Life Sciences

67. *Biology.* A yeast culture is growing at the rate of $W'(t) = 0.2e^{0.1t}$ grams per hour. If the starting culture weighs 2 grams, what will be the weight of the culture, $W(t)$, after t hours? After 8 hours?

68. *Medicine.* The rate of healing for a skin wound (in square centimeters per day) is approximated by $A'(t) = -0.9e^{-0.1t}$. If the initial wound has an area of 9 square centimeters, what will its area, $A(t)$, be after t days? After 5 days?

69. *Pollution.* A contaminated lake is treated with a bactericide. The rate of decrease in harmful bacteria t days after the treatment is given by

$$\frac{dN}{dt} = -\frac{2{,}000t}{1+t^2} \qquad 0 \leq t \leq 10$$

where $N(t)$ is the number of bacteria per milliliter of water. If the initial count was 5,000 bacteria per milliliter, find $N(t)$ and then find the bacteria count after 10 days.

70. *Pollution.* An oil tanker aground on a reef is losing oil and producing an oil slick that is radiating outward at a rate given approximately by

$$\frac{dR}{dt} = \frac{60}{\sqrt{t+9}} \qquad t \geq 0$$

where R is the radius (in feet) of the circular slick after t minutes. Find the radius of the slick after 16 minutes if the radius is 0 when $t = 0$.

Social Sciences

71. *Learning.* In a particular business college, it was found that an average student enrolled in an advanced typing class progressed at a rate of $N'(t) = 6e^{-0.1t}$ words per minute per week, t weeks after enrolling in a 15 week course. If at the beginning of the course a student could type 40 words per minute, how many words per minute, $N(t)$, would the student be expected to type t weeks into the course? After completing the course?

72. *Learning.* In the same business college, it was also found that an average student enrolled in a beginning shorthand class progressed at a rate of $N'(t) = 12e^{-0.06t}$ words per minute per week, t weeks after enrolling in a 15 week course. If at the beginning of the course a student could take dictation in shorthand at 0 words per minute, how many words per minute, $N(t)$, would the student be expected to handle t weeks into the course? After completing the course?

73. *College enrollment.* The projected rate of increase in enrollment in a new college is estimated by

$$\frac{dE}{dt} = 5{,}000(t+1)^{-3/2} \qquad t \geq 0$$

where $E(t)$ is the projected enrollment in t years. If enrollment is 2,000 now ($t = 0$), find the projected enrollment 15 years from now.

SECTION 6-3 Differential Equations—Growth and Decay

- DIFFERENTIAL EQUATIONS
- CONTINUOUS COMPOUND INTEREST REVISITED
- EXPONENTIAL GROWTH LAW
- POPULATION GROWTH; RADIOACTIVE DECAY; LEARNING
- A COMPARISON OF EXPONENTIAL GROWTH PHENOMENA

DIFFERENTIAL EQUATIONS

In the previous section, we considered equations of the form

$$\frac{dy}{dx} = 6x^2 - 4x \qquad p'(x) = -400e^{-0.04x}$$

These are examples of *differential equations*. In general, an equation is a **differential equation** if it involves an unknown function (often denoted by y) and one or more of its derivatives. Other examples of differential equations are

$$\frac{dy}{dx} = ky \qquad y'' - xy' + x^2 = 5$$

Finding solutions to different types of differential equations (functions that satisfy the equation) is the subject matter for whole books and courses on this topic. Here, we will consider only a few very special but very important types of equations that have immediate and significant applications. We start by considering the problem of continuous compound interest from another point of view, which will enable us to generalize the concept and apply the results to problems from a number of different fields.

◆ CONTINUOUS COMPOUND INTEREST REVISITED

Let P be the initial amount of money deposited in an account, and let A be the amount in the account at any time t. Instead of assuming that the money in the account earns a particular rate of interest, suppose we say that the rate of growth of the amount of money in the account at any time t is proportional to the amount present at that time. Since dA/dt is the rate of growth of A with respect to t, we have

$$\frac{dA}{dt} = rA \qquad A(0) = P \qquad A, P > 0 \tag{1}$$

where r is an appropriate constant. We would like to find a function $A = A(t)$ that satisfies these conditions. Multiplying both sides of equation (1) by $1/A$, we obtain

$$\frac{1}{A}\frac{dA}{dt} = r$$

Now we integrate each side with respect to t:

$$\int \frac{1}{A}\frac{dA}{dt}\,dt = \int r\,dt \qquad \frac{dA}{dt}\,dt = A'(t)\,dt = dA$$

$$\int \frac{1}{A}\,dA = \int r\,dt \qquad \text{Use formula 6 to evaluate the left side.}$$

$$\ln|A| = rt + C \qquad |A| = A \text{ since } A > 0$$

$$\ln A = rt + C$$

We convert this last equation into the equivalent exponential form

$$A = e^{rt+C} \qquad \text{Definition of logarithmic function: } y = \ln x \text{ if and only if } x = e^y$$

$$= e^C e^{rt} \qquad \text{Property of exponents: } b^m b^n = b^{m+n}$$

Since $A(0) = P$, we evaluate $A(t) = e^C e^{rt}$ at $t = 0$ and set it equal to P:

$$A(0) = e^C e^0 = e^C = P$$

Hence, $e^C = P$, and we can rewrite $A = e^C e^{rt}$ in the form

$$A = Pe^{rt}$$

This is the same continuous compound interest formula obtained in Section 5-1, where the principal P is invested at an annual nominal rate of r compounded continuously for t years.

◆ EXPONENTIAL GROWTH LAW

In general, if the rate of change with respect to time of a quantity Q is proportional to the amount present and $Q(0) = Q_0$, then proceeding in exactly the same way as above, we obtain the following:

Exponential Growth Law

If $\dfrac{dQ}{dt} = rQ$ and $Q(0) = Q_0$, then $Q = Q_0 e^{rt}$,

where

Q_0 = Amount at $t = 0$

r = Continuous compound growth rate (expressed as a decimal)

t = Time

Q = Quantity at time t

The constant r in the exponential growth law is sometimes called the **growth constant,** or the **growth rate.** This last term can be misleading, since the rate of growth of Q with respect to time is dQ/dt, not r. Notice that if $r < 0$, then $dQ/dt < 0$ and Q is decreasing. This type of growth is called **exponential decay.**

Once we know that the rate of growth of something is proportional to the amount present, then we know it has exponential growth and we can use the results summarized in the box without having to solve the differential equation each time. The exponential growth law applies not only to money invested at interest compounded continuously, but also to many other types of problems—population growth, radioactive decay, natural resource depletion, and so on.

◆ POPULATION GROWTH; RADIOACTIVE DECAY; LEARNING

The world population is growing at an ever-increasing rate, as illustrated in Figure 2 at the top of the next page. **Population growth** over certain periods of time often can be approximated by the exponential growth law described above.

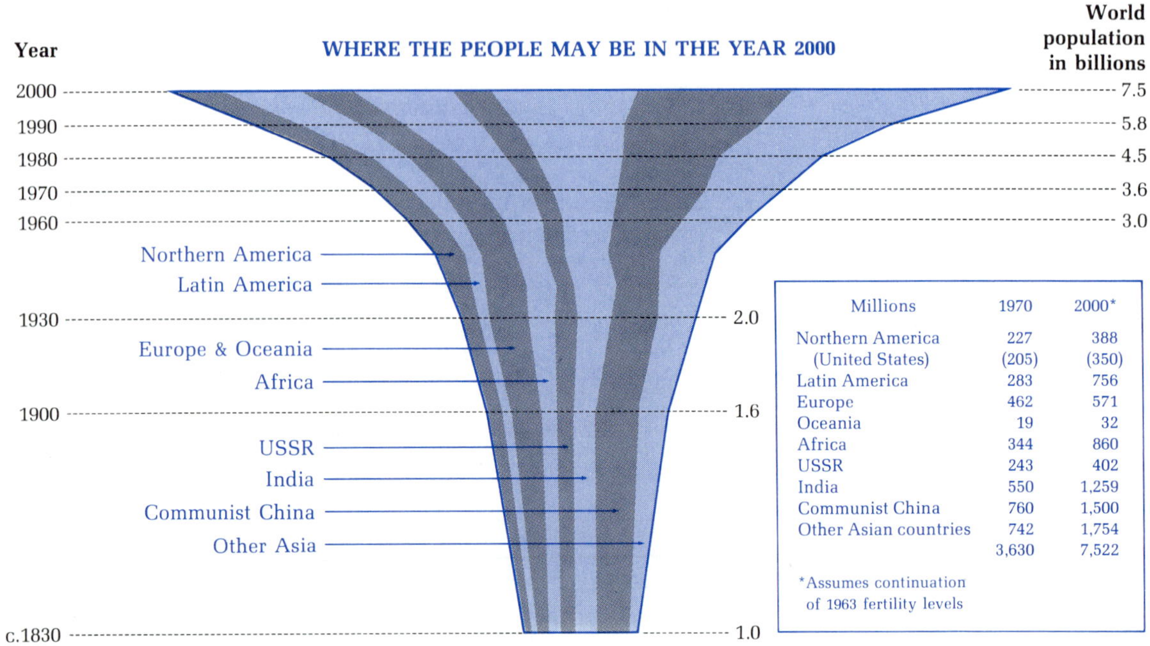

Millions	1970	2000*
Northern America	227	388
(United States)	(205)	(350)
Latin America	283	756
Europe	462	571
Oceania	19	32
Africa	344	860
USSR	243	402
India	550	1,259
Communist China	760	1,500
Other Asian countries	742	1,754
	3,630	7,522

*Assumes continuation of 1963 fertility levels

FIGURE 2
The population explosion
Source: U.S. State Department

◆ EXAMPLE 13 *Population Growth*

India had a population of 800 million people in 1987 ($t = 0$) and a growth rate of 3% per year (which we will assume is compounded continuously). If P is the population (in millions) t years after 1987 and the same growth rate continues, then

$$\frac{dP}{dt} = 0.03P \qquad P(0) = 800$$

Thus, using the exponential growth law, we obtain

$$P = 800e^{0.03t}$$

With this result, we can estimate the population of India in 2007 ($t = 20$) to be

$$\begin{aligned} P(20) &= 800e^{0.03(20)} \\ &\approx 1{,}458 \text{ million people} \end{aligned}$$

◆

PROBLEM 13 Assuming the same continuous compound growth rate as in Example 13, what will India's population be in the year 2012? ◆

◆ EXAMPLE 14 — *Population Growth*

If the exponential growth law applies to Canada's population growth, at what continuous compound growth rate will the population double over the next 100 years?

Solution

The problem is to find r, given $P = 2P_0$ and $t = 100$:

$$P = P_0 e^{rt}$$
$$2P_0 = P_0 e^{100r}$$
$$2 = e^{100r}$$

Take the natural logarithm of both sides and reverse the equation.

$$100r = \ln 2$$
$$r = \frac{\ln 2}{100}$$
$$\approx 0.0069 \quad \text{or} \quad 0.69\%$$

◆

PROBLEM 14

If the exponential growth law applies to population growth in Mexico, find the doubling time of the population if it continues to grow at 3.2% per year compounded continuously. ◆

We now turn to another type of exponential growth—**radioactive decay.** In 1946, Willard Libby (who later received a Nobel prize in chemistry) found that as long as a plant or animal is alive, radioactive carbon-14 is maintained at a constant level in its tissues. Once the plant or animal is dead, however, the radioactive carbon-14 diminishes by radioactive decay at a rate proportional to the amount present. Thus,

$$\frac{dQ}{dt} = rQ \qquad Q(0) = Q_0$$

and we have another example of the exponential growth law. The continuous compound rate of decay for radioactive carbon-14 has been found to be 0.000 123 8; thus, $r = -0.000\ 123\ 8$, since decay implies a negative continuous compound growth rate.

◆ EXAMPLE 15 — *Archaeology*

A piece of human bone was found at an archaeological site in Africa. If 10% of the original amount of radioactive carbon-14 was present, estimate the age of the bone.

Solution

Using the exponential growth law for

$$\frac{dQ}{dt} = -0.000\ 123\ 8Q \qquad Q(0) = Q_0$$

we find that

$$Q = Q_0 e^{-0.0001238t}$$

and our problem is to find t so that $Q = 0.1Q_0$ (since the amount of carbon-14 present now is 10% of the amount present, Q_0, at the death of the person). Thus,

$$\begin{aligned} 0.1Q_0 &= Q_0 e^{-0.0001238t} \\ 0.1 &= e^{-0.0001238t} \\ \ln 0.1 &= \ln e^{-0.0001238t} \\ t &= \frac{\ln 0.1}{-0.0001238} \approx 18{,}600 \text{ years} \end{aligned}$$

◆

PROBLEM 15 Estimate the age of the bone in Example 15 if 50% of the original amount of carbon-14 is present. ◆

In learning certain skills such as typing and swimming, a mathematical model often used is one that assumes there is a maximum skill attainable, say, M, and the rate of improving is proportional to the difference between that achieved, y, and that attainable, M. Mathematically,

$$\frac{dy}{dt} = k(M - y) \qquad y(0) = 0$$

We solve this type of problem using the same technique that was used to obtain the exponential growth law. First, multiply both sides of the first equation by $1/(M - y)$ to obtain

$$\frac{1}{M - y}\frac{dy}{dt} = k$$

and then integrate each side with respect to t:

$$\int \frac{1}{M - y}\frac{dy}{dt}\,dt = \int k\,dt$$

$$-\int \frac{1}{M - y}\left(-\frac{dy}{dt}\right)dt = \int k\,dt \qquad \text{Substitute } u = M - y \text{ and } du = -dy = -\frac{dy}{dt}\,dt.$$

$$-\int \frac{1}{u}\,du = \int k\,dt$$

$$-\ln|u| = kt + C \qquad \text{Substitute } M - y = u.$$

$$-\ln(M - y) = kt + C \qquad \text{Absolute value signs are not required. (Why?)}$$

$$\ln(M - y) = -kt - C$$

Change this last equation to equivalent exponential form:

$$M - y = e^{-kt-C}$$
$$M - y = e^{-C}e^{-kt}$$
$$y = M - e^{-C}e^{-kt}$$

Now, $y(0) = 0$; hence,

$$y(0) = M - e^{-C}e^{0} = 0$$

Solving for e^{-C}, we obtain

$$e^{-C} = M$$

and our final solution is

$$y = M - Me^{-kt} = M(1 - e^{-kt})$$

◆ EXAMPLE 16 *Learning*

For a particular person who is learning to swim, it is found that the distance y (in feet) the person is able to swim in 1 minute after t hours of practice is given approximately by

$$y = 50(1 - e^{-0.04t})$$

What is the rate of improvement after 10 hours of practice?

Solution

$$y = 50 - 50e^{-0.04t}$$
$$y'(t) = 2e^{-0.04t}$$
$$y'(10) = 2e^{-0.04(10)} \approx 1.34 \text{ feet per hour of practice}$$ ◆

PROBLEM 16 In Example 16, what is the rate of improvement after 50 hours of practice? ◆

◆ A COMPARISON OF EXPONENTIAL GROWTH PHENOMENA

The graphs and equations given in Table 1 (on the next page) compare several widely used growth models. These are divided basically into two groups: unlimited growth and limited growth. Following each equation and graph is a short (and necessarily incomplete) list of areas in which the models are used. This only touches on a subject that has been extensively developed and which you are likely to encounter in greater depth in the future.

TABLE 1

Exponential Growth

DESCRIPTION	MODEL	SOLUTION	GRAPH	USES
Unlimited growth: Rate of growth is proportional to the amount present	$\frac{dy}{dt} = ky$ $k, t > 0$ $y(0) = c$	$y = ce^{kt}$	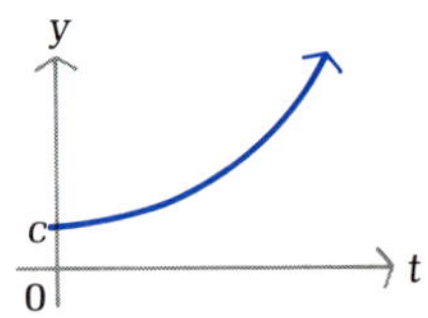	• Short-term population growth (people, bacteria, etc.) • Growth of money at continuous compound interest • Price–supply curves • Depletion of natural resources
Exponential decay: Rate of growth is proportional to the amount present	$\frac{dy}{dt} = -ky$ $k, t > 0$ $y(0) = c$	$y = ce^{-kt}$	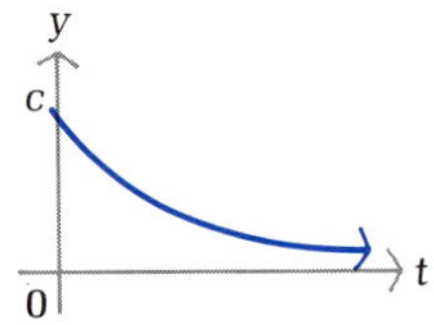	• Radioactive decay • Light absorption in water • Price–demand curves • Atmospheric pressure (t is altitude)
Limited growth: Rate of growth is proportional to the difference between the amount present and a fixed limit	$\frac{dy}{dt} = k(M - y)$ $k, t > 0$ $y(0) = 0$	$y = M(1 - e^{-kt})$	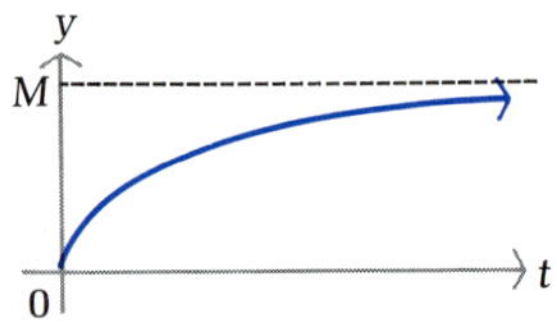	• Sales fads (e.g., skateboards) • Depreciation of equipment • Company growth • Learning
Logistic growth: Rate of growth is proportional to the amount present and to the difference between the amount present and a fixed limit	$\frac{dy}{dt} = ky(M - y)$ $k, t > 0$ $y(0) = \frac{M}{1 + c}$	$y = \frac{M}{1 + ce^{-kMt}}$	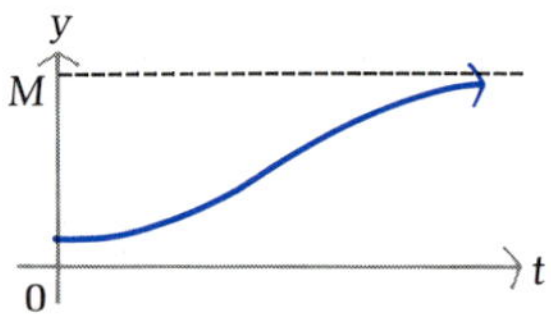	• Long-term population growth • Epidemics • Sales of new products • Rumor spread • Company growth

Answers to Matched Problems

13. 1,694 million people
14. Approx. 22 yr
15. Approx. 5,600 yr
16. Approx. 0.27 ft/hr

EXERCISE 6-3

APPLICATIONS

Business & Economics

1. *Continuous compound interest.* Find the amount A in an account after t years if

$$\frac{dA}{dt} = 0.08A \quad \text{and} \quad A(0) = 1{,}000$$

2. *Continuous compound interest.* Find the amount A in an account after t years if

$$\frac{dA}{dt} = 0.12A \quad \text{and} \quad A(0) = 5{,}250$$

3. *Continuous compound interest.* Find the amount A in an account after t years if

$$\frac{dA}{dt} = rA \qquad A(0) = 8{,}000 \qquad A(2) = 9{,}020$$

4. *Continuous compound interest.* Find the amount A in an account after t years if

$$\frac{dA}{dt} = rA \qquad A(0) = 5{,}000 \qquad A(5) = 7{,}460$$

5. *Price–demand.* If the marginal price dp/dx at x units of demand per week is proportional to the price p, and if at a price of \$100 there is no weekly demand [$p(0) = 100$] and at a price of \$77.88 there is a weekly demand of 5 units [$p(5) = 77.88$], find the price–demand equation.

6. *Price–supply.* If the marginal price dp/dx at x units of supply per day is proportional to the price p, and if at a price of \$10 there is no daily supply [$p(0) = 10$] and at a price of \$12.84 there is a daily supply of 50 units [$p(50) = 12.84$], find the price–supply equation.

7. *Advertising.* A company is trying to expose a new product to as many people as possible through television advertising. Suppose the rate of exposure to new people is proportional to the number of those who have not seen the product out of L possible viewers. If no one is aware of the product at the start of the campaign and after 10 days 40% of L are aware of the product, solve

$$\frac{dN}{dt} = k(L - N) \qquad N(0) = 0 \qquad N(10) = 0.4L$$

for $N = N(t)$, the number of people who are aware of the product after t days of advertising.

8. *Advertising.* Repeat Problem 7 for

$$\frac{dN}{dt} = k(L - N) \qquad N(0) = 0 \qquad N(10) = 0.1L$$

Life Sciences

9. *Ecology.* For relatively clear bodies of water, light intensity is reduced according to

$$\frac{dI}{dx} = -kI \qquad I(0) = I_0$$

where I is the intensity of light at x feet below the surface. For the Sargasso Sea off the West Indies, $k = 0.009\ 42$. Find I in terms of x, and find the depth at which the light is reduced to half of that at the surface.

10. *Blood pressure.* It can be shown under certain assumptions that blood pressure P in the largest artery in the human body (the aorta) changes between beats with respect to time t according to

$$\frac{dP}{dt} = -aP \qquad P(0) = P_0$$

where a is a constant. Find $P = P(t)$ that satisfies both conditions.

11. *Drug concentrations.* A single injection of a drug is administered to a patient. The amount Q in the body then decreases at a rate proportional to the amount present, and for this particular drug the rate is 4% per hour. Thus,

$$\frac{dQ}{dt} = -0.04Q \qquad Q(0) = Q_0$$

where t is time in hours. If the initial injection is 3 milliliters $[Q(0) = 3]$, find $Q = Q(t)$ that satisfies both conditions. How many milliliters of the drug are still in the body after 10 hours?

12. *Simple epidemic.* A community of 1,000 individuals is assumed to be homogeneously mixed. One individual who has just returned from another community has influenza. Assume the home community has not had influenza shots and all are susceptible. One mathematical model for an influenza epidemic assumes that influenza tends to spread at a rate in direct proportion to the number who have it, N, and to the number who have not contracted it, in this case, $1{,}000 - N$. Mathematically,

$$\frac{dN}{dt} = kN(1{,}000 - N) \qquad N(0) = 1$$

where N is the number of people who have contracted influenza after t days. For $k = 0.0004$, it can be shown that $N(t)$ is given by

$$N(t) = \frac{1{,}000}{1 + 999e^{-0.4t}}$$

See Table 1 (logistic growth) for the characteristic graph.

(A) How many people have contracted influenza after 10 days? After 20 days?

(B) How many days will it take until half the community has contracted influenza?

(C) Find $\lim_{t\to\infty} N(t)$.

Reuters/Bettmann Newsphotos

13. *Nuclear accident.* One of the dangerous radioactive isotopes detected after the Chernobyl nuclear accident in 1986 was cesium-137. If 93.3% of the cesium-137 emitted during the accident is still present 3 years later, find the continuous compound rate of decay of this isotope.

14. *Insecticides.* Many countries have banned the use of the insecticide DDT because of its long-term adverse effects. Five years after a particular country stopped using DDT, the amount of DDT in the ecosystem had declined to 75% of the amount present at the time of the ban. Find the continuous compound rate of decay of DDT.

Social Sciences

15. *Archaeology.* A skull from an ancient tomb was discovered and was found to have 5% of the original amount of radioactive carbon-14 present. Estimate the age of the skull. (See Example 15.)

16. *Learning.* For a particular person learning to type, it was found that the number of words per minute, N, the person was able to type after t hours of practice was given approximately by

$$N = 100(1 - e^{-0.02t})$$

See Table 1 (limited growth) for a characteristic graph. What is the rate of improvement after 10 hours of practice? After 40 hours of practice?

17. *Small group analysis.* In a study on small group dynamics, sociologists Stephan and Mischler found that, when the members of a discussion group of 10 were ranked according to the number of times each participated, the number of times $N(k)$ the kth-ranked person participated was given approximately by

$$N(k) = N_1 e^{-0.11(k-1)} \qquad 1 \leq k \leq 10$$

where N_1 is the number of times the 1st-ranked person participated in the discussion. If, in a particular discussion group of 10 people, $N_1 = 180$, estimate how many times the 6th-ranked person participated. The 10th-ranked person.

18. *Perception.* One of the oldest laws in mathematical psychology is the Weber–Fechner law (discovered in the middle of the nineteenth century). It concerns a person's sensed perception of various strengths of stimulation involving weights, sound, light, shock, taste, and so on. One form of the law states that the rate of change of sensed sensation S with respect to stimulus R is inversely proportional to the strength of the stimulus R. Thus,

$$\frac{dS}{dR} = \frac{k}{R}$$

where k is a constant. If we let R_0 be the threshold level at which the stimulus R can be detected (the least amount of sound, light, weight, and so on, that can be detected), then it is appropriate to write

$$S(R_0) = 0$$

Find a function S in terms of R that satisfies the above conditions.

19. *Rumor spread.* A group of 400 parents, relatives, and friends are waiting anxiously at Kennedy Airport for a student charter flight to return after a year in Europe. It is stormy and the plane is late. A particular parent thought he had heard that the plane's radio had gone out and related this news to some friends, who in turn passed it on to others, and so on. Sociologists have studied rumor propagation and have found that a rumor tends to spread at a rate in direct proportion to the number who have heard it, x, and to the number who have not, $P - x$, where P is the total population.

Mathematically, for our case, $P = 400$ and

$$\frac{dx}{dt} = 0.001x(400 - x) \qquad x(0) = 1$$

where t is time (in minutes). From this, it can be shown that

$$x(t) = \frac{400}{1 + 399e^{-0.4t}}$$

See Table 1 (logistic growth) for a characteristic graph.

(A) How many people have heard the rumor after 5 minutes? 20 minutes?
(B) Find $\lim_{t \to \infty} x(t)$.

20. Rumor spread. In Problem 19, how long (to the nearest minute) will it take for half of the group of 400 to have heard the rumor?

In Problems 21–28, use a graphic calculator or a computer to graph, as indicated, the various cases in Table 1.

21. *Unlimited growth:*
$y = 1{,}000e^{0.08t}$
t range: [0, 15]
y range: [0, 3,500]

22. *Unlimited growth:*
$y = 5{,}250e^{0.12t}$
t range: [0, 10]
y range: [0, 20,000]

23. *Exponential decay:*
$p = 100e^{-0.05x}$
x range: [0, 30]
p range: [0, 100]

24. *Exponential decay:*
$p = 1{,}000e^{-0.08x}$
x range: [0, 40]
p range: [0, 1,000]

25. *Limited growth:*
$N = 100(1 - e^{-0.05t})$
t range: [0, 100]
N range: [0, 100]

26. *Limited growth:*
$N = 1{,}000(1 - e^{-0.07t})$
t range: [0, 70]
N range: [0, 1,000]

27. *Logistic growth:*

$$N = \frac{1{,}000}{1 + 999e^{-0.4t}}$$

t range: [0, 40]
N range: [0, 1,000]

28. *Logistic growth:*

$$N = \frac{400}{1 + 99e^{-0.4t}}$$

t range: [0, 30]
N range: [0, 400]

SECTION 6-4

Area Under a Curve

- SUMMATION NOTATION
- AREAS UNDER CURVES

Finding areas bounded by various kinds of curves has intrigued people for thousands of years. In fact, this is one of the three basic problems of calculus stated at the beginning of Chapter 3. The approach we are going to take to solving this type of problem leads to the second kind of integral mentioned earlier, the *definite integral*. The definite integral has wide applications in many fields, as we will see.

Let us start by considering the area between the x axis and the curve $f(x) = x^2 + 3$ from $x = 1$ to $x = 5$, as indicated in Figure 3. Since the computation of the area of a rectangle is very easy, let us cover the indicated area with rectangles and use the sum of the areas of the rectangles as approximations of the desired area. Figure 4 shows three of many possibilities. Notice that the top of each rectangle touches the graph of $f(x) = x^2 + 3$, and in each case the bases of the rectangles are chosen to be equal.

It appears that if we continue to increase the number of rectangles, the rectangle approximation of the actual area will improve. In fact, it appears that the approximation will approach the exact area in question as a limit. A glance at the computation in Figure 4C might discourage you in this approach; however, we can tidy things up quite a bit by the introduction of *summation notation*. Let us digress for a moment to discuss summation notation and some of its basic properties before continuing the line of reasoning suggested in Figure 4.

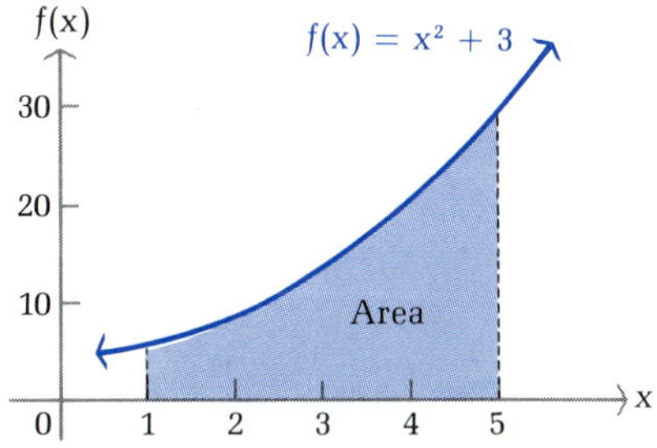

FIGURE 3

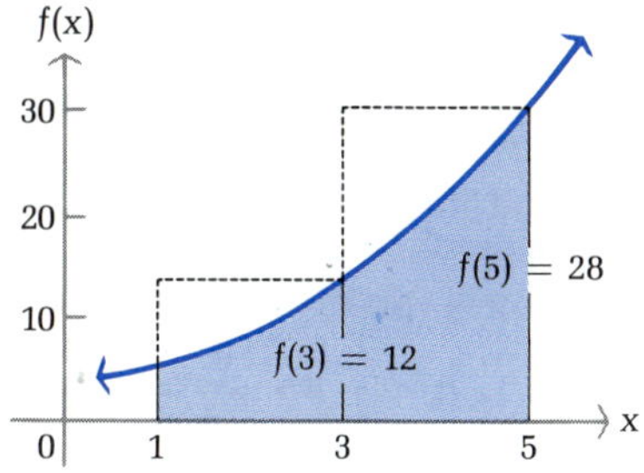

(A) Two-rectangle approximation

$$\begin{aligned} \text{Area} &\approx f(3)(2) + f(5)(2) \\ &= [f(3) + f(5)]2 \\ &= [12 + 28]2 \\ &= 80 \end{aligned}$$

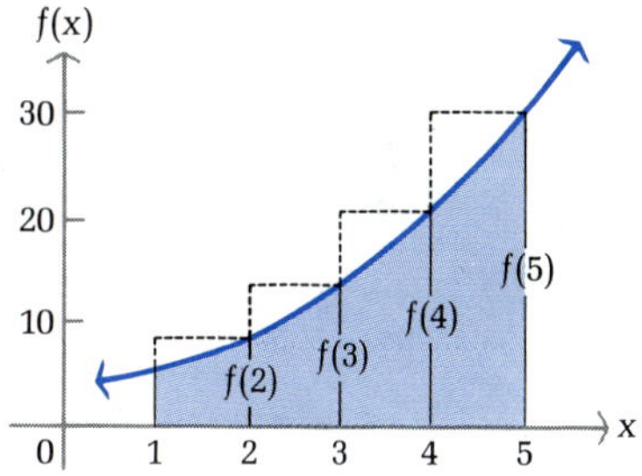

(B) Four-rectangle approximation

$$\begin{aligned} \text{Area} &\approx f(2)(1) + f(3)(1) + f(4)(1) + f(5)(1) \\ &= [f(2) + f(3) + f(4) + f(5)](1) \\ &= [7 + 12 + 19 + 28]1 \\ &= 66 \end{aligned}$$

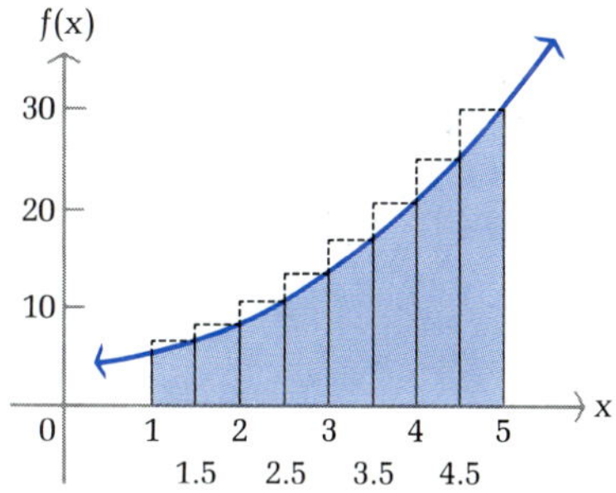

(C) Eight-rectangle approximation

$$\begin{aligned} \text{Area} &\approx f(1.5)(0.5) + f(2)(0.5) + f(2.5)(0.5) + f(3)(0.5) + f(3.5)(0.5) + f(4)(0.5) \\ &\qquad + f(4.5)(0.5) + f(5)(0.5) \\ &= [f(1.5) + f(2) + f(2.5) + f(3) + f(3.5) + f(4) + f(4.5) + f(5)](0.5) \\ &= (119)(0.5) \\ &= 59.5 \end{aligned}$$

FIGURE 4

◆ SUMMATION NOTATION

We need a compact way of expressing the sum of a large number of terms. The symbol Σ is used to denote a sum, and

$$\sum_{k=2}^{6} f(k)$$

means that we are to add $f(k)$ for successive integer values of k from 2 to 6. Thus, we write

$$\sum_{k=2}^{6} f(k) = f(2) + f(3) + f(4) + f(5) + f(6)$$

The **summing index *k*** may start at any integer and end at any integer greater than or equal to the starting integer. Consider the examples below.

◆ EXAMPLE 17

(A) $\sum_{k=4}^{6} g(k) = g(4) + g(5) + g(6)$

(B) $$\begin{aligned}\sum_{k=1}^{5} k^2 &= 1^2 + 2^2 + 3^2 + 4^2 + 5^2 \\ &= 1 + 4 + 9 + 16 + 25 = 55\end{aligned}$$

(C) $$\begin{aligned}\sum_{k=0}^{3} (1 + 2k) &= (1 + 2 \cdot 0) + (1 + 2 \cdot 1) + (1 + 2 \cdot 2) + (1 + 2 \cdot 3) \\ &= 1 + 3 + 5 + 7 = 16\end{aligned}$$

(D) $\sum_{k=1}^{5} 3 = 3 + 3 + 3 + 3 + 3 = 5(3) = 15$

(E) $\sum_{k=1}^{n} f(k) = f(1) + f(2) + \cdots + f(n)$ ◆

PROBLEM 17

Write each of the following in expanded form, as in Example 17. Find the values of the sums in parts B, C, and D.

(A) $\sum_{k=3}^{7} h(k)$ (B) $\sum_{k=0}^{4} (2k + 3)$ (C) $\sum_{k=1}^{3} k^3$

(D) $\sum_{k=1}^{3} 8$ (E) $\sum_{k=1}^{n} \frac{1}{k}$ ◆

Summation notation has several properties that greatly facilitate its use. For example,

$$\begin{aligned}\sum_{k=1}^{n} cf(k) &= cf(1) + cf(2) + \cdots + cf(n) \\ &= c[f(1) + f(2) + \cdots + f(n)] \\ &= c \sum_{k=1}^{n} f(k)\end{aligned}$$

Thus,

$$\sum_{k=1}^{5} 2\frac{k}{n} = \frac{2}{n}\sum_{k=1}^{5} k \qquad \text{Since } 2/n \text{ is a constant relative to the summing index } k$$

This property of moving a constant factor across the summation sign and two other properties are listed below for convenient reference.

Summation Notation Properties

1. $\sum_{k=1}^{n} c = nc \qquad c$ a constant
2. $\sum_{k=1}^{n} cf(k) = c\sum_{k=1}^{n} f(k)$
3. $\sum_{k=1}^{n} [f(k) \pm g(k)] = \sum_{k=1}^{n} f(k) \pm \sum_{k=1}^{n} g(k)$

In addition to these special properties, we will find the two special formulas below useful. These can be established by a process called *mathematical induction*, which we do not include here.

Special Formulas

1. $\sum_{k=1}^{n} k = 1 + 2 + 3 + \cdots + n = \frac{n(n+1)}{2}$
2. $\sum_{k=1}^{n} k^2 = 1^2 + 2^2 + 3^2 + \cdots + n^2 = \frac{n(n+1)(2n+1)}{6}$

◆ EXAMPLE 18

(A)
$$\begin{aligned}\sum_{k=1}^{29} (4k+2) &= \sum_{k=1}^{29} 4k + \sum_{k=1}^{29} 2 && \text{Property 3}\\ &= 4\sum_{k=1}^{29} k + \sum_{k=1}^{29} 2 && \text{Property 2}\\ &= 4\left[\frac{29(29+1)}{2}\right] + 2(29) && \text{Property 1 and formula 1}\\ &= 1{,}798\end{aligned}$$

(B)
$$\begin{aligned}\sum_{k=1}^{40} k(k+6) &= \sum_{k=1}^{40} (k^2+6k)\\ &= \sum_{k=1}^{40} k^2 + 6\sum_{k=1}^{40} k && \text{Properties 2 and 3}\\ &= \frac{40(41)(81)}{6} + 6\,\frac{40(41)}{2} && \text{Formulas 1 and 2}\\ &= 27{,}060\end{aligned}$$

◆

PROBLEM 18 Evaluate as in Example 18.

(A) $\sum_{k=1}^{20} (6k - 4)$ (B) $\sum_{k=1}^{60} (k^2 + 2k - 2)$ ◆

Now we will see how these summation properties and formulas can be used to compute areas bounded by certain curves.

◆ AREAS UNDER CURVES

Before returning to the area problem presented at the beginning of this section, let us consider a simpler area problem—one where the exact area is known. We will show that the method of rectangle approximations will lead to this known area.

◆ EXAMPLE 19 Use the method of rectangle approximations to determine the area between the x axis and $f(x) = x - 2$ from $x = 2$ to $x = 8$.

Solution The figure is a triangle with an exact area of 18, as shown below:

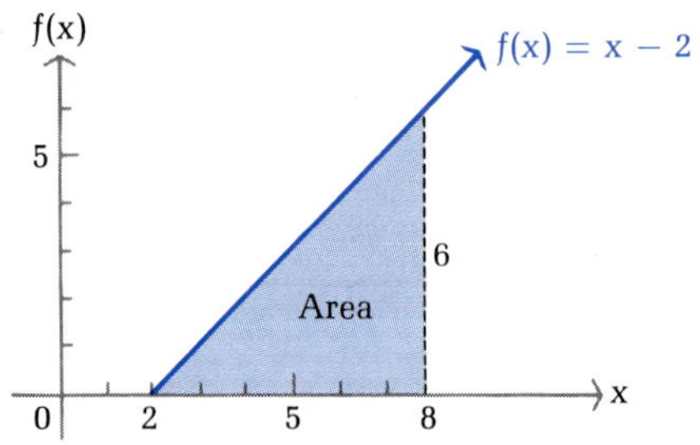

$$\text{Area of triangle} = \frac{(\text{Base})(\text{Height})}{2} = \frac{(8-2)(6)}{2} = 18$$

Let us approximate this area with rectangles and see what happens as we let the number of rectangles increase without bound:

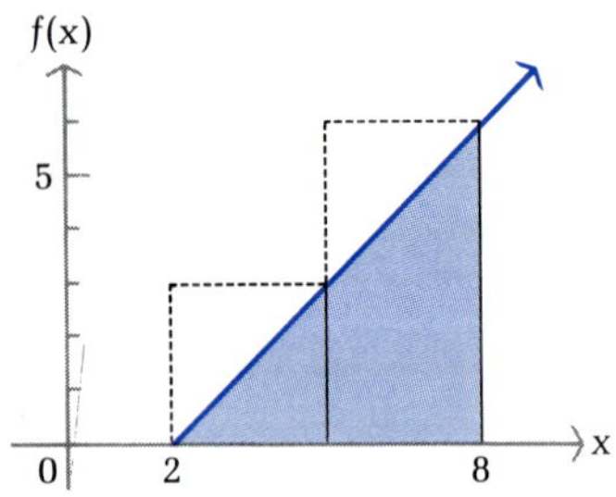

(A) Two rectangles

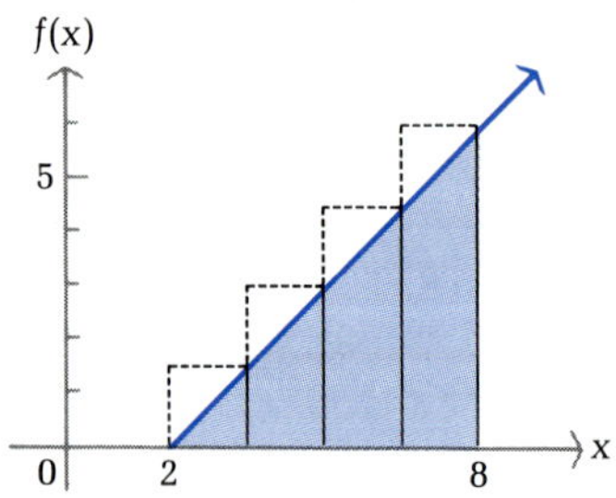

(B) Four rectangles

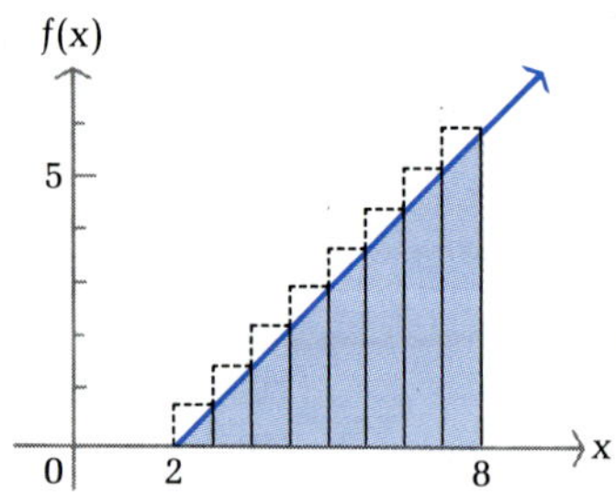

(C) Eight rectangles

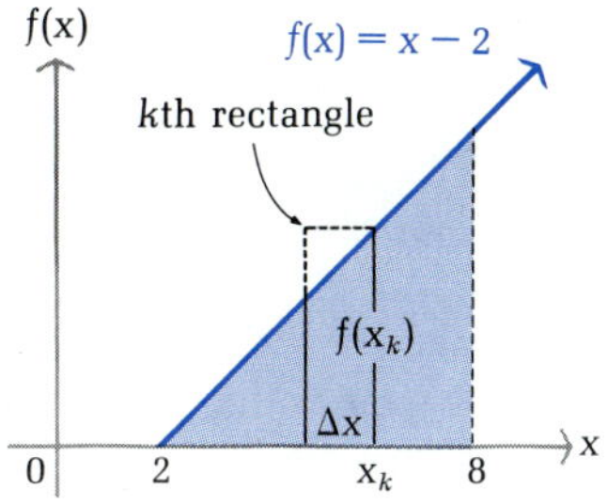

If we compute the sums of the areas of the rectangles in figures A, B, and C, we obtain *approximations* to the area of the triangle. To find the *exact area* by the **rectangle method,** we need an expression that represents the area of n rectangles for arbitrarily large n. Let us form n rectangles by dividing the interval $[2, 8]$ into n equal parts. In the figure in the margin we show only the kth typical rectangle, where the point x_k at the right end of the base is to be used to determine the height $f(x_k)$ of the kth rectangle. Actually, any point on each base may be used, but the right end point makes the computation simpler.

$$\text{Base} = \Delta x = \frac{8-2}{n} = \frac{6}{n}$$

$$\text{Right end point} = x_k = 2 + k\Delta x = 2 + k\frac{6}{n}$$

$$\text{Height} = f(x_k) = x_k - 2 = \left(2 + k\frac{6}{n}\right) - 2 = k\frac{6}{n}$$

$$\begin{aligned}\text{Area of } k\text{th rectangle} &= (\text{Height})(\text{Base})\\ &= f(x_k)\Delta x = k\frac{6}{n}\cdot\frac{6}{n}\end{aligned}$$

$$\text{Area of } n \text{ rectangles} = \sum_{k=1}^{n} f(x_k)\Delta x = \sum_{k=1}^{n} k\frac{6}{n}\cdot\frac{6}{n}$$

Now, we are interested in the limit of this sum (if it exists) as n tends to infinity. That is, we would like to be able to compute

$$\lim_{n\to\infty}\sum_{k=1}^{n} f(x_k)\Delta x = \lim_{n\to\infty}\sum_{k=1}^{n} k\frac{6}{n}\cdot\frac{6}{n} \tag{1}$$

And if this limit exists, we would like to call it the area of the triangle. But it is not possible to evaluate (1) in its present form. The difficulty lies in the fact that the number of terms in the sum is increasing as n increases, even though each term is approaching 0 (we are adding more and more areas of smaller and smaller rectangles). In order to use the familiar limit properties from Chapter 3, we must replace the summation in (1) with an equivalent expression that does not involve the sum of n terms. This is called finding a **closed form** for the summation. We use summation properties to find this closed form:

$$\begin{aligned}\sum_{k=1}^{n} f(x_k)\Delta x &= \sum_{k=1}^{n}\frac{36}{n^2}k\\ &= \frac{36}{n^2}\sum_{k=1}^{n} k && \text{Property 2}\\ &= \frac{36}{n^2}\cdot\frac{n(n+1)}{2} && \text{Special formula 1}\\ &= 18 + \frac{18}{n} && \text{Closed form}\end{aligned}$$

Now that we have a closed form for the summation, we can evaluate the limit:

$$\lim_{n\to\infty} \sum_{k=1}^{n} f(x_k)\Delta x = \lim_{n\to\infty} \left(18 + \frac{18}{n}\right)$$
$$= 18$$

This is the same result obtained by using the area formula, but with quite a bit more trouble. The advantage of the rectangle method, however, is that it can be used to find areas of figures that do not have simple area formulas. ◆

The key steps for using the rectangle method are summarized in the box.

The Rectangle Method for Finding Area—Key Steps

Let $f(x)$ be continuous and $f(x) \geq 0$ on the closed interval $[a, b]$.

Step 1. Sketch a typical kth rectangle and identify its critical parts (see the figure):

$$\text{Base} = \Delta x = \frac{b-a}{n}$$
$$\text{Right end point} = x_k = a + k\Delta x$$
$$\text{Height} = f(x_k)$$
$$\text{Area} = f(x_k)\Delta x$$

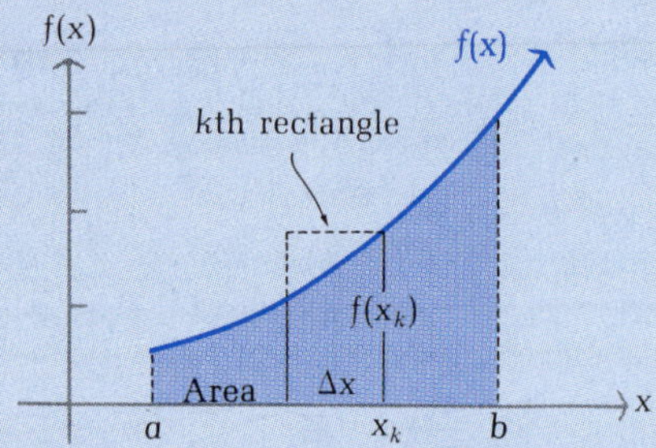

Step 2. Form the summation representing the area of n rectangles,

$$\sum_{k=1}^{n} f(x_k)\Delta x$$

and use summation properties and special formulas to find a closed form for the summation.

Step 3. Use the closed form from step 2 to find the exact area between the graph of $y = f(x)$ and the x axis from $x = a$ to $x = b$:

$$\text{Area} = \lim_{n\to\infty} \sum_{k=1}^{n} f(x_k)\Delta x$$

PROBLEM 19 Use the rectangle method to find the exact area between the x axis and the graph of $f(x) = x - 1$ from $x = 1$ to $x = 11$. ◆

We now return to the example discussed at the beginning of this section and use the rectangle method to find the exact area.

◆ EXAMPLE 20 Use the rectangle method to find the exact area between the x axis and the graph of $f(x) = x^2 + 3$ from $x = 1$ to $x = 5$.

Solution

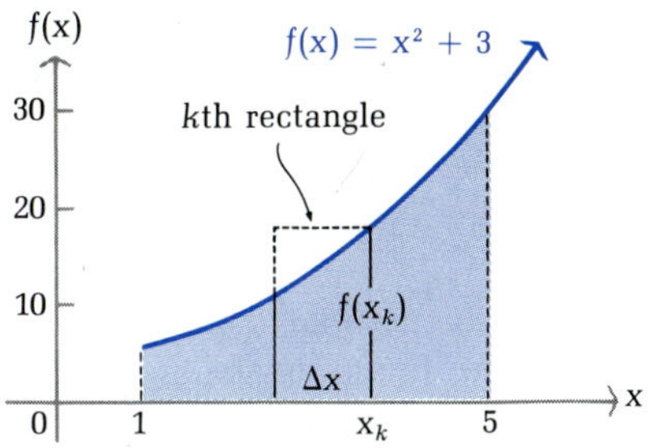

Step 1. Sketch a typical kth rectangle and identify its critical parts:

$$\text{Base} = \Delta x = \frac{5-1}{n} = \frac{4}{n}$$

$$\text{Right end point} = x_k = 1 + k\Delta x$$
$$= 1 + k\frac{4}{n}$$

$$\text{Height} = f(x_k) = x_k^2 + 3$$
$$= \left(1 + k\frac{4}{n}\right)^2 + 3$$

$$\text{Area} = f(x_k)\Delta x$$
$$= \left[\left(1 + k\frac{4}{n}\right)^2 + 3\right]\frac{4}{n}$$
Simplify and collect like terms.
$$= \frac{16}{n} + k\left(\frac{32}{n^2}\right) + k^2\left(\frac{64}{n^3}\right)$$
Area of one rectangle

Step 2. Form the summation representing the area of n rectangles, and find a closed form:

$$\sum_{k=1}^{n} f(x_k)\Delta x = \sum_{k=1}^{n}\left[\frac{16}{n} + k\left(\frac{32}{n^2}\right) + k^2\left(\frac{64}{n^3}\right)\right]$$

$$= \sum_{k=1}^{n}\frac{16}{n} + \sum_{k=1}^{n} k\left(\frac{32}{n^2}\right) + \sum_{k=1}^{n} k^2\left(\frac{64}{n^3}\right)$$
Property 3

$$= \frac{16}{n}\cdot n + \frac{32}{n^2}\sum_{k=1}^{n} k + \frac{64}{n^3}\sum_{k=1}^{n} k^2$$
Properties 1 and 2

$$= 16 + \frac{32}{n^2}\left[\frac{n(n+1)}{2}\right] + \frac{64}{n^3}\left[\frac{n(n+1)(2n+1)}{6}\right]$$
Formulas 1 and 2

$$= 16 + 16\left[\frac{n+1}{n}\right] + \frac{32}{3}\left[\frac{2n^2+3n+1}{n^2}\right]$$
$$= 16 + 16 + \frac{16}{n} + \frac{64}{3} + \frac{32}{n} + \frac{32}{3n^2}$$
Simplify and express as a sum of terms of the form A/Bn^p, $p \geq 0$.

$$= \frac{160}{3} + \frac{48}{n} + \frac{32}{3n^2}$$
Area of n rectangles in closed form

Step 3. Use the closed form to find the exact area:

$$\text{Area} = \lim_{n\to\infty}\sum_{k=1}^{n} f(x_k)\Delta x$$

$$= \lim_{n\to\infty}\left(\frac{160}{3} + \frac{48}{n} + \frac{32}{3n^2}\right)$$
$\lim_{n\to\infty}\frac{A}{Bn^p} = 0$, $p > 0$ (Theorem 3, Section 3-1)

$$= \frac{160}{3}$$
Exact area

Since the graph of f is not a straight line, we could not find this exact area by using simple formulas for the areas of triangles, rectangles, etc. Now you see the power of the rectangle method. ◆

PROBLEM 20 Use the rectangle method to find the exact area between the x axis and the graph of $f(x) = x^2 + 3$ from $x = 2$ to $x = 4$. ◆

To determine area by the rectangle method, we must be able to find a closed form for the summation. This can be very difficult for functions that are more complicated than the ones we have considered in this section. In Section 6-6 we will discuss the fundamental theorem of calculus. This incredible theorem provides a simple procedure for finding the limit of a summation without first finding a closed form. A firm understanding of the rectangle method will then become important in applications of this theorem (just as understanding the limit definition of the derivative was important in applications involving differentiation).

Answers to Matched Problems

17. (A) $h(3) + h(4) + h(5) + h(6) + h(7)$
    (B) $(2 \cdot 0 + 3) + (2 \cdot 1 + 3) + (2 \cdot 2 + 3) + (2 \cdot 3 + 3) + (2 \cdot 4 + 3) = 35$
    (C) $1^3 + 2^3 + 3^3 = 36$ (D) $8 + 8 + 8 = 24$
    (E) $\frac{1}{1} + \frac{1}{2} + \frac{1}{3} + \cdots + \frac{1}{n}$
18. (A) 1,180 (B) 77,350
19. Area = 50 20. Area = $\frac{74}{3}$

EXERCISE 6-4

A

Write Problems 1–4 in expanded form without summation notation and evaluate.

1. $\sum_{k=2}^{4} (2k - 1)$ 2. $\sum_{k=3}^{5} (3k - 8)$ 3. $\sum_{k=1}^{4} 2^k$ 4. $\sum_{k=0}^{3} 4^k$

Write Problems 5–8 in expanded form without summation notation. Do not simplify.

5. $\sum_{k=1}^{3} f(x_k)\Delta x$ 6. $\sum_{k=1}^{4} g(x_k)\Delta x$ 7. $\sum_{k=1}^{3} f(2 + k\Delta x)\Delta x$ 8. $\sum_{k=1}^{3} g(3 + k\Delta x)\Delta x$

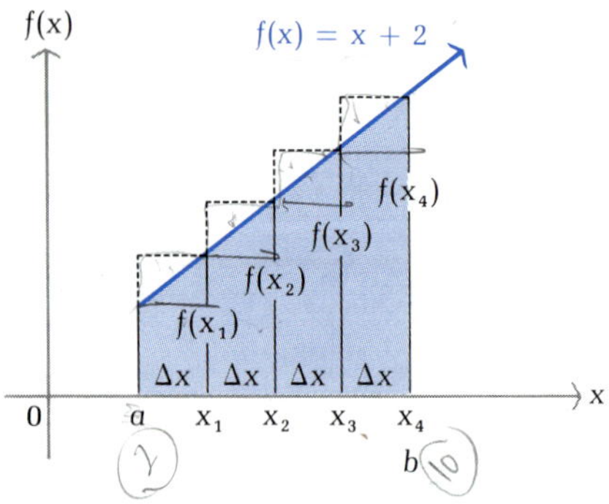

For the figure in the margin, find each of the following in Problems 9–12:

(A) $\Delta x = \frac{b - a}{n}$ (B) $x_k = a + k\Delta x$

(C) x_1, x_2, x_3, *and* x_4 *from part B* (D) $f(x_1), f(x_2), f(x_3), f(x_4)$

(E) $\sum_{k=1}^{4} f(x_k)\Delta x$

9. Use $a = 2$, $b = 10$, and $n = 4$.
10. Use $a = 1$, $b = 3$, and $n = 4$.
11. Use $a = 0$, $b = 6$, and $n = 4$.
12. Use $a = 6$, $b = 12$, and $n = 4$.

B *Use the properties of summation notation and the special summaton formulas to compute the quantities in Problems 13–24.*

13. $\sum_{k=1}^{99} 10$
14. $\sum_{k=1}^{49} (-10)$
15. $\sum_{k=1}^{99} 2k$
16. $\sum_{k=1}^{49} 6k$
17. $\sum_{k=1}^{99} (2k + 10)$
18. $\sum_{k=1}^{49} (6k - 10)$
19. $\sum_{k=1}^{20} 3k^2$
20. $\sum_{k=1}^{50} 12k^2$
21. $\sum_{k=1}^{10} k(k - 1)$
22. $\sum_{k=1}^{10} k(k + 1)$
23. $\sum_{k=1}^{20} (3k^2 - 2k + 5)$
24. $\sum_{k=1}^{40} (6k^2 - 4k - 5)$

In Problems 25–28, find a closed form for the summaton, and evaluate the limit.

25. $\lim_{n\to\infty} \sum_{k=1}^{n} \frac{1}{n^2} k$
26. $\lim_{n\to\infty} \sum_{k=1}^{n} \frac{4}{n^2} k$
27. $\lim_{n\to\infty} \sum_{k=1}^{n} \left(\frac{4}{n^2} k + \frac{2}{n}\right)$
28. $\lim_{n\to\infty} \sum_{k=1}^{n} \left(\frac{9}{n^2} k - \frac{3}{n}\right)$

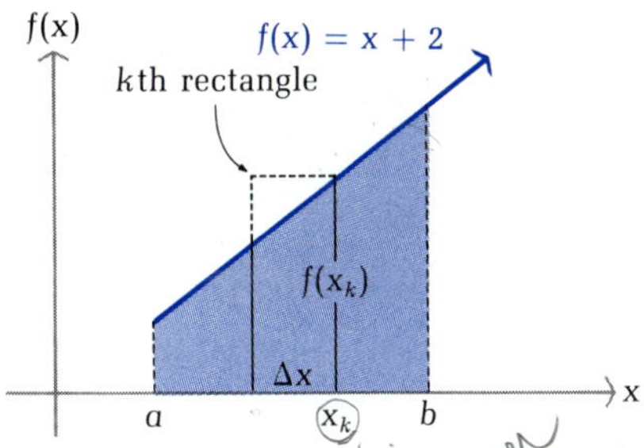

Problems 29–36 refer to the figure in the margin and the formulas below. [Note: Problems 29, 31, and 33 are related as are Problems 30, 32, and 34.]

$$\Delta x = \frac{b - a}{n} \qquad x_k = a + k\Delta x$$

29. For $a = 3$, $b = 7$, and n arbitrary, find:
    (A) Δx (B) x_k (C) $f(x_k)$ (D) $f(x_k)\Delta x$
30. For $a = 1$, $b = 6$, and n arbitrary, find:
    (A) Δx (B) x_k (C) $f(x_k)$ (D) $f(x_k)\Delta x$
31. Using $f(x_k)\Delta x$ obtained in Problem 29D, summation properties, and special summation formulas, show that

$$\sum_{k=1}^{n} f(x_k)\Delta x = 28 + \frac{8}{n}$$

32. Using $f(x_k)\Delta x$ obtained in Problem 30D, summation properties, and special summation formulas, show that

$$\sum_{k=1}^{n} f(x_k)\Delta x = \frac{55}{2} + \frac{25}{2n}$$

33. Use the results of Problem 31 to find

$$\lim_{n\to\infty} \sum_{k=1}^{n} f(x_k)\Delta x$$

which is the area between $f(x) = x + 2$ and the x axis from $x = 3$ to $x = 7$.

34. Use the results of Problem 32 to find

$$\lim_{n\to\infty} \sum_{k=1}^{n} f(x_k)\Delta x$$

which is the area between $f(x) = x + 2$ and the x axis from $x = 1$ to $x = 6$.

35. Use the rectangle method to find the area between the graph of $f(x) = x + 2$ and the x axis from $x = 5$ to $x = 7$.

36. Repeat Problem 35 for the interval [4, 8].

37. Use the rectangle method to find the area between the graph of $f(x) = 2x + 5$ and the x axis from $x = -2$ to $x = 1$.

38. Repeat Problem 37 for the interval [−1, 3].

C *In Problems 39–42, find a closed form for the summation, and evaluate the limit.*

39. $\lim_{n\to\infty} \sum_{k=1}^{n} \frac{27}{n^3} k^2$

40. $\lim_{n\to\infty} \sum_{k=1}^{n} \frac{8}{n^3} k^2$

41. $\lim_{n\to\infty} \sum_{k=1}^{n} \left(\frac{8}{n^3} k^2 - \frac{8}{n^2} k + \frac{2}{n}\right)$

42. $\lim_{n\to\infty} \sum_{k=1}^{n} \left(\frac{27}{n^3} k^2 - \frac{18}{n^2} k + \frac{3}{n}\right)$

Problems 43–50 refer to the figure and formulas below. [Note: Problems 43, 45, and 47 are related, as are Problems 44, 46, and 48.]

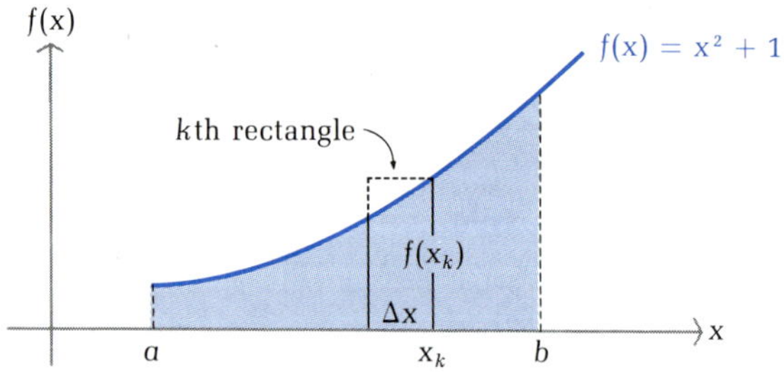

$$\Delta x = \frac{b-a}{n}$$

$$x_k = a + k\Delta x$$

43. For $a = 2$, $b = 6$, and n arbitrary, find:

(A) Δx (B) x_k (C) $f(x_k)$ (D) $f(x_k)\Delta x$

44. For $a = -1$, $b = 4$, and n arbitrary, find:

(A) Δx (B) x_k (C) $f(x_k)$ (D) $f(x_k)\Delta x$

45. Using $f(x_k)\Delta x$ obtained in Problem 43D, summation properties, and special summation formulas, show that

$$\sum_{k=1}^{n} f(x_k)\Delta x = \frac{220}{3} + \frac{64}{n} + \frac{32}{3n^2}$$

46. Using $f(x_k)\Delta x$ obtained in Problem 44D, summation properties, and special summation formulas, show that

$$\sum_{k=1}^{n} f(x_k)\Delta x = \frac{80}{3} + \frac{75}{2n} + \frac{125}{6n^2}$$

47. Use the results of Problem 45 to find

$$\lim_{n\to\infty} \sum_{k=1}^{n} f(x_k)\Delta x$$

which is the area between $f(x) = x^2 + 1$ and the x axis from $x = 2$ to $x = 6$.

48. Use the results of Problem 46 to find

$$\lim_{n\to\infty} \sum_{k=1}^{n} f(x_k)\Delta x$$

which is the area between $f(x) = x^2 + 1$ and the x axis from $x = -1$ to $x = 4$.

49. Use the rectangle method to find the area between the graph of $f(x) = x^2 + 1$ and the x axis from $x = -2$ to $x = 3$.

50. Repeat Problem 49 for the interval $[-3, 1]$.

51. Use the rectangle method to find the area between the graph of $f(x) = (x + 1)^2$ and the x axis from $x = 0$ to $x = 1$.

52. Repeat Problem 51 for the interval $[-1, 2]$.

SECTION 6-5

Definite Integrals

- DEFINITE INTEGRALS
- DIRECT EVALUATION OF A DEFINITE INTEGRAL
- RECOGNIZING A DEFINITE INTEGRAL—AVERAGE VALUE
- PROPERTIES OF DEFINITE INTEGRALS
- DEFINITE INTEGRALS OF TABULAR FUNCTIONS

DEFINITE INTEGRALS

Recall that the derivative of a function f was defined by

$$f'(x) = \lim_{h\to 0} \frac{f(x + h) - f(x)}{h}$$

This form is generally not easy to compute directly, but it is easy to recognize in certain practical problems (slope, instantaneous velocity, rates of change, etc.). Once we have recognized that we are dealing with a derivative, we can then try to compute it using derivative formulas and rules.

The area form introduced in the preceding section,

$$\lim_{n\to\infty} \sum_{k=1}^{n} f(x_k)\Delta x$$

can be generalized to make it applicable to problems that may not have anything at all to do with area. To do this, we introduce the idea of a *definite integral*. The development of the definite integral will parallel the development of the derivative. That is, we will define the concept through a form that is not easy to compute, but is easy to recognize in certain practical problems, and then we will introduce important properties that will enable us (in many cases) to compute the definite integral with relative ease. We start with a definition.

Definition of a Definite Integral

Let f be a continuous function defined on the closed interval $[a, b]$, and let:

1. $a = x_0 \leq x_1 \leq \cdots \leq x_{n-1} \leq x_n = b$
2. $\Delta x_k = x_k - x_{k-1}$ for $k = 1, 2, \ldots, n$
3. $\Delta x_k \to 0$ as $n \to \infty$
4. $x_{k-1} \leq c_k \leq x_k$ for $k = 1, 2, \ldots, n$

Then

$$\int_a^b f(x)\,dx = \lim_{n\to\infty} \sum_{k=1}^{n} f(c_k)\Delta x_k$$

is called the **definite integral** of f from a to b. The **integrand** is $f(x)$, the **upper limit** is b, and the **lower limit** is a.

In the definition of a definite integral, we divide the closed interval $[a, b]$ into n subintervals of arbitrary length in such a way that the length of each subinterval $\Delta x_k = x_k - x_{k-1}$ tends to 0 as n increases without bound. From each of the n subintervals we then select a point c_k and form the sum

$$\sum_{k=1}^{n} f(c_k)\Delta x_k$$

This is called a **Riemann sum,** named after the celebrated German mathematician Georg Riemann (1826–1866).

Under the conditions stated in the definition, it can be shown that the limit of a Riemann sum always exists, and it is a real number. The limit is independent of the nature of the subdivisions of $[a, b]$ as long as condition 3 holds, and it is independent of the choice of c_k as long as condition 4 holds. In particular, we can always divide $[a, b]$ into n equal parts so that $\Delta x_k = \Delta x = (b - a)/n$ for each k, and we can always choose c_k as the right end point of the kth interval so that $c_k = a + k\Delta x$ (or any other convenient choice, such as the left end point or midpoint).

Referring to the discussion of area in the preceding section, we have the following important interpretation of the definite integral:

Definite Integrals and Area

If f is continuous and $f(x) \geq 0$ over the interval $[a, b]$, then the area bounded by $y = f(x)$, the x axis ($y = 0$), and the vertical lines $x = a$ and $x = b$ is given exactly by

$$A = \int_a^b f(x)\, dx$$

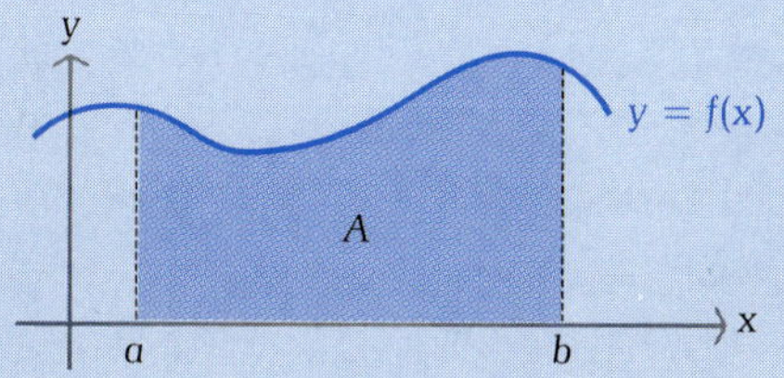

Remark

Do not confuse a definite integral with an indefinite integral. The definite integral $\int_a^b f(x)\, dx$ is a real number; the indefinite integral $\int f(x)\, dx$ is a whole set of functions—all the antiderivatives of $f(x)$. The integral sign $\int$ is used in both cases because of the fundamental relationship between definite and indefinite integrals discussed in the next section.

◆ DIRECT EVALUATION OF A DEFINITE INTEGRAL

◆ EXAMPLE 21

(A) Approximate $\int_0^3 (x^2 - 4)\, dx$ using three equal subdivisions of [0, 3].
(B) Evaluate $\int_0^3 (x^2 - 4)\, dx$ exactly using the definition of the definite integral.

Solutions

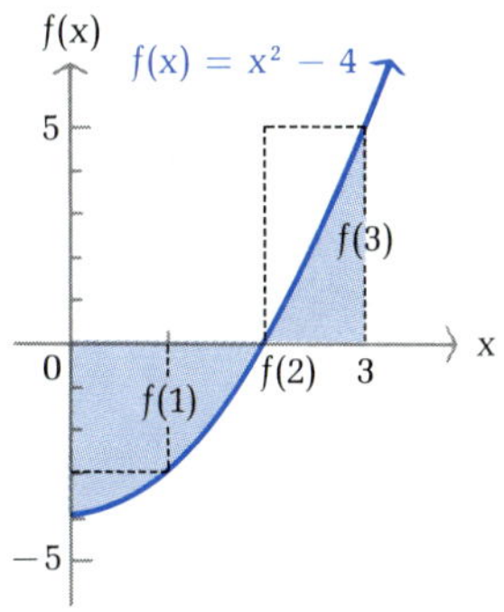

(A) We first sketch a graph of $f(x) = x^2 - 4$ and label important parts of the graph:

$$\Delta x = \frac{b - a}{n} = \frac{3 - 0}{3} = 1$$

$$c_k = a + k\Delta x = 0 + k(1) = k$$

$$f(c_k) = c_k^2 - 4 = k^2 - 4$$

Then

$$\begin{aligned}\int_0^3 f(x)\, dx &\approx \sum_{k=1}^{3} f(c_k)\Delta x \\ &= \sum_{k=1}^{3} (k^2 - 4)1 \\ &= (1^2 - 4)1 + (2^2 - 4)1 + (3^2 - 4)1 \\ &= -3 + 0 + 5 = 2\end{aligned}$$

Of course, this is a very rough approximation of $\int_0^3 (x^2 - 4)\,dx$, and we would expect the approximation to improve as we increase n, the number of subdivisions of the interval [0, 3].

(B) Let us now increase n without bound to obtain the exact value of $\int_0^3 (x^2 - 4)\,dx$:

$$\int_0^3 f(x)\,dx = \lim_{n\to\infty} \sum_{k=1}^{n} f(c_k)\Delta x_k$$

Divide [0, 3] into n equal parts:

$$\Delta x_k = \Delta x = \frac{b-a}{n} = \frac{3-0}{n} = \frac{3}{n}$$

Choose c_k as the right end point of each subinterval:

$$c_k = a + k\Delta x = 0 + k\frac{3}{n} = \frac{3}{n}k$$

We use the summation properties from the preceding section to find a closed form of the Riemann sum so that we can evaluate the limit:

$$\begin{aligned}
\sum_{k=1}^{n} f(c_k)\Delta x_k &= \sum_{k=1}^{n} (c_k^2 - 4)\Delta x_k \\
&= \sum_{k=1}^{n} \left[\left(\frac{3}{n}k\right)^2 - 4\right]\frac{3}{n} \\
&= \sum_{k=1}^{n} \left(\frac{27}{n^3}k^2 - \frac{12}{n}\right) \\
&= \frac{27}{n^3}\sum_{k=1}^{n} k^2 - \sum_{k=1}^{n} \frac{12}{n} \\
&= \frac{27}{n^3}\left[\frac{n(n+1)(2n+1)}{6}\right] - \left(\frac{12}{n}\right)n \\
&= -3 + \frac{27}{2n} + \frac{9}{2n^2}
\end{aligned}$$

Thus,

$$\begin{aligned}
\int_0^3 (x^2 - 4)\,dx &= \lim_{n\to\infty} \sum_{k=1}^{n} (c_k^2 - 4)\Delta x_k \\
&= \lim_{n\to\infty}\left(-3 + \frac{27}{2n} + \frac{9}{2n^2}\right) \\
&= -3
\end{aligned}$$

Note that -3 is *not* the area of the region between the graph of $y = f(x)$ and the x axis from $x = 0$ to $x = 3$ (look back at the figure). Remember, in order for $\int_a^b f(x)\,dx$ to represent an area, it must be true that $f(x) \geqslant 0$ on the interval $[a, b]$. We will discuss techniques for finding the area of the shaded region in the figure on page 407 in the next chapter. ◆

PROBLEM 21 (A) Approximate $\int_0^8 (x-6)\,dx$ using four equal subdivisions of [0, 8] and right end points of subintervals for c_k.

(B) Evaluate $\int_0^8 (x-6)\,dx$ exactly using the definition of the definite integral.

◆

◆ RECOGNIZING A DEFINITE INTEGRAL—AVERAGE VALUE

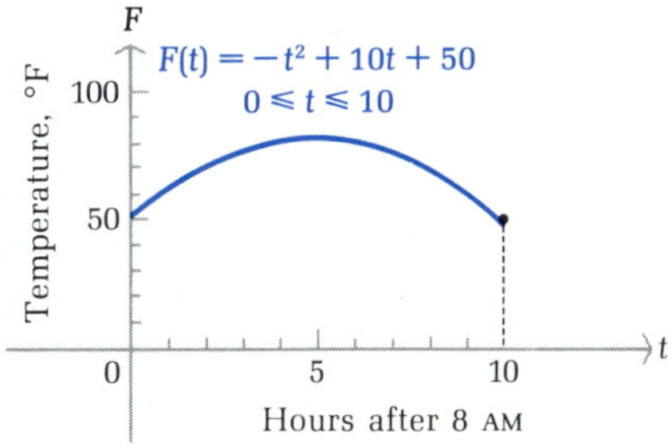

FIGURE 5

It is important to be able to recognize when definite integrals are involved in the solutions of applications. We will start with a simple example here and look at many more before our discussion of integrals is finished.

Suppose the temperature F (in degrees Fahrenheit) in the middle of a small lake from 8 AM ($t = 0$) to 6 PM ($t = 10$) during the month of May is given approximately as shown in Figure 5.

How can we compute the average temperature from 8 AM to 6 PM? We know that the average of a finite number of values $a_1, a_2, \ldots, a_n$ is given by

$$\text{Average} = \frac{a_1 + a_2 + \cdots + a_n}{n} = \frac{1}{n}\sum_{k=1}^{n} a_k$$

But how can we handle a continuous function with infinitely many values? It would seem reasonable to divide the time interval [0, 10] into n equal subintervals, compute the temperature at a point in each subinterval, and then use the average of these values as an approximation of the average value of the continuous function $F = F(t)$ over [0, 10]. We would expect the approximations to improve as n increases. In fact, we would be inclined to define the limit of the average for n values as $n \to \infty$ as *the average value of F* over [0, 10], if the limit exists. This is exactly what we will do:

$$\left(\begin{array}{c}\text{Average temperature}\\ \text{for } n \text{ values}\end{array}\right) = \frac{1}{n}\sum_{k=1}^{n} F(c_k) \qquad (1)$$

where c_k is a point in the kth subinterval. We will call the limit of (1) as $n \to \infty$ *the average temperature over the time interval* [0, 10].

The summation in (1) is similar to a Riemann sum, but there is no Δt_k. We take care of this by multiplying (1) by $(b - a)/(b - a)$, which will change the form of (1) without changing its value:

$$\begin{aligned}\frac{b-a}{b-a}\cdot\frac{1}{n}\sum_{k=1}^{n} F(c_k) &= \frac{1}{b-a}\cdot\frac{b-a}{n}\sum_{k=1}^{n} F(c_k)\\ &= \frac{1}{b-a}\sum_{k=1}^{n} F(c_k)\frac{b-a}{n} \qquad \Delta t_k = \Delta t = \frac{b-a}{n}\\ &= \frac{1}{b-a}\sum_{k=1}^{n} F(c_k)\Delta t\end{aligned}$$

Now the summation has the form of a Riemann sum. Thus,

$$\begin{pmatrix}\text{Average temperature}\\ \text{over } [a, b] = [0, 10]\end{pmatrix} = \lim_{n\to\infty}\left[\frac{1}{b-a}\sum_{k=1}^{n} F(c_k)\Delta t\right] \quad \text{Definition of average temperature}$$

$$= \frac{1}{b-a}\left[\lim_{n\to\infty}\sum_{k=1}^{n} F(c_k)\Delta t\right] \quad \text{Limit property}$$

$$= \frac{1}{b-a}\int_a^b F(t)\,dt \quad \text{Definition of a definite integral}$$

$$= \frac{1}{10}\int_0^{10} (-t^2 + 10t + 50)\,dt$$

We will not evaluate this definite integral, but in the next section we will solve similar problems using a method that is much easier than the direct evaluation method discussed earlier. For now, it is important to recognize how a limit of a sum was involved in defining the average value and that the result is a definite integral.

Proceeding as above for an arbitrary continuous function f over an interval $[a, b]$, we obtain the following general formula:

Average Value of a Continuous Function f Over $[a, b]$

$$\frac{1}{b-a}\int_a^b f(x)\,dx$$

◆ EXAMPLE 22 Use the results of Example 21 to find the average value of $f(x) = x^2 - 4$ over $[0, 3]$.

Solution From Example 21,

$$\int_0^3 (x^2 - 4)\,dx = -3$$

Thus,

$$\frac{1}{b-a}\int_a^b f(x)\,dx = \frac{1}{3-0}\int_0^3 (x^2 - 4)\,dx = \frac{1}{3}(-3) = -1$$ ◆

PROBLEM 22 Use the results of Problem 21 to find the average value of $f(x) = x - 6$ over $[0, 8]$. ◆

◆ PROPERTIES OF DEFINITE INTEGRALS

In some situations, we will encounter definite integrals where the lower limit is greater than the upper limit or even where both limits of integration are the same. We extend the definition to cover these cases as follows:

Definite Integral Definition—Extended

If f is continuous on $[a, b]$, then

$$\int_b^a f(x)\,dx = -\int_a^b f(x)\,dx \qquad \text{and} \qquad \int_a^a f(x)\,dx = 0$$

In the next box we state some useful properties of the definite integral. Note that properties 1 and 2 parallel properties 3 and 4 for the indefinite integral listed in Section 6-1.

Definite Integral Properties

If f and g are continuous functions over all indicated intervals and M is a constant, then:

1. $\displaystyle\int_a^b Mf(x)\,dx = M\int_a^b f(x)\,dx$

2. $\displaystyle\int_a^b [f(x) \pm g(x)]\,dx = \int_a^b f(x)\,dx \pm \int_a^b g(x)\,dx$

3. $\displaystyle\int_a^b f(x)\,dx = \int_a^c f(x)\,dx + \int_c^b f(x)\,dx$ where c is any real number

Note that it is not necessary for c to be between a and b or for a to be less than b.

These properties follow from limit and summation properties. For example,

$$\begin{aligned}
\int_a^b Mf(x)\,dx &= \lim_{n\to\infty} \sum_{k=1}^{n} Mf(c_k)\Delta x_k && \text{Definition}\\
&= \lim_{n\to\infty} M \sum_{k=1}^{n} f(c_k)\Delta x_k && \text{Summation property}\\
&= M\left[\lim_{n\to\infty} \sum_{k=1}^{n} f(c_k)\Delta x_k\right] && \text{Limit property}\\
&= M\int_a^b f(x)\,dx && \text{Definition}
\end{aligned}$$

These properties will be used extensively in the sections that follow. Example 23 illustrates how they can be used.

◆ EXAMPLE 23 If

$$\int_0^2 x\,dx = 2 \qquad \int_0^2 x^2\,dx = \frac{8}{3} \qquad \int_2^3 x^2\,dx = \frac{19}{3}$$

then:

(A) $\int_0^2 12x^2\,dx = 12\int_0^2 x^2\,dx = 12\left(\frac{8}{3}\right) = 32$

(B) $\int_0^2 (2x - 6x^2)\,dx = 2\int_0^2 x\,dx - 6\int_0^2 x^2\,dx = 2(2) - 6\left(\frac{8}{3}\right) = -12$

(C) $\int_3^2 x^2\,dx = -\int_2^3 x^2\,dx = -\frac{19}{3}$

(D) $\int_5^5 3x^2\,dx = 0$

(E) $\int_0^3 3x^2\,dx = 3\int_0^2 x^2\,dx + 3\int_2^3 x^2\,dx = 3\left(\frac{8}{3}\right) + 3\left(\frac{19}{3}\right) = 27$ ◆

PROBLEM 23 Using the same integral values given in Example 23, find:

(A) $\int_2^3 6x^2\,dx$ (B) $\int_0^2 (9x^2 - 4x)\,dx$ (C) $\int_2^0 3x\,dx$
(D) $\int_{-2}^{-2} 3x\,dx$ (E) $\int_0^3 12x^2\,dx$ ◆

◆ DEFINITE INTEGRALS OF TABULAR FUNCTIONS

Many applications involve **tabular functions**—that is, functions defined by tables rather than formulas. It is impossible to find the exact value of the definite integral of a tabular function, but it is always possible to approximate the definite integral with a Riemann sum. The following example illustrates this approach.

◆ EXAMPLE 24 *Real Estate* A developer is interested in estimating the area of the irregularly shaped property shown in Figure 6A. A surveyor used the straight horizontal road at the

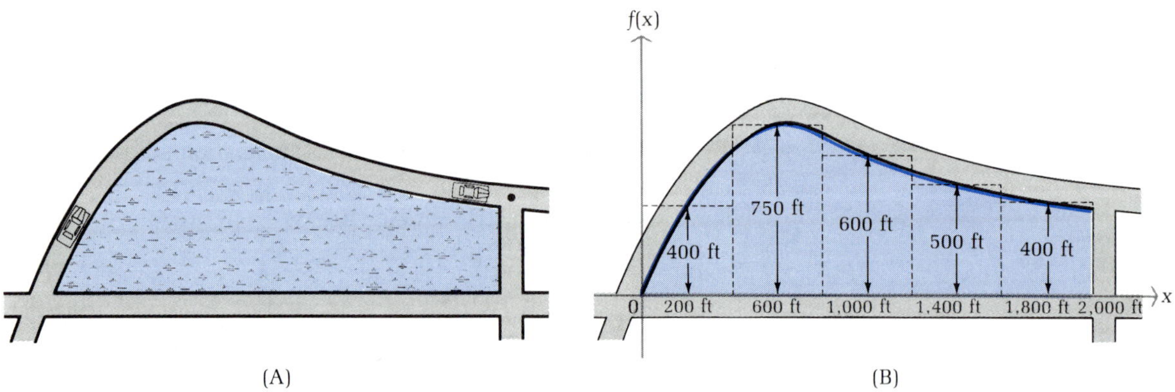

FIGURE 6

bottom of the property as the x axis and measured the vertical distance across the property at 400 foot intervals, starting at 200 (see Figure 6B). These distances can be viewed as the values of the continuous function f whose graph forms the top of the property. Use a Riemann sum to approximate the area of the property.

Solution

x	f(x)
200	400
600	750
1,000	600
1,400	500
1,800	400

We list the values of the function f in the table in the margin. Since $f(x) \geqslant 0$ on $[0, 2{,}000]$, the area of the property is given by the definite integral:

$$\text{Area} = \int_0^{2,000} f(x)\, dx$$

We are limited to using the values of f at the five values of x given in the table, so we want to divide the interval [0, 2,000] into five subintervals, each containing one of these values. We can accomplish this by choosing (see Figure 6B)

$$n = 5$$

$$\Delta x_k = \Delta x = \frac{2{,}000 - 0}{5} = 400$$

$$c_k = \text{Midpoint* of } k\text{th interval}$$

Using the values in the table to evaluate $f(c_k)$, we have

$$\begin{aligned}\text{Area} &\approx \sum_{k=1}^{5} f(c_k)\Delta x \\ &= f(200)\Delta x + f(600)\Delta x + f(1{,}000)\Delta x + f(1{,}400)\Delta x + f(1{,}800)\Delta x \\ &= 400 \cdot 400 + 750 \cdot 400 + 600 \cdot 400 + 500 \cdot 400 + 400 \cdot 400 \\ &= 1{,}060{,}000 \text{ square feet}\end{aligned}$$

We have used all the given values of the function f and cannot improve the accuracy of this approximation unless we obtain more values of f. This would require additional measurements by the surveyor (see Problem 24). ◆

PROBLEM 24 To obtain a more accurate approximation of the area of the property shown in Figure 6A, the surveyor measures the vertical distances at 200 foot intervals, starting at 100. The results are listed in the table. Use these values with $n = 10$ and $\Delta x = 200$ to compute a Riemann sum approximating the area of the property.

x	f(x)	x	f(x)
100	225	1,100	575
300	500	1,300	525
500	700	1,500	475
700	725	1,700	425
900	650	1,900	375

◆

* Remember, in the definition of a definite integral, we are free to choose any point in the kth subinterval. In approximation problems, the midpoint is often a convenient point to choose, because then the tops of the rectangles usually are above part of the graph and below part of the graph. This tends to cancel some of the error that occurs in the approximation.

Answers to Matched Problems

21. (A) $\int_0^8 (x-6)\,dx \approx -8$ (B) $\int_0^8 (x-6)\,dx = -16$
22. $\frac{1}{8}\int_0^8 (x-6)\,dx = -2$
23. (A) 38 (B) 16 (C) −6 (D) 0 (E) 108
24. 1,035,000 ft^2

EXERCISE 6-5

A

Use properties of definite integrals to evaluate Problems 1–10, given $\int_0^2 dx = 2$, $\int_0^2 x\,dx = 2$, $\int_0^2 x^2\,dx = \frac{8}{3}$, and $\int_2^4 x\,dx = 6$.

1. $\displaystyle\int_0^2 3x^2\,dx$ **2.** $\displaystyle\int_0^2 7\,dx$ **3.** $\displaystyle\int_0^2 (2x+3)\,dx$

4. $\displaystyle\int_0^2 (6x^2-5)\,dx$ **5.** $\displaystyle\int_4^2 3x\,dx$ **6.** $\displaystyle\int_2^0 12x^2\,dx$

7. $\displaystyle\int_0^4 3x\,dx$ **8.** $\displaystyle\int_4^0 2x\,dx$ **9.** $\displaystyle\int_0^2 (3x^2-2x+5)\,dx$

10. $\displaystyle\int_0^2 (6-8x-6x^2)\,dx$

In Problems 11–14, use a Riemann sum with $n = 4$ and the values of f in the given table to approximate the indicated definite integral. Choose c_k as the midpoint of each interval.

11. $\displaystyle\int_0^8 f(x)\,dx$

x	f(x)
1	4.5
3	3.2
5	2.4
7	1.6

12. $\displaystyle\int_1^9 f(x)\,dx$

x	f(x)
2	3.2
4	4.5
6	7.9
8	9.4

13. $\displaystyle\int_1^5 f(x)\,dx$

x	f(x)
1.5	12.5
2.5	16.7
3.5	15.4
4.5	10.7

14. $\displaystyle\int_0^4 f(x)\,dx$

x	f(x)
0.5	9.4
1.5	14.7
2.5	11.5
3.5	6.4

B

Write in the form $\int_a^b f(x)\,dx$.

15. $\displaystyle\lim_{n\to\infty}\sum_{k=1}^{n}(1-c_k^2)\frac{5-2}{n};\quad c_k = 2+k\frac{3}{n}$

16. $\displaystyle\lim_{n\to\infty}\sum_{k=1}^{n}(c_k^2-3)\frac{10-0}{n};\quad c_k = 0+k\frac{10}{n}$

17. $\displaystyle\lim_{n\to\infty}\sum_{k=1}^{n}(3c_k^2-2c_k+3)\frac{12-2}{n};\quad c_k = 2+k\frac{10}{n}$

18. $\lim_{n\to\infty} \sum_{k=1}^{n} (4c_k^3 - 3c_k^2 - 5)\frac{2-1}{n}$; $c_k = 1 + k\frac{1}{n}$

Write in the form

$$\lim_{n\to\infty} \sum_{k=1}^{n} f(c_k)\frac{b-a}{n}$$

where c_k is the right end point of the kth subinterval.

19. $\int_1^3 (2x - 3)\, dx$

20. $\int_1^3 (6x + 5)\, dx$

21. $\int_0^4 (3x^2 - 4)\, dx$

22. $\int_0^2 (6x^2 - 2x)\, dx$

Evaluate each definite integral using the definition of the definite integral and the summation properties from Section 6-4.

23. $\int_1^3 (2x - 3)\, dx$

24. $\int_1^3 (6x + 5)\, dx$

25. $\int_0^4 (3x^2 - 4)\, dx$

26. $\int_0^2 (6x^2 - 2x)\, dx$

27. $\int_{-3}^4 (4 - x^2)\, dx$

28. $\int_{-1}^2 (x^2 - 4x)\, dx$

Use the definite integrals evaluated in Problems 23–28 to find the average value of each function over the indicated interval.

29. $f(x) = 2x - 3$; $[1, 3]$
30. $f(x) = 6x + 5$; $[1, 3]$
31. $f(x) = 3x^2 - 4$; $[0, 4]$
32. $f(x) = 6x^2 - 2x$; $[0, 2]$
33. $f(x) = 4 - x^2$; $[-3, 4]$
34. $f(x) = x^2 - 4x$; $[-1, 2]$

C *Use the definition of the definite integral and the special summation formula given below to evaluate the definite integrals in Problems 35 and 36.*

$$\sum_{k=1}^{n} k^3 = \frac{n^2(n+1)^2}{4}$$

35. $\int_0^2 x^3\, dx$

36. $\int_0^1 x^3\, dx$

In Problems 37 and 38, use the definition of the definite integral, along with limit and summation properties, to show that each statement is true. (Assume $a < b$.)

37. $\int_a^b M\, dx = M(b - a)$, M a constant

38. $\int_a^b x\, dx = \frac{b^2 - a^2}{2}$

Business & Economics

39. *Cash reserves.* The cash reserves (in thousands of dollars) of a company x months after the first of the year are approximated by

$$C(x) = 1 + 12x - x^2 \qquad 0 \leq x \leq 12$$

Express the average cash reserves during the first quarter in terms of a definite integral. Do not evaluate.

40. *Cash reserves.* Repeat Problem 39 for the second quarter.

41. *Real estate.* A surveyor produced the table below by measuring the vertical distance (in feet) across a piece of real estate at 600 foot intervals, starting at 300 (see the figure). Use these values and a Riemann sum to estimate the area of the property.

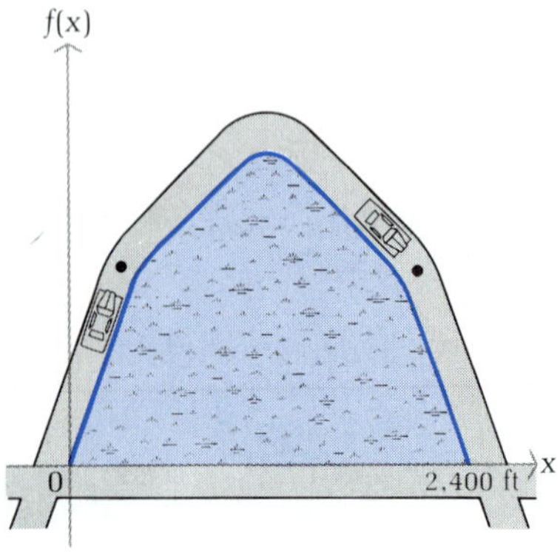

x	f(x)
300	900
900	1,700
1,500	1,700
2,100	900

42. *Real estate.* Repeat Problem 41 for the following table of measurements:

x	f(x)
200	600
600	1,400
1,000	1,800
1,400	1,800
1,800	1,400
2,200	600

Life Sciences

43. *Medicine—circulation.* An indicator dye is injected into the bloodstream of a patient at a constant rate for 2 minutes. The concentration of the dye (in milligrams per liter) after t minutes is approximated by

$$C(t) = 1 + \tfrac{1}{2}t + \tfrac{1}{4}t^2 \qquad 0 \leq t \leq 2$$

Express the average concentration during the first 30 seconds after the injection began as a definite integral. Do not evaluate.

44. *Medicine—circulation.* Repeat Problem 43 for the second 30 seconds after the injection began.

45. *Medicine—respiration.* Physiologists use a machine called a pneumotachograph to produce a graph of the rate of flow $R(t)$ of air into the lungs (inspiration) and out of the lungs (expiration). The figure gives the graph of the inspiration phase of the breathing cycle for an individual at rest. The

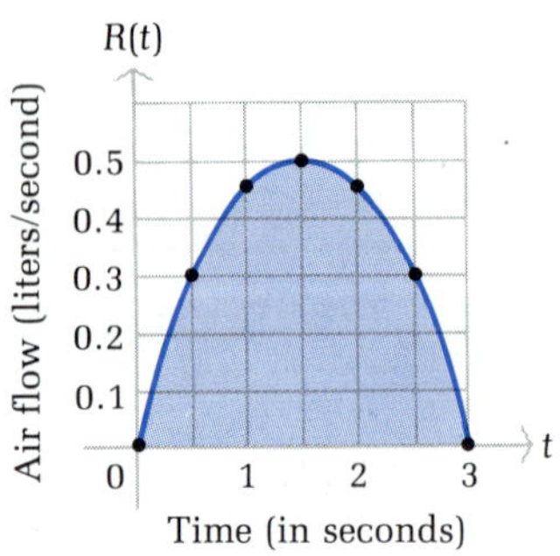

area under this graph represents the total volume of air inhaled during the inspiration phase. Use a Riemann sum with $n = 3$, $\Delta t = 1$, and c_k the midpoint of each subinterval to approximate the area under the graph. Estimate the necessary function values from the graph.

46. *Medicine—respiration.* Repeat Problem 45 using $n = 6$, $\Delta t = 0.5$, and c_k the right end point of each subinterval.

Social Sciences

47. *Population composition.* Because of various factors (such as birth rate expansion, then contraction; family flights from urban areas; etc.), the number of children in a large city was found to increase and then decrease rather drastically (see the figure). The number of children (in millions) was found to be given approximately by

$$N(t) = -\tfrac{1}{4}t^2 + t + 4 \qquad 0 \le t \le 6$$

Express the average number of children in the city during this 6 year period in terms of a definite integral. Do not evaluate.

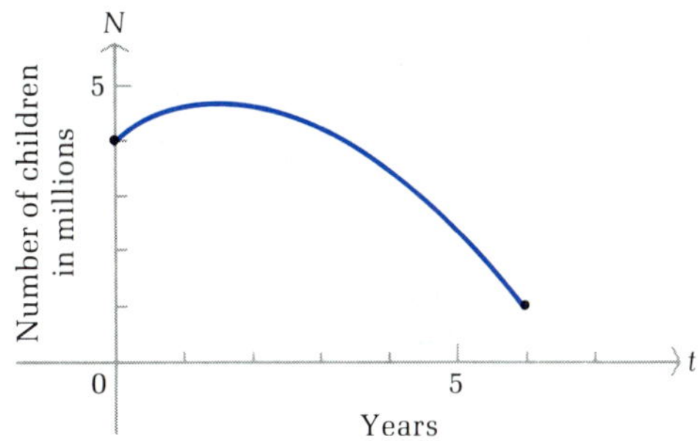

SECTION 6-6 The Fundamental Theorem of Calculus

- THE FUNDAMENTAL THEOREM OF CALCULUS
- USING THE FUNDAMENTAL THEOREM
- DEFINITE INTEGRALS AND SUBSTITUTION
- APPLICATIONS
- APPROXIMATING DEFINITE INTEGRALS

THE FUNDAMENTAL THEOREM OF CALCULUS

We now know what a definite integral is and how to recognize one in a practical problem by observing the presence of a certain kind of limit of a sum. Its exact evaluation by direct computation of the limit is a tedious process at best; and it is only because of special summation formulas that we are able to succeed at all. But now we introduce a theorem that will show us how a definite integral

$\int_a^b f(x)\,dx$ can be evaluated exactly if the indefinite integral $\int f(x)\,dx$ can be found.

Recall that

$$\int f(x)\,dx = F(x) + C$$

where $F(x)$ is any function such that

$$F'(x) = f(x)$$

That is, F is an antiderivative of f.

THEOREM 2

Fundamental Theorem of Calculus

If f is a continuous function on the closed interval $[a, b]$ and F is any antiderivative of f, then

$$\int_a^b f(x)\,dx = F(x)\Big|_a^b = F(b) - F(a) \qquad F'(x) = f(x)$$

The symbol $F(x)|_a^b$ is used to represent the **net change** in $F(x)$ from $x = a$ to $x = b$ and is included as a convenient intermediate step in the evaluation of a definite integral by the fundamental theorem.

To gain some insight into the fundamental theorem, we give a brief outline of the proof. A complete proof can be found in more advanced texts. For the purpose of this discussion, we will assume that $f(x) \geq 0$ on $[a, b]$ so that we can interpret the definite integral as area.

Proof We start by defining a function $A(x)$ called the *area function*. For $a \leq x \leq b$, let $A(x)$ be the area under the graph of f from a to x, as indicated in Figure 7:

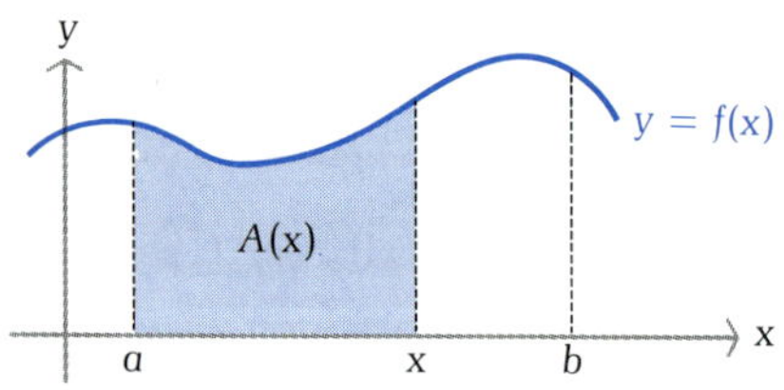

$A(x) =$ Area from a to x

$A(b) =$ Area from a to b

$A(a) =$ Area from a to $a = 0$

FIGURE 7

Note that $A(a) = 0$ and $A(b)$ is the area under the graph of f from $x = a$ to $x = b$.

But we have already seen that this area is given by the definite integral of f from $x = a$ to $x = b$ (Section 6-5). Thus,

$$\int_a^b f(x)\,dx = A(b)$$

The next step is to show that $A(x)$ is an antiderivative of f; that is, $A'(x) = f(x)$. To do this, we use the definition of derivative in increment form (Section 4-5) and write

$$A'(x) = \lim_{\Delta x \to 0} \frac{A(x + \Delta x) - A(x)}{\Delta x}$$

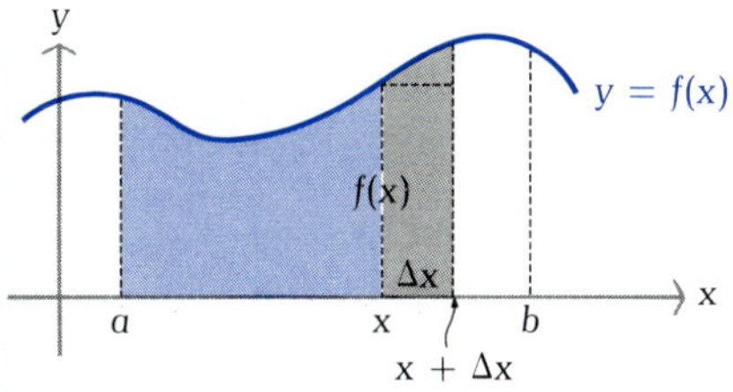

FIGURE 8

If we assume $\Delta x > 0$, then $A(x + \Delta x) - A(x)$ is the area from x to $x + \Delta x$ (see Figure 8). This area is given approximately by the area of the rectangle $\Delta x \cdot f(x)$, and the smaller Δx is, the better the approximation.

Using

$$A(x + \Delta x) - A(x) \approx \Delta x \cdot f(x)$$

and dividing both sides by Δx, we obtain

$$\frac{A(x + \Delta x) - A(x)}{\Delta x} \approx f(x)$$

Now, if we let $\Delta x \to 0$, then the left side has $A'(x)$ as a limit, which is equal to the right side. Hence,

$$A'(x) = f(x)$$

that is, $A(x)$ is an antiderivative of $f(x)$. Finally, let $F(x)$ be any antiderivative of $f(x)$. Then there exists a constant C so that (Theorem 1 in Section 6-1)

$$F(x) = A(x) + C$$

Thus,

$$\begin{aligned} F(b) - F(a) &= [A(b) + C] - [A(a) + C] && A(a) = 0 \\ &= A(b) && A(b) \text{ is the area from } x = a \text{ to } x = b. \\ &= \int_a^b f(x)\,dx \end{aligned}$$

which is the statement in the fundamental theorem.

◆ USING THE FUNDAMENTAL THEOREM

You now see why we discussed indefinite integration in the earlier sections of this chapter. To use the fundamental theorem, you must be able to find antiderivatives.

EXAMPLE 25 Use the fundamental theorem to evaluate $\int_0^3 (x^2 - 4)\, dx$.

Solution We choose the simplest antiderivative of $x^2 - 4$, namely $(x^3/3) - 4x$, since any antiderivative will do:

$$\int_0^3 (x^2 - 4)\, dx = \left(\frac{x^3}{3} - 4x\right)\Big|_0^3$$

Evaluate at $x = 3$ and at $x = 0$; then subtract.

$$= \left(\frac{3^3}{3} - 4 \cdot 3\right) - \left(\frac{0^3}{3} - 4 \cdot 0\right)$$

$$= -3 - 0 = -3$$

(Compare this process with the direct evaluation by the definition in Example 21, Section 6-5.) ◆

PROBLEM 25 Use the fundamental theorem to evaluate $\int_0^8 (x - 6)\, dx$. (Compare with Problem 21, Section 6-5.) ◆

EXAMPLE 26 Evaluate: $\displaystyle\int_1^2 \left(2x + 3e^x - \frac{4}{x}\right) dx$

Solution

$$\int_1^2 \left(2x + 3e^x - \frac{4}{x}\right) dx = (x^2 + 3e^x - 4\ln|x|)\Big|_1^2$$

$$= (2^2 + 3e^2 - 4\ln 2) - (1^2 + 3e^1 - 4\ln 1)$$

$$= 3 + 3e^2 - 3e - 4\ln 2 \approx 14.24$$ ◆

PROBLEM 26 Evaluate: $\displaystyle\int_1^3 \left(4x - 2e^x + \frac{5}{x}\right) dx$ ◆

DEFINITE INTEGRALS AND SUBSTITUTION

The evaluation of a definite integral is a two-step process: First, find an antiderivative, and then find the net change in that antiderivative. If substitution techniques are required to find the antiderivative, there are two different ways to proceed. The next example illustrates both methods.

EXAMPLE 27 Evaluate: $\displaystyle\int_0^5 \frac{x}{x^2 + 10}\, dx$

Solution We will solve this problem using substitution in two different ways:

Method 1. Use substitution in an indefinite integral to find an antiderivative as a function of x; then evaluate the definite integral:

$$\int \frac{x}{x^2+10}\,dx = \frac{1}{2}\int \frac{1}{x^2+10}\,2x\,dx \qquad \text{Substitute } u = x^2+10 \text{ and } du = 2x\,dx.$$

$$= \frac{1}{2}\int \frac{1}{u}\,du$$

$$= \tfrac{1}{2}\ln|u| + C$$

$$= \tfrac{1}{2}\ln(x^2+10) + C \qquad \text{Since } u = x^2+10 > 0$$

We choose $C = 0$ and use the antiderivative $\frac{1}{2}\ln(x^2+10)$ to evaluate the definite integral:

$$\int_0^5 \frac{x}{x^2+10}\,dx = \frac{1}{2}\ln(x^2+10)\Big|_0^5$$

$$= \tfrac{1}{2}\ln 35 - \tfrac{1}{2}\ln 10 \approx 0.626$$

Method 2. Substitute directly in the definite integral, changing both the variable of integration and the limits of integration: In the definite integral

$$\int_0^5 \frac{x}{x^2+10}\,dx$$

the upper limit is $x = 5$ and the lower limit is $x = 0$. When we make the substitution $u = x^2 + 10$ in this definite integral, we must change the limits of integration to the corresponding values of u:

$x = 5$	implies	$u = 5^2 + 10 = 35$	New upper limit
$x = 0$	implies	$u = 0^2 + 10 = 10$	New lower limit

Thus, we have

$$\int_0^5 \frac{x}{x^2+10}\,dx = \frac{1}{2}\int_0^5 \frac{1}{x^2+10}\,2x\,dx$$

$$= \frac{1}{2}\int_{10}^{35} \frac{1}{u}\,du$$

$$= \frac{1}{2}\left(\ln|u|\Big|_{10}^{35}\right)$$

$$= \tfrac{1}{2}(\ln 35 - \ln 10) \approx 0.626$$

◆

PROBLEM 27 Use both methods described in Example 27 to evaluate: $\int_0^1 \frac{1}{2x+4}\,dx$ ◆

◆ EXAMPLE 28 Use method 2 described in Example 27 to evaluate: $\int_{-4}^1 \sqrt{5-t}\,dt$

Solution If $u = 5 - t$, then $du = -dt$, and

$t = 1$	implies	$u = 5 - 1 = 4$	New upper limit
$t = -4$	implies	$u = 5 - (-4) = 9$	New lower limit

Notice that the lower limit for u is larger than the upper limit. Be careful not to reverse these two values when substituting in the definite integral.

$$\begin{aligned}\int_{-4}^{1} \sqrt{5-t}\, dt &= -\int_{-4}^{1} \sqrt{5-t}\,(-dt)\\ &= -\int_{9}^{4} \sqrt{u}\, du\\ &= -\int_{9}^{4} u^{1/2}\, du\\ &= -\left(\frac{u^{3/2}}{\frac{3}{2}}\Big|_9^4\right)\\ &= -\left[\tfrac{2}{3}(4)^{3/2} - \tfrac{2}{3}(9)^{3/2}\right]\\ &= -\left[\tfrac{16}{3} - \tfrac{54}{3}\right] = \tfrac{38}{3} \approx 12.667\end{aligned}$$

◆

PROBLEM 28 Use method 2 described in Example 27 to evaluate: $\displaystyle\int_{2}^{5} \frac{1}{\sqrt{6-t}}\, dt$ ◆

◆ APPLICATIONS

As we indicated earlier, in practical problems we can often recognize that a definite integral is involved in the solution by observing the presence of a particular kind of limit of a sum. If the form

$$\lim_{n\to\infty} \sum_{k=1}^{n} f(c_k)\Delta x_k$$

is involved in a problem, then we are dealing with a definite integral, providing f, c_k, and x_k meet the conditions in the definition of a definite integral. If f has an antiderivative F, then the limit can be determined using the fundamental theorem. In summary:

Definite Integral Definition and the Fundamental Theorem

Definition ↓   Fundamental theorem ↓

$$\lim_{n\to\infty} \sum_{k=1}^{n} f(c_k)\Delta x_k = \int_a^b f(x)\, dx = F(b) - F(a) \qquad F'(x) = f(x) \text{ on } [a, b]$$

◆ EXAMPLE 29 Let us reconsider Example 20 in Section 6-4 in light of the fundamental theorem. Find the exact area between the x axis and the graph of $f(x) = x^2 + 3$ from $x = 1$ to $x = 5$.

Solution

$$\Delta x_k = \frac{5-1}{n} \qquad c_k = 1 + k\Delta x_k$$

Think

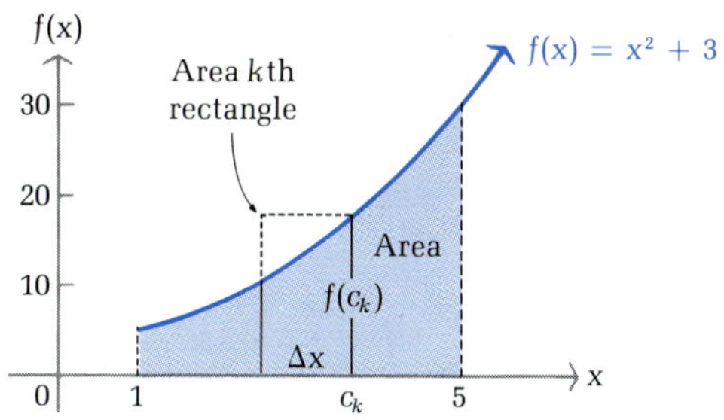

$$\text{Area} = \lim_{n\to\infty} \sum_{k=1}^{n} f(c_k)\Delta x_k = \int_1^5 f(x)\,dx$$

The area is given by the definite integral

$$\int_1^5 (x^2+3)\,dx$$

and this integral can be easily evaluated using the fundamental theorem. Thus, we have

$$\text{Area} = \int_1^5 (x^2+3)\,dx = \left(\frac{x^3}{3} + 3x\right)\Bigg|_1^5$$

$$= \left(\frac{5^3}{3} + 3\cdot 5\right) - \left(\frac{1^3}{3} + 3\cdot 1\right)$$

$$= \frac{160}{3}$$

◆

PROBLEM 29 Set up a definite integral for the area between the x axis and the graph of $f(x) = 3x^2 + 2x$ from $x = 2$ to $x = 4$. Evaluate it. ◆

We now consider an application related to average value, which was defined in Section 6-5.

◆ EXAMPLE 30 *Advertising*

A metropolitan newspaper currently has a daily circulation of 50,000 papers (weekdays and Sunday). The management of the paper decides to initiate an aggressive advertising campaign to increase circulation. Suppose that the daily circulation (in thousands of papers) t days after the beginning of the campaign is given by

$$S(t) = 100 - 50e^{-0.01t}$$

What is the average daily circulation during the first 30 days of the campaign?

Solution

$$\begin{pmatrix}\text{Average daily circulation}\\ \text{over } [a, b] = [0, 30]\end{pmatrix} = \frac{1}{b-a}\int_a^b S(t)\,dt$$

$$= \frac{1}{30}\int_0^{30} (100 - 50e^{-0.01t})\,dt$$

$$= \frac{1}{30}(100t + 5{,}000e^{-0.01t})\Big|_0^{30}$$

$$= \frac{1}{30}(3{,}000 + 5{,}000e^{-0.3} - 5{,}000)$$

$$\approx 56.8 \quad \text{or } 56{,}800 \text{ papers}$$

◆

PROBLEM 30 Refer to Example 30. Satisfied with the increase in circulation, management decides to terminate the advertising campaign. Suppose that the daily circulation (in thousands of papers) t days after the end of the advertising campaign is given by

$$S(t) = 65 + 8e^{-0.02t}$$

What is the average daily circulation during the first 30 days after the end of the campaign? ◆

In Section 6-1 we saw that given the instantaneous rate of change (derivative) of a function, antidifferentiation can be used to find the function. Now that the fundamental theorem has established a relationship between definite integrals and antiderivatives, we can use definite integrals to solve problems of this type.

◆ EXAMPLE 31 *Mining*

Using data from the first 5 years of production, the management of a mining company has determined that a gold mine produces ore at a rate given by

$$Q'(t) = 600 - 20t$$

where $Q(t)$ is the total amount of ore (in thousands of tons) produced after t years of operating the mine. Assuming that the mine continues to produce $Q'(t)$ thousands of tons per year, how much ore will be produced during the next 5 years of operation?

Solution The total production after 5 years is $Q(5)$, the total production after 10 years is $Q(10)$, and the total production during the second 5 years of operating the mine is given by $Q(10) - Q(5)$. Since we are given $Q'(t)$ and since $Q(t)$ is an antiderivative of $Q'(t)$ (think about this), we can use the fundamental theorem to find this quantity:

$$\begin{aligned} Q(10) - Q(5) &= \int_5^{10} Q'(t)\, dt \\ &= \int_5^{10} (600 - 20t)\, dt \\ &= (600t - 10t^2)\Big|_5^{10} \\ &= [600(10) - 10(10)^2] - [600(5) - 10(5^2)] \\ &= 5{,}000 - 2{,}750 = 2{,}250 \quad \text{or } 2{,}250{,}000 \text{ tons} \end{aligned}$$

◆

PROBLEM 31 A silver mine is producing ore at the rate of $Q'(t) = 900 - 30t$ thousand tons per year. Find the total amount of ore produced from the end of the tenth year to the end of the fifteenth year of production. ◆

Remark **Net Change in a Function**

In general, if $F'(x) = f(x)$ and the fundamental theorem is restated in terms of F' instead of f, we obtain

$$\int_a^b F'(x)\, dx = F(b) - F(a)$$

Thus, **the definite integral of the rate of change of a function (the derivative) from *a* to *b* gives the net change in the function from *a* to *b*.**

◆ APPROXIMATING DEFINITE INTEGRALS

The fundamental theorem of calculus is a powerful tool for evaluating definite integrals, but it has some limitations. In order to use this theorem, we must be able to find the antiderivative of the integrand. But suppose we cannot find an antiderivative (it may not even exist in a convenient or closed form). In this case, we simply return to the definition of the definite integral and use the Riemann sum with a sufficiently large n (and a computer) to approximate $\int_a^b f(x)\, dx$. This method is called **numerical integration,** and using the Riemann sum approximation

$$\sum_{k=1}^{n} f(c_k)\Delta x_k \approx \int_a^b f(x)\, dx \qquad n \text{ sufficiently large}$$

is usually referred to as the **rectangle rule for numerical integration.** The computer supplement for this text (see the Preface) contains a program that can be used to compute Riemann sums.

To illustrate this approach, consider the definite integral

$$\int_0^1 e^{-x^2}\,dx$$

Definite integrals involving the function $f(x) = e^{-x^2}$ are very important in statistics and always must be approximated since f does not have an antiderivative that can be expressed in a closed form in terms of elementary functions. The Riemann sums obtained by dividing the interval $[0, 1]$ into n equal subintervals and choosing c_k as the midpoint of each subinterval provide approximate values for $\int_0^1 e^{-x^2}\,dx$. Table 2 lists these Riemann sums for selected values of n. Examining the values in Table 2, we conclude that

$$\int_0^1 e^{-x^2}\,dx \approx 0.746\ 831$$

TABLE 2

Riemann Sum Approximations of $\int_0^1 e^{-x^2}\,dx$

n	RIEMANN SUM
8	0.747 303
16	0.746 943
32	0.746 854
64	0.746 831

The Riemann sums in Table 2 were calculated using the program in the computer supplement. If you have any programming experience, you should have no trouble programming a graphic calculator or a computer to calculate Riemann sums.

Common Errors

1. $$\int_0^2 e^x\,dx = e^x\Big|_0^2 \neq e^2$$

Do not forget to evaluate the antiderivative at both the upper and lower limits of integration, and do not assume that the antiderivative is 0 just because the lower limit is 0. The correct procedure for this integral is

$$\int_0^2 e^x\,dx = e^x\Big|_0^2 = e^2 - e^0 = e^2 - 1$$

2. $$\int_2^5 \frac{1}{2x+3}\,dx \neq \frac{1}{2}\int_2^5 \frac{1}{u}\,du \qquad u = 2x + 3,\ du = 2\,dx$$

If a substitution is made in a definite integral, the limits of integration also must be changed. The new limits are determined by the particular substitution used in the integral. The correct procedure for this integral is

$$\int_2^5 \frac{1}{2x+3}\,dx = \frac{1}{2}\int_7^{13} \frac{1}{u}\,du \qquad \begin{array}{l} x = 5 \text{ implies } u = 2(5) + 3 = 13 \\ x = 2 \text{ implies } u = 2(2) + 3 = 7 \end{array}$$

$$= \frac{1}{2}\left(\ln|u|\Big|_7^{13}\right)$$

$$= \tfrac{1}{2}(\ln 13 - \ln 7)$$

Answers to Matched Problems

25. -16 26. $16 - 2e^3 + 2e + 5 \ln 3 \approx -13.24$
27. $\frac{1}{2}(\ln 6 - \ln 4) \approx 0.203$ 28. 2 29. $\int_2^4 (3x^2 + 2x)\,dx = 68$
30. $(2{,}350 - 400e^{-0.6})/30 \approx 71.0$ or 71,000 papers 31. 2,625,000 tons

EXERCISE 6-6

A *Evaluate.*

1. $\int_2^3 2x\,dx$
2. $\int_1^2 3x^2\,dx$
3. $\int_3^4 5\,dx$
4. $\int_{12}^{20} dx$
5. $\int_1^3 (2x - 3)\,dx$
6. $\int_1^3 (6x + 5)\,dx$
7. $\int_0^4 (3x^2 - 4)\,dx$
8. $\int_0^2 (6x^2 - 2x)\,dx$
9. $\int_{-3}^4 (4 - x^2)\,dx$
10. $\int_{-1}^2 (x^2 - 4x)\,dx$
11. $\int_0^1 24x^{11}\,dx$
12. $\int_0^2 30x^5\,dx$
13. $\int_0^1 e^{2x}\,dx$
14. $\int_{-1}^1 e^{5x}\,dx$
15. $\int_1^{3.5} 2x^{-1}\,dx$
16. $\int_1^2 \frac{dx}{x}$

Find the area between the graph of $y = f(x)$ and the x axis over the indicated interval.

17. $f(x) = 4 - x^2;\quad -1 \leqslant x \leqslant 2$
18. $f(x) = 12 - 3x^2;\quad -2 \leqslant x \leqslant 1$
19. $f(x) = e^x;\quad -1 \leqslant x \leqslant 2$
20. $f(x) = e^{-x};\quad -2 \leqslant x \leqslant 1$
21. $f(x) = \frac{1}{x};\quad 0.5 \leqslant x \leqslant 1$
22. $f(x) = \frac{1}{x};\quad 0.1 \leqslant x \leqslant 1$

B *Evaluate.* substitution ?

23. $\int_1^2 (2x^{-2} - 3)\,dx$
24. $\int_1^2 (5 - 16x^{-3})\,dx$
25. $\int_1^4 3\sqrt{x}\,dx$
26. $\int_4^{25} \frac{2}{\sqrt{x}}\,dx$
27. $\int_2^3 12(x^2 - 4)^5 x\,dx$
28. $\int_0^1 32(x^2 + 1)^7 x\,dx$
29. $\int_3^9 \frac{1}{x - 1}\,dx$
30. $\int_2^8 \frac{1}{x + 1}\,dx$
31. $\int_{-5}^{10} e^{-0.05x}\,dx$

32. $\int_{-10}^{25} e^{-0.01x}\,dx$

33. $\int_{-6}^{0} \sqrt{4-2x}\,dx$

34. $\int_{-4}^{2} \frac{1}{\sqrt{8-2x}}\,dx$

35. $\int_{-1}^{7} \frac{x}{\sqrt{x+2}}\,dx$

36. $\int_{0}^{3} x\sqrt{x+1}\,dx$

37. $\int_{0}^{1} (e^{2x}-2x)^2(e^{2x}-1)\,dx$

38. $\int_{0}^{1} \frac{2e^{4x}-3}{e^{2x}}\,dx$

39. $\int_{-2}^{-1} (x^{-1}+2x)\,dx$

40. $\int_{-3}^{-1} (-3x^{-2}+x^{-1})\,dx$

Find the average value of each function over the indicated interval.

41. $f(x) = 500 - 50x$; $[0, 10]$

42. $g(x) = 2x + 7$; $[0, 5]$

43. $f(t) = 3t^2 - 2t$; $[-1, 2]$

44. $g(t) = 4t - 3t^2$; $[-2, 2]$

45. $f(x) = \sqrt[3]{x}$; $[1, 8]$

46. $g(x) = \sqrt{x+1}$; $[3, 8]$

47. $f(x) = 4e^{-0.2x}$; $[0, 10]$

48. $f(x) = 64e^{0.08x}$; $[0, 10]$

Write in the form $\int_a^b f(x)\,dx$ *and evaluate using the fundamental theorem of calculus.*

49. $\lim_{n\to\infty} \sum_{k=1}^{n} (1 - c_k^2)\frac{5-2}{n}$; $c_k = 2 + k\frac{3}{n}$

50. $\lim_{n\to\infty} \sum_{k=1}^{n} (c_k^2 - 3)\frac{10-0}{n}$; $c_k = 0 + k\frac{10}{n}$

51. $\lim_{n\to\infty} \sum_{k=1}^{n} (3c_k^2 - 2c_k + 3)\frac{12-2}{n}$; $c_k = 2 + k\frac{10}{n}$

52. $\lim_{n\to\infty} \sum_{k=1}^{n} (4c_k^3 + 3c_k^2 - 5)\frac{2-1}{n}$; $c_k = 1 + k\frac{1}{n}$

C

Evaluate.

53. $\int_{2}^{3} x\sqrt{2x^2-3}\,dx$

54. $\int_{0}^{1} x\sqrt{3x^2+2}\,dx$

55. $\int_{0}^{1} \frac{x-1}{x^2-2x+3}\,dx$

56. $\int_{1}^{2} \frac{x+1}{2x^2+4x+4}\,dx$

57. $\int_{-1}^{1} \frac{e^{-x}-e^{x}}{(e^{-x}+e^{x})^2}\,dx$

58. $\int_{6}^{7} \frac{\ln(t-5)}{t-5}\,dt$

59. Find the average value of $f'(x)$ over the interval $[a, b]$ for any differentiable function f.

60. Show that the average value of $f(x) = Ax + B$ over the interval $[a, b]$ is

$$f\left(\frac{a+b}{2}\right)$$

APPLICATIONS

Business & Economics

61. *Inventory.* A store orders 600 units of a product every 3 months. If the product is steadily depleted to 0 by the end of each 3 months, the inventory on hand, I, at any time t during the year is illustrated as follows:

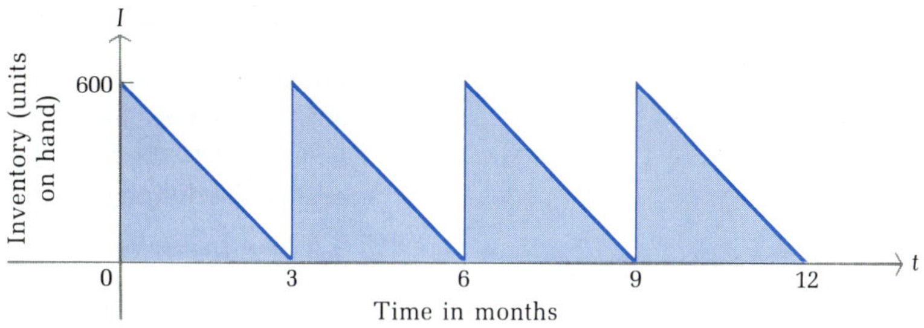

(A) Write an inventory function (assume it is continuous) for the first 3 months. [The graph is a straight line joining (0, 600) and (3, 0).]

(B) What is the average number of units on hand for a 3 month period?

62. *Inventory.* Repeat Problem 61 with an order of 1,200 units every 4 months.

63. *Salvage value.* A new piece of industrial equipment will depreciate in value rapidly at first, then less rapidly as time goes on. Suppose the rate (in dollars per year) at which the book value of a new milling machine changes is given approximately by

$$V'(t) = f(t) = 500(t - 12) \qquad 0 \leq t \leq 10$$

where $V(t)$ is the value of the machine after t years. What is the total loss in value of the machine in the first 5 years? In the second 5 years? Set up appropriate integrals and solve.

64. *Maintenance costs.* Maintenance costs for an apartment house generally increase as the building gets older. From past records, a managerial service determines that the rate of increase in maintenance costs (in dollars per year) for a particular apartment complex is given approximately by

$$M'(x) = f(x) = 90x^2 + 5{,}000$$

where x is the age of the apartment complex in years and $M(x)$ is the total (accumulated) cost of maintenance for x years. Write a definite integral that will give the total maintenance costs from 2 to 7 years after the apartment house was built, and evaluate it.

65. *Average cost.* The total cost (in dollars) of manufacturing x auto body frames is

$$C(x) = 60{,}000 + 300x$$

(A) Find the average cost per unit if 500 frames are produced. [*Hint:* Recall that $\overline{C}(x)$ is the average cost per unit.]
(B) Find the average value of the cost function over the interval [0, 500].

66. *Average cost.* The total cost (in dollars) of printing x dictionaries is

$$C(x) = 20{,}000 + 10x$$

(A) Find the average cost per unit if 1,000 dictionaries are produced.
(B) Find the average value of the cost function over the interval [0, 1,000].

67. *Marketing.* An automobile company is ready to introduce a new line of cars with a national sales campaign. After test marketing the line in a carefully selected city, the marketing research department estimates that sales (in millions of dollars) will increase at the monthly rate of

$$S'(t) = 10 - 10e^{-0.1t} \qquad 0 \leq t \leq 24$$

t months after the national campaign is started. What will be the approximate total sales during the first 12 months of the campaign? The second 12 months of the campaign?

68. *Marketing.* Repeat Problem 67 if the monthly rate of increase in sales is found to be approximated by

$$S'(t) = 20 - 20e^{-0.05t} \qquad 0 \leq t \leq 24$$

69. *Continuous compound interest.* If $100 is deposited in an account that earns interest at an annual nominal rate of 8% compounded continuously, find the amount in the account after 5 years and the average amount in the account during this 5 year period. (See Section 5-1 for a discussion of continuous compound interest.)

70. *Continuous compound interest.* If $500 is deposited in an account that earns interest at an annual nominal rate of 12% compounded continuously, find the amount in the account after 10 years and the average amount in the account during this 10 year period.

71. *Advertising.* The number of hamburgers (in thousands) sold each day by a chain of restaurants t days after the beginning of an advertising campaign is given by

$$S(t) = 20 - 10e^{-0.1t}$$

What is the average number of hamburgers sold each day during the first week of the advertising campaign? During the second week of the campaign?

72. *Advertising.* The number of hamburgers (in thousands) sold each day by another chain of restaurants t days after the end of an advertising campaign is given by

$$S(t) = 10 + 8e^{-0.2t}$$

What is the average number of hamburgers sold each day during the first week after the end of the advertising campaign? During the second week after the end of the campaign?

73. *Oil production.* Using data from the first 3 years of production as well as geological studies, the management of an oil company estimates that oil will be pumped from a producing field at a rate given by

$$R(t) = \frac{100}{t+1} + 5 \qquad 0 \leq t \leq 20$$

where $R(t)$ is the rate of production in thousands of barrels per year t years after pumping begins. Approximately how many barrels of oil will the field produce during the first 10 years of production? From the end of the 10th year to the end of the 20th year of production?

74. *Oil production.* In Problem 73, if the rate is found to be

$$R(t) = \frac{120t}{t^2+1} + 3 \qquad 0 \leq t \leq 20$$

approximately how many barrels of oil will the field produce during the first 5 years of production? The second 5 years of production?

Life Sciences

75. *Natural resource depletion.* The instantaneous rate of change of demand for wood in the United States since 1970 ($t = 0$) in billions of cubic feet per year is estimated to be given by

$$Q'(t) = 12 + 0.006t^2 \qquad 0 \leq t \leq 50$$

where $Q(t)$ is the total amount of wood consumed (in billions of cubic feet) t years after 1970. How many billions of cubic feet of wood will be consumed from 2000 to 2010?

76. *Medicine.* The rate of healing for a skin wound (in square centimeters per day) is approximated by $A'(t) = -0.9e^{-0.1t}$. The initial wound has an area of 9 square centimeters. How much will the area change during the first 5 days? The second 5 days?

77. *Temperature.* If the temperature $C(t)$ in an aquarium was made to change according to

$$C(t) = t^3 - 2t + 10 \qquad 0 \leq t \leq 2$$

(in degrees Celsius) over a 2 hour period, what is the average temperature over this period?

78. *Medicine.* A drug is injected into the bloodstream of a patient through her right arm. The concentration of the drug in the bloodstream of the left arm t hours after the injection is given by

$$C(t) = \frac{0.14t}{t^2 + 1}$$

What is the average concentration of the drug in the bloodstream of the left arm during the first hour after the injection? During the first 2 hours after the injection?

Social Sciences

79. *Learning.* In a particular business college, it was found that an average student enrolled in an advanced typing class progressed at a rate of $N'(t) = 6e^{-0.1t}$ words per minute per week, t weeks after enrolling in a 15 week course. At the beginning of the course an average student could type 40 words per minute. How much improvement would be expected during the first 5 weeks of the course? The second 5 weeks of the course? The last 5 weeks of the course?

80. *Learning.* In the same business college, it was also found that an average student enrolled in a beginning shorthand class progressed at a rate of $N'(t) = 12e^{-0.06t}$ words per minute per week, t weeks after enrolling in a 15 week course. At the beginning of the course none of the students could take any dictation by shorthand. How much improvement would be expected during the first 5 weeks of the course? The second 5 weeks of the course? The last 5 weeks of the course?

81. *Politics.* Public awareness of a Congressional candidate before and after a successful campaign was approximated by

$$P(t) = \frac{8.4t}{t^2 + 49} + 0.1 \qquad 0 \leq t \leq 24$$

where t is time in months after the campaign started and $P(t)$ is the fraction of people in the Congressional district who could recall the candidate's name. What is the average fraction of people who could recall the candidate's name during the first 7 months after the campaign began? During the first 2 years after the campaign began?

82. *Public transportation.* Bus ridership (in millions of riders) on a metropolitan transit system over a 4 year period was found to be given approximately by

$$R(t) = 2t^3 - 15t^2 + 22t + 50 \qquad 0 \leq t \leq 4$$

What was the average number of bus riders over this 4 year time period?

Use a graphic calculator or a computer for Problems 83–90. For a given problem, graph each pair of equations in the same viewing rectangle. The symbol $\bar{y}$ represents the average value of the first function over the indicated interval.

83. From Problem 69:
$y = 100e^{0.08x}$
$\bar{y} = 122.96$
$0 \le x \le 5$

84. From Problem 70:
$y = 500e^{0.12x}$
$\bar{y} = 966.72$
$0 \le x \le 10$

85. From Problem 71:
$y = 20 - 10e^{-0.1x}$
$\bar{y} = 12.8$
$0 \le x \le 7$

86. From Problem 72:
$y = 10 + 8e^{-0.2x}$
$\bar{y} = 14.3$
$0 \le x \le 7$

87. From Problem 77:
$y = x^3 - 2x + 10$
$\bar{y} = 10$
$0 \le x \le 2$

88. From Problem 78:
$$y = \frac{0.14x}{x^2 + 1}$$
$\bar{y} = 0.0485$
$0 \le x \le 1$

89. From Problem 81:
$$y = \frac{8.4x}{x^2 + 49} + 0.1$$
$\bar{y} = 0.516$
$0 \le x \le 7$

90. From Problem 82:
$y = 2x^3 - 15x^2 + 22x + 50$
$\bar{y} = 46$
$0 \le x \le 4$

SECTION 6-7 Chapter Review

Important Terms and Symbols

6-1 *Antiderivatives and Indefinite Integrals.* Antiderivative; indefinite integral; integral sign; integrand; constant of integration

$$\int f(x)\,dx$$

6-2 *Integration by Substitution.* General integral formulas; method of substitution

6-3 *Differential Equations—Growth and Decay.* Differential equation; continuous compound interest; exponential growth law; population growth; radioactive decay

$$\frac{dQ}{dt} = rQ; \quad Q = Q_0e^{rt}$$

6-4 *Area Under a Curve.* Summation notation; summing index; summation notation properties; special summation formulas; area under a curve; rectangle method for finding area; closed form

$$\sum_{k=1}^{n} f(k); \quad \sum_{k=1}^{n} k = \frac{n(n+1)}{2}; \quad \sum_{k=1}^{n} k^2 = \frac{n(n+1)(2n+1)}{6}$$

6-5 *Definite Integrals.* Definite integral; integrand; upper limit; lower limit; Riemann sum; average value of a continuous function; definite integral properties; tabular functions

$$\sum_{k=1}^{n} f(c_k)\Delta x_k; \quad \int_a^b f(x)\,dx$$

6-6 *The Fundamental Theorem of Calculus.* The fundamental theorem of calculus; net change in a function; numerical integration; rectangle rule

$$F(x)\Big|_a^b = F(b) - F(a)$$

Integration Formulas

$$\int k\,dx = kx + C$$

$$\int kf(x)\,dx = k\int f(x)\,dx$$

$$\int [f(x) \pm g(x)]\,dx = \int f(x)\,dx \pm \int g(x)\,dx$$

$$\int u^n\,du = \frac{u^{n+1}}{n+1} + C \qquad n \neq -1$$

$$\int e^u\,du = e^u + C$$

and

$$\int e^{au}\,du = \frac{1}{a}e^{au} + C \qquad a \neq 0$$

$$\int \frac{1}{u}\,du = \ln|u| + C \qquad u \neq 0$$

EXERCISE 6-7

Chapter Review

Work through all the problems in this chapter review and check your answers in the back of the book. (Answers to all review problems are there.) Where weaknesses show up, review appropriate sections in the text.

The following special summation formulas will be of use in some of the problems in this exercise set.

$$\sum_{k=1}^{n} c = cn \qquad \sum_{k=1}^{n} k = \frac{n(n+1)}{2} \qquad \sum_{k=1}^{n} k^2 = \frac{n(n+1)(2n+1)}{6}$$

A *Find each integral in Problems 1–6.*

1. $\int (3t^2 - 2t)\,dt$ **2.** $\int_2^5 (2x-3)\,dx$ **3.** $\int (3t^{-2} - 3)\,dt$

4. $\int_1^4 x\,dx$

5. $\int e^{-0.5x}\,dx$

6. $\int_1^5 \frac{2}{u}\,du$

7. Find a function $y = f(x)$ that satisfies both conditions:

$$\frac{dy}{dx} = 3x^2 - 2 \qquad f(0) = 4$$

8. Write in expanded form without summation notation and evaluate:

$$\sum_{k=1}^{5} (4k + 3)$$

9. For $f(x) = x^2 + 7$, $a = 1$, $b = 3$, $n = 4$, and $\Delta x = (b - a)/n$, find

$$\sum_{k=1}^{n} f(a + k\Delta x)\Delta x$$

x	f(x)
3	1.2
7	3.4
11	2.6
15	0.5

10. Use the table of values in the margin and a Riemann sum with $n = 4$ and c_k the midpoint of each subinterval to approximate $\int_1^{17} f(x)\,dx$.

11. Find the area between the graph of $f(x) = 3x^2 + 1$ and the x axis for $-1 \leqslant x \leqslant 2$.

12. Find the average value of $f(x) = 6x^2 + 2x$ over the interval $[-1, 2]$.

B *Find each integral in Problems 13–18.*

13. $\int \sqrt[3]{6x - 5}\,dx$

14. $\int_0^1 10(2x - 1)^4\,dx$

15. $\int \left(\frac{2}{x^2} - 2xe^{x^2}\right) dx$

16. $\int_0^4 \sqrt{x^2 + 4}\,x\,dx$

17. $\int (e^{-2x} + x^{-1})\,dx$

18. $\int_0^{10} 10e^{-0.02x}\,dx$

19. Find a function $y = f(x)$ that satisfies both conditions:

$$\frac{dy}{dx} = 3x^{-1} - x^{-2} \qquad f(1) = 5$$

20. Find the equation of the curve that passes through (2, 10) if its slope is given by

$$\frac{dy}{dx} = 6x + 1 \qquad \text{for each } x$$

21. Use properties of summation notation and special summation formulas to find

$$\sum_{k=1}^{100} (1 + k + k^2)$$

22. Find a closed form for the summation and evaluate the limit:

$$\lim_{n\to\infty} \sum_{k=1}^{n} \left(\frac{4}{n} + \frac{16}{n^2} k\right)$$

23. (A) Use the method of rectangles to find the area between the graph of $f(x) = 2x + 4$ and the x axis from $x = 0$ to $x = 4$.
(B) Use a definite integral and the fundamental theorem of calculus to find the area in part A.

24. Given that f and g are continuous on $[-1, 3]$, $\int_{-1}^{3} f(x)\,dx = 12$, and $\int_{-1}^{3} g(x)\,dx = 9$, use properties of the definite integral to find:

(A) $\int_{-1}^{3} 3f(x)\,dx$ (B) $\int_{3}^{-1} g(x)\,dx$ (C) $\int_{-1}^{3} [f(x) - g(x)]\,dx$

25. Find the area bounded by the graph of $f(x) = 6x^2 + 2$ and the x axis from $x = -1$ to $x = 2$.

26. Find the average value of $f(x) = 3x^{1/2}$ over the interval $[1, 9]$.

C

Find each integral in Problems 27–36.

27. $\displaystyle\int_0^3 \frac{x}{1 + x^2}\,dx$

28. $\displaystyle\int_0^3 \frac{x}{(1 + x^2)^2}\,dx$

29. $\displaystyle\int x^3(2x^4 + 5)^5\,dx$

30. $\displaystyle\int \frac{e^{-x}}{e^{-x} + 3}\,dx$

31. $\displaystyle\int \frac{e^x}{(e^x + 2)^2}\,dx$

32. $\displaystyle\int \frac{(\ln x)^2}{x}\,dx$

33. $\displaystyle\int x(x^3 - 1)^2\,dx$

34. $\displaystyle\int \frac{x}{\sqrt{6 - x}}\,dx$

35. $\displaystyle\int_0^7 x\sqrt{16 - x}\,dx$

36. $\displaystyle\int_{-1}^{1} x(x + 1)^4\,dx$

37. Find a function $y = f(x)$ that satisfies both conditions:

$$\frac{dy}{dx} = 9x^2 e^{x^3} \qquad f(0) = 2$$

38. Solve the differential equation:

$$\frac{dN}{dt} = 0.06N \qquad N(0) = 800,\ N > 0$$

39. (A) Use summation properties and special summation formulas to evaluate

$$\lim_{n\to\infty} \sum_{k=1}^{n} (3c_k^2 + 5)\,\frac{5 - 0}{n} \qquad c_k = \frac{5k}{n}$$

(B) Write the limit in part A as a definite integral and evaluate by using the fundamental theorem of calculus.

Business & Economics

40. *Profit function.* If the marginal profit for producing x units per day is given by

$$P'(x) = 100 - 0.02x \qquad P(0) = 0$$

where $P(x)$ is the profit in dollars, find the profit function P and the profit on 10 units of production per day.

41. *Resource depletion.* An oil well starts out producing oil at the rate of 60,000 barrels of oil per year, but the production rate is expected to decrease by 4,000 barrels per year. Thus, if $P(t)$ is the total production (in thousands of barrels) in t years, then

$$P'(t) = f(t) = 60 - 4t \qquad 0 \leq t \leq 15$$

Write a definite integral that will give the total production after 15 years of operation. Evaluate it.

42. *Profit and production.* The weekly marginal profit for an output of x units is given approximately by

$$P'(x) = 150 - \frac{x}{10} \qquad 0 \leq x \leq 40$$

What is the total change in profit for a production change from 10 units per week to 40 units? Set up a definite integral and evaluate it.

43. *Inventory.* Suppose the inventory of a certain item t months after the first of the year is given approximately by

$$I(t) = 10 + 36t - 3t^2 \qquad 0 \leq t \leq 12$$

What is the average inventory for the second quarter of the year?

44. *Advertising.* The number of customers (in thousands) each day at a major theme park t days after the beginning of an advertising campaign is given by

$$S(t) = 25 - 10e^{-0.1t}$$

What is the average number of customers each day during the first 30 days of the campaign?

45. *Marketing.* The market research department for an automobile company estimates that the sales (in millions of dollars) of a new automobile will increase at the monthly rate of

$$S'(t) = 4e^{-0.08t} \qquad 0 \leq t \leq 24$$

t months after the introduction of the automobile. What will be the total sales $S(t)$, t months after the automobile is introduced if we assume that there were 0 sales at the time the automobile entered the marketplace? What are the estimated total sales during the first 12 months after the introduction of the automobile? How long will it take for the total sales to reach $40 million?

Life Sciences

46. *Wound healing.* The area of a small, healing surface wound changes at a rate given approximately by

$$\frac{dA}{dt} = -5t^{-2} \qquad 1 \leq t \leq 5$$

where t is time in days and $A(1) = 5$ square centimeters. What will the area of the wound be in 5 days?

47. *Pollution.* An environmental protection agency estimates that the rate of seepage of toxic chemicals from a waste dump in gallons per year is given by

$$R(t) = \frac{1{,}000}{(1 + t)^2}$$

where t is time in years since the discovery of the seepage. Find the total amount of toxic chemicals that seep from the dump during the first 4 years after the seepage is discovered.

48. *Population.* The population of the United States in 1980 was approximately 226 million. The continuous compound growth rate for the decade from 1970 to 1980 was 1.1% (data from the 1980 census).

(A) If the population continued to grow at the same continuous compound growth rate, what is the predicted 1990 population?
(B) How long will it take the population to double at this continuous compound growth rate?

Social Sciences

49. *Archaeology.* The continuous compound rate of decay for carbon-14 is $r = -0.000\ 123\ 8$. A piece of animal bone found at an archaeological site contains 4% of the original amount of carbon-14. Estimate the age of the bone.

50. *Learning.* In a particular business college, it was found that an average student enrolled in a typing class progressed at a rate of $N'(t) = 7e^{-0.1t}$ words per minute t weeks after enrolling in a 15 week course. If at the beginning of the course a student could type 25 words per minute, how many words per minute, $N(t)$, would the student be expected to type t weeks into the course? After completing the course?

CHAPTER 7

Additional Integration Topics

CHAPTER 7

Contents

SECTION 7-1

Area between Curves

- AREA BETWEEN A CURVE AND THE X AXIS
- AREA BETWEEN TWO CURVES
- APPLICATION: DISTRIBUTION OF INCOME

AREA BETWEEN A CURVE AND THE X AXIS

From Chapter 6, we know that if $f(x)$ is continuous and greater than or equal to 0 on $[a, b]$, then the area between the graph of f and the x axis from $x = a$ to $x = b$ is given by

$$\text{Area} = \lim_{n\to\infty} \sum_{k=1}^{n} f(c_k)\Delta x_k = \int_a^b f(x)\,dx$$

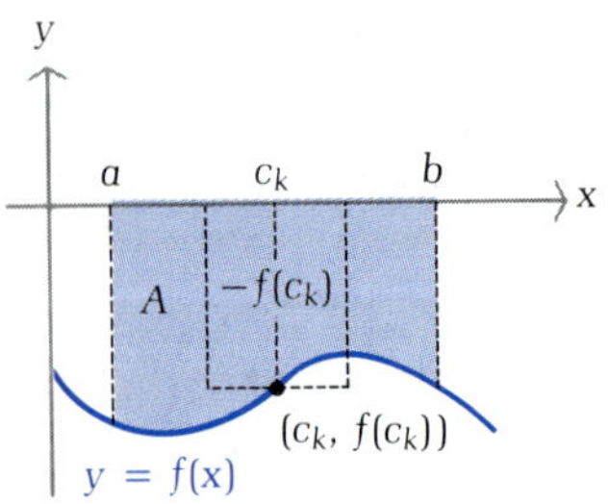

FIGURE 1

The requirement $f(x) \geq 0$ is necessary in order to interpret each term $f(c_k)\Delta x_k$ in the Riemann sum as the area of a rectangle (area is never negative). How can we find the area between the graph of f and the x axis if $f(x) \leq 0$ on $[a, b]$ or if $f(x)$ is both positive and negative on $[a, b]$? To begin, suppose $f(x) \leq 0$ and A is the area between the graph of f and the x axis for $a \leq x \leq b$, as illustrated in Figure 1.

If we divide the interval $[a, b]$ in Figure 1 into n equal subintervals and select a point c_k in each subinterval, then the y coordinate of the point where the base of the kth rectangle intersects the curve is the negative number $f(c_k)$, but the height of the rectangle is the positive number $-f(c_k)$ (see Figure 1):

kth rectangle
Width $= \Delta x$
Height $= -f(c_k)$
Area $= [-f(c_k)]\Delta x$

Thus,

$$A = \lim_{n\to\infty} \sum_{k=1}^{n} [-f(c_k)]\Delta x_k = \int_a^b [-f(x)]\,dx$$

That is, **the area between the graph of a negative function and the x axis is equal to the definite integral of the negative of the function.** Finally, if $f(x)$ is positive for some values of x and negative for others, the area between the graph of f and the x axis can be obtained by dividing $[a, b]$ into intervals over which f is always positive or always negative, finding the area over each interval, and then summing these areas.

◆ EXAMPLE 1

Find the area between the graph of $f(x) = x^2 - 2x$ and the x axis over the indicated intervals:

(A) $[1, 2]$ (B) $[-1, 1]$

Solutions

We begin by sketching the graph of f, as shown in the margin. (The solution of every area problem should begin with a sketch.)

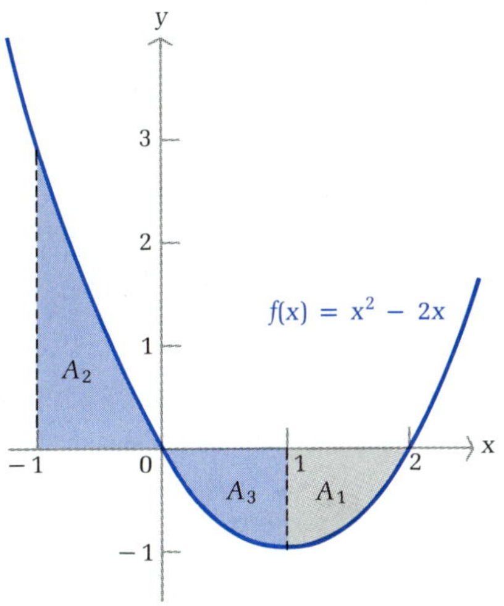

(A) From the graph, we see that $f(x) \le 0$ for $1 \le x \le 2$, so we integrate $-f(x)$:

$$\begin{aligned} A_1 &= \int_1^2 [-f(x)]\, dx \\ &= \int_1^2 (2x - x^2)\, dx \\ &= \left(x^2 - \frac{x^3}{3}\right)\Big|_1^2 \\ &= \left[(2)^2 - \frac{(2)^3}{3}\right] - \left[(1)^2 - \frac{(1)^3}{3}\right] \\ &= 4 - \tfrac{8}{3} - 1 + \tfrac{1}{3} = \tfrac{2}{3} \end{aligned}$$

(B) Since the graph shows that $f(x) \ge 0$ on $[-1, 0]$ and $f(x) \le 0$ on $[0, 1]$, the computation of this area will require two integrals:

$$\begin{aligned} A &= A_2 + A_3 \\ &= \int_{-1}^0 f(x)\, dx + \int_0^1 [-f(x)]\, dx \\ &= \int_{-1}^0 (x^2 - 2x)\, dx + \int_0^1 (2x - x^2)\, dx \\ &= \left(\frac{x^3}{3} - x^2\right)\Big|_{-1}^0 + \left(x^2 - \frac{x^3}{3}\right)\Big|_0^1 \\ &= \tfrac{4}{3} + \tfrac{2}{3} = 2 \end{aligned}$$

◆

PROBLEM 1

Find the area between the graph of $f(x) = x^2 - 9$ and the x axis over the indicated intervals:

(A) $[0, 2]$ (B) $[2, 4]$ ◆

◆ AREA BETWEEN TWO CURVES

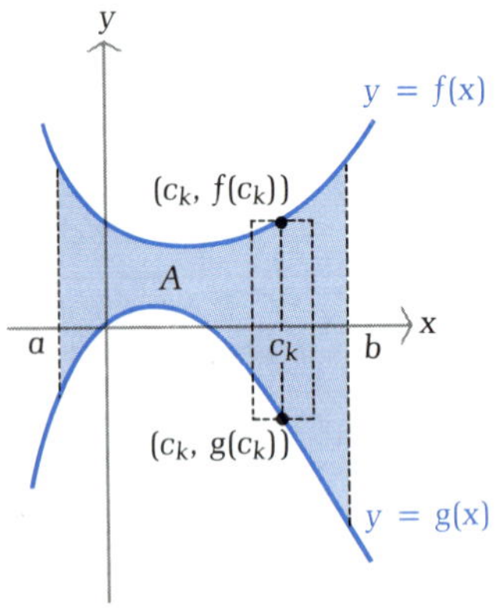

FIGURE 2

Let $f(x)$ and $g(x)$ be continuous functions satisfying $f(x) \geqslant g(x)$ for $a \leqslant x \leqslant b$, and let A be the area of the region bounded by the graphs of f and g for $a \leqslant x \leqslant b$, as indicated in Figure 2.

kth rectangle
Width $= \Delta x$
Height $= f(c_k) - g(c_k)$
Area $= [f(c_k) - g(c_k)]\Delta x$

To find A, we first divide $[a, b]$ into n equal subintervals, choose a point c_k in each subinterval, and draw a typical rectangle (see Figure 2). Since $f(x) \geqslant g(x)$ over the interval $[a, b]$, the top of the kth rectangle intersects the graph of f at $(c_k, f(c_k))$, and the bottom intersects the graph of g at $(c_k, g(c_k))$. Regardless of the signs of $f(c_k)$ and $g(c_k)$, the height of the kth rectangle is the difference* $f(c_k) - g(c_k)$, and

$$A = \lim_{n\to\infty} \sum_{k=1}^{n} [f(c_k) - g(c_k)]\Delta x_k$$

$$= \int_a^b [f(x) - g(x)]\,dx$$

Thus, **the area between the graphs of two functions is equal to the definite integral of the upper function minus the lower function.** This result is summarized in the box for convenient reference.

Area between Two Curves

If f and g are continuous and $f(x) \geqslant g(x)$ over the interval $[a, b]$, then the area bounded by $y = f(x)$ and $y = g(x)$ for $a \leqslant x \leqslant b$ is given exactly by

$$A = \int_a^b [f(x) - g(x)]\,dx$$

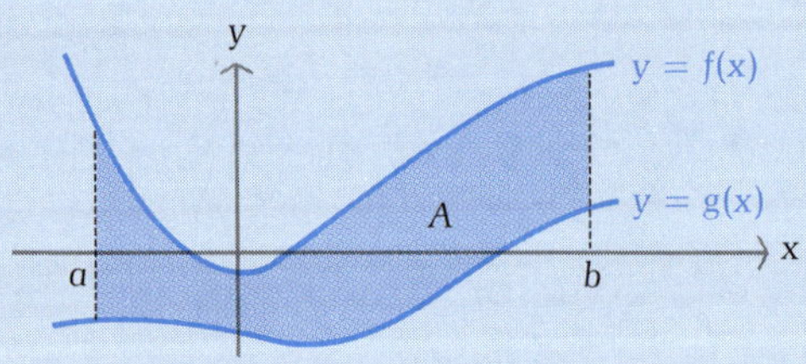

* Recall that if a and b are two points on a real number line and $a < b$, then $b - a$ is a positive number representing the distance between a and b.

◆ EXAMPLE 2 Find the area bounded by $f(x) = \frac{1}{2}x + 3$, $g(x) = -x^2 + 1$, $x = -2$, and $x = 1$.

Solution We first sketch the area, then set up and evaluate an appropriate definite integral:

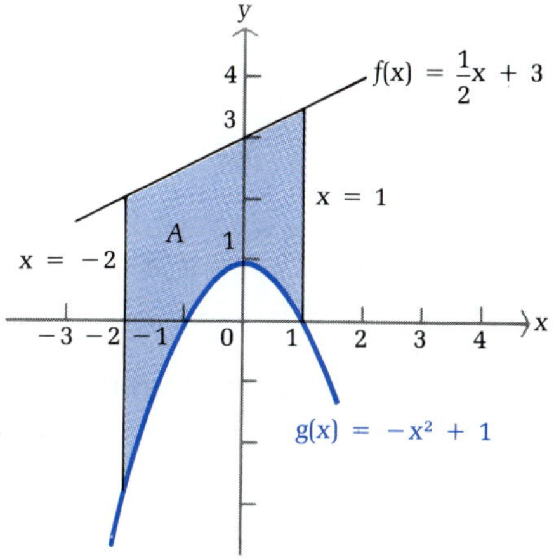

We observe from the graph that $f(x) \geqslant g(x)$ for $-2 \leqslant x \leqslant 1$, so

$$A = \int_{-2}^{1} [f(x) - g(x)]\,dx = \int_{-2}^{1} \left[\left(\frac{x}{2} + 3\right) - (-x^2 + 1)\right] dx$$

$$= \int_{-2}^{1} \left(x^2 + \frac{x}{2} + 2\right) dx$$

$$= \left(\frac{x^3}{3} + \frac{x^2}{4} + 2x\right)\Bigg|_{-2}^{1} = \left(\frac{1}{3} + \frac{1}{4} + 2\right) - \left(\frac{-8}{3} + \frac{4}{4} - 4\right) = \frac{33}{4}$$ ◆

PROBLEM 2 Find the area bounded by $f(x) = x^2 - 1$, $g(x) = -\frac{1}{2}x - 3$, $x = -1$, and $x = 2$. ◆

◆ EXAMPLE 3 Find the area bounded by $f(x) = 5 - x^2$ and $g(x) = 2 - 2x$.

Solution First, graph f and g on the same coordinate system, as shown in the margin. Since the statement of the problem does not include any limits on the values of x, we must determine the appropriate values from the graph. The graph of f is a parabola and the graph of g is a line, as shown in the figure. The area bounded by these two graphs extends from the intersection point on the left to the intersection point on the right. To find these intersection points, we solve the equation $f(x) = g(x)$ for x:

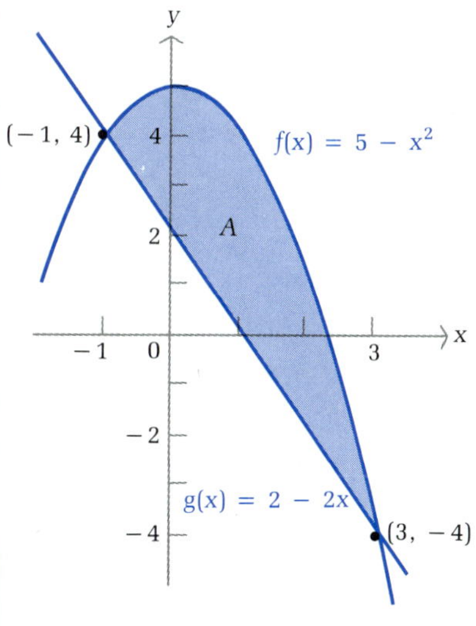

$$f(x) = g(x)$$
$$5 - x^2 = 2 - 2x$$
$$x^2 - 2x - 3 = 0$$
$$x = -1, 3$$

(Note that the area between the graphs for $x < -1$ is unbounded on the left, and the area between the graphs for $x > 3$ is unbounded on the right.) The figure shows that $f(x) \geq g(x)$ over the interval $[-1, 3]$, so we have

$$A = \int_{-1}^{3} [f(x) - g(x)]\,dx = \int_{-1}^{3} [5 - x^2 - (2 - 2x)]\,dx$$
$$= \int_{-1}^{3} (3 + 2x - x^2)\,dx$$
$$= \left(3x + x^2 - \frac{x^3}{3}\right)\Big|_{-1}^{3}$$
$$= \left[3(3) + (3)^2 - \frac{(3)^3}{3}\right] - \left[3(-1) + (-1)^2 - \frac{(-1)^3}{3}\right]$$
$$= 9 + 9 - 9 + 3 - 1 - \tfrac{1}{3} = \tfrac{32}{3}$$ ◆

PROBLEM 3 Find the area bounded by $f(x) = 6 - x^2$ and $g(x) = x$. ◆

◆ EXAMPLE 4 Find the area bounded by $f(x) = x^2 - x$ and $g(x) = 2x$ for $-2 \leq x \leq 3$.

Solution The graphs of f and g are shown in the figure. Examining the graph, we see that $f(x) \geq g(x)$ on the interval $[-2, 0]$, but $g(x) \geq f(x)$ on the interval $[0, 3]$. Thus, two integrals are required to compute this area:

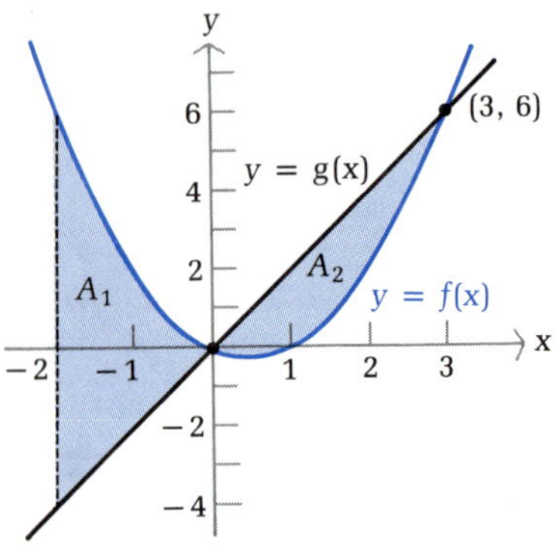

$$A_1 = \int_{-2}^{0} [f(x) - g(x)]\,dx \qquad f(x) \geq g(x) \text{ on } [-2, 0]$$
$$= \int_{-2}^{0} [x^2 - x - 2x]\,dx$$
$$= \int_{-2}^{0} (x^2 - 3x)\,dx$$
$$= \left(\frac{x^3}{3} - \frac{3}{2}x^2\right)\Big|_{-2}^{0}$$
$$= (0) - \left[\frac{(-2)^3}{3} - \frac{3}{2}(-2)^2\right]$$
$$= \tfrac{8}{3} + 6 = \tfrac{26}{3}$$

$$A_2 = \int_{0}^{3} [g(x) - f(x)]\,dx \qquad g(x) \geq f(x) \text{ on } [0, 3]$$
$$= \int_{0}^{3} [2x - (x^2 - x)]\,dx$$
$$= \int_{0}^{3} (3x - x^2)\,dx$$
$$= \left(\frac{3}{2}x^2 - \frac{x^3}{3}\right)\Big|_{0}^{3}$$
$$= \left[\frac{3}{2}(3)^2 - \frac{(3)^3}{3}\right] - (0)$$
$$= \tfrac{27}{2} - 9 = \tfrac{9}{2}$$

The total area between the two graphs is

$$A = A_1 + A_2 = \frac{26}{3} + \frac{9}{2} = \frac{79}{6}$$ ◆

PROBLEM 4 Find the area bounded by $f(x) = 2x^2$ and $g(x) = 4 - 2x$ for $-2 \leq x \leq 2$. ◆

◆ EXAMPLE 5 *Useful Life*

An amusement company maintains records for each video game it installs in an arcade. Suppose that $C(t)$ and $R(t)$ represent the total accumulated costs and revenues (in thousands of dollars), respectively, t years after a particular game has been installed. Also, suppose that

$$C'(t) = 2 \qquad \text{and} \qquad R'(t) = 9e^{-0.5t}$$

As long as $R'(t) \geq C'(t)$, the total accumulated profits from the game will continue to increase. The value of t for which $C'(t) = R'(t)$ is called the **useful life** of the game, and the area between the graphs of $C'(t)$ and $R'(t)$ over this time period represents the total accumulated profit for the useful life of the game. Find the useful life and the total profit.

Solution First, we find the useful life of the game:

$$\begin{aligned} R'(t) &= C'(t) \\ 9e^{-0.5t} &= 2 \\ e^{-0.5t} &= \tfrac{2}{9} \\ -0.5t &= \ln \tfrac{2}{9} \\ t &= -2 \ln \tfrac{2}{9} \approx 3 \text{ years} \end{aligned}$$

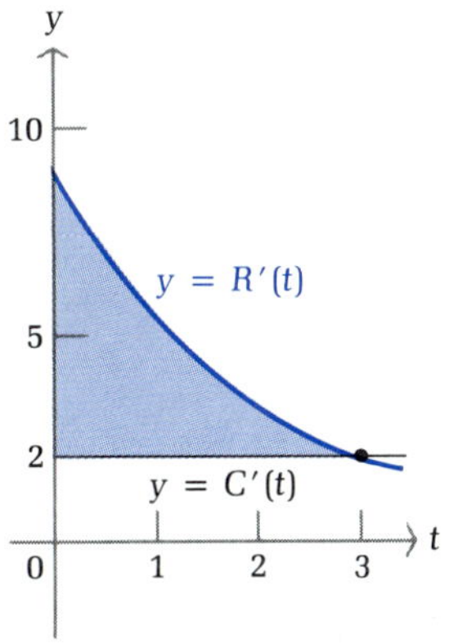

The graphs of C' and R' are shown in the figure in the margin. The total profit accumulated during the useful life of the game is

$$\begin{aligned} P(3) - P(0) &= \int_0^3 P'(t)\, dt \\ &= \int_0^3 [R'(t) - C'(t)]\, dt \qquad \text{The shaded area in the figure} \\ &= \int_0^3 (9e^{-0.5t} - 2)\, dt \\ &= (-18e^{-0.5t} - 2t)\Big|_0^3 \\ &= (-18e^{-1.5} - 6) - (-18e^0 - 0) \\ &= 12 - 18e^{-1.5} \approx 7.984 \quad \text{or } \$7{,}984 \end{aligned}$$ ◆

PROBLEM 5 Repeat Example 5 if $C'(t) = 1$ and $R'(t) = 7.5e^{-0.5t}$. ◆

TABLE 1

Income Distribution in the United States (1983)

INCOME LEVEL	x	y
Under \$10,000	0.16	0.03
Under \$15,000	0.28	0.08
Under \$25,000	0.51	0.24
Under \$35,000	0.71	0.44
Under \$50,000	0.88	0.69
Under \$75,000	0.97	0.88

Source: U.S. Bureau of the Census

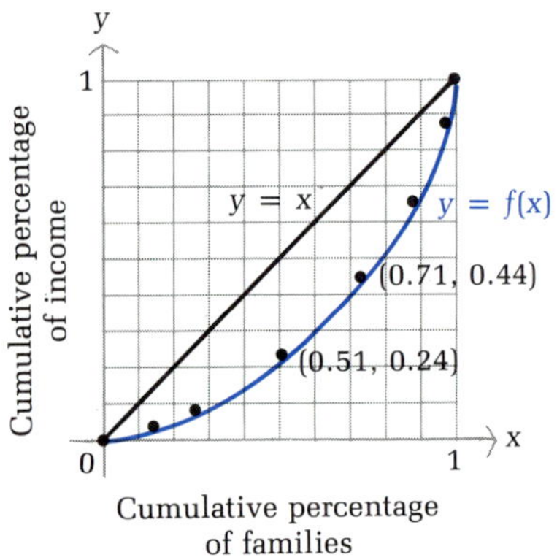

FIGURE 3
Lorenz curve, $y = f(x)$

◆ APPLICATION: DISTRIBUTION OF INCOME

Economists often use a graph called a **Lorenz curve** to provide a graphical description of the distribution of income among various groups of people. For example, the distribution of personal income in the United States given in Table 1 can be represented by the Lorenz curve shown in Figure 3. In both Table 1 and Figure 3, **x represents the cumulative percentage of families at or below a given income level** and **y represents the cumulative percentage of total personal income received by all these families.** For example, the point (0.51, 0.24) on the Lorenz curve in Figure 3 indicates that the bottom 51% of families (those with incomes under \$25,000) received 24% of the total income, the point (0.71, 0.44) indicates that the bottom 71% of families received 44% of the total income, and so on.

Absolute equality of income would occur if every family received the same income. That is, 20% of the families receive 20% of the income, 40% of the families receive 40% of the income, and so on. This distribution is represented by the graph of $y = x$ in Figure 3. The area between the Lorenz curve and the line $y = x$ can be used to measure how much the distribution of income differs from absolute equality.

More precisely, the **coefficient of inequality** of income distribution is defined to be the ratio of the area between the line $y = x$ and the Lorenz curve to the area between the line $y = x$ and the x axis. If we are given a function f whose graph is a Lorenz curve, then the area between the line $y = x$ and the Lorenz curve is $\int_0^1 [x - f(x)]\,dx$. Since the area between the graph of the line $y = x$ and the x axis from $x = 0$ to $x = 1$ is $\frac{1}{2}$, it follows that the coefficient of inequality is given by

$$\frac{\text{Area between } y = x \text{ and } y = f(x)}{\text{Area between } y = x \text{ and } x \text{ axis}} = \frac{\int_0^1 [x - f(x)]\,dx}{\frac{1}{2}} = 2\int_0^1 [x - f(x)]\,dx$$

Coefficient of Inequality

If $y = f(x)$ is the equation of a Lorenz curve, then

$$\text{Coefficient of inequality} = 2\int_0^1 [x - f(x)]\,dx$$

The coefficient of inequality is always a number between 0 and 1. If the coefficient is near 0, then the income distribution is close to absolute equality. The closer the coefficient is to 1, the greater the inequality (or disparity) of the distribution. The coefficient of inequality can be used to compare income distributions at various points in time, between different groups of people, before and after taxes are paid, between different countries, and so on.

◆ EXAMPLE 6

Distribution of Income

The Lorenz curve for the distribution of income in a certain country in 1990 is given by $f(x) = x^{2.6}$. Economists predict that the Lorenz curve for the country in the year 2010 will be given by $g(x) = x^{1.8}$. Find the coefficient of inequality for each curve, and interpret the results.

Solution The Lorenz curves are shown in the figure below:

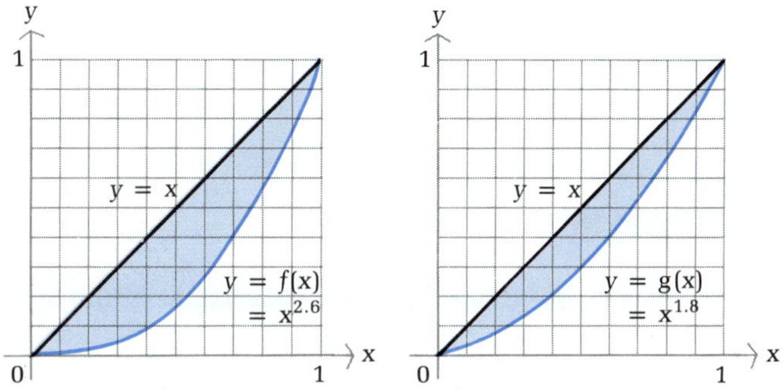

(A) Lorenz curve for 1990 (B) Projected Lorenz curve for 2010

The coefficient of inequality in 1990 is (see figure A)

$$\begin{aligned} 2\int_0^1 [x - f(x)]\,dx &= 2\int_0^1 [x - x^{2.6}]\,dx \\ &= 2\left(\frac{1}{2}x^2 - \frac{1}{3.6}x^{3.6}\right)\Big|_0^1 \\ &= 2\left(\frac{1}{2} - \frac{1}{3.6}\right) \approx 0.444 \end{aligned}$$

The projected coefficient of inequality in 2010 is (see figure B)

$$\begin{aligned} 2\int_0^1 [x - g(x)]\,dx &= 2\int_0^1 [x - x^{1.8}]\,dx \\ &= 2\left(\frac{1}{2}x^2 - \frac{1}{2.8}x^{2.8}\right)\Big|_0^1 \\ &= 2\left(\frac{1}{2} - \frac{1}{2.8}\right) \approx 0.286 \end{aligned}$$

If this projection is correct, the coefficient of inequality will decrease, and income will be more equally distributed in the year 2010 than in 1990. ◆

PROBLEM 6 Repeat Example 6 if the projected Lorenz curve in the year 2010 is given by $g(x) = x^{3.8}$. ◆

Answers to Matched Problems

1. (A) $A = \int_0^2 (9 - x^2)\,dx = \frac{46}{3}$
   (B) $A = \int_2^3 (9 - x^2)\,dx + \int_3^4 (x^2 - 9)\,dx = 6$
2. $A = \int_{-1}^{2} \left[(x^2 - 1) - \left(-\frac{x}{2} - 3 \right) \right] dx = \frac{39}{4}$
3. $A = \int_{-3}^{2} [(6 - x^2) - x]\,dx = \frac{125}{6}$
4. $A = \int_{-2}^{1} [(4 - 2x) - 2x^2]\,dx + \int_1^2 [2x^2 - (4 - 2x)]\,dx = \frac{38}{3}$
5. Useful life $= -2 \ln \frac{2}{15} \approx 4$ yr; Total profit $= 11 - 15e^{-2} \approx 8.970$ or \$8,970
6. Coefficient of inequality ≈ 0.583; income will be less equally distributed in 2010.

EXERCISE 7-1

A *Find the area between the graph of f and the x axis over the indicated interval.*

1. $f(x) = 1 - x;\quad 2 \le x \le 3$
2. $f(x) = 2 - 4x;\quad 1 \le x \le 2$
3. $f(x) = x^2 - 9;\quad -1 \le x \le 2$
4. $f(x) = x^2 - 5x;\quad 1 \le x \le 3$

Find the area between the graphs of f and g over the indicated interval.

5. $f(x) = 12,\quad g(x) = -2x + 8;\quad -1 \le x \le 2$
6. $f(x) = 2x + 6,\quad g(x) = 3;\quad -1 \le x \le 2$
7. $f(x) = x^2 + 1,\quad g(x) = 2x - 2;\quad -1 \le x \le 2$
8. $f(x) = x^2 - 1,\quad g(x) = x - 2;\quad -2 \le x \le 1$
9. $f(x) = e^{0.5x},\quad g(x) = -1/x;\quad 1 \le x \le 2$
10. $f(x) = 1/x,\quad g(x) = -e^x;\quad 0.5 \le x \le 1$

B *Find the area bounded by the graphs of the indicated equations.*

11. $y = -x,\quad y = 0;\quad -2 \le x \le 1$
12. $y = -x + 1,\quad y = 0;\quad -1 \le x \le 2$
13. $y = x^2 - 4,\quad y = 0;\quad 0 \le x \le 3$
14. $y = 4\sqrt[3]{x},\quad y = 0;\quad -1 \le x \le 8$
15. $y = 3x^2,\quad y = 12$
16. $y = x^2,\quad y = 9$
17. $y = 4 - x^2,\quad y = -5$
18. $y = x^2 - 1,\quad y = 3$
19. $y = x^2 + 2x + 3,\quad y = 2x + 4$
20. $y = 8 + 4x - x^2,\quad y = x^2 - 2x$
21. $y = x^2 - 4x - 10,\quad y = 14 - 2x - x^2$
22. $y = 6 + 6x - x^2,\quad y = 13 - 2x$

C

23. $y = 4/x,\quad y = x;\quad 1 \le x \le 4$
24. $y = 8/x,\quad y = x^2;\quad 1 \le x \le 4$
25. $y = x^3 - 2x^2 - 15x,\quad y = 0$
26. $y = x^3 - x^2 - 56x,\quad y = 0$
27. $y = x^3,\quad y = 4x$
28. $y = x^3 + 1,\quad y = x + 1$
29. $y = x^3 - 3x^2 - 9x + 12,\quad y = x + 12$
30. $y = x^3 - 6x^2 + 9x,\quad y = x$

Business & Economics

Photo courtesy of Shell Oil Company

31. *Oil production.* Using data from the first 3 years of production as well as geological studies, the management of an oil company estimates that oil will be pumped from a producing field at a rate given by

$$R(t) = \frac{100}{t+10} + 10 \qquad 0 \leq t \leq 15$$

where $R(t)$ is the rate of production (in thousands of barrels per year) t years after pumping begins. Find the area between the graph of R and the t axis over the interval $[5, 10]$ and interpret the results.

32. *Oil production.* In Problem 31, if the rate is found to be

$$R(t) = \frac{100t}{t^2+25} + 4 \qquad 0 \leq t \leq 25$$

find the area between the graph of R and the t axis over the interval $[5, 15]$ and interpret the results.

33. *Useful life.* The total accumulated costs $C(t)$ and revenues $R(t)$ (in thousands of dollars), respectively, for a coin-operated photocopying machine satisfy

$$C'(t) = 3 \qquad \text{and} \qquad R'(t) = 8e^{-0.5t}$$

where t is time in years. Find the useful life of the machine to the nearest year and the total profit accumulated during the useful life of the machine.

34. *Useful life.* The total accumulated costs $C(t)$ and revenues $R(t)$ (in thousands of dollars), respectively, for a coal mine satisfy

$$C'(t) = 3 \qquad \text{and} \qquad R'(t) = 15e^{-0.1t}$$

where t is time in years. Find the useful life of the mine to the nearest year and the total profit accumulated during the useful life of the mine.

35. *Income distribution.* As part of a study of the effects of World War II on the economy of the United States, an economist used data from the U.S. Bureau of the Census to produce the following Lorenz curves for distribution of income in the United States in 1935 and in 1947:

$f(x) = x^{2.4}$ Lorenz curve for 1935

$g(x) = x^{1.6}$ Lorenz curve for 1947

Find the coefficient of inequality for each Lorenz curve and inerpret the results.

36. *Income distribution.* Using data from the U.S. Bureau of the Census, an economist produced the following Lorenz curves for distribution of income in the United States in 1962 and in 1972:

$$f(x) = \tfrac{3}{10}x + \tfrac{7}{10}x^2 \qquad \text{Lorenz curve for 1962}$$
$$g(x) = \tfrac{1}{2}x + \tfrac{1}{2}x^2 \qquad \text{Lorenz curve for 1972}$$

Find the coefficient of inequality for each Lorenz curve and interpret the results.

37. *Distribution of wealth.* Lorenz curves also can be used to provide a relative measure of the distribution of the total assets of a country. Using data in a report by the U.S. Congressional Joint Economic Committee, an economist produced the following Lorenz curves for the distribution of total assets in the United States in 1963 and in 1983:

$$f(x) = x^{10} \qquad \text{Lorenz curve for 1963}$$
$$g(x) = x^{12} \qquad \text{Lorenz curve for 1983}$$

Find the coefficient of inequality for each Lorenz curve and interpret the results.

38. *Income distribution.* The government of a small country is planning sweeping changes in the tax structure in order to provide a more equitable distribution of income. The Lorenz curves for the current income distribution and for the projected income distribution after enactment of the tax changes are given below. Find the coefficient of inequality for each Lorenz curve. Will the proposed changes provide a more equitable income distribution?

$$f(x) = x^{2.3} \qquad \text{Current Lorenz curve}$$
$$g(x) = 0.4x + 0.6x^2 \qquad \text{Projected Lorenz curve after changes in tax laws}$$

Life Sciences

39. *Biology.* A yeast culture is growing at a rate of $W'(t) = 0.3e^{0.1t}$ grams per hour. Find the area between the graph of W' and the t axis over the interval $[0, 10]$ and interpret the results.

40. *Natural resource depletion.* The instantaneous rate of change of the demand for lumber in the United States since 1970 ($t = 0$) in billions of cubic feet per year is estimated to be given by

$$Q'(t) = 12 + 0.006t^2 \qquad 0 \le t \le 50$$

Find the area between the graph of Q' and the t axis over the interval $[15, 20]$ and inerpret the results.

41. *Environment.* The United States loses farmland to urban sprawl, highways, and other development at the rate of 1 million acres per year. Much of this loss is due to the large number of farmers that have been unable to meet their mortgage payments. Financial analysts estimate that providing financial aid to farmers would reduce the rate of loss of farmland to $R(t) = e^{-0.05t}$ million acres per year over the next 20 years, where t is time in years. Find

the area between the graphs of $f(t) = 1$ and $R(t) = e^{-0.05t}$ for $0 \leq t \leq 20$ and interpret the results.

42. *Environment.* The United States loses soil through erosion at the rate of 4 billion tons per year. Agronomists estimate that the enactment of a soil preservation bill would change this rate to $R(t) = 4e^{-0.1t}$ billion tons per year over the next 5 years, where t is time in years. find the area between the graphs of $f(t) = 4$ and $R(t) = 4e^{-0.1t}$ for $0 \leq t \leq 5$ and interpret the results.

Social Sciences

43. *Learning.* A beginning high school language class was chosen for an experiment in learning. Using a list of 50 words, the experiment involved measuring the rate of vocabulary memorization at different times during a continuous 5 hour study session. It was found that the average rate of learning for the whole class was inversely proportional to the time spent studying and was given approximately by

$$V'(t) = \frac{15}{t} \qquad 1 \leq t \leq 5$$

Find the area between the graph of V' and the t axis over the interval $[2, 4]$ and interpret the results.

44. *Learning.* Repeat Problem 43 if

$$V'(t) = \frac{13}{t^{1/2}}$$

and the interval $[2, 4]$ is changed to $[1, 4]$.

Problems 45–52 require the use of a graphic calculator or a computer. Graph the indicated equations, approximate x coordinates of intersection points correct to two decimal places, and find the area bounded by the graphs.

45. $y = \sqrt{x}, \quad y = x^2 - 4x + 3$
46. $y = \sqrt{x}, \quad y = -x^2 + 4x - 1$
47. $y = e^{2x}, \quad y = 2x + 3$
48. $y = e^{-4x}, \quad y = 2 - 4x$
49. $y = 2 + 6x - x^3, \quad y = 0$
50. $y = x^3 - 4x^2 + 7, \quad y = 0$
51. $y = \dfrac{6x}{x^2 + 1}, \quad y = x - 1$
52. $y = \dfrac{10x + 10}{x^2 + 2x + 3}, \quad y = x$

SECTION 7-2 Applications in Business and Economics

- CONTINUOUS INCOME STREAM
- FUTURE VALUE OF A CONTINUOUS INCOME STREAM
- CONSUMERS' AND PRODUCERS' SURPLUS

CONTINUOUS INCOME STREAM

We start this discussion with an example.

◆ EXAMPLE 7 The rate of change of the income produced by a vending machine (in dollars per year) is given by

$$f(t) = 5{,}000e^{0.04t}$$

where t is time in years since the installation of the machine. Find the total income produced by the machine during the first 5 years of operation.

Solution Since we have been given the rate of change of income, we can find the total income by using a definite integral:

$$\begin{aligned}\text{Total income} &= \int_0^5 5{,}000e^{0.04t}\,dt \\ &= 125{,}000e^{0.04t}\Big|_0^5 \\ &= 125{,}000e^{0.04(5)} - 125{,}000e^{0.04(0)} \\ &= 152{,}675 - 125{,}000 \\ &= \$27{,}675 \quad \text{Rounded to the nearest dollar}\end{aligned}$$

Thus, the vending machine produces a total income of $27,675 during the first 5 years of operation. ◆

PROBLEM 7 Refer to Example 7. Find the total income produced during the second 5 years of operation. ◆

In reality, income from a vending machine is not received as a single payment at the end of the 5 year period. Instead, the income is collected on a regular basis, perhaps daily or weekly. In problems of this type, it is convenient to assume that income is actually received in a **continuous stream;** that is, we assume that income is a continuous function of time. The rate of change of income is called the **rate of flow** of the continuous income stream. In general, we have the following:

Total Income for a Continuous Income Stream

If $f(t)$ is the rate of flow of a continuous income stream, then the **total income** produced during the time period from $t = a$ to $t = b$ is

$$\text{Total income} = \int_a^b f(t)\,dt$$

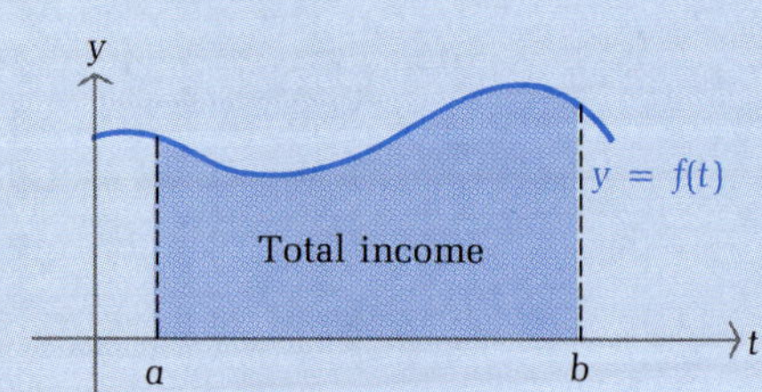

◆ FUTURE VALUE OF A CONTINUOUS INCOME STREAM

In Section 5-1, we discussed the continuous compound interest formula

$$A = Pe^{rt}$$

where P is the principal (or present value), A is the amount (or future value), r is the annual rate of continuous compounding (expressed as a decimal), and t is time in years. For example, if money is worth 12% compounded continuously, then the future value of a \$10,000 investment in 5 years is (to the nearest dollar)

$$A = 10{,}000e^{0.12(5)} = \$18{,}221$$

Now we want to apply the future value concept to the income produced by a continuous income stream. Suppose $f(t)$ is the rate of flow of a continuous income stream, and the income produced by this continuous income stream is invested as soon as it is received at a rate r, compounded continuously. We already know how to find the total income produced after T years, but how can we find the total of the income produced and the interest earned by this income? Since the income is received in a continuous flow, we cannot just use the formula $A = Pe^{rt}$. This formula is valid only for a single deposit P, not for a continuous flow of income. Instead, we use a Riemann sum approach that will allow us to apply the formula $A = Pe^{rt}$ repeatedly. To begin, we divide the time interval $[0, T]$ into n equal subintervals of length Δt and choose an arbitrary point c_k in each subinterval, as illustrated in Figure 4.

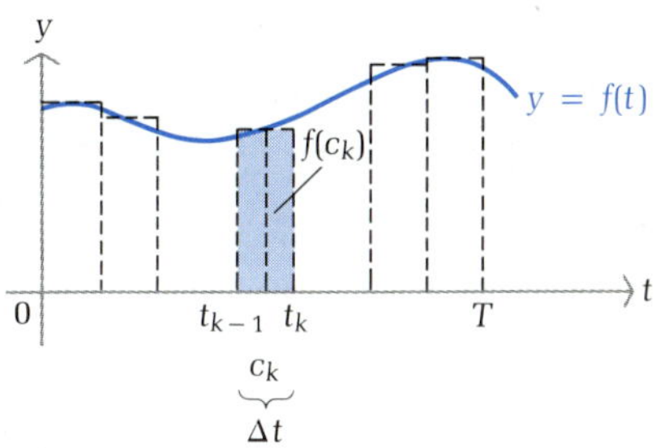

FIGURE 4

The total income produced during the time period from $t = t_{k-1}$ to $t = t_k$ is equal to the area under the graph of $f(t)$ over this subinterval and is approximately equal to $f(c_k)\Delta t$, the area of the shaded rectangle in Figure 4. The income received during this time period will earn interest for approximately $T - c_k$ years. Thus, using the future value formula $A = Pe^{rt}$ with $P = f(c_k)\Delta t$ and $t = T - c_k$, the future value of the income produced during the time period from $t = t_{k-1}$ to $t = t_k$ is approximately equal to

$$f(c_k)\Delta t e^{(T-c_k)r}$$

The total of these approximate future values over n subintervals is then

$$f(c_1)\Delta t e^{(T-c_1)r} + f(c_2)\Delta t e^{(T-c_2)r} + \cdots + f(c_n)\Delta t e^{(T-c_n)r}$$

This has the form of a Riemann sum, and the limit of this sum is a definite integral. (See the definition of definite integral in Section 6-5.) Thus, the *future value, FV*, of the income produced by the continuous income stream is given by

$$FV = \lim_{n\to\infty} \sum_{k=1}^{\infty} f(c_k)e^{r(T-c_k)}\Delta t = \int_0^T f(t)e^{r(T-t)}\,dt$$

Since r and T are constants, we also can write

$$FV = \int_0^T f(t)e^{rT}e^{-rt}\,dt = e^{rT}\int_0^T f(t)e^{-rt}\,dt \tag{1}$$

This last form is preferable, since the integral is usually easier to evaluate than the intermediate form in (1).

Future Value of a Continuous Income Stream

If $f(t)$ is the rate of flow of a continuous income stream, $0 \le t \le T$, and if the income is continuously invested at a rate r compounded continuously, then the **future value, *FV*,** at the end of T years is given by

$$FV = \int_0^T f(t)e^{r(T-t)}\,dt = e^{rT}\int_0^T f(t)e^{-rt}\,dt$$

The future value of a continuous income stream is the total value of all money produced by the continuous income stream (income and interest) at the end of T years.

◆ EXAMPLE 8 Let $f(t) = 5{,}000e^{0.04t}$ be the rate of flow of the income produced by the vending machine in Example 7. Find the future value of this income stream at 12% compounded continuously for 5 years, and find the total interest earned.

Solution Using the formula

$$FV = e^{rT}\int_0^T f(t)e^{-rt}\,dt$$

with $r = 0.12$, $T = 5$, and $f(t) = 5{,}000e^{0.04t}$, we have

$$\begin{aligned} FV &= e^{0.12(5)}\int_0^5 5{,}000e^{0.04t}e^{-0.12t}\,dt \\ &= 5{,}000e^{0.6}\int_0^5 e^{-0.08t}\,dt \\ &= 5{,}000e^{0.6}\left(\frac{e^{-0.08t}}{-0.08}\right)\Bigg|_0^5 \\ &= 5{,}000e^{0.6}(-12.5e^{-0.4} + 12.5) \\ &= \$37{,}545 \qquad \text{Rounded to the nearest dollar} \end{aligned}$$

In Example 7, we saw that the total income produced by this vending machine over a 5 year period was \$27,675. Since the future value is the total of the income produced and the interest earned, the difference between the future value and income is interest. Thus,

$$\$37{,}545 - \$27{,}675 = \$9{,}870$$

is the interest earned by the income produced by the vending machine during the 5 year period. ◆

PROBLEM 8 Repeat Example 8 if the interest rate is 9% compounded continuously. ◆

◆ CONSUMERS' AND PRODUCERS' SURPLUS

Let $p = D(x)$ be the price–demand equation for a product, where x is the number of units of the product that consumers will purchase at a price of \$$p$ per unit. Notice that we are expressing price $p = D(x)$ as a function of demand x (see Section 3-7). Suppose $\overline{p}$ is the current price and $\overline{x}$ is the number of units that can be sold at that price. The price-demand curve in Figure 5 shows that if the price is higher than $\overline{p}$, then the demand x is less than $\overline{x}$, but some consumers are still willing to pay the higher price. The consumers who were willing to pay more than $\overline{p}$ have saved money. We want to determine the total amount saved by all the consumers who were willing to pay a higher price for this product.

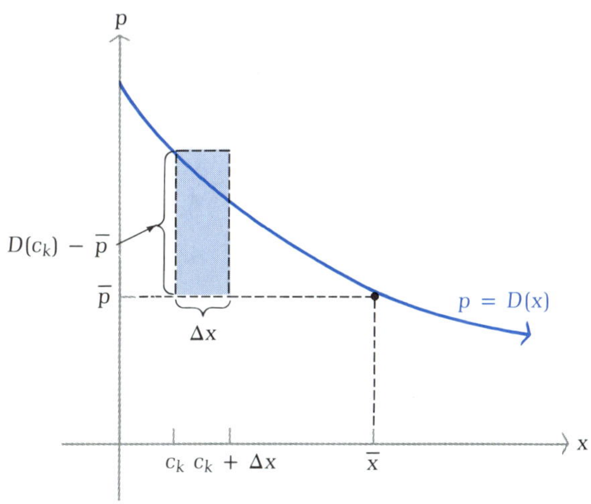

FIGURE 5

To do this, consider the interval $[c_k, c_k + \Delta x]$, where $c_k + \Delta x < \overline{x}$. If the price remained constant over this interval, then the savings on each unit would be the difference between $D(c_k)$, the price consumers are willing to pay, and $\overline{p}$, the price they actually pay. Since Δx represents the number of units purchased by consumers over the interval, the total savings to consumers over this interval is approximately equal to

$$[D(c_k) - \overline{p}]\Delta x \qquad \text{(Savings per unit)} \times \text{(Number of units)}$$

which is the area of the shaded rectangle shown in Figure 5. If we divide the interval $[0, \overline{x}]$ into n equal subintervals, then the total savings to consumers is approximately equal to

$$[D(c_1) - \overline{p}]\Delta x + [D(c_2) - \overline{p}]\Delta x + \cdots + [D(c_n) - \overline{p}]\Delta x$$

which we recognize as a Riemann sum. Thus, we define the *consumers' surplus* to be

$$CS = \lim_{n\to\infty} \sum_{k=1}^{n} [D(c_k) - \overline{p}]\Delta x = \int_0^{\overline{x}} [D(x) - \overline{p}]\, dx$$

Consumers' Surplus

If $(\overline{x}, \overline{p})$ is a point on the graph of the price–demand equation $p = D(x)$ for a particular product, then the **consumers' surplus, *CS*,** at a price level of $\overline{p}$ is

$$CS = \int_0^{\overline{x}} [D(x) - \overline{p}]\, dx$$

which is the area between $p = \overline{p}$ and $p = D(x)$ from $x = 0$ to $x = \overline{x}$.

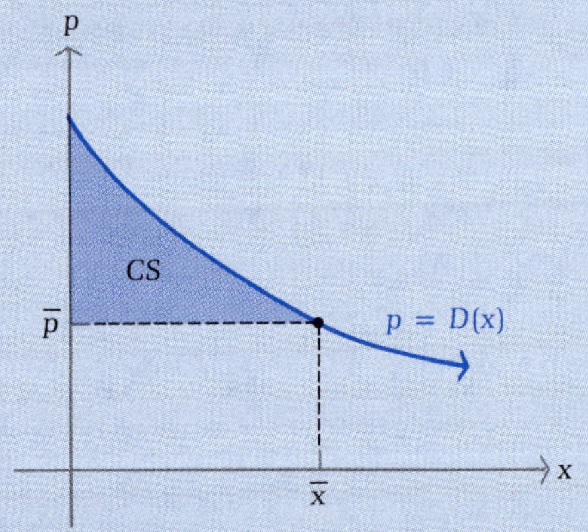

The consumers' surplus represents the total savings to consumers who are willing to pay a higher price for the product.

◆ **EXAMPLE 9** Find the consumers' surplus at a price level of \$8 for the price–demand equation

$$p = D(x) = 20 - \tfrac{1}{20}x$$

Solution **Step 1.** Find $\overline{x}$, the demand when the price is $\overline{p} = 8$:

$$\begin{aligned} \overline{p} &= 20 - \tfrac{1}{20}\overline{x} \\ 8 &= 20 - \tfrac{1}{20}\overline{x} \\ \tfrac{1}{20}\overline{x} &= 12 \\ \overline{x} &= 240 \end{aligned}$$

Step 2. Sketch a graph:

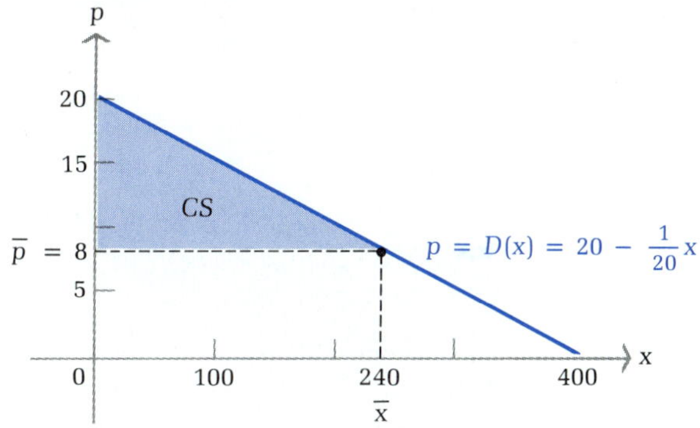

Step 3. Find the consumers' surplus (the shaded area in the graph):

$$\begin{aligned} CS &= \int_0^{\overline{x}} [D(x) - \overline{p}]\, dx \\ &= \int_0^{240} (20 - \tfrac{1}{20}x - 8)\, dx \\ &= \int_0^{240} (12 - \tfrac{1}{20}x)\, dx \\ &= (12x - \tfrac{1}{40}x^2)\Big|_0^{240} \\ &= 2{,}880 - 1{,}440 = \$1{,}440 \end{aligned}$$

Thus, the total savings to consumers who are willing to pay a higher price for the product is \$1,440. ◆

PROBLEM 9 Repeat Example 9 for a price level of \$4. ◆

If $p = S(x)$ is the price–supply equation for a product, $\overline{p}$ is the current price, and $\overline{x}$ is the current supply, then some suppliers are still willing to supply some units at a lower price. The additional money that these suppliers gain from the higher price is called the *producers' surplus* and can be expressed in terms of a definite integral (proceeding as we did for the consumers' surplus).

Producers' Surplus

If $(\bar{x}, \bar{p})$ is a point on the graph of the price–supply equation $p = S(x)$, then the **producers' surplus, *PS*,** at a price level of $\bar{p}$ is

$$PS = \int_0^{\bar{x}} [\bar{p} - S(x)]\, dx$$

which is the area between $p = \bar{p}$ and $p = S(x)$ from $x = 0$ to $x = \bar{x}$.

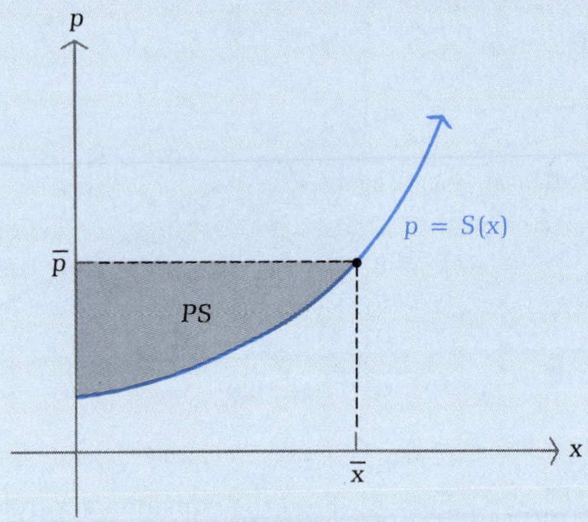

The producers' surplus represents the total gain to producers who are willing to supply units at a lower price.

◆ EXAMPLE 10 Find the producers' surplus at a price level of \$20 for the price–supply equation

$$p = S(x) = 2 + \tfrac{1}{5,000}x^2$$

Solution Step 1. Find $\bar{x}$, the supply when the price is $\bar{p} = 20$:

$$\bar{p} = 2 + \tfrac{1}{5,000}\bar{x}^2$$
$$20 = 2 + \tfrac{1}{5,000}\bar{x}^2$$
$$\tfrac{1}{5,000}\bar{x}^2 = 18$$
$$\bar{x}^2 = 90,000$$
$$\bar{x} = 300 \qquad \text{There is only one solution since } \bar{x} \geq 0.$$

Step 2. Sketch a graph:

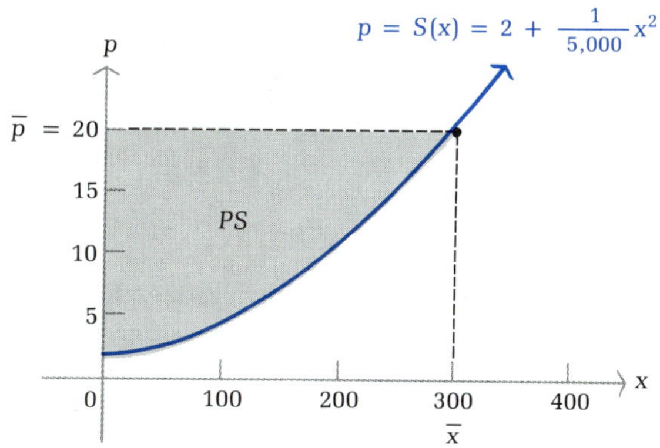

Step 3. Find the producers' surplus (the shaded area in the graph):

$$PS = \int_0^{\bar{x}} [\bar{p} - S(x)]\, dx$$
$$= \int_0^{300} [20 - (2 + \tfrac{1}{5{,}000}x^2)]\, dx$$
$$= \int_0^{300} (18 - \tfrac{1}{5{,}000}x^2)\, dx$$
$$= (18x - \tfrac{1}{15{,}000}x^3)\Big|_0^{300}$$
$$= 5{,}400 - 1{,}800 = \$3{,}600$$

Thus, the total gain to producers who are willing to supply units at a lower price is $3,600. ◆

PROBLEM 10 Repeat Example 10 for a price level of $4. ◆

In a free competitive market, the price of a product is determined by the relationship between supply and demand. If $p = D(x)$ and $p = S(x)$ are the price–demand and price–supply equations, respectively, for a product and if $(\bar{x}, \bar{p})$ is the point of intersection of these equations, then $\bar{p}$ is called the **equilibrium price** and $\bar{x}$ is called the **equilibrium quantity.** If the price stabilizes at the equilibrium price $\bar{p}$, then this is the price level that will determine both the consumers' surplus and the producers' surplus.

◆ EXAMPLE 11 Find the equilibrium price and then find the consumers' surplus and producers' surplus at the equilibrium price level if

$$p = D(x) = 20 - \tfrac{1}{20}x \qquad \text{and} \qquad p = S(x) = 2 + \tfrac{1}{5{,}000}x^2$$

Solution Step 1. Find the equilibrium point. Set $D(x)$ equal to $S(x)$ and solve:

$$
\begin{aligned}
D(x) &= S(x) \\
20 - \tfrac{1}{20}x &= 2 + \tfrac{1}{5{,}000}x^2 \\
\tfrac{1}{5{,}000}x^2 + \tfrac{1}{20}x - 18 &= 0 \\
x^2 + 250x - 90{,}000 &= 0 \\
x &= 200, -450
\end{aligned}
$$

Since x cannot be negative, the only solution is $x = 200$. The equilibrium price can be determined by using $D(x)$ or $S(x)$. We will use both to check our work:

$$
\begin{aligned}
\bar{p} &= D(200) \\
&= 20 - \tfrac{1}{20}(200) = 10
\end{aligned}
\qquad
\begin{aligned}
\bar{p} &= S(200) \\
&= 2 + \tfrac{1}{5{,}000}(200)^2 = 10
\end{aligned}
$$

Thus, the equilibrium price is $\bar{p} = 10$, and the equilibrium quantity is $\bar{x} = 200$.

Step 2. Sketch a graph:

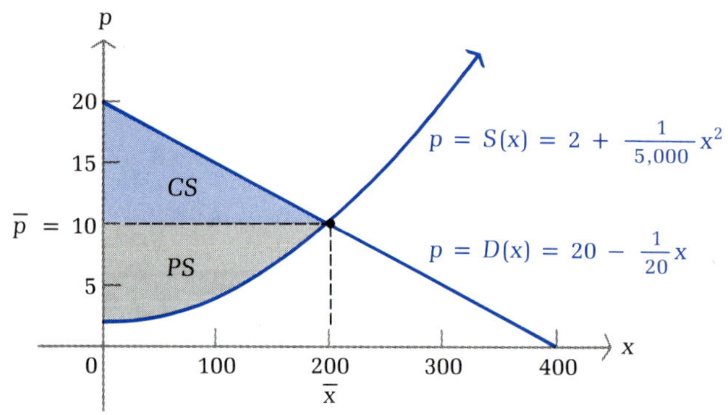

Step 3. Find the consumers' surplus:

$$
\begin{aligned}
CS &= \int_0^{\bar{x}} [D(x) - \bar{p}]\, dx \\
&= \int_0^{200} (20 - \tfrac{1}{20}x - 10)\, dx \\
&= \int_0^{200} (10 - \tfrac{1}{20}x)\, dx \\
&= (10x - \tfrac{1}{40}x^2)\Big|_0^{200} \\
&= 2{,}000 - 1{,}000 = \$1{,}000
\end{aligned}
$$

Step 4. Find the producers' surplus:

$$PS = \int_0^{\bar{x}} [\bar{p} - S(x)]\, dx$$

$$= \int_0^{200} [10 - (2 + \tfrac{1}{5{,}000}x^2)]\, dx$$

$$PS = \int_0^{200} (8 - \tfrac{1}{5{,}000}x^2)\, dx$$

$$= (8x - \tfrac{1}{15{,}000}x^3)\Big|_0^{200}$$

$$= 1{,}600 - \tfrac{1{,}600}{3} \approx \$1{,}067$$ Rounded to the nearest dollar ◆

PROBLEM 11 Repeat Example 11 for

$$p = D(x) = 25 - \tfrac{1}{1{,}000}x^2 \qquad \text{and} \qquad p = S(x) = 5 + \tfrac{1}{10}x$$ ◆

Answers to Matched Problems

7. \$33,803 8. FV = \$34,691; interest = \$7,016
9. \$2,560 10. \$133 11. $\bar{p} = 15$; CS = \$667; PS = \$500

EXERCISE 7-2

APPLICATIONS

Business & Economics

1. Find the total income produced by a continuous income stream in the first 5 years if the rate of flow is $f(t) = 2{,}500$.
2. Find the total income produced by a continuous income stream in the first 10 years if the rate of flow is $f(t) = 3{,}000$.
3. Find the total income produced by a continuous income stream in the first 3 years if the rate of flow is $f(t) = 400e^{0.05t}$.
4. Find the total income produced by a continuous income stream in the first 2 years if the rate of flow is $f(t) = 600e^{0.06t}$.
5. Find the future value at 8% interest compounded continuously for 10 years for the continuous income stream with rate of flow $f(t) = 800$.

6. Find the future value at 6% interest compounded continuously for 8 years for the continuous income stream with rate of flow $f(t) = 1{,}200$.

7. Find the future value at 10% interest compounded continuously for 4 years for the continuous income stream with rate of flow $f(t) = 1{,}500e^{-0.02t}$.

8. Find the future value at 7% interest compounded continuously for 6 years for the continuous income stream with rate of flow $f(t) = 2{,}000e^{0.06t}$.

9. Find the interest earned at 12% compounded continuously for 3 years by a continuous income stream with rate of flow $f(t) = 3{,}000$.

10. Find the interest earned at 14% compounded continuously for 2 years by a continuous income stream with rate of flow $f(t) = 7{,}000$.

11. An investor is presented with a choice of two investments, an established clothing store and a new computer store. Each choice requires the same initial investment and each produces a continuous income stream at 10% compounded continuously. The rate of flow of income from the clothing store is $f(t) = 12{,}000$, and the rate of flow of income from the computer store is expected to be $g(t) = 10{,}000e^{0.05t}$. Compare the future values of these investments to determine which is the better choice over the next 5 years.

12. Refer to Problem 11. Which investment is the better choice over the next 10 years?

13. An investor has \$10,000 to invest in either a bond that matures in 5 years or a business that will produce a continuous stream of income over the next 5 years with rate of flow $f(t) = 2{,}000$. If both the bond and the continuous income stream earn 8% compounded continuously, which is the better investment?

14. Refer to Problem 13. Which is the better investment if the rate of flow of the income from the business is $f(t) = 3{,}000$?

15. A business is planning to purchase a piece of equipment that will produce a continuous stream of income for 8 years with rate of flow $f(t) = 9{,}000$. If the continuous income stream earns 12% compounded continuously, what single deposit into an account earning the same interest rate will produce the same future value as the continuous income stream? (This deposit is called the **present value** of the continuous income stream.)

16. Refer to Problem 15. Find the present value of a continuous income stream at 8% compounded continuously for 12 years if the rate of flow is $f(t) = 1{,}000e^{0.03t}$.

17. Find the future value at a rate r compounded continuously for T years for a continuous income stream with rate of flow $f(t) = k$, where k is a constant.

18. Find the future value at a rate r compounded continuously for T years for a continuous income stream with rate of flow $f(t) = ke^{ct}$, where c and k are constants, $c \neq r$.

19. Find the consumers' surplus at a price level of $\overline{p} = \$150$ for the price–demand equation

$$p = D(x) = 400 - \tfrac{1}{20}x$$

20. Find the consumers' surplus at a price level of $\overline{p} = \$120$ for the price–demand equation

$$p = D(x) = 200 - \tfrac{1}{50}x$$

21. Find the producers' surplus at a price level of $\overline{p} = \$65$ for the price–supply equation

$$p = S(x) = 10 + \tfrac{1}{10}x + \tfrac{1}{3,600}x^2$$

22. Find the producers' surplus at a price level of $\overline{p} = \$55$ for the price–supply equation

$$p = S(x) = 15 + \tfrac{1}{10}x + \tfrac{3}{1,000}x^2$$

23. Find the consumers' surplus and the producers' surplus at the equilibrium price level for

$$p = D(x) = 50 - \tfrac{1}{10}x \qquad \text{and} \qquad p = S(x) = 11 + \tfrac{1}{20}x$$

24. Find the consumers' surplus and the producers' surplus at the equilibrium price level for

$$p = D(x) = 20 - \tfrac{1}{600}x^2 \qquad \text{and} \qquad p = S(x) = 2 + \tfrac{1}{300}x^2$$

25. Find the consumers' surplus and the producers' surplus at the equilibrium price level (rounded to the nearest dollar) for

$$p = D(x) = 80e^{-0.001x} \qquad \text{and} \qquad p = S(x) = 30e^{0.001x}$$

26. Find the consumers' surplus and the producers' surplus at the equilibrium price level (rounded to the nearest dollar) for

$$p = D(x) = 185e^{-0.005x} \qquad \text{and} \qquad p = S(x) = 25e^{0.005x}$$

27. Suppose that in a monopolistic market, the production level for a certain product is always chosen so that the profit is maximized. The price–demand equation for this product is

$$p = 100 - \tfrac{1}{10}x$$

and the cost equation is

$$C(x) = 5,000 + 60x$$

Find the consumers' surplus at the production level that maximizes profit.

28. Suppose that in a certain market, the production level for a product is always chosen so that the revenue is maximized. The price–demand equation for this product is

$$p = 100 - \tfrac{1}{10}x$$

Find the consumers' surplus at the production level that maximizes revenue.

Use a graphic calculator or a computer for Problems 29–32. For a given problem, graph each pair of equations in the same viewing rectangle, approximate the equilibrium price, and then add the graph of $\overline{p}$ = (Equilibrium price) to the graphs in the viewing rectangle.

29. From Problem 23:
$p = D(x) = 50 - \tfrac{1}{10}x$
$p = S(x) = 11 + \tfrac{1}{20}x$
x range: [0, 400]
y range: [0, 50]

30. From Problem 24:
$p = D(x) = 20 - \tfrac{1}{600}x^2$
$p = S(x) = 2 + \tfrac{1}{300}x^2$
x range: [0, 75]
y range: [0, 25]

31. From Problem 25:
$p = D(x) = 80e^{-0.001x}$
$p = S(x) = 30e^{0.001x}$
x range: [0, 750]
y range: [0, 100]

32. From Problem 26:
$p = D(x) = 185e^{-0.005x}$
$p = S(x) = 25e^{0.005x}$
x range: [0, 300]
y range: [0, 200]

SECTION 7-3 Integration by Parts

In Section 6-1, we said we would return later to the indefinite integral

$$\int \ln x \, dx$$

since none of the integration techniques considered up to that time could be used to find an antiderivative for ln x. We will now develop a very useful technique, called *integration by parts*, that will enable us to find not only the above integral, but also many others, including integrals such as

$$\int x \ln x \, dx \qquad \text{and} \qquad \int xe^x \, dx$$

The technique of integration by parts is also used to derive many integration formulas that are tabulated in mathematical handbooks.

The method of integration by parts is based on the product formula for derivatives. If f and g are differentiable functions, then

$$D_x[f(x)g(x)] = f(x)g'(x) + g(x)f'(x)$$

which can be written in the equivalent form

$$f(x)g'(x) = D_x[f(x)g(x)] - g(x)f'(x)$$

Integrating both sides, we obtain

$$\int f(x)g'(x)\,dx = \int D_x[f(x)g(x)]\,dx - \int g(x)f'(x)\,dx$$

The first integral to the right of the equal sign is $f(x)g(x) + C$. (Why?) We will leave out the constant of integration for now, since we can add it after integrating the second integral to the right of the equal sign. So we have

$$\int f(x)g'(x)\,dx = f(x)g(x) - \int g(x)f'(x)\,dx$$

This equation can be transformed into a more convenient form by letting $u = f(x)$ and $v = g(x)$; then $du = f'(x)\,dx$ and $dv = g'(x)\,dx$. Making these substitutions, we obtain the **integration by parts formula:**

Integration by Parts Formula

$$\int u\,dv = uv - \int v\,du$$

This formula can be very useful when the integral on the left is difficult to integrate using standard formulas. If u and dv are chosen with care, then the integral on the right side may be easier to integrate than the one on the left. Several examples will demonstrate the use of the formula.

◆ EXAMPLE 12 Find $\int xe^x\,dx$ using integration by parts.

Solution First, write the integration by parts formula:

$$\int u\,dv = uv - \int v\,du$$

Then, try to identify u and dv in $\int u\,dv$ (this is the key step) so that when $\int u\,dv$ is written in the form $uv - \int v\,du$, the new integral will be easier to evaluate.

Suppose we choose

$$u = x \quad \text{and} \quad dv = e^x\,dx \qquad u\,dv = xe^x\,dx$$

Then

$$du = (D_x\, x)\, dx = dx \qquad \text{and} \qquad v = \int e^x\, dx = e^x$$

Any constant may be added to v (we always choose 0 for simplicity). The general arbitrary constant of integration will be added at the end of the process.

Using the chosen u and dv and the corresponding du and v in the integration by parts formula, we obtain

$$\int u\, dv \;= uv \;- \int v\, du$$

$$\int xe^x\, dx = xe^x - \int e^x\, dx \qquad \text{This new integral is easy to evaluate.}$$

$$= xe^x - e^x + C$$

Check $D_x(xe^x - e^x + C) = xe^x + e^x - e^x = xe^x$ ◆

PROBLEM 12 Find: $\int xe^{2x}\, dx$ ◆

◆ EXAMPLE 13 Find: $\int x \ln x\, dx$

Solution We write the integration by parts formula,

$$\int u\, dv = uv - \int v\, du$$

and choose

$$u = \ln x \qquad \text{and} \qquad dv = x\, dx \qquad u\, dv = (\ln x)x\, dx = x \ln x\, dx$$

Then,

$$du = (D_x \ln x)\, dx = \frac{1}{x}\, dx \qquad \text{and} \qquad v = \int x\, dx = \frac{x^2}{2}$$

and we have

$$\int x \ln x\, dx = (\ln x)\left(\frac{x^2}{2}\right) - \int \left(\frac{x^2}{2}\right)\left(\frac{1}{x}\, dx\right)$$

$$= \frac{x^2}{2} \ln x - \int \frac{x}{2}\, dx \qquad \text{This integral is easy to evaluate.}$$

$$= \frac{x^2}{2} \ln x - \frac{x^2}{4} + C$$

Check $D_x\left(\frac{x^2}{2} \ln x - \frac{x^2}{4} + C\right) = x \ln x + \frac{x^2}{2}\left(\frac{1}{x}\right) - \frac{x}{2} = x \ln x$ ◆

PROBLEM 13 Find: $\int x \ln 2x \, dx$ ◆

Before considering additional examples, let us examine the process of selecting u and dv more closely. In the integral $\int xe^x \, dx$ (Example 12), suppose we had chosen $u = e^x$ and $dv = x \, dx$. Then $du = e^x \, dx$, $v = \frac{1}{2}x^2$, and the integration by parts formula gives

$$\int xe^x \, dx = e^x(\tfrac{1}{2}x^2) - \int \tfrac{1}{2}x^2e^x \, dx$$

$$= \tfrac{1}{2}x^2e^x - \tfrac{1}{2}\int x^2e^x \, dx$$

For this choice of u and dv, the new integral $\int v \, du$ is more complicated than the original integral $\int u \, dv$. This does not mean there is an error in these calculations or in the integration by parts formula. It simply means that this choice of u and dv does not change the original problem into one we can solve. When this happens, we must look for a different choice of u and dv. In some problems, it is possible that no choice will work. These observations and some guidelines for selecting u and dv are summarized below.

Integration by Parts: Selection of *u* and *dv*

1. The product $u \, dv$ must equal the original integrand.
2. It must be possible to integrate dv (preferably by using standard formulas or simple substitutions).
3. The new integral $\int v \, du$, should not be any more involved than the original integral $\int u \, dv$.
4. For integrals involving x^pe^{ax}, try

   $u = x^p$ and $dv = e^{ax} \, dx$

5. For integrals involving $x^p(\ln x)^q$, try

   $u = (\ln x)^q$ and $dv = x^p \, dx$

In some cases, repeated use of the integration by parts formula will lead to the evaluation of the original integral. The next example provides an illustration of such a case.

◆ EXAMPLE 14 Find: $\int x^2e^{-x} \, dx$

Solution Following suggestion 4 in the box, we choose

$$u = x^2 \qquad dv = e^{-x} \, dx$$

Then,

$$du = 2x \, dx \qquad v = -e^{-x}$$

and

$$\int x^2e^{-x}\,dx = x^2(-e^{-x}) - \int (-e^{-x})2x\,dx$$

$$= -x^2e^{-x} + 2\int xe^{-x}\,dx \qquad (1)$$

The new integral is not one we can evaluate by standard formulas, but it is simpler than the original integral. Applying the integration by parts formula to it will produce an even simpler integral. For the integral $\int xe^{-x}\,dx$, we choose

$$u = x \qquad dv = e^{-x}\,dx$$

Then,

$$du = dx \qquad v = -e^{-x}$$

and

$$\int xe^{-x}\,dx = x(-e^{-x}) - \int (-e^{-x})\,dx$$

$$= -xe^{-x} + \int e^{-x}\,dx$$

$$= -xe^{-x} - e^{-x} + C \qquad (2)$$

Substituting (2) into (1), we have

$$\int x^2e^{-x}\,dx = -x^2e^{-x} + 2(-xe^{-x} - e^{-x}) + C$$

$$= -x^2e^{-x} - 2xe^{-x} - 2e^{-x} + C$$

◆

PROBLEM 14 Find: $\int x^2e^{2x}\,dx$ ◆

◆ EXAMPLE 15 Find: $\int_1^e \ln x\,dx$

Solution First, we will find $\int \ln x\,dx$, and then return to the definite integral. Following suggestion 5 in the box (with $p = 0$), we choose

$$u = \ln x \qquad dv = dx$$

Then,

$$du = \frac{1}{x}\,dx \qquad v = x$$

Hence,

$$\int \ln x\,dx = (\ln x)(x) - \int (x)\frac{1}{x}\,dx$$

$$= x\ln x - x + C$$

This is the important result we mentioned at the beginning of this section.

Now, we have

$$\int_1^e \ln x\, dx = (x \ln x - x)\Big|_1^e$$
$$= (e \ln e - e) - (1 \ln 1 - 1)$$
$$= (e - e) - (0 - 1)$$
$$= 1$$

◆

PROBLEM 15 Find: $\int_1^2 \ln 3x\, dx$ ◆

Answers to Matched Problems

12. $\frac{x}{2} e^{2x} - \frac{1}{4} e^{2x} + C$
13. $\frac{x^2}{2} \ln 2x - \frac{x^2}{4} + C$
14. $\frac{x^2}{2} e^{2x} - \frac{x}{2} e^{2x} + \frac{1}{4} e^{2x} + C$
15. $2 \ln 6 - \ln 3 - 1 \approx 1.4849$

EXERCISE 7-3

A

Integrate using the method of integration by parts. Assume $x > 0$ whenever the natural logarithm function is involved.

1. $\int xe^{3x}\, dx$
2. $\int xe^{4x}\, dx$
3. $\int x^2 \ln x\, dx$
4. $\int x^3 \ln x\, dx$

B

Problems 5–18 are mixed—some require integration by parts and others can be solved using techniques we have considered earlier. Integrate as indicated, assuming $x > 0$ whenever the natural logarithm function is involved.

5. $\int xe^{-x}\, dx$
6. $\int (x - 1)e^{-x}\, dx$
7. $\int xe^{x^2}\, dx$
8. $\int xe^{-x^2}\, dx$
9. $\int_0^1 (x - 3)e^x\, dx$
10. $\int_0^2 (x + 5)e^x\, dx$
11. $\int_1^3 \ln 2x\, dx$
12. $\int_2^3 \ln 7x\, dx$
13. $\int \frac{2x}{x^2 + 1}\, dx$
14. $\int \frac{x^2}{x^3 + 5}\, dx$
15. $\int \frac{\ln x}{x}\, dx$
16. $\int \frac{e^x}{e^x + 1}\, dx$
17. $\int \sqrt{x} \ln x\, dx$
18. $\int \frac{\ln x}{\sqrt{x}}\, dx$

C

Problems 19–36 are mixed—some may require use of the integration by parts formula along with techniques we have considered earlier; others may require repeated use of the integration by parts formula. Assume $g(x) > 0$ whenever $\ln g(x)$ is involved.

19. $\int x^2 e^x\, dx$
20. $\int x^3 e^x\, dx$
21. $\int xe^{ax}\, dx, \quad a \neq 0$

22. $\int \ln ax\, dx, \quad a > 0$ 23. $\int_1^e \frac{\ln x}{x^2}\, dx$ 24. $\int_1^2 x^3 e^{x^2}\, dx$

25. $\int_0^2 \ln(x + 4)\, dx$ 26. $\int_0^2 \ln(4 - x)\, dx$ 27. $\int xe^{x-2}\, dx$

28. $\int xe^{x+1}\, dx$ 29. $\int x \ln(1 + x^2)\, dx$ 30. $\int x \ln(1 + x)\, dx$

31. $\int e^x \ln(1 + e^x)\, dx$ 32. $\int \frac{\ln(1 + \sqrt{x})}{\sqrt{x}}\, dx$ 33. $\int (\ln x)^2\, dx$

34. $\int x(\ln x)^2\, dx$ 35. $\int (\ln x)^3\, dx$ 36. $\int x(\ln x)^3\, dx$

Find the area bounded by the graphs of the indicated equations on the given intervals.

37. $y = x - \ln x$ and $y = 0, \quad 1 \le x \le e$
38. $y = 1 - \ln x$ and $y = 0, \quad 1 \le x \le 4$
39. $y = (x - 2)e^x$ and $y = 0, \quad 0 \le x \le 3$
40. $y = (3 - x)e^x$ and $y = 0, \quad 0 \le x \le 3$

Use a graphic calculator or a computer for Problems 41–44. Graph each equation over the indicated interval.

41. From Problem 37:
$y = x - \ln x, \quad [1, e]$

42. From Problem 38:
$y = 1 - \ln x, \quad [1, 4]$

43. From Problem 39:
$y = (x - 2)e^x, \quad [0, 3]$

44. From Problem 40:
$y = (3 - x)e^x, \quad [0, 3]$

APPLICATIONS

Business & Economics

45. Profit. If the rate of change of profit (in millions of dollars per year) is given by

$$P'(t) = 2t - te^{-t}$$

where t is time in years and the profit at time 0 is 0, find $P = P(t)$.

46. Production. An oil field is estimated to produce oil at a rate of $R(t)$ thousand barrels per month t months from now, as given by

$$R(t) = 10te^{-0.1t}$$

Estimate the total production in the first year of operation by use of an appropriate definite integral.

47. Continuous income stream. Find the future value at 8% compounded continuously for 5 years for the continuous income stream with rate of flow

$$f(t) = 1{,}000 - 200t$$

48. *Continuous income stream.* Find the interest earned at 10% compounded continuously for 4 years for the continuous income stream with rate of flow

$$f(t) = 1{,}000 - 250t$$

49. *Income distribution.* Find the coefficient of inequality for the Lorenz curve with equation

$$y = xe^{x-1}$$

50. *Producers' surplus.* Find the producers' surplus at a price level of $\bar{p} = \$26$ for the price–supply equation

$$p = S(x) = 5 \ln(x + 1)$$

Life Sciences

51. *Pollution.* The concentration of particulate matter (in parts per million) t hours after a factory ceases operation for the day is given by

$$C(t) = \frac{20 \ln(t + 1)}{(t + 1)^2}$$

Find the average concentration for the time period from $t = 0$ to $t = 5$.

52. *Medicine.* After a person takes a pill, the drug contained in the pill is assimilated into the bloodstream. The rate of assimilation t minutes after taking the pill is

$$R(t) = te^{-0.2t}$$

Find the total amount of the drug that is assimilated into the bloodstream during the first 10 minutes after the pill is taken.

Social Sciences

53. *Politics.* The number of voters (in thousands) in a certain city is given by

$$N(t) = 20 + 4t - 5te^{-0.1t}$$

where t is time in years. Find the average number of voters during the time period from $t = 0$ to $t = 5$.

SECTION 7-4 Integration Using Tables

- INTRODUCTION
- USING A TABLE OF INTEGRALS
- SUBSTITUTION AND INTEGRAL TABLES
- REDUCTION FORMULAS
- APPLICATION

INTRODUCTION

A **table of integrals** is a list of integration formulas used to evaluate integrals. People who frequently evaluate complex integrals may refer to tables that contain hundreds of formulas. Tables of this type are included in mathematical handbooks available in most college book stores. Table II of Appendix C contains a short list of integral formulas illustrating the types found in more extensive tables. Some of these formulas can be derived using the integration techniques discussed earlier, while others require techniques we have not considered. However, it is possible to verify each formula by differentiating the right side.

◆ USING A TABLE OF INTEGRALS

The formulas in Table II (and in larger integral tables) are organized by categories, such as "*Integrals Involving* $a + bu$," "*Integrals Involving* $\sqrt{u^2 - a^2}$," and so on. The variable u is the variable of integration. All other symbols represent constants. To use a table to evaluate an integral, you must first find the category that most closely agrees with the form of the integrand and then find a formula in that category that can be made to match the integrand exactly by assigning values to the constants in the formula. The following examples illustrate this process.

◆ EXAMPLE 16 Use Table II to find: $\int \frac{x}{(5 + 2x)(4 + 3x)}\, dx$

Solution Since the integrand

$$f(x) = \frac{x}{(5 + 2x)(4 + 3x)}$$

is a rational function involving terms of the form $a + bu$ and $c + du$, we examine formulas 15–20 in Table II to see if any of the integrands in these formulas can be made to match $f(x)$ exactly. Comparing the integrand in formula 16 with $f(x)$, we see that this integrand will match $f(x)$ if we let $a = 5$, $b = 2$, $c = 4$, and $d = 3$. Letting $u = x$ and substituting for a, b, c, and d in formula 16, we have

$$\int \frac{u}{(a + bu)(c + du)}\, du = \frac{1}{ad - bc}\left(\frac{a}{b}\ln|a + bu| - \frac{c}{d}\ln|c + du|\right) \qquad \text{Formula 16}$$

$$\int \frac{x}{(5 + 2x)(4 + 3x)}\, dx = \frac{1}{5 \cdot 3 - 2 \cdot 4}\left(\frac{5}{2}\ln|5 + 2x| - \frac{4}{3}\ln|4 + 3x|\right) + C$$

$$a \quad b \quad c \quad d \qquad\qquad a \cdot d - b \cdot c = 5 \cdot 3 - 2 \cdot 4 = 7$$

$$= \tfrac{5}{14}\ln|5 + 2x| - \tfrac{4}{21}\ln|4 + 3x| + C$$

Notice that the constant of integration C is not included in any of the formulas in Table II. However, you must still include C in all antiderivatives. ◆

PROBLEM 16 Use Table II to find: $\int \frac{1}{(5 + 3x)^2(1 + x)}\, dx$ ◆

◆ EXAMPLE 17 Evaluate: $\displaystyle\int_3^4 \frac{1}{x\sqrt{25 - x^2}}\,dx$

Solution First, we use Table II to find

$$\int \frac{1}{x\sqrt{25 - x^2}}\,dx$$

Since the integrand involves the expression $\sqrt{25 - x^2}$, we examine formulas 29–31 and select formula 29 with $a^2 = 25$ and $a = 5$:

$$\int \frac{1}{u\sqrt{a^2 - u^2}}\,du = -\frac{1}{a}\ln\left|\frac{a + \sqrt{a^2 - u^2}}{u}\right| \qquad \text{Formula 29}$$

$$\int \frac{1}{x\sqrt{25 - x^2}}\,dx = -\frac{1}{5}\ln\left|\frac{5 + \sqrt{25 - x^2}}{x}\right| + C$$

Thus,

$$\int_3^4 \frac{1}{x\sqrt{25 - x^2}}\,dx = -\frac{1}{5}\ln\left|\frac{5 + \sqrt{25 - x^2}}{x}\right|\Bigg|_3^4$$

$$= -\frac{1}{5}\ln\left|\frac{5 + 3}{4}\right| + \frac{1}{5}\ln\left|\frac{5 + 4}{3}\right|$$

$$= -\tfrac{1}{5}\ln 2 + \tfrac{1}{5}\ln 3 = \tfrac{1}{5}\ln 1.5 \approx 0.0811$$ ◆

PROBLEM 17 Evaluate: $\displaystyle\int_6^8 \frac{1}{x^2\sqrt{100 - x^2}}\,dx$ ◆

◆ SUBSTITUTION AND INTEGRAL TABLES

As Examples 16 and 17 illustrate, if the integral we want to evaluate can be made to match one in the table exactly, then evaluating the indefinite integral consists of simply substituting the correct values of the constants into the formula. What happens if we cannot match an integral with one of the formulas in the table? In many cases, a substitution will change the given integral into one that corresponds to a table entry. The following examples illustrate several frequently used substitutions.

◆ EXAMPLE 18 Find: $\displaystyle\int \frac{x^2}{\sqrt{16x^2 - 25}}\,dx$

Solution In order to relate this integral to one of the formulas involving $\sqrt{u^2 - a^2}$ (formulas 40–45), we observe that if $u = 4x$, then

$$u^2 = 16x^2 \qquad \text{and} \qquad \sqrt{16x^2 - 25} = \sqrt{u^2 - 25}$$

Thus, we will use the substitution $u = 4x$ to change this integral into one that appears in the table:

$$\int \frac{x^2}{\sqrt{16x^2 - 25}}\,dx = \frac{1}{4}\int \frac{\frac{1}{16}u^2}{\sqrt{u^2 - 25}}\,du$$

$$= \frac{1}{64}\int \frac{u^2}{\sqrt{u^2 - 25}}\,du$$

Substitution:
$u = 4x$, $du = 4\,dx$
$x = \frac{1}{4}u$

This last integral can be evaluated by using formula 44 with $a = 5$:

$$\int \frac{u^2}{\sqrt{u^2 - a^2}}\,du = \frac{1}{2}\left(u\sqrt{u^2 - a^2} + a^2 \ln|u + \sqrt{u^2 - a^2}|\right)$$ Formula 44

$$\int \frac{x^2}{\sqrt{16x^2 - 25}}\,dx = \frac{1}{64}\int \frac{u^2}{\sqrt{u^2 - 25}}\,du$$ Use formula 44 with $a = 5$.

$$= \frac{1}{128}\left(u\sqrt{u^2 - 25} + 25 \ln|u + \sqrt{u^2 - 25}|\right) + C$$ Substitute $u = 4x$.

$$= \frac{1}{128}\left(4x\sqrt{16x^2 - 25} + 25 \ln|4x + \sqrt{16x^2 - 25}|\right) + C$$ ◆

PROBLEM 18 Find: $\int \sqrt{9x^2 - 16}\,dx$ ◆

◆ EXAMPLE 19 Find: $\int \frac{x}{\sqrt{x^4 + 1}}\,dx$

Solution None of the formulas in the table involve fourth powers; however, if we let $u = x^2$, then

$$\sqrt{x^4 + 1} = \sqrt{u^2 + 1}$$

and this form does appear in formulas 32–39. Thus, we substitute $u = x^2$:

$$\int \frac{1}{\sqrt{x^4 + 1}}\,x\,dx = \frac{1}{2}\int \frac{1}{\sqrt{u^2 + 1}}\,du$$

Substitution:
$u = x^2$, $du = 2x\,dx$

We recognize the last integral as formula 36 with $a = 1$:

$$\int \frac{1}{\sqrt{u^2 + a^2}}\,du = \ln|u + \sqrt{u^2 + a^2}|$$ Formula 36

$$\int \frac{x}{\sqrt{x^4 + 1}}\,dx = \frac{1}{2}\int \frac{1}{\sqrt{u^2 + 1}}\,du$$ Use formula 36 with $a = 1$.

$$= \frac{1}{2}\ln|u + \sqrt{u^2 + 1}| + C$$ Substitute $u = x^2$.

$$= \frac{1}{2}\ln|x^2 + \sqrt{x^4 + 1}| + C$$ ◆

PROBLEM 19 Find: $\int x\sqrt{x^4+1}\,dx$ ◆

◆ REDUCTION FORMULAS

◆ EXAMPLE 20 Use Table II to find: $\int x^2e^{3x}\,dx$

Solution Since the integrand involves the function e^{3x}, we examine formulas 46–48 and conclude that formula 47 can be used for this problem. Letting $u = x$, $n = 2$, and $a = 3$ in formula 47, we have

$$\int u^n e^{au}\,du = \frac{u^n e^{au}}{a} - \frac{n}{a}\int u^{n-1}e^{au}\,du \qquad \text{Formula 47}$$

$$\int x^2e^{3x}\,dx = \frac{x^2e^{3x}}{3} - \frac{2}{3}\int xe^{3x}\,dx$$

Notice that the expression on the right still contains an integral, but the exponent of x has been reduced by 1. Formulas of this type are called **reduction formulas** and are designed to be applied repeatedly until an integral that can be evaluated is obtained. Applying formula 47 to $\int xe^{3x}\,dx$ with $n = 1$, we have

$$\begin{aligned}\int x^2e^{3x}\,dx &= \frac{x^2e^{3x}}{3} - \frac{2}{3}\left(\frac{xe^{3x}}{3} - \frac{1}{3}\int e^{3x}\,dx\right)\\ &= \frac{x^2e^{3x}}{3} - \frac{2xe^{3x}}{9} + \frac{2}{9}\int e^{3x}\,dx\end{aligned}$$

This last expression contains an integral that is easy to evaluate:

$$\int e^{3x}\,dx = \tfrac{1}{3}e^{3x}$$

After making a final substitution and adding a constant of integration, we have

$$\int x^2e^{3x}\,dx = \frac{x^2e^{3x}}{3} - \frac{2xe^{3x}}{9} + \frac{2}{27}e^{3x} + C$$ ◆

PROBLEM 20 Use Table II to find: $\int (\ln x)^2\,dx$ ◆

◆ APPLICATION

◆ EXAMPLE 21 Find the producers' surplus at a price level of $20 for the price–supply equation

Producers' Surplus

$$p = S(x) = \frac{5x}{500 - x}$$

Solution Step 1. Find $\bar{x}$, the supply when the price is $\bar{p} = 20$:

$$\bar{p} = \frac{5\bar{x}}{500 - \bar{x}}$$

$$20 = \frac{5\bar{x}}{500 - \bar{x}}$$

$$10{,}000 - 20\bar{x} = 5\bar{x}$$

$$10{,}000 = 25\bar{x}$$

$$\bar{x} = 400$$

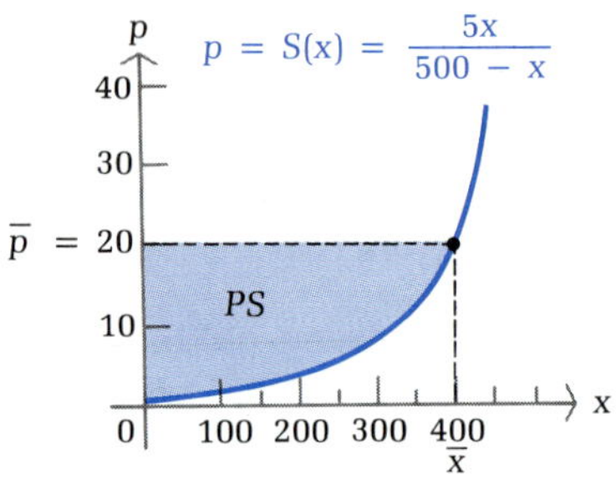

Step 2. Sketch a graph, as shown in the margin.

Step 3. Find the producers' surplus (the shaded area in the graph):

$$PS = \int_0^{\bar{x}} [\bar{p} - S(x)]\, dx$$

$$= \int_0^{400} \left(20 - \frac{5x}{500 - x}\right) dx$$

$$= \int_0^{400} \frac{10{,}000 - 25x}{500 - x}\, dx$$

Use formula 20 with $a = 10{,}000$, $b = -25$, $c = 500$, and $d = -1$:

$$\int \frac{a + bu}{c + du}\, du = \frac{bu}{d} + \frac{ad - bc}{d^2} \ln|c + du| \qquad \text{Formula 20}$$

$$PS = (25x + 2{,}500 \ln|500 - x|)\Big|_0^{400}$$

$$= 10{,}000 + 2{,}500 \ln|100| - 2{,}500 \ln|500|$$
$$\approx \$5{,}976$$ ◆

PROBLEM 21 Find the consumers' surplus at a price level of \$10 for the price–demand equation

$$p = D(x) = \frac{20x - 8{,}000}{x - 500}$$ ◆

Answers to Matched Problems

16. $\frac{1}{2}\left(\frac{1}{5 + 3x}\right) + \frac{1}{4} \ln\left|\frac{1 + x}{5 + 3x}\right| + C$ 17. $\frac{7}{1{,}200} \approx 0.0058$
18. $\frac{1}{6}(3x\sqrt{9x^2 - 16} - 16 \ln|3x + \sqrt{9x^2 - 16}|) + C$
19. $\frac{1}{4}(x^2\sqrt{x^4 + 1} + \ln|x^2 + \sqrt{x^4 + 1}|) + C$
20. $x(\ln x)^2 - 2x \ln x + 2x + C$
21. $3{,}000 + 2{,}000 \ln 200 - 2{,}000 \ln 500 \approx \$1{,}167$

EXERCISE 7-4

A *Use Table II to find each indefinite integral.*

1. $\int \frac{1}{x(1+x)}\,dx$

2. $\int \frac{1}{x^2(1+x)}\,dx$

3. $\int \frac{1}{(3+x)^2(5+2x)}\,dx$

4. $\int \frac{x}{(5+2x)^2(2+x)}\,dx$

5. $\int \frac{x}{\sqrt{16+x}}\,dx$

6. $\int \frac{1}{x\sqrt{16+x}}\,dx$

7. $\int \frac{1}{x\sqrt{x^2+4}}\,dx$

8. $\int \frac{1}{x^2\sqrt{x^2-16}}\,dx$

9. $\int x^2 \ln x\,dx$

10. $\int x^3 \ln x\,dx$

Evaluate each definite integral. Use Table II to find the antiderivative.

11. $\int_1^3 \frac{x^2}{3+x}\,dx$

12. $\int_2^6 \frac{x}{(6+x)^2}\,dx$

13. $\int_0^7 \frac{1}{(3+x)(1+x)}\,dx$

14. $\int_0^7 \frac{x}{(3+x)(1+x)}\,dx$

15. $\int_0^4 \frac{1}{\sqrt{x^2+9}}\,dx$

16. $\int_4^5 \sqrt{x^2-16}\,dx$

B *Use substitution techniques and Table II to find each indefinite integral.*

17. $\int \frac{\sqrt{4x^2+1}}{x^2}\,dx$

18. $\int x^2\sqrt{9x^2-1}\,dx$

19. $\int \frac{x}{\sqrt{x^4-16}}\,dx$

20. $\int x\sqrt{x^4-16}\,dx$

21. $\int x^2\sqrt{x^6+4}\,dx$

22. $\int \frac{x^2}{\sqrt{x^6+4}}\,dx$

23. $\int \frac{1}{x^3\sqrt{4-x^4}}\,dx$

24. $\int \frac{\sqrt{x^4+4}}{x}\,dx$

25. $\int \frac{e^x}{(2+e^x)(3+4e^x)}\,dx$

26. $\int \frac{e^x}{(4+e^x)^2(2+e^x)}\,dx$

27. $\int \frac{\ln x}{x\sqrt{4+\ln x}}\,dx$

28. $\int \frac{1}{x \ln x\sqrt{4+\ln x}}\,dx$

C *Use Table II to find each indefinite integral.*

29. $\int x^2e^{5x}\,dx$

30. $\int x^2e^{-4x}\,dx$

31. $\int x^3e^{-x}\,dx$

32. $\int x^3e^{2x}\,dx$

33. $\int (\ln x)^3\,dx$

34. $\int (\ln x)^4\,dx$

Problems 35–42 are mixed—some require the use of Table II and others can be solved using techniques we considered earlier.

35. $\int_3^5 x\sqrt{x^2-9}\,dx$ **36.** $\int_3^5 x^2\sqrt{x^2-9}\,dx$ **37.** $\int_2^4 \frac{1}{x^2-1}\,dx$

38. $\int_2^4 \frac{x}{(x^2-1)^2}\,dx$ **39.** $\int \frac{x+1}{x^2+2x}\,dx$ **40.** $\int \frac{x+1}{x^2+x}\,dx$

41. $\int \frac{x+1}{x^2+3x}\,dx$ **42.** $\int \frac{x^2+1}{x^2+3x}\,dx$

APPLICATIONS

Use Table II to evaluate all integrals involved in the solutions of Problems 43–54.

Business & Economics

43. Consumers' surplus. Find the consumers' surplus at a price level of $\bar{p} = \$15$ for the price–demand equation

$$p = D(x) = \frac{7{,}500 - 30x}{300 - x}$$

44. Producers' surplus. Find the producers' surplus at a price level of $\bar{p} = \$20$ for the price–supply equation

$$p = S(x) = \frac{10x}{300 - x}$$

45. Continuous income stream. Find the future value at 10% compounded continuously for 10 years for the continuous income stream with rate of flow $f(t) = 50t^2$.

46. Continuous income stream. Find the interest earned at 8% compounded continuously for 5 years for the continuous income stream with rate of flow $f(t) = 200t$.

47. Income distribution. Find the coefficient of inequality for the Lorenz curve with equation

$$y = \tfrac{1}{2}x\sqrt{1+3x}$$

48. Income distribution. Find the coefficient of inequality for the Lorenz curve with equation

$$y = \tfrac{1}{2}x^2\sqrt{1+3x}$$

49. Marketing. After test marketing a new high-fiber cereal, the market research department of a major food producer estimates that monthly sales (in millions of dollars) will grow at the monthly rate of

$$S'(t) = \frac{t^2}{(1+t)^2}$$

t months after the cereal is introduced. If we assume 0 sales at the time the cereal is first introduced, find the total sales, $S(t)$, t months after the cereal is introduced. Find the total sales during the first 2 years this cereal is on the market.

50. *Average price.* At a discount department store, the price–demand equation for premium motor oil is given by

$$p = p(x) = \frac{50}{\sqrt{100 + 6x}}$$

where x is the number of cans of oil that can be sold at a price of $\$p$. Find the average price over the demand interval [50, 250].

Life Sciences

51. *Pollution.* An oil tanker aground on a reef is losing oil and producing an oil slick that is radiating outward at a rate given approximately by

$$\frac{dR}{dt} = \frac{100}{\sqrt{t^2 + 9}} \qquad t \geqslant 0$$

where R is the radius (in feet) of the circular slick after t minutes. Find the radius of the slick after 4 minutes if the radius is 0 when $t = 0$.

52. *Pollution.* The concentration of particulate matter (in parts per million) during a 24 hour period is given approximately by

$$C(t) = t\sqrt{24 - t} \qquad 0 \leqslant t \leqslant 24$$

where t is time in hours. Find the average concentration during the time period from $t = 0$ to $t = 24$.

Social Sciences

53. *Learning.* A person learns N items at a rate given approximately by

$$N'(t) = \frac{60}{\sqrt{t^2 + 25}} \qquad t \geqslant 0$$

where t is the number of hours of continuous study. Determine the total number of items learned in the first 12 hours of continuous study.

54. *Politics.* The number of voters (in thousands) in a metropolitan area is given approximately by

$$f(t) = \frac{500}{2 + 3e^{-t}} \qquad t \geqslant 0$$

where t is time in years. Find the average number of voters during the time period from $t = 0$ to $t = 10$.

Use a graphic calculator or a computer for Problems 55 and 56. For a given problem, graph each pair of equations in the same viewing rectangle. Use approximation techniques to estimate $\bar{x}$ to the nearest unit for the given $\bar{p}$ price level.

55. From Problem 43:

$$p = D(x) = \frac{7{,}500 - 30x}{300 - x}$$

$\bar{p} = 15$
x range: [0, 250]
y range: [0, 30]

56. From Problem 44:

$$p = S(x) = \frac{10x}{300 - x}$$

$\bar{p} = 20$
x range: [0, 250]
y range: [0, 50]

SECTION 7-5 Improper Integrals

- IMPROPER INTEGRALS
- APPLICATION: CAPITAL VALUE
- PROBABILITY DENSITY FUNCTIONS

IMPROPER INTEGRALS

We are now going to consider an integral form that has wide application in probability studies as well as other areas. Earlier, when we introduced the idea of a definite integral,

$$\int_a^b f(x)\,dx \tag{1}$$

we required f to be continuous over a closed interval $[a, b]$. Now we are going to extend the meaning of (1) so that the interval $[a, b]$ may become infinite in length.

Let us investigate a particular example that will motivate several general definitions. What would be a reasonable interpretation for the following expression?

$$\int_1^\infty \frac{dx}{x^2}$$

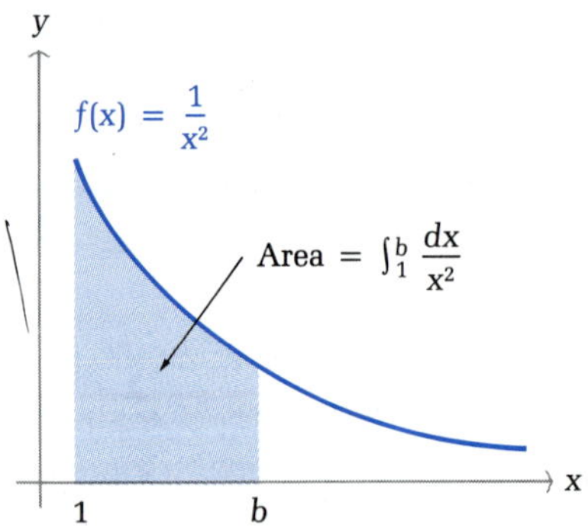

FIGURE 6

Sketching a graph of $f(x) = 1/x^2$, $x \geq 1$ (see Figure 6), we note that for any fixed $b > 1$, $\int_1^b f(x)\,dx$ is the area between the curve $y = 1/x^2$, the x axis, $x = 1$, and $x = b$.

Let us see what happens when we let $b \to \infty$; that is, when we compute the following limit:

$$\lim_{b\to\infty} \int_1^b \frac{dx}{x^2} = \lim_{b\to\infty} \left[(-x^{-1})\Big|_1^b \right]$$
$$= \lim_{b\to\infty} \left(-\frac{1}{b} + 1 \right) = 1$$

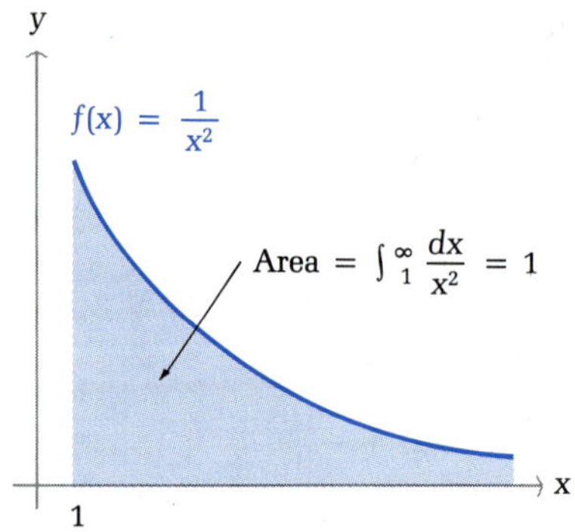

FIGURE 7

Did you expect this result? No matter how large b is taken, the area under the curve from $x = 1$ to $x = b$ never exceeds 1, and in the limit it is 1 (see Figure 7). This suggests that we write

$$\int_1^\infty \frac{dx}{x^2} = \lim_{b\to\infty} \int_1^b \frac{dx}{x^2} = 1$$

This integral is an example of an *improper integral.* In general, the forms

$$\int_{-\infty}^b f(x)\,dx \qquad \int_a^\infty f(x)\,dx \qquad \int_{-\infty}^\infty f(x)\,dx$$

where f is continuous over the indicated interval, are called **improper integrals.** (These integrals are "improper" because the interval of integration is unbounded, as indicated by the use of ∞ for one or both limits of integration. There are other types of improper integrals involving certain types of points of discontinuity within the interval of integration, but these will not be considered here.) Each type of improper integral above is formally defined in the following box:

Improper Integrals

If f is continuous over the indicated interval and the limit exists, then:

1. $\int_a^{\infty} f(x)\,dx = \lim_{b\to\infty} \int_a^b f(x)\,dx$
2. $\int_{-\infty}^{b} f(x)\,dx = \lim_{a\to-\infty} \int_a^b f(x)\,dx$
3. $\int_{-\infty}^{\infty} f(x)\,dx = \int_{-\infty}^{c} f(x)\,dx + \int_c^{\infty} f(x)\,dx$

   where c is any point on $(-\infty, \infty)$, provided *both* improper integrals on the right exist.

If the indicated limit exists, then the improper integral is said to exist or to **converge;** if the limit does not exist, then the improper integral is said not to exist or to **diverge** (and no value is assigned to it).

◆ EXAMPLE 22 Evaluate the following, if it converges: $\int_2^{\infty} \frac{dx}{x}$

Solution

$$\int_2^{\infty} \frac{dx}{x} = \lim_{b\to\infty} \int_2^b \frac{dx}{x}$$
$$= \lim_{b\to\infty} \left[(\ln x)\Big|_2^b \right]$$
$$= \lim_{b\to\infty} (\ln b - \ln 2)$$

Since $\ln b \to \infty$ as $b \to \infty$, the limit does not exist. Hence, the improper integral diverges. ◆

PROBLEM 22 Evaluate the following, if it converges: $\int_3^{\infty} \frac{dx}{(x-1)^2}$ ◆

◆ EXAMPLE 23 Evaluate the following, if it converges: $\int_{-\infty}^{2} e^x\,dx$

Solution

$$\begin{aligned}\int_{-\infty}^{2} e^x\,dx &= \lim_{a\to-\infty}\int_a^2 e^x\,dx\\ &= \lim_{a\to-\infty}\left(e^x\Big|_a^2\right)\\ &= \lim_{a\to-\infty}(e^2 - e^a) = e^2 - 0 = e^2\end{aligned}$$

The integral converges. ◆

PROBLEM 23 Evaluate the following, if it converges: $\int_{-\infty}^{-1} x^{-2}\,dx$ ◆

◆ EXAMPLE 24 Evaluate the following, if it converges: $\int_{-\infty}^{\infty} \frac{2x}{(1+x^2)^2}\,dx$

Solution

$$\begin{aligned}\int_{-\infty}^{\infty} \frac{2x}{(1+x^2)^2}\,dx &= \int_{-\infty}^{0}(1+x^2)^{-2}2x\,dx + \int_0^{\infty}(1+x^2)^{-2}2x\,dx\\ &= \lim_{a\to-\infty}\int_a^0(1+x^2)^{-2}2x\,dx + \lim_{b\to\infty}\int_0^b(1+x^2)^{-2}2x\,dx\\ &= \lim_{a\to-\infty}\left[\frac{(1+x^2)^{-1}}{-1}\Big|_a^0\right] + \lim_{b\to\infty}\left[\frac{(1+x^2)^{-1}}{-1}\Big|_0^b\right]\\ &= \lim_{a\to-\infty}\left[-1+\frac{1}{1+a^2}\right] + \lim_{b\to\infty}\left[-\frac{1}{1+b^2}+1\right]\\ &= -1+1 = 0\end{aligned}$$

The integral converges. ◆

PROBLEM 24 Evaluate the following, if it converges: $\int_{-\infty}^{\infty} \frac{dx}{e^x}$ ◆

◆ EXAMPLE 25 It is estimated that an oil well will produce oil at a rate of $R(t)$ million barrels per year t years from now, as given by

Oil Production

$$R(t) = te^{-0.1t}$$

Estimate the total amount of oil that will be produced by this well.

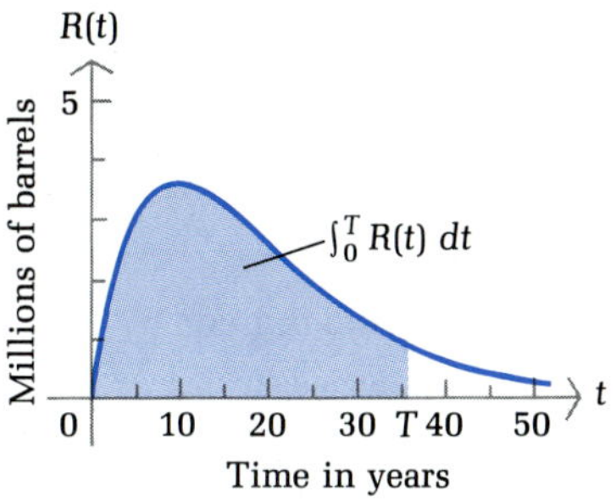

Solution The total amount of oil produced in T years of operation is $\int_0^T R(t)\,dt$ (see the figure in the margin). At some point in time, the annual production rate will become so low that it will no longer be economically feasible to operate the well. However, since we do not know when this will occur, it is convenient to assume that the well is operated indefinitely and use an improper integral. Thus, the total amount of oil produced is approximately

$$\int_0^\infty R(t)\,dt = \lim_{T\to\infty} \int_0^T R(t)\,dt$$

$$= \lim_{T\to\infty} \int_0^T te^{-0.1t}\,dt \quad \text{Use integration by parts or formula 47 in Table II.}$$

$$= \lim_{T\to\infty} \left[(-10te^{-0.1t} - 100e^{-0.1t}) \Big|_0^T \right]$$

$$= \lim_{T\to\infty} (-10Te^{-0.1T} - 100e^{-0.1T} + 100)$$

∞/∞ form

$$= \lim_{T\to\infty} \frac{-10T}{e^{0.1T}} - 0 + 100 \quad \text{Use L'Hôpital's rule.}$$

$$= \lim_{T\to\infty} \frac{-10}{0.1e^{0.1T}} + 100$$

$$= 100 \quad \text{or } 100{,}000{,}000 \text{ barrels}$$

The total production over any finite time period $[0, T]$ is less than 100,000,000 barrels, and the larger T is, the closer the total production will be to 100,000,000 barrels (look back at the figure). ◆

PROBLEM 25 The annual production rate (in millions of barrels) for an oil well is given by

$$R(t) = 10te^{-0.5t}$$

Assuming that the well is operated indefinitely, find the total production. ◆

◆ APPLICATION: CAPITAL VALUE

Recall that if money is invested at a rate r compounded continuously for T years, then the present value, PV, and the future value, FV, are related by the continuous compound interest formula (see Section 5-1)

$$FV = PVe^{rT}$$

which also can be written as

$$PV = e^{-rT}FV$$

This relationship is valid for both single deposits and amounts generated by continuous income streams. Thus, if $f(t)$ is a continuous income stream, the future value is given by (see Section 7-2)

$$FV = e^{rT} \int_0^T f(t)e^{-rt}\,dt \qquad \text{Future value}$$

and, using the compound interest formula, the present value is given by

$$PV = e^{-rT}FV = e^{-rT}e^{rT} \int_0^T f(t)e^{-rt}\,dt$$

Since $e^{-rT}e^{rT} = 1$, we have

$$PV = \int_0^T f(t)e^{-rt}dt \qquad \text{Present value}$$

If we let T approach ∞ in this formula for present value, we obtain an improper integral that represents the *capital value* of the income stream.

Capital Value of a Perpetual Income Stream

A continuous income stream is called **perpetual** if it never stops producing income. The **capital value, CV,** of a perpetual income stream $f(t)$ at a rate r compounded continuously is the present value over the time interval $[0, \infty)$. That is,

$$CV = \int_0^\infty f(t)e^{-rt}\,dt$$

Capital value provides a method for expressing the worth (in terms of today's dollars) of an investment that will produce income for an indefinite period of time.

◆ EXAMPLE 26 A family has leased the oil rights of a property to a petroleum company in return for a perpetual annual payment of $1,200. Find the capital value of this lease at 10% compounded continuously.

Solution The annual payments from the oil company produce a continuous income stream with rate of flow $f(t) = 1{,}200$ that continues indefinitely. (It is common practice to treat a sequence of equal periodic payments as a continuous income stream with a constant rate of flow, even if the income is received only at the end of each period.) Thus, the capital value is

$$CV = \int_0^\infty f(t)e^{-rt}\, dt$$

$$= \int_0^\infty 1{,}200e^{-0.1t}\, dt$$

$$= \lim_{T\to\infty} \int_0^T 1{,}200e^{-0.1t}\, dt$$

$$= \lim_{T\to\infty} \left[-12{,}000e^{-0.1t} \Big|_0^T \right]$$

$$= \lim_{T\to\infty} (-12{,}000e^{-0.1T} + 12{,}000) = \$12{,}000$$ ◆

PROBLEM 26 Repeat Example 26 if the interest rate is 8%, compounded continuously. ◆

◆ PROBABILITY DENSITY FUNCTIONS

We will now take a brief look at the use of improper integrals relative to *probability density functions.* The approach will be intuitive and informal.

Suppose an experiment is designed in such a way that any real number x on the interval $[a, b]$ is a possible outcome. For example, x may represent an IQ score, the height of a person in inches, or the life of a light bulb in hours.

In certain situations it is possible to find a function f with x as an independent variable that can be used to determine the probability that x will assume a value on a given subinterval of $(-\infty, \infty)$. Such a function, called a **probability density function,** must satisfy the following three conditions (see Figure 8):

1. $f(x) \geq 0$ for all $x \in (-\infty, \infty)$
2. $\int_{-\infty}^{\infty} f(x)\, dx = 1$
3. If $[c, d]$ is a subinterval of $(-\infty, \infty)$, then

$$\text{Probability}(c \leq x \leq d) = \int_c^d f(x)\, dx$$

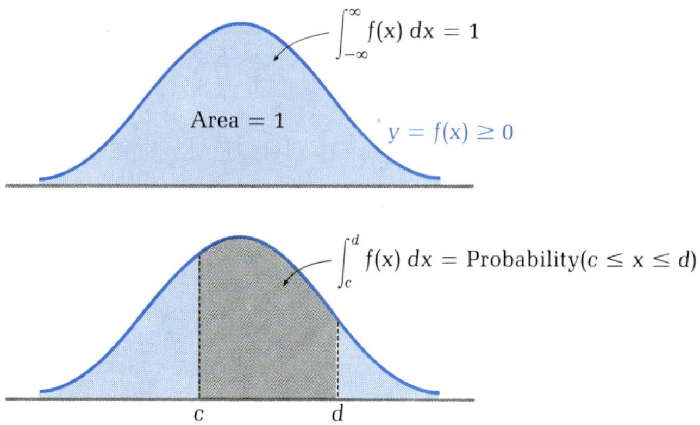

FIGURE 8

◆ EXAMPLE 27

Finish Time

A sailing club has a race over the same course twice a month. The races always start at 12 noon on Sunday, and the boats finish according to the probability density function (where x is hours after noon):

$$f(x) = \begin{cases} -\frac{x}{2} + 2 & \text{if } 2 \leq x \leq 4 \\ 0 & \text{otherwise} \end{cases}$$

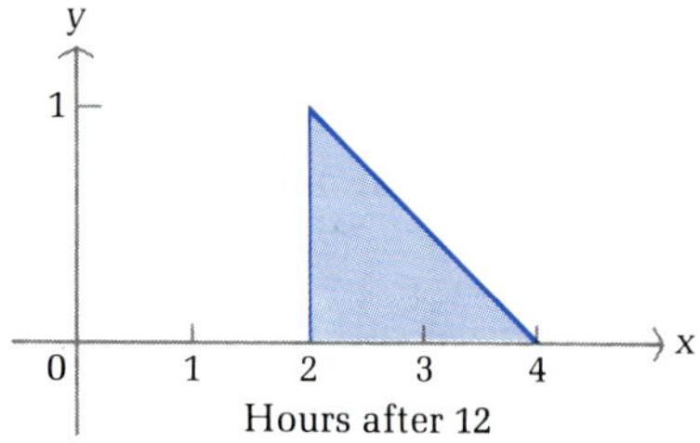

Note that $f(x) \geq 0$ and

$$\int_{-\infty}^{\infty} f(x)\,dx = \underbrace{\int_{-\infty}^{2} f(x)\,dx}_{f(x)=0} + \underbrace{\int_{2}^{4} f(x)\,dx}_{f(x) = -\frac{x}{2}+2} + \underbrace{\int_{4}^{\infty} f(x)\,dx}_{f(x)=0}$$

$$= 0 + \int_{2}^{4} \left(-\frac{x}{2} + 2\right) dx + 0$$

$$= \left(-\frac{x^2}{4} + 2x\right)\Big|_2^4 = 1$$

The probability that a boat selected at random from the sailing fleet will finish between 2 and 3 hours after the start is given by

$$\text{Probability}(2 \leq x \leq 3) = \int_2^3 \left(-\frac{x}{2} + 2\right) dx$$

$$= \left(-\frac{x^2}{4} + 2x\right)\Big|_2^3 = .75$$

which is the area under the curve from $x = 2$ to $x = 3$. ◆

PROBLEM 27

In Example 27, find the probability that a boat selected at random from the fleet will finish between 2:30 and 3:30 PM. ◆

◆ EXAMPLE 28

Duration of Telephone Calls

Suppose the length of telephone calls (in minutes) in a public telephone booth has the probability density function

$$f(x) = \begin{cases} \frac{1}{4}e^{-t/4} & \text{if } t \geq 0 \\ 0 & \text{otherwise} \end{cases}$$

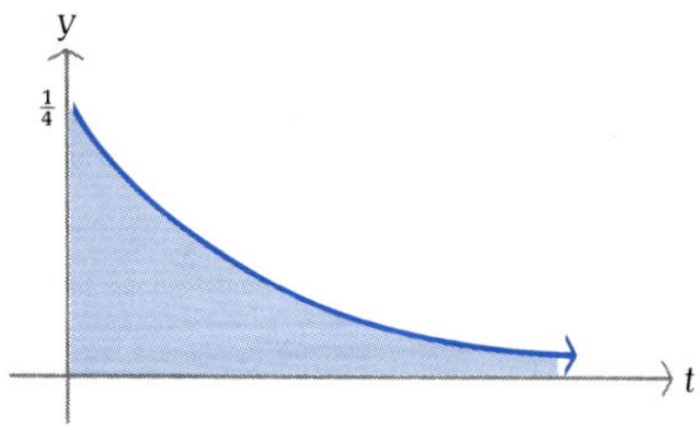

(A) Compute $\int_{-\infty}^{\infty} f(t)\, dt$.

(B) Determine the probability that a call selected at random will last between 2 and 3 minutes.

Solutions (A) $$\int_{-\infty}^{\infty} f(t)\, dx = \int_{-\infty}^{0} f(t)\, dt + \int_{0}^{\infty} f(t)\, dt$$

$$= 0 + \int_0^{\infty} \frac{1}{4} e^{-t/4}\, dt$$

$$= \lim_{b\to\infty} \int_0^{b} \frac{1}{4} e^{-t/4}\, dt$$

$$= \lim_{b\to\infty} \left(-e^{-t/4} \Big|_0^b \right)$$

$$= \lim_{b\to\infty} (-e^{-b/4} + e^0)$$

$$= \lim_{b\to\infty} \left(-\frac{1}{e^{b/4}} + 1 \right)$$

$$= 0 + 1 = 1$$

(B) $$\text{Probability}(2 \leqslant t \leqslant 3) = \int_2^3 \frac{1}{4} e^{-t/4}\, dt$$

$$= (-e^{-t/4}) \Big|_2^3$$

$$= -e^{-3/4} + e^{-1/2} \approx .13$$ ◆

PROBLEM 28 In Example 28, find the probability that a call selected at random will last longer than 4 minutes. ◆

The most important probability density function is the **normal probability density function** defined below and graphed in Figure 9.

$$f(x) = \frac{1}{\sigma\sqrt{2\pi}} e^{-(x-\mu)^2/2\sigma^2}$$ μ is the mean

σ is the standard deviation

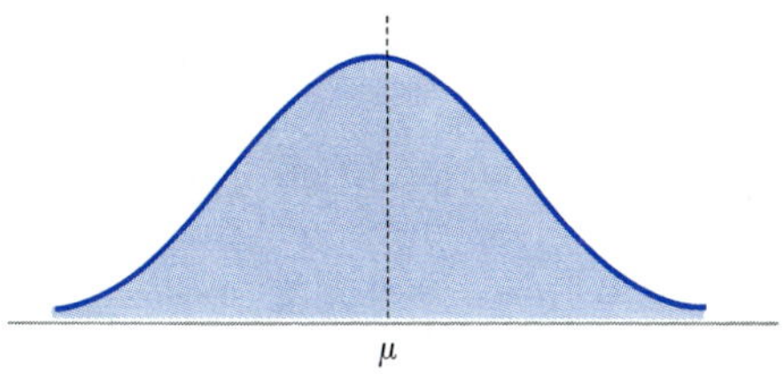

FIGURE 9
Normal curve

It can be shown, but not easily, that

$$\frac{1}{\sigma\sqrt{2\pi}} \int_{-\infty}^{\infty} e^{-(x-\mu)^2/2\sigma^2}\, dx = 1$$

Since $\int e^{-x^2}\,dx$ is nonintegrable in terms of elementary functions (that is, the antiderivative cannot be expressed as a finite combination of simple functions), probabilities such as

$$\text{Probability}(c \leq x \leq d) = \frac{1}{\sigma\sqrt{2\pi}} \int_c^d e^{-(x-\mu)^2/2\sigma^2}\,dx$$

cannot be evaluated by using the fundamental theorem of calculus. However, these probabilities can be approximated by Riemann sums or other numerical integration techniques. In Section 12-7 we will use a table to approximate area under the standard normal curve (that is, the normal curve with $\mu = 0$ and $\sigma = 1$) and we will see that this table can be used to approximate area under any normal curve. Many calculators also have the capability of computing areas under the standard normal curve.

Answers to Matched Problems

22. $\frac{1}{2}$ 23. 1 24. Diverges 25. 40,000,000 barrels
26. \$15,000 27. .5 28. $e^{-1} \approx .37$

EXERCISE 7-5

Find the value of each improper integral that converges.

A

1. $\int_1^\infty \frac{dx}{x^4}$
2. $\int_1^\infty \frac{dx}{x^3}$
3. $\int_0^\infty e^{-x/2}\,dx$
4. $\int_0^\infty e^{-x}\,dx$

B

5. $\int_1^\infty \frac{dx}{\sqrt{x}}$
6. $\int_1^\infty \frac{dx}{\sqrt[3]{x}}$
7. $\int_0^\infty \frac{dx}{(x+1)^2}$
8. $\int_0^\infty \frac{dx}{(x+1)^3}$
9. $\int_0^\infty \frac{dx}{(x+1)^{2/3}}$
10. $\int_0^\infty \frac{dx}{\sqrt{x+1}}$
11. $\int_1^\infty \frac{dx}{x^{0.99}}$
12. $\int_1^\infty \frac{dx}{x^{1.01}}$
13. $0.3\int_0^\infty e^{-0.3x}\,dx$
14. $0.01\int_0^\infty e^{-0.1x}\,dx$
15. In Example 27, find the probability that a randomly selected boat will finish before 3:30 PM.
16. In Example 27, find the probability that a randomly selected boat will finish after 2:30 PM.
17. In Example 28, find the probability that a telephone call selected at random will last longer than 1 minute.
18. In Example 28, find the probability that a telephone call selected at random will last less than 3 minutes.

C *Find the value of each improper integral that converges.*

19. $\int_0^\infty \frac{1}{k}e^{-x/k}\,dx,\quad k > 0$
20. $\int_1^\infty \frac{1}{x^p}\,dx,\quad p > 1$
21. $\int_0^\infty xe^{-x}\,dx$
22. $\int_0^\infty x^2e^{-x}\,dx$
23. $\int_{-\infty}^\infty \frac{x}{1+x^2}\,dx$
24. $\int_{-\infty}^\infty xe^{-x^2}\,dx$

25. $\int_0^\infty (e^{-x} - e^{-2x})\,dx$

26. $\int_0^\infty (e^{-x} + 2e^{x})\,dx$

27. $\int_{-\infty}^0 \frac{1}{\sqrt{1-x}}\,dx$

28. $\int_{-\infty}^0 \frac{1}{\sqrt[3]{(1-x)^4}}\,dx$

29. $\int_1^\infty \frac{\ln x}{x}\,dx$

30. $\int_1^\infty \frac{\ln x}{x^2}\,dx$

In Problems 31–38, find F(b), use a graphing calculator or a computer to graph F(b), and use the graph to estimate $\lim_{b\to\infty} F(b)$.

31. From Problem 1, $F(b) = \int_1^b \frac{dx}{x^4}$

32. From Problem 2, $F(b) = \int_1^b \frac{dx}{x^3}$

33. From Problem 3, $F(b) = \int_0^b e^{-x/2}\,dx$

34. From Problem 4, $F(b) = \int_0^b e^{-x}\,dx$

35. From Problem 5, $F(b) = \int_1^b \frac{dx}{\sqrt{x}}$

36. From Problem 6, $F(b) = \int_1^b \frac{dx}{\sqrt[3]{x}}$

37. From Problem 7, $F(b) = \int_0^b \frac{dx}{(x+1)^2}$

38. From Problem 8, $F(b) = \int_0^b \frac{dx}{(x+1)^3}$

APPLICATIONS

Business & Economics

39. *Capital value.* The perpetual annual rent for a property is \$6,000. Find the capital value at 12% compounded continuously.

40. *Capital value.* The perpetual annual rent for a property is \$10,000. Find the capital value at 5% compounded continuously.

41. *Capital value.* A trust fund produces a perpetual stream of income with rate of flow

$$f(t) = 1{,}500e^{0.04t}$$

Find the capital value at 9% compounded continuously.

42. *Capital value.* A trust fund produces a perpetual stream of income with rate of flow

$$f(t) = 1{,}000t$$

Find the capital value at 10% compounded continuously.

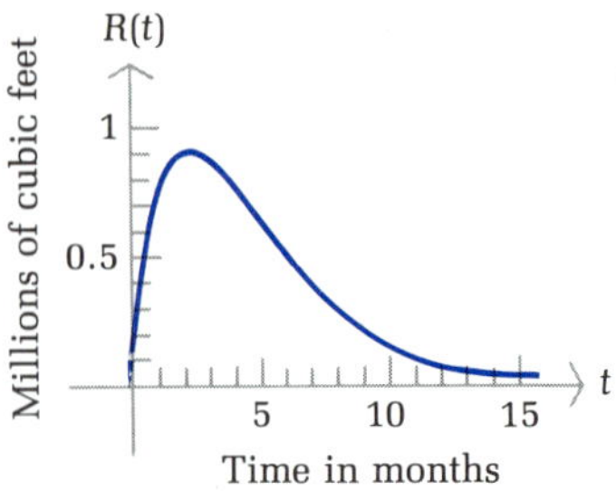

43. Production. The rate of production of a natural gas well (in millions of cubic feet per month) is given by

$$R(t) = te^{-0.4t}$$

See the figure in the margin. Assuming that the well is operated indefinitely, find the total production.

44. Production. Repeat Problem 43 for $R(t) = t^2e^{-0.4t}$.

45. Product life. The life expectancy (in years) of a certain brand of clock radios is a continuous random variable with probability density function

$$f(x) = \begin{cases} 2/(x+2)^2 & \text{if } x \geq 0 \\ 0 & \text{otherwise} \end{cases}$$

Find the probability that a randomly selected clock radio lasts:

(A) At most 6 years (B) From 6 to 12 years

46. Product life. The shelf life (in years) of a certain brand of flashlight batteries is a continuous random variable with probability density function

$$f(x) = \begin{cases} 1/(x+1)^2 & \text{if } x \geq 0 \\ 0 & \text{otherwise} \end{cases}$$

Find the probability that a randomly selected battery has a shelf life of:

(A) 3 years or less (B) From 3 to 9 years

47. Warranty. A manufacturer guarantees a product for 1 year. The time to failure of the product after it is sold is given by the probability density function

$$f(x) = \begin{cases} .01e^{-.01t} & \text{if } t \geq 0 \\ 0 & \text{otherwise} \end{cases}$$

where t is time in months. What is the probability that a buyer chosen at random will have a product failure:

(A) During the warranty period?
(B) During the second year after purchase?

48. Consumption. In a certain city, the daily use of water (in hundreds of gallons) per household is a continuous random variable with probability density function

$$f(x) = \begin{cases} .15e^{-.15x} & \text{if } x \geq 0 \\ 0 & \text{otherwise} \end{cases}$$

Find the probability that a household chosen at random will use:

(A) At most 400 gallons of water per day
(B) Between 300 and 600 gallons of water per day

49. Warranty. In Problem 47, what is the probability that the product will last at least 1 year?

50. Consumption. In Problem 48, what is the probability that a household will use more than 400 gallons of water per day?

Life Sciences

51. *Pollution.* It has been estimated that the rate of seepage of toxic chemicals from a waste dump is $R(t)$ gallons per year t years from now, where

$$R(t) = \frac{500}{(1 + t)^2}$$

Assuming that this seepage continues indefinitely, find the total amount of toxic chemicals that seep from the dump.

52. *Drug assimilation.* When a person takes a drug, the body does not assimilate all of the drug. One way to determine the amount of the drug that is assimilated is to measure the rate at which the drug is eliminated from the body. If the rate of elimination of the drug (in milliliters per minute) is given by

$$R(t) = te^{-0.2t}$$

where t is the time in minutes since the drug was administered, how much of the drug is eliminated from the body?

53. *Medicine.* If the length of stay for people in a hospital has a probability density function

$$g(t) = \begin{cases} .2e^{-.2t} & \text{if } t \geq 0 \\ 0 & \text{otherwise} \end{cases}$$

where t is time in days, find the probability that a patient chosen at random will stay in the hospital for at most 5 days.

54. *Medicine.* For a particular disease, the length of time in days for recovery has a probability density function of the form

$$R(t) = \begin{cases} .03e^{-.03t} & \text{if } t \geq 0 \\ 0 & \text{otherwise} \end{cases}$$

For a randomly selected person who contracts this disease, what is the probability that he or she will take at least 7 days to recover?

Social Sciences

55. *Politics.* In a particular election, the length of time each voter spent on campaigning for a candidate or issue was found to have a probability density function

$$F(x) = \begin{cases} \dfrac{1}{(x + 1)^2} & \text{if } x \geq 0 \\ 0 & \text{otherwise} \end{cases}$$

where x is time in minutes. For a voter chosen at random, what is the probability of his or her spending at least 9 minutes on the campaign?

56. *Psychology.* In an experiment on conditioning, pigeons were required to recognize on a light display one pattern of dots out of five possible patterns to receive a food pellet. After the ninth successful trial, it was found that the probability density function for the length of time in seconds until success on the tenth trial is given by

$$f(t) = \begin{cases} e^{-t} & \text{if } t \geq 0 \\ 0 & \text{otherwise} \end{cases}$$

What is the probability that a pigeon selected at random from those having successfully completed nine trials will take 2 or more seconds to complete the tenth trial successfully?

SECTION 7-6 Chapter Review

Important Terms and Symbols

7-1 *Area between Curves.* Area between a curve and the x axis; area between two curves; useful life; distribution of income; Lorenz curve; absolute equality; coefficient of inequality

7-2 *Applications in Business and Economics.* Continuous income stream; rate of flow; total income; future value; consumers' surplus; producers' surplus; equilibrium price; equilibrium quantity

7-3 *Integration by Parts.*

$$\int u\,dv = uv - \int v\,du$$

7-4 *Integration Using Tables.* Table of integrals; substitution and integral tables; reduction formulas

7-5 *Improper Integrals.* Improper integral; converge; diverge; perpetual income stream; capital value; probability density function; normal probability density function

$$\int_a^\infty f(x)\,dx = \lim_{b\to\infty}\int_a^b f(x)\,dx; \quad \int_{-\infty}^b f(x)\,dx = \lim_{a\to-\infty}\int_a^b f(x)\,dx;$$

$$\int_{-\infty}^\infty f(x)\,dx = \int_{-\infty}^c f(x)\,dx + \int_c^\infty f(x)\,dx$$

EXERCISE 7-6 Chapter Review

Work through all the problems in this chapter review and check your answers in the back of the book. (Answers to all review problems are there.) Where weaknesses show up, review appropriate sections in the text.

A *Evaluate the indicated integrals, if possible.*

1. $\int xe^{4x}\,dx$ **2.** $\int x \ln x\,dx$ **3.** $\int \frac{1}{x(1+x)^2}\,dx$

4. $\int \frac{1}{x^2\sqrt{1+x}}\,dx$ 5. $\int_0^{\infty} e^{-2x}\,dx$ 6. $\int_0^{\infty} \frac{1}{x+1}\,dx$

7. $\int_1^{\infty} \frac{16}{x^3}\,dx$

8. Find the area between the graph of $f(x) = x^2 - 16$ and the x axis for $-2 \leqslant x \leqslant 4$.

9. Find the area between the graphs of $f(x) = e^{-x}$ and $g(x) = 3$ for $0 \leqslant x \leqslant 2$.

B *Evaluate the indicated integrals.*

10. $\int_0^1 xe^x\,dx$ 11. $\int_{-\infty}^0 e^x\,dx$ 12. $\int_0^3 \frac{x^2}{\sqrt{x^2+16}}\,dx$

13. $\int \sqrt{9x^2-49}\,dx$ 14. $\int te^{-0.5t}\,dt$ 15. $\int x^2 \ln x\,dx$

16. $\int \frac{1}{1+2e^x}\,dx$ 17. $\int_0^{\infty} \frac{1}{(x+3)^2}\,dx$

18. Find the area between the graphs of $y = \ln x$ and $y = 0$ for $0.5 \leqslant x \leqslant e$.

19. Find the area between the graphs of $y = x^2 - 6x + 9$ and $y = 9 - x$.

C *Evaluate the indicated integrals.*

20. $\int \frac{(\ln x)^2}{x}\,dx$ 21. $\int x(\ln x)^2\,dx$ 22. $\int xe^{-2x^2}\,dx$

23. $\int x^2e^{-2x}\,dx$ 24. $\int \frac{x}{\sqrt{x^2-36}}\,dx$ 25. $\int \frac{x}{\sqrt{x^4-36}}\,dx$

26. $\int_0^4 x\ln(10-x)\,dx$ 27. $\int (\ln x)^2\,dx$ 28. $\int_{-\infty}^{\infty} \frac{x}{(1+x^2)^3}\,dx$

29. $\int_0^{\infty} (x+1)e^{-x}\,dx$ 30. $\int_1^{\infty} \frac{\ln x}{x^3}\,dx$

31. Find the area bounded by the graphs of $y = x^3 - 6x^2 + 9x$ and $y = x$.

APPLICATIONS

Business & Economics

32. *Useful life.* The total accumulated costs $C(t)$ and revenues $R(t)$ (in thousands of dollars), respectively, for a coal mine satisfy

$$C'(t) = 3 \quad \text{and} \quad R'(t) = 20e^{-0.1t}$$

where t is the number of years the mine has been in operation. Find the useful life of the mine to the nearest year. What is the total profit accumulated during the useful life of the mine?

33. *Income distribution.* An economist produced the following Lorenz curves for the current income distribution and the projected income distribution 10 years from now in a certain country:

$f(x) = \frac{1}{10}x + \frac{9}{10}x^2$ Current Lorenz curve

$g(x) = x^{1.5}$ Projected Lorenz curve

Find the coefficient of inequality for each Lorenz curve and interpret the results.

34. *Continuous income stream.* The rate of flow of a continuous income stream for a 5 year period is given by

$$f(t) = 2{,}500e^{0.05t} \qquad 0 \leq t \leq 5$$

(A) Find the total income over this 5 year period.
(B) Find the future value at the end of this 5 year period at 15% compounded continuously.
(C) Find the interest earned during this 5 year period.

35. *Consumers' and producers' surplus.* Given the price–demand and price–supply equations

$$p = D(x) = 70 - \tfrac{1}{5}x \qquad p = S(x) = 13 + \tfrac{3}{2{,}500}x^2$$

(A) Find the consumers' surplus at a price level of $\overline{p} = \$50$.
(B) Find the producers' surplus at a price level of $\overline{p} = \$25$.
(C) Find the equilibrium price, and then find the consumers' surplus and the producers' surplus at the equilibrium price level.

36. *Production.* An oil field is estimated to produce oil at the rate of $R(t)$ thousand barrels per month t months from now, as given by

$$R(t) = 25te^{-0.05t}$$

How much oil is produced during the first 2 years of operation? If the well is operated indefinitely, what is the total amount of oil produced?

37. *Capital value.* The perpetual annual rent for a property is $2,400. Find the capital value at 12% compounded continuously.

38. *Marketing.* The market research department of a major soft-drink producer estimates that the monthly sales (in millions of dollars) of their new diet soda will grow at the monthly rate of

$$S'(t) = \frac{2}{1 + e^{-0.2t}}$$

t months after the soda is introduced. If we assume 0 sales at the time the new soda was first introduced, find the total sales, $S(t)$, t months after the soda is first introduced. Find the total sales during the first 2 years the soda is on the market.

39. *Parts testing.* If in testing printed circuits for calculators, failures occur relative to time in hours according to the probability density function

$$F(t) = \begin{cases} .02e^{-.02t} & \text{if } t \geqslant 0 \\ 0 & \text{otherwise} \end{cases}$$

what is the probability that a circuit chosen at random will fail in the first hour of testing?

Life Sciences

40. *Drug assimilation.* The rate at which the body eliminates a drug (in milliliters per hour) is given by

$$R(t) = \frac{60t}{(t+1)^2(t+2)}$$

where t is the number of hours since the drug was administered. How much of the drug is eliminated in the first hour after it was administered? What is the total amount of the drug that is eliminated by the body?

41. *Medicine.* For a particular doctor, the length of time in hours spent with a patient per office visit has the probability density function

$$f(t) = \begin{cases} \dfrac{\frac{4}{3}}{(t+1)^2} & \text{if } 0 \leqslant t \leqslant 3 \\ 0 & \text{otherwise} \end{cases}$$

What is the probability that the doctor will spend more than 1 hour with a randomly selected patient?

Social Sciences

42. *Politics.* The rate of change of the voting population of a city with respect to time t in years is estimated to be

$$N'(t) = \frac{100t}{(1+t^2)^2}$$

where $N(t)$ is in thousands. If $N(0)$ is the current voting population, how much will this population increase during the next 3 years? If the population continues to grow at this rate indefinitely, what is the total increase in the voting population?

43. *Psychology.* Rats were trained to go through a maze by rewarding them with a food pellet upon successful completion. After the seventh successful run, it was found that the probability density function for length of time in minutes until success on the eighth trial is given by

$$f(t) = \begin{cases} .5e^{-.5t} & \text{if } t \geqslant 0 \\ 0 & \text{otherwise} \end{cases}$$

What is the probability that a rat selected at random after seven successful runs will take 2 or more minutes to complete the eighth run successfully?

CHAPTER 8

Multivariable Calculus

Contents

CHAPTER 8

SECTION 8-1 Functions of Several Variables

- FUNCTIONS OF TWO OR MORE INDEPENDENT VARIABLES
- EXAMPLES OF FUNCTIONS OF SEVERAL VARIABLES
- THREE-DIMENSIONAL COORDINATE SYSTEMS

◆ FUNCTIONS OF TWO OR MORE INDEPENDENT VARIABLES

In Section 1-2, we introduced the concept of a function with one independent variable. Now we will broaden the concept to include functions with more than one independent variable. We start with an example.

A small manufacturing company produces a standard type of surfboard and no other products. If fixed costs are \$500 per week and variable costs are \$70 per board produced, then the weekly cost function is given by

$$C(x) = 500 + 70x \tag{1}$$

where x is the number of boards produced per week. The cost function is a function of a single independent variable x. For each value of x from the domain of C there exists exactly one value of $C(x)$ in the range of C.

Now, suppose the company decides to add a high-performance competition board to its line. If the fixed costs for the competition board are \$200 per week and the variable costs are \$100 per board, then the cost function (1) must be modified to

$$C(x, y) = 700 + 70x + 100y \tag{2}$$

where $C(x, y)$ is the cost for weekly output of x standard boards and y competition boards. Equation (2) is an example of a function with two independent variables, x and y. Of course, as the company expands its product line even further, its weekly cost function must be modified to include more and more independent variables, one for each new product produced.

In general, an equation of the form

$$z = f(x, y)$$

describes a **function of two independent variables** if for each permissible ordered pair (x, y), there is one and only one value of z determined by $f(x, y)$. The variables x and y are **independent variables,** and the variable z is a **dependent variable.** The set of all ordered pairs of permissible values of x and y is the **domain** of the function, and the set of all corresponding values $f(x, y)$ is the **range** of the function. Unless otherwise stated, we will assume that the domain of a function specified by an equation of the form $z = f(x, y)$ is the set of all ordered pairs of real numbers (x, y) such that $f(x, y)$ is also a real number. It should be noted, however, that certain conditions in practical problems often lead to further restrictions of the domain of a function.

We can similarly define functions of three independent variables, $w = f(x, y, z)$; of four independent variables, $u = f(w, x, y, z)$; and so on. In this chapter, we will primarily concern ourselves with functions of two independent variables.

◆ **EXAMPLE 1** For the cost function $C(x, y) = 700 + 70x + 100y$ described earlier, find $C(10, 5)$.

Solution

$$C(10, 5) = 700 + 70(10) + 100(5) = \$1{,}900$$ ◆

PROBLEM 1 Find $C(20, 10)$ for the cost function in Example 1. ◆

◆ **EXAMPLE 2** For $f(x, y, z) = 2x^2 - 3xy + 3z + 1$, find $f(3, 0, -1)$.

Solution

$$f(3, 0, -1) = 2(3)^2 - 3(3)(0) + 3(-1) + 1 = 18 - 0 - 3 + 1 = 16$$ ◆

PROBLEM 2 Find $f(-2, 2, 3)$ for f in Example 2. ◆

◆ **EXAMPLE 3** *Revenue, Cost, and Profit Functions*

The surfboard company discussed at the beginning of this section has determined that the demand equations for the two types of boards they produce are given by

$$p = 210 - 4x + y$$
$$q = 300 + x - 12y$$

where p is the price of the standard board, q is the price of the competition board, x is the weekly demand for standard boards, and y is the weekly demand for competition boards.

(A) Find the weekly revenue function $R(x, y)$ and evaluate $R(20, 10)$.

(B) If the weekly cost function is

$$C(x, y) = 700 + 70x + 100y$$

find the weekly profit function $P(x, y)$ and evaluate $P(20, 10)$.

Solutions (A)

$$\text{Revenue} = \begin{pmatrix}\text{Demand for}\\ \text{standard}\\ \text{boards}\end{pmatrix} \times \begin{pmatrix}\text{Price of a}\\ \text{standard}\\ \text{board}\end{pmatrix} + \begin{pmatrix}\text{Demand for}\\ \text{competition}\\ \text{boards}\end{pmatrix} \times \begin{pmatrix}\text{Price of a}\\ \text{competition}\\ \text{board}\end{pmatrix}$$

$$\begin{aligned} R(x, y) &= xp + yq \\ &= x(210 - 4x + y) + y(300 + x - 12y) \\ &= 210x + 300y - 4x^2 + 2xy - 12y^2 \end{aligned}$$

$$\begin{aligned} R(20, 10) &= 210(20) + 300(10) - 4(20)^2 + 2(20)(10) - 12(10)^2 \\ &= \$4,800 \end{aligned}$$

(B)

$$\text{Profit} = \text{Revenue} - \text{Cost}$$

$$\begin{aligned} P(x, y) &= R(x, y) - C(x, y) \\ &= 210x + 300y - 4x^2 + 2xy - 12y^2 - 700 - 70x - 100y \\ &= 140x + 200y - 4x^2 + 2xy - 12y^2 - 700 \end{aligned}$$

$$\begin{aligned} P(20, 10) &= 140(20) + 200(10) - 4(20)^2 + 2(20)(10) - 12(10)^2 - 700 \\ &= \$1,700 \end{aligned}$$

◆

PROBLEM 3 Repeat Example 3 if the demand and cost equations are given by

$$\begin{aligned} p &= 220 - 6x + y \\ q &= 300 + 3x - 10y \\ C(x, y) &= 40x + 80y + 1,000 \end{aligned}$$

◆

◆ EXAMPLES OF FUNCTIONS OF SEVERAL VARIABLES

A number of concepts we have already considered can be thought of in terms of functions of two or more variables. We list a few of these below.

Area of a rectangle $A(x, y) = xy$

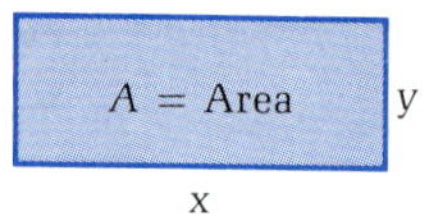

Volume of a box $V(x, y, z) = xyz$

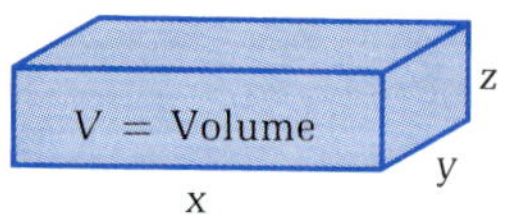

Volume of a right circular cylinder $V(r, h) = \pi r^2 h$

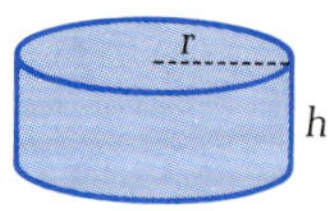

Simple interest	$A(P, r, t) = P(1 + rt)$	A = Amount P = Principal r = Annual rate t = Time in years
Compound interest	$A(P, r, t, n) = P\left(1 + \frac{r}{n}\right)^{nt}$	A = Amount P = Principal r = Annual rate t = Time in years n = Compound periods per year
IQ	$Q(M, C) = \frac{M}{C}(100)$	Q = IQ = Intelligence quotient M = MA = Mental age C = CA = Chronological age
Resistance for blood flow in a vessel (Poiseuille's law)	$R(L, r) = k\frac{L}{r^4}$	R = Resistance L = Length of vessel r = Radius of vessel k = Constant

◆ EXAMPLE 4

Package Design

A company uses a box with a square base and an open top for one of its products (see the figure). If x is the length (in inches) of each side of the base and y is the height (in inches), find the total amount of material $M(x, y)$ required to construct one of these boxes, and evaluate $M(5, 10)$.

Solution

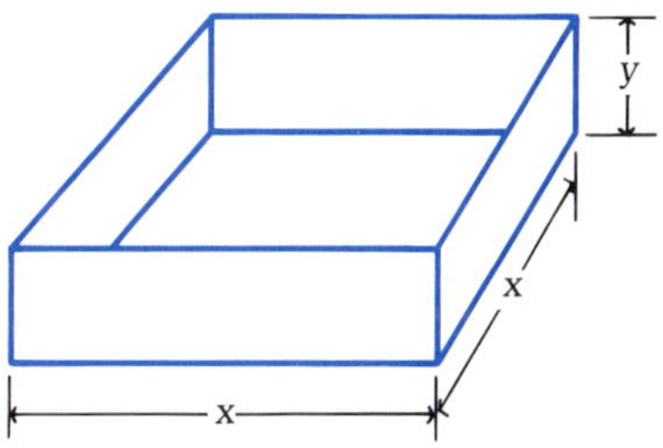

Area of base $= x^2$

Area of one side $= xy$

Total material = (Area of base) + 4(Area of one side)

$$M(x, y) = x^2 + 4xy$$

$$M(5, 10) = (5)^2 + 4(5)(10)$$
$$= 225 \text{ square inches}$$

◆

PROBLEM 4 For the box in Example 4, find the volume $V(x, y)$, and evaluate $V(5, 10)$. ◆

The next example concerns the **Cobb–Douglas production function,**

$$f(x, y) = kx^m y^n$$

where k, m, and n are positive constants with $m + n = 1$. Economists use this function to describe the number of units $f(x, y)$ produced from the utilization of x units of labor and y units of capital (for equipment such as tools, machinery, buildings, and so on). Cobb–Douglas production functions are also used to describe the productivity of a single industry, of a group of industries producing the same product, or even of an entire country.

◆ EXAMPLE 5 *Productivity*

The productivity of a steel manufacturing company is given approximately by the function

$$f(x, y) = 10x^{0.2}y^{0.8}$$

with the utilization of x units of labor and y units of capital. If the company uses 3,000 units of labor and 1,000 units of capital, how many units of steel will be produced?

Solution The number of units of steel produced is given by

$$f(3{,}000, 1{,}000) = 10(3{,}000)^{0.2}(1{,}000)^{0.8} \quad \text{Use a calculator.}$$
$$\approx 12{,}457 \text{ units}$$

◆

PROBLEM 5 Refer to Example 5. Find the steel production if the company uses 1,000 units of labor and 2,000 units of capital. ◆

◆ THREE-DIMENSIONAL COORDINATE SYSTEMS

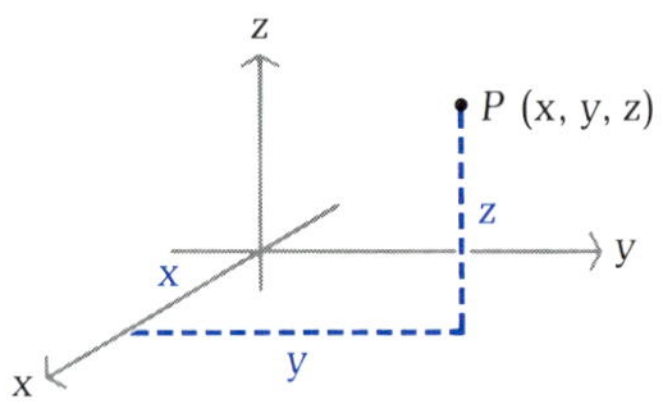

FIGURE 1
Rectangular coordinate system

We now take a brief look at some graphs of functions of two independent variables. Since functions of the form $z = f(x, y)$ involve two independent variables, x and y, and one dependent variable, z, we need a *three-dimensional coordinate system* for their graphs. A **three-dimensional coordinate system** is formed by three mutually perpendicular number lines intersecting at their origins. To represent a three-dimensional coordinate system on a piece of paper, we draw two of the number lines (usually the y and z axes) in the plane of the paper and draw the third (the x axis) as a diagonal line which we visualize as projecting out from the paper (see Fig. 1). In such a system, every ordered **triplet of numbers (x, y, z)** can be associated with a unique point, and conversely.

◆ EXAMPLE 6 Locate $(-3, 5, 2)$ in a rectangular coordinate system.

Solution

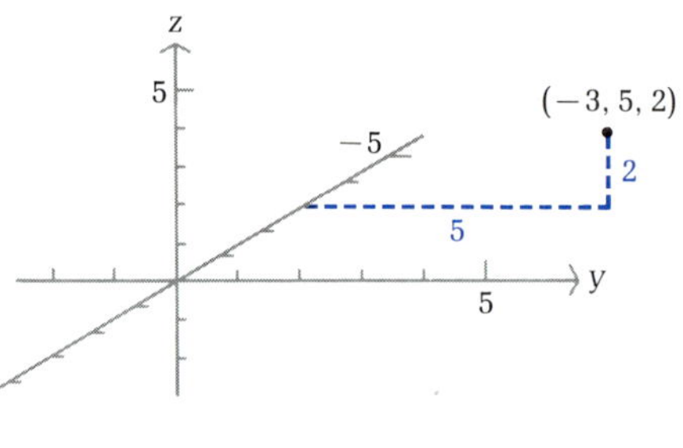

◆

PROBLEM 6 Find the coordinates of the corners A, C, G, and D of the rectangular box shown in the figure.

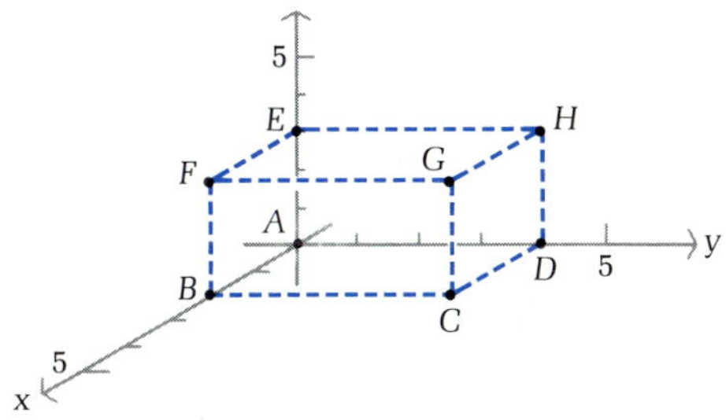

What does the graph of $z = x^2 + y^2$ look like? If we let $x = 0$ and graph $z = 0^2 + y^2 = y^2$ in the yz plane, we obtain a parabola; if we let $y = 0$ and graph $z = x^2 + 0^2 = x^2$ in the xz plane, we obtain another parabola. It can be shown that the graph of $z = x^2 + y^2$ is either one of these parabolas rotated around the z axis (see Fig. 2). This cup-shaped figure is a *surface* and is called a **paraboloid.**

In general, the graph of any function of the form $z = f(x, y)$ is called a **surface.** The graph of such a function is the graph of all ordered triplets of numbers (x, y, z) that satisfy the equation. Graphing functions of two independent variables is often a very difficult task, and the general process will not be dealt with in this book. We present only a few simple graphs to suggest extensions of earlier geometric interpretations of the derivative and local maxima and minima to functions of two variables. Note that $z = f(x, y) = x^2 + y^2$ appears (see Fig. 2) to have a local minimum at $(x, y) = (0, 0)$. Figure 3 shows a local maximum at $(x, y) = (0, 0)$.

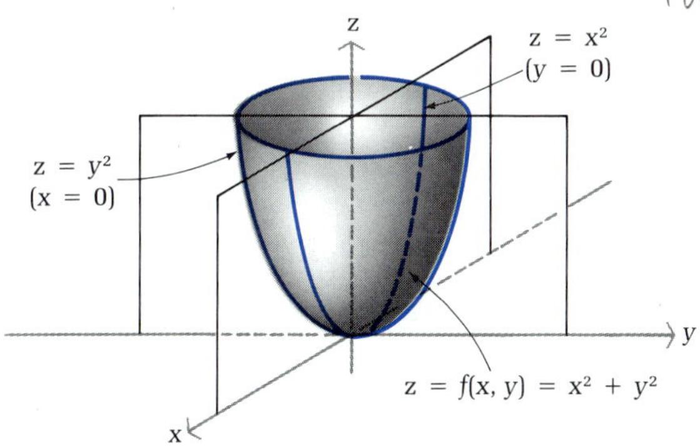

FIGURE 2
Paraboloid

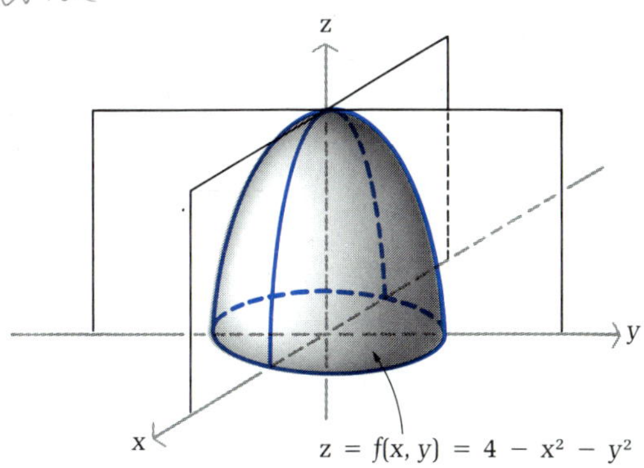

FIGURE 3
Local maximum: $f(0, 0) = 4$

Figure 4 shows a point at $(x, y) = (0, 0)$, called a **saddle point,** which is neither a local minimum nor a local maximum. Note that if $x = 0$, then the saddle point is a local minimum, and if $y = 0$, then the saddle point is a local maximum. More will be said about local maxima and minima in Section 8-3.

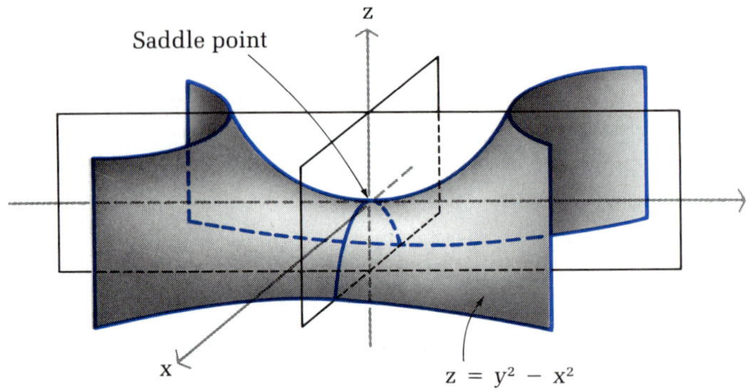

FIGURE 4
Saddle point at (0, 0, 0)

Answers to Matched Problems

1. \$3,100 2. 30
3. (A) $R(x, y) = 220x + 300y - 6x^2 + 4xy - 10y^2$; $R(20, 10) = \$4,800$
   (B) $P(x, y) = 180x + 220y - 6x^2 + 4xy - 10y^2 - 1,000$; $P(20, 10) = \$2,200$
4. $V(x, y) = x^2y$; $V(5, 10) = 250$ in.^{3} 5. 17,411 units
6. $A(0, 0, 0)$; $C(2, 4, 0)$; $G(2, 4, 3)$; $D(0, 4, 0)$

EXERCISE 8-1

A *For the functions*

$$f(x, y) = 10 + 2x - 3y \quad \text{and} \quad g(x, y) = x^2 - 3y^2$$

find each of the following:

1. $f(0, 0)$ **2.** $f(2, 1)$ **3.** $f(-3, 1)$ **4.** $f(2, -7)$
5. $g(0, 0)$ **6.** $g(0, -1)$ **7.** $g(2, -1)$ **8.** $g(-1, 2)$

B *Find each of the following:*

9. $A(2, 3)$ for $A(x, y) = xy$
10. $V(2, 4, 3)$ for $V(x, y, z) = xyz$
11. $Q(12, 8)$ for $Q(M, C) = \dfrac{M}{C}(100)$
12. $T(50, 17)$ for $T(V, x) = \dfrac{33V}{x + 33}$
13. $V(2, 4)$ for $V(r, h) = \pi r^2 h$
14. $S(4, 2)$ for $S(x, y) = 5x^2y^3$
15. $R(1, 2)$ for $R(x, y) = -5x^2 + 6xy - 4y^2 + 200x + 300y$
16. $P(2, 2)$ for $P(x, y) = -x^2 + 2xy - 2y^2 - 4x + 12y + 5$
17. $R(6, 0.5)$ for $R(L, r) = 0.002\dfrac{L}{r^4}$

18. $L(2{,}000, 50)$ for $L(w, v) = (1.25 \times 10^{-5})wv^2$

19. $A(100, 0.06, 3)$ for $A(P, r, t) = P + Prt$

20. $A(10, 0.04, 3, 2)$ for $A(P, r, t, n) = P\left(1 + \dfrac{r}{n}\right)^{tn}$

21. $A(100, 0.08, 10)$ for $A(P, r, t) = Pe^{rt}$

22. $A(1{,}000, 0.06, 8)$ for $A(P, r, t) = Pe^{rt}$

C

23. For the function $f(x, y) = x^2 + 2y^2$, find: $\dfrac{f(x + h, y) - f(x, y)}{h}$

24. For the function $f(x, y) = x^2 + 2y^2$, find: $\dfrac{f(x, y + k) - f(x, y)}{k}$

25. For the function $f(x, y) = 2xy^2$, find: $\dfrac{f(x + h, y) - f(x, y)}{h}$

26. For the function $f(x, y) = 2xy^2$, find: $\dfrac{f(x, y + k) - f(x, y)}{k}$

27. Find the coordinates of E and F in the figure for Problem 6 in the text.

28. Find the coordinates of B and H in the figure for Problem 6 in the text.

APPLICATIONS

Business & Economics

29. *Cost function.* A small manufacturing company produces two models of a surfboard: a standard model and a competition model. If the standard model is produced at a variable cost of \$70 each, the competition model at a variable cost of \$100 each, and the total fixed costs per month are \$2,000, then the monthly cost function is given by

$$C(x, y) = 2{,}000 + 70x + 100y$$

where x and y are the numbers of standard and competition models produced per month, respectively. Find $C(20, 10)$, $C(50, 5)$, and $C(30, 30)$.

30. *Advertising and sales.* A company spends \$x thousand per week on newspaper advertising and \$y thousand per week on television advertising. Its weekly sales are found to be given by

$$S(x, y) = 5x^2y^3$$

Find $S(3, 2)$ and $S(2, 3)$.

31. *Revenue function.* A supermarket sells two brands of coffee: brand A at \$p per pound and brand B at \$q per pound. The daily demand equations for brands A and B are, respectively,

$$x = 200 - 5p + 4q$$
$$y = 300 + 2p - 4q$$

(both in pounds). Find the daily revenue function $R(p, q)$. Evaluate $R(2, 3)$ and $R(3, 2)$.

32. *Revenue, cost, and profit functions.* A company manufactures ten-speed and three-speed bicycles. The weekly demand and cost equations are

$$p = 230 - 9x + y$$
$$q = 130 + x - 4y$$
$$C(x, y) = 200 + 80x + 30y$$

where $\$p$ is the price of a ten-speed bicycle, $\$q$ is the price of a three-speed bicycle, x is the weekly demand for ten-speed bicycles, y is the weekly demand for three-speed bicycles, and $C(x, y)$ is the cost function. Find the weekly revenue function $R(x, y)$ and the weekly profit function $P(x, y)$. Evaluate $R(10, 15)$ and $P(10, 15)$.

33. *Productivity.* The Cobb–Douglas production function for a petroleum company is given by

$$f(x, y) = 20x^{0.4}y^{0.6}$$

where x is the utilization of labor and y is the utilization of capital. If the company uses 1,250 units of labor and 1,700 units of capital, how many units of petroleum will be produced?

34. *Productivity.* The petroleum company in Problem 33 is taken over by another company that decides to double both the units of labor and the units of capital utilized in the production of petroleum. Use the Cobb–Douglas production function given in Problem 33 to find the amount of petroleum that will be produced by this increased utilization of labor and capital. What is the effect on productivity of doubling both the units of labor and the units of capital?

35. *Future value.* At the end of each year, \$2,000 is invested into an IRA earning 9% compounded annually. How much will be in the account at the end of 30 years? Use the annuity formula

$$F(P, i, n) = P\frac{(1 + i)^n - 1}{i}$$

where

P = Periodic payment	n = Number of payments (periods)
i = Rate per period	$F = FV$ = Future value

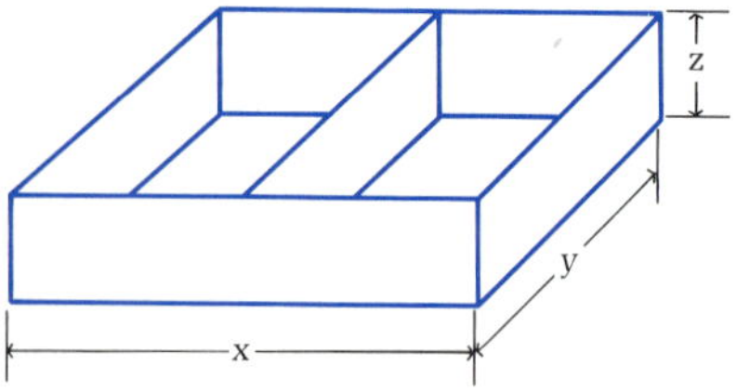

36. *Package design.* The packaging department in a company has been asked to design a rectangular box with no top and a partition down the middle (see the figure in the margin). If x, y, and z are the dimensions (in inches), find the total amount of material $M(x, y, z)$ used in constructing one of these boxes, and evaluate $M(10, 12, 6)$.

Life Sciences

37. *Marine biology.* In using scuba diving gear, a marine biologist estimates the time of a dive according to the equation

$$T(V, x) = \frac{33V}{x + 33}$$

where

T = Time of dive in minutes

V = Volume of air, at sea level pressure, compressed into tanks

x = Depth of dive in feet

Find $T(70, 47)$ and $T(60, 27)$.

38. *Blood flow.* Poiseuille's law states that the resistance, R, for blood flowing in a blood vessel varies directly as the length of the vessel, L, and inversely as the fourth power of its radius, r. Stated as an equation,

$$R(L, r) = k\frac{L}{r^4} \qquad k \text{ a constant}$$

Find $R(8, 1)$ and $R(4, 0.2)$.

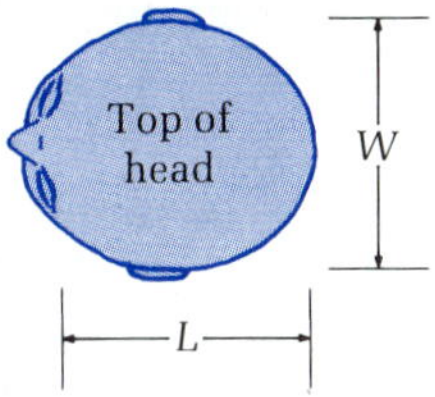

39. *Physical anthropology.* Anthropologists, in their study of race and human genetic groupings, often use an index called the *cephalic index*. The cephalic index, C, varies directly as the width, W, of the head, and inversely as the length, L, of the head (both viewed from the top). In terms of an equation,

$$C(W, L) = 100\frac{W}{L}$$

where

W = Width in inches $\qquad$ L = Length in inches

Find $C(6, 8)$ and $C(8.1, 9)$.

Social Sciences

40. *Safety research.* Under ideal conditions, if a person driving a car slams on the brakes and skids to a stop, the length of the skid marks (in feet) is given by the formula

$$L(w, v) = kwv^2$$

where

k = Constant $\qquad$ w = Weight of car in pounds

v = Speed of car in miles per hour

For $k = 0.000\ 013\ 3$, find $L(2{,}000, 40)$ and $L(3{,}000, 60)$.

41. *Psychology.* The intelligence quotient (IQ) is defined to be the ratio of mental age (MA), as determined by certain tests, and chronological age (CA), multiplied by 100. Stated as an equation,

$$Q(M, C) = \frac{M}{C} \cdot 100$$

where

$$Q = \text{IQ} \qquad M = \text{MA} \qquad C = \text{CA}$$

Find $Q(12, 10)$ and $Q(10, 12)$.

SECTION 8-2 Partial Derivatives

- PARTIAL DERIVATIVES
- SECOND-ORDER PARTIAL DERIVATIVES

PARTIAL DERIVATIVES

We know how to differentiate many kinds of functions of one independent variable and how to interpret the results. What about functions with two or more independent variables? Let us return to the surfboard example considered at the beginning of the chapter.

For the company producing only the standard board, the cost function was

$$C(x) = 500 + 70x$$

Differentiating with respect to x, we obtain the marginal cost function

$$C'(x) = 70$$

Since the marginal cost is constant, \$70 is the change in cost for a 1 unit increase in production at any output level.

For the company producing two types of boards, a standard model and a competition model, the cost function was

$$C(x, y) = 700 + 70x + 100y$$

Now suppose we differentiate with respect to x, holding y fixed, and denote this by $C_x(x, y)$; or suppose we differentiate with respect to y, holding x fixed, and denote this by $C_y(x, y)$. Differentiating in this way, we obtain

$$C_x(x, y) = 70 \qquad C_y(x, y) = 100$$

Each of these is called a **partial derivative,** and, in this example, each represents marginal cost. The first is the change in cost due to a 1 unit increase in production of the standard board with the production of the competition model held fixed. The second is the change in cost due to a 1 unit increase in production of the competition board with the production of the standard board held fixed.

In general, if $z = f(x, y)$, then the **partial derivative of f with respect to x,** denoted by $\partial z/\partial x$, f_x, or $f_x(x, y)$, is defined by

$$\frac{\partial z}{\partial x} = \lim_{h \to 0} \frac{f(x + h, y) - f(x, y)}{h}$$

provided the limit exists. We recognize this as the ordinary derivative of f with respect to x, holding y constant. Thus, we are able to continue to use all the derivative rules and properties discussed in Chapters 3–5 for partial derivatives.

Similarly, the **partial derivative of f with respect to y,** denoted by $\partial z/\partial y$, f_y, or $f_y(x, y)$, is defined by

$$\frac{\partial z}{\partial y} = \lim_{k \to 0} \frac{f(x, y + k) - f(x, y)}{k}$$

which is the ordinary derivative with respect to y, holding x constant.

Parallel definitions and interpretations hold for functions with three or more independent variables.

◆ EXAMPLE 7 For $z = f(x, y) = 2x^2 - 3x^2y + 5y + 1$, find:

(A) $\partial z/\partial x$ (B) $f_x(2, 3)$

Solutions (A) $z = 2x^2 - 3x^2y + 5y + 1$

Differentiating with respect to x, holding y constant (that is, treating y as a constant), we obtain

$$\frac{\partial z}{\partial x} = 4x - 6xy$$

(B) $f(x, y) = 2x^2 - 3x^2y + 5y + 1$

First, differentiate with respect to x (part A) to obtain

$$f_x(x, y) = 4x - 6xy$$

Then evaluate at (2, 3):

$$f_x(2, 3) = 4(2) - 6(2)(3) = -28$$ ◆

PROBLEM 7 For f in Example 7, find:

(A) $\partial z/\partial y$ (B) $f_y(2, 3)$ ◆

◆ EXAMPLE 8 For $z = f(x, y) = e^{x^2+y^2}$, find:

(A) $\partial z/\partial x$ (B) $f_y(2, 1)$

Solutions (A) Using the chain rule [thinking of $z = e^u$, $u = u(x)$; y is held constant], we obtain

$$\frac{\partial z}{\partial x} = e^{x^2+y^2} \frac{\partial(x^2 + y^2)}{\partial x}$$
$$= 2xe^{x^2+y^2}$$

(B) $$f_y(x, y) = e^{x^2+y^2} \frac{\partial(x^2 + y^2)}{\partial y} = 2ye^{x^2+y^2}$$
$$f_y(2, 1) = 2(1)e^{(2)^2+(1)^2}$$
$$= 2e^5$$

◆

PROBLEM 8 For $z = f(x, y) = (x^2 + 2xy)^5$, find:

(A) $\partial z/\partial y$ (B) $f_x(1, 0)$ ◆

◆ EXAMPLE 9 *Profit* The profit function for the surfboard company in Example 3 in Section 8-1 was

$$P(x, y) = 140x + 200y - 4x^2 + 2xy - 12y^2 - 700$$

Find $P_x(15, 10)$ and $P_x(30, 10)$, and interpret the results.

Solution

$$P_x(x, y) = 140 - 8x + 2y$$
$$P_x(15, 10) = 140 - 8(15) + 2(10) = 40$$
$$P_x(30, 10) = 140 - 8(30) + 2(10) = -80$$

At a production level of 15 standard and 10 competition boards per week, increasing the production of standard boards by 1 unit and holding the production of competition boards fixed at 10 will increase profit by approximately \$40. At a production level of 30 standard and 10 competition boards per week, increasing the production of standard boards by 1 unit and holding the production of competition boards fixed at 10 will decrease profit by approximately \$80. ◆

PROBLEM 9 For the profit function in Example 9, find $P_y(25, 10)$ and $P_y(25, 15)$, and interpret the results. ◆

◆ EXAMPLE 10 *Productivity* The productivity of a major computer manufacturer is given approximately by the Cobb–Douglas production function

$$f(x, y) = 15x^{0.4}y^{0.6}$$

with the utilization of x units of labor and y units of capital. The partial derivative $f_x(x, y)$ represents the rate of change of productivity with respect to labor and is called the **marginal productivity of labor.** The partial derivative $f_y(x, y)$ represents the rate of change of productivity with respect to capital and is called the **marginal productivity of capital.** If the company is currently utilizing 4,000 units of labor and 2,500 units of capital, find the marginal productivity of labor and the marginal productivity of capital. For the greatest increase in productivity, should the management of the company encourage increased use of labor or increased use of capital?

Solution

$$f_x(x, y) = 6x^{-0.6}y^{0.6}$$

$$f_x(4{,}000,\ 2{,}500) = 6(4{,}000)^{-0.6}(2{,}500)^{0.6} \approx 4.53 \qquad \text{Marginal productivity of labor}$$

$$f_y(x, y) = 9x^{0.4}y^{-0.4}$$

$$f_y(4{,}000,\ 2{,}500) = 9(4{,}000)^{0.4}(2{,}500)^{-0.4} \approx 10.86 \qquad \text{Marginal productivity of capital}$$

At the current level of utilization of 4,000 units of labor and 2,500 units of capital, each 1 unit increase in labor utilization (keeping capital utilization fixed at 2,500 units) will increase production by approximately 4.53 units and each 1 unit increase in capital utilization (keeping labor utilization fixed at 4,000 units) will increase production by approximately 10.86 units. Thus, the management of the company should encourage increased use of capital. ◆

PROBLEM 10 The productivity of an airplane manufacturing company is given approximately by the Cobb–Douglas production function

$$f(x, y) = 40x^{0.3}y^{0.7}$$

(A) Find $f_x(x, y)$ and $f_y(x, y)$.

(B) If the company is currently using 1,500 units of labor and 4,500 units of capital, find the marginal productivity of labor and the marginal productivity of capital.

(C) For the greatest increase in productivity, should the management of the company encourage increased use of labor or increased use of capital? ◆

Partial derivatives have simple geometric interpretations, as indicated in Figure 5. If we hold x fixed, say, $x = a$, then $f_y(a, y)$ is the slope of the curve obtained by intersecting the plane $x = a$ with the surface $z = f(x, y)$. A similar interpretation is given to $f_x(x, b)$.

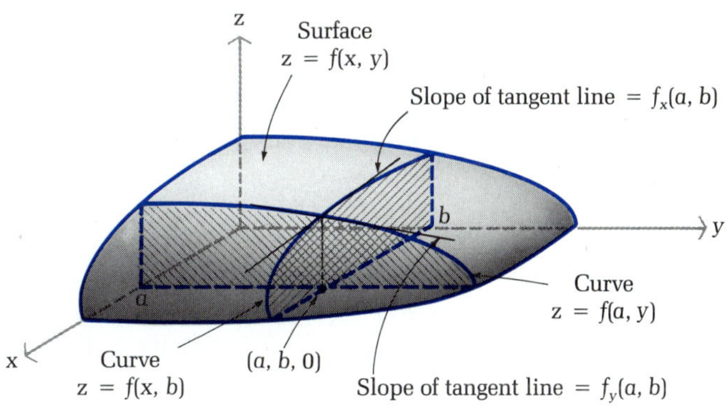

FIGURE 5

◆ SECOND-ORDER PARTIAL DERIVATIVES

The function

$$z = f(x, y) = x^4y^7$$

has two **first-order partial derivatives,**

$$\frac{\partial z}{\partial x} = f_x = f_x(x, y) = 4x^3y^7 \qquad \text{and} \qquad \frac{\partial z}{\partial y} = f_y = f_y(x, y) = 7x^4y^6$$

Each of these partial derivatives, in turn, has two partial derivatives which are called **second-order partial derivatives** of $z = f(x, y)$. Generalizing the various notations we have for first-order partial derivatives, the four second-order partial derivatives of $z = f(x, y) = x^4y^7$ are written as

Equivalent notations

$$f_{xx} = f_{xx}(x, y) = \frac{\partial^2 z}{\partial x^2} = \frac{\partial}{\partial x}\left(\frac{\partial z}{\partial x}\right) = \frac{\partial}{\partial x}(4x^3y^7) = 12x^2y^7$$

$$f_{xy} = f_{xy}(x, y) = \frac{\partial^2 z}{\partial y\,\partial x} = \frac{\partial}{\partial y}\left(\frac{\partial z}{\partial x}\right) = \frac{\partial}{\partial y}(4x^3y^7) = 28x^3y^6$$

$$f_{yx} = f_{yx}(x, y) = \frac{\partial^2 z}{\partial x\,\partial y} = \frac{\partial}{\partial x}\left(\frac{\partial z}{\partial y}\right) = \frac{\partial}{\partial x}(7x^4y^6) = 28x^3y^6$$

$$f_{yy} = f_{yy}(x, y) = \frac{\partial^2 z}{\partial y^2} = \frac{\partial}{\partial y}\left(\frac{\partial z}{\partial y}\right) = \frac{\partial}{\partial y}(7x^4y^6) = 42x^4y^5$$

In the mixed partial derivative $\partial^2 z/\partial y\,\partial x = f_{xy}$, we started with $z = f(x, y)$ and first differentiated with respect to x (holding y constant). Then we differentiated with respect to y (holding x constant). In the other mixed partial derivative, $\partial^2 z/\partial x\,\partial y = f_{yx}$, the order of differentiation was reversed; however, the final result was the same—that is, $f_{xy} = f_{yx}$. Although it is possible to find functions for which $f_{xy} \neq f_{yx}$, such functions rarely occur in applications involving partial derivatives. Thus, for all the functions in this text, we will assume that $f_{xy} = f_{yx}$.

In general, we have the following definitions:

Second-Order Partial Derivatives

If $z = f(x, y)$, then

$$f_{xx} = f_{xx}(x, y) = \frac{\partial^2 z}{\partial x^2} = \frac{\partial}{\partial x}\left(\frac{\partial z}{\partial x}\right)$$

$$f_{xy} = f_{xy}(x, y) = \frac{\partial^2 z}{\partial y\,\partial x} = \frac{\partial}{\partial y}\left(\frac{\partial z}{\partial x}\right)$$

$$f_{yx} = f_{yx}(x, y) = \frac{\partial^2 z}{\partial x\,\partial y} = \frac{\partial}{\partial x}\left(\frac{\partial z}{\partial y}\right)$$

$$f_{yy} = f_{yy}(x, y) = \frac{\partial^2 z}{\partial y^2} = \frac{\partial}{\partial y}\left(\frac{\partial z}{\partial y}\right)$$

◆ EXAMPLE 11 For $z = f(x, y) = 3x^2 - 2xy^3 + 1$, find:

(A) $\frac{\partial^2 z}{\partial x\, \partial y}$, $\frac{\partial^2 z}{\partial y\, \partial x}$ (B) $\frac{\partial^2 z}{\partial x^2}$ (C) $f_{yx}(2, 1)$

Solutions (A) First differentiate with respect to y and then with respect to x:

$$\frac{\partial z}{\partial y} = -6xy^2 \qquad \frac{\partial^2 z}{\partial x\, \partial y} = \frac{\partial}{\partial x}\left(\frac{\partial z}{\partial y}\right) = \frac{\partial}{\partial x}(-6xy^2) = -6y^2$$

First differentiate with respect to x and then with respect to y:

$$\frac{\partial z}{\partial x} = 6x - 2y^3 \qquad \frac{\partial^2 z}{\partial y\, \partial x} = \frac{\partial}{\partial y}\left(\frac{\partial z}{\partial x}\right) = \frac{\partial}{\partial y}(6x - 2y^3) = -6y^2$$

(B) Differentiate with respect to x twice:

$$\frac{\partial z}{\partial x} = 6x - 2y^3 \qquad \frac{\partial^2 z}{\partial x^2} = \frac{\partial}{\partial x}\left(\frac{\partial z}{\partial x}\right) = 6$$

(C) First find $f_{yx}(x, y)$; then evaluate at (2, 1). Again, remember that f_{yx} means to differentiate with respect to y first and then with respect to x. Thus,

$$f_y(x, y) = -6xy^2 \qquad f_{yx}(x, y) = -6y^2$$

and

$$f_{yx}(2, 1) = -6(1)^2 = -6$$ ◆

PROBLEM 11 For $z = f(x, y) = x^3y - 2y^4 + 3$, find:

(A) $\frac{\partial^2 z}{\partial y\, \partial x}$ (B) $\frac{\partial^2 z}{\partial y^2}$ (C) $f_{xy}(2, 3)$ (D) $f_{yx}(2, 3)$ ◆

Answers to Matched Problems

7. (A) $\partial z/\partial y = -3x^2 + 5$ (B) $f_y(2, 3) = -7$
8. (A) $10x(x^2 + 2xy)^4$ (B) 10
9. $P_y(25, 10) = 10$: at a production level of $x = 25$ and $y = 10$, increasing y by 1 unit and holding x fixed at 25 will increase profit by approx. \$10; $P_y(25, 15) = -110$: at a production level of $x = 25$ and $y = 15$, increasing y by 1 unit and holding x fixed at 25 will decrease profit by approx. \$110
10. (A) $f_x(x, y) = 12x^{-0.7}y^{0.7}$; $f_y(x, y) = 28x^{0.3}y^{-0.3}$
    (B) Marginal productivity of labor ≈ 25.89; marginal productivity of capital ≈ 20.14
    (C) Labor
11. (A) $3x^2$ (B) $-24y^2$ (C) 12 (D) 12

EXERCISE 8-2

A *For $z = f(x, y) = 10 + 3x + 2y$, find each of the following:*

1. $\partial z/\partial x$ **2.** $\partial z/\partial y$ **3.** $f_y(1, 2)$ **4.** $f_x(1, 2)$

For $z = f(x, y) = 3x^2 - 2xy^2 + 1$, *find each of the following:*

5. $\partial z/\partial y$ **6.** $\partial z/\partial x$ **7.** $f_x(2, 3)$ **8.** $f_y(2, 3)$

For $S(x, y) = 5x^2y^3$, *find each of the following:*

9. $S_x(x, y)$ **10.** $S_y(x, y)$ **11.** $S_y(2, 1)$ **12.** $S_x(2, 1)$

B For $C(x, y) = x^2 - 2xy + 2y^2 + 6x - 9y + 5$, *find each of the following:*

13. $C_x(x, y)$ **14.** $C_y(x, y)$ **15.** $C_x(2, 2)$ **16.** $C_y(2, 2)$
17. $C_{xy}(x, y)$ **18.** $C_{yx}(x, y)$ **19.** $C_{xx}(x, y)$ **20.** $C_{yy}(x, y)$

For $z = f(x, y) = e^{2x+3y}$, *find each of the following:*

21. $\dfrac{\partial z}{\partial x}$ **22.** $\dfrac{\partial z}{\partial y}$ **23.** $\dfrac{\partial^2 z}{\partial x\, \partial y}$ **24.** $\dfrac{\partial^2 z}{\partial y\, \partial x}$

25. $f_{xy}(1, 0)$ **26.** $f_{yx}(0, 1)$ **27.** $f_{xx}(0, 1)$ **28.** $f_{yy}(1, 0)$

Find $f_x(x, y)$ *and* $f_y(x, y)$ *for each function f given by:*

29. $f(x, y) = (x^2 - y^3)^3$ **30.** $f(x, y) = \sqrt{2x - y^2}$ **31.** $f(x, y) = (3x^2y - 1)^4$
32. $f(x, y) = (3 + 2xy^2)^3$ **33.** $f(x, y) = \ln(x^2 + y^2)$ **34.** $f(x, y) = \ln(2x - 3y)$

35. $f(x, y) = y^2e^{xy^2}$ **36.** $f(x, y) = x^3e^{x^2y}$ **37.** $f(x, y) = \dfrac{x^2 - y^2}{x^2 + y^2}$

38. $f(x, y) = \dfrac{2x^2y}{x^2 + y^2}$

Find $f_{xx}(x, y)$, $f_{xy}(x, y)$, $f_{yx}(x, y)$, *and* $f_{yy}(x, y)$ *for each function f given by:*

39. $f(x, y) = x^2y^2 + x^3 + y$ **40.** $f(x, y) = x^3y^3 + x + y^2$

41. $f(x, y) = \dfrac{x}{y} - \dfrac{y}{x}$ **42.** $f(x, y) = \dfrac{x^2}{y} - \dfrac{y^2}{x}$

43. $f(x, y) = xe^{xy}$ **44.** $f(x, y) = x\ln(xy)$

C **45.** For

$$P(x, y) = -x^2 + 2xy - 2y^2 - 4x + 12y - 5$$

find values of x and y such that

$$P_x(x, y) = 0 \quad \text{and} \quad P_y(x, y) = 0$$

simultaneously.

46. For

$$C(x, y) = 2x^2 + 2xy + 3y^2 - 16x - 18y + 54$$

find values of x and y such that

$$C_x(x, y) = 0 \quad \text{and} \quad C_y(x, y) = 0$$

simultaneously.

In Problems 47 and 48, show that the function f satisfies $f_{xx}(x, y) + f_{yy}(x, y) = 0$.

47. $f(x, y) = \ln(x^2 + y^2)$

48. $f(x, y) = x^3 - 3xy^2$

49. For $f(x, y) = x^2 + 2y^2$, find:

(A) $\lim_{h \to 0} \frac{f(x + h, y) - f(x, y)}{h}$ (B) $\lim_{k \to 0} \frac{f(x, y + k) - f(x, y)}{k}$

50. For $f(x, y) = 2xy^2$, find:

(A) $\lim_{h \to 0} \frac{f(x + h, y) - f(x, y)}{h}$ (B) $\lim_{k \to 0} \frac{f(x, y + k) - f(x, y)}{k}$

APPLICATIONS

Business & Economics

51. *Profit function.* A firm produces two types of calculators, x of type *A* and y of type *B* each week. The weekly revenue and cost functions (in dollars) are

$$R(x, y) = 80x + 90y + 0.04xy - 0.05x^2 - 0.05y^2$$
$$C(x, y) = 8x + 6y + 20{,}000$$

Find $P_x(1{,}200, 1{,}800)$ and $P_y(1{,}200, 1{,}800)$, and interpret the results.

52. *Advertising and sales.* A company spends \$x per week on newspaper advertising and \$y per week on television advertising. Its weekly sales were found to be given by

$$S(x, y) = 10x^{0.4}y^{0.8}$$

Find $S_x(3{,}000, 2{,}000)$ and $S_y(3{,}000, 2{,}000)$, and interpret the results.

53. *Demand equations.* A supermarket sells two brands of coffee, brand *A* at \$p per pound and brand *B* at \$q per pound. The daily demand equations for brands *A* and *B* are, respectively,

$$x = 200 - 5p + 4q$$
$$y = 300 + 2p - 4q$$

Find $\partial x/\partial p$ and $\partial y/\partial p$, and interpret the results.

54. *Revenue and profit functions.* A company manufactures ten-speed and three-speed bicycles. The weekly demand and cost functions are

$$p = 230 - 9x + y$$
$$q = 130 + x - 4y$$
$$C(x, y) = 200 + 80x + 30y$$

where \$p is the price of a ten-speed bicycle, \$q is the price of a three-speed bicycle, x is the weekly demand for ten-speed bicycles, y is the weekly demand for three-speed bicycles, and $C(x, y)$ is the cost function. Find $R_x(10, 5)$ and $P_x(10, 5)$, and interpret the results.

55. *Productivity.* The productivity of a certain third-world country is given approximately by the function

$$f(x, y) = 10x^{0.75}y^{0.25}$$

with the utilization of x units of labor and y units of capital.

(A) Find $f_x(x, y)$ and $f_y(x, y)$.

(B) If the country is now using 600 units of labor and 100 units of capital, find the marginal productivity of labor and the marginal productivity of capital.

(C) For the greatest increase in the country's productivity, should the government encourage increased use of labor or increased use of capital?

56. *Productivity.* The productivity of an automobile manufacturing company is given approximately by the function

$$f(x, y) = 50\sqrt{xy} = 50x^{0.5}y^{0.5}$$

with the utilization of x units of labor and y units of capital.

(A) Find $f_x(x, y)$ and $f_y(x, y)$.

(B) If the company is now using 250 units of labor and 125 units of capital, find the marginal productivity of labor and the marginal productivity of capital.

(C) For the greatest increase in the company's productivity, should the management encourage increased use of labor or increased use of capital?

Problems 57–60 refer to the following: If a decrease in demand for one product results in an increase in demand for another product, then the two products are said to be ***competitive,*** *or* ***substitute, products.*** *(Real whipping cream and imitation whipping cream are examples of competitive, or substitute, products.) If a decrease in demand for one product results in a decrease in demand for another product, then the two products are said to be* ***complementary products.*** *(Fishing boats and outboard motors are examples of complementary products.) Partial derivatives can be used to test whether two products are competitive, complementary, or neither. We start with demand functions for two products where the demand for either depends on the prices for both:*

$x = f(p, q)$ *Demand function for product A*

$y = g(p, q)$ *Demand function for product B*

The variables x and y represent the number of units demanded of products A and B, respectively, at a price p for 1 unit of product A and a price q for 1 unit of product B. Normally, if the price of A increases while the price of B is held constant, then the demand for A will decrease; that is, $f_p(p, q) < 0$. Then, if A and B are competitive products, the demand for B will increase; that is, $g_p(p, q) > 0$. Similarly, if the price of B increases while the price of A is held constant, then the *demand for B will decrease; that is, $g_q(p, q) < 0$. And if A and B are competitive products, then the demand for A will increase; that is, $f_q(p, q) > 0$. Reasoning similarly for complementary products, we arrive at the following test:*

Test for Competitive and Complementary Products

PARTIAL DERIVATIVES			PRODUCTS *A* AND *B*
$f_q(p, q) > 0$	and	$g_p(p, q) > 0$	Competitive (Substitute)
$f_q(p, q) < 0$	and	$g_p(p, q) < 0$	Complementary
$f_q(p, q) \geqslant 0$	and	$g_p(p, q) \leqslant 0$	Neither
$f_q(p, q) \leqslant 0$	and	$g_p(p, q) \geqslant 0$	Neither

Use this test in Problems 57–60 to determine whether the indicated products are competitive, complementary, or neither.

57. *Product demand.* The weekly demand equations for the sale of butter and margarine in a supermarket are

$$x = f(p, q) = 8{,}000 - 0.09p^2 + 0.08q^2 \qquad \text{Butter}$$
$$y = g(p, q) = 15{,}000 + 0.04p^2 - 0.3q^2 \qquad \text{Margarine}$$

58. *Product demand.* The daily demand equations for the sale of brand *A* coffee and brand *B* coffee in a supermarket are

$$x = f(p, q) = 200 - 5p + 4q \qquad \text{Brand } A \text{ coffee}$$
$$y = g(p, q) = 300 + 2p - 4q \qquad \text{Brand } B \text{ coffee}$$

59. *Product demand.* The monthly demand equations for the sale of skis and ski boots in a sporting goods store are

$$x = f(p, q) = 800 - 0.004p^2 - 0.003q^2 \qquad \text{Skis}$$
$$y = g(p, q) = 600 - 0.003p^2 - 0.002q^2 \qquad \text{Ski boots}$$

60. *Product demand.* The monthly demand equations for the sale of tennis rackets and tennis balls in a sporting goods store are

$$x = f(p, q) = 500 - 0.5p - q^2 \qquad \text{Tennis rackets}$$
$$y = g(p, q) = 10{,}000 - 8p - 100q^2 \qquad \text{Tennis balls (cans)}$$

Life Sciences

61. *Medicine.* The following empirical formula relates the surface area *A* (in square inches) of an average human body to its weight *w* (in pounds) and its height *h* (in inches):

$$A = f(w, h) = 15.64w^{0.425}h^{0.725}$$

Knowing the surface area of a human body is useful, for example, in studies pertaining to hypothermia (heat loss due to exposure).

(A) Find $f_w(w, h)$ and $f_h(w, h)$.

(B) For a 65 pound child who is 57 inches tall, find $f_w(65, 57)$ and $f_h(65, 57)$, and interpret the results.

62. *Blood flow.* Poiseuille's law states that the resistance, R, for blood flowing in a blood vessel varies directly as the length of the vessel, L, and inversely as the fourth power of its radius, r. Stated as an equation,

$$R(L, r) = k\frac{L}{r^4} \qquad k \text{ a constant}$$

Find $R_L(4, 0.2)$ and $R_r(4, 0.2)$, and interpret the results.

Social Sciences

63. *Physical anthropology.* Anthropologists, in their study of race and human genetic groupings, often use the cephalic index, C, which varies directly as the width, W, of the head, and inversely as the length, L, of the head (both viewed from the top). In terms of an equation,

$$C(W, L) = 100\frac{W}{L}$$

where

W = Width in inches $\qquad$ L = Length in inches

Find $C_W(6, 8)$ and $C_L(6, 8)$, and interpret the results.

64. *Safety research.* Under ideal conditions, if a person driving a car slams on the brakes and skids to a stop, the length of the skid marks (in feet) is given by the formula

$$L(w, v) = kwv^2$$

where

k = Constant $\qquad$ w = Weight of car in pounds

v = Speed of car in miles per hour

For $k = 0.000\ 013\ 3$, find $L_w(2{,}500, 60)$ and $L_v(2{,}500, 60)$, and interpret the results.

SECTION 8-3 Maxima and Minima

We are now ready to undertake a brief but useful analysis of local maxima and minima for functions of the type $z = f(x, y)$. Basically, we are going to extend the second-derivative test developed for functions of a single independent variable. To start, we assume that all second-order partial derivatives exist for the function f in some circular region in the xy plane. This guarantees that the surface $z = f(x, y)$ has no sharp points, breaks, or ruptures. In other words, we are dealing only with smooth surfaces with no edges (like the edge of a box); or breaks (like an earthquake fault); or sharp points (like the bottom point of a golf tee). See Figure 6.

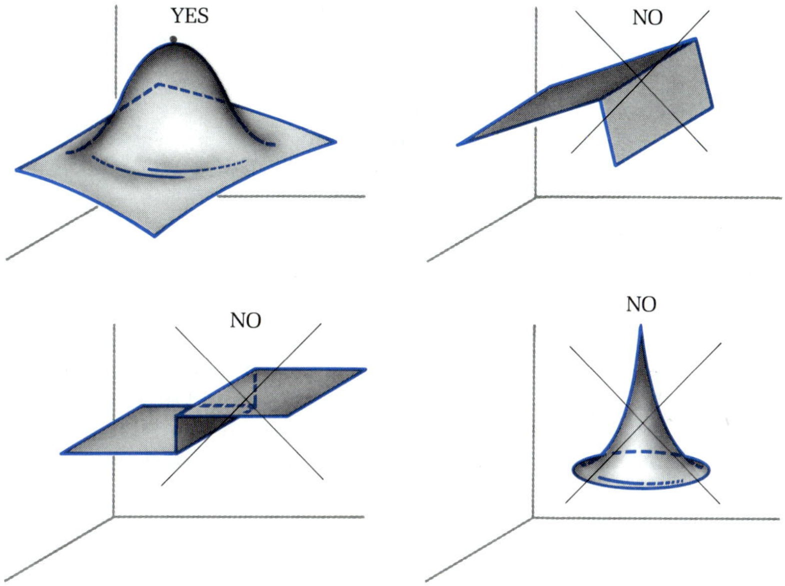

FIGURE 6

In addition, we will not concern ourselves with boundary points or absolute maxima–minima theory. In spite of these restrictions, the procedure we are now going to describe will help us solve a large number of useful problems.

What does it mean for $f(a, b)$ to be a local maximum or a local minimum? We say that ***f*(*a*, *b*) is a local maximum** if there exists a circular region in the domain of f with (a, b) as the center, such that

$$f(a, b) \geq f(x, y)$$

for all (x, y) in the region. Similarly, we say that ***f*(*a*, *b*) is a local minimum** if there exists a circular region in the domain of f with (a, b) as the center, such that

$$f(a, b) \leq f(x, y)$$

for all (x, y) in the region. Figure 7A illustrates a local maximum, Figure 7B a local minimum, and Figure 7C a **saddle point,** which is neither a local maximum nor a local minimum.

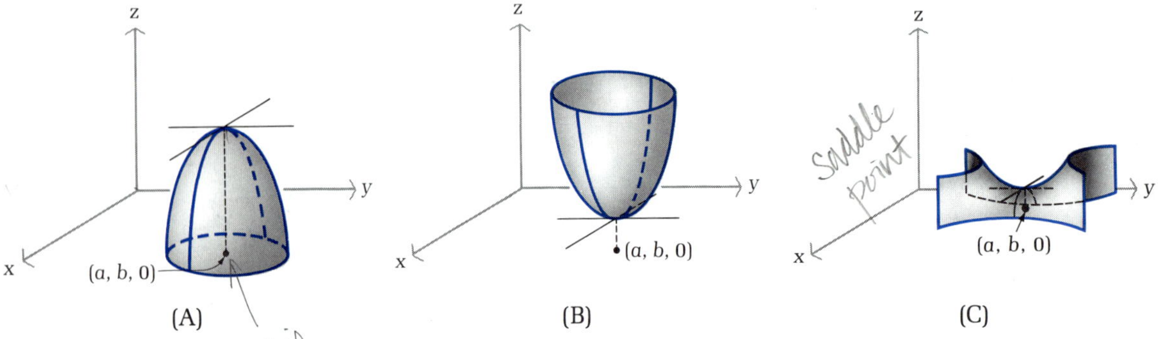

FIGURE 7

What happens to $f_x(a, b)$ and $f_y(a, b)$ if $f(a, b)$ is a local minimum or a local maximum and the partial derivatives of f exist in a circular region containing (a, b)? Figure 7 suggests that $f_x(a, b) = 0$ and $f_y(a, b) = 0$, since the tangent lines to the given curves are horizontal. Theorem 1 indicates that our intuitive reasoning is correct.

THEOREM 1

Let $f(a, b)$ be a local extremum (a local maximum or a local minimum) for the function f. If both f_x and f_y exist at (a, b), then

$$f_x(a, b) = 0 \qquad \text{and} \qquad f_y(a, b) = 0 \tag{1}$$

The converse of this theorem is false. That is, if $f_x(a, b) = 0$ and $f_y(a, b) = 0$, then $f(a, b)$ may or may not be a local extremum; for example, the point $(a, b, f(a, b))$ may be a saddle point (see Fig. 7C).

Theorem 1 gives us *necessary* (but not *sufficient*) conditions for $f(a, b)$ to be a local extremum. We thus find all points (a, b) such that $f_x(a, b) = 0$ and $f_y(a, b) = 0$ and test these further to determine whether $f(a, b)$ is a local extremum or a saddle point. Points (a, b) such that conditions (1) hold are called **critical points.** The next theorem, using second-derivative tests, gives us *sufficient* conditions for a critical point to produce a local extremum or a saddle point. (As was the case with Theorem 1, we state this theorem without proof.)

THEOREM 2

Second-Derivative Test for Local Extrema

Given

1. $z = f(x, y)$
2. $f_x(a, b) = 0$ and $f_y(a, b) = 0$ [(a, b) is a critical point]
3. All second-order partial derivatives of f exist in some circular region containing (a, b) as a center
4. $A = f_{xx}(a, b)$, $B = f_{xy}(a, b)$, $C = f_{yy}(a, b)$

Then:

Case 1. If $AC - B^2 > 0$ and $A < 0$, then $f(a, b)$ is a local maximum.

Case 2. If $AC - B^2 > 0$ and $A > 0$, then $f(a, b)$ is a local minimum.

Case 3. If $AC - B^2 < 0$, then f has a saddle point at (a, b).

Case 4. If $AC - B^2 = 0$, the test fails.

To illustrate the use of Theorem 2, we will first find the local extrema for a very simple function whose solution is almost obvious: $z = f(x, y) = x^2 + y^2 + 2$. From the function f itself and its graph (Fig. 8), it is clear that a local minimum is found at $(0, 0)$. Let us see how Theorem 2 confirms this observation.

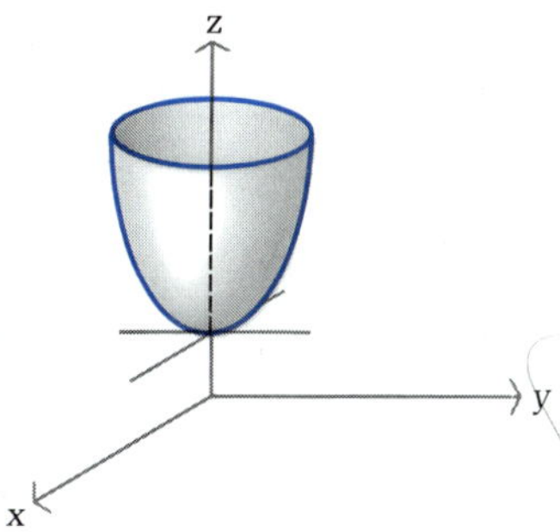

FIGURE 8

Step 1. Find critical points: Find (x, y) such that $f_x(x, y) = 0$ and $f_y(x, y) = 0$ simultaneously:

$$f_x(x, y) = 2x = 0 \qquad f_y(x, y) = 2y = 0$$
$$x = 0 \qquad\qquad y = 0$$

The only critical point is $(a, b) = (0, 0)$.

Step 2. Compute $A = f_{xx}(0, 0)$, $B = f_{xy}(0, 0)$, and $C = f_{yy}(0, 0)$:

$$f_{xx}(x, y) = 2 \quad \text{thus} \quad A = f_{xx}(0, 0) = 2$$
$$f_{xy}(x, y) = 0 \quad \text{thus} \quad B = f_{xy}(0, 0) = 0$$
$$f_{yy}(x, y) = 2 \quad \text{thus} \quad C = f_{yy}(0, 0) = 2$$

Step 3. Evaluate $AC - B^2$ and try to classify the critical point $(0, 0)$ using Theorem 2:

$$AC - B^2 = (2)(2) - (0)^2 = 4 > 0 \qquad \text{and} \qquad A = 2 > 0$$

Therefore, case 2 in Theorem 2 holds. That is, $f(0, 0) = 2$ is a local minimum.

We will now use Theorem 2 in the following examples to analyze extrema without the aid of graphs.

◆ EXAMPLE 12 Use Theorem 2 to find local extrema for: $f(x, y) = -x^2 - y^2 + 6x + 8y - 21$

Solution Step 1. Find critical points: Find (x, y) such that $f_x(x, y) = 0$ and $f_y(x, y) = 0$ simultaneously:

$$f_x(x, y) = -2x + 6 = 0 \qquad f_y(x, y) = -2y + 8 = 0$$
$$x = 3 \qquad y = 4$$

The only critical point is $(a, b) = (3, 4)$.

Step 2. Compute $A = f_{xx}(3, 4)$, $B = f_{xy}(3, 4)$, and $C = f_{yy}(3, 4)$:

$$f_{xx}(x, y) = -2 \quad \text{thus} \quad A = f_{xx}(3, 4) = -2$$
$$f_{xy}(x, y) = 0 \quad \text{thus} \quad B = f_{xy}(3, 4) = 0$$
$$f_{yy}(x, y) = -2 \quad \text{thus} \quad C = f_{yy}(3, 4) = -2$$

Step 3. Evaluate $AC - B^2$ and try to classify the critical point $(3, 4)$ using Theorem 2:

$$AC - B^2 = (-2)(-2) - (0)^2 = 4 > 0 \quad \text{and} \quad A = -2 < 0$$

Therefore, case 1 in Theorem 2 holds. That is, $f(3, 4) = 4$ is a local maximum. ◆

PROBLEM 12 Use Theorem 2 to find local extrema for: $f(x, y) = x^2 + y^2 - 10x - 2y + 36$ ◆

◆ EXAMPLE 13 Use Theorem 2 to find local extrema for: $f(x, y) = x^3 + y^3 - 6xy$

Solution Step 1. Find critical points for $f(x, y) = x^3 + y^3 - 6xy$:

$$f_x(x, y) = 3x^2 - 6y = 0 \qquad \text{Solve for } y.$$
$$6y = 3x^2$$
$$y = \tfrac{1}{2}x^2 \qquad (2)$$
$$f_y(x, y) = 3y^2 - 6x = 0$$
$$3y^2 = 6x \qquad \text{Use (2) to eliminate } y.$$
$$3(\tfrac{1}{2}x^2)^2 = 6x$$
$$\tfrac{3}{4}x^4 = 6x \qquad \text{Solve for } x.$$
$$3x^4 - 24x = 0$$
$$3x(x^3 - 8) = 0$$
$$x = 0 \quad \text{or} \quad x = 2$$
$$y = 0 \qquad y = \tfrac{1}{2}(2)^2 = 2$$

The critical points are $(0, 0)$ and $(2, 2)$. Since there are two critical points, steps 2 and 3 must be performed twice.

Test (0, 0) Step 2. Compute $A = f_{xx}(0, 0)$, $B = f_{xy}(0, 0)$, and $C = f_{yy}(0, 0)$:

$$f_{xx}(x, y) = 6x \quad \text{thus} \quad A = f_{xx}(0, 0) = 0$$
$$f_{xy}(x, y) = -6 \quad \text{thus} \quad B = f_{xy}(0, 0) = -6$$
$$f_{yy}(x, y) = 6y \quad \text{thus} \quad C = f_{yy}(0, 0) = 0$$

Step 3. Evaluate $AC - B^2$ and try to classify the critical point (0, 0) using Theorem 2:

$$AC - B^2 = (0)(0) - (-6)^2 = -36 < 0$$

Therefore, case 3 in Theorem 2 applies. That is, f has a saddle point at (0, 0).

Now we will consider the second critical point, (2, 2).

Test (2, 2) Step 2. Compute $A = f_{xx}(2, 2)$, $B = f_{xy}(2, 2)$, and $C = f_{yy}(2, 2)$:

$$f_{xx}(x, y) = 6x \quad \text{thus} \quad A = f_{xx}(2, 2) = 12$$
$$f_{xy}(x, y) = -6 \quad \text{thus} \quad B = f_{xy}(2, 2) = -6$$
$$f_{yy}(x, y) = 6y \quad \text{thus} \quad C = f_{yy}(2, 2) = 12$$

Step 3. Evaluate $AC - B^2$ and try to classify the critical point (2, 2) using Theorem 2:

$$AC - B^2 = (12)(12) - (-6)^2 = 108 > 0 \quad \text{and} \quad A = 12 > 0$$

Thus, case 2 in Theorem 2 applies, and $f(2, 2) = -8$ is a local minimum. ◆

PROBLEM 13 Use Theorem 2 to find local extrema for: $f(x, y) = x^3 + y^2 - 6xy$ ◆

◆ EXAMPLE 14

Profit

Suppose the surfboard company discussed earlier has developed the yearly profit equation

$$P(x, y) = -2x^2 + 2xy - y^2 + 10x - 4y + 107$$

where x is the number (in thousands) of standard surfboards produced per year, y is the number (in thousands) of competition surfboards produced per year, and P is profit (in thousands of dollars). How many of each type of board should be produced per year to realize a maximum profit? What is the maximum profit?

Solution Step 1. Find critical points:

$$P_x(x, y) = -4x + 2y + 10 = 0$$
$$P_y(x, y) = 2x - 2y - 4 = 0$$

Solving this system, we obtain (3, 1) as the only critical point.

Step 2. Compute $A = P_{xx}(3, 1)$, $B = P_{xy}(3, 1)$, and $C = P_{yy}(3, 1)$:

$$P_{xx}(x, y) = -4 \quad \text{thus} \quad A = P_{xx}(3, 1) = -4$$
$$P_{xy}(x, y) = 2 \quad \text{thus} \quad B = P_{xy}(3, 1) = 2$$
$$P_{yy}(x, y) = -2 \quad \text{thus} \quad C = P_{yy}(3, 1) = -2$$

Step 3. Evaluate $AC - B^2$ and try to classify the critical point (3, 1) using Theorem 2:

$$AC - B^2 = (-4)(-2) - (2)^2 = 8 - 4 = 4 > 0 \quad \text{and} \quad A = -4 < 0$$

Therefore, case 1 in Theorem 2 applies. That is, $P(3, 1) = \$120{,}000$ is a local maximum. This is obtained by producing 3,000 standard boards and 1,000 competition boards per year. ◆

PROBLEM 14 Repeat Example 14 with: $P(x, y) = -2x^2 + 4xy - 3y^2 + 4x - 2y + 77$ ◆

◆ EXAMPLE 15

Package Design

The packaging department in a company has been asked to design a rectangular box with no top and a partition down the middle. The box must have a volume of 48 cubic inches. Find the dimensions that will minimize the amount of material used to construct the box.

Solution

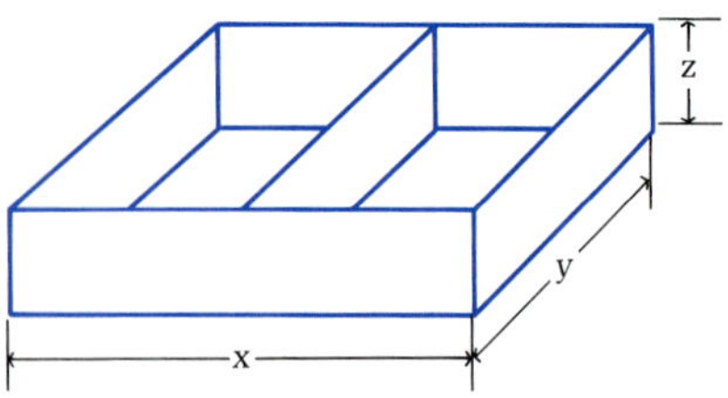

Refer to the figure in the margin. The amount of material used in constructing this box is

$$M = \underset{\text{Base}}{xy} + \underset{\text{Front, back}}{2xz} + \underset{\text{Sides, partition}}{3yz} \tag{3}$$

The volume of the box is

$$V = xyz = 48 \tag{4}$$

Since Theorem 2 applies only to functions with two independent variables, we must use (4) to eliminate one of the variables in (3):

$$\begin{aligned} M &= xy + 2xz + 3yz \qquad \text{Substitute } z = 48/xy. \\ &= xy + 2x\left(\frac{48}{xy}\right) + 3y\left(\frac{48}{xy}\right) \\ &= xy + \frac{96}{y} + \frac{144}{x} \end{aligned}$$

Thus, we must find the minimum value of

$$M(x, y) = xy + \frac{96}{y} + \frac{144}{x} \qquad x > 0 \quad \text{and} \quad y > 0$$

Step 1. Find critical points:

$$M_x(x, y) = y - \frac{144}{x^2} = 0$$

$$y = \frac{144}{x^2} \tag{5}$$

$$M_y(x, y) = x - \frac{96}{y^2} = 0$$

$$x = \frac{96}{y^2}$$ Solve for y^2.

$$y^2 = \frac{96}{x}$$ Use (5) to eliminate y and solve for x.

$$\left(\frac{144}{x^2}\right)^2 = \frac{96}{x}$$

$$\frac{20{,}736}{x^4} = \frac{96}{x}$$ Multiply both sides by $x^4/96$ (recall, $x > 0$).

$$x^3 = \tfrac{20{,}736}{96} = 216$$

$$x = 6$$ Use (5) to find y.

$$y = \tfrac{144}{36} = 4$$

Thus, (6, 4) is the only critical point.

Step 2. Compute $A = M_{xx}(6, 4)$, $B = M_{xy}(6, 4)$, and $C = M_{yy}(6, 4)$:

$$M_{xx}(x, y) = \frac{288}{x^3} \quad \text{thus} \quad A = M_{xx}(6, 4) = \tfrac{288}{216} = \tfrac{4}{3}$$

$$M_{xy}(x, y) = 1 \quad \text{thus} \quad B = M_{xy}(6, 4) = 1$$

$$M_{yy}(x, y) = \frac{192}{y^3} \quad \text{thus} \quad C = M_{yy}(6, 4) = \tfrac{192}{64} = 3$$

Step 3. Evaluate $AC - B^2$ and try to classify the critical point (6, 4) using Theorem 2:

$$AC - B^2 = (\tfrac{4}{3})(3) - (1)^2 = 3 > 0 \quad \text{and} \quad A = \tfrac{4}{3} > 0$$

Therefore, case 2 in Theorem 2 applies; $M(x, y)$ has a local minimum at (6, 4). If $x = 6$ and $y = 4$, then

$$z = \frac{48}{xy} = \frac{48}{(6)(4)} = 2$$

Thus, the dimensions that will require the minimum amount of material are 6 inches by 4 inches by 2 inches. ◆

PROBLEM 15 If the box in Example 15 must have a volume of 384 cubic inches, find the dimensions that will require the least amount of material. ◆

Answers to Matched Problems

12. $f(5, 1) = 10$ is a local minimum
13. f has a saddle point at (0, 0); $f(6, 18) = -108$ is a local minimum
14. Local maximum for $x = 2$ and $y = 1$; $P(2, 1) = \$80{,}000$
15. 12 in. by 8 in. by 4 in.

EXERCISE 8-3

Find local extrema using Theorem 2.

A

1. $f(x, y) = 6 - x^2 - 4x - y^2$
2. $f(x, y) = 3 - x^2 - y^2 + 6y$
3. $f(x, y) = x^2 + y^2 + 2x - 6y + 14$
4. $f(x, y) = x^2 + y^2 - 4x + 6y + 23$

B

5. $f(x, y) = xy + 2x - 3y - 2$
6. $f(x, y) = x^2 - y^2 + 2x + 6y - 4$
7. $f(x, y) = -3x^2 + 2xy - 2y^2 + 14x + 2y + 10$
8. $f(x, y) = -x^2 + xy - 2y^2 + x + 10y - 5$
9. $f(x, y) = 2x^2 - 2xy + 3y^2 - 4x - 8y + 20$
10. $f(x, y) = 2x^2 - xy + y^2 - x - 5y + 8$

C

11. $f(x, y) = e^{xy}$
12. $f(x, y) = x^2y - xy^2$
13. $f(x, y) = x^3 + y^3 - 3xy$
14. $f(x, y) = 2y^3 - 6xy - x^2$
15. $f(x, y) = 2x^4 + y^2 - 12xy$
16. $f(x, y) = 16xy - x^4 - 2y^2$
17. $f(x, y) = x^3 - 3xy^2 + 6y^2$
18. $f(x, y) = 2x^2 - 2x^2y + 6y^3$

APPLICATIONS

Business & Economics

19. *Product mix for maximum profit.* A firm produces two types of calculators, x thousand of type *A* and y thousand of type *B* per year. If the revenue and cost equations for the year are (in millions of dollars)

$$R(x, y) = 2x + 3y$$
$$C(x, y) = x^2 - 2xy + 2y^2 + 6x - 9y + 5$$

determine how many of each type of calculator should be produced per year to maximize profit. What is the maximum profit?

20. *Automation–labor mix for minimum cost.* The annual labor and automated equipment cost (in millions of dollars) for a company's production of television sets is given by

$$C(x, y) = 2x^2 + 2xy + 3y^2 - 16x - 18y + 54$$

where x is the amount spent per year on labor and y is the amount spent per year on automated equipment (both in millions of dollars). Determine how much should be spent on each per year to minimize this cost. What is the minimum cost?

21. *Maximizing profit.* A department store sells two brands of inexpensive calculators. The store pays $6 for each brand *A* calculator and $8 for each brand *B* calculator. The research department has estimated the following weekly demand equations for these two competitive products:

$x = 116 - 30p + 20q$ Demand equation for brand *A*

$y = 144 + 16p - 24q$ Demand equation for brand *B*

where p is the selling price for brand *A* and q is the selling price for brand *B*.

(A) Determine the demands x and y when $p = \$10$ and $q = \$12$; when $p = \$11$ and $q = \$11$.

(B) How should the store price each calculator to maximize weekly profits? What is the maximum weekly profit? [*Hint:* $C = 6x + 8y$, $R = px + qy$, and $P = R - C$.]

22. *Maximizing profit.* A store sells two brands of color print film. The store pays \$2 for each roll of brand *A* film and \$3 for each roll of brand *B* film. A consulting firm has estimated the following daily demand equations for these two competitive products:

$$x = 75 - 40p + 25q \qquad \text{Demand equation for brand } A$$
$$y = 80 + 20p - 30q \qquad \text{Demand equation for brand } B$$

where p is the selling price for brand *A* and q is the selling price for brand *B*.

(A) Determine the demands x and y when $p = \$4$ and $q = \$5$; when $p = \$4$ and $q = \$4$.

(B) How should the store price each brand of film to maximize daily profits? What is the maximum daily profit? [*Hint:* $C = 2x + 3y$, $R = px + qy$, and $P = R - C$.]

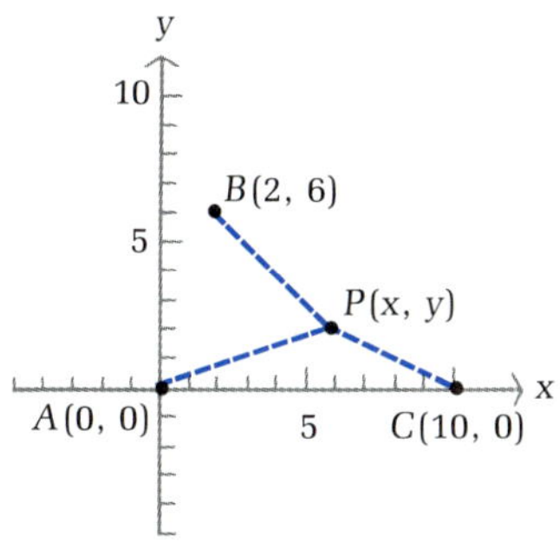

23. *Minimizing cost.* A satellite television reception station is to be located at P(x, y) so that the sum of the squares of the distances from P to the three towns A, B, and C is minimum (see the figure). Find the coordinates of P. This location will minimize the cost of providing satellite cable television for all three towns.

24. *Minimizing cost.* Repeat Problem 23 replacing the coordinates of B with B(6, 9) and the coordinates of C with C(9, 0).

25. *Minimum material.* A rectangular box with no top and two parallel partitions (see the figure below) is to be made to hold a volume of 64 cubic inches. Find the dimensions that will require the least amount of material.

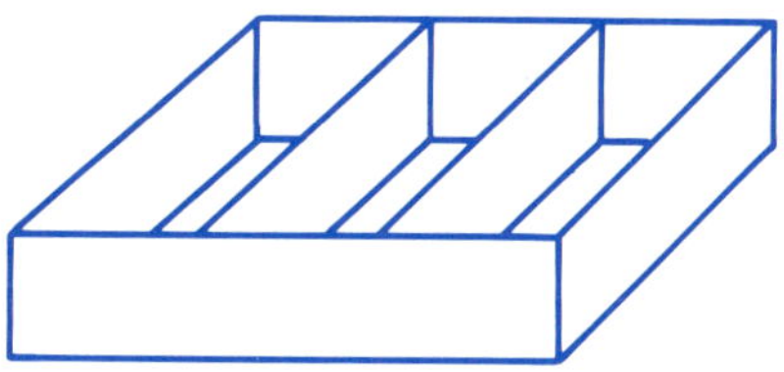

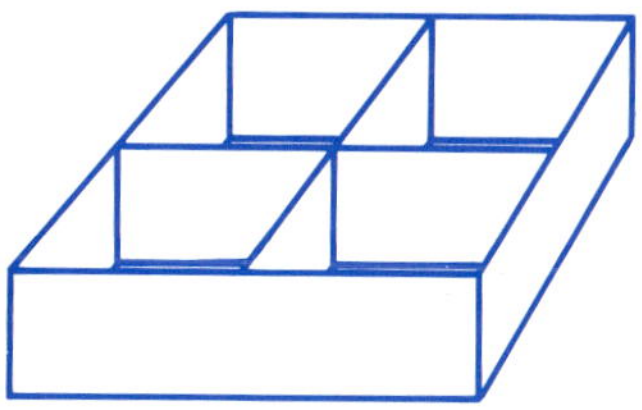

26. *Minimum material.* A rectangular box with no top and two intersecting partitions (see the figure in the margin) is to be made to hold a volume of 72 cubic inches. What should its dimensions be in order to use the least amount of material in its construction?

27. *Maximum volume.* A mailing service states that a rectangular package shall have the sum of the length and girth not to exceed 120 inches (see the figure). What are the dimensions of the largest (in volume) mailing carton that can be constructed meeting these restrictions?

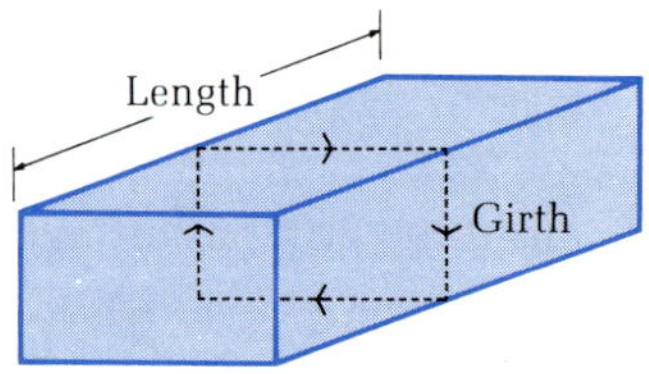

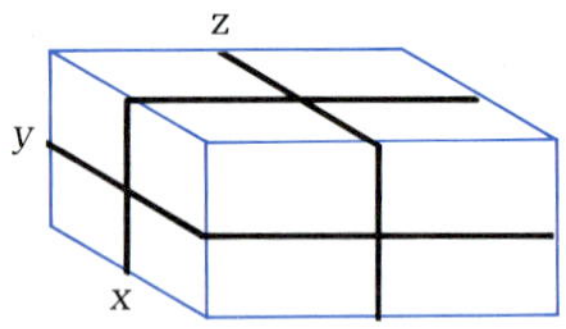

28. *Maximum shipping volume.* A shipping box is reinforced with steel bands in all three directions, as indicated in the figure in the margin. A total of 150 inches of steel tape are to be used, with 6 inches of waste because of a 2 inch overlap in each direction. Find the dimensions of the box with maximum volume that can be taped as indicated.

SECTION 8-4 Maxima and Minima Using Lagrange Multipliers

- FUNCTIONS OF TWO INDEPENDENT VARIABLES
- FUNCTIONS OF THREE INDEPENDENT VARIABLES

◆ FUNCTIONS OF TWO INDEPENDENT VARIABLES

We will now consider a particularly powerful method of solving a certain class of maxima–minima problems. The method is due to Joseph Louis Lagrange (1736–1813), an eminent eighteenth century French mathematician, and it is called the **method of Lagrange multipliers.** We introduce the method through an example; then we will formalize the discussion in the form of a theorem.

A rancher wants to construct two feeding pens of the same size along an existing fence (see Fig. 9). If the rancher has 720 feet of fencing materials available, how long should x and y be in order to obtain the maximum total area? What is the maximum area?

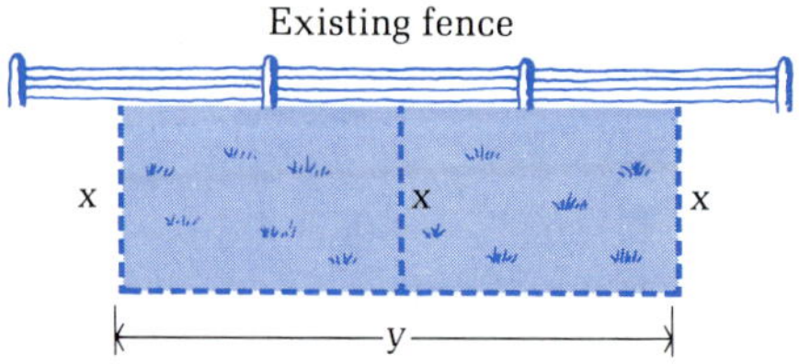

FIGURE 9

The total area is given by

$$f(x, y) = xy$$

which can be made as large as we like, providing there are no restrictions on x and y. But there are restrictions on x and y, since we have only 720 feet of fencing. That is, x and y must be chosen so that

$$3x + y = 720$$

This restriction on x and y, called a **constraint,** leads to the following maxima–minima problem:

$$\text{Maximize} \quad f(x, y) = xy \tag{1}$$

$$\text{Subject to} \quad 3x + y = 720 \qquad \text{or} \qquad 3x + y - 720 = 0 \tag{2}$$

This problem is a special case of a general class of problems of the form

$$\text{Maximize (or Minimize)} \quad z = f(x, y) \tag{3}$$

$$\text{Subject to} \quad g(x, y) = 0 \tag{4}$$

Of course, we could try to solve (4) for y in terms of x, or for x in terms of y, then substitute the result into (3), and use methods developed in Section 4-4 for functions of a single variable. But what if (4) were more complicated than (2), and solving for one variable in terms of the other was either very difficult or impossible? In the method of Lagrange multipliers, we will work with $g(x, y)$ directly and avoid having to solve (4) for one variable in terms of the other. In addition, the method generalizes to functions of arbitrarily many variables subject to one or more constraints.

Now, to the method. We form a new function F, using functions f and g in (3) and (4), as follows:

$$F(x, y, \lambda) = f(x, y) + \lambda g(x, y) \tag{5}$$

where λ (the Greek letter lambda) is called a **Lagrange multiplier.** Theorem 3 gives the basis for the method.

THEOREM 3

Any local maxima or minima of the function $z = f(x, y)$ subject to the constraint $g(x, y) = 0$ will be among those points (x_0, y_0) for which (x_0, y_0, λ_0) is a solution to the system

$$F_x(x, y, \lambda) = 0$$
$$F_y(x, y, \lambda) = 0$$
$$F_\lambda(x, y, \lambda) = 0$$

where $F(x, y, \lambda) = f(x, y) + \lambda g(x, y)$, provided all the partial derivatives exist.

We now solve the fence problem using the method of Lagrange multipliers.

Step 1. Formulate the problem in the form of equations (3) and (4):

$$\text{Maximize} \quad f(x, y) = xy$$
$$\text{Subject to} \quad g(x, y) = 3x + y - 720 = 0$$

Step 2. Form the function F, introducing the Lagrange multiplier λ:

$$\begin{aligned} F(x, y, \lambda) &= f(x, y) + \lambda g(x, y) \\ &= xy + \lambda(3x + y - 720) \end{aligned}$$

Step 3. Solve the system $F_x = 0$, $F_y = 0$, $F_\lambda = 0$. (The solutions are called **critical points** for F.)

$$F_x = y + 3\lambda = 0$$
$$F_y = x + \lambda = 0$$
$$F_\lambda = 3x + y - 720 = 0$$

From the first two equations, we see that

$$y = -3\lambda$$
$$x = -\lambda$$

Substitute these values for x and y into the third equation and solve for λ:

$$\begin{aligned} -3\lambda - 3\lambda &= 720 \\ -6\lambda &= 720 \\ \lambda &= -120 \end{aligned}$$

Thus,

$$y = -3(-120) = 360 \text{ feet}$$
$$x = -(-120) = 120 \text{ feet}$$

and $(x_0, y_0, \lambda_0) = (120, 360, -120)$ is the only critical point for F.

Step 4. According to Theorem 3, if the function $f(x, y)$, subject to the constraint $g(x, y) = 0$, has a local maximum or minimum, it must occur at $x = 120$, $y = 360$. Although it is possible to develop a test similar to Theorem 2 in Section 8-3 to determine the nature of this local extremum, we will not do so. [Note that Theorem 2 cannot be applied to $f(x, y)$ at (120, 360), since this point is not a critical point of the unconstrained function $f(x, y)$.] We will simply assume that the maximum value of $f(x, y)$ must occur for $x = 120$, $y = 360$. Thus,

$$\begin{aligned} \text{Max } f(x, y) &= f(120, 360) \\ &= (120)(360) = 43{,}200 \text{ square feet} \end{aligned}$$

The key steps in applying the method of Lagrange multipliers are listed in the following box:

Method of Lagrange Multipliers—Key Steps

Step 1. Formulate the problem in the form

Maximize (or Minimize) $z = f(x, y)$
Subject to $g(x, y) = 0$

Step 2. Form the function F:

$$F(x, y, \lambda) = f(x, y) + \lambda g(x, y)$$

Step 3. Find the critical points for F; that is, solve the system

$$F_x(x, y, \lambda) = 0$$
$$F_y(x, y, \lambda) = 0$$
$$F_\lambda(x, y, \lambda) = 0$$

Step 4. If (x_0, y_0, λ_0) is the only critical point of F, then we assume that (x_0, y_0) will always produce the solution to the problems we consider. If F has more than one critical point, then we evaluate $z = f(x, y)$ at (x_0, y_0) for each critical point (x_0, y_0, λ_0) of F. For the problems we consider, we assume that the largest of these values is the maximum value of $f(x, y)$, subject to the constraint $g(x, y) = 0$, and the smallest is the minimum value of $f(x, y)$, subject to the constraint $g(x, y) = 0$.

◆ EXAMPLE 16 Minimize $f(x, y) = x^2 + y^2$ subject to $x + y = 10$.

Solution Step 1. Minimize $f(x, y) = x^2 + y^2$
Subject to $g(x, y) = x + y - 10 = 0$

Step 2. $F(x, y, \lambda) = x^2 + y^2 + \lambda(x + y - 10)$

Step 3. $F_x = 2x + \lambda = 0$
$F_y = 2y + \lambda = 0$
$F_\lambda = x + y - 10 = 0$

From the first two equations,

$$x = -\frac{\lambda}{2} \qquad y = -\frac{\lambda}{2}$$

Substituting these into the third equation, we obtain

$$-\frac{\lambda}{2} - \frac{\lambda}{2} = 10$$
$$-\lambda = 10$$
$$\lambda = -10$$

The only critical point is $(x_0, y_0, \lambda_0) = (5, 5, -10)$.

Step 4. Since (5, 5, −10) is the only critical point for F, we conclude that (see step 4 in the box)

$$\text{Min } f(x, y) = f(5, 5) = (5)^2 + (5)^2 = 50$$ ◆

PROBLEM 16 Maximize $f(x, y) = 25 - x^2 - y^2$ subject to $x + y = 4$. ◆

Figures 10 and 11 illustrate the results obtained in Example 16 and Problem 16, respectively.

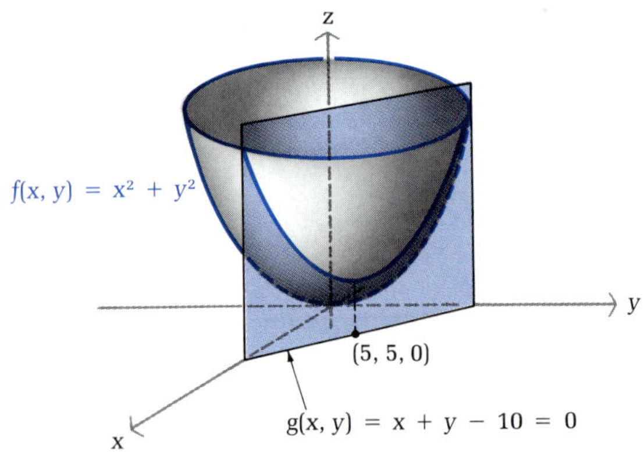

FIGURE 10

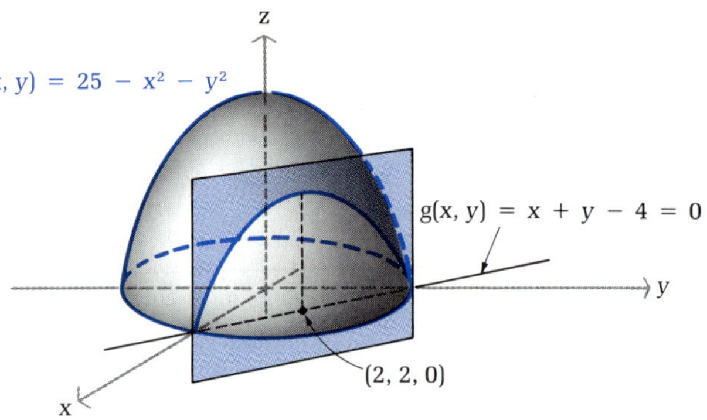

FIGURE 11

◆ EXAMPLE 17 The Cobb–Douglas production function for a new product is given by

Productivity

$$N(x, y) = 16x^{0.25}y^{0.75}$$

where x is the number of units of labor and y is the number of units of capital required to produce $N(x, y)$ units of the product. Each unit of labor costs \$50 and each unit of capital costs \$100. If \$500,000 has been budgeted for the production of this product, how should this amount be allocated between labor and capital in order to maximize production? What is the maximum number of units that can be produced?

Solution The total cost of using x units of labor and y units of capital is $50x + 100y$. Thus, the constraint imposed by the \$500,000 budget is

$$50x + 100y = 500{,}000$$

Step 1. Maximize $N(x, y) = 16x^{0.25}y^{0.75}$
Subject to $g(x, y) = 50x + 100y - 500{,}000 = 0$

Step 2. $F(x, y, \lambda) = 16x^{0.25}y^{0.75} + \lambda(50x + 100y - 500{,}000)$

Step 3. $F_x = 4x^{-0.75}y^{0.75} + 50\lambda = 0$
$F_y = 12x^{0.25}y^{-0.25} + 100\lambda = 0$
$F_\lambda = 50x + 100y - 500{,}000 = 0$

From the first two equations,

$$\lambda = -\tfrac{2}{25}x^{-0.75}y^{0.75} \qquad \text{and} \qquad \lambda = -\tfrac{3}{25}x^{0.25}y^{-0.25}$$

Thus,

$$\begin{aligned} -\tfrac{2}{25}x^{-0.75}y^{0.75} &= -\tfrac{3}{25}x^{0.25}y^{-0.25} \qquad &&\text{Multiply both sides by } x^{0.75}y^{0.25}. \\ -\tfrac{2}{25}y &= -\tfrac{3}{25}x &&\text{(We can assume } x \neq 0 \text{ and } y \neq 0.) \\ y &= \tfrac{3}{2}x \end{aligned}$$

Now, substitute for y in the third equation and solve for x:

$$\begin{aligned} 50x + 100(\tfrac{3}{2}x) - 500{,}000 &= 0 \\ 200x &= 500{,}000 \\ x &= 2{,}500 \end{aligned}$$

Thus,

$$y = \tfrac{3}{2}(2{,}500) = 3{,}750$$

and

$$\lambda = -\tfrac{2}{25}(2{,}500)^{-0.75}(3{,}750)^{0.75} \approx -0.1084$$

The only critical point of F is $(2{,}500, 3{,}750, -0.1084)$.

Step 4. Since F has only one critical point, we conclude that maximum productivity occurs when 2,500 units of labor and 3,750 units of capital are used (see step 4 in the box). Thus,

$$\begin{aligned} \text{Max } N(x, y) &= N(2{,}500, 3{,}750) \\ &= 16(2{,}500)^{0.25}(3{,}750)^{0.75} \\ &\approx 54{,}216 \text{ units} \end{aligned}$$

◆

The negative of the value of the Lagrange multiplier found in step 3 is called the **marginal productivity of money** and gives the approximate increase in production for each additional dollar spent on production. In Example 17, increasing the production budget from \$500,000 to \$600,000 would result in an approximate increase in production of

$$0.1084(100{,}000) = 10{,}840 \text{ units}$$

Note that simplifying the constraint equation

$$50x + 100y - 500{,}000 = 0$$

to

$$x + 2y - 10{,}000 = 0$$

before forming the function $F(x, y, \lambda)$ would make it difficult to interpret $-\lambda$ correctly. Thus, **in marginal productivity problems, the constraint equation should not be simplified.**

PROBLEM 17

The Cobb–Douglas production function for a new product is given by

$$N(x, y) = 20x^{0.5}y^{0.5}$$

where x is the number of units of labor and y is the number of units of capital required to produce $N(x, y)$ units of the product. Each unit of labor costs \$40 and each unit of capital costs \$120.

(A) If \$300,000 has been budgeted for the production of this product, how should this amount be allocated in order to maximize production? What is the maximum production?

(B) Find the marginal productivity of money in this case, and estimate the increase in production if an additional \$40,000 is budgeted for production.

◆

◆ FUNCTIONS OF THREE INDEPENDENT VARIABLES

We have indicated that the method of Lagrange multipliers can be extended to functions with arbitrarily many independent variables with one or more constraints. We now state a theorem for functions with three independent variables and one constraint, and consider an example that will demonstrate the advantage of the method of Lagrange multipliers over the method used in Section 8-3.

THEOREM 4

Any local maxima or minima of the function $w = f(x, y, z)$ subject to the constraint $g(x, y, z) = 0$ will be among the set of points (x_0, y_0, z_0) for which $(x_0, y_0, z_0, \lambda_0)$ is a solution to the system

$$F_x(x, y, z, \lambda) = 0$$
$$F_y(x, y, z, \lambda) = 0$$
$$F_z(x, y, z, \lambda) = 0$$
$$F_\lambda(x, y, z, \lambda) = 0$$

where $F(x, y, z, \lambda) = f(x, y, z) + \lambda g(x, y, z)$, provided all the partial derivatives exist.

◆ EXAMPLE 18

Package Design

A rectangular box with an open top and one partition is to be constructed from 162 square inches of cardboard. Find the dimensions that will result in a box with the largest possible volume.

Solution We must maximize

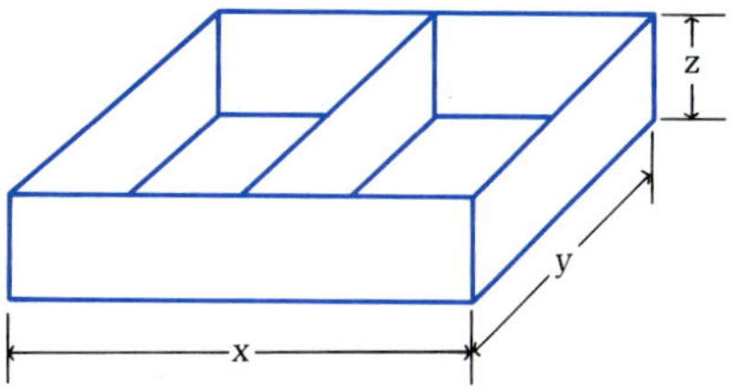

$$V(x, y, z) = xyz$$

subject to the constraint that the amount of material used is 162 square inches. Thus, x, y, and z must satisfy

$$xy + 2xz + 3yz = 162$$

Step 1. Maximize $V(x, y, z) = xyz$
Subject to $g(x, y, z) = xy + 2xz + 3yz - 162 = 0$

Step 2. $F(x, y, z, \lambda) = xyz + \lambda(xy + 2xz + 3yz - 162)$

Step 3.
$$\begin{aligned} F_x &= yz + \lambda(y + 2z) = 0 \\ F_y &= xz + \lambda(x + 3z) = 0 \\ F_z &= xy + \lambda(2x + 3y) = 0 \\ F_\lambda &= xy + 2xz + 3yz - 162 = 0 \end{aligned}$$

From the first two equations, we can write

$$\lambda = \frac{-yz}{y + 2z} \qquad \lambda = \frac{-xz}{x + 3z}$$

Eliminating λ, we have

$$\begin{aligned} \frac{-yz}{y + 2z} &= \frac{-xz}{x + 3z} \\ -xyz - 3yz^2 &= -xyz - 2xz^2 \\ 3yz^2 &= 2xz^2 \qquad \text{We can assume } z \neq 0. \\ 3y &= 2x \\ x &= \tfrac{3}{2}y \end{aligned}$$

From the second and third equations,

$$\lambda = \frac{-xz}{x + 3z} \qquad \lambda = \frac{-xy}{2x + 3y}$$

Eliminating λ, we have

$$\begin{aligned} \frac{-xz}{x + 3z} &= \frac{-xy}{2x + 3y} \\ -2x^2z - 3xyz &= -x^2y - 3xyz \\ 2x^2z &= x^2y \qquad \text{We can assume } x \neq 0. \\ 2z &= y \\ z &= \tfrac{1}{2}y \end{aligned}$$

Substituting $x = \frac{3}{2}y$ and $z = \frac{1}{2}y$ in the fourth equation, we have

$$(\tfrac{3}{2}y)y + 2(\tfrac{3}{2}y)(\tfrac{1}{2}y) + 3y(\tfrac{1}{2}y) - 162 = 0$$

$$\tfrac{3}{2}y^2 + \tfrac{3}{2}y^2 + \tfrac{3}{2}y^2 = 162$$

$$y^2 = 36 \qquad \text{We can assume } y > 0.$$

$$y = 6$$

$$x = \tfrac{3}{2}(6) = 9 \qquad \text{Using } x = \tfrac{3}{2}y$$

$$z = \tfrac{1}{2}(6) = 3 \qquad \text{Using } z = \tfrac{1}{2}y$$

and, finally,

$$\lambda = \frac{-(6)(3)}{6 + 2(3)} = -\frac{3}{2} \qquad \text{Using } \lambda = \frac{-yz}{y + 2z}$$

Thus, the only critical point of F with x, y, and z all positive is $(9, 6, 3, -\frac{3}{2})$.

Step 4. The box with maximum volume has dimensions 9 inches by 6 inches by 3 inches. ◆

PROBLEM 18 A box of the same type as described in Example 18 is to be constructed from 288 square inches of cardboard. Find the dimensions that will result in a box with the largest possible volume. ◆

Suppose we had decided to solve Example 18 by the method used in Section 8-3. First we would have to solve the material constraint for one of the variables, say, z:

$$z = \frac{162 - xy}{2x + 3y}$$

Then we would eliminate z in the volume function and maximize

$$V(x, y) = xy\,\frac{162 - xy}{2x + 3y}$$

Using the method of Lagrange multipliers allows us to avoid the formidable task of finding the partial derivatives of V.

Answers to Matched Problems

16. Max $f(x, y) = f(2, 2) = 17$ (see Fig. 11)
17. (A) 3,750 units of labor and 1,250 units of capital;
    Max $N(x, y) = N(3{,}750, 1{,}250) \approx 43{,}301$ units
    (B) Marginal productivity of money ≈ 0.1443; increase in production $\approx$ 5,774 units
18. 12 in. by 8 in. by 4 in.

EXERCISE 8-4

Use the method of Lagrange multipliers in the following problems:

A

1. Maximize $f(x, y) = 2xy$
   Subject to $x + y = 6$
2. Minimize $f(x, y) = 6xy$
   Subject to $y - x = 6$
3. Minimize $f(x, y) = x^2 + y^2$
   Subject to $3x + 4y = 25$
4. Maximize $f(x, y) = 25 - x^2 - y^2$
   Subject to $2x + y = 10$

B

5. Find the maximum and minimum of $f(x, y) = 2xy$ subject to $x^2 + y^2 = 18$.
6. Find the maximum and minimum of $f(x, y) = x^2 - y^2$ subject to $x^2 + y^2 = 25$.
7. Maximize the product of two numbers if their sum must be 10.
8. Minimize the product of two numbers if their difference must be 10.

C

9. Minimize $f(x, y, z) = x^2 + y^2 + z^2$
   Subject to $2x - y + 3z = -28$
10. Maximize $f(x, y, z) = xyz$
    Subject to $2x + y + 2z = 120$
11. Maximize and minimize $f(x, y, z) = x + y + z$
    Subject to $x^2 + y^2 + z^2 = 12$
12. Maximize and minimize $f(x, y, z) = 2x + 4y + 4z$
    Subject to $x^2 + y^2 + z^2 = 9$

APPLICATIONS

Business & Economics

13. *Budgeting for least cost.* A manufacturing company produces two models of a television set, x units of model A and y units of model B per week, at a cost (in dollars) of

$$C(x, y) = 6x^2 + 12y^2$$

If it is necessary (because of shipping considerations) that

$$x + y = 90$$

how many of each type of set should be manufactured per week to minimize cost? What is the minimum cost?

14. *Budgeting for maximum production.* A manufacturing firm has budgeted $60,000 per month for labor and materials. If $x thousand is spent on labor and $y thousand is spent on materials, and if the monthly output (in units) is given by

$$N(x, y) = 4xy - 8x$$

how should the $60,000 be allocated to labor and materials in order to maximize N? What is the maximum N?

15. *Productivity.* A consulting firm for a manufacturing company arrived at the following Cobb–Douglas production function for a particular product:

$$N(x, y) = 50x^{0.8}y^{0.2}$$

where x is the number of units of labor and y is the number of units of capital required to produce $N(x, y)$ units of the product. Each unit of labor costs \$40 and each unit of capital costs \$80.

(A) If \$400,000 is budgeted for production of the product, determine how this amount should be allocated to maximize production, and find the maximum production.

(B) Find the marginal productivity of money in this case, and estimate the increase in production if an additional \$50,000 is budgeted for the production of this product.

16. *Productivity.* The research department for a manufacturing company arrived at the following Cobb–Douglas production function for a particular product:

$$N(x, y) = 10x^{0.6}y^{0.4}$$

where x is the number of units of labor and y is the number of units of capital required to produce $N(x, y)$ units of the product. Each unit of labor costs \$30 and each unit of capital costs \$60.

(A) If \$300,000 is budgeted for production of the product, determine how this amount should be allocated to maximize production, and find the maximum production.

(B) Find the marginal productivity of money in this case, and estimate the increase in production if an additional \$80,000 is budgeted for the production of this product.

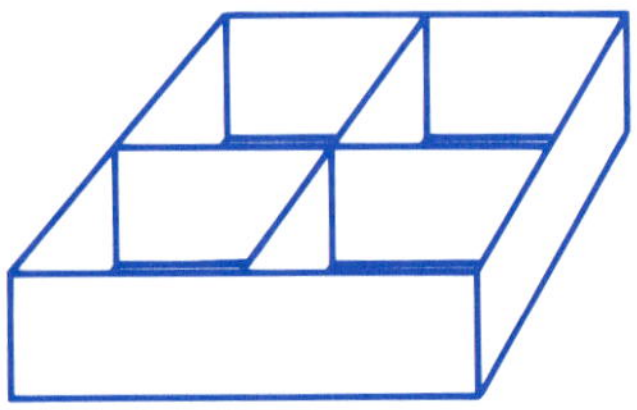

17. *Maximum volume.* A rectangular box with no top and two intersecting partitions is to be constructed from 192 square inches of cardboard (see the figure in the margin). Find the dimensions that will maximize the volume.

18. *Maximum volume.* A mailing service states that a rectangular package shall have the sum of the length and girth not to exceed 120 inches (see the figure). What are the dimensions of the largest (in volume) mailing carton that can be constructed meeting these restrictions?

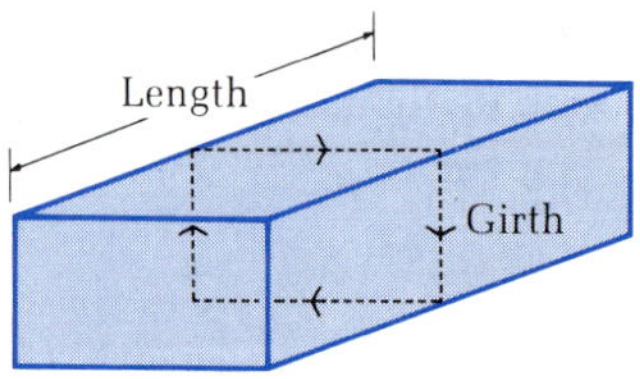

Life Sciences

19. *Agriculture.* Three pens of the same size are to be built along an existing fence (see the figure). If 400 feet of fencing are available, what length should x and y be to produce the maximum total area? What is the maximum area?

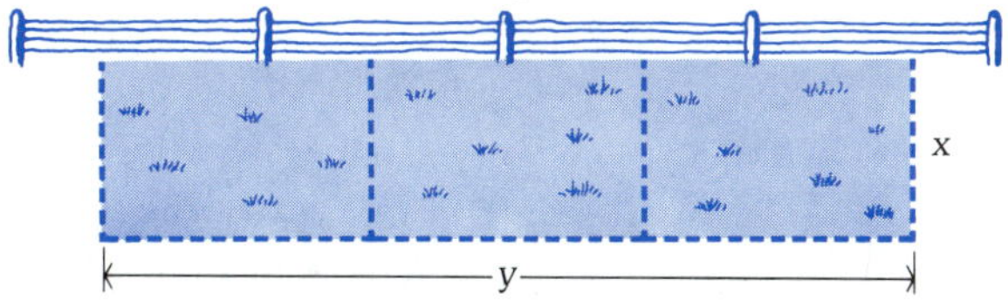

20. *Diet and minimum cost.* A group of guinea pigs is to receive 25,600 calories per week. Two available foods produce $200xy$ calories for a mixture of x kilograms of type M food and y kilograms of type N food. If type M costs \$1 per kilogram and type N costs \$2 per kilogram, how much of each type of food should be used to minimize weekly food costs? What is the minimum cost? [*Note:* $x \geq 0$, $y \geq 0$]

SECTION 8-5

Method of Least Squares

- LEAST SQUARES APPROXIMATION
- APPLICATIONS

LEAST SQUARES APPROXIMATION

In this section we will use the optimization techniques discussed in Section 8-3 to find the equation of a line that is a "best" approximation to a set of points in a rectangular coordinate system. This very popular method is known as **least squares approximation,** or **linear regression.** Let us begin by considering a specific case.

A manufacturer wants to approximate the cost function for a product. The value of the cost function has been determined for certain levels of production, as listed in Table 1. Although these points do not all lie on a line (see Fig. 12, page 540), they are very close to being linear. The manufacturer would like to approximate the cost function by a linear function; that is, determine values m and d so that the line

$$y = mx + d$$

is, in some sense, the "best" approximation to the cost function.

TABLE 1

NUMBER OF UNITS *x, in hundreds*	COST *y, in thousands of dollars*
2	4
5	6
6	7
9	8

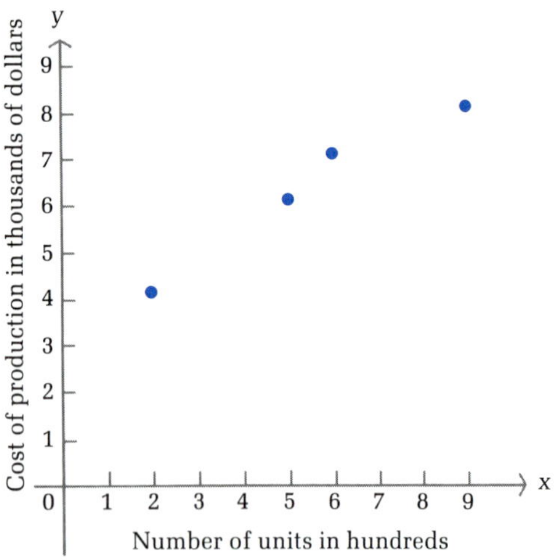

FIGURE 12

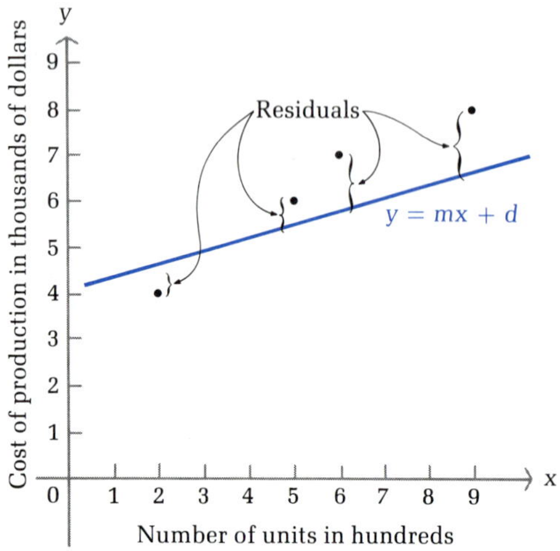

FIGURE 13

What do we mean by "best"? Since the line $y = mx + d$ will not go through all four points, it is reasonable to examine the differences between the y coordinates of the points listed in the table and the y coordinates of the corresponding points on the line. Each of these differences is called the **residual** at that point (see Fig. 13). For example, at $x = 2$, the point from Table 1 is (2, 4) and the point on the line is $(2, 2m + d)$, so the residual is

$$4 - (2m + d) = 4 - 2m - d$$

All the residuals are listed in Table 2.

TABLE 2

x	y	mx + d	Residual
2	4	2m + d	4 − 2m − d
5	6	5m + d	6 − 5m − d
6	7	6m + d	7 − 6m − d
9	8	9m + d	8 − 9m − d

Our criterion for the "best" approximation is the following: Determine the values of m and d that *minimize the sum of the squares* of the residuals. The resulting line is called the **least squares line,** or the **regression line.** To this end, we minimize

$$F(m, d) = (4 - 2m - d)^2 + (6 - 5m - d)^2 + (7 - 6m - d)^2 + (8 - 9m - d)^2$$

Step 1. Find critical points:

$$\begin{aligned} F_m(m, d) &= 2(4 - 2m - d)(-2) + 2(6 - 5m - d)(-5) \\ &\quad + 2(7 - 6m - d)(-6) + 2(8 - 9m - d)(-9) \\ &= -304 + 292m + 44d = 0 \end{aligned}$$

$$\begin{aligned} F_d(m, d) &= 2(4 - 2m - d)(-1) + 2(6 - 5m - d)(-1) \\ &\quad + 2(7 - 6m - d)(-1) + 2(8 - 9m - d)(-1) \\ &= -50 + 44m + 8d = 0 \end{aligned}$$

Solving the system

$$\begin{aligned} -304 + 292m + 44d &= 0 \\ -50 + 44m + 8d &= 0 \end{aligned}$$

we obtain $(m, d) = (0.58, 3.06)$ as the only critical point.

Step 2. Compute $A = F_{mm}(m, d)$, $B = F_{md}(m, d)$, and $C = F_{dd}(m, d)$:

$$F_{mm}(m, d) = 292 \quad \text{thus} \quad A = F_{mm}(0.58, 3.06) = 292$$
$$F_{md}(m, d) = 44 \quad \text{thus} \quad B = F_{md}(0.58, 3.06) = 44$$
$$F_{dd}(m, d) = 8 \quad \text{thus} \quad C = F_{dd}(0.58, 3.06) = 8$$

Step 3. Evaluate $AC - B^2$ and try to classify the critical point (m, d) using Theorem 2 in Section 8-3:

$$AC - B^2 = (292)(8) - (44)^2 = 400 > 0 \quad \text{and} \quad A = 292 > 0$$

Therefore, case 2 in Theorem 2 applies, and $F(m, d)$ has a local minimum at the critical point (0.58, 3.06).

Thus, the least squares line for the given data is

$$y = 0.58x + 3.06 \qquad \text{Least squares line}$$

The sum of the squares of the residuals is minimized for this choice of m and d (see Fig. 14).

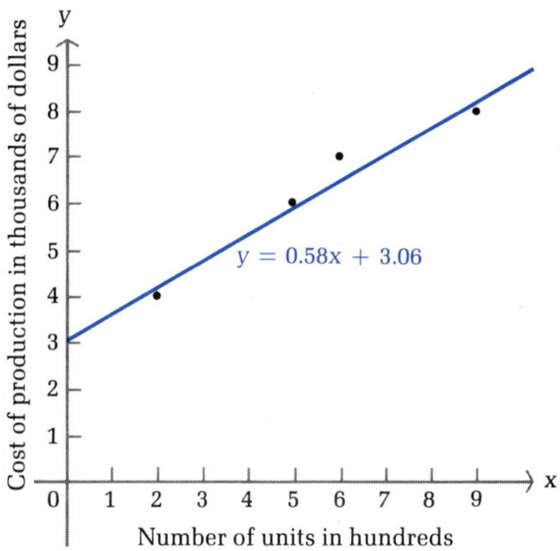

FIGURE 14

This linear function can now be used by the manufacturer to estimate any of the quantities normally associated with the cost function—such as costs, marginal costs, average costs, and so on. For example, the cost of producing 2,000 units is approximately

$$y = (0.58)(20) + 3.06 = 14.66 \quad \text{or} \quad \$14,660$$

The marginal cost function is

$$\frac{dy}{dx} = 0.58$$

The average cost function is

$$\bar{y} = \frac{0.58x + 3.06}{x}$$

In general, if we are given a set of n points $(x_1, y_1), (x_2, y_2), \ldots, (x_n, y_n)$, then proceeding as we did with the points in Table 1, it can be shown that the coefficients m and d of the least squares line $y = mx + d$ must satisfy the following system of *normal equations:*

$$\begin{aligned} \left(\sum_{k=1}^{n} x_k\right)m + nd &= \sum_{k=1}^{n} y_k \\ \left(\sum_{k=1}^{n} x_k^2\right)m + \left(\sum_{k=1}^{n} x_k\right)d &= \sum_{k=1}^{n} x_k y_k \end{aligned} \qquad (1)$$

Solving system (1) for m and d produces the formulas given in the box.

Least Squares Approximation

For a set of n points $(x_1, y_1), (x_2, y_2), \ldots, (x_n, y_n)$, the coefficients m and d of the least squares line

$$y = mx + d$$

are the solutions of the system of **normal equations**

$$\begin{aligned} \left(\sum_{k=1}^{n} x_k\right)m + nd &= \sum_{k=1}^{n} y_k \\ \left(\sum_{k=1}^{n} x_k^2\right)m + \left(\sum_{k=1}^{n} x_k\right)d &= \sum_{k=1}^{n} x_k y_k \end{aligned} \qquad (1)$$

and are given by the formulas

$$m = \frac{n\left(\sum_{k=1}^{n} x_k y_k\right) - \left(\sum_{k=1}^{n} x_k\right)\left(\sum_{k=1}^{n} y_k\right)}{n\left(\sum_{k=1}^{n} x_k^2\right) - \left(\sum_{k=1}^{n} x_k\right)^2} \qquad (2)$$

and

$$d = \frac{\sum_{k=1}^{n} y_k - m\left(\sum_{k=1}^{n} x_k\right)}{n} \qquad (3)$$

To find m and d, we can either solve system (1) directly or use formulas (2) and (3). If the formulas are used, note that the value of m is needed in formula (3); thus, the value of m always must be computed first.

◆ APPLICATIONS

◆ EXAMPLE 19 *Educational Testing*

Table 3 lists the midterm and final examination scores for 10 students in a calculus course.

TABLE 3

MIDTERM	FINAL	MIDTERM	FINAL
49	61	78	77
53	47	83	81
67	72	85	79
71	76	91	93
74	68	99	99

(A) Find the least squares line for the data given in the table.

(B) Use the least squares line to predict the final examination score for a student who scored 95 on the midterm examination.

(C) Graph the data and the least squares line on the same set of axes.

Solutions

(A) Table 4 shows a convenient way to compute all the sums in the formulas for m and d.

TABLE 4

	x_k	y_k	$x_k y_k$	x_k^2
	49	61	2,989	2,401
	53	47	2,491	2,809
	67	72	4,824	4,489
	71	76	5,396	5,041
	74	68	5,032	5,476
	78	77	6,006	6,084
	83	81	6,723	6,889
	85	79	6,715	7,225
	91	93	8,463	8,281
	99	99	9,801	9,801
Totals	750	753	58,440	58,496

Thus,

$$\sum_{k=1}^{10} x_k = 750 \qquad \sum_{k=1}^{10} y_k = 753 \qquad \sum_{k=1}^{10} x_k y_k = 58{,}440 \qquad \sum_{k=1}^{10} x_k^2 = 58{,}496$$

and the normal equations are

$$750m + \quad 10d = 753$$
$$58{,}496m + 750d = 58{,}440$$

We can either solve this system directly or use formulas (2) and (3). We choose to use the formulas:

$$m = \frac{10(58{,}440) - (750)(753)}{10(58{,}496) - (750)^2} = \frac{19{,}650}{22{,}460} \approx 0.875$$

$$d = \frac{753 - 0.875(750)}{10} = 9.675$$

The least squares line is given (approximately) by

$$y = 0.875x + 9.675$$

(B) If $x = 95$, then the predicted score on the final exam is

$$y = 0.875(95) + 9.675$$
$$\approx 93 \qquad \text{Assuming the score must be an integer}$$

(C)

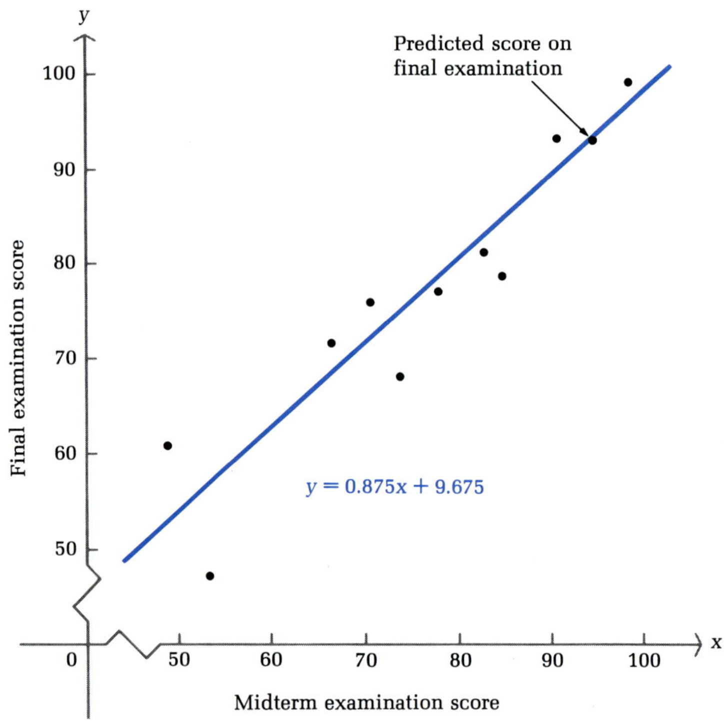

◆

PROBLEM 19 Repeat Example 19 for the scores listed in Table 5.

TABLE 5

MIDTERM	FINAL	MIDTERM	FINAL
54	50	84	80
60	66	88	95
75	80	89	85
76	68	97	94
78	71	99	86

◆

◆ EXAMPLE 20 *Wool Production*

Table 6 lists the annual production of wool throughout the world for the years 1970–1980. Use the data in the table to predict the worldwide wool production for 1981.

TABLE 6

World Wool Production

YEAR	MILLIONS OF POUNDS	YEAR	MILLIONS OF POUNDS
1970	6,107	1976	5,827
1971	5,972	1977	5,838
1972	5,560	1978	5,983
1973	5,474	1979	6,168
1974	5,769	1980	6,285
1975	5,911		

Solution Solving this problem by hand is certainly possible, but would require considerable effort. Instead, we used a computer to perform the necessary computations. A typical computer output is shown in Table 7.

TABLE 7

INPUT TO PROGRAM

```
* - LEAST SQUARES TECHNIQUE - *
              DATA
PT.  X COORDINATE  Y COORDINATE
===  ============  ============
 1              0          6107
 2              1          5972
 3              2          5560
 4              3          5474
 5              4          5769
 6              5          5911
 7              6          5827
 8              7          5838
 9              8          5983
 10             9          6168
 11            10          6285
```

OUTPUT FROM PROGRAM

```
* - LEAST SQUARES TECHNIQUE - *
          POLYNOMIAL

Y = 33.9X + 5729.95

* - LEAST SQUARES TECHNIQUE - *
      EVALUATE POLYNOMIAL

       X            CALCULATED Y
================================
      11                 6102.85
```

Notice that we used $x = 0$ for 1970, $x = 1$ for 1971, and so on. Examining the computer output in Table 7, we see that the least squares line is

$$y = 33.9x + 5{,}729.95$$

and the estimated worldwide wool production in 1981 is 6,102.85 million pounds. ◆

PROBLEM 20 Use the least squares line in Example 20 to estimate the worldwide wool production in 1982. ◆

Answers to Matched Problems

19. (A) $y = 0.85x + 9.47$ (B) 90.2

(C)

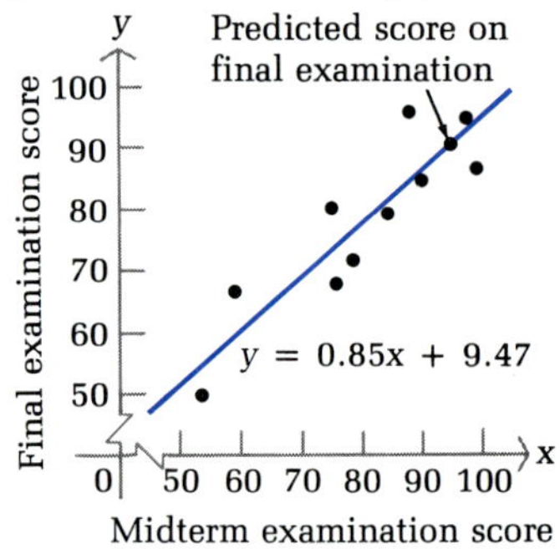

20. 6,136.75 million pounds

EXERCISE 8-5

A *Find the least squares line. Graph the data and the least squares line.*

1.

x	y
1	1
2	3
3	4
4	3

2.

x	y
1	−2
2	−1
3	3
4	5

3.

x	y
1	8
2	5
3	4
4	0

4.

x	y
1	20
2	14
3	11
4	3

5.

x	y
1	3
2	4
3	5
4	6

6.

x	y
1	2
2	3
3	3
4	2

B *Find the least squares line and use it to estimate y for the indicated value of x.*

7.

x	y
0	10
5	22
10	31
15	46
20	51

Estimate y when x = 25.

8.

x	y
−5	60
0	50
5	30
10	20
15	15

Estimate y when x = 20.

9.

x	y
−1	14
1	12
3	8
5	6
7	5

Estimate y when x = 2.

10.

x	y
2	−4
6	0
10	8
14	12
18	14

Estimate y when x = 15.

11.

x	y	x	y
0.5	25	9.5	12
2	22	11	11
3.5	21	12.5	8
5	21	14	5
6.5	18	15.5	1

Estimate y when x = 8.

12.

x	y	x	y
0	−15	12	11
2	−9	14	13
4	−7	16	19
6	−7	18	25
8	−1	20	33

Estimate y when x = 10.

C **13.** The method of least squares can be generalized to curves other than straight lines. To find the coefficients of the parabola

$$y = ax^2 + bx + c$$

that is the "best" fit for the points (1, 2), (2, 1), (3, 1), and (4, 3), minimize the sum of the squares of the residuals

$$F(a, b, c) = (a + b + c - 2)^2 + (4a + 2b + c - 1)^2 + (9a + 3b + c - 1)^2 + (16a + 4b + c - 3)^2$$

by solving the system

$$F_a(a, b, c) = 0 \qquad F_b(a, b, c) = 0 \qquad F_c(a, b, c) = 0$$

for a, b, and c. Graph the points and the parabola.

14. Repeat Problem 13 for the points (−1, −2), (0, 1), (1, 2), and (2, 0).

Problems 15 and 16 refer to the system of normal equations and the formulas for m and d stated on page 542.

15. Verify formulas (2) and (3) by solving the system of normal equations (1) for m and d.

16. If

$$\overline{x} = \frac{1}{n}\sum_{k=1}^{n} x_k \quad \text{and} \quad \overline{y} = \frac{1}{n}\sum_{k=1}^{n} y_k$$

are the averages of the x and y coordinates, respectively, show that the point $(\overline{x}, \overline{y})$ satisfies the equation of the least squares line $y = mx + d$.

APPLICATIONS

Business & Economics

17. *Income.* Data for the per capita personal income in the United States from 1980 to 1986 are given in the table below.

(A) Find the least squares line for the data using $x = 0$ for 1980.
(B) Use the least squares line to estimate the per capita income in 1990.

YEAR	INCOME Dollars	YEAR	INCOME Dollars
1980	9,910	1984	13,115
1981	10,940	1985	13,687
1982	11,470	1986	14,461
1983	12,093		

18. *Consumer price index.* Data for the consumer price index (CPI) in the United States (in 1967 dollars) from 1980 to 1985 are given in the table below.

(A) Find the least squares line for the data using $x = 0$ for 1980.
(B) Use the least squares line to estimate the consumer price index in 1990.

YEAR	CPI	YEAR	CPI
1980	246.8	1983	298.4
1981	272.4	1984	311.1
1982	289.1	1985	322.2

x	y
5.0	2.0
5.5	1.8
6.0	1.4
6.5	1.2
7.0	1.1

19. *Maximizing profit.* The market research department for a drug store chain chose two summer resort areas to test market a new sun screen lotion packaged in 4 ounce plastic bottles. After a summer of varying the selling price and recording the monthly demand, the research department arrived at the demand table given in the margin, where y is the number of bottles purchased per month (in thousands) at x dollars per bottle.

(A) Find a demand equation using the method of least squares.
(B) If each bottle of sun screen costs the drug store chain \$4, how should it be priced to achieve a maximum monthly profit? [*Hint:* Use the result of part A, with $C = 4y$, $R = xy$, and $P = R - C$.]

x	y
4.0	4.2
4.5	3.5
5.0	2.7
5.5	1.5
6.0	0.7

20. *Maximizing profit.* A market research consultant for a supermarket chain chose a large city to test market a new brand of mixed nuts packaged in 8 ounce cans. After a year of varying the selling price and recording the monthly demand, the consultant arrived at the demand table given in the margin, where y is the number of cans purchased per month (in thousands) at x dollars per can.

(A) Find a demand equation using the method of least squares.

(B) If each can of nuts costs the supermarket chain \$3, how should it be priced to achieve a maximum monthly profit?

Life Sciences

WATER TEMPERATURE	PULSE RATE REDUCTION
50	15
55	13
60	10
65	6
70	2

21. *Medicine.* If a person dives into cold water, a neural reflex response automatically shuts off blood circulation to the skin and muscles and reduces the pulse rate. A medical research team conducted an experiment using a group of ten 2-year-olds. A child's face was placed momentarily in cold water, and the corresponding reduction in pulse rate was recorded. The data for the average reduction in heart rate for each temperature are summarized in the table in the margin.

(A) If T is water temperature (in degrees Fahrenheit) and P is pulse rate reduction (in beats per minute), use the method of least squares to find a linear equation relating T and P.

(B) Use the equation found in part A to find P when $T = 57$.

22. *Biology.* In biology there is an approximate rule, called the *bioclimatic rule for temperate climates,* that has been known for a couple of hundred years. This rule states that in spring and early summer, periodic phenomena such as blossoming of flowers, appearance of insects, and ripening of fruit usually come about 4 days later for each 500 feet of altitude. Stated as a formula,

$$d = 8h \qquad 0 \leq h \leq 4$$

h	d
0	0
1	7
2	18
3	28
4	33

where d is the change in days and h is the altitude (in thousands of feet). To test this rule, an experiment was set up to record the difference in blossoming time of the same type of apple tree at different altitudes. A summary of the results is given in the table in the margin.

(A) Use the method of least squares to find a linear equation relating h and d. Does the bioclimatic rule, $d = 8h$, appear to be approximately correct?

(B) How much longer will it take this type of apple tree to blossom at 3.5 thousand feet than at sea level? [Use the linear equation found in part A.]

Social Sciences

23. *Political science.* Association of economic class and party affiliation did not start with Roosevelt's New Deal; it goes back to the time of Andrew Jackson (1767–1845). Paul Lazarsfeld of Columbia University published an article

Political Affiliations, 1836

WARD	AVERAGE ASSESSED VALUE PER PERSON *(in \$100)*	DEMOCRATIC VOTES *(%)*
12	1.7	51
3	2.1	49
1	2.3	53
5	2.4	36
2	3.6	65
11	3.7	35
10	4.7	29
4	6.2	40
6	7.1	34
9	7.4	29
8	8.7	20
7	11.9	23

in the November 1950 issue of *Scientific American* in which he discusses statistical investigations of the relationships between economic class and party affiliation. The data in the table are taken from this article.

(A) If A represents the average assessed value per person in a given ward in 1836 and D represents the percentage of people in that ward voting Democratic in 1836, use the method of least squares to find a linear equation relating A and D.

(B) If the average assessed value per person in a ward had been \$300, what is the predicted percentage of people in that ward that would have voted Democratic?

24. *Education.* The table lists the high school grade-point averages (GPA's) of 10 students, along with their grade-point averages after one semester of college.

HIGH SCHOOL GPA	COLLEGE GPA	HIGH SCHOOL GPA	COLLEGE GPA
2.0	1.5	3.0	2.3
2.2	1.5	3.1	2.5
2.4	1.6	3.3	2.9
2.7	1.8	3.4	3.2
2.9	2.1	3.7	3.5

(A) Find the least squares line for the data.

(B) Estimate the college GPA for a student with a high school GPA of 3.5.

(C) Estimate the high school GPA necessary for a college GPA of 2.7.

SECTION 8-6 Double Integrals Over Rectangular Regions

- INTRODUCTION
- DEFINITION OF THE DOUBLE INTEGRAL
- AVERAGE VALUE OVER RECTANGULAR REGIONS
- VOLUME AND DOUBLE INTEGRALS

◆ INTRODUCTION

We have generalized the concept of differentiation to functions with two or more independent variables. How can we do the same with integration, and how can we interpret the results? Let us first look at the operation of antidifferentiation. We can antidifferentiate a function of two or more variables with respect to one of the variables by treating all the other variables as though they were constants. Thus, this operation is the reverse operation of partial differentiation, just as ordinary antidifferentiation is the reverse operation of ordinary differentiation. We write $\int f(x, y)\,dx$ to indicate that we are to antidifferentiate $f(x, y)$ with respect to x, holding y fixed; we write $\int f(x, y)\,dy$ to indicate that we are to antidifferentiate $f(x, y)$ with respect to y, holding x fixed.

◆ EXAMPLE 21 Evaluate:

(A) $\int (6xy^2 + 3x^2)\,dy$ (B) $\int (6xy^2 + 3x^2)\,dx$

Solutions (A) Treating x as a constant and using the properties of antidifferentiation from Section 6-1, we have

$$\begin{aligned}\int (6xy^2 + 3x^2)\,dy &= \int 6xy^2\,dy + \int 3x^2\,dy \\ &= 6x \int y^2\,dy + 3x^2 \int dy \\ &= 6x\left(\frac{y^3}{3}\right) + 3x^2(y) + C(x) \\ &= 2xy^3 + 3x^2y + C(x)\end{aligned}$$

The dy tells us we are looking for the antiderivative of $6xy^2 + 3x^2$ with respect to y only, holding x constant.

Notice that the constant of integration actually can be *any function of x alone*, since, for any such function

$$\frac{\partial}{\partial y} C(x) = 0$$

We can verify that our answer is correct by using partial differentiation:

$$\begin{aligned}\frac{\partial}{\partial y}[2xy^3 + 3x^2y + C(x)] &= 6xy^2 + 3x^2 + 0 \\ &= 6xy^2 + 3x^2\end{aligned}$$

(B) Now we treat y as a constant:

$$\begin{aligned}\int (6xy^2 + 3x^2)\,dx &= \int 6xy^2\,dx + \int 3x^2\,dx \\ &= 6y^2 \int x\,dx + 3 \int x^2\,dx \\ &= 6y^2\left(\frac{x^2}{2}\right) + 3\left(\frac{x^3}{3}\right) + E(y) \\ &= 3x^2y^2 + x^3 + E(y)\end{aligned}$$

This time, the antiderivative contains an arbitrary function $E(y)$ of y alone.

Check:
$$\begin{aligned}\frac{\partial}{\partial x}[3x^2y^2 + x^3 + E(y)] &= 6xy^2 + 3x^2 + 0 \\ &= 6xy^2 + 3x^2\end{aligned}$$ ◆

PROBLEM 21 Evaluate:

(A) $\int (4xy + 12x^2y^3)\,dy$ (B) $\int (4xy + 12x^2y^3)\,dx$ ◆

Now that we have extended the concept of antidifferentiation to functions with two variables, we also can evaluate definite integrals of the form

$$\int_a^b f(x, y)\,dx \qquad \text{or} \qquad \int_c^d f(x, y)\,dy$$

◆ EXAMPLE 22 Evaluate, substituting the limits of integration in y if dy is used and in x if dx is used:

(A) $\int_0^2 (6xy^2 + 3x^2)\,dy$ (B) $\int_0^1 (6xy^2 + 3x^2)\,dx$

Solutions

(A) From Example 21A, we know that $\int (6xy^2 + 3x^2)\,dy = 2xy^3 + 3x^2y + C(x)$. According to the definition of the definite integral for a function of one variable, we can use any antiderivative to evaluate the definite integral. Thus, choosing $C(x) = 0$, we have

$$\begin{aligned}\int_0^2 (6xy^2 + 3x^2)\,dy &= (2xy^3 + 3x^2y)\Big|_{y=0}^{y=2}\\ &= [2x(2)^3 + 3x^2(2)] - [2x(0)^3 + 3x^2(0)]\\ &= 16x + 6x^2\end{aligned}$$

(B) From Example 21B, we know that $\int (6xy^2 + 3x^2)\,dx = 3x^2y^2 + x^3 + E(y)$. Thus, choosing $E(y) = 0$, we have

$$\begin{aligned}\int_0^1 (6xy^2 + 3x^2)\,dx &= (3x^2y^2 + x^3)\Big|_{x=0}^{x=1}\\ &= [3y^2(1)^2 + (1)^3] - [3y^2(0)^2 + (0)^3]\\ &= 3y^2 + 1\end{aligned}$$

◆

PROBLEM 22 Evaluate:

(A) $\int_0^1 (4xy + 12x^2y^3)\,dy$ (B) $\int_0^3 (4xy + 12x^2y^3)\,dx$ ◆

Notice that integrating and evaluating a definite integral with integrand $f(x, y)$ with respect to y produces a function of x alone (or a constant). Likewise, integrating and evaluating a definite integral with integrand $f(x, y)$ with respect to x produces a function of y alone (or a constant). Each of these results, involving at most one variable, can now be used as an integrand in a second definite integral.

◆ EXAMPLE 23 Evaluate:

(A) $\int_0^1 \left[\int_0^2 (6xy^2 + 3x^2)\, dy \right] dx$ (B) $\int_0^2 \left[\int_0^1 (6xy^2 + 3x^2)\, dx \right] dy$

Solutions (A) Example 22A showed that

$$\int_0^2 (6xy^2 + 3x^2)\, dy = 16x + 6x^2$$

Thus,

$$\begin{aligned}
\int_0^1 \left[\int_0^2 (6xy^2 + 3x^2)\, dy \right] dx &= \int_0^1 (16x + 6x^2)\, dx \\
&= (8x^2 + 2x^3)\Big|_{x=0}^{x=1} \\
&= [8(1)^2 + 2(1)^3] - [8(0)^2 + 2(0)^3] \\
&= 10
\end{aligned}$$

(B) Example 22B showed that

$$\int_0^1 (6xy^2 + 3x^2)\, dx = 3y^2 + 1$$

Thus,

$$\begin{aligned}
\int_0^2 \left[\int_0^1 (6xy^2 + 3x^2)\, dx \right] dy &= \int_0^2 (3y^2 + 1)\, dy \\
&= (y^3 + y)\Big|_{y=0}^{y=2} \\
&= [(2)^3 + 2] - [(0)^3 + 0] \\
&= 10
\end{aligned}$$

◆

PROBLEM 23 Evaluate:

(A) $\int_0^3 \left[\int_0^1 (4xy + 12x^2y^3)\, dy \right] dx$ (B) $\int_0^1 \left[\int_0^3 (4xy + 12x^2y^3)\, dx \right] dy$ ◆

◆ DEFINITION OF THE DOUBLE INTEGRAL

Notice that the answers in Examples 23A and 23B are identical. This is not an accident. In fact, it is this property that enables us to define the *double integral*, as follows:

Double Integral

The **double integral** of a function $f(x, y)$ over a rectangle $R = \{(x, y) | a \le x \le b, \quad c \le y \le d\}$ is

$$\iint_R f(x, y)\, dA = \int_a^b \left[\int_c^d f(x, y)\, dy \right] dx = \int_c^d \left[\int_a^b f(x, y)\, dx \right] dy$$

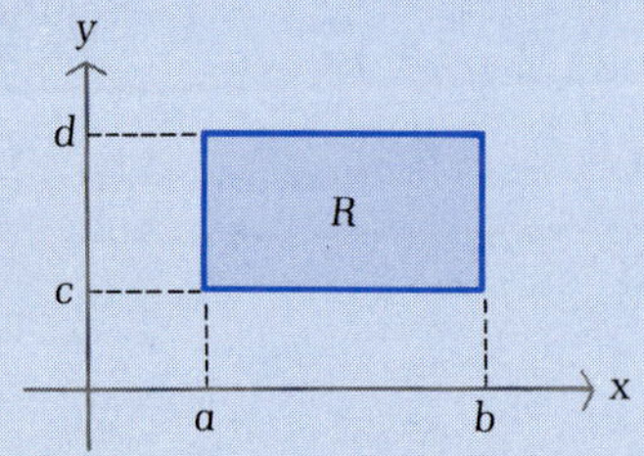

In the double integral $\iint_R f(x, y)\, dA$, $f(x, y)$ is called the **integrand** and R is called the **region of integration.** The expression dA indicates that this is an integral over a two-dimensional region. The integrals

$$\int_a^b \left[\int_c^d f(x, y)\, dy \right] dx \qquad \text{and} \qquad \int_c^d \left[\int_a^b f(x, y)\, dx \right] dy$$

are referred to as **iterated integrals** (the brackets are often omitted), and the order in which dx and dy are written indicates the order of integration. This is not the most general definition of the double integral over a rectangular region; however, it is equivalent to the general definition for all the functions we will consider.

◆ EXAMPLE 24 Evaluate:

$$\iint_R (x + y)\, dA \qquad \text{over} \qquad R = \{(x, y) | 1 \le x \le 3, \quad -1 \le y \le 2\}$$

Solution We can choose either order of iteration. As a check, we will evaluate the integral both ways:

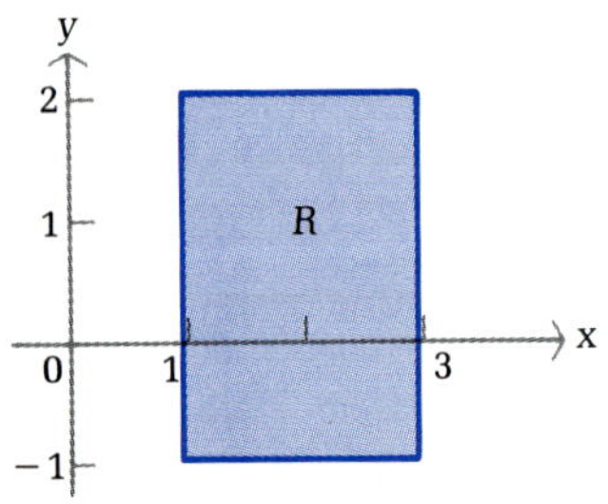

$$\begin{aligned}\iint_R (x + y)\, dA &= \int_1^3 \int_{-1}^2 (x + y)\, dy\, dx \\ &= \int_1^3 \left[\left(xy + \frac{y^2}{2} \right) \Big|_{y=-1}^{y=2} \right] dx \\ &= \int_1^3 [(2x + 2) - (-x + \tfrac{1}{2})]\, dx \\ &= \int_1^3 (3x + \tfrac{3}{2})\, dx\end{aligned}$$

$$\int_1^3 \left(3x + \tfrac{3}{2}\right) dx = \left(\tfrac{3}{2}x^2 + \tfrac{3}{2}x\right)\Big|_{x=1}^{x=3}$$

$$= \left(\tfrac{27}{2} + \tfrac{9}{2}\right) - \left(\tfrac{3}{2} + \tfrac{3}{2}\right)$$

$$= 18 - 3 = 15$$

$$\iint_R (x + y)\, dA = \int_{-1}^{2} \int_1^3 (x + y)\, dx\, dy$$

$$= \int_{-1}^{2} \left[\left(\frac{x^2}{2} + xy\right)\Big|_{x=1}^{x=3}\right] dy$$

$$= \int_{-1}^{2} \left[\left(\tfrac{9}{2} + 3y\right) - \left(\tfrac{1}{2} + y\right)\right] dy$$

$$= \int_{-1}^{2} (4 + 2y)\, dy$$

$$= (4y + y^2)\Big|_{y=-1}^{y=2}$$

$$= (8 + 4) - (-4 + 1)$$

$$= 12 - (-3) = 15 \quad \blacklozenge$$

PROBLEM 24 Evaluate both ways:

$$\iint_R (2x - y)\, dA \qquad \text{over} \qquad R = \{(x, y) \mid -1 \leqslant x \leqslant 5,\quad 2 \leqslant y \leqslant 4\} \quad \blacklozenge$$

◆ EXAMPLE 25 Evaluate:

$$\iint_R 2xe^{x^2+y}\, dA \qquad \text{over} \qquad R = \{(x, y) \mid 0 \leqslant x \leqslant 1,\quad -1 \leqslant y \leqslant 1\}$$

Solution

$$\iint_R 2xe^{x^2+y}\, dA = \int_{-1}^{1} \int_0^1 2xe^{x^2+y}\, dx\, dy$$

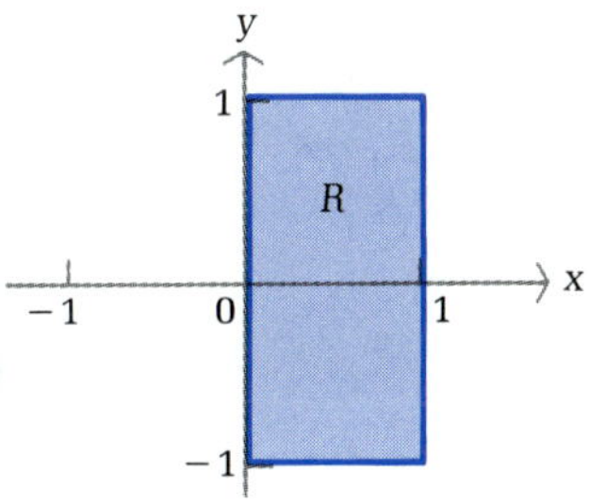

$$= \int_{-1}^{1} \left[\left(e^{x^2+y}\right)\Big|_{x=0}^{x=1}\right] dy$$

$$= \int_{-1}^{1} (e^{1+y} - e^y)\, dy$$

$$= (e^{1+y} - e^y)\Big|_{y=-1}^{y=1}$$

$$= (e^2 - e) - (e^0 - e^{-1})$$

$$= e^2 - e - 1 + e^{-1} \quad \blacklozenge$$

PROBLEM 25 Evaluate:

$$\iint_R \frac{x}{y^2} e^{x/y}\, dA \qquad \text{over} \qquad R = \{(x, y) \mid 0 \leqslant x \leqslant 1,\quad 1 \leqslant y \leqslant 2\} \quad \blacklozenge$$

◆ AVERAGE VALUE OVER RECTANGULAR REGIONS

In Section 6-5, the average value of a function $f(x)$ over an interval $[a, b]$ was defined as

$$\frac{1}{b-a}\int_a^b f(x)\,dx$$

This definition is easily extended to functions of two variables over rectangular regions, as shown in the box. Notice that the denominator in the expression given in the box, $(b-a)(d-c)$, is simply the area of the rectangle R.

Average Value over Rectangular Regions

The **average value** of the function $f(x, y)$ over the rectangle $R = \{(x, y) | a \le x \le b, \quad c \le y \le d\}$ is

$$\frac{1}{(b-a)(d-c)}\iint_R f(x, y)\,dA$$

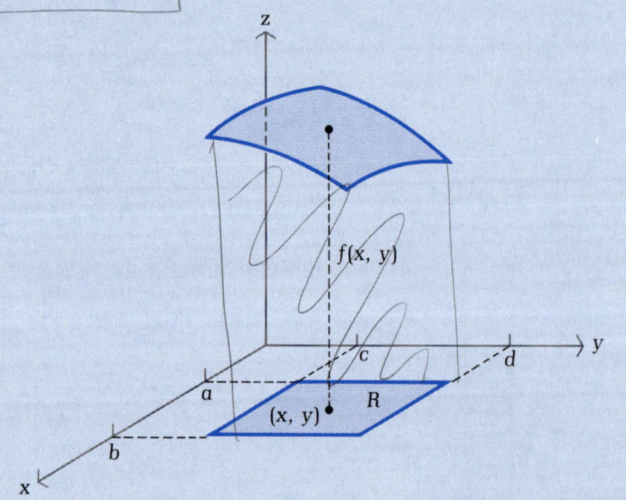

◆ EXAMPLE 26 Find the average value of $f(x, y) = 4 - \frac{1}{2}x - \frac{1}{2}y$ over the rectangle $R = \{(x, y) | 0 \le x \le 2, \ 0 \le y \le 2\}$.

Solution

$$\frac{1}{(b-a)(d-c)}\iint_R f(x, y)\,dA = \frac{1}{(2-0)(2-0)}\iint_R \left(4-\frac{1}{2}x-\frac{1}{2}y\right) dA$$

$$= \tfrac{1}{4}\int_0^2\int_0^2 (4-\tfrac{1}{2}x-\tfrac{1}{2}y)\,dy\,dx$$

$$= \tfrac{1}{4}\int_0^2 \left[(4y-\tfrac{1}{2}xy-\tfrac{1}{4}y^2)\Big|_{y=0}^{y=2}\right] dx$$

$$= \tfrac{1}{4}\int_0^2 (7-x)\,dx$$

$$= \tfrac{1}{4}(7x-\tfrac{1}{2}x^2)\Big|_{x=0}^{x=2}$$

$$= \tfrac{1}{4}(12) = 3$$

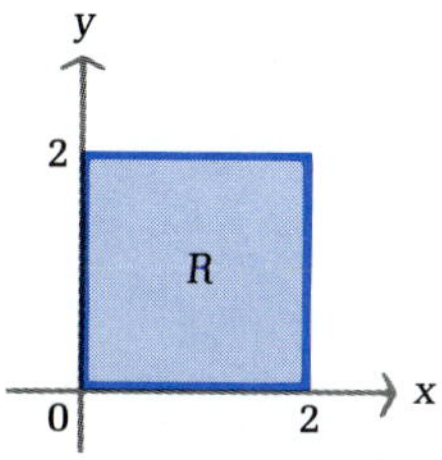

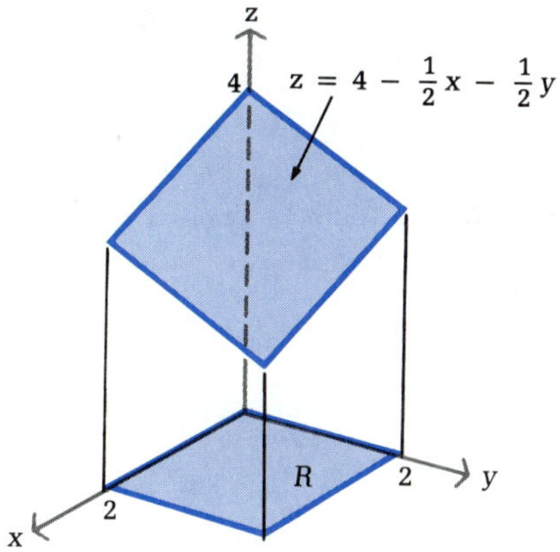

◆

PROBLEM 26 Find the average value of $f(x, y) = x + 2y$ over the rectangle $R = \{(x, y) | 0 \le x \le 2, \quad 0 \le y \le 1\}$. ◆

◆ VOLUME AND DOUBLE INTEGRALS

One application of the definite integral of a function with one variable is the calculation of areas, so it is not surprising that the definite integral of a function of two variables can be used to calculate volumes of solids.

Volume under a Surface

If $f(x, y) \geq 0$ over a rectangle $R = \{(x, y) | a \leq x \leq b, \quad c \leq y \leq d\}$, then the volume of the solid formed by graphing f over the rectangle R is given by

$$V = \iint_R f(x, y)\, dA$$

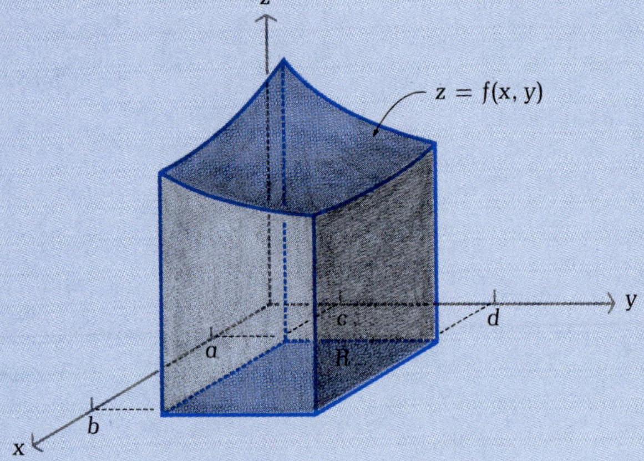

A proof of the statement in the box is left to a more advanced text.

◆ EXAMPLE 27 Find the volume of the solid under the graph of $f(x, y) = 1 + x^2 + y^2$ over the rectangle $R = \{(x, y) | 0 \leq x \leq 1, \quad 0 \leq y \leq 1\}$.

Solution

$$V = \iint_R (1 + x^2 + y^2)\, dA$$

$$= \int_0^1 \int_0^1 (1 + x^2 + y^2)\, dx\, dy$$

$$= \int_0^1 \left[(x + \tfrac{1}{3}x^3 + xy^2)\Big|_{x=0}^{x=1} \right] dy$$

$$= \int_0^1 (\tfrac{4}{3} + y^2)\, dy$$

$$= (\tfrac{4}{3}y + \tfrac{1}{3}y^3)\Big|_{y=0}^{y=1} = \tfrac{5}{3} \text{ cubic units}$$

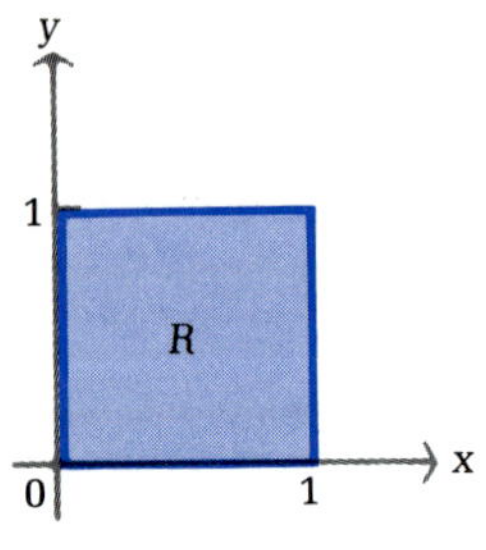

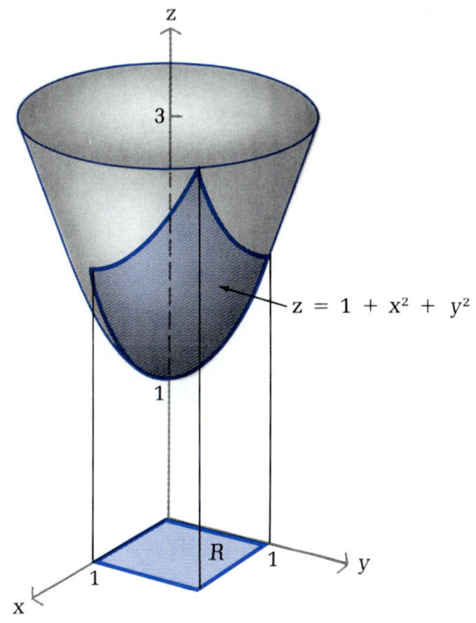

◆

PROBLEM 27 Find the volume of the solid under the graph of $f(x, y) = 1 + x + y$ over the rectangle $R = \{(x, y) | 0 \leq x \leq 1, \quad 0 \leq y \leq 2\}$. ◆

Answers to Matched Problems

21. (A) $2xy^2 + 3x^2y^4 + C(x)$ (B) $2x^2y + 4x^3y^3 + E(y)$
22. (A) $2x + 3x^2$ (B) $18y + 108y^3$
23. (A) 36 (B) 36
24. 12
25. $e - 2e^{1/2} + 1$
26. 2
27. 5 cubic units

EXERCISE 8-6

A *Find each antiderivative. Then use the antiderivative to evaluate the definite integral.*

1. (A) $\int 12x^2y^3\,dy$ (B) $\int_0^1 12x^2y^3\,dy$

2. (A) $\int 12x^2y^3\,dx$ (B) $\int_{-1}^2 12x^2y^3\,dx$

3. (A) $\int (4x + 6y + 5)\,dx$ (B) $\int_{-2}^3 (4x + 6y + 5)\,dx$

4. (A) $\int (4x + 6y + 5)\,dy$ (B) $\int_1^4 (4x + 6y + 5)\,dy$

5. (A) $\displaystyle\int \frac{x}{\sqrt{y + x^2}}\, dx$ (B) $\displaystyle\int_0^2 \frac{x}{\sqrt{y + x^2}}\, dx$

6. (A) $\displaystyle\int \frac{x}{\sqrt{y + x^2}}\, dy$ (B) $\displaystyle\int_1^5 \frac{x}{\sqrt{y + x^2}}\, dy$

B *Evaluate each iterated integral. (See the indicated problem for the evaluation of the inner integral.)*

7. $\displaystyle\int_{-1}^{2}\int_0^1 12x^2y^3\, dy\, dx$

(see Problem 1)

8. $\displaystyle\int_0^1\int_{-1}^{2} 12x^2y^3\, dx\, dy$

(see Problem 2)

9. $\displaystyle\int_1^4\int_{-2}^{3} (4x + 6y + 5)\, dx\, dy$

(see Problem 3)

10. $\displaystyle\int_{-2}^{3}\int_1^4 (4x + 6y + 5)\, dy\, dx$

(see Problem 4)

11. $\displaystyle\int_1^5\int_0^2 \frac{x}{\sqrt{y + x^2}}\, dx\, dy$

(see Problem 5)

12. $\displaystyle\int_0^2\int_1^5 \frac{x}{\sqrt{y + x^2}}\, dy\, dx$

(see Problem 6)

Use both orders of iteration to evaluate each double integral.

13. $\displaystyle\iint_R xy\, dA;\quad R = \{(x, y)|0 \leqslant x \leqslant 2,\quad 0 \leqslant y \leqslant 4\}$

14. $\displaystyle\iint_R \sqrt{xy}\, dA;\quad R = \{(x, y)|1 \leqslant x \leqslant 4,\quad 1 \leqslant y \leqslant 9\}$

15. $\displaystyle\iint_R (x + y)^5\, dA;\quad R = \{(x, y)|-1 \leqslant x \leqslant 1,\quad 1 \leqslant y \leqslant 2\}$

16. $\displaystyle\iint_R xe^y\, dA;\quad R = \{(x, y)|-2 \leqslant x \leqslant 3,\quad 0 \leqslant y \leqslant 2\}$

Find the average value of each function over the given rectangle.

17. $f(x, y) = (x + y)^2;\quad R = \{(x, y)|1 \leqslant x \leqslant 5,\quad -1 \leqslant y \leqslant 1\}$
18. $f(x, y) = x^2 + y^2;\quad R = \{(x, y)|-1 \leqslant x \leqslant 2,\quad 1 \leqslant y \leqslant 4\}$
19. $f(x, y) = x/y;\quad R = \{(x, y)|1 \leqslant x \leqslant 4,\quad 2 \leqslant y \leqslant 7\}$
20. $f(x, y) = x^2y^3;\quad R = \{(x, y)|-1 \leqslant x \leqslant 1,\quad 0 \leqslant y \leqslant 2\}$

Find the volume of the solid under the graph of each function over the given rectangle.

21. $f(x, y) = 2 - x^2 - y^2;\quad R = \{(x, y)|0 \leqslant x \leqslant 1,\quad 0 \leqslant y \leqslant 1\}$
22. $f(x, y) = 5 - x;\quad R = \{(x, y)|0 \leqslant x \leqslant 5,\quad 0 \leqslant y \leqslant 5\}$
23. $f(x, y) = 4 - y^2;\quad R = \{(x, y)|0 \leqslant x \leqslant 2,\quad 0 \leqslant y \leqslant 2\}$
24. $f(x, y) = e^{-x-y};\quad R = \{(x, y)|0 \leqslant x \leqslant 1,\quad 0 \leqslant y \leqslant 1\}$

C *Evaluate each double integral. Select the order of integration carefully—each problem is easy to do one way and difficult the other.*

25. $\iint\limits_R xe^{xy}\,dA; \quad R = \{(x, y)|0 \le x \le 1, \quad 1 \le y \le 2\}$

26. $\iint\limits_R xye^{x^2y}\,dA; \quad R = \{(x, y)|0 \le x \le 1, \quad 1 \le y \le 2\}$

27. $\iint\limits_R \frac{2y + 3xy^2}{1 + x^2}\,dA; \quad R = \{(x, y)|0 \le x \le 1, \quad -1 \le y \le 1\}$

28. $\iint\limits_R \frac{2x + 2y}{1 + 4y + y^2}\,dA; \quad R = \{(x, y)|1 \le x \le 3, \quad 0 \le y \le 1\}$

APPLICATIONS

Business & Economics

29. *Multiplier principle.* Suppose Congress enacts a one-time-only 10% tax rebate that is expected to infuse \$y billion, $5 \le y \le 7$, into the economy. If every individual and corporation is expected to spend a proportion x, $0.6 \le x \le 0.8$, of each dollar received, then by the **multiplier principle** in economics, the total amount of spending S (in billions of dollars) generated by this tax rebate is given by

$$S(x, y) = \frac{y}{1 - x}$$

What is the average total amount of spending for the indicated ranges of the values of x and y? Set up a double integral and evaluate.

30. *Multiplier principle.* Repeat Problem 29 if $6 \le y \le 10$ and $0.7 \le x \le 0.9$.

31. *Cobb–Douglas production function.* If an industry invests x thousand labor-hours, $10 \le x \le 20$, and \$y million, $1 \le y \le 2$, in the production of N thousand units of a certain item, then N is given by

$$N(x, y) = x^{0.75}y^{0.25}$$

What is the average number of units produced for the indicated ranges of x and y? Set up a double integral and evaluate.

32. *Cobb–Douglas production function.* Repeat Problem 31 for

$$N(x, y) = x^{0.5}y^{0.5}$$

where $10 \le x \le 30$ and $1 \le y \le 3$.

Life Sciences

33. *Population distribution.* In order to study the population distribution of a certain species of insects, a biologist has constructed an artificial habitat in the shape of a rectangle 16 feet long and 12 feet wide. The only food available to the insects in this habitat is located at its center. The biologist has determined that the concentration C of insects per square foot at a point d units from the food supply (see the figure) is given approximately by

$$C = 10 - \tfrac{1}{10}d^2$$

What is the average concentration of insects throughout the habitat? Express C as a function of x and y, set up a double integral, and evaluate.

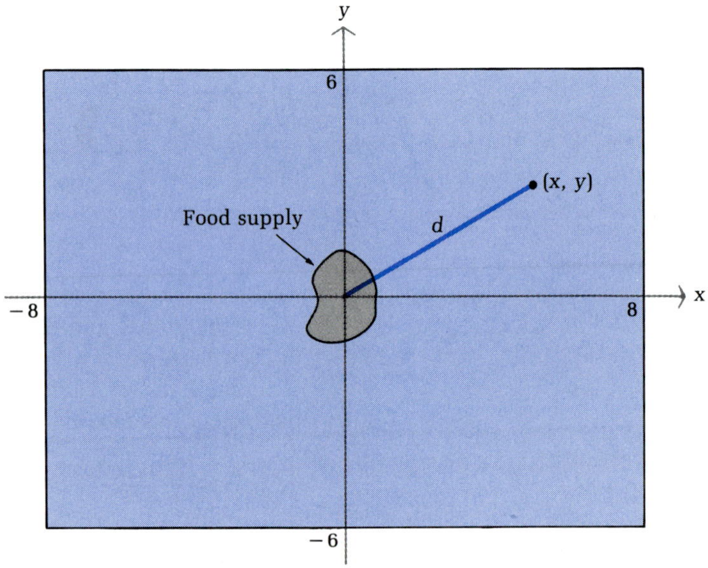

34. *Population distribution.* Repeat Problem 33 for a square habitat that measures 12 feet on each side, where the insect concentration is given by

$$C = 8 - \tfrac{1}{10}d^2$$

35. *Pollution.* A heavy industrial plant located in the center of a small town emits particulate matter into the atmosphere. Suppose the concentration of particulate matter (in parts per million) at a point d miles from the plant (see the figure at the top of the next page) is given by

$$C = 100 - 15d^2$$

If the boundaries of the town form a rectangle 4 miles long and 2 miles wide, what is the average concentration of particulate matter throughout the city? Express C as a function of x and y, set up a double integral, and evaluate.

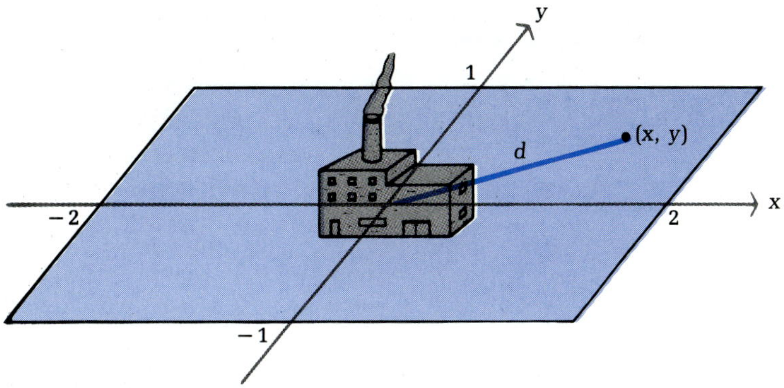

36. *Pollution.* Repeat Problem 35 if the boundaries of the town form a rectangle 8 miles long and 4 miles wide and the concentration of particulate matter is given by

$$C = 100 - 3d^2$$

Social Sciences

37. *Safety research.* Under ideal conditions, if a person driving a car slams on the brakes and skids to a stop, the length of the skid marks (in feet) is given by the formula

$$L = 0.000\ 013\ 3xy^2$$

where x is the weight of the car (in pounds) and y is the speed of the car (in miles per hour). What is the average length of the skid marks for cars weighing between 2,000 and 3,000 pounds and traveling at speeds between 50 and 60 miles per hour? Set up a double integral and evaluate.

38. *Safety research.* Repeat Problem 37 for cars weighing between 2,000 and 2,500 pounds and traveling at speeds between 40 and 50 miles per hour.

39. *Psychology.* The intelligence quotient Q for an individual with mental age x and chronological age y is given by

$$Q(x, y) = 100\frac{x}{y}$$

In a group of sixth graders, the mental age varies between 8 and 16 years and the chronological age varies between 10 and 12 years. What is the average intelligence quotient for this group? Set up a double integral and evaluate.

40. *Psychology.* Repeat Problem 39 for a group with mental ages between 6 and 14 years and chronological ages between 8 and 10 years.

SECTION 8-7 Double Integrals Over More General Regions

- REGULAR REGIONS
- DOUBLE INTEGRALS OVER REGULAR REGIONS
- REVERSING THE ORDER OF INTEGRATION
- VOLUME AND DOUBLE INTEGRALS

In this section we will extend the concept of double integration to nonrectangular regions. We begin with an example and some new terminology.

REGULAR REGIONS

Let R be the region graphed in Figure 15. We can describe R with the following inequalities:

$$R = \{(x, y) | x \leq y \leq 6x - x^2, \quad 0 \leq x \leq 5\}$$

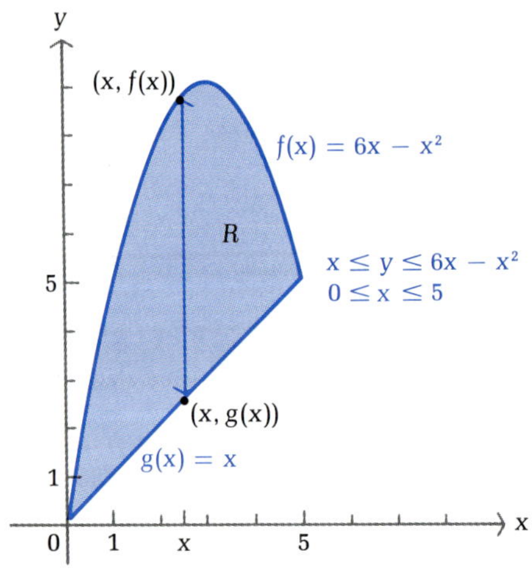

FIGURE 15

The region R can be viewed as a union of vertical line segments. For each x in the interval $[0, 5]$, the line segment from the point $(x, g(x))$ to the point $(x, f(x))$ lies in the region R. Any region that can be covered by vertical line segments in this manner is called a *regular x region*.

Now consider the region S in Figure 16. This is *not* a regular x region, but it can be described with the following inequalities:

$$S = \{(x, y) | y^2 \leq x \leq y + 2, \quad -1 \leq y \leq 2\}$$

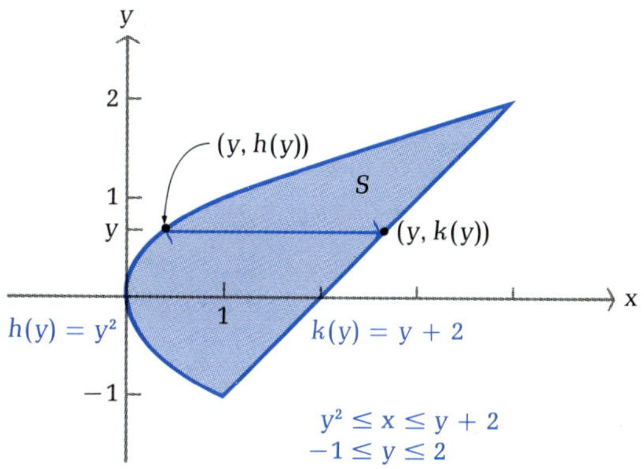

FIGURE 16

The region S can be viewed as a union of horizontal line segments going from the graph of $h(y) = y^2$ to the graph of $k(y) = y + 2$ on the interval $[-1, 2]$. Regions that can be described in this manner are called *regular y regions.*

In general, *regular regions* are defined as follows:

Regular Regions

A region R in the xy plane is a **regular x region** if there exist functions $f(x)$ and $g(x)$ and numbers a and b so that

$$R = \{(x, y) | g(x) \leq y \leq f(x), \quad a \leq x \leq b\}$$

A region R is a **regular *y* region** if there exist functions $h(y)$ and $k(y)$ and numbers c and d so that

$$R = \{(x, y) | h(y) \leq x \leq k(y), \quad c \leq y \leq d\}$$

See Figure 17 for a geometric interpretation.

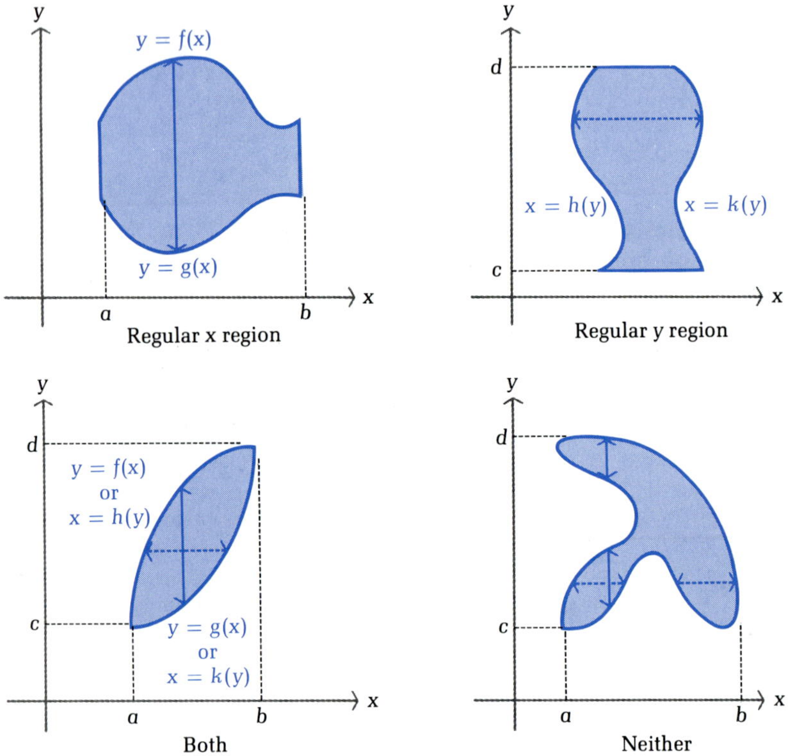

FIGURE 17

◆ EXAMPLE 28 The region R is bounded by the graphs of $y = 4 - x^2$ and $y = x - 2$, $x \geqslant 0$, and the y axis. Graph R and describe R as a regular x region, a regular y region, both, or neither. If possible, represent R in terms of set notation and double inequalities.

Solution

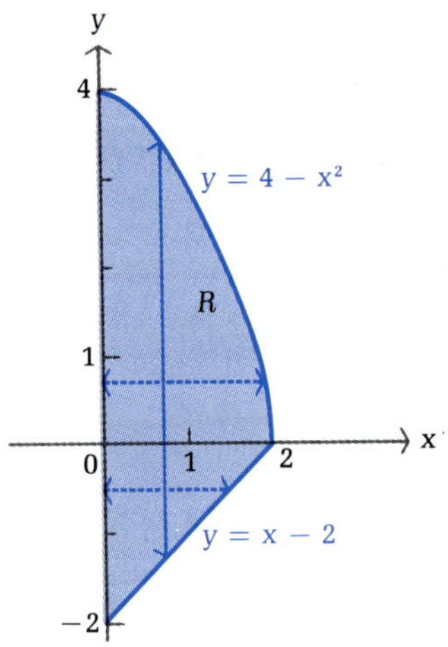

As the solid line in the figure indicates, R can be covered by vertical line segments that go from the graph of $y = x - 2$ to the graph of $y = 4 - x^2$. Thus, R is a regular x region. In terms of set notation and double inequalities, we can write

$$R = \{(x, y) | x - 2 \leq y \leq 4 - x^2, \quad 0 \leq x \leq 2\}$$

On the other hand, a horizontal line passing through a point in the interval $[-2, 0]$ on the y axis will intersect R in a line segment that goes from the y axis to the graph of $y = x - 2$, while one that passes through a point in the interval $[0, 4]$ on the y axis goes from the y axis to the graph of $y = 4 - x^2$. Two such segments are shown as dashed lines in the figure. Thus, the region is not a regular y region. ◆

PROBLEM 28 Repeat Example 28 for the region R bounded by the graphs of $x = 6 - y$, $x = y^2$, $y \geq 0$, and the x axis, as shown in the figure below:

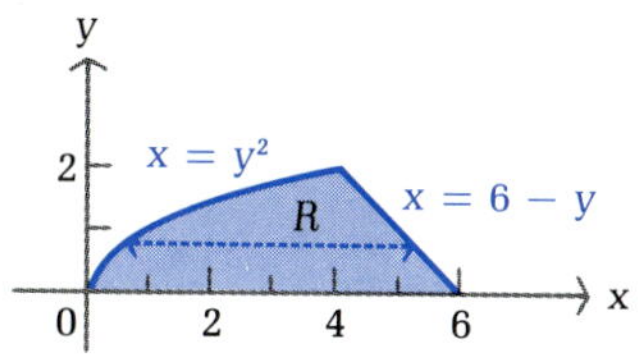

◆

◆ EXAMPLE 29 The region R is bounded by the graphs of $x + y^2 = 9$ and $x + 3y = 9$. Graph R and describe R as a regular x region, a regular y region, both, or neither. If possible, represent R using set notation and double inequalities.

Solution

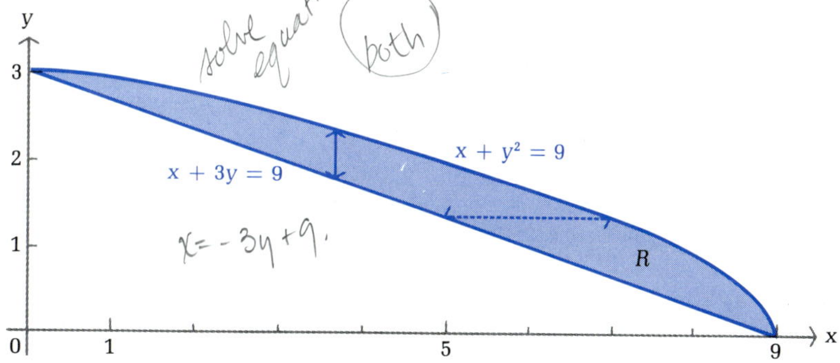

Test for x Region Region R can be covered by vertical line segments that go from the graph of $x + 3y = 9$ to the graph of $x + y^2 = 9$. Thus, R is a regular x region. In order to describe R with inequalities, we must solve each equation for y in terms of x:

$$x + 3y = 9$$
$$3y = 9 - x$$
$$y = 3 - \tfrac{1}{3}x$$

$$x + y^2 = 9$$
$$y^2 = 9 - x$$
$$y = \sqrt{9 - x}$$

We use the positive square root since the graph is in the first quadrant.

Thus,

$$R = \{(x, y) | 3 - \tfrac{1}{3}x \leqslant y \leqslant \sqrt{9 - x}, \quad 0 \leqslant x \leqslant 9\}$$

Test for y Region

Since region R also can be covered by horizontal line segments (see the dashed line in the figure) that go from the graph of $x + 3y = 9$ to the graph of $x + y^2 = 9$, it is a regular y region. Now we must solve each equation for x in terms of y:

$$x + 3y = 9$$
$$x = 9 - 3y$$

$$x + y^2 = 9$$
$$x = 9 - y^2$$

Thus,

$$R = \{(x, y) | 9 - 3y \leqslant x \leqslant 9 - y^2, \quad 0 \leqslant y \leqslant 3\}$$ ◆

PROBLEM 29 Repeat Example 29 for the region bounded by the graphs of $2y - x = 4$ and $y^2 - x = 4$, as shown in the following figure:

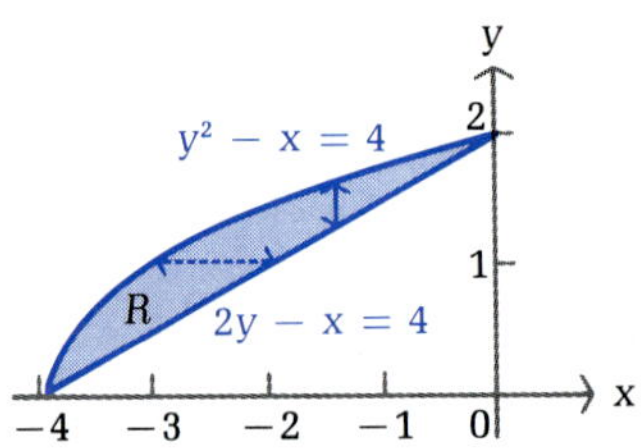

◆ DOUBLE INTEGRALS OVER REGULAR REGIONS

Now we want to extend the definition of double integration to include regular x regions and regular y regions.

Double Integration Over Regular Regions

REGULAR X REGION

If $R = \{(x, y) | g(x) \leq y \leq f(x), \quad a \leq x \leq b\}$, then

$$\iint_R F(x, y)\, dA = \int_a^b \left[\int_{g(x)}^{f(x)} F(x, y)\, dy \right] dx$$

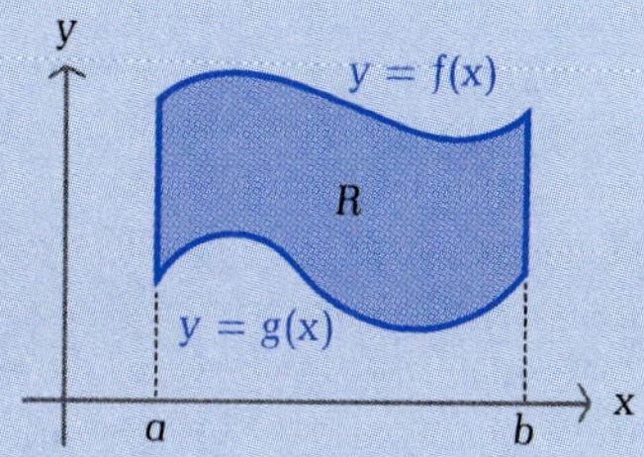

REGULAR y REGION

If $R = \{(x, y) | h(y) \leq x \leq k(y), \quad c \leq y \leq d\}$, then

$$\iint_R F(x, y)\, dA = \int_c^d \left[\int_{h(y)}^{k(y)} F(x, y)\, dx \right] dy$$

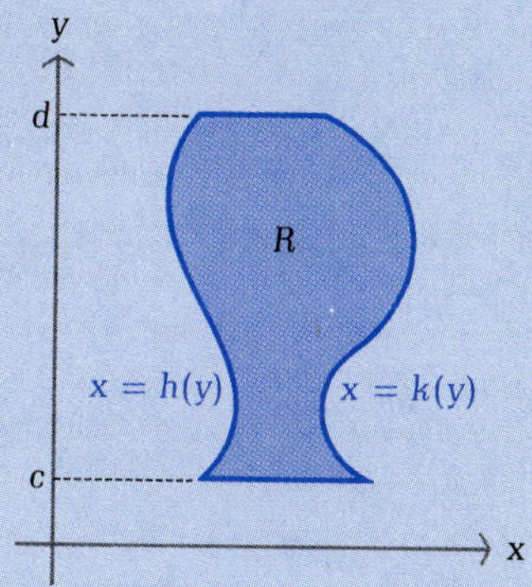

Notice that the order of integration now depends on the nature of the region R. If R is a regular x region, we integrate with respect to y first, while if R is a regular y region, we integrate with respect to x first.

It is also important to note that the variable limits of integration (when present) are always on the inner integral, and the constant limits of integration are always on the outer integral.

◆ EXAMPLE 30 Evaluate $\iint_R 2xy\, dA$, where R is the region bounded by the graphs of $y = -x$ and $y = x^2$, $x \geq 0$, and the graph of $x = 1$.

Solution From the graph we can see that R is a regular x region described by

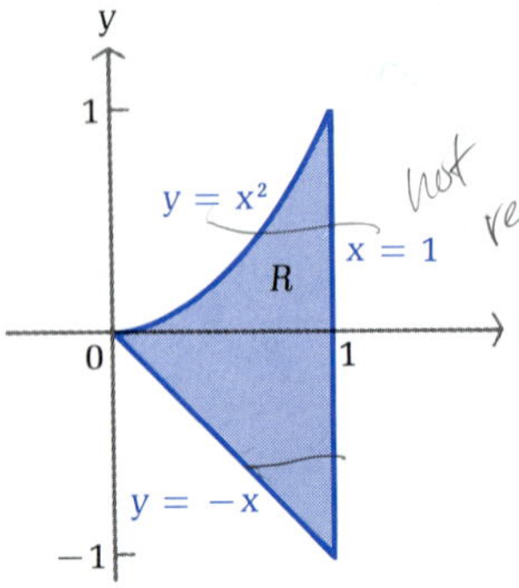

$$R = \{(x, y) | -x \leq y \leq x^2, \quad 0 \leq x \leq 1\}$$

Thus,

$$\iint_R 2xy\, dA = \int_0^1 \left[\int_{-x}^{x^2} 2xy\, dy \right] dx$$

$$= \int_0^1 \left[xy^2 \Big|_{y=-x}^{y=x^2} \right] dx$$

$$= \int_0^1 [x(x^2)^2 - x(-x)^2]\, dx$$

$$= \int_0^1 (x^5 - x^3)\, dx$$

$$= \left(\frac{x^6}{6} - \frac{x^4}{4} \right) \Big|_{x=0}^{x=1}$$

$$= (\tfrac{1}{6} - \tfrac{1}{4}) - (0 - 0) = -\tfrac{1}{12}$$ ◆

PROBLEM 30 Evaluate $\iint_R 3xy^2\, dA$, where R is the region in Example 30. ◆

◆ EXAMPLE 31 Evaluate $\iint_R (2x + y)\, dA$, where R is the region bounded by the graphs of $y = \sqrt{x}$, $x + y = 2$, and $y = 0$.

Solution From the graph we can see that R is a regular y region. After solving each equation for x, we can write

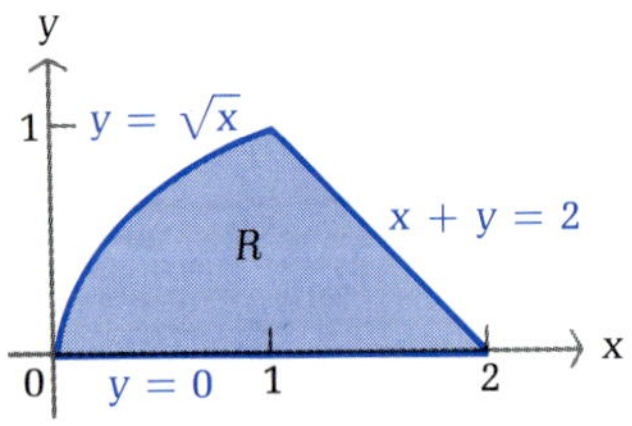

$$R = \{(x, y) | y^2 \leq x \leq 2 - y, \quad 0 \leq y \leq 1\}$$

Thus,

$$\iint_R (2x + y)\, dA = \int_0^1 \left[\int_{y^2}^{2-y} (2x + y)\, dx \right] dy$$

$$= \int_0^1 \left[(x^2 + yx) \Big|_{x=y^2}^{x=2-y} \right] dy$$

$$= \int_0^1 \{[(2-y)^2 + y(2-y)] - [(y^2)^2 + y(y^2)]\}\, dy$$

$$= \int_0^1 (4 - 2y - y^3 - y^4)\, dy$$

$$= (4y - y^2 - \tfrac{1}{4}y^4 - \tfrac{1}{5}y^5) \Big|_{y=0}^{y=1}$$

$$= (4 - 1 - \tfrac{1}{4} - \tfrac{1}{5}) - 0 = \tfrac{51}{20}$$ ◆

PROBLEM 31 Evaluate $\iint_R (y - 4x)\, dA$, where R is the region in Example 31. ◆

◆ EXAMPLE 32 The region R is bounded by the graphs of $y = \sqrt{x}$ and $y = \frac{1}{2}x$. Evaluate $\iint_R 4xy^3\, dA$ two different ways.

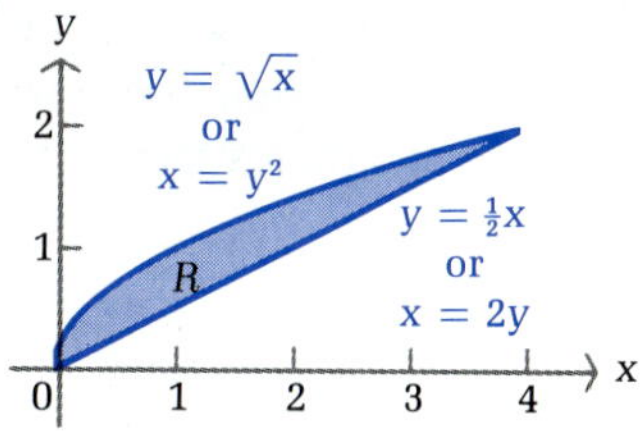

Solution Region R is both a regular x region and a regular y region:

$$R = \{(x, y) \mid \tfrac{1}{2}x \leq y \leq \sqrt{x}, \quad 0 \leq x \leq 4\} \qquad \text{Regular x region}$$

$$R = \{(x, y) \mid y^2 \leq x \leq 2y, \quad 0 \leq y \leq 2\} \qquad \text{Regular y region}$$

Using the first representation (a regular x region), we obtain

$$\begin{aligned}
\iint_R 4xy^3\,dA &= \int_0^4 \left[\int_{(1/2)x}^{\sqrt{x}} 4xy^3\,dy\right] dx \\
&= \int_0^4 \left[xy^4\Big|_{y=(1/2)x}^{y=\sqrt{x}}\right] dx \\
&= \int_0^4 [x(\sqrt{x})^4 - x(\tfrac{1}{2}x)^4]\,dx \\
&= \int_0^4 (x^3 - \tfrac{1}{16}x^5)\,dx \\
&= (\tfrac{1}{4}x^4 - \tfrac{1}{96}x^6)\Big|_{x=0}^{x=4} \\
&= (64 - \tfrac{128}{3}) - 0 = \tfrac{64}{3}
\end{aligned}$$

Using the second representation (a regular y region), we obtain

$$\begin{aligned}
\iint_R 4xy^3\,dA &= \int_0^2 \left[\int_{y^2}^{2y} 4xy^3\,dx\right] dy \\
&= \int_0^2 \left[2x^2y^3\Big|_{x=y^2}^{x=2y}\right] dy \\
&= \int_0^2 [2(2y)^2y^3 - 2(y^2)^2y^3]\,dy \\
&= \int_0^2 (8y^5 - 2y^7)\,dy \\
&= (\tfrac{4}{3}y^6 - \tfrac{1}{4}y^8)\Big|_{y=0}^{y=2} \\
&= (\tfrac{256}{3} - 64) - 0 = \tfrac{64}{3}
\end{aligned}$$

◆

PROBLEM 32 The region R is bounded by the graphs of $y = x$ and $y = \frac{1}{2}x^2$. Evaluate $\iint_R 4x^3y\,dA$ two different ways. ◆

◆ REVERSING THE ORDER OF INTEGRATION

Example 32 shows that

$$\iint_R 4xy^3\,dA = \int_0^4 \left[\int_{(1/2)x}^{\sqrt{x}} 4xy^3\,dy\right] dx = \int_0^2 \left[\int_{y^2}^{2y} 4xy^3\,dx\right] dy$$

In general, if R is both a regular x region and a regular y region, the two iterated integrals are equal. In rectangular regions, reversing the order of integration in an iterated integral was a simple matter. As Example 32 illustrates, the process is more complicated in nonrectangular regions. The next example illustrates how to start with an iterated integral and reverse the order of integration. Since we are interested in the reversal process and not in the value of either integral, the integrand will not be specified.

◆ EXAMPLE 33 Reverse the order of integration in $\int_1^3 \left[\int_0^{x-1} f(x, y)\, dy \right] dx$.

Solution The order of integration indicates that the region of integration is a regular x region:

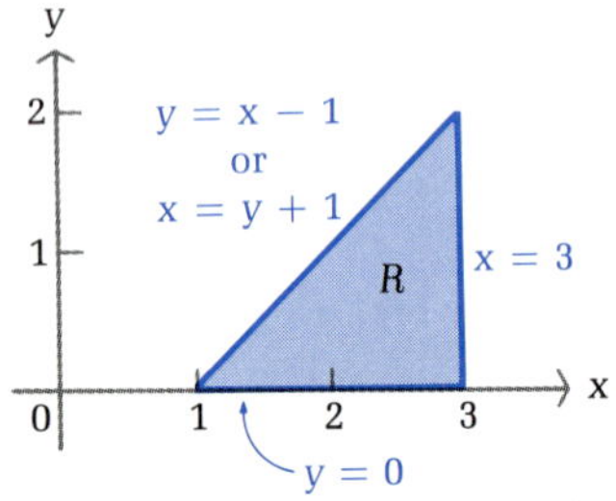

$$R = \{(x, y) | 0 \leq y \leq x - 1, \quad 1 \leq x \leq 3\}$$

Graph region R to determine whether it is also a regular y region. The graph shows that R is also a regular y region, and we can write

$$R = \{(x, y) | y + 1 \leq x \leq 3, \quad 0 \leq y \leq 2\}$$

Thus,

$$\int_1^3 \left[\int_0^{x-1} f(x, y)\, dy \right] dx = \int_0^2 \left[\int_{y+1}^3 f(x, y)\, dx \right] dy$$ ◆

PROBLEM 33 Reverse the order of integration in $\int_2^4 \left[\int_0^{4-x} f(x, y)\, dy \right] dx$. ◆

◆ VOLUME AND DOUBLE INTEGRALS

In Section 8-6 we used the double integral to calculate the volume of a solid with a rectangular base. In general, if a solid can be described by the graph of a positive function $f(x, y)$ over a regular region R (not necessarily a rectangle), then the double integral of the function f over the region R still represents the volume of the corresponding solid.

◆ EXAMPLE 34 The region R is bounded by the graphs of $x + y = 1$, $y = 0$, and $x = 0$. Find the volume of the solid under the graph of $z = 1 - x - y$ over the region R.

Solution

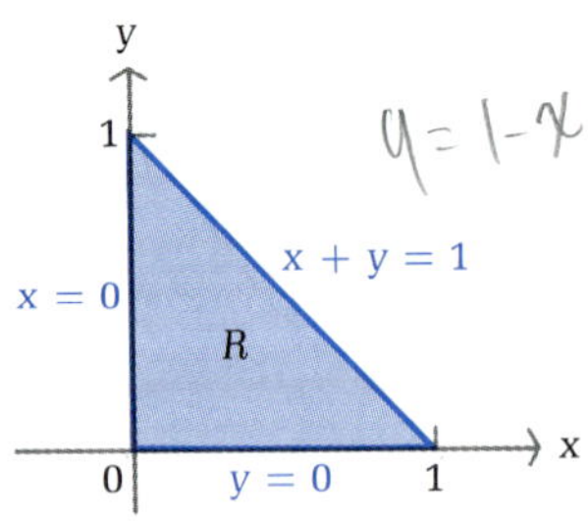

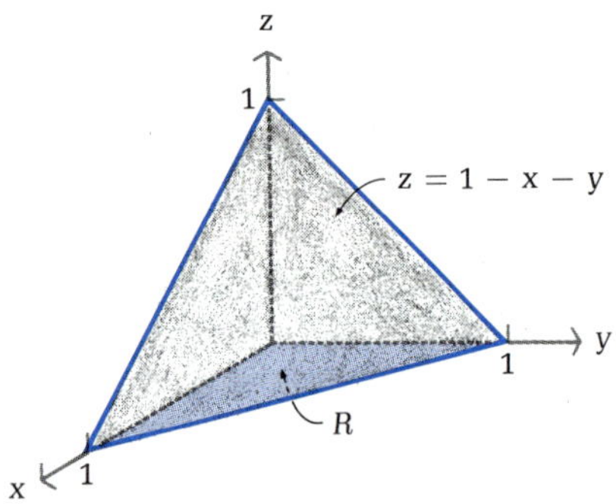

The graph of R indicates that R is both a regular x region and a regular y region.

We choose to use the regular x region description.

$$R = \{(x, y) | 0 \leqslant y \leqslant 1 - x, \quad 0 \leqslant x \leqslant 1\}$$

Thus, the volume of the solid is

$$V = \iint_R (1 - x - y)\, dA = \int_0^1 \left[\int_0^{1-x} (1 - x - y)\, dy \right] dx$$

$$= \int_0^1 \left[(y - xy - \tfrac{1}{2}y^2) \Big|_{y=0}^{y=1-x} \right] dx$$

$$= \int_0^1 [(1 - x) - x(1 - x) - \tfrac{1}{2}(1 - x)^2]\, dx$$

$$= \int_0^1 (\tfrac{1}{2} - x + \tfrac{1}{2}x^2)\, dx$$

$$= (\tfrac{1}{2}x - \tfrac{1}{2}x^2 + \tfrac{1}{6}x^3) \Big|_{x=0}^{x=1}$$

$$= (\tfrac{1}{2} - \tfrac{1}{2} + \tfrac{1}{6}) - 0 = \tfrac{1}{6}$$ ◆

PROBLEM 34 The region R is bounded by the graphs of $y + 2x = 2$, $y = 0$, and $x = 0$. Find the volume of the solid under the graph of $z = 2 - 2x - y$ over the region R. [*Hint:* Sketch the region first—the solid does not have to be sketched.] ◆

Answers to Matched Problems

28. $R = \{(x, y) | y^2 \leqslant x \leqslant 6 - y, \quad 0 \leqslant y \leqslant 2\}$ is a regular y region; R is not a regular x region
29. R is both a regular x region and a regular y region;

$$R = \{(x, y) | \tfrac{1}{2}x + 2 \leqslant y \leqslant \sqrt{x + 4}, \quad -4 \leqslant x \leqslant 0\}$$
$$= \{(x, y) | y^2 - 4 \leqslant x \leqslant 2y - 4, \quad 0 \leqslant y \leqslant 2\}$$

30. $\frac{13}{40}$ 31. $-\frac{77}{20}$ 32. $\frac{16}{3}$ 33. $\int_0^2 \int_2^{4-y} f(x, y)\, dx\, dy$ 34. $\frac{2}{3}$

EXERCISE 8-7

A

Graph the region R bounded by the graphs of the equations. Express R in terms of set notation and double inequalities that describe R as a regular x region, a regular y region, or both.

1. $y = 4 - x^2, \quad y = 0, \quad 0 \leqslant x \leqslant 2$
2. $y = x^2, \quad y = 9, \quad 0 \leqslant x \leqslant 3$
3. $y = x^3, \quad y = 12 - 2x, \quad x = 0$
4. $y = 5 - x, \quad y = 1 + x, \quad y = 0$
5. $y^2 = 2x, \quad y = x - 4$
6. $y = 4 + 3x - x^2, \quad x + y = 4$

Evaluate each integral.

7. $\int_0^1 \int_0^x (x + y)\, dy\, dx$
8. $\int_0^2 \int_0^y xy\, dx\, dy$
9. $\int_0^1 \int_{y^3}^{\sqrt{y}} (2x + y)\, dx\, dy$
10. $\int_1^4 \int_x^{x^2} (x^2 + 2y)\, dy\, dx$

B *Use the description of the region R to evaluate the indicated integral.*

11. $\iint_R (x^2 + y^2)\, dA; \quad R = \{(x, y) | 0 \leq y \leq 2x, \quad 0 \leq x \leq 2\}$

12. $\iint_R 2x^2y\, dA; \quad R = \{(x, y) | 0 \leq y \leq 9 - x^2, \quad -3 \leq x \leq 3\}$

13. $\iint_R (x + y - 2)^3\, dA; \quad R = \{(x, y) | 0 \leq x \leq y + 2, \quad 0 \leq y \leq 1\}$

14. $\iint_R (2x + 3y)\, dA; \quad R = \{(x, y) | y^2 - 4 \leq x \leq 4 - 2y, \quad 0 \leq y \leq 2\}$

15. $\iint_R e^{x+y}\, dA; \quad R = \{(x, y) | -x \leq y \leq x, \quad 0 \leq x \leq 2\}$

16. $\iint_R \frac{x}{\sqrt{x^2 + y^2}}\, dA; \quad R = \{(x, y) | 0 \leq x \leq \sqrt{4y - y^2}, \quad 0 \leq y \leq 2\}$

Graph the region R bounded by the graphs of the indicated equations. Describe R in set notation with double inequalities and evaluate the indicated integral.

17. $y = x + 1, \quad y = 0, \quad x = 0, \quad x = 1; \quad \iint_R \sqrt{1 + x + y}\, dA$

18. $y = x^2, \quad y = \sqrt{x}; \quad \iint_R 12xy\, dA$

19. $y = 4x - x^2, \quad y = 0; \quad \iint_R \sqrt{y + x^2}\, dA$

20. $x = 1 + 3y, \quad x = 1 - y, \quad y = 1; \quad \iint_R (x + y + 1)^3\, dA$

21. $y = 1 - \sqrt{x}, \quad y = 1 + \sqrt{x}, \quad x = 4; \quad \iint_R x(y - 1)^2\, dA$

22. $y = \frac{1}{2}x, \quad y = 6 - x, \quad y = 1; \quad \iint_R \frac{1}{x + y}\, dA$

Evaluate each integral. Graph the region of integration, reverse the order of integration, and then evaluate the integral with the order reversed.

23. $\int_0^3 \int_0^{3-x} (x + 2y)\, dy\, dx$

24. $\int_0^2 \int_0^y (y - x)^4\, dx\, dy$

25. $\int_0^1 \int_0^{1-x^2} x\sqrt{y}\, dy\, dx$

26. $\int_0^2 \int_{x^3}^{4x} (1 + 2y)\, dy\, dx$

27. $\int_0^4 \int_{x/4}^{\sqrt{x}/2} x\, dy\, dx$

28. $\int_0^4 \int_{y^2/4}^{2\sqrt{y}} (1 + 2xy)\, dx\, dy$

Find the volume of the solid under the graph of f(x, y) over the region R bounded by the graphs of the indicated equations. Sketch the region R—the solid does not have to be sketched.

29. $f(x, y) = 4 - x - y$; R is bounded by the graphs of $x + y = 4$, $y = 0$, $x = 0$

30. $f(x, y) = (x - y)^2$; R is the region bounded by the graphs of $y = x$, $y = 2$, $x = 0$

31. $f(x, y) = 4$; R is the region bounded by the graphs of $y = 1 - x^2$ and $y = 0$ for $0 \leq x \leq 1$

32. $f(x, y) = 4xy$; R is the region bounded by the graphs of $y = \sqrt{1 - x^2}$ and $y = 0$ for $0 \leq x \leq 1$

C

Reverse the order of integration for each integral. Evaluate the integral with the order reversed. Do not attempt to evaluate the integral in the original form.

33. $\int_0^2 \int_{x^2}^4 \frac{4x}{1 + y^2}\, dy\, dx$

34. $\int_0^1 \int_y^1 \sqrt{1 - x^2}\, dx\, dy$

35. $\int_0^1 \int_{y^2}^1 4ye^{x^2}\, dx\, dy$

36. $\int_0^4 \int_{\sqrt{x}}^2 \sqrt{3x + y^2}\, dy\, dx$

Problems 37–42 require the use of a graphic calculator or a computer. Graph the region R bounded by the graphs of the indicated equations. Use approximation techniques to find intersection points correct to two decimal places. Describe R in set notation with double inequalities, and evaluate the indicated integral correct to two decimal places.

37. $y = 1 + \sqrt{x}$, $y = x^2$, $x = 0$; $\iint_R x\, dA$

38. $y = 1 + \sqrt[3]{x}$, $y = x$, $x = 0$; $\iint_R x\, dA$

39. $y = \sqrt[3]{x}$, $y = 1 - x$, $y = 0$; $\iint_R 24xy\, dA$

40. $y = x^3$, $y = 1 - x$, $y = 0$; $\iint_R 48xy\, dA$

41. $y = e^{-x}$, $y = 3 - x$; $\iint_R 4y\, dA$

42. $y = e^x$, $y = 2 + x$; $\iint_R 8y\, dA$

SECTION 8-8 Chapter Review

Important Terms and Symbols

8-1 *Functions of Several Variables.* Functions of two independent variables; functions of several independent variables; Cobb-Douglas production function; three-dimensional coordinate system; triplet of numbers (x, y, z); surface; paraboloid; saddle point

$$z = f(x, y); \quad w = f(x, y, z)$$

8-2 *Partial Derivatives.* Partial derivative of f with respect to x; partial derivative of f with respect to y; marginal productivity of labor; marginal productivity of capital; second-order partial derivatives

$$\frac{\partial z}{\partial x}; \quad \frac{\partial z}{\partial y}; \quad f_x(x, y); \quad f_y(x, y); \quad \frac{\partial^2 z}{\partial x^2} = f_{xx}(x, y); \quad \frac{\partial^2 z}{\partial x\, \partial y} = f_{yx}(x, y);$$

$$\frac{\partial^2 z}{\partial y\, \partial x} = f_{xy}(x, y); \quad \frac{\partial^2 z}{\partial y^2} = f_{yy}(x, y)$$

8-3 *Maxima and Minima.* Local maximum; local minimum; saddle point; critical point; second-derivative test

8-4 *Maxima and Minima Using Lagrange Multipliers.* Constraint; Lagrange multiplier; critical point; method of Lagrange multipliers for functions of two variables; marginal productivity of money; method of Lagrange multipliers for functions of three variables

8-5 *Method of Least Squares.* Least squares approximation; linear regression; residual; least squares line; regression line; normal equations

8-6 *Double Integrals Over Rectangular Regions.* Double integral; integrand; region of integration; iterated integral; average value over rectangular region; volume under a surface

$$\iint_R f(x, y)\, dA = \int_a^b \left[\int_c^d f(x, y)\, dy\right] dx = \int_c^d \left[\int_a^b f(x, y)\, dx\right] dy$$

$$\frac{1}{(b - a)(d - c)} \iint_R f(x, y)\, dA$$

8-7 *Double Integrals Over More General Regions.* Regular x region; regular y region; reversing the order of integration; volume under a surface

$$\int_a^b \left[\int_{g(x)}^{f(x)} F(x, y)\, dy\right] dx; \quad \int_c^d \left[\int_{h(y)}^{k(y)} F(x, y)\, dx\right] dy$$

EXERCISE 8-8 Chapter Review

Work through all the problems in this chapter review and check your answers in the back of the book. (Answers to all review problems are there.) Where weaknesses show up, review appropriate sections in the text.

A

1. For $f(x, y) = 2{,}000 + 40x + 70y$, find $f(5, 10)$, $f_x(x, y)$, and $f_y(x, y)$.
2. For $z = x^3y^2$, find $\partial^2 z/\partial x^2$ and $\partial^2 z/\partial x\, \partial y$.
3. Evaluate: $\int (6xy^2 + 4y)\, dy$
4. Evaluate: $\int (6xy^2 + 4y)\, dx$
5. Evaluate: $\int_0^1 \int_0^1 4xy\, dy\, dx$
6. Evaluate: $\int_0^1 \int_0^x 4xy\, dy\, dx$

B

7. For $f(x, y) = 3x^2 - 2xy + y^2 - 2x + 3y - 7$, find $f(2, 3)$, $f_y(x, y)$, and $f_y(2, 3)$.
8. For $f(x, y) = -4x^2 + 4xy - 3y^2 + 4x + 10y + 81$, find

$$[f_{xx}(2, 3)][f_{yy}(2, 3)] - [f_{xy}(2, 3)]^2$$

9. If $f(x, y) = x + 3y$ and $g(x, y) = x^2 + y^2 - 10$, find the critical points of $F(x, y, \lambda) = f(x, y) + \lambda g(x, y)$.
10. Use the least squares line for the data in the table in the margin to estimate y when $x = 10$.

x	y
2	12
4	10
6	7
8	3

11. For $R = \{(x, y) | -1 \leqslant x \leqslant 1,\quad 1 \leqslant y \leqslant 2\}$, evaluate the following in two ways:

$$\iint_R (4x + 6y)\, dA$$

12. For R the region bounded by the graphs of $y = x^2$, $y = -x$, and $x = 2$, graph R, describe R in set notation with double inequalities, and evaluate the following:

$$\iint_R (4x + 5y)\, dA$$

13. For $R = \{(x, y) | 1 \leqslant y \leqslant x,\quad 1 \leqslant x \leqslant 3\}$, graph R and evaluate the following integral in two ways:

$$\iint_R 30x^2y\, dA$$

C

14. For $f(x, y) = e^{x^2+2y}$, find f_x, f_y, and f_{xy}.
15. For $f(x, y) = (x^2 + y^2)^5$, find f_x and f_{xy}.
16. Find all critical points and test for extrema for

$$f(x, y) = x^3 - 12x + y^2 - 6y$$

17. Use Lagrange multipliers to maximize $f(x, y) = xy$ subject to $2x + 3y = 24$.
18. Use Lagrange multipliers to minimize $f(x, y, z) = x^2 + y^2 + z^2$ subject to $2x + y + 2z = 9$.
19. Find the least squares line for the data in the table in the margin.

x	y	x	y
10	50	60	80
20	45	70	85
30	50	80	90
40	55	90	90
50	65	100	110

20. Find the average value of $f(x, y) = x^{2/3}y^{1/3}$ over the rectangle

$$R = \{(x, y) | -8 \leqslant x \leqslant 8,\quad 0 \leqslant y \leqslant 27\}$$

21. Find the volume of the solid under the graph of $z = 3x^2 + 3y^2$ over the rectangle

$$R = \{(x, y) | 0 \leqslant x \leqslant 1,\quad -1 \leqslant y \leqslant 1\}$$

22. Find the volume of the solid under the graph of $f(x, y) = 6 - x - y$ over the region R, where R is bounded by the graphs of $x + y = 6$, $x = 0$, and $y = 0$. Sketch the region R—the solid does not have to be sketched.

APPLICATIONS

Business & Economics

23. *Maximizing profit.* A company produces x units of product A and y units of product B (both in hundreds per month). The monthly profit equation (in thousand of dollars) is found to be

$$P(x, y) = -4x^2 + 4xy - 3y^2 + 4x + 10y + 81$$

(A) Find $P_x(1, 3)$ and interpret the results.
(B) How many of each product should be produced each month to maximize profit? What is the maximum profit?

24. *Minimizing material.* A rectangular box with no top and six compartments (see the figure) is to have a volume of 96 cubic inches. Find the dimensions that will require the least amount of material.

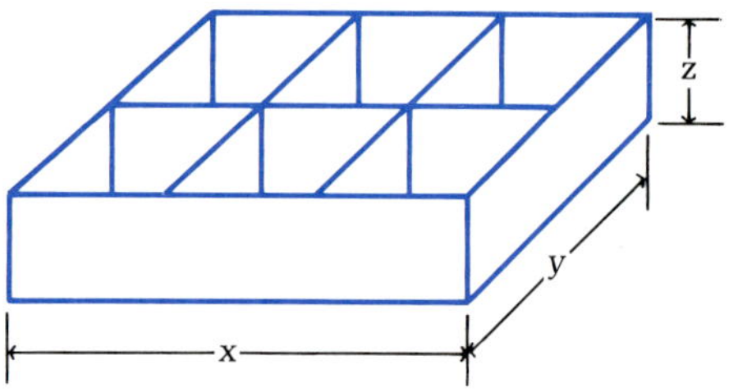

YEAR	PROFIT
1	2
2	2.5
3	3.1
4	4.2
5	4.3

25. *Profit.* A company's annual profits (in millions of dollars) over a 5 year period are given in the table. Use the least squares line to estimate the profit for the sixth year.

26. *Productivity.* The Cobb–Douglas production function for a product is

$$N(x, y) = 10x^{0.8}y^{0.2}$$

where x is the number of units of labor and y is the number of units of capital required to produce N units of the product.

(A) Find the marginal productivity of labor and the marginal productivity of capital at $x = 40$ and $y = 50$. For the greatest increase in productivity, should management encourage increased use of labor or increased use of capital?
(B) If each unit of labor costs \$100, each unit of capital costs \$50, and \$10,000 is budgeted for production of this product, use the method of Lagrange multipliers to determine the allocations of labor and capital that will maximize the number of units produced and find the maxi-

mum production. Find the marginal productivity of money and approximate the increase in production that would result from an increase of $2,000 in the amount budgeted for production.

(C) If $50 \leq x \leq 100$ and $20 \leq y \leq 40$, find the average number of units produced. Set up a definite integral and evaluate.

Life Sciences

27. *Marine biology.* The function used for timing dives with scuba gear is

$$T(V, x) = \frac{33V}{x + 33}$$

where T is the time of the dive in minutes, V is the volume of air (at sea level pressure) compressed into tanks, and x is the depth of the dive in feet. Find $T_x(70, 17)$ and interpret the results.

28. *Pollution.* A heavy industrial plant located in the center of a small town emits particulate matter into the atmosphere. Suppose the concentration of particular matter (in parts per million) at a point d miles from the plant is given by

$$C = 100 - 24d^2$$

If the boundaries of the town form a square 4 miles long and 4 miles wide, what is the average concentration of particulate matter throughout the town? Express C as a function of x and y, set up a double integral, and evaluate.

Social Sciences

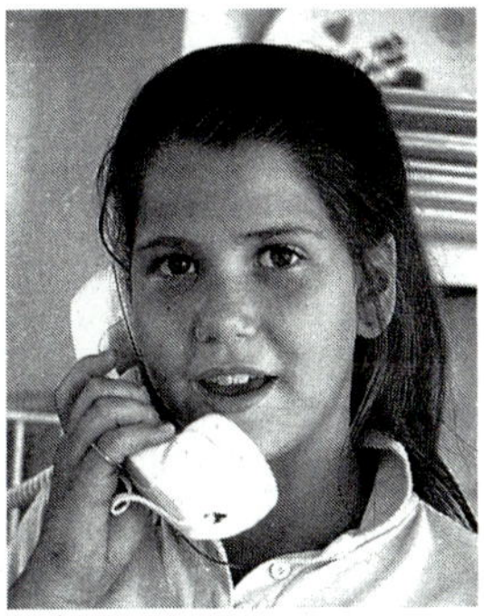

29. *Sociology.* Joseph Cavanaugh, a sociologist, found that the number of long-distance telephone calls, n, between two cities in a given period of time varied (approximately) jointly as the populations P_1 and P_2 of the two cities, and varied inversely as the distance, d, between the two cities. In terms of an equation for a time period of 1 week,

$$n(P_1, P_2, d) = 0.001\,\frac{P_1P_2}{d}$$

Find $n(100{,}000, 50{,}000, 100)$.

30. *Education.* At the beginning of the semester, students in a foreign language course are given a proficiency exam. The same exam is given at the end of the semester. The results for 5 students are given in the table. Use the least squares line to estimate the score on the second exam for a student who scored 40 on the first exam.

FIRST EXAM	SECOND EXAM
30	60
50	75
60	80
70	85
90	90

CHAPTER 9

Differential Equations

CHAPTER 9

Contents

In Section 6-3 we considered *differential equations* of the form

$$\frac{dy}{dt} = ky, \qquad \frac{dy}{dt} = k(M - y), \qquad \text{and} \qquad \frac{dy}{dt} = ky(M - y)$$

Each of these equations establishes a relationship between a quantity y and its rate of growth dy/dt. Their solutions provide models for several types of exponential growth laws (see Table 1, page 390). Many different relationships in business, economics, and the sciences can be stated in terms of differential equations. Unfortunately, there is no single method that will solve all the differential equations that may be encountered—even in very simple applications. In this chapter, after discussing some basic concepts in the first section, we will consider methods for solving several types of differential equations that have significant applications.

SECTION 9-1 Basic Concepts

- SOLUTIONS OF DIFFERENTIAL EQUATIONS
- IMPLICIT SOLUTIONS
- APPLICATION

In this section we introduce the concept of a *differential equation* and discuss what is meant by a *solution*, including explicit and implicit representations. Actual techniques for finding solutions of differential equations will be discussed in the following sections.

SOLUTIONS OF DIFFERENTIAL EQUATIONS

A **differential equation** is an equation involving an unknown function, usually denoted by y, and one or more of its derivatives. For example,

$$y' = 2xy \tag{1}$$

is a differential equation. Since only the first derivative of the unknown function y appears in this equation, it is called a **first-order** differential equation. In general, the **order** of a differential equation is the highest derivative of the unknown function present in the equation.

Now consider the function

$$y = 4e^{x^2}$$

whose derivative is

$$y' = 4e^{x^2}(2x) = 8xe^{x^2}$$

Substituting for y and y' in equation (1) gives

$$\begin{aligned} y' &= 2xy \\ 8xe^{x^2} &= 2x(4e^{x^2}) \\ 8xe^{x^2} &= 8xe^{x^2} \end{aligned}$$

which is certainly true for all values of x. This shows that the function $y = 4e^{x^2}$ is a **solution** of equation (1). But this function is not the only solution. In fact, if C is any constant, substituting $y = Ce^{x^2}$ and $y' = 2xCe^{x^2}$ in equation (1) yields the identity

$$\begin{aligned} y' &= 2xy \\ 2xCe^{x^2} &= 2xCe^{x^2} \end{aligned}$$

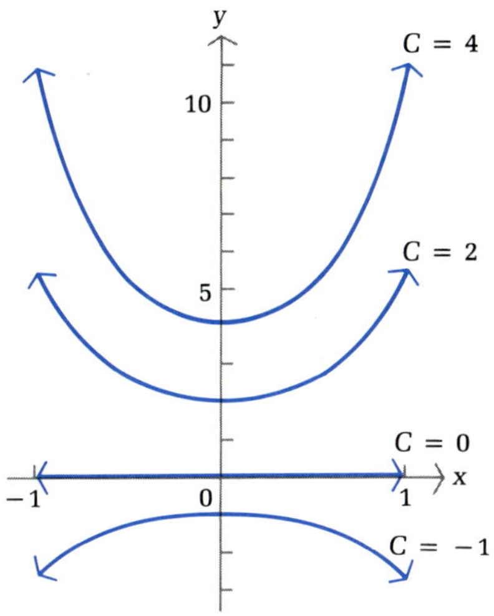

FIGURE 1
Particular solutions of $y' = 2xy$

It turns out that all solutions of $y' = 2xy$ can be obtained from $y = Ce^{x^2}$ by assigning C appropriate values; hence, $y = Ce^{x^2}$ is called the **general solution** of equation (1). The collection of all functions of the form $y = Ce^{x^2}$ is called the **family of solutions** of equation (1). The function $y = 4e^{x^2}$, obtained by letting $C = 4$ in the general solution, is called a **particular solution** of equation (1). Other particular solutions are

$y = -e^{x^2}$	$C = -1$
$y = 0$	$C = 0$
$y = 2e^{x^2}$	$C = 2$

These particular solutions are graphed in Figure 1.

Remark **Recognizing General Solutions**

Most, but not all, first-order differential equations have general solutions that involve one arbitrary constant. Whenever we find a function, such as $y = Ce^{x^2}$, that satisfies a given differential equation for any value of the constant C, we will assume this function is the general solution.

◆ EXAMPLE 1 Show that

$$y = Cx^2 + 1$$

is the general solution of the differential equation

$$xy' = 2y - 2$$

On the same set of axes, graph the particular solutions obtained by letting $C = -2, -1, 0, 1,$ and 2.

Solution Substituting $y = Cx^2 + 1$ and $y' = 2Cx$ in the differential equation, we have

$$\begin{aligned} xy' &= 2y - 2 \\ x(2Cx) &= 2(Cx^2 + 1) - 2 \\ 2Cx^2 &= 2Cx^2 + 2 - 2 \\ 2Cx^2 &= 2Cx^2 \end{aligned}$$

which shows that $y = Cx^2 + 1$ is the general solution. The particular solutions corresponding to $C = -2, -1, 0, 1$, and 2 are graphed in Figure 2.

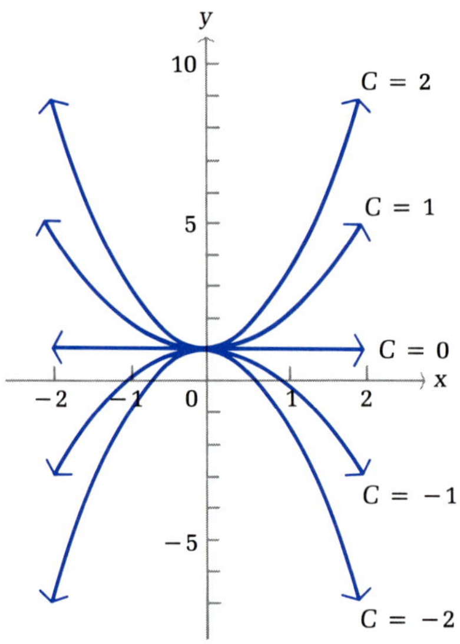

FIGURE 2
Particular solutions of $xy' = 2y - 2$ ◆

PROBLEM 1 Show that

$$y = Cx + 1$$

is the general solution of the differential equation

$$xy' = y - 1$$

On the same set of axes, graph the particular solutions obtained by letting $C = -2, -1, 0, 1$, and 2. ◆

In many applications, we will be interested in finding a particular solution $y(x)$ that satisfies an **initial condition** of the form $y(x_0) = y_0$. The value of the constant C in the general solution must then be selected so that this initial condition is satisfied; that is, so that the solution curve (from the family of solution curves) passes through (x_0, y_0).

◆ EXAMPLE 2 Using the general solution in Example 1, find a particular solution of the differential equation

$$xy' = 2y - 2$$

that satisfies the indicated initial condition, if such a solution exists.

(A) $y(1) = 3$ (B) $y(0) = 3$ (C) $y(0) = 1$

Solutions

(A) *Initial Condition $y(1) = 3$.* From Example 1, the general solution of the differential equation is

$$y = Cx^2 + 1$$

Substituting $x = 1$ and $y = 3$ in this general solution yields

$$\begin{aligned} 3 &= C(1)^2 + 1 \\ &= C + 1 \\ C &= 2 \end{aligned}$$

Thus, the particular solution satisfying the initial condition $y(1) = 3$ is

$$y = 2x^2 + 1$$

See the graph labeled $C = 2$ in Figure 2.

(B) *Initial Condition $y(0) = 3$.* Substituting $x = 0$ and $y = 3$ in the general solutions yields

$$3 = C(0)^2 + 1$$

No matter what value of C we select, $C(0)^2 = 0$ and this equation reduces to

$$3 = 1$$

which is *never* valid. Thus, we must conclude that there is no particular solution of this differential equation that will satisfy the initial condition $y(0) = 3$.

(C) *Initial Condition $y(0) = 1$.* Substituting $x = 0$ and $y = 1$ in the general solution, we have

$$1 = C(0)^2 + 1$$
$$1 = 1$$

This equation is valid for *all* values of C. Thus, all solutions of this differential equation satisfy the initial condition $y(0) = 1$ (see Figure 2). ◆

Remark **Nature of Particular Solutions**

The situation illustrated in Example 2A is the most common. Most first-order differential equations you will encounter will have exactly one particular solution that satisfies a given initial condition. However, we must always be aware of the possibility that a differential equation may not have any particular solutions that satisfy a given initial condition, or the possibility that it may have many particular solutions satisfying a given initial condition.

PROBLEM 2 Using the general solution in Problem 1, find a particular solution of the differential equation

$$xy' = y - 1$$

that satisfies the indicated initial condition, if such a solution exists.

(A) $y(1) = 2$ (B) $y(0) = 2$ (C) $y(0) = 1$ ◆

◆ IMPLICIT SOLUTIONS

In many cases, the solution of a differential equation is defined by an implicit equation rather than by an explicit formula. As with all implicitly defined functions, it may not be possible to derive an explicit formula for an implicitly defined solution.

◆ **EXAMPLE 3** If y is defined implicitly by the equation

$$y^3 + e^y - x^4 = C \tag{2}$$

show that y satisfies the differential equation

$$(3y^2 + e^y)y' = 4x^3$$

Solution We use implicit differentiation to show that y satisfies the given differential equation:

$$y^3 + e^y - x^4 = C$$
$$D_x(y^3 + e^y - x^4) = D_x\, C$$
$$D_x\, y^3 + D_x\, e^y - D_x\, x^4 = 0$$
$$3y^2y' + e^y y' - 4x^3 = 0$$
$$(3y^2 + e^y)y' = 4x^3$$

Remember, $D_x\, x^n = nx^{n-1}$ and $D_x\, e^x = e^x$, but $D_x\, y^n = ny^{n-1}y'$ and $D_x\, e^y = e^y y'$ since y is a function of x.

Since the last equation is the given differential equation, our calculations show that any function y defined implicitly by equation (2) is a solution of this differential equation. We cannot find an explicit formula for y, since none exists in terms of finite combinations of elementary functions. ◆

PROBLEM 3 If y is defined implicitly by the equation

$$y + e^{y^2} - x^2 = C$$

show that y satisfies the differential equation

$$(1 + 2ye^{y^2})y' = 2x$$ ◆

◆ **EXAMPLE 4** If y is defined implicitly by the equation

$$y^2 - x^2 = C$$

show that y satisfies the differential equation

$$yy' = x$$

Find an explicit expression for the particular solution that satisfies the initial condition $y(0) = 2$.

Solution Using implicit differentiation, we have

$$\begin{aligned} y^2 - x^2 &= C \\ D_x(y^2 - x^2) &= D_x C \\ 2yy' - 2x &= 0 \\ 2yy' &= 2x \\ yy' &= x \end{aligned}$$

which shows that y satisfies the given differential equation. Substituting $x = 0$ and $y = 2$ in $y^2 - x^2 = C$, we have

$$\begin{aligned} y^2 - x^2 &= C \\ (2)^2 - (0)^2 &= C \\ 4 &= C \end{aligned}$$

Thus, the particular solution satisfying $y(0) = 2$ is a solution of the equation

$$y^2 - x^2 = 4$$

or

$$y^2 = 4 + x^2$$

Solving the last equation for y yields two explicit solutions,

$$y_1(x) = \sqrt{4 + x^2} \qquad \text{and} \qquad y_2(x) = -\sqrt{4 + x^2}$$

The first of these two solutions satisfies

$$y_1(0) = \sqrt{4 + (0)^2} = \sqrt{4} = 2$$

while the second satisfies

$$y_2(0) = -\sqrt{4 + (0)^2} = -\sqrt{4} = -2$$

Thus, the particular solution of the differential equation $yy' = x$ that satisfies the initial condition $y(0) = 2$ is

$$y(x) = \sqrt{4 + x^2}$$

Check

$$\begin{aligned} y'(x) &= \frac{1}{2}(4 + x^2)^{-1/2}2x \\ &= \frac{x}{\sqrt{4 + x^2}} \end{aligned}$$

$$\begin{aligned} yy' &= x \\ \sqrt{4 + x^2}\left(\frac{x}{\sqrt{4 + x^2}}\right) &\stackrel{?}{=} x \\ x &\stackrel{\checkmark}{=} x \end{aligned}$$

◆

PROBLEM 4 If y is defined implicitly by the equation

$$y^2 - x = C$$

show that y satisfies the differential equation

$$2yy' = 1$$

Find an explicit expression for the particular solution that satisfies the initial condition $y(0) = 3$. ◆

◆ APPLICATION

In economics, the price of a product is often studied over a period of time, and so it is natural to view price as a function of time. Let $p(t)$ be the price of a particular product at time t. If $p(t)$ approaches a limiting value $\bar{p}$ as t approaches infinity, then the price for this product is said to be **dynamically stable** and $\bar{p}$ is referred to as the **equilibrium price.** (Later in this chapter, this definition of equilibrium price will be related to the one given in Section 7-2.) In order to study the behavior of price as a function of time, economists often assume that the price satisfies a differential equation. This approach is illustrated in the next example.

◆ EXAMPLE 5

Dynamic Price Stability

The price $p(t)$ of a product is assumed to satisfy the differential equation

$$\frac{dp}{dt} = 10 - 0.5p$$

Show that

$$p(t) = 20 - Ce^{-0.5t}$$

is the general solution of this differential equation and evaluate

$$\bar{p} = \lim_{t\to\infty} p(t)$$

On the same set of axes, graph the three particular solutions that satisfy the initial conditions $p(0) = 40$, $p(0) = 10$, and $p(0) = 20$.

Solution

$$p(t) = 20 - Ce^{-0.5t}$$
$$p'(t) = 0.5Ce^{-0.5t}$$

Substituting in the given differential equation, we have

$$\frac{dp}{dt} = 10 - 0.5p$$
$$\begin{aligned} 0.5Ce^{-0.5t} &= 10 - 0.5(20 - Ce^{-0.5t}) \\ &= 10 - 10 + 0.5Ce^{-0.5t} \\ &= 0.5Ce^{-0.5t} \end{aligned}$$

which shows that $p(t) = 20 - Ce^{-0.5t}$ is the general solution of this differential equation. To find the equilibrium price, we must evaluate $\lim_{t\to\infty} p(t)$.

$$\begin{aligned}\bar{p} &= \lim_{t\to\infty} p(t) \\ &= \lim_{t\to\infty} (20 - Ce^{-0.5t}) \\ &= 20 - C \lim_{t\to\infty} e^{-0.5t} \qquad \lim_{t\to\infty} e^{-0.5t} = 0 \\ &= 20 - C \cdot 0 \\ &= 20\end{aligned}$$

Before we can graph the three particular solutions, we must evaluate the constant C for each of the indicated initial conditions. In each case, we will make use of the equation

$$p(0) = 20 - Ce^0 = 20 - C$$

$p(0) = 40$	$p(0) = 10$	$p(0) = 20$
$20 - C = 40$	$20 - C = 10$	$20 - C = 20$
$C = -20$	$C = 10$	$C = 0$
$p(t) = 20 + 20e^{-0.5t}$	$p(t) = 20 - 10e^{-0.5t}$	$p(t) = 20$

Notice that the equilibrium price $\bar{p}$ does not depend on the constant C and consequently does not depend on the initial value of the price function $p(0)$ (see Figure 3). If $p(0) > \bar{p}$, then the price decreases and approaches $\bar{p}$ as a limit. If $p(0) < \bar{p}$, then the price increases and approaches $\bar{p}$ as a limit. If $p(0) = \bar{p}$, then the price remains constant for all t.

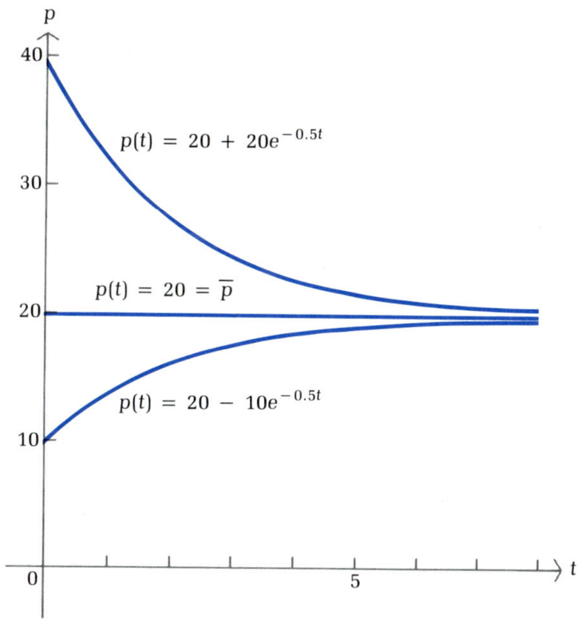

FIGURE 3

Particular solutions of $\dfrac{dp}{dt} = 10 - 0.5p$

◆

PROBLEM 5 Show that the price function

$$p(t) = 25 - Ce^{-0.2t}$$

is the general solution of the differential equation

$$\frac{dp}{dt} = 5 - 0.2p$$

Find the equilibrium price $\bar{p}$. On the same set of axes, graph the three particular solutions that satisfy the initial conditions $p(0) = 40$, $p(0) = 5$, and $p(0) = 25$. ◆

Answers to Matched Problems

1. 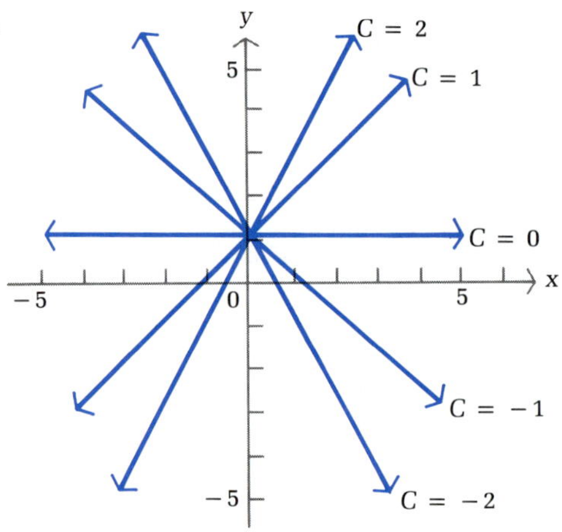

2. (A) $y = x + 1$
   (B) No particular solution exists
   (C) $y = Cx + 1$ for any C
3. $D_x\, y + D_x\, e^{y^2} - D_x\, x^2 = D_x\, C$
   $$y' + e^{y^2}2yy' - 2x = 0$$
   $$(1 + 2ye^{y^2})y' = 2x$$
4. $y(x) = \sqrt{9 + x}$
5. $\bar{p} = 25$

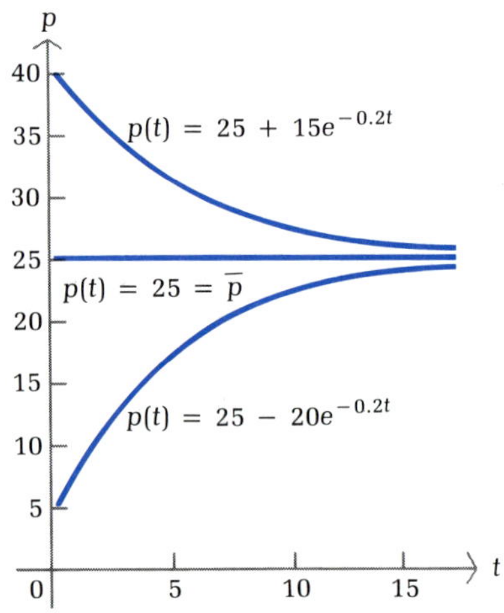

EXERCISE 9-1

A *Show that the given function y is the general solution of the indicated differential equation. On the same set of axes, graph the particular solutions obtained by letting $C = -2, -1, 0, 1$, and 2.*

1. $y = Cx^2$; $xy' = 2y$
2. $y = C(x - 1)^2$; $(x - 1)y' = 2y$
3. $y = \frac{C}{x}$; $xy' = -y$
4. $y = C(1 + x^2)$; $(1 + x^2)y' = 2xy$

B *Show that the given function y is the general solution of the indicated differential equation. Find a particular solution satisfying the given initial condition.*

5. $y = Ce^x - 5x - 5$; $y' = y + 5x$; $y(0) = 2$
6. $y = Ce^{-x} + 2x - 2$; $y' = 2x - y$; $y(0) = 3$
7. $y = e^x + Ce^{2x}$; $y' = 2y - e^x$; $y(0) = -1$
8. $y = e^x + Ce^{-x}$; $y' = 2e^x - y$; $y(0) = 1$
9. $y = x + \frac{C}{x}$; $xy' = 2x - y$; $y(2) = 3$
10. $y = Cx + \frac{2}{x}$; $x^2y' = xy - 4$; $y(2) = -3$

If y is defined implicitly by the given equation, use implicit differentiation to show that y satisfies the indicated differential equation.

11. $y^3 + xy - x^3 = C$; $(3y^2 + x)y' = 3x^2 - y$
12. $y + x^3y^3 - 2x = C$; $(1 + 3x^3y^2)y' = 2 - 3x^2y^3$
13. $xy + e^{y^2} - x^2 = C$; $(x + 2ye^{y^2})y' = 2x - y$
14. $y + e^{xy} - x = C$; $(1 + xe^{xy})y' = 1 - ye^{xy}$

If y is defined implicitly by the given equation, use implicit differentiation to show that y satisfies the indicated differential equation. Find an explicit expression for the particular solution that satisfies the given initial condition.

15. $y^2 + x^2 = C$; $yy' = -x$; $y(0) = 3$
16. $y^2 + 2x^2 = C$; $yy' = -2x$; $y(0) = -4$
17. $\ln(2 - y) = x + C$; $y' = y - 2$; $y(0) = 1$
18. $\ln(5 - y) = 2x + C$; $y' = 2(y - 5)$; $y(0) = 2$

Use the general solution y of each differential equation to find a particular solution that satisfies the indicated initial condition. Graph the particular solutions for $x \geqslant 0$.

19. $y = 2 + Ce^{-x}$; $y' = 2 - y$

    (A) $y(0) = 1$ (B) $y(0) = 2$ (C) $y(0) = 3$

20. $y = 4 + Ce^{-x}$; $y' = 4 - y$

    (A) $y(0) = 2$ (B) $y(0) = 4$ (C) $y(0) = 5$

21. $y = 2 + Ce^x$; $y' = y - 2$

(A) $y(0) = 1$ (B) $y(0) = 2$ (C) $y(0) = 3$

22. $y = 4 + Ce^x$; $y' = y - 4$

(A) $y(0) = 2$ (B) $y(0) = 4$ (C) $y(0) = 5$

C *Use the general solution y of each differential equation to find a particular solution that satisfies the indicated initial condition. Graph the particular solutions for $x \geqslant 0$.*

23. $y = \dfrac{10}{1 + Ce^{-x}}$; $y' = 0.1y(10 - y)$

(A) $y(0) = 1$ (B) $y(0) = 10$ (C) $y(0) = 20$

[*Hint:* The particular solution in part A has an inflection point at $x = \ln 9$.]

24. $y = \dfrac{5}{1 + Ce^{-x}}$; $y' = 0.2y(5 - y)$

(A) $y(0) = 1$ (B) $y(0) = 5$ (C) $y(0) = 10$

[*Hint:* The particular solution in part A has an inflection point at $x = \ln 4$.]

Use the general solution y of each differential equation to find a particular solution that satisfies the indicated initial condition.

25. $y = Cx^3 + 2$; $xy' = 3y - 6$

(A) $y(0) = 2$ (B) $y(0) = 0$ (C) $y(1) = 1$

26. $y = Cx^4 + 1$; $xy' = 4y - 4$

(A) $y(0) = 1$ (B) $y(0) = 0$ (C) $y(1) = 2$

Problems 27 and 28 require the use of a graphic calculator or a computer. In each problem, use $[-5, 5]$ for both the x range and the y range.

27. Given that $y = x + Ce^{-x}$ is the general solution of the differential equation $y' + y = 1 + x$:
   (A) In the same viewing rectangle, graph the particular solutions obtained by letting $C = 0, 1, 2,$ and 3.
   (B) What do the graphs of the solutions for $C > 0$ have in common?
   (C) In the same viewing rectangle, graph the particular solutions obtained by letting $C = 0, -1, -2,$ and -3.
   (D) What do the graphs of the solutions for $C < 0$ have in common?

28. Repeat Problem 27, given that $y = x + (C/x)$ is the general solution of the differential equation $xy' = 2x - y$.

Business & Economics

29. *Price stability.* The price $p(t)$ of a product is assumed to satisfy the differential equation

$$\frac{dp}{dt} = 0.5 - 0.1p$$

Show that

$$p(t) = 5 - Ce^{-0.1t}$$

is the general solution of this differential equation and evaluate

$$\bar{p} = \lim_{t \to \infty} p(t)$$

Graph the particular solutions that satisfy the initial conditions $p(0) = 1$ and $p(0) = 10$.

30. *Price stability.* The price $p(t)$ of a product is assumed to satisfy the differential equation

$$\frac{dp}{dt} = 0.8 - 0.2p$$

Show that

$$p(t) = 4 - Ce^{-0.2t}$$

is the general solution of this differential equation and evaluate

$$\bar{p} = \lim_{t \to \infty} p(t)$$

Graph the particular solutions that satisfy the initial conditions $p(0) = 2$ and $p(0) = 8$.

31. *Continuous compound interest.* If money is deposited at the continuous rate of \$200 per year into an account earning 8% compounded continuously, then the amount A in the account after t years satisfies the differential equation

$$\frac{dA}{dt} = 0.08A + 200$$

Show that

$$A = Ce^{0.08t} - 2{,}500$$

is the general solution of this differential equation. Graph the particular solutions satisfying $A(0) = 0$ and $A(0) = 1{,}000$.

32. *Continuous compound interest.* If money is withdrawn at the continuous rate of \$500 per year from an account earning 10% compounded continuously, then the amount A in the account after t years satisfies the differential equation

$$\frac{dA}{dt} = 0.1A - 500$$

Show that

$$A = 5{,}000 + Ce^{0.1t}$$

is the general solution of this differential equation. Graph the particular solutions satisfying $A(0) = 4{,}000$ and $A(0) = 6{,}000$.

Life Sciences

33. *Population growth—Verhulst growth law.* The number $N(t)$ of bacteria in a culture at time t is assumed to satisfy the Verhulst growth law

$$\frac{dN}{dt} = 100 - 0.5N$$

Show that

$$N(t) = 200 - Ce^{-0.5t}$$

is the general solution of this differential equation and evaluate

$$\overline{N} = \lim_{t \to \infty} N(t)$$

where $\overline{N}$ is the equilibrium size of the population. Graph the particular solutions that satisfy $N(0) = 50$ and $N(0) = 300$. (Note that growth can be "negative," that is, populations can decrease in size.)

Amwest

34. *Population growth—logistic growth.* The population $N(t)$ of a certain species of animal in a controlled habitat at time t is assumed to satisfy the logistic growth law

$$\frac{dN}{dt} = \frac{1}{500} N(1{,}000 - N)$$

Show that

$$N(t) = \frac{1{,}000}{1 + Ce^{-2t}}$$

is the general solution of this differential equation and evaluate

$$\overline{N} = \lim_{t \to \infty} N(t)$$

Graph the particular solution that satisfies the initial condition $N(0) = 200$. [*Hint:* The particular solution has an inflection point at $t = \ln 2$.]

Social Sciences

35. *Rumor spread—Gompertz growth law.* The rate of propagation of a rumor is assumed to satisfy the Gompertz growth law

$$\frac{dN}{dt} = Ne^{-0.5t}$$

where $N(t)$ is the number of individuals who have heard the rumor at time t. Show that

$$N(t) = Ce^{-2e^{-0.5t}}$$

is the general solution of this differential equation. Find the particular solution that satisfies the initial condition $N(0) = 200$, evaluate

$$\overline{N} = \lim_{t\to\infty} N(t)$$

and sketch the graph of this particular solution. [*Hint:* $N(t)$ has an inflection point at $t = \ln 4$.]

SECTION 9-2 Separation of Variables

- SEPARATION OF VARIABLES
- EXPONENTIAL GROWTH
- LIMITED GROWTH
- LOGISTIC GROWTH

SEPARATION OF VARIABLES

In this section we will develop a technique called *separation of variables*, which can be used to solve differential equations that can be expressed in the form

$$f(y)y' = g(x) \tag{1}$$

We have already used this technique informally in Section 6-3 to derive some important exponential growth laws. We will now present a more formal development of this method.

The solution of differential equations by the method of separating the variables is based on the substitution formula for indefinite integrals. If $y = y(x)$ is a differentiable function of x, then

$$\int f(y)\,dy = \int f[y(x)]y'(x)\,dx \tag{2}$$

Substitution formula for indefinite integrals

If $y(x)$ is also the solution of (1), then $y(x)$ and $y'(x)$ must satisfy (1). That is,

$$f[y(x)]y'(x) = g(x)$$

Substituting $g(x)$ for $f[y(x)]y'(x)$ in (2), we have

$$\int f(y)\,dy = \int g(x)\,dx$$

Thus, the solution $y(x)$ of (1) is given implicitly by this equation. Of course, for this method to be useful, we must be able to evaluate both of these indefinite integrals.

This discussion is summarized in Theorem 1:

THEOREM 1

Separation of Variables

The solution of the differential equation

$$f(y)y' = g(x)$$

is given implicitly by the equation

$$\int f(y)\,dy = \int g(x)\,dx$$

Several examples will demonstrate the use of this method.

◆ **EXAMPLE 6** Solve: $y' = 2xy^2$

Solution

$$y' = 2xy^2 \qquad \text{Multiply by } 1/y^2 \text{ to separate the variables.}$$

$$\frac{1}{y^2}y' = 2x \qquad \text{Convert to an equation involving indefinite integrals.}$$

$$\int \frac{1}{y^2}\,dy = \int 2x\,dx \qquad \text{Evaluate each indefinite integral.}$$

$$-\frac{1}{y} + C_1 = x^2 + C_2 \qquad \text{Combine the two constants of integration into a single arbitrary constant.}$$

$$-\frac{1}{y} = x^2 + C \qquad \text{Solve for } y.$$

$$y = -\frac{1}{x^2 + C} \qquad \text{General solution of } y' = 2xy^2$$

Check

$$y' = D_x\left(-\frac{1}{x^2 + C}\right) = \frac{2x}{(x^2 + C)^2}$$

$$y' = 2xy^2 \qquad \text{Substitute } y' = 2x/(x^2 + C)^2 \text{ and } y = -1/(x^2 + C).$$

$$\frac{2x}{(x^2 + C)^2} \stackrel{?}{=} 2x\left(-\frac{1}{x^2 + C}\right)^2$$

$$\frac{2x}{(x^2 + C)^2} \stackrel{\checkmark}{=} \frac{2x}{(x^2 + C)^2}$$

This verifies that our general solution is correct. (You should develop the habit of checking the solution of each differential equation you solve, as we have done here. From now on, we will leave it up to you to check most of the examples worked in the text.) ◆

PROBLEM 6 Solve: $y' = 4x^3y^2$ ◆

In some cases the general solution obtained by the technique of separation of variables does not include all the solutions to a differential equation. For example, the constant function $y = 0$ also satisfies the differential equation in Example 6 (verify this). Yet the solution $y = 0$ cannot be obtained from the expression

$$y = -\frac{1}{x^2 + C}$$

for any choice of the constant C. Solutions of this type are referred to as **singular solutions** and are usually discussed in more advanced courses. We will not attempt to find the singular solutions of any of the differential equations we consider.

◆ EXAMPLE 7 Find the general solution of

$$(1 + x^2)y' = 2x(y - 1)$$

Then find the particular solution that satisfies the initial condition $y(0) = 3$.

Solution First, we find the general solution:

$$(1 + x^2)y' = 2x(y - 1)$$

Multiply by $1/(1 + x^2)$ and $1/(y - 1)$ to separate the variables.

$$\frac{y'}{y - 1} = \frac{2x}{1 + x^2}$$

Convert to an equation involving indefinite integrals.

$$\int \frac{dy}{y - 1} = \int \frac{2x\,dx}{1 + x^2}$$

Evaluate each indefinite integral.

$$\ln|y - 1| = \ln(1 + x^2) + C$$

where C is an arbitrary constant. Notice that the two constants of integration always can be combined to form a single arbitrary constant. Also note that we can use $1 + x^2$ instead of $|1 + x^2|$, since $1 + x^2$ is always positive. In order to solve for y, we convert this last equation to exponential form:

$$\begin{aligned} |y - 1| &= e^{\ln(1+x^2)+C} \\ &= e^C e^{\ln(1+x^2)} \qquad \text{Use the property } e^{\ln r} = r. \\ &= e^C(1 + x^2) \end{aligned}$$

It can be shown that if we replace e^C with an arbitrary constant K, then we can omit the absolute value signs on the left side of the last equation.* The resulting

---

* In any problem involving separation of variables, you may assume that the equations

$$e^{\ln|f(y)|} = e^C g(x) \qquad \text{and} \qquad f(y) = Kg(x)$$

are equivalent (both C and K are arbitrary constants). Justification of this assumption involves properties of the absolute value and exponential functions.

equation can then be used to find the general solution to the original differential equation. Thus,

$$y - 1 = K(1 + x^2)$$
$$y = 1 + K(1 + x^2) \qquad \text{General solution}$$

To find the particular solution that satisfies $y(0) = 3$, we substitute $x = 0$ and $y = 3$ in the general solution and solve for K:

$$3 = 1 + K(1 + 0)$$
$$K = 2$$
$$y = 1 + 2(1 + x^2) = 3 + 2x^2 \qquad \text{Particular solution}$$

◆

PROBLEM 7 Find the general solution of

$$(2 + x^4)y' = 4x^3(y - 3)$$

Then find the particular solution that satisfies the initial condition $y(0) = 5$. ◆

◆ EXPONENTIAL GROWTH

We now return to the study of exponential growth laws first begun in Section 6-3. This time we will place more emphasis on determining the relevant growth law for a particular application and on using separation of variables to solve the corresponding differential equation in each problem. We begin with the familiar exponential growth law.

Exponential Growth Law

If the rate of change with respect to time t of a quantity y is proportional to the amount present, then y satisfies the differential equation

$$\frac{dy}{dt} = ky$$

Exponential growth includes both the case where y is increasing and the case where y is decreasing, or decaying.

◆ EXAMPLE 8

Product Analysis

A certain brand of mothballs evaporate at a rate proportional to their volume, losing half their volume every 4 weeks. If the volume of each mothball is initially 15 cubic centimeters and a mothball becomes ineffective when its volume reaches 1 cubic centimeter, how long will these mothballs be effective?

Solution

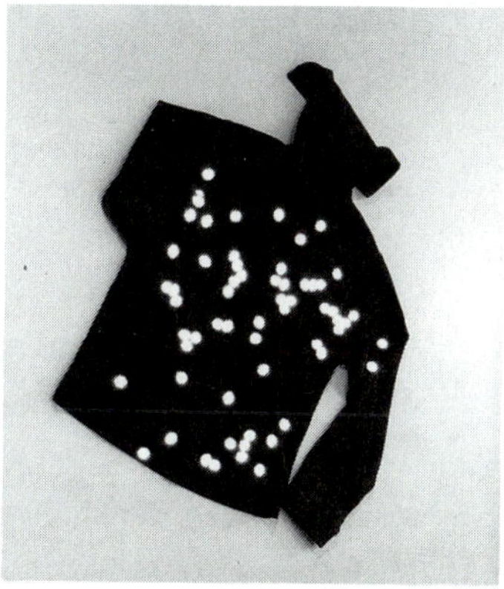

The volume of each mothball is decaying at a rate proportional to its volume. If V is the volume of a mothball after t weeks, then

$$\frac{dV}{dt} = kV$$

Since the initial volume is 15 cubic centimeters, we know that $V(0) = 15$. After 4 weeks, the volume will be half the original volume, so $V(4) = 7.5$. Summarizing these requirements, we have the following exponential decay model:

$$\frac{dV}{dt} = kV$$

$$V(0) = 15 \qquad V(4) = 7.5$$

We want to determine the value of t that satisfies the equation $V(t) = 1$. First, we use separation of variables to find the general solution of the differential equation:

$$\frac{dV}{dt} = kV$$

$$\frac{1}{V}\frac{dV}{dt} = k$$

$$\int \frac{dV}{V} = \int k\,dt$$

$$\ln V = kt + C \qquad \text{We can write } \ln V \text{ in place of } \ln|V|\text{, since } V > 0.$$

$$V = e^{kt+C} = e^C e^{kt} = Ae^{kt} \qquad \text{General solution}$$

where $A = e^C$ is a positive constant. Now we use the initial condition to determine the value of the constant A:

$$V(0) = Ae^0 = A = 15$$

$$V(t) = 15e^{kt}$$

Next, we apply the condition $V(4) = 7.5$ to determine the constant k:

$$V(4) = 15e^{4k} = 7.5$$

$$e^{4k} = \frac{7.5}{15} = 0.5$$

$$4k = \ln 0.5$$

$$k = \frac{\ln 0.5}{4} \approx -0.1733$$

$$V(t) = 15e^{(t/4)\ln 0.5} \qquad \text{Particular solution}$$

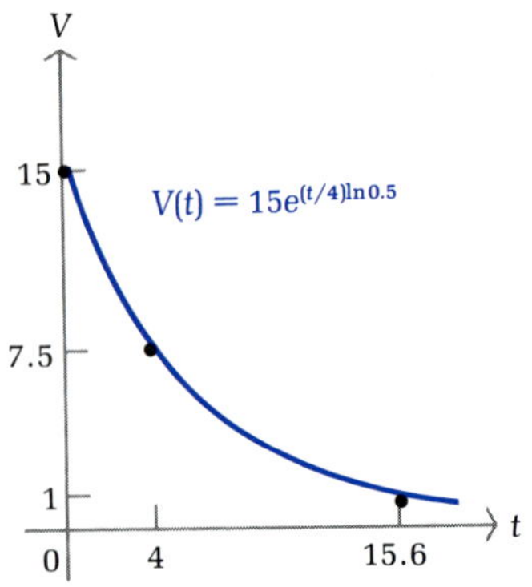

Exponential decay

The graph of $V(t)$ is shown in the figure in the margin. To determine how long the mothballs will be effective, we find t when $V = 1$:

$$V(t) = 1$$
$$15e^{(t/4)\ln 0.5} = 1$$
$$e^{(t/4)\ln 0.5} = \frac{1}{15}$$
$$\frac{t}{4}\ln 0.5 = \ln\frac{1}{15}$$
$$t = \frac{4\ln\frac{1}{15}}{\ln 0.5} \approx 15.6 \text{ weeks}$$

◆

PROBLEM 8 Repeat Example 8 if the mothballs lose half their volume every 5 weeks. ◆

◆ LIMITED GROWTH

In certain situations there is an upper limit (or a lower limit), say M, on the values a variable can assume. This limiting value leads to the limited growth law.

Limited Growth Law

If the rate of change with respect to time t of a quantity y is proportional to the difference between y and a limiting value M, then y satisfies the differential equation

$$\frac{dy}{dt} = k(M - y)$$

When we speak of limited growth, we will include both the case where y increases and approaches M from below and the case where y decreases and approaches M from above.

◆ EXAMPLE 9

Sales Growth

The annual sales of a new company are expected to grow at a rate proportional to the difference between the sales and an upper limit of \$20 million. The sales are 0 initially and \$4 million for the second year of operation.

(A) What should the company expect the sales to be during the tenth year?
(B) In what year should the sales be expected to reach \$15 million?

Solutions If S is the annual sales in millions of dollars during year t, then the model for this problem is

$$\frac{dS}{dt} = k(20 - S) \qquad (3)$$

$$S(0) = 0 \qquad S(2) = 4$$

This is a limited growth model. For part A we want to find $S(10)$, and for part B we want to solve $S(t) = 15$ for t. First, separating the variables in (3) and integrating both sides, we obtain

$$\int \frac{dS}{20 - S} = \int k\, dt$$

$$-\ln(20 - S) = kt + C$$

$$\ln(20 - S) = -kt - C$$

We can write $-\ln(20 - S)$ in place of $-\ln|20 - S|$, since $0 < S < 20$.

$$20 - S = e^{-kt-C} = e^{-C}e^{-kt} = Ae^{-kt} \qquad A = e^{-C}$$

$$S = 20 - Ae^{-kt} \qquad \text{General solution}$$

Now we use the conditions $S(0) = 0$ and $S(2) = 4$ to determine the constants A and k:

$$S(0) = 20 - Ae^0$$

$$= 20 - A = 0$$

$$A = 20$$

$$S(t) = 20 - 20e^{-kt}$$

$$S(2) = 20 - 20e^{-2k} = 4$$

$$20e^{-2k} = 16$$

$$e^{-2k} = \frac{16}{20} = 0.8$$

$$-2k = \ln 0.8$$

$$k = -\frac{\ln 0.8}{2} \approx 0.1116$$

$$S(t) = 20 - 20e^{(t/2)\ln 0.8} \qquad \text{Particular solution}$$

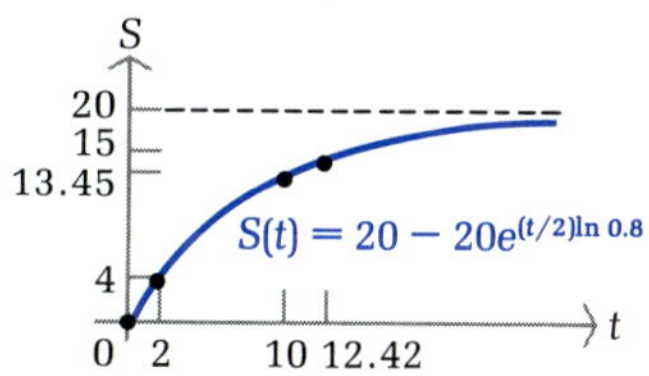

Limited growth

The graph of $S(t)$ is shown in the figure in the margin. Notice that the upper limit of \$20 million is a horizontal asymptote.

(A) $S(10) = 20 - 20e^{5\ln 0.8} \approx \$13.45$ million

(B) $S(t) = 20 - 20e^{(t/2)\ln 0.8} = 15$

$$20e^{(t/2)\ln 0.8} = 5$$

$$e^{(t/2)\ln 0.8} = \frac{5}{20} = 0.25$$

$$\frac{t}{2}\ln 0.8 = \ln 0.25$$

$$t = \frac{2\ln 0.25}{\ln 0.8} \approx 12.43 \text{ years}$$

The annual sales will exceed \$15 million in the thirteenth year. ◆

PROBLEM 9 Repeat Example 9 if the sales during the second year are \$3 million. ◆

◆ LOGISTIC GROWTH

If a quantity first begins to grow exponentially, but then starts to approach a limiting value, it is said to exhibit logistic growth. More formally, we have:

Logistic Growth Law

If the rate of change with respect to time t of a quantity y is proportional to both the amount present and the difference between y and a limiting value M, then y satisfies the differential equation

$$\frac{dy}{dt} = ky(M - y)$$

Theoretically, functions satisfying logistic growth laws can be increasing or decreasing, just as was the case for the exponential and limited growth laws. However, decreasing functions are seldom encountered in actual practice.

◆ EXAMPLE 10

Population Growth

In a study of ciliate protozoans, it has been shown that the rate of growth of the number of *Paramecium caudatum* in a medium with fixed volume is proportional to the product of the number present and the difference between an upper limit of 375 and the number present. The medium initially contains 25 paramecia. After 1 hour there are 125 paramecia. How many paramecia are present after 2 hours?

Solution

If P is the number of paramecia in the medium at time t, then the model for this problem is the following logistic growth model:

$$\frac{dP}{dt} = kP(375 - P)$$

$$P(0) = 25 \qquad P(1) = 125$$

We want to find $P(2)$. First, we separate the variables and convert to an equation involving indefinite integrals:

$$\frac{1}{P(375 - P)} \frac{dP}{dt} = k$$

$$\int \frac{1}{P(375 - P)}\, dP = \int k\, dt \tag{4}$$

The integral on the left side of (4) can be evaluated either by using an algebraic identity* or by using formula 9 in Table II. We will use formula 9 with $u = P$, $a = 375$, and $b = -1$:

$$\int \frac{1}{u(a + bu)}\,du = \frac{1}{a}\ln\left|\frac{u}{a + bu}\right| \qquad \text{Formula 9}$$

$$\int \frac{1}{P(375 - P)}\,dP = \frac{1}{375}\ln\left|\frac{P}{375 - P}\right|$$

$$= \frac{1}{375}\ln\left(\frac{P}{375 - P}\right) \qquad \text{Since } 0 < P < 375 \text{, the absolute value signs can be omitted.}$$

Returning to equation (4), we have

$$\frac{1}{375}\ln\left(\frac{P}{375 - P}\right) = \int k\,dt = kt + D \qquad D \text{ is a constant.}$$

$$\ln\left(\frac{P}{375 - P}\right) = 375kt + 375D$$

$$\frac{P}{375 - P} = e^{375kt + 375D} = e^{Bt}e^{C} \qquad B = 375k,\ C = 375D$$

$$P = 375e^{Bt}e^{C} - Pe^{Bt}e^{C}$$

$$P(e^{Bt}e^{C} + 1) = 375e^{Bt}e^{C}$$

$$P = \frac{375e^{Bt}e^{C}}{e^{Bt}e^{C} + 1} \qquad \text{Multiply numerator and denominator by } e^{-Bt}e^{-C} \text{ and let } A = e^{-C}.$$

$$= \frac{375}{1 + Ae^{-Bt}} \qquad \text{General solution}$$

Now we use the conditions $P(0) = 25$ and $P(1) = 125$ to evaluate the constants A and B:

$$P(0) = \frac{375}{1 + A} = 25$$

$$375 = 25 + 25A$$

$$A = 14$$

$$P(t) = \frac{375}{1 + 14e^{-Bt}}$$

---

* If you do not wish to use Table II to evaluate integrals, you can use the algebraic identity

$$\frac{1}{x(a - x)} = \frac{1}{a}\left[\frac{(a - x) + x}{x(a - x)}\right] = \frac{1}{a}\left(\frac{1}{x} + \frac{1}{a - x}\right)$$

to evaluate this integral and in all logistic growth problems in Exercise 9-2.

$$P(1) = \frac{375}{1 + 14e^{-B}} = 125$$

$$375 = 125 + 1{,}750e^{-B}$$

$$e^{-B} = \frac{250}{1{,}750} = \frac{1}{7}$$

$$-B = \ln\frac{1}{7} = -\ln 7$$

$$B = \ln 7$$

$$P(t) = \frac{375}{1 + 14e^{-t\ln 7}} \quad \text{Particular solution}$$

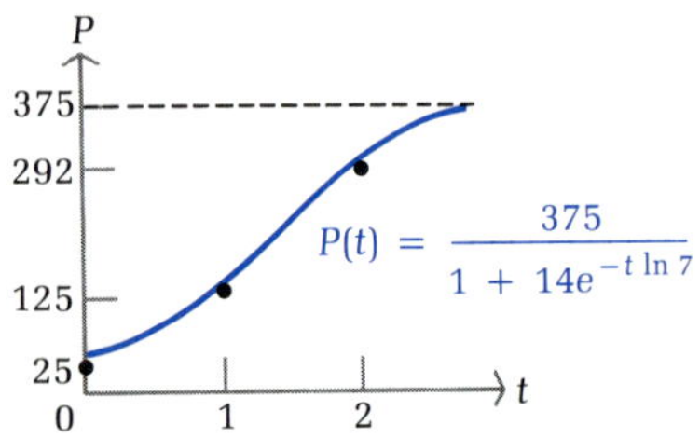

Logistic growth

To determine the population after 2 hours, we evaluate $P(2)$:

$$P(2) = \frac{375}{1 + 14e^{-2\ln 7}} \approx 292 \text{ paramecia}$$

The graph of $P(t)$ is shown in the figure in the margin. This S-shaped curve is typical of logistic growth functions. Notice that the upper limit of 375 is a horizontal asymptote. ◆

PROBLEM 10 Repeat Example 10 for 15 paramecia initially and 150 paramecia after 1 hour. ◆

Answers to Matched Problems

6. $y = -1/(x^4 + C)$
7. General solution: $y = 3 + K(2 + x^4)$; particular solution: $y = 5 + x^4$
8. 19.5 weeks
9. (A) \$11.13 million (B) Seventeenth year ($t \approx 17.06$ yr)
10. 343 paramecia

EXERCISE 9-2

A *Find the general solution for each differential equation. Then find the particular solution satisfying the initial condition.*

1. $y' = 1$; $y(0) = 2$
2. $y' = 2x$; $y(0) = 4$
3. $y' = \frac{1}{x^{1/2}}$; $y(1) = -2$
4. $y' = 3x^{1/2}$; $y(0) = 7$
5. $y' = y$; $y(0) = 10$
6. $y' = y - 10$; $y(0) = 15$
7. $y' = 25 - y$; $y(0) = 5$
8. $y' = 3x^2y$; $y(0) = \frac{1}{2}$
9. $y' = \frac{y}{x}$; $y(1) = 5$; $x > 0$
10. $y' = \frac{y}{x^2}$; $y(-1) = 2e$

B *Find the general solution for each differential equation. Then find the particular solution satisfying the initial condition.*

11. $y' = \dfrac{1}{y^2}$; $y(1) = 3$

12. $y' = \dfrac{x^2}{y^2}$; $y(0) = 2$

13. $y' = ye^x$; $y(0) = 3e$

14. $y' = -y^2e^x$; $y(0) = \frac{1}{2}$

15. $y' = \dfrac{e^x}{e^y}$; $y(0) = \ln 2$

16. $y' = y^2(2x + 1)$; $y(0) = -\frac{1}{5}$

17. $y' = xy + x$; $y(0) = 2$

18. $y' = (2x + 4)(y - 3)$; $y(0) = 2$

19. $y' = (2 - y)^2e^x$; $y(0) = 1$

20. $y' = \dfrac{x^{1/2}}{(y - 5)^2}$; $y(1) = 7$

Find the general solution for each differential equation. Do not attempt to find an explicit expression for the solution.

21. $y' = \dfrac{1 + x^2}{1 + y^2}$

22. $y' = \dfrac{6x - 9x^2}{2y + 4}$

23. $xyy' = (1 + x^2)(1 + y^2)$

24. $(xy^2 - x)y' = x^2y - x^2$

25. $x^2e^yy' = x^3 + x^3e^y$

26. $y' = \dfrac{xe^x}{\ln y}$

C *Find an explicit expression for the particular solution for each differential equation.*

27. $xyy' = \ln x$; $y(1) = 1$

28. $y' = \dfrac{xe^{x^2}}{y}$; $y(0) = 2$

29. $xy' = x\sqrt{y} + 2\sqrt{y}$; $y(1) = 4$

30. $y' = x(x - 1)^{1/2}(y - 1)^{1/2}$; $y(1) = 1$

31. $yy' = xe^{-y^2}$; $y(0) = 1$

32. $yy' = x(1 + y^2)$; $y(0) = 1$

APPLICATIONS

Business & Economics

33. Continuous compound interest. If $5,000 is invested at 12% compounded continuously, how much will be in the account at the end of 10 years?

34. Continuous compound interest. If a sum of money is invested at 9% compounded continuously, how long will it take for the sum to double?

35. Advertising. A company is using radio advertising to introduce a new product to a community of 100,000 people. Suppose the rate at which people learn about the new product is proportional to the number who have not yet heard of it. If no one is aware of the product at the start of the advertising campaign and after 7 days 20,000 people are aware of the product, how long will it take for 50,000 people to become aware of the product?

36. *Advertising.* Prior to the beginning of an advertising campaign, 10% of the potential users of a certain brand are aware of the brand name. After the first week of the campaign, 20% of the consumers are aware of the brand name. If the percentage of informed consumers is growing at a rate proportional to the product of the percentage of informed consumers and the percentage of uninformed consumers, what percentage of consumers will be aware of the brand name after 5 weeks of advertising?

37. *Product analysis.* A company wishes to analyze a new room deodorizer. The active ingredient evaporates at a rate proportional to the amount present. Half of the ingredient evaporates in the first 30 days after the deodorizer is installed. If the deodorizer becomes ineffective after 90% of the active ingredient has evaporated, how long will one of these deodorizers remain effective?

38. *Natural resources.* According to the U.S. Department of Agriculture Forest Service, the total consumption of wood in the United States was 11.6 billion cubic feet in 1960 and 13.4 billion cubic feet in 1970. If the consumption is increasing at a rate proportional to the total consumption, how much wood will be used in the year 2000?

39. *Sales growth.* The annual sales of a new company are expected to grow at a rate proportional to the difference between the sales and an upper limit of \$5 million. If the sales are 0 initially and \$1 million during the fourth year of operation, when will the sales reach \$4 million?

40. *Sales analysis.* The annual sales of a company have declined from \$8 million 2 years ago to \$6 million today. If the annual sales continue to decline at a rate proportional to the difference between the annual sales and a lower limit of \$3 million, find the annual sales 3 years from now.

Newton's law of cooling states that the rate of change of the temperature of an object is proportional to the difference between the temperature of the object and the temperature of the surrounding medium. Use this law to formulate models for Problems 41–44, and then solve using the techniques discussed in this section.

41. *Manufacturing.* As part of a manufacturing process, a metal bar is to be heated in an oven until its temperature reaches 500°F. The oven is maintained at a constant temperature of 800°F. Before it is placed in the oven, the temperature of the bar is 80°F. After 2 minutes in the oven, the temperature of the bar is 200°F. How long should the bar be left in the oven?

42. *Manufacturing.* The next step in the manufacturing process described in Problem 41 calls for the heated bar to be cooled in a vat of water until its temperature reaches 100°F. The water in the vat is maintained at a con-

stant temperature of 50°F. If the temperature of the bar is 500°F when it is first placed in the water and the bar has cooled to 400°F after 5 minutes in the water, how long should the bar be left in the water?

Life Sciences

43. *Food preparation.* A pie is removed from an oven where the temperature is 325°F and placed in a freezer with a constant temperature of 25°F. After 1 hour in the freezer, the temperature of the pie is 225°F. What is the temperature of the pie after 4 hours in the freezer?

44. *Food preparation.* A roast is taken from a freezer where the temperature is 20°F and placed in an oven with a constant temperature of 350°F. After 1 hour in the oven, the temperature of the roast is 185°F. What is the temperature of the roast after 3 hours in the oven?

45. *Population growth.* A culture of bacteria is growing at a rate proportional to the number present. The culture initially contains 100 bacteria. After 1 hour there are 140 bacteria in the culture.

(A) How many bacteria will be present after 5 hours?
(B) When will the culture contain 1,000 bacteria?

46. *Population growth.* A culture of bacteria is growing in a medium that can support a maximum of 1,100 bacteria. The rate of change of the number of bacteria is proportional to the product of the number present and the difference between 1,100 and the number present. The culture initially contains 100 bacteria. After 1 hour there are 140 bacteria.

(A) How many bacteria are present after 5 hours?
(B) When will the culture contain 1,000 bacteria?

47. *Simple epidemic.* An influenza epidemic has spread throughout a community of 50,000 people at a rate proportional to the product of the number of people who have been infected and the number who have not been infected. If 100 individuals were infected initially and 500 were infected 10 days later:

(A) How many people will be infected after 20 days?
(B) When will half the community be infected?

48. *Ecology.* A fish population in a large lake is declining at a rate proportional to the difference between the population and a lower limit of 5,000 fish. If the population has declined from 15,000 fish 3 years ago to 10,000 today, find the population 6 years from now.

Social Sciences

49. *Sensory perception.* A person is subjected to a physical stimulus that has a measurable magnitude, but the intensity of the resulting sensation is difficult to measure. If s is the magnitude of the stimulus and $I(s)$ is the intensity of sensation, experimental evidence suggests that

$$\frac{dI}{ds} = k\frac{I}{s}$$

for some constant k. Express I as a function of s.

50. *Learning.* The number of words per minute, N, a person can type increases with practice. Suppose the rate of change of N is proportional to the difference between N and an upper limit of 140. It is reasonable to assume that a beginner cannot type at all. Thus, $N = 0$ when $t = 0$. If a person can type 35 words per minute after 10 hours of practice:

    (A) How many words per minute can that individual type after 20 hours of practice?
    (B) How many hours must that individual practice to be able to type 105 words per minute?

51. *Rumor spread.* A rumor spreads through a population of 1,000 people at a rate proportional to the product of the number who have heard it and the number who have not heard it. If 5 people initiated a rumor and 10 people had heard it after 1 day:

    (A) How many people will have heard the rumor after 7 days?
    (B) How long will it take for 850 people to hear the rumor?

Problems 52–58 require the use of a graphic calculator or a computer.

52. *Rumor spread.* Refer to Problem 51. Graph the particular solution and use approximation techniques to determine how long it will take for 600 people to hear the rumor. Compute the answer to the nearest day.

Business & Economics

53. *Advertising.* Refer to Problem 35. Graph the particular solution and use approximation techniques to determine how long it will take for 70,000 people to become aware of the product. Compute the answer to the nearest day.

54. *Advertising.* Refer to Problem 36. Graph the particular solution and use approximation techniques to determine how long it will take for 60% of the consumers to become aware of the product. Compute the answer correct to one decimal place.

55. *Sales analysis.* A new company has 0 sales initially, sales of \$2 million during the first year, and sales of \$5 million during the third year. If the

annual sales S are assumed to be growing at a rate proportional to the difference between the sales and an unknown upper limit M, then by the limited growth law,

$$S(t) = M(1 - e^{kt})$$

where t is time (in years) and $S(t)$ represents sales (in millions of dollars). Use approximation techniques to find k to one decimal place and M to the nearest million. [*Note:* An extraneous solution must be discarded.]

56. *Sales analysis.* Refer to Problem 55. Approximate k to one decimal place and M to the nearest million if the sales during the third year are \$4 million and all other information is unchanged.

Life Sciences

57. *Simple epidemic.* Refer to Problem 47. Graph the particular solution and use approximation techniques to determine how long it will take for 40,000 people to be infected. Compute the answer to the nearest day.

58. *Ecology.* Refer to Problem 48. Graph the particular solution and use approximation techniques to determine how long it will take for the fish population to decline to 5,500. Compute the answer to the nearest year.

SECTION 9-3 First-Order Linear Differential Equations

- SOLUTION OF FIRST-ORDER LINEAR DIFFERENTIAL EQUATIONS
- APPLICATIONS

◆ SOLUTION OF FIRST-ORDER LINEAR DIFFERENTIAL EQUATIONS

A differential equation that can be expressed in the form

$$y' + f(x)y = g(x)$$

is called a **first-order linear differential equation.**

For example,

$$y' + \frac{2}{x}y = x \tag{1}$$

is a first-order linear differential equation with $f(x) = 2/x$ and $g(x) = x$. This equation cannot be solved by the method of separation of variables. (Try to separate the variables to convince yourself that this is true.) Instead, we will change the form of the equation by multiplying both sides by x^2:

$$x^2y' + 2xy = x^3$$

How was x^2 chosen? We will discuss that below. Let us first see how this choice leads to a solution of the problem.

Recall that the product rule for differentiation can be written as

$$FS' + F'S = (FS)'$$

Notice the similarity between the left-hand sides of the last two equations. In fact, if we equate F with x^2 and S with y, then the two expressions are identical:

$$x^2y' + 2xy = FS' + F'S = (FS)' = (x^2y)'$$

Thus, making use of the product rule, we can write the differential equation as

$$(x^2y)' = x^3$$

Now we can integrate both sides:

$$\int (x^2y)'\,dx = \int x^3\,dx$$

$$x^2y = \frac{x^4}{4} + C$$

Solving for y, we obtain the general solution

$$y = \frac{x^2}{4} + \frac{C}{x^2}$$

The function x^2, which we used to transform the original equation into one we could solve as illustrated, is called an *integrating factor*. It turns out that there is a specific formula for determining the integrating factor for any first-order linear differential equation. Furthermore, this integrating factor can then be used to find the solution of the differential equation, just as we used x^2 to find the solution to (1). The formula for the integrating factor and a step-by-step summary of the solution process are given in the box.

Solving First-Order Linear Differential Equations

Step 1. Write the equation in the **standard form:**

$$y' + f(x)y = g(x)$$

Step 2. Compute the **integrating factor:**

$$I(x) = e^{\int f(x)\,dx}$$

(When evaluating $\int f(x)\,dx$, choose 0 for the constant of integration.)

Step 3. Multiply both sides of the standard form by the integrating factor $I(x)$. The left side should now be in the form $[I(x)y]'$:

$$[I(x)y]' = I(x)g(x)$$

Step 4. Integrate both sides:

$$I(x)y = \int I(x)g(x)\,dx$$

(When evaluating $\int I(x)g(x)\,dx$, include an arbitrary constant of integration.)

Step 5. Solve for y to obtain the **general solution:**

$$y = \frac{1}{I(x)}\int I(x)g(x)\,dx$$

◆ EXAMPLE 11 Solve: $2xy' + y = 10x^2$

Solution Step 1. Multiply both sides by $1/(2x)$ to obtain the standard form:

$$y' + \frac{1}{2x}y = 5x \qquad f(x) = \frac{1}{2x} \text{ and } g(x) = 5x$$

Step 2. Find the integrating factor:

$$\begin{aligned} I(x) &= e^{\int f(x)\,dx} \\ &= e^{\int [1/(2x)]\,dx} && \text{Assume } x > 0 \text{ and choose 0 for the constant of integration.} \\ &= e^{(1/2)\ln x} && r \ln t = \ln t^r \\ &= e^{\ln x^{1/2}} && e^{\ln r} = r,\ r > 0 \\ &= x^{1/2} && \text{Integrating factor} \end{aligned}$$

Step 3. Multiply both sides of the standard form by the integrating factor:

$$x^{1/2}\left(y' + \frac{1}{2x}y\right) = x^{1/2}(5x)$$

$$x^{1/2}y' + \tfrac{1}{2}x^{-1/2}y = 5x^{3/2} \qquad \text{The left side should have the form } [I(x)y]'.$$

$$(x^{1/2}y)' = 5x^{3/2}$$

Step 4. Integrate both sides:

$$\int (x^{1/2}y)'\,dx = \int 5x^{3/2}\,dx \qquad \text{Include an arbitrary constant of integration on the right side.}$$

$$x^{1/2}y = 2x^{5/2} + C$$

Step 5. Solve for y:

$$y = \frac{1}{x^{1/2}}(2x^{5/2} + C)$$

$$= 2x^2 + \frac{C}{x^{1/2}} \qquad \text{General solution}$$ ◆

PROBLEM 11 Solve: $xy' + 3y = 4x$ ◆

In Example 11, notice that we assumed $x > 0$ to avoid introducing absolute value signs in the integrating factor. Many of the problems in this section will require the evaluation of expressions of the form $e^{\int [h'(x)/h(x)]\,dx}$. In order to avoid the complications caused by the introduction of absolute value signs, we will assume that the domain of $h(x)$ has been restricted so that $h(x) > 0$. This will simplify the solution process. We state the following familiar formulas for convenient reference:

> If the domain of $h(x)$ is restricted so that $h(x) > 0$, then
>
> $$\int \frac{h'(x)}{h(x)}\,dx = \ln h(x) \qquad \text{and} \qquad e^{\ln h(x)} = h(x)$$

If a first-order linear differential equation is written in standard form, then multiplying both sides of the equation by its integrating factor will always convert the left side of the equation into the derivative of $I(x)y$. Thus, it is possible to omit steps 3–4 and proceed directly to step 5. This approach is illustrated in the next example. You decide which is easier to use—the step-by-step procedure or the formula in step 5.

◆ EXAMPLE 12 Find the particular solution of the equation

$$y' + 2xy = 4x$$

satisfying the initial condition $y(0) = 5$.

Solution Since the equation is already in standard form, we begin by finding the integrating factor:

$$I(x) = e^{\int 2x\,dx} = e^{x^2} \qquad f(x) = 2x$$

Proceeding directly to step 5, we have

$$y = \frac{1}{I(x)} \int I(x)g(x)\,dx \qquad g(x) = 4x$$

$$= \frac{1}{e^{x^2}} \int e^{x^2}(4x)\,dx$$

$$= e^{-x^2} \int 4xe^{x^2}\,dx \qquad \text{Let } u = x^2,\ du = 2x\,dx.$$

$$= e^{-x^2}(2e^{x^2} + C)$$

$$= 2 + Ce^{-x^2} \qquad \text{General solution}$$

Substituting $x = 0$ and $y = 5$ in the general solution, we have

$$5 = 2 + C$$

$$C = 3$$

$$y = 2 + 3e^{-x^2} \qquad \text{Particular solution}$$

◆

PROBLEM 12 Find the particular solution of $y' + 3x^2y = 9x^2$ satisfying the initial condition $y(0) = 7$. ◆

Common Errors

1. When integrating both sides of an equation such as

$$(x^{1/2}y)' = 5x^{3/2}$$

remember that

$$x^{1/2}y = \int 5x^{3/2}\,dx \neq 2x^{5/2} \qquad \text{“+ C” is missing in the antiderivative.}$$

Remember: A constant of integration must be included when evaluating $\int I(x)g(x)\,dx$. If you omit this constant, you will not be able to find the general solution of the differential equation. See step 4 of Example 11 for the correct procedure.

2. $$\frac{1}{e^{x^2}} \int 4xe^{x^2}\,dx \neq \frac{1}{\cancel{e^{x^2}}} \int 4x\cancel{e^{x^2}}\,dx = \int 4x\,dx$$

Just as a variable factor cannot be moved across the integral sign, **a variable factor outside the integral sign cannot be used to cancel a factor inside the integral sign.** See Example 12 for the correct procedure.

◆ APPLICATIONS

If P is the initial amount deposited into an account earning $100r\%$ compounded continuously, and A is the amount in the account after t years, then we saw in

Section 6-3 that A satisfies the exponential growth equation

$$\frac{dA}{dt} = rA \qquad A(0) = P$$

Now suppose that money is continuously withdrawn from this account at a rate of \$$m$ per year. Then the amount A in the account at time t must satisfy

$$\begin{pmatrix}\text{Rate of change}\\ \text{of amount } A\end{pmatrix} = \begin{pmatrix}\text{Rate of growth}\\ \text{from continuous}\\ \text{compounding}\end{pmatrix} - \begin{pmatrix}\text{Rate of}\\ \text{withdrawal}\end{pmatrix}$$

$$\frac{dA}{dt} \quad = \quad rA \quad - \quad m$$

or

$$\frac{dA}{dt} - rA = -m$$

◆ EXAMPLE 13

Continuous Compound Interest

An initial deposit of \$10,000 is made into an account earning 8% compounded continuously. Money is then continuously withdrawn at a constant rate of \$1,000 a year until the account is depleted. Find the amount in the account at any time t. When will the amount be 0? What is the total amount withdrawn from this account?

Solution

The amount A in the account at any time t must satisfy

$$\frac{dA}{dt} - 0.08A = -1{,}000 \qquad A(0) = 10{,}000$$

The integrating factor for this equation is

$$I(t) = e^{\int -0.08\,dt} = e^{-0.08t} \qquad f(t) = -0.08$$

Multiplying both sides of the differential equation by $I(t)$ and following the step-by-step procedure, we have

$$\begin{aligned} e^{-0.08t}\frac{dA}{dt} - 0.08e^{-0.08t}A &= -1{,}000e^{-0.08t} \\ (e^{-0.08t}A)' &= -1{,}000e^{-0.08t} \\ e^{-0.08t}A &= \int -1{,}000e^{-0.08t}\,dt \\ &= 12{,}500e^{-0.08t} + C \\ A &= 12{,}500 + Ce^{0.08t} \qquad \text{General solution} \end{aligned}$$

Applying the initial condition $A(0) = 10{,}000$ yields

$$\begin{aligned} A(0) = 12{,}500 + C &= 10{,}000 \\ C &= -2{,}500 \\ A(t) = 12{,}500 - 2{,}500e^{0.08t} & \qquad \text{Amount in the account at any time } t \end{aligned}$$

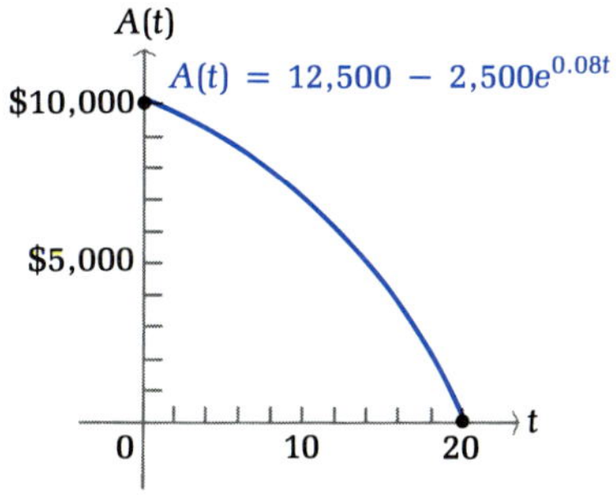

To determine when the amount in the account is 0, we solve $A(t) = 0$ for t:

$$\begin{aligned} A(t) &= 0 \\ 12{,}500 - 2{,}500e^{0.08t} &= 0 \\ 12{,}500 &= 2{,}500e^{0.08t} \\ 5 &= e^{0.08t} \\ t &= \frac{\ln 5}{0.08} \approx 20.118 \text{ years} \end{aligned}$$

Thus, the account is depleted after 20.118 years (see the figure in the margin). Since money is being withdrawn at the rate of $1,000 per year, the total amount withdrawn is

$$1{,}000(20.118) = \$20{,}118$$ ◆

PROBLEM 13 Repeat Example 13 if the account earns 5% compounded continuously. ◆

◆ EXAMPLE 14

Equilibrium Price

In economics, the supply S and the demand D for a commodity often can be considered as functions of both the price, $p(t)$, and the rate of change of the price, $p'(t)$. (Thus, S and D are ultimately functions of time t.) The **equilibrium price at time *t*** is the solution of the equation $S = D$. If $p(t)$ is the solution of this equation, then the **long-range equilibrium price** is

$$\bar{p} = \lim_{t\to\infty} p(t)$$

For example, if

$$\begin{aligned} D &= 50 - 2p(t) + 2p'(t) \\ S &= 20 + 4p(t) + 5p'(t) \end{aligned}$$

and $p(0) = 15$, then the equilibrium price at time t is the solution of the equation

$$50 - 2p(t) + 2p'(t) = 20 + 4p(t) + 5p'(t)$$

This simplifies to

$$p'(t) + 2p(t) = 10 \qquad f(t) = 2 \text{ and } g(t) = 10$$

which is a first-order linear equation with integrating factor

$$I(t) = e^{\int 2\,dt} = e^{2t}$$

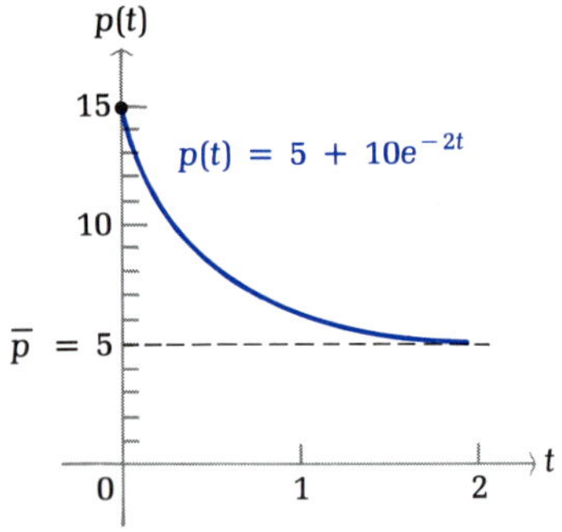

Proceeding directly to step 5, we have

$$p(t) = \frac{1}{e^{2t}} \int 10e^{2t}\, dt$$

$$= e^{-2t}(5e^{2t} + C)$$

$$p(t) = 5 + Ce^{-2t} \qquad \text{General solution}$$

$$p(0) = 5 + C = 15$$

$$C = 10$$

$$p(t) = 5 + 10e^{-2t} \qquad \text{Equilibrium price at time } t \text{ (see the figure in the margin)}$$

$$\bar{p} = \lim_{t\to\infty} (5 + 10e^{-2t}) = 5 \qquad \text{Long-range equilibrium price}$$

◆

PROBLEM 14 If $D = 70 + 2p(t) + 2p'(t)$, $S = 30 + 6p(t) + 3p'(t)$, and $p(0) = 25$, find the equilibrium price at time t and find the long-range equilibrium price. ◆

◆ EXAMPLE 15

Pollution Control

A company has a 1,000 gallon holding tank which is used to control the release of pollutants into a sewage system. Initially, the tank contains 500 gallons of water. Each gallon of water contains 2 pounds of pollutants. Additional polluted water containing 5 pounds of pollutants per gallon is pumped into the tank at the rate of 100 gallons per hour and is thoroughly mixed with the water already present in the tank. At the same time, the uniformly mixed water in the tank is released into the sewage system at a rate of 50 gallons per hour. This process continues for 5 hours. At the end of this 5 hour period, determine:

(A) The total amount of pollutants in the tank.
(B) The rate (in pounds per gallon) at which pollutants are being released into the sewage system.

Solutions

Let $p(t)$ be the total amount (in pounds) of pollutants in the tank t hours after this process begins. Since the tank initially contains 500 gallons of water and each gallon of water contains 2 pounds of pollutants,

$$p(0) = 2 \cdot 500 = 1{,}000 \qquad \text{Initial amount of pollutants in the tank}$$

Since polluted water is entering and leaving the tank at different rates and with different concentrations of pollutants, the rate of change of the amount of pollutants in the tank will depend on the rate at which pollutants enter the tank and the rate at which they leave the tank:

$$\begin{pmatrix}\text{Rate of change}\\ \text{of pollutants}\end{pmatrix} = \begin{pmatrix}\text{Rate pollutants}\\ \text{enter the tank}\end{pmatrix} - \begin{pmatrix}\text{Rate pollutants}\\ \text{leave the tank}\end{pmatrix}$$

Finding expressions for the two rates on the right side of this equation will produce a differential equation involving $p'(t)$.

Since water containing 5 pounds of pollutants per gallon is entering the tank at a constant rate of 100 gallons per hour, pollutants are entering the tank at a

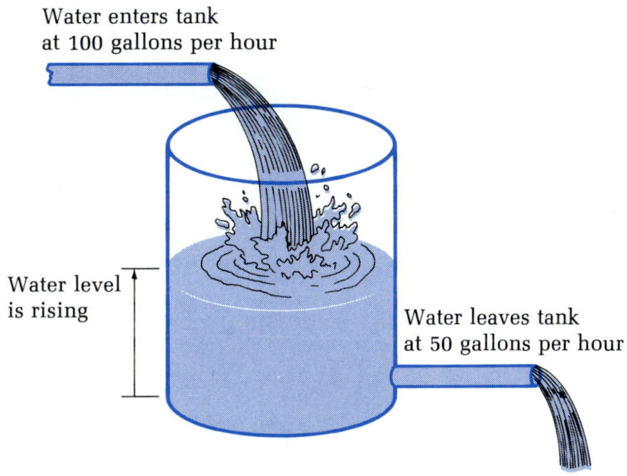

constant rate of

$$\begin{pmatrix}\text{5 pounds}\\ \text{per gallon}\end{pmatrix} \times \begin{pmatrix}\text{100 gallons}\\ \text{per hour}\end{pmatrix} = \text{500 pounds per hour}$$

How fast are the pollutants leaving the tank? Since the total amount of pollutants in the tank is increasing, the rate at which the pollutants leave the tank will depend on the amount of pollutants in the tank at time t and the amount of water in the tank at time t. Since 100 gallons of water enter the tank each hour and only 50 gallons leave each hour, the amount of water in the tank increases at the rate of 50 gallons per hour. Thus, the amount of water in the tank at time t is

$$\begin{pmatrix}\text{Initial amount}\\ \text{of water}\end{pmatrix} + \begin{pmatrix}\text{Gallons}\\ \text{per hour}\end{pmatrix} \times \begin{pmatrix}\text{Number}\\ \text{of hours}\end{pmatrix}$$

$$500 \qquad + \qquad 50t$$

The amount of pollutants in each gallon of water at any time t is

$$\frac{p(t)}{500 + 50t} \qquad \frac{\begin{pmatrix}\text{Total amount}\\ \text{of pollutants}\end{pmatrix}}{\begin{pmatrix}\text{Total amount}\\ \text{of water}\end{pmatrix}} = \text{Pollutants per gallon}$$

Since water is leaving the tank at the rate of 50 gallons per hour, the rate at which the pollutants are leaving the tank is

$$\frac{50p(t)}{500 + 50t} = \frac{p(t)}{10 + t}$$

Thus, the rate of change of $p(t)$ must satisfy

$$\begin{pmatrix}\text{Rate of change}\\ \text{of pollutants}\end{pmatrix} = \begin{pmatrix}\text{Rate pollutants}\\ \text{enter the tank}\end{pmatrix} - \begin{pmatrix}\text{Rate pollutants}\\ \text{leave the tank}\end{pmatrix}$$

$$p'(t) \qquad = \qquad 500 \qquad - \qquad \frac{p(t)}{10 + t}$$

This gives the following model for this problem:

$$p'(t) = 500 - \frac{p(t)}{10 + t} \qquad p(0) = 1{,}000$$

or

$$p'(t) + \frac{1}{10 + t}\,p(t) = 500$$ First-order linear equation with $f(t) = 1/(10 + t)$ and $g(t) = 500$

$$I(t) = e^{\int dt/(10+t)} = e^{\ln(10+t)} = 10 + t$$ Integrating factor

$$p(t) = \frac{1}{10 + t}\int 500(10 + t)\,dt$$ Proceeding directly to step 5

$$= \frac{1}{10 + t}\,[250(10 + t)^2 + C]$$

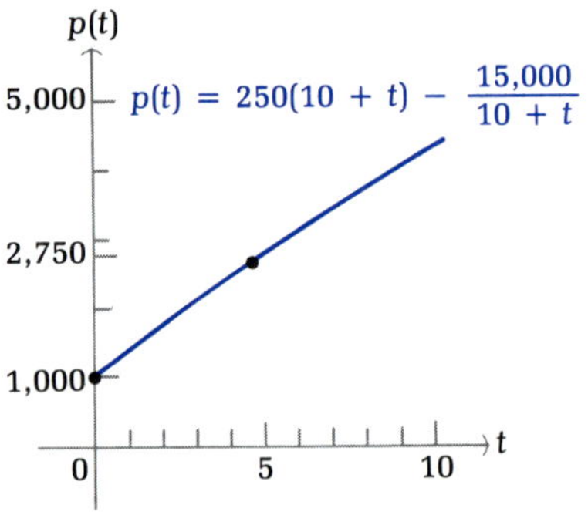

$$p(t) = 250(10 + t) + \frac{C}{10 + t}$$ General solution

$$1{,}000 = 250(10) + \frac{C}{10}$$ Initial condition: $p(0) = 1{,}000$

$$\frac{C}{10} = -1{,}500$$

$$C = -15{,}000$$

$$p(t) = 250(10 + t) - \frac{15{,}000}{10 + t}$$ Particular solution (see the figure in the margin)

(A) To find the total amount of pollutants after 5 hours, we evaluate $p(5)$:

$$p(5) = 250(15) - \frac{15{,}000}{15} = 2{,}750 \text{ pounds}$$

(B) After 5 hours, the tank contains 750 gallons of water. The rate at which pollutants are being released into the sewage system is

$$\frac{2{,}750}{750} \approx 3.67 \text{ pounds per gallon}$$ ◆

PROBLEM 15 Repeat Example 15 if water is released from the tank at the rate of 75 gallons per hour. ◆

Answers to Matched Problems

11. $I(x) = x^3$; $y = x + (C/x^3)$ 12. $I(x) = e^{x^3}$; $y = 3 + 4e^{-x^3}$
13. $A(t) = 20{,}000 - 10{,}000e^{0.05t}$; $A = 0$ when $t = (\ln 2)/0.05 \approx 13.863$ yr; total withdrawals = \$13,863
14. $p(t) = 10 + 15e^{-4t}$; $\bar{p} = 10$
15. $p(t) = 125(20 + t) - \dfrac{12{,}000{,}000}{(20 + t)^3}$ (A) 2,357 lb (B) Approx. 3.77 lb/gal

EXERCISE 9-3

A *In all problems, assume $h(x) > 0$ whenever $\ln h(x)$ is involved.*

In Problems 1–12, find the integrating factor, the general solution, and the particular solution satisfying the given initial condition.

1. $y' + 2y = 4;\quad y(0) = 1$

2. $y' - 3y = 3;\quad y(0) = -1$

3. $y' + y = e^{-2x};\quad y(0) = 3$

4. $y' - 2y = e^{3x};\quad y(0) = 2$

5. $y' - y = 2e^{x};\quad y(0) = -4$

6. $y' + 4y = 3e^{-4x};\quad y(0) = 5$

7. $y' + y = 9x^2e^{-x};\quad y(0) = 2$

8. $y' - 3y = 6\sqrt{x}\, e^{3x};\quad y(0) = -2$

9. $y' + \frac{1}{x}y = 2;\quad y(1) = 1$

10. $y' - \frac{3}{x}y = 4;\quad y(1) = 1$

11. $y' + \frac{2}{x}y = 10x^2;\quad y(2) = 8$

12. $y' - \frac{1}{x}y = \frac{9}{x^3};\quad y(2) = 3$

B *Find the integrating factor $I(x)$ for each equation, and then find the general solution.*

13. $y' + xy = 5x$

14. $y' - 2xy = 6x$

15. $y' - 2y = 4x$

16. $y' + y = x^2$

17. $y' + \frac{1}{x}y = e^{x}$

18. $y' + \frac{2}{x}y = e^{3x}$

19. $y' + \frac{1}{x}y = \ln x$

20. $y' - \frac{1}{x}y = \ln x$

C *In Problems 21–26, find the general solution two ways. First, use an integrating factor and then use separation of variables.*

21. $y' = \frac{1-y}{x}$

22. $y' = \frac{y+2}{x+1}$

23. $y' = \frac{2x+2xy}{1+x^2}$

24. $y' = \frac{4x+2xy}{4+x^2}$

25. $y' = 2x(y+1)$

26. $y' = 3x^2(y+2)$

27. Use an integrating factor to find the general solution of the unlimited growth model,

$$\frac{dy}{dt} = ky$$

[*Hint:* Remember, the antiderivative of the constant function 0 is an arbitrary constant C.]

28. Use an integrating factor to find the general solution of the limited growth model,

$$\frac{dy}{dt} = k(L - y)$$

Business & Economics

29. *Continuous compound interest.* An initial deposit of \$20,000 is made into an account that earns 4% compounded continuously. Money is then withdrawn at a constant rate of \$4,000 a year until the amount in the account is 0. Find the amount in the account at any time t. When is the amount 0? What is the total amount withdrawn from the account?

30. *Continuous compound interest.* An initial deposit of \$50,000 is made into an account that earns 10% compounded continuously. Money is then withdrawn at a constant rate of \$6,000 a year until the amount in the account is 0. Find the amount in the account at any time t. When is the amount 0? What is the total amount withdrawn from the account?

31. *Continuous compound interest.* An initial deposit of \$$P$ is made into an account that earns 5% compounded continuously. Money is then withdrawn at a constant rate of \$1,500 a year. After 10 years of continuous withdrawals, the amount in the account is 0. Find the initial deposit P.

32. *Continuous compound interest.* An initial deposit of \$$P$ is made into an account that earns 8% compounded continuously. Money is then withdrawn at a constant rate of \$3,000 a year. After 5 years of continuous withdrawals, the amount in the account is 0. Find the initial deposit P.

33. *Continuous compound interest.* An initial deposit of \$7,000 is made into an account earning 8% compounded continuously. Thereafter, money is deposited into the account at a constant rate of \$2,000 per year. Find the amount in this account at any time t. How much is in this account after 5 years?

34. *Continuous compound interest.* An initial deposit of \$10,000 is made into an account earning 6% compounded continuously. Thereafter, money is deposited into the account at a constant rate of \$6,000 per year. Find the amount in this account at any time t. How much is in this account after 10 years?

35. *Supply–demand.* The supply S and demand D for a certain commodity satisfy the equations

$$S = 35 - 2p(t) + 3p'(t) \qquad \text{and} \qquad D = 95 - 5p(t) + 2p'(t)$$

If $p(0) = 30$, find the equilibrium price at time t and the long-range equilibrium price.

36. *Supply–demand.* The supply S and demand D for a certain commodity satisfy the equations

$$S = 70 - 3p(t) + 2p'(t) \qquad \text{and} \qquad D = 100 - 5p(t) + p'(t)$$

If $p(0) = 5$, find the equilibrium price at time t and the long-range equilibrium price.

Life Sciences

37. *Pollution.* A 1,000 gallon holding tank contains 200 gallons of water. Initially, each gallon of water in the tank contains 2 pounds of pollutants. Water containing 3 pounds of pollutants per gallon enters the tank at a rate of 75 gallons per hour, and the uniformly mixed water is released from the tank at a rate of 50 gallons per hour. How many pounds of pollutants are in the tank after 2 hours? At what rate (in pounds per gallon) are the pollutants being released after 2 hours?

38. *Pollution.* Rework Problem 37 if water is entering the tank at the rate of 100 gallons per hour.

39. *Pollution.* Rework Problem 37 if water is entering the tank at the rate of 50 gallons per hour.

40. *Pollution.* Rework Problem 37 if water is entering the tank at the rate of 150 gallons per hour.

In a recent article in the College Mathematics Journal *(January 1987, 18:1), Arthur Segal proposes the following model for weight loss or gain:*

$$\frac{dw}{dt} + 0.005w = \frac{1}{3,500}C$$

where $w(t)$ is a person's weight (in pounds) after t days of consuming exactly C calories per day. Use this model to solve Problems 41–44.

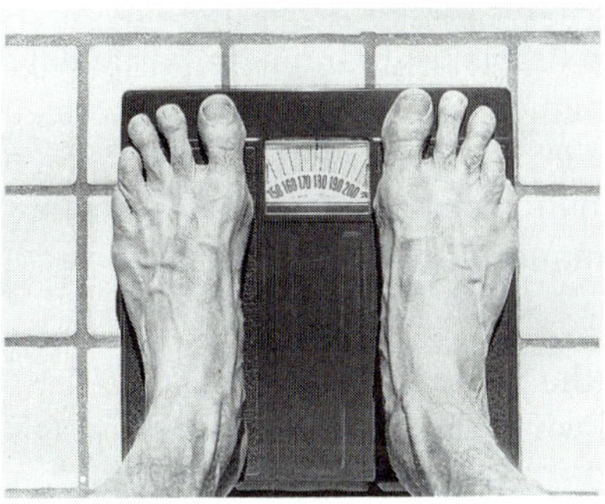

41. *Weight loss.* A person weighing 160 pounds goes on a diet of 2,100 calories per day. How much will this person weigh after 30 days on this diet? How long will it take this person to lose 10 pounds? Find $\lim_{t\to\infty} w(t)$ and interpret the results.

42. *Weight loss.* A person weighing 200 pounds goes on a diet of 2,800 calories per day. How much will this person weigh after 90 days on this diet? How long will it take this person to lose 25 pounds? Find $\lim_{t\to\infty} w(t)$ and interpret the results.

43. *Weight loss.* A person weighing 130 pounds would like to lose 5 pounds during a 30 day period. How many calories per day should this person consume to reach this goal?

44. *Weight loss.* A person weighing 175 pounds would like to lose 10 pounds during a 45 day period. How many calories per day should this person consume to reach this goal?

Social Sciences

In 1960, William K. Estes proposed the following model for measuring a student's performance in the classroom:

$$\frac{dk}{dt} + lk = \lambda l$$

where $k(t)$ is the student's knowledge after t weeks (expressed as a percentage and measured by performance on examinations), l is a constant called the coefficient of learning and representing the student's ability to learn (expressed as a per-

centage and determined by IQ or some similar general intelligence predictor), and λ is a constant representing the fraction of available time the student spends performing helpful acts that should increase knowledge of the subject (studying, going to class, and so on). Use this model to solve Problem 45.

45. *Learning theory.* Students enrolled in a beginning Spanish class are given a pretest the first day of class in order to determine their initial knowledge of the subject. The results of the pretest, the coefficient of learning, and the fraction of time spent performing helpful acts for two students in the class are given in the table. Use the Estes model to predict the knowledge of each student after 6 weeks in the class.

	SCORE ON PRETEST	COEFFICIENT OF LEARNING	FRACTION OF HELPFUL ACTS
STUDENT A	0.1 (10%)	0.8	0.9
STUDENT B	0.4 (40%)	0.8	0.7

Problems 46–50 require the use of a graphic calculator or a computer. Compute all approximations correct to one decimal place.

46. *Learning theory.* Refer to Problem 45. Graph both particular solutions in the same viewing rectangle and use approximation techniques to approximate the time when both students have the same level of knowledge.

Business & Economics

47. *Continuous compound interest.* Refer to Problem 33. Graph the particular solution and use approximation techniques to determine when the account will contain $50,000.

48. *Continuous compound interest.* Refer to Problem 34. Graph the particular solution and use approximation techniques to determine when the account will contain $200,000.

Life Sciences

49. *Pollution.* Refer to Problem 37. Graph the particular solution and use approximation techniques to determine when the tank will contain 1,000 pounds of pollutants.

50. *Pollution.* Refer to Problem 40. Graph the particular solution and use approximation techniques to determine when the tank will contain 1,500 pounds of pollutants.

SECTION 9-4 Second-Order Differential Equations

- HOMOGENEOUS SECOND-ORDER EQUATIONS
- NONHOMOGENEOUS SECOND-ORDER EQUATIONS

An equation that can be written in the form

$$a(x)y'' + b(x)y' + c(x)y = d(x) \tag{1}$$

is called a **second-order linear differential equation.** Unfortunately, there is no simple general procedure that can be used to solve every equation of this form. However, there are some special cases of (1) that are easy to solve. In this section we will consider two of these special cases: **homogeneous second-order equations** of the form

$$ay'' + by' + c = 0$$

and **nonhomogeneous second-order equations** of the form

$$ay'' + by' + c = d$$

where a, b, c, and d are constants.

◆ HOMOGENEOUS SECOND-ORDER EQUATIONS

The homogeneous first-order equation

$$ay' + by = 0$$

has a general solution of the form

$$y = Ce^{mx} \qquad m = -\frac{b}{a}$$

where C is an arbitrary constant (verify this). What should we expect the general solution of a homogeneous second-order equation to look like? It is not unreasonable to expect that the general solution of a second-order equation will involve two functions and two arbitrary constants. Theorem 2, which will form the basis for most of our work with second-order equations, states that this is the case.

THEOREM 2

General Solution of Homogeneous Second-Order Equations

If f and g are two functions that satisfy the differential equation

$$ay'' + by' + cy = 0$$

and if g is not a constant multiple of f, then the general solution of this differential equation is

$$y = C_1f(x) + C_2g(x)$$

where C_1 and C_2 are arbitrary constants.

The condition that g not be a constant multiple of f is necessary to ensure that y is in fact the general solution of this equation.

Theorem 2 provides a strategy for solving homogeneous second-order equations. First, we will find two different functions that satisfy the equation, and then we will combine these two functions to form the general solution.

◆ EXAMPLE 16 Solve: $y'' - 2y' - 3y = 0$

Solution Since the general solution of a homogeneous first-order equation involves a function of the form $y = e^{mx}$, we will try to determine whether this second-order equation has any solutions of this form. We begin by substituting $y = e^{mx}$, $y' = me^{mx}$, and $y'' = m^2e^{mx}$ in the given differential equation:

$$\begin{aligned} y'' \quad - \quad 2y' \quad - \quad 3y &= 0 \\ m^2e^{mx} - 2me^{mx} - 3e^{mx} &= 0 \\ e^{mx}(m^2 - 2m - 3) &= 0 \end{aligned}$$

Since e^{mx} is never 0, in order for this equation to be satisfied, we must have

$$\begin{aligned} m^2 - 2m - 3 &= 0 \\ (m - 3)(m + 1) &= 0 \end{aligned}$$

$$m_1 = 3 \qquad m_2 = -1$$

Thus, the functions

$$f(x) = e^{3x} \qquad \text{and} \qquad g(x) = e^{-x}$$

both satisfy the given differential equation, and e^{-x} is not a constant multiple of e^{3x}. According to Theorem 2, the general solution is

$$y = C_1e^{3x} + C_2e^{-x}$$

Check

$$\begin{aligned} y &= C_1e^{3x} + C_2e^{-x} \\ y' &= 3C_1e^{3x} - C_2e^{-x} \\ y'' &= 9C_1e^{3x} + C_2e^{-x} \end{aligned}$$

$$\begin{aligned} y'' - 2y' - 3y &= (9C_1e^{3x} + C_2e^{-x}) - 2(3C_1e^{3x} - C_2e^{-x}) - 3(C_1e^{3x} + C_2e^{-x}) \\ &= (9 - 6 - 3)C_1e^{3x} + (1 + 2 - 3)C_2e^{-x} \\ &\stackrel{\checkmark}{=} 0 \end{aligned}$$

◆

PROBLEM 16 Solve: $y'' + y' - 6y = 0$ ◆

The quadratic equation

$$m^2 - 2m - 3 = 0$$

which we obtained by substituting $y = e^{mx}$ in the differential equation

$$y'' - 2y' - 3y = 0$$

is called the *characteristic equation* for this differential equation.

The Characteristic Equation

The **characteristic equation** for

$$ay'' + by' + cy = 0$$

is

$$am^2 + bm + c = 0$$

If m is a real root of the characteristic equation, then

$$f(x) = e^{mx}$$

is a solution of the differential equation.

◆ EXAMPLE 17 Solve: $y'' - 16y = 0$

Solution

$m^2 - 16 = 0$	Characteristic equation
$(m - 4)(m + 4) = 0$	
$m_1 = 4 \qquad m_2 = -4$	Roots of characteristic equation
$f(x) = e^{4x} \qquad g(x) = e^{-4x}$	Two different solutions of the differential equation
$y = C_1e^{4x} + C_2e^{-4x}$	General solution ◆

PROBLEM 17 Solve: $y'' - y' - 2y = 0$ ◆

Since the general solution of a second-order equation involves two arbitrary constants, two conditions are required to determine a particular solution. If the value of y and the value of y' are both given for the same value of x, then both conditions are called initial conditions.

◆ EXAMPLE 18 Find the particular solution of $2y'' + 3y' - 2y = 0$ that satisfies the initial conditions $y(0) = 2$ and $y'(0) = -1$.

Solution

$2m^2 + 3m - 2 = 0$	Characteristic equation
$(2m - 1)(m + 2) = 0$	
$m_1 = \frac{1}{2} \qquad m_2 = -2$	Roots of characteristic equation
$y = C_1e^{x/2} + C_2e^{-2x}$	General solution
$y(0) = C_1 + C_2 = 2$	First initial condition
$y' = \frac{1}{2}C_1e^{x/2} - 2C_2e^{-2x}$	
$y'(0) = \frac{1}{2}C_1 - 2C_2 = -1$	Second initial condition

In order to determine the values of C_1 and C_2, we must solve the system of equations:

$$C_1 + C_2 = 2$$
$$\tfrac{1}{2}C_1 - 2C_2 = -1$$

$C_1 = 2 - C_2$ Solve the first equation for C_1.

$\tfrac{1}{2}(2 - C_2) - 2C_2 = -1$, $C_2 = \tfrac{4}{5}$ Substitute for C_1 in the second equation and solve for C_2.

$C_1 = 2 - \tfrac{4}{5} = \tfrac{6}{5}$ Substitute the value for C_2 and solve for C_1.

Thus, the particular solution satisfying the given initial conditions is

$$y = \tfrac{6}{5}e^{x/2} + \tfrac{4}{5}e^{-2x}$$

◆

PROBLEM 18 Find the particular solution of $3y'' - 7y' - 6y = 0$ that satisfies the initial conditions $y(0) = 1$ and $y'(0) = 2$. ◆

◆ EXAMPLE 19 Solve: $y'' - 2y' + y = 0$

Solution

$m^2 - 2m + 1 = 0$ Characteristic equation

$(m - 1)(m - 1) = 0$

$m = 1$ Single repeated root of characteristic equation

This characteristic equation has a single repeated root, $m = 1$, which provides us with one solution of the differential equation $f(x) = e^x$. How can we find a second solution? The answer to this question is given in Theorem 3:

THEOREM 3

Characteristic Equations with a Repeated Root

If the characteristic equation

$$am^2 + bm + c = 0$$

has a single repeated root m, then

$$g(x) = xe^{mx}$$

is a solution of the differential equation

$$ay'' + by' + cy = 0$$

Returning to the differential equation

$$y'' - 2y' + y = 0$$

whose characteristic equation has the single repeated root $m = 1$, we can now conclude that $f(x) = e^x$ and $g(x) = xe^x$ are two different solutions to this equation. Thus, the general solution is

$$y = C_1e^x + C_2xe^x$$

You should check this solution. ◆

PROBLEM 19 Solve: $y'' + 6y' + 9y = 0$ ◆

We now know how to find the solution of a homogeneous second-order equation if the characteristic equation has two distinct real roots or one repeated real root. But there is a third possibility. If the discriminant $b^2 - 4ac$ of the characteristic equation is negative, the characteristic equation has two imaginary roots. In this case, the general solution involves both exponential functions and trigonometric functions. For completeness, the imaginary roots case is included in the box (also, see Problems 27–30 in Exercise 9-4).

Homogeneous Second-Order Differential Equations

Differential Equation	Characteristic Equation
$ay'' + by' + cy = 0$	$am^2 + bm + c = 0$
Solution	**Nature of Roots**
$y = C_1e^{m_1x} + C_2e^{m_2x}$	Two distinct real roots, m_1 and m_2
$y = C_1e^{mx} + C_2xe^{mx}$	A single repeated root, m
$y = e^{px}(C_1 \cos qx + C_2 \sin qx)$	Two distinct imaginary roots, $m_1 = p + qi$ and $m_2 = p - qi$

◆ NONHOMOGENEOUS SECOND-ORDER EQUATIONS

Now we want to consider nonhomogeneous second-order equations of the form

$$ay'' + by' + cy = d$$

where d is a nonzero constant. The following example illustrates how to solve such an equation.

◆ EXAMPLE 20 Solve: $y'' - 2y' - 8y = 16$

Solution We use a three-step process to solve the equation.

Step 1. Find a constant function satisfying the nonhomogeneous equation. If $y = k$ is a constant function, then $y' = 0$ and $y'' = 0$. Substituting in the equation, we have

$$y'' - 2y' - 8y = 16$$
$$0 - 2(0) - 8k = 16$$
$$k = -2$$

Thus, $y = -2$ is a constant function satisfying the nonhomogeneous equation.

Step 2. Solve the associated homogeneous equation (the equation formed by replacing d with 0):

$$y'' - 2y' - 8y = 0 \quad \text{Associated homogeneous equation}$$

$$m^2 - 2m - 8 = 0 \quad \text{Characteristic equation}$$

$$(m - 4)(m + 2) = 0$$

$$m_1 = 4 \qquad m_2 = -2 \quad \text{Roots of characteristic equation}$$

$$y = C_1e^{4x} + C_2e^{-2x} \quad \text{Solution of homogeneous equation}$$

Step 3. Add together the solutions from steps 1 and 2 to form the general solution of the nonhomogeneous equation:

$$y = -2 + C_1e^{4x} + C_2e^{-2x} \quad \text{General solution of the nonhomogeneous equation}$$

You should check this solution. ◆

PROBLEM 20 Solve: $y'' + y' - 12y = 6$ ◆

The three-step process outlined in the solution of Example 20 can be used to solve any nonhomogeneous equation of the form

$$ay'' + by' + cy = d$$

provided $c \neq 0$. If $c = 0$, then there are no solutions of the form $y = k$. In this case, the first step consists of finding any solutions of the nonhomogeneous equation of the form $y = kx$, where k is a constant. The remaining steps are unchanged (see Problems 19–22 in Exercise 9-4).

This three-step process is a simple application of an important theorem studied in more advanced courses. This theorem states that if y_p is any particular solution of a nonhomogeneous differential equation and y_h is the general solution of the associated homogeneous equation, then $y = y_p + y_h$ is the general solution of the nonhomogeneous equation.

Answers to Matched Problems

16. $y = C_1e^{-3x} + C_2e^{2x}$
17. $y = C_1e^{2x} + C_2e^{-x}$
18. $y = \frac{8}{11}e^{3x} + \frac{3}{11}e^{-2x/3}$
19. $y = C_1e^{-3x} + C_2xe^{-3x}$
20. $y = C_1e^{-4x} + C_2e^{3x} - \frac{1}{2}$

EXERCISE 9-4

A *Find the general solution for each equation.*

1. $y'' + 3y' + 2y = 0$
2. $y'' - 6y' + 8y = 0$
3. $y'' + 2y' - 15y = 0$
4. $y'' - 25y = 0$
5. $y'' + 6y' = 0$
6. $y'' - 3y' = 0$
7. $y'' - 4y' + 4y = 0$
8. $y'' + 10y' + 25y = 0$

B *Find the particular solution for each equation that satisfies the initial conditions.*

9. $y'' - y = 0; \quad y(0) = 3; \quad y'(0) = 1$
10. $y'' - y' - 2y = 0; \quad y(0) = 1; \quad y'(0) = 2$
11. $3y'' - 10y' + 3y = 0; \quad y(0) = 1; \quad y'(0) = -1$
12. $y'' - 4y' = 0; \quad y(0) = 0; \quad y'(0) = 3$
13. $y'' + 2y' + y = 0; \quad y(0) = 2; \quad y'(0) = 4$
14. $y'' - 2y = 0; \quad y(0) = 0; \quad y'(0) = 1$

Find the general solution for each equation.

15. $y'' - 3y' - 4y = 12$
16. $y'' + 2y' + y = 5$

Find the particular solution that satisfies the initial conditions.

17. $y'' + y' - 2y = 6; \quad y(0) = 0; \quad y'(0) = 0$
18. $y'' - 3y' - 10y = 100; \quad y(0) = -10; \quad y'(0) = 0$

C *Use the following strategy to find the general solution in Problems 19–22:*

Step 1. *Find a solution of the nonhomogeneous equation of the form $y = kx$, k a constant.*

Step 2. *Find the general solution of the associated homogeneous equation.*

Step 3. *Add together the solutions from steps 1 and 2.*

19. $2y'' + y' = 1$
20. $y'' + y' = 2$
21. $y'' - 2y' = 3$
22. $2y'' - 4y' = 5$

If the solution of a second-order equation is required to satisfy two conditions of the form $y(a) = y_1$ and $y(b) = y_2$, then these conditions are referred to as boundary conditions. Find the particular solution for each equation that satisfies the given boundary conditions.

23. $y'' - 8y' + 16y = 0; \quad y(0) = 1; \quad y(1) = e^4$
24. $4y'' + 4y' + y = 0; \quad y(0) = 0; \quad y(1) = 2e^{-0.5}$
25. $2y'' - 5y' + 2y = 0; \quad y(0) = 0; \quad y(2) = e^4 - e$
26. $y'' - 3y' = 0; \quad y(0) = 3; \quad y(1) = 2 + e^3$

Problems 27–30 are optional (trigonometric functions are involved). Find the general solution for each equation.

27. $y'' + y = 0$
28. $y'' + 4y = 0$
29. $y'' - 4y' + 13y = 0$
30. $y'' + 2y' + 3y = 0$

APPLICATIONS

Business & Economics

31. Supply–demand. In earlier exercises, supply and demand were considered as functions of the price, $p(t)$, and the rate of change of the price, $p'(t)$. In studying certain markets, economists include the second derivative $p''(t)$ in the differential equation to reflect whether the rate of change of $p(t)$

is increasing or decreasing. Suppose that S and D satisfy the equations

$$S = 3 + 0.2p' - 0.05p - p'' \qquad \text{and} \qquad D = 2 + 0.8p' - 0.01p + p''$$

and that $p(0) = 75$ and $p'(0) = -15$. The equilibrium price at time t is the solution of the equation $S = D$. Find the equilibrium price at time t. Find the long-range equilibrium price.

32. *Public debt.* According to the Domar burden-of-debt model, the total public debt $D(t)$ can be modeled by the equation

$$D''(t) - \beta D(t) = 0$$

where β is the constant relative growth rate of income $(0 < \beta < 1)$.

(A) Find the general solution of this equation for any constant β.
(B) Find the particular solution satisfying $D(0) = 1$ and $D'(0) = -\sqrt{\beta}$.
(C) Find the limit of this particular solution as $t \to \infty$.

Social Sciences

33. *Learning theory.* The differential equation

$$y'' + 5y' + 4y = 8$$

is typical of the equations that occur in the study of learning curves of rats in certain types of psychological experiments. Find the particular solution satisfying $y(0) = 1$ and $y'(0) = 1$. Find the limit of this particular solution as $t \to \infty$.

SECTION 9-5 Systems of Differential Equations

- SOLUTION OF SYSTEMS OF DIFFERENTIAL EQUATIONS
- APPLICATIONS

◆ SOLUTION OF SYSTEMS OF DIFFERENTIAL EQUATIONS

In many applications we are interested in the relationship between two quantities, both of which are changing with respect to time. This often leads to a system of differential equations of the type illustrated in Example 21.

◆ EXAMPLE 21 Solve:

$$\frac{dx}{dt} = x + y \tag{1}$$

$$\frac{dy}{dt} = 4x - 2y \tag{2}$$

where $x = x(t)$ and $y = y(t)$ are functions of t.

Solution Equations (1) and (2) form a **first-order linear system of differential equations.** Systems of this form can be solved by eliminating one of the variables, in the same way systems of linear algebraic equations are solved.

$$\frac{dx}{dt} = x + y \qquad \text{Solve equation (1) for } y. \tag{1}$$

$$y = \frac{dx}{dt} - x \qquad \text{Differentiate with respect to } t. \tag{3}$$

$$\frac{dy}{dt} = \frac{d^2x}{dt^2} - \frac{dx}{dt} \tag{4}$$

$$\frac{dy}{dt} = 4x - 2y \tag{2}$$

Use equations (3) and (4) to substitute for dy/dt and y in equation (2).

$$\frac{d^2x}{dt^2} - \frac{dx}{dt} = 4x - 2\left(\frac{dx}{dt} - x\right)$$

Simplify and collect like terms to obtain a second-order homogeneous equation.

$$\frac{d^2x}{dt^2} + \frac{dx}{dt} - 6x = 0$$

Thus, we have completely eliminated y and dy/dt, so we now have a second-order differential equation involving x alone. The characteristic equation is $m^2 + m - 6 = 0$, which has roots $m_1 = -3$ and $m_2 = 2$. The general solution is

$$x = C_1e^{-3t} + C_2e^{2t}$$

To determine y, we must first compute dx/dt:

$$\frac{dx}{dt} = -3C_1e^{-3t} + 2C_2e^{2t}$$

Substituting for dx/dt and x in equation (3) gives

$$\begin{aligned} y &= -3C_1e^{-3t} + 2C_2e^{2t} - C_1e^{-3t} - C_2e^{2t} \\ &= -4C_1e^{-3t} + C_2e^{2t} \end{aligned}$$

Thus, the solution to this system is

$$x = C_1e^{-3t} + C_2e^{2t} \qquad \text{and} \qquad y = -4C_1e^{-3t} + C_2e^{2t}$$

You should check this solution. ◆

PROBLEM 21 Solve:

$$\frac{dx}{dt} = x + y$$

$$\frac{dy}{dt} = 3x - y$$

◆

Now, suppose we want to find a particular solution of the system in Example 21. Since there are two arbitrary constants in the general solution, this will require two initial conditions, one for x and one for y.

◆ EXAMPLE 22 Use the general solution in Example 21 to find the particular solution of the system

$$\frac{dx}{dt} = x + y$$

$$\frac{dy}{dt} = 4x - 2y$$

that satisfies the initial conditions $x(0) = 3$ and $y(0) = -2$.

Solution Substituting $x = 3$, $y = -2$, and $t = 0$ in the general solution from Example 21,

$$x = C_1e^{-3t} + C_2e^{2t}$$
$$y = -4C_1e^{-3t} + C_2e^{2t}$$

we obtain the following system of equations:

$$C_1 + C_2 = 3$$
$$-4C_1 + C_2 = -2$$

The solution to this system is $C_1 = 1$ and $C_2 = 2$. Thus, the particular solution we are seeking is

$$x = e^{-3t} + 2e^{2t} \qquad \text{and} \qquad y = -4e^{-3t} + 2e^{2t}$$ ◆

PROBLEM 22 Use the general solution in Problem 21 to find the particular solution of the system

$$\frac{dx}{dt} = x + y$$
$$\frac{dy}{dt} = 3x - y$$

that satisfies the initial conditions $x(0) = 5$ and $y(0) = -3$. ◆

◆ APPLICATIONS

In studying the relationship between the prices of two commodities in an interrelated market, economists often assume that the rate of change of each price is proportional to a linear combination of both prices. This leads to a system of first-order linear differential equations.

◆ EXAMPLE 23 *Interrelated Markets*

The prices p and q (in dollars) of two commodities in an interrelated market satisfy the following system of differential equations, where p and q are functions of time t (in years):

$$p' = p + 2q - 300 \qquad p(0) = 150 \tag{5}$$
$$q' = -4p - 5q + 900 \qquad q(0) = 25 \tag{6}$$

Find the solution to this system and analyze the long-term behavior of p and q, that is, the behavior of p and q as t gets larger and larger.

Solution

$$p' = p + 2q - 300 \qquad \text{Solve equation (5) for } q. \tag{5}$$
$$q = \tfrac{1}{2}p' - \tfrac{1}{2}p + 150 \qquad \text{Differentiate with respect to } t. \tag{7}$$
$$q' = \tfrac{1}{2}p'' - \tfrac{1}{2}p' \tag{8}$$

$$q' = -4p - 5q + 900 \qquad (6)$$

Use equations (7) and (8) to substitute for q' and q in equation (6).

$$\tfrac{1}{2}p'' - \tfrac{1}{2}p' = -4p - 5(\tfrac{1}{2}p' - \tfrac{1}{2}p + 150) + 900 \qquad \text{Simplify.}$$

$$p'' + 4p' + 3p = 300$$

We have eliminated q and q', but the resulting equation is nonhomogeneous. Using the method illustrated in Example 20, we find (details omitted) the general solution of this nonhomogeneous equation to be

$$p = C_1e^{-t} + C_2e^{-3t} + 100$$

Substituting for p and p' in (7) determines q:

$$q = -C_1e^{-t} - 2C_2e^{-3t} + 100$$

Applying the initial conditions $p = 150$ and $q = 25$ when $t = 0$ produces the following system of equations:

$$\begin{aligned} C_1 + C_2 &= 50 \\ C_1 + 2C_2 &= 75 \end{aligned}$$

The solution to this system is $C_1 = 25$ and $C_2 = 25$. Thus,

$$\begin{aligned} p &= 25e^{-t} + 25e^{-3t} + 100 \\ q &= -25e^{-t} - 50e^{-3t} + 100 \end{aligned}$$

To determine the behavior of p and q for large values of t, first note that

$$p' = -25e^{-t} - 75e^{-3t} < 0 \qquad \text{for all } t$$

and

$$q' = 25e^{-t} + 150e^{-3t} > 0 \qquad \text{for all } t$$

Thus, p is always decreasing and q is always increasing. Furthermore,

$$\begin{aligned} \lim_{t\to\infty} p &= \lim_{t\to\infty}(25e^{-t} + 25e^{-3t} + 100) \\ &= 0 + 0 + 100 = 100 \end{aligned}$$

and

$$\begin{aligned} \lim_{t\to\infty} q &= \lim_{t\to\infty}(-25e^{-t} - 50e^{-3t} + 100) \\ &= 0 + 0 + 100 = 100 \end{aligned}$$

The graph of $y = p(t)$ is always falling and approaches the line $y = 100$ from above, while the graph of $y = q(t)$ is always rising and approaches the line $y = 100$ from below (see the figure at the top of the next page).

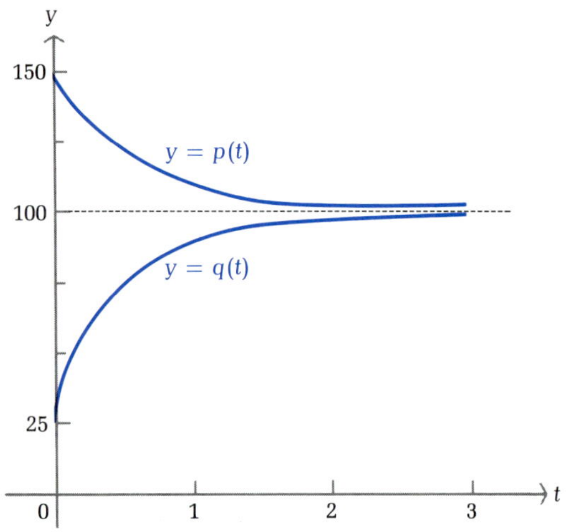

PROBLEM 23 Repeat Example 23 for the system

$$\begin{aligned} p' &= -3p + q + 225 \qquad & p(0) = 150 \\ q' &= 2p - 4q + 100 \qquad & q(0) = 35 \end{aligned}$$

In studying organs in the body, scientists often must deal with the relationship between different organs or different parts of the same organ. Each organ or part of an organ is considered to be a compartment. A given substance, such as a drug, may be able to enter or leave a compartment at certain rates and may be able to pass back and forth between adjacent compartments.

EXAMPLE 24 *Compartment Analysis*

Suppose an organ has two compartments separated by a membrane, and assume that a drug injected into compartment 1 can move back and forth between the compartments and can also leave the organ from compartment 2 (see the figure).

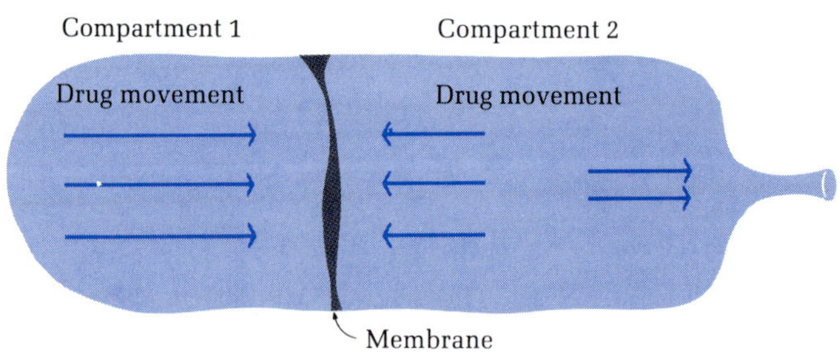

If x is the drug concentration in compartment 1 and y is the drug concentration in compartment 2 at any time t, experimental evidence suggests that x and y satisfy the following system of differential equations:

$$\frac{dx}{dt} = y - 6x$$

$$\frac{dy}{dt} = 6x - 7y$$

where x and y are functions of time t in hours. If there was no drug present in either compartment prior to the injection and if 40 units are injected into compartment 1, then we have the initial conditions $x(0) = 40$ and $y(0) = 0$. Find the solution of the system, and find the amount of the drug present in each compartment after 6 minutes.

Solution Proceeding as before (details omitted), we arrive at the second-order equation

$$\frac{d^2x}{dt^2} + 13\frac{dx}{dt} + 36x = 0$$

which has the solution

$$x = C_1e^{-9t} + C_2e^{-4t}$$

This gives

$$y = -3C_1e^{-9t} + 2C_2e^{-4t}$$

Applying the initial conditions produces the following system of equations:

$$\begin{aligned} C_1 + C_2 &= 40 \\ -3C_1 + 2C_2 &= 0 \end{aligned}$$

which has the solution $C_1 = 16$ and $C_2 = 24$. Thus, the amount of the drug present in each compartment at time t is

$$x = 16e^{-9t} + 24e^{-4t} \quad \text{Compartment 1}$$

$$y = -48e^{-9t} + 48e^{-4t} \quad \text{Compartment 2}$$

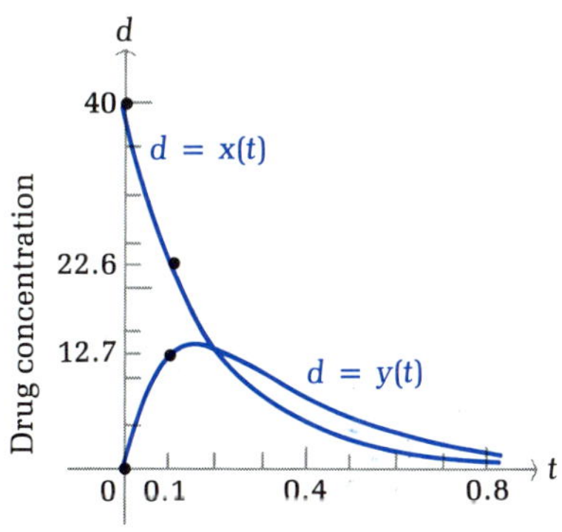

If t is measured in hours, then after 6 minutes, or 0.1 hour, the amount of the drug present in each compartment is (see the figure in the margin)

$$x(0.1) = 16e^{-0.9} + 24e^{-0.4} \approx 22.6 \text{ units}$$

$$y(0.1) = -48e^{-0.9} + 48e^{-0.4} \approx 12.7 \text{ units}$$

Of the original 40 units injected into compartment 1, 22.6 units are still present, 12.7 units are now in compartment 2, and so, 4.7 units have left the organ completely. ◆

PROBLEM 24 If the drug concentrations x and y are governed by the system of equations

$$\frac{dx}{dt} = y - 2x$$

$$\frac{dy}{dt} = 2x - 3y$$

where t is measured in hours and the initial concentration is 30 units in compartment 1 and none in compartment 2, find the concentration in each compartment after 6 minutes. After 30 minutes. After 1 hour. ◆

In Section 6-3 we saw that the population growth of many different species over short periods of time is often governed by the exponential growth law $dx/dt = kx$. If we consider the situation where two species compete for the same food supply, it is reasonable to assume that the rate of growth of each species will be affected by the size of the populations of both species. If x and y are the populations of the two species, then the following system of differential equations can be used as a model for this situation:

$$\frac{dx}{dt} = \quad ax - by$$

$$\frac{dy}{dt} = -cx + dy$$

where a, b, c, and d are positive constants. Notice that the first equation indicates that the rate of growth of x increases as x increases but decreases as y increases. The term $-by$ introduces the competition between the two species. The second equation may be interpreted in a similar manner.

◆ EXAMPLE 25

Growth of Interacting Species

The populations x and y (in thousands) of two species satisfy the system of differential equations

$$\frac{dx}{dt} = 0.125x - 0.05y$$

$$\frac{dy}{dt} = -0.05x + 0.05y$$

where t is measured in years and the initial conditions are $x(0) = 90$ and $y(0) = 130$. Find the solution to this system and analyze the long-term growth of each species.

Solution Eliminating y and dy/dt (details omitted) leads to the second-order differential equation

$$\frac{d^2x}{dt^2} - 0.175\frac{dx}{dt} + 0.00375x = 0$$

which has the solution

$$x = C_1e^{0.15t} + C_2e^{0.025t}$$

Substituting for dx/dt and x in the first equation and solving for y gives

$$y = -0.5C_1e^{0.15t} + 2C_2e^{0.025t}$$

Applying the initial conditions gives the system

$$\begin{aligned} C_1 + \ C_2 &= \ 90 \\ -\tfrac{1}{2}C_1 + 2C_2 &= 130 \end{aligned}$$

which has the solution $C_1 = 20$ and $C_2 = 70$. Thus,

$$x = 20e^{0.15t} + 70e^{0.025t} \qquad \text{and} \qquad y = -10e^{0.15t} + 140e^{0.025t}$$

For example, the population of each species after 10 years is

$$\begin{aligned} x(10) &= 20e^{1.5} + 70e^{0.25} \approx 180 \\ y(10) &= -10e^{1.5} + 140e^{0.25} \approx 135 \end{aligned}$$

We see that x has increased from 90 to 180 thousand and y has increased from 130 to 135 thousand. Since x is the sum of two increasing functions, the first species will continue to increase. On the other hand, y is the difference between two increasing functions. Is it possible that the second species will die out? That is, is y ever 0? To find out, we set the solution for y equal to 0 and try to solve for t:

$$\begin{aligned} y = -10e^{0.15t} + 140e^{0.025t} &= 0 \\ 10e^{0.15t} &= 140e^{0.025t} \\ \frac{e^{0.15t}}{e^{0.025t}} &= \frac{140}{10} \\ e^{0.125t} &= 14 \\ 0.125t &= \ln 14 \\ t &= \frac{\ln 14}{0.125} \approx 21 \end{aligned}$$

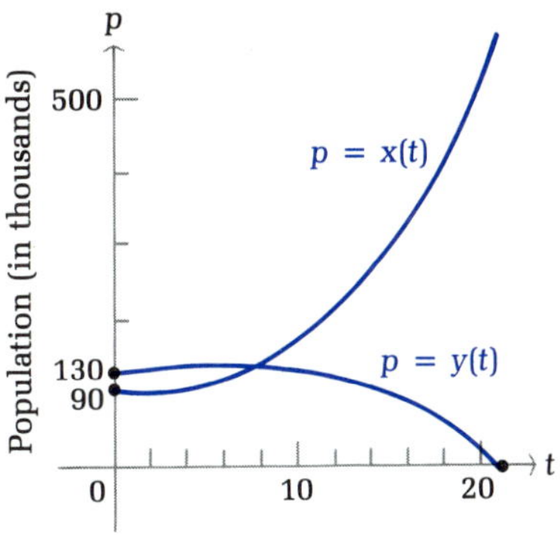

This indicates that the second species will die out after 21 years (see the figure in the margin), and since negative populations do not make any sense, the solution to this system should not be used for larger values of t. ◆

PROBLEM 25 Repeat Example 25 for the system

$$\begin{aligned} \frac{dx}{dt} &= \ \ 0.04x - 0.02y \qquad x(0) = 50 \\ \frac{dy}{dt} &= -0.01x + 0.03y \qquad y(0) = 35 \end{aligned}$$

◆

Answers to Matched Problems

21. $x = C_1e^{2t} + C_2e^{-2t}$; $y = C_1e^{2t} - 3C_2e^{-2t}$
22. $x = 3e^{2t} + 2e^{-2t}$; $y = 3e^{2t} - 6e^{-2t}$
23. $p = 20e^{-2t} + 30e^{-5t} + 100$, $q = 20e^{-2t} - 60e^{-5t} + 75$; $\lim_{t\to\infty} p = 100$, $\lim_{t\to\infty} q = 75$
24. Solution to the system: $x = 10e^{-4t} + 20e^{-t}$, $y = -20e^{-4t} + 20e^{-t}$ Concentrations are (approx.) 24.8 units in compartment 1 and 4.7 units in compartment 2 after 6 min, 13.5 units in compartment 1 and 9.4 units in compartment 2 after 30 min, 7.5 units in compartment 1 and 7.0 units in compartment 2 after 1 hr
25. $x(t) = 40e^{0.02t} + 10e^{0.05t}$, $y(t) = 40e^{0.02t} - 5e^{0.05t}$; the first species grows without bound; the second species will die out after approx. 69 yr

EXERCISE 9-5

A *Find the general solution for each system of differential equations, and find the particular solution when initial conditions are given.*

1. $x' = -x + y$; $y' = 2x$
2. $x' = 3x + y$; $y' = -2x$
3. $x' = 2x - y$; $y' = 3x - 2y$; $x(0) = 1$; $y(0) = -1$
4. $x' = 5x + 2y$; $y' = 2x + 2y$; $x(0) = 3$; $y(0) = -1$
5. $x' = 2x + y$; $y' = 2x + y$; $x(0) = 2$; $y(0) = -1$
6. $x' = x - y$; $y' = x - y$; $x(0) = 1$; $y(0) = 2$

B *Find the general solution for each system of differential equations, and find the particular solution when initial conditions are given.*

7. $x' = -2x + y$; $y' = -3x + 2y$
8. $x' = 3x - y$; $y' = 4x - y$
9. $x' = 5x - 3y + 2$; $y' = 6x - 4y + 4$
10. $x' = y - 3$; $y' = 4x - 16$; $x(0) = 3$; $y(0) = 1$

APPLICATIONS

Business & Economics

11. Interrelated markets. The prices p and q (in dollars) of two commodities in an interrelated market satisfy the following system of equations:

$$p' = -4p + q + 260 \qquad p(0) = 100$$
$$q' = -2p - q + 250 \qquad q(0) = 100$$

Find the solution to this system and analyze the long-tem behavior of p and q.

12. Interrelated markets. Repeat Problem 11 for the following system:

$$p' = -26p - 40q + 1{,}320 \qquad p(0) = 33$$
$$q' = 15p + 23q - 760 \qquad q(0) = 12$$

Life Sciences

13. Compartment analysis. An organ has two compartments separated by a membrane. The drug concentration in the first compartment is represented by $x(t)$ and in the second compartment by $y(t)$ at any time t (measured in hours). Initially, there are no traces of a drug in either compartment. Fifty units of a drug are injected into compartment 1. The flow of the drug is governed by the equations

$$x' = y - 5x$$
$$y' = x - 5y$$

Find the concentration in each compartment after 6 minutes. After 30 minutes. After 1 hour.

14. Population growth. The populations x and y (in thousands) of two competing species satisfy the system of differential equations

$$\frac{dx}{dt} = 0.09x - 0.02y \qquad x(0) = 200$$

$$\frac{dy}{dt} = -0.02x + 0.06y \qquad y(0) = 150$$

where t is measured in years. Find the solution of this system and analyze the long-term growth of each species.

In Problems 15–18, use a graphic calculator or a computer to graph the particular solution of the indicated system in the given viewing rectangle.

Business & Economics

15. Interrelated markets. System from Problem 11
t range: [0, 2]
p and q range: [75, 100]

16. Interrelated markets. System from Problem 12
t range: [0, 3]
p and q range: [0, 40]

Life Sciences

17. Compartment analysis. System from Problem 13
t range: [0, 1]
x and y range: [0, 50]

18. Population growth. System from Problem 14
t range: [0, 30]
x and y range: [0, 2,000]

SECTION 9-6

Chapter Review

Important Terms and Symbols

9-1 Basic Concepts. Differential equation; order; solution; general solution; family of solutions; particular solution; initial condition; implicit solution; dynamically stable price; equilibrium price

9-2 *Separation of Variables.* Separation of variables; singular solution; growth laws

$$\int f(y)\,dy = \int g(x)\,dx$$

9-3 *First-Order Linear Differential Equations.* First-order linear differential equation; standard form; integrating factor; general solution; equilibrium price at time t; long-range equilibrium price

$$y' + f(x)y = g(x); \quad I(x) = e^{\int f(x)\,dx}; \quad y = \frac{1}{I(x)} \int I(x)g(x)\,dx$$

9-4 *Second-Order Differential Equations.* Second-order linear differential equation; homogeneous second-order equation; nonhomogeneous second-order equation; characteristic equation; distinct roots; repeated root

$$ay'' + by' + c = 0; \quad ay'' + by' + c = d$$

9-5 *Systems of Differential Equations.* First-order linear system of differential equations; interrelated markets; compartment analysis; interacting species

EXERCISE 9-6

Chapter Review

Work through all the problems in this chapter review and check your answers in the back of the book. (Answers to all review problems are there.) Where weaknesses show up, review appropriate sections in the text.

A

Show that the function y is the general solution of the differential equation. On the same set of axes, graph the particular solutions obtained by letting $C = -2$, -1, 0, 1, and 2.

1. $y = C\sqrt{x}; \quad 2xy' = y$

2. $y = 1 + Ce^{-x}; \quad y' + y = 1$

Find the general solution.

3. $y' = -\dfrac{4y}{x}$

4. $y' = -\dfrac{4y}{x} + x$

5. $y' = 3x^2y^2$

6. $y' = 2y - e^x$

7. $y' = \dfrac{5}{x}y + x^6$

8. $y' = \dfrac{3 + y}{2 + x}; \quad x > -2$

9. $y'' - 4y' - 21y = 0$

10. $y'' + 12y' + 36y = 0$

B

Find the particular solution that satisfies the given condition(s).

11. $y' = 10 - y; \quad y(0) = 0$

12. $y' + y = x; \quad y(0) = 0$

13. $y' = 2ye^{-x}; \quad y(0) = 1$

14. $y' = \dfrac{2x - y}{x + 4}; \quad y(0) = 1$

15. $y' = \frac{x}{y + 4}$; $y(0) = 0$

16. $y' + \frac{2}{x} y = \ln x$; $y(1) = 2$

17. $yy' = \frac{x(1 + y^2)}{1 + x^2}$; $y(0) = 1$

18. $y' + 2xy = 2e^{-x^2}$; $y(0) = 1$

19. $y'' + 4y' = 0$; $y(0) = 1$; $y'(0) = -2$

20. $y'' - 16y = 16$; $y(0) = 1$; $y'(0) = 0$

C *Find the particular solution that satisfies the initial conditions.*

21. $x' = -2x + y$
$y' = -4x + 2y$
$x(0) = 1$; $y(0) = 1$

22. $x' = -2x + 2y + 6$
$y' = 4x - 8$
$x(0) = 2$; $y(0) = 2$

APPLICATIONS

Business & Economics

23. *Depreciation.* A refrigerator costs \$500 when it is new. The value of the refrigerator depreciates to \$25 over a 20 year period. If the rate of change of the value is proportional to the value, find the value of the refrigerator 5 years after it was purchased.

24. *Sales growth.* A new company has sales of \$50,000 during the first year of operation. The rate of growth of the annual sales s is proportional to the difference between s and an upper limit of \$200,000. Assuming $s = 0$ at $t = 0$, how long will it take for the annual sales to reach \$150,000?

25. *Supply–demand.* The supply S and demand D for a certain commodity satisfy the equations

$$S = 100 + p + p' \quad \text{and} \quad D = 200 - p' - p$$

If $p = 75$ when $t = 0$, find the equilibrium price at time t, and find the long-range equilibrium price.

26. *Continuous compound interest.* An initial deposit of \$60,000 is made into an account that earns 5% compounded continuously. Money is then withdrawn at a constant rate of \$5,000 a year until the amount in the account is 0. Find the amount in the account at any time t. When is the amount 0? What is the total amount withdrawn from the account?

27. *Interrelated markets.* The prices p and q (in dollars) of two commodities of an interrelated market satisfy the following system of equations:

$$p' = -2p + q + 125 \qquad p(0) = 50$$
$$q' = -2p - 5q + 575 \qquad q(0) = 150$$

Find the solution of this system and analyze the long-term behavior of p and q.

Life Sciences

28. *Crop yield.* The yield per acre, $y(t)$, of a corn crop satisfies the equation

$$\frac{dy}{dt} = 100 + e^{-t} - y$$

If $y(0) = 0$, find y at any time t. (The yield is in bushels per acre.)

29. *Pollution.* A 1,000 gallon holding tank contains 100 gallons of unpolluted water. Water containing 2 pounds of pollutant per gallon is pumped into the tank at the rate of 100 gallons per hour. The uniformly mixed water is released from the tank at 50 gallons per hour. Find the total amount of pollutants in the tank after 2 hours.

30. *Population growth.* The populations x and y (in thousands) of two competing species satisfy the following system of differential equations:

$$\frac{dx}{dt} = 0.03x - 0.01y \qquad x(0) = 75$$

$$\frac{dy}{dt} = -0.02x + 0.02y \qquad y(0) = 75$$

where t is measured in years. Find the solution of this system. Analyze the long-term behavior of each species.

Social Sciences

31. *Rumor spread.* A single individual starts a rumor in a community of 200 people. The rumor spreads at a rate proportional to the number of people who have not yet heard the rumor. After 2 days, 10 people have heard the rumor.

(A) How many people will have heard the rumor after 5 days?
(B) How long will it take for the rumor to spread to 100 people?

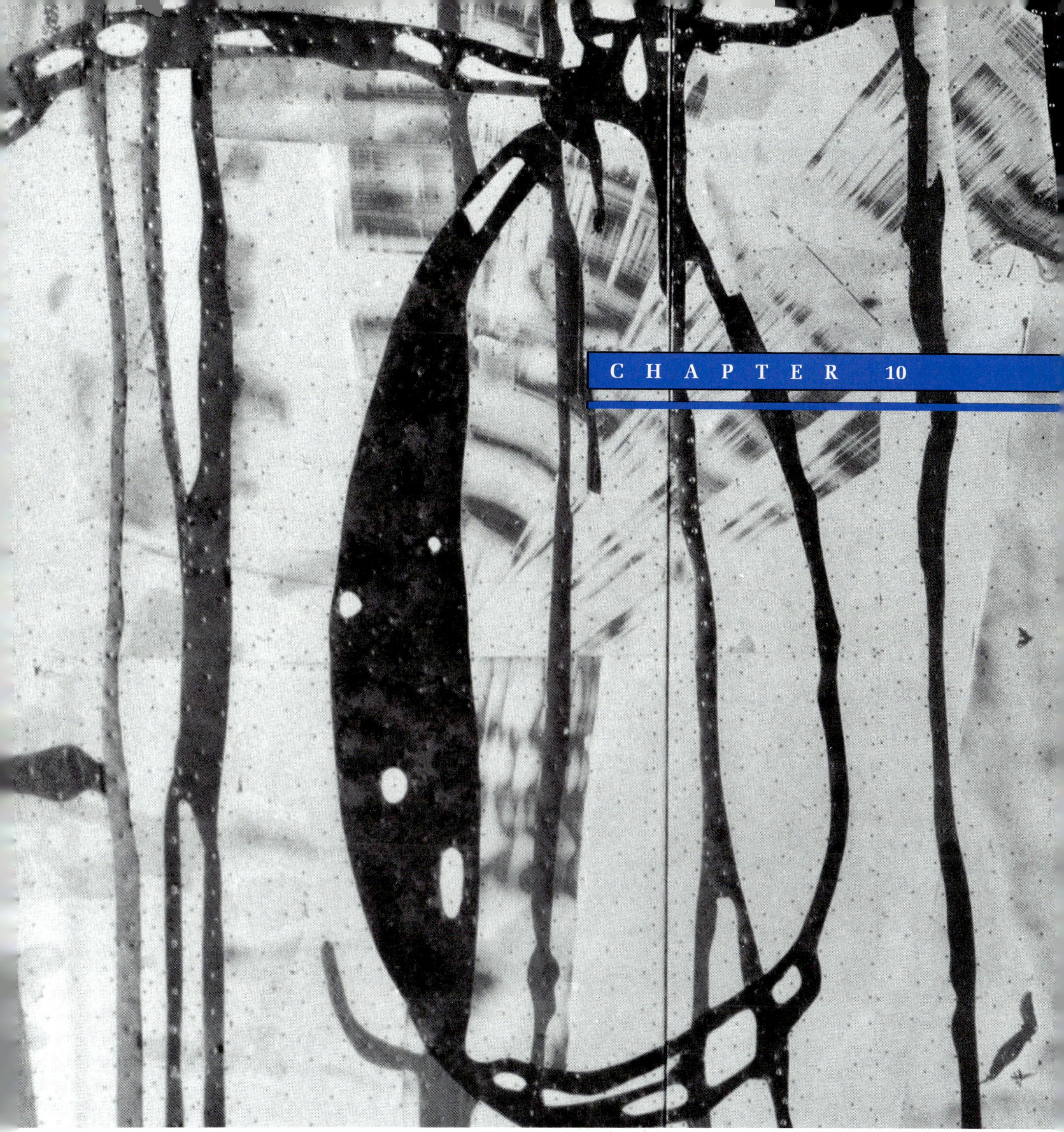

CHAPTER 10

Taylor Polynomials and Infinite Series

CHAPTER 10

Contents

The circuits inside a calculator are capable only of performing the basic operations of addition, subtraction, multiplication, and division. Yet, many calculators have keys that allow you to evaluate functions such as e^x, ln x, and sin x. How is this done? In most cases, the values of these functions are *approximated* by using a carefully selected polynomial, and, of course, polynomials can be evaluated by using the basic arithmetic operations. Thus, the approximating polynomials give the calculator the capability of evaluating nonpolynomial functions. In this chapter we will study one type of approximating polynomial, called a *Taylor polynomial* after the English mathematician Brook Taylor (1685–1731). For a given function f, we will determine the coefficients of the Taylor polynomial, the values of x where the Taylor polynomial approximates $f(x)$, and the accuracy of this approximation. We will also consider various applications involving Taylor polynomials, including approximation of definite integrals.

SECTION 10-1 Taylor Polynomials

- HIGHER-ORDER DERIVATIVES
- APPROXIMATING e^x WITH POLYNOMIALS
- TAYLOR POLYNOMIALS AT 0
- TAYLOR POLYNOMIALS AT a
- APPLICATION

HIGHER-ORDER DERIVATIVES

Up to this point, we have considered only the first and second derivatives of a function. In the work that follows, we will need to find higher-order derivatives. For example, if we start with the function f defined by

$$f(x) = x^{-1}$$

then the first derivative of f is

$$f'(x) = -x^{-2}$$

and the second derivative of f is

$$f''(x) = D_x f'(x) = D_x(-x^{-2}) = 2x^{-3}$$

Additional higher-order derivatives are found by successive differentiation. Thus, the third derivative is

$$f^{(3)}(x) = D_x f''(x) = D_x(2x^{-3}) = -6x^{-4}$$

the fourth derivative is

$$f^{(4)}(x) = D_x f^{(3)}(x) = D_x(-6x^{-4}) = 24x^{-5}$$

and so on.

In general, the symbol $f^{(n)}$ is used to represent the **nth derivative of the function f.** When stating formulas involving a function and its higher-order derivatives, it is convenient to let $f^{(0)}$ represent the function f (that is, the zeroth derivative of a function is just the function itself).

The order of the derivative must be enclosed in parentheses. In most contexts, $f^n(x)$ is interpreted to mean the nth power of f, not the nth derivative. Thus, $f^2(x) = [f(x)]^2 = f(x)f(x)$, while $f^{(2)}(x) = f''(x)$.

Finding a particular higher-order derivative is a routine calculation. However, finding a formula for the nth derivative for arbitrary n requires careful observation of the patterns that develop as each successive derivative is found. Study the next example carefully. Many problems in this chapter will involve similar concepts.

◆ EXAMPLE 1 Find the nth derivative of: $f(x) = \dfrac{1}{1+x} = (1+x)^{-1}$

Solution We begin by finding the first four derivatives of f:

$$\begin{aligned}
f(x) &= (1+x)^{-1} \\
f'(x) &= (-1)(1+x)^{-2} \\
f''(x) &= (-1)(-2)(1+x)^{-3} \\
f^{(3)}(x) &= (-1)(-2)(-3)(1+x)^{-4} \\
f^{(4)}(x) &= (-1)(-2)(-3)(-4)(1+x)^{-5}
\end{aligned}$$

Notice that we did not multiply out the coefficient of $(1+x)^{-k}$ in each derivative. Our objective is to look for a pattern in the form of each derivative. Multiplying out these coefficients would tend to obscure any pattern that is developing. Instead, we observe that each coefficient is a product of successive negative integers that can be written as follows:

$$\begin{aligned}
(-1) &= (-1)^1(1) \\
(-1)(-2) &= (-1)^2(1)(2) \\
(-1)(-2)(-3) &= (-1)^3(1)(2)(3) \\
(-1)(-2)(-3)(-4) &= (-1)^4(1)(2)(3)(4)
\end{aligned}$$

Next we note that the product of natural numbers in each expression on the right can be written in terms of a factorial:*

$$(-1) = (-1)^1 1! \qquad 1! = 1$$
$$(-1)(-2) = (-1)^2 2! \qquad 2! = 1 \cdot 2$$
$$(-1)(-2)(-3) = (-1)^3 3! \qquad 3! = 1 \cdot 2 \cdot 3$$
$$(-1)(-2)(-3)(-4) = (-1)^4 4! \qquad 4! = 1 \cdot 2 \cdot 3 \cdot 4$$

Substituting these last expressions in the derivatives of f, we have

$$f'(x) = (-1)^1 1!(1+x)^{-2} \qquad n = 1$$
$$f''(x) = (-1)^2 2!(1+x)^{-3} \qquad n = 2$$
$$f^{(3)}(x) = (-1)^3 3!(1+x)^{-4} \qquad n = 3$$
$$f^{(4)}(x) = (-1)^4 4!(1+x)^{-5} \qquad n = 4$$

This suggests that

$$f^{(n)}(x) = (-1)^n n!(1+x)^{-(n+1)} \qquad \text{Arbitrary } n$$ ◆

PROBLEM 1 Find the nth derivative of: $f(x) = \ln x$ ◆

◆ APPROXIMATING e^x WITH POLYNOMIALS

We have already seen that the irrational number e and the exponential function e^x play important roles in many applications, including continuous compound interest, population growth, and exponential decay, to name a few. Now, given the function

$$f(x) = e^x$$

we would like to construct a polynomial function p whose values are close to the values of f, at least for some values of x. If we are successful, then we can use the values of p (which are easily computed) to approximate the values of f. We begin by trying to approximate f for values of x near 0 with a first-degree polynomial of the form

$$p_1(x) = a_0 + a_1 x \tag{1}$$

We want to place conditions on p_1 that will enable us to determine the unknown coefficients a_0 and a_1. Since we want to approximate f for values near 0, it is reasonable to require that f and p_1 agree at 0. Thus,

$$a_0 = p_1(0) = f(0) = e^0 = 1$$

---

* For n a natural number, $n! = n(n-1)! = n(n-1)(n-2) \cdot \cdots \cdot 2 \cdot 1$ and $0! = 1$. (See Appendix A-12.)

This determines the value of a_0. To determine the value of a_1, we require that both functions have the same slope at 0. Since $p_1'(x) = a_1$ and $f'(x) = e^x$, this implies that

$$a_1 = p_1'(0) = f'(0) = e^0 = 1$$

Thus, after substituting $a_0 = 1$ and $a_1 = 1$ into (1), we obtain

$$p_1(x) = 1 + x$$

which is a first-degree polynomial satisfying

$$p_1(0) = f(0) \qquad \text{and} \qquad p_1'(0) = f'(0)$$

How well does $1 + x$ approximate e^x? Examining the graph in Figure 1, it appears that $1 + x$ is a good approximation to e^x for x very close to 0. However, as x moves away from 0, in either direction, the distance between the values of $1 + x$ and e^x increases and the accuracy of the approximation decreases.

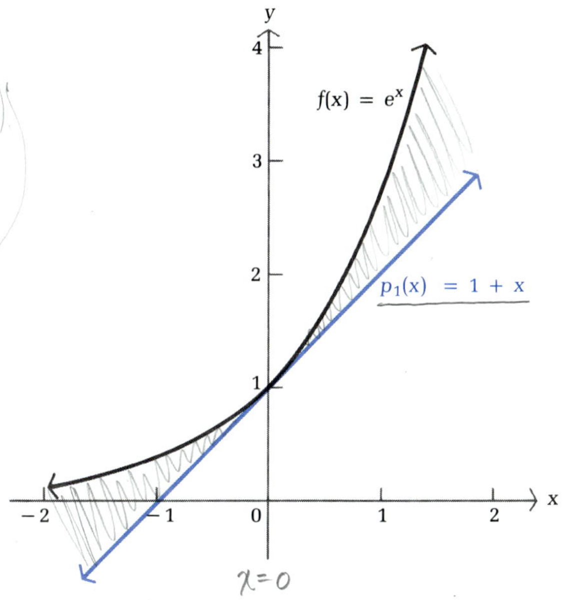

FIGURE 1

Now we will try to improve this approximation by using a second-degree polynomial of the form

$$p_2(x) = a_0 + a_1x + a_2x^2 \tag{2}$$

We need three conditions to determine the coefficients a_0, a_1, and a_2. We still require that $p_2(0) = f(0)$ and $p_2'(0) = f'(0)$ and add the condition that $p_2''(0) = f''(0)$. This ensures that the graphs of p_2 and f have the same concavity at $x = 0$.

Proceeding as before, we compute the first and second derivatives of p_2 and f and apply these conditions:

$$\begin{aligned} p_2(x) &= a_0 + a_1x + a_2x^2 \qquad & f(x) &= e^x \\ p_2'(x) &= a_1 + 2a_2x & f'(x) &= e^x \\ p_2''(x) &= 2a_2 & f''(x) &= e^x \end{aligned}$$

Thus,

$$\begin{aligned} a_0 &= p_2(0) = f(0) = e^0 = 1 \quad & &\text{implies} \quad & a_0 &= 1 \\ a_1 &= p_2'(0) = f'(0) = e^0 = 1 & &\text{implies} & a_1 &= 1 \\ 2a_2 &= p_2''(0) = f''(0) = e^0 = 1 & &\text{implies} & a_2 &= \tfrac{1}{2} \end{aligned}$$

and, substituting in (2), we obtain

$$p_2(x) = 1 + x + \tfrac{1}{2}x^2$$

The graph is shown in Figure 2.

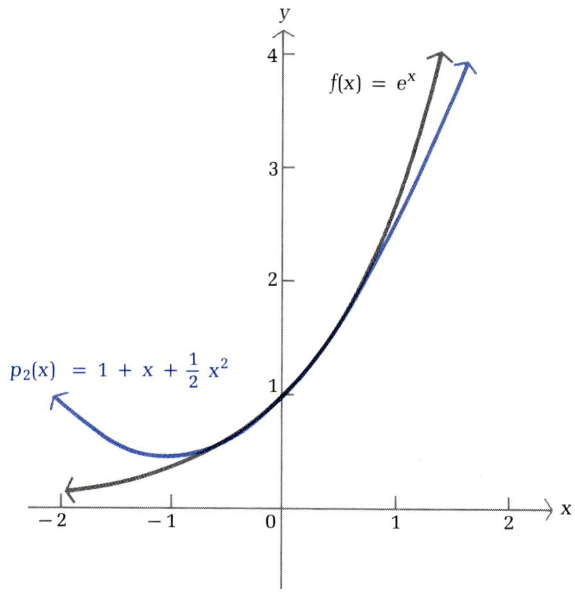

FIGURE 2

Comparing Figures 1 and 2, we see that for values of x near 0, the graph of $p_2(x) = 1 + x + \frac{1}{2}x^2$ is closer to the graph of $f(x) = e^x$ than was the graph of $p_1(x) = 1 + x$. Thus, the polynomial $1 + x + \frac{1}{2}x^2$ can be used to approximate e^x over a larger range of x values than the polynomial $1 + x$.

It seems reasonable to assume that a third-degree polynomial would yield a still better approximation. Using the notation for higher-order derivatives discussed earlier in this section, we can state the required condition as:

Find

$$p_3(x) = a_0 + a_1x + a_2x^2 + a_3x^3 \tag{3}$$

Satisfying

$$p_3^{(k)}(0) = f^{(k)}(0) \qquad k = 0, 1, 2, 3$$

As before, we obtain the value of the additional coefficient a_3 by adding the requirement that

$$p_3^{(3)}(0) = f^{(3)}(0)$$

Thus,

$$\begin{array}{llll} p_3(x) = a_0 + a_1x + a_2x^2 + a_3x^3 & & & f(x) = e^x \\ p_3'(x) = a_1 + 2a_2x + 3a_3x^2 & & & f'(x) = e^x \\ p_3''(x) = 2a_2 + 6a_3x & & & f''(x) = e^x \\ p_3^{(3)}(x) = 6a_3 & & & f^{(3)}(x) = e^x \end{array}$$

Applying the conditions $p_3^{(k)}(0) = f^{(k)}(0)$, we have

$$\begin{array}{lll} a_0 = p_3(0) = f(0) = e^0 = 1 & \text{implies} & a_0 = 1 \\ a_1 = p_3'(0) = f'(0) = e^0 = 1 & \text{implies} & a_1 = 1 \\ 2a_2 = p_3''(0) = f''(0) = e^0 = 1 & \text{implies} & a_2 = \frac{1}{2} \\ 6a_3 = p_3^{(3)}(0) = f^{(3)}(0) = e^0 = 1 & \text{implies} & a_3 = \frac{1}{6} \end{array}$$

Substituting in (3), we obtain

$$p_3(x) = 1 + x + \tfrac{1}{2}x^2 + \tfrac{1}{6}x^3$$

Figure 3 (on page 650) indicates that the approximations provided by p_3 are an improvement over those provided by p_2 and p_1.

In order to compare all three approximations, Table 1 lists the values of p_1, p_2, p_3, and e^x at selected values of x and Table 2 lists the absolute value of the

TABLE 1

$p_1(x) = 1 + x$, $p_2(x) = 1 + x + \frac{1}{2}x^2$, $p_3(x) = 1 + x + \frac{1}{2}x^2 + \frac{1}{6}x^3$

x	$p_1(x)$	$p_2(x)$	$p_3(x)$	e^x
−0.2	0.8	0.820	0.818 667	0.818 731
−0.1	0.9	0.905	0.904 833	0.904 837
0	1	1	1	1
0.1	1.1	1.105	1.105 167	1.105 171
0.2	1.2	1.220	1.221 333	1.221 403

TABLE 2

x	$\lvert p_1(x) - e^x \rvert$	$\lvert p_2(x) - e^x \rvert$	$\lvert p_3(x) - e^x \rvert$
−0.2	0.018 731	0.001 269	0.000 064
−0.1	0.004 837	0.000 163	0.000 004
0	0	0	0
0.1	0.005 171	0.000 171	0.000 004
0.2	0.021 403	0.001 403	0.000 069

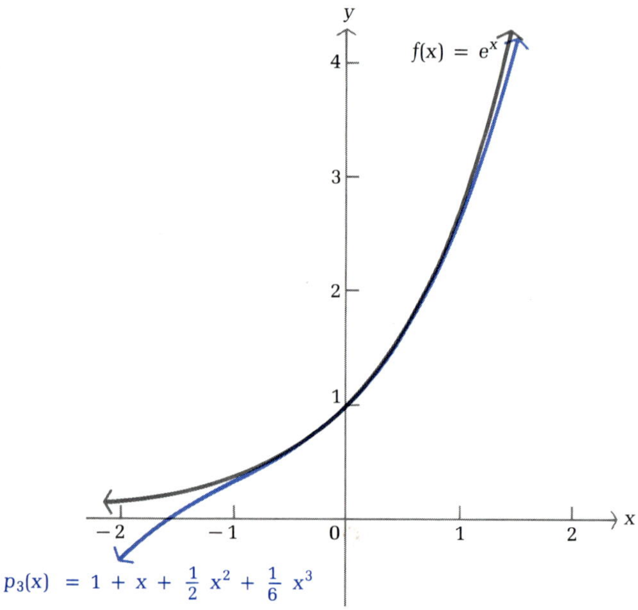

FIGURE 3

difference between these polynomials and e^x for the same x values. The values of e^x were obtained by using a calculator and are rounded to six decimal places. Comparing the columns in Table 2, we see that increasing the degree of the approximating polynomial decreases the difference between the polynomial and e^x. Thus, p_2 provides a better approximation than p_1, and p_3 provides a better approximation than p_2. We will have more to say about the accuracy of these approximations later in this chapter.

◆ TAYLOR POLYNOMIALS AT 0

The process we have used to determine p_1, p_2, and p_3 can be continued. Given any positive integer n, we define

$$p_n(x) = a_0 + a_1x + a_2x^2 + \cdots + a_nx^n$$

and require that

$$p_n^{(k)}(0) = f^{(k)}(0) \qquad k = 0, 1, 2, \ldots, n$$

The polynomial p_n is called a *Taylor polynomial*. Before determining p_n for $f(x) = e^x$, it will be convenient to make some general statements concerning the relationship between a_k, $p_n^{(k)}(0)$, and $f^{(k)}(0)$ for an arbitrary function f. First, $p_n(x)$ is differentiated n times to obtain the following relationships:

$$\begin{aligned}
p_n(x) &= a_0 + a_1x + a_2x^2 + a_3x^3 + \cdots + a_nx^n \\
p_n'(x) &= a_1 + 2a_2x + 3a_3x^2 + \cdots + na_nx^{n-1} \\
p_n''(x) &= 2a_2 + (3)(2)a_3x + \cdots + n(n-1)a_nx^{n-2} \\
p_n^{(3)}(x) &= (3)(2)a_3 + \cdots + n(n-1)(n-2)a_nx^{n-3} \\
&\vdots \\
p_n^{(n)}(x) &= n(n-1)(n-2) \cdot \cdots \cdot (1)a_n = n!a_n
\end{aligned}$$

Evaluating each derivative at 0 and applying the requirement that $p_n^{(k)}(0) = f^{(k)}(0)$ leads to the following equations:

$$\begin{aligned}
a_0 &= p_n(0) = f(0) \\
a_1 &= p_n'(0) = f'(0) \\
2a_2 &= p_n''(0) = f''(0) \\
(3)(2)a_3 &= p_n^{(3)}(0) = f^{(3)}(0) \\
&\vdots \\
n!a_n &= p_n^{(n)}(0) = f^{(n)}(0)
\end{aligned}$$

Solving each equation for a_k, we have

$$a_k = \frac{p_n^{(k)}(0)}{k!} = \frac{f^{(k)}(0)}{k!}$$

This relationship enables us to state the general definition of a Taylor polynomial.

Taylor Polynomial at 0*

If f has n derivatives at 0, then the **nth-degree Taylor polynomial for *f* at 0** is

$$p_n(x) = a_0 + a_1x + a_2x^2 + \cdots + a_nx^n = \sum_{k=0}^{n} a_kx^k$$

where

$$p_n^{(k)}(0) = f^{(k)}(0) \quad \text{and} \quad a_k = \frac{f^{(k)}(0)}{k!} \qquad k = 0, 1, 2, \ldots, n$$

* Taylor polynomials at 0 are also often referred to as *Maclaurin polynomials*, but we will not use this terminology in this book.

This result can be stated in a form that is more readily remembered, as follows:

Taylor Polynomial at 0: Concise Form

The Taylor polynomial of degree n for f at 0 is

$$p_n(x) = f(0) + f'(0)x + \frac{f''(0)}{2!}x^2 + \cdots + \frac{f^{(n)}(0)}{n!}x^n = \sum_{k=0}^{n} \frac{f^{(k)}(0)}{k!}x^k$$

provided f has n derivatives at 0.

In each of the preceding definitions, notice that we have used both the **expanded notation**

$$\begin{aligned} p_n(x) &= a_0 + a_1x + a_2x^2 + \cdots + a_nx^n \\ &= f(0) + f'(0)x + \frac{f''(0)}{2!}x^2 + \cdots + \frac{f^{(n)}(0)}{n!}x^n \end{aligned}$$

and the more compact **summation notation**

$$p_n(x) = \sum_{k=0}^{n} a_kx^k = \sum_{k=0}^{n} \frac{f^{(k)}(0)}{k!}x^k$$

to represent a finite sum. Recall that we used the summation notation extensively in Chapters 6 and 7. In this chapter we will place more emphasis on the expanded notation, since it is usually easier to visualize a polynomial written in this notation, but we will continue to include the summation notation where appropriate.

Returning to our original function $f(x) = e^x$, it is now an easy matter to find the nth-degree Taylor polynomial for this function. Since $D_x\, e^x = e^x$, it follows that

$$f^{(k)}(x) = e^x \qquad f^{(k)}(0) = e^0 = 1 \qquad a_k = \frac{f^{(k)}(0)}{k!} = \frac{1}{k!}$$

for all values of k. Thus, for any n, the nth-degree Taylor polynomial for e^x is

$$p_n(x) = 1 + x + \frac{1}{2!}x^2 + \frac{1}{3!}x^3 + \cdots + \frac{1}{n!}x^n = \sum_{k=0}^{n} \frac{1}{k!}x^k$$

◆ EXAMPLE 2 Find the third-degree Taylor polynomial at 0 for $f(x) = \sqrt{x+4}$. Use p_3 to approximate $\sqrt{5}$.

Solution Step 1. Find the derivatives:

$$f(x) = (x + 4)^{1/2}$$
$$f'(x) = \tfrac{1}{2}(x + 4)^{-1/2}$$
$$f''(x) = -\tfrac{1}{4}(x + 4)^{-3/2}$$
$$f^{(3)}(x) = \tfrac{3}{8}(x + 4)^{-5/2}$$

Step 2. Evaluate the derivatives at 0:

$$f(0) = 4^{1/2} = 2$$
$$f'(0) = \tfrac{1}{2}(4^{-1/2}) = \tfrac{1}{4}$$
$$f''(0) = -\tfrac{1}{4}(4^{-3/2}) = -\tfrac{1}{32}$$
$$f^{(3)}(0) = \tfrac{3}{8}(4^{-5/2}) = \tfrac{3}{256}$$

Step 3. Find the coefficients of the Taylor polynomial:

$$a_0 = \frac{f(0)}{0!} = f(0) = 2$$
$$a_1 = \frac{f'(0)}{1!} = f'(0) = \frac{1}{4}$$
$$a_2 = \frac{f''(0)}{2!} = \frac{-\frac{1}{32}}{2} = -\frac{1}{64}$$
$$a_3 = \frac{f^{(3)}(0)}{3!} = \frac{\frac{3}{256}}{6} = \frac{1}{512}$$

Step 4. Write down the Taylor polynomial:

$$p_3(x) = 2 + \tfrac{1}{4}x - \tfrac{1}{64}x^2 + \tfrac{1}{512}x^3 \qquad \text{Taylor polynomial}$$

To use p_3 to approximate $\sqrt{5}$, we must first determine the appropriate value of x:

$$f(x) = \sqrt{x + 4} = \sqrt{5} \qquad \text{Square both sides.}$$
$$x + 4 = 5 \qquad \text{Solve for } x.$$
$$x = 1 \qquad \text{As can easily be checked.}$$

Thus,

$$\sqrt{5} = f(1) \approx p_3(1) = 2 + \tfrac{1}{4} - \tfrac{1}{64} + \tfrac{1}{512} \approx 2.236\ 328\ 1$$

[*Note:* The value obtained by using a calculator is $\sqrt{5} \approx 2.236\ 068$.] ◆

PROBLEM 2 Find the second-degree Taylor polynomial at 0 for $f(x) = \sqrt{x + 9}$. Use $p_2(x)$ to approximate $\sqrt{10}$. ◆

Many of the problems in this section involve approximating the values of functions that also can be evaluated with the use of a calculator. In such problems you should compare the Taylor polynomial approximation with the calculator value (which is also an approximation). This will give you some indication of the accuracy of the Taylor polynomial approximation. Later in this chapter we will discuss methods for determining the accuracy of any Taylor polynomial approximation.

◆ EXAMPLE 3 Find the nth-degree Taylor polynomial at 0 for $f(x) = e^{2x}$.

Solution

Step 1	Step 2	Step 3
$f(x) = e^{2x}$	$f(0) = 1$	$a_0 = f(0) = 1$
$f'(x) = 2e^{2x}$	$f'(0) = 2$	$a_1 = f'(0) = 2$
$f''(x) = 2^2e^{2x}$	$f''(0) = 2^2$	$a_2 = \frac{f''(0)}{2!} = \frac{2^2}{2!}$
$f^{(3)}(x) = 2^3e^{2x}$	$f^{(3)}(0) = 2^3$	$a_3 = \frac{f^{(3)}(0)}{3!} = \frac{2^3}{3!}$
$\vdots$	$\vdots$	$\vdots$
$f^{(n)}(x) = 2^ne^{2x}$	$f^{(n)}(0) = 2^n$	$a_n = \frac{f^{(n)}(0)}{n!} = \frac{2^n}{n!}$

Step 4. Write down the Taylor polynomial:

$$p_n(x) = 1 + 2x + \frac{2^2}{2!}x^2 + \frac{2^3}{3!}x^3 + \cdots + \frac{2^n}{n!}x^n = \sum_{k=0}^{n} \frac{2^k}{k!}x^k$$ ◆

PROBLEM 3 Find the nth-degree Taylor polynomial at 0 for $f(x) = e^{x/3}$. ◆

◆ TAYLOR POLYNOMIALS AT a

Suppose we want to approximate the function $f(x) = \sqrt[4]{x}$ with a polynomial. Since $f^{(k)}(0)$ is undefined for $k = 1, 2, \ldots$, this function does not have a Taylor polynomial at 0. However, all the derivatives of f exist at any $x > 0$. How can we generalize the definition of the Taylor polynomial in order to approximate functions such as this one? Before answering this question, we need to review a basic property of polynomials.

We are used to expressing polynomials in powers of x. However, it is also possible to express any polynomial in powers of $x - a$ for an arbitrary number a. For example, the following three expressions all represent the same polynomial:

$$\begin{aligned} p(x) &= 3 + 2x + x^2 && \text{Powers of } x \\ &= 6 + 4(x-1) + (x-1)^2 && \text{Powers of } x-1 \\ &= 3 - 2(x+2) + (x+2)^2 && \text{Powers of } x+2 \end{aligned}$$

To verify this statement, we expand the second and third expressions:

$$\begin{aligned} 6 + 4(x-1) + (x-1)^2 &= 6 + 4x - 4 + x^2 - 2x + 1 \\ &= 3 + 2x + x^2 \end{aligned}$$

$$\begin{aligned} 3 - 2(x+2) + (x+2)^2 &= 3 - 2x - 4 + x^2 + 4x + 4 \\ &= 3 + 2x + x^2 \end{aligned}$$

Now we return to the problem of generalizing the definition of the Taylor polynomial. Proceeding as we did before, given a function f with n derivatives at a number a, we want to find an nth-degree polynomial p_n with the property that

$$p^{(k)}(a) = f^{(k)}(a) \qquad k = 0, 1, \ldots, n$$

That is, we require that p_n and its first n derivatives agree with f and its first n derivatives at the number a. It turns out that it is much easier to find p_n when it is expressed in powers of $x - a$. The general expression for an nth-degree polynomial in powers of $x - a$ and its first n derivatives are listed below.

$$\begin{aligned}
p_n(x) &= a_0 + a_1(x-a) + a_2(x-a)^2 + a_3(x-a)^3 + \cdots + a_n(x-a)^n \\
p_n'(x) &= a_1 + 2a_2(x-a) + 3a_3(x-a)^2 + \cdots + na_n(x-a)^{n-1} \\
p_n''(x) &= 2a_2 + (3)(2)a_3(x-a) + \cdots + n(n-1)a_n(x-a)^{n-2} \\
p_n^{(3)}(x) &= (3)(2)(1)a_3 + \cdots + n(n-1)(n-2)a_n(x-a)^{n-3} \\
&\vdots \\
p_n^{(n)}(x) &= n(n-1) \cdot \cdots \cdot (1)a_n = n!a_n
\end{aligned}$$

Now we evaluate each function at a and apply the appropriate condition:

$$\begin{aligned}
a_0 &= p_n(a) = f(a) && k = 0 \\
a_1 &= p_n'(a) = f'(a) && k = 1 \\
2a_2 &= p_n''(a) = f''(a) && k = 2 \\
(3)(2)(1)a_3 &= p_n^{(3)}(a) = f^{(3)}(a) && k = 3 \\
&\vdots && \vdots \\
n!a_n &= p_n^{(n)}(a) = f^{(n)}(a) && k = n
\end{aligned}$$

Thus, each coefficient of p_n satisfies

$$a_k = \frac{f^{(k)}(a)}{k!} \qquad k = 0, 1, 2, \ldots, n$$

instead of 0

Taylor Polynomial at *a*

The ***n*th-degree Taylor polynomial at *a*** for a function f is

$$p_n(x) = f(a) + f'(a)(x-a) + \frac{f''(a)}{2!}(x-a)^2 + \cdots + \frac{f^{(n)}(a)}{n!}(x-a)^n$$
$$= \sum_{k=0}^{n} \frac{f^{(k)}(a)}{k!}(x-a)^k$$

provided f has n derivatives at a.

Theoretically, a function f has an nth-degree Taylor polynomial at any value a where it has n derivatives. In practice, we usually choose a so that f and its derivatives are easy to evaluate at $x = a$. For example, if $f(x) = \sqrt{x}$, then good choices for a are 1, 4, 9, 16, and so on.

◆ EXAMPLE 4 Find the third-degree Taylor polynomial at $a = 1$ for $f(x) = \sqrt[4]{x}$. Use $p_3(x)$ to approximate $\sqrt[4]{2}$.

Solution Step 1. Find the derivatives:

$$f(x) = x^{1/4}$$
$$f'(x) = \tfrac{1}{4}x^{-3/4}$$
$$f''(x) = -\tfrac{3}{16}x^{-7/4}$$
$$f^{(3)}(x) = \tfrac{21}{64}x^{-11/4}$$

Step 2. Evaluate the derivatives at $a = 1$:

$$f(1) = 1$$
$$f'(1) = \tfrac{1}{4}$$
$$f''(1) = -\tfrac{3}{16}$$
$$f^{(3)}(1) = \tfrac{21}{64}$$

Step 3. Find the coefficients of the Taylor polynomial:

$$a_0 = f(1) = 1$$
$$a_1 = f'(1) = \tfrac{1}{4}$$
$$a_2 = \frac{f''(1)}{2!} = \frac{-\frac{3}{16}}{2} = -\frac{3}{32}$$
$$a_3 = \frac{f^{(3)}(1)}{3!} = \frac{\frac{21}{64}}{6} = \frac{7}{128}$$

Step 4. Write down the Taylor polynomial:

$$p_3(x) = 1 + \tfrac{1}{4}(x - 1) - \tfrac{3}{32}(x - 1)^2 + \tfrac{7}{128}(x - 1)^3$$

Now we use the Taylor polynomial to approximate $\sqrt[4]{2}$:

$$\sqrt[4]{2} = f(2) \approx p_3(2) = 1 + \tfrac{1}{4} - \tfrac{3}{32} + \tfrac{7}{128} = 1.210\ 937\ 5$$

[*Note:* The value obtained by using a calculator is $\sqrt[4]{2} \approx 1.189\ 207\ 1$.] ◆

PROBLEM 4 Find the second-degree Taylor polynomial at $a = 8$ for $f(x) = \sqrt[3]{x}$. Use $p_2(x)$ to approximate $\sqrt[3]{9}$. ◆

◆ APPLICATION

◆ **EXAMPLE 5** *Average Price*

Given the demand function

$$p = D(x) = \sqrt{2{,}500 - x^2}$$

use the second-degree Taylor polynomial at 0 to approximate the average price (in dollars) over the demand interval [10, 40].

Solution The average price over the demand interval [10, 40] is given by

$$\text{Average price} = \frac{1}{30}\int_{10}^{40} \sqrt{2{,}500 - x^2}\,dx$$

This integral cannot be evaluated by any of the techniques we have discussed. However, we can use a Taylor polynomial to approximate the value of the integral. Omitting the details (which you should supply), the second-degree Taylor polynomial at 0 for $D(x) = \sqrt{2{,}500 - x^2}$ is

$$p_2(x) = 50 - \tfrac{1}{100}x^2$$

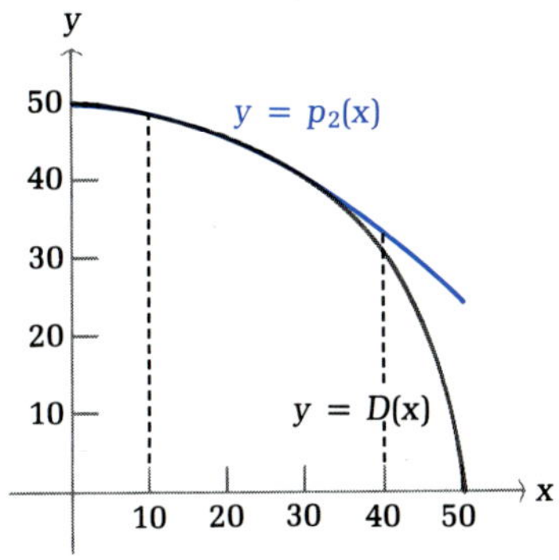

Assuming that $D(x) \approx p_2(x)$ for $10 \le x \le 40$ (see the figure in the margin), we have

$$\begin{aligned}\text{Average price} &\approx \frac{1}{30}\int_{10}^{40}\left(50 - \frac{1}{100}x^2\right)dx\\ &= \frac{1}{30}\left(50x - \frac{1}{300}x^3\right)\Bigg|_{10}^{40}\\ &= \$43\end{aligned}$$

◆

PROBLEM 5 Given the demand function

$$p = D(x) = \sqrt{400 - x^2}$$

use the second-degree Taylor polynomial at 0 to approximate the average price (in dollars) over the demand interval [5, 15]. ◆

Example 5 illustrates the convenience of using Taylor polynomials to approximate more complicated functions, but it also raises several important questions. First, how can we determine whether it is reasonable to use the Taylor polynomial to approximate a function on a given interval? Second, is it possible to find the Taylor polynomial without resorting to successive differentiation of the function? Finally, can we determine the accuracy of an approximation such as the approximate average price found in Example 5? These questions will be discussed in the next three sections.

Answers to Matched Problems

1. $f^{(n)}(x) = (-1)^{n-1}(n-1)!x^{-n}$
2. $p_2(x) = 3 + \frac{1}{6}x - \frac{1}{216}x^2$; $\sqrt{10} \approx 3.162\ 037$
3. $p_n(x) = 1 + \dfrac{1}{3}x + \dfrac{1}{3^2}\dfrac{1}{2!}x^2 + \dfrac{1}{3^3}\dfrac{1}{3!}x^3 + \cdots + \dfrac{1}{3^n}\dfrac{1}{n!}x^n = \displaystyle\sum_{k=0}^{n}\frac{1}{3^k}\frac{1}{k!}x^k$

4. $p_2(x) = 2 + \frac{1}{12}(x - 8) - \frac{1}{288}(x - 8)^2$; $\sqrt[3]{9} \approx 2.079\ 861\ 1$

5. $\frac{1}{10}\int_5^{15}\left(20 - \frac{x^2}{40}\right)dx = \frac{415}{24} \approx \$17.29$

EXERCISE 10-1

A *In Problems 1–4, find $f^{(4)}(x)$.*

1. $f(x) = \frac{1}{2 - x}$
2. $f(x) = \ln(3 + x)$
3. $f(x) = e^{-2x}$
4. $f(x) = \sqrt{x}$

In Problems 5–10, find the indicated Taylor polynomial at 0.

5. $f(x) = e^{-x}$; $p_4(x)$
6. $f(x) = e^{4x}$; $p_3(x)$
7. $f(x) = \ln(1 + 2x)$; $p_3(x)$
8. $f(x) = \ln(1 + \frac{1}{2}x)$; $p_4(x)$
9. $f(x) = \sqrt[3]{x + 1}$; $p_3(x)$
10. $f(x) = \sqrt[4]{x + 16}$; $p_2(x)$

In Problems 11–14, find the indicated Taylor polynomial at the given value of a.

11. $f(x) = e^{x-1}$; $p_4(x)$ at 1
12. $f(x) = e^{2x}$; $p_3(x)$ at $\frac{1}{2}$
13. $f(x) = \ln(3x)$; $p_3(x)$ at $\frac{1}{3}$
14. $f(x) = \ln(2 - x)$; $p_4(x)$ at 1

15. Use the third-degree Taylor polynomial at 0 for $f(x) = e^{-2x}$ and $x = 0.25$ to approximate $e^{-0.5}$.
16. Use the fourth-degree Taylor polynomial at 0 for $f(x) = \ln(1 - x)$ and $x = 0.1$ to approximate $\ln 0.9$.
17. Use the second-degree Taylor polynomial at 0 for $f(x) = \sqrt{x + 16}$ and $x = 1$ to approximate $\sqrt{17}$.
18. Use the third-degree Taylor polynomial at 0 for $f(x) = \sqrt{(x + 4)^3}$ and $x = 1$ to approximate $\sqrt{125}$.
19. Use the fourth-degree Taylor polynomial at 1 for $f(x) = \sqrt{x}$ and $x = 1.2$ to approximate $\sqrt{1.2}$.
20. Use the third-degree Taylor polynomial at 4 for $f(x) = \sqrt{x}$ and $x = 3.95$ to approximate $\sqrt{3.95}$.

B *Find $f^{(n)}(x)$.*

21. $f(x) = e^{5x}$
22. $f(x) = e^{-3x}$
23. $f(x) = \ln(5 + x)$
24. $f(x) = \ln(2 - x)$

Find the nth-degree Taylor polynomial at 0. Write the answer in expanded notation.

25. $f(x) = \ln(1 + x)$
26. $f(x) = e^{-x}$
27. $f(x) = \frac{1}{1 - x}$
28. $f(x) = \ln(1 + 2x)$
29. $f(x) = \ln(3 + 4x)$
30. $f(x) = \frac{x}{2 - x}$

Find the nth-degree Taylor polynomial at the indicated value of a. Write the answer in expanded notation.

31. $f(x) = e^x$ at 3 **32.** $f(x) = \dfrac{1}{3+x}$ at -2 **33.** $f(x) = \ln x$ at $\frac{1}{2}$

34. $f(x) = \dfrac{2+x}{x}$ at 2 **35.** $f(x) = e^{-x}$ at 10 **36.** $f(x) = x \ln x$ at 1

C

37. Find the first three Taylor polynomials at 0 for $f(x) = \ln(1 + x)$, and use these polynomials to complete the two tables below. Round all table entries to six decimal places.

x	$p_1(x)$	$p_2(x)$	$p_3(x)$	$f(x)$
-0.2				
-0.1				
0				
0.1				
0.2				

x	$\lvert p_1(x) - f(x)\rvert$	$\lvert p_2(x) - f(x)\rvert$	$\lvert p_3(x) - f(x)\rvert$
-0.2			
-0.1			
0			
0.1			
0.2			

38. Repeat Problem 37 for $f(x) = \sqrt{1 + x}$.

Problems 39 and 40 require the use of a graphic calculator or a computer.

39. Refer to Problem 37. Graph $f(x) = \ln(1 + x)$ and its first three Taylor polynomials in the same viewing rectangle. Use $[-3, 3]$ for both the x range and the y range.

40. Refer to Problem 38. Graph $f(x) = \sqrt{1 + x}$ and its first three Taylor polynomials in the same viewing rectangle. Use $[-3, 3]$ for both the x range and the y range.

In Problems 41–44, write the answer in both expanded notation and summation notation.

41. Find the nth-degree Taylor polynomial for $f(x) = e^x$ at any point a.

42. Find the nth-degree Taylor polynomial for $f(x) = \ln x$ at any $a > 0$.

43. Find the nth-degree Taylor polynomial for $f(x) = 1/x$ at any $a \neq 0$.

44. Find the nth-degree Taylor polynomial at 0 for $f(x) = (x + c)^n$, where c is a constant.

Business & Economics

45. *Average price.* Given the demand equation

$$p = D(x) = \tfrac{1}{10}\sqrt{10{,}000 - x^2}$$

use the second-degree Taylor polynomial at 0 to approximate the average price (in dollars) over the demand interval [0, 30].

46. *Average price.* Given the demand equation

$$p = D(x) = \tfrac{1}{5}\sqrt{1{,}600 - x^2}$$

use the second-degree Taylor polynomial at 0 to approximate the average price (in dollars) over the demand interval [0, 15].

47. *Average price.* Refer to Problem 45. Use the second-degree Taylor polynomial at $a = 60$ to approximate the average price over the demand interval [60, 80].

48. *Average price.* Refer to Problem 46. Use the second-degree Taylor polynomial at $a = 24$ to approximate the average price over the demand interval [24, 32].

49. *Production.* The rate of production of a mine (in millions of dollars per year) is given by

$$R(t) = 2 + 8e^{-0.1t^2}$$

Use the second-degree Taylor polynomial at 0 to approximate the total production during the first 2 years of operation of the mine.

50. *Production.* The rate of production of an oil well (in millions of dollars per year) is given by

$$R(t) = 5 + 10e^{-0.05t^2}$$

Use the second-degree Taylor polynomial at 0 to approximate the total production during the first 3 years of operation of the well.

Life Sciences

51. *Medicine.* The rate of healing for a skin wound (in square centimeters per day) is given by

$$A'(t) = \frac{-75}{t^2 + 25}$$

The initial wound has an area of 12 square centimeters. Use the second-degree Taylor polynomial at 0 for $A'(t)$ to approximate the area of the wound after 2 days.

52. *Medicine.* Rework Problem 51 for

$$A'(t) = \frac{-60}{t^2 + 20}$$

53. *Pollution.* On an average summer day in a particular large city the air pollution level (in parts per million) is given by

$$P(x) = 20\sqrt{3x^2 + 25} - 80$$

where x is the number of hours elapsed since 8:00 AM. Use the second-degree Taylor polynomial for $P(x)$ at $a = 5$ to approximate the average pollution level during the 10 hour period from 8:00 AM to 6:00 PM.

54. *Pollution.* On an average summer day in another large city the air pollution level (in parts per million) is given by

$$P(x) = 10\sqrt{4x^2 + 36} - 50$$

where x is the number of hours elapsed since 8:00 AM. Use the second-degree Taylor polynomial for $P(x)$ at $a = 4$ to approximate the average pollution level during the 8 hour period from 8:00 AM to 4:00 PM.

Social Sciences

55. *Learning.* In a particular business college, it was found that an average student enrolled in an advanced typing class progresses at a rate of $N'(t) = 6e^{-0.01t^2}$ words per minute per week, t weeks after enrolling in a 15 week course. At the beginning of the course an average student could type 40 words per minute. Use the second-degree Taylor polynomial at 0 for $N'(t)$ to approximate the improvement in typing after 5 weeks in the course.

56. *Learning.* In the same business college, it was also found that an average student enrolled in a beginning shorthand class progressed at a rate of $N'(t) = 12e^{-0.005t^2}$ words per minute per week, t weeks after enrolling in a 15 week course. At the beginning of the course none of the students could take any dictation by shorthand. Use the second-degree Taylor polynomial at 0 for $N'(t)$ to approximate the improvement after 5 weeks in the course.

In Problems 57–64, use a graphic calculator or a computer to graph the given function and its second-degree Taylor polynomial at a in the indicated viewing rectangle.

Business & Economics

57. *Average price.* From Problem 45:
$p = D(x) = \frac{1}{10}\sqrt{10{,}000 - x^2}$
$a = 0$
x range: [0, 100]
p range: [0, 10]

58. *Average price.* From Problem 46:
$p = D(x) = \frac{1}{5}\sqrt{1{,}600 - x^2}$
$a = 0$
x range: [0, 40]
p range: [0, 8]

59. *Production.* From Problem 49:
$y = R(t) = 2 + 8e^{-0.1t^2}$
$a = 0$
t range: [0, 5]
y range: [0, 10]

60. *Production.* From Problem 50:
$y = R(t) = 5 + 10e^{-0.05t^2}$
$a = 0$
t range: [0, 5]
y range: [0, 15]

Life Sciences

61. *Pollution.* From Problem 53:
$y = P(x) = 20\sqrt{3x^2 + 25} - 80$
$a = 5$
x range: [0, 10]
y range: [0, 300]

62. *Pollution.* From Problem 54:
$y = P(x) = 10\sqrt{4x^2 + 36} - 50$
$a = 4$
x range: [0, 10]
y range: [0, 200]

Social Sciences

63. *Learning.* From Problem 55:
$y = N'(t) = 6e^{-0.01t^2}$
$a = 0$
t range: [0, 10]
y range: [0, 6]

64. *Learning.* From Problem 56:
$y = N'(t) = 12e^{-0.005t^2}$
$a = 0$
t range: [0, 15]
y range: [0, 12]

SECTION 10-2 Taylor Series

- INTRODUCTION
- TAYLOR SERIES
- REPRESENTATION OF FUNCTIONS BY TAYLOR SERIES

INTRODUCTION

If f is a function with derivatives of all order at a point a, then we can construct the Taylor polynomial p_n at a for any integer n. Now we are interested in the relationship between the original function f and the corresponding Taylor polynomial p_n as n assumes larger and larger values. If the Taylor polynomial is to be a useful tool for approximating functions, then the accuracy of the approximation should improve as we increase the size of n. Indeed, Figures 1, 2, and 3 and Tables 1 and 2 in the preceding section indicate that this is the case for the function e^x.

For a given value of x, we would like to know whether we can make $p_n(x)$ arbitrarily close to $f(x)$ by making n sufficiently large. In other words, we want to know whether

$$\lim_{n\to\infty} p_n(x) = f(x) \tag{1}$$

It turns out that for most functions with derivatives of all order at a point, there is a set of values of x for which equation (1) is valid. We will begin by considering a specific example to illustrate this.

Let

$$f(x) = \frac{1}{1-x} \qquad \text{and} \qquad a = 0$$

First, we find p_n:

Step 1	Step 2	Step 3
$f(x) = (1-x)^{-1}$	$f(0) = 1$	$a_0 = f(0) \quad = 1$
$f'(x) = (-1)(1-x)^{-2}(-1) = (1-x)^{-2}$	$f'(0) = 1$	$a_1 = f'(0) \quad = 1$
$f''(x) = (-2)(1-x)^{-3}(-1) = 2(1-x)^{-3}$	$f''(0) = 2$	$a_2 = \dfrac{f''(0)}{2!} = \dfrac{2}{2!} = 1$
$f^{(3)}(x) = (-3)(2)(1-x)^{-4}(-1) = 3!(1-x)^{-4}$	$f^{(3)}(0) = 3!$	$a_3 = \dfrac{f^{(3)}(0)}{3!} = \dfrac{3!}{3!} = 1$
$\vdots$	$\vdots$	$\vdots$
$f^{(n)}(x) = n!(1-x)^{-n-1}$	$f^{(n)}(0) = n!$	$a_n = \dfrac{f^{(n)}(0)}{n!} = \dfrac{n!}{n!} = 1$

Step 4. Write down the Taylor polynomial:

$$p_n(x) = 1 + x + x^2 + x^3 + \cdots + x^n \qquad (2)$$
$$= \sum_{k=0}^{n} x^k$$

Now we want to evaluate

$$\lim_{n\to\infty} p_n(x) = \lim_{n\to\infty} \sum_{k=0}^{n} x^k$$

As we saw in Chapter 6, it is not possible to evaluate limits written in this form. The difficulty lies in the fact that the number of terms is increasing as n increases. First, we must find a closed form for the summation that does not involve a sum of n terms. [You may recognize that $p_n(x)$ is a finite geometric series with common ratio x.]

If we multiply both sides of (2) by x and subtract this new equation from the original, we obtain an equation that we can solve for $p_n(x)$. Notice how many terms drop out when the subtraction is performed.

$$\begin{aligned}
p_n(x) &= 1 + x + x^2 + \cdots + x^{n-1} + x^n \\
xp_n(x) &= \phantom{1 +{}} x + x^2 + \cdots + x^{n-1} + x^n + x^{n+1} \\
\hline
p_n(x) - xp_n(x) &= 1 \qquad\qquad\qquad\qquad\qquad\qquad - x^{n+1} \\
p_n(x)(1-x) &= 1 - x^{n+1} \\
p_n(x) &= \frac{1 - x^{n+1}}{1-x} \qquad x \neq 1
\end{aligned}$$

Since we had to exclude the value $x = 1$ to avoid division by 0, we must consider $x = 1$ as a special case:

$$p_n(1) = \overbrace{1 + 1 + 1^2 + \cdots + 1^n}^{n+1 \text{ terms}}$$
$$= n + 1$$

Thus,

$$p_n(x) = \begin{cases} \dfrac{1 - x^{n+1}}{1 - x} & \text{if } x \neq 1 \\ n + 1 & \text{if } x = 1 \end{cases}$$

If $x = 1$, it is clear that

$$\lim_{n\to\infty} p_n(1) = \lim_{n\to\infty} (n + 1) = \infty$$

That is, $\lim_{n\to\infty} p_n(1)$ does not exist. What happens to this limit for other values of x? It can be shown that

$$\lim_{n\to\infty} x^{n+1} = \begin{cases} 0 & \text{if } -1 < x < 1 \\ 1 & \text{if } x = 1 \\ \text{Does not exist} & \text{if } x \leq -1 \text{ or } x > 1 \end{cases}$$

Experimenting with a calculator for various values of x and large values of n should convince you that this statement is valid.

Thus,

$$\lim_{n\to\infty} p_n(x) = \begin{cases} \dfrac{1}{1 - x} & \text{if } -1 < x < 1 & \lim_{n\to\infty} x^{n+1} = 0 \\ \text{Does not exist} & \text{if } x = 1 & \lim_{n\to\infty} (n + 1) \text{ does not exist} \\ \text{Does not exist} & \text{if } x \leq -1 \text{ or } x > 1 & \lim_{n\to\infty} x^{n+1} \text{ does not exist} \end{cases}$$

We can now conclude that

$$f(x) = \frac{1}{1 - x}$$
$$= \lim_{n\to\infty} p_n(x) \qquad -1 < x < 1$$
$$= \lim_{n\to\infty} \sum_{k=0}^{n} x^k$$

and that the limit does not exist for any other values of x. This is often abbreviated by writing

$$\frac{1}{1-x} = 1 + x + x^2 + \cdots + x^n + \cdots \qquad -1 < x < 1$$
$$= \sum_{k=0}^{\infty} x^k$$

The expression $\Sigma_{k=0}^{\infty} x^k$ is called the *Taylor series for f at 0*, and the expanded notation

$$1 + x + x^2 + \cdots + x^n + \cdots$$

is just another way to write this series. The Taylor series is said to *converge at x* if $\lim_{n\to\infty} p_n(x)$ exists and to *diverge at x* if this limit fails to exist. Thus, $\Sigma_{k=0}^{\infty} x^k$ converges for $-1 < x < 1$ and diverges for $x \leqslant -1$ or $x \geqslant 1$. The set of values $\{x|-1 < x < 1\} = (-1, 1)$ where this Taylor series converges is called the *interval of convergence*. Since $f(x)$ is equal to the Taylor series for any x in the interval of convergence, we say that f is represented by its Taylor series throughout this interval. That is,

$$f(x) = \frac{1}{1-x}$$
$$= 1 + x + x^2 + \cdots + x^n + \cdots \qquad -1 < x < 1$$

There can be no relationship between f and its Taylor seies outside the interval of convergence since the series is not defined there (see Figure 4).

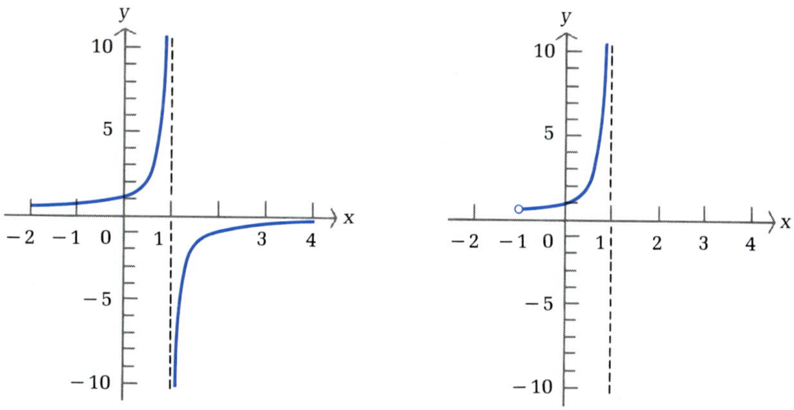

$f(x) = \dfrac{1}{1-x}$
Defined for $x \neq 1$

$1 + x + x^2 + \ldots + x^n + \ldots$
Defined for $-1 < x < 1$

FIGURE 4

◆ TAYLOR SERIES

We now generalize the concepts introduced in the preceding discussion to arbitrary functions.

Taylor Series

If f is a function with derivatives of all order at a point a, $p_n(x)$ is the nth-degree Taylor polynomial at a for f, and $a_n = f^{(n)}(a)/n!$, $n = 0, 1, 2, \ldots$, then

$$\sum_{k=0}^{\infty} a_k(x-a)^k = a_0 + a_1(x-a) + a_2(x-a)^2 + \cdots + a_n(x-a)^n + \cdots$$

is the **Taylor series at *a* for *f*.** The Taylor series **converges at x** if

$$\lim_{n\to\infty} p_n(x) = \lim_{n\to\infty} \sum_{k=0}^{n} a_k(x-a)^k$$

exists and **diverges at x** if this limit does not exist. The set of values of x for which this limit exists is called the **interval of convergence.**

We must emphasize that we are not really adding up an infinite number of terms in a Taylor series, nor is a Taylor series an "infinite" polynomial. Rather, for x in the interval of convergence,

$$\sum_{k=0}^{\infty} a_k(x-a)^k = \lim_{n\to\infty} \sum_{k=0}^{n} a_k(x-a)^k$$

Thus, when we write

$$\frac{1}{1-x} = 1 + x + x^2 + \cdots + x^n + \cdots \qquad -1 < x < 1$$

we mean

$$\frac{1}{1-x} = \lim_{n\to\infty}(1 + x + x^2 + \cdots + x^n) \qquad -1 < x < 1$$

In general, it is very difficult to find the interval of convergence by directly evaluating

$$\lim_{n\to\infty} \sum_{k=0}^{n} a_k(x-a)^k$$

For many Taylor series, Theorem 1 (on the next page) can be used to find the interval of convergence.

THEOREM 1

Interval of Convergence

Let f be a function with derivatives of all order at a point a, let $a_n = f^{(n)}(a)/n!$, $n = 0, 1, 2, \ldots$, and let

$$\sum_{k=0}^{\infty} a_k(x-a)^k = a_0 + a_1(x-a) + a_2(x-a)^2 + \cdots + a_n(x-a)^n + \cdots$$

be the Taylor series at a for f. If $a_n \neq 0$ for $n \geqslant n_0$, then:

Case 1. If

$$\lim_{n\to\infty}\left|\frac{a_{n+1}}{a_n}\right| = L > 0$$

and $R = 1/L$, then the series converges for $|x-a| < R$ and diverges for $|x-a| > R$.

Case 2. If

$$\lim_{n\to\infty}\left|\frac{a_{n+1}}{a_n}\right| = 0$$

then the series converges for all values of x.

Case 3. If

$$\lim_{n\to\infty}\left|\frac{a_{n+1}}{a_n}\right|$$

does not exist, then the series converges only at $x = a$.

The various possibilities for the interval of convergence are illustrated in Figure 5. In case 1, the series may or may not converge at the end points $x = a - R$ and $x = a + R$. (Determination of the behavior of a series at the end points of the interval of convergence requires techniques that we will not discuss.)

Case 1. $\lim_{n\to\infty}\left|\frac{a_{n+1}}{a_n}\right| = L > 0$

$R = 1/L$

Case 2. $\lim_{n\to\infty}\left|\frac{a_{n+1}}{a_n}\right| = 0$

Case 3. $\lim_{n\to\infty}\left|\frac{a_{n+1}}{a_n}\right|$

does not exist

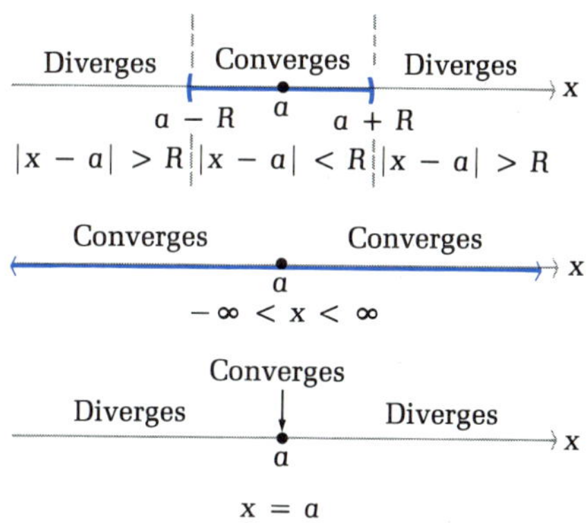

FIGURE 5
Intervals of convergence

Notice that the set of values where a series converges can be expressed in interval notation as $(a - R, a + R)$ in case 1 and as $(-\infty, \infty)$ in case 2. We were anticipating this result when we used the term *interval of convergence* to describe the set of points where a Taylor series converges.

Finally, the requirement that all the coefficients be nonzero from some point on is necessary to ensure that the ratio a_{n+1}/a_n is well-defined.

The next two examples illustrate the application of Theorem 1. The details of finding the Taylor series are omitted. In the next section we will discuss methods that greatly simplify the process of finding Taylor series. For now, we want to concentrate on understanding the significance of the interval of convergence.

◆ EXAMPLE 6 The Taylor series at 0 for $f(x) = e^{2x}$ is

$$1 + 2x + \frac{2^2}{2!}x^2 + \cdots + \frac{2^n}{n!}x^n + \cdots$$

Find the interval of convergence.

Solution

$$a_n = \frac{2^n}{n!}$$ Notice that $a_n > 0$ for $n \geq 0$.

$$a_{n+1} = \frac{2^{n+1}}{(n+1)!}$$

$$\frac{a_{n+1}}{a_n} = \frac{\dfrac{2^{n+1}}{(n+1)!}}{\dfrac{2^n}{n!}}$$ Form the ratio a_{n+1}/a_n and simplify.

$$= \frac{2^{n+1}}{(n+1)!} \cdot \frac{n!}{2^n}$$ $\dfrac{n!}{(n+1)!} = \dfrac{n!}{(n+1)n!} = \dfrac{1}{n+1}$

$$= \frac{2}{n+1}$$

$$\lim_{n\to\infty}\left|\frac{a_{n+1}}{a_n}\right| = \lim_{n\to\infty}\left|\frac{2}{n+1}\right|$$ Absolute value signs can be dropped since $a_{n+1}/a_n > 0$.

$$= \lim_{n\to\infty}\frac{2}{n+1} = 0$$

Case 2 in Theorem 1 applies, and the series converges for all values of x. ◆

PROBLEM 6 The Taylor series at 0 for $f(x) = e^{x/3}$ is

$$1 + \frac{1}{3}x + \frac{1}{3^2 2!}x^2 + \cdots + \frac{1}{3^n n!}x^n + \cdots$$

Find the interval of convergence. ◆

◆ EXAMPLE 7 The Taylor series at 0 for $f(x) = 1/(1+5x)^2$ is

$$1 - 2 \cdot 5x + 3 \cdot 5^2x^2 - \cdots + (-1)^n(n+1)5^nx^n + \cdots$$

Find the interval of convergence.

Solution

$$a_n = (-1)^n(n+1)5^n$$

Notice that $a_n \neq 0$ for $n \geqslant 0$.

$$a_{n+1} = (-1)^{n+1}(n+2)5^{n+1}$$

$$\frac{a_{n+1}}{a_n} = \frac{(-1)^{n+1}(n+2)5^{n+1}}{(-1)^n(n+1)5^n} = \frac{-5(n+2)}{n+1}$$

$$\lim_{n\to\infty}\left|\frac{a_{n+1}}{a_n}\right| = \lim_{n\to\infty}\left|\frac{-5(n+2)}{n+1}\right| = \lim_{n\to\infty}\frac{5(n+2)}{n+1} = \lim_{n\to\infty}\frac{5n+10}{n+1} = \lim_{n\to\infty}\frac{5+\frac{10}{n}}{1+\frac{1}{n}} = 5$$

Divide numerator and denominator by n.

Applying case 1 in Theorem 1, we have $L = 5$, $R = \frac{1}{5}$, and the series converges for $-\frac{1}{5} < x < \frac{1}{5}$. ◆

PROBLEM 7 The Taylor series at 0 for $f(x) = 4/(2+x)^2$ is

$$1 - \frac{2}{2}x + \frac{3}{2^2}x^2 - \cdots + \frac{(-1)^n(n+1)}{2^n}x^n + \cdots$$

Find the interval of convergence. ◆

◆ REPRESENTATION OF FUNCTIONS BY TAYLOR SERIES

Referring to the function $f(x) = 1/(1-x)$ discussed earlier in this section, our calculations showed that the Taylor series at 0 for f is

$$\lim_{n\to\infty} p_n(x) = 1 + x + x^2 + \cdots + x^n + \cdots$$

with interval of convergence $(-1, 1)$. Furthermore, we showed that f is represented by its Taylor series throughout the interval of convergence. That is,

$$f(x) = \frac{1}{1-x} = \lim_{n\to\infty} p_n(x) = 1 + x + x^2 + \cdots + x^n + \cdots \qquad -1 < x < 1$$

A similar statement can be made for most functions. However, there do exist functions that cannot be represented by their Taylor series. Since it is very unlikely that we will ever encounter any of these functions, we will make the following assumption for the functions we consider.

Representation of Functions by Taylor Series

If f is a function with derivatives of all order at a point a, *then* **f is represented by its Taylor series throughout the interval of convergence of the series.** Thus, if $p_n(x)$ is the nth-degree Taylor polynomial at a for f and $a_n = f^{(n)}(a)/n!$, $n = 0, 1, 2, \ldots$, then

$$f(x) = \lim_{n\to\infty} p_n(x) = \sum_{k=0}^{\infty} a_k(x-a)^k$$

for any x in the interval of convergence of the Taylor series at a for f.

Consequently, **the interval of convergence determines the values of x for which the Taylor polynomials $p_n(x)$ can be used to approximate the values of the function f.**

Referring to Examples 6 and 7, we can now write

$$e^{2x} = 1 + 2x + \frac{2^2}{2!}x^2 + \cdots + \frac{2^n}{n!}x^n + \cdots \qquad -\infty < x < \infty$$

and

$$\frac{1}{(1+5x)^2} = 1 - 2 \cdot 5x + 3 \cdot 5^2x^2 - \cdots + (-1)^n(n+1)5^nx^n + \cdots$$

$$-\tfrac{1}{5} < x < \tfrac{1}{5}$$

◆ EXAMPLE 8 Let $f(x) = \ln x$.

(A) Find the nth-degree Taylor polynomial at $a = 1$ for f.
(B) Find the Taylor series at $a = 1$ for f.
(C) Determine the values of x for which $f(x) = \lim_{n\to\infty} p_n(x)$.

Solutions (A) We use the four-step process from the preceding section to find the nth-degree Taylor polynomial:

Step 1	Step 2	Step 3
$f(x) = \ln x$	$f(1) = \ln 1 = 0$	$a_0 = f(1) = 0$
$f'(x) = \frac{1}{x} = x^{-1}$	$f'(1) = 1$	$a_1 = f'(1) = 1$
$f''(x) = (-1)x^{-2}$	$f''(1) = -1$	$a_2 = \frac{f''(1)}{2!} = -\frac{1}{2}$
$f^{(3)}(x) = (-1)(-2)x^{-3} = (-1)^2 2! x^{-3}$	$f^{(3)}(1) = (-1)^2 2!$	$a_3 = \frac{f^{(3)}(1)}{3!} = \frac{(-1)^2 2!}{3!} = \frac{1}{3}$
$f^{(4)}(x) = (-1)(-2)(-3)x^{-4} = (-1)^3 3! x^{-4}$	$f^{(4)}(1) = (-1)^3 3!$	$a_4 = \frac{f^{(4)}(1)}{4!} = \frac{(-1)^3 3!}{4!} = \frac{-1}{4}$
$\vdots$	$\vdots$	$\vdots$
$f^{(n)}(x) = (-1)^{n-1}(n-1)! x^{-n}$	$f^{(n)}(1) = (-1)^{n-1}(n-1)!$	$a_n = \frac{f^{(n)}(1)}{n!} = \frac{(-1)^{n-1}(n-1)!}{n!} = \frac{(-1)^{n-1}}{n}$

Step 4. The nth-degree Taylor polynomial at $a = 1$ for $f(x) = \ln x$ is

$$p_n(x) = (x-1) - \frac{1}{2}(x-1)^2 + \frac{1}{3}(x-1)^3 - \cdots + \frac{(-1)^{n-1}}{n}(x-1)^n$$

(B) Once the nth-degree Taylor polynomial for a function has been determined, finding the Taylor series is simply a matter of using the notation correctly. Using the Taylor polynomial from part A, the Taylor series at $a = 1$ for $f(x) = \ln x$ can be written as

$$\lim_{n\to\infty} p_n(x) = (x-1) - \frac{1}{2}(x-1)^2 + \frac{1}{3}(x-1)^3 - \cdots + \frac{(-1)^{n-1}}{n}(x-1)^n + \cdots$$

or, using summation notation,* as

$$\lim_{n\to\infty} p_n(x) = \sum_{k=0}^{\infty} \frac{(-1)^{k-1}}{k}(x-1)^k$$

(C) To determine the values for which $f(x) = \lim_{n\to\infty} p_n(x)$, we use Theorem 1 to find the interval of convergence of the Taylor series in part B:

$$a_n = \frac{(-1)^{n-1}}{n}$$

$$a_{n+1} = \frac{(-1)^n}{n+1}$$

* From this point on, we will use the expanded notation exclusively when writing Taylor series.

$$\frac{a_{n+1}}{a_n} = \frac{\dfrac{(-1)^n}{n+1}}{\dfrac{(-1)^{n-1}}{n}} = \frac{(-1)^n}{n+1} \cdot \frac{n}{(-1)^{n-1}} = \frac{-n}{n+1}$$

$$\lim_{n\to\infty}\left|\frac{a_{n+1}}{a_n}\right| = \lim_{n\to\infty}\left|\frac{-n}{n+1}\right|$$

$$= \lim_{n\to\infty}\frac{n}{n+1} \qquad \text{Divide numerator and denominator by } n.$$

$$= \lim_{n\to\infty}\frac{1}{1+\dfrac{1}{n}} = 1$$

From case 1 of Theorem 1, $L = 1$, $R = 1/L = 1$, and the series converges for $|x - 1| < 1$. Converting this to double inequalities, we have

$$|x - 1| < 1 \qquad |y| < c \text{ is equivalent to } -c < y < c.$$

$$-1 < x - 1 < 1 \qquad \text{Add 1 to each side.}$$

$$0 < x < 2 \qquad \text{Interval of convergence}$$

Thus, $f(x) = \lim_{n\to\infty} p_n(x)$ for $0 < x < 2$ or, equivalently,

$$\ln x = (x-1) - \frac{1}{2}(x-1)^2 + \frac{1}{3}(x-1)^3 - \cdots + \frac{(-1)^{n-1}}{n}(x-1)^n + \cdots \qquad 0 < x < 2$$

◆

PROBLEM 8 Repeat Example 8 for $f(x) = 1/x$. ◆

Table 3 displays the values of the Taylor polynomials for $f(x) = \ln x$ at $x = 1.5$ and at $x = 3$ for $n = 1, 2, \ldots, 10$ (see Example 8). Notice that $x = 1.5$ is in the

TABLE 3

$f(x) = \ln x$

$$p_n(x) = (x-1) - \frac{1}{2}(x-1)^2 + \frac{1}{3}(x-1)^3 - \cdots + \frac{(-1)^{n-1}}{n}(x-1)^n$$

n	$p_n(1.5)$	$p_n(3)$
1	0.5	2
2	0.375	0
3	0.416 667	2.666 667
4	0.401 042	−1.333 333
5	0.407 292	5.066 667
6	0.404 688	−5.6
7	0.405 804	12.685 714
8	0.405 315	−19.314 286
9	0.405 532	37.574 603
10	0.405 435	−64.825 397
	$\ln 1.5 = 0.405\ 465$	$\ln 3 = 1.098\ 612$

interval of convergence of the Taylor series at 1 for $f(x)$ and that the values of $p_n(1.5)$ are approaching $\ln 1.5$ as n increases. On the other hand, $x = 3$ is not in the interval of convergence and the values of $p_n(3)$ do not approach $\ln 3$. These calculations illustrate the importance of the interval of convergence.

The values of x must be in the interval of convergence in order to use Taylor polynomials to approximate the values of a function.

Answers to Matched Problems

6. $-\infty < x < \infty$ 7. $-2 < x < 2$
8. (A) $1 - (x - 1) + (x - 1)^2 - \cdots + (-1)^n(x - 1)^n$
   (B) $1 - (x - 1) + (x - 1)^2 - \cdots + (-1)^n(x - 1)^n + \cdots$
   (C) $0 < x < 2$

EXERCISE 10-2

Find the interval of convergence of the given Taylor series representation.

A

1. $\dfrac{2}{1 - x} = 2 + 2x + 2x^2 + \cdots + 2x^n + \cdots$
2. $\dfrac{1}{1 + x} = 1 - x + x^2 - \cdots + (-1)^n x^n + \cdots$
3. $\dfrac{1}{1 - 2x} = 1 + 2x + 2^2x^2 + \cdots + 2^n x^n + \cdots$
4. $\dfrac{1}{1 - 3x} = 1 + 3x + 3^2x^2 + \cdots + 3^n x^n + \cdots$
5. $\ln(1 + x) = x - \dfrac{1}{2}x^2 + \dfrac{1}{3}x^3 - \cdots + \dfrac{(-1)^{n-1}}{n}x^n + \cdots$
6. $\ln(1 - x) = -x - \dfrac{1}{2}x^2 - \dfrac{1}{3}x^3 - \cdots - \dfrac{1}{n}x^n - \cdots$
7. $\dfrac{1}{(1 + x)^2} = 1 - 2x + 3x^2 - \cdots + (-1)^n(n + 1)x^n + \cdots$
8. $\dfrac{x}{(1 - x)^2} = x + 2x^2 + 3x^3 + \cdots + nx^n + \cdots$
9. $e^x = 1 + x + \dfrac{1}{2!}x^2 + \cdots + \dfrac{1}{n!}x^n + \cdots$
10. $e^{-x} = 1 - x + \dfrac{1}{2!}x^2 - \cdots + \dfrac{(-1)^n}{n!}x^n + \cdots$

B

11. $\dfrac{1}{5 - x} = 1 + (x - 4) + (x - 4)^2 + \cdots + (x - 4)^n + \cdots$
12. $\dfrac{1}{3 - x} = 1 + (x - 2) + (x - 2)^2 + \cdots + (x - 2)^n + \cdots$
13. $\dfrac{3}{x + 4} = 1 - \dfrac{1}{3}(x + 1) + \dfrac{1}{3^2}(x + 1)^2 - \cdots + \dfrac{(-1)^n}{3^n}(x + 1)^n + \cdots$

14. $\frac{4}{x+2} = 1 - \frac{1}{4}(x-2) + \frac{1}{4^2}(x-2)^2 - \cdots + \frac{(-1)^n}{4^n}(x-2)^n + \cdots$

15. $e^{-3x} = 1 - 3x + \frac{3^2}{2!}x^2 - \cdots + \frac{(-1)^n 3^n}{n!}x^n + \cdots$

16. $e^{x/5} = 1 + \frac{1}{5}x + \frac{1}{5^2 2!}x^2 + \cdots + \frac{1}{5^n n!}x^n + \cdots$

17. $\ln(5-4x) = -4(x-1) - \frac{4^2}{2}(x-1)^2 - \frac{4^3}{3}(x-1)^3 - \cdots - \frac{4^n}{n}(x-1)^n - \cdots$

18. $\ln(6+5x) = 5(x+1) - \frac{5^2}{2}(x+1)^2 + \frac{5^3}{3}(x+1)^3 - \cdots + \frac{(-1)^{n-1}5^n}{n}(x+1)^n + \cdots$

C *In Problems 19–24, find the nth-degree Taylor polynomial at 0 for f, find the Taylor series at 0 for f, and determine the values of x for which $f(x) = \lim_{n\to\infty} p_n(x)$.*

19. $f(x) = e^{4x}$

20. $f(x) = e^{-x/2}$

21. $f(x) = \ln(1 + 2x)$

22. $f(x) = \ln(1 - 4x)$

23. $f(x) = \frac{2}{2-x}$

24. $f(x) = \frac{3}{3+x}$

In Problems 25–28, find the nth-degree Taylor polynomial at the indicated value of a for f, find the Taylor series at a for f, and determine the values of x for which $f(x) = \lim_{n\to\infty} p_n(x)$.

25. $f(x) = \ln(2 - x)$; $a = 1$

26. $f(x) = \ln(2 + x)$; $a = -1$

27. $f(x) = \frac{1}{1-x}$; $a = 2$

28. $f(x) = \frac{1}{4-x}$; $a = 3$

SECTION 10-3 Operations on Taylor Series

- BASIC TAYLOR SERIES
- ADDITION AND MULTIPLICATION
- DIFFERENTIATION AND INTEGRATION
- SUBSTITUTION

BASIC TAYLOR SERIES

In the preceding sections, we used repeated differentiation and the formula $a_n = f^{(n)}(a)/n!$ to find Taylor polynomials and Taylor series for a variety of simple functions. In this section we will see how to use the series for these simple functions and some basic properties of Taylor series to find the Taylor series for more complicated functions. Table 4 (on the next page) lists the series we will use in this process for convenient reference.

TABLE 4

Basic Taylor Series

FUNCTION $f(x)$	TAYLOR SERIES AT 0	INTERVAL OF CONVERGENCE
$\frac{1}{1-x}$	$= 1 + x + x^2 + \cdots + x^n + \cdots$	$-1 < x < 1$
$\frac{1}{1+x}$	$= 1 - x + x^2 - \cdots + (-1)^n x^n + \cdots$	$-1 < x < 1$
$\ln(1-x)$	$= -x - \frac{1}{2}x^2 - \frac{1}{3}x^3 - \cdots - \frac{1}{n}x^n - \cdots$	$-1 < x < 1$
$\ln(1+x)$	$= x - \frac{1}{2}x^2 + \frac{1}{3}x^3 - \cdots + \frac{(-1)^{n-1}}{n}x^n + \cdots$	$-1 < x < 1$
e^x	$= 1 + x + \frac{1}{2!}x^2 + \cdots + \frac{1}{n!}x^n + \cdots$	$-\infty < x < \infty$
e^{-x}	$= 1 - x + \frac{1}{2!}x^2 - \cdots + \frac{(-1)^n}{n!}x^n + \cdots$	$-\infty < x < \infty$

Most of the discussion in this section involves Taylor series at 0. At the end of the section we will see that Taylor series at points other than 0 can be obtained from Taylor series at 0 by using a simple substitution process.

◆ ADDITION AND MULTIPLICATION

Property 1: Addition

Two Taylor series at 0 can be added term by term:

If

$$f(x) = a_0 + a_1x + a_2x^2 + \cdots + a_nx^n + \cdots$$

and

$$g(x) = b_0 + b_1x + b_2x^2 + \cdots + b_nx^n + \cdots$$

then

$$f(x) + g(x) = (a_0 + b_0) + (a_1 + b_1)x + (a_2 + b_2)x^2 + \cdots + (a_n + b_n)x^n + \cdots$$

This operation is valid in the intersection of the intervals of convergence of the series for f and g.

◆ EXAMPLE 9 Find the Taylor series at 0 for $f(x) = e^x + \ln(1 + x)$, and find the interval of convergence.

Solution From Table 4,

$$e^x = 1 + x + \frac{1}{2!}x^2 + \frac{1}{3!}x^3 + \cdots + \frac{1}{n!}x^n + \cdots \qquad -\infty < x < \infty$$

and

$$\ln(1 + x) = x - \frac{1}{2}x^2 + \frac{1}{3}x^3 - \cdots + \frac{(-1)^{n-1}}{n}x^n + \cdots \qquad -1 < x < 1$$

Adding the coefficients of corresponding powers of x, we have

$$\begin{aligned} f(x) &= e^x + \ln(1 + x) \\ &= 1 + 2x + \left(\frac{1}{2!} - \frac{1}{2}\right)x^2 + \left(\frac{1}{3!} + \frac{1}{3}\right)x^3 + \cdots + \left[\frac{1}{n!} + \frac{(-1)^{n-1}}{n}\right]x^n + \cdots \end{aligned}$$

The series for e^x converges for all x; however, the series for $\ln(1 + x)$ converges only for $-1 < x < 1$. Thus, the combined series converges for $-1 < x < 1$, the intersection of these two intervals of convergence. ◆

PROBLEM 9 Find the Taylor series at 0 for $f(x) = e^x + 1/(1 - x)$, and find the interval of convergence. ◆

Property 2: Multiplication

A Taylor series at 0 can be multiplied term by term by an expression of the form cx^r, where c is a constant and r is a nonnegative integer:

If

$$f(x) = a_0 + a_1x + a_2x^2 + \cdots + a_nx^n + \cdots$$

then

$$cx^rf(x) = ca_0x^r + ca_1x^{r+1} + ca_2x^{r+2} + \cdots + ca_nx^{r+n} + \cdots$$

The Taylor series for $cx^rf(x)$ has the same interval of convergence as the Taylor series for f.

◆ EXAMPLE 10 Find the Taylor series at 0 for $f(x) = 2x^2e^{-x}$, and find the interval of convergence.

Solution From Table 4,

$$e^{-x} = 1 - x + \frac{1}{2!}x^2 - \cdots + \frac{(-1)^n}{n!}x^n + \cdots \qquad -\infty < x < \infty$$

Multiplying each term of this series by $2x^2$, the series for $f(x)$ is

$$\begin{aligned} f(x) &= 2x^2e^{-x} \\ &= 2x^2\left[1 - x + \frac{1}{2!}x^2 - \cdots + \frac{(-1)^n}{n!}x^n + \cdots\right] \\ &= 2x^2 - 2x^3 + \frac{2}{2!}x^4 - \cdots + \frac{2(-1)^n}{n!}x^{n+2} + \cdots \end{aligned}$$

Since the series for e^{-x} converges for all x, the series for $f(x)$ also converges for $-\infty < x < \infty$. ◆

PROBLEM 10 Find the Taylor series at 0 for $f(x) = 3x^3 \ln(1 - x)$, and find the interval of convergence. ◆

◆ DIFFERENTIATION AND INTEGRATION

Property 3: Differentiation

A Taylor series at 0 can be differentiated term by term:

If

$$f(x) = a_0 + a_1x + a_2x^2 + a_3x^3 + \cdots + a_nx^n + \cdots$$

then

$$f'(x) = a_1 + 2a_2x + 3a_3x^2 + \cdots + na_nx^{n-1} + \cdots$$

The Taylor series for f' has the same interval of convergence as the Taylor series for f.

◆ EXAMPLE 11 Find the Taylor series at 0 for $f(x) = 1/(1 - x)^2$, and find the interval of convergence.

Solution We want to relate f to the derivative of one of the functions in Table 4, so that we can apply property 3. Since

$$\begin{aligned} D_x\left(\frac{1}{1-x}\right) &= D_x(1 - x)^{-1} \\ &= -(1 - x)^{-2}(-1) \\ &= \frac{1}{(1 - x)^2} \\ &= f(x) \end{aligned}$$

we can apply property 3 to the series for $1/(1-x)$. From Table 4,

$$\frac{1}{1-x} = 1 + x + x^2 + x^3 + \cdots + x^n + \cdots \qquad -1 < x < 1$$

Differentiating both sides of this equation,

$$D_x\left(\frac{1}{1-x}\right) = D_x(1 + x + x^2 + x^3 + \cdots + x^n + \cdots)$$

$$\frac{1}{(1-x)^2} = D_x\,1 + D_x\,x + D_x\,x^2 + D_x\,x^3 + \cdots + D_x\,x^n + \cdots$$

$$f(x) = 1 + 2x + 3x^2 + \cdots + nx^{n-1} + \cdots$$

Since the series for $1/(1-x)$ converges for $-1 < x < 1$, the series for $f(x)$ also converges for $-1 < x < 1$. [Verify this statement by applying Theorem 1 in the preceding section to the series for $f(x)$.] ◆

PROBLEM 11 Find the Taylor series at 0 for $f(x) = 1/(1+x)^2$, and find the interval of convergence. ◆

Property 4: Integration

A Taylor series at 0 can be integrated term by term:

If

$$f(x) = a_0 + a_1x + a_2x^2 + \cdots + a_nx^n + \cdots$$

then

$$\int f(x)\,dx = C + a_0x + \frac{1}{2}a_1x^2 + \frac{1}{3}a_2x^3 + \cdots + \frac{1}{n+1}a_nx^{n+1} + \cdots$$

where C is the constant of integration. The Taylor series for $\int f(x)\,dx$ has the same interval of convergence as the Taylor series for f.

◆ EXAMPLE 12 If f is a function that satisfies $f'(x) = x^2e^x$ and $f(0) = 2$, find the Taylor series at 0 for f. Find the interval of convergence.

Solution Using the series for e^x from Table 4 and property 2, the series for f' is

$$\begin{aligned} f'(x) &= x^2e^x \\ &= x^2\left(1 + x + \frac{1}{2!}x^2 + \cdots + \frac{1}{n!}x^n + \cdots\right) \qquad -\infty < x < \infty \\ &= x^2 + x^3 + \frac{1}{2!}x^4 + \cdots + \frac{1}{n!}x^{n+2} + \cdots \end{aligned}$$

Integrating term by term produces a series for f:

$$
\begin{aligned}
f(x) &= \int f'(x)\,dx \\
&= \int \left(x^2 + x^3 + \frac{1}{2!}x^4 + \cdots + \frac{1}{n!}x^{n+2} + \cdots\right) dx \\
&= \int x^2\,dx + \int x^3\,dx + \frac{1}{2!}\int x^4\,dx + \cdots + \frac{1}{n!}\int x^{n+2}\,dx + \cdots \\
&= C + \frac{1}{3}x^3 + \frac{1}{4}x^4 + \frac{1}{(5)2!}x^5 + \cdots + \frac{1}{(n+3)n!}x^{n+3} + \cdots
\end{aligned}
$$

Now we use the condition $f(0) = 2$ to evaluate the constant of integration C:

$$
\begin{aligned}
2 &= f(0) \\
&= C + 0 + 0 + \cdots + 0 + \cdots \\
&= C
\end{aligned}
$$

Thus,

$$
f(x) = 2 + \frac{1}{3}x^3 + \frac{1}{4}x^4 + \cdots + \frac{1}{(n+3)n!}x^{n+3} + \cdots
$$

Since the series for f' converges for $-\infty < x < \infty$, the series for f converges for $-\infty < x < \infty$. ◆

PROBLEM 12 If f is a function that satisfies $f'(x) = x\ln(1+x)$ and $f(0) = 4$, find the Taylor series at 0 for f. Find the interval of convergence. ◆

◆ SUBSTITUTION

◆ EXAMPLE 13 Find the Taylor series at 0 for $f(x) = e^{-x^2}$, and find the interval of convergence.

Solution Suppose we try to solve this problem by finding the general form of the nth derivative of f:

$$
\begin{aligned}
f(x) &= e^{-x^2} \\
f'(x) &= -2xe^{-x^2} \\
f''(x) &= -2e^{-x^2} + 4x^2e^{-x^2} \\
f^{(3)}(x) &= 4xe^{-x^2} + 8xe^{-x^2} - 8x^3e^{-x^2} \\
&= 12xe^{-x^2} - 8x^3e^{-x^2} \\
f^{(4)}(x) &= 12e^{-x^2} - 24x^2e^{-x^2} - 24x^2e^{-x^2} + 16x^4e^{-x^2} \\
&= 12e^{-x^2} - 48x^2e^{-x^2} + 16x^4e^{-x^2}
\end{aligned}
$$

Since the higher-order derivatives are becoming very complicated and no general pattern is emerging, we will try another approach. How can we relate f to one of the functions in Table 4? If we let $g(x) = e^{-x}$, then f and g are related by

$$f(x) = e^{-x^2} = g(x^2)$$

From Table 4,

$$\begin{aligned} g(x) &= e^{-x} \\ &= 1 - x + \frac{1}{2!}x^2 - \frac{1}{3!}x^3 + \cdots + \frac{(-1)^n}{n!}x^n + \cdots \qquad -\infty < x < \infty \end{aligned}$$

Substituting x^2 for x in the series for g, we have

$$\begin{aligned} f(x) &= g(x^2) \\ &= 1 - x^2 + \frac{1}{2!}(x^2)^2 - \frac{1}{3!}(x^2)^3 + \cdots + \frac{(-1)^n}{n!}(x^2)^n + \cdots \\ &= 1 - x^2 + \frac{1}{2!}x^4 - \frac{1}{3!}x^6 + \cdots + \frac{(-1)^n}{n!}x^{2n} + \cdots \end{aligned}$$

Since the series for g converges for all values of x, the series for f must also converge for all values of x. ◆

PROBLEM 13 Find the Taylor series at 0 for $f(x) = e^{x^3}$, and find the interval of convergence. ◆

Making a substitution in a known Taylor series in order to obtain a new series is a very useful technique. Since there are many different substitutions that can be used, it is difficult to make a general statement concerning the effect of a substitution on the interval of convergence. The following examples illustrate some of the possibilities that may occur.

◆ EXAMPLE 14 Find the Taylor series at 0 for $f(x) = 1/(4 - x)$, and find the inteval of convergence.

Solution If we factor a 4 out of the denominator of f, we can establish a relationship between $f(x)$ and $g(x) = 1/(1 - x)$. Thus,

$$\begin{aligned} f(x) &= \frac{1}{4 - x} \\ &= \left(\frac{1}{4}\right)\frac{1}{1 - (x/4)} \qquad g(x) = \frac{1}{1 - x} \\ &= \frac{1}{4}g\left(\frac{x}{4}\right) \end{aligned}$$

From Table 4,

$$g(x) = \frac{1}{1-x}$$
$$= 1 + x + x^2 + \cdots + x^n + \cdots \qquad -1 < x < 1$$

Substituting $x/4$ for x in this series and multiplying by $\frac{1}{4}$, we have

$$f(x) = \frac{1}{4} g\left(\frac{x}{4}\right)$$
$$= \frac{1}{4}\left[1 + \left(\frac{x}{4}\right) + \left(\frac{x}{4}\right)^2 + \cdots + \left(\frac{x}{4}\right)^n + \cdots\right]$$
$$= \frac{1}{4} + \frac{1}{4^2}x + \frac{1}{4^3}x^2 + \cdots + \frac{1}{4^{n+1}}x^n + \cdots$$

Since the original series for g converges for $-1 < x < 1$ and we substituted $x/4$ for x in that series, the series for f converges for

$$-1 < \frac{x}{4} < 1 \qquad \text{Multiply each member by 4.}$$
$$-4 < x < 4$$

◆

PROBLEM 14 Find the Taylor series at 0 for $f(x) = 1/(3 + x)$, and find the interval of convergence. ◆

◆ EXAMPLE 15 Find the Taylor series at 0 for $f(x) = \ln(1 + 4x^2)$, and find the interval of convergence.

Solution If we let $g(x) = \ln(1 + x)$, then

$$f(x) = \ln(1 + 4x^2)$$
$$= g(4x^2)$$

From Table 4, the series for g is

$$g(x) = x - \frac{1}{2}x^2 + \frac{1}{3}x^3 - \cdots + \frac{(-1)^{n-1}}{n}x^n + \cdots \qquad -1 < x < 1$$

Substituting $4x^2$ for x in this series, we have

$$f(x) = g(4x^2)$$
$$= 4x^2 - \frac{1}{2}(4x^2)^2 + \frac{1}{3}(4x^2)^3 - \cdots + \frac{(-1)^{n-1}}{n}(4x^2)^n + \cdots$$
$$= 4x^2 - \frac{4^2}{2}x^4 + \frac{4^3}{3}x^6 - \cdots + \frac{(-1)^{n-1}4^n}{n}x^{2n} + \cdots$$

The series for g converges for $-1 < x < 1$. Since we substituted $4x^2$ for x, the series for f converges for

$$-1 < 4x^2 < 1$$

Inequalities of this type are easier to solve if we use absolute value notation:

$$-1 < 4x^2 < 1 \qquad \text{Change to absolute value notation.}$$
$$|4x^2| < 1 \qquad \text{Multiply by } \tfrac{1}{4}.$$
$$|x^2| < \tfrac{1}{4}$$
$$|x|^2 < \tfrac{1}{4}$$
$$|x| < \tfrac{1}{2} \qquad \text{Convert to double inequalities.}$$
$$-\tfrac{1}{2} < x < \tfrac{1}{2}$$

Thus, the series for f converges for $-\frac{1}{2} < x < \frac{1}{2}$. ◆

PROBLEM 15 Find the Taylor series at 0 for $f(x) = 1/(1 + 9x^2)$, and find the interval of convergence. ◆

Up to this point in this section we have restricted our attention to Taylor series at 0. How can we use the techniques discussed to find a Taylor series at a point $a \neq 0$? Properties 1–4 could be stated in terms of Taylor series at an arbitrary point a; however, there is an easier way to proceed. The method of substitution allows us to use Taylor series at 0 to find Taylor series at other points. The following example illustrates this technique.

◆ EXAMPLE 16 Find the Taylor series at 1 for $f(x) = 1/(2 - x)$, and find the interval of convergence.

Solution In order to find a Taylor series for f in powers of $x - 1$, we will use the substitution $t = x - 1$ to express f as a function of t. If we find the Taylor series at 0 for this new function and then replace t with $x - 1$, we will have obtained the Taylor series at 1 for f.

$$t = x - 1 \qquad \text{Solve for x.}$$
$$x = t + 1 \qquad \text{Substitute for x in } f(x) = 1/(2 - x).$$
$$\frac{1}{2 - x} = \frac{1}{2 - (t + 1)}$$
$$= \frac{1}{1 - t} \qquad \text{Find the Taylor series at 0 for this function of } t.$$
$$= 1 + t + t^2 + \cdots + t^n + \cdots \qquad \text{Substitute } x - 1 \text{ for } t.$$
$$-1 < t < 1$$
$$= 1 + (x - 1) + (x - 1)^2 + \cdots + (x - 1)^n + \cdots$$

Since $t = x - 1$ and the series in powers of t converges for $-1 < t < 1$, the series in powers of $x - 1$ converges for

$$-1 < x - 1 < 1 \qquad \text{Add 1 to each member.}$$
$$0 < x < 2$$ ◆

PROBLEM 16 Find the Taylor series at 1 for $f(x) = 1/x$, and find the interval of convergence. ◆

Answers to Matched Problems

9. $2 + 2x + \left(\frac{1}{2!} + 1\right)x^2 + \cdots + \left(\frac{1}{n!} + 1\right)x^n + \cdots, -1 < x < 1$
10. $-3x^4 - \frac{3}{2}x^5 - x^6 - \cdots - \frac{3}{n}x^{n+3} - \cdots, -1 < x < 1$
11. $1 - 2x + 3x^2 - \cdots + (-1)^{n+1}nx^{n-1} + \cdots, -1 < x < 1$
12. $4 + \frac{1}{3}x^3 - \frac{1}{8}x^4 + \frac{1}{15}x^5 - \cdots + \frac{(-1)^{n-1}}{n(n+2)}x^{n+2} + \cdots, -1 < x < 1$
13. $1 + x^3 + \frac{1}{2!}x^6 + \cdots + \frac{1}{n!}x^{3n} + \cdots, -\infty < x < \infty$
14. $\frac{1}{3} - \frac{1}{3^2}x + \frac{1}{3^3}x^2 - \cdots + \frac{(-1)^n}{3^{n+1}}x^n + \cdots, -3 < x < 3$
15. $1 - 9x^2 + 9^2x^4 - \cdots + (-1)^n 9^n x^{2n} + \cdots, -\frac{1}{3} < x < \frac{1}{3}$
16. $1 - (x - 1) + (x - 1)^2 - \cdots + (-1)^n(x - 1)^n + \cdots, 0 < x < 2$

EXERCISE 10-3

Solve all the problems in this exercise by performing operations on the Taylor series in Table 4. State the interval of convergence for each series you find.

A *In Problems 1–14, find the Taylor series at 0.*

1. $f(x) = \frac{1}{1 - x} + \ln(1 + x)$
2. $f(x) = \frac{1}{1 - x} + \ln(1 - x)$
3. $f(x) = \frac{x^3}{1 - x}$
4. $f(x) = x^3e^x$
5. $f(x) = xe^{x^2}$
6. $f(x) = xe^{-2x^2}$
7. $f(x) = \frac{1}{1 - x^3}$
8. $f(x) = \ln(1 + x^3)$

B

9. $f(x) = \frac{1}{2 - x}$
10. $f(x) = \frac{1}{5 + x}$
11. $f(x) = \frac{1}{1 - 8x^3}$
12. $f(x) = \ln(1 + 16x^2)$

13. $f(x) = \dfrac{1}{4 + x^2}$

14. $f(x) = \dfrac{1}{9 - x^2}$

15. Find the Taylor series at 0 for:

(A) $f(x) = \dfrac{1}{1 - x^2}$ (B) $g(x) = \dfrac{2x}{(1 - x^2)^2}$

[*Hint:* Compare $f'(x)$ and $g(x)$.]

16. Find the Taylor series at 0 for:

(A) $f(x) = \dfrac{x}{1 - x^2}$ (B) $g(x) = \dfrac{1 + x^2}{(1 - x^2)^2}$

[*Hint:* Compare $f'(x)$ and $g(x)$.]

17. If $f(x)$ satisfies $f'(x) = 1/(1 + x^2)$ and $f(0) = 0$, find the Taylor series at 0 for $f(x)$.

18. If $f(x)$ satisfies $f'(x) = \ln(1 + x^2)$ and $f(0) = 1$, find the Taylor series at 0 for $f(x)$.

Use the substitution $t = x - a$ to find the Taylor series at the indicated value of a.

19. $f(x) = \dfrac{1}{4 - x}$ at 3

20. $f(x) = \dfrac{1}{3 + x}$ at -2

21. $f(x) = \ln x$ at 1

22. $f(x) = \ln(3 - x)$ at 2

23. $f(x) = \dfrac{1}{4 - 3x}$ at 1

24. $f(x) = \dfrac{1}{5 - 2x}$ at 2

C

25. Use the Taylor series at 0 for $1/(1 - x)$ and repeated applications of property 3 to find the Taylor series at 0 for

$$f(x) = \frac{1}{(1 - x)^3}$$

26. Use the Taylor series at 0 for $1/(1 + x)$ and repeated applications of property 3 to find the Taylor series at 0 for

$$f(x) = \frac{1}{(1 + x)^4}$$

27. Find the Taylor series at 0 for: $f(x) = \dfrac{1 + x}{1 - x}$

$$\left[\textit{Note:}\quad \frac{1 + x}{1 - x} = \frac{1}{1 - x} + \frac{x}{1 - x}\right]$$

28. Find the Taylor series at 0 for: $f(x) = \dfrac{1 - 2x}{1 + x}$

$$\left[\textit{Note:}\quad \frac{1 - 2x}{1 + x} = \frac{1}{1 + x} - \frac{2x}{1 + x}\right]$$

29. Find the Taylor series at 0 for: $f(x) = \frac{1}{2}\ln\left(\frac{1+x}{1-x}\right)$

$\left[\textit{Note:}\ \ln\left(\frac{1+x}{1-x}\right) = \ln(1+x) - \ln(1-x)\right]$

30. Find the Taylor series at 0 for: $f(x) = \frac{e^x - e^{-x}}{2}$

31. Find the Taylor series at 0 for: $f(x) = \frac{e^x + e^{-x}}{2}$

32. Find the Taylor series at any a for $f(x) = e^x$.
33. Find the Taylor series at any $a > 0$ for $f(x) = \ln x$.
34. If a and b are constants ($a \neq b$), find the Taylor series at a for

$$f(x) = \frac{1}{b - x}$$

SECTION 10-4 Approximations Using Taylor Series

- THE REMAINDER
- TAYLOR'S FORMULA FOR THE REMAINDER
- TAYLOR SERIES WITH ALTERNATING TERMS
- APPROXIMATING DEFINITE INTEGRALS

THE REMAINDER

Now that we can find Taylor series for a variety of functions, we return to our original goal: approximating the values of a function.

If x is in the interval of convergence of the Taylor series for a function f and $p_n(x)$ is the nth-degree Taylor polynomial, then

$$f(x) = \lim_{n\to\infty} p_n(x)$$

and $p_n(x)$ can be used to approximate $f(x)$. We want to consider two questions:

1. If we select a particular value of n, how accurate is the approximation $f(x) \approx p_n(x)$?
2. If we want the approximation $f(x) \approx p_n(x)$ to have a specified accuracy, how do we select the proper value of n?

It turns out that both of these questions can be answered by examining the difference between $f(x)$ and $p_n(x)$. This difference is called the *remainder* and is defined in the box on the next page.

The Remainder of a Taylor Series

If p_n is the nth-degree Taylor polynomial for f, then the **remainder** is

$$R_n(x) = f(x) - p_n(x)$$

The **error** in the approximation $f(x) \approx p_n(x)$ is

$$|f(x) - p_n(x)| = |R_n(x)|$$

If the Taylor series at 0 for f is

$$f(x) = \overbrace{a_0 + a_1x + a_2x^2 + \cdots + a_nx^n}^{p_n(x)} + \overbrace{a_{n+1}x^{n+1} + \cdots}^{R_n(x)}$$

then

$$p_n(x) = a_0 + a_1x + a_2x^2 + \cdots + a_nx^n$$

and

$$R_n(x) = a_{n+1}x^{n+1} + a_{n+2}x^{n+2} + \cdots$$

Similar statements can be made for Taylor series at a.

It follows from our basic assumption (see page 670) for the functions we consider that

$$\lim_{n\to\infty} R_n(x) = \lim_{n\to\infty} [f(x) - p_n(x)] = f(x) - f(x) = 0$$

if and only if x is in the interval of convergence of f.*

In general, it is difficult to find the exact value of $R_n(x)$. In fact, since $f(x) = p_n(x) + R_n(x)$, this is equivalent to finding the exact value of $f(x)$. Instead, we will discuss two methods for *estimating* the value of $R_n(x)$. The first method works in all cases, but can be difficult to apply, while the second method is easy to apply, but does not work in all cases.

◆ TAYLOR'S FORMULA FOR THE REMAINDER

The first method for estimating the remainder of a Taylor series is based on *Taylor's formula for the remainder.*

---

* In more advanced texts, the statement $\lim_{x\to\infty} R_n(x) = 0$ for x in the interval of convergence is actually proved for functions such as e^x, making our basic assumption unnecessary.

Taylor's Formula for the Remainder

If f has derivatives of all order at 0, then

$$R_n(x) = \frac{f^{(n+1)}(t)x^{n+1}}{(n+1)!}$$

where t is a number between 0 and x.

A similar formula can be stated for the remainder of a Taylor series at an arbitrary point a. (We will not use this formula in this text.)

In most applications of the remainder formula, it is not possible to find the value of t. However, if we can find a number M satisfying

$$|f^{(n+1)}(t)| < M \qquad \text{for } t \text{ between 0 and } x$$

then we can estimate $R_n(x)$ as follows:

$$|R_n(x)| = \left|\frac{f^{(n+1)}(t)x^{n+1}}{(n+1)!}\right| < \frac{M|x|^{n+1}}{(n+1)!}$$

This technique is illustrated in the next example.

◆ EXAMPLE 17 Use the second-degree Taylor polynomial at 0 for $f(x) = e^x$ to approximate $e^{0.1}$. Use Taylor's formula for the remainder to estimate the error in this approximation.

Solution Since $f^{(n)}(x) = e^x$ for any n, we can write

$$f(x) = e^x = p_2(x) + R_2(x)$$

$$= \underbrace{1 + x + \frac{1}{2!}x^2}_{p_2(x)} + \underbrace{\frac{e^t x^3}{3!}}_{R_2(x)} \qquad f^{(3)}(t) = e^t$$

where t is a number between 0 and x. Thus,

$$\begin{aligned} f(0.1) &= e^{0.1} \\ &= p_2(0.1) + R_2(0.1) \\ &\approx p_2(0.1) \\ &= 1 + 0.1 + \tfrac{1}{2}(0.1)^2 \\ &= 1.105 \end{aligned}$$

To estimate the error in this approximation, we must estimate

$$|R_2(0.1)| = \left|\frac{e^t(0.1)^3}{3!}\right| = \frac{e^t}{6{,}000}$$

where $0 \le t \le 0.1$. In order to estimate $|R_2(0.1)|$, we must estimate $g(t) = e^t$ for $0 \le t \le 0.1$. Since e^t is always increasing, $e^t \le e^{0.1}$ for $0 \le t \le 0.1$. However, $e^{0.1}$ is

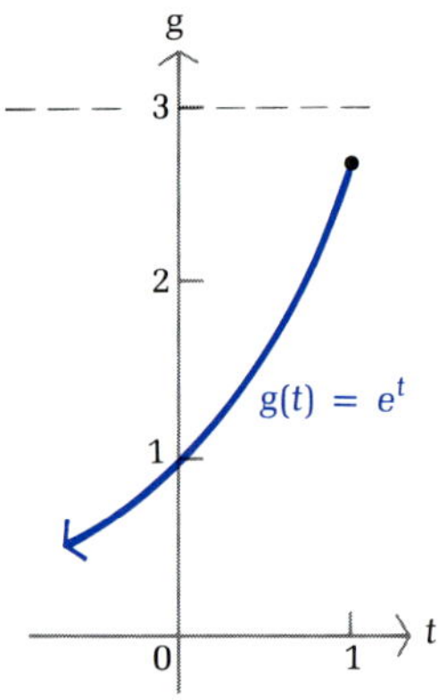

the number we are trying to approximate. We do not want to use this number in our estimate of the error. (This situation occurs frequently in approximation problems involving the exponential function.) Instead, we will use the following rough estimate for e^t (see the figure):

$$\text{If} \quad t \leq 1, \quad \text{then} \quad e^t \leq 3. \tag{1}$$

Since (1) holds for $t \leq 1$, it certainly holds for $0 \leq t \leq 0.1$. Thus,

$$|R_2(0.1)| = \frac{e^t}{6{,}000} \leq \frac{3}{6{,}000} = 0.0005$$

and we can conclude that the approximate value 1.105 is within ± 0.0005 of the exact value of $e^{0.1}$. ◆

PROBLEM 17 Use the second-degree Taylor polynomial at 0 for $f(x) = e^x$ to approximate $e^{0.2}$. Use Taylor's formula for the remainder to estimate the error in the approximation. ◆

◆ TAYLOR SERIES WITH ALTERNATING TERMS

Taylor's formula for the remainder can be difficult to apply for large values of n. For most functions, the formula for the nth derivative becomes very complicated as n increases. For a certain type of problem, there is another method that does not require estimation of the nth derivative. If the series of numbers that is formed by evaluating a Taylor series at a given number x_0 *alternates in sign and decreases in absolute value,* then the remainder can be estimated by simply examining the numbers in this series. Series of numbers whose terms alternate in sign are called **alternating series.** The estimate for the remainder in an alternating series is given in Theorem 2.

THEOREM 2 **Error Estimation for Alternating Series**

If x_0 is a number in the interval of convergence for

$$f(x) = a_0 + a_1x + a_2x^2 + \cdots + a_kx^k + \cdots$$

and the terms in the series

$$f(x_0) = a_0 + a_1x_0 + a_2x_0^2 + \cdots + a_kx_0^k + \cdots$$

are alternating in sign and decreasing in absolute value, then the error in the approximation

$$f(x_0) \approx a_0 + a_1x_0 + a_2x_0^2 + \cdots + a_nx_0^n$$

is strictly less than the absolute value of the next term. That is,

$$|R_n(x_0)| < |a_{n+1}x_0^{n+1}|$$

◆ EXAMPLE 18 Use the Taylor series at 0 for $f(x) = e^{-x}$ to approximate $e^{-0.3}$ with an error of no more than 0.0005.

Solution From Table 4, the Taylor series at 0 for $f(x) = e^{-x}$ is

$$e^{-x} = 1 - x + \frac{1}{2!}x^2 - \frac{1}{3!}x^3 + \cdots + \frac{(-1)^k}{k!}x^k + \cdots$$

If we substitute $x = 0.3$ in this series, we obtain

$$\begin{aligned} f(0.3) &= e^{-0.3} \\ &= 1 - 0.3 + \tfrac{1}{2}(0.3)^2 - \tfrac{1}{6}(0.3)^3 + \tfrac{1}{24}(0.3)^4 - \cdots \\ &= 1 - 0.3 + 0.045 - 0.0045 + 0.000\ 337\ 5 - \cdots \end{aligned}$$

Since the terms in this series are alternating in sign and decreasing in absolute value, Theorem 2 applies. If we use the first four terms in this series to approximate $e^{-0.3}$, then the error in this approximation is less than the absolute value of the fifth term. That is,

$$|R_3(0.3)| < 0.000\ 337\ 5 < 0.0005$$

Thus,

$$e^{-0.3} \approx 1 - 0.3 + 0.045 - 0.0045 = 0.7405$$

and the error in this approximation is less than the specified accuracy of 0.0005. ◆

PROBLEM 18 Use the Taylor series at 0 for $f(x) = e^{-x}$ to approximate $e^{-0.1}$ with an eror of no more than 0.0005. ◆

As Example 18 illustrates, Theorem 2 is much easier to use than Taylor's formula for the remainder. Notice that we did not have to find an estimate for $f^{(n+1)}(t)$. However, it is very important to understand that Theorem 2 can be applied only if the terms in the series alternate in sign *after they have been evaluated at* x_0. For example, if we try to use the series for e^{-x} to approximate $e^{0.3}$ by substituting $x_0 = -0.3$, we have

$$\begin{aligned} e^{0.3} &= 1 - (-0.3) + \tfrac{1}{2}(-0.3)^2 - \tfrac{1}{6}(-0.3)^3 + \cdots \\ &= 1 + 0.3 + \tfrac{1}{2}(0.3)^2 + \tfrac{1}{6}(0.3)^3 + \cdots \end{aligned}$$

Since these numbers do not alternate in sign, Theorem 2 does not apply. Taylor's formula for the remainder would have to be used to estimate the error in approximations obtained from this series.

◆ APPROXIMATING DEFINITE INTEGRALS

In order to find the exact value of a definite integral $\int_a^b f(x)\,dx$, we must first find an antiderivative of the function f. But suppose we cannot find an antiderivative of f (it may not even exist in a convenient form). In Section 6-6 we saw that the rectangle rule can be used to approximate definite integrals. Taylor series techniques provide an alternative method for approximating definite integrals that is often more efficient than the rectangle rule and, in the case of alternating series, automatically determines the accuracy of the approximation.

◆ EXAMPLE 19 Approximate $\int_0^1 e^{-x^2}\,dx$ with a maximum error of 0.005.

Solution If $F(x)$ is an antiderivative of e^{-x^2}, then

$$\int_0^1 e^{-x^2}\,dx = F(1) - F(0)$$

It is not possible to express $F(x)$ as a finite combination of simple functions; however, it is possible to find a Taylor series for F. This series can be used to approximate the values of F and, consequently, of the definite integral.

Step 1. Find a Taylor series for the integrand. From Table 4,

$$e^x = 1 + x + \frac{1}{2!}x^2 + \cdots + \frac{1}{n!}x^n + \cdots \qquad -\infty < x < \infty$$

Thus,

$$\begin{aligned} e^{-x^2} &= 1 + (-x^2) + \frac{1}{2!}(-x^2)^2 + \cdots + \frac{1}{n!}(-x^2)^n + \cdots \\ &= 1 - x^2 + \frac{1}{2!}x^4 - \cdots + \frac{(-1)^n}{n!}x^{2n} + \cdots \qquad -\infty < x < \infty \end{aligned}$$

Step 2. Find the Taylor series for the antiderivative. Integrating term by term, we have

$$\begin{aligned} F(x) &= \int e^{-x^2}\,dx \\ &= \int \left[1 - x^2 + \frac{1}{2!}x^4 - \cdots + \frac{(-1)^n}{n!}x^{2n} + \cdots\right] dx \\ &= C + x - \frac{1}{3}x^3 + \frac{1}{10}x^5 - \cdots + \frac{(-1)^n}{n!(2n+1)}x^{2n+1} + \cdots \qquad -\infty < x < \infty \end{aligned}$$

Step 3. Approximate the definite integral. If we choose $C = 0$, then $F(0) = 0$ and

$$\begin{aligned} \int_0^1 e^{-x^2}\,dx &= F(1) - F(0) \\ &= \left[x - \frac{1}{3}x^3 + \frac{1}{10}x^5 - \cdots + \frac{(-1)^n}{n!(2n+1)}x^{2n+1} + \cdots\right]\Bigg|_0^1 \\ &= \left[1 - \frac{1}{3} + \frac{1}{10} - \cdots + \frac{(-1)^n}{n!(2n+1)} + \cdots\right] - 0 \\ &= 1 - \frac{1}{3} + \frac{1}{10} - \frac{1}{42} + \frac{1}{216} - \cdots \\ &= 1 - 0.333\ 333 + 0.1 - 0.023\ 809 + 0.004\ 630 - \cdots \end{aligned}$$

Since the Taylor series for $F(x)$ converges for all values of x, we can use this last series of numbers to approximate $F(1)$. Notice that the terms in this series are alternating in sign and decreasing in absolute value. Applying Theorem 2, we conclude that the error introduced by approximating $F(1)$ with the first four terms of this series will be no more than 0.004 630, the absolute value of the fifth term. Since this is less than the specified error of 0.005, we have

$$\begin{aligned}\int_0^1 e^{-x^2}\,dx &= F(1) - F(0)\\ &\approx 1 - 0.333\ 333 + 0.1 - 0.023\ 809\\ &\approx 0.743 \qquad \text{Rounded to three decimal places}\end{aligned}$$

◆

PROBLEM 19 Approximate $\int_0^{0.5} e^{-x^2}\,dx$ with a maximum error of 0.005. ◆

◆ EXAMPLE 20 *Income Distribution* Approximate the coefficient of inequality for the Lorenz curve given by

$$f(x) = \frac{11x^4}{10 + x^2}$$

with an error of no more than 0.005.

Solution Referring to Section 7-1, the coefficient of inequality for a Lorenz curve is twice the area between the graph of the Lorenz curve and the graph of the line $y = x$ (see the figure). Thus, we must evaluate the integral

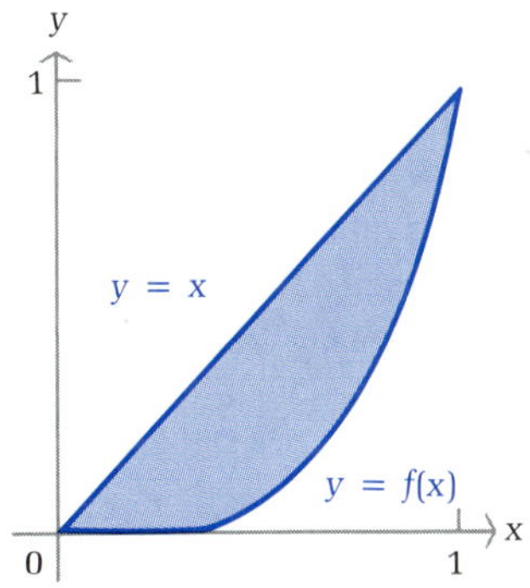

$$\begin{aligned}2\int_0^1 [x - f(x)]\,dx &= 2\int_0^1 x\,dx - 2\int_0^1 f(x)\,dx\\ &= \int_0^1 2x\,dx - \int_0^1 \frac{22x^4}{10 + x^2}\,dx \qquad (2)\end{aligned}$$

The first integral in (2) is easy to evaluate:

$$\int_0^1 2x\,dx = x^2\Big|_0^1 = 1 - 0 = 1$$

Since the second integral in (2) cannot be evaluated by any of the techniques we have discussed, we will use a Taylor series to approximate this integral.

Step 1. Find a Taylor series for the integrand:

$$\frac{22x^4}{10 + x^2} = \frac{22x^4}{10}\left[\frac{1}{1 + (x^2/10)}\right] \qquad \text{Substitute } x^2/10 \text{ for } x \text{ in the series for } 1/(1+x).$$

$$= 2.2x^4\left[1 - \frac{x^2}{10} + \left(\frac{x^2}{10}\right)^2 - \cdots + (-1)^n\left(\frac{x^2}{10}\right)^n + \cdots\right] \qquad -1 < \frac{x^2}{10} < 1$$

$$= 2.2x^4 - \frac{2.2}{10}x^6 + \frac{2.2}{10^2}x^8 - \cdots + \frac{2.2(-1)^n}{10^n}x^{2n+4} + \cdots$$

To find the interval of convergence, we solve $-1 < x^2/10 < 1$ for x:

$-1 < \frac{x^2}{10} < 1$	Change to absolute value notation.
$\left\lvert\frac{x^2}{10}\right\rvert < 1$	Multiply by 10.
$\lvert x^2\rvert < 10$	Take the square root of both sides.
$\lvert x\rvert < \sqrt{10}$	Convert to double inequalities.
$-\sqrt{10} < x < \sqrt{10}$	Interval of convergence

Step 2. Find the Taylor series for the antiderivative. Using property 4,

$$\int \frac{22x^4}{10+x^2}\,dx = \int \left[2.2x^4 - \frac{2.2}{10}x^6 + \frac{2.2}{10^2}x^8 - \cdots + \frac{2.2(-1)^n}{10^n}x^{2n+4} + \cdots\right] dx$$
$$= C + \frac{2.2}{5}x^5 - \frac{2.2}{7\cdot 10}x^7 + \frac{2.2}{9\cdot 10^2}x^9 - \cdots + \frac{2.2(-1)^n}{(2n+5)10^n}x^{2n+5} + \cdots$$

This series also converges for $-\sqrt{10} < x < \sqrt{10}$.

Step 3. Approximate the definite integral. Choosing $C = 0$ in the antiderivative, we have

$$\int_0^1 \frac{22x^4}{10+x^2}\,dx = \left[\frac{2.2}{5}x^5 - \frac{2.2}{7\cdot 10}x^7 + \frac{2.2}{9\cdot 10^2}x^9 - \cdots + \frac{2.2(-1)^n}{(2n+5)10^n}x^{2n+5} + \cdots\right]\Bigg|_0^1$$
$$= \left[\frac{2.2}{5} - \frac{2.2}{7\cdot 10} + \frac{2.2}{9\cdot 10^2} - \cdots + \frac{2.2(-1)^n}{(2n+5)10^n} + \cdots\right] - 0$$
$$= 0.44 - 0.031\ 429 + 0.002\ 444 - \cdots$$

Since the limits of integration, 0 and 1, are within the interval of convergence of the Taylor series for the antiderivative, we can use this series to approximate the value of the definite integral. Notice that the numbers in this series are alternating in sign and decreasing in absolute value. Since the absolute value of the third term is less than 0.005, Theorem 2 implies that we can use the first two terms of this series to approximate the definite integral to the specified accuracy. Thus,

$$\int_0^1 \frac{22x^4}{10+x^2}\,dx \approx 0.44 - 0.031\ 429$$
$$\approx 0.409 \qquad \text{Rounded to three decimal places}$$

Returning to (2), we have

$$2\int_0^1 [x - f(x)]\,dx = 2\int_0^1 x\,dx - 2\int_0^1 f(x)\,dx$$
$$\approx 1 - 0.409$$
$$= 0.591 \qquad \text{Coefficient of inequality}$$

◆

PROBLEM 20 Repeat Example 20 for: $f(x) = \dfrac{11x^5}{10 + x^2}$ ◆

Answers to Matched Problems

17. $e^{0.2} \approx 1 + 0.2 + \frac{1}{2}(0.2)^2 = 1.22$; $|R_2(0.2)| < 0.004$
18. $e^{-0.1} \approx 1 - 0.1 + 0.005 = 0.905$ within $\pm 0.000\ 167$
19. $\int_0^{0.5} e^{-x^2}\, dx \approx 0.5 - \frac{1}{3}(0.5)^3 \approx 0.458$ within $\pm 0.003\ 125$
20. $\int_0^1 \dfrac{22x^5}{10 + x^2}\, dx \approx \dfrac{2.2}{6} - \dfrac{2.2}{80} \approx 0.339$ within ± 0.0022; coefficient of inequality ≈ 0.661

EXERCISE 10-4

In Problems 1–30, use Theorem 2 to perform the indicated error estimations.

A

Evaluating the Taylor series at 0 for $f(x) = e^{-x}$ at $x = 0.6$ produces the following series:

$$e^{-0.6} = 1 - 0.6 + 0.18 - 0.036 + 0.0054 - 0.000\ 648 + \cdots$$

In Problems 1–4, use the indicated number of terms in this series to approximate $e^{-0.6}$, and then estimate the error in this approximation.

1. Two terms **2.** Three terms **3.** Four terms **4.** Five terms

Evaluating the Taylor series at 0 for $f(x) = \ln(1 + x)$ at $x = 0.3$ produces the following series:

$$\ln 1.3 = 0.3 - 0.045 + 0.009 - 0.002\ 025 + 0.000\ 486 - 0.000\ 121\ 5 + \cdots$$

In Problems 5–8, use the indicated number of terms in this series to approximate $\ln 1.3$, and then estimate the error in this approximation.

5. Two terms **6.** Three terms **7.** Four terms **8.** Five terms

Use the third-degree Taylor polynomial at 0 for $f(x) = e^{-x}$ to approximate each of the following, and then estimate the error in the approximation.

9. $e^{-0.2}$ **10.** $e^{-0.5}$ **11.** $e^{-0.03}$ **12.** $e^{-0.06}$

Use the third-degree Taylor polynomial at 0 for $f(x) = \ln(1 + x)$ to approximate each of the following, and then estimate the error in the approximation.

13. $\ln 1.6$ **14.** $\ln 1.8$ **15.** $\ln 1.06$ **16.** $\ln 1.08$

B

Use a Taylor polynomial at 0 to approximate each of the following with an error of no more than 0.0005. Select the polynomial of lowest degree that can be used to obtain this accuracy and state the degree of this polynomial.

17. $e^{-0.4}$ **18.** $e^{-0.8}$ **19.** $e^{-0.04}$ **20.** $e^{-0.08}$
21. $\ln 1.1$ **22.** $\ln 1.2$ **23.** $\ln 1.01$ **24.** $\ln 1.02$

Use a Taylor series at 0 to approximate each integral with an error of no more than 0.0005.

25. $\int_0^{0.2} \frac{1}{1+x^2}\,dx$ **26.** $\int_0^{0.5} \frac{x}{1+x^4}\,dx$ **27.** $\int_0^{0.6} \ln(1+x^2)\,dx$

28. $\int_0^{0.7} x\ln(1+x^4)\,dx$ **29.** $\int_0^{0.4} x^2e^{-x^2}\,dx$ **30.** $\int_0^{0.8} x^4e^{-x^2}\,dx$

C *Use the second-degree Taylor polynomial at 0 for $f(x) = e^x$ to approximate each of the following. Use Taylor's formula for the remainder to estimate the error in each approximation.*

31. $e^{0.3}$ **32.** $e^{0.4}$ **33.** $e^{0.05}$ **34.** $e^{0.01}$

35. The Taylor series at 0 for $f(x) = \sqrt{16+x}$ converges for $-16 < x < 16$. Use the second-degree Taylor polynomial at 0 to approximate $\sqrt{17}$. Use Taylor's formula for the remainder to estimate the error in this approximation.

36. The Taylor series at 0 for $f(x) = \sqrt{(4+x)^3}$ converges for $-4 < x < 4$. Use the third-degree Taylor polynomial at 0 to approximate $\sqrt{125}$. Use Taylor's formula for the remainder to estimate the error in this approximation.

APPLICATIONS

In Problems 37–48, use Theorem 2 to perform the indicated error estimations.

Business & Economics

37. Income distribution. The income distribution for a certain country is represented by the Lorenz curve with the equation

$$f(x) = \frac{5x^6}{4+x^2}$$

Approximate the coefficient of inequality to within ± 0.005.

38. Income distribution. Repeat Problem 37 for

$$f(x) = \frac{10x^4}{9+x^2}$$

39. Marketing. A soft drink manufacturer is ready to introduce a new diet soda by a national sales campaign. After test marketing the soda in a carefully selected city, the market research department estimates that sales (in millions of dollars) will increase at the monthly rate of

$$S'(t) = 10 - 10e^{-0.01t^2} \qquad 0 \le t \le 12$$

t months after the national campaign is started. Use the fourth-degree Taylor polynomial at 0 for $S'(t)$ to approximate the total sales during the first 4 months of the campaign, and estimate the error in this approximation.

40. *Marketing.* Repeat Problem 39 if the monthly rate of increase in sales is given by

$$S'(t) = 10 - 10e^{-0.005t^2} \qquad 0 \leq t \leq 12$$

41. *Useful life.* A computer store rents time on desktop publishing systems. The total accumulated costs $C(t)$ and revenues $R(t)$ (in thousands of dollars) from a particular system satisfy

$$C'(t) = 4 \qquad \text{and} \qquad R'(t) = \frac{80}{16 + t^2}$$

where t is the time in years that the system has been available for rental. Find the useful life of the system, and approximate the total profit during the useful life to within ± 0.005.

42. *Average price.* Given the demand equation

$$p = D(x) = 10 - 20 \ln\left(1 + \frac{x^2}{2{,}500}\right) \qquad 0 \leq x \leq 40$$

approximate the average price (in dollars) over the demand interval $[0, 20]$ to within ± 0.005.

Life Sciences

43. *Temperature.* The temperature (in degrees Celsius) in an artificial habitat is made to change according to the equation

$$C(t) = 20 + 800 \ln\left(1 + \frac{t^2}{100}\right) \qquad 0 \leq t \leq 2$$

Use a Taylor series at 0 to approximate the average temperature over the time interval $[0, 2]$ to within ± 0.005.

44. *Temperature.* Repeat Problem 43 for

$$C(t) = 10 + 200 \ln\left(1 + \frac{t^2}{50}\right) \qquad 0 \leq t \leq 2$$

45. *Medicine.* The rate of healing for a skin wound (in square centimeters per day) is given by

$$A'(t) = \frac{-75}{t^2 + 25}$$

The initial wound has an area of 12 square centimeters. Use the second-degree Taylor polynomial at 0 for $A'(t)$ to approximate the area of the wound after 2 days, and estimate the error in this approximation.

46. *Medicine.* Repeat Problem 45 for

$$A'(t) = \frac{-60}{t^2 + 20}$$

Social Sciences

47. *Learning.* In a particular business college, it was found that an average student enrolled in an advanced typing class progresses at a rate of $N'(t) = 6e^{-0.01t^2}$ words per minute per week, t weeks after enrolling in a 15 week course. At the beginning of the course an average student could type 40 words per minute. Use the second-degree Taylor polynomial at 0 for $N'(t)$ to approximate the improvement in typing after 5 weeks in the course, and estimate the error in this approximation.

48. *Learning.* In the same business college, it was also found that an average student enrolled in a beginning shorthand class progressed at a rate of $N'(t) = 12e^{-0.005t^2}$ words per minute per week, t weeks after enrolling in a 15 week course. At the beginning of the course none of the students could take any dictation by shorthand. Use the second-degree Taylor polynomial at 0 for $N'(t)$ to approximate the improvement after 5 weeks in the course, and estimate the error in this approximation.

SECTION 10-5 Chapter Review

Important Terms and Symbols

10-1 *Taylor Polynomials.* Higher-order derivatives; nth derivative; Taylor polynomial at 0; expanded notation; summation notation; Taylor polynomial at a

$$p_n(x) = \sum_{k=0}^{n} \frac{f^{(k)}(0)}{k!} x^k$$

$$= f(0) + f'(0)x + \frac{f''(0)}{2!} x^2 + \cdots + \frac{f^{(n)}(0)}{n!} x^n$$

$$p_n(x) = \sum_{k=0}^{n} \frac{f^{(k)}(a)}{k!} (x-a)^k$$

$$= f(a) + f'(a)(x-a) + \frac{f''(a)}{2!} (x-a)^2 + \cdots + \frac{f^{(n)}(a)}{n!} (x-a)^n$$

10-2 *Taylor Series.* Taylor series at a; converge; diverge; interval of convergence; using $\lim_{n\to\infty}|a_{n+1}/a_n|$ to find the interval of convergence; representation of functions by Taylor series

$$f(x) = \sum_{k=0}^{\infty} \frac{f^{(k)}(a)}{k!} (x-a)^k, \text{ for } x \text{ in the interval of convergence}$$

10-3 *Operations on Taylor Series.* Basic Taylor series; addition; multiplication; differentiation; integration; substitution

10-4 *Approximations Using Taylor Series.* Remainder; error; Taylor's formula for the remainder; alternating series; error estimation; approximating definite integrals

$$R_n(x) = f(x) - p_n(x)$$

EXERCISE 10-5 Chapter Review

Work through all the problems in this chapter review and check your answers in the back of the book. (Answers to all review problems are there.) Where weaknesses show up, review appropriate sections in the text.

Unless directed otherwise, use Theorem 2 in all problems involving error estimation.

A

1. Find $f^{(4)}(x)$ for $f(x) = \ln(x + 5)$.
2. Use the third-degree Taylor polynomial at 0 for $f(x) = \sqrt[3]{1 + x}$ and $x = 0.01$ to approximate $\sqrt[3]{1.01}$.
3. Use the third-degree Taylor polynomial at $a = 3$ for $f(x) = \sqrt{1 + x}$ and $x = 2.9$ to approximate $\sqrt{3.9}$.
4. Use the second-degree Taylor polynomial at 0 for $f(x) = \sqrt{9 + x^2}$ and $x = 0.1$ to approximate $\sqrt{9.01}$.

Use Theorem 1 to find the interval of convergence of each Taylor series representation given in Problems 5–8.

5. $\dfrac{1}{1 - 4x} = 1 + 4x + 4^2x^2 + \cdots + 4^nx^n + \cdots$

6. $\dfrac{5}{x - 1} = 1 - \dfrac{1}{5}(x - 6) + \dfrac{1}{5^2}(x - 6)^2 - \cdots + \dfrac{(-1)^n}{5^n}(x - 6)^n + \cdots$

7. $\dfrac{2x}{(1 - x)^3} = 1 \cdot 2x + 2 \cdot 3x^2 + 3 \cdot 4x^3 + \cdots + n(n + 1)x^n + \cdots$

8. $e^{10x} = 1 + 10x + \dfrac{10^2}{2!}x^2 + \cdots + \dfrac{10^n}{n!}x^n + \cdots$

9. Find the nth derivative of $f(x) = e^{-9x}$.

B

Use the formula $a_n = f^{(n)}(a)/n!$ to find the Taylor series at the indicated value of a. Use Theorem 1 to find the interval of convergence.

10. $f(x) = \dfrac{1}{7 - x}$ at 0
11. $f(x) = \ln x$ at 2

Use Table 4 and the properties of Taylor series to find the Taylor series of each function at the indicated value of a. Find the interval of convergence.

12. $f(x) = \dfrac{1}{10 + x}$ at 0
13. $f(x) = \dfrac{x^2}{4 - x^2}$ at 0
14. $f(x) = x^2e^{3x}$ at 0
15. $f(x) = x\ln(e + x)$ at 0
16. $f(x) = \dfrac{1}{4 - x}$ at 2

Find the Taylor series at 0 for f(x) and use the relationship $g(x) = f'(x)$ to find the Taylor series at 0 for g(x). Find the interval of convergence of both series.

17. $f(x) = \dfrac{1}{2 - x}$; $g(x) = \dfrac{1}{(2 - x)^2}$

18. $f(x) = \dfrac{x^2}{1 + x^2}$; $g(x) = \dfrac{2x}{(1 + x^2)^2}$

Find the Taylor series at 0 and find the interval of convergence.

19. $f(x) = \displaystyle\int_0^x \frac{t^2}{9 + t^2}\,dt$

20. $f(x) = \displaystyle\int_0^x \frac{t^4}{16 - t^2}\,dt$

Use the second-degree Taylor polynomial at 0 for $f(x) = e^x$ to approximate each of the following. Use Taylor's formula for the remainder to estimate the error in the approximaton.

21. $e^{0.6}$

22. $e^{0.06}$

Use a Taylor polynomial at 0 for $f(x) = \ln(1 + x)$ to estimate each of the following to within ± 0.0005. Give the degree of the Taylor polynomial of lowest degree that will provide this accuracy.

23. ln 1.3

24. ln 1.03

C *Approximate each integral to within ± 0.0005.*

25. $\displaystyle\int_0^1 \frac{1}{16 + x^2}\,dx$

26. $\displaystyle\int_0^1 x^2 e^{-0.1x^2}\,dx$

APPLICATIONS

Business & Economics

27. Average price. Given the demand equation

$$p = D(x) = \tfrac{1}{10}\sqrt{2{,}500 - x^2}$$

use the second-degree Taylor polynomial at 0 to approximate the average price (in dollars) over the demand interval [0, 15].

28. Production. The rate of production of an oil well (in thousands of dollars per year) is given by

$$R(t) = 6 + 3e^{-0.01t^2}$$

Use the second-degree Taylor polynomial at 0 to approximate the total production during the first 10 years of operation of the well.

29. Income distribution. The income distribution for a certain country is represented by the Lorenz curve with the equation

$$f(x) = \frac{9x^3}{8 + x^2}$$

Approximate the coefficient of inequality to within ± 0.005.

30. *Marketing.* A cereal manufacturer is ready to introduce a new high-fiber cereal by a national sales campaign. After test marketing the cereal in a carefully selected city, the market research department estimates that sales (in millions of dollars) will increase at the monthly rate of

$$S'(t) = 20 - 20e^{-0.001t^2} \qquad 0 \le t \le 12$$

t months after the national campaign is started. Use the fourth-degree Taylor polynomial at 0 for $S'(t)$ to approximate the total sales during the first 8 months of the campaign, and estimate the error in this approximation.

Life Sciences

31. *Medicine.* The rate of healing for a skin wound (in square centimeters per day) is given by

$$A'(t) = \frac{-100}{t^2 + 40}$$

The initial wound has an area of 15 square centimeters. Use the second-degree Taylor polynomial at 0 for $A'(t)$ to approximate the area of the wound after 2 days, and estimate the error in this approximation.

32. *Medicine.* A large injection of insulin is administered to a patient. The level of insulin in the bloodstream t minutes after the injection is given approximately by

$$L(t) = \frac{5{,}000t^2}{10{,}000 + t^4}$$

Express the average insulin level over the time interval $[0, 5]$ as a definite integral and use a Taylor series at 0 to approximate this integral to within ± 0.005.

Social Sciences

33. *Politics.* In a newly incorporated city, the number of voters (in thousands) t years after incorporation is given by

$$N(t) = 10 + 2t - 5e^{-0.01t^2} \qquad 0 \le t \le 5$$

Express the average number of voters over the time interval $[0, 5]$ as a definite integral and use a Taylor series at 0 to approximate this integral to within ± 0.05.

CHAPTER 11

Numerical Techniques

Contents

CHAPTER 11

Calculus is a powerful tool for solving a wide variety of problems, as evidenced by all the applications we have discussed in this text. However, many real-world applications of calculus cannot be solved by the techniques we have discussed so far. For instance, not all equations can be solved exactly, not all definite integrals can be evaluated by finding an antiderivative, and not all differential equations have solutions that can be expressed in terms of familiar functions. In such cases, numerical methods often are used to approximate solutions that cannot be found exactly. In this chapter we will develop methods for approximating the roots of an equation, the values of a function defined by a table, the value of a definite integral, and a particular solution of a differential equation. In actual practice, computers are usually used to perform the calculations involved in these approximations. *Visual Calculus*, the software that accompanies this text (see the Preface), can be used to produce these approximations, as can a number of other commonly available programs (such as Matlab, Mathematica, Maple, and Derive). Certain models of graphic calculators also have some built-in approximation routines and can be programmed to perform other routines. However, it is instructive for the beginning student to perform some of these calculations directly using nothing more than an ordinary scientific calculator. We have used a scientific calculator to perform our calculations, and you will need a calculator with similar capabilities in order to work the problems and exercises in this chapter.

Calculator Variation

Do not be alarmed if your calculator results do not agree exactly with those printed in this chapter or in the answer section. The large variety of scientific calculators on the market do not all perform these operations in exactly the same way, so you can expect slight variations in the last two or three decimal places when the calculators are used to their full capacities. Slight variations in results may even occur in the same calculator when a given calculation is performed in two different ways.

SECTION 11-1 Newton's Method for Approximating Roots

- ZEROS, ROOTS, AND INTERCEPTS
- NEWTON'S METHOD FOR APPROXIMATING ROOTS
- APPLICATIONS
- INTERNAL RATE OF RETURN

◆ ZEROS, ROOTS, AND INTERCEPTS

A number r is a **zero** of a function f, or a **root** of the equation $f(x) = 0$, if $f(r) = 0$. If $f(x) = ax^2 + bx + c$ is any second-degree polynomial, then its zeros are given by the *quadratic formula,*

$$r = \frac{-b \pm \sqrt{b^2 - 4ac}}{2a}$$

◆ EXAMPLE 1 Find the zeros of $f(x) = x^2 - 8x + 13$.

Solution Since f is a second-degree polynomial and $a = 1$, $b = -8$, $c = 13$, the zeros are

$$r_1 = \frac{8 - \sqrt{64 - 52}}{2} = \frac{8 - 2\sqrt{3}}{2} = 4 - \sqrt{3} \approx 2.267\ 949\ 2$$

$$r_2 = \frac{8 + \sqrt{64 - 52}}{2} = 4 + \sqrt{3} \approx 5.732\ 050\ 8$$ ◆

PROBLEM 1 Use the quadratic formula to find the zeros of $f(x) = x^2 - 6x + 4$. ◆

Notice the use of the symbols = and ≈ in Example 1. The *exact* zero r_1 of f is the irrational number $4 - \sqrt{3}$. However, when we use a calculator to evaluate the irrational number $4 - \sqrt{3}$, we obtain the finite decimal 2.267 949 2, which is a rational number that *approximates* r_1.

If we graph $y = f(x)$, then the zeros are often referred to as the **x intercepts,** since these are the points where the graph crosses the x axis. Figure 1 illustrates this for the function in Example 1.

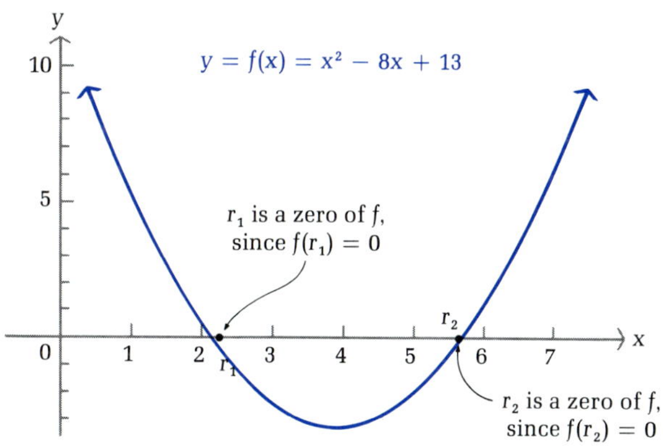

FIGURE 1

Using algebraic methods to find the exact zeros of more complicated functions can be very difficult, if not impossible. Many methods have been developed to approximate the zeros of a function. For example, computers and calculators with graphing capabilities can be used to approximate the intercepts of a function by repeatedly expanding the graph near the intercept and displaying the coordinates of the point on the screen that is closest to the actual intercept. Figure 2 illustrates this process for the function $f(x)$ in Example 1.

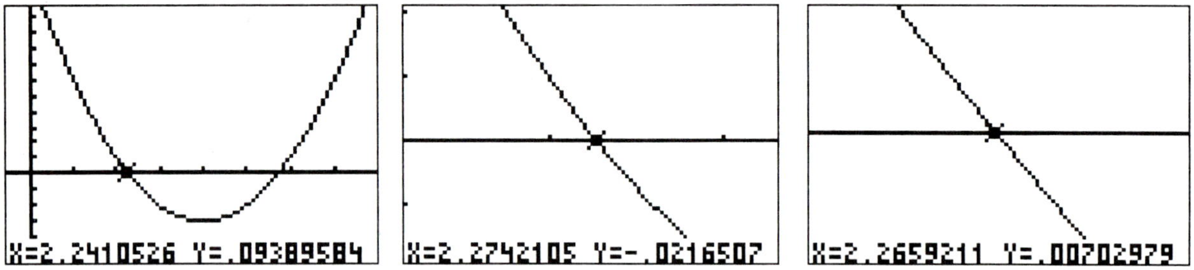

(A) Initial graph (B) First expansion (C) Second expansion

FIGURE 2
Graphical approximation of the smaller intercept of $f(x) = x^2 - 8x + 13$

The coordinates of the point at the location of the cursor are displayed at the bottom of each graph. The graph in Figure 1(C) indicates that $f(x)$ has an intercept at approximately $x = 2.265\ 921\ 1$. Comparing this result with r_1 in Example 1, we see that this approximation is accurate to two decimal places. Repeated expansions of the graph will improve the accuracy of this approximation.

Graphical approximation techniques are easy to understand and to apply, but they are not practical for all purposes. They require user interaction at each step, and the number of steps can be large. For example, eight more expansions of the graph in Figure 2 are required to approximate the intercept to seven decimal places. Numerical methods that can be performed automatically by a computer or programmable calculator are generally used in actual practice. One of these—usually called *Newton's method*—produces very accurate approximations in relatively few steps.

◆ NEWTON'S METHOD FOR APPROXIMATING ROOTS

Given a differentiable function f, suppose we are able to determine that f has a zero at some number r, but we are unable to find the exact value of r by algebraic methods. In order to approximate the value of r, we begin by selecting an initial value x_1, which we suspect is close to r. (We will have more to say about the selection of this initial value later.) If $f'(x_1) \neq 0$, then the line tangent to the graph of $y = f(x)$ at the point $(x_1, f(x_1))$ must intersect the x axis (see Figure 3).

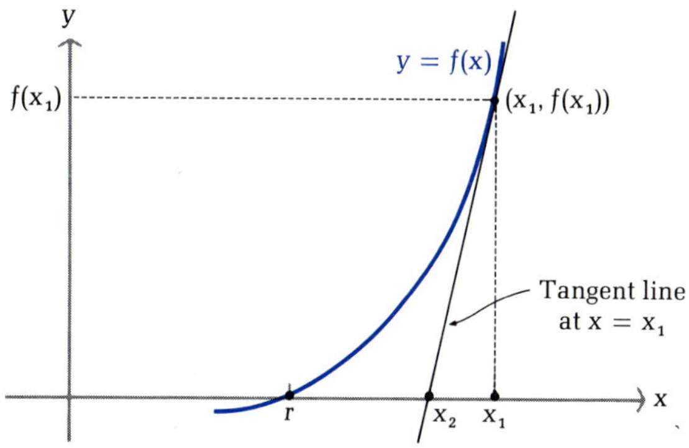

FIGURE 3

Let x_2 be the x intercept of this tangent line. Examining Figure 3, it appears that x_2 is closer to r than x_1 is. How can we find the value of x_2? The equation for the tangent line at $(x_1, f(x_1))$ is

$$y - f(x_1) = f'(x_1)(x - x_1) \tag{1}$$

If x_2 is the x intercept of this line, then the point $(x_2, 0)$ must satisfy the equation of the line. Substituting x_2 for x and 0 for y in (1) and solving for x_2, we have

$$0 - f(x_1) = f'(x_1)(x_2 - x_1)$$

$$x_2 - x_1 = -\frac{f(x_1)}{f'(x_1)} \qquad f'(x_1) \neq 0$$

$$x_2 = x_1 - \frac{f(x_1)}{f'(x_1)} \tag{2}$$

Notice that if $f'(x_1) = 0$, then the tangent line at $(x_1, f(x_1))$ is horizontal and there is no x intercept. To avoid this case, we will assume that $f'(x) \neq 0$ for all values of x near the zero r. Thus, given an initial approximation x_1, we can use (2) to compute x_2, a second approximation to r. For most functions, x_2 will be a better approximation than x_1. However, x_2 still might not be a sufficiently accurate approximation. If we repeat this process, beginning now with the tangent line at the point $(x_2, f(x_2))$, we should obtain an even better approximation to r (see Figure 4 at the top of the next page).

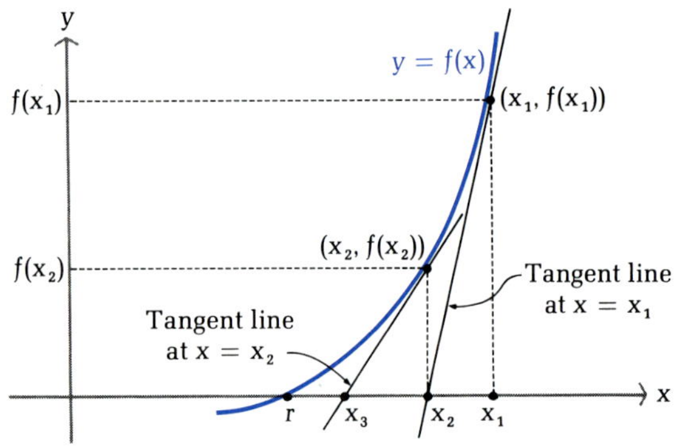

FIGURE 4

Let x_3 denote the x intercept of the tangent line at $(x_2, f(x_2))$. The equation of this line is

$$y - f(x_2) = f'(x_2)(x - x_2)$$

Substituting x_3 for x and 0 for y and solving for x_3, we have

$$0 - f(x_2) = f'(x_2)(x_3 - x_2)$$

$$x_3 = x_2 - \frac{f(x_2)}{f'(x_2)} \tag{3}$$

Notice the similarity between the expression for x_2 in equation (2) and the one for x_3 in equation (3). Continued repetition of this process will produce a sequence of numbers $x_1, x_2, \ldots, x_n, \ldots$, where each number in the sequence (after x_1) is obtained by using the preceding number in the **recursion formula**

$$x_n = x_{n-1} - \frac{f(x_{n-1})}{f'(x_{n-1})} \qquad n > 1$$

If the numbers $x_1, x_2, \ldots, x_n, \ldots$ approach a limit r, then it can be shown that $f(r) = 0$. Hence, r is a zero of f. The process described above is referred to as *Newton's method for approximating roots.*

Newton's Method

Given the function f and the initial approximation x_1, define

$$x_n = x_{n-1} - \frac{f(x_{n-1})}{f'(x_{n-1})} \qquad n > 1, \quad f'(x_{n-1}) \neq 0$$

If $\lim_{n\to\infty} x_n$ exists, then

$$r = \lim_{n\to\infty} x_n$$

is a zero of f.

◆ EXAMPLE 2 Use Newton's method to approximate the smaller zero of $f(x) = x^2 - 8x + 13$. Compare the result with that obtained by use of the quadratic formula in Example 1.

Solution Step 1. Find the recursion formula for x_n:

$$f(x) = x^2 - 8x + 13$$

$$f'(x) = 2x - 8$$

$$\begin{aligned} x_n &= x_{n-1} - \frac{f(x_{n-1})}{f'(x_{n-1})} \\ &= x_{n-1} - \frac{x_{n-1}^2 - 8x_{n-1} + 13}{2x_{n-1} - 8} \\ &= \frac{x_{n-1}(2x_{n-1} - 8) - (x_{n-1}^2 - 8x_{n-1} + 13)}{2x_{n-1} - 8} \end{aligned}$$

$$x_n = \frac{x_{n-1}^2 - 13}{2x_{n-1} - 8} \tag{4}$$

Step 2. Approximate the root using equation (4). Examining the graph of $y = f(x)$ in Figure 1 (and ignoring the fact that we know the exact root), we see that $x_1 = 2$ is a reasonable first approximation to the root r_1. Using equation (4) and $x_1 = 2$ as a first approximation, we obtain the following successive approximations for r_1:

$$x_1 = 2$$

$$x_2 = \frac{(\mathbf{2})^2 - 13}{2(\mathbf{2}) - 8} = 2.25$$

$$x_3 = \frac{(\mathbf{2.25})^2 - 13}{2(\mathbf{2.25}) - 8} \approx 2.267\ 857\ 1$$

$$x_4 = \frac{(\mathbf{2.267\ 857\ 1})^2 - 13}{2(\mathbf{2.267\ 857\ 1}) - 8} \approx 2.267\ 949\ 2$$

$$x_5 = \frac{(\mathbf{2.267\ 949\ 2})^2 - 13}{2(\mathbf{2.267\ 949\ 2}) - 8} \approx 2.267\ 949\ 2$$

Since $x_4 = x_5$ in the display in our calculator, we have reached the limits of accuracy for our calculator and assume $r_1 \approx 2.267\ 949\ 2$. Notice that this is the *same approximation to the exact zero* that we obtained in Example 1 using the quadratic formula and our calculator directly. ◆

PROBLEM 2 Use Newton's method to approximate the larger zero of $f(x) = x^2 - 8x + 13$. First examine Figure 1 to obtain a first approximation to the larger zero of f. Then use Newton's method to approximate the larger root to the limit of accuracy of your calculator—that is, until two successive approximations are the same in the display of your calculator. ◆

In Example 2, it took us only three calculations to approximate the zero correct to seven decimal places. We were able to obtain this much accuracy with so few computations because the initial approximation was reasonably close to the actual zero. It is always important to try to select a good first approximation. Figures 5 and 6 illustrate two situations that may occur if x_1 is too far away from r. In Figure 5, each x_n is larger than the preceding value and further away from r. This type of behavior is very common if the initial approximation x_1 is not selected carefully. In Figure 6, the approximating values oscillate between x_1 and x_2, never getting any closer to r. If either of these situations occur when you are using Newton's method, you must select a better initial value for x_1.

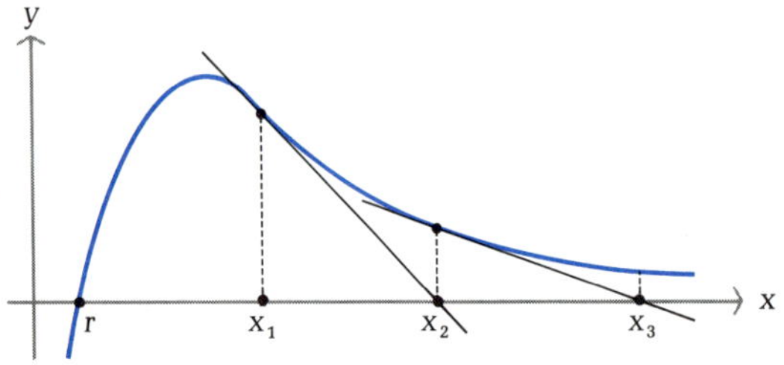

FIGURE 5

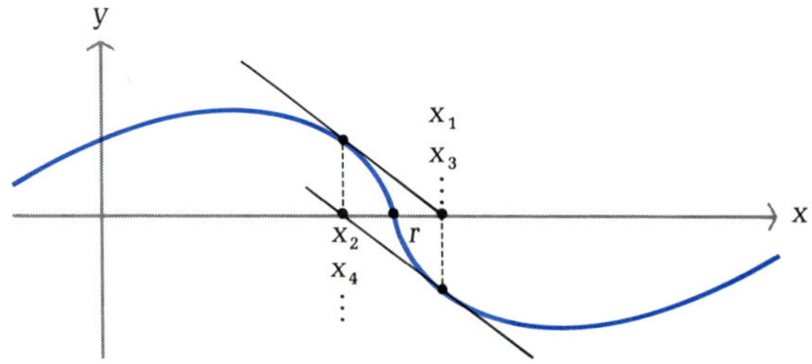

FIGURE 6

◆ EXAMPLE 3 Sketch the graph of $f(x) = x^3 - 9x^2 + 15x + 10$ and approximate the x intercepts.

Solution Step 1. Sketch the graph of f:

$$f'(x) = 3x^2 - 18x + 15 = 3(x - 5)(x - 1)$$

Critical values: $x = 1, 5$

The graph of f rises for x in the intervals $(-\infty, 1)$ and $(5, \infty)$ and falls for x in the interval $(1, 5)$. The second derivative,

$$f''(x) = 6x - 18 = 6(x - 3)$$

tells us the graph of f is concave downward in the interval $(-\infty, 3)$ and concave upward in the interval $(3, \infty)$. Using this information and point-by-point plotting to sketch the graph of f, as shown in the figure at the top of the next page, we conclude that f has three zeros: r_1 in $(-1, 0)$, r_2 in $(3, 4)$, and r_3 in $(6, 7)$. Notice that the intervals containing intercepts also can be identified by examining the table in the margin for changes in the sign of $f(x)$. Drawing the graph assures us that there is only one zero in each of these intervals and that there are no zeros anywhere else.

Step 2. Write the recursion formula for x_n:

$$x_n = x_{n-1} - \frac{f(x_{n-1})}{f'(x_{n-1})} = x_{n-1} - \frac{x_{n-1}^3 - 9x_{n-1}^2 + 15x_{n-1} + 10}{3x_{n-1}^2 - 18x_{n-1} + 15} \qquad (5)$$

This time we will not simplify the recursion formula.

x	f(x)	
−1	−15	Sign change
0	10	
1	17	
2	12	
3	1	Sign change
4	−10	
5	−15	
6	−8	Sign change
7	17	

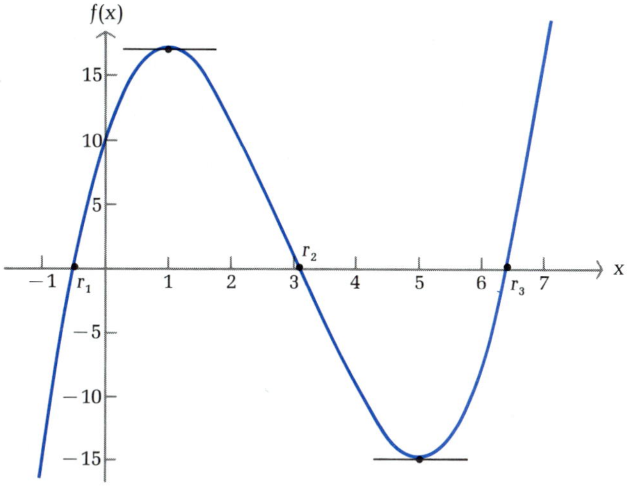

Step 3. Approximate r_1: We will use $x_1 = -1$ for an initial approximation and use a calculator to repeatedly evaluate the expression in (5). At first you may want to evaluate $f(x_{n-1})$ and $f'(x_{n-1})$ separately, recording each value, and then compute x_n. To assist in this process, we have listed both partial and final results of these calculations in the table below. After some practice, you should be able to store x_{n-1} in the memory of your calculator and evaluate (5) directly, recalling the stored value of x_{n-1} each time you need it. It is good practice to record each value of x_n as you compute it so that you can reenter a previous value in case of an error in a subsequent calculation. If you are using a graphic calculator, you can enter the recursion formula (5) (omitting the subscripts) in the function definition screen of the calculator and repeatedly evaluate this expression for successive values of x.

n	x_{n-1}	$f(x_{n-1})$	$f'(x_{n-1})$	$x_n = x_{n-1} - \dfrac{f(x_{n-1})}{f'(x_{n-1})}$
2	−1	−15	36	$-1 - \dfrac{-15}{36} \approx -0.583\ 333\ 33$
3	−0.583 333 33	−2.010 995 3	26.520 833	$-0.583\ 333\ 33 - \dfrac{-2.010\ 995\ 3}{26.520\ 833} \approx -0.507\ 506\ 33$
4	−0.507 506 33	−0.061 373 72	24.907 803	$-0.507\ 506\ 33 - \dfrac{-0.061\ 373\ 72}{24.907\ 803} \approx -0.505\ 042\ 29$
5	−0.505 042 29	−0.000 063 77	24.855 964	$-0.505\ 042\ 29 - \dfrac{-0.000\ 063\ 77}{24.855\ 964} \approx -0.505\ 039\ 72$
6	−0.505 039 72	0.000 000 11	24.855 910	$-0.505\ 039\ 72 - \dfrac{0.000\ 000\ 11}{24.855\ 910} \approx -0.505\ 039\ 73$

Since x_5 and x_6 are very nearly the same, we conclude that $r_1 \approx -0.505\ 039\ 7$ is a good approximation.

Step 4. Approximate r_2: Initial approximation: $x_1 = 3$.

n	x_{n-1}	$f(x_{n-1})$	$f'(x_{n-1})$	$x_n = x_{n-1} - \dfrac{f(x_{n-1})}{f'(x_{n-1})}$
2	3	1	−12	3.083 333 3
3	3.083 333 3	0.000 578 77	−11.979 167	3.083 381 6
4	3.083 381 6	0.000 000 54	−11.979 143	3.083 381 6

Thus, $r_2 \approx 3.083\ 381\ 6$.

Step 5. Approximate r_3: Initial approximation: $x_1 = 6$.

n	x_{n-1}	$f(x_{n-1})$	$f'(x_{n-1})$	$x_n = x_{n-1} - \dfrac{f(x_{n-1})}{f'(x_{n-1})}$
2	6	−8	15	6.533 333 3
3	6.533 333 3	2.711 703 6	25.453 333	6.426 797
4	6.426 797	0.119 100 63	23.228 814	6.421 669 8
5	6.421 669 8	0.000 270 04	23.123 472	6.421 658 1
6	6.421 658 1	0.000 000 47	23.123 232	6.421 658 1

Thus, $r_3 \approx 6.421\ 658\ 1$. ◆

PROBLEM 3 Use Newton's method to approximate the zeros of $f(x) = x^4 - 4x^3 + 10$. (Sketch a graph of f to obtain initial approximations and to be certain all zeros have been located.) ◆

◆ APPLICATIONS

Many applications in earlier sections involved solving equations in order to find quantities such as the value of x that produces a given functional value, the points of intersection of two graphs, or the critical values of a function. All the problems we considered were carefully selected so that the solutions could be found by algebraic methods. Now that we can use Newton's method to find the roots of equations that can be written in the form $f(x) = 0$, we can consider problems involving equations that do not lend themselves to algebraic solutions. Since Newton's method generally produces an accurate approximation in a relatively small number of steps, most problems can be solved with a scientific calculator. However, performing all these calculations does become tedious, and it is convenient to use a graphic calculator or a computer if one is available. The tables in Examples 4, 5, and 6 were produced using a graphic calculator. All the values in these tables were rounded to seven decimal places.

◆ EXAMPLE 4 *Pollution*

The pollution level (in parts per million) in a river t days after an industrial accident near the river is given by

$$P(t) = 50te^{-0.1t}$$

The water is considered to be contaminated as long as the pollution level is above 125 parts per million. Graph $P(t)$ and determine the time period during which the river will be contaminated. Round values of t to one decimal place.

Solution The graph of $P(t)$ is shown in the figure (graphing details omitted):

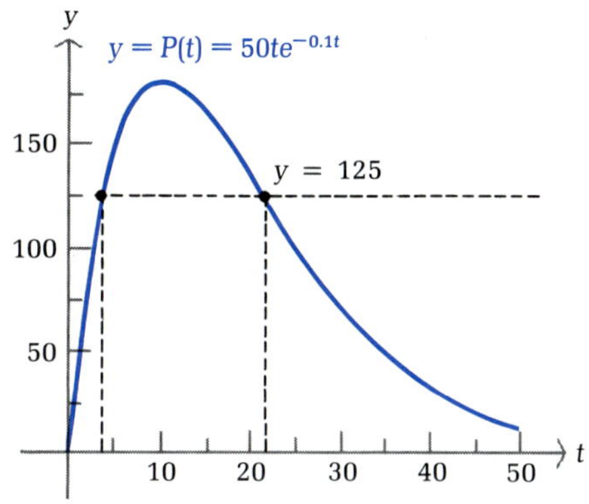

To determine the time interval where $P(t) > 125$ we must first solve the equation

$$P(t) = 125$$

for t. The graph shows that this equation has two solutions, one near 5 and another near 20. Since Newton's method can be used only to solve equations of the form $F(t) = 0$, we define

$$\begin{aligned} F(t) &= P(t) - 125 \\ &= 50te^{-0.1t} - 125 \end{aligned}$$

and apply Newton's method to $F(t)$, first with an initial approximation of 5 and then with an initial approximation of 20. The table gives the results:

Approximation of the Zeros of $F(t)$

(A) NEAR $t = 5$		(B) NEAR $t = 20$	
n	t_n	n	t_n
1	5	1	20
2	3.243 606 4	2	21.527 359 8
3	3.560 726 3	3	21.532 923 4
4	3.574 007	4	21.532 923 6
5	3.574 029 6		

According to the table, $F(t) = 0$ at $t \approx 3.6$ and $t \approx 21.5$ (rounded to one decimal place). Examining the figure, we see that the graph of $P(t)$ is above the (dashed) horizontal line $y = 125$ for $3.6 < t < 21.5$. Thus,

$$P(t) > 125 \qquad \text{for} \qquad 3.6 < t < 21.5$$

and the river is contaminated for t in the interval $(3.6, 21.5)$. ◆

PROBLEM 4 Repeat Example 4 if the river is considered to be contaminated when the pollution level is above 160 parts per million. ◆

◆ EXAMPLE 5 *Equilibrium Price*

Find the equilibrium price correct to two decimal places, given the demand and supply functions

$$p = D(x) = 100 - 10\sqrt{x} \qquad \text{and} \qquad p = S(x) = 10 + \tfrac{1}{50}x^2$$

where p is the price in dollars and x is the number of units in thousands.

Solution Recall that the equilibrium price $\bar{p}$ is the common value of $S(x)$ and $D(x)$ at the point where their graphs intersect. Thus, we must first solve the equation $S(x) = D(x)$ for x. Proceeding as before, we define

$$F(x) = D(x) - S(x) = 90 - 10\sqrt{x} - \tfrac{1}{50}x^2$$

The zeros of F will be the x coordinates of the points of intersection of the graphs of D and S. The figure at the top of the next page shows that the graphs of D and S have one point of intersection and that the x coordinate of this point is close to 40. We apply Newton's method to F with $x_1 = 40$.

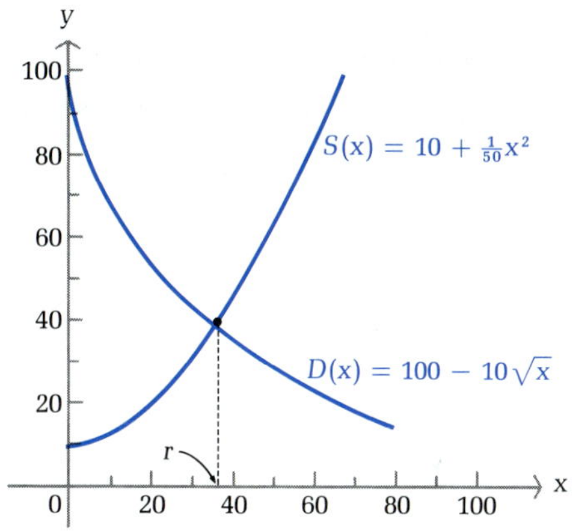

n	x_n
1	40
2	37.805 730 6
3	37.774 841 5
4	37.774 835 5

According to the table, the zero of F is approximately 37.775. We find the equilibrium price (to two decimal places) by using a calculator to evaluate either the demand function or the supply function at $x = 37.775$:

$$D(37.775) \approx 38.54 \approx S(37.775)$$

Thus, the equilibrium price is $\bar{p} \approx \$38.54$. ◆

PROBLEM 5 Repeat Example 5 if

$$p = D(x) = 36 - \tfrac{1}{16}x^2 \qquad \text{and} \qquad p = S(x) = \sqrt{100 + x^2}$$ ◆

◆ INTERNAL RATE OF RETURN

If a \$10,000 investment returns a single payment of \$13,000 at the end of 3 years, then the annual rate of return r on this investment can be found by using the compound interest formula (see Section 2-1):

$$A = P\left(1 + \frac{r}{m}\right)^{mt}$$

Substituting 10,000 for the present value P, 13,000 for the future value A, 1 for the number of compounding periods m in a year, and 3 for the number of years t, we obtain an equation we can solve for r:

$$\begin{aligned} 13{,}000 &= 10{,}000(1 + r)^3 \\ 1.3 &= (1 + r)^3 \\ \sqrt[3]{1.3} &= 1 + r \\ r &= \sqrt[3]{1.3} - 1 \approx 0.0914 \end{aligned}$$

Thus, the rate of return is 9.14% compounded annually.

In the next example we consider an investment that returns a sequence of payments, resulting in a more complicated equation for the rate of return r, and we use Newton's method to solve the equation.

EXAMPLE 6 *Internal Rate of Return*

A $10,000 investment returns payments of $5,000 at the end of the first year, $4,000 at the end of the second year, and a final payment of $3,000 at the end of the third year. The present values of these three payments are

$$P_1 = \frac{5{,}000}{1+r} \qquad P_2 = \frac{4{,}000}{(1+r)^2} \qquad P_3 = \frac{3{,}000}{(1+r)^3}$$

The **internal rate of return** of this investment is the annual rate r that produces three present values whose sum equals the original investment of $10,000. Find the internal rate of return. Express the answer as a percentage correct to two decimal places.

Solution The internal rate of return r is the solution of the equation

$$\begin{pmatrix}\text{Initial}\\ \text{investment}\end{pmatrix} = \begin{pmatrix}\text{Sum of present values}\\ \text{of all payments}\end{pmatrix}$$

$$10{,}000 = \frac{5{,}000}{1+r} + \frac{4{,}000}{(1+r)^2} + \frac{3{,}000}{(1+r)^3}$$

or, equivalently, the root of the equation

$$10{,}000 - \frac{5{,}000}{1+r} - \frac{4{,}000}{(1+r)^2} - \frac{3{,}000}{(1+r)^3} = 0$$

Multiplying both sides of this equation by 0.001, we have

$$10 - \frac{5}{1+r} - \frac{4}{(1+r)^2} - \frac{3}{(1+r)^3} = 0$$

If we let $x = 1 + r$ and use negative exponents, then this last equation can be written more simply as

$$f(x) = 10 - 5x^{-1} - 4x^{-2} - 3x^{-3} = 0$$

Since r represents an interest rate, we expect its value to be between 0 and 1. Hence, we expect to find a zero of $f(x)$ between $x = 1$ and $x = 2$. Evaluating f at these values, we have

$$f(1) = -2 \qquad \text{and} \qquad f(2) = 6.125$$

Thus, f does have a zero between $x = 1$ and $x = 2$. Furthermore, since

$$f'(x) = 5x^{-2} + 8x^{-3} + 9x^{-4} > 0$$

for $x > 0$, f is increasing on $(0, \infty)$ and can have only one positive zero. We apply Newton's method to $f(x)$ with $x_1 = 1$ to find this zero.

n	x_n
1	1
2	1.090 909 1
3	1.106 174 5
4	1.106 516 6
5	1.106 516 8

The table indicates that the zero of f (to four decimal places) is $x = 1.1065$. Thus, $r = x - 1 = 0.1065$ and the internal rate of return is 10.65% compounded annually. ◆

PROBLEM 6 An investment of \$15,000 returns payments of \$5,000 at the end of the first year, \$6,000 at the end of the second year, and a final payment of \$7,000 at the end of the third year. Find the internal rate of return. Express the answer as a percentage correct to two decimal places. ◆

Remarks

1. The internal rate of return is used by businesses to help make decisions involving long-term capital investments such as equipment replacement, facility expansion, mergers, or new product development.
2. As long as all the payments are nonnegative and their sum exceeds the initial investment, the internal rate of return will be a nonnegative number that can be determined using Newton's method as illustrated in Example 6. Situations where some of the payments are negative (indicating an additional amount was invested) or where the sum of the payments is less than the initial investment (indicating a loss to the investor) require a more detailed analysis. We will not consider these situations.
3. This procedure also can be applied to investments involving payment periods other than annual payments.

Answers to Matched Problems

1. $r_1 = 3 + \sqrt{5} \approx 5.236\ 068$; $r_2 = 3 - \sqrt{5} \approx 0.763\ 932\ 02$
2. $r_2 \approx 5.732\ 050\ 8$ 3. $r_1 \approx 1.611\ 793\ 4$; $r_2 \approx 3.820\ 704\ 4$
4. $5.6 < t < 16.2$ 5. \$19.20 6. 9.15%

EXERCISE 11-1

A *In Problems 1–10, find and simplify the recursion formula for x_n. Use this formula and the given value of x_1 to compute x_n for the indicated value of n.*

1. $f(x) = x^2 - 4$; $x_1 = 1$; $n = 5$
2. $f(x) = x^2 - 2$; $x_1 = 1$; $n = 5$
3. $f(x) = x^3 - 8$; $x_1 = 3$; $n = 5$
4. $f(x) = x^3 - 2$; $x_1 = 4$; $n = 7$
5. $f(x) = e^x + x$; $x_1 = 0$; $n = 5$
6. $f(x) = e^x + 2x$; $x_1 = 1$; $n = 5$
7. $f(x) = \ln x + x$; $x_1 = 1$; $n = 5$
8. $f(x) = \ln x + 2x - 8$; $x_1 = 10$; $n = 4$
9. $f(x) = \ln x + x^2$; $x_1 = 2$; $n = 5$
10. $f(x) = \ln x - e^{-x}$; $x_1 = 3$; $n = 7$

B *In Problems 11–18, use Newton's method to find all the zeros of each function. Sketch the graph of the function to obtain initial approximations and to be certain that all the zeros have been located.*

11. $f(x) = x^2 - 7x + 2$
12. $f(x) = x^2 + 5x + 3$
13. $f(x) = x^3 + 4x + 10$
14. $f(x) = x^3 - 6x^2 + 12x - 4$

15. $f(x) = x^3 - 12x^2 + 22$

16. $f(x) = x^3 - 18x^2 + 60x - 11$

17. $f(x) = x^4 - 4x^3 - 8x^2 + 4$

18. $f(x) = x^4 - 4x^3 + 4x^2 + 1$

In Problems 19–24, graph f and g on the same set of axes and use Newton's method to find the x coordinate of all points of intersection of the two graphs.

19. $f(x) = x^3$; $g(x) = x + 4$

20. $f(x) = \sqrt{x + 1}$; $g(x) = x^2 - 4$

21. $f(x) = x^{1/3}$; $g(x) = 12 - x$

22. $f(x) = x^3$; $g(x) = \dfrac{1}{1 + x^2}$

23. $f(x) = e^x$; $g(x) = x + 2$

24. $f(x) = \ln x$; $g(x) = x^2 - 4$

C

25. Newton's algorithm for approximating the square root of a positive number A is often stated as

$$x_n = \frac{x_{n-1}}{2} + \frac{A}{2x_{n-1}}$$

Show that this recursion formula can be derived by applying Newton's method to the function $f(x) = x^2 - A$.

26. Apply Newton's method to the function $f(x) = x^3 - A$ and derive a recursion formula for approximating cube roots.

27. Apply Newton's method to the function $f(x) = x^p - A$, where p is a positive integer, and derive a recursion formula for approximating the pth root of A.

28. Apply Newton's method to the function

$$f(x) = \frac{1}{x} - A$$

and derive a recursion formula for approximating reciprocals without the use of division.

29. Apply Newton's method to $f(x) = x/\sqrt{1 + x^2}$ with $x_1 = 1$. Why does Newton's method fail in this case?

30. Apply Newton's method to $f(x) = 17 + 8x^2 - x^4$ with $x_1 = 1$. Why does Newton's method fail in this case?

Problems 31 and 32 require the use of a graphic calculator or a computer.

31. Refer to Problem 29. Find the equations of the lines tangent to the graph of $f(x) = x/\sqrt{1 + x^2}$ at $x = 1$ and at $x = -1$. Graph f and both tangent lines in the same viewing rectangle. Use $[-2, 2]$ for both the x range and the y range.

32. Refer to Problem 30. Find the equations of the lines tangent to the graph of $f(x) = 17 + 8x^2 - x^4$ at $x = 1$ and at $x = -1$. Graph f and both tangent lines in the same viewing rectangle. Use $[-5, 5]$ for the x range and $[-35, 35]$ for the y range.

Business & Economics

33. *Revenue.* The revenue (in thousands of dollars) from the sale of x thousand units of a food processor is given by

$$R(x) = 100xe^{-0.05x}$$

How many food processors must be sold to generate a revenue of \$600,000?

34. *Revenue.* The revenue (in thousands of dollars) from the sale of x thousand units of a compact disk is given by

$$R(x) = 20xe^{-0.01x}$$

How many disks must be sold to generate a revenue of \$500,000?

35. *Supply–demand.* The supply and demand equations for premium motor oil in a particular market are given by

$$p = S(x) = e^{0.02x} \qquad \text{and} \qquad p = D(x) = 20 - e^{0.08x}$$

where p is the price of a can of oil (in dollars) and x is the number of cans of motor oil (in thousands). Find the equilibrium price (correct to two decimal places).

36. *Supply–demand.* The supply and demand equations for 20 pound bags of dog food in a certain city are given by

$$p = S(x) = e^{0.1x} \qquad \text{and} \qquad p = D(x) = 10 - e^{0.05x}$$

where p is the price of a bag of dog food (in dollars) and x is the number of bags of dog food (in thousands). Find the equilibrium price (correct to two decimal places).

Marty Haigney

37. *Internal rate of return.* A motion picture distributor pays \$6 million for exclusive rights to a film for 2 years, after which all rights revert back to the film's producer. The distributor realizes a profit of \$5 million for the first year of distributing the film and \$3 million for the second year. Find the internal rate of return of the distributor's investment. Express the answer as a percentage correct to two decimal places.

38. *Internal rate of return.* A library buys a coin-operated photocopying machine for \$5,000. The copier generates annual profits of \$2,000 for the library over a 3 year period, after which the copier is sold for \$1,000. Find the internal rate of return of this investment. Express the answer as a percentage correct to two decimal places.

39. *Internal rate of return.* An individual pays \$160,000 for a two-family housing unit which generates annual profits of \$10,000. After 4 years, the property is sold for \$200,000. Find the internal rate of return of this investment. Express the answer as a percentage correct to two decimal places.

YEAR	PROFIT
1	\$50,000
2	\$70,000
3	\$80,000
4	\$60,000
5	\$30,000

40. *Internal rate of return.* A company spends \$200,000 on a piece of equipment with a 5 year life span. The annual profits contributed by the equipment (including any salvage value at the end of the fifth year) are shown in the table in the margin. Find the internal rate of return of the investment in this equipment. Express the answer as a percentage correct to two decimal places.

Life Sciences

Marty Haigney

41. *Pollution.* A swimming pool is treated with a bactericide each morning. The number of bacteria per milliliter of water t hours after the treatment is given by

$$N(t) = 1{,}750 - 800te^{-0.2t}$$

The water is considered safe for swimming as long as the bacteria count is less than 600 bacteria per milliliter. Determine the time period during which this water is safe for swimming. Round values of t to one decimal place.

42. *Pollution.* In a certain city the amount of a pollutant (in parts per million) in the air t hours after 6 AM is given by

$$P(t) = 100te^{-0.2t}$$

The air quality is classified as poor whenever the amount of pollutant is above 150 parts per million. Determine the time interval (in hours after 6 AM) during which the air quality is poor. Round values of t to one decimal place.

43. *Drug concentration.* The concentration of a drug in the bloodstream t hours after injection is given approximately by

$$C(t) = 10e^{-0.1t} + 15e^{-0.2t}$$

where $C(t)$ is concentration in milligrams per milliliter. The initial concentration is $C(0) = 25$ milligrams per milliliter. A second injection is to be given when the concentration reaches 15 milligrams per milliliter. Find the time (correct to one decimal place) when the second injection should be administered.

44. *Drug sensitivity.* One hour after x milligrams of a particular drug is given to a person, the change in body temperature $T(x)$ in degrees Fahrenheit is given by

$$T(x) = x^2\left(1 - \frac{x}{9}\right) \qquad 0 \leq x \leq 6$$

Find the dosage (correct to two decimal places) that would produce a temperature change of 4°F.

Social Sciences

45. *Learning.* The relationship between the number of units $N(t)$ a worker can assemble in 1 day and the number of hours t of training the worker has received is given by

$$N(t) = t^3 - 6t^2 + 25t$$

How many hours (correct to one decimal place) of training are required in order for a worker to be able to assemble 500 units a day?

46. *Urban growth.* The population of a newly incorporated city (in thousands) t years after incorporation is given by

$$N(t) = 12 + t^2 - \tfrac{1}{15}t^3 \qquad 0 \leq t \leq 10$$

The city will qualify for certain types of federal aid when its population reaches 40,000. When will this city qualify for federal aid? Round the value of t to one decimal place.

SECTION 11-2 Interpolating Polynomials and Divided Differences

- INTRODUCTION
- THE INTERPOLATING POLYNOMIAL
- DIVIDED DIFFERENCE TABLES
- APPLICATION

In the preceding chapter, we used the Taylor polynomial at a to approximate the values of a function for values of x near a. In order to find the Taylor polynomial of degree n, we had to evaluate the function and its first n derivatives at a. This required that the function be defined by an equation, such as $f(x) = e^x$. But many important applications involve functions that are defined by tables instead of equations. For example, a retail sales firm may have defined the demand equation for a certain item by examining their past sales records and forming a table that lists the number of items sold at various prices. A biologist may have defined the relationship between the yield of a crop and the amount of a nutrient in the soil by conducting a series of experiments involving different amounts of the nutrient and listing the results in a table. Functions defined by tables cannot be approximated by Taylor polynomials, since it is not possible to compute the derivatives of such functions. In this section we will see how to use another type of polynomial to approximate functions that are defined by tables.

INTRODUCTION

We will begin by considering an example to illustrate the basic concepts and then proceed to more general observations.

EXAMPLE 7 *Revenue*

A manufacturing company has defined the revenue function for one of its products by examining past records and listing the revenue (in thousands of dollars) for certain levels of production (in thousands of units). Use the revenue function defined by Table 1 (on the next page) to estimate the revenue if 3,000 units are produced and if 7,000 units are produced.

Solution One way to approximate values of a function defined by a table is to use a **piecewise linear approximation.** To form the piecewise linear approximation for Table 1, we simply use the point–slope formula to find the equation of the line joining each successive pair of points in the table (see Figure 7).

TABLE 1

Revenue R Defined as a Function of Production x by a Table

x	1	4	6	8
R(x)	65	80	40	16

This type of approximation is very useful in certain applications, but it has several disadvantages. First, the piecewise linear approximation usually has a sharp corner at each point in the table and thus is not differentiable at these points. Second, the piecewise linear approximation requires the use of a different formula between each successive pair of points in the table (see Figure 7).

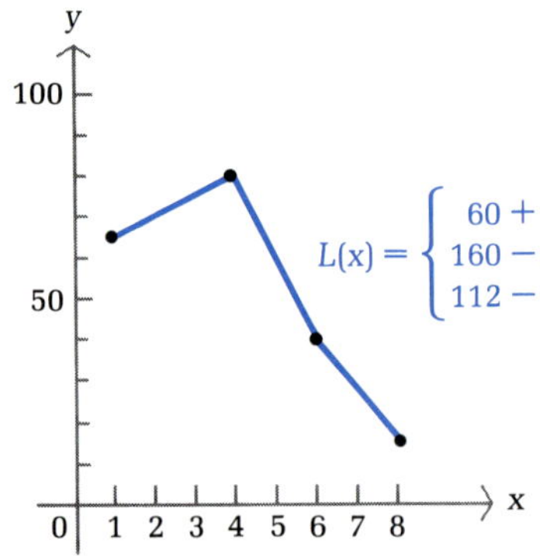

$$L(x) = \begin{cases} 60 + 5x & \text{if } 1 \leq x \leq 4 \\ 160 - 20x & \text{if } 4 < x \leq 6 \\ 112 - 12x & \text{if } 6 < x \leq 8 \end{cases}$$

FIGURE 7
$L(x)$ is the piecewise linear approximation for $R(x)$

Instead of using the piecewise linear approximation, we will outline a method that will produce a polynomial whose values agree with $R(x)$ at each point in Table 1. This will provide us with a differentiable function given by a single formula that can be used to approximate $R(x)$ for any value of x between 1 and 8.

Suppose $p(x)$ is a polynomial whose values agree with the values of $R(x)$ at the four x values given in Table 1. Instead of expressing $p(x)$ in terms of powers of x, the standard method for writing polynomial forms, we use the first three x values in the table to write

$$p(x) = a_0 + a_1(x-1) + a_2(x-1)(x-4) + a_3(x-1)(x-4)(x-6)$$

As we will see, writing $p(x)$ in this special form will greatly simplify our work.

Since $p(x)$ is to agree with $R(x)$ at each x value in Table 1, we can compute the coefficients a_0, a_1, a_2, and a_3 as follows:

$$R(1) = p(1)$$
$$65 = a_0 + a_1(0) + a_2(0)(-3) + a_3(0)(-3)(-5)$$

Notice that all terms after the first are 0. Thus,

$$a_0 = 65$$

Now we use the second point in Table 1 to determine a_1:

$$\begin{aligned} R(4) &= p(4) \\ 80 &= a_0 + a_1(3) + a_2(3)(0) + a_3(3)(0)(-2) && \text{Substitute } a_0 = 65. \\ 80 &= 65 + 3a_1 \\ 15 &= 3a_1 \end{aligned}$$

Thus,

$$a_1 = 5$$

Once again, notice that all terms after the term containing a_1 are 0. Now you begin to see the reason for writing $p(x)$ in this special form. Using the third point in Table 1, we have

$$\begin{aligned} R(6) &= p(6) \\ 40 &= a_0 + a_1(5) + a_2(5)(2) + a_3(5)(2)(0) && \text{Substitute } a_0 = 65,\ a_1 = 5. \\ 40 &= 65 + 25 + 10a_2 \\ -50 &= 10a_2 \end{aligned}$$

Thus,

$$a_2 = -5$$

Finally, we use the fourth point in Table 1 to determine a_3:

$$\begin{aligned} R(8) &= p(8) \\ 16 &= a_0 + a_1(7) + a_2(7)(4) + a_3(7)(4)(2) && \text{Substitute } a_0 = 65, \\ & && a_1 = 5,\ a_2 = -5. \\ 16 &= 65 + 35 - 140 + 56a_3 \\ 56 &= 56a_3 \end{aligned}$$

Thus,

$$a_3 = 1$$

and

$$p(x) = 65 + 5(x - 1) - 5(x - 1)(x - 4) + (x - 1)(x - 4)(x - 6)$$

The polynomial $p(x)$ agrees with $R(x)$ at each x value in Table 1 (verify this) and can be used to approximate $R(x)$ for values of x between 1 and 8 (see Figure 8).

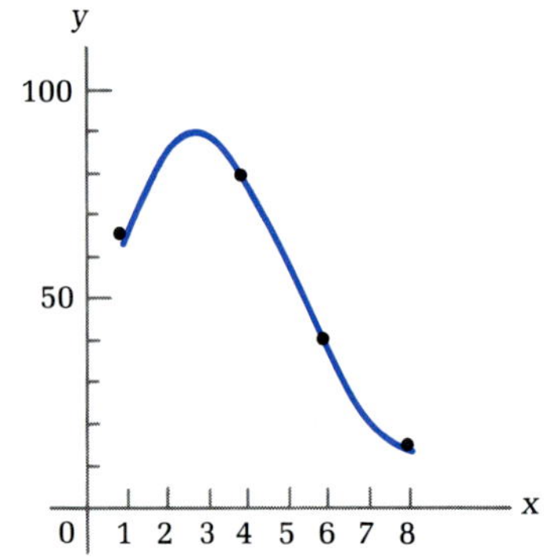

FIGURE 8
$p(x) = 65 + 5(x - 1) - 5(x - 1)(x - 4) + (x - 1)(x - 4)(x - 6)$

TABLE 1

x	1	4	6	8
$R(x)$	65	80	40	16

If 3,000 units are produced, then the revenue can be approximated by evaluating $p(3)$:

$$\begin{aligned} R(3) \approx p(3) &= 65 + 5(2) - 5(2)(-1) + (2)(-1)(-3) \\ &= 91 \text{ or } \$91{,}000 \end{aligned}$$

If 7,000 units are produced, then

$$\begin{aligned} R(7) \approx p(7) &= 65 + 5(6) - 5(6)(3) + (6)(3)(1) \\ &= 23 \text{ or } \$23{,}000 \end{aligned}$$

◆

PROBLEM 7 Refer to Example 7. Approximate the revenue if 2,000 units are produced and if 5,000 units are produced. ◆

Since the revenue function in Example 7 was defined by a table, we have no information about this function for any value of x other than those listed in the table. Thus, we cannot say anything about the accuracy of the approximations obtained by using $p(x)$. As we mentioned earlier, the piecewise linear approximation might provide a better approximation in some cases. The primary advantage of using $p(x)$ is that we have a differentiable function that is defined by a single equation and agrees with the revenue function at every value of x in the table.

◆ THE INTERPOLATING POLYNOMIAL

The procedure we used to find a polynomial approximation for the revenue function in Example 7 can be applied to any function that is defined by a table. The polynomial that is obtained in this way is referred to as the *interpolating polynomial*. The basic concepts are summarized in the next box.

The Interpolating Polynomial

If $f(x)$ is the function defined by the following table of $n + 1$ points:

x	x_0	x_1	$\cdots$	x_n
$f(x)$	y_0	y_1	$\cdots$	y_n

then the **interpolating polynomial** for $f(x)$ is the polynomial $p(x)$ of degree less than or equal to n that satisfies

$$\begin{aligned} p(x_0) &= y_0 = f(x_0) \\ p(x_1) &= y_1 = f(x_1) \\ &\vdots \\ p(x_n) &= y_n = f(x_n) \end{aligned}$$

The **general form** of the interpolating polynomial is

$$p(x) = a_0 + a_1(x - x_0) + a_2(x - x_0)(x - x_1) + \cdots + a_n(x - x_0)(x - x_1) \cdot \cdots \cdot (x - x_{n-1})$$

Notice that if we graph the points in the defining table and the interpolating polynomial $p(x)$ on the same set of axes, then the graph of $p(x)$ will pass through every point given in the table (see Figure 8). Is it possible to find a polynomial that is different from $p(x)$ and also has a graph that passes through all the points in the table? In more advanced texts, it is shown that:

The interpolating polynomial is the only polynomial of degree less than or equal to *n* whose graph will pass through every point in the table.

Any other polynomial whose graph goes through all these points must be of degree greater than n. The steps we used in finding the interpolating polynomial are summarized in the box.

Steps for Finding the Intepolating Polynomial

Write the general form of $p(x)$ and proceed as follows:

Step 1. Use the condition $p(x_0) = y_0$ to find a_0.
Step 2. Use the condition $p(x_1) = y_1$ and the value of a_0 determined in the preceding step to find a_1.
Step 3. Use the condition $p(x_2) = y_2$ and the values of a_0 and a_1 determined in the preceding steps to find a_2.

$\vdots$

Step $n + 1$. Use the condition $p(x_n) = y_n$ and the values of $a_0, a_1, \ldots, a_{n-1}$ determined in the preceding steps to find a_n.

◆ EXAMPLE 8 Find the interpolating polynomial for the function defined by the following table:

x	0	1	2	3
$f(x)$	5	4	−3	−4

Solution The general form of $p(x)$ for this table is

$$p(x) = a_0 + a_1x + a_2x(x-1) + a_3x(x-1)(x-2)$$

Step 1. $f(0) = p(0)$

$5 = a_0$ All other terms in $p(0)$ are 0.

Step 2. $f(1) = p(1)$

$4 = a_0 + a_1(1)$ Substitute $a_0 = 5$.

$4 = 5 + a_1$

$-1 = a_1$

Step 3. $f(2) = p(2)$

$-3 = a_0 + a_1(2) + a_2(2)(1)$ Substitute $a_0 = 5$, $a_1 = -1$.

$-3 = 5 - 2 + 2a_2$

$-6 = 2a_2$

$-3 = a_2$

Step 4. $f(3) = p(3)$

$-4 = a_0 + a_1(3) + a_2(3)(2) + a_3(3)(2)(1)$ Substitute $a_0 = 5$, $a_1 = -1$, $a_2 = -3$.

$-4 = 5 - 3 - 18 + 6a_3$

$12 = 6a_3$

$2 = a_3$

Since there are no more points in the table, we stop and conclude that the interpolating polynomial for this table is

$$p(x) = 5 - x - 3x(x-1) + 2x(x-1)(x-2)$$

This form of $p(x)$ is suitable for evaluating the polynomial. For other operations, such as differentiation, integration, or graphing, it is preferable to perform the indicated multiplications, collect like terms, and express $p(x)$ in the standard polynomial form

$$p(x) = 2x^3 - 9x^2 + 6x + 5$$

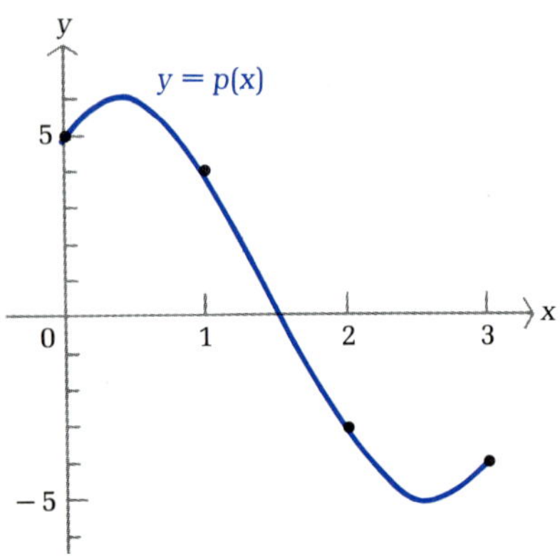

The graph of this polynomial is shown in the figure in the margin. (Graphing details are omitted.) ◆

PROBLEM 8 Find the interpolating polynomial for the function defined by the following table:

x	-1	0	1	2
$f(x)$	5	3	3	11

◆

TABLE 1

x	1	4	6	8
$f(x)$	65	80	40	16

◆ DIVIDED DIFFERENCE TABLES

We now present a simple computational procedure for finding the coefficients $a_0, a_1, \ldots, a_n$ in the general form of an interpolating polynomial. To introduce this method, we return to Table 1 in Example 7, which we restate here.

The coefficients in the general form of the interpolating polynomial for this table were $a_0 = 65$, $a_1 = 5$, $a_2 = -5$, and $a_3 = 1$. We will now construct a table, called a *divided difference table*, which will produce these coefficients with a minimum of computation. To begin, we place the x and y values in the first two columns of a new table. Then we compute the ratio of the change in y to the change in x for each successive pair of points in the table, and place the result on the line between the two points (see Table 2). These ratios are called the **first divided differences.**

TABLE 2
First Divided Differences

x_k	y_k	FIRST DIVIDED DIFFERENCE
1	65	
		$\frac{80-65}{4-1} = \frac{15}{3} = 5$
4	80	
		$\frac{40-80}{6-4} = \frac{-40}{2} = -20$
6	40	
		$\frac{16-40}{8-6} = \frac{-24}{2} = -12$
8	16	

To form the next column in the table, we repeat this process, using the change in the first divided differences in the numerator and the change in *two* successive values of x in the denominator. These ratios are called the **second divided differences** and are placed on the line between the corresponding first divided differences (see Table 3).

TABLE 3

Second Divided Differences

x_k	y_k	FIRST DIVIDED DIFFERENCE	SECOND DIVIDED DIFFERENCE
1	65		
		5	
4	80		$\frac{-20-5}{6-1}=\frac{-25}{5}=-5$
		-20	
6	40		$\frac{-12-(-20)}{8-4}=\frac{8}{4}=2$
		-12	
8	16		

To form the next column of the table, we form the ratio of the change in the second divided differences to the change in *three* successive values of x. These ratios are called the **third divided differences** and are placed on the line between the corresponding second divided differences (see Table 4). Since our table has only two second divided differences, there is only one third divided difference and this process is now complete.

TABLE 4

Third Divided Differences

x_k	y_k	FIRST DIVIDED DIFFERENCE	SECOND DIVIDED DIFFERENCE	THIRD DIVIDED DIFFERENCE
1	65			
		5		
4	80		-5	
		-20		$\frac{2-(-5)}{8-1}=\frac{7}{7}=1$
6	40		2	
		-12		
8	16			

We have presented each step in constructing the divided difference table here in a separate table to clearly illustrate this process. In applications of this technique, these steps are combined into a single table. With a little practice, you should be able to proceed quickly from the defining table for the function (Table 1) to the final form of the divided difference table (Table 5).

TABLE 5

Divided Difference Table—Final Form

x_k	y_k	FIRST DIVIDED DIFFERENCE	SECOND DIVIDED DIFFERENCE	THIRD DIVIDED DIFFERENCE
1	**65**			
		$\frac{80-65}{4-1} =$ **5**		
4	80		$\frac{-20-5}{6-1} =$ **−5**	
		$\frac{40-80}{6-4} = -20$		$\frac{2-(-5)}{8-1} =$ **1**
6	40		$\frac{-12-(-20)}{8-4} = 2$	
		$\frac{16-40}{8-6} = -12$		
8	16			

Now that we have computed the divided difference table, how do we use it? If we write the first number from each column of the divided difference table, beginning with the second column:

$$65 \qquad 5 \qquad -5 \qquad 1$$

we see that these numbers are the coefficients of the interpolating polynomial for Table 1 (see Example 7). Thus, Table 5 contains all the information we need to write the interpolating polynomial:

$$p(x) = 65 + 5(x-1) - 5(x-1)(x-4) + (x-1)(x-4)(x-6)$$

The divided difference table provides an alternate method for finding interpolating polynomials that generally requires fewer computations and can be implemented easily on a computer. The ideas introduced in the preceding discussion are summarized in the box on the next page.

Divided Difference Tables and Interpolating Polynomials

Given the defining table for a function $f(x)$ with $n + 1$ points,

x	x_0	x_1	$\cdots$	x_n
$f(x)$	y_0	y_1	$\cdots$	y_n

where $x_0 < x_1 < \cdots < x_n$, then the **divided difference table** is computed as follows:

Column 1: x values from the defining table

Column 2: y values from the defining table

Column 3: First divided differences computed using columns 1 and 2

Column 4: Second divided differences computed using columns 1 and 3

$\vdots$

Column n + 2: nth divided differences computed using columns 1 and $n + 1$

The coefficients in the general form of the interpolating polynomial,

$$p(x) = a_0 + a_1(x - x_0) + a_2(x - x_0)(x - x_1) + \cdots + a_n(x - x_0)(x - x_1) \cdot \cdots \cdot (x - x_{n-1})$$

are the first numbers in each column of the divided difference table, beginning with column 2.

Remarks

1. The points in the defining table must be arranged with increasing x values before computing the divided difference table. If the x values are out of order, then the divided difference table will not contain the coefficients of the general form of the interpolating polynomial.
2. Since each column in the divided difference table uses all the values in the preceding column, it is necessary to compute all the numbers in every column, even though we are interested only in the first number in each column.
3. Other methods can be used to find interpolating polynomials. Referring to Table 1, we could write $p(x)$ in standard polynomial notation

$$p(x) = b_3x^3 + b_2x^2 + b_1x + b_0$$

and use the points in the table to write the following system of linear equations:

$$\begin{aligned}
p(1) &= b_3 + b_2 + b_1 + b_0 = 65 \\
p(4) &= 64b_3 + 16b_2 + 4b_1 + b_0 = 80 \\
p(6) &= 216b_3 + 36b_2 + 6b_1 + b_0 = 40 \\
p(8) &= 512b_3 + 64b_2 + 8b_1 + b_0 = 16
\end{aligned}$$

The computations required to solve this system of equations are far more complicated than those involved in finding the divided difference table.

◆ EXAMPLE 9 Given the table:

x	0	1	2	3
$f(x)$	5	4	−3	−4

(A) Find the divided difference table.
(B) Use the divided difference table to find the interpolating polynomial. Compare the result with the answer in Example 8.

Solutions (A) The divided difference table is given below:

x_k	y_k	FIRST DIVIDED DIFFERENCE	SECOND DIVIDED DIFFERENCE	THIRD DIVIDED DIFFERENCE
0	**5**			
		$\frac{4-5}{1-0} =$ **−1**		
1	4		$\frac{-7-(-1)}{2-0} =$ **−3**	
		$\frac{-3-4}{2-1} = -7$		$\frac{3-(-3)}{3-0} =$ **2**
2	−3		$\frac{-1-(-7)}{3-1} = 3$	
		$\frac{-4-(-3)}{3-2} = -1$		
3	−4			

(B) The general form of the interpolating polynomial is

$$p(x) = a_0 + a_1x + a_2x(x-1) + a_3x(x-1)(x-2)$$

Substituting the values from the divided difference table for the coefficients in this general form, we have

$$p(x) = 5 - x - 3x(x - 1) + 2x(x - 1)(x - 2)$$

which is the same polynomial we obtained in Example 8. ◆

PROBLEM 9 Given the table:

x	−1	0	1	2
$f(x)$	5	3	3	11

(A) Find the divided difference table.
(B) Use the divided difference table to find the interpolating polynomial. Compare the result with the answer in Problem 8. ◆

◆ EXAMPLE 10 Given the table:

x	−1	0	2	5	10
$f(x)$	53	20	−28	−55	20

(A) Find the divided difference table.
(B) Use the divided difference table to find the interpolating polynomial.

Solutions (A) The divided difference table is given below:

x_k	y_k	FIRST DIVIDED DIFFERENCE	SECOND DIVIDED DIFFERENCE	THIRD DIVIDED DIFFERENCE	FOURTH DIVIDED DIFFERENCE
−1	**53**				
		−33			
0	20		**3**		
		−24		**0**	
2	−28		3		**0**
		−9		0	
5	−55		3		
		15			
10	20				

(B) The general form of the interpolating polynomial is

$$p(x) = a_0 + a_1(x + 1) + a_2(x + 1)x + a_3(x + 1)x(x - 2) + a_4(x + 1)x(x - 2)(x - 5)$$

Substituting the values from the divided difference table for the coefficients in the general form, we have

$$p(x) = 53 - 33(x + 1) + 3(x + 1)x + 0(x + 1)x(x - 2) + 0(x + 1)x(x - 2)(x - 5)$$
$$= 53 - 33(x + 1) + 3(x + 1)x$$

Check

$$p(-1) = 53 - 0 + 0 \stackrel{\checkmark}{=} 53$$
$$p(0) = 53 - 33 + 0 \stackrel{\checkmark}{=} 20$$
$$p(2) = 53 - 99 + 18 \stackrel{\checkmark}{=} -28$$
$$p(5) = 53 - 198 + 90 \stackrel{\checkmark}{=} -55$$
$$p(10) = 53 - 363 + 330 \stackrel{\checkmark}{=} 20$$

◆

PROBLEM 10 Given the table:

x	-2	-1	1	4	8
$f(x)$	-32	-11	19	34	-2

(A) Find the divided difference table.
(B) Use the divided difference table to find the interpolating polynomial. ◆

Remark

Notice that the table in Example 10 has five points, but the interpolating polynomial is only a second-degree polynomial. If a table has $n + 1$ points, then the degree of the interpolating polynomial is always less than or equal to n and may be strictly less than n.

◆ APPLICATION

◆ EXAMPLE 11 *Inventory*

A store orders 8,000 units of a new product. The inventory I on hand t weeks after the order arrived is given in the table:

t	0	2	4	6	8
$I(t)$	8,000	5,952	3,744	1,568	0

Use the interpolating polynomial to approximate the inventory after 5 weeks and the average inventory during the first 5 weeks after the order arrived.

Solution The divided difference table is shown below:

t_k	y_k	FIRST DIVIDED DIFFERENCE	SECOND DIVIDED DIFFERENCE	THIRD DIVIDED DIFFERENCE	FOURTH DIVIDED DIFFERENCE
0	8,000				
		−1,024			
2	5,952		−20		
		−1,104		4	
4	3,744		4		1
		−1,088		12	
6	1,568		76		
		−784			
8	0				

The interpolating polynomial is

$$p(t) = 8{,}000 - 1{,}024t - 20t(t-2) + 4t(t-2)(t-4) + t(t-2)(t-4)(t-6)$$

or, after simplifying,

$$p(t) = t^4 - 8t^3 - 1{,}000t + 8{,}000$$

The inventory after 5 weeks is given approximately by

$$p(5) = 5^4 - 8(5^3) - 1{,}000(5) + 8{,}000 = 2{,}625 \text{ units}$$

The average inventory during the first 5 weeks is given approximately by

$$\begin{aligned}\frac{1}{5}\int_0^5 p(t)\,dt &= \frac{1}{5}\int_0^5 (t^4 - 8t^3 - 1{,}000t + 8{,}000)\,dt \\ &= \frac{1}{5}\left(\frac{1}{5}t^5 - 2t^4 - 500t^2 + 8{,}000t\right)\Big|_0^5 \\ &= \frac{1}{5}(625 - 1{,}250 - 12{,}500 + 40{,}000) - \frac{1}{5}(0) \\ &= 5{,}375 \text{ units}\end{aligned}$$

◆

PROBLEM 11 Refer to Example 11. Approximate the inventory after 7 weeks and the average inventory during the first 7 weeks. ◆

Answers to Matched Problems

7. $p(2) = 88$ or \$88,000; $p(5) = 61$ or \$61,000
8. $p(x) = 5 - 2(x+1) + (x+1)x + (x+1)x(x-1)$

9. (A)

x_k	y_k	FIRST DIVIDED DIFFERENCE	SECOND DIVIDED DIFFERENCE	THIRD DIVIDED DIFFERENCE
−1	5			
		−2		
0	3		1	
		0		1
1	3		4	
		8		
2	11			

(B) $p(x) = 5 - 2(x + 1) + (x + 1)x + (x + 1)x(x - 1)$

10. (A)

x_k	y_k	FIRST DIVIDED DIFFERENCE	SECOND DIVIDED DIFFERENCE	THIRD DIVIDED DIFFERENCE	FOURTH DIVIDED DIFFERENCE
−2	−32				
		21			
−1	−11		−2		
		15		0	
1	19		−2		0
		5		0	
4	34		−2		
		−9			
8	−2				

(B) $p(x) = -32 + 21(x + 2) - 2(x + 2)(x + 1)$

11. 657 units; 4,294.2 units

EXERCISE 11-2

A *Use the method outlined in the box on page 723 to find the interpolating polynomial for each table.*

1.

x	1	3	4
$f(x)$	2	6	11

2.

x	−1	1	2
$f(x)$	1	3	7

3.

x	−1	0	2	4
$f(x)$	6	5	15	−39

4.

x	−1	0	2	3
$f(x)$	5	1	5	1

Find the divided difference table and then find the interpolating polynomial.

5.

x	1	2	3
$f(x)$	4	8	14

6.

x	1	2	3
$f(x)$	1	3	7

7.

x	-1	0	1	2
$f(x)$	-3	1	3	9

8.

x	-1	0	1	2
$f(x)$	5	6	3	2

9.

x	-2	1	2	4
$f(x)$	25	10	17	13

10.

x	-1	0	3	5
$f(x)$	17	10	25	5

B *Use the interpolating polynomial to approximate the value of the function defined by the table at the indicated values of x.*

11.

x	-4	0	4	8
$f(x)$	-64	32	0	224

(A) $f(2) \approx ?$ (B) $f(6) \approx ?$

12.

x	-5	0	5	10
$f(x)$	250	50	100	-350

(A) $f(-3) \approx ?$ (B) $f(8) \approx ?$

13.

x	-1	0	1	4
$f(x)$	0	0	0	15

(A) $f(2) \approx ?$ (B) $f(3) \approx ?$

14.

x	-2	0	2	6
$f(x)$	0	0	0	-96

(A) $f(1) \approx ?$ (B) $f(4) \approx ?$

15.

x	-4	-2	0	2	4
$f(x)$	24	2	0	-6	8

(A) $f(-3) \approx ?$ (B) $f(1) \approx ?$

16.

x	-6	-2	0	2	6
$f(x)$	19	3	10	3	19

(A) $f(1) \approx ?$ (B) $f(5) \approx ?$

17.

x	-3	-2	-1	1	2	3
$f(x)$	-24	-6	0	0	6	24

(A) $f(-0.5) \approx ?$ (B) $f(2.5) \approx ?$

18.

x	-3	-2	-1	0	1	2	3
$f(x)$	40	0	0	4	0	0	40

(A) $f(-2.5) \approx ?$ (B) $f(1.5) \approx ?$

Find the interpolating polynomial. Graph the interpolating polynomial and the points in the given table on the same set of axes.

19.

x	−2	0	2
f(x)	2	0	2

20.

x	−2	0	2
f(x)	2	0	−2

21.

x	0	1	2
f(x)	−4	−2	0

22.

x	0	1	2
f(x)	−4	−3	0

23.

x	−1	0	2	3
f(x)	0	2	0	−4

24.

x	−3	−1	0	1
f(x)	0	4	3	0

25.

x	−2	−1	0	1	2
f(x)	1	5	3	1	5

26.

x	−2	−1	0	1	2
f(x)	−8	0	2	4	12

27.

x	−2	−1	0	1	2
f(x)	−3	0	5	0	−3

28.

x	−1	0	1	2	3
f(x)	6	2	0	−6	2

C

29. The following table was obtained from the function $f(x) = \sqrt{x}$:

x	1	4	9
f(x)	1	2	3

Find the interpolating polynomial for this table. Compare the values of the interpolating polynomial p(x) and the original function $f(x) = \sqrt{x}$ by completing the table below. Use a calculator to evaluate $\sqrt{x}$ and round each value to one decimal place.

x	1	2	3	4	5	6	7	8	9
p(x)	1			2					3
$\sqrt{x}$	1			2					3

30. The following table was obtained from the function $f(x) = 6/\sqrt{x}$:

x	1	4	9
$f(x)$	6	3	2

Find the interpolating polynomial for this table. Compare the values of the interpolating polynomial $p(x)$ and the original function $f(x) = 6/\sqrt{x}$ by completing the table below. Use a calculator to evaluate $6/\sqrt{x}$ and round each value to one decimal place.

x	1	2	3	4	5	6	7	8	9
$p(x)$	6			3					2
$6/\sqrt{x}$	6			3					2

31. The following table was obtained from the function $f(x) = 10x/(1 + x^2)$:

x	-2	-1	0	1	2
$f(x)$	-4	-5	0	5	4

Find the interpolating polynomial $p(x)$ for this table. Graph $p(x)$ and $f(x)$ on the same set of axes.

32. The table below was obtained from the function $f(x) = (9 - x^2)/(1 + x^2)$:

x	-2	-1	0	1	2
$f(x)$	1	4	9	4	1

Find the interpolating polynomial $p(x)$ for this table. Graph $p(x)$ and $f(x)$ on the same set of axes.

33. Find the equation of the parabola whose graph passes through the points $(-x_1, y_1)$, $(0, y_2)$, and (x_1, y_1), where $x_1 > 0$ and $y_1 \neq y_2$.

34. Find the equation of the parabola whose graph passes through the points $(0, 0)$, (x_1, y_1), and $(2x_1, 0)$, where $x_1 > 0$ and $y_1 \neq 0$.

Business & Economics

35. *Cash reserves.* Suppose cash reserves (in thousands of dollars) are given by the following table, where t is the number of months after the first of the year:

t	0	4	8	12
$C(t)$	2	32	38	20

Find the interpolating polynomial for this table. Approximate the cash reserves after 6 months. Approximate the average cash reserves for the first quarter.

36. *Inventory.* A store orders 147 units of a product. The inventory I on hand t months after the order arrived is given in the table. Find the interpolating polynomial for this table. Approximate the average number of units on hand for this 3 month period.

t	0	1	2	3
$I(t)$	147	66	19	0

37. *Income distribution.* The income distribution for a certain country is represented by the Lorenz curve $y = f(x)$, where $f(x)$ is given in the table. Find the interpolating polynomial for this table and approximate the coefficient of inequality.

x	0	0.2	0.8	1
$f(x)$	0	0.1	0.4	1

38. *Income distribution.* Repeat Problem 37 for the Lorenz curve defined by the following table:

x	0	0.2	0.4	1
$f(x)$	0	0.12	0.16	1

39. *Maximum revenue.* The revenue R (in thousands of dollars) from the sale of x thousand units of a product is given in the table. Find the interpolating polynomial for this table. Approximate the revenue if 5,000 units are produced. Approximate the production level that will maximize the revenue.

x	2	4	6
$R(x)$	24.4	36	34.8

40. *Minimum average cost.* The cost C (in hundreds of dollars) of producing x hundred units of a product is given in the table. Find the interpolating polynomial for this table. Approximate the cost of producing 400 units. Approximate the production level that will minimize the average cost.

x	1	3	5
C(x)	215	535	1,055

Life Sciences

41. *Temperature.* The temperature C (in degrees Celsius) in an artificial habitat after t hours is given in the table. Find the interpolating polynomial for this table. Approximate the average temperature over this 4 hour period.

t	0	1	2	3	4
C(t)	14	13	16	17	10

42. *Drug concentration.* The concentration $C(t)$ in milligrams per cubic centimeter of a particular drug in a patient's bloodstream t hours after the drug is taken is given in the table. Find the interpolating polynomial for this table. Approximate the number of hours it will take for the drug concentration to reach its maximum level.

t	0	1	2	3	4
C(t)	0	0.032	0.036	0.024	0.008

43. *Bacterial control.* A lake that is used for recreational swimming is treated periodically to control harmful bacteria growth. The concentration, C, of bacteria per cubic centimeter t days after a treatment is given in the table. Find the interpolating polynomial for this table. Approximate the number of days it will take for the bacteria concentration to reach its minimal level.

t	0	2	4	6
C(t)	450	190	90	150

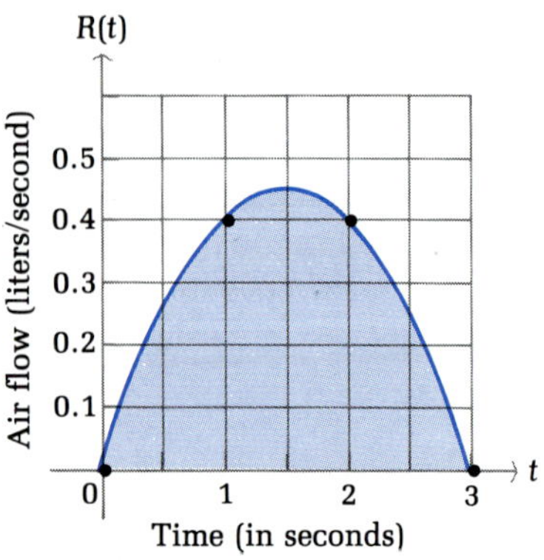

44. *Medicine—respiration.* Physiologists use a machine called a pneumotachograph to produce a graph of the rate of flow $R(t)$ of air into the lungs (inspiration) and out (expiration). The figure in the margin gives the graph of the inspiration phase of the breathing cycle of an individual at rest. The area under this graph represents the total volume of air inhaled during the inspiration phrase. Use the values given by the graph at $t = 0, 1, 2,$ and 3 to find the interpolating polynomial for $R(t)$. Then use the interpolating polynomial to approximate the total volume of air inhaled.

Social Sciences

45. *Voter registration.* The number N of registered voters over a 30 year period is given in the table. Find the interpolating polynomial for this table. Approximate the average number of voters over the first 20 years of this period.

t	0	10	20	30
$N(t)$	10,000	13,500	20,000	23,500

46. *Politics.* The voting population N in a city over a 10 year period is given in the table. Find the interpolating polynomial for this table. Approximate the year t when the rate of increase in the voting population is most rapid.

t	0	4	6	10
$N(t)$	15,000	18,800	22,200	26,000

SECTION 11-3 Numerical Integration

- INTRODUCTION
- THE TRAPEZOID RULE
- SIMPSON'S RULE
- APPLICATIONS
- ERROR ESTIMATES

INTRODUCTION

If f is continuous on the interval $[a, b]$ and F is an antiderivative of f, then the definite integral of f from $x = a$ to $x = b$ is given by

$$\int_a^b f(x)\,dx = F(b) - F(a)$$

However, the antiderivative of f may be difficult to find, or it may not even exist in a convenient closed form. In this case, numerical integration can be used to approximate $\int_a^b f(x)\,dx$.

In Section 6-6 we discussed the rectangle rule for numerical integration, which uses Riemann sums to approximate definite integrals. That is,

$$\int_a^b f(x)\,dx \approx \sum_{k=1}^{n} f(c_k)\Delta x_k \qquad n \text{ sufficiently large}$$

Recall that if $f(x) \geq 0$ for $a \leq x \leq b$, then each term in the Riemann sum can be interpreted as the area of a rectangle with width Δx and altitude $f(c_k)$; hence the name "rectangle rule" (see Figure 9 at the top of the next page).

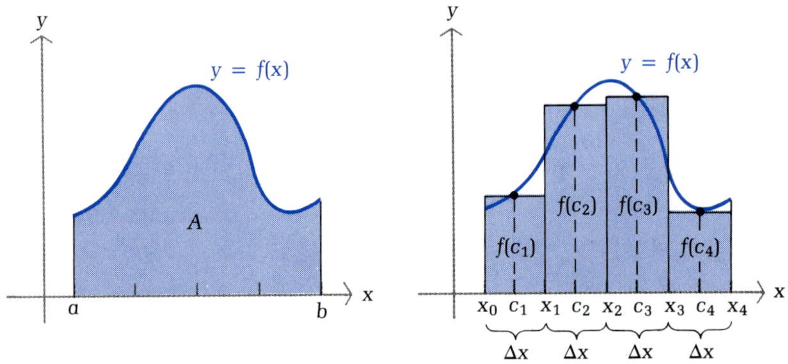

(A) $\int_a^b f(x)\,dx = A$

(B) $\int_a^b f(x)\,dx \approx f(c_1)\Delta x + f(c_2)\Delta x + f(c_3)\Delta x + f(c_4)\Delta x$

FIGURE 9

In this section we will develop two other important methods for approximating definite integrals, the *trapezoid rule* and *Simpson's rule*. Each of these methods is based on using a sum of areas of nonrectangular regions to approximate the area under a curve. Both methods are used extensively to approximate definite integrals on computers. Some calculators also have the capability to approximate integrals using one or the other of these methods.

◆ THE TRAPEZOID RULE

In the rectangle rule, the area under the graph of a positive function $f(x)$ over the interval $[x_{k-1}, x_k]$ was approximated by the area of the rectangle with width $\Delta x = x_k - x_{k-1}$ and altitude $f(c_k)$. Now we will approximate the area under the graph of $f(x)$ from $x = x_{k-1}$ to $x = x_k$ by using the area of a trapezoid. Recall that the area of a trapezoid with altitude h and bases b_1 and b_2 (see Figure 10A) is given by

$$A = \frac{h}{2}(b_1 + b_2) \tag{1}$$

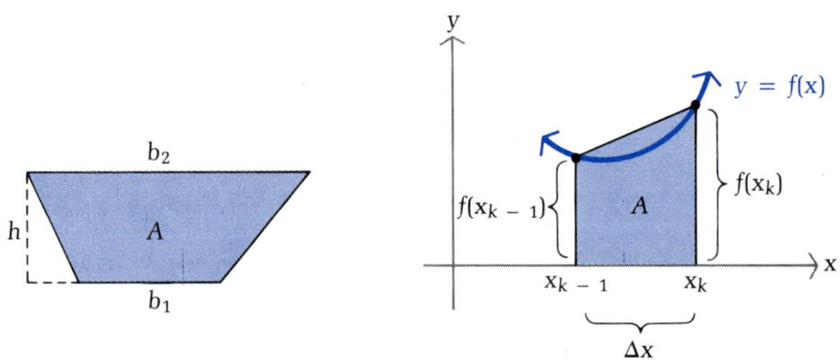

(A) $A = \frac{h}{2}(b_1 + b_2)$

(B) $A = \frac{\Delta x}{2}[f(x_{k-1}) + f(x_k)]$

FIGURE 10

If the points $(x_{k-1}, f(x_{k-1}))$ and $(x_k, f(x_k))$ on the graph of $y = f(x)$ are connected with a straight line segment, a trapezoid is formed (see Figure 10B). The area of this trapezoid is given by

$$A = \frac{\Delta x}{2}[f(x_{k-1}) + f(x_k)] \qquad (2)$$

Let $h = \Delta x$, $b_1 = f(x_{k-1})$, and $b_2 = f(x_k)$ in (1).

Connecting the points $(x_0, f(x_0))$, $(x_1, f(x_1))$, . . . , $(x_n, f(x_n))$ with straight line segments forms n trapezoids (as illustrated in Figure 11 for $n = 4$). If T_n denotes the sum of the areas of these trapezoids, then T_n is an approximation to the area under the graph of f.

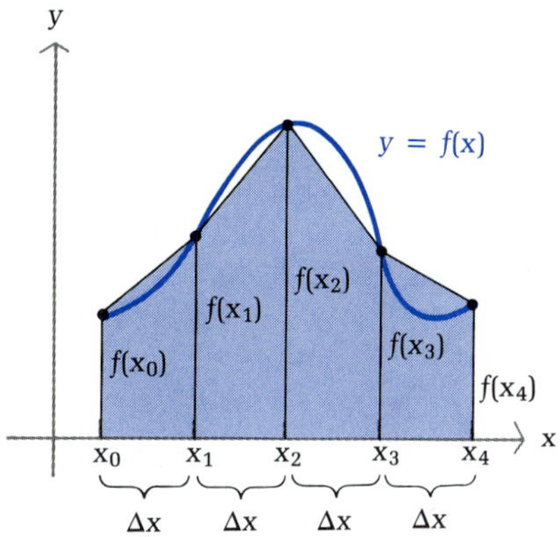

FIGURE 11

$$\int_a^b f(x)\,dx \approx T_4$$

$$= \frac{\Delta x}{2}[f(x_0) + f(x_1)] + \frac{\Delta x}{2}[f(x_1) + f(x_2)] + \frac{\Delta x}{2}[f(x_2) + f(x_3)] + \frac{\Delta x}{2}[f(x_3) + f(x_4)]$$

Using formula (2) for the area of each trapezoid, we have

$$T_n = \frac{\Delta x}{2}[f(x_0) + f(x_1)] + \frac{\Delta x}{2}[f(x_1) + f(x_2)] + \frac{\Delta x}{2}[f(x_2) + f(x_3)] + \cdots + \frac{\Delta x}{2}[f(x_{n-1}) + f(x_n)]$$

Factor out $\Delta x/2$.

$$= \frac{\Delta x}{2}[f(x_0) + f(x_1) + f(x_1) + f(x_2) + f(x_2) + f(x_3) + \cdots + f(x_{n-1}) + f(x_n)]$$

Note: Excluding $f(x_0)$ and $f(x_n)$, each term $f(x_k)$ occurs twice in this sum.

$$= \frac{\Delta x}{2}[f(x_0) + 2f(x_1) + 2f(x_2) + \cdots + 2f(x_{n-1}) + f(x_n)]$$

Although this discussion is based on approximating the area under the graph of a positive function $f(x)$, it can be shown that

$$\lim_{n\to\infty} T_n = \int_a^b f(x)\,dx$$

for any function $f(x)$ that is continuous on $[a, b]$. This leads to the *trapezoid rule for approximating definite integrals.*

Trapezoid Rule

Divide the interval from $x = a$ to $x = b$ into n equal subintervals of length $\Delta x = (b - a)/n$. Let $a = x_0 < x_1 < \cdots < x_n = b$ be the end points of these subintervals. Then

$$\int_a^b f(x)\,dx \approx T_n = \frac{\Delta x}{2}\,[f(x_0) + 2f(x_1) + \cdots + 2f(x_{n-1}) + f(x_n)]$$

◆ EXAMPLE 12 Use the trapezoid rule with the indicated values of n to approximate

$$\int_1^3 \ln x\,dx$$

Round each approximation to three decimal places.

(A) $n = 4$ (B) $n = 10$

Solutions (A) *Step 1.* Compute Δx:

$$\Delta x = \frac{b - a}{n} = \frac{3 - 1}{4} = 0.5$$

Step 2. Find the end points of the subintervals:

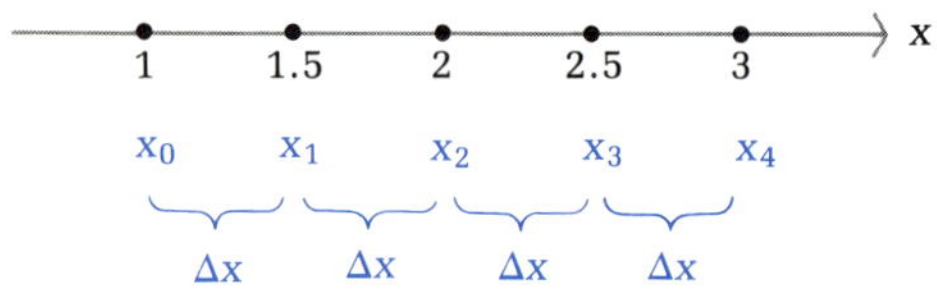

Step 3. Compute T_4:

$$\begin{aligned} T_4 &= \frac{\Delta x}{2}\,[f(x_0) + 2f(x_1) + 2f(x_2) + 2f(x_3) + f(x_4)] \\ &= \frac{0.5}{2}\,[f(1) + 2f(1.5) + 2f(2) + 2f(2.5) + f(3)] \\ &= 0.25[(\ln 1) + 2(\ln 1.5) + 2(\ln 2) + 2(\ln 2.5) + (\ln 3)] \\ &\approx 0.25[5.128\ 418\ 3] \approx 1.282\ 104\ 6 \end{aligned}$$

We used a calculator to evaluate T_4. Rounding this value to three decimal places, we can conclude that

$$\int_1^3 \ln x\,dx \approx T_4 \approx 1.282$$

(B) *Step 1.* Compute Δx:

$$\Delta x = \frac{3 - 1}{10} = 0.2$$

Step 2. Find the end points of the subintervals:

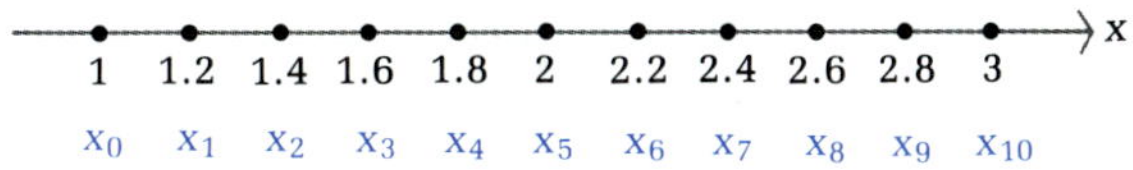

Step 3. Compute T_{10}:

$$\begin{aligned} T_{10} &= \frac{\Delta x}{2}[f(x_0) + 2f(x_1) + \cdots + 2f(x_9) + f(x_{10})] \\ &= \frac{0.2}{2}[f(1) + 2f(1.2) + \cdots + 2f(2.8) + f(3)] \\ &= 0.1[(\ln 1) + 2(\ln 1.2) + \cdots + 2(\ln 2.8) + (\ln 3)] \\ &\approx 0.1[12.936\ 189] \approx 1.293\ 618\ 9 \end{aligned}$$

Thus,

$$\int_1^3 \ln x\, dx \approx T_{10} \approx 1.294$$

◆

PROBLEM 12 Use the trapezoid rule with the indicated values of n to approximate

$$\int_2^4 \frac{1}{x}\, dx$$

Round each approximation to three decimal places.

(A) $n = 4$ (B) $n = 10$ ◆

Referring to Example 12, which approximation is more accurate, T_4 or T_{10}? In order to be able to answer this question, we purposely selected an integral that can be evaluated directly. (In general, we apply approximation methods to integrals that cannot be evaluated directly.) Using integration by parts or a table of integrals to find an antiderivative for ln x, we have

$$\begin{aligned} \int_1^3 \ln x\, dx &= (x \ln x - x)\Big|_1^3 \\ &= 3 \ln 3 - 2 \\ &\approx 1.296 \qquad \text{Rounded to three decimal places} \end{aligned}$$

Comparing this value with the approximations obtained in Example 12, we have

$$\int_1^3 \ln x\, dx - T_4 \approx 1.296 - 1.282 = 0.014$$

and

$$\int_1^3 \ln x\,dx - T_{10} \approx 1.296 - 1.294 = 0.002$$

Thus, T_{10} provides the more accurate approximation. In most cases, increasing the value of n increases the accuracy of the approximation provided by the trapezoid rule. We will have more to say about the error in this approximation later in this section.

◆ SIMPSON'S RULE

In the development of the trapezoid rule, we connected each pair of points $(x_{k-1}, f(x_{k-1}))$ and $(x_k, f(x_k))$ with a straight line segment and used the area under the graph of this line segment to approximate the area under the graph of $f(x)$ from $x = x_{k-1}$ to $x = x_k$.

In order to obtain a more accurate approximation, we will now use the area under the graph of a parabola that passes through three successive points on the graph of f. First, we need a formula for the area under the graph of a parabola passing through three noncollinear points. The formula is surprisingly simple.

Let (c, y_0), $(c + h, y_1)$, and $(c + 2h, y_2)$ be three points on the graph of a parabola P (see Figure 12). Notice that the x coordinates of these points are equally spaced. Using integration, it can be shown that the area under the graph of P from $x = c$ to $x = c + 2h$ is

$$A = \frac{h}{3}(y_0 + 4y_1 + y_2) \tag{3}$$

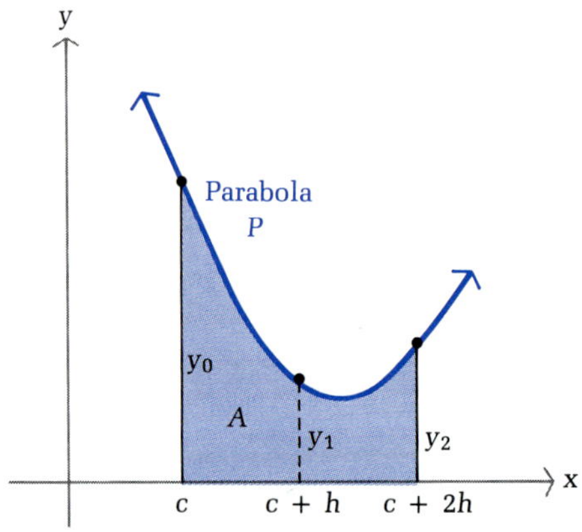

FIGURE 12

$A = \frac{h}{3}(y_0 + 4y_1 + y_2)$

If the graph of the parabola P passes through the points $(x_{k-1}, f(x_{k-1}))$, $(x_k, f(x_k))$, and $(x_{k+1}, f(x_{k+1}))$, then the area under the graph of P from $x = x_{k-1}$ to $x = x_{k+1}$ is

$$A = \frac{\Delta x}{3}[f(x_{k-1}) + 4f(x_k) + f(x_{k+1})] \qquad (4)$$

Let $h = \Delta x$, $y_0 = f(x_{k-1})$, $y_1 = f(x_k)$, and $y_2 = f(x_{k+1})$ in (3).

Since A is the area under a parabola over two subintervals of $[a, b]$, we must assume n is even to use this process. If S_n denotes the sum of the areas under the parabolas over the intervals $[x_0, x_2]$, $[x_2, x_4]$, . . . , $[x_{n-2}, x_n]$, then S_n is an approximation to the area under the graph of f. This is illustrated in Figure 13 using $n = 4$ and two parabolas, P_1 and P_2.

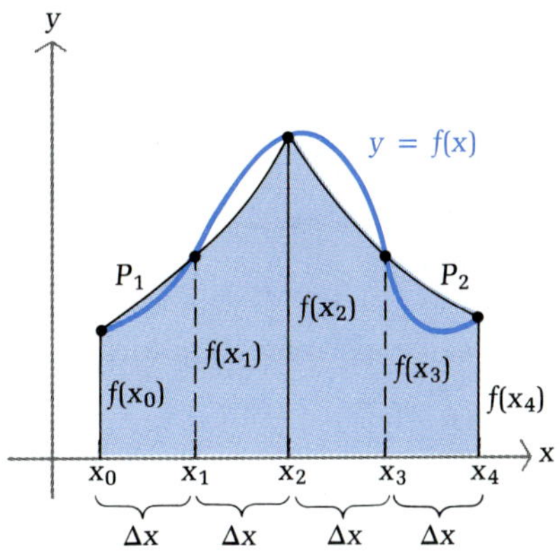

FIGURE 13

$$\int_a^b f(x)\,dx \approx S_4 = \underbrace{\frac{\Delta x}{3}[f(x_0) + 4f(x_1) + f(x_2)]}_{\text{Area under } P_1} + \underbrace{\frac{\Delta x}{3}[f(x_2) + 4f(x_3) + f(x_4)]}_{\text{Area under } P_2}$$

For arbitrary n, using formula (4) for the area under each parabola, we have

$$S_n = \frac{\Delta x}{3}[f(x_0) + 4f(x_1) + f(x_2)] + \frac{\Delta x}{3}[f(x_2) + 4f(x_3) + f(x_4)]$$
$$+ \frac{\Delta x}{3}[f(x_4) + 4f(x_5) + f(x_6)] + \cdots + \frac{\Delta x}{3}[f(x_{n-2}) + 4f(x_{n-1}) + f(x_n)]$$

Excluding $f(x_0)$ and $f(x_n)$, each term $f(x_k)$ with an *even* subscript occurs twice in this sum. Factoring out $\Delta x/3$ and combining like terms, we can write

$$S_n = \frac{\Delta x}{3}[f(x_0) + 4f(x_1) + 2f(x_2) + 4f(x_3) + 2f(x_4) + \cdots + 2f(x_{n-2}) + 4f(x_{n-1}) + f(x_n)]$$

If f is continuous on $[a, b]$, it can be shown that

$$\int_a^b f(x)\,dx = \lim_{n\to\infty} S_n$$

and S_n can be used to approximate the defnite integral. The formula for S_n is referred to as *Simpson's rule.*

Simpson's Rule

Given an even integer n, divide the interval from $x = a$ to $x = b$ into n equal subintervals of length $\Delta x = (b - a)/n$. Let $a = x_0 < x_1 < \cdots < x_n = b$ be the end points of these subintervals. Then

$$\int_a^b f(x)\,dx \approx S_n = \frac{\Delta x}{3}[f(x_0) + 4f(x_1) + 2f(x_2) + 4f(x_3) + \cdots + 2f(x_{n-2}) + 4f(x_{n-1}) + f(x_n)]$$

◆ EXAMPLE 13 Use Simpson's rule with the indicated values of n to approximate

$$\int_1^3 \ln x\,dx$$

Round each approximation to five decimal places.

(A) $n = 4$ (B) $n = 10$

Solutions (A) *Step 1.* Compute Δx:

$$\Delta x = \frac{3-1}{4} = 0.5$$

Step 2. Find the end points of each subinterval:

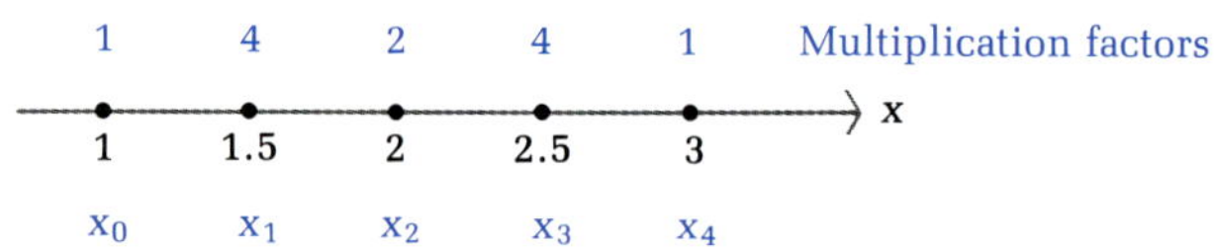

Placing the correct multiplication factor (1, 2, or 4) for $f(x_k)$ above each end point will help avoid errors in computing S_n.

Step 3. Compute S_4:

$$S_4 = \frac{\Delta x}{3}[f(x_0) + 4f(x_1) + 2f(x_2) + 4f(x_3) + f(x_4)]$$

$$= \frac{0.5}{3}[(\ln 1) + 4(\ln 1.5) + 2(\ln 2) + 4(\ln 2.5) + (\ln 3)]$$

$$\approx \frac{0.5}{3}[7.771\ 93] \approx 1.295\ 321\ 7$$

Thus,

$$\int_1^3 \ln x\, dx \approx S_4 \approx 1.295\ 32 \qquad \text{Rounded to five decimal places}$$

(B) Step 1. Compute Δx:

$$\Delta x = \frac{3-1}{10} = 0.2$$

Step 2. Find the end points of the subintervals:

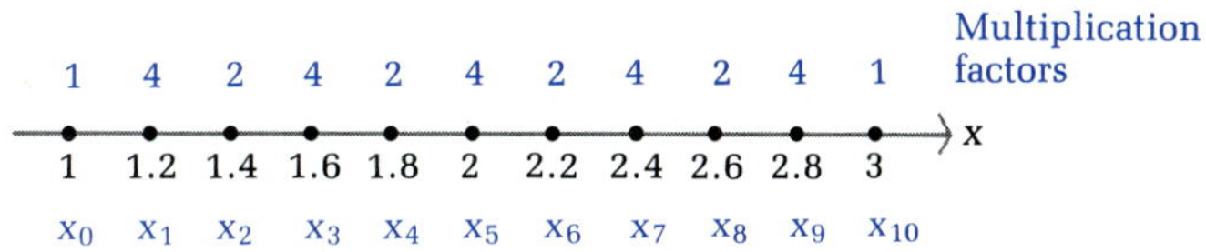

Step 3. Compute S_{10}:

$$S_{10} = \frac{\Delta x}{3}[f(x_0) + 4f(x_1) + 2f(x_2) + 4f(x_3) + 2f(x_4) + 4f(x_5) + 2f(x_6) + 4f(x_7) + 2f(x_8) + 4f(x_9) + f(x_{10})]$$

$$= \frac{0.2}{3}[(\ln 1) + 4(\ln 1.2) + 2(\ln 1.4) + 4(\ln 1.6) + 2(\ln 1.8) + 4(\ln 2) + 2(\ln 2.2) + 4(\ln 2.4) + 2(\ln 2.6) + 4(\ln 2.8) + (\ln 3)]$$

$$\approx \frac{0.2}{3}[19.437\ 31] \approx 1.295\ 820\ 7$$

Thus,

$$\int_1^3 \ln x\, dx \approx S_{10} \approx 1.295\ 82 \qquad \text{Rounded to five decimal places}$$ ◆

PROBLEM 13 Use Simpson's rule with the indicated values of n to approximate

$$\int_2^4 \frac{1}{x}\, dx$$

Round each approximation to five decimal places.

(A) $n = 4$ (B) $n = 10$ ◆

Referring to the discussion following Example 12, the value of $\int_1^3 \ln x\, dx$ rounded to five decimal places is

$$\int_1^3 \ln x\, dx = 3 \ln 3 - 2 \approx 1.295\ 84$$

Comparing this value with the approximations obtained in Example 13, we have

$$\int_1^3 \ln x\, dx - S_4 \approx 1.295\ 84 - 1.295\ 32 = 0.000\ 52$$

and

$$\int_1^3 \ln x\,dx - S_{10} \approx 1.295\ 84 - 1.295\ 82 = 0.000\ 02$$

Once again, using a larger value of n provides a more accurate approximation. Later in this section we will compare in more detail the accuracy of the approximations obtained by using Simpson's rule with those obtained from the trapezoid rule.

◆ APPLICATIONS

In the preceding two examples, we purposely selected a definite integral that could be evaluated directly so that we could compare approximate and exact values. The next two examples involve integrals that cannot be evaluated directly and must be approximated.

◆ **EXAMPLE 14** *Consumers' and Producers' Surplus*

Find the equilibrium price and then use Simpson's rule with $n = 4$ to approximate the consumers' surplus at the equilibrium price level for

$$p = D(x) = \sqrt{100 - 8x^3} \qquad \text{and} \qquad p = S(x) = \sqrt{4 + 4x^3}$$

Solution To find the equilibrium price, set $\sqrt{100 - 8x^3}$ equal to $\sqrt{4 + 4x^3}$ and solve for x:

$$\sqrt{100 - 8x^3} = \sqrt{4 + 4x^3} \qquad \text{Square both sides.}$$

$$100 - 8x^3 = 4 + 4x^3$$

$$-12x^3 = -96$$

$$x^3 = 8$$

$$x = 2$$

Check

$$D(2) = \sqrt{100 - 8(2^3)} = \sqrt{36} = 6$$

$$S(2) = \sqrt{4 + 4(2^3)} = \sqrt{36} = 6$$

$$D(2) \stackrel{\checkmark}{=} S(2)$$

Thus, the equilibrium price is $\bar{p} = 6$, and the corresponding supply and demand is $\bar{x} = 2$. Sketch a graph:

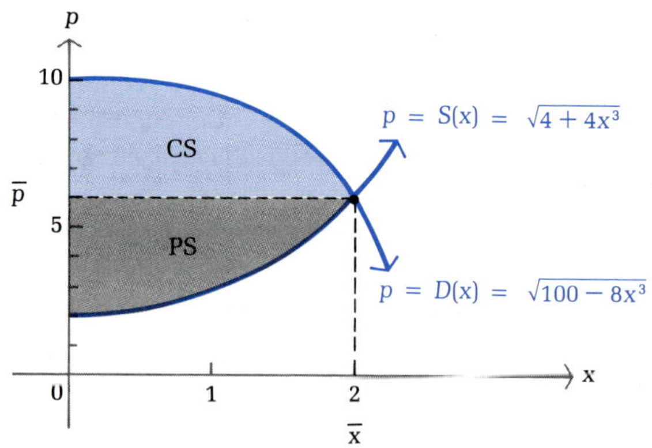

Now,

$$CS = \int_0^2 [D(x) - 6]\, dx$$

$$= \int_0^2 (\sqrt{100 - 8x^3} - 6)\, dx$$

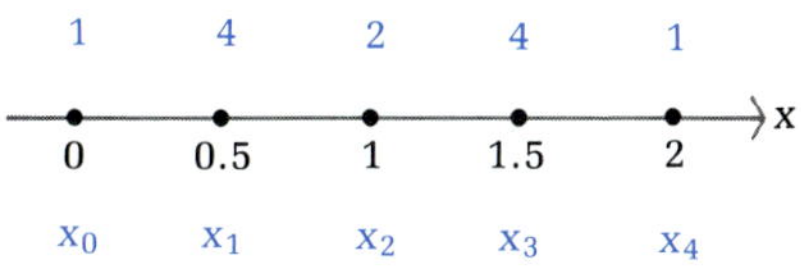

Use Simpson's rule with $n = 4$ to approximate this integral:

$$\Delta x = \frac{2 - 0}{4} = 0.5$$

$$S_4 = \frac{\Delta x}{3}[f(0) + 4f(0.5) + 2f(1) + 4f(1.5) + f(2)]$$

$$= \frac{0.5}{3}[(\sqrt{100} - 6) + 4(\sqrt{99} - 6) + 2(\sqrt{92} - 6) + 4(\sqrt{73} - 6) + (\sqrt{36} - 6)]$$

$$\approx \frac{0.5}{3}[37.158\ 84] \approx 6.193\ 14$$

Thus,

$$CS = \int_0^2 (\sqrt{100 - 8x^3} - 6)\, dx$$

$$\approx 6.193\ 14$$

◆

PROBLEM 14 Use Simpson's rule with $n = 4$ to approximate the producers' surplus at the equilibrium price level for Example 14. ◆

Remark

The integral in Example 14 cannot be evaluated directly. The function $D(x) = \sqrt{100 - 8x^3}$ does not have an antiderivative that can be expressed in terms of any of the functions we have studied. You are not expected to recognize functions that do not have antiderivatives. Instead, if the instructions for a problem call for the use of an approximation method, you should assume that the integral may be one that cannot be evaluated directly and you should not spend time trying to find an antiderivative.

◆ EXAMPLE 15 *Biology*

A biologist studying the aquatic life in a small pond needs to determine the area of the surface of the pond, so the distance across the pond is measured at 10 foot intervals. Use the data in the figure at the top of the next page and the trapezoid rule to approximate the area of the pond.

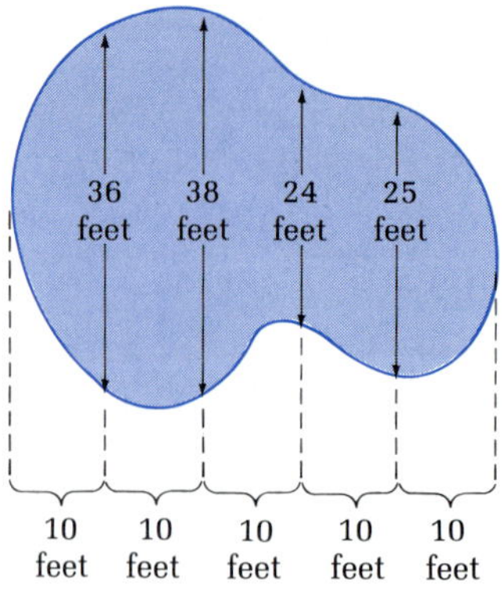

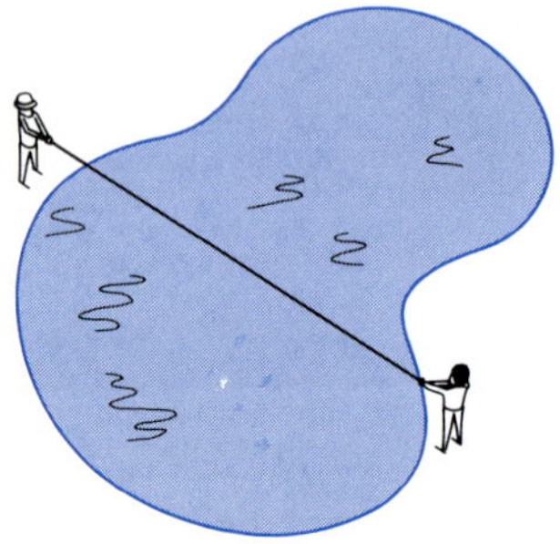

Solution

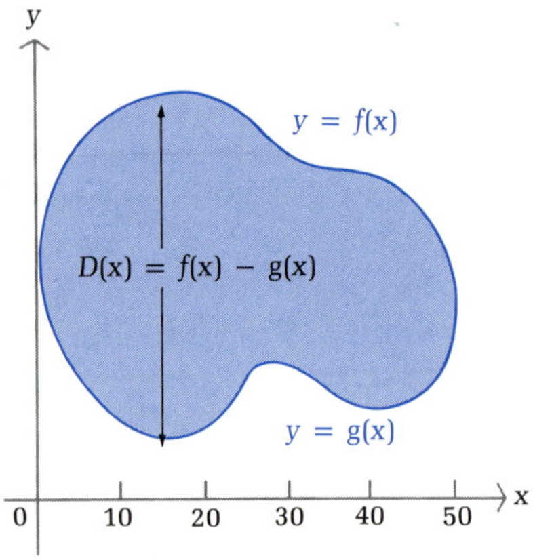

We begin by introducing a coordinate system, as shown in the figure in the margin. If the pond is bounded by the graphs of $y = f(x)$ and $y = g(x)$, then $D(x) = f(x) - g(x)$ is the distance across the pond. The surface area A of the pond is given by

$$A = \int_0^{50} [f(x) - g(x)]\, dx = \int_0^{50} D(x)\, dx$$

Since there are no equations defining f, g, or D, we cannot evaluate this integral directly. However, we can use the measurements taken by the biologist to approximate the integral. We consider D to be defined by the following table:

x	0	10	20	30	40	50
$D(x)$	0	36	38	24	25	0

Applying the trapezoid rule with $n = 5$ and $\Delta x = 10$, we have

$$\begin{aligned} T_5 &= \tfrac{10}{2}[D(0) + 2D(10) + 2D(20) + 2D(30) + 2D(40) + D(50)] \\ &= 5[0 + 72 + 76 + 48 + 50 + 0] \\ &= 1{,}230 \end{aligned}$$

Thus, the surface area of the pond is approximately 1,230 square feet. ◆

PROBLEM 15 Repeat Example 15 for the following measurements:

x	0	10	20	30	40	50
$D(x)$	0	26	42	33	17	0

◆

◆ ERROR ESTIMATES

In many actual applications involving numerical integration, it is necessary to approximate the definite integral to a specified accuracy. The difference between the exact value of a definite integral and the approximate value obtained by a numerical method is called the **error in the approximation.**

Error in Approximations

If T_n and S_n are approximations to $\int_a^b f(x)\,dx$ obtained by using the trapezoid rule and Simpson's rule, respectively, then

TRAPEZOID RULE ERROR $\quad \text{Err}(T_n) = \int_a^b f(x)\,dx - T_n$

SIMPSON'S RULE ERROR $\quad \text{Err}(S_n) = \int_a^b f(x)\,dx - S_n$

Earlier in this section, we computed four different approximations to $\int_1^3 \ln x\,dx$ and compared the approximations to the exact value $3 \ln 3 - 2$. These results are summarized in Table 6.

TABLE 6

Approximations to $\int_1^3 \ln x\,dx$

n	T_n	ERR(T_n)	S_n	ERR(S_n)
4	1.282	0.014	1.295 32	0.000 52
10	1.294	0.002	1.295 82	0.000 02

Examining the column for $\text{Err}(T_n)$ in Table 6, we see that increasing n decreases the error in the approximation. This is also true for $\text{Err}(S_n)$. Although exceptions may occur, in most applications of either the trapezoid rule or Simpson's rule, **increasing *n* decreases the error in the approximation.** Next, comparing $\text{Err}(T_n)$ and $\text{Err}(S_n)$ in each row of Table 6, we see that $\text{Err}(S_n)$ is much smaller than $\text{Err}(T_n)$. In fact, $\text{Err}(S_4)$ is even smaller than $\text{Err}(T_{10})$. Again, exceptions may occur, but in most cases, **Simpson's rule provides a more accurate approximation than the trapezoid rule does (and with very little extra work).**

The values of $\text{Err}(T_n)$ and $\text{Err}(S_n)$ in Table 6 were obtained by using the exact value of the integral being approximated. If we apply an approximation method to a definite integral whose exact value is not known, then we cannot expect to compute the actual error in the approximation. Instead, we must *estimate* the error in the approximation. Theorem 1 provides estimates for these errors.

THEOREM 1

Error Estimates for the Trapezoid Rule and Simpson's Rule

Given the definite integral $\int_a^b f(x)\,dx$:

ERROR ESTIMATE FOR THE TRAPEZOID RULE

If $|f''(x)| \leq M_1$ for $a \leq x \leq b$, then $|\text{Err}(T_n)| \leq \dfrac{(b-a)^3 M_1}{12n^2}$

ERROR ESTIMATE FOR SIMPSON'S RULE

If $|f^{(4)}(x)| \leq M_2$ for $a \leq x \leq b$, then $|\text{Err}(S_n)| \leq \dfrac{(b-a)^5 M_2}{180n^4}$

◆ EXAMPLE 16 Use Theorem 1 to estimate $\text{Err}(T_n)$ and $\text{Err}(S_n)$ for the definite integral

$$\int_1^3 \ln x\,dx$$

if $n = 10, 20, 30, 40$, or 50

Solution To use Theorem 1, we must maximize the absolute value of the second and fourth derivatives of $f(x) = \ln x$ for $1 \leq x \leq 3$. Thus, we find

$$f(x) = \ln x \qquad f'(x) = \frac{1}{x} \qquad f''(x) = -\frac{1}{x^2}$$

$$f^{(3)}(x) = \frac{2}{x^3} \qquad f^{(4)}(x) = -\frac{6}{x^4}$$

Examining the graphs in figures A and B, we see that

$$M_1 = \text{Max}|f''(x)| = |f''(1)| = 1 \qquad 1 \leq x \leq 3$$

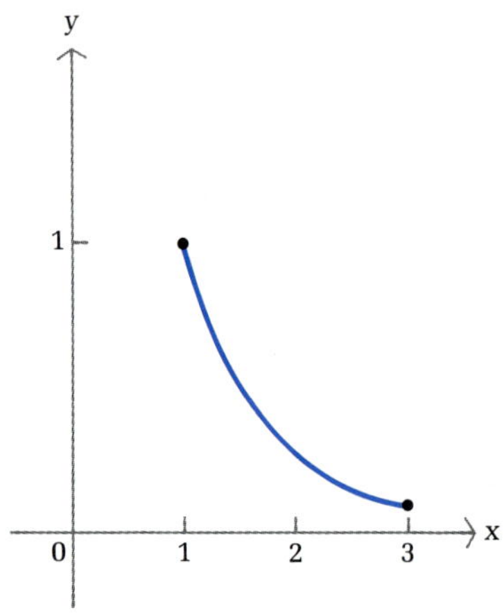

(A) $y = |f''(x)| = \dfrac{1}{x^2}$

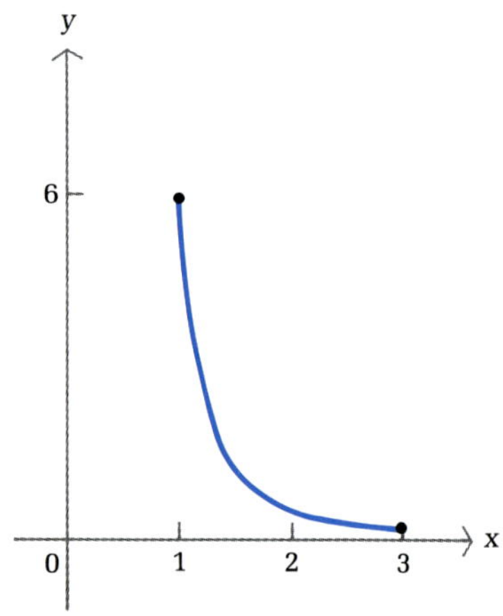

(B) $y = |f^{(4)}(x)| = \dfrac{6}{x^4}$

and

$$M_2 = \text{Max}|f^{(4)}(x)| = |f^{(4)}(1)| = 6 \qquad 1 \leq x \leq 3$$

Thus, according to Theorem 1,

$$|\text{Err}(T_n)| \leq \frac{(b-a)^3 M_1}{12n^2} = \frac{(3-1)^3 \cdot 1}{12n^2} = \frac{2}{3n^2}$$

and

$$|\text{Err}(S_n)| \leq \frac{(b-a)^5 M_2}{180n^4} = \frac{(3-1)^5 \cdot 6}{180n^4} = \frac{16}{15n^4}$$

The estimates for specific values of n are listed in the table in the margin. All values in this table have been rounded to the first nonzero decimal place. ◆

Error Estimates for $\int_1^3 \ln x\, dx$

n	ESTIMATE FOR ERR(T_n) $2/(3n^2)$	ESTIMATE FOR ERR(S_n) $16/(15n^4)$
10	0.007	0.000 1
20	0.002	0.000 007
30	0.0007	0.000 001
40	0.0004	0.000 000 4
50	0.0003	0.000 000 2

PROBLEM 16 Repeat Example 16 for: $\int_2^4 \frac{1}{x}\, dx$ ◆

In theory, the estimates in Theorem 1 can be used to determine the value of n required for either rule to provide an approximation of a specified accuracy. For example, the table in Example 16 indicates that using Simpson's rule with $n = 30$ will provide an approximation correct to within $\pm 0.000\ 001$. We were able to obtain this estimate because the function $f(x) = \ln x$ has a very simple fourth derivative. In most applications, maximizing the absolute value of the second or fourth derivative of the integrand can be a formidable task. For example, to estimate the error in using Simpson's rule to approximate

$$\int_0^2 f(x)\, dx = \int_0^2 (\sqrt{100 - 8x^3} - 6)\, dx \tag{5}$$

(see Example 14), we would have to maximize the absolute value of

$$f^{(4)}(x) = \frac{2{,}304x^2(x^6 - 700x^3 - 12{,}500)}{(100 - 8x^3)^{7/2}}$$

In actual practice, computers are used to compute approximations to definite integrals. Theorem 1 can be used to devise a variety of numerical techniques for estimating the error in computer-generated approximations to definite integrals. One of the simplest techniques is to continue to double the number of subintervals until the difference between S_n and S_{2n} is less than the desired accuracy. Computer-generated values of S_n for the integral in (5) are listed in Table 7 for $n = 4, 8, \ldots, 128$. Each value is rounded to six decimal places. Since S_{64} and S_{128} are the same, it is reasonable to conclude that the approximate value of the integral in (5) is 6.204 489, correct to six decimal places.

TABLE 7

Computer Approximation of $\int_0^2 (\sqrt{100 - 8x^3} - 6)\, dx$

```
* - SIMPSON'S RULE - *
A = 0              B = 2
F(X)=SQR(100-8*X^3)-6
----------------------
N               S(N)
----------------------
4               6.19314
8               6.203338
16              6.204398
32              6.204483
64              6.204489
128             6.204489
----------------------
```

Thus, in practice, it is possible to compute approximations by Simpson's rule to a specified accuracy without maximizing the absolute value of the fourth derivative of the integrand.

Answers to Matched Problems

12. (A) 0.697 (B) 0.694 13. (A) 0.693 25 (B) 0.693 15
14. 5.520 78 15. Approx. 1,180 ft^2

16.

n	$1/(6n^2)$	$2/(15n^4)$
10	0.002	0.000 01
20	0.000 4	0.000 000 8
30	0.000 2	0.000 000 2
40	0.000 1	0.000 000 05
50	0.000 07	0.000 000 02

EXERCISE 11-3

A *Write out all the terms in the trapezoid rule for the indicated value of n.*

1. $n = 5$

2. $n = 6$

Write out all the terms in Simpson's rule for the indicated value of n.

3. $n = 6$

4. $n = 8$

Given a, b, and n, divide the interval [a, b] into n equal subintervals and graph the end points of these subintervals on a number line. Use the values of these end points to write a formula for T_n.

5. $a = 0, \quad b = 2, \quad n = 5$

6. $a = 0, \quad b = 3, \quad n = 4$

Given a, b, and n, divide the interval [a, b] into n equal subintervals and graph the end points of these subintervals. Use the values of these end points to write a formula for S_n.

7. $a = 0, \quad b = 3, \quad n = 4$

8. $a = 0, \quad b = 9, \quad n = 6$

B *Use the trapezoid rule with n = 5 to approximate each integral. Round each result to four decimal places.*

9. $\int_0^1 \sqrt{1 + x^3}\, dx$

10. $\int_0^1 \sqrt{1 + x^4}\, dx$

11. $\int_2^4 \ln(1 + x^2)\, dx$

12. $\int_2^4 \ln(1 + x^3)\, dx$

Use Simpson's rule with n = 4 to approximate each definite integral. Round each result to four decimal places.

13. $\int_0^1 e^{x^2}\,dx$

14. $\int_0^1 e^{-x^2}\,dx$

15. $\int_{-1}^1 \frac{1}{\sqrt{4+x^3}}\,dx$

16. $\int_{-1}^1 \frac{1}{\sqrt{4+x^4}}\,dx$

Use the values in each table and the trapezoid rule to approximate the area bounded by the graph of the function and the x axis over the interval of x values in the table.

17.

x	0	0.5	1	1.5	2
f(x)	5	7	6	4	2

18.

x	0	0.5	1	1.5	2
f(x)	4	3	4	6	9

19.

x	−3	−2	−1	0	1	2	3
f(x)	10	12	14	15	13	11	9

20.

x	−6	−4	−2	0	2	4	6
f(x)	25	20	15	5	10	25	35

Use Simpson's rule with the indicated value of n to approximate the area of the region bounded by the graphs of the given equations. Round each result to five decimal places.

21. $y = \frac{10}{4+x^2};\quad y = 0,\quad 0 \leq x \leq 4,\quad n = 8$

22. $y = \frac{10}{1+x^2};\quad y = 0,\quad -2 \leq x \leq 2,\quad n = 8$

23. $y = \sqrt{x - x^2};\quad y = 0,\quad 0 \leq x \leq 1,\quad n = 10$

24. $y = \sqrt{2x - x^2};\quad y = 0,\quad 0 \leq x \leq 2,\quad n = 10$

C *Find the exact value of each integral; then construct a table containing T_n, $Err(T_n)$, S_n, and $Err(S_n)$ for n = 4 and n = 10. Round all values in the table to six decimal places.*

25. $\int_0^1 e^x\,dx$

26. $\int_1^2 \frac{1}{\sqrt{x}}\,dx$

In Problems 27 and 28, estimate $Err(T_n)$ and $Err(S_n)$ for each integral. Construct a table containing the values of these estimates rounded to the first nonzero decimal place for n = 10, 20, 30, 40, and 50.

27. $\int_1^3 (x \ln x)\,dx$

28. $\int_0^4 \sqrt{x+4}\,dx$

29. Show that the trapezoid rule gives the exact value of

$$\int_a^b (Ax + B)\, dx$$

for any constants A and B. [*Hint:* Use Theorem 1 to estimate $\text{Err}(T_n)$ for $f(x) = Ax + B$.]

30. Show that Simpson's rule gives the exact value of

$$\int_a^b (Ax^3 + Bx^2 + Cx + D)\, dx$$

for any constants A, B, C, and D. [*Hint:* Use Theorem 1 to estimate $\text{Err}(S_n)$ for $f(x) = Ax^3 + Bx^2 + Cx + D$.]

Problems 31–38 are related to Problems 9–16. In each problem, use a numerical integration routine on a calculator or a computer to approximate the integral to four decimal places.

31. $\int_0^1 \sqrt{1 + x^3}\, dx$

32. $\int_0^1 \sqrt{1 + x^4}\, dx$

33. $\int_2^4 \ln(1 + x^2)\, dx$

34. $\int_2^4 \ln(1 + x^3)\, dx$

35. $\int_0^1 e^{x^2}\, dx$

36. $\int_0^1 e^{-x^2}\, dx$

37. $\int_{-1}^1 \frac{1}{\sqrt{4 + x^3}}\, dx$

38. $\int_{-1}^1 \frac{1}{\sqrt{4 + x^4}}\, dx$

APPLICATIONS

In Problems 39–52, use Simpson's rule with $n = 4$ to approximate the required definite integrals. Round each result to two decimal places.

Business & Economics

39. Consumers' and producers' surplus. Find the equilibrium price (in dollars), and then find the consumers' and producers' surplus for

$$p = D(x) = \sqrt{400 - 4x^3} \qquad \text{and} \qquad p = S(x) = \sqrt{16 + 2x^3}$$

40. Oil production. The management of an oil company estimates that oil will be pumped from a producing field at a rate given by

$$R(t) = \frac{100}{\sqrt{t^2 + 1}} \qquad 0 \le t \le 8$$

where $R(t)$ is the rate of production in thousands of barrels per year t years after pumping begins. Approximately how much oil will the field produce during the first 8 years of production?

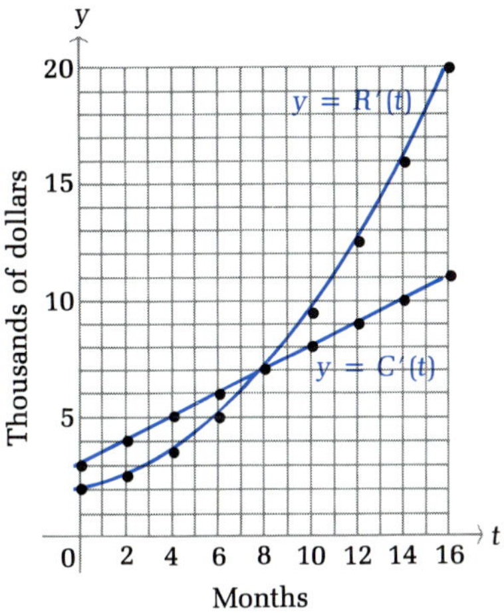

41. *Marginal revenue.* The total accumulated revenues and costs (in thousands of dollars) t months after the opening of a new restaurant are denoted by $R(t)$ and $C(t)$, respectively. The graphs of $R'(t)$ and $C'(t)$ are shown in the figure. Typically, newly opened restaurants lose money until they develop a clientele. How much did this restaurant lose during the first 8 months of operation? Set up an appropriate definite integral and use the graph to estimate the necessary values of $R'(t)$ and $C'(t)$.

42. *Marginal revenue.* Refer to Problem 41. How much profit did the restaurant make during the second 8 months of operation? What was the total profit for the first 16 months of operation?

The results of surveying two irregularly shaped pieces of riverfront property are given in the figure. Use these measurements in Problems 43 and 44.

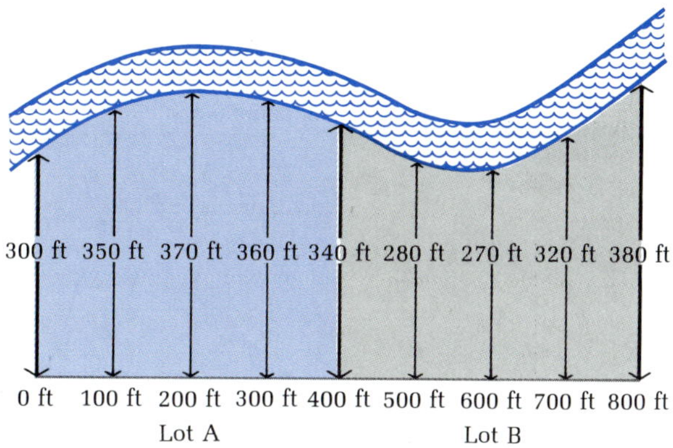

43. *Real estate.* Find the area of lot A.

44. *Real estate.* Find the area of lot B.

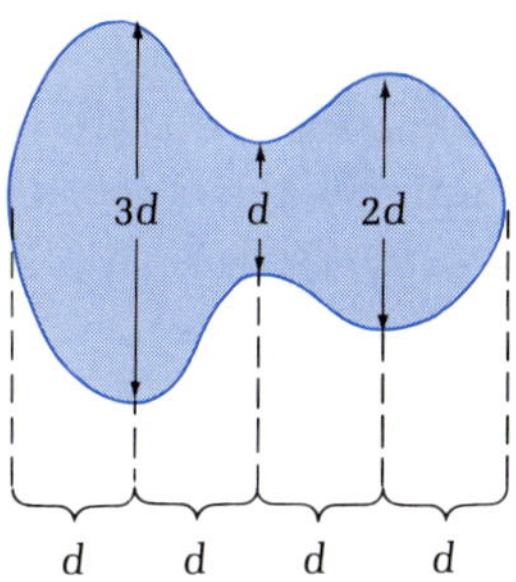

45. *Construction.* A city park commission is planning to construct ponds in several parks. All the ponds will have the same basic shape, but the surface area will depend on the location selected for the pond. The city architect designed the basic shape of the pond using an unspecified dimension d (see the figure in the margin). Determine the value of d for a pond with a surface area of 1,650 square meters.

46. *Construction.* Repeat Problem 45 for a pond with a surface area of 2,400 square meters whose basic shape is given in the figure below.

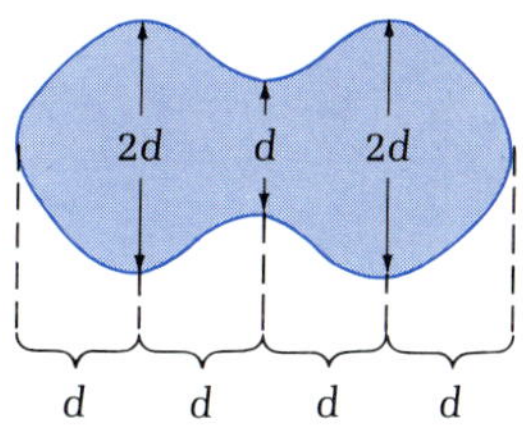

Life Sciences

47. Medicine. The body assimilates a drug at a rate given by

$$R'(t) = 7 - 3\ln(t^2 - 6t + 10) \qquad 0 \le t \le 3$$

where t is the time (in hours) since the drug was administered. Find the total amount of the drug assimilated in the first 3 hours.

48. Medicine. The level of concentration of a certain drug in the bloodstream t minutes after it is administered is given by

$$L(t) = 10 - 6\ln(t^2 - 4t + 5) \qquad 0 \le t \le 4$$

Find the average level of concentration for t in the interval [0, 4].

49. Temperature. The temperature C (in degrees Celsius) in an artificial habitat was graphed by a recording device over a 2 hour period (see the figure). What was the average temperature during this period? Set up an appropriate definite integral and use the graph below to estimate the necessary values of C.

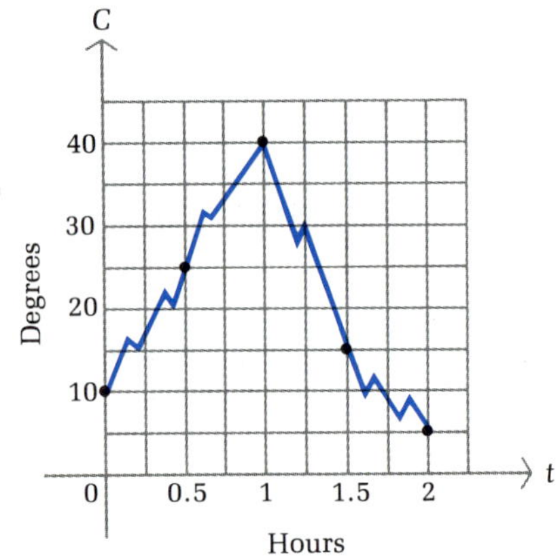

50. Ecology. A biologist is studying the ecosystem on a small island. In order to determine the area of the island, a grid was imposed over an aerial photograph of the island (see the figure). Find the area of the island. Set up an appropriate definite integral and use the graph to estimate the necessary distances.

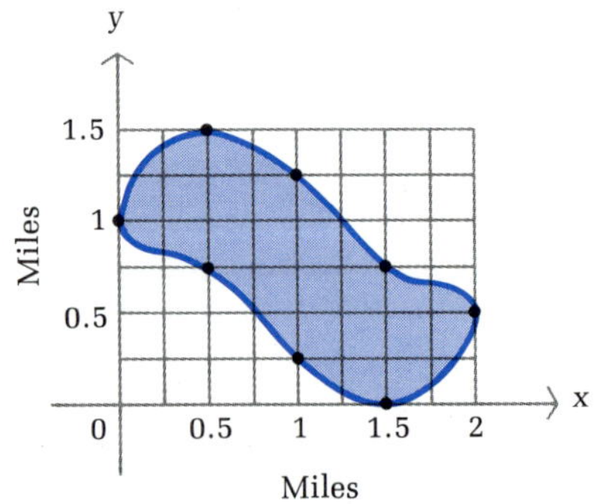

Social Sciences

51. *Learning.* A person learns N items at a rate given by

$$N'(t) = \sqrt{24 + \frac{1}{t^2}} \qquad 1 \leq t \leq 6$$

where t is the number of hours of continuous study. Find the total number of items N learned from $t = 1$ to $t = 6$ hours of study.

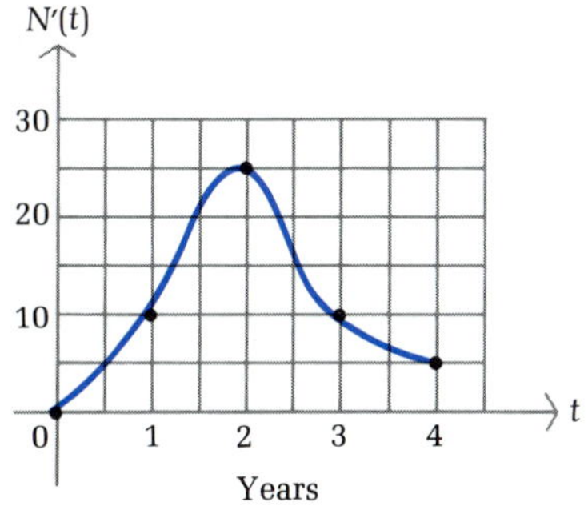

52. *Politics.* In a newly incorporated city, the rate of change of the voting population, $N'(t)$, with respect to time t (in years) is shown in the figure, where $N(t)$ is in thousands. What is the total increase in the voting population during the first 4 years? Set up an appropriate definite integral and estimate the necessary values from the graph.

SECTION 11-4 Euler's Method

- APPROXIMATE SOLUTIONS
- EULER'S METHOD
- COMPARISON OF EXACT AND APPROXIMATE SOLUTIONS
- APPLICATION
- EULER'S METHOD FOR SYSTEMS OF DIFFERENTIAL EQUATIONS

APPROXIMATE SOLUTIONS

In Chapter 9 we developed techniques for finding the solutions to certain types of differential equations. But it is just a fortunate accident when a differential equation has a simple, neat solution expressed in terms of a finite combination of familiar functions. Most do not. When a differential equation does not have such a nice solution, it is necessary to settle for an approximation to the solution.

The problem of finding the particular solution to a differential equation that satisfies a given initial condition is called an **initial value problem.** In this section we develop a numerical technique for approximating the solution to initial value problems of the form

$$y' = f(x, y) \qquad y(a) = C \qquad a \leq x \leq b \tag{1}$$

To avoid confusion between the actual solution of an initial value problem and an approximation to that solution, we will often refer to the actual solution as the *exact solution*. That is, a function $y = y(x)$ is the **exact solution** of (1) if

$$y'(x) = f(x, y(x)) \qquad a \leq x \leq b \qquad \text{and} \qquad y(a) = C$$

Note that an exact solution is always a particular solution. General solutions involving arbitrary constants cannot be approximated numerically.

Given any initial value problem of the form (1), we want to find a function $p(x)$ whose values can be used to approximate the values of the exact solution $y(x)$ for $a \leq x \leq b$. Furthermore, we want to be able to find this approximate solution $p(x)$ without first finding the exact solution $y(x)$. The procedure we will use to find an approximate solution is outlined in the box.

Approximating the Solution of:

$$y' = f(x, y) \qquad y(a) = C \qquad a \leq x \leq b$$

Step 1. Divide the interval $a \leq x \leq b$ into n equal subintervals of length $\Delta x = (b - a)/n$ with end points

$$a = x_0 < x_1 < \cdots < x_n = b$$

Step 2. Let $y_0 = y(a) = C$ and find a sequence of values $y_1, y_2, \ldots, y_n$ that are approximations to the values of the exact solution $y(x)$ at $x = x_1, x_2, \ldots, x_n$. That is,

$$y(x_k) \approx y_k \qquad k = 1, 2, \ldots, n$$

Step 3. Connect the points $(x_0, y_0), (x_1, y_1), \ldots, (x_n, y_n)$ with straight line segments to form a continuous piecewise linear function $p(x)$ that approximates $y(x)$ for $a \leq x \leq b$.

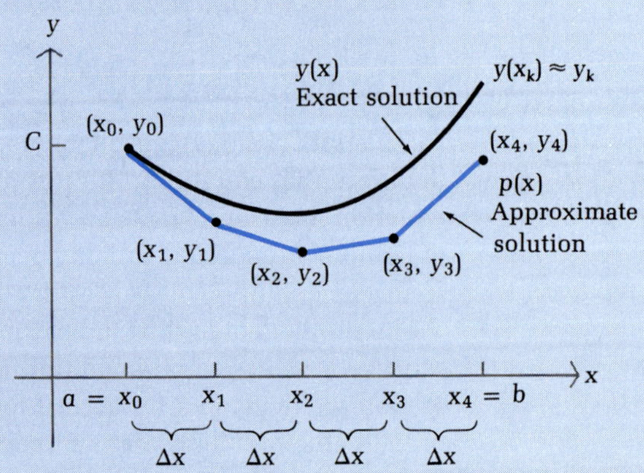

Many different methods can be used to find the approximate values $y_1, y_2, \ldots, y_n$ mentioned in step 2 in the box. One of the oldest and easiest to use is *Euler's method.*

◆ EULER'S METHOD

Euler's method of approximating exact solutions applies only to initial value problems of the form

$$y' = f(x, y) \qquad y(a) = C \qquad a \leq x \leq b \tag{1}$$

However, this type of problem occurs frequently in applications, and in most cases the problem does not have a simple exact solution. Note that $y' = f(x, y)$ gives us the slope of a tangent line to a solution curve at a point (x, y) on the curve. Euler's method for computing the approximate values $y_1, y_2, \ldots, y_n$ is based on the use of actual and approximate tangent lines to the exact solution curve. The slopes of these tangent lines are determined by $y' = f(x, y)$.

The method starts by finding the line tangent to the graph of the exact solution $y(x)$ at $x = x_0$. The equation of this tangent line is

$$y = y(x_0) + y'(x_0)(x - x_0) \qquad \text{Tangent line at } x = x_0 \tag{2}$$

Since $x_0 = a$, the initial condition in (1) implies that $y(x_0) = y(a) = C$. In the box outlining the approximation procedure, we have denoted this initial value by y_0 instead of C for consistency in notation.

Now, how can we determine the value of $y'(x_0)$? Remember that $y(x)$ is the solution of (1) and must satisfy

$$y'(x) = f(x, y(x))$$

Substituting x_0 for x in this equation, we have

$$\begin{aligned} y'(x_0) &= f(x_0, y(x_0)) \qquad y(x_0) = y_0 \\ &= f(x_0, y_0) \end{aligned}$$

Substituting y_0 for $y(x_0)$ and $f(x_0, y_0)$ for $y'(x_0)$ in (2) enables us to write

$$y = y_0 + f(x_0, y_0)(x - x_0) \qquad \text{Tangent line at } x = x_0$$

This line can be used to approximate $y(x)$ for values of x near x_0. In particular, we can use this line to approximate $y(x_1)$.

Let

$$\begin{aligned} y_1 &= y_0 + f(x_0, y_0)(x_1 - x_0) \qquad \text{Tangent line at } x_0 \text{ evaluated at } x = x_1 \\ &= y_0 + f(x_0, y_0)\Delta x \end{aligned}$$

As Figure 14 (p. 762) illustrates, y_1 can be used to approximate $y(x_1)$. That is,

$$y(x_1) \approx y_1 = y_0 + f(x_0, y_0)\Delta x$$

Notice that y_1 is computed by using the given function $f(x, y)$, the initial values $x_0 = a$ and $y_0 = C$, and $\Delta x = x_1 - x_0$. In particular, we do not need to know the actual solution $y(x)$ in order to compute y_1.

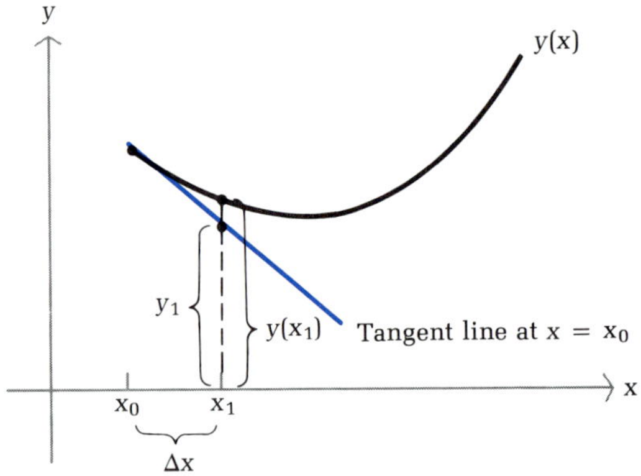

FIGURE 14
$y(x_1) \approx y_1 = y_0 + f(x_0, y_0)\Delta x$

Suppose we try to use the tangent at $x = x_1$ to find an approximate value for $y(x_2)$. The equation of this tangent line is

$$y = y(x_1) + y'(x_1)(x - x_1) \qquad \text{Tangent line at } x = x_1$$

Since we do not know the exact values of $y(x_1)$ or $y'(x_1)$, we cannot use this line to approximate $y(x_2)$. However, we have already concluded that $y(x_1) \approx y_1$. Furthermore, since $y(x)$ satisfies the differential equation in (1) for $a \leq x \leq b$, we know that $y'(x_1) = f(x_1, y(x_1))$. It is reasonable to assume that $y(x_1) \approx y_1$ implies

$$y'(x_1) = f(x_1, y(x_1)) \approx f(x_1, y_1)$$

Thus, we can use the approximations $y(x_1) \approx y_1$ and $y'(x_1) \approx f(x_1, y_1)$ to find the equation of a line that approximates the tangent line at x_1. That is,

$$y = \underbrace{y(x_1)}_{\approx} + \underbrace{y'(x_1)}_{\approx}(x - x_1) \qquad \text{Tangent line at } x_1$$

$$y = \overbrace{y_1} + \overbrace{f(x_1, y_1)}(x - x_1) \qquad \text{Approximate tangent line at } x_1$$

Now we can use this approximate tangent line to approximate $y(x_2)$.

Let

$$\begin{aligned} y_2 &= y_1 + f(x_1, y_1)(x_2 - x_1) \qquad \text{Approximate tangent line evaluated at } x = x_2 \\ &= y_1 + f(x_1, y_1)\Delta x \end{aligned}$$

Referring to Figure 15, we see that

$$y(x_2) \approx y_2 = y_1 + f(x_1, y_1)\Delta x$$

Notice that the calculation of y_2 depends on the given function $f(x, y)$, the known values of x_1 and Δx, and the previously computed approximate value y_1.

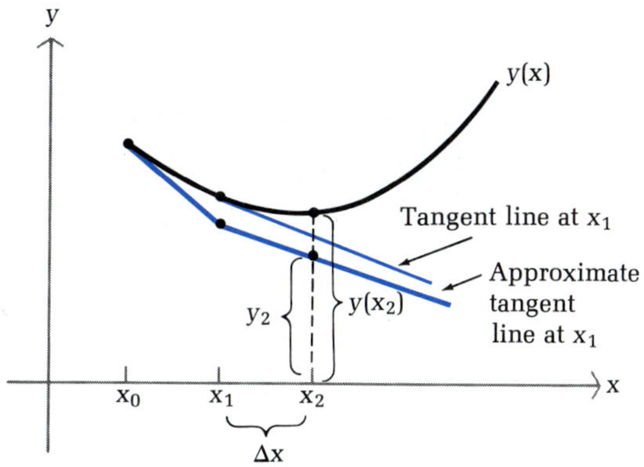

FIGURE 15
$y(x_2) \approx y_2 = y_1 + f(x_1, y_1)\Delta x$

If we find the approximate tangent line at x_2 and use this line to approximate $y(x_3)$ (see Figure 16), we obtain

$$y(x_3) \approx y_3 = y_2 + f(x_2, y_2)\Delta x$$

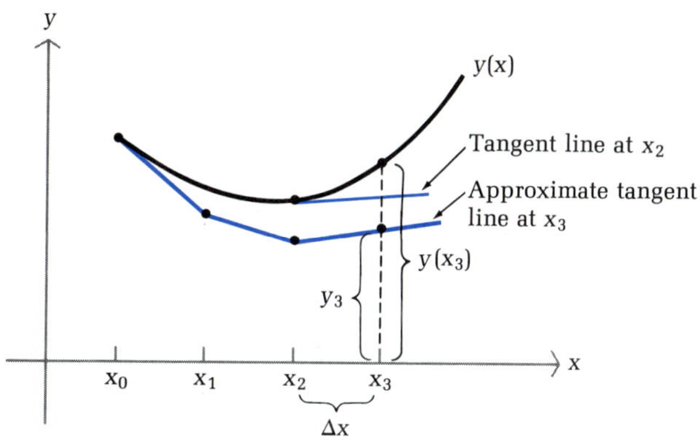

FIGURE 16
$y(x_3) \approx y_3 = y_2 + f(x_2, y_2)\Delta x$

Continuing this process, we are led to the **recursion formula** (each approximate value is calculated in terms of the previous one)

$$y(x_k) \approx y_k = y_{k-1} + f(x_{k-1}, y_{k-1})\Delta x \qquad k = 1, 2, \ldots, n$$

which is ideal for calculator or computer application. This recursion formula is the heart of Euler's method.

Each calculation in using this formula is called a **step** and $\Delta x = x_k - x_{k-1}$ is called the **step size.** (It is convenient, but not necessary, to use equal step sizes, as we have done here.) We summarize *Euler's method* for computing the approximate values of the solution of an initial value problem in the box.

Euler's Method

If $y(x)$ is the (exact) solution to the initial value problem

$$y' = f(x, y) \qquad y(x_0) = y_0$$

$x_0, x_1, \ldots, x_n$ is the sequence of values defined by

$$x_k = x_{k-1} + \Delta x \qquad k = 1, 2, \ldots, n$$

and $y_1, y_2, \ldots, y_n$ is the sequence of values defined by

$$y_k = y_{k-1} + f(x_{k-1}, y_{k-1})\Delta x \qquad k = 1, 2, \ldots, n$$

then

$$y(x_k) \approx y_k \qquad k = 1, 2, \ldots, n$$

◆ EXAMPLE 17 Use Euler's method with a step size of $\Delta x = 0.2$ to approximate the solution of

$$y' = 10\sqrt{y} - 12x \qquad y(1) = 2$$

for $1 \leq x \leq 2$. Round each approximate value to two decimal places and graph the approximate solution.

Solution Let $x_0 = 1$, $y_0 = 2$, $\Delta x = 0.2$, and

$$f(x, y) = 10\sqrt{y} - 12x$$

Then Euler's formula for y_k is

$$\begin{aligned} y_k &= y_{k-1} + f(x_{k-1}, y_{k-1})\Delta x \\ &= y_{k-1} + (10\sqrt{y_{k-1}} - 12x_{k-1})(0.2) \end{aligned}$$

We use this recursion formula to obtain the sequence of approximate values listed in the table:

Approximate Values for $y' = 10\sqrt{y} - 12x$, $y(1) = 2$

k	x_k	$y_k = y_{k-1} + (10\sqrt{y_{k-1}} - 12x_{k-1})(0.2)$
0	1	2
1	1.2	$2 + [10\sqrt{2} - 12(1)](0.2) \approx 2.428\ 427\ 1$
2	1.4	$2.428\ 427\ 1 + [10\sqrt{2.428\ 427\ 1} - 12(1.2)](0.2) \approx 2.665\ 109\ 4$
3	1.6	$2.665\ 109\ 4 + [10\sqrt{2.665\ 109\ 4} - 12(1.4)](0.2) \approx 2.570\ 142$
4	1.8	$2.570\ 142 + [10\sqrt{2.570\ 142} - 12(1.6)](0.2) \approx 1.936\ 474\ 4$
5	2	$1.936\ 474\ 4 + [10\sqrt{1.936\ 474\ 4} - 12(1.8)](0.2) \approx 0.399\ 619\ 75$

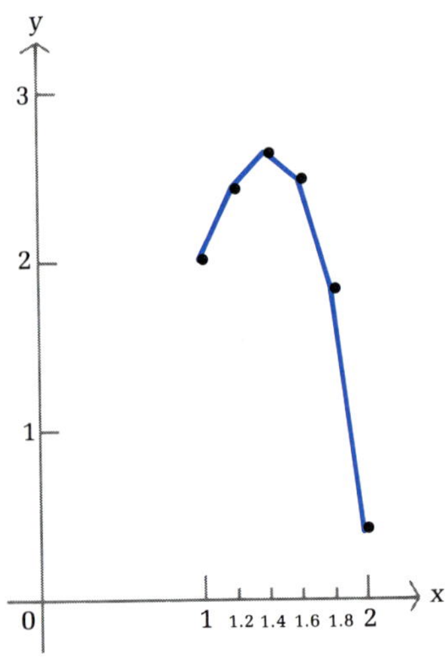

Rounding the values of y_k from this table to 2.43, 2.67, 2.57, 1.94, and 0.40, we graph the approximate solution, as shown in the figure. ◆

PROBLEM 17 Use Euler's method with a step size of $\Delta x = 0.2$ to approximate the solution of

$$y' = 10x - 9\sqrt{y} \qquad y(1) = 3$$

for $1 \leqslant x \leqslant 2$. Round each approximate value to two decimal places and graph the approximate solution. ◆

A calculator with a memory will be very helpful in computing the approximate values $y_1, \ldots, y_n$. In most problems, each value of y_k can be stored in the memory and used to compute the next value without reentering the previously computed value. We used this procedure in Example 17 to calculate the values of y_k in the table (to the limit of accuracy provided by our calculator) and then rounded these values to two decimal places for graphing purposes. Unless otherwise noted, we will follow this procedure in all the examples and exercises in this section.

◆ COMPARISON OF EXACT AND APPROXIMATE SOLUTIONS

It is very difficult to make any estimates concerning the error in approximations obtained by using Euler's method and we will not do so in this text. However, it is instructive to apply Euler's method to a problem whose solution is known, so that we can compare the graphs of the exact and approximate solutions.

◆ EXAMPLE 18 Approximate the solution of

$$y' = y + 7x - 3 \qquad y(0) = 1$$

for $0 \leqslant x \leqslant 1$ using Euler's method with a step size of:

(A) $\Delta x = 0.2$ (B) $\Delta x = 0.1$

(C) The exact solution of this equation is

$$y = 5e^x - 7x - 4$$

Graph the exact solution and both approximate solutions from parts A and B on the same set of axes.

Solutions (A) $x_0 = 0$, $y_0 = 1$, $\Delta x = 0.2$, $f(x, y) = y + 7x - 3$, and

$$y_k = y_{k-1} + (y_{k-1} + 7x_{k-1} - 3)(0.2)$$

k	x_k	$y_k = y_{k-1} + (y_{k-1} + 7x_{k-1} - 3)(0.2)$
0	0	1
1	0.2	0.6
2	0.4	0.4
3	0.6	0.44
4	0.8	$0.768 \approx 0.77$
5	1	$1.4416 \approx 1.44$

(B) $x_0 = 0$, $y_0 = 1$, $\Delta x = 0.1$, $f(x, y) = y + 7x - 3$, and
$y_k = y_{k-1} + (y_{k-1} + 7x_{k-1} - 3)(0.1)$

k	x_k	$y_k = y_{k-1} + (y_{k-1} + 7x_{k-1} - 3)(0.1)$
0	0	1
1	0.1	0.8
2	0.2	0.65
3	0.3	0.555 ≈ 0.56
4	0.4	0.520 5 ≈ 0.52
5	0.5	0.552 55 ≈ 0.55
6	0.6	0.657 805 ≈ 0.66
7	0.7	0.843 585 5 ≈ 0.84
8	0.8	1.117 944 1 ≈ 1.12
9	0.9	1.489 738 5 ≈ 1.49
10	1	1.968 712 3 ≈ 1.97

(C) In order to make an accurate comparison, we used point-by-point plotting to graph $y(x) = 5e^x - 7x - 4$ in the following figure:

x	$y(x)$
0	1
0.1	0.83
0.2	0.71
0.3	0.65
0.4	0.66
0.5	0.74
0.6	0.91
0.7	1.17
0.8	1.53
0.9	2.00
1	2.59

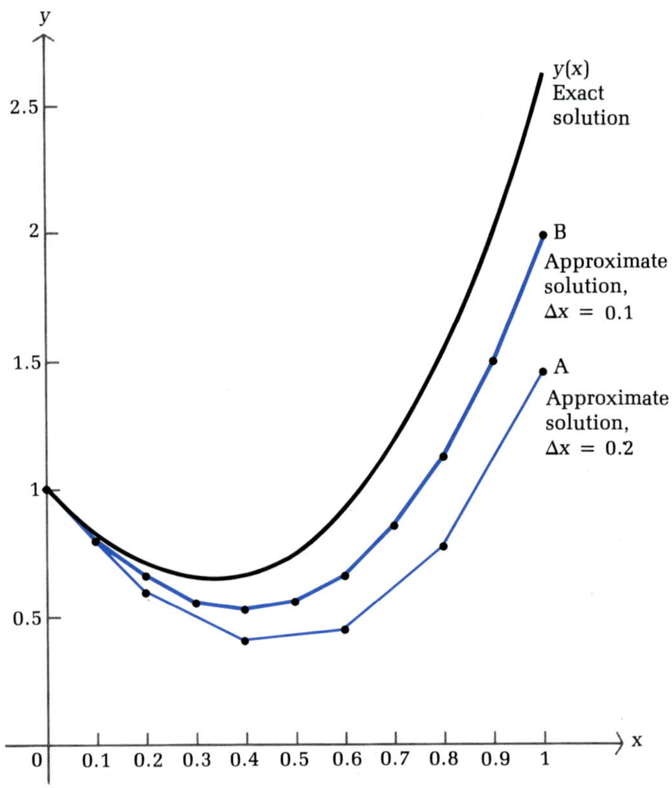

PROBLEM 18 Approximate the solution of

$$y' = y - 8x + 1 \qquad y(0) = 1$$

for $0 \leqslant x \leqslant 1$ using Euler's method with a step size of:

(A) $\Delta x = 0.2$ (B) $\Delta x = 0.1$

(C) The exact solution of this equation is

$$y = -6e^x + 8x + 7$$

Graph the exact solution and both approximate solutions from parts A and B on the same set of axes. ◆

The figure in Example 18 illustrates fairly typical behavior for approximate solutions obtained by Euler's method. Notice that the distance between the graph of the actual solution and the graph of each approximate solution increases with each step. For most initial value problems, **the error in the approximation increases with each step.**

Comparing the graphs of the two approximate solutions, we see that the graph of the approximate solution for $\Delta x = 0.1$ is closer to the graph of the exact solution than is the graph of the approximate solution for $\Delta x = 0.2$. Thus, **decreasing the step size usually increases the accuracy in the approximation.**

◆ APPLICATION

Up to this point, we have been concerned with approximating the solution of an initial value problem over an interval $a \leqslant x \leqslant b$. In many applications, we are interested in approximating the value of the solution only at $x = b$. Euler's method is a step-by-step process, with each approximate value depending on the previous one. Thus, to approximate the solution at $x = b$, we must compute all the intermediate approximate values.

◆ EXAMPLE 19 *Sales Growth*

The annual sales S (in millions of dollars) of a new company are expected to satisfy the growth equation

$$\frac{dS}{dt} = S(2 - \sqrt{S})$$

where t is the number of years the company has been in operation. If the annual sales for the first year ($t = 1$) are \$1.5 million, approximate the annual sales for the third year by using Euler's method with a step size of:

(A) $\Delta t = 1$ (B) $\Delta t = 0.5$ (C) $\Delta t = 0.25$

Solutions If $S(t)$ is the solution of

$$\frac{dS}{dt} = S(2 - \sqrt{S}) \qquad S(1) = 1.5$$

we must approximate $S(3)$. Using successively smaller step sizes will give us some information about the accuracy of our approximations.

(A) $\Delta t = 1$, $t_0 = 1$, $S_0 = 1.5$, $f(t, S) = S(2 - \sqrt{S})$, and
$S_k = S_{k-1} + [S_{k-1}(2 - \sqrt{S_{k-1}})](1)$

k	t_k	$S_k = S_{k-1} + [S_{k-1}(2 - \sqrt{S_{k-1}})](1)$
0	1	1.5
1	2	2.662 882 7
2	3	3.643 265 2

(B) $\Delta t = 0.5$

k	t_k	$S_k = S_{k-1} + [S_{k-1}(2 - \sqrt{S_{k-1}})](0.5)$
0	1	1.5
1	1.5	2.081 441 3
2	2	2.661 414
3	2.5	3.151 933 8
4	3	3.505 945 4

(C) $\Delta t = 0.25$

k	t_k	$S_k = S_{k-1} + [S_{k-1}(2 - \sqrt{S_{k-1}})](0.25)$
0	1	1.5
1	1.25	1.790 720 7
2	1.5	2.087 005 2
3	1.75	2.376 761 3
4	2	2.649 093 7
5	2.25	2.895 721 9
6	2.5	3.111 683 3
7	2.75	3.295 275 4
8	3	3.447 443 5

The last line of each of these tables contains an approximation to $S(3)$. These values, rounded to one decimal place, are displayed in the table below. Since decreasing the step size did not produce a large change in the approximate value of $S(3)$, we can conclude that $S(3) \approx 3.4$ is a reasonable approximation. Thus, we would expect the annual sales for the third year to be approximately $3.4 million.

Approximations to $S(3)$

Δt	1	0.5	0.25
Approximation	3.6	3.5	3.4

◆

PROBLEM 19 Repeat Example 19 if the annual sales during the first year are $2.5 million. ◆

Refer to the table in Example 19, where the approximations to $S(3)$ are summarized. How can we be sure that 3.4 is close to the exact value of $S(3)$? Perhaps the approximations keep decreasing as the step size decreases and the exact value of $S(3)$ is nowhere near 3.4. One way to answer this question is to continue to compute approximate values of $S(3)$, using successively smaller step sizes. A common procedure is to halve the step size each time, as we did in Example 19. Since halving the step size doubles the number of steps and the number of intermediate values that must be calculated, it is convenient to use a computer to carry out these calculations.

Table 8 contains the approximate values found in Example 19, along with the approximate values obtained by using a computer program with step sizes of 0.125 and 0.0625, all rounded to one decimal place. (*Visual Calculus*, the software that accompanies this text, contains a routine that will calculate these approximations; see the Preface.) After examining the values in Table 8, we can feel more confident in concluding that $S(3) \approx 3.4$.

TABLE 8

Approximations to $S(3)$

Δt	1	0.5	0.25	0.125	0.0625
Approximation	3.6	3.5	3.4	3.4	3.4

◆ EULER'S METHOD FOR SYSTEMS OF DIFFERENTIAL EQUATIONS

Euler's method is easily modified to approximate the solution to a system of differential equations. We simply compute approximate values for each variable in the system, as described in the box at the top of the next page.

Euler's Method for a System of Differential Equations

If $x = x(t)$ and $y = y(t)$ is the solution to the system of differential equations

$$\frac{dx}{dt} = f(x, y) \qquad x(t_0) = x_0$$

$$\frac{dy}{dt} = g(x, y) \qquad y(t_0) = y_0$$

$t_0, t_1, \ldots, t_n$ is the sequence of values defined by

$$t_k = t_{k-1} + \Delta t \qquad k = 1, 2, \ldots, n$$

and $x_1, x_2, \ldots, x_n$ and $y_1, y_2, \ldots, y_n$ are defined by

$$x_k = x_{k-1} + f(x_{k-1}, y_{k-1})\Delta t$$
$$y_k = y_{k-1} + g(x_{k-1}, y_{k-1})\Delta t \qquad k = 1, 2, \ldots, n$$

then

$$x(t_k) \approx x_k \qquad \text{and} \qquad y(t_k) \approx y_k \qquad k = 1, 2, \ldots, n$$

◆ EXAMPLE 20

Interrelated Prices

The prices (in dollars) of two interrelated commodities are p and q. The rate of change of each price is related to both prices by

$$\frac{dp}{dt} = -0.15p + 0.1q + 10$$

$$\frac{dq}{dt} = 0.1p - 0.2q + 15$$

where t is time in months. Initially, $p = 150$ and $q = 200$. Use Euler's method with a step size of $\Delta t = 1$ and five steps to approximate the solution to this system. Round the approximate values to one decimal place after each step and use these rounded values in the next step.

Solution Let $\Delta t = 1$, $t_0 = 0$, $p_0 = 150$, $q_0 = 200$,

$$f(p, q) = -0.15p + 0.1q + 10$$

and

$$g(p, q) = 0.1p - 0.2q + 15$$

Then

$$p_k = p_{k-1} + f(p_{k-1}, q_{k-1})(1)$$

and

$$q_k = q_{k-1} + g(p_{k-1}, q_{k-1})(1)$$

Notice that we cannot compute these approximations independently. Instead, we must compute p_1 and q_1, then p_2 and q_2, and so on. Also note that we can omit the factor Δt, since $\Delta t = 1$ for this approximation.

$$p_k = p_{k-1} + f(p_{k-1}, q_{k-1}) \qquad q_k = q_{k-1} + g(p_{k-1}, q_{k-1})$$

$$p_0 = 150 \qquad q_0 = 200$$

$$\begin{aligned} p_1 &= 150 + f(150, 200) \\ &= 150 + 7.5 = 157.5 \end{aligned} \qquad \begin{aligned} q_1 &= 200 + g(150, 200) \\ &= 200 - 10 = 190 \end{aligned}$$

$$\begin{aligned} p_2 &= 157.5 + f(157.5, 190) \\ &= 157.5 + 5.375 \approx 162.9 \end{aligned} \qquad \begin{aligned} q_2 &= 190 + g(157.5, 190) \\ &= 190 - 7.25 \approx 182.8 \end{aligned}$$

$$\begin{aligned} p_3 &= 162.9 + f(162.9, 182.8) \\ &= 162.9 + 3.845 \approx 166.7 \end{aligned} \qquad \begin{aligned} q_3 &= 182.8 + g(162.9, 182.8) \\ &= 182.8 - 5.27 \approx 177.5 \end{aligned}$$

$$\begin{aligned} p_4 &= 166.7 + f(166.7, 177.5) \\ &= 166.7 + 2.745 \approx 169.4 \end{aligned} \qquad \begin{aligned} q_4 &= 177.5 + g(166.7, 177.5) \\ &= 177.5 - 3.83 \approx 173.7 \end{aligned}$$

$$\begin{aligned} p_5 &= 169.4 + f(169.4, 173.7) \\ &= 169.4 + 1.96 \approx 171.4 \end{aligned} \qquad \begin{aligned} q_5 &= 173.7 + g(169.4, 173.7) \\ &= 173.7 - 2.8 \approx 170.9 \end{aligned}$$

In summary:

p_k	150	157.5	162.9	166.7	169.4	171.4
q_k	200	190	182.8	177.5	173.7	170.9

Thus, over a 5 month period, the price p increases from \$150 to approximately \$171.40 and the price q decreases from \$200 to approximately \$170.90. ◆

PROBLEM 20 Repeat Example 20 if the initial values are $p = 200$ and $q = 150$. ◆

Answers to Matched Problems

17.

k	x_k	y_k
0	1	3
1	1.2	1.88
2	1.4	1.81
3	1.6	2.19
4	1.8	2.73
5	2	3.35

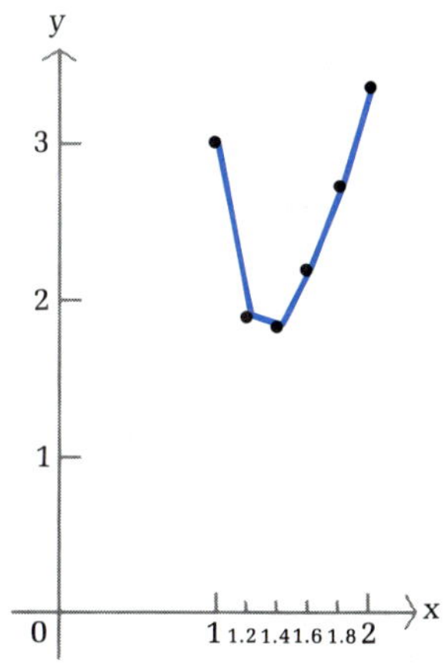

18. (A)

k	x_k	y_k
0	0	1
1	0.2	1.4
2	0.4	1.56
3	0.6	1.43
4	0.8	0.96
5	1	0.07

(B)

k	x_k	y_k
0	0	1
1	0.1	1.2
2	0.2	1.34
3	0.3	1.41
4	0.4	1.42
5	0.5	1.34
6	0.6	1.17
7	0.7	0.91
8	0.8	0.54
9	0.9	0.05
10	1	−0.56

(C)

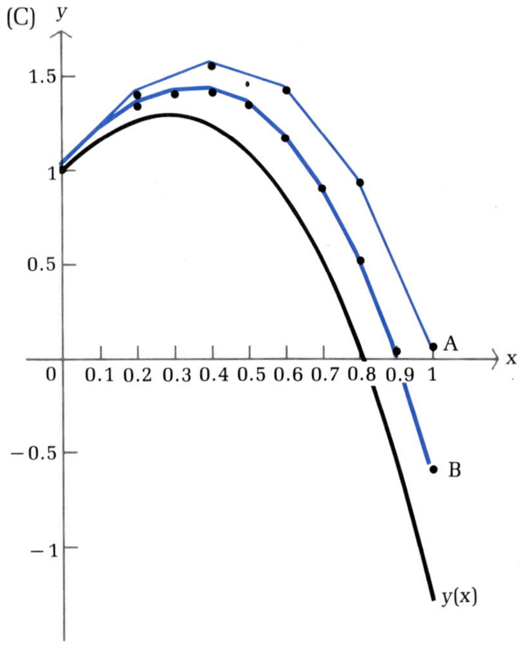

19.

Δt	1	0.5	0.25
Approximation	4.0	3.8	3.8

Approximately \$3.8 million

20.

p_k	200	195	191.3	188.5	186.3	184.4
q_k	150	155	158.5	160.9	162.6	163.7

EXERCISE 11-4

A *In Problems 1–6, use Euler's method with a step size of $\Delta x = 0.2$ to approximate the solution for $0 \le x \le 1$. Round each approximate value to two decimal places and graph the approximate solution.*

1. $y' = x + y;\quad y(0) = 2$

2. $y' = x - y;\quad y(0) = 2$

3. $y' = y;\quad y(0) = 1$

4. $y' = \dfrac{1}{2y};\quad y(0) = 1$

5. $y' = 2\sqrt{y} - 8x;\quad y(0) = 1$

6. $y' = x - 2y^2;\quad y(0) = 1$

B *In Problems 7–10, use Euler's method with a step size of $\Delta x = 0.2$ to approximate the solution for $0 \leqslant x \leqslant 1$. Round each approximate value to two decimal places. Graph the approximate solution and the exact solution on the same set of axes.*

7. $y' = -y$; $y(0) = 1$; exact solution: $y(x) = e^{-x}$
8. $y' = -y^2$; $y(0) = 1$; exact solution: $y(x) = 1/(1 + x)$
9. $y' = 1 + y$; $y(0) = 0$; exact solution: $y(x) = e^x - 1$
10. $y' = 1 - y$; $y(0) = 0$; exact solution: $y(x) = 1 - e^{-x}$

In Problems 11 and 12, use Euler's method with a step size of $\Delta x = 0.2$ and $\Delta x = 0.1$ to find two approximate solutions for $0 \leqslant x \leqslant 1$. Round each approximate value to two decimal places. Graph both approximate solutions and the exact solution on the same set of axes.

11. $y' = y - 6x$; $y(0) = 1$; exact solution: $y(x) = -5e^x + 6x + 6$
12. $y' = y - 7x + 1$; $y(0) = 1$; exact solution: $y(x) = -5e^x + 7x + 6$

In Problems 13–20, approximate the indicated value of y by using Euler's method with the given step size. Round the approximate value to two decimal places.

13. $y' = 1 + y^2$; $y(0) = 0$; use $\Delta x = 0.2$ to approximate $y(1)$
14. $y' = \sqrt{1 + y^2}$; $y(0) = 0$; use $\Delta x = 0.2$ to approximate $y(1)$
15. $y' = y^2 - x^2$; $y(1) = -1$; use $\Delta x = 0.4$ to approximate $y(3)$
16. $y' = \sqrt{x^2 + y^2}$; $y(0) = 0$; use $\Delta x = 0.1$ to approximate $y(0.5)$
17. $y' = x + e^{-y}$; $y(0) = 0$; use $\Delta x = 0.1$ to approximate $y(1)$
18. $y' = x + \ln y$; $y(0) = 1$; use $\Delta x = 0.3$ to approximate $y(1.2)$
19. $y' = \ln(x + y)$; $y(1) = 0$; use $\Delta x = 0.2$ to approximate $y(3)$
20. $y' = e^{-xy}$; $y(0) = 0$; use $\Delta x = 0.5$ to approximate $y(3)$

In Problems 21–24, use Euler's method with $\Delta t = 0.2$ and five steps to approximate the solution. Round the approximate values to two decimal places after each step and use these rounded values in the next step.

21. $\dfrac{dx}{dt} = 2x - y$, $\dfrac{dy}{dt} = 3x - 2y$; $x(0) = 1$, $y(0) = -1$

22. $\dfrac{dx}{dt} = 5x + 2y$, $\dfrac{dy}{dt} = 2x + 2y$; $x(0) = 3$, $y(0) = -1$

23. $\dfrac{dx}{dt} = 2x + y$, $\dfrac{dy}{dt} = 2x + y$; $x(0) = 2$, $y(0) = -1$

24. $\dfrac{dx}{dt} = x - y$, $\dfrac{dy}{dt} = x - y$; $x(0) = 1$, $y(0) = 2$

C 25. Use Euler's method with a step size of $\Delta x = 0.2$ to approximate the solution of

$$y' = 5x - 2\sqrt{y}$$

on the interval $0 \leq x \leq 1$ if:

(A) $y(0) = 1$ (B) $y(0) = 2$ (C) $y(0) = 3$

Round all approximate values to two decimal places and graph all three approximate solutions on the same set of axes.

26. Repeat Problem 25 for the equation $y' = 4x - \sqrt{y}$.

27. Use step sizes of $\Delta x = 1$, $\Delta x = 0.5$, and $\Delta x = 0.25$ to approximate $y(3)$ if $y(x)$ is the solution of

$$y' = \sqrt{y}(2 - \sqrt{y}) \qquad y(1) = 2$$

Round each approximate value to one decimal place.

28. Repeat Problem 27 for

$$y' = \sqrt{y}(4 - \sqrt{y}) \qquad y(1) = 2$$

APPLICATIONS

Business & Economics

29. *Advertising.* A company is using an extensive newspaper advertising campaign to introduce a new product. The number of people N (in thousands) who have heard of the product satisfies

$$\frac{dN}{dt} = 0.1N(10 - \sqrt{N})$$

If $N(0) = 1$, use Euler's method with $\Delta t = 1$ to approximate $N(5)$.

30. *Sales growth.* The annual sales s (in millions of dollars) of a company satisfy

$$\frac{ds}{dt} = 0.01s(50 - s^{1.5})$$

If the annual sales now ($t = 0$) are \$2 million, use Euler's method with a step size of 1 to approximate the annual sales 5 years from now.

31. *Interrelated markets.* The prices p and q (in dollars) of two commodities in an interrelated market satisfy the following system of differential equations:

$$\frac{dp}{dt} = 2p - q$$

$$\frac{dq}{dt} = -p + q$$

where t is time in years. If $p(0) = 100$ and $q(0) = 160$, use Euler's method with a step size of 0.5 to approximate $p(2)$ and $q(2)$. Round the approximate values to one decimal place after each step and use the rounded values in the next step.

32. *Interrelated markets.* Repeat Problem 31 for the system

$$\frac{dp}{dt} = p - q \qquad p(0) = 150$$

$$\frac{dq}{dt} = p - 2q \qquad q(0) = 100$$

Life Sciences

33. *Epidemic.* The rate at which a disease spreads through a community is given by

$$\frac{dN}{dt} = 0.1\sqrt{N}(10 - \sqrt{N})$$

where N is the total number (in thousands) of infected individuals at time t. If $N(0) = 1$, use Euler's method with a step size of 1 to approximate $N(4)$.

34. *Population growth.* The population y (in tens of thousands) of a certain species of bird satisfies the growth equation

$$\frac{dy}{dt} = -y + y^2 - 0.1y^3$$

where t is measured in years. If the initial population is 5,000, use Euler's method with a step size of 0.5 to approximate the population after 2 years.

35. *Population growth.* If an ecological system contains two species (say, rabbits and foxes) that interact with each other in a predator–prey relationship, then the rate of growth of each species is related to the current population of both species. Suppose that the number of rabbits, y (in thousands), and the number of foxes, x (in thousands), satisfy the system of equations

$$\frac{dy}{dt} = y(0.8 - 0.04x)$$

$$\frac{dx}{dt} = x(0.006y - 0.3)$$

where t is time in years. If $y(0) = 55$ and $x(0) = 15$, use Euler's method with a step size of 1 to approximate $y(4)$ and $x(4)$. Round the approximate values to one decimal place after each step and use the rounded values in the next step.

36. *Population growth.* Repeat Problem 35 for the system

$$\frac{dy}{dt} = y(0.5 - 0.02x) \qquad y(0) = 60$$

$$\frac{dx}{dt} = x(0.01y - 0.5) \qquad x(0) = 20$$

Social Sciences

37. *Rumor spread.* A rumor spreads through a community at a rate given by

$$\frac{dN}{dt} = 0.1N^{2/3}(100 - N^{1/3})$$

where N is the total number of individuals who have heard the rumor at time t. If $N(0) = 1{,}000$, use Euler's method with a step size of 1 to approximate $N(4)$.

SECTION 11-5 Chapter Review

Important Terms and Symbols

11-1 *Newton's Method for Approximating Roots.* Zero; root; x intercept; Newton's method; recursion formula for x_n; internal rate of return

$$x_n = x_{n-1} - \frac{f(x_{n-1})}{f'(x_{n-1})}$$

11-2 *Interpolating Polynomials and Divided Differences.* Piecewise linear approximation; interpolating polynomial; general form; divided difference; divided difference table

$$p(x) = a_0 + a_1(x - x_0) + a_2(x - x_0)(x - x_1) + \cdots + a_n(x - x_0)(x - x_1) \cdot \cdots \cdot (x - x_{n-1})$$

11-3 *Numerical Integration.* Trapezoid rule; Simpson's rule; error estimates; $\text{Err}(T_n)$; $\text{Err}(S_n)$

$$T_n = \frac{\Delta x}{2}[f(x_0) + 2f(x_1) + \cdots + 2f(x_{n-1}) + f(x_n)]$$

$$S_n = \frac{\Delta x}{3}[f(x_0) + 4f(x_1) + 2f(x_2) + 4f(x_3) + \cdots + 2f(x_{n-2}) + 4f(x_{n-1}) + f(x_n)]$$

11-4 *Euler's Method.* Initial value problem; exact solution; approximate solution; recursion formula; step; step size; Euler's method; Euler's method for a system of differential equations

$$y_k = y_{k-1} + f(x_{k-1}, y_{k-1})\Delta x$$

EXERCISE 11-5 Chapter Review

Work through all the problems in this chapter review and check your answers in the back of the book. (Answers to all review problems are there.) Where weaknesses show up, review appropriate sections in the text.

A

In Problems 1 and 2, find and simplify Newton's recursion formula for x_n. Use this expression and the given value of x_1 to compute x_n for the indicated value of n.

1. $f(x) = x^2 - 10; \quad x_1 = 3, \quad n = 4$
2. $f(x) = \ln x + x + 1; \quad x_1 = 0.5, \quad n = 5$

3. Find the interpolating polynomial for:

x	1	2	5
$f(x)$	11	12	3

4. Find the divided difference table and then find the interpolating polynomial for:

x	−1	0	1	2
$f(x)$	1	6	5	10

5. Use the trapezoid rule with $n = 5$ to approximate

$$\int_0^2 e^{-x^2}\,dx$$

Round the result to three decimal places.

6. Use Simpson's rule with $n = 4$ to approximate

$$\int_0^3 \sqrt{4 + x^3}\,dx$$

Round the result to three decimal places.

In Problems 7 and 8, use Euler's method with a step size of $\Delta x = 0.2$ to approximate the solution for $0 \leq x \leq 1$. Round each approximate value to two decimal places and graph the approximate solution.

7. $y' = -2y;\quad y(0) = 3$

8. $y' = y - 5x;\quad y(0) = 1$

B *In Problems 9 and 10, sketch the graph of the function and use Newton's method to find the x intercept(s).*

9. $f(x) = x^3 - 12x + 3$

10. $f(x) = x^3 - 9x^2 + 15x + 30$

In Problems 11 and 12, graph both functions on the same set of axes and find their point(s) of intersection.

11. $f(x) = e^{-x};\quad g(x) = x^3$

12. $f(x) = \dfrac{1}{x^3};\quad g(x) = x + 1$

In Problems 13 and 14, use the interpolating polynomial to approximate the value of the function defined by the table at the indicated value of x.

13.

x	1	2	3	5
$f(x)$	16	10	8	40

$f(4) \approx ?$

14.

x	−4	−2	0	2	4
$f(x)$	20	40	124	80	100

$f(1) \approx ?$

In Problems 15 and 16, graph the interpolating polynomial and the points in the table on the same set of axes.

15.

x	0	2	5
f(x)	10	2	5

16.

x	−3	−1	1	3
f(x)	16	12	32	−20

17. Use the trapezoid rule with $n = 10$ to approximate

$$\int_0^2 \frac{1}{e^x + e^{-x}}\,dx$$

Round the result to five decimal places.

18. Use Simpson's rule with $n = 8$ to approximate

$$\int_1^3 [\ln(1 + x^2)]^2\,dx$$

Round the result to five decimal places.

19. Use the values in the table and the trapezoid rule to approximate the area bounded by the graph of the function f and the x axis over the interval of x values in the table.

x	−3	−2	−1	0	1	2	3
f(x)	5	7	8	12	11	6	4

20. Use Simpson's rule with $n = 10$ to approximate the area of the region bounded by the graphs of

$$y = \frac{1}{\ln x} \qquad \text{and} \qquad y = 0 \qquad 2 \leqslant x \leqslant 3$$

Round the result to five decmal places.

21. Use Euler's method with $\Delta x = 0.2$ to approximate the solution to

$$y' = -\frac{x}{y} \qquad y(0) = 1$$

for $0 \leqslant x \leqslant 1$. Round each approximate value to two decimal places. The exact solution to this equation is

$$y = \sqrt{1 - x^2}$$

Graph the approximate solution and the exact solution on the same set of axes.

In Problems 22 and 23, use Euler's method and the given step size to approximate the indicated value of the solution.

22. $y' = x + \sqrt{y}$; $y(0) = 0$; use $\Delta x = 0.2$ to approximate $y(1)$

23. $y' = \frac{1}{y} + \frac{1}{x}$; $y(1) = 1$; use $\Delta x = 0.4$ to approximate $y(3)$

24. Use Euler's method with $\Delta t = 0.2$ and five steps to approximate the solution to

$$\frac{dx}{dt} = x + 2y \qquad x(0) = 1$$

$$\frac{dy}{dt} = y + 4x \qquad y(0) = 0$$

Round the approximate values to one decimal place after each step and use the rounded values in the next step.

C 25. Apply Newton's method to $f(x) = ax^2 + bx + c$ and derive a recursion formula that approximates the roots of a quadratic polynomial.

26. The table below was obtained from the function $f(x) = 10x^2/(1 + x^2)$:

x	-2	-1	0	1	2
$f(x)$	8	5	0	5	8

Find the interpolating polynomial $p(x)$ for this table. Graph $p(x)$ and $f(x)$ on the same set of axes.

27. Find the exact value of

$$\int_0^2 x^4\, dx$$

Then construct a table containing the values of T_n, $\mathrm{Err}(T_n)$, S_n, and $\mathrm{Err}(S_n)$ for $n = 4$ and $n = 8$. Round all values in the table to three decimal places.

28. Use Euler's method with step sizes of $\Delta x = 0.2$ and $\Delta x = 0.1$ to find two approximate solutions to

$$y' = 2y - 10x + 1 \qquad y(0) = 1$$

for $0 \leqslant y \leqslant 1$. Round each approximate value to two decimal places. The exact solution is

$$y = -e^{2x} + 5x + 2$$

Graph both approximate solutions and the exact solution on the same set of axes.

Business & Economics

29. *Break-even point.* The cost and revenue functions for a product are

$$C(x) = 10 + 0.5x \qquad \text{and} \qquad R(x) = 2x + \frac{10x}{1 + x^2} \qquad x \geqslant 1$$

Use Newton's method to approximate the break-even point.

30. *Internal rate of return.* A small business pays \$20,000 to lease a delivery van for 3 years. The annual profits generated by this van during the leasing period are \$8,000, \$10,000, and \$9,000, respectively. Find the internal rate of return of this investment. Express the answer as a percentage correct to two decimal places.

31. *Revenue.* The revenue (in thousands of dollars) from the sale of x thousand units of a product is given in the table. Find the interpolating polynomial for this table. Approximate the revenue if 3,000 units are produced. Approximate the production level that will maximize revenue.

x	0	2	4	6
R(x)	0	44	112	108

32. *Consumers' and producers' surplus.* Find the equilibrium price (in dollars) and then find the consumers' and producers' surplus for

$$p = D(x) = \sqrt{144 - 4x^3} \qquad \text{and} \qquad p = S(x) = \sqrt{9 + x^3}$$

Use Simpson's rule with $n = 4$, and round each result to two decimal places.

33. *Marginal analysis.* The marginal cost and revenue functions for a machine are

$$C'(t) = \frac{2t^2}{4 + t^2} \qquad \text{and} \qquad R'(t) = \frac{8}{4 + t^2}$$

where $C(t)$ and $R(t)$ are the total cost and revenue functions (in thousands of dollars) t years after the machine has been put into use. Find the useful life of the machine, and use Simpson's rule with $n = 4$ to approximate the total profit earned by the machine during its useful life. Round the result to the nearest hundred dollars.

34. *Sales growth.* The annual sales s (in thousands of dollars) of a company satisfy the equation

$$\frac{ds}{dt} = 0.1\sqrt{s}(100 - \sqrt{s})$$

If $s(0) = 50$, use Euler's method with $\Delta t = 1$ to approximate $s(5)$. Round the result to the nearest thousand dollars.

Life Sciences

35. Medicine. A patient is given a drug containing a particular hormone. The level of the hormone in the bloodstream (in milligrams per milliliter) t hours after the drug was administered is given by

$$L(t) = 100te^{-0.4t}$$

Approximate the time interval during which the hormone level is greater than 50 milligrams per milliliter. Round the values of t to one decimal place.

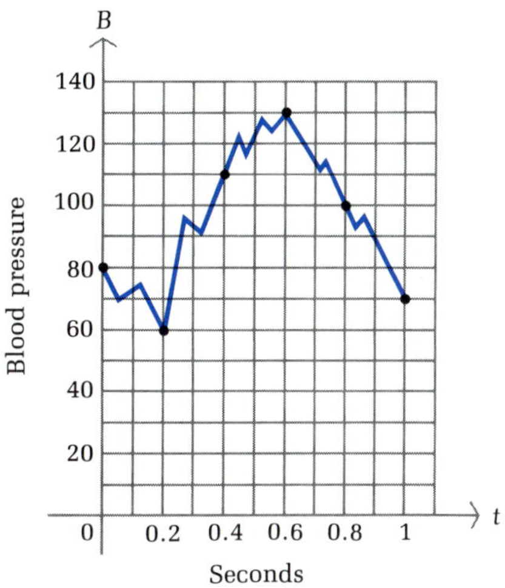

36. Medicine. The blood pressure B of an individual was graphed by a recording device over a 1 second interval (see the figure). What was the average blood pressure during this time period? Set up an appropriate definite integral, use the trapezoid rule with $n = 5$ to approximate this integral, and use the graph to estimate the necessary values of B, where B is expressed in a unit of measurement for pressure called a *torr*.

37. Population growth. Suppose the number of foxes, x (in thousands), and the number of rabbits, y (in thousands), in an ecological system satisfy

$$\frac{dy}{dt} = y(0.9 - 0.05x) \qquad y(0) = 50$$

$$\frac{dx}{dt} = x(0.005y - 0.2) \qquad x(0) = 15$$

where t is measured in years. Use Euler's method with $\Delta t = 1$ to approximate the population of each species after 5 years. Round the approximate values to one decimal place after each step and use the rounded values in the next step.

Social Sciences

38. Psychology. The rate at which workers can assemble electronic components t hours after they begin work is given by

$$N'(t) = \frac{4t^2 + 4t + 15}{4t^2 + 4t + 5}$$

Use Simpson's rule with $n = 4$ to approximate the total number of components a worker will assemble during the first 4 hours at work. Round the result to the nearest integer.

CHAPTER 12

Probability and Calculus

CHAPTER 12

Contents

In this chapter we discuss some basic probability concepts, with an emphasis on probability distributions. We first consider some important classes of probability distributions associated with discrete sample spaces and then discuss several different probability distributions associated with continuous sample spaces. As we will see, calculus techniques play an important role in extending probability concepts from discrete to continuous sample spaces.

SECTION 12-1

Finite Probability Models

- SAMPLE SPACE
- RANDOM VARIABLE; PROBABILITY DISTRIBUTION
- EXPECTED VALUE OF A RANDOM VARIABLE
- MEAN AND STANDARD DEVIATION

This section provides a relatively brief and informal introduction to probability. More detailed and formal treatments can be found in books and courses devoted entirely to the subject.

◆ SAMPLE SPACE

Probability studies are based on experiments that do not yield the same results each time they are performed, no matter how carefully they are repeated under the same conditions. These experiments are called **random experiments.** Familiar examples of random experiments are flipping coins, rolling dice, observing the frequency of defective items from an assembly line, or observing the frequency of deaths in a certain age group. Probability theory is a branch of mathematics that has been developed to deal with outcomes of random experiments, both real and conceptual. In the work that follows, we will simply use the word **experiment** to mean a random experiment.

If we identify a set S of outcomes of an experiment in such a way that in each trial of the experiment one and only one of the outcomes in the set will occur, then we call the set S a **sample space** for the experiment. Each element in S is called a **simple outcome.** For example, if 3 distinct coins are tossed, then a sample space for this experiment is

$$S = \{HHH, HHT, HTH, HTT, THH, THT, TTH, TTT\}$$

where H is heads and T is tails.

An **event** E is defined to be any subset of the sample space (including the empty set, ø, and the entire sample space S). An event is called a **simple event** if it contains only one element and a **compound event** if it contains more than one element. We say that **an event E occurs** if any of the simple outcomes in E occurs. Referring to the sample space S given above, the event that all 3 coins turn up heads and the event that all 3 coins turn up the same can be described, respectively, as

$$E_1 = \{HHH\} \qquad \text{and} \qquad E_2 = \{HHH, TTT\}$$

Event E_1 is a simple event,* while event E_2 is a compound event.

If each of these 3 coins is fair (a head is as likely to occur as a tail), then each simple event in the sample space is **equally likely** to occur. Since there are 8 different simple events in S, the probability of any one of these simple events occurring is

$$P(e) = \tfrac{1}{8} \qquad e \in S$$

More generally, we have the following result.

THEOREM 1

Probability in Equally Likely Sample Spaces

If we assume each simple event in sample space S is as likely as any other to occur, then the probability of an arbitrary event E in S is given by

$$P(E) = \frac{\text{Number of elements in } E}{\text{Number of elements in } S} = \frac{n(E)}{n(S)}$$

Let's apply this theorem to the events E_1 and E_2 in the coin-tossing example discussed above:

$$P(E_1) = \frac{n(E_1)}{n(S)} = \frac{1}{8} \qquad E_1 = \{HHH\}$$

$$P(E_2) = \frac{n(E_2)}{n(S)} = \frac{2}{8} = \frac{1}{4} \qquad E_2 = \{HHH, TTT\}$$

* Technically, there is a logical distinction between the simple outcome HHH (a single element in the sample space) and the simple event {HHH} (a subset of the sample space consisting of a single element). But we will just keep this in mind and use the terms *simple outcome* and *simple event* interchangeably.

◆ RANDOM VARIABLE; PROBABILITY DISTRIBUTION

When performing a random experiment, a sample space S is selected in such a way that all probability problems of interest relative to the experiment can be solved. In many situations we may not be interested in each simple event in the sample space S but in some numerical value associated with the event. For example, if 3 coins are tossed, we may be interested in the number of heads that turn up rather than in the particular pattern that turns up. Or, in selecting a random sample of students, we may be interested in the proportion that are women rather than which particular students are women. In the same way, a "craps" player is usually interested in the sum of the dots on the showing faces rather than the pattern of dots on each face.

In each of these examples, we have a rule that assigns to each simple event in S a single real number. Mathematically speaking, we are dealing with a function (see Section 1-2). Historically, this particular type of function has been called a "random variable."

Random Variable

A **random variable** is a function that assigns a numerical value to each simple event in a sample space S.

The term *random variable* is an unfortunate choice, since it is neither random nor a variable—it is a function with a numerical value and it is defined on a sample space. But the terminology has stuck and is now standard, so we shall have to live with it. Capital letters, such as X, are used to represent random variables.

TABLE 1

Number of Heads in the Toss of 3 Coins

SAMPLE SPACE S	NUMBER OF HEADS $X(e_i)$
e_1: TTT	0
e_2: TTH	1
e_3: THT	1
e_4: HTT	1
e_5: THH	2
e_6: HTH	2
e_7: HHT	2
e_8: HHH	3

Let us return to the experiment of tossing 3 coins. A sample space S of equally likely simple events is indicated in Table 1. Suppose we are interested in the number of heads (0, 1, 2, or 3) appearing on each toss of the 3 coins and the probability of each of these events. We introduce a random variable X (a function) that indicates the number of heads for each simple event in S (see the second column in Table 1). For example, $X(e_1) = 0$, $X(e_2) = 1$, and so on. The random variable X assigns a numerical value to each simple event in the sample space S.

We are interested in the probability of the occurrence of each image value of X; that is, in the probability of the occurrence of 0 heads, 1 head, 2 heads, or 3 heads in the single toss of 3 coins. We indicate this probability by

$$P(x) \qquad \text{where} \qquad x \in \{0, 1, 2, 3\}$$

The function P is called the **probability function* of the random variable X.**

What is $P(2)$, the probability of getting exactly 2 heads on the single toss of 3 coins? "Exactly 2 heads occur" is the event

$$E = \{\text{THH, HTH, HHT}\}$$

Thus,

$$P(2) = \frac{n(E)}{n(S)} = \frac{3}{8}$$

TABLE 2

Probability Distribution

NUMBER OF HEADS x	0	1	2	3
PROBABILITY $P(x)$	$\frac{1}{8}$	$\frac{3}{8}$	$\frac{3}{8}$	$\frac{1}{8}$

Proceeding similarly for $P(0)$, $P(1)$, and $P(3)$, we obtain the results in Table 2. This table is called a **probability distribution for the random variable X.** Probability distributions are also represented graphically, as shown in Figure 1. The graph of a probability distribution is often called a **histogram.**

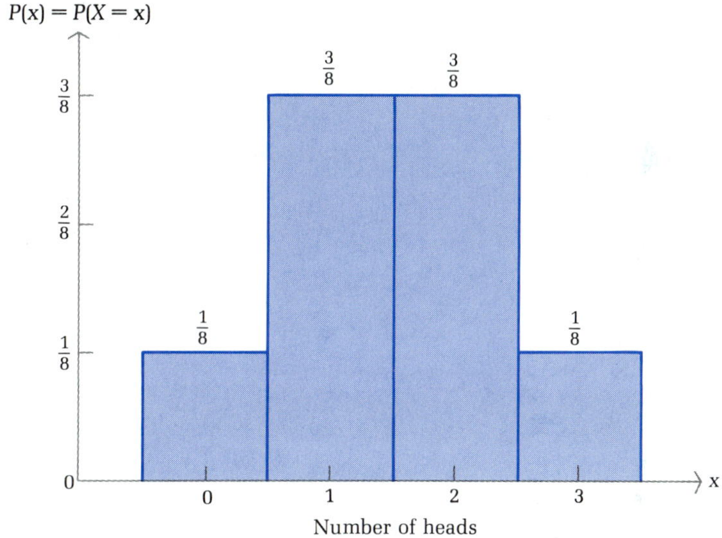

FIGURE 1
Histogram for a probability distribution

Note from Table 2 or Figure 1 that

1. $0 \le P(x) \le 1, \quad x \in \{0, 1, 2, 3\}$
2. $P(0) + P(1) + P(2) + P(3) = \frac{1}{8} + \frac{3}{8} + \frac{3}{8} + \frac{1}{8} = 1$

These are general properties that any probability distribution of a random variable X associated with a finite sample space must have.

---

* Formally, the probability function P of the random variable X is defined by $P(x) = P(\{e_i \in S | X(e_i) = x\})$, which, because of its cumbersome nature, is usually simplified to $P(X = x)$ or, simply, $P(x)$. We will use the simplified notation.

Probability Distribution of a Random Variable X

A probability function $P(X = x) = P(x)$ is a **probability distribution of the random variable X** if

1. $0 \le P(x) \le 1, \quad x \in \{x_1, x_2, \ldots, x_n\}$
2. $P(x_1) + P(x_2) + \cdots + P(x_n) = 1$

where $\{x_1, x_2, \ldots, x_n\}$ are the (range) values of X (see Figure 2).

Figure 2 illustrates the process of forming a probability distribution of a random variable.

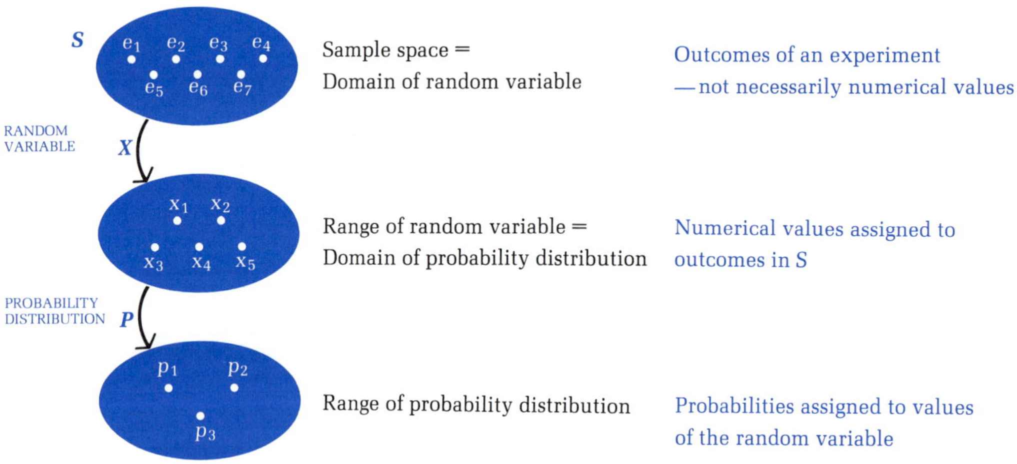

FIGURE 2
Probability distribution of a random variable for a finite sample space

◆ EXAMPLE 1 A spinner is numbered from 1 to 4, and each number is as likely to turn up as any other. The random variable X represents the sum of the numbers obtained in two consecutive spins of the pointer. Find and graph the probability distribution of X.

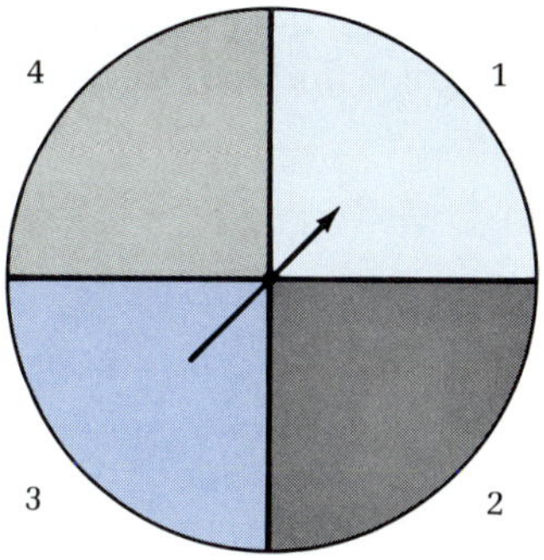

Solution The possible values of X are 2, 3, 4, 5, 6, 7, and 8. In order to determine the probability of each of these sums occurring, we consider the sample space S consisting of the sixteen possible outcomes listed in the following table:

		SECOND SPIN			
		1	2	3	4
	1	(1, 1)	(1, 2)	(1, 3)	(1, 4)
FIRST	2	(2, 1)	(2, 2)	(2, 3)	(2, 4)
SPIN	3	(3, 1)	(3, 2)	(3, 3)	(3, 4)
	4	(4, 1)	(4, 2)	(4, 3)	(4, 4)

In this sample space, the simple outcome (3, 4) is to be distinguished from the simple outcome (4, 3) even though both produce the same sum. The former indicates that the pointer stopped at 3 on the first spin and at 4 on the second spin, while the latter indicates that the pointer stopped at 4 on the first spin and at 3 on the second spin. Since each of the four possible outcomes of a single spin are equally likely, pairing all possible outcomes from the first spin with all possible outcomes from the second spin produces an equally likely sample space of combined outcomes. Thus, we can use Theorem 1 to compute the probability distribution of the random variable X, which assumes the values 2, 3, 4, 5, 6, 7, and 8:

$$P(2) = P(X = 2) = P(\{(1, 1)\}) = \tfrac{1}{16}$$
$$P(3) = P(X = 3) = P(\{(1, 2), (2, 1)\}) = \tfrac{2}{16}$$
$$P(4) = P(X = 4) = P(\{(1, 3), (2, 2), (3, 1)\}) = \tfrac{3}{16}$$

and so on. The probability distribution for X is

x	2	3	4	5	6	7	8
P(x)	$\frac{1}{16}$	$\frac{2}{16}$	$\frac{3}{16}$	$\frac{4}{16}$	$\frac{3}{16}$	$\frac{2}{16}$	$\frac{1}{16}$

And the graph is

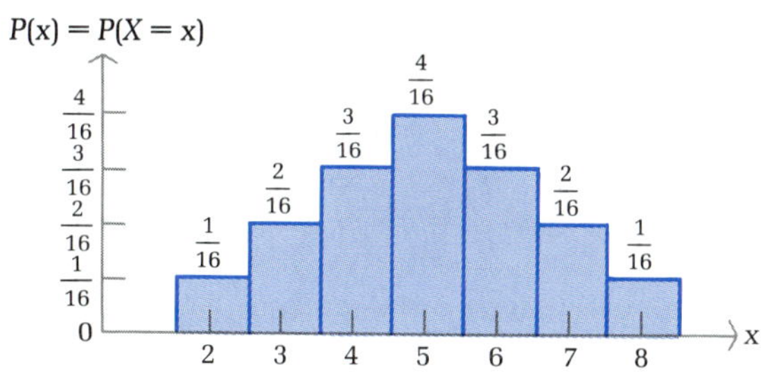

◆

PROBLEM 1 A spinner is marked from 1 to 3, and each number is as likely to come up as any other. The random variable X represents the sum of the numbers obtained in two consecutive spins of the pointer. Find and graph the probability distribution of X.

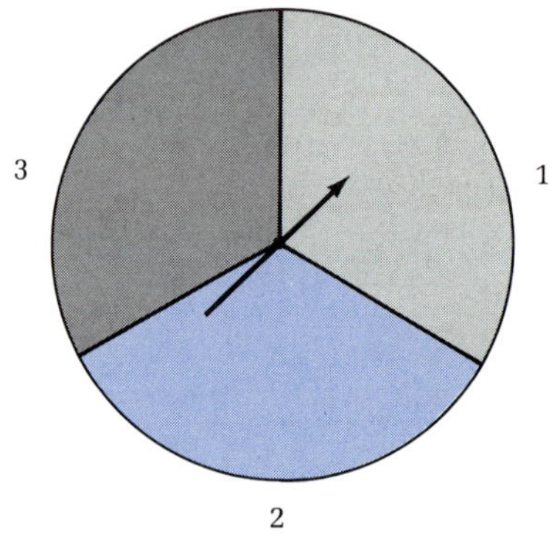

◆ EXPECTED VALUE OF A RANDOM VARIABLE

If the experiment of tossing 3 coins were repeated 1,000 times, then, based on the probability distribution in Table 2, we would expect to obtain the following results:

0 heads approximately $\frac{1}{8}(1{,}000) = 125$ times
1 head approximately $\frac{3}{8}(1{,}000) = 375$ times
2 heads approximately $\frac{3}{8}(1{,}000) = 375$ times
3 heads approximately $\frac{1}{8}(1{,}000) = 125$ times

Of course, it is not likely that we would obtain these exact values, but the actual results should be close to those predicted by the probabilities. Now suppose we were interested in the average number of heads per toss (the total number of heads in all tosses divided by the total number of tosses). Using the values listed above, we would expect the average number of heads per toss of the 3 coins, or the *expected value* $E(X)$, to be given by

$$E(X) = \frac{0 \cdot 125 + 1 \cdot 375 + 2 \cdot 375 + 3 \cdot 125}{1{,}000} \qquad \frac{\text{Total number of heads}}{\text{Total number of tosses}}$$

$$= 0\left(\frac{125}{1{,}000}\right) + 1\left(\frac{375}{1{,}000}\right) + 2\left(\frac{375}{1{,}000}\right) + 3\left(\frac{125}{1{,}000}\right)$$

$$= 0\left(\frac{1}{8}\right) + 1\left(\frac{3}{8}\right) + 2\left(\frac{3}{8}\right) + 3\left(\frac{1}{8}\right) = \frac{12}{8} = 1.5$$

It is important to note that the expected value is not a value that will necessarily occur in a single experiment (1.5 heads cannot occur in the toss of 3 coins), but it is an average of what occurs over a large number of experiments. Sometimes, we will toss more than 1.5 heads and sometimes less, but if the experiment were repeated many times, the average number of heads per experiment would approach 1.5.

We now make the above discussion more precise through the following definition of expected value:

Expected Value of a Random Variable X

Given the probability distribution for the random variable X,

x_i	x_1	x_2	$\cdots$	x_m
p_i	p_1	p_2	$\cdots$	p_m

where $p_i = P(x_i)$, we define the **expected value of X,** denoted **$E(X)$,** by the formula

$$E(X) = x_1p_1 + x_2p_2 + \cdots + x_mp_m$$

We again emphasize that the expected value is not to be expected to occur in a single experiment; it is a long-run average of repeated experiments—it is the weighted average of the possible outcomes, each weighted by its probability.

Steps for Computing the Expected Value of a Random Variable X

Step 1. Form the probability distribution of the random variable X.

Step 2. Multiply each image value of X, x_i, by its corresponding probability of occurence p_i; then add the results.

◆ EXAMPLE 2 What is the expected value (long-run average) of the number of dots facing up for the roll of a single die?

Solution If we choose

$$S = \{1, 2, 3, 4, 5, 6\}$$

as our sample space, then each simple event is a numerical outcome reflecting our interest, and each is equally likely. The random variable X in this case is just the identity function (each number is associated with itself). Thus, the probability distribution for X is

x_i	1	2	3	4	5	6
p_i	$\frac{1}{6}$	$\frac{1}{6}$	$\frac{1}{6}$	$\frac{1}{6}$	$\frac{1}{6}$	$\frac{1}{6}$

Hence,

$$\begin{aligned} E(X) &= 1(\tfrac{1}{6}) + 2(\tfrac{1}{6}) + 3(\tfrac{1}{6}) + 4(\tfrac{1}{6}) + 5(\tfrac{1}{6}) + 6(\tfrac{1}{6}) \\ &= \tfrac{21}{6} = 3.5 \end{aligned}$$ ◆

PROBLEM 2 Suppose the die in Example 2 is not fair and we obtain (empirically) the following probability distribution for X:

x_i	1	2	3	4	5	6
p_i	.14	.13	.18	.20	.11	.24

[Note: Sum = 1.]

What is the expected value of X? ◆

◆ EXAMPLE 3 A spinner device is numbered from 0 to 5, and each of the 6 numbers is as likely to come up as any other. A player who bets \$1 on any given number wins \$4 (and gets the bet back) if the pointer comes to rest on the chosen number; otherwise, the \$1 bet is lost. What is the expected value of the game (long-run average gain or loss per game)?

Solution The sample space of equally likely events is

$$S = \{0, 1, 2, 3, 4, 5\}$$

Each simple outcome occurs with a probability of $\frac{1}{6}$. The random variable X assigns \$4 to the winning number and −\$1 to each of the remaining numbers. Thus, the probability distribution for X, called a **payoff table,** is as shown below:

Payoff Table (Probability Distribution for X)

x_i	\$4	−\$1
p_i	$\frac{1}{6}$	$\frac{5}{6}$

The probability of winning \$4 is $\frac{1}{6}$ and of losing \$1 is $\frac{5}{6}$. We can now compute the expected value of the game:

$$E(X) = \$4(\tfrac{1}{6}) + (-\$1)(\tfrac{5}{6}) = -\$\tfrac{1}{6} \approx -\$0.1667 \approx -17¢ \text{ per game}$$

Thus, in the long run the player will lose an average of about 17¢ per game. ◆

In general, a game is said to be **fair** if $E(X) = 0$. The game in Example 3 is not fair—the "house" has an advantage, on the average, of about 17¢ per game.

PROBLEM 3 Repeat Example 3 with the player winning $5 instead of $4 if the chosen number turns up. The loss is still $1 if any other number turns up. Is this now a fair game? ◆

◆ EXAMPLE 4 Suppose you are interested in insuring a car stereo system for $500 against theft. An insurance company charges a premium of $60 for coverage for 1 year, claiming an empirically determined probability of .1 that the stereo will be stolen some time during the year. What is your expected return from the insurance company if you take out this insurance?

Solution This is actually a game of chance in which your stake is $60. You have a .1 chance of receiving $440 from the insurance company ($500 minus your stake of $60) and a .9 chance of losing your stake of $60. What is the expected value of this "game"? We form a payoff table (the probability distribution for X) as shown below:

Payoff Table

x_i	$440	−$60
p_i	.1	.9

Then we compute the expected value as follows:

$$E(X) = (\$440)(.1) + (-\$60)(.9) = -\$10$$

This means that if you insure with this company over many years and the circumstances remain the same, you would have an average net loss to the insurance company of $10 per year. ◆

PROBLEM 4 Find the expected value in Example 4 from the insurance company's point of view. ◆

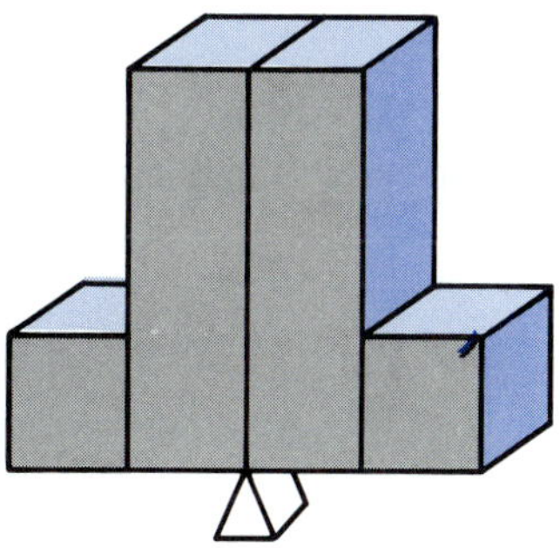

FIGURE 3
The balance point on the histogram is $\mu = 1.5$

◆ MEAN AND STANDARD DEVIATION

Since the expected value of a random variable represents the long-run average of repeated experiments, it is often referred to as the **arithmetic average,** or **mean.** Traditionally, the Greek letter μ is used to denote the mean. Thus,

$$\mu = E(X) = x_1p_1 + x_2p_2 + \cdots + x_mp_m$$

is the mean of the random variable X. Geometrically, the mean is the center of the values of X and is often referred to as a *measure of central tendency*. For example, if the histogram in Figure 1 were drawn on a piece of wood of uniform thickness and the wood cut around the outside of the figure, then the resulting object would balance on a wedge placed at the mean $\mu = 1.5$ (see Figure 3).

Another numerical quantity that is used to describe the properties of a random variable is the *standard deviation*. This quantity gives a *measure of the dispersion*, or *spread*, of the random variable X about the mean μ.

Standard Deviation of a Random Variable X

Given the probability distribution for the random variable X,

x_i	x_1	x_2	$\cdots$	x_m
p_i	p_1	p_2	$\cdots$	p_m

and the mean

$$\mu = x_1p_1 + x_2p_2 + \cdots + x_mp_m$$

we define the **variance of *X*,** denoted by ***V*(*X*)**, by the formula

$$V(X) = (x_1 - \mu)^2p_1 + (x_2 - \mu)^2p_2 + \cdots + (x_m - \mu)^2p_m$$

and the **standard deviation of *X*,** denoted by σ (Greek letter sigma), by the formula

$$\sigma = \sqrt{V(X)}$$

In other words, the variance is the expected value of the squares of the distances from each value of X to the mean. Standard deviation is defined using the square root so that it will be expressed in the same units as the values of X.

Returning to the coin-tossing experiment, we have already shown that $\mu = E(X) = 1.5$. Thus,

$$\begin{aligned} V(X) &= (0 - 1.5)^2(\tfrac{1}{8}) + (1 - 1.5)^2(\tfrac{3}{8}) + (2 - 1.5)^2(\tfrac{3}{8}) + (3 - 1.5)^2(\tfrac{1}{8}) \\ &= .75 \end{aligned}$$

and

$$\sigma = \sqrt{V(X)} = \sqrt{.75} \approx .866$$

◆ EXAMPLE 5 Find the variance and standard deviation for the random variable in Example 2.

Solution

$$\begin{aligned} \mu &= E(X) = 3.5 \\ V(X) &= (1 - 3.5)^2(\tfrac{1}{6}) + (2 - 3.5)^2(\tfrac{1}{6}) + (3 - 3.5)^2(\tfrac{1}{6}) \\ &\qquad + (4 - 3.5)^2(\tfrac{1}{6}) + (5 - 3.5)^2(\tfrac{1}{6}) + (6 - 3.5)^2(\tfrac{1}{6}) \\ &= \tfrac{35}{12} \\ \sigma &= \sqrt{V(X)} = \sqrt{\tfrac{35}{12}} \approx 1.708 \end{aligned}$$

◆

PROBLEM 5 Find the variance and standard deviation for the random variable in Problem 2.

◆

The standard deviation is often used to compare different probability distributions. Figure 4 shows four different probability distributions and their graphs. Notice the relationship between the standard deviation σ and the dispersion of the probability distribution about the mean. The tighter the cluster of the probability distribution about the mean, the smaller the standard deviation.

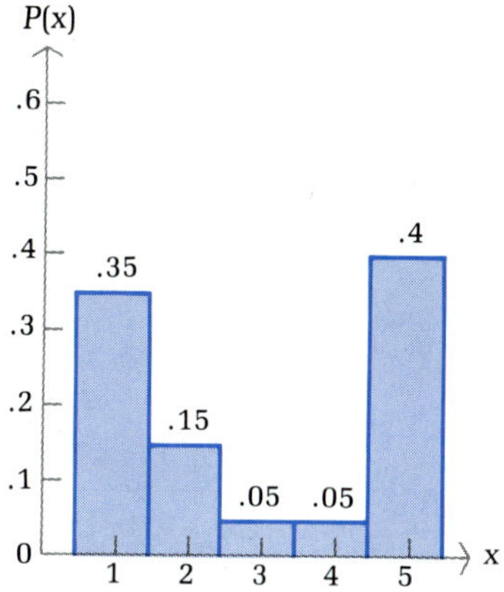

$\mu = 3,\quad V = 3.2,\quad \sigma \approx 1.7889$

x_i	1	2	3	4	5
p_i	.35	.15	.05	.05	.4

(A)

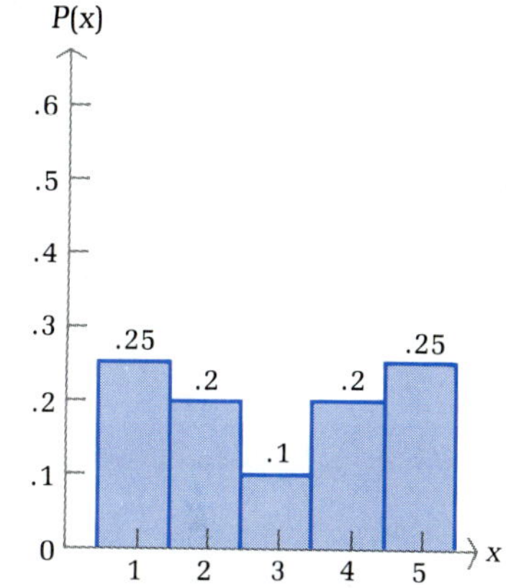

$\mu = 3,\quad V = 2.4,\quad \sigma \approx 1.5492$

x_i	1	2	3	4	5
p_i	.25	.2	.1	.2	.25

(B)

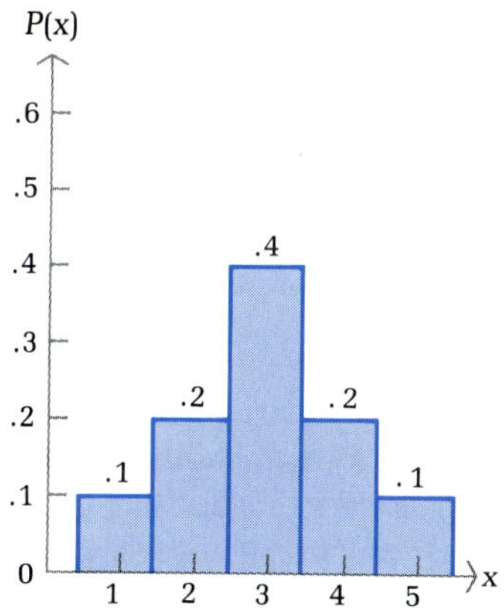

$\mu = 3,\quad V = 1.2,\quad \sigma \approx 1.0954$

x_i	1	2	3	4	5
p_i	.1	.2	.4	.2	.1

(C)

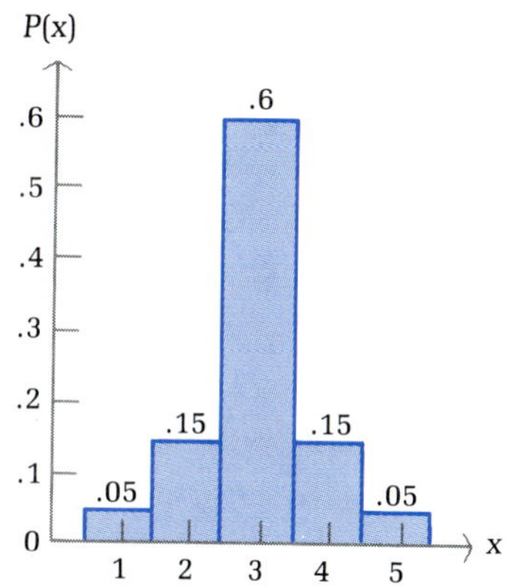

$\mu = 3,\quad V = .7\quad \sigma \approx .8367$

x_i	1	2	3	4	5
p_i	.05	.15	.6	.15	.05

(D)

FIGURE 4

Answers to Matched Problems

1.

x	2	3	4	5	6
P(x)	$\frac{1}{9}$	$\frac{2}{9}$	$\frac{3}{9}$	$\frac{2}{9}$	$\frac{1}{9}$

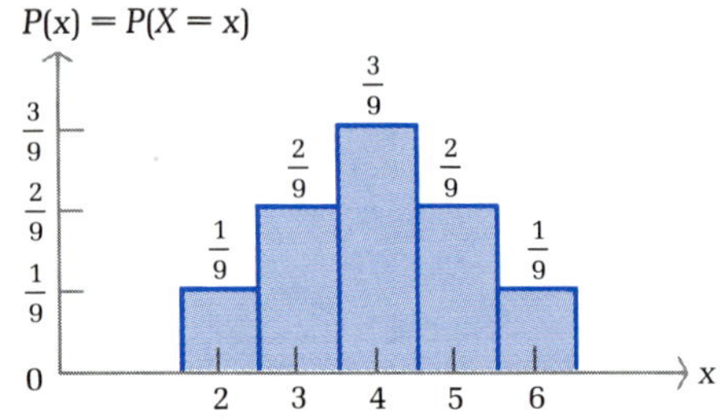

2. $E(X) = 3.73$ 3. $E(X) = \$0$; the game is fair
4. $E(X) = (-\$440)(.1) + (\$60)(.9) = \$10$; this amount, of course, is necessary to cover expenses and profit
5. $V(X) = 2.9571$; $\sigma \approx 1.720$

EXERCISE 12-1

A

In Problems 1–4, graph the probability distribution, and find the mean and standard deviation of the random variable X.

1.

x_i	−2	−1	0	1	2
p_i	.1	.2	.4	.2	.1

2.

x_i	−2	−1	0	1	2
p_i	.1	.1	.2	.2	.4

3.

x_i	−2	−1	0	1	2
p_i	.5	.2	.1	.1	.1

4.

x_i	−2	−1	0	1	2
p_i	.3	.1	.1	.1	.4

A spinner is marked from 1 to 10 and each number is as likely to turn up as any other. Problems 5–10 refer to this experiment.

5. Find a sample space S consisting of equally likely simple events.
6. The random variable X represents the number that turns up on the spinner. Find the probability distribution for X.
7. What is the probability of obtaining an even number?
8. What is the probability of obtaining a number that is exactly divisible by 3?
9. What is the expected value of X?
10. What is the standard deviation of X?

B

11. In tossing 2 fair coins once, what is the expected number of heads?
12. In a family with 2 children, excluding multiple births and assuming a boy is as likely as a girl at each birth, what is the expected number of boys?

An experiment consists of tossing a coin 4 times in succession. Answer the questions in Problems 13–18 regarding this experiment.

13. The random variable X represents the number of heads that occur in 4 tosses. Find and graph the probability distribution of X.
14. What is the probability of getting 2 or more heads?
15. What is the probability of getting an even number of heads?
16. What is the probability of getting more heads than tails?
17. Find the expected number of heads.
18. Find the standard deviation of X.

An experiment consists of rolling 2 fair dice. Answer the questions in Problems 19–24 regarding this experiment.

19. The random variable X represents the sum of the dots on the 2 up faces of the dice. Find and graph the probability distribution of X.
20. What is the probability that the sum is 7 or 11?
21. What is the probability that the sum is less than 6?
22. What is the probability that the sum is an even number?
23. Find the expected value of the sum of the dots.
24. Find the standard deviation of X.

C

25. After you pay \$4 to play a game, a single fair die is rolled and you are paid back the number of dollars equal to the number of dots facing up. For example, if 5 dots turn up, \$5 is returned to you for a net gain of \$1. If 1 dot turns up, \$1 is returned to you for a net gain, or payoff, of −\$3; and so on. If X is the random variable that represents net gain, or payoff, what is the expected value of X?
26. Repeat Problem 25 with the same game costing \$3.50 for each play.
27. A player tosses 2 coins and wins \$3 if 2 heads appear and \$1 if 1 head appears, but loses \$6 if 2 tails appear. If X is the random variable representing the player's net gain, what is the expected value of X?
28. Repeat Problem 27 if the player wins \$5 if 2 heads appear and \$2 if 1 head appears, but loses \$7 if 2 tails appear.
29. Roulette wheels in the United States generally have 38 equally spaced slots numbered 00, 0, 1, 2, . . . , 36. A player who bets \$1 on any given number wins \$35 (and gets the bet back) if the ball comes to rest on the chosen number; otherwise, the \$1 bet is lost. If X is the random variable that represents the player's net gain, what is the expected value of X?
30. In roulette (see Problem 29) the numbers from 1 to 36 are evenly divided between red and black. A player who bets \$1 on black, wins \$1 (and gets the bet back) if the ball comes to rest on black; otherwise (if the ball lands on red, 0, or 00), the \$1 bet is lost. If X is the random variable that represents the player's net gain, what is the expected value of X?

31. One thousand raffle tickets are sold at $10 each. A ticket will be drawn at random, and the ticket holder will be paid $5,000. Suppose you buy 5 tickets.

(A) Create a payoff table for the random variable X representing your net gain in this raffle.
(B) What is the expected value of X?

32. Repeat Problem 31 with the purchase of 10 tickets.

APPLICATIONS

Business & Economics

33. *Insurance.* The annual premium for a $5,000 insurance policy against the theft of a painting is $150. If the (empirical) probability that the painting will be stolen during the year is .01, what is your expected return from the insurance company if you take out this insurance?

34. *Insurance.* Repeat Problem 33 from the point of view of the insurance company.

35. *Decision analysis.* An oil company, after careful testing and analysis, is considering drilling in two different sites. It is estimated that site A will net $30 million if successful (probability .2) and lose $3 million if not (probability .8); site B will net $70 million if successful (probability .1) and lose $4 million if not (probability .9). Which site should the company choose according to the expected return from each site?

36. *Decision analysis.* Repeat Problem 35, assuming additional analysis caused the estimated probability of success in field B to be changed from .1 to .11.

Life Sciences

37. *Genetics.* Suppose that, at each birth, having a girl is not as likely as having a boy. The probability assignments for the number of boys in a 3-child family are approximated empirically from past records and are given in the table below. What is the expected number of boys in a 3-child family?

NUMBER OF BOYS x_i	p_i
0	.12
1	.36
2	.38
3	.14

38. *Genetics.* A pink-flowering plant is of genotype RW. If two such plants are crossed, we obtain a red plant (RR) with probability .25, a pink plant (RW or WR) with probability .50, and a white plant (WW) with probability .25, as shown in the table at the top of the next page. What is the expected number of W genes present in a crossing of this type?

NUMBER OF W GENES PRESENT x_i	p_i
0	.25
1	.50
2	.25

Social Sciences

39. *Politics.* A money drive is organized by a campaign committee for a candidate running for public office. Two approaches are considered:

A_1: A general mailing with a followup mailing

A_2: Door-to-door solicitation with followup telephone calls

From campaign records of previous committees, average donations and their corresponding probabilities are estimated to be:

A_1		A_2	
x_i (return per person)	p_i	x_i (return per person)	p_i
$10	.3	$15	.3
5	.2	3	.1
0	.5	0	.6
	1.0		1.0

What are the expected returns? Which course of action should be taken according to the expected returns?

SECTION 12-2 Binomial Distributions

- BERNOULLI TRIALS
- BINOMIAL FORMULA (BRIEF REVIEW)
- BINOMIAL DISTRIBUTION
- APPLICATION

BERNOULLI TRIALS

If we toss a coin, either a head occurs or it does not. If we roll a die, either a 3 shows or it fails to show. If you are vaccinated for smallpox, either you contract smallpox or you do not. What do all these situations have in common? All can be classified as experiments with two possible outcomes, each the complement of the other. An experiment for which there are only two possible outcomes, E or E', is called a **Bernoulli experiment,** or **trial,** after Jacob Bernoulli (1654–1705), a Swiss scientist and mathematician who was one of the first people to study systematically the probability problems related to a two-outcome experiment.

In a Bernoulli experiment or trial, it is customary to refer to one of the two outcomes as a **success** S and to the other as a **failure** F. If we designate the probability of success by

$$P(S) = p$$

then the probability of failure will be

$$P(F) = 1 - p = q \qquad \textit{Note:} \quad p + q = 1$$

◆ EXAMPLE 6 We roll a fair die and ask for the probability of a 6 turning up. This can be viewed as a Bernoulli trial by identifying a success with a 6 turning up and a failure with any of the other numbers turning up. Thus,

$$p = \tfrac{1}{6} \qquad \text{and} \qquad q = 1 - \tfrac{1}{6} = \tfrac{5}{6}$$ ◆

PROBLEM 6 Identify p and q for a single roll of a fair die where a success is a number divisible by 3 turning up. ◆

Now, suppose a Bernoulli experiment is repeated 5 times. How can we compute the probability of the outcome $SSFFS$? In order to answer this question, we must make two basic assumptions about the trials in a sequence of Bernoulli experiments. First, we will assume that the probability of a success remains the same from trial to trial. Second, we will assume that the trials are **independent;** that is, the outcome of one trial has no effect on the outcome of any of the other trials. With these two assumptions, it can be shown that *the probability of a sequence of events is equal to the product of the probability of each event in the sequence.* Thus,

$$\begin{aligned} P(SSFFS) &= P(S)P(S)P(F)P(F)P(S) \\ &= ppqqp \\ &= p^3q^2 \end{aligned}$$

In general, we define a sequence of Bernoulli trials as follows:

Bernoulli Trials

A sequence of experiments is called a **sequence of Bernoulli trials** if:

1. Only two outcomes are possible on each trial.
2. The probability of success p for each trial is a constant (probability of failure is then $q = 1 - p$).
3. All trials are independent.

In simple Bernoulli experiments, such as tossing a coin or rolling a die, it seems very reasonable to assume that the trials are independent. In more complicated situations, it can be very difficult to determine whether the trials are actually independent. We will assume that all the Bernoulli experiments we consider in this book have independent trials.

◆ EXAMPLE 7 If we roll a fair die 5 times and identify a success in a single roll with a 1 turning up, what is the probability of the sequence *SFFSS* occurring?

Solution

$p = \frac{1}{6} \qquad q = 1 - p = \frac{5}{6}$

$$P(SFFSS) = pqqpp = p^3q^2$$
$$= (\tfrac{1}{6})^3(\tfrac{5}{6})^2 \approx .003$$ ◆

PROBLEM 7 In Example 7, find the probability of the outcome *FSSSF*. ◆

In most applications involving sequences of Bernoulli trials, we will be interested in the number of successes, rather than in a specific outcome of the form *SSFFS*. If X is the random variable associated with the number of successes in a sequence of Bernoulli trials, we would like to find the probability distribution for X. Since this probability distribution is closely related to the *binomial formula*, it will be helpful if we first review this important formula. (A more detailed discussion of this formula and factorial notation can be found in Appendix A.)

◆ BINOMIAL FORMULA (BRIEF REVIEW)

To start, let us calculate directly the first five natural number powers of $(a + b)^x$:

$$(a + b)^1 = a + b$$
$$(a + b)^2 = a^2 + 2ab + b^2$$
$$(a + b)^3 = a^3 + 3a^2b + 3ab^2 + b^3$$
$$(a + b)^4 = a^4 + 4a^3b + 6a^2b^2 + 4ab^3 + b^4$$
$$(a + b)^5 = a^5 + 5a^4b + 10a^3b^2 + 10a^2b^3 + 5ab^4 + b^5$$

In general, it can be shown that a binomial expansion is given by the well-known **binomial formula:**

Binomial Formula

For n a natural number,

$$(a + b)^n = C_{n,0}a^n + C_{n,1}a^{n-1}b + C_{n,2}a^{n-2}b^2 + \cdots + C_{n,n}b^n$$

where

$$C_{n,r} = \frac{n!}{r!(n-r)!} \qquad n \geq r \geq 0$$

◆ EXAMPLE 8 Use the binomial formula to expand $(q + p)^3$.

Solution

$$(q + p)^3 = C_{3,0}\, q^3 + C_{3,1}\, q^2p + C_{3,2}\, qp^2 + C_{3,3}\, p^3$$
$$= q^3 + 3q^2p + 3qp^2 + p^3$$ ◆

PROBLEM 8 Use the binomial formula to expand $(q + p)^4$. ◆

◆ EXAMPLE 9 Use the binomial formula to find the fifth term in the expansion of $(q + p)^6$.

Solution The fifth term is given by

$$C_{6,4}\, q^2p^4 = \frac{6!}{4!(6-4)!}\, q^2p^4 = 15q^2p^4$$ ◆

PROBLEM 9 Use the binomial formula to find the third term in the expansion of $(q + p)^7$. ◆

◆ BINOMIAL DISTRIBUTION

We now generalize the discussion of Bernoulli trials to *binomial distributions*. We start by considering a sequence of three Bernoulli trials. Let the random variable X_3 represent the number of successes in three trials, 0, 1, 2, or 3. We are interested in the probability distribution for this random variable.

Which outcomes of an experiment consisting of a sequence of three Bernoulli trials lead to the random variable values 0, 1, 2, and 3, and what are the probabilities associated with these values? Table 3 answers these questions completely.

TABLE 3

SIMPLE EVENT	PROBABILITY OF SIMPLE EVENT	X_3 x successes in 3 trials	$P(X_3 = x)$
FFF	$qqq = q^3$	0	q^3
FFS	$qqp = q^2p$		
FSF	$qpq = q^2p$	1	$3q^2p$
SFF	$pqq = q^2p$		
FSS	$qpp = qp^2$		
SFS	$pqp = qp^2$	2	$3qp^2$
SSF	$ppq = qp^2$		
SSS	$ppp = p^3$	3	p^3

The terms in the last column of Table 3 are the terms in the binomial expansion of $(q + p)^3$, as we saw in Example 8. The last two columns in Table 3 provide a probability distribution for the random variable X_3. Note that both conditions for a probability distribution (see Section 12-1) are met:

1. $0 \leq P(X_3 = x) \leq 1, \quad x \in \{0, 1, 2, 3\}$
2. $1 = 1^3 = (q + p)^3$ — Recall that $q + p = 1$.

$$\begin{aligned} &= C_{3,0}q^3 + C_{3,1}q^2p + C_{3,2}qp^2 + C_{3,3}p^3 \\ &= q^3 + 3q^2p + 3qp^2 + p^3 \\ &= P(X_3 = 0) + P(X_3 = 1) + P(X_3 = 2) + P(X_3 = 3) \end{aligned}$$

Reasoning in the same way for the general case, we see why the probability distribution of a random variable associated with the number of successes in a sequence of n Bernoulli trials is called a *binomial distribution*—the probability of each number is a term in the binomial expansion of $(q + p)^n$. For this reason, a sequence of Bernoulli trials is often referred to as a *binomial experiment*. In terms of a formula, we have Theorem 2:

THEOREM 2

Binomial Distribution

$$\begin{aligned} P(X_n = x) &= P(x \text{ successes in } n \text{ trials}) \\ &= C_{n,x}p^x q^{n-x} \qquad x \in \{0, 1, 2, \ldots, n\} \end{aligned}$$

where p is the probability of success and q is the probability of failure on each trial.

Informally, we will write $P(x)$ in place of $P(X_n = x)$.

◆ **EXAMPLE 10** If a fair coin is tossed 4 times, what is the probability of tossing:

(A) Exactly 2 heads? (B) At least 2 heads?

Solutions (A) Use Theorem 2 with $n = 4$, $x = 2$, $p = \frac{1}{2}$, and $q = \frac{1}{2}$:

$$\begin{aligned} P(2) &= P(X_4 = 2) \\ &= C_{4,2}\left(\frac{1}{2}\right)^2\left(\frac{1}{2}\right)^2 \\ &= \frac{4!}{2!2!}\left(\frac{1}{2}\right)^4 = .375 \end{aligned}$$

(B) Notice how this problem differs from part A. Here we have

$$\begin{aligned} P(X_4 \geq 2) &= P(2) + P(3) + P(4) \\ &= C_{4,2}\left(\frac{1}{2}\right)^2\left(\frac{1}{2}\right)^2 + C_{4,3}\left(\frac{1}{2}\right)^3\left(\frac{1}{2}\right)^1 + C_{4,4}\left(\frac{1}{2}\right)^4\left(\frac{1}{2}\right)^0 \\ &= \frac{4!}{2!2!}\left(\frac{1}{2}\right)^4 + \frac{4!}{3!1!}\left(\frac{1}{2}\right)^4 + \frac{4!}{4!0!}\left(\frac{1}{2}\right)^4 \\ &= .375 + .25 + .0625 = .6875 \end{aligned}$$

◆

PROBLEM 10 If a fair coin is tossed 4 times, what is the probability of tossing:

(A) Exactly 1 head? (B) At most 1 head?

◆

◆ EXAMPLE 11 Suppose a fair die is rolled 3 times and a success on a single roll is considered to be rolling a number divisible by 3.

(A) Write the probability function for the binomial distribution.
(B) Construct a table for this binomial distribution.
(C) Draw a histogram for this binomial distribution.

Solutions (A) $p = \frac{1}{3}$ — Since two numbers out of six are divisible by 3

$q = 1 - p = \frac{2}{3}$

$n = 3$

Hence,

$$P(x) = P(x \text{ successes in 3 trials}) = C_{3,x}\left(\tfrac{1}{3}\right)^x\left(\tfrac{2}{3}\right)^{3-x}$$

(B)

x	P(x)
0	$C_{3,0}(\frac{1}{3})^0(\frac{2}{3})^3 \approx .30$
1	$C_{3,1}(\frac{1}{3})^1(\frac{2}{3})^2 \approx .44$
2	$C_{3,2}(\frac{1}{3})^2(\frac{2}{3})^1 \approx .22$
3	$C_{3,3}(\frac{1}{3})^3(\frac{2}{3})^0 \approx .04$
	1.00

(C)

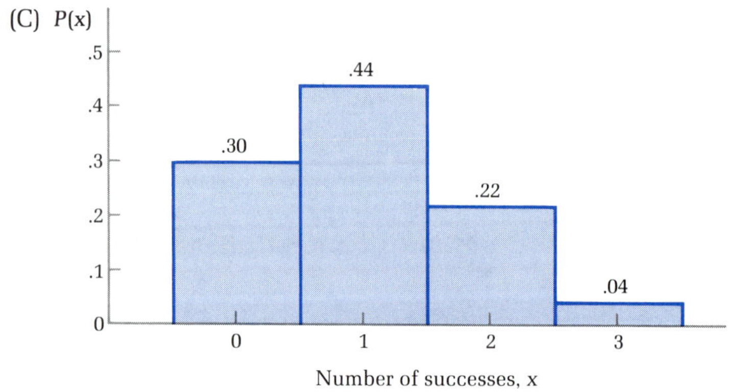

◆

If we actually performed the binomial experiment described in Example 11 a large number of times with a fair die, we would find that we would roll no number divisible by 3 in 3 rolls of a die about 30% of the time, one number divisible by 3 in 3 rolls about 44% of the time, two numbers divisible by 3 in 3 rolls about 22% of the time, and three numbers divisible by 3 in 3 rolls only 4% of the time. Note that the sum of all the probabilities is 1, as it should be.

PROBLEM 11 Repeat Example 11 where the binomial experiment consists of 2 rolls of a die instead of 3 rolls. ◆

We close our discussion of binomial distributions by stating (without proof) formulas for the mean and standard deviation of the random variable associated with the distribution.

Mean and Standard Deviation (Random Variable in a Binomial Distribution)

MEAN $\mu = np$

STANDARD DEVIATION $\sigma = \sqrt{npq}$

◆ EXAMPLE 12 Compute the mean and standard deviation for the random variable in Example 11.

Solution

$$n = 3 \qquad p = \tfrac{1}{3} \qquad q = 1 - \tfrac{1}{3} = \tfrac{2}{3}$$

$$\mu = np = 3(\tfrac{1}{3}) = 1 \qquad \sigma = \sqrt{npq} = \sqrt{3(\tfrac{1}{3})(\tfrac{2}{3})} \approx .82$$ ◆

PROBLEM 12 Compute the mean and standard deviation for the random variable in Problem 11 above. ◆

◆ APPLICATION

Binomial experiments are associated with a wide variety of practical problems: industrial sampling, drug testing, genetics, epidemics, medical diagnosis, opinion polls, analysis of social phenomena, qualifying tests, and so on. Several types of applications are included in Exercise 12-2. We will now consider one application in detail.

◆ EXAMPLE 13 *Medicine* The probability of recovering after a particular type of operation is .5. Let us investigate the binomial distribution involving 8 patients undergoing this operation.

(A) Write the function defining this distribution.
(B) Construct a table for the distribution.
(C) Construct a histogram for the distribution.
(D) Find the mean and standard deviation for the distribution.

Solutions

(A) Letting a recovery be a success, we have

$$p = .5 \qquad q = 1 - p = .5 \qquad n = 8$$

Hence,

$$P(x) = P(\text{Exactly } x \text{ successes in 8 trials}) = C_{8,x}(.5)^x(.5)^{8-x} = C_{8,x}(.5)^8$$

(B)

x	$P(x)$
0	$C_{8,0}(.5)^8 \approx .004$
1	$C_{8,1}(.5)^8 \approx .031$
2	$C_{8,2}(.5)^8 \approx .109$
3	$C_{8,3}(.5)^8 \approx .219$
4	$C_{8,4}(.5)^8 \approx .273$
5	$C_{8,5}(.5)^8 \approx .219$
6	$C_{8,6}(.5)^8 \approx .109$
7	$C_{8,7}(.5)^8 \approx .031$
8	$C_{8,8}(.5)^8 \approx .004$
	$.999 \approx 1$

(C)

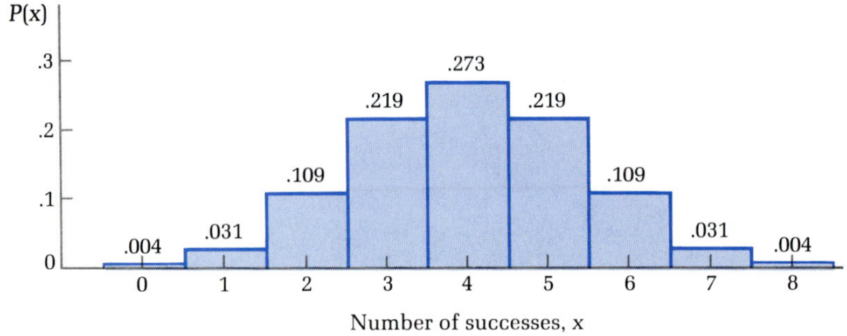

(D) $\mu = np = 8(.5) = 4 \qquad \sigma = \sqrt{npq} = \sqrt{8(.5)(.5)} \approx 1.41$ ◆

PROBLEM 13 Repeat Example 13 for 4 patients. ◆

Answers to Matched Problems

6. $p = \frac{1}{3}, q = \frac{2}{3}$ 7. $p^3q^2 = (\frac{1}{6})^3(\frac{5}{6})^2 \approx .003$

8. $C_{4,0}q^4 + C_{4,1}q^3p + C_{4,2}q^2p^2 + C_{4,3}qp^3 + C_{4,4}p^4$
$= q^4 + 4q^3p + 6q^2p^2 + 4qp^3 + p^4$ 9. $C_{7,2}q^5p^2 = 21q^5p^2$

10. (A) $P(1) = C_{4,1}(\frac{1}{2})^1(\frac{1}{2})^3 = .25$ (B) $P(X_4 \leq 1) = P(0) + P(1)$
$= C_{4,0}(\frac{1}{2})^0(\frac{1}{2})^4 + C_{4,1}(\frac{1}{2})^1(\frac{1}{2})^3$
$= .3125$

11. (A) $P(x) = P(x \text{ successes in 2 trials}) = C_{2,x}(\frac{1}{3})^x(\frac{2}{3})^{2-x}, x \in \{0, 1, 2\}$

(B)

x	$P(x)$
0	$\frac{4}{9} \approx .44$
1	$\frac{4}{9} \approx .44$
2	$\frac{1}{9} \approx .11$

(C)

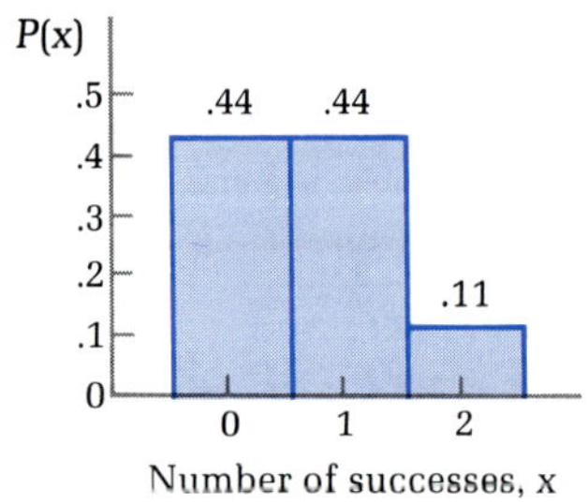

12. $\mu \approx .67$; $\sigma \approx .67$

13. (A) $P(x) = P(\text{Exactly } x \text{ successes in 4 trials}) = C_{4,x}(.5)^4$

(B)

x	P(x)
0	.06
1	.25
2	.38
3	.25
4	.06
	1.00

(C)

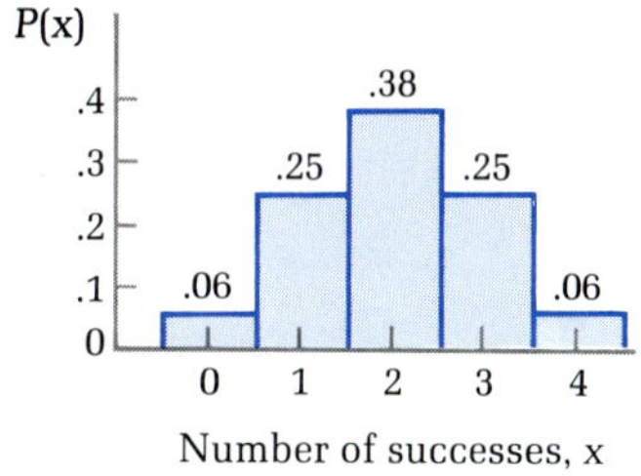

(D) $\mu = 2$; $\sigma = 1$

EXERCISE 12-2

A *Evaluate $C_{n,x}p^xq^{n-x}$ for the following values of n, x, and p:*

1. $n = 3, \quad x = 2, \quad p = \frac{1}{2}$

2. $n = 3, \quad x = 1, \quad p = \frac{1}{2}$

3. $n = 3, \quad x = 0, \quad p = \frac{1}{2}$

4. $n = 3, \quad x = 3, \quad p = \frac{1}{2}$

5. $n = 5, \quad x = 3, \quad p = .4$

6. $n = 5, \quad x = 0, \quad p = .4$

A fair coin is tossed 3 times. What is the probability of obtaining:

7. Exactly 2 heads?

8. Exactly 1 head?

9. 0 heads?

10. 3 heads?

11. At least 2 heads?

12. At least 1 head?

Construct a histogram for each of the binomial distributions in Problems 13–16. Compute the mean and standard deviation for each distribution.

13. $P(x) = C_{2,x}(.3)^x(.7)^{2-x}$

14. $P(x) = C_{2,x}(.7)^x(.3)^{2-x}$

15. $P(x) = C_{4,x}(.5)^x(.5)^{4-x}$

16. $P(x) = C_{6,x}(.5)^x(.5)^{6-x}$

B *A fair die is rolled 4 times. What is the probability of rolling:*

17. Exactly three 2's?

18. Exactly two 3's?

19. No 1's?

20. All 4's?

21. At least one 6?

22. At least one 4?

23. If a baseball player has a batting average of .350, what is the probability that the player will get the following number of hits in the next 4 times at bat?

(A) Exactly 2 hits (B) At least 2 hits

24. If a true–false test with 10 questions is given, what is the probability of scoring:

(A) Exactly 70% just by guessing? (B) 70% or better just by guessing?

Construct a histogram for each of the binomial distributions in Problems 25–28. Compute the mean and standard deviation for each distribution.

25. $P(x) = C_{6,x}(.4)^x(.6)^{6-x}$

26. $P(x) = C_{6,x}(.6)^x(.4)^{6-x}$

27. $P(x) = C_{8,x}(.3)^x(.7)^{8-x}$

28. $P(x) = C_{8,x}(.7)^x(.3)^{8-x}$

C

In Problems 29 and 30, a coin is weighted so that the probability of a head occurring on a single toss is $\frac{3}{4}$. In 5 tosses of the coin, what is the probability of getting:

29. All heads or all tails?

30. Exactly 2 heads or exactly 2 tails?

APPLICATIONS

Business & Economics

31. *Management training.* Each year a company selects a number of employees for a management training program given by a nearby university. On the average, 70% of those sent complete the program. Out of 7 people sent by the company, what is the probability that:

(A) Exactly 5 complete the program?
(B) 5 or more complete the program?

32. *Employee turnover.* If the probability of a new employee in a fast-food chain still being with the company at the end of 1 year is .6, what is the probability that out of 8 newly hired people:

(A) 5 will still be with the company after 1 year?
(B) 5 or more will still be with the company after 1 year?

33. *Quality control.* A manufacturing process produces, on the average, 6 defective items out of 100. To control quality, each day a sample of 10 completed items is selected at random and inspected. If the sample produces more than 2 defective items, then the whole day's output is inspected and the manufacturing process is reviewed. What is the probability of this happening, assuming that the process is still producing 6% defective items?

34. *Guarantees.* A manufacturing process produces, on the average, 3% defective items. The company ships 10 items in each box and wishes to guarantee no more than 1 defective item per box. If this guarantee accompanies each box, what is the probability that the box will fail to satisfy the guarantee?

35. *Quality control.* A manufacturing process produces, on the average, 5 defective items out of 100. To control quality, each day a random sample of 6 completed items is selected and inspected. If a success on a single trial (inspection of 1 item) is finding the item defective, then the inspection of each of the 6 items in the sample constitutes a binomial experiment, which has a binomial distribution.

(A) Write the function defining the distribution.
(B) Construct a table for the distribution.
(C) Draw a histogram.
(D) Compute the mean and standard deviation.

36. *Management training.* Each year a company selects 5 employees for a management training program given at a nearby university. On the average, 40% of those sent complete the course in the top 10% of their class. If we consider an employee finishing in the top 10% of the class a success in a binomial experiment, then for the 5 employees entering the program there exists a binomial distribution involving $P(x$ successes out of 5).

(A) Write the function defining the distribution.
(B) Construct a table for the distribution.
(C) Draw a histogram.
(D) Compute the mean and standard deviation.

Life Sciences

37. *Medical diagnosis.* A person with tuberculosis is given a chest x ray. Four tuberculosis x-ray specialists examine each x ray independently. If each specialist can detect tuberculosis 80% of the time when it is present, what is the probability that at least 1 of the specialists will detect tuberculosis in this person?

38. *Harmful side effects of drugs.* A pharmaceutical laboratory claims that a drug it produces causes serious side effects in 20 people out of 1,000, on the average. To check this claim, a hospital administers the drug to 10 randomly chosen patients and finds that 3 suffer from serious side effects. If the laboratory's claims are correct, what is the probability of the hospital obtaining these results?

39. *Genetics.* The probability that brown-eyed parents, both with the recessive gene for blue, will have a child with brown eyes is .75. If such parents have 5 children, what is the probability that they will have:

(A) All blue-eyed children?
(B) Exactly 3 children with brown eyes?
(C) At least 3 children with brown eyes?

40. *Gene mutations.* The probability of gene mutation under a given level of radiation is 3×10^{-5}. What is the probability of the occurrence of at least 1 gene mutation if 10^5 genes are exposed to this level of radiation?

41. *Epidemics.* If the probability of a person contracting influenza on exposure is .6, consider the binomial distribution for a family of 6 that has been exposed.

(A) Write the function defining the distribution.
(B) Construct a table for the distribution.
(C) Draw a histogram.
(D) Compute the mean and standard deviation.

42. *Side effects of drugs.* The probability that a given drug will produce a serious side effect in a person using the drug is .02. In the binomial distribution for 450 people using the drug, what are the mean and standard deviation?

Social Sciences

43. *Testing.* A multiple-choice test is given with 5 choices for each of 10 questions. What is the probability of passing the test with a grade of 70% or better just by guessing?

44. *Opinion polls.* An opinion poll based on a small sample can be unrepresentative of the population. To see why, let us assume that 40% of the electorate favors a certain candidate. If a random sample of 7 is asked their preference, what is the probability that a majority will favor the candidate?

45. *Testing.* A multiple-choice test is given with 5 choices for each of 5 questions. Answering each of the 5 questions by guessing constitutes a binomial experiment with an associated binomial distribution.

(A) Write the function defining the distribution.
(B) Construct a table for the distribution.
(C) Draw a histogram.
(D) Compute the mean and standard deviation.

46. *Sociology.* The probability that a marriage will end in divorce within 10 years is .4. What are the mean and standard deviation for the binomial distribution involving 1,000 marriages?

47. *Sociology.* If the probability is .60 that a marriage will end in divorce within 20 years after its start, what is the probability that out of 6 couples just married, in the next 20 years:

(A) None will be divorced? (B) All will be divorced?
(C) Exactly 2 will be divorced? (D) At least 2 will be divorced?

SECTION 12-3 Poisson Distributions

- POISSON APPROXIMATION TO THE BINOMIAL DISTRIBUTION
- POISSON DISTRIBUTIONS IN GENERAL

Among the most important probability distributions encountered throughout probability studies are the binomial distribution, the *Poisson distribution*, and the *normal distribution*. The Poisson distribution will be investigated in this section and the normal distribution in Section 12-7. The Poisson distribution was first introduced by the French mathematician Siméon Denis Poisson (1781–1840) in a book on applications of probability theory to lawsuits and criminal trials published in 1837.

POISSON APPROXIMATION TO THE BINOMIAL DISTRIBUTION

Many important binomial applications involve very large values for n and very small values for p. As n increases and p decreases, computations of binomial values become increasingly tedious. The Poisson distribution provides an excellent approximation to the binomial distribution for large n and small p. To appreciate this point, consider the following example. If the probability of being dealt a royal flush (10, jack, queen, king, and ace in one suit) in a 5-card poker hand is $\frac{1}{649{,}740}$, what is the probability of being dealt 2 royal flushes in 649,740 hands of 5 cards? Since each deal is a Bernoulli trial, we have a binomial experiment, and the problem can be solved using the binomial distribution:

$$P(\text{2 royal flushes in 649,740 deals}) = C_{649{,}740,\,2}\left(\frac{1}{649{,}740}\right)^2\left(\frac{649{,}739}{649{,}740}\right)^{649{,}738}$$

$$\approx .1839^*$$

* If your calculator cannot compute $C_{649{,}740,\,2}$ using the $C_{n,x}$ key because of the large value of n, then use the formula

$$C_{649{,}740,\,2} = \frac{649{,}740!}{2!649{,}738!} = \frac{649{,}740 \cdot 649{,}739}{2}$$

Compare this solution with the following Poisson approximation:

$$P(2 \text{ royal flushes in } 649{,}740 \text{ deals}) \approx e^{-1}\left(\frac{1^2}{2!}\right)$$
$$\approx .1839$$

This excellent aproximation is not a coincidence. To understand why the approximation is so good, we start with the binomial distribution and let n increase without bound as we hold np constant at $np = \lambda$ (the Greek letter lambda is customarily used for this constant). To facilitate the process, we convert the formula for the binomial distribution into a more convenient form:

$$\begin{aligned} P(X_n = x) &= C_{n,x}p^x q^{n-x} = \frac{n!}{x!(n-x)!}\,p^x(1-p)^{n-x} \\ &= \frac{n!}{x!(n-x)!}\left(\frac{\lambda}{n}\right)^x\left(1-\frac{\lambda}{n}\right)^{n-x} \qquad \text{Since } p = \lambda/n \\ &= \frac{n(n-1)\cdot\ \cdots\ \cdot(n-x+1)}{n^x}\left(\frac{\lambda^x}{x!}\right)\frac{\left(1-\frac{\lambda}{n}\right)^n}{\left(1-\frac{\lambda}{n}\right)^x} \end{aligned} \qquad (1)$$

To find the limit of this last expression as $n \to \infty$, we proceed by finding the limit of each key part (remembering that x and λ are constant):

1. $$\begin{aligned} \lim_{n\to\infty}\frac{n(n-1)\cdot\ \cdots\ \cdot(n-x+1)}{n^x} &= \lim_{n\to\infty}\left(\frac{n}{n}\cdot\frac{n-1}{n}\cdot\ \cdots\ \cdot\frac{n-x+1}{n}\right) \\ &= \lim_{n\to\infty}\left[\frac{n}{n}\cdot\left(1-\frac{1}{n}\right)\cdot\ \cdots\ \cdot\left(1-\frac{x-1}{n}\right)\right] \\ &= 1 \end{aligned}$$

2. $$\lim_{n\to\infty}\left(1-\frac{\lambda}{n}\right)^x = 1^x = 1$$

3. $$\begin{aligned} \lim_{n\to\infty}\left(1-\frac{\lambda}{n}\right)^n &= \lim_{n\to\infty}\left[\left(1-\frac{\lambda}{n}\right)^{n/\lambda}\right]^\lambda && \text{Let } s = -\lambda/n. \\ &= \lim_{s\to 0}[(1+s)^{-1/s}]^\lambda && n\to\infty \text{ implies } s = -\lambda/n \to 0. \\ &= \lim_{s\to 0}\left[\frac{1}{(1+s)^{1/s}}\right]^\lambda && \text{See Section 5-1.} \\ &= \left[\frac{1}{e}\right]^\lambda = e^{-\lambda} \end{aligned}$$

Using limits 1–3 and expression (1), it follows that if $np = \lambda$, λ a positive constant, then

$$\lim_{n\to\infty} C_{n,x}p^x q^{n-x} = \frac{\lambda^x}{x!}e^{-\lambda} \qquad x = 0, 1, 2, \ldots$$

From the property of limits, we thus have the classic Poisson approximation to the binomial distribution:

Poisson Approximation to the Binomial Distribution

For n large and p small, and $\lambda = np$,

$$C_{n,x}p^x q^{n-x} \approx \frac{\lambda^x}{x!}e^{-\lambda} \qquad x \in \{0, 1, \ldots, n\}$$

RULE-OF-THUMB

Excellent approximations can be obtained when $n \geqslant 100$ and $np \leqslant 10$.

Table 4 and Figure 5 illustrate the Poisson approximation for the binomial distribution when $n = 100$ and $p = .02$. For the Poisson approximation we use $\lambda = np = 100(.02) = 2$.

TABLE 4

Poisson Approximation of the Binomial

x	BINOMIAL $C_{100,x}(.02)^x(.98)^{100-x}$	POISSON $(2^x/x!)e^{-2}$
0	.1326	.1353
1	.2707	.2707
2	.2734	.2707
3	.1823	.1804
4	.0902	.0902
5	.0353	.0361
6	.0114	.0120
7	.0031	.0034
8	.0007	.0009
9	.0002	.0002
⋮	⋮	⋮
100		

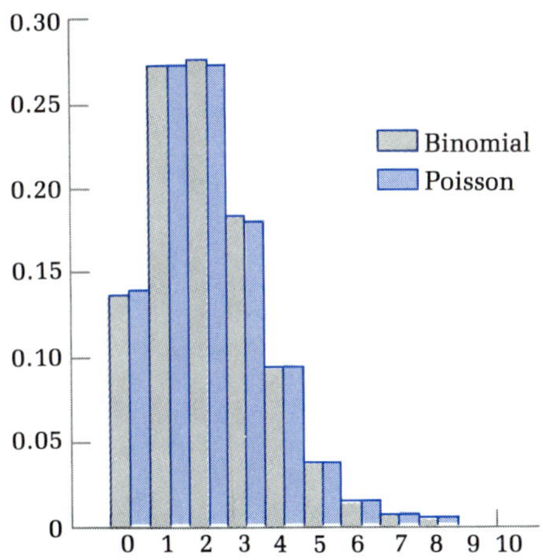

FIGURE 5
Poisson approximation to the binomial distribution

◆ EXAMPLE 14

Quality Control

A company manufactures inexpensive disposable flashlights. The management accepts a manufacturing process that yields, on the average, 2 defective flashlights per 100 produced. In an order for 200 flashlights, what is the probability that:

(A) The order will contain exactly 6 defective flashlights?
(B) The order will contain 6 or more defective flashlights?

Solutions

This problem can be solved using a binomial distribution with $p = .02$ and $n = 200$. But since n is large (≥ 100) and $np = 4 \leq 10$, we can use the Poisson approximation and the calculations will be less tedious:

$$\frac{\lambda^x}{x!}e^{-\lambda} \qquad \text{with } \lambda = np = 4$$

(A) $P(\text{Exactly 6 defective flashlights}) = P(6) \approx \frac{4^6}{6!}e^{-4} \approx .1042$

[*Note:* The direct use of the binomial formula produces $P(6) = C_{200,6}(.02)^6(.98)^{194} \approx .1047$.]

(B) $P(\text{6 or more defective flashlights})$

$$\begin{aligned}
&= P(6 \leq X \leq 200) \\
&= 1 - P(0 \leq X < 6) \qquad \text{Since } P(0) + P(1) + \cdots + P(200) = 1 \\
&= 1 - [P(0) + P(1) + P(2) + P(3) + P(4) + P(5)] \\
&\approx 1 - \left[\frac{4^0}{0!}e^{-4} + \frac{4^1}{1!}e^{-4} + \cdots + \frac{4^5}{5!}e^{-4}\right] \qquad \text{Poisson approximation} \\
&= 1 - e^{-4}\left[\frac{4^0}{0!} + \frac{4^1}{1!} + \frac{4^2}{2!} + \frac{4^3}{3!} + \frac{4^4}{4!} + \frac{4^5}{5!}\right] \\
&\approx 1 - .7851 = .2149
\end{aligned}$$

◆

PROBLEM 14

A company manufactures inexpensive solar credit-card size calculators with company logos for business promotions. The management has accepted a manufacturing process that, on the average, produces 1 defective calculator per 100 produced. For an order of 300 calculators, what is the probability that:

(A) The order will contain no defective calculators?
(B) The order will contain at least 5 defective calculators?

(Compute answers to four decimal places.)

◆

◆ POISSON DISTRIBUTIONS IN GENERAL

We now turn to the Poisson distribution as a probability distribution in its own right. In fact, the Poisson distribution can be applied to many problems that have no direct connection to the binomial distribution. A **Poisson experiment** usually consists of counting the number of times a random event occurs in a given unit of measure such as time, distance, area, volume, weight, and so on. The possible values of the Poisson random variable are 0, 1, 2, ... Some applications that often involve Poisson distributions are the following:

Business & Economics

The number of items taken from inventory per hour

The number of telephone calls arriving at a switchboard per minute

The number of death claims arriving at an insurance company per day

The number of industrial accidents occurring at a company per month

The number of typesetting errors per page

The number of flaws per 1,000 feet of video tape

Life Sciences

The number of parts per million of a toxic substance found in 1 milliliter of water

The number of births or deaths per day in a city

The number of yeast cells suspended in a small drop of liquid

The number of diseased plants per acre

Social Sciences

The number of suicides per month from the Golden Gate Bridge

The number of accidents occurring on a certain stretch of highway per week

The number of automobiles arriving at a toll booth per hour

The number of unscheduled arrivals at a hospital emergency room per day

Three conditions (listed below) must be satisfied for ***X* to be a Poisson random variable.** (To simplify our statements of these conditions, we restrict these statements to time. Similar statements may be made for distance, area, volume, and other units of measure.)

time.

1. The number of occurrences of an event over one unit of time is independent of the number of occurrences in any other nonoverlapping unit of time.
2. The expected or average number of occurrences of an event over any time interval is proportional to the length of the time interval. (For example, if on the average 5 calls arrive at a switchboard per minute, then 10 calls will arrive on the average over a 2 minute period.)
3. Two events cannot occur at exactly the same time.

A Poisson random variable X has the following probability distribution:

Poisson Distribution with Mean and Standard Deviation

$$P(X = x) = P(x \text{ occurrences per unit of measure})$$
$$= \frac{\lambda^x}{x!} e^{-\lambda} \qquad x \in \{0, 1, 2, \ldots\}$$

MEAN $\mu = \lambda$
STANDARD DEVIATION $\sigma = \sqrt{\lambda}$

Informally, we will write $P(x)$ in place of $P(X = x)$.

Most Poisson distributions of practical importance are fairly tightly clustered about the mean. In fact, most of the distribution is often found within 2–3 units of the mean (see Figure 5, where the mean is 2 and the standard deviation is $\sqrt{2}$).

Binomial and Poisson random variables are **discrete random variables.** A random variable is said to be discrete if its set of possible values (range) contains a finite number of elements or infinitely many elements that can be arranged into a sequence of the form $e_1, e_2, e_3, \ldots$. A binomial random variable has a finite discrete range, and the range for a Poisson random variable is discrete and infinite. In Sections 12-4–12-7, we will study *continuous* random variables. Probabilities of events involving discrete random variables are obtained by additions, whereas integrations are required for continuous random variables.

To show that the Poisson distribution is actually a probability distribution, we must show that

1. $0 \leqslant P(X = x) \leqslant 1, \quad x = 0, 1, 2, \ldots$
2. $P(0) + P(1) + P(2) + \cdots = 1$

Since λ^x, $x!$, and $e^{-\lambda}$ are all nonnegative, and $x!$ is never 0, we see that

$$P(X = x) = \frac{\lambda^x}{x!} e^{-\lambda} \geqslant 0$$

which establishes the left-hand part of the inequality in 1. (We will establish the right-hand part of the inequality in 1 after establishing 2.)

To establish 2, we use the following property of the exponential function:

$$e^{\lambda} = \lim_{n\to\infty} \left(\frac{\lambda^0}{0!} + \frac{\lambda^1}{1!} + \frac{\lambda^2}{2!} + \cdots + \frac{\lambda^n}{n!}\right)$$
$$= \frac{\lambda^0}{0!} + \frac{\lambda^1}{1!} + \frac{\lambda^2}{2!} + \cdots$$

Using this fact, we have

$$P(0) + P(1) + P(2) + \cdots = \frac{\lambda^0}{0!}e^{-\lambda} + \frac{\lambda^1}{1!}e^{-\lambda} + \frac{\lambda^2}{2!}e^{-\lambda} + \cdots$$
$$= e^{-\lambda}\left(\frac{\lambda^0}{0!} + \frac{\lambda^1}{1!} + \frac{\lambda^2}{2!} + \cdots\right) = e^{-\lambda}e^{\lambda} = 1$$

To complete the task at hand, we need to establish the right-hand part of the inequality in 1. Since (from 2)

$$P(0) + P(1) + P(2) + \cdots = 1$$

and each term is nonnegative, it follows that $P(X = x) \leq 1$ for each x. Thus, the Poisson distribution is a probability distribution.

We now turn to some examples illustrating the use of the Poisson distribution in interesting applications.

◆ EXAMPLE 15

Arrival Rates

During a business day the switchboard in a large legal firm receives 180 calls per hour, on the average. Assuming that the calls are coming in randomly during the business day, compute the probability that during a particular minute:

(A) No calls will come in. (B) 10 calls will come in.

Solutions

The three conditions of a Poisson random variable are met. Since, on the average, 180 calls arrive per hour (and the calls are coming in randomly), then the average number of calls per minute is $\frac{180}{60} = 3 = \lambda$. We use the Poisson distribution with $\lambda = 3$:

$$P(x) = \frac{\lambda^x}{x!}e^{-\lambda} = \frac{3^x}{x!}e^{-3} \qquad x = 0, 1, 2, \ldots$$

(A) $P(0) = \frac{3^0}{0!}e^{-3} = e^{-3} \approx .0498$

(B) $P(10) = \frac{3^{10}}{10!}e^{-3} \approx .0008$ ◆

PROBLEM 15

In Example 15, compute the probability of the switchboard receiving between 5 and 7 calls, inclusive, in a particular period of 2 minutes. ◆

◆ EXAMPLE 16

Quality Control

Flaws occur randomly in the manufacturing of a certain brand of blank audio cassette tapes, at the average rate of 2 flaws per 1,125 feet of tape produced (60 minutes of tape played at 3.75 inches per second). This flaw rate is acceptable to management. Random samples of 1,125 feet of tape are periodically inspected, and the management has agreed to shut down the manufacturing process for servicing if more than 4 flaws are found in a sample. Assuming the manufacturing process is operating normally:

(A) What is the probability that the plant will be shut down after an inspection?
(B) What is the probability that if you purchase a 60 minute tape, it will have no flaws?
(C) Compute the mean and standard deviation for the distribution.

Solutions The three conditions for a Poisson random variable are satisfied, so we use the Poisson distribution with $\lambda = 2$ (the average number of flaws per 1,125 feet of tape produced):

$$P(x) = \frac{\lambda^x}{x!} e^{-\lambda} = \frac{2^x}{x!} e^{-2} \qquad x = 0, 1, 2, \ldots$$

(A) $P(X > 4) = 1 - P(0 \leq X \leq 4)$ Since $P(0) + P(1) + \cdots = 1$

$$= 1 - [P(0) + P(1) + P(2) + P(3) + P(4)]$$
$$= 1 - e^{-2}\left(\frac{2^0}{0!} + \frac{2^1}{1!} + \frac{2^2}{2!} + \frac{2^3}{3!} + \frac{2^4}{4!}\right)$$
$$\approx 1 - .9473 = .0527$$

Thus, if the manufacturing process is operating normally, then the probability that the plant will be (unnecessarily) shut down for service is .0527. [*Note:* The management's policy of shutting down the manufacturing process for service when more than 4 flaws occur in a sample means that they have chosen to reject the assumption that the manufacturing process is functioning normally when more than 4 flaws occur in a sample. Under this policy, they will make the wrong decision about 5% of the time when the manufacturing process is operating normally.]

(B) $P(0) = \frac{2^0}{0!} e^{-2} = e^{-2} \approx .1353$

(C) Mean: $\mu = \lambda = 2$ Standard deviation: $\sigma = \sqrt{\lambda} = \sqrt{2} \approx 1.4142$ ◆

PROBLEM 16 In Example 16, find the probability of a 60 minute tape chosen at random containing no more than 4 flaws. ◆

Answers to Matched Problems

14. (A) $P(0) \approx \frac{3^0}{0!} e^{-3} \approx .0498$

(B) $P(X \geq 5) \approx 1 - e^{-3}\left(\frac{3^0}{0!} + \frac{3^1}{1!} + \frac{3^2}{2!} + \frac{3^3}{3!} + \frac{3^4}{4!}\right) \approx 1 - .8153 = .1847$

15. $P(5 \le X \le 7) = e^{-6}\left(\frac{6^5}{5!} + \frac{6^6}{6!} + \frac{6^7}{7!}\right) \approx .4589$

16. $e^{-2}\left(\frac{2^0}{0!} + \frac{2^1}{1!} + \frac{2^2}{2!} + \frac{2^3}{3!} + \frac{2^4}{4!}\right) \approx .9473$

(The accepted manufacturing process will produce 60 minute tapes containing no more than 4 flaws about 95% of the time.)

EXERCISE 12-3

Unless instructed to the contrary, compute all answers to four decimal places.

A *In Problems 1–8, compute each Poisson probability.*

1. $P(0) = \frac{4^0}{0!} e^{-4}$

2. $P(0) = \frac{(0.5)^0}{0!} e^{-0.5}$

3. $P(2) = \frac{(0.25)^2}{2!} e^{-0.25}$

4. $P(3) = \frac{1^3}{3!} e^{-1}$

5. $P(X < 3) = e^{-2}\left(\frac{2^0}{0!} + \frac{2^1}{1!} + \frac{2^2}{2!}\right)$

6. $P(X \le 2) = e^{-3}\left(\frac{3^0}{0!} + \frac{3^1}{1!} + \frac{3^2}{2!}\right)$

7. $P(X \ge 3) = 1 - e^{-2}\left(\frac{2^0}{0!} + \frac{2^1}{1!} + \frac{2^2}{2!}\right)$

8. $P(X > 2) = 1 - e^{-3}\left(\frac{3^0}{0!} + \frac{3^1}{1!} + \frac{3^2}{2!}\right)$

B *In Problems 9–12, compute P(3) for both the binomial distribution and the Poisson approximation to the binomial for each of the indicated values of n and p.*

9. $n = 200, \quad p = .01$

10. $n = 1{,}000, \quad p = .003$

11. $n = 10{,}000, \quad p = .000\ 05$

12. $n = 5{,}000, \quad p = .000\ 02$

In Problems 13 and 14, use the Poisson approximation to the binomial distribution to determine the indicated probabilities.

13. The probability of rolling a pair of 1's (snake eyes) in a single roll of a pair of fair dice is $\frac{1}{36}$. What is the probability of:

(A) Rolling no snake eyes in 108 rolls of a pair of fair dice?
(B) Rolling 5 snake eyes in 108 rolls of a pair of fair dice?

14. The probability of drawing the ace of spades from a standard 52-card deck is $\frac{1}{52}$. What is the probability of:

(A) Not drawing an ace of spaces in 104 single-card draws (with replacement) from a 52-card deck?
(B) Drawing an ace of spades 4 times in 104 single-card draws (with replacement) from a 52-card deck?

C *In Problems 15 and 16, find N to the nearest integer so that:*

15. $P(0) = \frac{(N/2{,}500)^0}{0!} e^{-N/2{,}500} = .01$

16. $P(0) = \frac{(N/500)^0}{0!} e^{-N/500} = .05$

Business & Economics

17. *Manufacturing.* A small bottling company bottles a popular brand of mineral water using automated equipment. The process produces, on the average, 5 defective (improperly filled) bottles per 1,000 produced. In an order for 100 bottles, what is the probability of the order:

(A) Containing no defective bottles?
(B) Containing at most 2 defective bottles?
(C) Containing more than 2 defective bottles?

Approximate these binomial probabilities using Poisson approximations.

18. *Manufacturing.* A pencil manufacturing company produces, on the average, 2 defective pencils per 1,000 produced. In an order for 500 pencils, what is the probability that:

(A) There will be no defective pencils?
(B) There will be 4 defective pencils?
(C) There will be at least 4 defective pencils?

Approximate these binomial probabilities using Poisson approximations.

19. *Hospital scheduling.* A community hospital wishes to schedule its delivery room facilities to take care of new births effectively without an excess of staff. On the average, 21 births occur randomly per week. During an 8 hour work shift, what is the probability that:

(A) No birth will occur? (B) 4 births will occur?
(C) 4 or more births will occur?

20. *Loading dock arrivals.* Over a period of several months it was observed that, on the average, 16 trucks arrived randomly at a store's loading dock in an 8 hour business day. Compute the probability that during a given business hour:

(A) No trucks will arrive. (B) 5 trucks will arrive.
(C) 5 or more trucks will arrive.

21. *Quality control.* In manufacturing a certain brand of carpeting, flaws occur randomly at the acceptable average rate of 3 flaws per 10 square yards of carpet. Random samples of 1 square yard of carpet are inspected periodically, and the management has a policy of shutting down and servicing the equipment if an inspection reveals more than 1 flaw in a 1 square yard sample. Assuming the manufacturing process is operating normally:

(A) What is the probability that the plant will be shut down after an inspection?
(B) What is the probability that a given 1 square yard sample will have no flaws?

22. *Quality control.* Wire is produced with an acceptable manufacturing flaw rate of 2 per 1,000 feet of wire produced. Random samples of 100 feet of wire are inspected periodically, and the manufacturing process is shut down for servicing if an inspection reveals more than 1 flaw. Assuming the manufacturing process is operating normally:

(A) What is the probability that the process will be shut down after an inspection?
(B) What is the probability of a given sample having no flaws?

23. *Equipment failure.* A company's records show that their computer is down on the average of about twice per week (5 business days). Compute the probability that during a given business day:

(A) The computer will not be down.
(B) The computer will be down twice.
(C) The computer will be down more than once.

24. *Equipment failure.* The elevator in a large old hotel breaks down, on the average, about 13 times a year. What is the probability that during a given week:

(A) The elevator will not break down?
(B) The elevator will break down once?
(C) The elevator will break down more than once?

25. *Typesetting.* A typesetting company produces, on the average, 1 error for every 15 pages of typesetting. In 30 pages, what is the probability of having:

(A) No typesetting errors? (B) At most 4 typesetting errors?
(C) More than 4 typesetting errors?

26. *Industrial accidents.* The average number of industrial accidents per day in a manufacturing company is 1.2. If accidents occur randomly over the working time, what is the probability that in 1 work week (5 days), there will be:

(A) 10 accidents? (B) At most 3 accidents?
(C) At least 3 accidents?

Life Sciences

27. *Vaccine testing.* In a certain city, on the average, 4% of the adult population will contract a certain strain of influenza during any given year. To test a newly discovered vaccine, 200 healthy adults are randomly selected and are inoculated with the new vaccine. Assuming the vaccine has no effect at all, what is the probability that during the test year:

(A) 2 out of the 200 inoculated adults contract influenza?
(B) No more than 2 out of the 200 inoculated adults contract influenza?

Approximate these binomial probabilities using Poisson approximations.

28. *Life expectancy.* In a stable community where there is little change in mortality rates or size, it is found that at any given time (on the average)

0.4% of the population are over 100 years old. Assuming there has not been a change in size or mortality rates, what is the probability that in a random sample of 500 people from the community:

(A) 5 people are over 100?
(B) More than 4 people are over 100?

Approximate these binomial probabilities using Poisson approximations.

29. *Organic pesticides.* An organic pesticide contains, on the average, 300,000 organisms per liter that are randomly distributed throughout the solution. The pesticide will be sprayed over a crop with a drop size of about $\frac{1}{100{,}000}$ liter.

(A) What is the probability that a single drop will have no organisms in it?
(B) How many organisms must be present per liter (on the average) so that the probability of a single drop not having any organisms will be .01? (Round your answer to the nearest thousand.)

30. *Plant nutrition.* Small slow-release nutrition pellets are randomly mixed throughout a dry planting mix. On the average, the mix contains 1,728 pellets per cubic foot.

(A) What is the probability that 1 cubic inch of the mix will contain no pellets?
(B) How many pellets must be present per cubic foot of mix (on the average) so that the probability of 1 cubic inch of the mix not containing any pellets is .1? (Round your answer to the nearest thousand.)

Social Sciences

31. *Traffic control.* A small river lock, accommodating one pleasure boat at a time, operates daily. It is found that, on the average, 5 boats arrive randomly per day. Compute the probabilities that on a given day:

(A) No boat will arrive. (B) From 4 to 6 boats will arrive.
(C) Fewer than 3 boats will arrive.

32. *Auto safety.* A major freeway interchange in Walnut Creek, California, handles about 280,000 vehicles per day, and, on the average, 1 car crashes every 30 hours. Compute the probability that:

(A) No car will crash in a given 24 hour period.
(B) 3 or more cars will crash in a 24 hour period.

33. *Emergency medical services.* A hospital wants to staff its emergency room from midnight to 6 AM so that adequate service is available most of the time. On the average, it is found that 18 emergency patients arrive randomly during these 6 hours. What is the probability that:

(A) No more than 5 emergency patients arrive in 1 hour?
(B) More than 5 emergency patients arrive in 1 hour?
(C) No more than 2 emergency patients arrive over a 2 hour period?

34. *Suicides.* On the average, about 4 people per year jump from the Golden Gate Bridge to their deaths. The suicides occur randomly over time. What is the probability that:

(A) No more than 2 suicides occur during a given year?
(B) More than 6 suicides occur during a given year?
(C) No more than 4 suicides occur over a 2 year period?

SECTION 12-4 Continuous Random Variables

- CONTINUOUS RANDOM VARIABLES
- PROBABILITY DENSITY FUNCTION
- COMPARING PROBABILITY DISTRIBUTION FUNCTIONS AND PROBABILITY DENSITY FUNCTIONS
- CUMULATIVE PROBABILITY DISTRIBUTION FUNCTION

◆ CONTINUOUS RANDOM VARIABLES

Suppose that during an 8 hour business day the switchboard of a large law firm receives an average of 3 calls per minute. As we saw in the preceding section, the number of calls received per minute is a Poisson random variable with the following discrete set of possible values (range):

$$S = \{0, 1, 2, \ldots\}$$

Now we want to consider a random variable X representing the time interval (in minutes) between incoming calls. What are the values that this random variable can assume? It is reasonable to assume that all the values of X must lie in the interval $[0, 480]$, since there are 480 minutes in an 8 hour business day. However, we cannot use the discrete set

$$S = \{0, 1, 2, \ldots, 480\}$$

for the range of this random variable, because this set excludes noninteger values such as 2.5 minutes or $\sqrt{35}$ minutes. Although it may be very unlikely that X will be close to 0 or to 480, theoretically there is no reason to exclude any of the values in the interval $[0, 480]$ as a possibility for the time between incoming calls. Thus, we assume that the range of X is the interval of real numbers $[0, 480]$, and we call X a *continuous random variable.*

Continuous Random Variable

A **continuous random variable** is a random variable with a set of possible values (range) that is an interval of real numbers. This interval may be open or closed, and it may be bounded or unbounded.

The term *continuous* is not used in the same sense here as it was used in Section 3-1. In this case, it refers to the fact that the values of the random variable form a continuous set of real numbers, such as $[0, \infty)$, rather than a discrete set, such as $\{5, 9, 17\}$ or $\{1, 3, 5, \ldots\}$.

◆ PROBABILITY DENSITY FUNCTION

If X is a discrete random variable, then the probability that X lies in a given interval can be computed by addition. For example, if X is a binomial or Poisson random variable, then

$$P(0 \le X \le 2) = P(X = 0) + P(X = 1) + P(X = 2)$$

The probability distribution function for X is used to evaluate each of the probabilities in the sum on the right and the results are added.

If X is a continuous random variable, the same approach will not work. Since X can now assume any real number in the interval $[0, 2]$, it is impossible to write $P(0 \le X \le 2)$ as a finite or even an infinite sum. (What would be the second term in such a sum? That is, what is the "next" real number after 0? Think about this.) Instead, we introduce a new type of function, called a *probability density function,* and use integrals involving this function to compute the probability that a continuous random variable X lies in a given interval. For example, we might use a probability density function to find the probability that the actual amount of soda in a 12 ounce can is between 11.9 and 12.1 ounces, or that the speed of a car involved in an accident was between 60 and 65 miles per hour.

For convenience in stating definitions and formulas, we will assume that the value of a continuous random variable can be any real number; that is, the range is $(-\infty, \infty)$.

Probability Density Function

The function $f(x)$ is a **probability density function** for a continuous random variable X if:

1. $f(x) \geqslant 0$ for all $x \in (-\infty, \infty)$
2. $\int_{-\infty}^{\infty} f(x)\, dx = 1$
3. The probability that X lies in the interval $[c, d]$ is given by

$$P(c \leqslant X \leqslant d) = \int_c^d f(x)\, dx$$

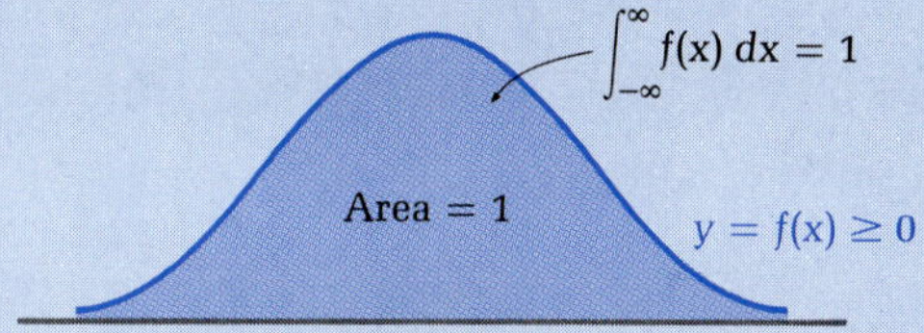

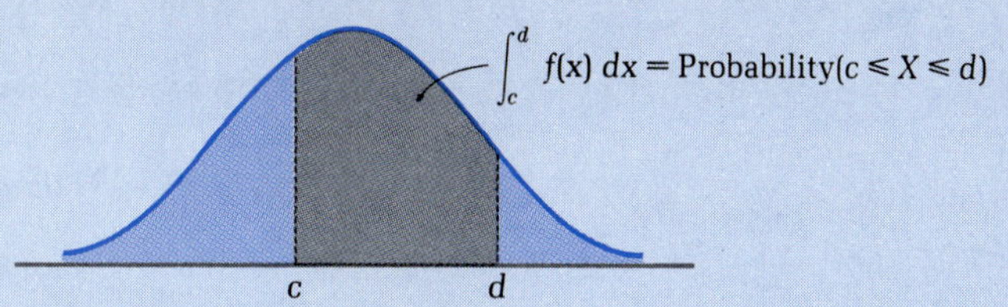

Range of $X = (-\infty, \infty) =$ Domain of f

◆ EXAMPLE 17 Let: $f(x) = \begin{cases} 12x^2 - 12x^3 & \text{if } 0 \leqslant x \leqslant 1 \\ 0 & \text{otherwise} \end{cases}$

(A) Verify that f satisfies the first two conditions for a probability density function.

(B) Compute $P(\frac{1}{4} \leqslant X \leqslant \frac{3}{4})$, $P(X \leqslant \frac{1}{2})$, $P(X \geqslant \frac{2}{3})$, and $P(X = \frac{1}{3})$.

Solutions (A) For $0 \leqslant x \leqslant 1$, we have $f(x) = 12x^2 - 12x^3 = 12x^2(1 - x) \geqslant 0$. Since $f(x) = 0$ for all other values of x, it follows that $f(x) \geqslant 0$ for all x. Also,

$$\int_{-\infty}^{\infty} f(x)\, dx = \int_0^1 (12x^2 - 12x^3)\, dx = (4x^3 - 3x^4)\Big|_0^1 = (4 - 3) - (0) = 1$$

(B) $P(\frac{1}{4} \leqslant X \leqslant \frac{3}{4}) = \int_{1/4}^{3/4} f(x)\, dx$

$$= \int_{1/4}^{3/4} (12x^2 - 12x^3)\, dx$$

$$= (4x^3 - 3x^4)\Big|_{1/4}^{3/4}$$

$$= \tfrac{189}{256} - \tfrac{13}{256} = \tfrac{11}{16}$$

$$P(X \leq \tfrac{1}{2}) = \int_{-\infty}^{1/2} f(x)\,dx \qquad \text{Note that } f(x) = 0 \text{ for } x < 0.$$

$$= \int_0^{1/2} (12x^2 - 12x^3)\,dx$$

$$= (4x^3 - 3x^4)\Big|_0^{1/2}$$

$$= \tfrac{5}{16}$$

$$P(X \geq \tfrac{2}{3}) = \int_{2/3}^{\infty} f(x)\,dx \qquad \text{Note that } f(x) = 0 \text{ for } x > 1.$$

$$= \int_{2/3}^{1} (12x^2 - 12x^3)\,dx$$

$$= (4x^3 - 3x^4)\Big|_{2/3}^{1}$$

$$= 1 - \tfrac{16}{27} = \tfrac{11}{27}$$

$$P(X = \tfrac{1}{3}) = \int_{1/3}^{1/3} f(x)\,dx = 0 \qquad \text{Definition, page 411}$$

◆

PROBLEM 17 Let: $f(x) = \begin{cases} 6x - 6x^2 & \text{if } 0 \leq x \leq 1 \\ 0 & \text{otherwise} \end{cases}$

(A) Verify that f satisfies the first two conditions for a probability density function.

(B) Compute $P(\frac{1}{3} \leq X \leq \frac{2}{3})$, $P(X \leq \frac{1}{5})$, $P(X \geq \frac{1}{2})$, $P(X = \frac{1}{4})$. ◆

◆ COMPARING PROBABILITY DISTRIBUTION FUNCTIONS AND PROBABILITY DENSITY FUNCTIONS

The last probability in Example 17 illustrates a fundamental difference between discrete and continuous random variables. In the discrete case, there is a *probability distribution* $P(x)$ that gives the probability of each possible value of the random variable. Thus, if c is one of the values of the random variable, then $P(X = c) = P(c)$. In the continuous case, the *integral of the probability density function* $f(x)$ gives the probability that the outcome lies in a certain interval. If c is any real number, then the probability that the outcome is *exactly* c is

$$P(X = c) = P(c \leq X \leq c) = \int_c^c f(x)\,dx = 0$$

Thus, $P(X = c) = 0$ for *any number* c and, since $f(c)$ is certainly not 0 for all values of c, we see that $f(x)$ does not play the same role for a continuous random variable as $P(x)$ does for a discrete random variable.

The fact that $P(X = c) = 0$ also implies that excluding either end point from an interval does not change the probability that the random variable lies in that interval; that is,

$$\begin{aligned} P(a < X < b) &= P(a < X \leq b) = P(a \leq X < b) = P(a \leq X \leq b) \\ &= \int_a^b f(x)\,dx \end{aligned}$$

◆ EXAMPLE 18 Use the probability density function in Example 17 to compute $P(.1 < X \leq .2)$ and $P(X > .9)$.

Solution

$$f(x) = \begin{cases} 12x^2 - 12x^3 & \text{if } 0 \leq x \leq 1 \\ 0 & \text{otherwise} \end{cases}$$

$$\begin{aligned} P(.1 < X \leq .2) &= \int_{.1}^{.2} f(x)\,dx \\ &= \int_{.1}^{.2} (12x^2 - 12x^3)\,dx \\ &= (4x^3 - 3x^4)\Big|_{.1}^{.2} \\ &= .0272 - .0037 = .0235 \end{aligned}$$

$$\begin{aligned} P(X > .9) &= \int_{.9}^{\infty} f(x)\,dx \\ &= \int_{.9}^{1} (12x^2 - 12x^3)\,dx \\ &= (4x^3 - 3x^4)\Big|_{.9}^{1} \\ &= 1 - .9477 = .0523 \end{aligned}$$

◆

PROBLEM 18 Use the probability density function in Problem 17 to compute $P(.2 \leq X < .4)$ and $P(X < .8)$. ◆

If $f(x)$ is the probability density function in Example 17, notice that $f(\frac{2}{3}) = \frac{16}{9} > 1$. Thus, a probability density function can assume values larger than 1. This illustrates another difference between probability density functions and probability distribution functions. In terms of inequalities, a probability distribution function must always satisfy $0 \leq P(x) \leq 1$, while a probability density function need only satisfy $f(x) \geq 0$. Despite these differences, we shall see that there are many similarities in the application of probability distribution functions and probability density functions.

◆ EXAMPLE 19 *Shelf-Life*

The shelf-life (in months) of a certain drug is a continuous random variable with probability density function (see the figure)

$$f(x) = \begin{cases} 50/(x+50)^2 & \text{if } x \geq 0 \\ 0 & \text{otherwise} \end{cases}$$

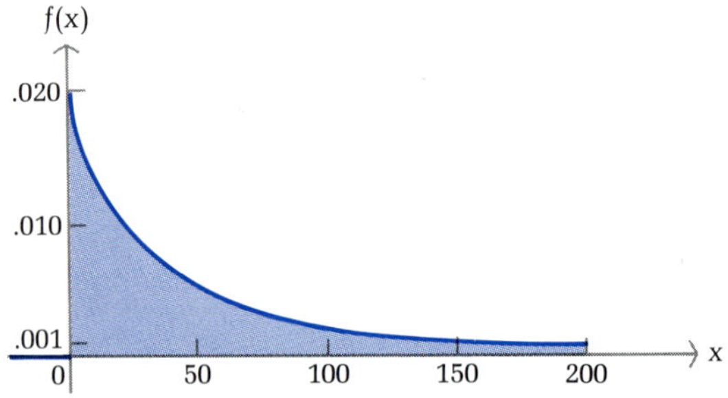

Find the probability that the drug has a shelf-life of:

(A) Between 10 and 20 months (B) At most 30 months
(C) Over 25 months

Solutions

(A) $$P(10 \leq X \leq 20) = \int_{10}^{20} f(x)\,dx = \int_{10}^{20} \frac{50}{(x+50)^2}\,dx = \frac{-50}{x+50}\Bigg|_{10}^{20}$$
$$= \left(-\tfrac{50}{70}\right) - \left(-\tfrac{50}{60}\right) = \tfrac{5}{42}$$

(B) $$P(X \leq 30) = \int_{-\infty}^{30} f(x)\,dx = \int_{0}^{30} \frac{50}{(x+50)^2}\,dx = \frac{-50}{x+50}\Bigg|_{0}^{30}$$
$$= \left(-\tfrac{50}{80}\right) - (-1) = \tfrac{3}{8}$$

(C) $$P(X > 25) = \int_{25}^{\infty} f(x)\,dx = \int_{25}^{\infty} \frac{50}{(x+50)^2}\,dx$$

This improper integral can be evaluated directly using the techniques discussed in Section 7-5. However, there is another method that does not involve evaluation of any improper integrals. Since f is a probability density function, we can write

$$1 = \int_{-\infty}^{\infty} f(x)\,dx$$
$$= \int_{0}^{\infty} \frac{50}{(x+50)^2}\,dx \qquad \int_a^b f(x)\,dx = \int_a^c f(x)\,dx + \int_c^b f(x)\,dx$$
$$= \int_{0}^{25} \frac{50}{(x+50)^2}\,dx + \int_{25}^{\infty} \frac{50}{(x+50)^2}\,dx$$

Solving this last equation for $\int_{25}^{\infty} [50/(x+50)^2]\, dx$, we have

$$\int_{25}^{\infty} \frac{50}{(x+50)^2}\, dx = 1 - \int_{0}^{25} \frac{50}{(x+50)^2}\, dx$$

$$= 1 - \left.\frac{-50}{x+50}\right|_{0}^{25}$$

$$= 1 - \left(\tfrac{-50}{75}\right) - 1 = \tfrac{2}{3}$$

Thus, $P(X > 25) = \frac{2}{3}$. ◆

PROBLEM 19 In Example 19 find the probability that the drug has a shelf-life of:

(A) Between 50 and 100 months (B) At most 20 months
(C) Over 10 months ◆

◆ CUMULATIVE PROBABILITY DISTRIBUTION FUNCTION

Each time we compute the probability for a continuous random variable, we must find the antiderivative of the probability density function. This antiderivative is used so often that it is convenient to give it a name.

Cumulative Probability Distribution Function

If f is a probability density function, then the associated **cumulative probability distribution function *F*** is defined by

$$F(x) = P(X \le x) = \int_{-\infty}^{x} f(t)\, dt$$

Furthermore,

$$P(c \le X \le d) = F(d) - F(c)$$

Figure 6 (page 830) gives a geometric interpretation of these ideas.

Notice that $F(x) = \int_{-\infty}^{x} f(t)\, dt$ is a function of x, the upper limit of integration, not t, the variable in the integrand. We state some important properties of cumulative probability distribution functions in the next box. These properties follow directly from the fact that $F(x)$ can be interpreted geometrically as the area under the graph of $y = f(t)$ from $-\infty$ to x (see Figure 6B).

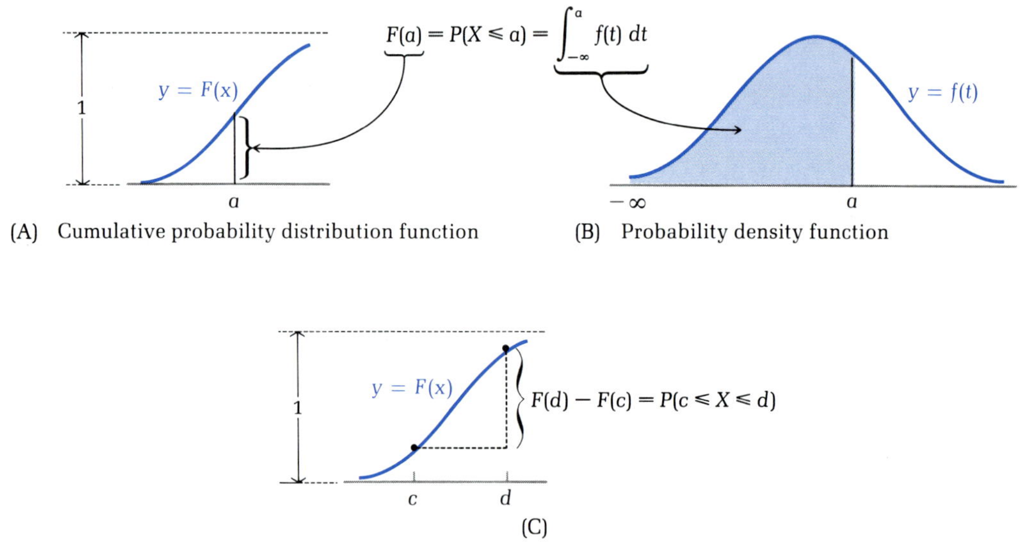

FIGURE 6

Properties of Cumulative Probability Distribution Functions

If f is a probability density function and

$$F(x) = \int_{-\infty}^{x} f(t)\, dt$$

is the associated cumulative probability distribution function, then:

1. $F'(x) = f(x)$ wherever f is continuous
2. $0 \leq F(x) \leq 1, \quad -\infty < x < \infty$
3. $F(x)$ is nondecreasing on $(-\infty, \infty)$*

◆ EXAMPLE 20 Find the cumulative probability distribution function for the probability density function in Example 17, and use it to compute $P(.1 \leq X \leq .9)$.

Solution If $x < 0$, then

$$F(x) = \int_{-\infty}^{x} f(t)\, dt \qquad f(x) = \begin{cases} 12x^2 - 12x^3 & \text{if } 0 \leq x \leq 1 \\ 0 & \text{otherwise} \end{cases}$$

$$= \int_{-\infty}^{x} 0\, dt = 0$$

* A function $F(x)$ is nondecreasing on (a, b) if $F(x_1) \leq F(x_2)$ for $a < x_1 < x_2 < b$.

If $0 \leq x \leq 1$, then

$$F(x) = \int_{-\infty}^{x} f(t)\,dt = \int_{-\infty}^{0} f(t)\,dt + \int_{0}^{x} f(t)\,dt$$

$$= 0 + \int_{0}^{x} (12t^2 - 12t^3)\,dt = (4t^3 - 3t^4)\Big|_0^x$$

$$= 4x^3 - 3x^4$$

If $x > 1$, then

$$F(x) = \int_{-\infty}^{x} f(t)\,dt = \int_{-\infty}^{0} f(t)\,dt + \int_{0}^{1} f(t)\,dt + \int_{1}^{x} f(t)\,dt$$

$$= 0 + 1 + 0 = 1$$

Thus,

$$F(x) = \begin{cases} 0 & \text{if } x < 0 \\ 4x^3 - 3x^4 & \text{if } 0 \leq x \leq 1 \\ 1 & \text{if } x > 1 \end{cases}$$

And

$$P(.1 \leq X \leq .9) = F(.9) - F(.1) = .9477 - .0037 = .944$$

See the figure.

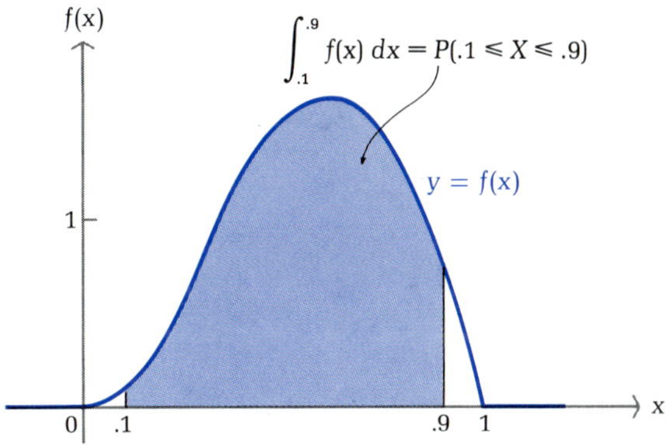

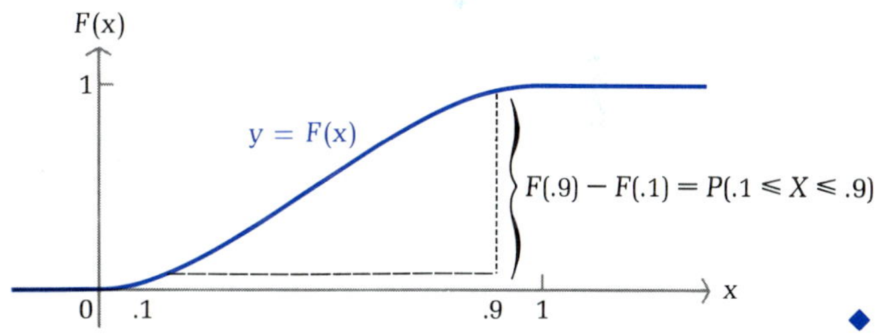

◆

PROBLEM 20 Find the cumulative probability distribution function for the probability density function in Problem 17, and use it to compute $P(.3 \leq X \leq .7)$. ◆

◆ EXAMPLE 21 *Shelf-Life*

Returning to the discussion of the shelf-life of a drug in Example 19, suppose a pharmacist wants to be 95% certain that the drug is still good when it is sold. How long is it safe to leave the drug on the shelf?

Solution Let x be the number of months the drug has been on the shelf when it is sold. The probability that the shelf-life of the drug is less than the number of months it has been sitting on the shelf is $P(0 \leq X \leq x)$. The pharmacist wants this probability to be .05. Thus, we must solve the equation $P(0 \le X \le x) = .05$ for x. First, we will find the cumulative probability distribution function F. For $x < 0$, we see that $F(x) = 0$. For $x \ge 0$,

$$F(x) = \int_0^x \frac{50}{(50 + t)^2}\,dt = \frac{-50}{50 + t}\bigg|_0^x = \frac{-50}{50 + x} - (-1) = 1 - \frac{50}{50 + x}$$
$$= \frac{x}{50 + x}$$

Thus,

$$F(x) = \begin{cases} 0 & \text{if } x < 0 \\ x/(50 + x) & \text{if } x \ge 0 \end{cases}$$

Now, to solve the equation $P(0 \le X \le x) = .05$, we solve

$$F(x) - F(0) = .05 \qquad F(0) = 0$$
$$\frac{x}{50 + x} = .05$$
$$x = 2.5 + .05x$$
$$.95x = 2.5$$
$$x \approx 2.6$$

If the drug is sold during the first 2.6 months it is on the shelf, then the probability that it is still good is .95. ◆

PROBLEM 21 Repeat Example 21 if the pharmacist wants the probability that the drug is still good to be .99. ◆

Answers to Matched Problems

17. (B) $\frac{13}{27}$; $\frac{13}{125}$; $\frac{1}{2}$; 0 18. .248; .896 19. (A) $\frac{1}{6}$ (B) $\frac{2}{7}$ (C) $\frac{5}{6}$

20. $F(x) = \begin{cases} 0 & \text{if } x < 0 \\ 3x^2 - 2x^3 & \text{if } 0 \le x \le 1 \\ 1 & \text{if } x > 1 \end{cases}$

$P(.3 \le X \le .7) = .568$

21. Approx. $\frac{1}{2}$ month, or 15 days

EXERCISE 12-4

A

In Problems 1 and 2, graph f, and show that f satisfies the first two conditions for a probability density function.

1. $f(x) = \begin{cases} \frac{1}{8}x & \text{if } 0 \le x \le 4 \\ 0 & \text{otherwise} \end{cases}$

2. $f(x) = \begin{cases} \frac{1}{9}x^2 & \text{if } 0 \le x \le 3 \\ 0 & \text{otherwise} \end{cases}$

3. Use the function in Problem 1 to find the indicated probabilities. Illustrate each probability with a graph.

(A) $P(1 \le X \le 3)$ (B) $P(X \le 2)$ (C) $P(X > 3)$

4. Use the function in Problem 2 to find the indicated probabilities. Illustrate each probability with a graph.

(A) $P(1 \leq X \leq 2)$ (B) $P(X \geq 1)$ (C) $P(X < 2)$

5. Use the function in Problem 1 to find the indicated probabilities.

(A) $P(X = 1)$ (B) $P(X > 5)$ (C) $P(X < 5)$

6. Use the function in Problem 2 to find the indicated probabilities.

(A) $P(X = 2)$ (B) $P(X > 4)$ (C) $P(X < 4)$

7. Find and graph the cumulative probability distribution function associated with the function in Problem 1.

8. Find and graph the cumulative probability distribution function associated with the function in Problem 2.

9. Use the cumulative probability distribution function from Problem 7 to find the indicated probabilities.

(A) $P(2 \leq X \leq 4)$ (B) $P(0 < X < 2)$

10. Use the cumulative probability distribution function from Problem 8 to find the indicated probabilities.

(A) $P(0 \leq X \leq 1)$ (B) $P(2 < X < 3)$

11. Use the cumulative probability distribution function from Problem 7 to find the value of x that satisfies each equation.

(A) $P(0 \leq X \leq x) = \frac{1}{4}$ (B) $P(0 \leq X \leq x) = \frac{1}{9}$

12. Use the cumulative probability distribution function from Problem 8 to find the value of x that satisfies each equation.

(A) $P(0 \leq X \leq x) = \frac{1}{8}$ (B) $P(0 \leq X \leq x) = \frac{1}{64}$

B

In Problems 13 and 14, graph f, and show that f satisfies the first two conditions for a probability density function.

13. $f(x) = \begin{cases} 2/(1+x)^3 & \text{if } x \geq 0 \\ 0 & \text{otherwise} \end{cases}$

14. $f(x) = \begin{cases} 2/(2+x)^2 & \text{if } x \geq 0 \\ 0 & \text{otherwise} \end{cases}$

15. Use the function in Problem 13 to find the indicated probabilities.

(A) $P(1 \leq X \leq 4)$ (B) $P(X > 3)$ (C) $P(X \leq 2)$

16. Use the function in Problem 14 to find the indicated probabilities.

(A) $P(2 \leq X \leq 8)$ (B) $P(X \geq 3)$ (C) $P(X < 1)$

17. Find and graph the cumulative probability distribution function associated with the function in Problem 13.

18. Find and graph the cumulative probability distribution function associated with the function in Problem 14.

19. Use the cumulative probability distribution function from Problem 17 to find the value of x that satisfies each equation.

(A) $P(0 \leqslant X \leqslant x) = \frac{3}{4}$ (B) $P(X \geqslant x) = \frac{1}{16}$

20. Use the cumulative probability distribution function from Problem 18 to find the value of x that satisfies each equation.

(A) $P(0 \leqslant X \leqslant x) = \frac{3}{4}$ (B) $P(X > x) = \frac{1}{3}$

In Problems 21–24, find the associated cumulative probability distribution function. Graph both functions (on separate sets of axes).

21. $f(x) = \begin{cases} \frac{3}{2}x - \frac{3}{4}x^2 & \text{if } 0 \leqslant x \leqslant 2 \\ 0 & \text{otherwise} \end{cases}$

22. $f(x) = \begin{cases} \frac{3}{4} - \frac{3}{4}x^2 & \text{if } -1 \leqslant x \leqslant 1 \\ 0 & \text{otherwise} \end{cases}$

23. $f(x) = \begin{cases} \frac{1}{2} + \frac{1}{2}x^3 & \text{if } -1 \leqslant x \leqslant 1 \\ 0 & \text{otherwise} \end{cases}$

24. $f(x) = \begin{cases} \frac{3}{4} - \frac{3}{8}\sqrt{x} & \text{if } 0 \leqslant x \leqslant 4 \\ 0 & \text{otherwise} \end{cases}$

In Problems 25–28, use a graphic calculator or a computer to approximate (to two decimal places) the value of x that satisfies the given equation for the indicated cumulative probability distribution function F(x).

25. $P(0 \leqslant X \leqslant x) = .2$ for F(x) from Problem 21
26. $P(-1 \leqslant X \leqslant x) = .4$ for F(x) from Problem 22
27. $P(-1 \leqslant X \leqslant x) = .6$ for F(x) from Problem 23
28. $P(0 \leqslant X \leqslant x) = .7$ for F(x) from Problem 24

In Problems 29–32, find the associated cumulative probability function, and use it to find the indicated probability.

29. Find $P(1 \leqslant X \leqslant 2)$ for

$$f(x) = \begin{cases} \ln x & \text{if } 1 \leqslant x \leqslant e \\ 0 & \text{otherwise} \end{cases}$$

30. Find $P(1 \leqslant X \leqslant 2)$ for

$$f(x) = \begin{cases} 3x/(8\sqrt{1+x}) & \text{if } 0 \leqslant x \leqslant 3 \\ 0 & \text{otherwise} \end{cases}$$

31. Find $P(X \geqslant 1)$ for

$$f(x) = \begin{cases} xe^{-x} & \text{if } x \geqslant 0 \\ 0 & \text{otherwise} \end{cases}$$

32. Find $P(X \geqslant e)$ for

$$f(x) = \begin{cases} (\ln x)/x^2 & \text{if } x \geqslant 1 \\ 0 & \text{otherwise} \end{cases}$$

C *In Problems 33–36, F(x) is the cumulative probability distribution function for a continuous random variable X. Find the probability density function f(x) associated with each F(x).*

33. $F(x) = \begin{cases} 0 & \text{if } x < 0 \\ x^2 & \text{if } 0 \leqslant x \leqslant 1 \\ 1 & \text{if } x > 1 \end{cases}$

34. $F(x) = \begin{cases} 0 & \text{if } x < 1 \\ \frac{1}{2}x - \frac{1}{2} & \text{if } 1 \leqslant x \leqslant 3 \\ 1 & \text{if } x > 3 \end{cases}$

35. $F(x) = \begin{cases} 0 & \text{if } x < 0 \\ 6x^2 - 8x^3 + 3x^4 & \text{if } 0 \leqslant x \leqslant 1 \\ 1 & \text{if } x > 1 \end{cases}$

36. $F(x) = \begin{cases} 1 - (1/x^3) & \text{if } x \geqslant 1 \\ 0 & \text{otherwise} \end{cases}$

In Problems 37 and 38, find the associated cumulative distribution function.

37. $f(x) = \begin{cases} x & \text{if } 0 \leq x \leq 1 \\ 2 - x & \text{if } 1 < x \leq 2 \\ 0 & \text{otherwise} \end{cases}$

38. $f(x) = \begin{cases} \frac{1}{4} & \text{if } 0 \leq x \leq 1 \\ \frac{1}{2} & \text{if } 1 < x \leq 2 \\ \frac{1}{4} & \text{if } 2 < x \leq 3 \\ 0 & \text{otherwise} \end{cases}$

APPLICATIONS

Business & Economics

39. *Electricity consumption.* The daily demand for electricity (in millions of kilowatt-hours) in a large city is a continuous random variable with probability density function

$$f(x) = \begin{cases} .2 - .02x & \text{if } 0 \leq x \leq 10 \\ 0 & \text{otherwise} \end{cases}$$

(A) What is the probability that the daily demand for electricity is less than 8 million kilowatt-hours?

(B) What is the probability that 5 million kilowatt-hours will not be sufficient to meet the daily demand?

40. *Gasoline consumption.* The daily demand for gasoline (in millions of gallons) in a large city is a continuous random variable with probability density function

$$f(x) = \begin{cases} .4 - .08x & \text{if } 0 \leq x \leq 5 \\ 0 & \text{otherwise} \end{cases}$$

(A) What is the probability that the daily demand is less than 2 million gallons?

(B) What is the probability that 3 million gallons will not be sufficient to meet the daily demand?

41. *Time-sharing.* In a computer time-sharing network, the time it takes (in seconds) to respond to a user's request is a continuous random variable with probability density function given by

$$f(x) = \begin{cases} \frac{1}{10}e^{-x/10} & \text{if } x \geq 0 \\ 0 & \text{otherwise} \end{cases}$$

(A) What is the probability that the computer responds within 1 second?

(B) What is the probability that a user must wait over 4 seconds for a response?

42. *Waiting time.* The time (in minutes) a customer must wait in line at a bank is a continuous random variable with probability density function given by

$$f(x) = \begin{cases} \frac{1}{2}e^{-x/2} & \text{if } x \geq 0 \\ 0 & \text{otherwise} \end{cases}$$

(A) What is the probability that a customer waits less than 3 minutes?

(B) What is the probability that a customer waits more than 5 minutes?

43. *Demand.* The weekly demand for hamburger (in thousands of pounds) for a chain of supermarkets is a continuous random variable with probability density function given by

$$f(x) = \begin{cases} .003x\sqrt{100 - x^2} & \text{if } 0 \leq x \leq 10 \\ 0 & \text{otherwise} \end{cases}$$

(A) What is the probability that more than 4,000 pounds of hamburger are demanded?

(B) The manager of the meat department orders 8,000 pounds of hamburger. What is the probability that the demand will not exceed this amount?

(C) The manager wants the probability that the demand does not exceed the amount ordered to be .9. How much hamburger meat should be ordered?

44. *Demand.* The demand for a weekly sports magazine (in thousands of copies) in a certain city is a continuous random variable with probability density function given by

$$f(x) = \begin{cases} \frac{3}{125}x\sqrt{25 - x^2} & \text{if } 0 \leq x \leq 5 \\ 0 & \text{otherwise} \end{cases}$$

(A) The magazine's distributor in this city orders 3,000 copies of the magazine. What is the probability that the demand exceeds this number?

(B) What is the probability that the demand does not exceed 4,000 copies?

(C) If the distributor wants to be 95% certain that the demand does not exceed the number ordered, how many copies should be ordered?

Life Sciences

45. *Life expectancy.* The life expectancy (in minutes) of a certain microscopic organism is a continuous random variable with probability density function given by

$$f(x) = \begin{cases} \frac{1}{5{,}000}(10x^3 - x^4) & \text{if } 0 \leq x \leq 10 \\ 0 & \text{otherwise} \end{cases}$$

(A) What is the probability that an organism lives for at least 7 minutes?

(B) What is the probability that an organism lives for at most 5 minutes?

46. *Life expectancy.* The life expectancy (in months) of plants of a certain species is a continuous random variable with probability density function given by

$$f(x) = \begin{cases} \frac{1}{36}(6x - x^2) & \text{if } 0 \leq x \leq 6 \\ 0 & \text{otherwise} \end{cases}$$

(A) What is the probability that one of these plants survives for at least 4 months?

(B) What is the probability that one of these plants survives for at most 5 months?

47. *Shelf-life.* The shelf-life (in days) of a perishable drug is a continuous random variable with probability density function given by

$$f(x) = \begin{cases} 800x/(400 + x^2)^2 & \text{if } x \geq 0 \\ 0 & \text{otherwise} \end{cases}$$

(A) What is the probability that the drug has a shelf-life of at most 20 days?
(B) What is the probability that the shelf-life exceeds 15 days?
(C) If the user wants the probability that the drug is still good to be .8, when is the last time it should be used?

48. *Shelf-life.* Repeat Problem 47 if

$$f(x) = \begin{cases} 200x/(100 + x^2)^2 & \text{if } x \geq 0 \\ 0 & \text{otherwise} \end{cases}$$

Social Sciences

49. *Learning.* The number of words per minute a beginner can type after 1 week of practice is a continuous random variable with probability density function given by

$$f(x) = \begin{cases} \frac{1}{20}e^{-x/20} & \text{if } x \geq 0 \\ 0 & \text{otherwise} \end{cases}$$

(A) What is the probability that a beginner can type at least 30 words per minute after 1 week of practice?
(B) What is the probability that a beginner can type at least 80 words per minute after 1 week of practice?

50. *Learning.* The number of hours it takes a chimpanzee to learn a new task is a continuous random variable with probability density function given by

$$f(x) = \begin{cases} \frac{4}{9}x^2 - \frac{4}{27}x^3 & \text{if } 0 \leq x \leq 3 \\ 0 & \text{otherwise} \end{cases}$$

(A) What is the probability that the chimpanzee learns the task in the first hour?
(B) What is the probability that the chimpanzee does not learn the task in the first 2 hours?

SECTION 12-5

Expected Value, Standard Deviation, and Median of Continuous Random Variables

- EXPECTED VALUE AND STANDARD DEVIATION
- ALTERNATE FORMULA FOR VARIANCE
- MEDIAN

EXPECTED VALUE AND STANDARD DEVIATION

Earlier, we used finite sums and the probability distribution function to define the expected value or mean, the variance, and the standard deviation of a discrete random variable. In much the same way, we now use integration and the probability density function to define these quantities for a continuous random variable. Compare the formulas below with those in Section 12-1.

Expected Value and Standard Deviation for a Continuous Random Variable

Let $f(x)$ be the probability density function for a continuous random variable X. The **expected value, or mean, of X** is

$$\mu = E(X) = \int_{-\infty}^{\infty} xf(x)\,dx$$

The **variance** is

$$V(X) = \int_{-\infty}^{\infty} (x - \mu)^2 f(x)\,dx$$

and the **standard deviation** is

$$\sigma = \sqrt{V(X)}$$

Just as in the discrete case, the mean of a continuous random variable is a measure of central tendency for the variable, and the standard deviation is a measure of the dispersion of the variable about the mean. This is illustrated in Figure 7. The probability density function in Figure 7A has a standard deviation of 1. Most of the area under the curve is near the mean. In Figure 7B, the standard deviation is four times as large, and the area under the graph is much more spread out.

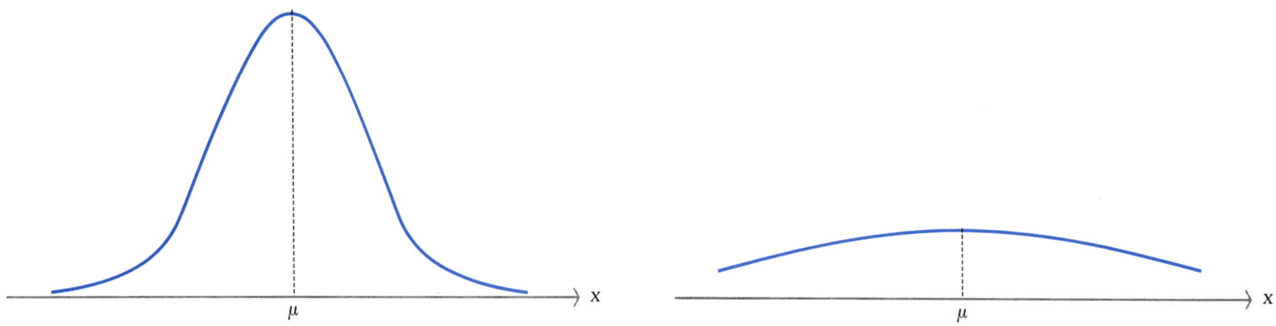

(A) $\sigma = 1$ (B) $\sigma = 4$

FIGURE 7

◆ EXAMPLE 22 Find the mean, variance, and standard deviation for

$$f(x) = \begin{cases} 12x^2 - 12x^3 & \text{if } 0 \le x \le 1 \\ 0 & \text{otherwise} \end{cases}$$

Solution

$$\mu = E(X) = \int_{-\infty}^{\infty} xf(x)\,dx = \int_0^1 x(12x^2 - 12x^3)\,dx = \int_0^1 (12x^3 - 12x^4)\,dx$$

$$= (3x^4 - \tfrac{12}{5}x^5)\Big|_0^1 = \tfrac{3}{5}$$

$$V(X) = \int_{-\infty}^{\infty} (x - \mu)^2 f(x)\,dx = \int_0^1 (x - \tfrac{3}{5})^2(12x^2 - 12x^3)\,dx$$

$$= \int_0^1 (x^2 - \tfrac{6}{5}x + \tfrac{9}{25})(12x^2 - 12x^3)\,dx$$

$$= \int_0^1 (\tfrac{108}{25}x^2 - \tfrac{468}{25}x^3 + \tfrac{132}{5}x^4 - 12x^5)\,dx$$

$$= (\tfrac{36}{25}x^3 - \tfrac{117}{25}x^4 + \tfrac{132}{25}x^5 - 2x^6)\Big|_0^1 = \tfrac{1}{25}$$

$$\sigma = \sqrt{V(X)} = \sqrt{\tfrac{1}{25}} = \tfrac{1}{5}$$ ◆

PROBLEM 22 Find the mean, variance, and standard deviation for

$$f(x) = \begin{cases} 6x - 6x^2 & \text{if } 0 \le x \le 1 \\ 0 & \text{otherwise} \end{cases}$$ ◆

The graph of the probability density function considered in Example 22 is shown in Figure 8, along with the indicated areas (computations omitted). Just as in the discrete case (see Figure 3), geometrically the mean represents the balance point for the region formed by the graph of f and the x axis. Notice that a vertical line through the mean does not divide the area under the graph of f into two regions with equal area, as you might expect. The value of x that does this is the *median*, which we will discuss later in this section.

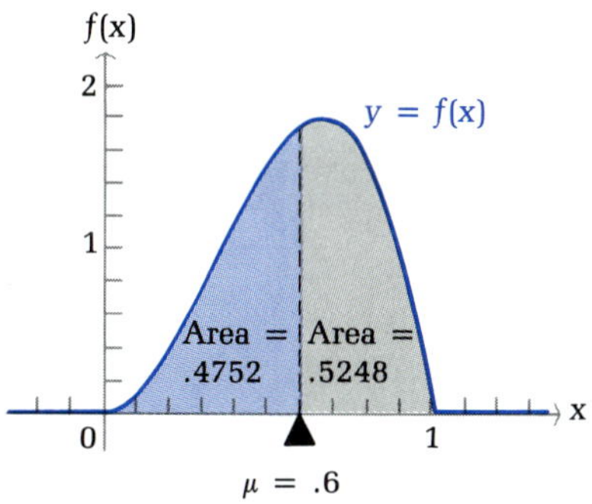

FIGURE 8
Mean of a continuous random variable

◆ EXAMPLE 23 *Life Expectancy*

The life expectancy (in hours) for a particular brand of light bulbs is a continuous random variable with probability density function

$$f(x) = \begin{cases} \frac{1}{100} - \frac{1}{20,000}x & \text{if } 0 \le x \le 200 \\ 0 & \text{otherwise} \end{cases}$$

(A) What is the average life expectancy of one of these light bulbs?
(B) What is the probability that a bulb will last longer than this average?

Solutions

(A) Since the value of this random variable is the number of hours a bulb lasts, the average life expectancy is just the expected value of the random variable. Thus,

$$E(X) = \int_{-\infty}^{\infty} xf(x)\,dx = \int_0^{200} x\left(\tfrac{1}{100} - \tfrac{1}{20,000}x\right) dx$$

$$= \int_0^{200} \left(\tfrac{1}{100}x - \tfrac{1}{20,000}x^2\right) dx = \left(\tfrac{1}{200}x^2 - \tfrac{1}{60,000}x^3\right)\Big|_0^{200}$$

$$= \tfrac{200}{3} \quad \text{or } 66\tfrac{2}{3} \text{ hours}$$

(B) The probability that a bulb lasts longer than $66\frac{2}{3}$ hours is

$$P\left(X > \tfrac{200}{3}\right) = \int_{200/3}^{\infty} f(x)\,dx = \int_{200/3}^{200} \left(\tfrac{1}{100} - \tfrac{1}{20,000}x\right) dx$$

$$= \left(\tfrac{1}{100}x - \tfrac{1}{40,000}x^2\right)\Big|_{200/3}^{200}$$

$$= 1 - \tfrac{5}{9} = \tfrac{4}{9}$$

◆

PROBLEM 23

Repeat Example 23 if the probability density function is

$$f(x) = \begin{cases} \frac{1}{200} - \frac{1}{90,000}x & \text{if } 0 \le x \le 300 \\ 0 & \text{otherwise} \end{cases}$$

◆

◆ ALTERNATE FORMULA FOR VARIANCE

The term $(x - \mu)^2$ in the formula for $V(X)$ introduces some complicated algebraic manipulations in the evaluation of the integral. We can use the properties of the definite integral to simplify this formula. Thus,

$$V(X) = \int_{-\infty}^{\infty} (x - \mu)^2 f(x)\,dx \qquad \text{Expand } (x-\mu)^2.$$

$$= \int_{-\infty}^{\infty} (x^2 - 2x\mu + \mu^2) f(x)\,dx \qquad \text{Multiply by } f(x).$$

$$= \int_{-\infty}^{\infty} [x^2 f(x) - 2x\mu f(x) + \mu^2 f(x)]\,dx \qquad \text{Use property 2, page 411.}$$

$$= \int_{-\infty}^{\infty} x^2 f(x)\,dx - \int_{-\infty}^{\infty} 2x\mu f(x)\,dx + \int_{-\infty}^{\infty} \mu^2 f(x)\,dx \qquad \text{Use property 1, page 411.}$$

$$V(X) = \int_{-\infty}^{\infty} x^2 f(x)\,dx - 2\mu \int_{-\infty}^{\infty} xf(x)\,dx + \mu^2 \int_{-\infty}^{\infty} f(x)\,dx \qquad \int_{-\infty}^{\infty} xf(x)\,dx = \mu, \quad \int_{-\infty}^{\infty} f(x)\,dx = 1$$

$$= \int_{-\infty}^{\infty} x^2 f(x)\,dx - 2\mu(\mu) + \mu^2(1)$$

$$= \int_{-\infty}^{\infty} x^2 f(x)\,dx - \mu^2$$

In general, it will be easier to evaluate $\int_{-\infty}^{\infty} x^2 f(x)\,dx$ than to evaluate $\int_{-\infty}^{\infty}(x-\mu)^2 f(x)\,dx$.

THEOREM 3

Alternate Formula for Variance

$$V(X) = \int_{-\infty}^{\infty} x^2 f(x)\,dx - \mu^2$$

◆ EXAMPLE 24 Use the alternate formula for variance (Theorem 3) to compute the variance in Example 22.

Solution From Example 22, we have $\mu = \int_{-\infty}^{\infty} xf(x)\,dx = \frac{3}{5}$. Thus,

$$\int_{-\infty}^{\infty} x^2 f(x)\,dx = \int_0^1 x^2(12x^2 - 12x^3)\,dx \qquad f(x) = \begin{cases} 12x^2 - 12x^3 & \text{if } 0 \le x \le 1 \\ 0 & \text{otherwise} \end{cases}$$

$$= \int_0^1 (12x^4 - 12x^5)\,dx$$

$$= \left(\tfrac{12}{5}x^5 - \tfrac{12}{6}x^6\right)\Big|_0^1 = \tfrac{2}{5}$$

$$V(X) = \int_{-\infty}^{\infty} x^2 f(x)\,dx - \mu^2 = \tfrac{2}{5} - \left(\tfrac{3}{5}\right)^2 = \tfrac{1}{25}$$

◆

PROBLEM 24 Use the alternate formula for variance (Theorem 3) to compute the variance in Problem 22. ◆

◆ EXAMPLE 25 Find the mean, variance, and standard deviation for

$$f(x) = \begin{cases} 3/x^4 & \text{if } x \ge 1 \\ 0 & \text{otherwise} \end{cases}$$

Solution

$$\mu = \int_{-\infty}^{\infty} xf(x)\,dx = \int_1^{\infty} x\frac{3}{x^4}\,dx = \lim_{R\to\infty}\int_1^R \frac{3}{x^3}\,dx$$

$$= \lim_{R\to\infty}\left[-\frac{3}{2}\left(\frac{1}{x^2}\right)\right]\Bigg|_1^R = \lim_{R\to\infty}\left[-\frac{3}{2}\left(\frac{1}{R^2}\right) + \frac{3}{2}\right] = \frac{3}{2}$$

$$\int_{-\infty}^{\infty} x^2 f(x)\,dx = \int_1^{\infty} x^2\frac{3}{x^4}\,dx = \lim_{R\to\infty}\int_1^R \frac{3}{x^2}\,dx = \lim_{R\to\infty}\left(-\frac{3}{x}\right)\Bigg|_1^R$$

$$= \lim_{R\to\infty}\left(-\frac{3}{R} + 3\right) = 3$$

$$V(X) = \int_{-\infty}^{\infty} x^2 f(x)\,dx - \mu^2 = 3 - \left(\frac{3}{2}\right)^2 = \frac{3}{4}$$

$$\sigma = \sqrt{V(X)} = \sqrt{\frac{3}{4}} = \frac{\sqrt{3}}{2} \approx .8660$$

PROBLEM 25 Find the mean, variance, and standard deviation for

$$f(x) = \begin{cases} 24/x^4 & \text{if } x \geqslant 2 \\ 0 & \text{otherwise} \end{cases}$$

MEDIAN

Another measurement often used to describe the properties of a random variable is the median. The **median** is the value of the random variable that divides the area under the graph of the probability density function into two equal parts (see Figure 9). If x_m is the median, then x_m must satisfy

$$P(X \leqslant x_m) = \tfrac{1}{2}$$

Generally, this equation is solved by first finding the cumulative probability distribution function.

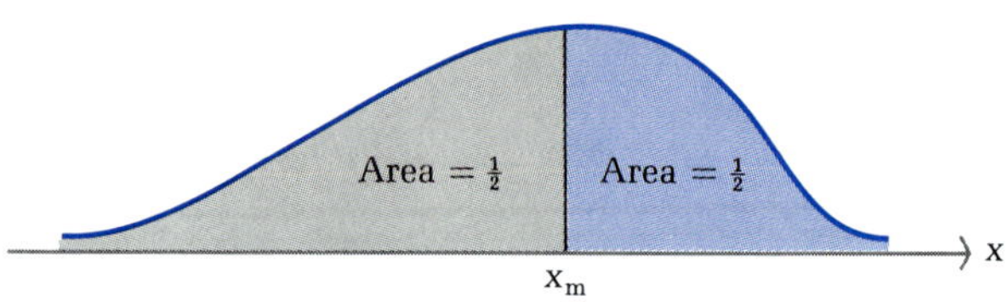

FIGURE 9

EXAMPLE 26 Find the median of the continuous random variable with probability density function

$$f(x) = \begin{cases} 3/x^4 & \text{if } x \geqslant 1 \\ 0 & \text{otherwise} \end{cases}$$

Solution Step 1. Find the cumulative probability distribution function. For $x < 1$, we have $F(x) = 0$. If $x \geqslant 1$, then

$$F(x) = \int_{-\infty}^{x} f(t)\,dt = \int_1^x \frac{3}{t^4}\,dt = -\frac{1}{t^3}\bigg|_1^x = -\frac{1}{x^3} + 1 = 1 - \frac{1}{x^3}$$

Step 2. Solve the equation $P(X \leqslant x_m) = \frac{1}{2}$ for x_m.

$$F(x_m) = P(X \leqslant x_m)$$

$$1 - \frac{1}{x_m^3} = \frac{1}{2}$$

$$\frac{1}{2} = \frac{1}{x_m^3}$$
$$x_m^3 = 2$$
$$x_m = \sqrt[3]{2}$$

Thus, the median is $\sqrt[3]{2} \approx 1.26$. ◆

PROBLEM 26 Find the median of the continuous random variable with probability density function

$$f(x) = \begin{cases} 24/x^4 & \text{if } x \geq 2 \\ 0 & \text{otherwise} \end{cases}$$ ◆

◆ EXAMPLE 27 *Life Expectancy* In Example 23, find the median life expectancy of a light bulb.

Solution Step 1. Find the cumulative probability distribution function. If $x < 0$, we have $F(x) = 0$. If $0 \leq x \leq 200$, then

$$F(x) = \int_{-\infty}^{x} f(t)\,dt \qquad f(x) = \begin{cases} \frac{1}{100} - \frac{1}{20{,}000}x & \text{if } 0 \leq x \leq 200 \\ 0 & \text{otherwise} \end{cases}$$
$$= \int_{0}^{x} \left(\frac{1}{100} - \frac{1}{20{,}000}t\right) dt$$
$$= \left(\frac{1}{100}t - \frac{1}{40{,}000}t^2\right)\Big|_0^x$$
$$= \frac{1}{100}x - \frac{1}{40{,}000}x^2$$

If $x > 200$, then

$$F(x) = \int_{-\infty}^{x} f(t)\,dt = \int_{-\infty}^{0} f(t)\,dt + \int_{0}^{200} f(t)\,dt + \int_{200}^{x} f(t)\,dt$$
$$= 0 + 1 + 0 = 1$$

Thus,

$$F(x) = \begin{cases} 0 & \text{if } x < 0 \\ \frac{1}{100}x - \frac{1}{40{,}000}x^2 & \text{if } 0 \leq x \leq 200 \\ 1 & \text{if } x > 200 \end{cases}$$

Step 2. Solve the equation $P(X \leq x_m) = \frac{1}{2}$ for x_m:

$$F(x_m) = P(X \leq x_m) = \tfrac{1}{2}$$
$$\frac{1}{100}x_m - \frac{1}{40{,}000}x_m^2 = \frac{1}{2}$$
$$x_m^2 - 400x_m + 20{,}000 = 0 \qquad \text{The solution must occur for } 0 \leq x_m \leq 200.$$

This quadratic equation has two solutions, $200 + 100\sqrt{2}$ and $200 - 100\sqrt{2}$. Since x_m must lie in the interval [0, 200], the second root is the correct answer.

Thus, the median life expectancy is $200 - 100\sqrt{2} \approx 58.58$ hours. ◆

PROBLEM 27 In Problem 23, find the median life expectancy of a light bulb. ◆

If you compare Examples 23 and 27, and Examples 25 and 26, you will see that the mean and the median generally are not equal (see Figure 10).

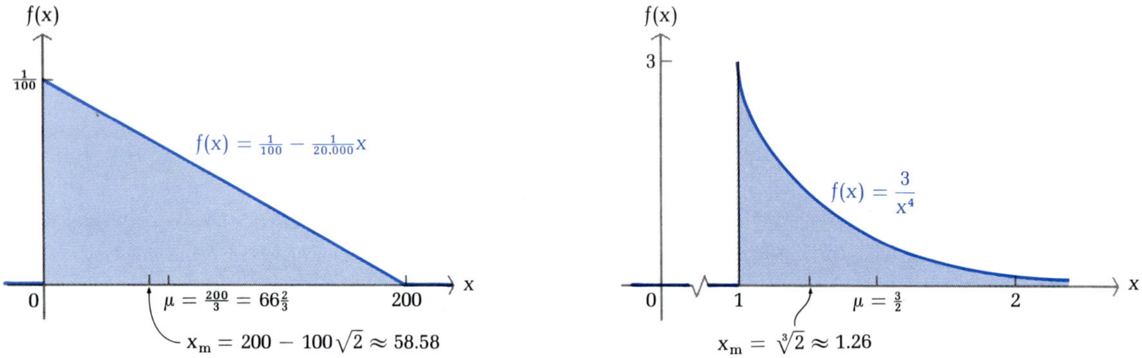

FIGURE 10

Answers to Matched Problems

22. $\mu = \frac{1}{2}$; $V(X) = \frac{1}{20}$; $\sigma \approx .2236$ 23. (A) 125 hr (B) $\frac{133}{288}$
24. $\frac{1}{20}$ 25. $\mu = 3$; $V(X) = 3$; $\sigma = \sqrt{3} \approx 1.732$
26. $x_m = \sqrt[3]{16} = 2\sqrt[3]{2} \approx 2.52$ 27. $x_m = 450 - 150\sqrt{5} \approx 114.59$ hr

EXERCISE 12-5

A *In Problems 1–6, find the mean, variance, and standard deviation.*

1. $f(x) = \begin{cases} \frac{1}{2}x & \text{if } 0 \leqslant x \leqslant 2 \\ 0 & \text{otherwise} \end{cases}$

2. $f(x) = \begin{cases} 3x^2 & \text{if } 0 \leqslant x \leqslant 1 \\ 0 & \text{otherwise} \end{cases}$

3. $f(x) = \begin{cases} \frac{1}{3} & \text{if } 2 \leqslant x \leqslant 5 \\ 0 & \text{otherwise} \end{cases}$

4. $f(x) = \begin{cases} \frac{1}{2} & \text{if } 1 \leqslant x \leqslant 3 \\ 0 & \text{otherwise} \end{cases}$

5. $f(x) = \begin{cases} 4 - 2x & \text{if } 1 \leqslant x \leqslant 2 \\ 0 & \text{otherwise} \end{cases}$

6. $f(x) = \begin{cases} 2 - \frac{1}{2}x & \text{if } 2 \leqslant x \leqslant 4 \\ 0 & \text{otherwise} \end{cases}$

In Problems 7–12, find the median.

7. $f(x) = \begin{cases} 2x & \text{if } 0 \leqslant x \leqslant 1 \\ 0 & \text{otherwise} \end{cases}$

8. $f(x) = \begin{cases} \frac{1}{8}x & \text{if } 0 \leqslant x \leqslant 4 \\ 0 & \text{otherwise} \end{cases}$

9. $f(x) = \begin{cases} \frac{1}{6}x & \text{if } 2 \leqslant x \leqslant 4 \\ 0 & \text{otherwise} \end{cases}$

10. $f(x) = \begin{cases} \frac{1}{4}x & \text{if } 1 \leqslant x \leqslant 3 \\ 0 & \text{otherwise} \end{cases}$

11. $f(x) = \begin{cases} \frac{1}{2} - \frac{1}{8}x & \text{if } 0 \leqslant x \leqslant 4 \\ 0 & \text{otherwise} \end{cases}$

12. $f(x) = \begin{cases} 2 - 2x & \text{if } 0 \leqslant x \leqslant 1 \\ 0 & \text{otherwise} \end{cases}$

B *In Problems 13–16, find the mean, variance, and standard deviation.*

13. $f(x) = \begin{cases} 4/x^5 & \text{if } x \geqslant 1 \\ 0 & \text{otherwise} \end{cases}$

14. $f(x) = \begin{cases} 5/x^6 & \text{if } x \geqslant 1 \\ 0 & \text{otherwise} \end{cases}$

15. $f(x) = \begin{cases} 64/x^5 & \text{if } x \geq 2 \\ 0 & \text{otherwise} \end{cases}$

16. $f(x) = \begin{cases} 81/x^4 & \text{if } x \geq 3 \\ 0 & \text{otherwise} \end{cases}$

In Problems 17–24, find the median.

17. $f(x) = \begin{cases} 1/x & \text{if } 1 \leq x \leq e \\ 0 & \text{otherwise} \end{cases}$

18. $f(x) = \begin{cases} 1/(2x) & \text{if } 0 \leq x \leq e^2 \\ 0 & \text{otherwise} \end{cases}$

19. $f(x) = \begin{cases} 4/(2+x)^2 & \text{if } 0 \leq x \leq 2 \\ 0 & \text{otherwise} \end{cases}$

20. $f(x) = \begin{cases} 2/(1+x)^2 & \text{if } 0 \leq x \leq 1 \\ 0 & \text{otherwise} \end{cases}$

21. $f(x) = \begin{cases} 1/(1+x)^2 & \text{if } x \geq 0 \\ 0 & \text{otherwise} \end{cases}$

22. $f(x) = \begin{cases} 3/(3+x)^2 & \text{if } x \geq 0 \\ 0 & \text{otherwise} \end{cases}$

23. $f(x) = \begin{cases} 2e^{-2x} & \text{if } x \geq 0 \\ 0 & \text{otherwise} \end{cases}$

24. $f(x) = \begin{cases} e^{-x} & \text{if } x \geq 0 \\ 0 & \text{otherwise} \end{cases}$

C

In Problems 25 and 26, $f(x)$ is a continuous probability density function with mean μ and standard deviation σ; a and b are constants. Evaluate each integral, expressing the result in terms of a, b, μ, and σ.

25. $\int_{-\infty}^{\infty} (ax + b)f(x)\,dx$

26. $\int_{-\infty}^{\infty} (x - a)^2 f(x)\,dx$

*The **quartile points** for a probability density function are the values x_1, x_2, x_3 that divide the area under the graph of the function into four equal parts. Find the quartile points for the probability density functions in Problems 27–30.*

27. $f(x) = \begin{cases} \frac{1}{2}x & \text{if } 0 \leq x \leq 2 \\ 0 & \text{otherwise} \end{cases}$

28. $f(x) = \begin{cases} 3x^2 & \text{if } 0 \leq x \leq 1 \\ 0 & \text{otherwise} \end{cases}$

29. $f(x) = \begin{cases} 3/(3+x)^2 & \text{if } x \geq 0 \\ 0 & \text{otherwise} \end{cases}$

30. $f(x) = \begin{cases} 1/(1+x)^2 & \text{if } x \geq 0 \\ 0 & \text{otherwise} \end{cases}$

In Problems 31–34, use a graphic calculator or a computer to approximate the median of the indicated probability density function f to two decimal places.

31. $f(x) = \begin{cases} 4x - 4x^3 & \text{if } 0 \leq x \leq 1 \\ 0 & \text{otherwise} \end{cases}$

32. $f(x) = \begin{cases} 3x - 3x^5 & \text{if } 0 \leq x \leq 1 \\ 0 & \text{otherwise} \end{cases}$

33. $f(x) = \begin{cases} \ln x & \text{if } 1 \leq x \leq e \\ 0 & \text{otherwise} \end{cases}$

34. $f(x) = \begin{cases} xe^{-x} & \text{if } x \geq 0 \\ 0 & \text{otherwise} \end{cases}$

APPLICATIONS

Business & Economics

35. *Profit.* A building contractor's profit (in thousands of dollars) on each unit in a subdivision is a continuous random variable with probability density function given by

$$f(x) = \begin{cases} \frac{1}{8}(10 - x) & \text{if } 6 \leq x \leq 10 \\ 0 & \text{otherwise} \end{cases}$$

(A) Find the contractor's expected profit.
(B) Find the median profit.

36. *Electricity consumption.* The daily consumption of electricity (in millions of kilowatt-hours) in a large city is a continuous random variable with probability density function

$$f(x) = \begin{cases} .2 - .02x & \text{if } 0 \leqslant x \leqslant 10 \\ 0 & \text{otherwise} \end{cases}$$

(A) Find the expected daily consumption of electricity.
(B) Find the median daily consumption of electricity.

37. *Waiting time.* The time (in minutes) a customer must wait in line at a bank is a continuous random variable with probability density function given by

$$f(x) = \begin{cases} \frac{1}{3}e^{-x/3} & \text{if } x \geqslant 0 \\ 0 & \text{otherwise} \end{cases}$$

Find the median waiting time.

38. *Product life.* The life expectancy (in years) of an automobile battery is a continuous random variable with probability density function given by

$$f(x) = \begin{cases} \frac{1}{2}e^{-x/2} & \text{if } x \geqslant 0 \\ 0 & \text{otherwise} \end{cases}$$

Find the median life expectancy.

39. *Water consumption.* The daily consumption of water (in millions of gallons) in a small city is a continuous random variable with probability density function given by

$$f(x) = \begin{cases} 1/(1 + x^2)^{3/2} & \text{if } x \geqslant 0 \\ 0 & \text{otherwise} \end{cases}$$

Find the expected daily consumption.

40. *Gasoline consumption.* The daily consumption of gasoline (in millions of gallons) in a large city is a continuous random variable with probability density function

$$f(x) = \begin{cases} 4/(4 + x^2)^{3/2} & \text{if } x \geqslant 0 \\ 0 & \text{otherwise} \end{cases}$$

Find the expected daily consumption of gasoline.

Life Sciences

41. *Life expectancy.* The life expectancy of a certain microscopic organism (in minutes) is a continuous random variable with probability density function given by

$$f(x) = \begin{cases} \frac{1}{5{,}000}(10x^3 - x^4) & \text{if } 0 \leqslant x \leqslant 10 \\ 0 & \text{otherwise} \end{cases}$$

Find the mean life expectancy of one of these organisms.

42. *Life expectancy.* The life expectancy (in months) of plants of a certain species is a continuous random variable with probability density function given by

$$f(x) = \begin{cases} \frac{1}{36}(6x - x^2) & \text{if } 0 \leq x \leq 6 \\ 0 & \text{otherwise} \end{cases}$$

Find the mean life expectancy of one of these plants.

43. *Shelf-life.* The shelf-life (in days) of a perishable drug is a continuous random variable with probability density function given by

$$f(x) = \begin{cases} 800x/(400 + x^2)^2 & \text{if } x \geq 0 \\ 0 & \text{otherwise} \end{cases}$$

Find the median shelf-life.

44. *Shelf life.* Repeat Problem 43 if

$$f(x) = \begin{cases} 200x/(100 + x^2)^2 & \text{if } x \geq 0 \\ 0 & \text{otherwise} \end{cases}$$

Social Sciences

45. *Learning.* The number of hours it takes a chimpanzee to learn a new task is a continuous random variable with probability density function given by

$$f(x) = \begin{cases} \frac{4}{9}x^2 - \frac{4}{27}x^3 & \text{if } 0 \leq x \leq 3 \\ 0 & \text{otherwise} \end{cases}$$

What is the expected number of hours it will take a chimpanzee to learn the task?

46. *Voter turnout.* The number of registered voters (in thousands) who vote in an off-year election in a small town is a continuous random variable with probability density function given by

$$f(x) = \begin{cases} 1 - \frac{1}{8}x & \text{if } 4 \leq x \leq 8 \\ 0 & \text{otherwise} \end{cases}$$

(A) Find the expected voter turnout.
(B) Find the median voter turnout.

SECTION 12-6 Uniform, Beta, and Exponential Distributions

- UNIFORM DISTRIBUTION
- BETA DISTRIBUTION
- EXPONENTIAL DISTRIBUTION

In this section we will examine several important probability density functions. In actual practice, we do not usually construct a probability density function for each experiment. Instead, we try to select a known probability density function that seems to give a reasonable description of the experiment. Thus, it is important to be familiar with the properties and applications of a variety of probability density functions.

◆ UNIFORM DISTRIBUTION

We begin with an example that will lead to some general observations.

◆ EXAMPLE 28

Waiting Time

A bus arrives every 30 minutes at a particular bus stop. If an individual arrives at the bus stop at a random time (that is, with no knowledge of the bus schedule), then the random variable X, representing the time this individual must spend waiting for the next bus, is said to be *uniformly distributed* on the interval $[0, 30]$. This means that the probability that X lies in a small interval of fixed length is independent of the location of that interval within $[0, 30]$. Thus, the probability of waiting between 0 and 5 minutes is the same as the probability of waiting between 5 and 10 minutes or the probability of waiting between 18 and 23 minutes. It can be shown that the probability density function for this uniformly distributed random variable is

$$f(x) = \begin{cases} \frac{1}{30} & \text{if } 0 \leqslant x \leqslant 30 \\ 0 & \text{otherwise} \end{cases}$$

Thus,

$$P(0 \leqslant X \leqslant 5) = \int_0^5 \frac{1}{30}\,dx = \frac{x}{30}\bigg|_0^5 = \frac{5}{30} - \frac{0}{30} = \frac{1}{6}$$

$$P(5 \leqslant X \leqslant 10) = \int_5^{10} \frac{1}{30}\,dx = \frac{x}{30}\bigg|_5^{10} = \frac{10}{30} - \frac{5}{30} = \frac{1}{6}$$

$$P(18 \leqslant X \leqslant 23) = \int_{18}^{23} \frac{1}{30}\,dx = \frac{x}{30}\bigg|_{18}^{23} = \frac{23}{30} - \frac{18}{30} = \frac{1}{6}$$

Each of these probabilities is represented as an area under the graph of f in the figure.

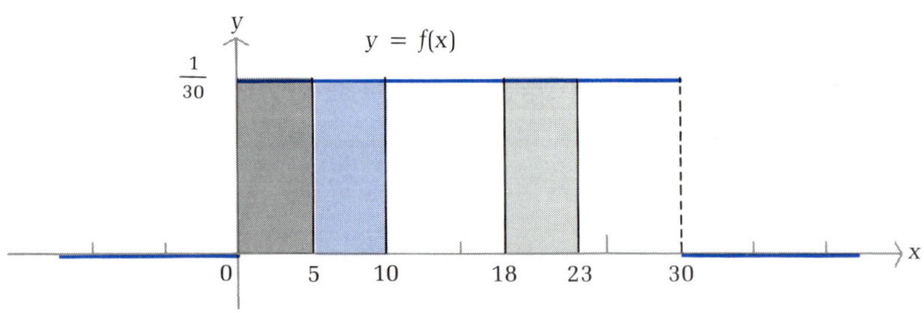

◆

PROBLEM 28 Use the probability density function given in Example 28 to find the probability that the individual waits:

(A) Between 0 and 10 minutes (B) Between 10 and 20 minutes
(C) Between 17 and 27 minutes ◆

In general, if the outcomes of an experiment lie in an interval $[a, b]$ and if the probability of the outcome lying in a small interval of fixed length is independent of the location of this small interval within $[a, b]$, then we say that the continuous random variable for this experiment is **uniformly distributed** on the interval $[a, b]$. The **uniform probability density function** is

$$f(x) = \begin{cases} \dfrac{1}{b-a} & \text{if } a \leq x \leq b \\ 0 & \text{otherwise} \end{cases}$$

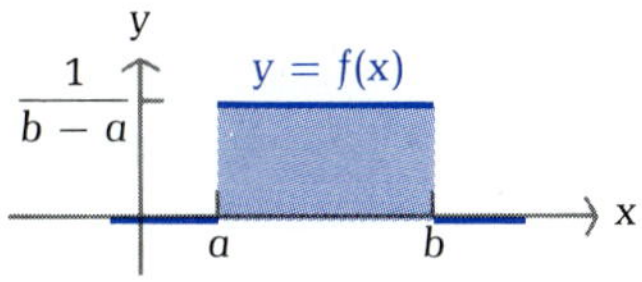

FIGURE 11

See Figure 11. Since $f(x) \geq 0$ and

$$\int_{-\infty}^{\infty} f(x)\,dx = \int_a^b \frac{1}{b-a}\,dx = \frac{x}{b-a}\Big|_a^b = \frac{b}{b-a} - \frac{a}{b-a} = 1$$

f satisfies the necessary conditions for a probability density function.

If F is the associated cumulative probability distribution function, then for $x < a$, $F(x) = 0$. For $a \leq x \leq b$, we have

$$F(x) = \int_{-\infty}^{x} f(t)\,dt = \int_a^x \frac{1}{b-a}\,dt = \frac{t}{b-a}\Big|_a^x$$

$$= \frac{x}{b-a} - \frac{a}{b-a} = \frac{x-a}{b-a}$$

For $x > b$, $F(x) = 1$.

Using the techniques discussed in the preceding section, it can be shown that (see Problems 25–28 in Exercise 12-6)

$$\mu = x_m = \frac{1}{2}(a+b) \qquad \text{and} \qquad \sigma = \frac{1}{\sqrt{12}}(b-a)$$

These properties are summarized in the box on the next page.

Uniform Probability Density Function

PROBABILITY DENSITY FUNCTION

$$f(x) = \begin{cases} \dfrac{1}{b-a} & \text{if } a \le x \le b \\ 0 & \text{otherwise} \end{cases}$$

CUMULATIVE PROBABILITY DISTRIBUTION

$$F(x) = \begin{cases} 0 & \text{if } x < a \\ \dfrac{x-a}{b-a} & \text{if } a \le x \le b \\ 1 & \text{if } x > b \end{cases}$$

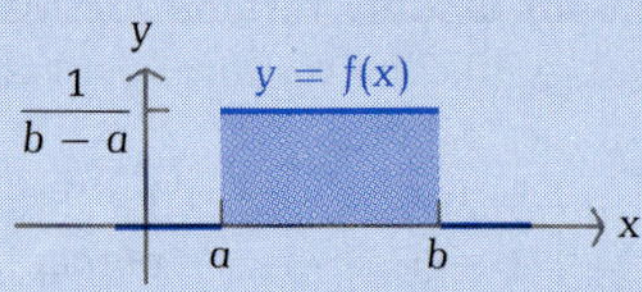

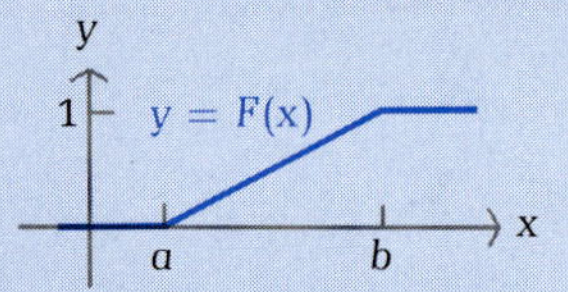

MEAN $\mu = \frac{1}{2}(a+b)$ MEDIAN $x_m = \frac{1}{2}(a+b)$

STANDARD DEVIATION $\sigma = \frac{1}{\sqrt{12}}(b-a)$

◆ EXAMPLE 29 *Electrical Current*

Standard electrical current is uniformly distributed between 110 and 120 volts. What is the probability that the current is between 113 and 118 volts?

Solution Since we are told that the current is uniformly distributed on the interval [110, 120], we choose the uniform probability density function

$$f(x) = \begin{cases} \frac{1}{10} & \text{if } 110 \le x \le 120 \\ 0 & \text{otherwise} \end{cases}$$

Then

$$P(113 \le X \le 118) = \int_{113}^{118} \frac{1}{10}\,dx = \frac{x}{10}\Big|_{113}^{118} = \frac{118}{10} - \frac{113}{10} = \frac{1}{2}$$ ◆

PROBLEM 29 In Example 29, what is the probability that the current is at least 116 volts? ◆

◆ BETA DISTRIBUTION

An area of applied statistics called *Bayesian inference*, named after the Presbyterian minister Thomas Bayes (1702–1763), has received a great deal of attention in recent years. Effective use of this method requires experience with a variety of random variables with values that can be expressed as fractions or percentages. One particular random variable that has been used extensively in this area of statistics is the *beta random variable*. Typical applications of beta random variables include the percentage of fast-food restaurants that make a profit during their first year of operation, the percentage of time each year a manufacturing plant is shut down, the consumption of natural gas as a percentage of capacity, and the percentage of job application forms that contain errors.

A continuous random variable has a **beta distribution*** and is referred to as a **beta random variable** if its probability density function is the **beta probability density function**

$$f(x) = \begin{cases} (\beta + 1)(\beta + 2)x^{\beta}(1 - x) & \text{if } 0 \leqslant x \leqslant 1 \\ 0 & \text{otherwise} \end{cases}$$

where β is a constant, $\beta \geqslant 0$. The value of β is usually determined by examining the results of a particular experiment. The values of a beta random variable can be expressed as fractions or percentages; however, percentages should be converted to fractions before performing calculations involving a beta random variable.

First, we show that f satisfies the requirements for a probability density function:

$$f(x) = (\beta + 1)(\beta + 2)x^{\beta}(1 - x) \geqslant 0 \qquad 0 \leqslant x \leqslant 1$$

$$\begin{aligned}
\int_{-\infty}^{\infty} f(x)\,dx &= \int_0^1 (\beta + 1)(\beta + 2)x^{\beta}(1 - x)\,dx \\
&= \int_0^1 (\beta + 1)(\beta + 2)(x^{\beta} - x^{\beta+1})\,dx \\
&= (\beta + 1)(\beta + 2)\left(\frac{x^{\beta+1}}{\beta + 1} - \frac{x^{\beta+2}}{\beta + 2}\right)\Bigg|_0^1 \\
&= (\beta + 1)(\beta + 2)\left(\frac{1}{\beta + 1} - \frac{1}{\beta + 2}\right) \\
&= (\beta + 2) - (\beta + 1) = 1
\end{aligned}$$

Thus, f is a probability density function.

If $F(x)$ is the associated cumulative probability distribution function, then for $x < 0$, $F(x) = 0$. For $0 \leqslant x \leqslant 1$, we have

$$\begin{aligned}
F(x) &= \int_{-\infty}^{x} f(t)\,dt = \int_0^x (\beta + 1)(\beta + 2)t^{\beta}(1 - t)\,dt \\
&= (\beta + 1)(\beta + 2)\left(\frac{t^{\beta+1}}{\beta + 1} - \frac{t^{\beta+2}}{\beta + 2}\right)\Bigg|_0^x \\
&= (\beta + 1)(\beta + 2)\left(\frac{x^{\beta+1}}{\beta + 1} - \frac{x^{\beta+2}}{\beta + 2}\right) \\
&= (\beta + 2)x^{\beta+1} - (\beta + 1)x^{\beta+2}
\end{aligned}$$

* There is a more general definition of a beta distribution, but we will not consider it here.

And for $x > 1$,

$$F(x) = 1$$

In general, it is not possible to solve the equation $F(x_m) = \frac{1}{2}$ for x_m. Thus, we will not discuss the median of a beta random variable. By straightforward (but tedious) integration we can show that

$$\mu = \frac{\beta + 1}{\beta + 3} \qquad \text{and} \qquad \sigma = \sqrt{\frac{2(\beta + 1)}{(\beta + 4)(\beta + 3)^2}}$$

The calculations are not included here. The above results are summarized in the box.

Beta Probability Density Function

$$f(x) = \begin{cases} (\beta + 1)(\beta + 2)x^{\beta}(1 - x) & \text{if } 0 \leq x \leq 1 \\ 0 & \text{otherwise} \end{cases} \qquad \text{where } \beta \geq 0$$

$$F(x) = \begin{cases} 0 & \text{if } x < 0 \\ (\beta + 2)x^{\beta+1} - (\beta + 1)x^{\beta+2} & \text{if } 0 \leq x \leq 1 \\ 1 & \text{if } x > 1 \end{cases}$$

MEAN $\mu = \dfrac{\beta + 1}{\beta + 3}$ STANDARD DEVIATION $\sigma = \sqrt{\dfrac{2(\beta + 1)}{(\beta + 4)(\beta + 3)^2}}$

Figure 12 shows the graphs of $f(x)$ for some typical values of β.

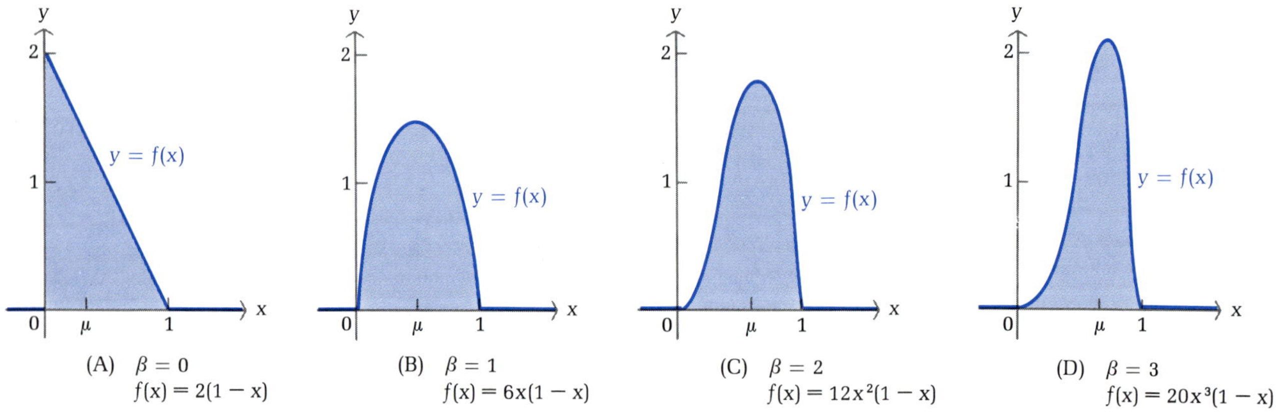

FIGURE 12
Beta probability density functions

◆ EXAMPLE 30 The annual percentage of correct income tax forms filed with the Internal Revenue Service is a beta random variable with $\beta = 8$.

Income Tax

(A) What is the probability that at least half the returns filed are correct?
(B) What is the expected percentage of correct returns?

Solutions Substituting $\beta = 8$ in the definition of the beta probability density function, we have

$$f(x) = \begin{cases} 90x^8(1-x) & \text{if } 0 \leqslant x \leqslant 1 \\ 0 & \text{otherwise} \end{cases}$$

$$\text{(A)}\quad P(X \geqslant \tfrac{1}{2}) = \int_{1/2}^{1} 90x^8(1-x)\,dx = \int_{1/2}^{1} (90x^8 - 90x^9)\,dx$$

$$= (10x^9 - 9x^{10})\Big|_{1/2}^{1} = 1 - \frac{11}{2^{10}} \approx .989$$

$$\text{(B)}\quad \mu = E(X) = \frac{\beta+1}{\beta+3} = \frac{8+1}{8+3} = \frac{9}{11} \approx .818$$

Thus, we expect approximately 82% of the returns to be correct. ◆

PROBLEM 30 In Example 30, what is the probability that at least 90% of the returns are correct? ◆

◆ EXAMPLE 31 A psychologist is studying the learning abilities of children in a certain age group by giving them 5 minutes to learn to perform a particular task. Repeated trials of the experiment have led to the conclusions that the percentage of children who can learn to perform this task in 5 minutes is a beta random variable and that the average percentage of children who learn the task is 75%. What is the appropriate value of β for this experiment?

Learning

Solution Since the average percentage of children who learned the task is 75% and the mean for any beta distribution is $(\beta + 1)/(\beta + 3)$, the value of β must satisfy

$$\frac{\beta+1}{\beta+3} = .75 \qquad \text{Convert 75\% to the decimal fraction .75.}$$

Solving this equation, we obtain $\beta = 5$. ◆

PROBLEM 31 In Example 31, what is the probability that at least 75% of the children will learn the task in 5 minutes? ◆

◆ EXPONENTIAL DISTRIBUTION

A continuous random variable has an **exponential distribution** and is referred to as an **exponential random variable** if its probability density function is the **exponential probability density function**

$$f(x) = \begin{cases} (1/\lambda)e^{-x/\lambda} & \text{if } x \geq 0 \\ 0 & \text{otherwise} \end{cases}$$

where λ is a positive constant. Exponential random variables are used in a variety of applications, including studies of the length of telephone conversations, the time customers spend waiting in line at a bank, and the life expectancy of a machine part.

Since $f(x) \geq 0$ and

$$\begin{aligned} \int_{-\infty}^{\infty} f(x)\,dx &= \int_0^{\infty} \frac{1}{\lambda} e^{-x/\lambda}\,dx \\ &= \lim_{R\to\infty} \int_0^R \frac{1}{\lambda} e^{-x/\lambda}\,dx \\ &= \lim_{R\to\infty} (-e^{-x/\lambda})\Big|_0^R \\ &= \lim_{R\to\infty} (-e^{-R/\lambda} + 1) = 1 \end{aligned}$$

f satisfies the conditions for a probability density function. If F is the cumulative distribution function, we see that $F(x) = 0$ for $x < 0$. For $x \geq 0$, we have

$$\begin{aligned} F(x) = \int_{-\infty}^{x} f(t)\,dt &= \int_0^x \frac{1}{\lambda} e^{-t/\lambda}\,dt \\ &= -e^{-t/\lambda}\Big|_0^x = 1 - e^{-x/\lambda} \end{aligned}$$

To find the median, we solve

$$F(x_m) = P(X \leq x_m) = \frac{1}{2}$$

$$1 - e^{-x_m/\lambda} = \frac{1}{2}$$

$$\frac{1}{2} = e^{-x_m/\lambda}$$

$$\ln\frac{1}{2} = -\frac{x_m}{\lambda}$$

$$x_m = -\lambda \ln\frac{1}{2} = \lambda \ln 2 \qquad \textit{Note:} \ \ln\tfrac{1}{2} = -\ln 2$$

Integration by parts can be used to show that $\mu = \lambda$ and $\sigma = \lambda$. The calculations are not included here. These results are summarized in the box at the top of the next page.

Exponential Probability Density Function

$$f(x) = \begin{cases} (1/\lambda)e^{-x/\lambda} & \text{if } x \geq 0 \\ 0 & \text{otherwise} \end{cases} \qquad F(x) = \begin{cases} 1 - e^{-x/\lambda} & \text{if } x \geq 0 \\ 0 & \text{otherwise} \end{cases}$$

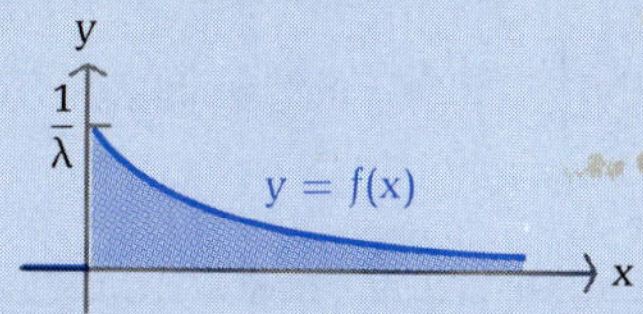

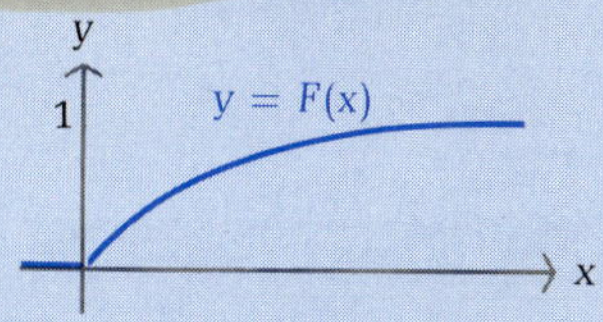

MEAN $\mu = \lambda$ MEDIAN $x_m = \lambda \ln 2$

STANDARD DEVIATION $\sigma = \lambda$

◆ EXAMPLE 32

Arrival Rates

The length of time between calls received by the switchboard in a large legal firm is an exponential random variable. The average length of time between calls is 20 seconds. If a call has just been received, what is the probability that no calls are received in the next 30 seconds?

Solution

Let X be the random variable that represents the length of time between calls received (in seconds). Since the average length of time is 20 seconds, we have $\mu = \lambda = 20$. Thus, the probability density function for X is

$$f(x) = \begin{cases} \frac{1}{20}e^{-x/20} & \text{if } x \geq 0 \\ 0 & \text{otherwise} \end{cases}$$

and

$$P(X \geq 30) = \int_{30}^{\infty} \tfrac{1}{20}e^{-x/20}\,dx \qquad \int_0^{30} f(x)\,dx + \int_{30}^{\infty} f(x)\,dx = 1$$

$$= 1 - \int_0^{30} \tfrac{1}{20}e^{-x/20}\,dx = 1 - (-e^{-x/20})\Big|_0^{30} = e^{-1.5} \approx .223$$ ◆

PROBLEM 32

In Example 32, if a call has just been received, what is the probability that no calls are received in the next 10 seconds? ◆

Referring to Example 32, if the average length of time between calls received by the switchboard is 20 seconds, then the switchboard must be receiving an average of 3 calls per minute [3(20 seconds) = 60 seconds, or 1 minute]. But in Section 12-3 we saw that the number of calls received each minute is a Poisson random variable. A moment's thought should convince you that this is not a coincidence. After all, the following two statements really say the same thing:

1. The average number of calls per minute is 3.
2. The average time between calls is $\frac{1}{3}$ minute, or 20 seconds.

In general, a Poisson random variable is concerned with the number of times an event occurs, which is a discrete concept. Associated with each Poisson random variable is an exponential random variable involving the length of time between occurrences of this event, which is a continuous concept. Thus, we see that there are applications that involve both discrete and continuous random variables.

Answers to Matched Problems

28. (A) $\frac{1}{3}$ (B) $\frac{1}{3}$ (C) $\frac{1}{3}$ 29. $\frac{2}{5}$ 30. .264
31. .555 32. $e^{-0.5} \approx .607$

EXERCISE 12-6

A

In Problems 1–6, find the probability density function f and the associated cumulative distribution function F for the continuous random variable X if:

1. X is uniformly distributed on $[0, 2]$.
2. X is uniformly distributed on $[3, 6]$.
3. X is a beta random variable with $\beta = 3$.
4. X is a beta random variable with $\beta = 5$.
5. X is an exponential random variable with $\lambda = \frac{1}{2}$.
6. X is an exponential random variable with $\lambda = \frac{1}{4}$.

In Problems 7–10, find the mean, median, and standard deviation of the continuous random variable X if:

7. X is uniformly distributed on $[1, 5]$.
8. X is uniformly distributed on $[2, 8]$.
9. X is an exponential random variable with $\lambda = 5$.
10. X is an exponential random variable with $\lambda = 3$.

In Problems 11 and 12, find the mean and standard deviation of the continuous random variable X if:

11. X is a beta random variable with $\beta = \frac{1}{2}$.
12. X is a beta random variable with $\beta = \frac{1}{3}$.

B

In Problems 13–18, X is a continuous random variable with mean μ. Find μ, and then find $P(X \leq \mu)$ if:

13. X is uniformly distributed on $[0, 4]$.
14. X is uniformly distributed on $[0, 10]$.
15. X is a beta random variable with $\beta = 0$.
16. X is a beta random variable with $\beta = 1$.
17. X is an exponential random variable with $\lambda = 1$.
18. X is an exponential random variable with $\lambda = 2$.

In Problems 19–22, X is a continuous random variable with mean μ and standard deviation σ. Find μ and σ, and then find $P(\mu - \sigma \leq X \leq \mu + \sigma)$ if:

19. X is uniformly distributed on $[-5, 5]$.
20. X is uniformly distributed on $[-2, 2]$.

21. X is an exponential random variable with $x_m = 6 \ln 2$.
22. X is an exponential random variable with $x_m = 4 \ln 2$.

In Problems 23 and 24, X is a beta random variable with mean μ. Find β, and then find $P(X \leq \mu)$ if:

23. $\mu = \frac{3}{5}$

24. $\mu = \frac{5}{7}$

C

Problems 25–28 refer to the uniformly distributed random variable X with probability density function

$$f(x) = \begin{cases} \dfrac{1}{b-a} & \text{if } a \leq x \leq b \\ 0 & \text{otherwise} \end{cases}$$

25. Show that $\mu = (a + b)/2$.

26. Show that $x_m = (a + b)/2$.

27. Show that $\int_{-\infty}^{\infty} x^2 f(x)\, dx = (b^2 + ab + a^2)/3$.

28. Show that $V(X) = (b - a)^2/12$.

Problems 29–32 require the use of a graphic calculator or a computer. For each value of β, approximate to two decimal places the median of the corresponding beta random variable.

29. $\beta = 2$

30. $\beta = 3$

31. $\beta = 4$

32. $\beta = 5$

APPLICATIONS

Business & Economics

33. *Waiting time.* The time (in minutes) applicants must wait for an officer to give them a driver's examination is uniformly distribued on the interval [0, 40]. What is the probability that an applicant must wait more than 25 minutes?

34. *Waiting time.* The time (in minutes) passengers must wait for a commuter plane in a large airport is uniformly distributed on the interval [0, 60]. What is the probability that a passenger waits less than 20 minutes?

35. *Business failures.* The percentage of restaurants that fail during the first year of operation is a beta random variable with $\beta = 2$.

    (A) What is the expected percentage of failures?
    (B) What is the probability that over 80% of the restaurants fail during the first year?

36. *Business failures.* The percentage of computer hobby stores that fail during the first year of operation is a beta random variable with $\beta = 4$.

    (A) What is the expected percentage of failures?
    (B) What is the probability that over 50% of the stores fail during the first year?

37. *Absenteeism.* The percentage of assembly line workers that are absent one Monday each month is a beta random variable. The mean percentage is 50%.

(A) What is the appropriate value of β?
(B) What is the probability that no more than 75% of the workers will be absent on one Monday each month?

38. *Insurance.* The percentage of insurance claims that contain errors is a beta random variable. The expected percentage of erroneous claims is 40%.

(A) What is the appropriate value of β?
(B) What is the probability that fewer than 25% of the claims are erroneous?

Marty Haigney

39. *Communication.* The length of time for telephone conversations (in minutes) is exponentially distributed. The average (mean) length of a conversation is 3 minutes. What is the probability that a conversation lasts less than 2 minutes?

40. *Waiting time.* The waiting time (in minutes) for customers at a drive-in bank is an exponential random variable. The average (mean) time a customer waits is 4 minutes. What is the probability that a customer waits more than 5 minutes?

41. *Service time.* The time between failures of a photo copier is an exponential random variable. Half the copiers require service during the first 2 years of operation. What is the probability that a copier requires service during the first year of operation?

42. *Component failure.* The life expectancy (in years) of a component in a microcomputer is an exponential random variable. Half the components fail in the first 3 years. The company that manufactures the component offers a 1 year warranty. What is the probability that a component will fail during the warranty period?

Life Sciences

43. *Nutrition.* The percentage of the daily requirement of vitamin D present in an 8 ounce serving of milk is a beta random variable with $\beta = .2$.

(A) What is the expected percentage of vitamin D per serving?
(B) What is the probability that a serving contains at least 50% of the daily requirement?

44. *Germination.* The percentage of a certain type of flower seeds that will germinate is a beta random variable with $\beta = 47$. What percentage of these seeds can be expected to germinate?

45. *Medical research.* A new test has been developed to detect a particular disease. The percentage of correct diagnoses obtained by using this test is a beta random variable with mean $\mu = .95$.

(A) What is the value of β?
(B) What is the probability that the percentage of correct diagnoses is greater than 90%?

46. *Medicine.* A scientist is measuring the percentage of a drug present in the bloodstream 10 minutes after an injection. The results indicate that the percentage of the drug present is a beta random variable with mean $\mu = .75$.

(A) What is the value of β?
(B) What is the probability that no more than 25% of the drug is present 10 minutes after an injection?

47. *Survival time.* The time of death (in years) after patients have contracted a certain disease is exponentially distributed. The probability that a patient dies within 1 year is .3.

(A) What is the expected time of death?
(B) what is the probability that a patient survives longer than the expected time of death?

48. *Survival time.* Repeat Problem 47 if the probability that a patient dies within 1 year is .5.

Social Sciences

49. *Education.* The percentage of entering students who complete the first year of college is a beta random variable with $\beta = 17$.

(A) What is the expected percentage of students who complete the first year?
(B) What is the probability that more than 95% of the students complete the first year?

50. *Voter turnout.* The percentage of registered voters who vote in a presidential election is a beta random variable with $\beta = 2.5$.

(A) What is the expected voter turnout?
(B) What is the probability that the voter turnout is greater than 50%?

51. *Learning.* The time (in minutes) it takes an adult to memorize a sequence of random digits is an exponential random variable. The average (mean) time is 2 minutes. What is the probability that it takes an adult over 5 minutes to memorize the digits?

52. *Psychology.* The time (in seconds) it takes rats to find their way through a maze is exponentially distributed. The average (mean) time is 30 seconds. What is the probability that it takes a rat over 1 minute to find a path through the maze?

SECTION 12-7

Normal Distributions

- NORMAL PROBABILITY DENSITY FUNCTIONS
- THE STANDARD NORMAL CURVE
- AREAS UNDER ARBITRARY NORMAL CURVES
- APPROXIMATING A BINOMIAL DISTRIBUTION WITH A NORMAL DISTRIBUTION

We will now consider the most important of all the probability density functions, the *normal probability density function.* This function is at the heart of a great deal of statistical theory, and it is also a useful tool in its own right for solving problems. We will see that the normal probability density function also can be used to provide a good approximation to the binomial distribution.

NORMAL PROBABILITY DENSITY FUNCTIONS

A continuous random variable X has a **normal distribution** and is referred to as a **normal random variable** if its probability density function is the **normal probability density function**

$$f(x) = \frac{1}{\sigma\sqrt{2\pi}} e^{-(x-\mu)^2/2\sigma^2}$$

where μ is any constant and σ is any positive constant. It can be shown, but not easily, that

$$\int_{-\infty}^{\infty} f(x)\,dx = 1$$

$$E(X) = \int_{-\infty}^{\infty} xf(x)\,dx = \mu$$

and

$$V(X) = \int_{-\infty}^{\infty} (x-\mu)^2 f(x)\,dx = \sigma^2$$

Thus, μ is the mean of the normal probability density function and σ is the standard deviation. The graph of $f(x)$ is always a bell-shaped curve called a **normal curve.** Figure 13 illustrates three normal curves for different values of μ and σ.

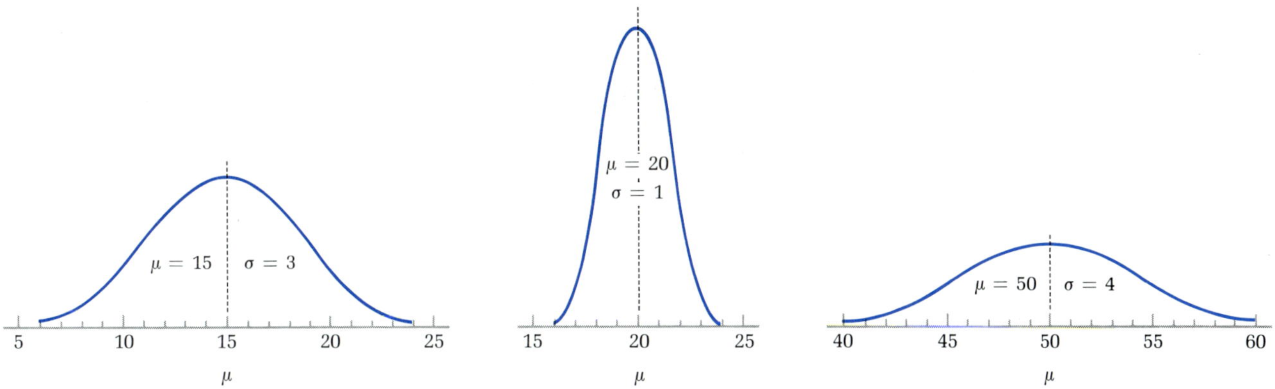

FIGURE 13
Normal probability distributions

The standard deviation measures the dispersion of the normal probability density function about the mean. A small standard deviation indicates a tight clustering about the mean and thus a tall, narrow curve; a large standard deviation indicates a large deviation from the mean and thus a broad, flat curve. Notice that each of the normal curves in Figure 13 is symmetric about a vertical line through the mean. This is true for any normal curve. Thus, the line $x = \mu$ divides the region under a normal curve into two regions with equal area. Since the total area under a normal curve is always 1, the area of each of these regions is .5. This implies that the median of a normal random variable is always equal to the mean (see Figure 14).

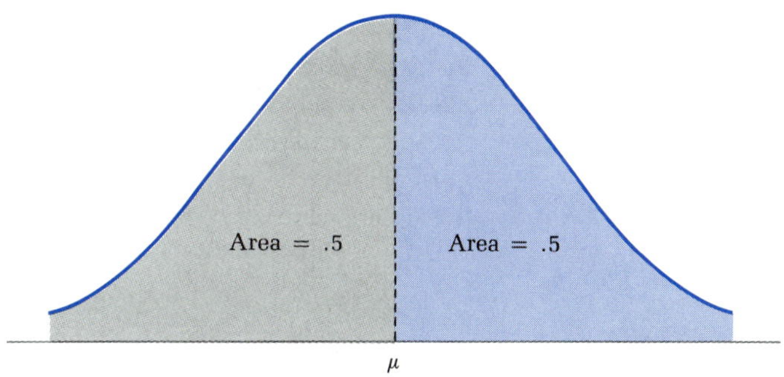

FIGURE 14
The mean and median of a normal random variable

The properties of the normal probability density function are summarized in the box on the next page for ease of reference.

Normal Probability Density Function

$$f(x) = \frac{1}{\sigma\sqrt{2\pi}} e^{-(x-\mu)^2/2\sigma^2} \qquad \sigma > 0$$

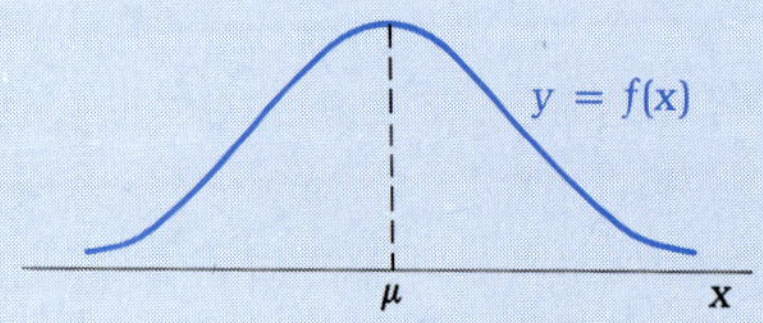

MEAN μ

MEDIAN μ

STANDARD DEVIATION σ

The graph of $f(x)$ is symmetric with respect to the line $x = \mu$.

The cumulative distribution function for a normal random variable is given formally by

$$F(x) = \frac{1}{\sigma\sqrt{2\pi}} \int_{-\infty}^{x} e^{-(t-\mu)^2/2\sigma^2}\, dt$$

It is not possible to express $F(x)$ as a finite combination of the functions we are familiar with. Furthermore, we cannot use antidifferentiation to evaluate probabilities such as

$$P(c \le X \le d) = \frac{1}{\sigma\sqrt{2\pi}} \int_{c}^{d} e^{-(x-\mu)^2/2\sigma^2}\, dx$$

Instead, we will use a table to approximate probabilities of this type. Fortunately, we can use the same table for all normal probability density functions, irrespective of the values of μ and σ. It is a remarkable fact that the area under a normal curve between the mean and a given number of standard deviations to the right (or left) of μ is the same regardless of the values of μ and σ (see Figure 15).

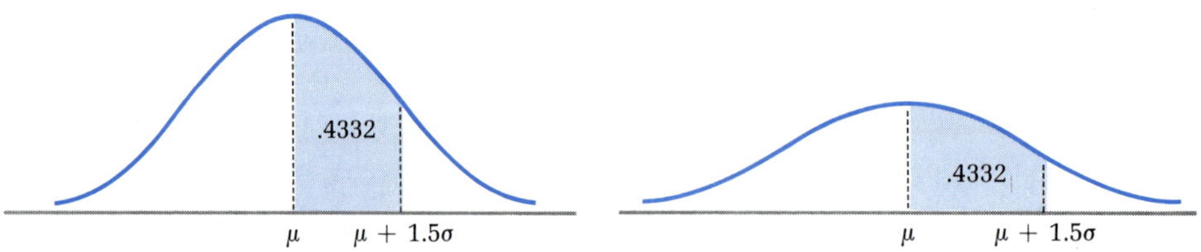

FIGURE 15

◆ THE STANDARD NORMAL CURVE

It is convenient to relate the area under an arbitrary normal curve to the area under a particular normal curve called the *standard normal curve.*

Standard Normal Curve

The normal random variable Z with mean $\mu = 0$ and standard deviation $\sigma = 1$ is called the **standard normal random variable,** and the graph of its probability density function is called the **standard normal curve.**

Table III in Appendix C gives the area under the standard normal curve from 0 to z for values of z in the range $0 \leq z \leq 3.99$. The values in this table, together with the familiar properties of area under a curve, can be used to compute probabilities involving the standard normal random variable.

◆ EXAMPLE 33 Use Table III to compute the following probabilities for the standard normal random variable Z:

(A) $P(0 \leq Z \leq .88)$ (B) $P(Z \leq 1.45)$ (C) $P(.3 \leq Z \leq 2.73)$

Solutions

(A) From Table III, the area under the standard normal curve from $z = 0$ to $z = .88$ is .3106. Thus,

$$P(0 \leq Z \leq .88) = .3106$$

(B) $P(Z \leq 1.45)$ is the area under the standard normal curve over the interval $(-\infty, 1.45]$. Since Table III gives only the area over intervals of the form $[0, z_0]$, we must divide this region into two parts. Let A_1 be the area of the region over the interval $(-\infty, 0]$, and let A_2 be the area of the region over the interval $[0, 1.45]$, as shown in the figure. The median of the standard normal random variable is 0; thus, $A_1 = .5$. From Table III, $A_2 = .4265$. Adding these areas we have

$$P(Z \leq 1.45) = A_1 + A_2 = .5 + .4265 = .9265$$

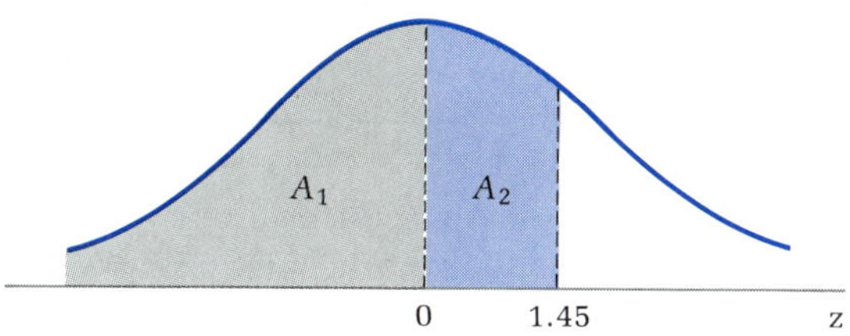

(C) This time we let A_1 be the area of the region from 0 to 2.73, and let A_2 be the area of the region from 0 to .3, as shown in figures below. Using the appropriate values from Table III, we have

$$P(.3 \leq Z \leq 2.73) = A_1 - A_2 = .4968 - .1179 = .3789$$

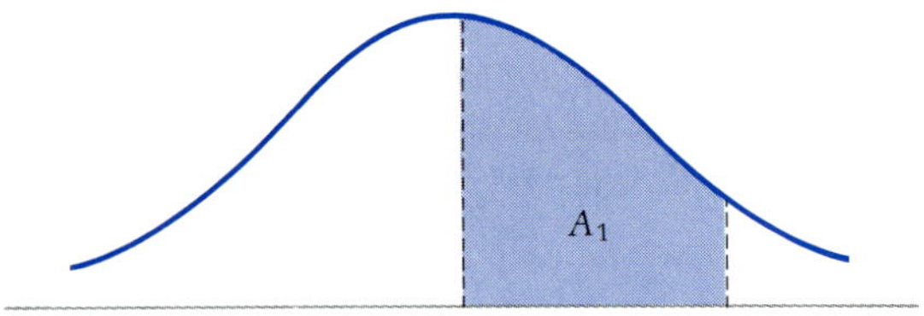

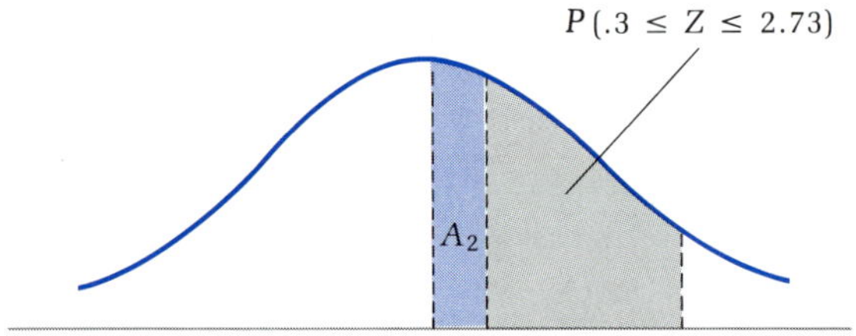

◆

PROBLEM 33 Use Table III to find the following probabilities for the standard normal random variable Z:

(A) $P(0 \leq Z \leq 2.15)$ (B) $P(Z \leq .75)$ (C) $P(.7 \leq Z \leq 3.2)$ ◆

◆ AREAS UNDER ARBITRARY NORMAL CURVES

Now that we have seen how to use Table III to determine probabilities involving the standard normal random variable, we want to consider the more general case. Theorem 4 relates areas under any normal curve to corresponding areas under the standard normal curve. This will enable us to use Table III to find areas under any normal curve, regardless of the values of μ and σ.

THEOREM 4

If X is a normal random variable with mean μ and standard deviation σ, Z is the standard random variable, and

$$z_i = \frac{x_i - \mu}{\sigma} \qquad i = 1, 2 \tag{1}$$

then

$$P(x_1 \leq X \leq x_2) = P(z_1 \leq Z \leq z_2) \tag{2}$$

$$P(x_1 \leq X) = P(z_1 \leq Z) \tag{3}$$

$$P(X \leq x_2) = P(Z \leq z_2) \tag{4}$$

◆ EXAMPLE 34 A manufacturing process produces light bulbs with life expectancies that are normally distributed with a mean of 500 hours and a standard deviation of 100 hours. What percentage of the light bulbs can be expected to last between 500 and 670 hours?

Solution Since the total area under a normal curve is 1, the percentage of light bulbs that can be expected to last between 500 and 670 hours is the same as the area under the curve from 500 to 670 (see the figure).

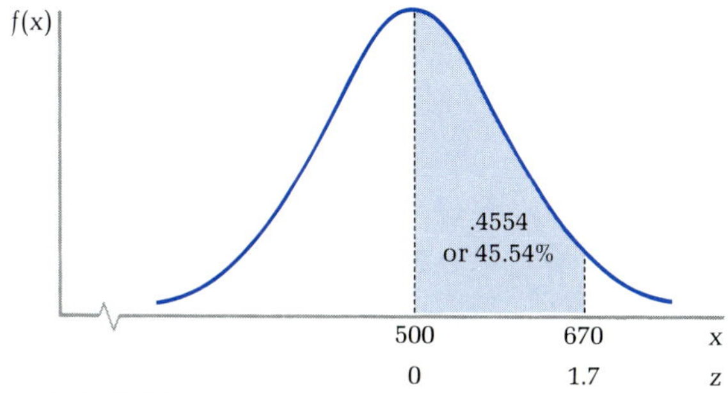

Light bulb life expectancy

If X is the random variable associated with the life expectancy of a light bulb, then we must find

$$P(500 \leq X \leq 670)$$

First we use equation (1) in Theorem 4 to find the corresponding z values:

$$z_1 = \frac{x_1 - \mu}{\sigma} = \frac{500 - 500}{100} = 0 \qquad z_2 = \frac{x_2 - \mu}{\sigma} = \frac{670 - 500}{100} = 1.7$$

Next we use equation (2) in Theorem 4 to write

$$P(500 \leq X \leq 670) = P(0 \leq Z \leq 1.7)$$

Finally, we use Table III to find the area under the standard normal curve from $z = 0$ to $z = 1.7$. This area is .4554. Thus,

$$\begin{aligned} P(500 \leq X \leq 670) &= P(0 \leq Z \leq 1.7) \\ &= .4554 \end{aligned}$$

and we conclude that 45.54% of the light bulbs produced will last between 500 and 670 hours. ◆

PROBLEM 34 Refer to Example 34. What percentage of the light bulbs can be expected to last between 500 and 750 hours? ◆

◆ EXAMPLE 35 Refer to Example 34. From all light bulbs produced, what is the probability that a light bulb chosen at random lasts between 380 and 500 hours?

Solution The corresponding z values are

$$z_1 = \frac{x_1 - \mu}{\sigma} = \frac{380 - 500}{100} = -1.2 \qquad z_2 = \frac{x_2 - \mu}{\sigma} = \frac{500 - 500}{100} = 0$$

Thus,

$$P(380 \leqslant X \leqslant 500) = P(-1.2 \leqslant Z \leqslant 0)$$

Table III does not include negative values of z, but because normal curves are symmetric with respect to a vertical line through the mean, we simply use the absolute value of z in Table III (see the figure).

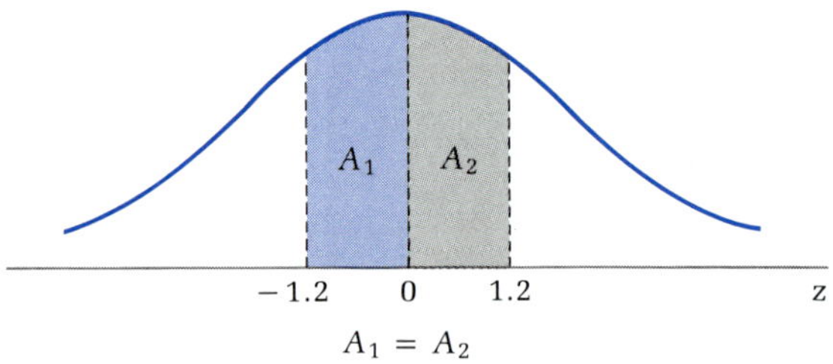

Area under the standard normal curve for negative z

The area under the standard normal curve from $z = -1.2$ to $z = 0$ is the same as the area from $z = 0$ to $z = 1.2$, which is .3849. Thus,

$P(380 \leqslant X \leqslant 500) = P(-1.2 \leqslant Z \leqslant 0)$	Theorem 4
$= P(0 \leqslant Z \leqslant 1.2)$	Symmetry property of the normal curve
$= .3849$	Table III ◆

PROBLEM 35 Refer to Example 35. What is the probability that a light bulb selected at random lasts between 400 and 500 hours? ◆

◆ APPROXIMATING A BINOMIAL DISTRIBUTION WITH A NORMAL DISTRIBUTION

In Section 12-3 we saw that a Poisson distribution can be used to approximate a binomial distribution when n is large:

Binomial distribution	Poisson approximation

$$C_{n,x}p^x q^{n-x} \approx \frac{\lambda^x}{x!}e^{-\lambda} \qquad \lambda = np, \quad x = 0, 1, \ldots, n$$

Although this approximation is quite useful in some situations, it still requires a great deal of computation in problems involving the sum of a large number of terms. For example, the computation of $P(1 \leqslant X \leqslant 100)$ for a binomial random variable X with a large n value (say, $n > 200$) requires a sum of 100 terms, whether you use the binomial distribution or a Poisson approximation. Furthermore, since a table of values for Poisson approximations requires a separate section for each value of λ, no single table can be used for all possible Poisson distributions.

Fortunately, it turns out that probabilities involving the binomial distribution also can be approximated by an appropriately selected normal distribution and evaluated easily using Table III. To clarify ideas and relationships, let us consider an example of a binomial distribution with a relatively small n value. Then we will consider an example with a large n value.

◆ EXAMPLE 36

Market Research

A credit card company claims that their card is used by 40% of the people buying gasoline in a particular city. A random sample of 20 gasoline purchasers is made. If the company's claim is correct, what is the probability that;

(A) From 6 to 12 people in the sample use the card?
(B) Fewer than 4 people in the sample use the card?

Solutions

Before we start, it is useful to look at a histogram of the binomial distribution with a normal distribution fitted to it using the same mean and standard deviation. The mean and standard deviation of the binomial distribution are

$$\mu = np = (20)(.4) = 8 \qquad n = \text{Sample size}$$
$$\sigma = \sqrt{npq} = \sqrt{(20)(.4)(.6)} \approx 2.19 \qquad p = .4 \text{ (from the 40\% claim)}$$

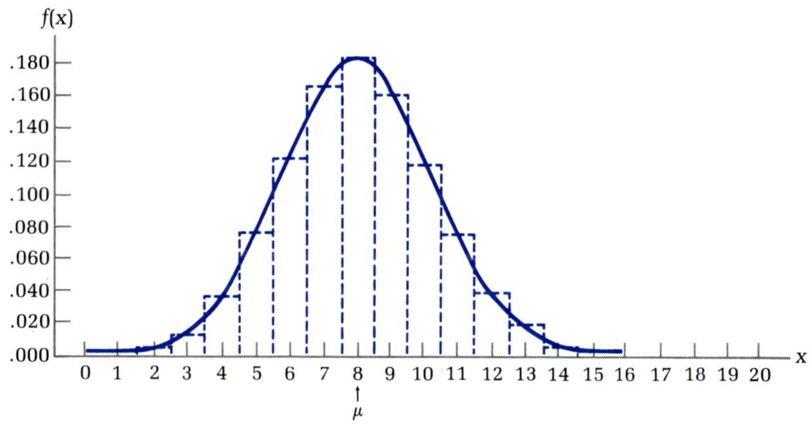

(A) To approximate the probability that 6 to 12 people in the sample use the credit card, we find the area under the normal curve from 5.5 to 12.5. We use 5.5 rather than 6, because the rectangle in the histogram corresponding to 6 extends from 5.5 to 6.5; and, reasoning in the same way, we use 12.5 instead of 12. To use Table III, we split the area into two parts: A_1 to the left of the mean and A_2 to the right of the mean. A sketch is helpful:

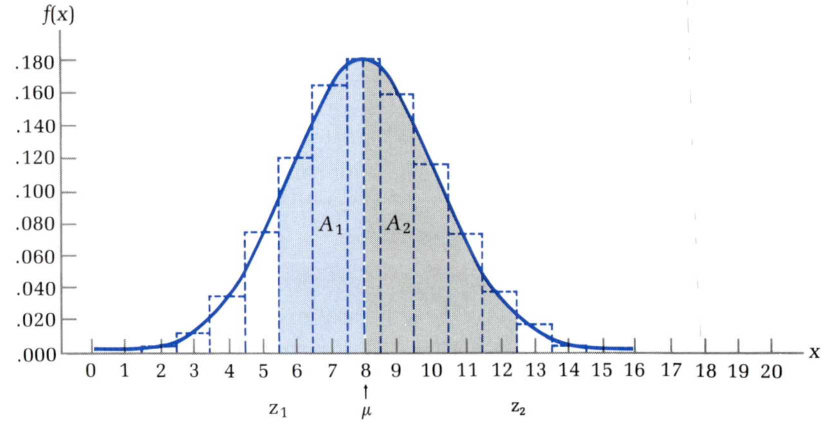

Areas A_1 and A_2 are found as follows:

$$z_1 = \frac{x - \mu}{\sigma} = \frac{5.5 - 8}{2.19} \approx -1.14 \qquad A_1 = .3729$$

$$z_2 = \frac{x - \mu}{\sigma} = \frac{12.5 - 8}{2.19} \approx 2.05 \qquad A_2 = .4798$$

$$\text{Total area} = A_1 + A_2 = .8527$$

Thus, the approximate probability that the sample will contain between 6 and 12 users of the credit card is .85 (assuming the firm's claim is correct).

(B) To use the normal curve to approximate the probability that the sample contains fewer than 4 users of the credit card, we must find the area A_1 under the normal curve to the left of 3.5. Again, a sketch is useful:

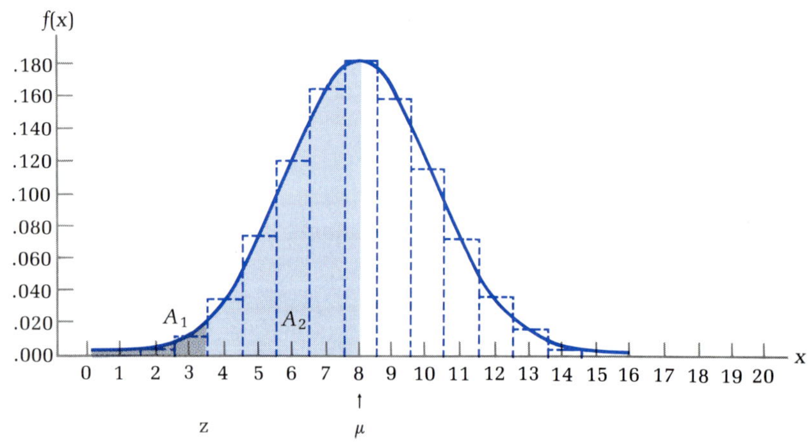

Since the total area under either half of the normal curve is .5, we first use Table III to find the area A_2 under the normal curve from 3.5 to the mean 8, and then subtract A_2 from .5:

$$z = \frac{x - \mu}{\sigma} = \frac{3.5 - 8}{2.19} \approx -2.05 \qquad A_2 = .4798$$

$$A_1 = .5 - A_2 = .5 - .4798 = .0202$$

Thus, the approximate probability that the sample contains fewer than 4 users of the credit card is .02 (assuming the company's claim is correct). ◆

PROBLEM 36 In Example 36 use the normal curve to approximate the probability that in the sample there are:

(A) From 5 to 9 users of the credit card
(B) More than 10 users of the card ◆

You no doubt are wondering how large n should be before a normal distribution provides an adequate approximation for a binomial distribution. Without getting too involved, the following rule-of-thumb provides a good test:

Rule-of-Thumb Test

Use a normal distribution to approximate a binomial distribution only if the interval $[\mu - 3\sigma, \mu + 3\sigma]$ lies entirely in the interval from 0 to n.

Note that in Example 36 the interval $[\mu - 3\sigma, \mu + 3\sigma] = [1.43, 14.57]$ lies entirely within the interval from 0 to 20; hence, the use of the normal distribution was justified.

◆ EXAMPLE 37 *Quality Control* A company manufactures 50,000 ballpoint pens each day. The manufacturing process produces 50 defective pens per 1,000, on the average. A random sample of 400 pens is selected from each day's production and tested. What is the probability that the sample contains:

(A) At least 14 and no more than 25 defective pens?
(B) 33 or more defective pens?

Solutions Is it appropriate to use a normal distribution to approximate this binomial distribution? The answer is yes, since the rule-of-thumb test passes with ease:

$$\mu = np = 400(.05) = 20 \qquad p = \tfrac{50}{1,000} = .05$$
$$\sigma = \sqrt{npq} = \sqrt{400(.05)(.95)} \approx 4.36$$
$$[\mu - 3\sigma, \mu + 3\sigma] = [6.92, 33.08]$$

This interval is well within the interval from 0 to 400.

(A) To find the approximate probability of the number of defective pens in a sample being at least 14 and not more than 25, we find the area under the normal curve from 13.5 to 25.5. To use Table III, we split the area into an area to the left of the mean and an area to the right of the mean, as shown:

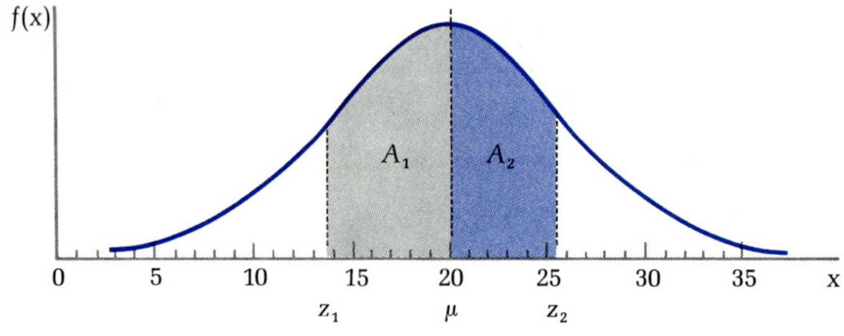

$$z_1 = \frac{x - \mu}{\sigma} = \frac{13.5 - 20}{4.36} \approx -1.49 \qquad A_1 = .4319$$

$$z_2 = \frac{x - \mu}{\sigma} = \frac{25.5 - 20}{4.36} \approx 1.26 \qquad A_2 = .3962$$

$$\text{Total area} = A_1 + A_2 = .8281$$

Thus, the approximate probability of the number of defective pens in the sample being at least 14 and not more than 25 is .83.

(B)

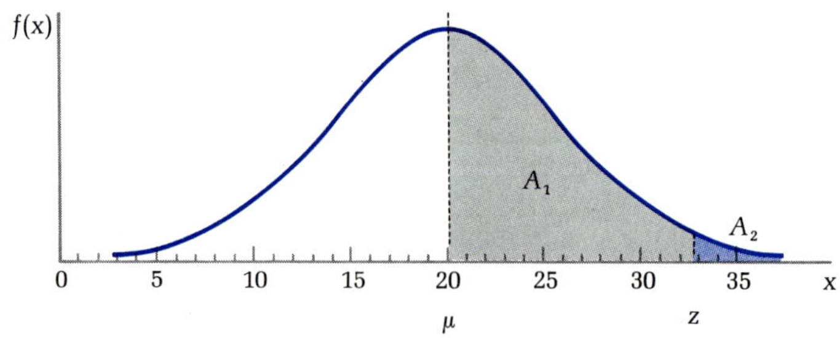

Since the total area under a normal curve from the mean on is .5, we find the area A_1 from Table III and subtract it from .5 to obtain A_2:

$$z = \frac{x - \mu}{\sigma} = \frac{32.5 - 20}{4.36} \approx 2.87 \qquad A_1 = .4979$$

$$A_2 = .5 - A_1 = .5 - .4979 = .0021 \approx .002$$

Thus, the approximate probability of finding 33 or more defective pens in the sample is .002. If a random sample of 400 included more than 33 defective pens, then the management would conclude that either a rare event has happened and the manufacturing process is still producing only 50 defective pens per 1,000, on the average, or something is wrong with the manufacturing process and it is producing more than 50 defective pens per 1,000, on the average. The company might very well have a policy of checking the manufacturing process whenever 33 or more defective pens are found in a sample rather than believing a rare event has happened and that the manufacturing process is still running smoothly. ◆

PROBLEM 37 Suppose in Example 37 that the manufacturing process produces 40 defective pens per 1,000, on the average. What is the approximate probability that in the sample of pens there are:

(A) At least 10 and no more than 20 defective pens?
(B) 27 or more defective pens? ◆

When to Use the .5 Adjustment

If we are assuming a normal probability distribution for a continuous random variable (such as that associated with heights or weights of people), then we find $P(a \leq x \leq b)$, where a and b are real numbers, by finding the area under the corresponding normal curve from a to b (see Example 34). However, if we use a normal probability distribution to approximate a binomial probability distribution, then we find $P(a \leq x \leq b)$, where a and b are nonnegative integers, by finding the area under the corresponding normal curve from $a - .5$ to $b + .5$ (see Examples 36 and 37).

Answers to Matched Problems

33. (A) .4842 (B) .7734 (C) .2413 34. 49.38% 35. .3413
36. (A) .70 (B) .13 37. (A) .83 (B) .004

EXERCISE 12-7

A *Use Table III to find the area under the standard normal curve from 0 to the indicated value of z.*

1. 1 **2.** 2 **3.** -3 **4.** -1
5. .9 **6.** -1.7 **7.** 2.47 **8.** -1.96

Given a normal distribution with mean 50 and standard deviation 10, use Theorem 4 and Table III to find the area under this normal curve from the mean to the indicated measurement.

9. 65 **10.** 75 **11.** 83 **12.** 79
13. 45 **14.** 38 **15.** 42 **16.** 26

B *In Problems 17–24, find the indicated probability for the standard normal random variable Z.*

17. $P(-1.7 \leqslant Z \leqslant .6)$ **18.** $P(-.4 \leqslant Z \leqslant 2)$ **19.** $P(.45 \leqslant Z \leqslant 2.25)$

20. $P(1 \leqslant Z \leqslant 2.75)$ **21.** $P(Z \geqslant .75)$ **22.** $P(Z \leqslant -1.5)$

23. $P(Z \leqslant 1.88)$ **24.** $P(Z \geqslant -.66)$

Given a normal random variable X with mean 70 and standard deviation 8, find the indicated probabilities.

25. $P(60 \leqslant X \leqslant 80)$ **26.** $P(50 \leqslant X \leqslant 90)$ **27.** $P(62 \leqslant X \leqslant 74)$

28. $P(66 \leqslant X \leqslant 78)$ **29.** $P(X \geqslant 88)$ **30.** $P(X \geqslant 90)$

31. $P(X \leqslant 60)$ **32.** $P(X \leqslant 56)$

In Problems 33–40, use the rule-of-thumb test to check whether a normal distribution (with the same mean and standard deviation as the binomial distribution) is a suitable approximation for the binomial distribution with:

33. $n = 15, \quad p = .7$ **34.** $n = 12, \quad p = .6$

35. $n = 15, \quad p = .4$ **36.** $n = 20, \quad p = .6$

37. $n = 100, \quad p = .05$ **38.** $n = 200, \quad p = .03$

39. $n = 500, \quad p = .05$ **40.** $n = 400, \quad p = .08$

A binomial experiment consists of 500 trials with the probability of success for each trial .4. What is the probability of obtaining the number of successes indicated in Problems 41–48? Approximate these probabilities to two decimal places using a normal curve. (This binomial experiment easily passes the rule-of-thumb test, as you can check. When computing the probabilities, adjust the intervals as in Examples 36 and 37.)

41. 185–220 **42.** 190–205 **43.** 210–220 **44.** 175–185

45. 225 or more **46.** 212 or more **47.** 175 or less **48.** 188 or less

To graph Problems 49–52, use a graphic calculator or a computer and refer to the normal probability distribution function with mean μ and standard deviation σ:

$$f(x) = \frac{1}{\sigma\sqrt{2\mu}} e^{-(x-\mu)^2/2\sigma^2} \tag{1}$$

49. Graph equation (1) with $\sigma = 5$ and:

(A) $\mu = 10$ (B) $\mu = 15$ (C) $\mu = 20$

Graph all three in the same viewing rectangle with the x range $= [-10, 40]$ and the y range $= [0, 0.1]$.

50. Graph equation (1) with $\sigma = 4$ and:

(A) $\mu = 8$ (B) $\mu = 12$ (C) $\mu = 16$

Graph all three in the same viewing rectangle with the x range $= [-5, 30]$ and the y range $= [0, 0.1]$.

51. Graph equation (1) with $\mu = 20$ and:

(A) $\sigma = 2$ (B) $\sigma = 4$

Graph both in the same viewing rectangle with the x range = [0, 40] and the y range = [0, 0.2].

52. Graph equation (1) with $\mu = 18$ and:

(A) $\sigma = 3$ (B) $\sigma = 6$

Graph both in the same viewing rectangle with the x range = [0, 40] and the y range = [0, 0.2].

APPLICATIONS

In problems involving random samples of a population, use normal distributions to approximate the appropriate binomial distributions. In all other problems, assume normal distributions.

Business & Economics

53. *Sales.* Salespeople for a business machine company have average annual sales of $200,000, with a standard deviation of $20,000. What percentage of the salespeople would be expected to make annual sales of $240,000 or more?

54. *Guarantees.* The average lifetime for a car battery of a certain brand is 170 weeks, with a standard deviation of 10 weeks. If the company guarantees the battery for 3 years, what percentage of the batteries sold would be expected to be returned before the end of the warranty period?

55. *Quality control.* A manufacturing process produces a critical part of average length 100 millimeters, with a standard deviation of 2 millimeters. All parts deviating by more than 5 millimeters from the mean must be rejected. What percentage of the parts must be rejected, on the average?

56. *Quality control.* An automated manufacturing process produces a component with an average width of 7.55 centimeters, with a standard deviation of 0.02 centimeter. All components deviating by more than 0.05 centimeter from the mean must be rejected. What percentage of the parts must be rejected, on the average?

57. *Marketing claims.* A company claims that 60% of the households in a given community use their product. A competitor surveys the community, using a random sample of 40 households, and finds only 15 households out of the 40 in the sample using the product. If the company's claim is correct, what is the probability of 15 or fewer households using the product in a sample of 40? What can you conclude?

58. *Labor relations.* A union representative claims 60% of the union membership will vote in favor of a particular settlement. A random sample of 100 members is polled, and out of these, 47 favor the settlement. What is the approximate probability of 47 or fewer in a sample of 100 favoring the settlement when 60% of all the membership favor the settlement? What can you conclude?

Life Sciences

59. *Medicine.* The average healing time of a certain type of incision is 240 hours, with standard deviation of 20 hours. What percentage of the people having this incision would heal in 8 days or less?

60. *Agriculture.* The average height of a hay crop is 38 inches, with a standard deviation of 1.5 inches. What percentage of the crop will be 40 inches or more?

61. *Genetics.* In a family with 2 children, the probability that both children are girls is approximately .25. In a random sample of 1,000 families with 2 children, what is the approximate probability that 220 or fewer will have 2 girls?

62. *Genetics.* In Problem 61, what is the approximate probability of the number of families with 2 girls in the sample being at least 225 and not more than 275?

Social Sciences

63. *Testing.* Scholastic Aptitude Tests are scaled so that the mean score is 500 and the standard deviation is 100. What percentage of the students taking this test should score 700 or more?

64. *Politics.* Candidate Harkins claims that a private poll indicates she will receive 52% of the vote for governor. Her opponent, Mankey, secures the services of another pollster, who finds that 470 out of a random sample of 1,000 registered voters favor Harkins. If Harkins' claim is correct, what is the probability that only 470 or fewer will favor her in a random sample of 1,000? What can you conclude?

65. *Grading on a curve.* An instructor grades on a curve by assuming the grades on a test are normally distributed. If the average grade is 70 and the standard deviation is 8, find the test scores for each grade interval if the instructor wishes to assign grades as follows: 10% A's, 20% B's, 40% C's, 20% D's, and 10% F's.

66. *Psychology.* A test devised to measure aggressive–passive personalities was standardized on a large group of people. The scores were normally distributed, with a mean of 50 and a standard deviation of 10. If we want to designate the highest 10% as aggressive, the next 20% as moderately aggressive, the middle 40% as average, the next 20% as moderately passive, and the lowest 10% as passive, what ranges of scores will be covered by these five designations?

SECTION 12-8 Chapter Review

Important Terms and Symbols

12-1 *Finite Probability Models.* Random experiment; sample space; simple outcome; event; simple event; compound event; event E occurs; equally likely; random variable; probability function; probability distribution of a random variable; histogram; expected value; payoff table; fair game; mean; variance; standard deviation

Probability distribution: $P(x) = P(X = x)$

Mean: $\mu = E(X) = x_1p_1 + x_2p_2 + \cdots + x_mp_m$

Variance: $V(X) = (x_1 - \mu)^2p_1 + (x_2 - \mu)^2p_2 + \cdots + (x_m - \mu)^2p_m$

Standard deviation: $\sigma = \sqrt{V(X)}$

12-2 *Binomial Distributions.* Bernoulli trial; success; failure; independent trials; sequence of Bernoulli trials; binomial formula; binomial distribution

$$P(x) = P(x \text{ successes in } n \text{ trials}) = C_{n,x}p^xq^{n-x},\ x \in \{0, 1, \ldots, n\}$$

Mean: $\mu = np$; Standard deviation: $\sigma = \sqrt{npq}$

12-3 *Poisson Distributions.* Approximation to the binomial distribution; rule-of-thumb; Poisson experiment; Poisson random variable; Poisson probability distribution; discrete random variable

$$C_{n,x}p^xq^{n-x} \approx \frac{\lambda^x}{x!}e^{-\lambda},\ \lambda = np,\ x \in \{0, 1, \ldots, n\}$$

$$P(x) = P(x \text{ occurrences per unit of measure}) = \frac{\lambda^x}{x!}e^{-\lambda},\ x \in \{0, 1, 2, \ldots\}$$

Mean: $\mu = \lambda$; Standard deviation: $\sigma = \sqrt{\lambda}$

12-4 *Continuous Random Variables.* Continuous random variable; probability density function; cumulative probability distribution function

$$P(c \leq X \leq d) = \int_c^d f(x)\,dx; \quad F(x) = P(X \leq x) = \int_{-\infty}^x f(t)\,dt$$

12-5 *Expected Value, Standard Deviation, and Median of Continuous Random Variables.* Expected value; mean; variance; standard deviation; alternate formula for variance; median

Mean: $\mu = E(X) = \int_{-\infty}^{\infty} xf(x)\,dx$

Variance: $V(x) = \int_{-\infty}^{\infty} (x - \mu)^2f(x)\,dx = \int_{-\infty}^{\infty} x^2f(x)\,dx - \mu^2$

Standard deviation: $\sigma = \sqrt{V(X)}$; Median: Solve $P(X \leq x_m) = \frac{1}{2}$ for x_m

12-6 *Uniform, Beta, and Exponential Distributions.*

Uniform probability density function: $f(x) = \begin{cases} \dfrac{1}{b-a} & \text{if } a \le x \le b \\ 0 & \text{otherwise} \end{cases}$

Beta probability density function: $f(x) = \begin{cases} (\beta+1)(\beta+2)x^{\beta}(1-x) & \text{if } 0 \le x \le 1 \\ 0 & \text{otherwise} \end{cases}$

Exponential probability density function: $f(x) = \begin{cases} (1/\lambda)e^{-x/\lambda} & \text{if } x \ge 0 \\ 0 & \text{otherwise} \end{cases}$

12-7 *Normal Distributions.* Normal distribution; normal random variable; normal probability density function; normal curve; standard normal random variable; standard normal curve; approximating binomial distributions with normal distributions; rule-of-thumb test

$$z = (x - \mu)/\sigma$$

EXERCISE 12-8

Chapter Review

Work through all the problems in this chapter review and check your answers in the back of the book. (Answers to all review problems are there.) Where weaknesses show up, review appropriate sections in the text.

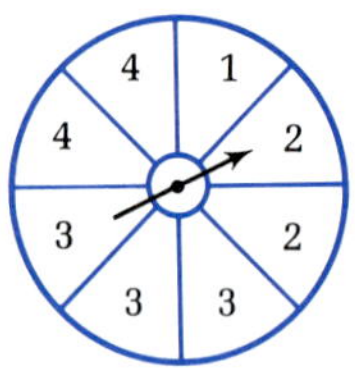

A *A spinner can land on any one of eight different sectors, and each sector is as likely to turn up as any other. The sectors are numbered in the figure in the margin. An experiment consists of spinning the dial once and recording the number in the sector that the spinner lands on. Problems 1–3 refer to this experiment.*

1. Find a sample space and probability distribution for this experiment.
2. What is the probability that the spinner stops on an even-numbered sector?
3. Find the expected value, variance, and standard deviation for the probability distribution found in Problem 1.

4. If a fair coin is tossed eight times, what is the probability of obtaining exactly 2 heads?

5. (A) Draw a histogram for the binomial distribution

$$P(x) = C_{3,x}(.4)^x(.6)^{3-x}$$

(B) What are the mean and standard deviation?

6. Find P(3) for the binomial distribution with $n = 500$ and $p = .004$, and then find the Poisson approximation to this probability.
7. Find $P(X \le 1)$ and $P(X > 2)$ for the Poisson random variable X with mean $\mu = 4$.

Problems 8–11 refer to the continuous random variable X with probability density function

$$f(x) = \begin{cases} 1 - \frac{1}{2}x & \text{if } 0 \leq x \leq 2 \\ 0 & \text{otherwise} \end{cases}$$

8. Find $P(0 \leq X \leq 1)$ and illustrate with a graph.
9. Find the mean, variance, and standard deviation.
10. Find and graph the associated cumulative probability distribution function.
11. Find the median.

12. If Z is the standard normal random variable, find $P(0 \leq Z \leq 2.5)$.
13. If X is a normal random variable with a mean of 100 and a standard deviation of 10, find $P(100 \leq X \leq 118)$.

B

14. (A) Construct a histogram for the binomial distribution

$$P(x) = C_{6,x}(.5)^x(.5)^{6-x}$$

(B) What are the mean and standard deviation?

15. The probability of drawing a king from a standard 52-card deck is $\frac{1}{13}$. Use the Poisson approximation to the binomial distribution to find the probability of drawing 5 kings in 117 single-card draws (with replacement) from a 52-card deck.

Problems 16–19 refer to the continuous random variable X with probability density function

$$f(x) = \begin{cases} \frac{5}{2}x^{-7/2} & \text{if } x \geq 1 \\ 0 & \text{otherwise} \end{cases}$$

16. Find $P(1 \leq X \leq 4)$ and illustrate with a graph.
17. Find the mean, variance, and standard deviation.
18. Find and graph the associated cumulative probability distribution function.
19. Find the median.

Problems 20–23 refer to a beta random variable X with $\beta = 5$.

20. Find and graph the probability density function.
21. Find $P(\frac{1}{4} \leq X \leq \frac{3}{4})$.
22. Find and graph the associated cumulative probability distribution function.
23. Find the mean and standard deviation.

Problems 24–27 refer to an exponentially distributed random variable X.

24. If $P(4 \leq X) = e^{-2}$, find the probability density function.
25. Find $P(0 \leq X \leq 2)$.
26. Find the associated cumulative probability distribution function.
27. Find the mean, standard deviation, and median.

28. What are the mean and standard deviation for a binomial distribution with $p = .6$ and $n = 1{,}000$?

29. If the probability of success in a single trial of a binomial experiment with 1,000 trials is .6, what is the probability of obtaining at least 550 and no more than 650 successes in 1,000 trials? [*Hint:* Approximate with a normal distribution.]

30. Given a normal distribution with mean 50 and standard deviation 6, find the area under the normal curve:

(A) Between 41 and 62 (B) From 59 on

C

31. If X is a beta random variable with mean $\mu = .8$, what is the value of β?

32. Find the mean and the median of the continuous random variable with probability density function

$$f(x) = \begin{cases} 50/(x+5)^3 & \text{if } x \geqslant 0 \\ 0 & \text{otherwise} \end{cases}$$

33. If $f(x)$ is a continuous probability density function with mean μ and standard deviation σ and a, b, and c are constants, evaluate the integral given below. Express the result in terms of μ, σ, a, b, and c.

$$\int_{-\infty}^{\infty} (ax^2 + bx + c)f(x)\, dx$$

APPLICATIONS

Business & Economics

34. *Quality control.* A manufacturing process produces, on the average, 6 defective items out of 100. To control quality, each day a sample of 10 completed items is selected at random and is inspected. If the sample produces more than 2 defective items, then the whole day's output is inspected and the manufacturing process is reviewed. What is the probability of this happening, assuming that the process is still producing 6% defective items and that the number of defective items is a binomial random variable?

35. *Equipment failure.* A company's records show that a printing press breaks down, on the average, 26 times a year. Assuming that the number of times the press breaks down per week is a Poisson random variable, find the probability that during a particular week:

(A) The printing press will not break down.
(B) The printing press will break down once.
(C) The printing press will break down more than 2 times.

36. *Demand.* The manager of a movie theater has determined that the weekly demand for popcorn (in pounds) is a continuous random variable with probability density function

$$f(x) = \begin{cases} \frac{1}{50}(1 - .01x) & \text{if } 0 \leq x \leq 100 \\ 0 & \text{otherwise} \end{cases}$$

(A) If the manager has 50 pounds of popcorn on hand at the beginning of the week, what is the probability that this will be enough to meet the weekly demand?

(B) If the manager wants the probability that the supply on hand exceeds the weekly demand to be .96, how much popcorn must be on hand at the beginning of the week?

37. *Credit applications.* The percentage of applications for a national credit card that are processed on the same day they are received is a beta random variable with $\beta = 1$.

(A) What is the probability that at least 20% of the applications received are processed the same day they arrive?

(B) What is the expected percentage of applications processed the same day they arrive?

38. *Computer failure.* A computer manufacturer has determined that the time between failures for its computers is an exponentially distributed random variable with a mean failure time of 4,000 hours. Suppose a particular computer has just been repaired.

(A) What is the probability that the computer operates for the next 4,000 hours without a failure?

(B) What is the probability that the computer fails in the next 1,000 hours?

39. *Waiting time.* Each business day the switchboard for a commercial airline receives an average of 120 calls per hour.

(A) Assume that the number of calls per minute is a Poisson random variable. What is the probability that 4 calls are received during a particular minute?

(B) Assume that the time between calls is an exponential random variable. If a call has just arrived, what is the probability that no other calls arrive during the next 45 seconds?

40. *Radial tire failure.* The life expectancy (in miles) of a certain brand of radial tire is a normal random variable with a mean of 35,000 and a standard deviation of 5,000. What is the probability that a tire fails during the first 25,000 miles of use?

41. *Personnel screening.* The scores on a screening test for new technicians are normally distributed with mean 100 and standard deviation 10. Find the approximate percent of applicants taking the test scoring:

(A) Between 92 and 108 (B) 115 or higher

42. *Market research.* A newspaper publisher claims that 70% of the people in a community read their newspaper. Doubting the assertion, a competitor randomly surveys 200 people in the community. Based on the publisher's claim (and assuming a binomial distribution):

(A) Compute the mean and standard deviation of the binomial distribution.
(B) Determine whether the rule-of-thumb test warrants the use of a normal distribution to approximate this binomial distribution.
(C) Calculate the approximate probability of finding at least 130 and no more than 155 readers in the sample.
(D) Determine the approximate probability of finding 125 or fewer readers in the sample.

Life Sciences

43. *Medicine.* The shelf-life (in months) of a certain drug is a continuous random variable with probability density function

$$f(x) = \begin{cases} 10/(x+10)^2 & \text{if } x \geqslant 0 \\ 0 & \text{otherwise} \end{cases}$$

(A) What is the probability that the drug is still usable after 5 months?
(B) What is the median shelf-life?

44. *Life expectancy.* The life expectancy (in months) after dogs have contracted a certain disease is an exponentially distributed random variable. The probability of surviving more than 1 month is e^{-2}. After contracting this disease:

(A) What is the probability of the dog surviving more than 2 months?
(B) What is the mean life expectancy?

45. *Harmful side effects of drugs.* A drug causes harmful side effects in 25% of the patients treated with the drug. If the drug is administered to 100 patients, what is the probability that 30 or more of these patients will suffer from the side effects? (Use a normal distribution to approximate this binomial distribution.)

46. *Measles epidemic.* Measles are found to be spreading through a high-rise college dormitory. A public health official determines that the number of infected students per floor is a Poisson random variable with mean $\mu = 3$. Find the probability that a particular floor has:

(A) No infected students (B) 1 infected student
(C) More than 3 infected students

Social Sciences

47. *Testing.* The percentage of correct answers on a college entrance examination is a beta random variable. The mean score is 75%. What is the probability that a student answers over 50% of the questions correctly?

48. *Testing.* The IQ scores for 6-year-old children in a certain area are normally distributed with a mean of 108 and a standard deviation of 12. What

percentage of the children can be expected to have IQ scores of 135 or more?

49. *Safety research.* The intersection of 76th Street and Good Hope Road has the highest accident rate of all the intersections in a certain city. A study of the traffic flow at this intersection shows that, on an average weekday, 2,880 vehicles arrive at this intersection at random times during the afternoon rush hour (from 4:00 PM to 5:00 PM). If the number of vehicle arrivals per second during this rush hour is a Poisson random variable, find the probability that during a particular second:

    (A) No vehicles arrive at the intersection.
    (B) 1 vehicle arrives at the intersection.
    (C) 5 or more vehicles arrive at the intersection.

APPENDIX A

CHAPTER 5 SYSTEMS OF LINEAR EQUATIONS; MATRICES FROM COLLEGE MATHEMATICS, SIXTH EDITION

APPENDIX B

CHAPTER 6 LINEAR INEQUALITIES AND LINEAR PROGRAMMING FROM COLLEGE MATHEMATICS, SIXTH EDITION

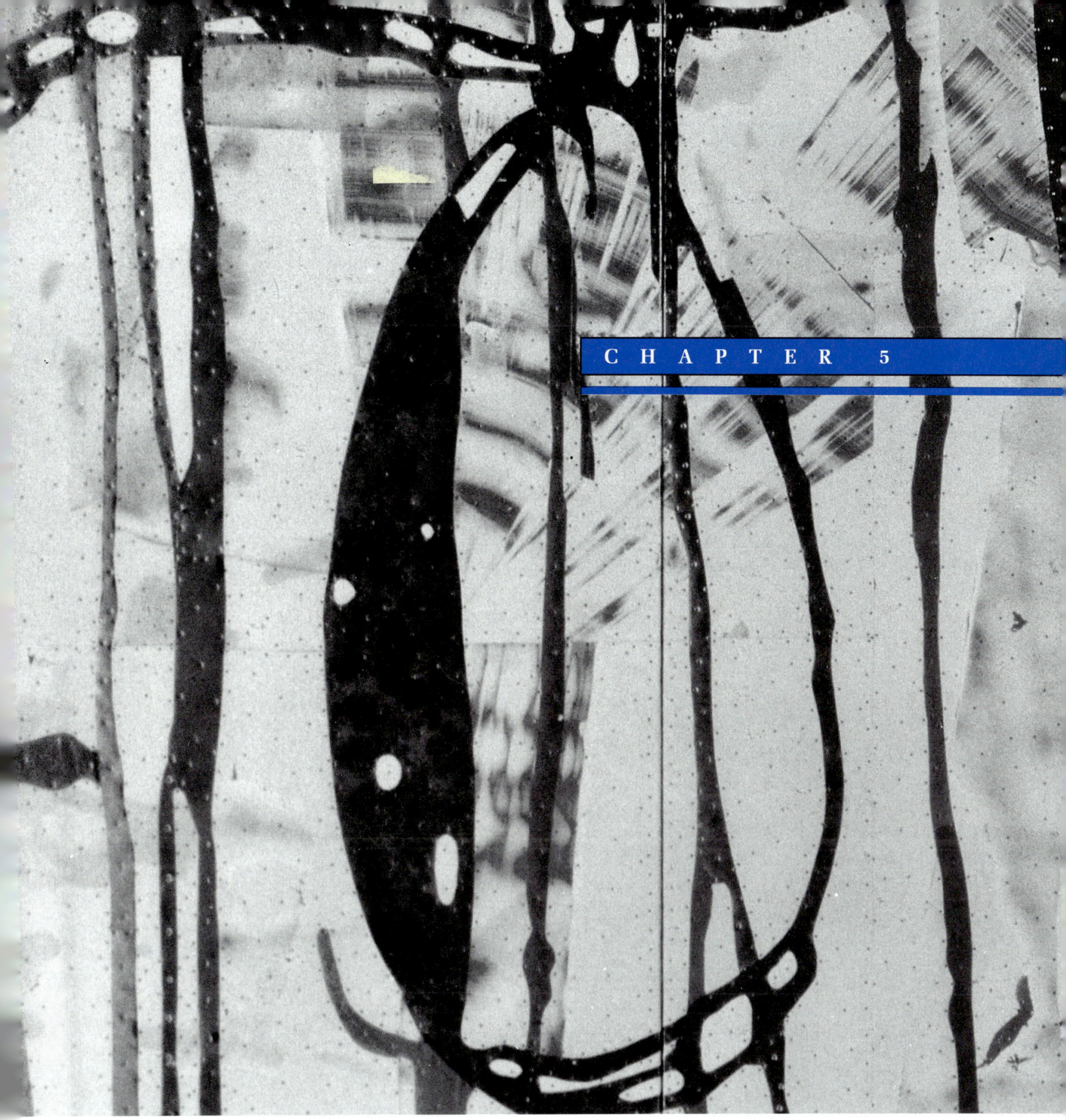

CHAPTER 5

Systems of Linear Equations; Matrices

CHAPTER 5

Contents

In this chapter we first review how systems of linear equations involving two variables are solved using techniques learned in elementary algebra. Because these techniques are not suitable for linear systems involving larger numbers of equations and variables, we then turn to a different method of solution involving the concept of an *augmented matrix*, which arises quite naturally when dealing with larger linear systems. We then study *matrices* and *matrix operations* in their own right as a new mathematical form. With this new tool, we return to systems of equations from a fresh point of view. The matrix techniques introduced in this chapter form the basis for computer solutions of large-scale systems.

SECTION 5-1

Review: Systems of Linear Equations in Two Variables

- ◆ SYSTEMS IN TWO VARIABLES
- ◆ APPLICATIONS

◆ SYSTEMS IN TWO VARIABLES

To establish basic concepts, consider the following simple example: If 2 adult tickets and 1 child ticket cost \$8, and if 1 adult ticket and 3 child tickets cost \$9, what is the price of each?

Let

x = Price of adult ticket

y = Price of child ticket

Then

$$\begin{aligned} 2x + y &= 8 \\ x + 3y &= 9 \end{aligned}$$

We now have a system of two linear equations in two variables. To solve this system, we find all ordered pairs of real numbers that satisfy both equations at the same time. In general, we are interested in solving linear systems of the type

$$ax + by = h$$
$$cx + dy = k$$

where a, b, c, d, h, and k are real constants. A pair of numbers $x = x_0$ and $y = y_0$ [also written as an ordered pair (x_0, y_0)] is a **solution** of this system if each equation is satisfied by the pair. The set of all such ordered pairs of numbers is called the **solution set** for the system. To **solve** a system is to find its solution set. We will consider three methods of solving such systems: *graphing*, *substitution*, and *elimination by addition*. Each method has certain advantages, depending on the situation.

Solution by Graphing

To solve the ticket problem above by graphing, we graph both equations in the same coordinate system. The coordinates of any points that the graphs have in common must be solutions to the system, since they must satisfy both equations.

◆ EXAMPLE 1 Solve the ticket problem by graphing:

$$2x + y = 8$$
$$x + 3y = 9$$

Solution Graph each equation in the same rectangular coordinate system (both graphs are straight lines). Then estimate the coordinates of any common points on the two lines.

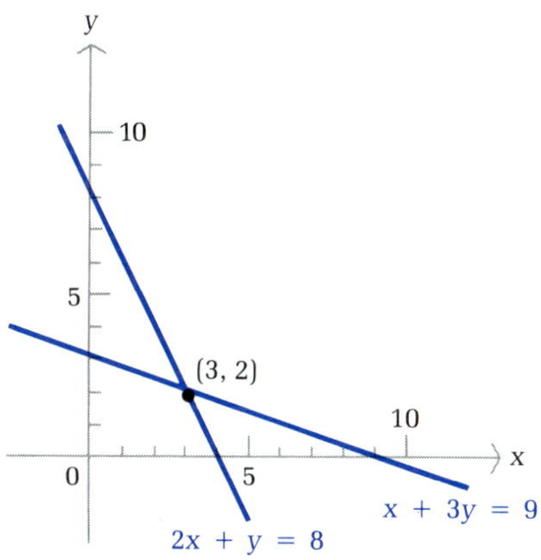

$x = \$3$ Adult ticket
$y = \$2$ Child ticket

Check

$2x + y = 8$	$x + 3y = 9$
$2(3) + 2 \overset{?}{=} 8$	$3 + 3(2) \overset{?}{=} 9$
$8 \overset{\checkmark}{=} 8$	$9 \overset{\checkmark}{=} 9$

(3, 2) must satisfy each original equation for a complete check. ◆

PROBLEM 1 Solve by graphing and check:

$$2x - y = -3$$
$$x + 2y = -4$$

◆

It is clear that Example 1 has exactly one solution, since the lines have exactly one point of intersection. In general, lines in a rectangular coordinate system are related to each other in one of the three ways illustrated in the next example.

◆ EXAMPLE 2 Solve each of the following systems by graphing:

(A) $x - 2y = 2$, $x + y = 5$ (B) $x + 2y = -4$, $2x + 4y = 8$ (C) $2x + 4y = 8$, $x + 2y = 4$

Solutions

(A)

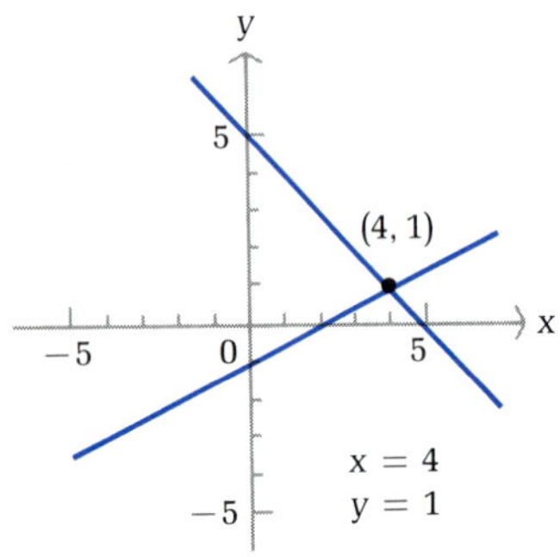

Intersection at one point only—exactly one solution

(B)

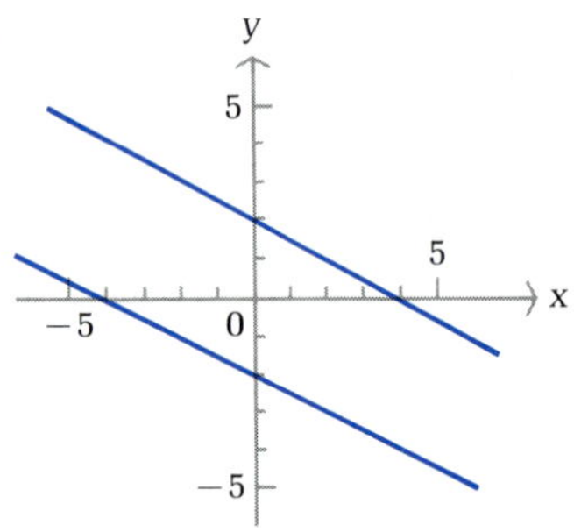

Lines are parallel (each has slope $-\frac{1}{2}$)—no solutions

(C)

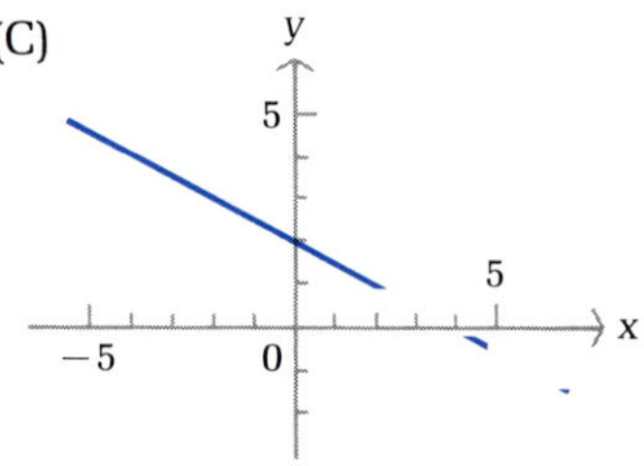

Lines coincide—infinite number of solutions

◆

PROBLEM 2 Solve each of the following systems by graphing:

(A) $x + y = 4$, $2x - y = 2$ (B) $6x - 3y = 9$, $2x - y = 3$ (C) $2x - y = 4$, $6x - 3y = -18$

◆

We now define some terms that we can use to describe the different types of solutions to systems of equations that we will encounter. A system of linear equations is **consistent** if it has one or more solutions and **inconsistent** if no solutions exist. Furthermore, a consistent system is said to be **independent** if it has exactly one solution (often referred to as the **unique solution**) and **dependent** if it has more than one solution. Referring to the three systems in Example 2, the system in part A is a consistent and independent system with the unique solution $x = 4$ and $y = 1$. The system in part B is inconsistent. And the system in part C is consistent and dependent.

By geometrically interpreting a system of two linear equations in two variables, we gain useful information about what to expect in the way of solutions to the system. In general, any two lines in a coordinate plane must intersect in exactly one point, be parallel, or coincide (have identical graphs). Thus, the systems in Example 2 illustrate the only three possible types of solutions for systems of two linear equations in two variables. These ideas are summarized in Theorem 1.

THEOREM 1

Possible Solutions to a Linear System

The linear system

$$ax + by = h$$
$$cx + dy = k$$

must have:

(A) Exactly one solution — CONSISTENT AND INDEPENDENT

Or:

(B) No solution — INCONSISTENT

Or:

(C) Infinitely many solutions — CONSISTENT AND DEPENDENT

There are no other possibilities.

In addition, graphs frequently reveal relationships in problems that might otherwise be hidden. Generally, however, graphic methods give us only rough approximations of solutions. The methods of substitution and elimination by addition, which we consider next, will yield exact solutions—assuming solutions exist.

Solution by Substitution

In this method, we choose one of two equations in a system and solve for one variable in terms of the other. (We make a choice that avoids fractions, if possible.) Then, we substitute the result into the other equation and solve the resulting linear equation in one variable. Finally, we substitute this result back into either of the original equations to find the second variable. An example should make the process clear.

◆ EXAMPLE 3

Solve by substitution:

$$5x + y = 4$$
$$2x - 3y = 5$$

Solution

Solve either equation for one variable in terms of the other; then substitute into the remaining equation. In this problem we can avoid fractions by choosing the first equation and solving for y in terms of x:

$$
\begin{aligned}
5x + y &= 4 && \text{Solve the first equation for } y \text{ in terms of } x. \\
y &= 4 - 5x && \text{Substitute into second equation.} \\
2x - 3y &= 5 && \text{Second equation} \\
2x - 3(4 - 5x) &= 5 && \text{Solve for } x. \\
2x - 12 + 15x &= 5 \\
17x &= 17 \\
x &= 1
\end{aligned}
$$

Now, replace x with 1 in $y = 4 - 5x$ to find y:

$$
\begin{aligned}
y &= 4 - 5x \\
y &= 4 - 5(1) \\
y &= -1
\end{aligned}
$$

Check

$$
\begin{aligned}
5x + y &= 4 & \qquad 2x - 3y &= 5 \\
5(1) + (-1) &\stackrel{?}{=} 4 & \qquad 2(1) - 3(-1) &\stackrel{?}{=} 5 \\
4 &\stackrel{\checkmark}{=} 4 & \qquad 5 &\stackrel{\checkmark}{=} 5
\end{aligned}
$$

◆

PROBLEM 3 Solve by substitution:

$$
\begin{aligned}
3x + 2y &= -2 \\
2x - y &= -6
\end{aligned}
$$

◆

Solution Using Elimination by Addition

Now we turn to **elimination by addition.** This is probably the most important method of solution, since it is readily generalized to higher-order systems. The method involves replacing systems of equations with simpler *equivalent systems* by performing appropriate operations, until we obtain a system with an obvious solution. **Equivalent systems** of equations are, as you would expect, systems that have exactly the same solution set. Theorem 2 lists the operations that produce equivalent systems.

THEOREM 2 **Operations That Produce Equivalent Systems**

A system of linear equations is transformed into an equivalent system if:

(A) Two equations are interchanged.
(B) An equation is multiplied by a nonzero constant.
(C) A constant multiple of one equation is added to another equation.

Any one of the three operations in Theorem 2 can be used to produce an equivalent system, but the operations in parts B and C will be of most use to us now. Part A becomes useful when we apply the theorem to larger systems. The use of Theorem 2 is best illustrated by examples.

◆ EXAMPLE 4 Solve the following system using elimination by addition:

$$\begin{aligned} 3x - 2y &= 8 \\ 2x + 5y &= -1 \end{aligned}$$

Solution We use Theorem 2 to eliminate one of the variables, thus obtaining a system with an obvious solution:

$$\begin{aligned} 3x - 2y &= 8 \\ 2x + 5y &= -1 \end{aligned}$$

Multiply the top equation by 5 and the bottom equation by 2 (Theorem 2B).

$$\begin{aligned} 5(3x - 2y) &= 5(8) \\ 2(2x + 5y) &= 2(-1) \end{aligned}$$

$$\begin{aligned} 15x - 10y &= 40 \\ 4x + 10y &= -2 \\ \hline 19x \qquad &= 38 \end{aligned}$$

Add the top equation to the bottom equation (Theorem 2C).

Divide both sides by 19, which is the same as multiplying the equation by $\frac{1}{19}$ (Theorem 2B).

$$\mathbf{x = 2}$$

This equation paired with either of the two original equations produces a system equivalent to the original system.

Knowing that $x = 2$, we substitute this number back into either of the two original equations (we choose the second) to solve for y:

$$\begin{aligned} 2(2) + 5y &= -1 \\ 5y &= -5 \\ \mathbf{y} &\mathbf{= -1} \end{aligned}$$

Check

$$\begin{aligned} 3x - 2y &= 8 & \qquad 2x + 5y &= -1 \\ 3(2) - 2(-1) &\stackrel{?}{=} 8 & 2(2) + 5(-1) &\stackrel{?}{=} -1 \\ 8 &\stackrel{\checkmark}{=} 8 & -1 &\stackrel{\checkmark}{=} -1 \end{aligned}$$

◆

PROBLEM 4 Solve the following system using elimination by addition:

$$\begin{aligned} 5x - 2y &= 12 \\ 2x + 3y &= 1 \end{aligned}$$

◆

Let us see what happens in the elimination process when a system has either no solution or infinitely many solutions. Consider the following system:

$$\begin{aligned} 2x + 6y &= -3 \\ x + 3y &= 2 \end{aligned}$$

Multiplying the second equation by -2 and adding, we obtain

$$\begin{array}{r} 2x + 6y = -3 \\ -2x - 6y = -4 \\ \hline 0 = -7 \end{array} \quad \text{Not possible}$$

We have obtained a contradiction. The assumption that the original system has solutions must be false (otherwise, we have proved that $0 = -7$). Thus, the system has no solutions and its solution set is the empty set. The graphs of the equations are parallel and the system is inconsistent.

Now consider the system

$$\begin{aligned} x - \tfrac{1}{2}y &= 4 \\ -2x + y &= -8 \end{aligned}$$

If we multiply the top equation by 2 and add the result to the bottom equation, we obtain

$$\begin{array}{r} 2x - y = 8 \\ -2x + y = -8 \\ \hline 0 = 0 \end{array}$$

Obtaining $0 = 0$ by addition implies that the equations are equivalent; that is, their graphs coincide and the system is dependent. If we let $x = k$, where k is any real number, and solve either equation for y, we obtain $y = 2k - 8$. Thus, $(k, 2k - 8)$ is a solution for any real number k. The variable k is called a **parameter,** and replacing k with a real number produces a **particular solution** to the system. For example, some particular solutions to this system are

$k = -1$	$k = 2$	$k = 5$	$k = 9.4$
$(-1, -10)$	$(2, -4)$	$(5, 2)$	$(9.4, 10.8)$

◆ APPLICATIONS

Many real-world problems are readily solved by applying two-equation–two-variable methods. We shall discuss two applications in detail.

◆ EXAMPLE 5

Diet

A dietitian in a hospital is to arrange a special diet comprised of two foods, M and N. Each ounce of food M contains 8 units of calcium and 2 units of iron. Each ounce of food N contains 5 units of calcium and 4 units of iron. How many ounces of foods M and N should be used to obtain a food mix that has exactly 74 units of calcium and 35 units of iron?

Solution It is convenient to first summarize the quantities involved in a table, as shown in the margin.

	FOOD M	FOOD N	TOTAL NEEDED
CALCIUM	8	5	74
IRON	2	4	35

Let

x = Number of ounces of food M

y = Number of ounces of food N

$$\begin{pmatrix}\text{Calcium in}\\ x \text{ oz of food } M\end{pmatrix} + \begin{pmatrix}\text{Calcium in}\\ y \text{ oz of food } N\end{pmatrix} = \begin{pmatrix}\text{Total calcium}\\ \text{needed}\end{pmatrix}$$

$$\begin{pmatrix}\text{Iron in } x \text{ oz}\\ \text{of food } M\end{pmatrix} + \begin{pmatrix}\text{Iron in } y \text{ oz}\\ \text{of food } N\end{pmatrix} = \begin{pmatrix}\text{Total iron}\\ \text{needed}\end{pmatrix}$$

$$8x + 5y = 74 \qquad \text{Units of calcium}$$

$$2x + 4y = 35 \qquad \text{Units of iron}$$

Solve using elimination by addition:

$$\begin{aligned} 8x + 5y &= 74 \\ -8x - 16y &= -140 \\ \hline -11y &= -66 \end{aligned}$$

$$y = 6 \text{ oz of food } N$$

$$2x + 4(6) = 35$$

$$2x = 11$$

$$x = 5.5 \text{ oz of food } M$$

Check

$$8x + 5y = 74 \qquad\qquad 2x + 4y = 35$$

$$8(5.5) + 5(6) \stackrel{?}{=} 74 \qquad 2(5.5) + 4(6) \stackrel{?}{=} 35$$

$$74 \stackrel{\checkmark}{=} 74 \qquad\qquad 35 \stackrel{\checkmark}{=} 35$$

◆

PROBLEM 5

Repeat Example 5 given that each ounce of food M contains 10 units of calcium and 4 units of iron, each ounce of food N contains 6 units of calcium and 4 units of iron, and the mix of M and N must have exactly 92 units of calcium and 44 units of iron. ◆

◆ EXAMPLE 6

Supply and Demand

The quantity of a product that people are willing to buy during some period of time depends on its price. Generally, the higher the price, the less the demand; the lower the price, the greater the demand. Similarly, the quantity of a product that a supplier is willing to sell during some period of time also depends on the price. Generally, a supplier will be willing to supply more of a product at higher prices and less of a product at lower prices. The simplest supply and demand model is a linear model where the graphs of a demand equation and a supply equation are straight lines.

Suppose that in a city on a particular day, the supply and demand equations for cherries are

$$p = -0.2q + 4 \qquad \text{Demand equation (consumer)}$$

$$p = 0.07q + 0.76 \qquad \text{Supply equation (supplier)}$$

where q represents the quantity in thousands of pounds and p represents the price in dollars. For example, we see that consumers will purchase 10 thousand pounds ($q = 10$) when the price is $p = -0.2(10) + 4 = \$2$ per pound. On the other hand, suppliers will be willing to supply 17.714 thousand pounds of cherries at \$2 per pound (solve $2 = 0.07q + 0.76$). Thus, at \$2 per pound the suppliers are willing to supply more cherries than consumers are willing to purchase. The supply exceeds the demand at that price and the price will come down. At what price will cherries stabilize for the day? That is, at what price will supply equal demand? This price, if it exists, is called the **equilibrium price,** and the quantity sold at that price is called the **equilibrium quantity.** How do we find these quantities? We solve the linear system

$$p = -0.2q + 4 \qquad \text{Demand equation}$$
$$p = 0.07q + 0.76 \qquad \text{Supply equation}$$

We solve this system using substitution (substituting $p = -0.2q + 4$ into the second equation):

$$-0.2q + 4 = 0.07q + 0.76$$
$$-0.27q = -3.24$$
$$\boldsymbol{q = 12 \text{ thousand pounds}} \qquad \text{Equilibrium quantity}$$

Now substitute $q = 12$ back into either of the original equations in the system and solve for p (we choose the first equation):

$$p = -0.2(12) + 4$$
$$\boldsymbol{p = \$1.60 \text{ per pound}} \qquad \text{Equilibrium price}$$

These results are interpreted geometrically in the figure.

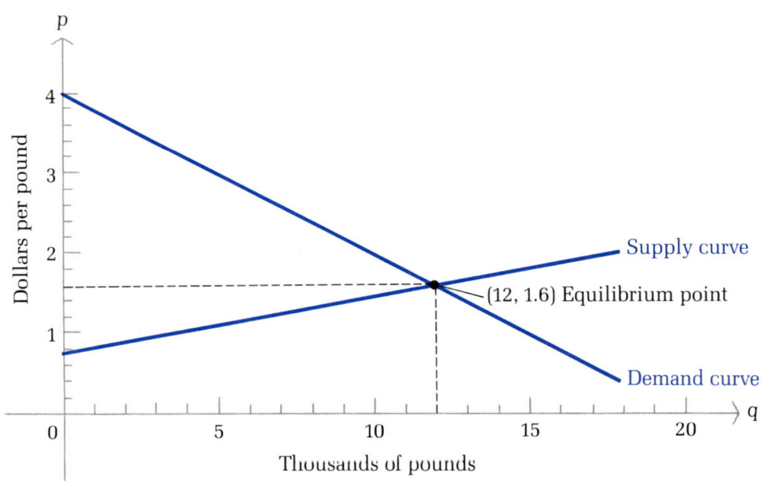

If the price was above the equilibrium price of \$1.60 per pound, the supply would exceed the demand and the price would come down. If the price was below the equilibrium price of \$1.60 per pound, the demand would exceed the supply and the price would rise. Thus, the price would reach equilibrium at \$1.60. At this price, suppliers would supply 12 thousand pounds of cherries and consumers would purchase 12 thousand pounds. ◆

PROBLEM 6 Repeat Example 6 (including drawing the graph) given:

$$p = -0.1q + 3 \qquad \text{Demand equation}$$
$$p = 0.08q + 0.66 \qquad \text{Supply equation}$$
◆

Answers to Matched Problems

1. $x = -2$, $y = -1$

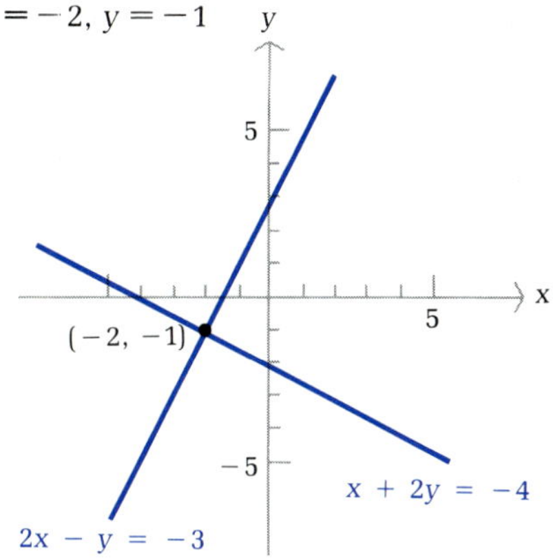

Check:
$$2x - y = -3$$
$$2(-2) - (-1) \stackrel{?}{=} -3$$
$$-3 \stackrel{\checkmark}{=} -3$$
$$x + 2y = -4$$
$$(-2) + 2(-1) \stackrel{?}{=} -4$$
$$-4 \stackrel{\checkmark}{=} -4$$

2. (A) $x = 2$, $y = 2$ (B) Infinitely many solutions (C) No solution
3. $x = -2$, $y = 2$
4. $x = 2$, $y = -1$
5. 6.5 oz of food M, 4.5 oz of food N
6. Equilibrium quantity = 13 thousand pounds; Equilibrium price = \$1.70 per pound

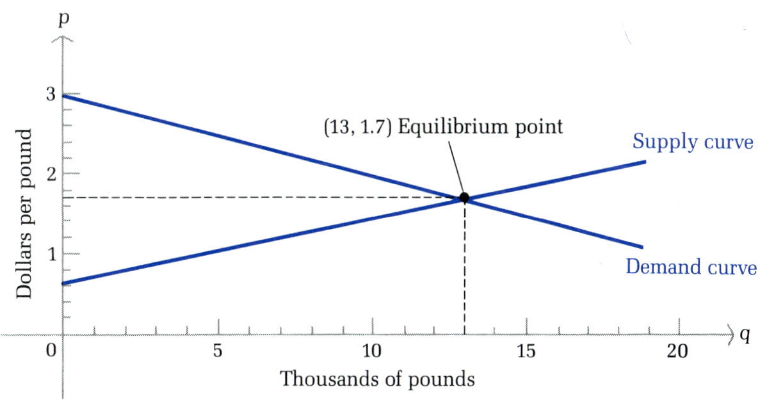

EXERCISE 5-1

A *Solve by graphing.*

1. $x + y = 5$, $x - y = 1$

2. $x - y = 2$, $x + y = 6$

3. $3x - y = 2$, $x + 2y = 10$

4. $3x - 2y = 12$, $7x + 2y = 8$

5. $m + 2n = 4$, $2m + 4n = -8$

6. $3u + 5v = 15$, $6u + 10v = -30$

Solve using substitution.

7. $y = 2x - 3$, $x + 2y = 14$

8. $y = x - 4$, $x + 3y = 12$

9. $2x + y = 6$, $x - y = -3$

10. $3x - y = 7$, $2x + 3y = 1$

Solve using elimination by addition.

11. $3u - 2v = 12$, $7u + 2v = 8$

12. $2x - 3y = -8$, $5x + 3y = 1$

13. $2m - n = 10$, $m - 2n = -4$

14. $2x + 3y = 1$, $3x - y = 7$

B *Solve using substitution or elimination by addition.*

15. $9x - 3y = 24$, $11x + 2y = 1$

16. $4x + 3y = 26$, $3x - 11y = -7$

17. $2x - 3y = -2$, $-4x + 6y = 7$

18. $3x - 6y = -9$, $-2x + 4y = 12$

19. $3x + 8y = 4$, $15x + 10y = -10$

20. $7m + 12n = -1$, $5m - 3n = 7$

21. $-6x + 10y = -30$, $3x - 5y = 15$

22. $2x + 4y = -8$, $x + 2y = 4$

23. $y = 0.07x$, $y = 80 + 0.05x$

24. $y = 0.08x$, $y = 100 + 0.04x$

25. $x + y = 1$, $0.3x - 0.4y = 0$

26. $x + y = 1$, $0.5x - 0.4y = 0$

27. $0.2x - 0.5y = 0.07$, $0.8x - 0.3y = 0.79$

28. $0.3u - 0.6v = 0.18$, $0.5u + 0.2v = 0.54$

29. $\frac{2}{5}x + \frac{3}{2}y = 2$, $\frac{7}{3}x - \frac{5}{4}y = -5$

30. $\frac{7}{2}x - \frac{5}{6}y = 10$, $\frac{2}{5}x + \frac{4}{3}y = 6$

C *In Problems 31–36, graph all three equations in the same coordinate system, and then find the coordinates of any points where two or more lines intersect.*

31. $x - 2y = -6$, $2x + y = 8$, $x + 2y = -2$

32. $x + y = 3$, $x + 3y = 15$, $3x - y = 5$

33. $x + y = 1$, $x - 2y = -8$, $3x + y = -3$

34. $x - y = 6$, $x - 2y = 8$, $x + 4y = -4$

35. $4x - 3y = -24$, $2x + 3y = 12$, $8x - 6y = 24$

36. $2x + 3y = 18$, $2x - 6y = -6$, $4x + 6y = -24$

Business & Economics

37. *Supply and demand.* Suppose the supply and demand equations for printed T-shirts in a resort town for a particular week are

$$p = 0.7q + 3 \qquad \text{Supply equation}$$
$$p = -1.7q + 15 \qquad \text{Demand equation}$$

where p is the price in dollars and q is the quantity in hundreds.

(A) Find the equilibrium price and quantity.
(B) Graph the two equations in the same coordinate system and identify the equilibrium point, supply curve, and demand curve.

38. *Supply and demand.* Repeat Problem 37 with the following supply and demand equations:

$$p = 0.4q + 3.2 \qquad \text{Supply equation}$$
$$p = -1.9q + 17 \qquad \text{Demand equation}$$

39. *Supply and demand.* At \$0.60 per bushel, the daily supply for wheat is 450 bushels, and the daily demand is 570 bushels. When the price is raised to \$0.75 per bushel, the daily supply increases to 600 bushels, and the daily demand decreases to 495 bushels. Assume that the supply and demand equations are linear.

(A) Find the supply equation. [*Hint:* Write the supply equation in the form $p = aq + b$ and solve for a and b.]
(B) Find the demand equation.
(C) Find the equilibrium price and quantity.
(D) Graph the two equations in the same coordinate system and identify the equilibrium point, supply curve, and demand curve.

40. *Supply and demand.* At \$1.40 per bushel, the daily supply for oats is 850 bushels, and the daily demand is 580 bushels. When the price falls to \$1.20 per bushel, the daily supply decreases to 350 bushels, and the daily demand increases to 980 bushels. Assume that the supply and demand equations are linear.

(A) Find the supply equation.
(B) Find the demand equation.
(C) Find the equilibrium price and quantity.
(D) Graph the two equations in the same coordinate system and identify the equilibrium point, supply curve, and demand curve.

41. *Break-even analysis.* A small company manufactures portable home computers. The plant has fixed costs (leases, insurance, and so on) of \$48,000 per month and variable costs (labor, materials, and so on) of \$1,400 per unit produced. The computers are sold for \$1,800 each. Thus, the cost and revenue equations are

$$y = 48{,}000 + 1{,}400x \qquad \text{Cost equation}$$
$$y = 1{,}800x \qquad \text{Revenue equation}$$

where x is the total number of computers produced and sold each month, and the monthly costs and revenue are in dollars.

(A) How many units must be manufactured and sold each month for the company to break even?

(B) Graph both equations in the same coordinate system and show the break-even point. Interpret the regions between the lines to the left and to the right of the break-even point.

42. *Break-even analysis.* Repeat Problem 41 with the cost and revenue equations

$$y = 65{,}000 + 1{,}100x \qquad \text{Cost equation}$$
$$y = 1{,}600x \qquad \text{Revenue equation}$$

43. *Break-even analysis.* A mail-order company markets videotapes that sell for \$19.95, including shipping and handling. The monthly fixed costs (advertising, rent, and so on) are \$24,000, and the variable costs (materials, shipping, and so on) are \$7.45 per tape.

(A) How many tapes must be sold each month for the company to break even?

(B) Graph both equations in the same coordinate system and show the break-even point. Interpret the regions between the lines to the left and to the right of the break-even point.

44. *Break-even analysis.* Repeat Problem 43 if the monthly fixed costs increase to \$27,200, the variable costs increase to \$9.15, and the company raises the selling price of the tapes to \$21.95.

Life Sciences

45. *Nutrition.* Animals in an experiment are to be kept under a strict diet. Each animal is to receive, among other things, 20 grams of protein and 6 grams of fat. The laboratory technician is able to purchase two food mixes of the following compositions: Mix *A* has 10% protein and 6% fat; mix *B* has 20% protein and 2% fat. How many grams of each mix should be used to obtain the right diet for a single animal?

46. *Nutrition—plants.* A fruit grower can use two types of fertilizer in an orange grove, brand *A* and brand *B*. Each bag of brand *A* contains 8 pounds of nitrogen and 4 pounds of phosphoric acid. Each bag of brand *B* contains 7 pounds of nitrogen and 6 pounds of phosphoric acid. Tests indicate that the grove needs 720 pounds of nitrogen and 500 pounds of phosphoric acid. How many bags of each brand should be used to provide the required amounts of nitrogen and phosphoric acid?

Social Sciences

47. *Psychology—approach and avoidance.* People often approach certain situations with "mixed emotions." For example, public speaking often brings forth the positive response of recognition and the negative response of

failure. Which dominates? J. S. Brown, in an experiment on approach and avoidance, trained rats by feeding them from a goal box. Then the rats received mild electric shocks from the same goal box. This established an approach–avoidance conflict relative to the goal box. Using appropriate apparatus, Brown arrived at the following relationships:

$$p = -\tfrac{1}{5}d + 70 \qquad \text{Approach equation}$$
$$p = -\tfrac{4}{3}d + 230 \qquad \text{Avoidance equation}$$

where $30 \leq d \leq 172.5$. The approach equation gives the pull (in grams) toward the food goal box when the rat is placed d centimeters away from it. The avoidance equation gives the pull (in grams) away from the shock goal box when the rat is placed d centimeters from it.

(A) Graph the approach equation and the avoidance equation in the same coordinate system.
(B) Find the value of d for the point of intersection of these two equations.
(C) What do you think the rat would do when placed the distance d from the box found in part B?

(For additional discussion of this phenomenon, see J. S. Brown, "Gradients of Approach and Avoidance Responses and Their Relation to Motivation," *Journal of Comparative and Physiological Psychology*, 1948, 41:450–465.

SECTION 5-2 Systems of Linear Equations and Augmented Matrices

- AUGMENTED MATRICES
- SOLVING LINEAR SYSTEMS USING AUGMENTED MATRICES
- SUMMARY

Most linear systems of any consequence involve large numbers of equations and variables. In the past these systems were solved on large computers, since hand methods were impractical. Now a wide array of readily available hardware and software can be used to solve linear systems. This ranges from graphic calculators, such as the Texas Instrument TI-81, to sophisticated mathematical software for personal computers, such as Matlab, Mathematica, and Maple. In addition, spreadsheet programs such as Excel and Lotus 1-2-3 can solve linear systems. All these various approaches have one thing in common: **The user is expected to be familar with the solution techniques used to solve large linear systems.**

In the remainder of this chapter, we will develop several methods for solving large systems that are commonly used with computers. It is important to keep in mind that we are not presenting these as more efficient methods for solving linear systems manually—there are none. Instead, we discuss methods that will help you effectively use a variety of software to solve systems. At the same time, you will gain experience in formulating problems in terms of linear systems and in interpreting the results, two activities that computers cannot perform for you.

◆ AUGMENTED MATRICES

In solving systems of equations using elimination by addition, the coefficients of the variables and the constant terms played a central role. The process can be made more efficient for generalization and computer work by the introduction of a mathematical form called a *matrix*. A **matrix** is a rectangular array of numbers written within brackets. Some examples are

$$\begin{bmatrix} 3 & 5 \\ 0 & -2 \end{bmatrix} \quad \begin{bmatrix} 2 \\ -3 \\ 0 \end{bmatrix} \quad \begin{bmatrix} 1 & -1 & 0 & 5 \end{bmatrix}$$

$$\begin{bmatrix} -1 & 2 & -5 & 0 \\ 0 & 3 & 2 & 1 \end{bmatrix} \quad \begin{bmatrix} 1 & 0 & 0 \\ 0 & 1 & 0 \\ 0 & 0 & 1 \end{bmatrix}$$

Each number in a matrix is called an **element** of the matrix.

Associated with the system

$$\begin{aligned} 2x - 3y &= 5 \\ x + 2y &= -3 \end{aligned} \tag{1}$$

is the *augmented matrix*

$$\left[\begin{array}{cc|c} 2 & -3 & 5 \\ 1 & 2 & -3 \end{array}\right]$$

which contains the essential parts of the system—namely, the coefficients of the variables and the constant terms. (The vertical bar is included only to separate the coefficients of the variables from the constant terms.)

For ease of generalization to the larger systems in the following sections, we are now going to change the notation for the variables in (1) to a subscript form (we would soon run out of letters, but we will not run out of subscripts). That is, in place of x and y, we will use x_1 and x_2, respectively, and (1) will be written as

$$\begin{aligned} 2x_1 - 3x_2 &= 5 \\ x_1 + 2x_2 &= -3 \end{aligned}$$

In general, associated with each linear system of the form

$$\begin{aligned} a_1x_1 + b_1x_2 &= k_1 \\ a_2x_1 + b_2x_2 &= k_2 \end{aligned} \tag{2}$$

where x_1 and x_2 are variables, is the **augmented matrix** of the system:

$$\begin{bmatrix} a_1 & b_1 & | & k_1 \\ a_2 & b_2 & | & k_2 \end{bmatrix} \begin{matrix} \leftarrow \text{Row 1 } (R_1) \\ \leftarrow \text{Row 2 } (R_2) \end{matrix}$$

Column 1 (C_1) points to the first column, Column 2 (C_2) to the second column, Column 3 (C_3) to the third column.

This matrix contains the essential parts of system (2). Our objective is to learn how to manipulate augmented matrices in order to solve system (2), if a solution exists. The manipulative process is a direct outgrowth of the elimination process discussed in Section 5-1.

Recall that two linear systems are said to be **equivalent** if they have exactly the same solution set. How did we transform linear systems into equivalent linear systems? We used Theorem 2, which we restate here.

THEOREM 2

Operations That Produce Equivalent Systems

A system of linear equations is transformed into an equivalent system if:

(A) Two equations are interchanged.
(B) An equation is multiplied by a nonzero constant.
(C) A constant multiple of one equation is added to another equation.

Paralleling the earlier discussion, we say that two augmented matrices are **row-equivalent,** denoted by the symbol ~ placed between the two matrices, if they are augmented matrices of equivalent systems of equations. (Think about this.) How do we transform augmented matrices into row-equivalent matrices? We use Theorem 3, which is a direct consequence of Theorem 2.

THEOREM 3

Operations That Produce Row-Equivalent Matrices

An augmented matrix is transformed into a row-equivalent matrix by performing any of the following **row operations:**

(A) Two rows are interchanged ($R_i \leftrightarrow R_j$).
(B) A row is multiplied by a nonzero constant ($kR_i \rightarrow R_i$).
(C) A constant multiple of one row is added to another row ($R_i + kR_j \rightarrow R_i$).

[*Note:* The arrow $\rightarrow$ means "replaces."]

◆ SOLVING LINEAR SYSTEMS USING AUGMENTED MATRICES

The use of Theorem 3 in solving systems in the form of (2) is best illustrated by examples.

◆ EXAMPLE 7 Solve using augmented matrix methods:

$$\begin{aligned} 3x_1 + 4x_2 &= 1 \\ x_1 - 2x_2 &= 7 \end{aligned} \tag{3}$$

Solution We start by writing the augmented matrix corresponding to (3)

$$\left[\begin{array}{rr|r} 3 & 4 & 1 \\ 1 & -2 & 7 \end{array}\right] \tag{4}$$

Our objective is to use row operations from Theorem 3 to try to transform (4) into the form

$$\left[\begin{array}{rr|r} 1 & 0 & m \\ 0 & 1 & n \end{array}\right] \tag{5}$$

where m and n are real numbers. The solution to system (3) will then be obvious, since matrix (5) will be the augmented matrix of the following system (a row in an augmented matrix always corresponds to an equation in a linear system):

$$\begin{aligned} x_1 &= m \qquad & x_1 + 0x_2 &= m \\ x_2 &= n \qquad & 0x_1 + x_2 &= n \end{aligned}$$

We now proceed to use row operations to transform (4) into form (5).

Step 1. To get a 1 in the upper left corner, we interchange Rows 1 and 2 (Theorem 3A):

$$\left[\begin{array}{rr|r} 3 & 4 & 1 \\ 1 & -2 & 7 \end{array}\right] \quad \overset{R_1 \leftrightarrow R_2}{\sim} \quad \left[\begin{array}{rr|r} 1 & -2 & 7 \\ 3 & 4 & 1 \end{array}\right]$$

Now you see why we wanted Theorem 2A!

Step 2. To get a 0 in the lower left corner, we multiply R_1 by (-3) and add to R_2 (Theorem 3C)—this changes R_2 but not R_1. Some people find it useful to write $(-3)R_1$ outside the matrix to help reduce errors in arithmetic, as shown:

$$\left[\begin{array}{rr|r} 1 & -2 & 7 \\ 3 & 4 & 1 \end{array}\right] \quad R_2 + (-3)R_1 \to R_2 \quad \left[\begin{array}{rr|r} 1 & -2 & 7 \\ 0 & 10 & -20 \end{array}\right]$$

$$-3 \quad 6 \quad -21$$

Step 3. To get a 1 in the second row, second column, we multiply R_2 by $\frac{1}{10}$ (Theorem 3B):

$$\left[\begin{array}{rr|r} 1 & -2 & 7 \\ 0 & 10 & -20 \end{array}\right] \quad \tfrac{1}{10}R_2 \to R_2 \quad \left[\begin{array}{rr|r} 1 & -2 & 7 \\ 0 & 1 & -2 \end{array}\right]$$

Step 4. To get a 0 in the first row, second column, we multiply R_2 by 2 and add the result to R_1 (Theorem 3C)—this changes R_1 but not R_2:

$$\begin{matrix} 0 & 2 & -4 \end{matrix}$$

$$\left[\begin{array}{cc|c} 1 & -2 & 7 \\ 0 & 1 & -2 \end{array}\right] \quad R_1 + 2R_2 \to R_1 \quad \sim \left[\begin{array}{cc|c} 1 & 0 & 3 \\ 0 & 1 & -2 \end{array}\right]$$

We have accomplished our objective! The last matrix is the augmented matrix for the system

$$\begin{aligned} x_1 &= 3 \qquad & x_1 + 0x_2 &= 3 \\ x_2 &= -2 \qquad & 0x_1 + x_2 &= -2 \end{aligned} \tag{6}$$

Since system (6) is equivalent to system (3), our starting system, we have solved (3); that is, $x_1 = 3$ and $x_2 = -2$.

Check

$$\begin{aligned} 3x_1 + 4x_2 &= 1 \qquad & x_1 - 2x_2 &= 7 \\ 3(3) + 4(-2) &\overset{?}{=} 1 \qquad & 3 - 2(-2) &\overset{?}{=} 7 \\ 1 &\overset{\checkmark}{=} 1 \qquad & 7 &\overset{\checkmark}{=} 7 \end{aligned}$$

◆

The above process may be written more compactly as follows:

Step 1: Need a 1 here.

$$\left[\begin{array}{cc|c} 3 & 4 & 1 \\ 1 & -2 & 7 \end{array}\right] \quad R_1 \leftrightarrow R_2$$

Step 2: Need a 0 here.

$$\sim \left[\begin{array}{cc|c} 1 & -2 & 7 \\ 3 & 4 & 1 \end{array}\right] \quad R_2 + (-3)R_1 \to R_2$$

$$\begin{matrix} -3 & 6 & -21 \end{matrix}$$

Step 3: Need a 1 here.

$$\sim \left[\begin{array}{cc|c} 1 & -2 & 7 \\ 0 & 10 & -20 \end{array}\right] \quad \tfrac{1}{10}R_2 \to R_2$$

Step 4: Need a 0 here.

$$\begin{matrix} 0 & 2 & -4 \end{matrix}$$

$$\sim \left[\begin{array}{cc|c} 1 & -2 & 7 \\ 0 & 1 & -2 \end{array}\right] \quad R_1 + 2R_2 \to R_1$$

$$\sim \left[\begin{array}{cc|c} 1 & 0 & 3 \\ 0 & 1 & -2 \end{array}\right]$$

Therefore, $x_1 = 3$ and $x_2 = -2$.

PROBLEM 7 Solve using augmented matrix methods:

$$\begin{aligned} 2x_1 - x_2 &= -7 \\ x_1 + 2x_2 &= 4 \end{aligned}$$

◆

◆ EXAMPLE 8 Solve using augmented matrix methods:

$$2x_1 - 3x_2 = 6$$
$$3x_1 + 4x_2 = \tfrac{1}{2}$$

Solution

Step 1: Need a 1 here.

$$\left[\begin{array}{cc|c} 2 & -3 & 6 \\ 3 & 4 & \tfrac{1}{2} \end{array}\right] \qquad \tfrac{1}{2}R_1 \to R_1$$

Step 2: Need a 0 here.

$$\sim \left[\begin{array}{cc|c} 1 & -\tfrac{3}{2} & 3 \\ 3 & 4 & \tfrac{1}{2} \end{array}\right] \qquad R_2 + (-3)R_1 \to R_2$$
$$\begin{array}{ccc} -3 & \tfrac{9}{2} & -9 \end{array}$$

Step 3: Need a 1 here.

$$\sim \left[\begin{array}{cc|c} 1 & -\tfrac{3}{2} & 3 \\ 0 & \tfrac{17}{2} & -\tfrac{17}{2} \end{array}\right] \qquad \tfrac{2}{17}R_2 \to R_2$$
$$\begin{array}{ccc} 0 & \tfrac{3}{2} & -\tfrac{3}{2} \end{array}$$

Step 4: Need a 0 here.

$$\sim \left[\begin{array}{cc|c} 1 & -\tfrac{3}{2} & 3 \\ 0 & 1 & -1 \end{array}\right] \qquad R_1 + \tfrac{3}{2}R_2 \to R_1$$

$$\sim \left[\begin{array}{cc|c} 1 & 0 & \tfrac{3}{2} \\ 0 & 1 & -1 \end{array}\right]$$

Thus, $x_1 = \tfrac{3}{2}$ and $x_2 = -1$. The check is left to the reader. ◆

PROBLEM 8 Solve using augmented matrix methods:

$$5x_1 - 2x_2 = 11$$
$$2x_1 + 3x_2 = \tfrac{5}{2}$$

◆

◆ EXAMPLE 9 Solve using augmented matrix methods:

$$\begin{aligned} 2x_1 - x_2 &= 4 \\ -6x_1 + 3x_2 &= -12 \end{aligned} \tag{7}$$

Solution

$$\left[\begin{array}{cc|c} 2 & -1 & 4 \\ -6 & 3 & -12 \end{array}\right] \qquad \begin{array}{l} \tfrac{1}{2}R_1 \to R_1 \text{ (to get a 1 in the upper left corner)} \\ \tfrac{1}{3}R_2 \to R_2 \text{ (this simplifies } R_2\text{)} \end{array}$$

$$\sim \left[\begin{array}{cc|c} 1 & -\tfrac{1}{2} & 2 \\ -2 & 1 & -4 \end{array}\right] \qquad R_2 + 2R_1 \to R_2 \text{ (to get a 0 in the lower left corner)}$$
$$\begin{array}{ccc} 2 & -1 & 4 \end{array}$$

$$\sim \left[\begin{array}{cc|c} 1 & -\tfrac{1}{2} & 2 \\ 0 & 0 & 0 \end{array}\right]$$

The last matrix corresponds to the system

$$\begin{aligned} x_1 - \tfrac{1}{2}x_2 &= 2 \qquad & x_1 - \tfrac{1}{2}x_2 &= 2 \\ 0 &= 0 \qquad & 0x_1 + 0x_2 &= 0 \end{aligned} \tag{8}$$

This system is equivalent to the original system. Geometrically, the graphs of the two original equations coincide and there are infinitely many solutions. In general, if we end up with a row of zeros in an augmented matrix for a two-equation–two-variable system, the system is dependent and there are infinitely many solutions.

We represent the infinitely many solutions using the same method that was used in Section 5-1; that is, by introducing a parameter. We start by solving $x_1 - \frac{1}{2}x_2 = 2$, the first equation in (8), for either variable in terms of the other. We choose to solve for x_1 in terms of x_2 because it is easier:

$$x_1 = \tfrac{1}{2}x_2 + 2 \tag{9}$$

Now we introduce a parameter t (we can use other letters, such as k, s, p, q, and so on, to represent a parameter just as well). If we let $x_2 = t$, then for t any real number,

$$\begin{aligned} x_1 &= \tfrac{1}{2}t + 2 \\ x_2 &= t \end{aligned} \tag{10}$$

represents a solution of system (7). Using ordered pair notation, we may also write: For any real number t,

$$(\tfrac{1}{2}t + 2,\ t) \tag{11}$$

is a solution of system (7). And more formally, we may write:

$$\text{Solution set} = \{(\tfrac{1}{2}t + 2,\ t) \mid t \in R\} \tag{12}$$

We will generally use the less formal forms (10) and (11) to represent the solution set for problems of this type.

Check The following is a check that (10) provides a solution for system (7) for any real number t:

$$\begin{aligned} 2x_1 - x_2 &= 4 \\ 2(\tfrac{1}{2}t + 2) - t &\stackrel{?}{=} 4 \\ t + 4 - t &\stackrel{?}{=} 4 \\ 4 &\stackrel{\checkmark}{=} 4 \end{aligned} \qquad \begin{aligned} -6x_1 + 3x_2 &= -12 \\ -6(\tfrac{1}{2}t + 2) + 3t &\stackrel{?}{=} -12 \\ -3t - 12 + 3t &\stackrel{?}{=} -12 \\ -12 &\stackrel{\checkmark}{=} -12 \end{aligned}$$

◆

PROBLEM 9 Solve using augmented matrix methods:

$$\begin{aligned} -2x_1 + 6x_2 &= 6 \\ 3x_1 - 9x_2 &= -9 \end{aligned}$$

◆

◆ EXAMPLE 10 Solve using augmented matrix methods:

$$\begin{aligned} 2x_1 + 6x_2 &= -3 \\ x_1 + 3x_2 &= 2 \end{aligned}$$

Solution

$$\left[\begin{array}{cc|c} 2 & 6 & -3 \\ 1 & 3 & 2 \end{array}\right] \quad R_1 \leftrightarrow R_2$$

$$\sim\left[\begin{array}{cc|c} 1 & 3 & 2 \\ 2 & 6 & -3 \end{array}\right] \quad R_2 + (-2)R_1 \rightarrow R_2$$

$$\begin{array}{ccc} -2 & -6 & -4 \end{array}$$

$$\sim\left[\begin{array}{cc|c} 1 & 3 & 2 \\ 0 & 0 & -7 \end{array}\right] \quad R_2 \text{ implies the contradiction } 0 = -7.$$

This is the augmented matrix of the system

$$\begin{aligned} x_1 + 3x_2 &= 2 \qquad & x_1 + 3x_2 &= 2 \\ 0 &= -7 \qquad & 0x_1 + 0x_2 &= -7 \end{aligned}$$

The second equation is not satisfied by any ordered pair of real numbers. Hence, the original system is inconsistent and has no solution—otherwise, we have proved that $0 = -7$! Thus, if in a row of an augmented matrix we obtain all zeros to the left of the vertical bar and a nonzero number to the right, then the system is inconsistent and there are no solutions. ◆

PROBLEM 10 Solve using augmented matrix methods:

$$\begin{aligned} 2x_1 - x_2 &= 3 \\ 4x_1 - 2x_2 &= -1 \end{aligned}$$

◆

◆ SUMMARY

Summary

Form 1 A Unique Solution (Consistent and Independent)	Form 2 Infinitely Many Solutions (Consistent and Dependent)	Form 3 No Solution (Inconsistent)
$\left[\begin{array}{cc\|c} 1 & 0 & m \\ 0 & 1 & n \end{array}\right]$	$\left[\begin{array}{cc\|c} 1 & m & n \\ 0 & 0 & 0 \end{array}\right]$	$\left[\begin{array}{cc\|c} 1 & m & n \\ 0 & 0 & p \end{array}\right]$

m, n, p real numbers; $p \neq 0$

The process of solving systems of equations described in this section is referred to as **Gauss–Jordan elimination.** We will formalize this method in the next section so that it will apply to systems of any size, including systems where the number of equations and the number of variables are not the same.

Answers to Matched Problems

7. $x_1 = -2,\ x_2 = 3$ 8. $x_1 = 2,\ x_2 = -\frac{1}{2}$
9. The system is dependent. For t any real number, a solution is

$$x_1 = 3t - 3$$
$$x_2 = t$$

10. Inconsistent—no solution

EXERCISE 5-2

A

Perform each of the indicated row operations on the following matrix:

$$\left[\begin{array}{cc|c} 1 & -3 & 2 \\ 4 & -6 & -8 \end{array}\right]$$

1. $R_1 \leftrightarrow R_2$
2. $\frac{1}{2}R_2 \rightarrow R_2$
3. $-4R_1 \rightarrow R_1$
4. $-2R_1 \rightarrow R_1$
5. $2R_2 \rightarrow R_2$
6. $-1R_2 \rightarrow R_2$
7. $R_2 + (-4)R_1 \rightarrow R_2$
8. $R_1 + (-\frac{1}{2})R_2 \rightarrow R_1$
9. $R_2 + (-2)R_1 \rightarrow R_2$
10. $R_2 + (-3)R_1 \rightarrow R_2$
11. $R_2 + (-1)R_1 \rightarrow R_2$
12. $R_2 + (1)R_1 \rightarrow R_2$

Solve using augmented matrix methods.

13. $\begin{aligned} x_1 + x_2 &= 5 \\ x_1 - x_2 &= 1 \end{aligned}$
14. $\begin{aligned} x_1 - x_2 &= 2 \\ x_1 + x_2 &= 6 \end{aligned}$

B *Solve using augmented matrix methods.*

15. $\begin{aligned} x_1 - 2x_2 &= 1 \\ 2x_1 - x_2 &= 5 \end{aligned}$
16. $\begin{aligned} x_1 + 3x_2 &= 1 \\ 3x_1 - 2x_2 &= 14 \end{aligned}$
17. $\begin{aligned} x_1 - 4x_2 &= -2 \\ -2x_1 + x_2 &= -3 \end{aligned}$
18. $\begin{aligned} x_1 - 3x_2 &= -5 \\ -3x_1 - x_2 &= 5 \end{aligned}$
19. $\begin{aligned} 3x_1 - x_2 &= 2 \\ x_1 + 2x_2 &= 10 \end{aligned}$
20. $\begin{aligned} 2x_1 + x_2 &= 0 \\ x_1 - 2x_2 &= -5 \end{aligned}$
21. $\begin{aligned} x_1 + 2x_2 &= 4 \\ 2x_1 + 4x_2 &= -8 \end{aligned}$
22. $\begin{aligned} 2x_1 - 3x_2 &= -2 \\ -4x_1 + 6x_2 &= 7 \end{aligned}$
23. $\begin{aligned} 2x_1 + x_2 &= 6 \\ x_1 - x_2 &= -3 \end{aligned}$
24. $\begin{aligned} 3x_1 - x_2 &= -5 \\ x_1 + 3x_2 &= 5 \end{aligned}$
25. $\begin{aligned} 3x_1 - 6x_2 &= -9 \\ -2x_1 + 4x_2 &= 6 \end{aligned}$
26. $\begin{aligned} 2x_1 - 4x_2 &= -2 \\ -3x_1 + 6x_2 &= 3 \end{aligned}$
27. $\begin{aligned} 4x_1 - 2x_2 &= 2 \\ -6x_1 + 3x_2 &= -3 \end{aligned}$
28. $\begin{aligned} -6x_1 + 2x_2 &= 4 \\ 3x_1 - x_2 &= -2 \end{aligned}$
29. $\begin{aligned} 2x_1 + x_2 &= 1 \\ 4x_1 - x_2 &= -7 \end{aligned}$
30. $\begin{aligned} 2x_1 - x_2 &= -8 \\ 2x_1 + x_2 &= 8 \end{aligned}$
31. $\begin{aligned} 4x_1 - 6x_2 &= 8 \\ -6x_1 + 9x_2 &= -10 \end{aligned}$
32. $\begin{aligned} 2x_1 - 4x_2 &= -4 \\ -3x_1 + 6x_2 &= 4 \end{aligned}$
33. $\begin{aligned} -4x_1 + 6x_2 &= -8 \\ 6x_1 - 9x_2 &= 12 \end{aligned}$
34. $\begin{aligned} -2x_1 + 4x_2 &= 4 \\ 3x_1 - 6x_2 &= -6 \end{aligned}$

C *Solve using augmented matrix methods.*

35. $\begin{aligned} 3x_1 - x_2 &= 7 \\ 2x_1 + 3x_2 &= 1 \end{aligned}$
36. $\begin{aligned} 2x_1 - 3x_2 &= -8 \\ 5x_1 + 3x_2 &= 1 \end{aligned}$
37. $\begin{aligned} 3x_1 + 2x_2 &= 4 \\ 2x_1 - x_2 &= 5 \end{aligned}$
38. $\begin{aligned} 4x_1 + 3x_2 &= 26 \\ 3x_1 - 11x_2 &= -7 \end{aligned}$
39. $\begin{aligned} 0.2x_1 - 0.5x_2 &= 0.07 \\ 0.8x_1 - 0.3x_2 &= 0.79 \end{aligned}$
40. $\begin{aligned} 0.3x_1 - 0.6x_2 &= 0.18 \\ 0.5x_1 - 0.2x_2 &= 0.54 \end{aligned}$

SECTION 5-3 Gauss–Jordan Elimination

- REDUCED MATRICES
- SOLVING SYSTEMS BY GAUSS–JORDAN ELIMINATION
- APPLICATION

Now that you have had some experience with row operations on simple augmented matrices, we will consider systems involving more than two variables. In addition, we will not require that a system have the same number of equations as variables. It turns out that the results for two-variable, two-equation linear systems, stated in Theorem 1, Section 5-1, actually hold for linear systems of any size.

Possible Solutions to a Linear System

It can be shown that any linear system must have exactly one solution, no solution, or an infinite number of solutions, regardless of the number of equations or number of variables in the system. The terms *unique solution, consistent, inconsistent, dependent,* and *independent* are used to describe these solutions, just as in the two-variable case.

REDUCED MATRICES

Our objective is to start with the augmented matrix of a linear system and transform it by using row operations from Theorem 3 (in the preceding section) into a simple form where the solution can be read by inspection. The simple form so obtained is called the *reduced form,* and we define it as follows:

Reduced Matrix

A matrix is in **reduced form** if:

1. Each row consisting entirely of zeros is below any row having at least one nonzero element.
2. The leftmost nonzero element in each row is 1.
3. All other elements in the column containing the leftmost 1 of a given row are zeros.
4. The leftmost 1 in any row is to the right of the leftmost 1 in the row above.

The following matrices are in reduced form. Check each one carefully to convince yourself that the conditions in the definition are met.

$$\left[\begin{array}{cc|c} 1 & 0 & 2 \\ 0 & 1 & -3 \end{array}\right] \quad \left[\begin{array}{ccc|c} 1 & 0 & 0 & 2 \\ 0 & 1 & 0 & -1 \\ 0 & 0 & 1 & 3 \end{array}\right] \quad \left[\begin{array}{cc|c} 1 & 0 & 3 \\ 0 & 1 & -1 \\ 0 & 0 & 0 \end{array}\right]$$

$$\left[\begin{array}{cccc|c} 1 & 4 & 0 & 0 & -3 \\ 0 & 0 & 1 & 0 & 2 \\ 0 & 0 & 0 & 1 & 6 \end{array}\right] \quad \left[\begin{array}{ccc|c} 1 & 0 & 4 & 0 \\ 0 & 1 & 3 & 0 \\ 0 & 0 & 0 & 1 \end{array}\right]$$

◆ EXAMPLE 11 The matrices below are not in reduced form. Indicate which condition in the definition is violated for each matrix.

(A) $\left[\begin{array}{cc|c} 0 & 1 & -2 \\ 1 & 0 & 3 \end{array}\right]$ (B) $\left[\begin{array}{ccc|c} 1 & 2 & -2 & 3 \\ 0 & 0 & 1 & -1 \end{array}\right]$

(C) $\left[\begin{array}{cc|c} 0 & 0 & 0 \\ 1 & 0 & -3 \\ 0 & 1 & -2 \end{array}\right]$ (D) $\left[\begin{array}{ccc|c} 1 & 0 & 0 & -1 \\ 0 & 2 & 0 & 3 \\ 0 & 0 & 1 & -5 \end{array}\right]$

Solutions

(A) Condition 4 is violated: The leftmost 1 in Row 2 is not to the right of the leftmost 1 in Row 1.

(B) Condition 3 is violated: The column containing the leftmost 1 in Row 2 has a nonzero element above the 1.

(C) Condition 1 is violated: The first row contains all zeros and it is not below any row having at least one nonzero element.

(D) Condition 2 is violated: The leftmost nonzero element in Row 2 is not 1. ◆

PROBLEM 11 The matrices below are not in reduced form. Indicate which condition in the definition is violated for each matrix.

(A) $\left[\begin{array}{cc|c} 1 & 0 & 2 \\ 0 & 3 & -6 \end{array}\right]$ (B) $\left[\begin{array}{ccc|c} 1 & 5 & 4 & 3 \\ 0 & 1 & 2 & -1 \\ 0 & 0 & 0 & 0 \end{array}\right]$

(C) $\left[\begin{array}{ccc|c} 0 & 1 & 0 & -3 \\ 1 & 0 & 0 & 0 \\ 0 & 0 & 1 & 2 \end{array}\right]$ (D) $\left[\begin{array}{ccc|c} 1 & 2 & 0 & 3 \\ 0 & 0 & 0 & 0 \\ 0 & 0 & 1 & 4 \end{array}\right]$ ◆

Example 12 illustrates methods of writing solutions to *reduced systems*. A **reduced system** is a linear system whose augmented coefficient matrix is in reduced form.

◆ EXAMPLE 12 Write the linear system corresponding to each reduced augmented matrix and solve.

(A) $\left[\begin{array}{ccc|c} 1 & 0 & 0 & 2 \\ 0 & 1 & 0 & -1 \\ 0 & 0 & 1 & 3 \end{array}\right]$ (B) $\left[\begin{array}{ccc|c} 1 & 0 & 4 & 0 \\ 0 & 1 & 3 & 0 \\ 0 & 0 & 0 & 1 \end{array}\right]$

(C) $\left[\begin{array}{ccc|c} 1 & 0 & 2 & -3 \\ 0 & 1 & -1 & 8 \\ 0 & 0 & 0 & 0 \end{array}\right]$ (D) $\left[\begin{array}{ccccc|c} 1 & 4 & 0 & 0 & 3 & -2 \\ 0 & 0 & 1 & 0 & -2 & 0 \\ 0 & 0 & 0 & 1 & 2 & 4 \end{array}\right]$

Solutions

(A)

$$\begin{aligned} x_1 \quad\quad &= 2 \\ x_2 \quad &= -1 \\ x_3 &= 3 \end{aligned}$$

The solution is obvious: $x_1 = 2$, $x_2 = -1$, $x_3 = 3$.

(B)

$$\begin{aligned} x_1 \quad\quad + 4x_3 &= 0 \\ x_2 + 3x_3 &= 0 \\ 0x_1 + 0x_2 + 0x_3 &= 1 \end{aligned}$$

The last equation implies $0 = 1$, which is a contradiction. Hence, the system is inconsistent and has no solution.

(C) The system corresponding to the reduced augmented matrix is

$$\begin{aligned} x_1 \quad + 2x_3 &= -3 \\ x_2 - \quad x_3 &= \quad 8 \end{aligned}$$

We discard the equation corresponding to the third (all 0) row in the matrix, since it is satisfied by all values of x_1, x_2, and x_3.

Note that the leftmost variable in each equation appears in one and only one equation. We solve for the leftmost variables x_1 and x_2 in terms of the remaining variable x_3:

$$\begin{aligned} x_1 &= -2x_3 - 3 \\ x_2 &= x_3 \quad + 8 \end{aligned}$$

If we let $x_3 = t$, then for any real number t,

$$\begin{aligned} x_1 &= -2t - 3 \\ x_2 &= t + 8 \\ x_3 &= t \end{aligned}$$

is a solution. Some particular solutions are

$t = 0$	$t = -2$	$t = 3.5$
$(-3, 8, 0)$	$(1, 6, -2)$	$(-10, 11.5, 3.5)$

In general, **when a reduced system has more variables than equations and contains no contradictions, the system is dependent and has infinitely many solutions.** Describing the solution set to this type of system is not difficult. In a reduced system, the **leftmost variables** correspond to the leftmost 1's in the corresponding reduced augmented matrix. The definition of reduced form for an augmented matrix ensures that each leftmost variable in the corresponding reduced system appears in one and only one equation of the system. Thus, for a reduced system with more variables than equations and no contradictions, solving for each leftmost variable in terms of the remaining variables and writing a general solution to the system is usually easy. (Part D illustrates a slightly more involved case.)

(D)

$$\begin{aligned} x_1 + 4x_2 \quad\quad + 3x_5 &= -2 \\ x_3 \quad - 2x_5 &= \quad 0 \\ x_4 + 2x_5 &= \quad 4 \end{aligned}$$

Solve for x_1, x_3, and x_4 (leftmost variables) in terms of x_2 and x_5 (remaining variables):

$$x_1 = -4x_2 - 3x_5 - 2$$
$$x_3 = 2x_5$$
$$x_4 = -2x_5 + 4$$

If we let $x_2 = s$ and $x_5 = t$, then for any real numbers s and t,

$$x_1 = -4s - 3t - 2$$
$$x_2 = s$$
$$x_3 = 2t$$
$$x_4 = -2t + 4$$
$$x_5 = t$$

is a solution. The system is dependent and has infinitely many solutions. ◆

PROBLEM 12 Write the linear system corresponding to each reduced augmented matrix and solve.

(A) $\left[\begin{array}{ccc|c} 1 & 0 & 0 & -5 \\ 0 & 1 & 0 & 3 \\ 0 & 0 & 1 & 6 \end{array}\right]$ (B) $\left[\begin{array}{ccc|c} 1 & 2 & -3 & 0 \\ 0 & 0 & 0 & 1 \\ 0 & 0 & 0 & 0 \end{array}\right]$

(C) $\left[\begin{array}{ccc|c} 1 & 0 & -2 & 4 \\ 0 & 1 & 3 & -2 \\ 0 & 0 & 0 & 0 \end{array}\right]$ (D) $\left[\begin{array}{cccc|c} 1 & 0 & 3 & 2 & 5 \\ 0 & 1 & -2 & -1 & 3 \\ 0 & 0 & 0 & 0 & 0 \end{array}\right]$ ◆

◆ SOLVING SYSTEMS BY GAUSS–JORDAN ELIMINATION

We are now ready to outline the Gauss–Jordan elimination method for solving systems of linear equations. The method systematically transforms an augmented matrix into a reduced form from which we can write the solution to the original system by inspection, if a solution exists. The method will also reveal when a solution fails to exist (see Example 15).

The Gauss–Jordan elimination method is named after the German mathematician Carl Friedrich Gauss (1777–1855) and the French mathematician Camille Jordan (1838–1922). Gauss, one of the greatest mathematicians of all time, used a method of solving systems of equations that was later generalized by Jordan to the method we present here.

◆ EXAMPLE 13 Solve by Gauss–Jordan elimination:

$$\begin{aligned} 2x_1 - 2x_2 + x_3 &= 3 \\ 3x_1 + x_2 - x_3 &= 7 \\ x_1 - 3x_2 + 2x_3 &= 0 \end{aligned}$$

Solution Write the augmented matrix and follow the steps indicated at the right.

Need a 1 here.

$$\left[\begin{array}{rrr|r} 2 & -2 & 1 & 3 \\ 3 & 1 & -1 & 7 \\ 1 & -3 & 2 & 0 \end{array}\right] \quad R_1 \leftrightarrow R_3$$

Step 1. Choose the leftmost nonzero column and get a 1 at the top.

Need 0's here.

$$\sim \left[\begin{array}{rrr|r} 1 & -3 & 2 & 0 \\ 3 & 1 & -1 & 7 \\ 2 & -2 & 1 & 3 \end{array}\right] \quad \begin{array}{l} R_2 + (-3)R_1 \rightarrow R_2 \\ R_3 + (-2)R_1 \rightarrow R_3 \end{array}$$

Step 2. Use multiples of the first row to get zeros below the 1 obtained in step 1.

Need a 1 here.

$$\sim \left[\begin{array}{rrr|r} 1 & -3 & 2 & 0 \\ 0 & 10 & -7 & 7 \\ 0 & 4 & -3 & 3 \end{array}\right] \quad \tfrac{1}{10}R_2 \rightarrow R_2$$

Step 3. Repeat steps 1 and 2 with the **submatrix** formed by (mentally) deleting the top row and first column.

Continue the above process (steps 1–3) until it is not possible to go further; then proceed with step 4.

Need a 0 here.

$$\sim \left[\begin{array}{rrr|r} 1 & -3 & 2 & 0 \\ 0 & 1 & -\frac{7}{10} & \frac{7}{10} \\ 0 & 4 & -3 & 3 \end{array}\right] \quad R_3 + (-4)R_2 \rightarrow R_3$$

Need a 1 here.

$$\sim \left[\begin{array}{rrr|r} 1 & -3 & 2 & 0 \\ 0 & 1 & -\frac{7}{10} & \frac{7}{10} \\ 0 & 0 & -\frac{1}{5} & \frac{1}{5} \end{array}\right] \quad (-5)R_3 \rightarrow R_3$$

Mentally delete the second row and the second column.

Need 0's here.

$$\sim \left[\begin{array}{rrr|r} 1 & -3 & 2 & 0 \\ 0 & 1 & -\frac{7}{10} & \frac{7}{10} \\ 0 & 0 & 1 & -1 \end{array}\right] \quad \begin{array}{l} R_1 + (-2)R_3 \rightarrow R_1 \\ R_2 + \frac{7}{10}R_3 \rightarrow R_2 \end{array}$$

Since steps 1–3 cannot be carried further, proceed to step 4.

Step 4. Consider the whole matrix. Begin with the bottom nonzero row and use appropriate multiples of it to get zeros above the leftmost 1. Continue the process, moving up row by row, until the matrix is in reduced form.

Need a 0 here.

$$\sim \left[\begin{array}{rrr|r} 1 & -3 & 0 & 2 \\ 0 & 1 & 0 & 0 \\ 0 & 0 & 1 & -1 \end{array}\right] \quad R_1 + 3R_2 \rightarrow R_1$$

$$\sim \left[\begin{array}{rrr|r} 1 & 0 & 0 & 2 \\ 0 & 1 & 0 & 0 \\ 0 & 0 & 1 & -1 \end{array}\right]$$

The matrix is in reduced form, and we can write the solution to the original system by inspection.

Solution: $x_1 = 2$, $x_2 = 0$, $x_3 = -1$. It is left to the reader to check this solution.

◆

Steps 1–4 outlined in the solution of Example 13 are referred to as *Gauss–Jordan elimination*. The steps are summarized below for easy reference:

Gauss–Jordan Elimination

Step 1. Choose the leftmost nonzero column and use appropriate row operations to get a 1 at the top.

Step 2. Use multiples of the first row to get zeros in all places below the 1 obtained in step 1.

Step 3. Delete (mentally) the top row and first column of the matrix. Repeat steps 1 and 2 with the **submatrix** (the matrix that remains after deleting the top row and first column). Continue this process (steps 1–3) until it is not possible to go further.

Step 4. Now consider the whole matrix. Begin with the bottom nonzero row and use appropriate multiples of it to get zeros above the leftmost 1. Continue this process, moving up row by row, until the matrix is finally in reduced form.

[*Note:* If at any point in this process we obtain a row with all zeros to the left of the vertical line and a nonzero number n to the right, we can stop, since we will have a contradiction: $0 = n$, $n \neq 0$. We can then conclude that the system has no solution.]

Remark

The sequence of steps (algorithm) presented here for transforming a matrix into a reduced form is not unique. That is, other sequences of steps (using row operations) can produce a reduced matrix. (For example, it is possible to use row operations in such a way that fractions can be avoided.) But we emphasize again that we are not interested in the most efficient hand methods for transforming a matrix into a reduced form. Our main interest is in giving you a little experience with a method that is suitable for computer use with large-scale systems.

PROBLEM 13 Solve by Gauss–Jordan elimination:

$$\begin{aligned} 3x_1 + x_2 - 2x_3 &= 2 \\ x_1 - 2x_2 + x_3 &= 3 \\ 2x_1 - x_2 - 3x_3 &= 3 \end{aligned}$$

◆

◆ EXAMPLE 14 Solve by Gauss–Jordan elimination:

$$\begin{aligned} 2x_1 - x_2 + 4x_3 &= -2 \\ 3x_1 + 2x_2 - x_3 &= 1 \end{aligned}$$

Solution

Need a 1 here.

$$\left[\begin{array}{rrr|r} 2 & -1 & 4 & -2 \\ 3 & 2 & -1 & 1 \end{array}\right] \quad \tfrac{1}{2}R_1 \to R_1$$

Need a 0 here.

$$\sim \left[\begin{array}{rrr|r} 1 & -\tfrac{1}{2} & 2 & -1 \\ 3 & 2 & -1 & 1 \end{array}\right] \quad R_2 + (-3)R_1 \to R_2$$

Need a 1 here.

$$\sim \left[\begin{array}{rrr|r} 1 & -\tfrac{1}{2} & 2 & -1 \\ 0 & \tfrac{7}{2} & -7 & 4 \end{array}\right] \quad \tfrac{2}{7}R_2 \to R_2$$

Need a 0 here.

$$\sim \left[\begin{array}{rrr|r} 1 & -\tfrac{1}{2} & 2 & -1 \\ 0 & 1 & -2 & \tfrac{8}{7} \end{array}\right] \quad R_1 + \tfrac{1}{2}R_2 \to R_1$$

$$\sim \left[\begin{array}{rrr|r} 1 & 0 & 1 & -\tfrac{3}{7} \\ 0 & 1 & -2 & \tfrac{8}{7} \end{array}\right]$$

The matrix is now in reduced form. Write the corresponding system and the solution.

$$\begin{aligned} x_1 + x_3 &= -\tfrac{3}{7} \\ x_2 - 2x_3 &= \tfrac{8}{7} \end{aligned}$$

Solve for the leftmost variables x_1 and x_2 in terms of the remaining variable x_3:

$$\begin{aligned} x_1 &= -x_3 - \tfrac{3}{7} \\ x_2 &= 2x_3 + \tfrac{8}{7} \end{aligned}$$

If $x_3 = t$, then for t any real number,

$$\begin{aligned} x_1 &= -t - \tfrac{3}{7} \\ x_2 &= 2t + \tfrac{8}{7} \\ x_3 &= t \end{aligned}$$

is a solution. Checking that $(-t - \tfrac{3}{7}, 2t + \tfrac{8}{7}, t)$ is a solution for any real t is left to the reader. ◆

PROBLEM 14 Solve by Gauss–Jordan elimination:

$$\begin{aligned} 3x_1 + 6x_2 - 3x_3 &= 2 \\ 2x_1 - x_2 + 2x_3 &= -1 \end{aligned}$$

◆

◆ EXAMPLE 15 Solve by Gauss–Jordan elimination:

$$\begin{aligned} 2x_1 - x_2 &= -4 \\ 2x_1 + 4x_2 &= 6 \\ 3x_1 - x_2 &= -1 \end{aligned}$$

Solution

$$\left[\begin{array}{cc|c} 2 & -1 & -4 \\ 2 & 4 & 6 \\ 3 & -1 & -1 \end{array}\right] \quad R_1 \leftrightarrow R_2$$

Interchanging R_1 and R_2 and then multiplying by $\frac{1}{2}$ will avoid fractions and simplify calculations.

$$\sim \left[\begin{array}{cc|c} 2 & 4 & 6 \\ 2 & -1 & -4 \\ 3 & -1 & -1 \end{array}\right] \quad \tfrac{1}{2}R_1 \rightarrow R_1$$

$$\sim \left[\begin{array}{cc|c} 1 & 2 & 3 \\ 2 & -1 & -4 \\ 3 & -1 & -1 \end{array}\right] \quad \begin{array}{l} R_2 + (-2)R_1 \rightarrow R_2 \\ R_3 + (-3)R_1 \rightarrow R_3 \end{array}$$

$$\sim \left[\begin{array}{cc|c} 1 & 2 & 3 \\ 0 & -5 & -10 \\ 0 & -7 & -10 \end{array}\right] \quad -\tfrac{1}{5}R_2 \rightarrow R_2$$

$$\sim \left[\begin{array}{cc|c} 1 & 2 & 3 \\ 0 & 1 & 2 \\ 0 & -7 & -10 \end{array}\right] \quad R_3 + 7R_2 \rightarrow R_3$$

$$\sim \left[\begin{array}{cc|c} 1 & 2 & 3 \\ 0 & 1 & 2 \\ 0 & 0 & 4 \end{array}\right]$$

We stop the Gauss–Jordan elimination, even though the matrix is not in a reduced form, since the last row produces a contradiction.

The last row implies $0 = 4$, which is a contradiction; therefore, the system has no solution. ◆

PROBLEM 15 Solve by Gauss–Jordan elimination:

$$\begin{aligned} 3x_1 + x_2 &= 5 \\ 2x_1 + 3x_2 &= 1 \\ 2x_1 - 2x_2 &= 6 \end{aligned}$$

◆

◆ APPLICATION

◆ EXAMPLE 16

Production Scheduling

A casting company produces three different bronze sculptures. The casting department has available a maximum of 350 labor-hours per week, and the finishing department has a maximum of 150 labor-hours available per week. Sculpture *A* requires 30 hours for casting and 10 hours for finishing; sculpture *B* requires 10 hours for casting and 10 hours for finishing; and sculpture *C* requires 10 hours for casting and 30 hours for finishing. If the plant is to operate at maximum capacity, how many of each sculpture should be produced each week?

Solution First, we summarize the relevant manufacturing data in a table:

	LABOR-HOURS PER SCULPTURE A	B	C	MAXIMUM LABOR-HOURS AVAILABLE PER WEEK
CASTING DEPARTMENT	30	10	10	350
FINISHING DEPARTMENT	10	10	30	150

Let

x_1 = Number of sculpture *A* produced per week

x_2 = Number of sculpture *B* produced per week

x_3 = Number of sculpture *C* produced per week

Then

$$30x_1 + 10x_2 + 10x_3 = 350 \qquad \text{Casting department}$$
$$10x_1 + 10x_2 + 30x_3 = 150 \qquad \text{Finishing department}$$

Now we can form the augmented matrix of the system and solve by using Gauss–Jordan elimination:

$$\left[\begin{array}{ccc|c} 30 & 10 & 10 & 350 \\ 10 & 10 & 30 & 150 \end{array}\right] \qquad \begin{array}{l} \frac{1}{10}R_1 \to R_1 \\ \frac{1}{10}R_2 \to R_2 \end{array} \qquad \text{Simplify each row.}$$

$$\sim \left[\begin{array}{ccc|c} 3 & 1 & 1 & 35 \\ 1 & 1 & 3 & 15 \end{array}\right] \qquad R_1 \leftrightarrow R_2$$

$$\sim \left[\begin{array}{ccc|c} 1 & 1 & 3 & 15 \\ 3 & 1 & 1 & 35 \end{array}\right] \qquad R_2 + (-3)R_1 \to R_2$$

$$\sim \left[\begin{array}{ccc|c} 1 & 1 & 3 & 15 \\ 0 & -2 & -8 & -10 \end{array}\right] \qquad -\tfrac{1}{2}R_2 \to R_2$$

$$\sim \left[\begin{array}{ccc|c} 1 & 1 & 3 & 15 \\ 0 & 1 & 4 & 5 \end{array}\right] \qquad R_1 + (-1)R_2 \to R_1$$

$$\sim \left[\begin{array}{ccc|c} 1 & 0 & -1 & 10 \\ 0 & 1 & 4 & 5 \end{array}\right] \qquad \text{Matrix is in reduced form.}$$

$$\begin{aligned} x_1 \quad - x_3 &= 10 \\ x_2 + 4x_3 &= 5 \end{aligned} \qquad \text{or} \qquad \begin{aligned} x_1 &= x_3 + 10 \\ x_2 &= -4x_3 + 5 \end{aligned}$$

Let $x_3 = t$. Then for t any real number,

$$x_1 = t + 10$$
$$x_2 = -4t + 5$$
$$x_3 = t$$

is a solution — or is it? We cannot produce a negative number of sculptures. If we also assume that we cannot produce a fractional number of sculptures, then t

must be a nonnegative whole number. And because of the middle equation ($x_2 = -4t + 5$), t can only assume the values 0 and 1. Thus, for $t = 0$, we have $x_1 = 10$, $x_2 = 5$, $x_3 = 0$; and for $t = 1$, we have $x_1 = 11$, $x_2 = 1$, $x_3 = 1$. These are the only possible production schedules that utilize the full capacity of the plant. ◆

PROBLEM 16 Repeat Example 16 given a casting capacity of 400 labor-hours per week and a finishing capacity of 200 labor-hours per week. ◆

Answers to Matched Problems

11. (A) Condition 2 is violated: The 3 in the second row should be a 1.
    (B) Condition 3 is violated: In the second column, the 5 should be a 0.
    (C) Condition 4 is violated: The leftmost 1 in the first row is not to the right of the leftmost 1 in the first row.
    (D) Condition 1 is violated: The all-zero second row should be at the bottom.

12. (A)
$$\begin{aligned} x_1 \qquad\quad &= -5 \\ x_2 \qquad &= 3 \\ x_3 &= 6 \end{aligned}$$
Solution: $x_1 = -5$, $x_2 = 3$, $x_3 = 6$

(B)
$$\begin{aligned} x_1 + 2x_2 - 3x_3 &= 0 \\ 0x_1 + 0x_2 + 0x_3 &= 1 \\ 0x_1 + 0x_2 + 0x_3 &= 0 \end{aligned}$$
Inconsistent; no solution

(C)
$$\begin{aligned} x_1 \quad - 2x_3 &= 4 \\ x_2 + 3x_3 &= -2 \end{aligned}$$
Dependent: Let $x_3 = t$. Then for any real t,
$$\begin{aligned} x_1 &= 2t + 4 \\ x_2 &= -3t - 2 \\ x_3 &= t \end{aligned}$$
is a solution.

(D)
$$\begin{aligned} x_1 \quad + 3x_3 + 2x_4 &= 5 \\ x_2 - 2x_3 - \quad x_4 &= 3 \end{aligned}$$
Dependent: Let $x_3 = s$ and $x_4 = t$. Then for any real s and t,
$$\begin{aligned} x_1 &= -3s - 2t + 5 \\ x_2 &= 2s + t + 3 \\ x_3 &= s \\ x_4 &= t \end{aligned}$$
is a solution.

13. $x_1 = 1$, $x_2 = -1$, $x_3 = 0$
14. $x_1 = -\frac{3}{5}t - \frac{4}{15}$, $x_2 = \frac{4}{5}t + \frac{7}{15}$, $x_3 = t$, t any real number
15. $x_1 = 2$, $x_2 = -1$
16. $x_1 = t + 10$, $x_2 = -4t + 10$, $x_3 = t$, where $t = 0, 1, 2$; that is, $(x_1, x_2, x_3) = (10, 10, 0), (11, 6, 1)$, or $(12, 2, 2)$

EXERCISE 5-3

A *Indicate whether each matrix is in reduced form.*

1. $\left[\begin{array}{cc|c} 1 & 0 & 2 \\ 0 & 1 & -1 \end{array}\right]$

2. $\left[\begin{array}{cc|c} 0 & 1 & 2 \\ 1 & 0 & -1 \end{array}\right]$

3. $\left[\begin{array}{ccc|c} 1 & 0 & 2 & 3 \\ 0 & 0 & 0 & 0 \\ 0 & 1 & -1 & 4 \end{array}\right]$

4. $\left[\begin{array}{ccc|c} 1 & 0 & 0 & -2 \\ 0 & 1 & 0 & 0 \\ 0 & 0 & 1 & 1 \end{array}\right]$

5. $\left[\begin{array}{ccc|c} 0 & 1 & 0 & 2 \\ 0 & 0 & 3 & -1 \\ 0 & 0 & 0 & 0 \end{array}\right]$

6. $\left[\begin{array}{ccc|c} 1 & 3 & 0 & 0 \\ 0 & 0 & 1 & 0 \\ 0 & 0 & 0 & 1 \end{array}\right]$

7. $\left[\begin{array}{cccc|c} 1 & 2 & 0 & 3 & 2 \\ 0 & 0 & 1 & -1 & 0 \end{array}\right]$

8. $\left[\begin{array}{ccc|c} 0 & 1 & 2 & 1 \\ 1 & 0 & -3 & 2 \end{array}\right]$

Write the linear system corresponding to each reduced augmented matrix and solve.

9. $\left[\begin{array}{ccc|c} 1 & 0 & 0 & -2 \\ 0 & 1 & 0 & 3 \\ 0 & 0 & 1 & 0 \end{array}\right]$

10. $\left[\begin{array}{cccc|c} 1 & 0 & 0 & 0 & -2 \\ 0 & 1 & 0 & 0 & 0 \\ 0 & 0 & 1 & 0 & 1 \\ 0 & 0 & 0 & 1 & 3 \end{array}\right]$

11. $\left[\begin{array}{ccc|c} 1 & 0 & -2 & 3 \\ 0 & 1 & 1 & -5 \\ 0 & 0 & 0 & 0 \end{array}\right]$

12. $\left[\begin{array}{ccc|c} 1 & -2 & 0 & -3 \\ 0 & 0 & 1 & 5 \\ 0 & 0 & 0 & 0 \end{array}\right]$

13. $\left[\begin{array}{cc|c} 1 & 0 & 0 \\ 0 & 1 & 0 \\ 0 & 0 & 1 \end{array}\right]$

14. $\left[\begin{array}{cc|c} 1 & 0 & 5 \\ 0 & 1 & -3 \\ 0 & 0 & 0 \end{array}\right]$

15. $\left[\begin{array}{cccc|c} 1 & -2 & 0 & -3 & -5 \\ 0 & 0 & 1 & 3 & 2 \end{array}\right]$

16. $\left[\begin{array}{cccc|c} 1 & 0 & -2 & 3 & 4 \\ 0 & 1 & -1 & 2 & -1 \end{array}\right]$

B *Use row operations to change each matrix to reduced form.*

17. $\left[\begin{array}{cc|c} 1 & 2 & -1 \\ 0 & 1 & 3 \end{array}\right]$

18. $\left[\begin{array}{cc|c} 1 & 3 & 1 \\ 0 & 2 & -4 \end{array}\right]$

19. $\left[\begin{array}{ccc|c} 1 & 0 & -3 & 1 \\ 0 & 1 & 2 & 0 \\ 0 & 0 & 3 & -6 \end{array}\right]$

20. $\left[\begin{array}{ccc|c} 1 & 0 & 4 & 0 \\ 0 & 1 & -3 & -1 \\ 0 & 0 & -2 & 2 \end{array}\right]$

21. $\left[\begin{array}{ccc|c} 1 & 2 & -2 & -1 \\ 0 & 3 & -6 & 1 \\ 0 & -1 & 2 & -\frac{1}{3} \end{array}\right]$

22. $\left[\begin{array}{ccc|c} 0 & -2 & 8 & 1 \\ 2 & -2 & 6 & -4 \\ 0 & -1 & 4 & \frac{1}{2} \end{array}\right]$

Solve using Gauss–Jordan elimination.

23. $\begin{aligned} 2x_1 + 4x_2 - 10x_3 &= -2 \\ 3x_1 + 9x_2 - 21x_3 &= 0 \\ x_1 + 5x_2 - 12x_3 &= 1 \end{aligned}$

24. $\begin{aligned} 3x_1 + 5x_2 - x_3 &= -7 \\ x_1 + x_2 + x_3 &= -1 \\ 2x_1 + 11x_3 &= 7 \end{aligned}$

25. $\begin{aligned} 3x_1 + 8x_2 - x_3 &= -18 \\ 2x_1 + x_2 + 5x_3 &= 8 \\ 2x_1 + 4x_2 + 2x_3 &= -4 \end{aligned}$

26. $\begin{aligned} 2x_1 + 7x_2 + 15x_3 &= -12 \\ 4x_1 + 7x_2 + 13x_3 &= -10 \\ 3x_1 + 6x_2 + 12x_3 &= -9 \end{aligned}$

27. $\begin{aligned} 2x_1 - x_2 - 3x_3 &= 8 \\ x_1 - 2x_2 &= 7 \end{aligned}$

28. $\begin{aligned} 2x_1 + 4x_2 - 6x_3 &= 10 \\ 3x_1 + 3x_2 - 3x_3 &= 6 \end{aligned}$

29. $\begin{aligned} 2x_1 + 3x_2 - x_3 &= 1 \\ x_1 - 2x_2 + 2x_3 &= -2 \end{aligned}$

30. $\begin{aligned} x_1 - 3x_2 + 2x_3 &= -1 \\ 3x_1 + 2x_2 - x_3 &= 2 \end{aligned}$

31.
$$\begin{aligned} 2x_1 + 2x_2 &= 2 \\ x_1 + 2x_2 &= 3 \\ -3x_2 &= -6 \end{aligned}$$

32.
$$\begin{aligned} 2x_1 - x_2 &= 0 \\ 3x_1 + 2x_2 &= 7 \\ x_1 - x_2 &= -1 \end{aligned}$$

33.
$$\begin{aligned} 2x_1 - x_2 &= 0 \\ 3x_1 + 2x_2 &= 7 \\ x_1 - x_2 &= -2 \end{aligned}$$

34.
$$\begin{aligned} x_1 - 3x_2 &= 5 \\ 2x_1 + x_2 &= 3 \\ x_1 - 2x_2 &= 5 \end{aligned}$$

35.
$$\begin{aligned} 3x_1 - 4x_2 - x_3 &= 1 \\ 2x_1 - 3x_2 + x_3 &= 1 \\ x_1 - 2x_2 + 3x_3 &= 2 \end{aligned}$$

36.
$$\begin{aligned} 3x_1 + 7x_2 - x_3 &= 11 \\ x_1 + 2x_2 - x_3 &= 3 \\ 2x_1 + 4x_2 - 2x_3 &= 10 \end{aligned}$$

37.
$$\begin{aligned} 3x_1 - 2x_2 + x_3 &= -7 \\ 2x_1 + x_2 - 4x_3 &= 0 \\ x_1 + x_2 - 3x_3 &= 1 \end{aligned}$$

38.
$$\begin{aligned} 2x_1 + 3x_2 + 5x_3 &= 21 \\ x_1 - x_2 - 5x_3 &= -2 \\ 2x_1 + x_2 - x_3 &= 11 \end{aligned}$$

39.
$$\begin{aligned} 2x_1 + 4x_2 - 2x_3 &= 2 \\ -3x_1 - 6x_2 + 3x_3 &= -3 \end{aligned}$$

40.
$$\begin{aligned} 3x_1 - 9x_2 + 12x_3 &= 6 \\ -2x_1 + 6x_2 - 8x_3 &= -4 \end{aligned}$$

C *Solve using Gauss–Jordan elimination.*

41.
$$\begin{aligned} 2x_1 - 3x_2 + 3x_3 &= -15 \\ 3x_1 + 2x_2 - 5x_3 &= 19 \\ 5x_1 - 4x_2 - 2x_3 &= -2 \end{aligned}$$

42.
$$\begin{aligned} 3x_1 - 2x_2 - 4x_3 &= -8 \\ 4x_1 + 3x_2 - 5x_3 &= -5 \\ 6x_1 - 5x_2 + 2x_3 &= -17 \end{aligned}$$

43.
$$\begin{aligned} 5x_1 - 3x_2 + 2x_3 &= 13 \\ 2x_1 + 4x_2 - 3x_3 &= -9 \\ 4x_1 - 2x_2 + 5x_3 &= 13 \end{aligned}$$

44.
$$\begin{aligned} 4x_1 - 2x_2 + 3x_3 &= 0 \\ 3x_1 - 5x_2 - 2x_3 &= -12 \\ 2x_1 + 4x_2 - 3x_3 &= -4 \end{aligned}$$

45.
$$\begin{aligned} x_1 + 2x_2 - 4x_3 - x_4 &= 7 \\ 2x_1 + 5x_2 - 9x_3 - 4x_4 &= 16 \\ x_1 + 5x_2 - 7x_3 - 7x_4 &= 13 \end{aligned}$$

46.
$$\begin{aligned} 2x_1 + 4x_2 + 5x_3 + 4x_4 &= 8 \\ x_1 + 2x_2 + 2x_3 + x_4 &= 3 \end{aligned}$$

APPLICATIONS

Solve all the following problems using Gauss–Jordan elimination.

Business & Economics

47. Production scheduling. A small manufacturing plant makes three types of inflatable boats: one-person, two-person, and four-person models. Each boat requires the services of three departments, as listed in the table. The cutting, assembly, and packaging departments have available a maximum of 380, 330, and 120 labor-hours per week, respectively. How many boats of each type must be produced each week for the plant to operate at full capacity?

	ONE-PERSON BOAT	TWO-PERSON BOAT	FOUR-PERSON BOAT
CUTTING DEPARTMENT	0.5 hr	1.0 hr	1.5 hr
ASSEMBLY DEPARTMENT	0.6 hr	0.9 hr	1.2 hr
PACKAGING DEPARTMENT	0.2 hr	0.3 hr	0.5 hr

48. Production scheduling. Repeat Problem 47 assuming the cutting, assembly, and packaging departments have available a maximum of 350, 330, and 115 labor-hours per week, respectively.

49. *Production scheduling.* Work Problem 47 assuming the packaging department is no longer used.

50. *Production scheduling.* Work Problem 48 assuming the packaging department is no longer used.

51. *Production scheduling.* Work Problem 47 assuming the four-person boat is no longer produced.

52. *Production scheduling.* Work Problem 48 assuming the four-person boat is no longer produced.

53. *Purchasing.* A chemical manufacturer wants to purchase a fleet of 24 railroad tank cars with a combined carrying capacity of 250,000 gallons. Tank cars with three different carrying capacities are available: 6,000 gallons, 8,000 gallons, and 18,000 gallons. How many of each type of tank car should be purchased?

54. *Purchasing.* A commuter airline wants to purchase a fleet of 30 airplanes with a combined carrying capacity of 960 passengers. The three available types of planes carry 18, 24, and 42 passengers, respectively. How many of each type of plane should be purchased?

55. *Income tax.* A company has a taxable income of $1,664,000. The federal income tax is 25% of the portion of the income that remains after the state and local taxes have been deducted. The state income tax is 10% of the portion of the income that remains after the federal and local taxes have been deducted, and the local tax is 5% of the portion of the income that remains after the federal and state taxes have been deducted. Find the company's federal, state, and local income taxes.

56. *Income tax.* Repeat Problem 55 if local taxes are not deducted before computing the federal and state taxes.

Life Sciences

57. *Nutrition.* A dietitian in a hospital is to arrange a special diet composed of three basic foods. The diet is to include exactly 340 units of calcium, 180 units of iron, and 220 units of vitamin A. The number of units per ounce of each special ingredient for each of the foods is indicated in the table. How many ounces of each food must be used to meet the diet requirements?

	UNITS PER OUNCE		
	Food A	*Food B*	*Food C*
CALCIUM	30	10	20
IRON	10	10	20
VITAMIN A	10	30	20

58. *Nutrition.* Repeat Problem 57 if the diet is to include exactly 400 units of calcium, 160 units of iron, and 240 units of vitamin A.

59. *Nutrition.* Solve Problem 57 with the assumption that food C is no longer available.

60. *Nutrition.* Solve Problem 58 with the assumption that food C is no longer available.

61. *Nutrition.* Solve Problem 57 with the assumption that the vitamin A requirement is deleted.

62. *Nutrition.* Solve Problem 58 with the assumption that the vitamin A requirement is deleted.

63. *Nutrition—plants.* A farmer can buy four types of plant food. Each barrel of mix *A* contains 30 pounds of phosphoric acid, 50 pounds of nitrogen, and 30 pounds of potash; each barrel of mix *B* contains 30 pounds of phosphoric acid, 75 pounds of nitrogen, and 20 pounds of potash; each barrel of mix C

contains 30 pounds of phosphoric acid, 25 pounds of nitrogen, and 20 pounds of potash; and each barrel of mix *D* contains 60 pounds of phosphoric acid, 25 pounds of nitrogen, and 50 pounds of potash. Soil tests indicate that a particular field needs 900 pounds of phosphoric acid, 750 pounds of nitrogen, and 700 pounds of potash. How many barrels of each type of food should the farmer mix together to supply the necessary nutrients for the field?

64. *Nutrition—animals.* In a laboratory experiment, rats are to be fed 5 packets of food containing a total of 80 units of vitamin E. There are four different brands of food packets that can be used. A packet of brand *A* contains 5 units of vitamin E, a packet of brand *B* contains 10 units of vitamin E, a packet of brand *C* contains 15 units of vitamin E, and a packet of brand *D* contains 20 units of vitamin E. How many packets of each brand should be mixed and fed to the rats?

Social Sciences

65. *Sociology.* Two sociologists have grant money to study school busing in a particular city. They wish to conduct an opinion survey using 600 telephone contacts and 400 house contacts. Survey company *A* has personnel to do 30 telephone and 10 house contacts per hour; survey company *B* can handle 20 telephone and 20 house contacts per hour. How many hours should be scheduled for each firm to produce exactly the number of contacts needed?

66. *Sociology.* Repeat Problem 65 if 650 telephone contacts and 350 house contacts are needed.

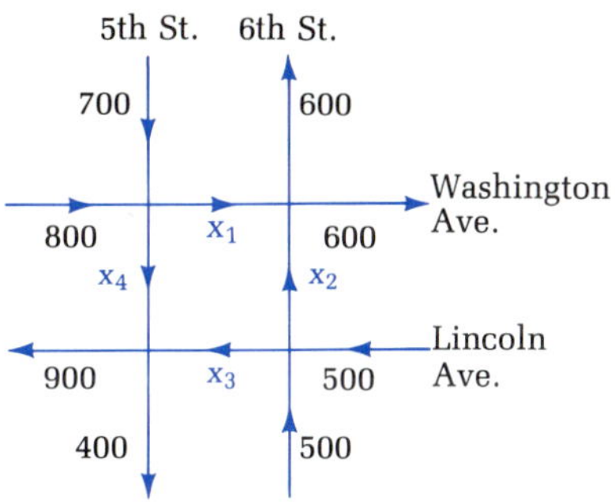

67. *Traffic flow.* The rush-hour traffic flow for a network of four one-way streets in a city is shown in the figure. The numbers next to each street indicate the number of vehicles per hour that enter and leave the network on that street. The variables x_1, x_2, x_3, and x_4 represent the flow of traffic between the four intersections in the network.

(A) For a smooth traffic flow, the number of vehicles entering each intersection should always equal the number leaving. For example, since 1,500 vehicles enter the intersection of 5th Street and Washington Avenue each hour and $x_1 + x_4$ vehicles leave this intersection, we see that $x_1 + x_4 = 1{,}500$. Find the equations determined by the traffic flow at each of the other three intersections.

(B) Find the solution to the system in part A.

(C) What is the maximum number of vehicles that can travel from Washington Avenue to Lincoln Avenue on 5th Street? What is the minimum number?

(D) If traffic lights are adjusted so that 1,000 vehicles per hour travel from Washington Avenue to Lincoln Avenue on 5th Street, determine the flow around the rest of the network.

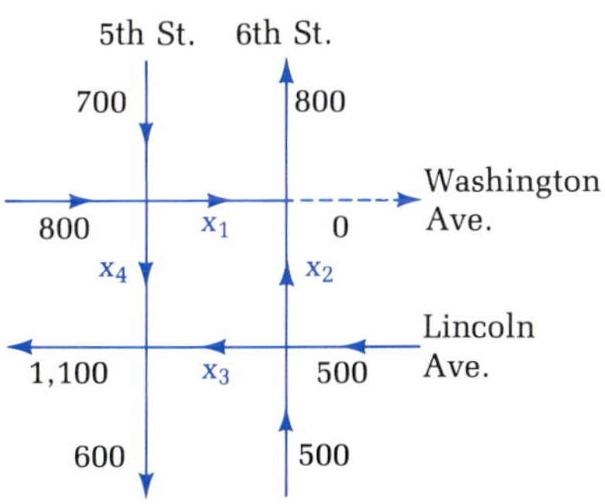

68. *Traffic flow.* Refer to Problem 67. Closing Washington Avenue east of 6th Street for construction changes the traffic flow for the network as indicated in the figure. Repeat parts A–D of Problem 67 for this traffic flow.

SECTION 5-4

Matrices — Addition and Multiplication by a Number

- DIMENSIONS OF A MATRIX
- ADDITION AND SUBTRACTION
- PRODUCT OF A NUMBER k AND A MATRIX M
- APPLICATION
- DOUBLE SUBSCRIPT NOTATION

In the last two sections we introduced the important new idea of matrices. In this and the following sections, we shall develop this concept further. Matrices are both a very ancient and a very current mathematical concept. References to matrices and systems of equations can be found in Chinese manuscripts dating back to around 200 B.C. More recently, the advent of personal and large-scale computers has made matrices a very useful tool for a wide variety of applications.

◆ DIMENSIONS OF A MATRIX

Recall that we defined a **matrix** as any rectangular array of real numbers enclosed within brackets and that each number in the array is called an **element** of the matrix. The **size** or **dimension of a matrix** is important to operations on matrices. We define an $\boldsymbol{m \times n}$ **matrix** (read "m by n matrix") to be one with m rows and n columns. It is important to note that the number of rows is always given first. If a matrix has the same number of rows and columns, it is called a **square matrix.** A matrix with only one column is called a **column matrix,** and a matrix with only one row is called a **row matrix.** These definitions are illustrated by the following:

$$\underset{3\times 2}{\begin{bmatrix} -2 & 5 \\ 0 & -2 \\ 3 & 6 \end{bmatrix}} \quad \underset{\text{Square matrix}}{\overset{3\times 3}{\begin{bmatrix} 0.5 & 0.2 & 1.0 \\ 0.0 & 0.3 & 0.5 \\ 0.7 & 0.0 & 0.2 \end{bmatrix}}} \quad \underset{\text{Column matrix}}{\overset{4\times 1}{\begin{bmatrix} 3 \\ -2 \\ 1 \\ 0 \end{bmatrix}}} \quad \underset{\text{Row matrix}}{\overset{1\times 4}{\begin{bmatrix} 2 & \frac{1}{2} & 0 & -\frac{2}{3} \end{bmatrix}}}$$

Two matrices are **equal** if they have the same dimension and their corresponding elements are equal. For example,

$$\overset{2\times 3}{\begin{bmatrix} a & b & c \\ d & e & f \end{bmatrix}} = \overset{2\times 3}{\begin{bmatrix} u & v & w \\ x & y & z \end{bmatrix}} \quad \text{if and only if} \quad \begin{matrix} a=u & b=v & c=w \\ d=x & e=y & f=z \end{matrix}$$

◆ ADDITION AND SUBTRACTION

The **sum of two matrices of the same dimension** is the matrix with elements that are the sum of the corresponding elements of the two given matrices.

Addition is not defined for matrices with different dimensions.

◆ EXAMPLE 17

(A) $\begin{bmatrix} a & b \\ c & d \end{bmatrix} + \begin{bmatrix} w & x \\ y & z \end{bmatrix} = \begin{bmatrix} (a+w) & (b+x) \\ (c+y) & (d+z) \end{bmatrix}$

(B) $\begin{bmatrix} 2 & -3 & 0 \\ 1 & 2 & -5 \end{bmatrix} + \begin{bmatrix} 3 & 1 & 2 \\ -3 & 2 & 5 \end{bmatrix} = \begin{bmatrix} 5 & -2 & 2 \\ -2 & 4 & 0 \end{bmatrix}$

(C) $\begin{bmatrix} 5 & 0 & -2 \\ 1 & -3 & 8 \end{bmatrix} + \begin{bmatrix} -1 & 7 \\ 0 & 6 \\ -2 & 8 \end{bmatrix}$ Not defined ◆

PROBLEM 17 Add: $\begin{bmatrix} 3 & 2 \\ -1 & -1 \\ 0 & 3 \end{bmatrix} + \begin{bmatrix} -2 & 3 \\ 1 & -1 \\ 2 & -2 \end{bmatrix}$ ◆

Because we add two matrices by adding their corresponding elements, it follows from the properties of real numbers that matrices of the same dimension are commutative and associative relative to addition. That is, if A, B, and C are matrices of the same dimension, then

COMMUTATIVE	$A + B = B + A$
ASSOCIATIVE	$(A + B) + C = A + (B + C)$

A matrix with elements that are all zeros is called a **zero matrix.** For example,

$$[0 \quad 0 \quad 0] \qquad \begin{bmatrix} 0 & 0 \\ 0 & 0 \end{bmatrix} \qquad \begin{bmatrix} 0 \\ 0 \\ 0 \\ 0 \end{bmatrix} \qquad \begin{bmatrix} 0 & 0 & 0 & 0 \\ 0 & 0 & 0 & 0 \\ 0 & 0 & 0 & 0 \end{bmatrix}$$

are zero matrices of different dimensions. [*Note:* "0" is often used to denote the zero matrix of an arbitrary dimension.] The **negative of a matrix *M*,** denoted by $-M$, is a matrix with elements that are the negatives of the elements in M. Thus, if

$$M = \begin{bmatrix} a & b \\ c & d \end{bmatrix} \qquad \text{then} \qquad -M = \begin{bmatrix} -a & -b \\ -c & -d \end{bmatrix}$$

Note that $M + (-M) = 0$ (a zero matrix).

If A and B are matrices of the same dimension, then we define **subtraction** as follows:

$$A - B = A + (-B)$$

Thus, to subtract matrix B from matrix A, we simply add the negative of B to A.

◆ EXAMPLE 18

$\begin{bmatrix} 3 & -2 \\ 5 & 0 \end{bmatrix} - \begin{bmatrix} -2 & 2 \\ 3 & 4 \end{bmatrix} = \begin{bmatrix} 3 & -2 \\ 5 & 0 \end{bmatrix} + \begin{bmatrix} 2 & -2 \\ -3 & -4 \end{bmatrix} = \begin{bmatrix} 5 & -4 \\ 2 & -4 \end{bmatrix}$ ◆

PROBLEM 18 Subtract: $[2 \quad -3 \quad 5] - [3 \quad -2 \quad 1]$ ◆

◆ PRODUCT OF A NUMBER k AND A MATRIX M

Finally, the **product of a number *k* and a matrix *M*,** denoted by ***kM***, is a matrix formed by multiplying each element of M by k. This definition is partly motivated by the fact that if M is a matrix, then we would like $M + M$ to equal $2M$.

◆ EXAMPLE 19

$$-2\begin{bmatrix} 3 & -1 & 0 \\ -2 & 1 & 3 \\ 0 & -1 & -2 \end{bmatrix} = \begin{bmatrix} -6 & 2 & 0 \\ 4 & -2 & -6 \\ 0 & 2 & 4 \end{bmatrix}$$ ◆

PROBLEM 19

Find: $10\begin{bmatrix} 1.3 \\ 0.2 \\ 3.5 \end{bmatrix}$ ◆

◆ APPLICATION

◆ EXAMPLE 20

Sales Commissions

Ms. Smith and Mr. Jones are salespeople in a new-car agency that sells only two models. August was the last month for this year's models, and next year's models were introduced in September. Gross dollar sales for each month are given in the following matrices:

$$\begin{array}{l} \\ \text{Ms. Smith} \\ \text{Mr. Jones} \end{array} \overset{\text{August sales}}{\overset{\text{Compact} \quad \text{Luxury}}{\begin{bmatrix} \$54{,}000 & \$88{,}000 \\ \$126{,}000 & 0 \end{bmatrix}}} = A \qquad \overset{\text{September sales}}{\overset{\text{Compact} \quad \text{Luxury}}{\begin{bmatrix} \$228{,}000 & \$368{,}000 \\ \$304{,}000 & \$322{,}000 \end{bmatrix}}} = B$$

(For example, Ms. Smith had \$54,000 in compact sales in August, and Mr. Jones had \$322,000 in luxury car sales in September.)

(A) What were the combined dollar sales in August and September for each person and each model?
(B) What was the increase in dollar sales from August to September?
(C) If both salespeople receive 5% commissions on gross dollar sales, compute the commission for each person for each model sold in September.

Solutions

(A) $$A + B = \overset{\text{Compact} \quad \text{Luxury}}{\begin{bmatrix} \$282{,}000 & \$456{,}000 \\ \$430{,}000 & \$322{,}000 \end{bmatrix}} \begin{array}{l} \text{Ms. Smith} \\ \text{Mr. Jones} \end{array}$$

(B) $$B - A = \overset{\text{Compact} \quad \text{Luxury}}{\begin{bmatrix} \$174{,}000 & \$280{,}000 \\ \$178{,}000 & \$322{,}000 \end{bmatrix}} \begin{array}{l} \text{Ms. Smith} \\ \text{Mr. Jones} \end{array}$$

(C) $$0.05B = \begin{bmatrix} (0.05)(\$228{,}000) & (0.05)(\$368{,}000) \\ (0.05)(\$304{,}000) & (0.05)(\$322{,}000) \end{bmatrix}$$

$$= \begin{bmatrix} \$11{,}400 & \$18{,}400 \\ \$15{,}200 & \$16{,}100 \end{bmatrix} \begin{array}{l} \text{Ms. Smith} \\ \text{Mr. Jones} \end{array}$$ ◆

PROBLEM 20 Repeat Example 20 with

$$A = \begin{bmatrix} \$45{,}000 & \$77{,}000 \\ \$106{,}000 & \$22{,}000 \end{bmatrix} \quad \text{and} \quad B = \begin{bmatrix} \$190{,}000 & \$345{,}000 \\ \$266{,}000 & \$276{,}000 \end{bmatrix}$$ ◆

Example 20 involved an agency with only 2 salespeople and 2 models. A more realistic problem might involve 20 salespeople and 15 models. Problems of this size are often solved with the aid of an electronic spreadsheet program on a personal computer. Figure 1 illustrates such a computer solution for Example 20.

	1	2	3	4	5	6	7
1		August Sales		September Sales		September Commissions	
2		Compact	Luxury	Compact	Luxury	Compact	Luxury
3	Smith	$54,000	$88,000	$228,000	$368,000	$11,400	$18,400
4	Jones	$126,000	$0	$304,000	$322,000	$15,200	$16,100
5		Combined Sales		Sales Increase			
6	Smith	$282,000	$456,000	$174,000	$280,000		
7	Jones	$430,000	$322,000	$178,000	$322,000		

FIGURE 1

◆ DOUBLE SUBSCRIPT NOTATION

In more advanced treatments of topics involving matrices, **double subscript notation** is used in writing general matrix forms. For example, a general 2 × 3 matrix would be written as

$$A = \begin{bmatrix} a_{11} & a_{12} & a_{13} \\ a_{21} & a_{22} & a_{23} \end{bmatrix}$$

The subscript is essentially the coordinates of the position of the element in the rectangular array of numbers—the first subscript number indicates the row in which the element appears, and the second subscript number indicates the column. Thus, a_{21} is the element in the second row and first column, and a_{12} is the element in the first row and second column. Note that a_{12} is read "a sub one two," *not* "a sub twelve." In the matrix

$$B = \begin{bmatrix} 7 & 9 & -5 \\ 3 & -6 & 4 \end{bmatrix}$$

$a_{23} = 4$, $a_{13} = -5$, and $a_{22} = -6$. The elements $a_{11} = 7$ and $a_{22} = -6$ make up the *principal diagonal* of B. In general, the **principal diagonal** of a matrix A consists of the elements $a_{11}, a_{22}, a_{33}, \ldots$. We will refer to the principal diagonal of a matrix in several of the sections that follow.

So that you can concentrate on key concepts and procedures without getting distracted by formal mathematical notation, we will make only light use of double subscript notation in developments that follow.

Answers to Matched Problems

17. $\begin{bmatrix} 1 & 5 \\ 0 & -2 \\ 2 & 1 \end{bmatrix}$ 18. $[-1 \quad -1 \quad 4]$ 19. $\begin{bmatrix} 13 \\ 2 \\ 35 \end{bmatrix}$

20. (A) $\begin{bmatrix} \$235{,}000 & \$422{,}000 \\ \$372{,}000 & \$298{,}000 \end{bmatrix}$ (B) $\begin{bmatrix} \$145{,}000 & \$268{,}000 \\ \$160{,}000 & \$254{,}000 \end{bmatrix}$

(C) $\begin{bmatrix} \$9{,}500 & \$17{,}250 \\ \$13{,}300 & \$13{,}800 \end{bmatrix}$

EXERCISE 5-4

A *Problems 1–20 refer to the following matrices:*

$$A = \begin{bmatrix} 2 & -1 \\ 3 & 0 \end{bmatrix} \qquad B = \begin{bmatrix} -3 & 1 \\ 2 & -3 \end{bmatrix} \qquad C = \begin{bmatrix} 2 \\ -3 \\ 0 \end{bmatrix}$$

$$D = \begin{bmatrix} 1 \\ 3 \\ 5 \end{bmatrix} \qquad E = [-4 \quad 1 \quad 0 \quad -2] \qquad F = \begin{bmatrix} 2 & -3 \\ -2 & 0 \\ 1 & 2 \\ 3 & 5 \end{bmatrix}$$

1. What are the dimensions of B? Of E?
2. What are the dimensions of F? Of D?
3. For the matrix F, find f_{12}, f_{22}, and f_{42}.
4. For the matrix F, find f_{11}, f_{21}, and f_{32}.
5. Write a zero matrix of the same dimension as B.
6. Write a zero matrix of the same dimension as E.
7. Identify all column matrices.
8. Identify all row matrices.
9. Identify all square matrices.
10. How many additional columns would F have to have to be a square matrix?
11. Find $A + B$. **12.** Find $C + D$. **13.** Find $E + F$. **14.** Find $B + C$.
15. Write the negative of matrix C.
16. Write the negative of matrix B.
17. Find $D - C$. **18.** Find $A - A$. **19.** Find $5B$. **20.** Find $-2E$.

B *In Problems 21–28, perform the indicated operations.*

21. $\begin{bmatrix} 3 & -2 & 0 & 1 \\ 2 & -3 & -1 & 4 \\ 0 & 2 & -1 & 6 \end{bmatrix} + \begin{bmatrix} -2 & 5 & -1 & 0 \\ -3 & -2 & 8 & -2 \\ 4 & 6 & 1 & -8 \end{bmatrix}$

22. $\begin{bmatrix} 4 & -2 & 8 \\ 0 & -1 & -4 \\ -6 & 5 & 2 \\ 1 & 3 & -6 \end{bmatrix} + \begin{bmatrix} -6 & -2 & -3 \\ 5 & 2 & 4 \\ 8 & 3 & -4 \\ 1 & -5 & 0 \end{bmatrix}$

23. $\begin{bmatrix} 1.3 & 2.5 & -6.1 \\ 8.3 & -1.4 & 6.7 \end{bmatrix} - \begin{bmatrix} -4.1 & 1.8 & -4.3 \\ 0.7 & 2.6 & -1.2 \end{bmatrix}$

24. $\begin{bmatrix} 2.6 & 3.8 \\ -1.9 & 7.3 \\ 5.6 & -0.4 \end{bmatrix} - \begin{bmatrix} 4.8 & -2.1 \\ 3.2 & 5.9 \\ -1.5 & 2.2 \end{bmatrix}$

25. $1{,}000 \begin{bmatrix} 0.25 & 0.36 \\ 0.04 & 0.35 \end{bmatrix}$

26. $100 \begin{bmatrix} 0.32 & 0.05 & 0.17 \\ 0.22 & 0.03 & 0.21 \end{bmatrix}$

27. $0.08 \begin{bmatrix} 24{,}000 & 35{,}000 \\ 12{,}000 & 24{,}000 \end{bmatrix} + 0.03 \begin{bmatrix} 12{,}000 & 22{,}000 \\ 14{,}000 & 13{,}000 \end{bmatrix}$

28. $0.05 \begin{bmatrix} 430 & 212 \\ 210 & 165 \\ 435 & 315 \end{bmatrix} + 0.07 \begin{bmatrix} 234 & 436 \\ 160 & 212 \\ 410 & 136 \end{bmatrix}$

29. Find a, b, c, and d so that

$$\begin{bmatrix} a & b \\ c & d \end{bmatrix} + \begin{bmatrix} 2 & -3 \\ 0 & 1 \end{bmatrix} = \begin{bmatrix} 1 & -2 \\ 3 & -4 \end{bmatrix}$$

30. Find w, x, y, and z so that

$$\begin{bmatrix} 4 & -2 \\ -3 & 0 \end{bmatrix} + \begin{bmatrix} w & x \\ y & z \end{bmatrix} = \begin{bmatrix} 2 & -3 \\ 0 & 5 \end{bmatrix}$$

31. Find x and y so that

$$\begin{bmatrix} 2x & 4 \\ -3 & 5x \end{bmatrix} + \begin{bmatrix} 3y & -2 \\ -2 & -y \end{bmatrix} = \begin{bmatrix} -5 & 2 \\ -5 & 13 \end{bmatrix}$$

32. Find x and y so that

$$\begin{bmatrix} 5 & 3x \\ 2x & -4 \end{bmatrix} + \begin{bmatrix} 1 & -4y \\ 7y & 4 \end{bmatrix} = \begin{bmatrix} 6 & -7 \\ 5 & 0 \end{bmatrix}$$

C *In Problems 33–36, is it possible to find 2×2 matrices A, B, and C and a real number k such that the indicated equation fails to hold? If so, state an example.*

33. $A + B = B + A$

34. $(A + B) + C = A + (B + C)$

35. $k(A + B) = kA + kB$

36. $k(A - B) = kA - kB$

APPLICATIONS

Business & Economics

37. *Cost analysis.* A company with two different plants manufactures guitars and banjos. Its production costs for each instrument are given in the following matrices:

Plant X

$$\begin{matrix} & \text{Guitar} & \text{Banjo} \\ \text{Materials} & \$30 & \$25 \\ \text{Labor} & \$60 & \$80 \end{matrix} = A$$

Plant Y

$$\begin{matrix} \text{Guitar} & \text{Banjo} \\ \$36 & \$27 \\ \$54 & \$74 \end{matrix} = B$$

Find $\frac{1}{2}(A + B)$, the average cost of production for the two plants.

38. *Cost analysis.* If both labor and materials at plant X in Problem 37 are increased by 20%, find $\frac{1}{2}(1.2A + B)$, the new average cost of production for the two plants.

39. *Markup.* An import car dealer sells three models of a car. The retail prices and the current dealer invoice prices (costs) for the basic models and indicated options are given in the following two matrices (where "Air" means air-conditioning):

Retail price

	Basic car	Air	AM/FM radio	Cruise control	
Model A	\$10,900	\$683	\$253	\$195	
Model B	\$13,000	\$738	\$382	\$206	$= M$
Model C	\$16,300	\$867	\$537	\$225	

Dealer invoice price

	Basic car	Air	AM/FM radio	Cruise control	
Model A	\$9,400	\$582	\$195	\$160	
Model B	\$11,500	\$621	\$295	\$171	$= N$
Model C	\$14,100	\$737	\$420	\$184	

We define the markup matrix to be $M - N$ (**markup** is the difference between the retail price and the dealer invoice price). Suppose the value of the dollar has had a sharp decline and the dealer invoice price is to have an across-the-board 15% increase next year. In order to stay competitive with domestic cars, the dealer increases the retail prices only 10%. Calculate a markup matrix for next year's models and the indicated options. (Compute results to the nearest dollar.)

40. *Markup.* Referring to Problem 39, what is the markup matrix resulting from a 20% increase in dealer invoice prices and an increase in retail prices of 15%? (Compute results to the nearest dollar.)

Life Sciences

41. *Heredity.* Gregor Mendel (1822–1884), an Austrian monk and botanist, made discoveries that revolutionized the science of genetics. In one experiment, he crossed dihybrid yellow round peas (yellow and round are dominant characteristics; the peas also contained genes for the recessive characteristics green and wrinkled) and obtained 560 peas of the types indicated in the matrix:

$$\begin{array}{c} \\ \text{Yellow} \\ \text{Green} \end{array} \begin{array}{c} \begin{array}{cc} \text{Round} & \text{Wrinkled} \end{array} \\ \begin{bmatrix} 319 & 101 \\ 108 & 32 \end{bmatrix} \end{array} = M$$

Suppose he carried out a second experiment of the same type and obtained 640 peas of the types indicated in this matrix:

$$\begin{array}{l} \\ \text{Yellow} \\ \text{Green} \end{array} \begin{array}{c} \begin{array}{cc} \text{Round} & \text{Wrinkled} \end{array} \\ \begin{bmatrix} 370 & 124 \\ 110 & 36 \end{bmatrix} \end{array} = N$$

If the results of the two experiments are combined, write the resulting matrix $M + N$. Compute the percentage of the total number of peas (1,200) in each category of the combined results.

Social Sciences

42. *Psychology.* Two psychologists independently carried out studies on the relationship between height and aggressive behavior in women over 18 years of age. The results of the studies are summarized in the following matrices:

Professor Aldquist

$$\begin{array}{l} \\ \text{Passive} \\ \text{Aggressive} \end{array} \begin{array}{c} \begin{array}{ccc} \text{Under 5 ft} & 5\text{–}5\tfrac{1}{2}\text{ ft} & \text{Over } 5\tfrac{1}{2}\text{ ft} \end{array} \\ \begin{bmatrix} 70 & 122 & 20 \\ 30 & 118 & 80 \end{bmatrix} \end{array} = A$$

Professor Kelley

$$\begin{array}{l} \\ \text{Passive} \\ \text{Aggressive} \end{array} \begin{array}{c} \begin{array}{ccc} \text{Under 5 ft} & 5\text{–}5\tfrac{1}{2}\text{ ft} & \text{Over } 5\tfrac{1}{2}\text{ ft} \end{array} \\ \begin{bmatrix} 65 & 160 & 30 \\ 25 & 140 & 75 \end{bmatrix} \end{array} = B$$

The two psychologists decided to combine their results and publish a joint paper. Write the matrix $A + B$ illustrating their combined results. Compute the percentage of the total sample in each category of the combined study.

SECTION 5-5 Matrix Multiplication

- MATRIX PRODUCT
- MULTIPLICATION PROPERTIES
- APPLICATION

In this section, we are going to introduce a matrix multiplication that will seem rather strange at first. Despite this apparent strangeness, this operation is well-founded in the general theory of matrices and, as we will see, is extremely useful in many practical problems.

Historically, matrix multiplication was introduced by the English mathematician Arthur Cayley (1821–1895) in studies of systems of linear equations and linear transformations. In Section 5-7, you will see how matrix multiplication is central to the process of expressing systems of linear equations as matrix equations and to the process of solving matrix equations. Matrix equations and their solutions provide us with an alternate method of solving linear systems with the same number of variables as equations.

◆ MATRIX PRODUCT

We start by defining the product of two special matrices, a row matrix and a column matrix.

Product of a Row Matrix and a Column Matrix

The **product** of a $1 \times n$ row matrix and an $n \times 1$ column matrix is a 1×1 matrix given by

$$\overset{1 \times n}{[a_1 \quad a_2 \quad \cdots \quad a_n]} \overset{n \times 1}{\begin{bmatrix} b_1 \\ b_2 \\ \cdot \\ \cdot \\ \cdot \\ b_n \end{bmatrix}} = [a_1b_1 + a_2b_2 + \cdots + a_nb_n]$$

Note that the number of elements in the row matrix and in the column matrix must be the same for the product to be defined.

◆ EXAMPLE 21

$$[2 \quad -3 \quad 0]\begin{bmatrix} -5 \\ 2 \\ -2 \end{bmatrix} = [(2)(-5) + (-3)(2) + (0)(-2)] = [-10 - 6 + 0] = [-16]$$ ◆

PROBLEM 21

$$[-1 \quad 0 \quad 3 \quad 2]\begin{bmatrix} 2 \\ 3 \\ 4 \\ -1 \end{bmatrix} = ?$$ ◆

Refer to Example 21. The distinction between the real number -16 and the 1×1 matrix $[-16]$ is a technical one, and it is common to see 1×1 matrices written as real numbers without brackets. In the work that follows, we will frequently refer to 1×1 matrices as real numbers and omit the brackets whenever it is convenient to do so.

◆ EXAMPLE 22

Labor Costs

A factory produces a slalom water ski that requires 4 labor-hours in the fabricating department and 1 labor-hour in the finishing department. Fabricating personnel receive \$8 per hour and finishing personnel receive \$6 per hour. Total labor cost per ski is given by the product:

$$[4 \quad 1]\begin{bmatrix} 8 \\ 6 \end{bmatrix} = [(4)(8) + (1)(6)] = [32 + 6] = [38] \quad \text{or \$38 per ski}$$ ◆

PROBLEM 22 If the factory in Example 22 also produces a trick water ski that requires 6 labor-hours in the fabricating department and 1.5 labor-hours in the finishing department, write a product between appropriate row and column matrices that will give the total labor cost for this ski. Compute the cost. ◆

We now use the product of a $1 \times n$ row matrix and an $n \times 1$ column matrix to extend the definition of matrix product to more general matrices.

Matrix Product

If A is an $m \times p$ matrix and B is a $p \times n$ matrix, then the **matrix product** of A and B, denoted AB, is an $m \times n$ matrix whose element in the ith row and jth column is the real number obtained from the product of the ith row of A and the jth column of B. If the number of columns in A does not equal the number of rows in B, then the matrix product AB is **not defined.**

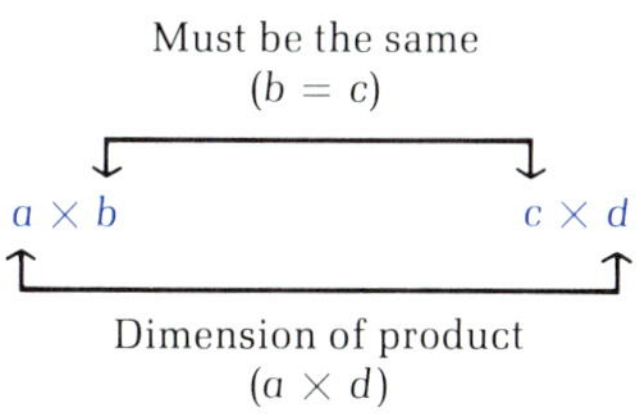

FIGURE 2

It is important to check dimensions before starting the multiplication process. If matrix A has dimension $a \times b$ and matrix B has dimension $c \times d$, then if $b = c$, the product AB will exist and will have dimension $a \times d$. This is shown schematically in Figure 2. The definition is not as complicated as it might first seem. An example should help to clarify the process. For

$$A = \begin{bmatrix} 2 & 3 & -1 \\ -2 & 1 & 2 \end{bmatrix} \quad \text{and} \quad B = \begin{bmatrix} 1 & 3 \\ 2 & 0 \\ -1 & 2 \end{bmatrix}$$

A is 2×3, B is 3×2, so AB is 2×2. To find the first row of AB, we take the product of the first row of A with every column of B and write each result as a real number, not as a 1×1 matrix. The second row of AB is computed in the same manner. The four products of row and column matrices used to produce the four elements in AB are shown in the following dashed box. These products are usually calculated mentally, or with the aid of a calculator, and need not be written out. The shaded portions highlight the steps involved in computing the element in the first row and second column of AB.

$$\underset{2 \times 3}{\begin{bmatrix} 2 & 3 & -1 \\ -2 & 1 & 2 \end{bmatrix}} \underset{3 \times 2}{\begin{bmatrix} 1 & 3 \\ 2 & 0 \\ -1 & 2 \end{bmatrix}} = \begin{bmatrix} [2 \quad 3 \quad -1]\begin{bmatrix} 1 \\ 2 \\ -1 \end{bmatrix} & [2 \quad 3 \quad -1]\begin{bmatrix} 3 \\ 0 \\ 2 \end{bmatrix} \\ [-2 \quad 1 \quad 2]\begin{bmatrix} 1 \\ 2 \\ -1 \end{bmatrix} & [-2 \quad 1 \quad 2]\begin{bmatrix} 3 \\ 0 \\ 2 \end{bmatrix} \end{bmatrix} = \underset{2 \times 2}{\begin{bmatrix} 9 & 4 \\ -2 & -2 \end{bmatrix}}$$

◆ EXAMPLE 23

(A) $\underset{3 \times 2}{\begin{bmatrix} 2 & 1 \\ 1 & 0 \\ -1 & 2 \end{bmatrix}} \underset{2 \times 4}{\begin{bmatrix} 1 & -1 & 0 & 1 \\ 2 & 1 & 2 & 0 \end{bmatrix}} = \underset{3 \times 4}{\begin{bmatrix} 4 & -1 & 2 & 2 \\ 1 & -1 & 0 & 1 \\ 3 & 3 & 4 & -1 \end{bmatrix}}$

(B) $\underset{2 \times 4}{\begin{bmatrix} 1 & -1 & 0 & 1 \\ 2 & 1 & 2 & 0 \end{bmatrix}} \underset{3 \times 2}{\begin{bmatrix} 2 & 1 \\ 1 & 0 \\ -1 & 2 \end{bmatrix}}$
Not defined

(C) $\begin{bmatrix} 2 & 6 \\ -1 & -3 \end{bmatrix} \begin{bmatrix} 1 & 2 \\ 3 & 6 \end{bmatrix} = \begin{bmatrix} 20 & 40 \\ -10 & -20 \end{bmatrix}$

(D) $\begin{bmatrix} 1 & 2 \\ 3 & 6 \end{bmatrix} \begin{bmatrix} 2 & 6 \\ -1 & -3 \end{bmatrix} = \begin{bmatrix} 0 & 0 \\ 0 & 0 \end{bmatrix}$

(E) $[2 \quad -3 \quad 0] \begin{bmatrix} -5 \\ 2 \\ -2 \end{bmatrix} = [-16]$

(F) $\begin{bmatrix} -5 \\ 2 \\ -2 \end{bmatrix} [2 \quad -3 \quad 0] = \begin{bmatrix} -10 & 15 & 0 \\ 4 & -6 & 0 \\ -4 & 6 & 0 \end{bmatrix}$ ◆

PROBLEM 23 Find each product, if it is defined:

(A) $\begin{bmatrix} -1 & 0 & 3 & -2 \\ 1 & 2 & 2 & 0 \end{bmatrix} \begin{bmatrix} -1 & 1 \\ 2 & 3 \\ 1 & 0 \end{bmatrix}$

(B) $\begin{bmatrix} -1 & 1 \\ 2 & 3 \\ 1 & 0 \end{bmatrix} \begin{bmatrix} -1 & 0 & 3 & -2 \\ 1 & 2 & 2 & 0 \end{bmatrix}$

(C) $\begin{bmatrix} 1 & 2 \\ -1 & -2 \end{bmatrix} \begin{bmatrix} -2 & 4 \\ 1 & -2 \end{bmatrix}$

(D) $\begin{bmatrix} -2 & 4 \\ 1 & -2 \end{bmatrix} \begin{bmatrix} 1 & 2 \\ -1 & -2 \end{bmatrix}$

(E) $[3 \quad -2 \quad 1] \begin{bmatrix} 4 \\ 2 \\ 3 \end{bmatrix}$

(F) $\begin{bmatrix} 4 \\ 2 \\ 3 \end{bmatrix} [3 \quad -2 \quad 1]$ ◆

◆ MULTIPLICATION PROPERTIES

In the arithmetic of real numbers it does not matter in which order we multiply; for example, $5 \times 7 = 7 \times 5$. In matrix multiplication, however, it does make a difference. That is, AB does not always equal BA, even if both multiplications are defined (see Examples 23C and 23D). Thus, **matrix multiplication is not commutative.** Also, AB may be zero with neither A nor B equal to zero (see Example 23D). Thus, **the zero property does not hold for matrix multiplication** (see Appendix A-1 for a discussion of the zero property for real numbers).

Matrix multiplication does have general properties, many of which are similar to the properties of real numbers. Some of these are listed in Theorem 4. In Section 5-7 we will use these properties to solve matrix equations. Notice that since matrix multiplication is not commutative, there are two distributive properties and two multiplication properties.

THEOREM 4

Properties of Matrix Multiplication

Assuming all products and sums are defined for the indicated matrices A, B, and C, then for k a real number:

ASSOCIATIVE PROPERTY:	1. $A(BC) = (AB)C$
LEFT DISTRIBUTIVE PROPERTY:	2. $A(B + C) = AB + AC$
RIGHT DISTRIBUTIVE PROPERTY:	3. $(B + C)A = BA + CA$
LEFT MULTIPLICATION PROPERTY:	4. If $A = B$, then $CA = CB$.
RIGHT MULTIPLICATION PROPERTY:	5. If $A = B$, then $AC = BC$.
	6. $k(AB) = (kA)B = A(kB)$

◆ APPLICATION

The next example illustrates the use of the matrix product in a business application.

◆ EXAMPLE 24

Labor Costs

Let us combine the time requirements for slalom and trick water skis discussed in Example 22 and Problem 22 into one matrix:

$$\begin{array}{r} \\ \\ \text{Trick ski} \\ \text{Slalom ski} \end{array} \begin{array}{c} \text{Fabricating} \quad \text{Finishing} \\ \text{department} \quad \text{department} \\ \begin{bmatrix} 6\text{ hr} & 1.5\text{ hr} \\ 4\text{ hr} & 1\text{ hr} \end{bmatrix} = A \end{array}$$

Now suppose the company has two manufacturing plants, X and Y, in different parts of the country and that their hourly rates for each department are given in the following matrix:

$$\begin{array}{r} \\ \text{Fabricating department} \\ \text{Finishing department} \end{array} \begin{array}{c} \text{Plant } X \quad \text{Plant } Y \\ \begin{bmatrix} \$8 & \$7 \\ \$6 & \$4 \end{bmatrix} = B \end{array}$$

To find the total labor costs for each ski at each factory, we multiply A and B:

$$AB = \underset{2 \times 2}{\begin{bmatrix} 6 & 1.5 \\ 4 & 1 \end{bmatrix}} \underset{2 \times 2}{\begin{bmatrix} 8 & 7 \\ 6 & 4 \end{bmatrix}} = \begin{array}{c} X \quad\ \ Y \\ \begin{bmatrix} \$57 & \$48 \\ \$38 & \$32 \end{bmatrix} \end{array} \begin{array}{l} \text{Trick ski} \\ \text{Slalom ski} \end{array}$$

Notice that the product of the first row matrix of A and the first column matrix of B gives us the labor costs, \$57, for a trick ski manufactured at plant X. The product of the second row matrix of A and the second column matrix of B gives us the labor costs, \$32, for manufacturing a slalom ski at plant Y. And so on. ◆

Example 24 is, of course, over simplified. Companies manufacturing many different items in many different plants deal with matrices that have very large numbers of rows and columns.

PROBLEM 24 Repeat Example 24 with

$$A = \begin{bmatrix} 7\text{ hr} & 2\text{ hr} \\ 5\text{ hr} & 1.5\text{ hr} \end{bmatrix} \quad \text{and} \quad B = \begin{bmatrix} \$10 & \$8 \\ \$6 & \$4 \end{bmatrix}$$ ◆

Answers to Matched Problems

21. 8 22. $[6 \quad 1.5]\begin{bmatrix} 8 \\ 6 \end{bmatrix} = [57]$, or \$57

23. (A) Not defined (B) $\begin{bmatrix} 2 & 2 & -1 & 2 \\ 1 & 6 & 12 & -4 \\ -1 & 0 & 3 & -2 \end{bmatrix}$ (C) $\begin{bmatrix} 0 & 0 \\ 0 & 0 \end{bmatrix}$

(D) $\begin{bmatrix} -6 & -12 \\ 3 & 6 \end{bmatrix}$ (E) [11] (F) $\begin{bmatrix} 12 & -8 & 4 \\ 6 & -4 & 2 \\ 9 & -6 & 3 \end{bmatrix}$

24. $\begin{matrix} & X & Y & \\ & \begin{bmatrix} \$82 & \$64 \\ \$59 & \$46 \end{bmatrix} & \begin{matrix} \text{Trick} \\ \text{Slalom} \end{matrix} \end{matrix}$

EXERCISE 5-5

A *Find the products.*

1. $[2 \quad 4]\begin{bmatrix} 3 \\ 1 \end{bmatrix}$ **2.** $[3 \quad 1]\begin{bmatrix} 2 \\ 4 \end{bmatrix}$ **3.** $[-3 \quad 2]\begin{bmatrix} -1 \\ -2 \end{bmatrix}$

4. $[3 \quad -2]\begin{bmatrix} -4 \\ -1 \end{bmatrix}$ **5.** $[2 \quad 5]\begin{bmatrix} 1 & -1 \\ 2 & 3 \end{bmatrix}$ **6.** $[1 \quad 3]\begin{bmatrix} 2 & 3 \\ 1 & -4 \end{bmatrix}$

7. $\begin{bmatrix} 3 & 4 \\ -1 & -2 \end{bmatrix}\begin{bmatrix} -1 \\ 2 \end{bmatrix}$ **8.** $\begin{bmatrix} -1 & 1 \\ 2 & -3 \end{bmatrix}\begin{bmatrix} 4 \\ -2 \end{bmatrix}$ **9.** $\begin{bmatrix} 2 & -3 \\ 1 & 2 \end{bmatrix}\begin{bmatrix} 1 & -1 \\ 0 & -2 \end{bmatrix}$

10. $\begin{bmatrix} -3 & 2 \\ 4 & -1 \end{bmatrix}\begin{bmatrix} -2 & 5 \\ -1 & 3 \end{bmatrix}$ **11.** $\begin{bmatrix} 1 & -1 \\ 0 & -2 \end{bmatrix}\begin{bmatrix} 2 & -3 \\ 1 & 2 \end{bmatrix}$ **12.** $\begin{bmatrix} -2 & 5 \\ -1 & 3 \end{bmatrix}\begin{bmatrix} -3 & 2 \\ 4 & -1 \end{bmatrix}$

13. $[5 \quad -2]\begin{bmatrix} -3 \\ -4 \end{bmatrix}$ **14.** $[-4 \quad 3]\begin{bmatrix} -2 \\ 1 \end{bmatrix}$ **15.** $\begin{bmatrix} -3 \\ -4 \end{bmatrix}[5 \quad -2]$

16. $\begin{bmatrix} -2 \\ 1 \end{bmatrix}[-4 \quad 3]$

B *Find the products.*

17. $\begin{bmatrix} -1 & -2 & 2 \end{bmatrix} \begin{bmatrix} 2 \\ -1 \\ 3 \end{bmatrix}$

18. $\begin{bmatrix} -2 & 4 & 0 \end{bmatrix} \begin{bmatrix} -1 \\ -3 \\ 2 \end{bmatrix}$

19. $\begin{bmatrix} -1 & -3 & 0 & 5 \end{bmatrix} \begin{bmatrix} 4 \\ -3 \\ -1 \\ 2 \end{bmatrix}$

20. $\begin{bmatrix} -1 & 2 & 3 & -2 \end{bmatrix} \begin{bmatrix} 3 \\ -2 \\ 0 \\ 4 \end{bmatrix}$

Find each matrix product, if it is defined.

21. $\begin{bmatrix} 2 & -1 & 1 \\ 1 & 3 & -2 \end{bmatrix} \begin{bmatrix} 1 & 3 \\ 0 & -1 \\ -2 & 2 \end{bmatrix}$

22. $\begin{bmatrix} -1 & -4 & 3 \\ 2 & 0 & 1 \end{bmatrix} \begin{bmatrix} 2 & -3 \\ 1 & 2 \\ 0 & -1 \end{bmatrix}$

23. $\begin{bmatrix} 1 & 3 \\ 0 & -1 \\ -2 & 2 \end{bmatrix} \begin{bmatrix} 2 & -1 & 1 \\ 1 & 3 & -2 \end{bmatrix}$

24. $\begin{bmatrix} 2 & -3 \\ 1 & 2 \\ 0 & -1 \end{bmatrix} \begin{bmatrix} -1 & -4 & 3 \\ 2 & 0 & 1 \end{bmatrix}$

25. $\begin{bmatrix} 3 & -2 & -4 \end{bmatrix} \begin{bmatrix} 1 \\ 2 \\ -3 \end{bmatrix}$

26. $\begin{bmatrix} 1 & -2 & 2 \end{bmatrix} \begin{bmatrix} 2 \\ -1 \\ 1 \end{bmatrix}$

27. $\begin{bmatrix} 1 \\ 2 \\ -3 \end{bmatrix} \begin{bmatrix} 3 & -2 & -4 \end{bmatrix}$

28. $\begin{bmatrix} 2 \\ -1 \\ 1 \end{bmatrix} \begin{bmatrix} 1 & -2 & 2 \end{bmatrix}$

29. $\begin{bmatrix} 1 & 2 \\ 2 & -1 \\ -3 & 1 \end{bmatrix} \begin{bmatrix} 3 & -2 & -4 \end{bmatrix}$

30. $\begin{bmatrix} 1 \\ 2 \\ -3 \end{bmatrix} \begin{bmatrix} 3 & -2 & 4 \\ 1 & -2 & 2 \end{bmatrix}$

31. $\begin{bmatrix} 1 & 2 & -1 \\ 3 & -1 & 4 \\ 2 & -4 & 5 \end{bmatrix} \begin{bmatrix} 4 \\ 5 \\ 7 \end{bmatrix}$

32. $\begin{bmatrix} 2 & 1 & -1 \\ 1 & 3 & 1 \\ -4 & 2 & 5 \end{bmatrix} \begin{bmatrix} -1 \\ 2 \\ -3 \end{bmatrix}$

33. $\begin{bmatrix} 2 & -1 & 3 & 0 \\ -3 & 4 & 2 & -1 \\ 0 & -2 & 1 & 4 \end{bmatrix} \begin{bmatrix} 2 & -3 & -2 \\ 1 & 0 & 1 \\ -1 & 2 & 0 \\ 2 & -2 & -3 \end{bmatrix}$

34. $\begin{bmatrix} 2 & -3 & -2 \\ 1 & 0 & 1 \\ -1 & 2 & 0 \\ 2 & -2 & -3 \end{bmatrix} \begin{bmatrix} 2 & -1 & 3 & 0 \\ -3 & 4 & 2 & -1 \\ 0 & -2 & 1 & 4 \end{bmatrix}$

C *In Problems 35–38, find $A^2 = AA$.*

35. $\begin{bmatrix} 6 & 9 \\ -4 & -6 \end{bmatrix}$

36. $\begin{bmatrix} -2 & 4 \\ -1 & 2 \end{bmatrix}$

37. $\begin{bmatrix} \frac{1}{3} & \frac{1}{3} \\ \frac{2}{3} & \frac{2}{3} \end{bmatrix}$

38. $\begin{bmatrix} \frac{3}{4} & \frac{3}{4} \\ \frac{1}{4} & \frac{1}{4} \end{bmatrix}$

In Problems 39–44, verify each statement by using the following matrices:

$$A = \begin{bmatrix} 1 & 2 \\ 0 & 1 \end{bmatrix} \qquad B = \begin{bmatrix} 1 & 1 \\ 2 & 3 \end{bmatrix} \qquad C = \begin{bmatrix} -3 & 1 \\ -1 & 2 \end{bmatrix}$$

39. $AB \neq BA$

40. $(AB)C = A(BC)$

41. $A(B + C) = AB + AC$

42. $(B + C)A = BA + CA$

43. $A^2 - B^2 \neq (A - B)(A + B)$

44. $A^2 + 2AB + B^2 \neq (A + B)(A + B)$

APPLICATIONS

Business & Economics

45. *Labor costs.* A company with manufacturing plants located in different parts of the country has labor-hour and wage requirements for the manufacturing of three types of inflatable boats as given in the following two matrices:

Labor-hours per boat

	Cutting department	Assembly department	Packaging department	
$M =$	0.6 hr	0.6 hr	0.2 hr	One-person boat
	1.0 hr	0.9 hr	0.3 hr	Two-person boat
	1.5 hr	1.2 hr	0.4 hr	Four-person boat

Hourly wages

	Plant I	Plant II	
$N =$	\$8	\$9	Cutting department
	\$10	\$12	Assembly department
	\$5	\$6	Packaging department

(A) Find the labor costs for a one-person boat manufactured at plant I. That is, find the product

$$[0.6 \quad 0.6 \quad 0.2] \begin{bmatrix} 8 \\ 10 \\ 5 \end{bmatrix}$$

(B) Find the labor costs for a four-person boat manufactured at plant II. Set up a product as in part A and multiply.

(C) What is the dimension of MN?

(D) Find MN and interpret.

46. *Inventory value.* A personal computer retail company sells five different computer models through three stores located in a large metropolitan area. The inventory of each model on hand in each store is summarized in matrix M. Wholesale (W) and retail (R) values of each model computer are summarized in matrix N.

$$\begin{array}{c} \text{Model} \\ \begin{array}{cc} & \begin{array}{ccccc} A & B & C & D & E \end{array} \\ M = & \begin{bmatrix} 4 & 2 & 3 & 7 & 1 \\ 2 & 3 & 5 & 0 & 6 \\ 10 & 4 & 3 & 4 & 3 \end{bmatrix} \begin{array}{l} \text{Store 1} \\ \text{Store 2} \\ \text{Store 3} \end{array} \end{array} \end{array}$$

$$\begin{array}{cc} & \begin{array}{cc} W & R \end{array} \\ N = & \begin{bmatrix} \$700 & \$840 \\ \$1{,}400 & \$1{,}800 \\ \$1{,}800 & \$2{,}400 \\ \$2{,}700 & \$3{,}300 \\ \$3{,}500 & \$4{,}900 \end{bmatrix} \begin{array}{c} A \\ B \\ C \\ D \\ E \end{array} \end{array}$$

(A) What is the retail value of the inventory at store 2?
(B) What is the wholesale value of the inventory at store 3?
(C) Compute MN and interpret.
(D) Multiply M by $[1 \quad 1 \quad 1]$ and interpret. (The multiplication makes sense in only one direction.)
(E) Multiply MN by $[1 \quad 1 \quad 1]$ and interpret. (The multiplication makes sense in only one direction.)

Although not essential, a graphic calculator or a computer will greatly simplify the computation of the matrix products and sums in Problems 47–49.

47. *Air freight.* A nationwide air freight service has connecting flights between five cities as illustrated in the diagram. To represent this schedule in matrix form, we construct a 5×5 **incidence matrix** A where the rows represent the origin of each flight and the columns represent the destinations. We place a 1 in the ith row and jth column of this matrix if there is a connecting flight from the ith city to the jth city; otherwise, insert a 0. We also place 0's on the principal diagonal, because a connecting flight with the same origin and destination makes no sense. With the schedule represented in the mathematical form of a matrix, we can perform operations on this matrix to obtain information about the schedule.

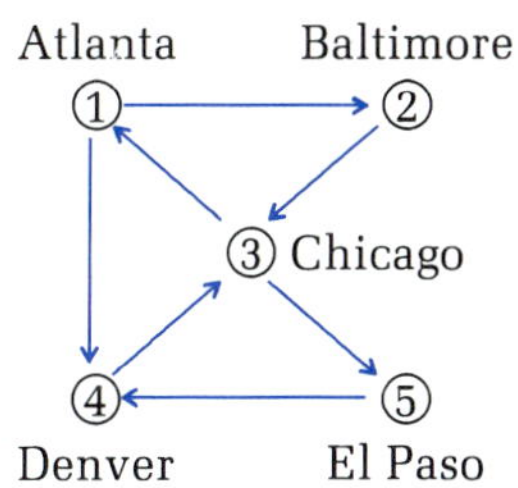

$$\begin{array}{c} \text{Destination} \\ \begin{array}{ccc} & & \begin{array}{ccccc} 1 & 2 & 3 & 4 & 5 \end{array} \\ \text{Origin} & \begin{array}{c} 1 \\ 2 \\ 3 \\ 4 \\ 5 \end{array} & \begin{bmatrix} 0 & 1 & 0 & 1 & 0 \\ 0 & 0 & 1 & 0 & 0 \\ 1 & 0 & 0 & 0 & 1 \\ 0 & 0 & 1 & 0 & 0 \\ 0 & 0 & 0 & 1 & 0 \end{bmatrix} = A \end{array} \end{array}$$

(A) Find A^2. What does the 1 in row 2 and column 1 of A^2 indicate about the schedule? What does the 2 in row 1 and column 3 indicate about the schedule? In general, how would you interpret each element not on the principal diagonal of A^2? [*Hint:* Examine the diagram for possible connections between the ith city and the jth city.]

(B) Find A^3. What does the 1 in row 4 and column 2 of A^3 indicate about the schedule? What does the 2 in row 1 and column 5 indicate about the schedule? In general, how would you interpret each element not on the principal diagonal of A^3?

48. *Air freight.* Refer to Problem 47. Compute A, $A + A^2$, $A + A^2 + A^3$, . . . , until you obtain a matrix with no 0 elements (except possibly on the principal diagonal), and interpret.

49. *Air freight.* Find the incidence matrix A for the flight schedule illustrated in the diagram. Compute A, $A + A^2$, $A + A^2 + A^3$, . . . , until you obtain a matrix with no 0 elements (except possibly on the principal diagonal), and interpret.

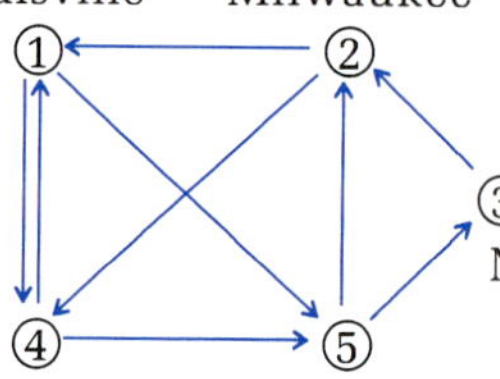

Life Sciences

50. *Nutrition.* A nutritionist for a cereal company blends two cereals in three different mixes. The amounts of protein, carbohydrate, and fat (in grams per ounce) in each cereal are given by matrix M. The amounts of each cereal used in the three mixes are given by matrix N.

$$M = \begin{array}{c} \text{Cereal } A \quad \text{Cereal } B \\ \begin{bmatrix} 4 \text{ g/oz} & 2 \text{ g/oz} \\ 20 \text{ g/oz} & 16 \text{ g/oz} \\ 3 \text{ g/oz} & 1 \text{ g/oz} \end{bmatrix} \end{array} \begin{array}{l} \\ \text{Protein} \\ \text{Carbohydrate} \\ \text{Fat} \end{array}$$

$$N = \begin{array}{c} \text{Mix } X \quad \text{Mix } Y \quad \text{Mix } Z \\ \begin{bmatrix} 15 \text{ oz} & 10 \text{ oz} & 5 \text{ oz} \\ 5 \text{ oz} & 10 \text{ oz} & 15 \text{ oz} \end{bmatrix} \end{array} \begin{array}{l} \\ \text{Cereal } A \\ \text{Cereal } B \end{array}$$

(A) Find the amount of protein in mix X by computing the product

$$[4 \quad 2]\begin{bmatrix} 15 \\ 5 \end{bmatrix}$$

(B) Find the amount of fat in mix Z. Set up a product as in part A and multiply.

(C) What is the dimension of MN?

(D) Find MN and interpret.

(E) Find $\frac{1}{20}MN$ and interpret.

Social Sciences

51. *Politics.* In a local California election, a group hired a public relations firm to promote its candidate in three ways: telephone calls, house calls, and letters. The cost per contact is given in matrix M:

$$M = \begin{bmatrix} \$0.40 \\ \$0.75 \\ \$0.25 \end{bmatrix} \begin{matrix} \text{Telephone call} \\ \text{House call} \\ \text{Letter} \end{matrix}$$

(column heading: Cost per contact)

The number of contacts of each type made in two adjacent cities is given in matrix N:

$$\begin{matrix} & \text{Telephone call} & \text{House call} & \text{Letter} & \\ N = & \begin{bmatrix} 1{,}000 & 500 & 5{,}000 \\ 2{,}000 & 800 & 8{,}000 \end{bmatrix} & & & \begin{matrix} \text{Berkeley} \\ \text{Oakland} \end{matrix} \end{matrix}$$

(A) Find the total amount spent in Berkeley by computing the product

$$[1{,}000 \quad 500 \quad 5{,}000] \begin{bmatrix} \$0.40 \\ \$0.75 \\ \$0.25 \end{bmatrix}$$

(B) Find the total amount spent in Oakland by computing the product of appropriate matrices.

(C) Compute NM and interpret.

(D) Multiply N by the matrix $[1 \quad 1]$ and interpret.

SECTION 5-6 Inverse of a Square Matrix

- IDENTITY MATRIX FOR MULTIPLICATION
- INVERSE OF A SQUARE MATRIX
- APPLICATION: CRYPTOLOGY

◆ IDENTITY MATRIX FOR MULTIPLICATION

We know that

$$1a = a1 = a \qquad \text{for all real numbers } a$$

The number 1 is called the **identity** for real number multiplication. Does the set of all matrices of a given dimension have an identity element for multiplication? That is, if M is an arbitrary $m \times n$ matrix, does M have an identity element I such that $IM = MI = M$? The answer, in general, is no. However, the set of all **square**

matrices of order n (dimension $n \times n$) does have an identity, and it is given as follows:

> The **identity element for multiplication** for the set of all square matrices of order n is the square matrix of order n, denoted by I, with 1's along the principal diagonal (from the upper left corner to the lower right) and 0's elsewhere.

For example,

$$\begin{bmatrix} 1 & 0 \\ 0 & 1 \end{bmatrix} \quad \text{and} \quad \begin{bmatrix} 1 & 0 & 0 \\ 0 & 1 & 0 \\ 0 & 0 & 1 \end{bmatrix}$$

are the identity matrices for all square matrices of order 2 and 3, respectively.

◆ EXAMPLE 25

(A) $\begin{bmatrix} 1 & 0 & 0 \\ 0 & 1 & 0 \\ 0 & 0 & 1 \end{bmatrix}\begin{bmatrix} a & b & c \\ d & e & f \\ g & h & i \end{bmatrix} = \begin{bmatrix} a & b & c \\ d & e & f \\ g & h & i \end{bmatrix}$

(B) $\begin{bmatrix} a & b & c \\ d & e & f \\ g & h & i \end{bmatrix}\begin{bmatrix} 1 & 0 & 0 \\ 0 & 1 & 0 \\ 0 & 0 & 1 \end{bmatrix} = \begin{bmatrix} a & b & c \\ d & e & f \\ g & h & i \end{bmatrix}$

(C) $\begin{bmatrix} 1 & 0 \\ 0 & 1 \end{bmatrix}\begin{bmatrix} a & b & c \\ d & e & f \end{bmatrix} = \begin{bmatrix} a & b & c \\ d & e & f \end{bmatrix}$

(D) $\begin{bmatrix} a & b & c \\ d & e & f \end{bmatrix}\begin{bmatrix} 1 & 0 & 0 \\ 0 & 1 & 0 \\ 0 & 0 & 1 \end{bmatrix} = \begin{bmatrix} a & b & c \\ d & e & f \end{bmatrix}$ ◆

PROBLEM 25

Multiply:

(A) $\begin{bmatrix} 1 & 0 \\ 0 & 1 \end{bmatrix}\begin{bmatrix} 2 & -3 \\ 5 & 7 \end{bmatrix}$ and $\begin{bmatrix} 2 & -3 \\ 5 & 7 \end{bmatrix}\begin{bmatrix} 1 & 0 \\ 0 & 1 \end{bmatrix}$

(B) $\begin{bmatrix} 1 & 0 & 0 \\ 0 & 1 & 0 \\ 0 & 0 & 1 \end{bmatrix}\begin{bmatrix} 4 & 2 \\ 3 & -5 \\ 6 & 8 \end{bmatrix}$ and $\begin{bmatrix} 4 & 2 \\ 3 & -5 \\ 6 & 8 \end{bmatrix}\begin{bmatrix} 1 & 0 \\ 0 & 1 \end{bmatrix}$ ◆

In general, we can show that if M is a square matrix of order n and I is the identity matrix of order n, then

$$IM = MI = M$$

If M is an $m \times n$ matrix that is not square ($m \neq n$), it is not possible to find a *single* identity matrix I satisfying both $IM = M$ and $MI = M$ (see Examples 25C and 25D). For this reason we restrict our attention in this section to square matrices.

◆ INVERSE OF A SQUARE MATRIX

In the set of real numbers, we know that for each real number a (except 0) there exists a real number a^{-1} such that

$$a^{-1}a = 1$$

The number a^{-1} is called the **inverse** of the number a relative to multiplication, or the **multiplicative inverse** of a. For example, 2^{-1} is the multiplicative inverse of 2, since $2^{-1} \cdot 2 = 1$. We use this idea to define the inverse of a square matrix.

Inverse of a Square Matrix

Let M be a square matrix of order n and I be the identity matrix of order n. If there exists a matrix M^{-1} (read "M inverse") such that

$$M^{-1}M = MM^{-1} = I$$

then M^{-1} is called the **multiplicative inverse of M** or, more simply, the **inverse of *M*.**

The multiplicative inverse of a nonzero real number a can also be written as $1/a$. This notation is not used for matrix inverses.

Let us use the above definition to find M^{-1} for

$$M = \begin{bmatrix} 2 & 3 \\ 1 & 2 \end{bmatrix}$$

We are looking for

$$M^{-1} = \begin{bmatrix} a & c \\ b & d \end{bmatrix}$$

such that

$$MM^{-1} = M^{-1}M = I$$

Thus, we write

$$\overset{M}{\begin{bmatrix} 2 & 3 \\ 1 & 2 \end{bmatrix}} \overset{M^{-1}}{\begin{bmatrix} a & c \\ b & d \end{bmatrix}} = \overset{I}{\begin{bmatrix} 1 & 0 \\ 0 & 1 \end{bmatrix}}$$

and try to find a, b, c, and d so that the product of M and M^{-1} is the identity matrix I. Multiplying M and M^{-1} on the left side, we obtain

$$\begin{bmatrix} (2a+3b) & (2c+3d) \\ (a+2b) & (c+2d) \end{bmatrix} = \begin{bmatrix} 1 & 0 \\ 0 & 1 \end{bmatrix}$$

which is true only if

$$\begin{aligned} 2a + 3b &= 1 \qquad & 2c + 3d &= 0 \\ a + 2b &= 0 \qquad & c + 2d &= 1 \end{aligned}$$

Solving these two systems, we find that $a = 2$, $b = -1$, $c = -3$, and $d = 2$. Thus,

$$M^{-1} = \begin{bmatrix} 2 & -3 \\ -1 & 2 \end{bmatrix}$$

as is easily checked:

$$\overset{M}{\begin{bmatrix} 2 & 3 \\ 1 & 2 \end{bmatrix}} \overset{M^{-1}}{\begin{bmatrix} 2 & -3 \\ -1 & 2 \end{bmatrix}} = \overset{I}{\begin{bmatrix} 1 & 0 \\ 0 & 1 \end{bmatrix}} = \overset{M^{-1}}{\begin{bmatrix} 2 & -3 \\ -1 & 2 \end{bmatrix}} \overset{M}{\begin{bmatrix} 2 & 3 \\ 1 & 2 \end{bmatrix}}$$

Unlike nonzero real numbers, inverses do not always exist for square matrices. For example, if

$$N = \begin{bmatrix} 2 & 1 \\ 4 & 2 \end{bmatrix}$$

then, proceeding as above, we are led to the systems

$$\begin{aligned} 2a + b &= 1 \qquad & 2c + d &= 0 \\ 4a + 2b &= 0 \qquad & 4c + 2d &= 1 \end{aligned}$$

These are both inconsistent and have no solution. Hence, N^{-1} does not exist.

Being able to find inverses, when they exist, leads to direct and simple solutions to many practical problems. In the next section, we shall show how inverses can be used to solve systems of linear equations.

The method outlined above for finding M^{-1}, if it exists, gets very involved for matrices of order larger than 2. Now that we know what we are looking for, we can introduce the idea of the augmented matrix (considered in Sections 5-2 and 5-3) to make the process more efficient.

◆ EXAMPLE 26 Find the inverse, if it exists, of the matrix

$$M = \begin{bmatrix} 1 & -1 & 1 \\ 0 & 2 & -1 \\ 2 & 3 & 0 \end{bmatrix}$$

Solution We start as before and write

$$\overset{M}{\begin{bmatrix} 1 & -1 & 1 \\ 0 & 2 & -1 \\ 2 & 3 & 0 \end{bmatrix}} \overset{M^{-1}}{\begin{bmatrix} a & d & g \\ b & e & h \\ c & f & i \end{bmatrix}} = \overset{I}{\begin{bmatrix} 1 & 0 & 0 \\ 0 & 1 & 0 \\ 0 & 0 & 1 \end{bmatrix}}$$

which is true only if

$$\begin{aligned} a - b + c &= 1 \\ 2b - c &= 0 \\ 2a + 3b &= 0 \end{aligned} \qquad \begin{aligned} d - e + f &= 0 \\ 2e - f &= 1 \\ 2d + 3e &= 0 \end{aligned} \qquad \begin{aligned} g - h + i &= 0 \\ 2h - i &= 0 \\ 2g + 3h &= 1 \end{aligned}$$

Now we write augmented matrices for each of the three systems:

$$\overset{\text{First}}{\left[\begin{array}{rrr|r} 1 & -1 & 1 & 1 \\ 0 & 2 & -1 & 0 \\ 2 & 3 & 0 & 0 \end{array}\right]} \qquad \overset{\text{Second}}{\left[\begin{array}{rrr|r} 1 & -1 & 1 & 0 \\ 0 & 2 & -1 & 1 \\ 2 & 3 & 0 & 0 \end{array}\right]} \qquad \overset{\text{Third}}{\left[\begin{array}{rrr|r} 1 & -1 & 1 & 0 \\ 0 & 2 & -1 & 0 \\ 2 & 3 & 0 & 1 \end{array}\right]}$$

Since each matrix to the left of the vertical bar is the same, exactly the same row operations can be used on each augmented matrix to transform it into a reduced form. We can speed up the process substantially by combining all three augmented matrices into the single augmented matrix form below:

$$\left[\begin{array}{rrr|rrr} 1 & -1 & 1 & 1 & 0 & 0 \\ 0 & 2 & -1 & 0 & 1 & 0 \\ 2 & 3 & 0 & 0 & 0 & 1 \end{array}\right] = [M|I] \tag{1}$$

We now try to perform row operations on matrix (1) until we obtain a row-equivalent matrix that looks like matrix (2):

$$\left[\begin{array}{ccc|ccc} \multicolumn{3}{c|}{I} & \multicolumn{3}{c}{B} \\ 1 & 0 & 0 & a & d & g \\ 0 & 1 & 0 & b & e & h \\ 0 & 0 & 1 & c & f & i \end{array}\right] = [I|B] \tag{2}$$

If this can be done, then the new matrix B to the right of the vertical bar will be M^{-1}! Now let us try to transform (1) into a form like (2). We will follow the same sequence of steps as we did in the solution of linear systems by Gauss–Jordan elimination (see Section 5-3).

$$\begin{array}{c} \qquad M \qquad\qquad\qquad I \\ \left[\begin{array}{rrr|rrr} 1 & -1 & 1 & 1 & 0 & 0 \\ 0 & 2 & -1 & 0 & 1 & 0 \\ 2 & 3 & 0 & 0 & 0 & 1 \end{array}\right] \end{array} \quad R_3 + (-2)R_1 \to R_3$$

$$\sim \left[\begin{array}{rrr|rrr} 1 & -1 & 1 & 1 & 0 & 0 \\ 0 & 2 & -1 & 0 & 1 & 0 \\ 0 & 5 & -2 & -2 & 0 & 1 \end{array}\right] \quad \tfrac{1}{2}R_2 \to R_2$$

$$\sim \left[\begin{array}{rrr|rrr} 1 & -1 & 1 & 1 & 0 & 0 \\ 0 & 1 & -\frac{1}{2} & 0 & \frac{1}{2} & 0 \\ 0 & 5 & -2 & -2 & 0 & 1 \end{array}\right] \quad R_3 + (-5)R_2 \to R_3$$

$$\sim \left[\begin{array}{rrr|rrr} 1 & -1 & 1 & 1 & 0 & 0 \\ 0 & 1 & -\frac{1}{2} & 0 & \frac{1}{2} & 0 \\ 0 & 0 & \frac{1}{2} & -2 & -\frac{5}{2} & 1 \end{array}\right] \quad 2R_3 \to R_3$$

$$\sim \left[\begin{array}{rrr|rrr} 1 & -1 & 1 & 1 & 0 & 0 \\ 0 & 1 & -\frac{1}{2} & 0 & \frac{1}{2} & 0 \\ 0 & 0 & 1 & -4 & -5 & 2 \end{array}\right] \quad \begin{array}{l} R_1 + (-1)R_3 \to R_1 \\ R_2 + \frac{1}{2}R_3 \to R_2 \end{array}$$

$$\sim \left[\begin{array}{rrr|rrr} 1 & -1 & 0 & 5 & 5 & -2 \\ 0 & 1 & 0 & -2 & -2 & 1 \\ 0 & 0 & 1 & -4 & -5 & 2 \end{array}\right] \quad R_1 + R_2 \to R_1$$

$$\begin{array}{c} \qquad I \qquad\qquad\qquad B \\ \sim \left[\begin{array}{rrr|rrr} 1 & 0 & 0 & 3 & 3 & -1 \\ 0 & 1 & 0 & -2 & -2 & 1 \\ 0 & 0 & 1 & -4 & -5 & 2 \end{array}\right] = [I|B] \end{array}$$

Converting back to systems of equations equivalent to our three original systems (we will not have to do this step in practice), we have

$$\begin{array}{lll} a = 3 & d = 3 & g = -1 \\ b = -2 & e = -2 & h = 1 \\ c = -4 & f = -5 & i = 2 \end{array}$$

And these are just the elements of M^{-1} that we are looking for! Hence,

$$M^{-1} = \begin{bmatrix} 3 & 3 & -1 \\ -2 & -2 & 1 \\ -4 & -5 & 2 \end{bmatrix}$$

Note that this is the matrix to the right of the vertical line in the last augmented matrix. That is, $M^{-1} = B$.

Since the definition of matrix inverse requires that

$$M^{-1}M = I \qquad \text{and} \qquad MM^{-1} = I \tag{3}$$

it appears that we must compute both $M^{-1}M$ and MM^{-1} to check our work. However, it can be shown that if one of the equations in (3) is satisfied, then the other is also satisfied. Thus, for checking purposes, it is sufficient to compute either $M^{-1}M$ or MM^{-1}; we do not need to do both.

Check

$$M^{-1}M = \begin{bmatrix} 3 & 3 & -1 \\ -2 & -2 & 1 \\ -4 & -5 & 2 \end{bmatrix} \begin{bmatrix} 1 & -1 & 1 \\ 0 & 2 & -1 \\ 2 & 3 & 0 \end{bmatrix} = \begin{bmatrix} 1 & 0 & 0 \\ 0 & 1 & 0 \\ 0 & 0 & 1 \end{bmatrix} = I$$

◆

PROBLEM 26 Let: $M = \begin{bmatrix} 3 & -1 & 1 \\ -1 & 1 & 0 \\ 1 & 0 & 1 \end{bmatrix}$

(A) Form the augmented matrix $[M|I]$.
(B) Use row operations to transform $[M|I]$ into $[I|B]$.
(C) Verify by multiplication that $B = M^{-1}$ (that is, show that $BM = I$). ◆

The procedure shown in Example 26 can be used to find the inverse of any square matrix, if the inverse exists, and will also indicate when the inverse does not exist. These ideas are summarized in Theorem 5.

THEOREM 5

Inverse of a Square Matrix *M*

If $[M|I]$ is transformed by row operations into $[I|B]$, then the resulting matrix B is M^{-1}. However, if we obtain 0's in one or more rows to the left of the vertical line, then M^{-1} does not exist.

◆ EXAMPLE 27 Find M^{-1}, given: $M = \begin{bmatrix} 3 & -1 \\ -4 & 2 \end{bmatrix}$

Solution

$$\left[\begin{array}{cc|cc} 3 & -1 & 1 & 0 \\ -4 & 2 & 0 & 1 \end{array}\right] \quad \tfrac{1}{3}R_1 \to R_1$$

$$\sim \left[\begin{array}{cc|cc} 1 & -\frac{1}{3} & \frac{1}{3} & 0 \\ -4 & 2 & 0 & 1 \end{array}\right] \quad R_2 + 4R_1 \to R_2$$

$$\sim \left[\begin{array}{cc|cc} 1 & -\frac{1}{3} & \frac{1}{3} & 0 \\ 0 & \frac{2}{3} & \frac{4}{3} & 1 \end{array}\right] \quad \tfrac{3}{2}R_2 \to R_2$$

$$\sim \left[\begin{array}{cc|cc} 1 & -\frac{1}{3} & \frac{1}{3} & 0 \\ 0 & 1 & 2 & \frac{3}{2} \end{array}\right] \quad R_1 + \tfrac{1}{3}R_2 \to R_1$$

$$\sim \left[\begin{array}{cc|cc} 1 & 0 & 1 & \frac{1}{2} \\ 0 & 1 & 2 & \frac{3}{2} \end{array}\right]$$

Thus,

$$M^{-1} = \begin{bmatrix} 1 & \frac{1}{2} \\ 2 & \frac{3}{2} \end{bmatrix}$$ Check by showing that $M^{-1}M = I$. ◆

PROBLEM 27 Find M^{-1}, given: $M = \begin{bmatrix} 2 & -6 \\ 1 & -2 \end{bmatrix}$ ◆

◆ EXAMPLE 28 Find M^{-1}, given: $M = \begin{bmatrix} 2 & -4 \\ -3 & 6 \end{bmatrix}$

Solution

$$\left[\begin{array}{rr|rr} 2 & -4 & 1 & 0 \\ -3 & 6 & 0 & 1 \end{array}\right] \quad \tfrac{1}{2}R_1 \to R_1$$

$$\sim \left[\begin{array}{rr|rr} 1 & -2 & \frac{1}{2} & 0 \\ -3 & 6 & 0 & 1 \end{array}\right] \quad R_2 + 3R_1 \to R_2$$

$$\sim \left[\begin{array}{rr|rr} 1 & -2 & \frac{1}{2} & 0 \\ 0 & 0 & \frac{3}{2} & 1 \end{array}\right]$$

We have all 0's in the second row to the left of the vertical bar; therefore, the inverse does not exist. ◆

PROBLEM 28 Find M^{-1}, given: $M = \begin{bmatrix} -6 & 3 \\ -4 & 2 \end{bmatrix}$ ◆

◆ APPLICATION: CRYPTOLOGY

Matrix inverses can provide a simple and effective procedure for encoding and decoding messages. To begin, assign the numbers 1–26 to the letters in the alphabet, as shown below. Also assign the number 27 to a blank to provide for space between words. (A more sophisticated code could include both capital and lowercase letters and punctuation symbols.)

A	B	C	D	E	F	G	H	I	J	K	L	M	N	O	P	Q	R	S	T	U	V	W	X	Y	Z	Blank
1	2	3	4	5	6	7	8	9	10	11	12	13	14	15	16	17	18	19	20	21	22	23	24	25	26	27

Thus, the message "SECRET CODE" corresponds to the sequence

19 5 3 18 5 20 27 3 15 4 5

Any matrix whose elements are positive integers and whose inverse exists can be used as an **encoding matrix.** For example, to use the 2 × 2 matrix

$$A = \begin{bmatrix} 4 & 3 \\ 1 & 1 \end{bmatrix}$$

to encode the above message, first divide the numbers in the sequence into groups of two and use these groups as the columns of a matrix with 2 rows. (Notice that we added an extra blank at the end of the message to make the columns come out even.) Then multiply this matrix on the left by A:

$$\begin{bmatrix} 4 & 3 \\ 1 & 1 \end{bmatrix} \begin{bmatrix} 19 & 3 & 5 & 27 & 15 & 5 \\ 5 & 18 & 20 & 3 & 4 & 27 \end{bmatrix} = \begin{bmatrix} 91 & 66 & 80 & 117 & 72 & 101 \\ 24 & 21 & 25 & 30 & 19 & 32 \end{bmatrix}$$

The coded message is

91 24 66 21 80 25 117 30 72 19 101 32

This message can be decoded simply by putting it back into matrix form and multiplying on the left by the **decoding matrix** A^{-1}. Since A^{-1} is easily determined if A is known, the encoding matrix A is the only key needed to decode messages encoded in this manner. Although simple in concept, codes of this type can be very difficult to crack.

Answers to Matched Problems

25. (A) $\begin{bmatrix} 2 & -3 \\ 5 & 7 \end{bmatrix}$ (B) $\begin{bmatrix} 4 & 2 \\ 3 & -5 \\ 6 & 8 \end{bmatrix}$

26. (A) $\left[\begin{array}{rrr|rrr} 3 & -1 & 1 & 1 & 0 & 0 \\ -1 & 1 & 0 & 0 & 1 & 0 \\ 1 & 0 & 1 & 0 & 0 & 1 \end{array}\right]$ (B) $\left[\begin{array}{rrr|rrr} 1 & 0 & 0 & 1 & 1 & -1 \\ 0 & 1 & 0 & 1 & 2 & -1 \\ 0 & 0 & 1 & -1 & -1 & 2 \end{array}\right]$

(C) $\begin{bmatrix} 1 & 1 & -1 \\ 1 & 2 & -1 \\ -1 & -1 & 2 \end{bmatrix}\begin{bmatrix} 3 & -1 & 1 \\ -1 & 1 & 0 \\ 1 & 0 & 1 \end{bmatrix} = \begin{bmatrix} 1 & 0 & 0 \\ 0 & 1 & 0 \\ 0 & 0 & 1 \end{bmatrix}$

27. $\begin{bmatrix} -1 & 3 \\ -\frac{1}{2} & 1 \end{bmatrix}$ 28. Does not exist

EXERCISE 5-6

A *Perform the indicated operations.*

1. $\begin{bmatrix} 1 & 0 \\ 0 & 1 \end{bmatrix}\begin{bmatrix} 2 & -3 \\ 4 & 5 \end{bmatrix}$

2. $\begin{bmatrix} 1 & 0 \\ 0 & 1 \end{bmatrix}\begin{bmatrix} -1 & 6 \\ 0 & 2 \end{bmatrix}$

3. $\begin{bmatrix} 2 & -3 \\ 4 & 5 \end{bmatrix}\begin{bmatrix} 1 & 0 \\ 0 & 1 \end{bmatrix}$

4. $\begin{bmatrix} -1 & 6 \\ 0 & 2 \end{bmatrix}\begin{bmatrix} 1 & 0 \\ 0 & 1 \end{bmatrix}$

5. $\begin{bmatrix} 1 & 0 & 0 \\ 0 & 1 & 0 \\ 0 & 0 & 1 \end{bmatrix}\begin{bmatrix} -2 & 1 & 3 \\ 2 & 4 & -2 \\ 5 & 1 & 0 \end{bmatrix}$

6. $\begin{bmatrix} 1 & 0 & 0 \\ 0 & 1 & 0 \\ 0 & 0 & 1 \end{bmatrix}\begin{bmatrix} 3 & -4 & 0 \\ 1 & 2 & -5 \\ 6 & -3 & -1 \end{bmatrix}$

7. $\begin{bmatrix} -2 & 1 & 3 \\ 2 & 4 & -2 \\ 5 & 1 & 0 \end{bmatrix}\begin{bmatrix} 1 & 0 & 0 \\ 0 & 1 & 0 \\ 0 & 0 & 1 \end{bmatrix}$

8. $\begin{bmatrix} 3 & -4 & 0 \\ 1 & 2 & -5 \\ 6 & -3 & -1 \end{bmatrix}\begin{bmatrix} 1 & 0 & 0 \\ 0 & 1 & 0 \\ 0 & 0 & 1 \end{bmatrix}$

In Problems 9–14, show that the two matrices are inverses of each other by showing that their product is the identity matrix I.

9. $\begin{bmatrix} 3 & -4 \\ -2 & 3 \end{bmatrix}; \begin{bmatrix} 3 & 4 \\ 2 & 3 \end{bmatrix}$

10. $\begin{bmatrix} 5 & -7 \\ -2 & 3 \end{bmatrix}; \begin{bmatrix} 3 & 7 \\ 2 & 5 \end{bmatrix}$

11. $\begin{bmatrix} -5 & 2 \\ -8 & 3 \end{bmatrix}; \begin{bmatrix} 3 & -2 \\ 8 & -5 \end{bmatrix}$

12. $\begin{bmatrix} 7 & 4 \\ -5 & -3 \end{bmatrix}; \begin{bmatrix} 3 & 4 \\ -5 & -7 \end{bmatrix}$

13. $\begin{bmatrix} 1 & -1 & 1 \\ 0 & 2 & -1 \\ 2 & 3 & 0 \end{bmatrix}$; $\begin{bmatrix} 3 & 3 & -1 \\ -2 & -2 & 1 \\ -4 & -5 & 2 \end{bmatrix}$

14. $\begin{bmatrix} 3 & 3 & -1 \\ -2 & -2 & 1 \\ -4 & -5 & 2 \end{bmatrix}$; $\begin{bmatrix} 1 & -1 & 1 \\ 0 & 2 & -1 \\ 2 & 3 & 0 \end{bmatrix}$

B *Given M in Problems 15–24, find M^{-1} and show that $M^{-1}M = I$.*

15. $\begin{bmatrix} -1 & 0 \\ -3 & 1 \end{bmatrix}$ 16. $\begin{bmatrix} 6 & 1 \\ -1 & 0 \end{bmatrix}$ 17. $\begin{bmatrix} 1 & 2 \\ 1 & 3 \end{bmatrix}$

18. $\begin{bmatrix} 2 & 1 \\ 5 & 3 \end{bmatrix}$ 19. $\begin{bmatrix} 1 & 3 \\ 2 & 7 \end{bmatrix}$ 20. $\begin{bmatrix} 2 & 1 \\ 1 & 1 \end{bmatrix}$

21. $\begin{bmatrix} 1 & -3 & 0 \\ 0 & 3 & 1 \\ 2 & -1 & 2 \end{bmatrix}$ 22. $\begin{bmatrix} 2 & 9 & 0 \\ 1 & 2 & 3 \\ 0 & -1 & 1 \end{bmatrix}$ 23. $\begin{bmatrix} 1 & 1 & 0 \\ 0 & 3 & -1 \\ 1 & 0 & 1 \end{bmatrix}$

24. $\begin{bmatrix} 1 & 0 & -1 \\ 2 & -1 & 0 \\ 1 & 1 & 1 \end{bmatrix}$

Find the inverse of each matrix in Problems 25–30, if it exists.

25. $\begin{bmatrix} 3 & 2 \\ -4 & -3 \end{bmatrix}$ 26. $\begin{bmatrix} -4 & 3 \\ -5 & 4 \end{bmatrix}$ 27. $\begin{bmatrix} 3 & 9 \\ 2 & 6 \end{bmatrix}$

28. $\begin{bmatrix} 2 & -4 \\ -3 & 6 \end{bmatrix}$ 29. $\begin{bmatrix} 3 & 1 \\ 4 & 2 \end{bmatrix}$ 30. $\begin{bmatrix} -5 & 3 \\ 2 & -2 \end{bmatrix}$

C *Find the inverse of each matrix in Problems 31–36, if it exists.*

31. $\begin{bmatrix} -5 & -2 & -2 \\ 2 & 1 & 0 \\ 1 & 0 & 1 \end{bmatrix}$ 32. $\begin{bmatrix} 2 & 0 & 1 \\ 1 & 1 & 0 \\ 1 & 0 & 1 \end{bmatrix}$ 33. $\begin{bmatrix} 2 & 1 & 1 \\ 1 & 1 & 0 \\ -1 & -1 & 0 \end{bmatrix}$

34. $\begin{bmatrix} 1 & -1 & 0 \\ 2 & -1 & 1 \\ 0 & 1 & 1 \end{bmatrix}$ 35. $\begin{bmatrix} -1 & -2 & 2 \\ 4 & 2 & 0 \\ 4 & 0 & 4 \end{bmatrix}$ 36. $\begin{bmatrix} 3 & 0 & 2 \\ 4 & 2 & 0 \\ 5 & 0 & 5 \end{bmatrix}$

37. Show that $(A^{-1})^{-1} = A$ for: $A = \begin{bmatrix} 3 & 4 \\ 2 & 3 \end{bmatrix}$

38. Show that $(AB)^{-1} = B^{-1}A^{-1}$ for: $A = \begin{bmatrix} 3 & 4 \\ 2 & 3 \end{bmatrix}$ and $B = \begin{bmatrix} 3 & 7 \\ 2 & 5 \end{bmatrix}$

In Problems 39–42, use a graphic calculator or a computer to find the inverse of each matrix, if it exists.

39. $\begin{bmatrix} 6 & 2 & 0 & 4 \\ 5 & 3 & 2 & 1 \\ 0 & -1 & 1 & -2 \\ 2 & -3 & 1 & 0 \end{bmatrix}$

40. $\begin{bmatrix} -2 & 4 & 0 & -1 \\ 2 & -1 & 2 & 5 \\ 0 & 2 & -1 & 7 \\ 2 & -3 & 0 & 5 \end{bmatrix}$

41. $\begin{bmatrix} 3 & 2 & 3 & 4 & 4 \\ 5 & 4 & 3 & 2 & 1 \\ -1 & -1 & 2 & -2 & 3 \\ 3 & -3 & 1 & 0 & 1 \\ 1 & 1 & 2 & 0 & 2 \end{bmatrix}$

42. $\begin{bmatrix} 1 & 2 & 3 & 4 & 5 \\ 2 & 6 & 4 & 5 & 6 \\ -1 & -2 & -1 & 2 & 3 \\ 1 & 6 & 1 & 6 & 4 \\ 1 & -4 & 3 & -7 & -4 \end{bmatrix}$

APPLICATIONS

Problems 43–46 refer to the encoding matrix: $A = \begin{bmatrix} 1 & 2 \\ 1 & 3 \end{bmatrix}$

Social Sciences

43. Cryptography. Encode the message "THE SUN ALSO RISES" using matrix A.

44. Cryptography. Encode the message "THE GRAPES OF WRATH" using matrix A.

45. Cryptography. The following message was encoded with matrix A. Decode this message:

37 52 24 29 73 96 49 69 62 89 36 44 59 86 41 50 22 26

46. Cryptography. The following message was encoded with matrix A. Decode this message:

9 13 40 49 29 34 29 30 22 26 33 36 43 57 29 34 54 74

Problems 47–50 require the use of a graphic calculator or a computer. To use the 5×5 encoding matrix B given below, form a matrix with 5 rows and as many columns as necessary to accommodate each message.

$$B = \begin{bmatrix} 1 & 0 & 1 & 0 & 1 \\ 0 & 1 & 1 & 0 & 3 \\ 2 & 1 & 1 & 1 & 1 \\ 0 & 0 & 1 & 0 & 2 \\ 1 & 1 & 1 & 2 & 1 \end{bmatrix}$$

47. Cryptography. Encode the message "THE BEST YEARS OF OUR LIVES" using matrix B.

48. Cryptography. Encode the message "THE BRIDGE ON THE RIVER KWAI" using matrix B.

49. *Cryptography.* The following message was encoded with matrix B. Decode this message:

32 34 87 19 94 24 21 67 11 69 54 71 112 43 112 56 92 109 55 109 48 66 98 41 89 62 135 124 81 143

50. *Cryptography.* The following message was encoded with matrix B. Decode this message:

28 22 83 11 90 57 108 102 69 105 57 78 91 57 89 66 111 112 79 125

SECTION 5-7

Matrix Equations and Systems of Linear Equations

- MATRIX EQUATIONS
- MATRIX EQUATIONS AND SYSTEMS OF LINEAR EQUATIONS
- APPLICATION

The identity matrix and inverse matrix discussed in the last section can be put to immediate use in the solution of certain simple matrix equations. Being able to solve a matrix equation gives us another important method of solving systems of equations, provided the system is independent and has the same number of variables as equations. If the system is dependent or if it has either fewer or more variables than equations, we must return to the Gauss–Jordan method of elimination.

MATRIX EQUATIONS

Solving simple matrix equations follows very much the same procedures used in solving real number equations. We have, however, less freedom with matrix equations, because matrix multiplication is not commutative. In solving matrix equations, we will be guided by Theorem 4 from Section 5-5 and some results from the last section, all of which are restated here for convenient reference.

THEOREM 4

Properties of Matrix Multiplication

Assuming all products and sums are defined for the indicated matrices A, B, and C, then for k a real number:

ASSOCIATIVE PROPERTY:	1. $A(BC) = (AB)C$
LEFT DISTRIBUTIVE PROPERTY:	2. $A(B + C) = AB + AC$
RIGHT DISTRIBUTIVE PROPERTY:	3. $(B + C)A = BA + CA$
LEFT MULTIPLICATION PROPERTY:	4. If $A = B$, then $CA = CB$.
RIGHT MULTIPLICATION PROPERTY:	5. If $A = B$, then $AC = BC$.
	6. $k(AB) = (kA)B = A(kB)$

Properties of Identity and Inverse Matrices

If A is a square matrix of order n and I is the identity matrix of order n, then

$$IA = AI = A$$

If, in addition, A^{-1} exists, then

$$A^{-1}A = AA^{-1} = I$$

The use of these properties in the solution of matrix equations is best illustrated by an example.

◆ EXAMPLE 29 Given an $n \times n$ matrix A and $n \times 1$ column matrices B and X, solve $AX = B$ for X. Assume all necessary inverses exist.

Solution We are interested in finding a column matrix X that satisfies the matrix equation $AX = B$. To solve this equation, we multiply both sides on the left by A^{-1}, assuming it exists, to isolate X on the left side.

$AX = B$	Use the left multiplication property.
$A^{-1}(AX) = A^{-1}B$	Use the associative property.
$(A^{-1}A)X = A^{-1}B$	$A^{-1}A = I$
$IX = A^{-1}B$	$IX = X$
$X = A^{-1}B$	

◆

PROBLEM 29 Given an $n \times n$ matrix A and $n \times 1$ column matrices B, C, and X, solve $AX + C = B$ for X. Assume all necessary inverses exist. ◆

Common Error

Do not mix the left multiplication property and the right multiplication property. If $AX = B$, then

$$A^{-1}(AX) \neq BA^{-1}$$

◆ MATRIX EQUATIONS AND SYSTEMS OF LINEAR EQUATIONS

We now show how independent systems of linear equations with the same number of variables as equations can be solved. First, convert the system into a matrix equation of the form $AX = B$, and then use $X = A^{-1}B$ as obtained in Example 29.

◆ EXAMPLE 30 Solve the system

$$\begin{aligned} x_1 - x_2 + x_3 &= k_1 \\ 2x_2 - x_3 &= k_2 \\ 2x_1 + 3x_2 &= k_3 \end{aligned} \tag{1}$$

for:

(A) $k_1 = 1, \quad k_2 = 1, \quad k_3 = 1$ (B) $k_1 = 3, \quad k_2 = 1, \quad k_3 = 4$
(C) $k_1 = -5, \quad k_2 = 2, \quad k_3 = -3$

Solutions The inverse of the coefficient matrix

$$A = \begin{bmatrix} 1 & -1 & 1 \\ 0 & 2 & -1 \\ 2 & 3 & 0 \end{bmatrix}$$

can be used to solve parts A, B, and C very easily. To see how, we convert system (1) into the following matrix equation:

$$\overset{A}{\begin{bmatrix} 1 & -1 & 1 \\ 0 & 2 & -1 \\ 2 & 3 & 0 \end{bmatrix}} \overset{X}{\begin{bmatrix} x_1 \\ x_2 \\ x_3 \end{bmatrix}} = \overset{B}{\begin{bmatrix} k_1 \\ k_2 \\ k_3 \end{bmatrix}} \tag{2}$$

Check that matrix equation (2) is equivalent to system (1) by finding the product of the left side and then equating corresponding elements on the left with those on the right. Now you see another important reason for defining matrix multiplication as we did.

We are interested in finding a column matrix X that satisfies the matrix equation $AX = B$. In Example 29 we found that if $AX = B$ and if A^{-1} exists, then

$$X = A^{-1}B$$

The inverse of A was found in Example 26 to be

$$A^{-1} = \begin{bmatrix} 3 & 3 & -1 \\ -2 & -2 & 1 \\ -4 & -5 & 2 \end{bmatrix}$$

Thus,

$$\overset{X}{\begin{bmatrix} x_1 \\ x_2 \\ x_3 \end{bmatrix}} = \overset{A^{-1}}{\begin{bmatrix} 3 & 3 & -1 \\ -2 & -2 & 1 \\ -4 & -5 & 2 \end{bmatrix}} \overset{B}{\begin{bmatrix} k_1 \\ k_2 \\ k_3 \end{bmatrix}}$$

Now, to solve parts A, B, and C, we simply replace k_1, k_2, and k_3 with the given values and multiply.

(A) $$\begin{bmatrix} x_1 \\ x_2 \\ x_3 \end{bmatrix} = \begin{bmatrix} 3 & 3 & -1 \\ -2 & -2 & 1 \\ -4 & -5 & 2 \end{bmatrix} \begin{bmatrix} 1 \\ 1 \\ 1 \end{bmatrix} = \begin{bmatrix} 5 \\ -3 \\ -7 \end{bmatrix}$$

Thus, $x_1 = 5$, $x_2 = -3$, and $x_3 = -7$.

(B) $$\begin{bmatrix} x_1 \\ x_2 \\ x_3 \end{bmatrix} = \begin{bmatrix} 3 & 3 & -1 \\ -2 & -2 & 1 \\ -4 & -5 & 2 \end{bmatrix} \begin{bmatrix} 3 \\ 1 \\ 4 \end{bmatrix} = \begin{bmatrix} 8 \\ -4 \\ -9 \end{bmatrix}$$

Thus, $x_1 = 8$, $x_2 = -4$, and $x_3 = -9$.

(C) $$\begin{bmatrix} x_1 \\ x_2 \\ x_3 \end{bmatrix} = \begin{bmatrix} 3 & 3 & -1 \\ -2 & -2 & 1 \\ -4 & -5 & 2 \end{bmatrix} \begin{bmatrix} -5 \\ 2 \\ -3 \end{bmatrix} = \begin{bmatrix} -6 \\ 3 \\ 4 \end{bmatrix}$$

Thus, $x_1 = -6$, $x_2 = 3$, and $x_3 = 4$. ◆

PROBLEM 30 Solve the system

$$\begin{aligned} 3x_1 - x_2 + x_3 &= k_1 \\ -x_1 + x_2 &= k_2 \\ x_1 + x_3 &= k_3 \end{aligned}$$

for:

(A) $k_1 = 1$, $k_2 = 3$, $k_3 = 2$ (B) $k_1 = 3$, $k_2 = -3$, $k_3 = 2$
(C) $k_1 = -5$, $k_2 = 1$, $k_3 = -4$

[*Note:* The inverse of the coefficient matrix was found in Problem 26 in Section 5-6.] ◆

As Example 30 illustrates, using an inverse matrix to solve a system of equations is a great advantage because, once the inverse is found, it can be used to solve any new system formed by changing only the constant terms. Furthermore, computer software for finding matrix inverses is readily available and easy to use.* Thus, the matrix inverse method is an important tool for solving linear systems. However, this method cannot be used in all cases (see the box at the top of the next page).

---

* See the Preface (under "Student Aids" and "Instructor Aids") for a description of the computer and graphic calculator manuals that accompany this book. Both provide solutions to matrix inverse problems.

Using Inverse Matrices to Solve Systems of Linear Equations

If the number of equations in a system equals the number of variables and the coefficient matrix has an inverse, then the system will always have a unique solution that can be found by using the inverse of the coefficient matrix. What happens if the coefficient matrix does not have an inverse? In this case, it can be shown that the system does not have a unique solution and is either dependent or inconsistent. Gauss–Jordan elimination must be used to determine which is the case. Also, as we mentioned earlier, Gauss–Jordan elimination always must be used if the number of variables is not the same as the number of equations.

◆ APPLICATION

The following application will illustrate the usefulness of the inverse matrix method for solving systems of equations.

◆ EXAMPLE 31

Investment Analysis

An investment advisor currently has two types of investments available for clients: a conservative investment A that pays 10% per year and an investment B of higher risk that pays 20% per year. Clients may divide their investments between the two to achieve any total return desired between 10% and 20%. However, the higher the desired return, the higher the risk. How should each client listed in the table invest to achieve the indicated return?

	CLIENT			
	1	2	3	k
TOTAL INVESTMENT	\$20,000	\$50,000	\$10,000	k_1
ANNUAL RETURN DESIRED	\$ 2,400 (12%)	\$ 7,500 (15%)	\$ 1,300 (13%)	k_2

Solution

We will solve the problem for an arbitrary client k by finding an inverse matrix. Then we will apply the result to the three specific clients.

Let

x_1 = Amount invested in A

x_2 = Amount invested in B

Then

$$x_1 + x_2 = k_1 \quad \text{Total invested}$$
$$0.1x_1 + 0.2x_2 = k_2 \quad \text{Total annual return desired}$$

Write as a matrix equation:

$$\overset{A}{\begin{bmatrix} 1 & 1 \\ 0.1 & 0.2 \end{bmatrix}} \overset{X}{\begin{bmatrix} x_1 \\ x_2 \end{bmatrix}} = \overset{B}{\begin{bmatrix} k_1 \\ k_2 \end{bmatrix}}$$

If A^{-1} exists, then

$$X = A^{-1}B$$

We now find A^{-1} by starting with the augmented matrix $[A|I]$ and proceeding as discussed in Section 5-6:

$$\left[\begin{array}{cc|cc} 1 & 1 & 1 & 0 \\ 0.1 & 0.2 & 0 & 1 \end{array}\right] \quad 10R_2 \to R_2$$

$$\sim \left[\begin{array}{cc|cc} 1 & 1 & 1 & 0 \\ 1 & 2 & 0 & 10 \end{array}\right] \quad R_2 + (-1)R_1 \to R_2$$

$$\sim \left[\begin{array}{cc|cc} 1 & 1 & 1 & 0 \\ 0 & 1 & -1 & 10 \end{array}\right] \quad R_1 + (-1)R_2 \to R_1$$

$$\sim \left[\begin{array}{cc|cc} 1 & 0 & 2 & -10 \\ 0 & 1 & -1 & 10 \end{array}\right]$$

Thus,

$$A^{-1} = \begin{bmatrix} 2 & -10 \\ -1 & 10 \end{bmatrix} \qquad \textit{Check:} \quad \overset{A^{-1}}{\begin{bmatrix} 2 & -10 \\ -1 & 10 \end{bmatrix}} \overset{A}{\begin{bmatrix} 1 & 1 \\ 0.1 & 0.2 \end{bmatrix}} = \overset{I}{\begin{bmatrix} 1 & 0 \\ 0 & 1 \end{bmatrix}}$$

and

$$\overset{X}{\begin{bmatrix} x_1 \\ x_2 \end{bmatrix}} = \overset{A^{-1}}{\begin{bmatrix} 2 & -10 \\ -1 & 10 \end{bmatrix}} \overset{B}{\begin{bmatrix} k_1 \\ k_2 \end{bmatrix}}$$

To solve each client's investment problem, we replace k_1 and k_2 with appropriate values from the table and multiply by A^{-1}:

Client 1

$$\begin{bmatrix} x_1 \\ x_2 \end{bmatrix} = \begin{bmatrix} 2 & -10 \\ -1 & 10 \end{bmatrix} \begin{bmatrix} 20{,}000 \\ 2{,}400 \end{bmatrix} = \begin{bmatrix} 16{,}000 \\ 4{,}000 \end{bmatrix}$$

Solution: $x_1 = \$16{,}000$ in A, $x_2 = \$4{,}000$ in B

Client 2

$$\begin{bmatrix} x_1 \\ x_2 \end{bmatrix} = \begin{bmatrix} 2 & -10 \\ -1 & 10 \end{bmatrix} \begin{bmatrix} 50{,}000 \\ 7{,}500 \end{bmatrix} = \begin{bmatrix} 25{,}000 \\ 25{,}000 \end{bmatrix}$$

Solution: $x_1 = \$25{,}000$ in A, $x_2 = \$25{,}000$ in B

Client 3

$$\begin{bmatrix} x_1 \\ x_2 \end{bmatrix} = \begin{bmatrix} 2 & -10 \\ -1 & 10 \end{bmatrix} \begin{bmatrix} 10{,}000 \\ 1{,}300 \end{bmatrix} = \begin{bmatrix} 7{,}000 \\ 3{,}000 \end{bmatrix}$$

Solution: $x_1 = \$7{,}000$ in A, $x_2 = \$3{,}000$ in B ◆

PROBLEM 31 Repeat Example 31 with investment A paying 8% and investment B paying 24%. ◆

Answers to Matched Problems

29.

$$AX + C = B$$

$$(AX + C) - C = B - C$$
$$AX + (C - C) = B - C$$
$$AX + 0 = B - C$$

$$AX = B - C$$

$$A^{-1}(AX) = A^{-1}(B - C)$$
$$(A^{-1}A)X = A^{-1}(B - C)$$
$$IX = A^{-1}(B - C)$$

$$X = A^{-1}(B - C)$$

30. (A) $x_1 = 2$, $x_2 = 5$, $x_3 = 0$ (B) $x_1 = -2$, $x_2 = -5$, $x_3 = 4$
(C) $x_1 = 0$, $x_2 = 1$, $x_3 = -4$

31. $A^{-1} = \begin{bmatrix} 1.5 & -6.25 \\ -0.5 & 6.25 \end{bmatrix}$; Client 1: \$15,000 in A and \$5,000 in B; Client 2: \$28,125 in A and \$21,875 in B; Client 3: \$6,875 in A and \$3,125 in B

EXERCISE 5-7

A *Write Problems 1–4 as systems of linear equations without matrices.*

1. $\begin{bmatrix} 3 & 1 \\ 2 & -1 \end{bmatrix} \begin{bmatrix} x_1 \\ x_2 \end{bmatrix} = \begin{bmatrix} 5 \\ -4 \end{bmatrix}$

2. $\begin{bmatrix} -2 & 1 \\ -3 & 4 \end{bmatrix} \begin{bmatrix} x_1 \\ x_2 \end{bmatrix} = \begin{bmatrix} -5 \\ 7 \end{bmatrix}$

3. $\begin{bmatrix} -3 & 1 & 0 \\ 2 & 0 & 1 \\ -1 & 3 & -2 \end{bmatrix} \begin{bmatrix} x_1 \\ x_2 \\ x_3 \end{bmatrix} = \begin{bmatrix} 3 \\ -4 \\ 2 \end{bmatrix}$

4. $\begin{bmatrix} 2 & -1 & 0 \\ -2 & 3 & -1 \\ 4 & 0 & 3 \end{bmatrix} \begin{bmatrix} x_1 \\ x_2 \\ x_3 \end{bmatrix} = \begin{bmatrix} 6 \\ -4 \\ 7 \end{bmatrix}$

Write each system in Problems 5–8 as a matrix equation of the form $AX = B$.

5. $$\begin{aligned} 3x_1 - 4x_2 &= 1 \\ 2x_1 + x_2 &= 5 \end{aligned}$$

6. $$\begin{aligned} 2x_1 + x_2 &= 8 \\ -5x_1 + 3x_2 &= -4 \end{aligned}$$

7. $$\begin{aligned} x_1 - 3x_2 + 2x_3 &= -3 \\ -2x_1 + 3x_2 &= 1 \\ x_1 + x_2 + 4x_3 &= -2 \end{aligned}$$

8. $$\begin{aligned} 3x_1 + 2x_3 &= 9 \\ -x_1 + 4x_2 + x_3 &= -7 \\ -2x_1 + 3x_2 &= 6 \end{aligned}$$

Find x_1 and x_2.

9. $\begin{bmatrix} x_1 \\ x_2 \end{bmatrix} = \begin{bmatrix} 3 & -2 \\ 1 & 4 \end{bmatrix} \begin{bmatrix} -2 \\ 1 \end{bmatrix}$

10. $\begin{bmatrix} x_1 \\ x_2 \end{bmatrix} = \begin{bmatrix} -2 & 1 \\ -1 & 2 \end{bmatrix} \begin{bmatrix} 3 \\ -2 \end{bmatrix}$

11. $\begin{bmatrix} x_1 \\ x_2 \end{bmatrix} = \begin{bmatrix} -2 & 3 \\ 2 & -1 \end{bmatrix} \begin{bmatrix} 3 \\ 2 \end{bmatrix}$

12. $\begin{bmatrix} x_1 \\ x_2 \end{bmatrix} = \begin{bmatrix} 3 & -1 \\ 0 & 2 \end{bmatrix} \begin{bmatrix} -2 \\ 1 \end{bmatrix}$

B Write each system as a matrix equation and solve by using inverses. [*Note:* The inverses were found in Problems 17–24 in Exercise 5-6.]

13. $$\begin{aligned} x_1 + 2x_2 &= k_1 \\ x_1 + 3x_2 &= k_2 \end{aligned}$$

(A) $k_1 = 1, \quad k_2 = 3$
(B) $k_1 = 3, \quad k_2 = 5$
(C) $k_1 = -2, \quad k_2 = 1$

14. $$\begin{aligned} 2x_1 + x_2 &= k_1 \\ 5x_1 + 3x_2 &= k_2 \end{aligned}$$

(A) $k_1 = 2, \quad k_2 = 13$
(B) $k_1 = 2, \quad k_2 = 4$
(C) $k_1 = 1, \quad k_2 = -3$

15. $$\begin{aligned} x_1 + 3x_2 &= k_1 \\ 2x_1 + 7x_2 &= k_2 \end{aligned}$$

(A) $k_1 = 2, \quad k_2 = -1$
(B) $k_1 = 1, \quad k_2 = 0$
(C) $k_1 = 3, \quad k_2 = -1$

16. $$\begin{aligned} 2x_1 + x_2 &= k_1 \\ x_1 + x_2 &= k_2 \end{aligned}$$

(A) $k_1 = -1, \quad k_2 = -2$
(B) $k_1 = 2, \quad k_2 = 3$
(C) $k_1 = 2, \quad k_2 = 0$

17. $$\begin{aligned} x_1 - 3x_2 &= k_1 \\ 3x_2 + x_3 &= k_2 \\ 2x_1 - x_2 + 2x_3 &= k_3 \end{aligned}$$

(A) $k_1 = 1, \quad k_2 = 0, \quad k_3 = 2$
(B) $k_1 = -1, \quad k_2 = 1, \quad k_3 = 0$
(C) $k_1 = 2, \quad k_2 = -2, \quad k_3 = 1$

18. $$\begin{aligned} 2x_1 + 9x_2 &= k_1 \\ x_1 + 2x_2 + 3x_3 &= k_2 \\ - x_2 + x_3 &= k_3 \end{aligned}$$

(A) $k_1 = 0, \quad k_2 = 2, \quad k_3 = 1$
(B) $k_1 = -2, \quad k_2 = 0, \quad k_3 = 1$
(C) $k_1 = 3, \quad k_2 = 1, \quad k_3 = 0$

19. $$\begin{aligned} x_1 + x_2 &= k_1 \\ 3x_2 - x_3 &= k_2 \\ x_1 + x_3 &= k_3 \end{aligned}$$

(A) $k_1 = 2, \quad k_2 = 0, \quad k_3 = 4$
(B) $k_1 = 0, \quad k_2 = 4, \quad k_3 = -2$
(C) $k_1 = 4, \quad k_2 = 2, \quad k_3 = 0$

20. $$\begin{aligned} x_1 - x_3 &= k_1 \\ 2x_1 - x_2 &= k_2 \\ x_1 + x_2 + x_3 &= k_3 \end{aligned}$$

(A) $k_1 = 4, \quad k_2 = 8, \quad k_3 = 0$
(B) $k_1 = 4, \quad k_2 = 0, \quad k_3 = -4$
(C) $k_1 = 0, \quad k_2 = 8, \quad k_3 = -8$

C For $n \times n$ matrices A and B, and $n \times 1$ column matrices C, D, and X, solve each matrix equation in Problems 21–26 for X. Assume all necessary inverses exist.

21. $AX - BX = C$ **22.** $AX + BX = C$ **23.** $AX + X = C$
24. $AX - X = C$ **25.** $AX - C = D - BX$ **26.** $AX + C = BX + D$

In Problems 27–30, write each system as a matrix equation and solve by using the inverse coefficient matrix. Use a graphic calculator or a computer to perform the necessary calculations.

27.
$$\begin{aligned} x_1 + 8x_2 + 7x_3 &= 135 \\ 6x_1 + 6x_2 + 8x_3 &= 155 \\ 3x_1 + 4x_2 + 6x_3 &= 75 \end{aligned}$$

28.
$$\begin{aligned} 5x_1 + 3x_2 - 2x_3 &= 112 \\ 7x_1 + 5x_2 \qquad &= 70 \\ 3x_1 + x_2 - 9x_3 &= 96 \end{aligned}$$

29.
$$\begin{aligned} 6x_1 + 9x_2 + 7x_3 + 5x_4 &= 250 \\ 6x_1 + 4x_2 + 7x_3 + 3x_4 &= 195 \\ 4x_1 + 5x_2 + 3x_3 + 2x_4 &= 145 \\ 4x_1 + 3x_2 + 8x_3 + 2x_4 &= 125 \end{aligned}$$

30.
$$\begin{aligned} 3x_1 + 3x_2 + 6x_3 + 5x_4 &= 10 \\ 4x_1 + 5x_2 + 8x_3 + 2x_4 &= 15 \\ 3x_1 + 6x_2 + 7x_3 + 4x_4 &= 30 \\ 4x_1 + x_2 + 6x_3 + 3x_4 &= 25 \end{aligned}$$

APPLICATIONS

Solve using systems of equations and matrix inverses.

Business & Economics

31. Resource allocation. A concert hall has 10,000 seats. If tickets are \$4 and \$8, how many of each type of ticket should be sold (assuming all seats can be sold) to bring in each of the returns indicated in the table?

	CONCERT 1	2	3
TICKETS SOLD	10,000	10,000	10,000
RETURN REQUIRED	\$56,000	\$60,000	\$68,000

32. Parking receipts. Parking fees at a municipal zoo are \$5.00 for local residents and \$7.50 for all others. At the end of each day, the total number of vehicles parked that day and the gross receipts for the day are recorded, but the number of vehicles in each category is not. The following table contains the relevant information for a recent 4 day period. How many vehicles in each category used the zoo's parking facilities each day?

	DAY 1	2	3	4
VEHICLES PARKED	1,200	1,550	1,740	1,400
GROSS RECEIPTS	\$7,125	\$9,825	\$11,100	\$8,650

PLANT	CAR FRAMES	TRUCK FRAMES
A	10	5
B	8	8

33. Production scheduling. A supplier for the automobile industry manufactures car and truck frames at two different plants. The production rates in frames per hour for each plant are given in the table in the margin. How many hours should each plant be scheduled to operate to exactly fill each of the orders in the following table?

	ORDERS		
	1	2	3
CAR FRAMES	3,000	2,800	2,600
TRUCK FRAMES	1,600	2,000	2,200

GUITAR MODEL	LABOR COST	MATERIAL COST
A	$30	$20
B	$40	$30

34. *Production scheduling.* Labor and material costs for manufacturing two guitar models are given in the table in the margin. If a total of $3,000 a week is allowed for labor and material, how many of each model should be produced each week to use exactly each of the allocations of the $3,000 indicated in the following table?

	WEEKLY ALLOCATION		
	1	2	3
LABOR	$1,800	$1,750	$1,720
MATERIAL	$1,200	$1,250	$1,280

Life Sciences

MIX	PROTEIN (%)	FAT (%)
A	20	2
B	10	6

35. *Diets.* A biologist has available two commercial food mixes containing the percentages of protein and fat given in the table in the margin. How many ounces of each mix should be used to prepare each of the diets listed in the following table?

	DIET		
	1	2	3
PROTEIN	20 oz	10 oz	10 oz
FAT	6 oz	4 oz	6 oz

Social Sciences

	RANK	
	Lecturer	*Instructor*
SECTIONS TAUGHT	3	4
ANNUAL SALARY *Thousand $*	20	25

36. *Education—resource allocation.* A state university system is planning to hire new faculty at the rank of lecturer or instructor for several of its 2 year community colleges. The number of sections taught and the annual salary (in thousands of dollars) for each rank are given in the table in the margin. The number of sections that must be taught by the new faculty and the amount budgeted for salaries (in thousands of dollars) at each of the colleges are given in the following table. How many faculty of each rank should be hired at each college to exactly meet the demand for sections and completely exhaust the salary budget?

	COMMUNITY COLLEGE		
	1	2	3
DEMAND FOR SECTIONS	30	33	35
SALARY BUDGET *Thousand $*	200	210	220

In Problems 37–42, use a graphic calculator or a computer to solve each system. Solve by the method that is easiest to use with the software available to you.

Business & Economics

37. *Incentive plan.* A small company provides an incentive plan for its top executives. Each executive receives as a bonus a percentage of the portion of the annual profit that remains after the bonuses for the other executives have been deducted (see the table). If the company has an annual profit of \$2 million, find the bonus for each executive. Round each bonus to the nearest hundred dollars.

OFFICER	BONUS
President	3%
Executive vice president	2.5%
Associative vice president	2%
Assistant vice president	1.5%

38. *Incentive plan.* Repeat Problem 37 if the company decides to include a 1% bonus for the sales manager in the incentive plan.

39. *Taxable income.* As a result of several mergers and acquisitions, stock in four companies has been distributed among the companies. The following table gives the percentage of stock held by each company and the annual net income of each company (in millions of dollars):

	PERCENTAGE OF STOCK HELD				ANNUAL NET INCOME
COMPANY	*A*	*B*	*C*	*D*	*Million \$*
A	82	8	3	7	3.2
B	12	64	11	13	2.6
C	11	9	72	8	3.8
D	6	2	14	78	4.4

Thus, company *A* holds 82% of its own stock, 8% of the stock in company *B*, 3% of the stock in company *C*, and so on. For the purpose of assessing a state tax on corporate income, the taxable income of each company is defined to be its share of its own annual net income plus its share of the taxable income of each of the other companies, as determined by the percentages in the table. What is the taxable income of each company to the nearest hundred thousand dollars? What is the difference between the total of the taxable incomes of the four companies and the total of their net incomes?

40. *Taxable income.* Repeat Problem 39 if tax law is changed so that the taxable income of a company is defined to be all of its own annual net income plus its share of the taxable income of each of the other companies.

41. *Operating costs.* A durable-goods manufacturer has two production departments, freezers and refrigerators, and two service departments, accounting and maintenance. The monthly direct cost of operating each department is given in the table, along with the percentage of service time each department receives from each service department:

DEPARTMENT	DIRECT COSTS *Monthly*	PERCENTAGE OF SERVICE TIME RECEIVED FROM *Accounting*	*Maintenance*
FREEZERS	\$260,000	38	42
REFRIGERATORS	\$190,000	34	28
ACCOUNTING	\$55,000	16	13
MAINTENANCE	\$95,000	12	17

The total cost of operating each department consists of its direct costs plus the cost of services received, which are determined by the percentages in the table. For example,

$$\begin{aligned}\text{Total costs for freezer department} &= \text{Direct costs for freezer department}\\ &\quad + 0.38\,(\text{Total costs for accounting department})\\ &\quad + 0.42\,(\text{Total costs for maintenance department})\end{aligned}$$

(A) Find the total cost of operating each department. Round each amount to the nearest thousand dollars.

(B) Compare the sum of the direct costs of all four departments with the sum of the total costs of the two production departments determined in part A, and interpret.

42. *Operating costs.* The manufacturer in Problem 41 adds another department that produces stoves. The relevant information is given in the table.

DEPARTMENT	DIRECT COSTS *Monthly*	PERCENTAGE OF SERVICE TIME RECEIVED FROM *Accounting*	*Maintenance*
FREEZERS	\$260,000	24	28
REFRIGERATORS	\$190,000	26	23
STOVES	\$210,000	22	19
ACCOUNTING	\$75,000	16	13
MAINTENANCE	\$115,000	12	17

(A) Find the total cost of operating each department. Round each amount to the nearest thousand dollars.

(B) Compare the sum of the direct costs of all five departments with the sum of the total costs of the three production departments determined in part A, and interpret.

SECTION 5-8 Leontief Input–Output Analysis

- INTRODUCTION
- TWO-INDUSTRY MODEL

INTRODUCTION

A very important application of matrices and their inverses is found in the branch of applied mathematics called **input–output analysis.** Wassily Leontief, the primary force behind these new developments, was awarded the Nobel prize in economics in 1973 because of the significant impact his work had on economic planning for industrialized countries. Among other things, he conducted a comprehensive study of how 500 sectors of the American economy interacted with each other. Of course, large-scale computers played an important role in this analysis.

Our investigation will be more modest. In fact, we will start with an economy comprised of only two industries. From these humble beginnings, ideas and definitions will evolve that can be readily generalized for more realistic economies. Input–output analysis attempts to establish equilibrium conditions under which industries in an economy have just enough output to satisfy each other's demands in addition to final (outside) demands. Given the internal demands within the industries for each other's output, the problem is to determine output levels that will meet various levels of final (outside) demands.

TWO-INDUSTRY MODEL

To make the problem concrete, let us start with a hypothetical economy with only two industries—electric company E and water company W. Output for both companies is measured in dollars. The electric company uses both electricity and water (input) in the production of electricity (output), and the water company uses both electricity and water (input) in the production of water (output). Suppose that the production of each dollar's worth of electricity requires \$0.10 worth of electricity and \$0.30 worth of water, and the production of each dollar's worth of water requires \$0.40 worth of electricity and \$0.20 worth of water. If the final demand from the outside sector of the economy (the demand from all other users of electricity and water) is

$d_1 = \$12$ million for electricity

$d_2 = \$6$ million for water

how much electricity and water should be produced to meet this final demand?

To begin, suppose the electricity company produces \$12 million worth of electricity and the water company produces \$6 million worth of water (the final demand). Then the production processes of the companies would require

Electricity required to produce electricity		Electricity required to produce water	
0.1(12)	+	0.4(6)	= \$3.6 million of electricity

and

Water required to produce electricity		Water required to produce water	
0.3(12)	+	0.2(6)	= \$4.8 million of water

leaving only \$8.4 million of electricity and \$1.2 million of water left to satisfy the final demand of the outside sector. Thus, in order to meet the internal demands of both companies and to end up with enough electricity for the final outside demand, both companies must produce more than just the amount demanded by the outside sector. In fact, they must produce exactly enough to meet their own internal demands plus that demanded by the outside sector.

We now state the main problem of input–output analysis.

Basic Input–Output Problem

Given the internal demands for each industry's output, determine output levels for the various industries that will meet a given final (outside) level of demand as well as the internal demand.

If

x_1 = Total output from electric company

x_2 = Total output from water company

then reasoning as above, the internal demands are

$0.1x_1 + 0.4x_2$ Internal demand for electricity

$0.3x_1 + 0.2x_2$ Internal demand for water

Combining the internal demand with the final demand produces the following system of equations:

Total output	Internal demand	Final demand

$$\begin{aligned} x_1 &= 0.1x_1 + 0.4x_2 + d_1 \\ x_2 &= 0.3x_1 + 0.2x_2 + d_2 \end{aligned} \qquad (1)$$

or, in matrix form,

$$\begin{bmatrix} x_1 \\ x_2 \end{bmatrix} = \begin{bmatrix} 0.1 & 0.4 \\ 0.3 & 0.2 \end{bmatrix} \begin{bmatrix} x_1 \\ x_2 \end{bmatrix} + \begin{bmatrix} d_1 \\ d_2 \end{bmatrix}$$

or

$$X = MX + D \qquad (2)$$

where

$$D = \begin{bmatrix} d_1 \\ d_2 \end{bmatrix} \qquad \text{Final demand matrix}$$

$$X = \begin{bmatrix} x_1 \\ x_2 \end{bmatrix} \qquad \text{Output matrix}$$

$$M = \begin{matrix} \\ E \\ W \end{matrix} \begin{matrix} \begin{matrix} E & \;\; W \end{matrix} \\ \begin{bmatrix} 0.1 & 0.4 \\ 0.3 & 0.2 \end{bmatrix} \end{matrix} \qquad \text{Technology matrix}$$

The **technology matrix** is the heart of input–output analysis. The elements in the technology matrix are determined as follows (read from left to right and then up):

$$\begin{matrix} & & \text{Output} & \\ & & E \qquad\qquad\qquad\qquad W & \\ & & \uparrow & \\ \text{Input} & \begin{matrix} E \rightarrow \\ \\ W \end{matrix} & \begin{bmatrix} \begin{pmatrix} \text{Input from } E \\ \text{to produce \$1} \\ \text{of electricity} \end{pmatrix} & \begin{pmatrix} \text{Input from } E \\ \text{to produce \$1} \\ \text{of water} \end{pmatrix} \\ \begin{pmatrix} \text{Input from } W \\ \text{to produce \$1} \\ \text{of electricity} \end{pmatrix} & \begin{pmatrix} \text{Input from } W \\ \text{to produce \$1} \\ \text{of water} \end{pmatrix} \end{bmatrix} & = M \end{matrix}$$

Now our problem is to solve equation (2) for X. We proceed as in the preceding section:

$$\begin{aligned} X &= MX + D \\ X - MX &= D \\ IX - MX &= D \qquad\qquad I = \begin{bmatrix} 1 & 0 \\ 0 & 1 \end{bmatrix} \\ (I - M)X &= D \\ X &= (I - M)^{-1}D \qquad \text{Assuming } I - M \text{ has an inverse} \end{aligned} \qquad (3)$$

Since

$$I - M = \begin{bmatrix} 0.9 & -0.4 \\ -0.3 & 0.8 \end{bmatrix}$$

and

$$(I - M)^{-1} = \begin{bmatrix} \frac{4}{3} & \frac{2}{3} \\ \frac{1}{2} & \frac{3}{2} \end{bmatrix}$$

We convert decimals to fractions in this example to work with exact forms.

we have

$$\begin{bmatrix} x_1 \\ x_2 \end{bmatrix} = \begin{bmatrix} \frac{4}{3} & \frac{2}{3} \\ \frac{1}{2} & \frac{3}{2} \end{bmatrix} \begin{bmatrix} d_1 \\ d_2 \end{bmatrix} \tag{4}$$

$$= \begin{bmatrix} \frac{4}{3} & \frac{2}{3} \\ \frac{1}{2} & \frac{3}{2} \end{bmatrix} \begin{bmatrix} 12 \\ 6 \end{bmatrix} = \begin{bmatrix} 20 \\ 15 \end{bmatrix}$$

Therefore, the electric company must have an output of \$20 million and the water company an output of \$15 million so that each company can meet both internal and final demands.

Check We use equation (2) to check our work:

$$X = MX + D$$

$$\begin{bmatrix} 20 \\ 15 \end{bmatrix} \stackrel{?}{=} \begin{bmatrix} 0.1 & 0.4 \\ 0.3 & 0.2 \end{bmatrix} \begin{bmatrix} 20 \\ 15 \end{bmatrix} + \begin{bmatrix} 12 \\ 6 \end{bmatrix}$$

$$\begin{bmatrix} 20 \\ 15 \end{bmatrix} \stackrel{?}{=} \begin{bmatrix} 8 \\ 9 \end{bmatrix} + \begin{bmatrix} 12 \\ 6 \end{bmatrix}$$

$$\begin{bmatrix} 20 \\ 15 \end{bmatrix} \stackrel{\checkmark}{=} \begin{bmatrix} 20 \\ 15 \end{bmatrix}$$

Actually, (4) solves the original problem for arbitrary final demands d_1 and d_2. This is very useful, since (4) gives a quick solution not only for the final demands stated but also to the original problem for various other projected final demands. If we had solved system (1) by elimination, then we would have to start over for each new set of final demands.

Suppose in the original problem that the projected final demands 5 years from now were $d_1 = 18$ and $d_2 = 12$. Determine each company's output for this projection. We simply substitute these values into (4) and multiply:

$$\begin{bmatrix} x_1 \\ x_2 \end{bmatrix} = \begin{bmatrix} \frac{4}{3} & \frac{2}{3} \\ \frac{1}{2} & \frac{3}{2} \end{bmatrix} \begin{bmatrix} 18 \\ 12 \end{bmatrix} = \begin{bmatrix} 32 \\ 27 \end{bmatrix}$$

We summarize these results for convenient reference.

Solution to a Two-Company Input–Output Problem

Given two companies, C_1 and C_2, with

Technology matrix

$$M = \begin{matrix} C_1 \\ C_2 \end{matrix} \overset{\begin{matrix} C_1 & \quad C_2 \end{matrix}}{\begin{bmatrix} a_{11} & a_{12} \\ a_{21} & a_{22} \end{bmatrix}}$$

Output matrix, Final demand matrix

$$X = \begin{bmatrix} x_1 \\ x_2 \end{bmatrix} \qquad D = \begin{bmatrix} d_1 \\ d_2 \end{bmatrix}$$

where a_{ij} is the input required from C_i to produce a dollar's worth of output for C_j, the solution to the input–output matrix equation

$$\underset{\text{Total output}}{X} = \underset{\text{Internal demand}}{MX} + \underset{\text{Final demand}}{D} \tag{2}$$

is

$$X = (I - M)^{-1}D \tag{3}$$

assuming $I - M$ has an inverse.

Another consequence of expressing the input–output problem in matrix form (2) is that matrix equation (2) and its solution (3) are the same for a three-industry economy, a four-industry economy, or an economy with n industries (where n is any natural number), since the steps we took going from (2) to (3) hold for arbitrary matrices as long as they are dimensionally correct and $(I - M)^{-1}$ exists. The next example illustrates this for a three-industry economy.

◆ EXAMPLE 32 An economy is based on three sectors, agriculture, energy, and manufacturing. Production of a dollar's worth of agriculture requires an input of \$0.20 from the agriculture sector and \$0.40 from the energy sector. Production of a dollar's worth of energy requires an input of \$0.20 from the energy sector and \$0.40 from the manufacturing sector. Production of a dollar's worth of manufacturing requires an input of \$0.10 from the agriculture sector, \$0.10 from the energy sector, and \$0.30 from the manufacturing sector. Find the output from each sector that is needed to satisfy a final demand of \$20 billion for agriculture, \$10 billion for energy, and \$30 billion for manufacturing.

Solution Since this is a three-industry problem, the technology matrix will be a 3×3 matrix, and the output and final demand matrices will be 3×1 column matrices. Using the information given in the problem, we can write

$$\begin{array}{c} \text{Technology matrix} \\ \begin{array}{cc} & \begin{array}{ccc} A & E & M \end{array} \\ M = \begin{array}{c} A \\ E \\ M \end{array} & \begin{bmatrix} 0.2 & 0 & 0.1 \\ 0.4 & 0.2 & 0.1 \\ 0 & 0.4 & 0.3 \end{bmatrix} \end{array} \end{array} \qquad \begin{array}{c} \text{Final demand} \\ \text{matrix} \\ D = \begin{bmatrix} 20 \\ 10 \\ 30 \end{bmatrix} \end{array} \qquad \begin{array}{c} \text{Output} \\ \text{matrix} \\ X = \begin{bmatrix} x_1 \\ x_2 \\ x_3 \end{bmatrix} \end{array}$$

where M, X, and D satisfy the input–output equation $X = MX + D$. Since the solution to this equation is $X = (I - M)^{-1}D$, we must first find $I - M$ and then $(I - M)^{-1}$. Omitting the details of the calculations, we have

$$I - M = \begin{bmatrix} 0.8 & 0 & -0.1 \\ -0.4 & 0.8 & -0.1 \\ 0 & -0.4 & 0.7 \end{bmatrix}$$

and

$$(I - M)^{-1} = \begin{bmatrix} 1.3 & 0.1 & 0.2 \\ 0.7 & 1.4 & 0.3 \\ 0.4 & 0.8 & 1.6 \end{bmatrix}$$

Thus, the output matrix X is given by

$$\overset{X}{\begin{bmatrix} x_1 \\ x_2 \\ x_3 \end{bmatrix}} = \overset{(I-M)^{-1}}{\begin{bmatrix} 1.3 & 0.1 & 0.2 \\ 0.7 & 1.4 & 0.3 \\ 0.4 & 0.8 & 1.6 \end{bmatrix}} \overset{D}{\begin{bmatrix} 20 \\ 10 \\ 30 \end{bmatrix}} = \begin{bmatrix} 33 \\ 37 \\ 64 \end{bmatrix}$$

An output of \$33 billion for agriculture, \$37 billion for energy, and \$64 billion for manufacturing will meet the given final demands. You should check this result in equation (2). ◆

PROBLEM 32 An economy is based on three sectors, coal, oil, and transportation. Production of a dollar's worth of coal requires an input of \$0.20 from the coal sector and \$0.40 from the transportation sector. Production of a dollar's worth of oil requires an input of \$0.10 from the oil sector and \$0.20 from the transportation sector. Production of a dollar's worth of transportation requires an input of \$0.40 from the coal sector, \$0.20 from the oil sector, and \$0.20 from the transportation sector.

(A) Find the technology matrix M.
(B) Find $(I - M)^{-1}$.
(C) Find the output from each sector that is needed to satisfy a final demand of \$30 billion for coal, \$10 billion for oil, and \$20 billion for transportation. ◆

Answers to Matched Problem

32. (A) $\begin{bmatrix} 0.2 & 0 & 0.4 \\ 0 & 0.1 & 0.2 \\ 0.4 & 0.2 & 0.2 \end{bmatrix}$ (B) $\begin{bmatrix} 1.7 & 0.2 & 0.9 \\ 0.2 & 1.2 & 0.4 \\ 0.9 & 0.4 & 1.8 \end{bmatrix}$

(C) \$71 billion for coal, \$26 billion for oil, and \$67 billion for transportation

EXERCISE 5-8

A

Problems 1–6 pertain to the following input–output model: Assume an economy is based on two industrial sectors, agriculture (A) and energy (E). The technology matrix M and final demand matrices are (in billions of dollars)

$$\begin{array}{c} \\ A \\ E \end{array} \begin{array}{c} \begin{array}{cc} A & E \end{array} \\ \begin{bmatrix} 0.4 & 0.2 \\ 0.2 & 0.1 \end{bmatrix} \end{array} = M \qquad D_1 = \begin{bmatrix} 6 \\ 4 \end{bmatrix} \qquad D_2 = \begin{bmatrix} 8 \\ 5 \end{bmatrix} \qquad D_3 = \begin{bmatrix} 12 \\ 9 \end{bmatrix}$$

1. How much input from A and E are required to produce a dollar's worth of output for A?
2. How much input from A and E are required to produce a dollar's worth of output for E?
3. Find $I - M$ and $(I - M)^{-1}$.
4. Find the output for each sector that is needed to satisfy the final demand D_1.
5. Repeat Problem 4 for D_2.
6. Repeat Problem 4 for D_3.

B

Problems 7–12 pertain to the following input–output model: Assume an economy is based on three industrial sectors, agriculture (A), building (B), and energy (E). The technology matrix M and final demand matrices (in billions of dollars) are

$$\begin{array}{c} \\ A \\ B \\ E \end{array} \begin{array}{c} \begin{array}{ccc} A & B & E \end{array} \\ \begin{bmatrix} 0.3 & 0.2 & 0.2 \\ 0.1 & 0.1 & 0.1 \\ 0.2 & 0.1 & 0.1 \end{bmatrix} \end{array} = M \qquad D_1 = \begin{bmatrix} 5 \\ 10 \\ 15 \end{bmatrix} \qquad D_2 = \begin{bmatrix} 20 \\ 15 \\ 10 \end{bmatrix}$$

7. How much input from A, B, and E is required to produce a dollar's worth of output for B?
8. How much of each of B's output dollars is required as input for each of the three sectors?
9. Show that

$$I - M = \begin{bmatrix} 0.7 & -0.2 & -0.2 \\ -0.1 & 0.9 & -0.1 \\ -0.2 & -0.1 & 0.9 \end{bmatrix}$$

10. Given

$$(I-M)^{-1} = \begin{bmatrix} 1.6 & 0.4 & 0.4 \\ 0.22 & 1.18 & 0.18 \\ 0.38 & 0.22 & 1.22 \end{bmatrix}$$

show that $(I-M)^{-1}(I-M) = I$.

11. Use $(I-M)^{-1}$ in Problem 10 to find the output for each sector that is needed to satisfy the final demand D_1.

12. Repeat Problem 11 for D_2.

In Problems 13–16, find $(I-M)^{-1}$ and X.

13. $M = \begin{bmatrix} 0.2 & 0.2 \\ 0.3 & 0.3 \end{bmatrix}$; $D = \begin{bmatrix} 10 \\ 25 \end{bmatrix}$

14. $M = \begin{bmatrix} 0.4 & 0.1 \\ 0.2 & 0.3 \end{bmatrix}$; $D = \begin{bmatrix} 15 \\ 20 \end{bmatrix}$

C

15. $M = \begin{bmatrix} 0.3 & 0.1 & 0.3 \\ 0.2 & 0.1 & 0.2 \\ 0.1 & 0.1 & 0.1 \end{bmatrix}$; $D = \begin{bmatrix} 20 \\ 5 \\ 10 \end{bmatrix}$

16. $M = \begin{bmatrix} 0.3 & 0.2 & 0.3 \\ 0.1 & 0.1 & 0.1 \\ 0.1 & 0.2 & 0.1 \end{bmatrix}$; $D = \begin{bmatrix} 10 \\ 25 \\ 15 \end{bmatrix}$

APPLICATIONS

Business & Economics

17. An economy is based on two industrial sectors, coal and steel. Production of a dollar's worth of coal requires an input of $0.10 from the coal sector and $0.20 from the steel sector. Production of a dollar's worth of steel requires an input of $0.20 from the coal sector and $0.40 from the steel sector. Find the output for each sector that is needed to satisfy a final demand of $20 billion for coal and $10 billion for steel.

18. An economy is based on two sectors, transportation and manufacturing. Production of a dollar's worth of transportation requires $0.10 of input from each sector and production of a dollar's worth of manufacturing requires an input of $0.40 from each sector. Find the output for each sector that is needed to satisfy a final demand of $5 billion for transportation and $20 billion for manufacturing.

19. The economy of a small island nation is based on two sectors, agriculture and tourism. Production of a dollar's worth of agriculture requires an input of $0.20 from agriculture and $0.15 from tourism. Production of a dollar's worth of tourism requires an input of $0.40 from agriculture and $0.30 from tourism. Find the output from each sector that is needed to satisfy a final demand of $60 million for agriculture and $80 million for tourism.

20. The economy of a country is based on two sectors, agriculture and oil. Production of a dollar's worth of agriculture requires an input of $0.40 from agriculture and $0.35 from oil. Production of a dollar's worth of oil requires

an input of $0.20 from agriculture and $0.05 from oil. Find the output from each sector that is needed to satisfy a final demand of $40 million for agriculture and $250 million for oil.

21. An economy is based on three sectors, agriculture, manufacturing, and energy. Production of a dollar's worth of agriculture requires inputs of $0.20 from agriculture, $0.20 from manufacturing, and $0.20 from energy. Production of a dollar's worth of manufacturing requires inputs of $0.40 from agriculture, $0.10 from manufacturing, and $0.10 from energy. Production of a dollar's worth of energy requires inputs of $0.30 from agriculture, $0.10 from manufacturing, and $0.10 from energy. Find the output for each sector that is needed to satisfy a final demand of $10 billion for agriculture, $15 billion for manufacturing, and $20 billion for energy.

22. A large energy company produces electricity, natural gas, and oil. The production of a dollar's worth of electricity requires inputs of $0.30 from electricity, $0.10 from natural gas, and $0.20 from oil. Production of a dollar's worth of natural gas requires inputs of $0.30 from electricity, $0.10 from natural gas, and $0.20 from oil. Production of a dollar's worth of oil requires inputs of $0.10 from each sector. Find the output for each sector that is needed to satisfy a final demand of $25 billion for electricity, $15 billion for natural gas, and $20 billion for oil.

Use a graphic calculator or a computer to solve Problems 23 and 24.

23. An economy is based on four sectors, agriculture (*A*), energy (*E*), labor (*L*), and manufacturing (*M*). The table gives the input requirements for a dollar's worth of output for each sector, along with the projected final demand (in billions of dollars) for a 3 year period. Find the output for each sector that is needed to satisfy each of these final demands. Round answers to the nearest billion dollars.

		OUTPUT				FINAL DEMAND		
		A	*E*	*L*	*M*	*1*	*2*	*3*
INPUT	*A*	0.05	0.17	0.23	0.09	23	32	55
	E	0.07	0.12	0.15	0.19	41	48	62
	L	0.25	0.08	0.03	0.32	18	21	25
	M	0.11	0.19	0.28	0.16	31	33	35

24. Repeat Problem 23 with the following table.

		OUTPUT				FINAL DEMAND		
		A	*E*	*L*	*M*	*1*	*2*	*3*
INPUT	*A*	0.07	0.09	0.27	0.12	18	22	37
	E	0.14	0.07	0.21	0.24	26	31	42
	L	0.17	0.06	0.02	0.21	12	19	28
	M	0.15	0.13	0.31	0.19	41	45	49

SECTION 5-9 Chapter Review

Important Terms and Symbols

5-1 *Review: Systems of Linear Equations in Two Variables.* Solution of a system; solution set; graphing method; consistent; inconsistent; independent; unique solution; dependent; substitution method; elimination by addition; equivalent systems; parameter; particular solution; equilibrium price; equilibrium quantity

5-2 *Systems of Linear Equations and Augmented Matrices.* Matrix; element; augmented matrix; column; row; equivalent systems; row-equivalent matrices; row operations; Gauss–Jordan elimination

$$R_i \leftrightarrow R_j; \quad kR_i \rightarrow R_i; \quad R_i + kR_j \rightarrow R_i$$

5-3 *Gauss–Jordan Elimination.* Reduced form; reduced system; leftmost variable; submatrix; Gauss–Jordan elimination

5-4 *Matrices—Addition and Multiplication by a Number.* Size or dimension of a matrix; $m \times n$ matrix; square matrix; column matrix; row matrix; equal matrices; sum of two matrices; zero matrix; negative of a matrix M; subtraction of matrices; product of a number k and a matrix M; double subscript notation; principal diagonal

5-5 *Matrix Multiplication.* Matrix product; associative property; left and right distributive properties; left and right multiplication properties; incidence matrix

5-6 *Inverse of a Square Matrix.* Identity element for multiplication; principal diagonal; multiplicative inverse; encoding matrix; decoding matrix

$$M^{-1}$$

5-7 *Matrix Equations and Systems of Linear Equations.* Matrix equation

5-8 *Leontief Input–Output Analysis.* Input–output analysis; technology matrix; final demand matrix; output matrix

EXERCISE 5-9 Chapter Review

Work through all the problems in this chapter review and check your answers in the back of the book. (Answers to all review problems are there.) Where weaknesses show up, review appropriate sections in the text.

A

1. Solve the following system by graphing:

$$\begin{aligned} 2x - y &= 4 \\ x - 2y &= -4 \end{aligned}$$

2. Solve the system in Problem 1 by substitution.

In Problems 3–11, perform the operations that are defined, given the following matrices:

$$A = \begin{bmatrix} 1 & 2 \\ 3 & 1 \end{bmatrix} \qquad B = \begin{bmatrix} 2 & 1 \\ 1 & 1 \end{bmatrix} \qquad C = [2 \quad 3] \qquad D = \begin{bmatrix} 1 \\ 2 \end{bmatrix}$$

3. $A + B$
4. $B + D$
5. $A - 2B$
6. AB
7. AC
8. AD
9. DC
10. CD
11. $C + D$
12. Find the inverse of the matrix A given below by appropriate row operations on $[A|I]$. Show that $A^{-1}A = I$.

$$A = \begin{bmatrix} 3 & 2 \\ 4 & 3 \end{bmatrix}$$

13. Solve the following system using elimination by addition:

$$3x_1 + 2x_2 = 3$$
$$4x_1 + 3x_2 = 5$$

14. Solve the system in Problem 13 by performing appropriate row operations on the augmented matrix of the system.
15. Solve the system in Problem 13 by writing the system as a matrix equation and using the inverse of the coefficient matrix (see Problem 12). Also, solve the system if the constants 3 and 5 are replaced by 7 and 10, respectively. By 4 and 2, respectively.

B *In Problems 16–21, perform the specified operations, given the following matrices:*

$$A = \begin{bmatrix} 2 & -2 \\ 1 & 0 \\ 3 & 2 \end{bmatrix} \qquad B = \begin{bmatrix} -1 \\ 2 \\ 3 \end{bmatrix} \qquad C = [2 \quad 1 \quad 3]$$

$$D = \begin{bmatrix} 3 & -2 & 1 \\ -1 & 1 & 2 \end{bmatrix} \qquad E = \begin{bmatrix} 3 & -4 \\ -1 & 0 \end{bmatrix}$$

16. $A + D$
17. $E + DA$
18. $DA - 3E$
19. BC
20. CB
21. $AD - BC$
22. Find the inverse of the matrix A given below by appropriate row operations on $[A|I]$. Show that $A^{-1}A = I$.

$$A = \begin{bmatrix} 1 & 2 & 3 \\ 2 & 3 & 4 \\ 1 & 2 & 1 \end{bmatrix}$$

23. Solve by Gauss–Jordan elimination:

(A)
$$\begin{aligned} x_1 + 2x_2 + 3x_3 &= 1 \\ 2x_1 + 3x_2 + 4x_3 &= 3 \\ x_1 + 2x_2 + x_3 &= 3 \end{aligned}$$

(B)
$$\begin{aligned} x_1 + 2x_2 - x_3 &= 2 \\ 2x_1 + 3x_2 + x_3 &= -3 \\ 3x_1 + 5x_2 &= -1 \end{aligned}$$

24. Solve the system in Problem 23A by writing the system as a matrix equation and using the inverse of the coefficient matrix (see Problem 22). Also, solve the system if the constants 1, 3, and 3 are replaced by 0, 0, and −2, respectively. By −3, −4, and 1, respectively.

25. Given the technology matrix M and the final demand matrix D (in billions of dollars), find $(I - M)^{-1}$ and the output matrix X:

$$M = \begin{bmatrix} 0.2 & 0.15 \\ 0.4 & 0.3 \end{bmatrix} \qquad D = \begin{bmatrix} 30 \\ 20 \end{bmatrix}$$

C

26. Find the inverse of the matrix A given below. Show that $A^{-1}A = I$.

$$A = \begin{bmatrix} 4 & 5 & 6 \\ 4 & 5 & -6 \\ 1 & 1 & 1 \end{bmatrix}$$

27. Solve the system

$$\begin{aligned} 0.04x_1 + 0.05x_2 + 0.06x_3 &= 360 \\ 0.04x_1 + 0.05x_2 - 0.06x_3 &= 120 \\ x_1 + x_2 + x_3 &= 7{,}000 \end{aligned}$$

by writing as a matrix equation and using the inverse of the coefficient matrix. (Before starting, multiply the first two equations by 100 to eliminate decimals. Also, see Problem 26.)

28. Solve Problem 27 by Gauss–Jordan elimination.

29. Given the technology matrix M and the final demand matrix D (in billions of dollars), find $(I - M)^{-1}$ and the output matrix X:

$$M = \begin{bmatrix} 0.2 & 0 & 0.4 \\ 0.1 & 0.3 & 0.1 \\ 0 & 0.4 & 0.2 \end{bmatrix} \qquad D = \begin{bmatrix} 40 \\ 20 \\ 30 \end{bmatrix}$$

APPLICATIONS

Business & Economics

ORE	NICKEL (%)	COPPER (%)
A	1	2
B	2	5

30. *Resource allocation.* A mining company has two mines with ore composition as given in the table. How many tons of each ore should be used to obtain exactly 4.5 tons of nickel and 10 tons of copper? Set up a system of equations and solve using Gauss–Jordan elimination.

31. (A) Set up Problem 30 as a matrix equation and solve using the inverse of the coefficient matrix.
    (B) Solve Problem 30 (as in part A) if 2.3 tons of nickel and 5 tons of copper are needed.

32. *Purchasing.* A soft-drink distributor has budgeted \$300,000 for the purchase of 12 new delivery trucks. If a model *A* truck costs \$18,000, a model *B* truck costs \$22,000, and a model *C* truck costs \$30,000, how many trucks of each model should the distributor purchase to use exactly all the budgeted funds?

33. *Material costs.* A manufacturer wishes to make two different bronze alloys in a metal foundry. The quantities of copper, tin, and zinc needed are indicated in matrix M. The costs for these materials (in dollars per pound) from two suppliers are summarized in matrix N. The company must choose one supplier or the other.

$$M = \begin{array}{c} \text{Copper} \quad \text{Tin} \quad \text{Zinc} \\ \begin{bmatrix} 4{,}800 \text{ lb} & 600 \text{ lb} & 300 \text{ lb} \\ 6{,}000 \text{ lb} & 1{,}400 \text{ lb} & 700 \text{ lb} \end{bmatrix} \end{array} \begin{array}{l} \text{Alloy 1} \\ \text{Alloy 2} \end{array}$$

$$N = \begin{array}{c} \text{Supplier } A \quad \text{Supplier } B \\ \begin{bmatrix} \$0.75 & \$0.70 \\ \$6.50 & \$6.70 \\ \$0.40 & \$0.50 \end{bmatrix} \end{array} \begin{array}{l} \text{Copper} \\ \text{Tin} \\ \text{Zinc} \end{array}$$

(A) Find MN and interpret. (B) Find $[1 \quad 1]MN$ and interpret.

34. *Labor costs.* A company with manufacturing plants in California and Texas has labor-hour and wage requirements for the manufacture of two inexpensive calculators as given in matrices M and N below:

Labor-hours per calculator

$$M = \begin{array}{c} \begin{array}{ccc} \text{Fabricating department} & \text{Assembly department} & \text{Packaging department} \end{array} \\ \begin{bmatrix} 0.15 \text{ hr} & 0.10 \text{ hr} & 0.05 \text{ hr} \\ 0.25 \text{ hr} & 0.20 \text{ hr} & 0.05 \text{ hr} \end{bmatrix} \end{array} \begin{array}{l} \text{Model } A \\ \text{Model } B \end{array}$$

Hourly wages

$$N = \begin{array}{c} \text{California plant} \quad \text{Texas plant} \\ \begin{bmatrix} \$15 & \$12 \\ \$12 & \$10 \\ \$\ 4 & \$\ 4 \end{bmatrix} \end{array} \begin{array}{l} \text{Fabricating department} \\ \text{Assembly department} \\ \text{Packaging department} \end{array}$$

(A) What is the labor cost for producing one model B calculator in California? Set up a matrix product and multiply.
(B) Find MN and interpret.

35. *Investment analysis.* A person has $5,000 to invest, part at 5% and the rest at 10%. How much should be invested at each rate to yield $400 per year? Set up a system of equations and solve using augmented matrix methods.

36. *Investment analysis.* Solve Problem 35 by using a matrix equation and the inverse of the coefficient matrix.

37. *Resource allocation.* An outdoor amphitheater has 25,000 seats. Ticket prices are $8, $12, and $20, and the number of tickets priced at $8 must equal the number priced at $20. How many tickets of each type should be

sold (assuming all seats can be sold) to bring in each of the returns indicated in the table? Set up a system of equations and solve using the inverse of the coefficient matrix.

	CONCERT		
	1	*2*	*3*
TICKETS SOLD	25,000	25,000	25,000
RETURN REQUIRED	$320,000	$330,000	$340,000

38. *Input–output analysis.* An economy is based on two industrial sectors, agriculture and fabrication. Production of a dollar's worth of agriculture requires an input of $0.30 from the agriculture sector and $0.20 from the fabricating sector. Production of a dollar's worth of fabrication requires $0.10 from the agriculture sector and $0.40 from the fabrication sector.

(A) Find the output for each sector that is needed to satisfy a final demand of $50 billion for agriculture and $20 billion for fabrication.

(B) Find the output for each sector that is needed to satisfy a final demand of $80 billion for agriculture and $60 billion for fabrication.

CHAPTER 6

Linear Inequalities and Linear Programming

Contents

CHAPTER 6

In this chapter we will discuss linear inequalities in two and more variables. In addition, we will introduce a relatively new and powerful mathematical tool called *linear programming*, which will be used to solve a variety of interesting practical problems. The row operations on matrices introduced in Chapter 5 will be particularly useful in Sections 6-4, 6-5, and 6-6.

SECTION 6-1

Systems of Linear Inequalities in Two Variables

- GRAPHING LINEAR INEQUALITIES IN TWO VARIABLES
- SOLVING SYSTEMS OF LINEAR INEQUALITIES GRAPHICALLY
- APPLICATIONS

Many applications of mathematics involve systems of inequalities rather than systems of equations. A graph is often the most convenient way to represent the solutions of a system of linear inequalities in two variables. In this section we discuss techniques for graphing both a single linear inequality in two variables and a system of linear inequalities in two variables.

◆ GRAPHING LINEAR INEQUALITIES IN TWO VARIABLES

We know how to graph first-degree equations such as

$$y = 2x - 3 \qquad \text{and} \qquad 2x - 3y = 5$$

but how do we graph first-degree inequalities such as the following?

$$y \leqslant 2x - 3 \qquad \text{and} \qquad 2x - 3y > 5$$

We will find that graphing these inequalities is almost as easy as graphing the equations, but first, we must discuss some important subsets of a plane with a rectangular coordinate system.

A line divides the plane into two halves called **half-planes.** A vertical line divides it into **left** and **right half-planes;** a nonvertical line divides it into **upper** and **lower half-planes** (Fig. 1).

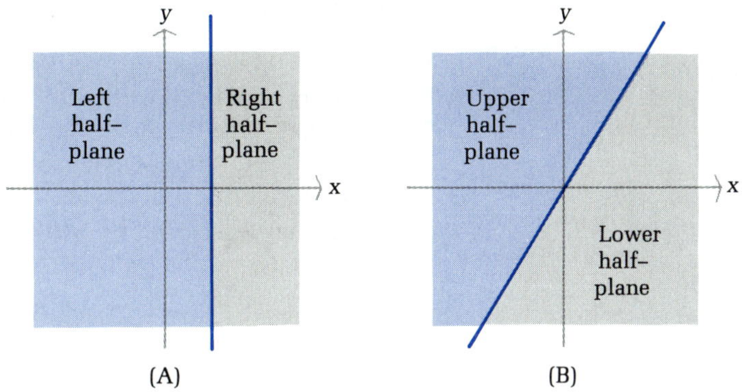

FIGURE 1

To investigate the half-planes determined by a linear equation such as $y - x = -2$, we rewrite the equation as $y = x - 2$. For any given value of x, there is exactly one value for y such that (x, y) lies on the line. For example, for $x = 4$, we have $y = 4 - 2 = 2$. For the same x and smaller values of y, the point (x, y) will lie below the line, since $y < x - 2$. Thus, the lower half-plane corresponds to the solution of the inequality $y < x - 2$. Similarly, the upper half-plane corresponds to $y > x - 2$, as shown in Figure 2.

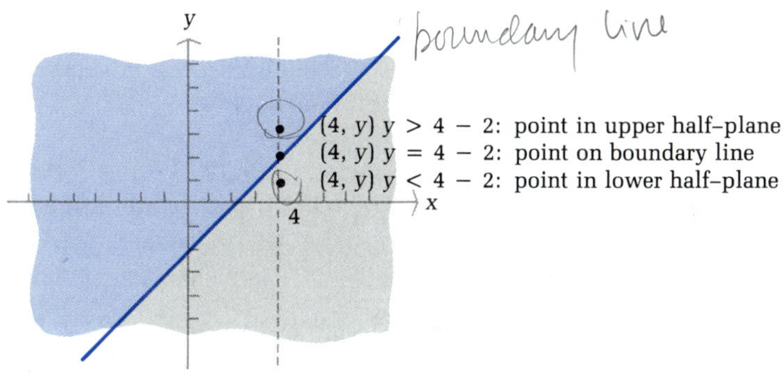

FIGURE 2

The four inequalities formed from $y = x - 2$ by replacing the = sign by $>$, $\geqslant$, $<$, and $\leqslant$, respectively, are

$$y > x - 2 \qquad y \geqslant x - 2 \qquad y < x - 2 \qquad y \leqslant x - 2$$

The graph of each is a half-plane, excluding the boundary line for $<$ and $>$, and including the boundary line for $\leqslant$ and $\geqslant$. In Figure 3 (page 290) the half-planes are indicated with small arrows on the graph of $y = x - 2$ and then graphed as shaded regions. Excluded boundary lines are shown as dashed lines, and included boundary lines are shown as solid lines.

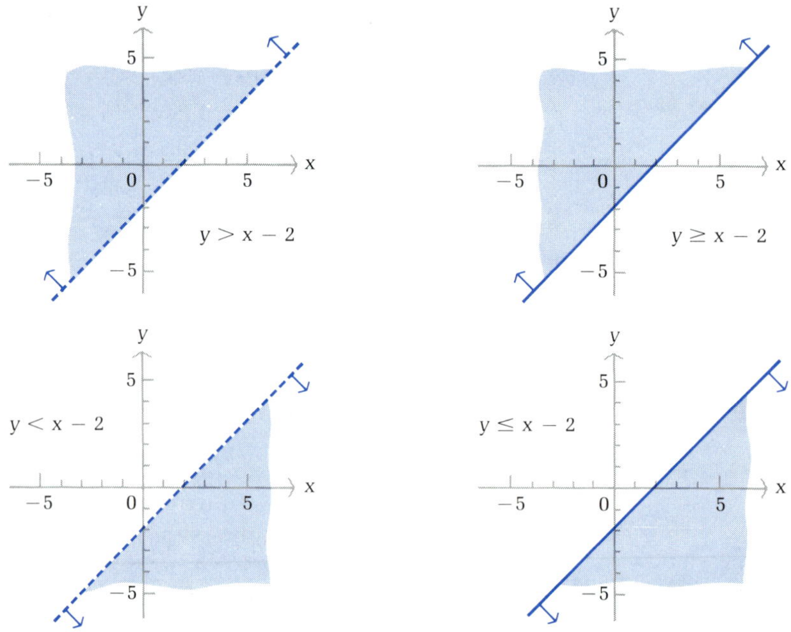

FIGURE 3

The preceding discussion suggests the following theorem, which is stated without proof:

THEOREM 1

Graphs of Linear Inequalities

The graph of the linear inequality

$$Ax + By < C \qquad \text{or} \qquad Ax + By > C$$

with $B \neq 0$, is either the upper half-plane or the lower half-plane (but not both) determined by the line $Ax + By = C$.

If $B = 0$, the graph of

$$Ax < C \qquad \text{or} \qquad Ax > C$$

is either the left half-plane or the right half-plane (but not both) determined by the line $Ax = C$.

As a consequence of this theorem, we state a simple and fast mechanical procedure for graphing linear inequalities.

Procedure for Graphing Linear Inequalities

Step 1. First graph $Ax + By = C$ as a dashed line if equality is not included in the original statement or as a solid line if equality is included.

Step 2. Choose a test point anywhere in the plane not on the line [the origin (0, 0) often requires the least computation] and substitute the coordinates into the inequality.

Step 3. The graph of the original inequality includes the half-plane containing the test point if the inequality is satisfied by that point or the half-plane not containing the test point if the inequality is not satisfied by that point.

◆ EXAMPLE 1 Graph $2x - 3y \leq 6$.

Solution *Step 1.* Graph $2x - 3y = 6$ as a solid line, since equality is included in the original statement.

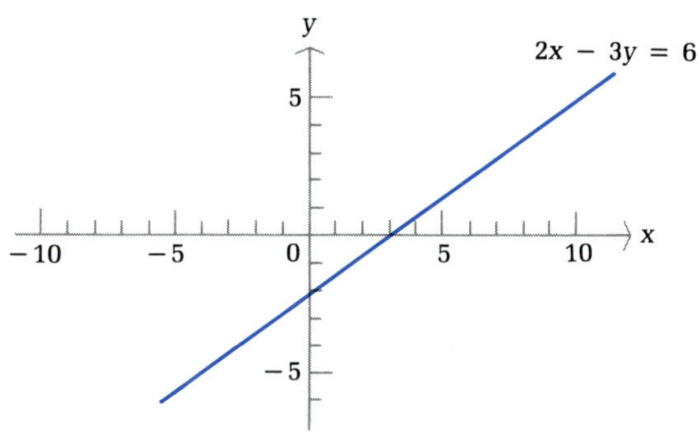

Step 2. Pick a convenient test point above or below the line. The origin (0, 0) requires the least computation, so, substituting (0, 0) into the inequality, we get

$$2x - 3y \leq 6$$
$$2(0) - 3(0) = 0 \leq 6$$

This is a true statement; therefore, the point (0, 0) is in the solution set.

Step 3. The line $2x - 3y = 6$ and the half-plane containing the origin form the graph of $2x - 3y \leq 6$, as shown at the top of the next page.

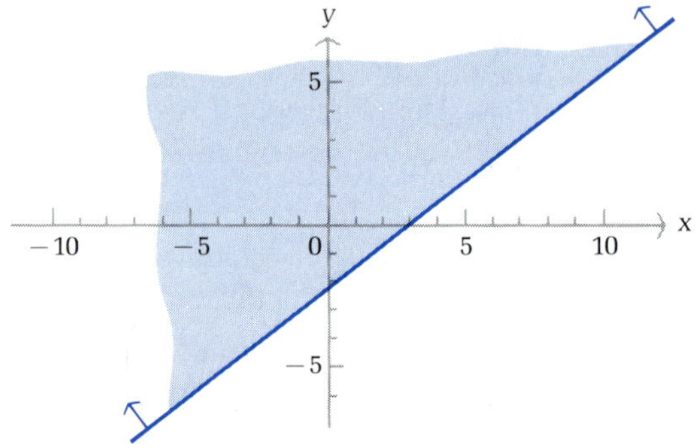

PROBLEM 1 Graph $6x - 3y > 18$.

EXAMPLE 2 Graph: (A) $y > -3$ (B) $2x \leq 5$ (C) $x \leq 3y$

Solutions

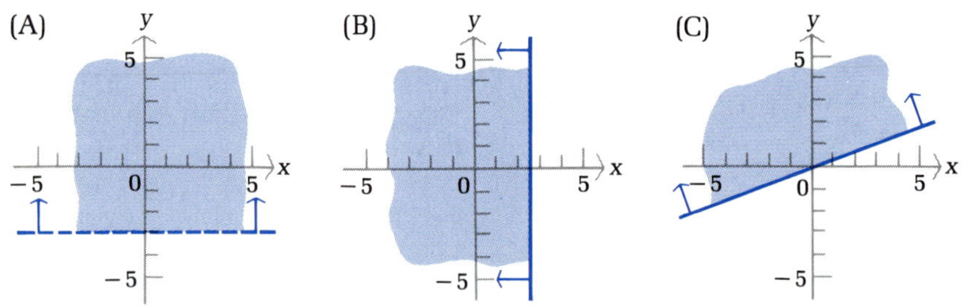

PROBLEM 2 Graph: (A) $y < 4$ (B) $4x \geq -9$ (C) $3x \geq 2y$

SOLVING SYSTEMS OF LINEAR INEQUALITIES GRAPHICALLY

We now consider systems of linear inequalities such as

$$\begin{aligned} x + y &\geq 6 \\ 2x - y &\geq 0 \end{aligned} \qquad \text{and} \qquad \begin{aligned} 2x + y &\leq 22 \\ x + y &\leq 13 \\ 2x + 5y &\leq 50 \\ x &\geq 0 \\ y &\geq 0 \end{aligned}$$

We wish to **solve** such systems **graphically;** that is, to find the graph of all ordered pairs of real numbers (x, y) that simultaneously satisfy all the inequalities in the system. The graph is called the **solution region** for the system. (In many applications, the solution region is also called the **feasible region.**) To find the solution region, we graph each inequality in the system and then take the intersection of all the graphs. To simplify the discussion that follows, **we will consider only systems of linear inequalities where equality is included in each statement in the system.**

◆ EXAMPLE 3 Solve the following system of linear inequalities graphically:

$$x + y \geqslant 6$$
$$2x - y \geqslant 0$$

Solution Graph the line $x + y = 6$ and shade the region that satisfies the inequality $x + y \geqslant 6$. This region is shaded in gray in figure A. Next, graph the line $2x - y = 0$ and shade the region that satisfies the inequality $2x - y \geqslant 0$. This region is shaded in color in figure A. The solution region for the system of inequalities is the intersection of these two regions. This is the region shaded in both gray and color in figure A and redrawn in figure B with only the solution region shaded. The coordinates of any point in the shaded region of figure B specify a solution to the system. For example, the points (2, 4), (6, 3), and (7.43, 8.56) are three of infinitely many solutions, as can be easily checked. The intersection point (2, 4) is obtained by solving the equations $x + y = 6$ and $2x - y = 0$ simultaneously.

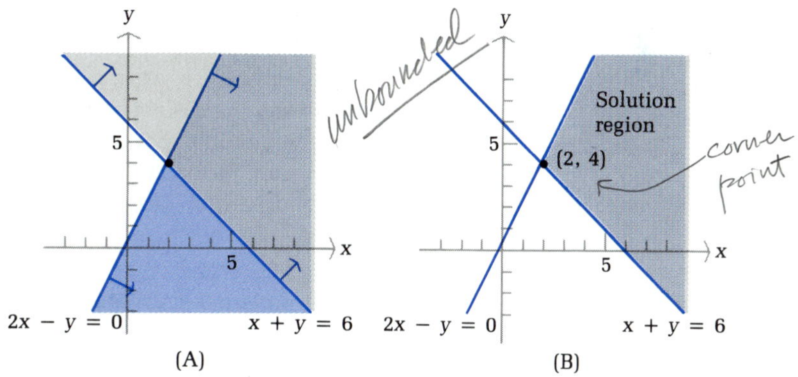

◆

PROBLEM 3 Solve the following system of linear inequalities graphically:

$$3x + \ \ y \leqslant 21$$
$$x - 2y \leqslant 0$$

◆

The points of intersection of the lines that form the boundary of a solution region will play a fundamental role in the solution of linear programming problems, which are discussed in the next section.

Corner Point

A **corner point** of a solution region is a point in the solution region that is the intersection of two boundary lines.

For example, the point (2, 4) is the only corner point of the solution region in Example 3.

◆ EXAMPLE 4 Solve the following system of linear inequalities graphically, and find the corner points:

$$2x + y \leq 22$$
$$x + y \leq 13$$
$$2x + 5y \leq 50$$
$$x \geq 0$$
$$y \geq 0$$

1st quadrant

Solution The inequalities $x \geq 0$ and $y \geq 0$ indicate that the solution region will lie in the first quadrant.* Thus, we can restrict our attention to that portion of the plane. First, we graph the lines

$$2x + y = 22$$
$$x + y = 13$$
$$2x + 5y = 50$$

Find the x and y intercepts of each line; then sketch the line through these points.

Next, choosing (0, 0) as a test point, we see that the graph of each of the first three inequalities in the system consists of its corresponding line and the half-plane lying below the line, as indicated by the small arrows in the figure. Thus, the solution region of the system consists of the points in the first quadrant that simultaneously lie on or below all three of these lines (see the shaded region in the figure).

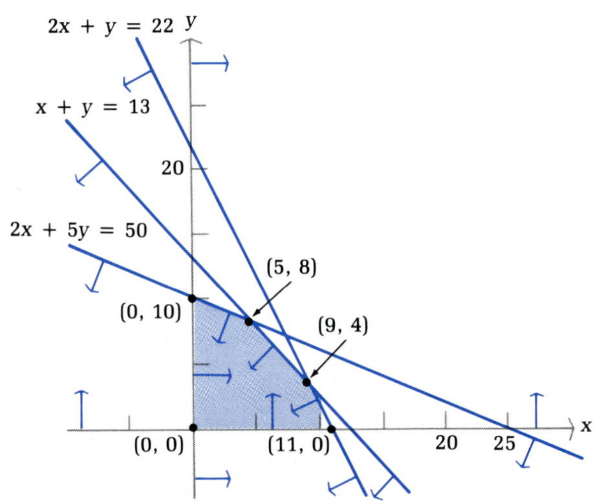

* The inequalities $x \geq 0$ and $y \geq 0$ will occur frequently in applications involving systems of inequalities, since x and y will often represent quantities that cannot be negative (number of units produced, number of hours worked, etc.).

The corner points (0, 0), (0, 10), and (11, 0) can be determined from the graph. The other two corner points are determined as follows:

Solve the system

$$2x + 5y = 50$$
$$x + y = 13$$

to obtain (5, 8).

Solve the system

$$2x + y = 22$$
$$x + y = 13$$

to obtain (9, 4).

Note that the lines $2x + 5y = 50$ and $2x + y = 22$ also intersect, but the intersection point is not part of the solution region, and hence, is not a corner point. ◆

PROBLEM 4 Solve the following system of linear inequalities graphically, and find the corner points:

$$5x + y \geqslant 20$$
$$x + y \geqslant 12$$
$$x + 3y \geqslant 18$$
$$x \geqslant 0$$
$$y \geqslant 0$$

◆

If we compare the solution regions of Example 3 and Example 4, we see that there is a fundamental difference between these two regions. We can draw a circle around the solution region in Example 4; however, it is impossible to include all the points in the solution region in Example 3 in any circle, no matter how large we draw it. This leads to the following definition:

Bounded and Unbounded Solution Regions

A solution region of a system of linear inequalities is **bounded** if it can be enclosed within a circle. If it cannot be enclosed within a circle, then it is **unbounded.**

Thus, the solution region for Example 4 is bounded and the solution region for Example 3 is unbounded. This definition will be important in the next section.

APPLICATIONS

EXAMPLE 5 *Medicine*

A patient in a hospital is required to have at least 84 units of drug A and 120 units of drug B each day (assume that an overdosage of either drug is harmless). Each gram of substance M contains 10 units of drug A and 8 units of drug B, and each gram of substance N contains 2 units of drug A and 4 units of drug B. How many grams of substances M and N can be mixed to meet the minimum daily requirements?

Solution To clarify relationships, we summarize the information in the following table:

	AMOUNT OF DRUG PER GRAM *Substance M*	*Substance N*	MINIMUM DAILY REQUIREMENT
DRUG A	10 units	2 units	84 units
DRUG B	8 units	4 units	120 units

Let

x = Number of grams of substance M used

y = Number of grams of substance N used

Then

$10x$ = Number of units of drug A in x grams of substance M

$2y$ = Number of units of drug A in y grams of substance N

$8x$ = Number of units of drug B in x grams of substance M

$4y$ = Number of units of drug B in y grams of substance N

The following conditions must be satisfied to meet daily requirements:

$$\begin{pmatrix}\text{Number of units of}\\ \text{drug } A\\ \text{in } x \text{ grams of substance } M\end{pmatrix} + \begin{pmatrix}\text{Number of units of}\\ \text{drug } A\\ \text{in } y \text{ grams of substance } N\end{pmatrix} \geqslant 84$$

$$\begin{pmatrix}\text{Number of units of}\\ \text{drug } B\\ \text{in } x \text{ grams of substance } M\end{pmatrix} + \begin{pmatrix}\text{Number of units of}\\ \text{drug } B\\ \text{in } y \text{ grams of substance } N\end{pmatrix} \geqslant 120$$

$$(\text{Number of grams of substance } M \text{ used}) \geqslant 0$$

$$(\text{Number of grams of substance } N \text{ used}) \geqslant 0$$

Converting these verbal statements into symbolic statements by using the variables x and y introduced above, we obtain the following system of linear inequalities:

$10x + 2y \geq 84$	Drug *A* restriction
$8x + 4y \geq 120$	Drug *B* restriction
$x \geq 0$	Cannot use a negative amount of *M*
$y \geq 0$	Cannot use a negative amount of *N*

Graphing this system of linear inequalities, we obtain the set of feasible solutions, or the feasible region (solution region), as shown in the figure. Thus, any point in the shaded area (including the straight-line boundaries) will meet the daily requirements; any point outside the shaded area will not. For example, 4 units of drug *M* and 23 units of drug *N* will meet the daily requirements, but 4 units of drug *M* and 21 units of drug *N* will not. (Note that the feasible region is unbounded.)

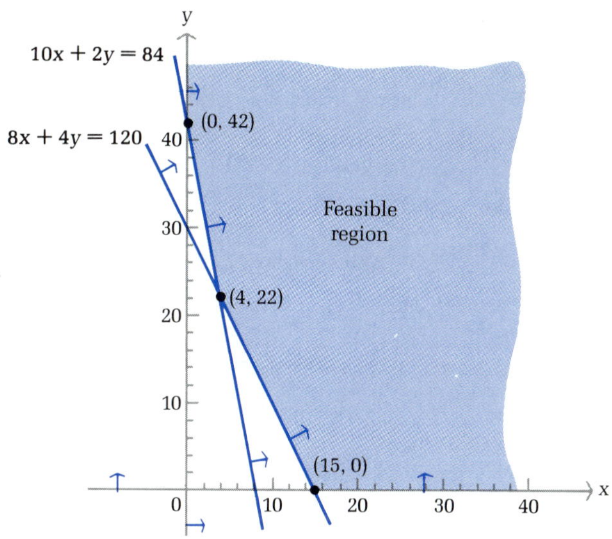

◆

PROBLEM 5

Resource Allocation

A manufacturing plant makes two types of inflatable boats, a two-person boat and a four-person boat. Each two-person boat requires 0.9 labor-hour in the cutting department and 0.8 labor-hour in the assembly department. Each four-person boat requires 1.8 labor-hours in the cutting department and 1.2 labor-hours in the assembly department. The maximum labor-hours available each month in the cutting and assembly departments are 864 and 672, respectively.

(A) Summarize this information in a table.

(B) If x two-person boats and y four-person boats are manufactured each month, write a system of linear inequalities that reflect the conditions indicated. Find the set of feasible solutions graphically. ◆

Answers to Matched Problems

1. Graph $6x - 3y = 18$ as a dashed line (since equality is not included). Choosing the origin (0, 0) as a test point, we see that $6(0) - 3(0) > 18$ is a false statement; thus, the lower half-plane determined by $6x - 3y = 18$ is the graph of $6x - 3y > 18$.

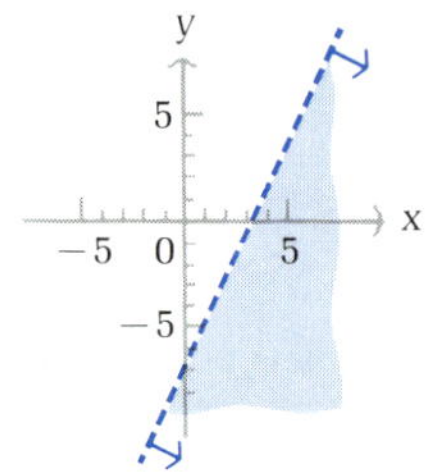

2. 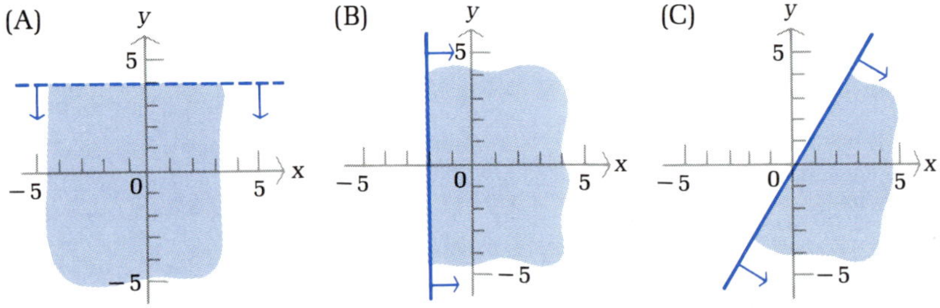

3. 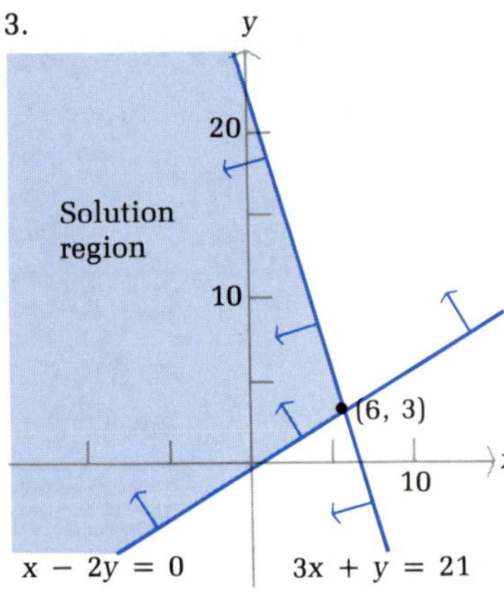

4. 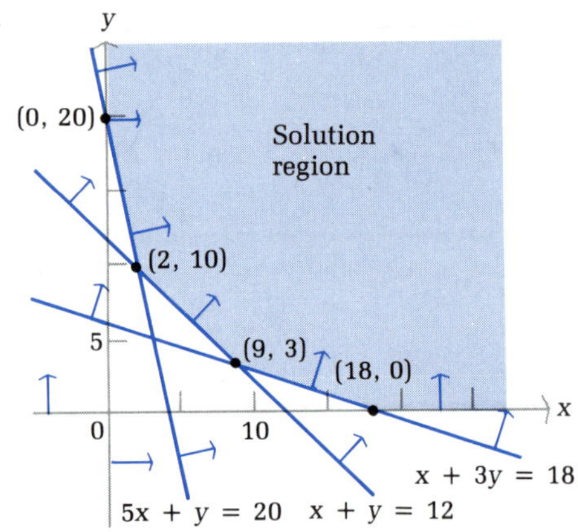

5. (A)

	LABOR-HOURS REQUIRED Two-Person Boat	Four-Person Boat	MAXIMUM LABOR-HOURS AVAILABLE PER MONTH
CUTTING DEPARTMENT	0.9	1.8	864
ASSEMBLY DEPARTMENT	0.8	1.2	672

(B) $0.9x + 1.8y \leq 864$

$0.8x + 1.2y \leq 672$

$x \geq 0$

$y \geq 0$

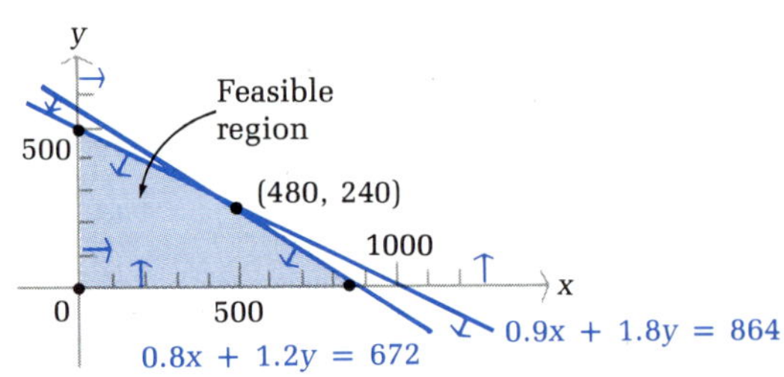

EXERCISE 6-1

A *Graph each inequality.*

1. $y \leq x - 1$
2. $y > x + 1$
3. $3x - 2y > 6$
4. $2x - 5y \leq 10$
5. $x \geq -4$
6. $y < 5$
7. $-4 \leq y < 4$
8. $0 \leq x < 6$
9. $6x + 4y \geq 24$
10. $4x + 8y \geq 32$
11. $5x \leq -2y$
12. $6x \geq 4y$

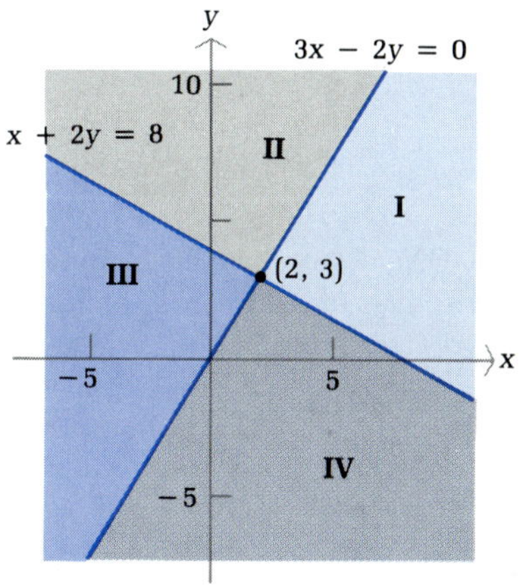

In Problems 13–16, match the solution region of each system of linear inequalities with one of the four regions shown in the figure in the margin.

13. $x + 2y \leq 8$, $3x - 2y \geq 0$
14. $x + 2y \geq 8$, $3x - 2y \leq 0$
15. $x + 2y \geq 8$, $3x - 2y \geq 0$
16. $x + 2y \leq 8$, $3x - 2y \leq 0$

Solve each system of linear inequalities graphically.

17. $3x + y \geq 6$, $x \leq 4$
18. $3x + 4y \leq 12$, $y \geq -3$
19. $x - 2y \leq 12$, $2x + y \geq 4$
20. $2x + 5y \leq 20$, $x - 5y \geq -5$

B *In Problems 21–24, match the solution region of each system of linear inequalities with one of the four regions shown in the figure in the margin. Identify the corner points of each solution region.*

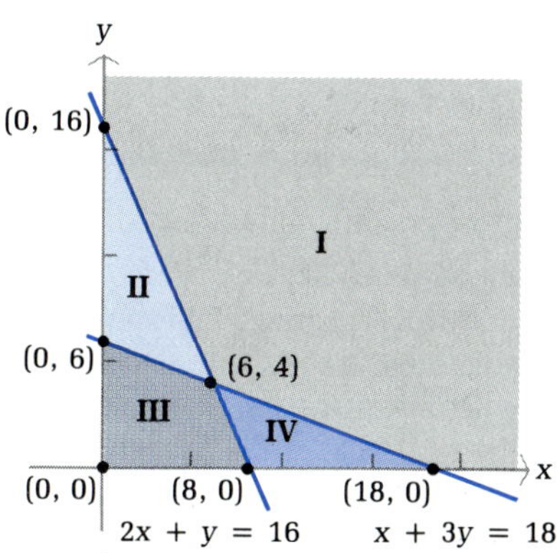

21. $x + 3y \leq 18$, $2x + y \geq 16$, $x \geq 0$, $y \geq 0$
22. $x + 3y \leq 18$, $2x + y \leq 16$, $x \geq 0$, $y \geq 0$
23. $x + 3y \geq 18$, $2x + y \geq 16$, $x \geq 0$, $y \geq 0$
24. $x + 3y \geq 18$, $2x + y \leq 16$, $x \geq 0$, $y \geq 0$

Solve the following systems graphically, and indicate whether each solution region is bounded or unbounded. Find the coordinates of each corner point.

25. $2x + 3y \leq 12$, $x \geq 0$, $y \geq 0$
26. $3x + 4y \leq 24$, $x \geq 0$, $y \geq 0$
27. $2x + y \leq 10$, $x + 2y \leq 8$, $x \geq 0$, $y \geq 0$
28. $6x + 3y \leq 24$, $3x + 6y \leq 30$, $x \geq 0$, $y \geq 0$
29. $2x + y \geq 10$, $x + 2y \geq 8$, $x \geq 0$, $y \geq 0$
30. $4x + 3y \geq 24$, $3x + 4y \geq 8$, $x \geq 0$, $y \geq 0$

31. $2x + y \le 10$
$x + y \le 7$
$x + 2y \le 12$
$x \ge 0$
$y \ge 0$

32. $3x + y \le 21$
$x + y \le 9$
$x + 3y \le 21$
$x \ge 0$
$y \ge 0$

33. $2x + y \ge 16$
$x + y \ge 12$
$x + 2y \ge 14$
$x \ge 0$
$y \ge 0$

34. $3x + y \ge 24$
$x + y \ge 16$
$x + 3y \ge 30$
$x \ge 0$
$y \ge 0$

C *Solve the following systems graphically, and indicate whether each solution region is bounded or unbounded. Find the coordinates of each corner point.*

35. $x + 4y \le 32$
$3x + y \le 30$
$4x + 5y \ge 51$

36. $x + y \le 11$
$x + 5y \ge 15$
$2x + y \ge 12$

37. $4x + 3y \le 48$
$2x + y \ge 24$
$x \le 9$

38. $2x + 3y \ge 24$
$x + 3y \le 15$
$y \ge 4$

39. $x - y \le 0$
$2x - y \le 4$
$0 \le x \le 8$

40. $2x + 3y \ge 12$
$-x + 3y \le 3$
$0 \le y \le 5$

41. $-x + 3y \ge 1$
$5x - y \ge 9$
$x + y \le 9$
$x \le 5$

42. $x + y \le 10$
$5x + 3y \ge 15$
$-2x + 3y \le 15$
$2x - 5y \le 6$

APPLICATIONS

Business & Economics

43. *Manufacturing—resource allocation.* A manufacturing company makes two types of water skis, a trick ski and a slalom ski. The trick ski requires 6 labor-hours for fabricating and 1 labor-hour for finishing. The slalom ski requires 4 labor-hours for fabricating and 1 labor-hour for finishing. The maximum labor-hours available per day for fabricating and finishing are 108 and 24, respectively. If x is the number of trick skis and y is the number of slalom skis produced per day, write a system of linear inequalities that indicates appropriate restraints on x and y. Find the set of feasible solutions graphically for the number of each type of ski that can be produced.

44. *Manufacturing—resource allocation.* A furniture manufacturing company manufactures dining room tables and chairs. A table requires 8 labor-hours for assembling and 2 labor-hours for finishing. A chair requires 2 labor-hours for assembling and 1 labor-hour for finishing. The maximum labor-hours available per day for assembly and finishing are 400 and 120, respectively. If x is the number of tables and y is the number of chairs produced per day, write a system of linear inequalities that indicates appropriate restraints on x and y. Find the set of feasible solutions graphically for the number of tables and chairs that can be produced.

Life Sciences

45. *Nutrition—plants.* A farmer can buy two types of plant food, mix *A* and mix *B*. Each cubic yard of mix *A* contains 20 pounds of phosphoric acid, 30 pounds of nitrogen, and 5 pounds of potash. Each cubic yard of mix *B* contains 10 pounds of phosphoric acid, 30 pounds of nitrogen, and 10 pounds of potash. The minimum monthly requirements are 460 pounds of phosphoric acid, 960 pounds of nitrogen, and 220 pounds of potash. If x is the number of cubic yards of mix *A* used and y is the number of cubic yards of mix *B* used, write a system of linear inequalities that indicates appropriate restraints on x and y. Find the set of feasible solutions graphically for the amounts of mix *A* and mix *B* that can be used.

46. *Nutrition—people.* A dietitian in a hospital is to arrange a special diet using two foods. Each ounce of food *M* contains 30 units of calcium, 10 units of iron, and 10 units of vitamin A. Each ounce of food *N* contains 10 units of calcium, 10 units of iron, and 30 units of vitamin A. The minimum requirements in the diet are 360 units of calcium, 160 units of iron, and 240 units of vitamin A. If x is the number of ounces of food *M* used and y is the number of ounces of food *N* used, write a system of linear inequalities that reflects the conditions indicated. Find the set of feasible solutions graphically for the amount of each kind of food that can be used.

Social Sciences

47. *Psychology.* In an experiment on conditioning, a psychologist uses two types of Skinner (conditioning) boxes with mice and rats. Each mouse spends 10 minutes per day in box *A* and 20 minutes per day in box *B*. Each rat spends 20 minutes per day in box *A* and 10 minutes per day in box *B*. The total maximum time available per day is 800 minutes for box *A* and 640 minutes for box *B*. We are interested in the various numbers of mice and rats that can be used in the experiment under the conditions stated. If we let x be the number of mice used and y the number of rats used, write a system of linear inequalities that indicates appropriate restrictions on x and y. Find the set of feasible solutions graphically.

SECTION 6-2 Linear Programming in Two Dimensions—A Geometric Approach

- A LINEAR PROGRAMMING PROBLEM
- LINEAR PROGRAMMING—A GENERAL DESCRIPTION
- GEOMETRIC SOLUTION OF LINEAR PROGRAMMING PROBLEMS
- APPLICATIONS

Several problems discussed in the last section are related to a more general type of problem called a *linear programming problem.* Linear programming is a mathematical process that has been developed to help management in decision-making, and it has become one of the most widely used and best-known tools of management science. We will introduce this topic by considering an example in detail, using an intuitive geometric approach. Insight gained from this approach will prove invaluable when we later consider an algebraic approach that is less intuitive but necessary in solving most real-world problems.

Notation Change

For efficiency of generalization in later sections, we will now change variable notation from letters such as x and y to subscript forms such as x_1 and x_2.

◆ A LINEAR PROGRAMMING PROBLEM

We begin our discussion with a concrete example. The geometric method of solution will suggest two important theorems and a simple general geometric procedure for solving linear programming problems in two variables.

◆ EXAMPLE 6

Production Scheduling

A manufacturer of lightweight mountain tents makes a standard model and an expedition model for national distribution. Each standard tent requires 1 labor-hour from the cutting department and 3 labor-hours from the assembly department. Each expedition tent requires 2 labor-hours from the cutting department and 4 labor-hours from the assembly department. The maximum labor-hours available per day in the cutting department and the assembly department are 32 and 84, respectively. If the company makes a profit of $50 on each standard tent and $80 on each expedition tent, how many tents of each type should be manufactured each day to maximize the total daily profit (assuming all tents can be sold)?

Solution This is an example of a linear programming problem. To see relationships more clearly, we summarize the manufacturing requirements, objectives, and restrictions in Table 1.

TABLE 1

	LABOR-HOURS PER TENT		MAXIMUM
	Standard Model	*Expedition Model*	LABOR-HOURS AVAILABLE PER DAY
CUTTING DEPARTMENT	1	2	32
ASSEMBLY DEPARTMENT	3	4	84
PROFIT PER TENT	$50	$80	

We now proceed to formulate a mathematical model for the problem and then to solve it by using geometric methods.

Objective Function

The *objective* of management is to *decide* how many of each tent model should be produced each day so as to maximize profit. Let

x_1 = Number of standard tents produced per day
x_2 = Number of expedition tents produced per day
} Decision variables

The following equation gives the total profit for x_1 standard tents and x_2 expedition tents manufactured each day, assuming all tents manufactured are sold:

$$\boxed{P = 50x_1 + 80x_2} \qquad \text{Objective function}$$

Mathematically, the manufacturer needs to decide on values for the **decision variables** (x_1, x_2) that achieve its objective—that is, maximize the **objective function** (profit) $P = 50x_1 + 80x_2$. (Note that P is a function of two independent variables.) It appears that the profit can be made as large as we like by manufacturing more and more tents—or can it?

Constraints

Any manufacturing company, no matter how large or small, has manufacturing limits imposed by available resources, plant capacity, demand, and so forth. These limits are referred to as **problem constraints.**

Cutting Department Constraint

$$\left(\begin{array}{c}\text{Daily cutting}\\ \text{time for } x_1\\ \text{standard tents}\end{array}\right) + \left(\begin{array}{c}\text{Daily cutting}\\ \text{time for } x_2\\ \text{expedition tents}\end{array}\right) \le \left(\begin{array}{c}\text{Maximum labor-}\\ \text{hours available}\\ \text{per day}\end{array}\right)$$

$$1x_1 \quad + \quad 2x_2 \quad \le \quad 32$$

Assembly Department Constraint

$$\left(\begin{array}{c}\text{Daily assembly}\\ \text{time for } x_1\\ \text{standard tents}\end{array}\right) + \left(\begin{array}{c}\text{Daily assembly}\\ \text{time for } x_2\\ \text{expedition tents}\end{array}\right) \le \left(\begin{array}{c}\text{Maximum labor-}\\ \text{hours available}\\ \text{per day}\end{array}\right)$$

$$3x_1 \quad + \quad 4x_2 \quad \le \quad 84$$

Nonnegative Constraints

It is not possible to manufacture a negative number of tents; thus, we have the **nonnegative constraints**

$$x_1 \ge 0$$
$$x_2 \ge 0$$

which we usually write in the form

$$x_1, x_2 \ge 0$$

Mathematical Model

We now have a **mathematical model** for the problem under consideration:

$$\begin{aligned} &\text{Maximize} \quad P = 50x_1 + 80x_2 && \text{Objective function} \\ &\text{Subject to} \quad \left.\begin{aligned} x_1 + 2x_2 &\le 32 \\ 3x_1 + 4x_2 &\le 84 \end{aligned}\right\} && \text{Problem constraints} \\ &\qquad\qquad\quad x_1, x_2 \ge 0 && \text{Nonnegative constraints} \end{aligned}$$

Graphical Solution

Solving the set of linear inequality constraints **graphically** (see the last section), we obtain the feasible region for production schedules (Fig. 4).

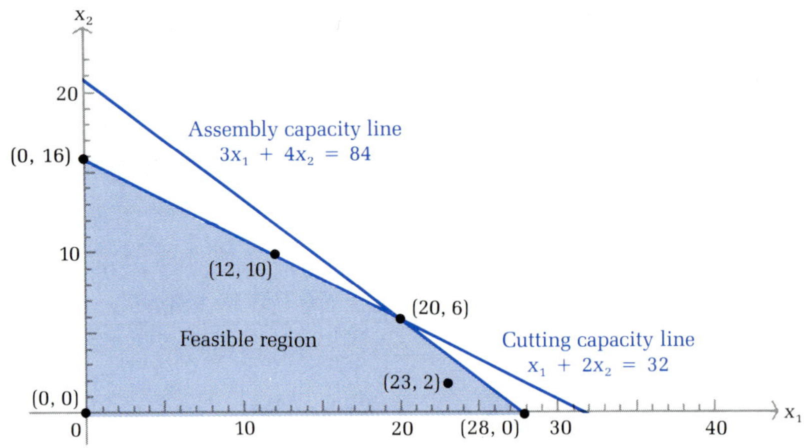

FIGURE 4

By choosing a production schedule (x_1, x_2) from the feasible region, a profit can be determined using the objective function

$$P = 50x_1 + 80x_2$$

For example, if $x_1 = 12$ and $x_2 = 10$, then the profit for the day would be

$$\begin{aligned} P &= 50(12) + 80(10) \\ &= \$1{,}400 \end{aligned}$$

Or if $x_1 = 23$ and $x_2 = 2$, then the profit for the day would be

$$\begin{aligned} P &= 50(23) + 80(2) \\ &= \$1{,}310 \end{aligned}$$

The question is, out of all possible production schedules (x_1, x_2) from the feasible region, which schedule(s) produces the maximum profit? Thus, we have a **maximization problem.** Since point by point checking is impossible (there are infinitely many points to check), we must find another way.

By assigning P in $P = 50x_1 + 80x_2$ a particular value and plotting the resulting equation in Figure 4, we obtain a **constant-profit line (isoprofit line).** Every point in the feasible region on this line represents a production schedule that will produce the same profit. By doing this for a number of values for P, we obtain a family of constant-profit lines (Fig. 5) that are parallel to each other, since they all have the same slope. To see this, we write $P = 50x_1 + 80x_2$ in the slope–intercept form

$$x_2 = -\frac{5}{8}x_1 + \frac{P}{80}$$

and note that for any profit P, the constant-profit line has slope $-\frac{5}{8}$. We also observe that as the profit P increases, the x_2 intercept $(P/80)$ increases, and the line moves away from the origin.

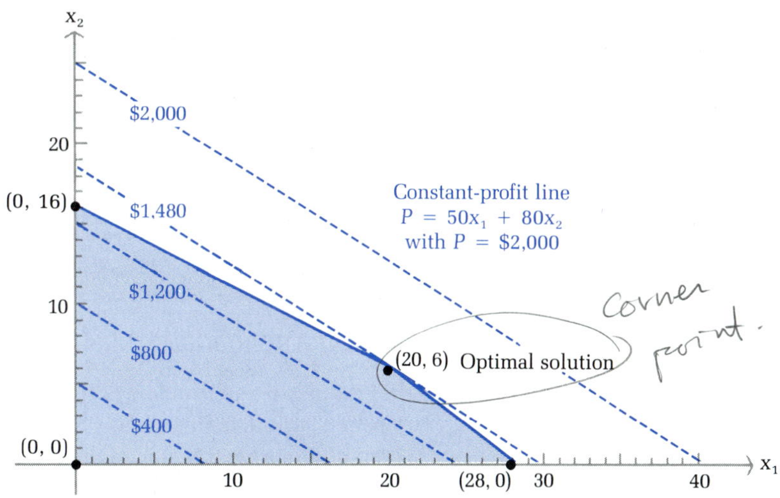

FIGURE 5
Constant-profit lines

Thus, the maximum profit occurs at a point where a constant-profit line is the farthest from the origin but still in contact with the feasible region. In this example, this occurs at (20, 6), as is seen in Figure 5. Thus, if the manufacturer makes 20 standard tents and 6 expedition tents per day, the profit will be maximized at

$$P = 50(20) + 80(6) \\ = \$1{,}480$$

The point (20, 6) is called an **optimal solution** to the problem, because it maximizes the objective (profit) function and is in the feasible region. In general, it appears that a maximum profit occurs at one of the corner points. We also note that the minimum profit ($P = 0$) occurs at the corner point (0, 0). ◆

PROBLEM 6 A manufacturing plant makes two types of inflatable boats, a two-person boat and a four-person boat. Each two-person boat requires 0.9 labor-hour from the cutting department and 0.8 labor-hour from the assembly department. Each four-person boat requires 1.8 labor-hours from the cutting department and 1.2 labor-hours from the assembly department. The maximum labor-hours available per month in the cutting department and the assembly department are 864 and 672, respectively. The company makes a profit of \$25 on each two-person boat and \$40 on each four-person boat.

(A) Summarize the relevant material in a table similar to Table 1 in Example 6.
(B) Identify the decision variables.
(C) Write the objective function P.
(D) Write the problem constraints and the nonnegative constraints.
(E) Graph the feasible region. Include graphs of the objective function for $P =$ \$5,000, $P =$ \$10,000, $P =$ \$15,000, and $P =$ \$21,600.
(F) From the graph and constant-profit lines, determine how many boats should be manufactured each month to maximize the profit. What is the maximum profit? ◆

◆ LINEAR PROGRAMMING—A GENERAL DESCRIPTION

In Example 6 and Problem 6, the optimal solution occurs at a corner point of the feasible region. Is this always the case? The answer is a qualified yes, as will be seen in Theorem 2, stated at the top of the next page. First, we give a few general definitions.

A **linear programming problem** is one that is concerned with finding the **optimal value** (maximum or minimum value) of a linear **objective function** of the form

$$z = c_1x_1 + c_2x_2 + \cdots + c_nx_n$$

where the **decision variables** $x_1, x_2, \ldots, x_n$ are subject to **problem constraints** in the form of linear inequalities and equations. In addition, the decision variables must satisfy the **nonnegative constraints** $x_i \geq 0, i = 1, 2, \ldots, n$. The set of points satisfying both the problem constraints and the nonnegative constraints is called the **feasible region** for the problem. Any point in the feasible region that produces the optimal value of the objective function over the feasible region is called an **optimal solution.**

THEOREM 2 — Fundamental Theorem of Linear Programming

If the optimal value of the objective function in a linear programming problem exists, then that value must occur at one (or more) of the corner points of the feasible region.

Theorem 2 provides a simple procedure for solving a linear programming problem, *provided the problem has a solution—not all do.* In order to use Theorem 2, we must know that the problem under consideration has a solution. Theorem 3 provides some conditions that will ensure that a linear programming problem has a solution.

THEOREM 3 — Existence of Solutions

(A) If the feasible region for a linear programming problem is bounded, then both the maximum value and the minimum value of the objective function always exist.

(B) If the feasible region is unbounded and the coefficients of the objective function are positive, then the minimum value of the objective function exists, but the maximum value does not.

(C) If the feasible region is empty (that is, there are no points that satisfy all the constraints), then both the maximum value and the minimum value of the objective function do not exist.

Theorem 3 does not cover all possibilities. For example, what happens if the feasible region is unbounded and one (or both) of the coefficients of the objective function are negative? Problems of this type require special techniques that we will not discuss. Since virtually all applied problems satisfy one of the conditions listed in Theorem 3, we will consider only problems of this type.

◆ GEOMETRIC SOLUTION OF LINEAR PROGRAMMING PROBLEMS

The discussion above leads to the following procedure for the geometric solution of linear programming problems with two decision variables:

Geometric Solution of a Linear Programming Problem with Two Decision Variables

Step 1. For an applied problem, summarize relevant material in table form (see Table 1 in Example 6).

Step 2. Form a mathematical model for the problem:

(A) Introduce decision variables, and write a linear objective function.
(B) Write problem constraints using linear inequalities and/or equations.
(C) Write nonnegative constraints.

Step 3. Graph the feasible region. Then, if according to Theorem 3 an optimal solution exists, find the coordinates of each corner point.

Step 4. Make a table listing the value of the objective function at each corner point.

Step 5. Determine the optimal solution(s) from the table in step 4.

Step 6. For an applied problem, interpret the optimal solution(s) in terms of the original problem.

Before we consider additional applications, let us use this procedure to solve some linear programming problems where the model has already been determined.

◆ EXAMPLE 7

(A) Minimize and maximize

$$z = 3x_1 + x_2$$

Subject to

$$\begin{aligned} 2x_1 + x_2 &\le 20 \\ 10x_1 + x_2 &\ge 36 \\ 2x_1 + 5x_2 &\ge 36 \\ x_1, x_2 &\ge 0 \end{aligned}$$

(B) Minimize and maximize

$$z = 10x_1 + 20x_2$$

Subject to

$$\begin{aligned} 6x_1 + 2x_2 &\ge 36 \\ 2x_1 + 4x_2 &\ge 32 \\ x_2 &\le 20 \\ x_1, x_2 &\ge 0 \end{aligned}$$

Solutions (A) We begin with step 3, since step 1 does not apply and step 2 has already been done for us.

Step 3. Graph the feasible region S. Then, after checking Theorem 3 to determine that an optimal solution exists, find the coordinates of each corner point. Since S is bounded, z will have both a maximum and a minimum on S (Theorem 3A) and these will both occur at corner points (Theorem 2).

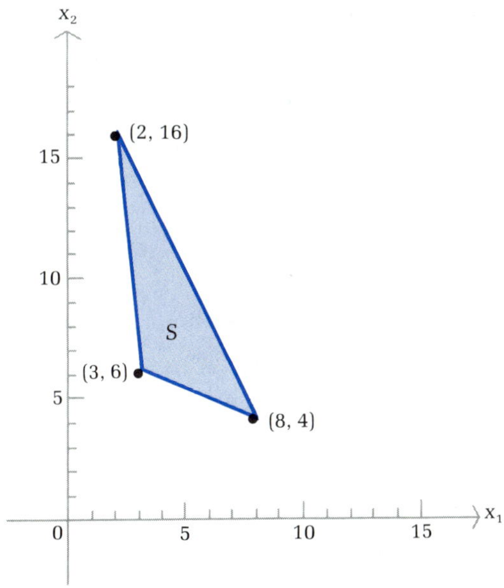

CORNER POINT (x_1, x_2)	$z = 3x_1 + x_2$
(3, 6)	15
(2, 16)	22
(8, 4)	28

Step 4. Evaluate the objective function at each corner point, as shown in the table in the margin.

Step 5. Determine the optimal solutions from step 4. Examining the values in the table, we see that the minimum value of z is 15 at (3, 6) and the maximum value of z is 28 at (8, 4).

(B) Again, we can begin with step 3.

Step 3. Graph the feasible region S, as shown on the next page. Then, after checking Theorem 3 to determine that an optimal solution exists, find the coordinates of each corner point. Since S is unbounded and the coefficients of the objective function are positive, z has a minimum value on S but no maximum value (Theorem 3B).

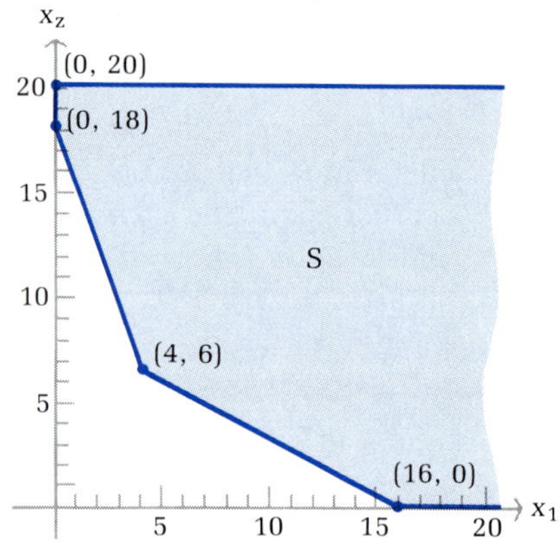

CORNER POINT (x_1, x_2)	$z = 10x_1 + 20x_2$
(0, 20)	400
(0, 18)	360
(4, 6)	160
(16, 0)	160

Step 4. Evaluate the objective function at each corner point, as shown in the table in the margin.

Step 5. Determine the optimal solution from step 4. The minimum value of z is 160 at (4, 6) and at (16, 0). ◆

This is a **multiple optimal solution.**

In general, if two corner points are both optimal solutions to a linear programming problem, then any point on the line segment joining them is also an optimal solution.

This is the only time that optimal solutions also occur at noncorner points.

PROBLEM 7

(A) Maximize and minimize $z = 4x_1 + 2x_2$ subject to the constraints given in Example 7A.

(B) Maximize and minimize $z = 20x_1 + 5x_2$ subject to the constraints given in Example 7B. ◆

For an illustration of Theorem 3C, consider the following:

$$\begin{aligned} \text{Maximize} \quad & P = 2x_1 + 3x_2 \\ \text{Subject to} \quad & x + x_2 \geq 8 \\ & x_1 + 2x_2 \leq 8 \\ & 2x_1 + x_2 \leq 10 \\ & x_1, x_2 \geq 0 \end{aligned}$$

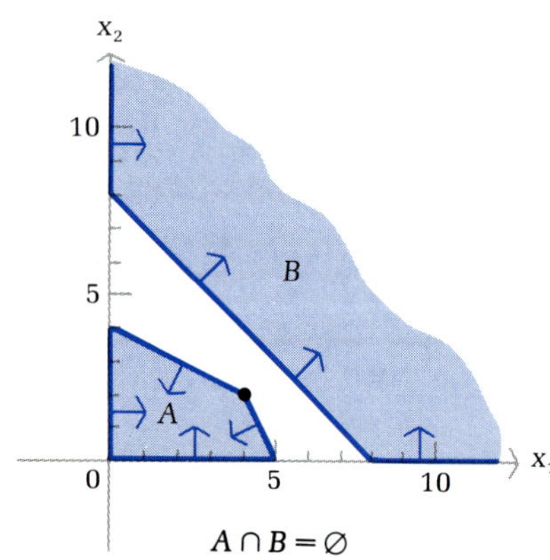

The intersection of the graphs of the constraint inequalities is the empty set; hence, the feasible region is empty. If this happens, then the problem should be reexamined to see if it has been formulated properly. If it has, then the management may have to reconsider items such as labor-hours, overtime, budget, and supplies allocated to the project in order to obtain a nonempty feasible region and a solution to the original problem.

◆ APPLICATIONS

◆ EXAMPLE 8 *Medicine*

We now convert Example 5 (in the preceding section) into a linear programming problem. A patient in a hospital is required to have at least 84 units of drug D_1 and 120 units of drug D_2 each day (assume that an overdosage of either drug is harmless). Each gram of substance M contains 10 units of drug D_1 and 8 units of drug D_2, and each gram of substance N contains 2 units of drug D_1 and 4 units of drug D_2. Now, suppose both M and N contain an undesirable drug D_3, 3 units per gram in M and 1 unit per gram in N. How many grams of each of substances M and N should be mixed to meet the minimum daily requirements and at the same time minimize the intake of drug D_3? How many units of the undesirable drug D_3 will be in this mixture?

Solution Step 1. Summarize relevant material in table form:

	AMOUNT OF DRUG PER GRAM		MINIMUM DAILY
	Substance M	*Substance N*	REQUIREMENT
DRUG D_1	10 units	2 units	84 units
DRUG D_2	8 units	4 units	120 units
DRUG D_3	3 units	1 unit	

Step 2. Form a mathematical model for the problem. Let

x_1 = Number of grams of substance M used
x_2 = Number of grams of substance N used
} Decision variables

We form the linear objective function

$$C = 3x_1 + x_2$$

which gives the amount of the undesirable drug D_3 in x_1 grams of M and x_2 grams of N. Proceeding as in Example 5, we formulate the following mathematical model for the problem:

Minimize	$C = 3x_1 + x_2$	Objective function
Subject to	$10x_1 + 2x_2 \geq 84$	Drug D_1 constraint
	$8x_1 + 4x_2 \geq 120$	Drug D_2 constraint
	$x_1, x_2 \geq 0$	Nonnegative constraints

Step 3. Graph the feasible region. Then, after checking Theorem 3 to determine that an optimal solution exists, find the coordinates of each corner point. Solving the system of constraint inequalities graphically, we obtain the feasible region shown in Figure 6. Since the feasible region is unbounded and the coefficients of the objective function are positive, this minimization problem has a solution.

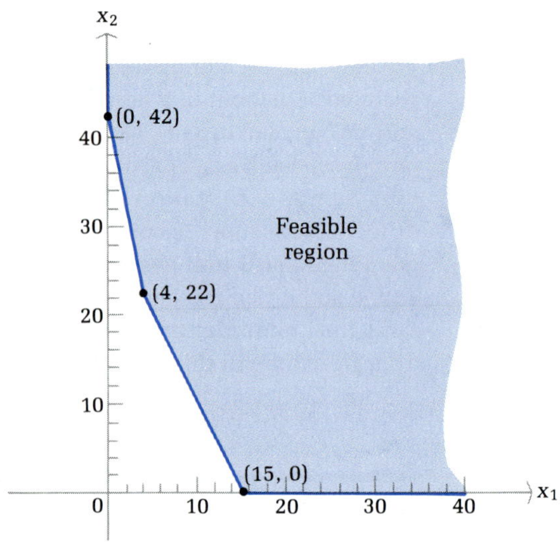

FIGURE 6

Step 4. Evaluate the objective function at each corner point, as shown in the table.

CORNER POINT (x_1, x_2)	$C = 3x_1 + x_2$
(0, 42)	42
(4, 22)	34
(15, 0)	45

Step 5. Determine the optimal solution from step 4. The optimal solution is $C = 34$ at the corner point (4, 22).

Step 6. Interpret the optimal solution in terms of the original problem. If we use 4 grams of substance M and 22 grams of substance N, we will supply the minimum daily requirements for drugs D_1 and D_2 and minimize the intake of the undesirable drug D_3 at 34 units. (Any other combination of M and N from the feasible region will result in a larger amount of the undesirable drug D_3.) ◆

PROBLEM 8

Agriculture

A chicken farmer can buy a special food mix A at 20¢ per pound and a special food mix B at 40¢ per pound. Each pound of mix A contains 3,000 units of nutrient N_1 and 1,000 units of nutrient N_2; each pound of mix B contains 4,000 units of nutrient N_1 and 4,000 units of nutrient N_2. If the minimum daily requirements for the chickens collectively are 36,000 units of nutrient N_1 and 20,000 units of nutrient N_2, how many pounds of each food mix should be used each day to minimize daily food costs while meeting (or exceeding) the minimum daily nutrient requirements? What is the minimum daily cost? ◆

Answers to Matched Problems

6. (A)

	LABOR-HOURS REQUIRED Two-Person Boat	Four-Person Boat	MAXIMUM LABOR-HOURS AVAILABLE PER MONTH
CUTTING DEPARTMENT	0.9	1.8	864
ASSEMBLY DEPARTMENT	0.8	1.2	672
PROFIT PER BOAT	$25	$40	

(B) x_1 = Number of two-person boats produced each month
x_2 = Number of four-person boats produced each month

(C) $P = 25x_1 + 40x_2$ (D) $0.9x_1 + 1.8x_2 \leq 864$
$0.8x_1 + 1.2x_2 \leq 672$
$x_1, x_2 \geq 0$

(E)

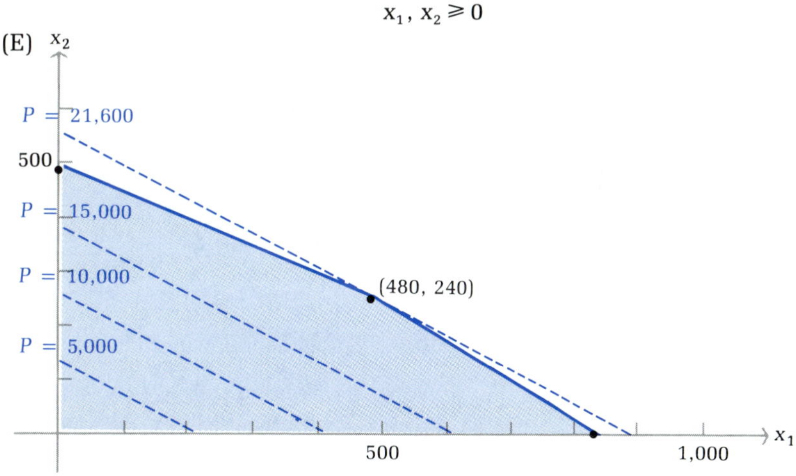

(F) 480 two-person boats, 240 four-person boats; Max $P = \$21,600$ per month

7. (A) Min $z = 24$ at (3, 6); Max $z = 40$ at (2, 16) and (8, 4) (multiple optimal solution)
(B) Min $z = 90$ at (0, 18); no maximum value

8. 8 lb of mix A, 3 lb of mix B; Min $C = \$2.80$ per day

EXERCISE 6-2

A *Find the maximum value of each objective function over the feasible region S shown in the figure.*

1. $z = x + y$

2. $z = 4x + y$

3. $z = 3x + 7y$

4. $z = 9x + 3y$

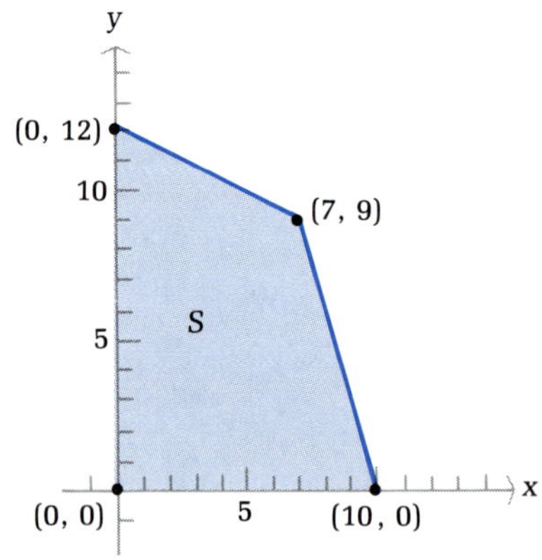

Find the minimum value of each objective function over the feasible region T shown in the figure.

5. $z = 7x + 4y$

6. $z = 7x + 9y$

7. $z = 3x + 8y$

8. $z = 5x + 4y$

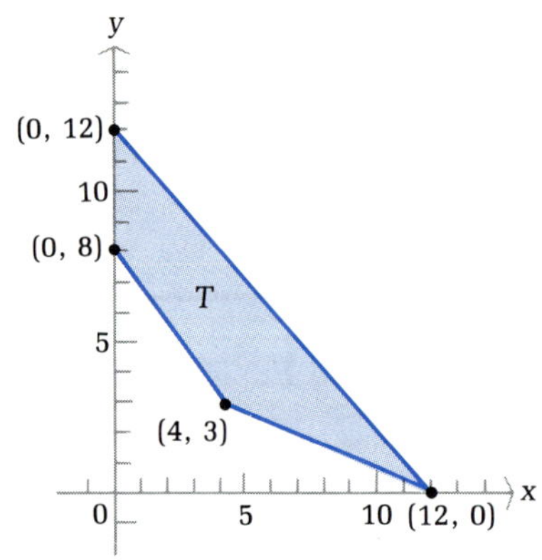

B *Solve the following linear programming problems:*

9. Maximize $P = 5x_1 + 5x_2$
Subject to $2x_1 + x_2 \leq 10$
$x_1 + 2x_2 \leq 8$
$x_1, x_2 \geq 0$

10. Maximize $P = 3x_1 + 2x_2$
Subject to $6x_1 + 3x_2 \leq 24$
$3x_1 + 6x_2 \leq 30$
$x_1, x_2 \geq 0$

11. Minimize and maximize

$$z = 2x_1 + 3x_2$$

Subject to $2x_1 + x_2 \geq 10$
$x_1 + 2x_2 \geq 8$
$x_1, x_2 \geq 0$

12. Minimize and maximize

$$z = 8x_1 + 7x_2$$

Subject to $4x_1 + 3x_2 \geq 24$
$3x_1 + 4x_2 \geq 8$
$x_1, x_2 \geq 0$

13. Maximize $P = 30x_1 + 40x_2$
Subject to
$$\begin{aligned} 2x_1 + x_2 &\leqslant 10 \\ x_1 + x_2 &\leqslant 7 \\ x_1 + 2x_2 &\leqslant 12 \\ x_1, x_2 &\geqslant 0 \end{aligned}$$

14. Maximize $P = 20x_1 + 10x_2$
Subject to
$$\begin{aligned} 3x_1 + x_2 &\leqslant 21 \\ x_1 + x_2 &\leqslant 9 \\ x_1 + 3x_2 &\leqslant 21 \\ x_1, x_2 &\geqslant 0 \end{aligned}$$

15. Minimize and maximize

$$z = 10x_1 + 30x_2$$

Subject to
$$\begin{aligned} 2x_1 + x_2 &\geqslant 16 \\ x_1 + x_2 &\geqslant 12 \\ x_1 + 2x_2 &\geqslant 14 \\ x_1, x_2 &\geqslant 0 \end{aligned}$$

16. Minimize and maximize

$$z = 400x_1 + 100x_2$$

Subject to
$$\begin{aligned} 3x_1 + x_2 &\geqslant 24 \\ x_1 + x_2 &\geqslant 16 \\ x_1 + 3x_2 &\geqslant 30 \\ x_1, x_2 &\geqslant 0 \end{aligned}$$

17. Minimize and maximize

$$P = 30x_1 + 10x_2$$

Subject to
$$\begin{aligned} 2x_1 + 2x_2 &\geqslant 4 \\ 6x_1 + 4x_2 &\leqslant 36 \\ 2x_1 + x_2 &\leqslant 10 \\ x_1, x_2 &\geqslant 0 \end{aligned}$$

18. Minimize and maximize

$$P = 2x_1 + x_2$$

Subject to
$$\begin{aligned} x_1 + x_2 &\geqslant 2 \\ 6x_1 + 4x_2 &\leqslant 36 \\ 4x_1 + 2x_2 &\leqslant 20 \\ x_1, x_2 &\geqslant 0 \end{aligned}$$

19. Minimize and maximize

$$P = 3x_1 + 5x_2$$

Subject to
$$\begin{aligned} x_1 + 2x_2 &\leqslant 6 \\ x_1 + x_2 &\leqslant 4 \\ 2x_1 + 3x_2 &\geqslant 12 \\ x_1, x_2 &\geqslant 0 \end{aligned}$$

20. Minimize and maximize

$$P = -x_1 + 3x_2$$

Subject to
$$\begin{aligned} 2x_1 - x_2 &\geqslant 4 \\ -x_1 + 2x_2 &\leqslant 4 \\ x_2 &\leqslant 6 \\ x_1, x_2 &\geqslant 0 \end{aligned}$$

21. Minimize and maximize

$$P = 20x_1 + 10x_2$$

Subject to
$$\begin{aligned} 2x_1 + 3x_2 &\geqslant 30 \\ 2x_1 + x_2 &\leqslant 26 \\ -2x_1 + 5x_2 &\leqslant 34 \\ x_1, x_2 &\geqslant 0 \end{aligned}$$

22. Minimize and maximize

$$P = 12x_1 + 14x_2$$

Subject to
$$\begin{aligned} -2x_1 + x_2 &\geqslant 6 \\ x_1 + x_2 &\leqslant 15 \\ 3x_1 - x_2 &\geqslant 0 \\ x_1, x_2 &\geqslant 0 \end{aligned}$$

23. Maximize $P = 20x_1 + 30x_2$
Subject to
$$\begin{aligned} 0.6x_1 + 1.2x_2 &\leqslant 960 \\ 0.03x_1 + 0.04x_2 &\leqslant 36 \\ 0.3x_1 + 0.2x_2 &\leqslant 270 \\ x_1, x_2 &\geqslant 0 \end{aligned}$$

24. Minimize $C = 30x_1 + 10x_2$
Subject to
$$\begin{aligned} 1.8x_1 + 0.9x_2 &\geqslant 270 \\ 0.3x_1 + 0.2x_2 &\geqslant 54 \\ 0.01x_1 + 0.03x_2 &\geqslant 3.9 \\ x_1, x_2 &\geqslant 0 \end{aligned}$$

C 25. The corner points for the bounded feasible region determined by the system of linear inequalities

$$x_1 + 2x_2 \leqslant 10$$
$$3x_1 + x_2 \leqslant 15$$
$$x_1, x_2 \geqslant 0$$

are $O = (0, 0)$, $A = (0, 5)$, $B = (4, 3)$, and $C = (5, 0)$. If $P = ax_1 + bx_2$ and $a, b > 0$, determine conditions on a and b that will ensure that the maximum value of P occurs:

(A) Only at A (B) Only at B (C) Only at C
(D) At both A and B (E) At both B and C

26. The corner points for the feasible region determined by the system of linear inequalities

$$x_1 + 4x_2 \geqslant 30$$
$$3x_1 + x_2 \geqslant 24$$
$$x_1, x_2 \geqslant 0$$

are $A = (0, 24)$, $B = (6, 6)$, $D = (30, 0)$. If $C = ax_1 + bx_2$ and $a, b > 0$, determine conditions on a and b that will ensure that the minimum value of C occurs:

(A) Only at A (B) Only at B (C) Only at D
(D) At both A and B (E) At both B and D

APPLICATIONS

Business & Economics

27. *Manufacturing—resource allocation.* A manufacturing company makes two types of water skis, a trick ski and a slalom ski. The relevant manufacturing data are given in the table. How many of each type of ski should be manufactured each day to realize a maximum profit? What is the maximum profit?

	LABOR-HOURS PER SKI		MAXIMUM LABOR-HOURS AVAILABLE PER DAY
	Trick Ski	*Slalom Ski*	
FABRICATING DEPARTMENT	6	4	108
FINISHING DEPARTMENT	1	1	24
PROFIT PER SKI	$40	$30	

28. *Manufacturing—resource allocation.* A furniture manufacturing company manufactures dining room tables and chairs. The relevant manufacturing data are given in the accompanying table.

(A) How many tables and chairs should be manufactured each day to realize a maximum profit? What is the maximum profit?

(B) Repeat part A if the marketing department of the company has decided that the number of chairs produced should be at least four times the number of tables produced.

	LABOR-HOURS PER UNIT *Table*	*Chair*	MAXIMUM LABOR-HOURS AVAILABLE PER DAY
ASSEMBLY DEPARTMENT	8	2	400
FINISHING DEPARTMENT	2	1	120
PROFIT PER UNIT	\$90	\$25	

29. *Manufacturing—production scheduling.* A furniture company has two plants that produce the lumber used in manufacturing tables and chairs. In 1 day of operation, plant A can produce the lumber required to manufacture 20 tables and 60 chairs, and plant B can produce the lumber required to manufacture 25 tables and 50 chairs. The company needs enough lumber to manufacture at least 200 tables and 500 chairs.

(A) If it costs \$1,000 to operate plant A for 1 day and \$900 to operate plant B for 1 day, how many days should each plant be operated in order to produce a sufficient amount of lumber at a minimum cost? What is the minimum cost?

(B) Repeat part A if the daily cost of operating plant A is reduced to \$600.

(C) Repeat part A if the daily cost of operating plant B is reduced to \$800.

30. *Manufacturing—resource allocation.* An electronics firm manufactures two types of personal computers, a standard model and a portable model. The production of a standard computer requires a capital expenditure of \$400 and 40 hours of labor. The production of a portable computer requires a capital expenditure of \$250 and 30 hours of labor. The firm has \$20,000 capital and 2,160 labor-hours available for production of standard and portable computers.

(A) What is the maximum number of computers the company is capable of producing?

(B) If each standard computer contributes a profit of \$320 and each portable computer contributes a profit of \$220, how many of each type of computer should the firm produce in order to maximize profit? What is the maximum profit?

31. *Transportation.* The officers of a high school senior class are planning to rent buses and vans for a class trip. Each bus can transport 40 students, requires 3 chaperones, and costs \$1,200 to rent. Each van can transport 8 students, requires 1 chaperone, and costs \$100 to rent. Since there are 400 students in the senior class that may be eligible to go on the trip, the officers must plan to accommodate at least 400 students. Since only 36 parents have volunteered to serve as chaperones, the officers must plan to use at most 36 chaperones. How many vehicles of each type should the officers rent in order to minimize the transportation costs? What are the minimal transportation costs?

32. *Investment.* An investor has \$24,000 to invest in bonds of AAA and B qualities. The AAA bonds yield on the average 6% and the B bonds yield 10%. The investor's policy requires that she invest at least three times as much money in AAA bonds as in B bonds. How much should she invest in each type of bond to maximize her return? What is the maximum return?

33. *Pollution control.* Because of new federal regulations on pollution, a chemical plant introduced a new, more expensive process to supplement or replace an older process used in the production of a particular chemical. The older process emitted 15 grams of sulfur dioxide and 40 grams of particulate matter into the atmosphere for each gallon of chemical produced. The new process emits 5 grams of sulfur dioxide and 20 grams of particulate matter for each gallon produced. The company makes a profit of 30¢ per gallon and 20¢ per gallon on the old and new processes, respectively. If the government allows the plant to emit no more than 10,500 grams of sulfur dioxide and no more than 30,000 grams of particulate matter daily, how many gallons of the chemical should be produced by each process to maximize daily profit? What is the maximum profit?

34. *Capital expansion.* A fast-food chain plans to expand by opening several new restaurants. The chain operates two types of restaurants, drive-through and full-service. A drive-through restaurant costs \$100,000 to construct, requires 5 employees, and has an expected annual revenue of \$200,000. A full-service restaurant costs \$150,000 to construct, requires 15 employees, and has an expected annual revenue of \$500,000. The chain has \$2,400,000 in capital available for expansion. Labor contracts require that they hire no more than 210 employees, and licensing restrictions require that they open no more than 20 new restaurants. How many restaurants of each type should the chain open in order to maximize the expected revenue? What is the maximum expected revenue? How much of their capital will they use and how many employees will they hire?

Life Sciences

35. *Nutrition—plants.* A fruit grower can use two types of fertilizer in his orange grove, brand A and brand B. The amounts (in pounds) of nitrogen, phosphoric acid, and chlorine in a bag of each brand are given in the accompanying table. Tests indicate that the grove needs at least 1,000 pounds of phosphoric acid and at most 400 pounds of chlorine.

    (A) If the grower wants to maximize the amount of nitrogen added to the grove, how many bags of each mix should be used? How much nitrogen will be added?

(B) If the grower wants to minimize the amount of nitrogen added to the grove, how many bags of each mix should be used? How much nitrogen will be added?

	POUNDS PER BAG	
	Brand A	*Brand B*
NITROGEN	8	3
PHOSPHORIC ACID	4	4
CHLORINE	2	1

36. *Nutrition—people.* A dietitian in a hospital is to arrange a special diet composed of two foods, M and N. Each ounce of food M contains 30 units of calcium, 10 units of iron, 10 units of vitamin A, and 8 units of cholesterol. Each ounce of food N contains 10 units of calcium, 10 units of iron, 30 units of vitamin A, and 4 units of cholesterol. If the minimum daily requirements are 360 units of calcium, 160 units of iron, and 240 units of vitamin A, how many ounces of each food should be used to meet the minimum requirements and at the same time minimize the cholesterol intake? What is the minimum cholesterol intake?

37. *Nutrition—plants.* A farmer can buy two types of plant food, mix A and mix B. Each cubic yard of mix A contains 20 pounds of phosphoric acid, 30 pounds of nitrogen, and 5 pounds of potash. Each cubic yard of mix B contains 10 pounds of phosphoric acid, 30 pounds of nitrogen, and 10 pounds of potash. The minimum monthly requirements are 460 pounds of phosphoric acid, 960 pounds of nitrogen, and 220 pounds of potash. If mix A costs \$30 per cubic yard and mix B costs \$35 per cubic yard, how many cubic yards of each mix should the farmer blend to meet the minimum monthly requirements at a minimal cost? What is this cost?

38. *Nutrition—animals.* A laboratory technician in a medical research center is asked to formulate a diet from two commercially packaged foods, food A and food B, for a group of animals. Each ounce of food A contains 8 units of fat, 16 units of carbohydrate, and 2 units of protein. Each ounce of food B contains 4 units of fat, 32 units of carbohydrate, and 8 units of protein. The minimum daily requirements are 176 units of fat, 1,024 units of carbohydrate, and 384 units of protein. If food A costs 5¢ per ounce and food B costs 5¢ per ounce, how many ounces of each food should be used to meet the minimum daily requirements at the least cost? What is the cost for this amount of food?

Social Sciences

39. *Psychology.* In an experiment on conditioning, a psychologist uses two types of Skinner boxes with mice and rats. The amount of time (in minutes) each mouse and each rat spends in each box per day is given in the table.

What is the maximum number of mice and rats that can be used in this experiment? How many mice and how many rats produce this maximum?

	TIME		MAXIMUM TIME
	Mice	*Rats*	AVAILABLE PER DAY
SKINNER BOX A	10 min	20 min	800 min
SKINNER BOX B	20 min	10 min	640 min

40. *Sociology.* A city council voted to conduct a study on inner-city community problems. A nearby university was contacted to provide sociologists and research assistants. Allocation of time and costs per week are given in the accompanying table. How many sociologists and how many research assistants should be hired to minimize the cost and meet the weekly labor-hour requirements? What is the minimum weekly cost?

	LABOR-HOURS		MINIMUM
		Research	LABOR-HOURS
	Sociologist	*Assistant*	NEEDED PER WEEK
FIELDWORK	10	30	180
RESEARCH CENTER	30	10	140
COSTS PER WEEK	$500	$300	

SECTION 6-3 A Geometric Introduction to the Simplex Method

- STANDARD MAXIMIZATION PROBLEMS IN STANDARD FORM
- SLACK VARIABLES
- BASIC AND NONBASIC VARIABLES; BASIC SOLUTIONS AND BASIC FEASIBLE SOLUTIONS
- BASIC FEASIBLE SOLUTIONS AND THE SIMPLEX METHOD

The geometric method of solving linear programming problems provides us with an overview of the subject and some useful terminology. But, practically speaking, the method is useful only for problems involving two decision variables and relatively few problem constraints. What happens when we need more decision variables and more problem constraints? We use an algebraic method called the *simplex method,* which was developed by George B. Dantzig in 1947 while on assignment to the U.S. Department of the Air Force. Ideally suited to computer use, the method is used routinely on applied problems involving hundreds and even thousands of variables and problem constraints.

The algebraic procedures utilized in the simplex method require the problem constraints to be written as equations rather than inequalities. This new form of the linear programming problem also prompts the use of some new terminology.

We introduce this new form and associated terminology through a simple example and an appropriate geometric interpretation. From this example we can illustrate what the simplex method does geometrically before we immerse ourselves in the algebraic details of the process.

◆ STANDARD MAXIMIZATION PROBLEMS IN STANDARD FORM

We now return to the tent production problem in Example 6 from the last section. Recall the mathematical model for the problem:

Maximize	$P = 50x_1 + 80x_2$	Objective function	
Subject to	$x_1 + 2x_2 \leq 32$	Cutting department constraint	(1)
	$3x_1 + 4x_2 \leq 84$	Assembly department constraint	
	$x_1, x_2 \geq 0$	Nonnegative constraints	

where the decision variables x_1 and x_2 are the number of standard and expedition tents, respectively, produced each day.

Notice that the problem constraints involve $\leq$ inequalities with positive constants to the right of the inequality. Maximization problems that satisfy this condition are called *standard maximization problems.* In this and the next section we will restrict our attention to standard maximization problems.

A Standard Maximization Problem in Standard Form

A linear programming problem is said to be a **standard maximization problem in standard form** if its mathematical model is of the following form:

Maximize the objective function

$$P = c_1x_1 + c_2x_2 + \cdots + c_nx_n$$

Subject to problem constraints of the form

$$a_1x_1 + a_2x_2 + \cdots + a_nx_n \leq b \qquad b \geq 0$$

With nonnegative constraints

$$x_1, x_2, \ldots, x_n \geq 0$$

[*Note:* Mathematical model (1) above is a standard maximization problem in standard form. Also note that the coefficients of the objective function can be any real numbers.]

◆ SLACK VARIABLES

To adapt a linear programming problem to the matrix methods used in the simplex process (as discussed in the next section), we convert the problem constraint inequalities into a system of linear equations by using a simple device called a *slack variable*. In particular, to convert the system of problem constraint inequalities from (1),

$$\begin{aligned} x_1 + 2x_2 &\le 32 \quad \text{Cutting department constraint} \\ 3x_1 + 4x_2 &\le 84 \quad \text{Assembly department constraint} \end{aligned} \tag{2}$$

into a system of equations, we add variables s_1 and s_2 to the left sides of (2) to obtain

$$\begin{aligned} x_1 + 2x_2 + s_1 &= 32 \\ 3x_1 + 4x_2 + s_2 &= 84 \end{aligned} \tag{3}$$

The variables s_1 and s_2 are called **slack variables** because each makes up the difference (takes up the slack) between the left and right sides of the inequalities in (2). For example, if we produced 20 standard tents ($x_1 = 20$) and 5 expedition tents ($x_2 = 5$), then the number of labor-hours used in the cutting department would be $20 + 2(5) = 30$, leaving a slack of 2 unused labor-hours out of the 32 available. Thus, s_1 would have the value of 2.

Notice that if the decision variables x_1 and x_2 satisfy the system of constraint inequalities (2), then the slack variables s_1 and s_2 are nonnegative. We will have more to say about this later in this discussion.

◆ BASIC AND NONBASIC VARIABLES; BASIC SOLUTIONS AND BASIC FEASIBLE SOLUTIONS

Observe that system (3) has infinitely many solutions—just solve for s_1 and s_2 in terms of x_1 and x_2, and then assign x_1 and x_2 arbitrary values. Certain solutions of system (3), called *basic solutions*, are related to the intersection points of the (extended) boundary lines of the feasible region in Figure 7.

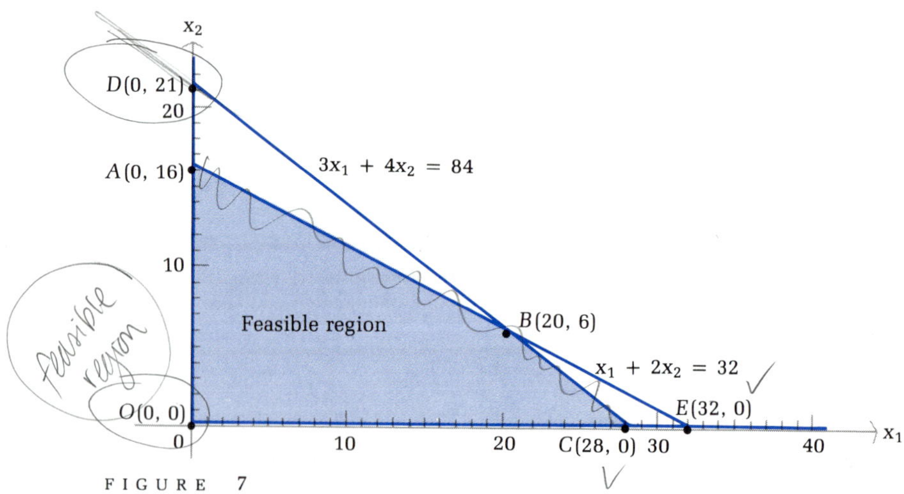

FIGURE 7

How are basic solutions to system (3) determined? System (3) involves four variables and two equations. We divide the four variables into two groups, called *basic variables* and *nonbasic variables*, as follows: Basic variables are selected arbitrarily with the restriction that there be as many basic variables as there are equations. The remaining variables are nonbasic variables.

Since system (3) has two equations, we can select any two of the four variables as basic variables. The remaining two variables are then nonbasic variables. A solution found by setting the two nonbasic variables equal to 0 and solving for the two basic variables is a basic solution. [Note that setting two variables equal to 0 in system (3) results in a system of two equations with two variables, which has (from Chapter 5) exactly one solution, infinitely many solutions, or no solution.]

◆ EXAMPLE 9

(A) Find two basic solutions for system (3) by first selecting s_1 and s_2 as basic variables, and then by selecting x_2 and s_1 as basic variables.

(B) Associate each basic solution found in part A with an intersection point of the (extended) boundary lines of the feasible region in Figure 7, and indicate which boundary lines produce each intersection point.

(C) Indicate which of the intersection points found in part B are in the feasible region.

Solutions

(A) Since s_1 and s_2 were selected as basic variables, x_1 and x_2 are nonbasic variables. A basic solution is found by setting the nonbasic variables equal to 0 and solving for the basic variables. If $x_1 = 0$ and $x_2 = 0$, then system (3) becomes

$$\begin{array}{ll} s_1 = 32 & \overset{0}{x_1} + \overset{0}{2x_2} + s_1 \qquad = 32 \\ s_2 = 84 & 3x_1 + 4x_2 \qquad + s_2 = 84 \end{array}$$

and the basic solution is

$$x_1 = 0, \quad x_2 = 0, \quad s_1 = 32, \quad s_2 = 84 \tag{4}$$

If we select x_2 and s_1 as basic variables, then x_1 and s_2 are nonbasic variables. Setting the nonbasic variables equal to 0, system (3) becomes

$$\begin{array}{rl} 2x_2 + s_1 = 32 & \overset{0}{x_1} + 2x_2 + s_1 \qquad \overset{0}{} = 32 \\ 4x_2 = 84 & 3x_1 + 4x_2 \qquad + s_2 = 84 \end{array}$$

Solving, we see that $x_2 = 21$ and $s_1 = -10$, and the basic solution is

$$x_1 = 0, \quad x_2 = 21, \quad s_1 = -10, \quad s_2 = 0 \tag{5}$$

(B) Basic solution (4)—since $x_1 = 0$ and $x_2 = 0$—corresponds to the origin $O(0, 0)$ in Figure 7, which is the intersection of the boundary lines $x_1 = 0$ and $x_2 = 0$. Basic solution (5)—since $x_1 = 0$ and $x_2 = 21$—corresponds to the intersection point $D(0, 21)$, which is the intersection of the boundary lines $x_1 = 0$ and $3x_1 + 4x_2 = 84$.

(C) The intersection point $O(0, 0)$ is in the feasible region; hence, it is natural to call the corresponding basic solution a *basic feasible solution*. The intersection point $D(0, 21)$ is not in the feasible region; hence, the corresponding basic solution is not feasible. ◆

PROBLEM 9 (A) Find two basic solutions for system (3) by first selecting x_1 and s_1 as basic variables, and then by selecting x_1 and s_2 as basic variables.

(B) Associate each basic solution found in part A with an intersection point of the (extended) boundary lines of the feasible region in Figure 7, and indicate which boundary lines produce each intersection point.

(C) Indicate which of the intersection points found in part B are in the feasible region. ◆

Proceeding systematically as in Example 9 and Problem 9, we can obtain all basic solutions to system (3). The results are summarized in Table 2, which also includes geometric interpretations of the basic solutions relative to Figure 7. Figure 8 summarizes these interpretations. A careful study of Table 2 and Figure 8 is very worthwhile. (Note that to be sure we have listed all possible basic solutions in Table 2, it is convenient to organize the table in terms of the zero values of the nonbasic variables.)

TABLE 2

Basic Solutions

BASIC SOLUTIONS x_1	x_2	s_1	s_2	INTERSECTION POINT	INTERSECTING BOUNDARY LINES	FEASIBLE
0	0	32	84	$O(0, 0)$	$x_1 = 0$ $x_2 = 0$	Yes
0	16	0	20	$A(0, 16)$	$x_1 = 0$ $x_1 + 2x_2 = 32$	Yes
0	21	−10	0	$D(0, 21)$	$x_1 = 0$ $3x_1 + 4x_2 = 84$	No
32	0	0	−12	$E(32, 0)$	$x_2 = 0$ $x_1 + 2x_2 = 32$	No
28	0	4	0	$C(28, 0)$	$x_2 = 0$ $3x_1 + 4x_2 = 84$	Yes
20	6	0	0	$B(20, 6)$	$x_1 + 2x_2 = 32$ $3x_1 + 4x_2 = 84$	Yes

Observations from Table 2 and Figure 8 Important to the Development of the Simplex Method

1. In Table 2, observe that a basic solution that is not feasible includes at least one negative value and that a basic feasible solution does not include any negative values. That is, we can determine the feasibility of a basic solution simply by examining the sign of all the variables in the solution.
2. In Table 2 and Figure 8, observe that basic feasible solutions are associated with the corner points of the feasible region, which include the optimal solution to the original linear programming problem.

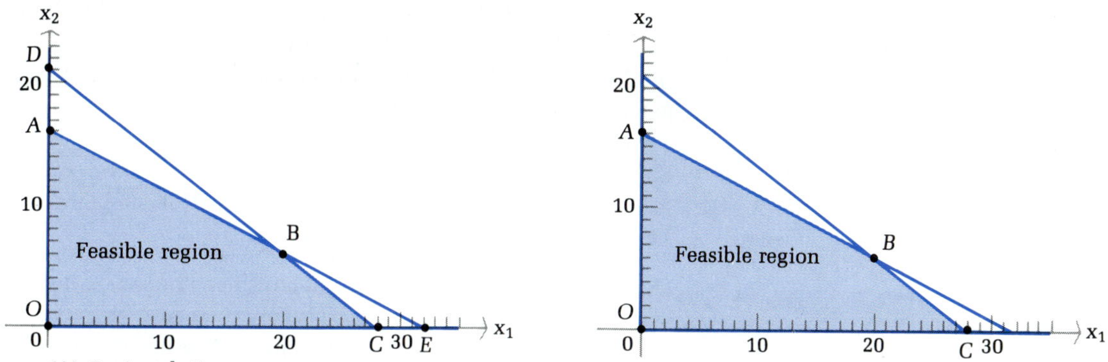

(A) Basic solutions: *O, A, B, C, D, E* (intersection points of all extended boundary lines)

(B) Basic feasible solutions: *O, A, B, C* (corner points of the feasible region)

FIGURE 8

Before proceeding further, let us formalize the definitions alluded to in the discussion above so that they apply to standard maximization problems in general, without any reference to geometric forms.

Basic Variables and Nonbasic Variables; Basic Solutions and Basic Feasible Solutions

Given a system of linear equations associated with a linear programming problem (such a system will always have more variables than equations):

The variables are divided into two (mutually exclusive) groups, as follows: **Basic variables** are selected arbitrarily with the one restriction that there be as many basic variables as there are equations. The remaining variables are called **nonbasic variables.**

A solution found by setting the nonbasic variables equal to 0 and solving for the basic variables is called a **basic solution.** If a basic solution has no negative values, it is a **basic feasible solution.**

◆ EXAMPLE 10 Suppose there is a system of three problem constraint equations with eight (slack and decision) variables associated with a standard maximization problem.

(A) How many basic variables and how many nonbasic variables are associated with the system?

(B) Setting the nonbasic variables equal to 0 will result in a system of how many linear equations with how many variables?

(C) If a basic solution has all nonnegative elements, is it feasible or not feasible?

Solutions

(A) Since there are three equations in the system, there should be three basic variables and five nonbasic variables.

(B) Three equations with three variables

(C) Feasible ◆

PROBLEM 10 Suppose there is a system of five problem constraint equations with eleven (slack and decision) variables associated with a standard maximization problem.

(A) How many basic variables and how many nonbasic variables are associated with the system?

(B) Setting the nonbasic variables equal to 0 will result in a system of how many linear equations with how many variables?

(C) If a basic solution has one or more negative elements, is it feasible or not feasible? ◆

◆ BASIC FEASIBLE SOLUTIONS AND THE SIMPLEX METHOD

The following important theorem, which is equivalent to the fundamental theorem (Theorem 2 in the preceding section), is stated without proof:

THEOREM 4

> If the optimal value of the objective function in a linear programming problem exists, then that value must occur at one (or more) of the basic feasible solutions.

corner points

Now you can understand why the concepts of basic and nonbasic variables and basic solutions and basic feasible solutions are so important—the concepts are central to the process of finding optimal solutions to linear programming problems.

We have taken the first step toward finding a general method of solving linear programming problems involving any number of variables and problem constraints. That is, **we have found a method of identifying all the corner points (basic feasible solutions) of a feasible region without drawing its graph.** This is a critical step if we want to consider problems with more than two decision variables. Unfortunately, the number of corner points increases dramatically as the number of variables and constraints increases. In real-world problems, it is not practical to find all the corner points in order to find the optimal solution. Thus, the next step is to find a method of locating the optimal solution without finding every corner point. The procedure for doing this is the simplex method mentioned at the beginning of this section.

The simplex method, using a special matrix and row operations, automatically moves from one basic feasible solution to another—that is, from one corner point of the feasible region to another—each time getting closer to an optimal solution (if one exists), until an optimal solution is reached. Then the process stops. The **simplex method** is an iterative algorithm (a repetitive procedure) that is shown schematically in Figure 9. A remarkable property of the simplex method is that in large linear programming problems it usually arrives at an optimal solution (if one exists) by testing relatively few of the large number of basic feasible solutions (corner points) available.

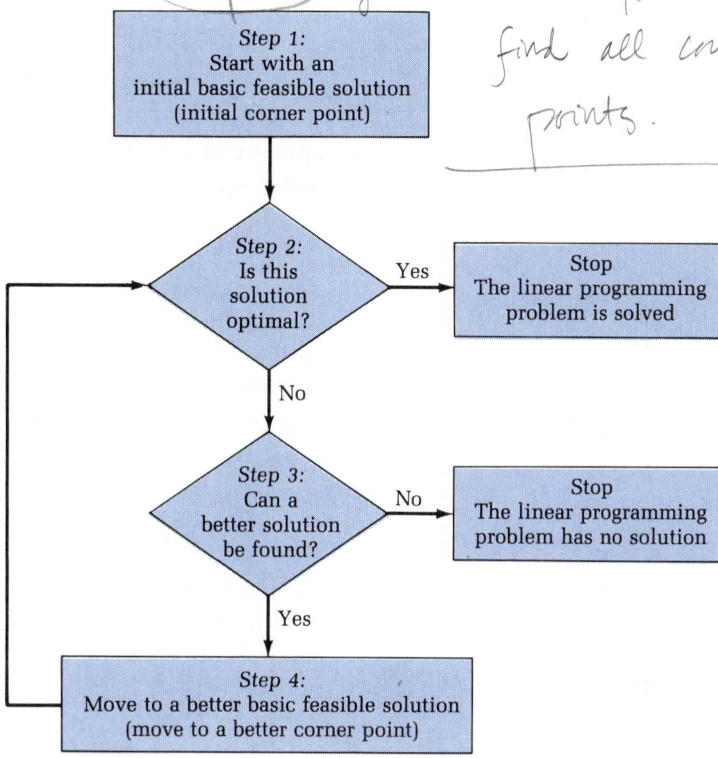

FIGURE 9
Simplex algorithm

With this background, we are now ready to discuss the algebraic details of the simplex method in the next section.

Answers to Matched Problems

9. (A) Basic solution corresponding to basic variables x_1 and s_1: $x_1 = 28$, $x_2 = 0$, $s_1 = 4$, $s_2 = 0$. Basic solution corresponding to basic variables x_1 and s_2: $x_1 = 32$, $x_2 = 0$, $s_1 = 0$, $s_2 = -12$.
   (B) The first basic solution corresponds to $C(28, 0)$, which is the intersection of the boundary lines $x_2 = 0$ and $3x_1 + 4x_2 = 84$. The second basic

solution corresponds to $E(32, 0)$, which is the intersection of the boundary lines $x_2 = 0$ and $x_1 + 2x_2 = 32$.

(C) $C(28, 0)$ is in the feasible region (hence, the corresponding basic solution is a basic feasible solution); $E(32, 0)$ is not in the feasible region (hence, the corresponding basic solution is not feasible).

10. (A) Five basic variables and six nonbasic variables
(B) Five equations and five variables (C) Not feasible

EXERCISE 6-3

A

1. Associated with a standard maximization problem is a system of problem constraint equations with nine variables, including five slack variables.
   (A) How many constraint equations are in the system?
   (B) How many decision variables are in the system?
   (C) How many basic variables and how many nonbasic variables are associated with the system?
   (D) Setting the nonbasic variables equal to 0 will result in a system of how many linear equations with how many variables?

2. Associated with a standard maximization problem is a system of problem constraint equations with ten variables, including four decision variables.
   (A) How many slack variables are in the system?
   (B) How many constraint equations are in the system?
   (C) How many basic variables and how many nonbasic variables are associated with the system?
   (D) Setting the nonbasic variables equal to 0 will result in a system of how many linear equations with how many variables?

3. Listed in the table below are all the basic solutions for the system

$$2x_1 + 3x_2 + s_1 \quad = 24$$
$$4x_1 + 3x_2 \quad + s_2 = 36$$

For each basic solution, identify the nonbasic variables and the basic variables. Then classify each basic solution as feasible or not feasible.

	x_1	x_2	s_1	s_2
(A)	0	0	24	36
(B)	0	8	0	12
(C)	0	12	−12	0
(D)	12	0	0	−12
(E)	9	0	6	0
(F)	6	4	0	0

4. Repeat Problem 3 for the system

$$\begin{aligned} 2x_1 + \ x_2 + s_1 \qquad &= 30 \\ x_1 + 5x_2 \qquad + s_2 &= 60 \end{aligned}$$

whose basic solutions are given in the following table:

	x_1	x_2	s_1	s_2
(A)	0	0	30	60
(B)	0	30	0	−90
(C)	0	12	18	0
(D)	15	0	0	45
(E)	60	0	−90	0
(F)	10	10	0	0

5. Listed in the table below are all the possible choices of nonbasic variables for the system

$$\begin{aligned} 2x_1 + \ x_2 + s_1 \qquad &= 50 \\ x_1 + 2x_2 \qquad + s_2 &= 40 \end{aligned}$$

In each case, find the values of the basic variables and determine whether the basic solution is feasible.

	x_1	x_2	s_1	s_2
(A)	0	0	?	?
(B)	0	?	0	?
(C)	0	?	?	0
(D)	?	0	0	?
(E)	?	0	?	0
(F)	?	?	0	0

6. Repeat Problem 5 for the system

$$\begin{aligned} x_1 + 2x_2 + s_1 \qquad &= 12 \\ 3x_1 + 2x_2 \qquad + s_2 &= 24 \end{aligned}$$

B *Graph the following systems of inequalities. Introduce slack variables to convert each system of inequalities to a system of equations, and find all the basic solutions of the system. Construct a table (similar to Table 2) listing each basic* *solution, the corresponding point on the graph, and whether the basic solution is feasible. (You do not need to list the intersecting lines.)*

7. $x_1 + x_2 \leq 16$
$2x_1 + x_2 \leq 20$
$x_1, x_2 \geq 0$

8. $5x_1 + x_2 \leq 35$
$4x_1 + x_2 \leq 32$
$x_1, x_2 \geq 0$

9. $2x_1 + x_2 \leq 22$
$x_1 + x_2 \leq 12$
$x_1 + 2x_2 \leq 20$
$x_1, x_2 \geq 0$

10. $4x_1 + x_2 \leq 28$
$2x_1 + x_2 \leq 16$
$x_1 + x_2 \leq 13$
$x_1, x_2 \geq 0$

SECTION 6-4

The Simplex Method: Maximization with Problem Constraints of the Form ≤

- INITIAL SYSTEM
- THE SIMPLEX TABLEAU
- THE PIVOT OPERATION
- INTERPRETING THE SIMPLEX PROCESS GEOMETRICALLY
- THE SIMPLEX METHOD SUMMARIZED
- APPLICATION

We are now ready to develop the simplex method for a standard maximization problem. Specific details in the presentation of the method generally vary from one person to another or from one book to another, even though the underlying process is the same. The presentation developed here emphasizes concept development and understanding.

As was pointed out in the last section, the simplex method is most efficient when used with computers. Consequently, it is not intended that you become expert in manually solving linear programming problems using the simplex method. It is important that you become proficient in setting up linear programming problems so that they can be solved using computers, and it is also important that you develop skill in interpreting the results. The best way to gain this proficiency and interpretive skill is to set up and manually solve a number of fairly simple linear programming problems using the simplex method, and this is the purpose of the remainder of this chapter. (The computer supplement for this book contains a computer program for solving linear programming problems—see the Preface.)

◆ INITIAL SYSTEM

We will introduce the concepts and procedures involved in the simplex method through an example—the tent production example we have discussed earlier.

We restate the problem here in standard form for convenient reference:

$$
\begin{aligned}
&\text{Maximize} \quad & P = 50x_1 + 80x_2 \quad & \text{Objective function} \\
&\text{Subject to} \quad & \left.\begin{aligned} x_1 + 2x_2 &\le 32 \\ 3x_1 + 4x_2 &\le 84 \end{aligned}\right\} \quad & \text{Problem constraints} \\
& & x_1, x_2 \ge 0 \quad & \text{Nonnegative constraints}
\end{aligned}
\qquad (1)
$$

Introducing slack variables s_1 and s_2, we convert the problem constraint inequalities in (1) into the system of problem constraint equations:

$$
\begin{aligned}
x_1 + 2x_2 + s_1 \qquad &= 32 \\
3x_1 + 4x_2 \qquad + s_2 &= 84 \\
x_1, x_2, s_1, s_2 &\ge 0
\end{aligned}
\qquad (2)
$$

Since a basic solution of (2) is not feasible if it contains any negative values, we have also included the nonnegative constraints for both the decision variables x_1 and x_2 and the slack variables s_1 and s_2. From our earlier discussion (in the last section), we know that out of the infinitely many solutions to system (2), an optimal solution is among the basic feasible solutions, which correspond to the corner points of the feasible region.

As part of the simplex method we add the objective function equation in the form $-50x_1 - 80x_2 + P = 0$ to system (2) to create what is called the **initial system:**

$$
\begin{aligned}
x_1 + 2x_2 + s_1 \qquad\qquad &= 32 \\
3x_1 + 4x_2 \qquad + s_2 \qquad &= 84 \\
-50x_1 - 80x_2 \qquad\qquad + P &= 0 \\
x_1, x_2, s_1, s_2 &\ge 0
\end{aligned}
\qquad (3)
$$

When we add the objective function equation to system (2), we must slightly modify the earlier definitions of basic solution and basic feasible solution so that they apply to the initial system (3).

Basic Solutions and Basic Feasible Solutions for Initial Systems

1. The objective function variable P is always selected as a basic variable and is never selected as a nonbasic variable.
2. Note that a basic solution of system (3) is also a basic solution of system (2) after P is deleted.
3. If a basic solution of system (3) is a basic feasible solution of system (2) after deleting P, then the basic solution of (3) is called a **basic feasible solution** of (3).
4. A basic feasible solution of system (3) can contain a negative number, but only if it is the value of P, the objective function variable.

These changes lead to a small change in Theorem 4:

THEOREM 5

> If the optimal value of the objective function in a linear programming problem exists, then that value must occur at one (or more) of the basic feasible solutions of the initial system.

With these adjustments understood, we start the simplex process with a basic feasible solution of the initial system (3), which we will refer to as an **initial basic feasible solution.** An initial basic feasible solution that is easy to find is the one associated with the origin.

Since system (3) has three equations and five variables, it has three basic variables and two nonbasic variables. Looking at the system, we see that x_1 and x_2 appear in all equations and s_1, s_2, and P each appears only once and each in a different equation. A basic solution can be found by inspection by selecting s_1, s_2, and P as the basic variables (remember, P is always selected as a basic variable) and x_1 and x_2 as the nonbasic variables to be set equal to 0. Setting x_1 and x_2 equal to 0 and solving for the basic variables, we obtain the basic solution:

$$x_1 = 0, \quad x_2 = 0, \quad s_1 = 32, \quad s_2 = 84, \quad P = 0$$

This basic solution is feasible since none of the variables (excluding P) are negative. Thus, this is the initial basic feasible solution we seek.

Now you can see why we wanted to add the objective function equation to system (2): A basic feasible solution of (3) not only includes a basic feasible solution of (2), but, in addition, it includes the value of P for that basic feasible solution of (2).

The initial basic feasible solution we just found is associated with the origin. Of course, if we do not produce any tents, we do not expect a profit, so $P = \$0$. Starting with this easily obtained initial basic feasible solution, the simplex process moves through each iteration (each repetition) to another basic feasible solution, each time improving the profit, and the process continues until the maximum profit is reached. Then the process stops.

◆ THE SIMPLEX TABLEAU

To facilitate the search for the optimal solution, we now turn to matrix methods discussed in Chapter 5. Our first step is to write the augmented matrix for the initial system (3). This matrix is called the **initial simplex tableau,** and it is simply a tabulation of the coefficients in system (3).

$$\begin{array}{c} \\ s_1 \\ s_2 \\ P \end{array} \begin{array}{c} \begin{array}{cccccc} x_1 & x_2 & s_1 & s_2 & P & \end{array} \\ \left[\begin{array}{ccccc|c} 1 & 2 & 1 & 0 & 0 & 32 \\ 3 & 4 & 0 & 1 & 0 & 84 \\ \hline -50 & -80 & 0 & 0 & 1 & 0 \end{array}\right] \end{array} \quad \text{Initial simplex tableau} \qquad (4)$$

In tableau (4), the row below the dashed line always corresponds to the objective function. Each of the basic variables we selected above, s_1, s_2, and P, is also placed on the left of the tableau so that the intersection element in its row and column is not 0. For example, we place the basic variable s_1 on the left so that the intersection element of the s_1 row and the s_1 column is 1 and not 0. The basic variable s_2 is similarly placed. The objective function variable P is always placed at the bottom. The reason for writing the basic variables on the left in this way, is that this placement makes it possible to read certain basic feasible solutions directly from the tableau. If $x_1 = 0$ and $x_2 = 0$, then the basic variables on the left of tableau (4) are lined up with their corresponding values, 32, 84, and 0, to the right of the vertical line.

Looking at tableau (4) relative to the choice of s_1, s_2, and P as basic variables, we see that each basic variable is above a column that has all 0 elements except for a single 1 and that no two such columns contain 1's in the same row. These observations lead to a formalization of the process of selecting basic and nonbasic variables that is an important part of the simplex method:

Selecting Basic and Nonbasic Variables for the Simplex Method

Given a simplex tableau:

Step 1. Determine the number of basic variables and the number of nonbasic variables. These numbers do not change during the simplex process.

Step 2. *Selecting basic variables:* A variable can be selected as a basic variable only if it corresponds to a column in the tableau that has exactly one nonzero element (usually 1) and the nonzero element in the column is not in the same row as the nonzero element in the column of another basic variable. (This procedure always selects P as a basic variable, since the P column never changes during the simplex process.)

Step 3. *Selecting nonbasic variables:* After the basic variables are selected in step 2, the remaining variables are selected as the nonbasic variables. (The tableau columns under the nonbasic variables will usually contain more than one nonzero element.)

The earlier selection of s_1, s_2, and P as basic variables and x_1 and x_2 as nonbasic variables conforms to this prescribed convention of selecting basic and nonbasic variables for the simplex process.

◆ THE PIVOT OPERATION

The simplex method will now switch one of the nonbasic variables, x_1 or x_2, for one of the basic variables, s_1 or s_2 (but not P), as a step toward improving the profit. For a nonbasic variable to be classified as a basic variable we need to perform appropriate row operations on the tableau so that the newly selected basic variable will end up with exactly one nonzero element in its column. In this process, the old basic variable will usually gain additional nonzero elements in its column as it becomes nonbasic.

Which nonbasic variable should we select to become basic? It makes sense to select the nonbasic variable that will increase the profit the most per unit change in that variable. Looking at the objective function

$$P = 50x_1 + 80x_2$$

we see that if x_1 stays a nonbasic variable (set equal to 0) and if x_2 becomes a new basic variable, then

$$P = 50(0) + 80x_2 = 80x_2$$

and for each unit increase in x_2, P will increase \$80. If x_2 stays a nonbasic variable and x_1 becomes a new basic variable, then (reasoning in the same way) for each unit increase in x_1, P will increase only \$50. So, we select the nonbasic variable x_2 to enter the set of basic variables, and call it the **entering variable.** (The basic variable leaving the set of basic variables to become a nonbasic variable is called the **exiting variable.** Exiting variables will be discussed shortly.)

We call the column corresponding to the entering variable the **pivot column.** Looking at the bottom row in tableau (4)—the objective function row below the dashed line—we see that the pivot column is associated with the column to the left of the P column that has the most negative bottom element. In general, the most negative element in the bottom row to the left of the P column *indicates* the variable above it that will produce the greatest increase in P for a unit increase in that variable. For this reason, we call the elements in the bottom row of the tableau to the left of the P column **indicators.**

We illustrate the indicators, the pivot column, the entering variable, and the initial basic feasible solution below:

Entering variable ↓

$$\begin{array}{c} \\ s_1 \\ s_2 \\ P \end{array}\begin{array}{c} \begin{array}{ccccc} x_1 & x_2 & s_1 & s_2 & P \end{array} \\ \left[\begin{array}{rrrrr|r} 1 & 2 & 1 & 0 & 0 & 32 \\ 3 & 4 & 0 & 1 & 0 & 84 \\ \hdashline -50 & -80 & 0 & 0 & 1 & 0 \end{array}\right] \end{array} \qquad (5)$$

Initial simplex tableau

Indicators are shown in color.

↑ Pivot column

$$x_1 = 0, \quad x_2 = 0, \quad s_1 = 32, \quad s_2 = 84, \quad P = 0$$

Initial basic feasible solution

Now that we have chosen the nonbasic variable x_2 as the entering variable (the nonbasic variable to become basic), which of the two basic variables, s_1 or s_2, should we choose as the exiting variable (the basic variable to become nonbasic)? We saw above that for $x_1 = 0$, each unit increase in the entering variable x_2 results in an increase of \$80 for P. Can we increase x_2 without limit? No! A limit is imposed by the nonnegative requirements for s_1 and s_2. (Remember that if any of the basic variables except P become negative, we no longer have a feasible solution.) So we rephrase the question and ask: How much can x_2 be increased when $x_1 = 0$ without causing s_1 or s_2 to become negative? To see how much x_2 can be increased, we refer to tableau (5) or system (3) and write the two problem constraint equations with $x_1 = 0$:

$$2x_2 + s_1 = 32$$
$$4x_2 + s_2 = 84$$

Solving for s_1 and s_2, we have

$$s_1 = 32 - 2x_2$$
$$s_2 = 84 - 4x_2$$

For s_1 and s_2 to be nonnegative, x_2 must be chosen so that both $32 - 2x_2$ and $84 - 4x_2$ are nonnegative. That is, so that

$$\begin{aligned} 32 - 2x_2 &\geq 0 \\ -2x_2 &\geq -32 \\ x_2 &\leq \tfrac{32}{2} = 16 \end{aligned} \qquad \text{and} \qquad \begin{aligned} 84 - 4x_2 &\geq 0 \\ -4x_2 &\geq -84 \\ x_2 &\leq \tfrac{84}{4} = 21 \end{aligned}$$

For both inequalities to be satisfied, x_2 must be less than or equal to the smaller of the values, which is 16. Thus, x_2 can increase to 16 without either s_1 or s_2 becoming negative. Now, observe how each value (16 and 21) can be obtained directly from tableau (6) below:

Entering variable ↓

$$\begin{array}{c} \\ s_1 \\ s_2 \\ P \end{array} \begin{array}{c} \begin{array}{ccccc|c} x_1 & x_2 & s_1 & s_2 & P & \end{array} \\ \left[\begin{array}{ccccc|c} 1 & 2 & 1 & 0 & 0 & 32 \\ 3 & 4 & 0 & 1 & 0 & 84 \\ \hdashline -50 & -80 & 0 & 0 & 1 & 0 \end{array}\right] \end{array} \qquad \begin{array}{l} \tfrac{32}{2} = 16 \text{ (smallest)} \\ \tfrac{84}{4} = 21 \end{array} \tag{6}$$

↑ Pivot column

From (6) we can determine the amount the entering variable can increase by choosing the smallest of the quotients obtained by dividing each element in the last column above the dashed line by the corresponding *positive* element in the pivot column. **If the corresponding element in the pivot column is 0 or nega-**

tive, no value is computed. The row with the smallest quotient is called the **pivot row,** and the variable to the left of the pivot row is the exiting variable. In this case, s_1 will be the exiting variable, and the roles of x_2 and s_1 will be interchanged. The element at the intersection of the pivot column and the pivot row, 2, is called the **pivot element,** and we circle this element for ease of recognition. Tableau (7) illustrates these ideas, which are also summarized in the box below.

Entering variable ↓ (x_2); Pivot element (2)

	x_1	x_2	s_1	s_2	P		
Exiting variable → s_1	1	(2)	1	0	0	32	← Pivot row
s_2	3	4	0	1	0	84	(7)
P	−50	−80	0	0	1	0	

↑ Pivot column

Selecting the Pivot Element

Step 1. Locate the most negative indicator in the bottom row of the tableau to the left of the P column (the negative number with the largest absolute value). The column containing this element is the *pivot column.* If there is a tie for the most negative, choose either.

Step 2. Divide each *positive* element in the pivot column above the dashed line into the corresponding element in the last column. The *pivot row* is the row corresponding to the smallest quotient obtained. If there is a tie for the smallest quotient, choose either. If the pivot column above the dashed line has no positive elements, then there is no solution, and we stop.

Step 3. The *pivot* (or *pivot element*) is the element in the intersection of the pivot column and pivot row. [*Note:* The pivot element is always positive and is never in the bottom row.]

[*Remember:* The entering variable is at the top of the pivot column and the exiting variable is at the left of the pivot row.]

In order for x_2 to be classified as a basic variable, we perform row operations on tableau (7) so that the pivot element is transformed into 1 and all other elements in the column into 0's. This procedure for transforming a nonbasic variable into a basic variable is called a *pivot operation,* or *pivoting,* and is summarized in the box.

Performing a Pivot Operation

A **pivot operation,** or **pivoting,** consists of performing row operations as follows:

Step 1. Multiply the pivot row by the reciprocal of the pivot element to transform the pivot element into a 1. (If the pivot element is already a 1, omit this step.)

Step 2. Add multiples of the pivot row to other rows in the tableau to transform all other nonzero elements in the pivot column into 0's.

[*Note:* Rows are not to be interchanged while performing a pivot operation. The only way the (positive) pivot element can be transformed into 1 (if it is not a 1 already) is for the pivot row to be multiplied by the reciprocal of the pivot element.]

Performing a pivot operation has the following effects:

1. The (entering) nonbasic variable becomes a basic variable.
2. The (exiting) basic variable becomes a nonbasic variable.
3. The value of the objective function is increased, or, in some cases, remains the same.

We now carry out the pivot operation on tableau (7). (To facilitate the process, we do not repeat the variables after the first tableau, and we use "Enter" and "Exit" for "Entering variable" and "Exiting variable," respectively.)

$$
\begin{array}{r}
\\ \\ \text{Exit} \rightarrow s_1 \\ s_2 \\ P
\end{array}
\begin{array}{c}
\qquad\quad \text{Enter} \\ \qquad\quad \downarrow \\
\begin{array}{ccccc} x_1 & x_2 & s_1 & s_2 & P \end{array} \\
\left[\begin{array}{rrrrr|r}
1 & \textcircled{2} & 1 & 0 & 0 & 32 \\
3 & 4 & 0 & 1 & 0 & 84 \\ \hdashline
-50 & -80 & 0 & 0 & 1 & 0
\end{array}\right]
\end{array}
\quad \tfrac{1}{2}R_1 \rightarrow R_1
$$

$$
\sim \left[\begin{array}{rrrrr|r}
\frac{1}{2} & \textcircled{1} & \frac{1}{2} & 0 & 0 & 16 \\
3 & 4 & 0 & 1 & 0 & 84 \\ \hdashline
-50 & -80 & 0 & 0 & 1 & 0
\end{array}\right]
\quad \begin{array}{l} R_2 + (-4)R_1 \rightarrow R_2 \\ R_3 + 80R_1 \rightarrow R_3 \end{array}
$$

$$
\sim \left[\begin{array}{rrrrr|r}
\frac{1}{2} & 1 & \frac{1}{2} & 0 & 0 & 16 \\
1 & 0 & -2 & 1 & 0 & 20 \\ \hdashline
-10 & 0 & 40 & 0 & 1 & 1{,}280
\end{array}\right]
$$

We have completed the pivot operation, and now we must insert appropriate variables for this new tableau. Observing that x_2, s_2, and P are now the basic variables, we write the new basic feasible solution by setting the nonbasic variables x_1 and s_1 equal to 0 and solving for the basic variables by inspection.

(Remember, the values of the basic variables listed on the left are the corresponding numbers to the right of the vertical line.)

$$\begin{array}{c|rrrrr|r|}
 & x_1 & x_2 & s_1 & s_2 & P & \\
\hline
x_2 & \frac{1}{2} & 1 & \frac{1}{2} & 0 & 0 & 16 \\
s_2 & 1 & 0 & -2 & 1 & 0 & 20 \\
\hdashline
P & -10 & 0 & 40 & 0 & 1 & 1{,}280
\end{array}$$

$x_1 = 0, \quad x_2 = 16, \quad s_1 = 0, \quad s_2 = 20, \quad P = 1{,}280$

A profit of \$1,280 is a marked improvement over the \$0 profit produced by the initial basic feasible solution. But we can improve P still further, since a negative indicator still remains in the bottom row. To see why, we write out the objective function:

$$-10x_1 + 40s_1 + P = 1{,}280$$

or

$$P = 10x_1 - 40s_1 + 1{,}280$$

If s_1 stays a nonbasic variable (set equal to 0) and x_1 becomes a new basic variable, then

$$P = 10x_1 - 40(0) + 1{,}280 = 10x_1 + 1{,}280$$

and for each unit increase in x_1, P will increase \$10.

We now go through another iteration of the simplex process (that is, we repeat the above sequence of steps) using another pivot element. The pivot element and the entering and exiting variables are shown in the following tableau:

$$\begin{array}{rc|rrrrr|r|l}
 & & \text{Enter } \downarrow & & & & & & \\
 & & x_1 & x_2 & s_1 & s_2 & P & & \\
\hline
 & x_2 & \frac{1}{2} & 1 & \frac{1}{2} & 0 & 0 & 16 & \frac{16}{1/2} = 32 \\
\text{Exit} \rightarrow & s_2 & \textcircled{1} & 0 & -2 & 1 & 0 & 20 & \frac{20}{1} = 20 \\
\hdashline
 & P & -10 & 0 & 40 & 0 & 1 & 1{,}280 &
\end{array}$$

We now pivot on (the circled) 1. That is, we perform a pivot operation using this 1 as the pivot element. Since the pivot element is 1, we do not need to perform the first step in the pivot operation, so we proceed to the second step to get 0's above and below the pivot element 1. As before, to facilitate the process, we omit writing the variables, except for the first tableau.

$$\begin{array}{rc|rrrrr|r|l}
 & & \text{Enter } \downarrow & & & & & & \\
 & & x_1 & x_2 & s_1 & s_2 & P & & \\
\hline
 & x_2 & \frac{1}{2} & 1 & \frac{1}{2} & 0 & 0 & 16 & R_1 + (-\frac{1}{2})R_2 \rightarrow R_1 \\
\text{Exit} \rightarrow & s_2 & \textcircled{1} & 0 & -2 & 1 & 0 & 20 & \\
\hdashline
 & P & -10 & 0 & 40 & 0 & 1 & 1{,}280 & R_3 + 10R_2 \rightarrow R_3
\end{array}$$

$$\sim \left[\begin{array}{rrrrr|r}
0 & 1 & \frac{3}{2} & -\frac{1}{2} & 0 & 6 \\
1 & 0 & -2 & 1 & 0 & 20 \\
\hdashline
0 & 0 & 20 & 10 & 1 & 1{,}480
\end{array}\right]$$

Since there are no more negative indicators in the bottom row, we are through. Let us insert the appropriate variables for this last tableau and write the corresponding basic feasible solution. The basic variables are now x_1, x_2, and P, so to get the corresponding basic feasible solution, we set the nonbasic variables s_1 and s_2 equal to 0 and solve for the basic variables by inspection.

$$\begin{array}{c} \\ x_2 \\ x_1 \\ P \end{array} \begin{array}{c} \begin{array}{ccccc} x_1 & x_2 & s_1 & s_2 & P \end{array} \\ \left[\begin{array}{ccccc|c} 0 & 1 & \frac{3}{2} & -\frac{1}{2} & 0 & 6 \\ 1 & 0 & -2 & 1 & 0 & 20 \\ \hdashline 0 & 0 & 20 & 10 & 1 & 1{,}480 \end{array}\right] \end{array}$$

$$x_1 = 20, \quad x_2 = 6, \quad s_1 = 0, \quad s_2 = 0, \quad P = 1{,}480$$

To see why this is the maximum, we rewrite the objective function from the bottom row:

$$20s_1 + 10s_2 + P = 1{,}480$$
$$P = 1{,}480 - 20s_1 - 10s_2$$

Since s_1 and s_2 cannot be negative, any increase of either from 0 will make the profit smaller.

Finally, returning to our original problem, we conclude that a production schedule of 20 standard tents and 6 expedition tents will produce a maximum profit of $1,480 per day, which is the same as the geometric solution obtained in Section 6-2. The fact that the slack variables are both 0 means that for this production schedule, the plant will operate at full capacity—there is no slack in either the cutting department or the assembly department.

◆ INTERPRETING THE SIMPLEX PROCESS GEOMETRICALLY

We can now interpret the simplex process just completed geometrically in terms of the feasible region graphed in the preceding section. Table 3 lists the three basic feasible solutions we just found using the simplex method (in the order they were found). The table also includes the corresponding corner points of the feasible region illustrated in Figure 10 (at the top of the next page).

TABLE 3

Basic Feasible Solutions (Obtained Above)

x_1	x_2	s_1	s_2	P	CORNER POINT
0	0	32	84	$ 0	$O(0, 0)$
0	16	0	20	$1,280	$A(0, 16)$
20	6	0	0	$1,480	$B(20, 6)$

Looking at Table 3 and Figure 10, we see that the simplex process started at the origin, moved to the adjacent corner point $A(0, 16)$, then to the optimal

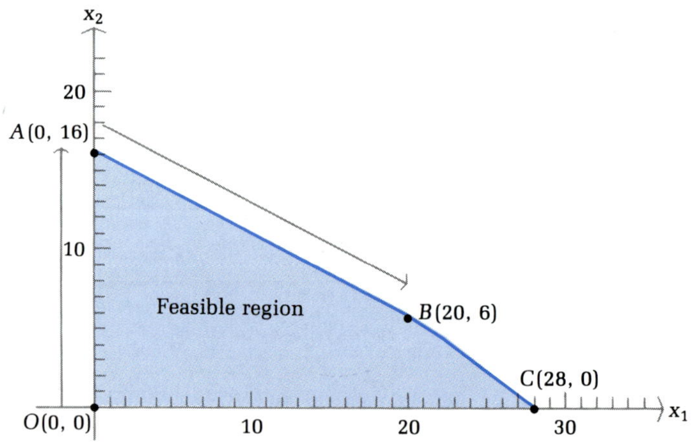

FIGURE 10

solution $B(20, 6)$ at the next adjacent corner point. This is typical of the simplex process.

◆ THE SIMPLEX METHOD SUMMARIZED

Before commencing with further examples, we summarize the important parts of the simplex method:

Simplex Method

KEY STEPS FOR STANDARD MAXIMIZATION PROBLEMS

(Problem constraints are of the $\leq$ form with nonnegative constants on the right. The coefficients of the objective function can be any real numbers.)

Step 1. Start with a standard maximization problem written in standard form.

Step 2. Introduce slack variables and write the initial system.

Step 3. Write the simplex tableau associated with the initial system.

Step 4. Determine the pivot element (if it exists) and the entering and exiting variables.

Step 5. Perform the pivot operation.

Step 6. Repeat steps 4 and 5 until all indicators in the bottom row are nonnegative. When this occurs, we stop the process and read the optimal solution.

Remarks

There are two different reasons for stopping the simplex process:

1. If we cannot select a new pivot column, we stop because the optimal solution has been found (see step 6 above).
2. If we select a new pivot column and then are unable to select a new pivot row, we stop because the problem has no solution (see step 4 above and step 2 on selecting the pivot element, page 336).

◆ EXAMPLE 11 Solve the following linear programming problem using the simplex method:

$$\begin{aligned} \text{Maximize} \quad & P = 10x_1 + 5x_2 \\ \text{Subject to} \quad & 6x_1 + 2x_2 \leqslant 36 \\ & 2x_1 + 4x_2 \leqslant 32 \\ & x_1, x_2 \geqslant 0 \end{aligned}$$

Solution Introduce slack variables s_1 and s_2 and write the initial system:

$$\begin{aligned} 6x_1 + 2x_2 + s_1 \qquad &= 36 \\ 2x_1 + 4x_2 \qquad + s_2 \qquad &= 32 \\ -10x_1 - 5x_2 \qquad\qquad + P &= 0 \\ x_1, x_2, s_1, s_2 &\geqslant 0 \end{aligned}$$

Write the simplex tableau and identify the first pivot element and the entering and exiting variables:

Enter ↓ (x_1)

$$\begin{matrix} & x_1 & x_2 & s_1 & s_2 & P & \\ \text{Exit} \rightarrow s_1 & \textcircled{6} & 2 & 1 & 0 & 0 & 36 \\ s_2 & 2 & 4 & 0 & 1 & 0 & 32 \\ P & -10 & -5 & 0 & 0 & 1 & 0 \end{matrix} \qquad \begin{matrix} \frac{36}{6} = 6 \\ \frac{32}{2} = 16 \end{matrix}$$

Perform the pivot operation:

Enter ↓ (x_1)

$$\begin{matrix} & x_1 & x_2 & s_1 & s_2 & P & \\ \text{Exit} \rightarrow s_1 & \textcircled{6} & 2 & 1 & 0 & 0 & 36 \\ s_2 & 2 & 4 & 0 & 1 & 0 & 32 \\ P & -10 & -5 & 0 & 0 & 1 & 0 \end{matrix} \qquad \tfrac{1}{6}R_1 \rightarrow R_1$$

$$\sim \left[\begin{array}{ccccc|c} \textcircled{1} & \frac{1}{3} & \frac{1}{6} & 0 & 0 & 6 \\ 2 & 4 & 0 & 1 & 0 & 32 \\ \hdashline -10 & -5 & 0 & 0 & 1 & 0 \end{array}\right] \qquad \begin{matrix} R_2 + (-2)R_1 \rightarrow R_2 \\ R_3 + 10R_1 \rightarrow R_3 \end{matrix}$$

$$\sim \begin{matrix} x_1 \\ s_2 \\ P \end{matrix} \left[\begin{array}{ccccc|c} 1 & \frac{1}{3} & \frac{1}{6} & 0 & 0 & 6 \\ 0 & \frac{10}{3} & -\frac{1}{3} & 1 & 0 & 20 \\ \hdashline 0 & -\frac{5}{3} & \frac{5}{3} & 0 & 1 & 60 \end{array}\right]$$

Since there still is a negative indicator in the last row, we repeat the process by finding a new pivot element:

$$
\begin{array}{r}
\\
x_1 \\
\text{Exit} \rightarrow s_2 \\
P
\end{array}
\begin{array}{c}
\qquad\text{Enter } \downarrow \text{ (on } x_2\text{)} \\
\begin{array}{ccccc|c}
x_1 & x_2 & s_1 & s_2 & P & \\
\end{array}\\
\left[\begin{array}{ccccc|c}
1 & \frac{1}{3} & \frac{1}{6} & 0 & 0 & 6 \\
0 & \textcircled{\frac{10}{3}} & -\frac{1}{3} & 1 & 0 & 20 \\
\hdashline
0 & -\frac{5}{3} & \frac{5}{3} & 0 & 1 & 60
\end{array}\right]
\end{array}
\qquad
\begin{array}{l}
6 \div \frac{1}{3} = 18 \\
20 \div \frac{10}{3} = 6
\end{array}
$$

Performing the pivot operation, we obtain

$$
\begin{array}{r}
x_1 \\
\text{Exit} \rightarrow s_2 \\
P
\end{array}
\left[\begin{array}{ccccc|c}
1 & \frac{1}{3} & \frac{1}{6} & 0 & 0 & 6 \\
0 & \textcircled{\frac{10}{3}} & -\frac{1}{3} & 1 & 0 & 20 \\
\hdashline
0 & -\frac{5}{3} & \frac{5}{3} & 0 & 1 & 60
\end{array}\right]
\qquad \tfrac{3}{10}R_2 \rightarrow R_2
$$

(Enter ↓ on column x_2; columns x_1, x_2, s_1, s_2, P)

$$
\sim \left[\begin{array}{ccccc|c}
1 & \frac{1}{3} & \frac{1}{6} & 0 & 0 & 6 \\
0 & \textcircled{1} & -\frac{1}{10} & \frac{3}{10} & 0 & 6 \\
\hdashline
0 & -\frac{5}{3} & \frac{5}{3} & 0 & 1 & 60
\end{array}\right]
\qquad
\begin{array}{l}
R + (-\frac{1}{3})R_2 \rightarrow R_1 \\
\\
R_3 + \frac{5}{3}R_2 \rightarrow R_3
\end{array}
$$

$$
\begin{array}{r}
x_1 \\
\sim x_2 \\
P
\end{array}
\left[\begin{array}{ccccc|c}
1 & 0 & \frac{1}{5} & -\frac{1}{10} & 0 & 4 \\
0 & 1 & -\frac{1}{10} & \frac{3}{10} & 0 & 6 \\
\hdashline
0 & 0 & \frac{3}{2} & \frac{1}{2} & 1 & 70
\end{array}\right]
$$

Since all the indicators in the last row are nonnegative, we stop and read the optimal solution:

$$\text{Max } P = 70 \quad \text{at} \quad x_1 = 4, \quad x_2 = 6, \quad s_1 = 0, \quad s_2 = 0$$

(If this still is not clear, write the objective function corresponding to the last row to see what happens to P when you try to increase s_1 or s_2.) ◆

PROBLEM 11 Solve the following linear programming problem using the simplex method:

$$
\begin{array}{ll}
\text{Maximize} & P = 2x_1 + x_2 \\
\text{Subject to} & 4x_1 + x_2 \leqslant 8 \\
& 2x_1 + 2x_2 \leqslant 10 \\
& x_1, x_2 \geqslant 0
\end{array}
$$

◆

◆ EXAMPLE 12 Solve using the simplex method:

$$
\begin{array}{ll}
\text{Maximize} & P = 6x_1 + 3x_2 \\
\text{Subject to} & -2x_1 + 3x_2 \leqslant 9 \\
& -x_1 + 3x_2 \leqslant 12 \\
& x_1, x_2 \geqslant 0
\end{array}
$$

Solution Write the initial system using the slack variables s_1 and s_2:

$$\begin{aligned} -2x_1 + 3x_2 + s_1 \qquad\quad &= 9 \\ -x_1 + 3x_2 \qquad + s_2 \quad &= 12 \\ -6x_1 - 3x_2 \qquad\qquad + P &= 0 \end{aligned}$$

Write the simplex tableau and identify the first pivot element:

$$\begin{matrix} & x_1 & x_2 & s_1 & s_2 & P & \\ s_1 & -2 & 3 & 1 & 0 & 0 & 9 \\ s_2 & -1 & 3 & 0 & 1 & 0 & 12 \\ P & -6 & -3 & 0 & 0 & 1 & 0 \end{matrix}$$

↑ Pivot column

Since both elements in the pivot column above the dashed line are negative, we are unable to select a pivot row. We stop and conclude that there is no solution. ◆

Notice that we do not try to continue with the simplex method by using a different column for the pivot column. The pivot column must correspond to the most negative indicator in the bottom row.

Once the pivot column has been selected, either there is a pivot row and the simplex method can be continued, or there is no pivot row and the simplex method stops, and we conclude that there is no solution.

PROBLEM 12 Solve using the simplex method:

Maximize $P = 2x_1 + 3x_2$

Subject to $-3x_1 + 4x_2 \leq 12$

$x_2 \leq 9$

$x_1, x_2 \geq 0$ ◆

◆ APPLICATION

◆ EXAMPLE 13 *Agriculture* A farmer owns a 100 acre farm and plans to plant at most three crops. The seed for crops *A*, *B*, and *C* costs \$40, \$20, and \$30 per acre, respectively. A maximum of \$3,200 can be spent on seed. Crops *A*, *B*, and *C* require 1, 2, and 1 workdays per acre, respectively, and there are a maximum of 160 workdays available. If the farmer can make a profit of \$100 per acre on crop *A*, \$300 per acre on crop *B*, and \$200 per acre on crop *C*, how many acres of each crop should be planted to maximize profit?

Solution Let

$$x_1 = \text{Number of acres of crop } A$$
$$x_2 = \text{Number of acres of crop } B$$
$$x_3 = \text{Number of acres of crop } C$$
$$P = \text{Total profit}$$

Then we have the following linear programming problem:

$$\begin{array}{lll} \text{Maximize} & P = 100x_1 + 300x_2 + 200x_3 & \text{Objective function} \\ \text{Subject to} & \left.\begin{array}{r} x_1 + x_2 + x_3 \leq 100 \\ 40x_1 + 20x_2 + 30x_3 \leq 3{,}200 \\ x_1 + 2x_2 + x_3 \leq 160 \end{array}\right\} & \text{Problem constraints} \\ & x_1, x_2, x_3 \geq 0 & \text{Nonnegative constraints} \end{array}$$

Next, we introduce slack variables and form the initial system:

$$\begin{aligned} x_1 + x_2 + x_3 + s_1 \qquad\qquad &= 100 \\ 40x_1 + 20x_2 + 30x_3 + s_2 \qquad &= 3{,}200 \\ x_1 + 2x_2 + x_3 + s_3 \quad &= 160 \\ -100x_1 - 300x_2 - 200x_3 + P &= 0 \\ x_1, x_2, x_3, s_1, s_2, s_3 &\geq 0 \end{aligned}$$

Notice that the initial system has $7 - 4 = 3$ nonbasic variables and 4 basic variables. Now we form the simplex tableau and solve by the simplex method:

$$\begin{array}{c} \\ \\ s_1 \\ s_2 \\ \text{Exit} \rightarrow s_3 \\ P \end{array} \begin{array}{c} \quad\text{Enter} \\ \quad\downarrow \\ \begin{array}{ccccccc|c} x_1 & x_2 & x_3 & s_1 & s_2 & s_3 & P & \\ 1 & 1 & 1 & 1 & 0 & 0 & 0 & 100 \\ 40 & 20 & 30 & 0 & 1 & 0 & 0 & 3{,}200 \\ 1 & \textcircled{2} & 1 & 0 & 0 & 1 & 0 & 160 \\ \hline -100 & -300 & -200 & 0 & 0 & 0 & 1 & 0 \end{array} \end{array} \quad 0.5R_3 \rightarrow R_3$$

$$\sim \left[\begin{array}{ccccccc|c} 1 & 1 & 1 & 1 & 0 & 0 & 0 & 100 \\ 40 & 20 & 30 & 0 & 1 & 0 & 0 & 3{,}200 \\ 0.5 & \textcircled{1} & 0.5 & 0 & 0 & 0.5 & 0 & 80 \\ \hline -100 & -300 & -200 & 0 & 0 & 0 & 1 & 0 \end{array}\right] \quad \begin{array}{l} R_1 + (-1)R_3 \rightarrow R_1 \\ R_2 + (-20)R_3 \rightarrow R_2 \\ \\ R_4 + 300R_3 \rightarrow R_4 \end{array}$$

$$
\begin{array}{c}
\\
\\
\text{Exit} \rightarrow \begin{array}{c} s_1 \\ s_2 \\ x_2 \\ P \end{array}
\end{array}
\begin{array}{c}
\qquad\qquad\quad \text{Enter} \\
\qquad\qquad\quad \downarrow \\
\begin{array}{ccccccc|c} x_1 & x_2 & x_3 & s_1 & s_2 & s_3 & P & \\ \end{array} \\
\left[\begin{array}{ccccccc|c}
0.5 & 0 & \textcircled{0.5} & 1 & 0 & -0.5 & 0 & 20 \\
30 & 0 & 20 & 0 & 1 & -10 & 0 & 1{,}600 \\
0.5 & 1 & 0.5 & 0 & 0 & 0.5 & 0 & 80 \\
\hdashline
50 & 0 & -50 & 0 & 0 & 150 & 1 & 24{,}000
\end{array}\right]
\end{array}
\quad 2R_1 \rightarrow R_1
$$

$$
\sim \left[\begin{array}{ccccccc|c}
1 & 0 & \textcircled{1} & 2 & 0 & -1 & 0 & 40 \\
30 & 0 & 20 & 0 & 1 & -10 & 0 & 1{,}600 \\
0.5 & 1 & 0.5 & 0 & 0 & 0.5 & 0 & 80 \\
\hdashline
50 & 0 & -50 & 0 & 0 & 150 & 1 & 24{,}000
\end{array}\right]
\quad
\begin{array}{l}
\\
R_2 + (-20)R_1 \rightarrow R_2 \\
R_3 + (-0.5)R_1 \rightarrow R_3 \\
R_4 + 50R_1 \rightarrow R_4
\end{array}
$$

$$
\sim \begin{array}{c} x_3 \\ s_2 \\ x_2 \\ P \end{array}
\left[\begin{array}{ccccccc|c}
1 & 0 & 1 & 2 & 0 & -1 & 0 & 40 \\
10 & 0 & 0 & -40 & 1 & 10 & 0 & 800 \\
0 & 1 & 0 & -1 & 0 & 1 & 0 & 60 \\
\hdashline
100 & 0 & 0 & 100 & 0 & 100 & 1 & 26{,}000
\end{array}\right]
$$

All indicators in the bottom row are nonnegative, and we can now read the optimal solution:

$$x_1 = 0, \quad x_2 = 60, \quad x_3 = 40, \quad s_1 = 0, \quad s_2 = 800, \quad s_3 = 0, \quad P = \$26{,}000$$

Thus, if the farmer plants 60 acres in crop *B*, 40 acres in crop *C*, and no crop *A*, the maximum profit of \$26,000 will be realized. The fact that $s_2 = 800$ tells us (look at the second row in the equations at the start) that this maximum profit is reached by using only \$2,400 of the \$3,200 available for seed; that is, we have a slack of \$800 that can be used for some other purpose. ◆

PROBLEM 13 Repeat Example 13 modified as follows:

	INVESTMENT PER ACRE Crop *A*	Crop *B*	Crop *C*	MAXIMUM AVAILABLE
SEED COST	\$24	\$40	\$30	\$3,600
WORKDAYS	1	2	2	160
PROFIT	\$140	\$200	\$160	

◆

Remarks

1. It can be shown that the feasible region for the linear programming problem in Example 13 has eight corner points, yet the simplex method found the solution in only two steps. Now you begin to see the power of the simplex method. In larger problems, the difference between the total number of corner points and the number of steps required by the simplex method is

even more dramatic. A feasible region may have hundreds or even thousands of corner points, yet the simplex method will often find the optimal solution in 10 or 15 steps.

2. Refer to the second problem constraint in the model for Example 13:

$$40x_1 + 20x_2 + 30x_3 \leq 3{,}200$$

Multiplying both sides of this inequality by $\frac{1}{10}$ before introducing a slack variable simplifies subsequent calculations. However, performing this operation has a side effect—it changes the units of the slack variable from dollars to tens of dollars. Compare the equations

$$40x_1 + 20x_2 + 30x_3 + s_2 = 3{,}200 \qquad s_2 \text{ represents dollars}$$

and

$$4x_1 + 2x_2 + 3x_3 + s_2' = 320 \qquad s_2' \text{ represents tens of dollars}$$

to see why this happens. In general, if you multiply a problem constraint by a number, remember to take this into account when you interpret the value of the slack variable for that constraint.

3. It is important to realize that in order to keep this introduction as simple as possible, we have purposely avoided certain degenerate cases that lead to difficulties. Discussion and resolution of these problems is left to a more advanced treatment of the subject.

Answers to Matched Problems

11. Max $P = 6$ when $x_1 = 1$ and $x_2 = 4$ 12. No solution
13. 40 acres of crop *A*, 60 acres of crop *B*, no crop *C*; Max $P = \$17{,}600$ (since $s_2 = 240$, \$240 out of the \$3,600 will not be spent)

EXERCISE 6-4

A

For the simplex tableaux in Problems 1–4:

(A) Identify the basic and nonbasic variables.
(B) Find the corresponding basic feasible solution.
(C) Determine whether the optimal solution has been found, an additional pivot is required, or the problem has no solution.

1.

$$\begin{array}{c} \begin{matrix} x_1 & x_2 & s_1 & s_2 & P \end{matrix} \\ \left[\begin{array}{rrrrr|r} 2 & 1 & 0 & 3 & 0 & 12 \\ 3 & 0 & 1 & -2 & 0 & 15 \\ \hline -4 & 0 & 0 & 4 & 1 & 20 \end{array}\right] \end{array}$$

2.

$$\begin{array}{c} \begin{matrix} x_1 & x_2 & s_1 & s_2 & P \end{matrix} \\ \left[\begin{array}{rrrrr|r} 1 & 4 & -2 & 0 & 0 & 10 \\ 0 & 2 & 3 & 1 & 0 & 25 \\ \hline 0 & 5 & 6 & 0 & 1 & 35 \end{array}\right] \end{array}$$

3.

x_1	x_2	x_3	s_1	s_2	s_3	P	
−2	0	1	3	1	0	0	5
0	1	0	−2	0	0	0	15
−1	0	0	4	1	1	0	12
−4	0	0	2	4	0	1	45

4.

x_1	x_2	x_3	s_1	s_2	s_3	P	
0	2	−1	1	4	0	0	5
0	1	2	0	−2	1	0	2
1	3	0	0	5	0	0	11
0	−5	4	0	−3	0	1	27

In Problems 5–8, find the pivot element, identify the entering and exiting variables, and perform one pivot operation.

5.

x_1	x_2	s_1	s_2	P	
1	4	1	0	0	4
3	5	0	1	0	24
−8	−5	0	0	1	0

6.

x_1	x_2	s_1	s_2	P	
1	6	1	0	0	36
3	1	0	1	0	5
−1	−2	0	0	1	0

7.

x_1	x_2	s_1	s_2	s_3	P	
2	1	1	0	0	0	4
3	0	1	1	0	0	8
0	0	2	0	1	0	2
−4	0	−3	0	0	1	5

8.

x_1	x_2	s_1	s_2	s_3	P	
0	0	2	1	1	0	2
1	0	−4	0	1	0	3
0	1	5	0	2	0	11
0	0	−6	0	−5	1	18

In Problems 9–12:

(A) Using slack variables, write the initial system for each linear programming problem.
(B) Write the simplex tableau, circle the first pivot, and identify the entering and exiting variables.
(C) Use the simplex method to solve the problem.

9. Maximize $P = 15x_1 + 10x_2$
Subject to $2x_1 + x_2 \leq 10$
$x_1 + 2x_2 \leq 8$
$x_1, x_2 \geq 0$

10. Maximize $P = 3x_1 + 2x_2$
Subject to $6x_1 + 3x_2 \leq 24$
$3x_1 + 6x_2 \leq 30$
$x_1, x_2 \geq 0$

11. Repeat Problem 9 with the objective function changed to $P = 30x_1 + x_2$.
12. Repeat Problem 10 with the objective function changed to $P = x_1 + 3x_2$.

B *Solve the following linear programming problems using the simplex method:*

13. Maximize $P = 30x_1 + 40x_2$
Subject to $2x_1 + x_2 \leq 10$
$x_1 + x_2 \leq 7$
$x_1 + 2x_2 \leq 12$
$x_1, x_2 \geq 0$

14. Maximize $P = 20x_1 + 10x_2$
Subject to $3x_1 + x_2 \leq 21$
$x_1 + x_2 \leq 9$
$x_1 + 3x_2 \leq 21$
$x_1, x_2 \geq 0$

15. Maximize $P = 2x_1 + 3x_2$
Subject to $-2x_1 + x_2 \leq 2$
$-x_1 + x_2 \leq 5$
$x_2 \leq 6$
$x_1, x_2 \geq 0$

16. Repeat Problem 15 with $P = -x_1 + 3x_2$.

17. Maximize $P = -x_1 + 2x_2$
Subject to $-x_1 + x_2 \leq 2$
$-x_1 + 3x_2 \leq 12$
$x_1 - 4x_2 \leq 4$
$x_1, x_2 \geq 0$

18. Repeat Problem 17 with $P = x_1 + 2x_2$.

19. Maximize $P = 5x_1 + 2x_2 - x_3$
Subject to $x_1 + x_2 - x_3 \leq 10$
$2x_1 + 4x_2 + x_3 \leq 30$
$x_1, x_2, x_3 \geq 0$

20. Maximize $P = 4x_1 - 3x_2 + 2x_3$
Subject to $x_1 + 2x_2 - x_3 \leq 5$
$3x_1 + 2x_2 + 2x_3 \leq 22$
$x_1, x_2, x_3 \geq 0$

21. Maximize $P = 2x_1 + 3x_2 + 4x_3$
Subject to $x_1 + x_3 \leq 4$
$x_2 + x_3 \leq 3$
$x_1, x_2, x_3 \geq 0$

22. Maximize $P = x_1 + x_2 + 2x_3$
Subject to $x_1 - 2x_2 + x_3 \leq 9$
$2x_1 + x_2 + 2x_3 \leq 28$
$x_1, x_2, x_3 \geq 0$

23. Maximize $P = 4x_1 + 3x_2 + 2x_3$
Subject to $3x_1 + 2x_2 + 5x_3 \leq 23$
$2x_1 + x_2 + x_3 \leq 8$
$x_1 + x_2 + 2x_3 \leq 7$
$x_1, x_2, x_3 \geq 0$

24. Maximize $P = 4x_1 + 2x_2 + 3x_3$
Subject to $x_1 + x_2 + x_3 \leq 11$
$2x_1 + 3x_2 + x_3 \leq 20$
$x_1 + 3x_2 + 2x_3 \leq 20$
$x_1, x_2, x_3 \geq 0$

C

25. Maximize $P = 20x_1 + 30x_2$
Subject to $0.6x_1 + 1.2x_2 \leq 960$
$0.03x_1 + 0.04x_2 \leq 36$
$0.3x_1 + 0.2x_2 \leq 270$
$x_1, x_2 \geq 0$

26. Repeat Problem 25 with $P = 20x_1 + 20x_2$.

27. Maximize $P = x_1 + 2x_2 + 3x_3$
Subject to $2x_1 + 2x_2 + 8x_3 \leq 600$
$x_1 + 3x_2 + 2x_3 \leq 600$
$3x_1 + 2x_2 + x_3 \leq 400$
$x_1, x_2, x_3 \geq 0$

28. Maximize $P = 10x_1 + 50x_2 + 10x_3$
Subject to $3x_1 + 3x_2 + 3x_3 \leq 66$
$6x_1 - 2x_2 + 4x_3 \leq 48$
$3x_1 + 6x_2 + 9x_3 \leq 108$
$x_1, x_2, x_3 \geq 0$

In Problems 29 and 30, first solve the linear programming problem by the simplex method, keeping track of the basic feasible solutions at each step. Then graph the feasible region and illustrate the path to the optimal solution determined by the simplex method.

29. Maximize $P = 2x_1 + 5x_2$

Subject to
$$x_1 + 2x_2 \le 40$$
$$x_1 + 3x_2 \le 48$$
$$x_1 + 4x_2 \le 60$$
$$x_2 \le 14$$
$$x_2, x_2 \ge 0$$

30. Maximize $P = 5x_1 + 3x_2$

Subject to
$$5x_1 + 4x_2 \le 100$$
$$2x_1 + x_2 \le 28$$
$$4x_1 + x_2 \le 42$$
$$x_1 \le 10$$
$$x_1, x_2 \ge 0$$

APPLICATIONS

Formulate each of the following as a linear programming problem. Then solve the problem using the simplex method.

Business & Economics

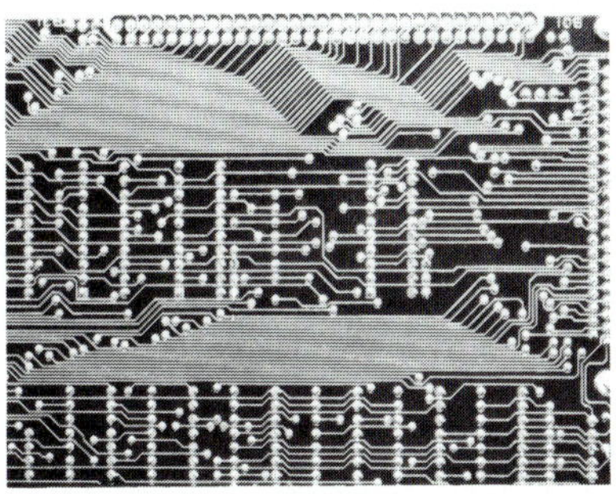

31. *Manufacturing—resource allocation.* A small company manufactures three different electronic components for computers. Component A requires 2 hours of fabrication and 1 hour of assembly; component B requires 1 hour of fabrication and 2 hours of assembly; and component C requires 2 hours of fabrication and 2 hours of assembly. The company has up to 1,000 labor-hours of fabrication time and 800 labor-hours of assembly time available per week. The profit on each component, A, B, and C, is \$7, \$9, and \$10, respectively. How many components of each type should the company manufacture each week in order to maximize its profit (assuming all components that it manufactures can be sold)? What is the maximum profit?

32. *Manufacturing—resource allocation.* Repeat Problem 31 under the additional assumption that the combined total number of components produced by the company cannot exceed 550.

33. *Investment.* An investor has at most \$100,000 to invest in government bonds, mutual funds, and money market funds. The average yields for government bonds, mutual funds, and money market funds are 8%, 13%, and 15%, respectively. The investor's policy requires that the total amount invested in mutual and money market funds not exceed the amount invested in government bonds. How much should be invested in each type of investment in order to maximize the return? What is the maximum return?

34. *Investment.* Repeat Problem 33 under the additional assumption that no more than \$30,000 can be invested in money market funds.

35. *Advertising.* A department store chain has up to \$20,000 to spend on television advertising for a sale. All ads will be placed with one television station, where a 30-second ad costs \$1,000 on daytime TV and is viewed by 14,000 potential customers, \$2,000 on prime-time TV and is viewed by 24,000 potential customers, and \$1,500 on late-night TV and is viewed by 18,000 potential customers. The television station will not accept a total of more than 15 ads in all three time periods. How many ads should be placed in each time period in order to maximize the number of potential customers who will see the ads? How many potential customers will see the ads? (Ignore repeated viewings of the ad by the same potential customer.)

36. *Advertising.* Repeat Problem 35 if the department store increases its budget to \$24,000 and requires that at least half of the ads be placed in prime-time shows.

37. *Construction — resource allocation.* A contractor is planning to build a new housing development consisting of colonial, split-level, and ranch-style houses. A colonial house requires $\frac{1}{2}$ acre of land, \$60,000 capital, and 4,000 labor-hours to construct, and returns a profit of \$20,000. A split-level house requires $\frac{1}{2}$ acre of land, \$60,000 capital, and 3,000 labor-hours to construct, and returns a profit of \$18,000. A ranch house requires 1 acre of land, \$80,000 capital, and 4,000 labor-hours to construct, and returns a profit of \$24,000. The contractor has available 30 acres of land, \$3,200,000 capital, and 180,000 labor-hours. How many houses of each type should be constructed to maximize the contractor's profit? What is the maximum profit?

38. *Manufacturing — resource allocation.* A company manufactures three-speed, five-speed, and ten-speed bicycles. Each bicycle passes through three departments, fabrication, painting & plating, and final assembly. The relevant manufacturing data are given in the table. How many bicycles of each type should the company manufacture per day in order to maximize its profit? What is the maximum profit?

	LABOR-HOURS PER BICYCLE			MAXIMUM
	Three-Speed	*Five-Speed*	*Ten-Speed*	LABOR-HOURS AVAILABLE PER DAY
FABRICATION	3	4	5	120
PAINTING & PLATING	5	3	5	130
FINAL ASSEMBLY	4	3	5	120
PROFIT PER BICYCLE	80	70	100	

39. *Packaging — product mix.* A candy company makes three types of candy, solid-center, fruit-filled, and cream-filled, and packages these candies in three different assortments. A box of assortment I contains 4 solid-center, 4 fruit-filled, and 12 cream-filled candies, and sells for \$9.40. A box of assortment II contains 12 solid-center, 4 fruit-filled, and 4 cream-filled candies, and sells for \$7.60. A box of assortment III contains 8 solid-center, 8 fruit-filled, and 8 cream-filled candies, and sells for \$11.00. The manufacturing costs per piece of candy are \$0.20 for solid-center, \$0.25 for fruit-filled, and \$0.30 for cream-filled. The company can manufacture 4,800 solid-center, 4,000 fruit-filled, and 5,600 cream-filled candies weekly. How many boxes of each type should they produce each week in order to maximize their profit? What is the maximum profit?

40. *Scheduling—resource allocation.* A small accounting firm prepares tax returns for three types of customers: individual, commercial, and industrial. The tax preparation process begins with a 1 hour interview with the customer. The data collected during this interview are entered into a time-sharing computer system, which produces the customer's tax return. It takes 1 hour to enter the data for an individual customer, 2 hours for a commercial customer, and $1\frac{1}{2}$ hours for an industrial customer. It takes 10 minutes of computer time to process an individual return, 25 minutes to process a commercial return, and 20 minutes to process an industrial return. The firm has one employee who conducts the initial interview and two who enter the data into the computer. The interviewer can work a maximum of 50 hours a week, and each of the data-entry employees can work a maximum of 40 hours a week. The computer is available for a maximum of 1,025 minutes a week. The firm makes a profit of $50 on each individual customer, $65 on each commercial customer, and $60 on each industrial customer. How many customers of each type should the firm schedule each week in order to maximize its profit? What is the maximum profit?

Life Sciences

41. *Nutrition—animals.* The natural diet of a certain animal consists of three foods, *A*, *B*, and *C*. The number of units of calcium, iron, and protein in 1 gram of each food and the average daily intake are given in the table. A scientist wants to investigate the effect of increasing the protein in the animal's diet while not allowing the units of calcium and iron to exceed their average daily intakes. How many grams of each food should be used to maximize the amount of protein in the diet? What is the maximum amount of protein?

	UNITS PER GRAM			AVERAGE DAILY INTAKE
	Food A	Food B	Food C	(UNITS)
CALCIUM	1	3	2	30
IRON	2	1	1	24
PROTEIN	3	3	5	60

42. *Nutrition—animals.* Repeat Problem 41 if the scientist wants to maximize the daily calcium intake while not allowing the intake of iron or protein to exceed the average daily intake.

Social Sciences

43. *Opinion survey.* A political scientist has received a grant to fund a research project involving voting trends. The budget of the grant includes $1,620 for conducting door-to-door interviews the day before an election. Undergraduate students, graduate students, and faculty members will be hired to

conduct the interviews. Each undergraduate student will conduct 18 interviews and be paid $60. Each graduate student will conduct 25 interviews and be paid $90. Each faculty member will conduct 30 interviews and be paid $120. Due to limited transportation facilities, no more than 20 interviewers can be hired. How many undergraduate students, graduate students, and faculty members should be hired in order to maximize the number of interviews that will be conducted? What is the maximum number of interviews?

44. *Opinion survey.* Repeat Problem 43 if one of the requirements of the grant is that at least 50% of the interviewers be undergraduate students.

SECTION 6-5

The Dual; Minimization with Problem Constraints of the Form $\geq$

- FORMATION OF THE DUAL PROBLEM
- SOLUTION OF MINIMIZATION PROBLEMS
- APPLICATION: A TRANSPORTATION PROBLEM
- SUMMARY OF PROBLEM TYPES AND SOLUTION METHODS

In the last section we restricted ourselves to standard maximization problems (problem constraints of the form $\leq$, with nonnegative constants on the right and objective function coefficients any real numbers). Now we will consider minimization problems with $\geq$ problem constraints. These two types of problems turn out to be very closely related.

FORMATION OF THE DUAL PROBLEM

Associated with each minimization problem is a maximization problem called the **dual problem.** To illustrate the procedure for finding the dual problem, we will use Example 8 from Section 6-2. There we solved the following linear programming problem geometrically. Now we will form its dual, and later we will solve the problem using the dual and the simplex method.

$$\begin{aligned} \text{Minimize} \quad & C = 3x_1 + x_2 \\ \text{Subject to} \quad & 10x_1 + 2x_2 \geq 84 \\ & 8x_1 + 4x_2 \geq 120 \\ & x_1, x_2 \geq 0 \end{aligned} \tag{1}$$

The first step in forming the dual problem is to construct a matrix by using the problem constraints and the objective function written in the following form:

$$\begin{aligned} 10x_1 + 2x_2 &\geq 84 \\ 8x_1 + 4x_2 &\geq 120 \\ 3x_1 + x_2 &= C \end{aligned} \qquad A = \left[\begin{array}{cc|c} 10 & 2 & 84 \\ 8 & 4 & 120 \\ \hline 3 & 1 & 1 \end{array}\right]$$

Be careful not to confuse this matrix with the simplex tableau. We use a solid horizontal line in the matrix to help distinguish the dual matrix from the simplex tableau. No slack variables are involved in matrix A, and the coefficient of C is in the same column as the constants from the problem constraints.

Now we will form a second matrix B by using the rows of A as the columns of B. [Technically, B is called the *transpose of A*. In general, the **transpose** of a given matrix is formed by interchanging its rows and corresponding columns (first row with first column, second row with second column, and so on).]

$$A = \left[\begin{array}{cc|c} 10 & 2 & 84 \\ 8 & 4 & 120 \\ \hline 3 & 1 & 1 \end{array}\right]$$

R_1 in $A = C_1$ in B

R_2 in $A = C_2$ in B

R_3 in $A = C_3$ in B

B is the transpose of A

$$\left[\begin{array}{cc|c} 10 & 8 & 3 \\ 2 & 4 & 1 \\ \hline 84 & 120 & 1 \end{array}\right] = B$$

Finally, we use the rows of B to define a new linear programming problem. This new problem will always be a maximization problem with $\leqslant$ problem constraints. To avoid confusion, we shall use different variables in this new problem:

$$\begin{aligned} 10y_1 + 8y_2 &\leqslant 3 \\ 2y_1 + 4y_2 &\leqslant 1 \\ 84y_1 + 120y_2 &= P \end{aligned} \qquad B = \left[\begin{array}{cc|c} y_1 & y_2 & \\ 10 & 8 & 3 \\ 2 & 4 & 1 \\ \hline 84 & 120 & 1 \end{array}\right]$$

The dual of the minimization problem (1) is:

$$\begin{aligned} \text{Maximize} \quad & P = 84y_1 + 120y_2 \\ \text{Subject to} \quad & 10y_1 + 8y_2 \leqslant 3 \\ & 2y_1 + 4y_2 \leqslant 1 \\ & y_1, y_2 \geqslant 0 \end{aligned}$$

This procedure is summarized in the box below:

Formation of the Dual Problem

Given a minimization problem with $\geqslant$ problem constraints:

Step 1. Use the coefficients of the problem constraints and the objective function to form a matrix A with the coefficients of the objective function in the last row.

Step 2. Use the *rows* of the matrix A as the *columns* of a second matrix B (matrix B is the transpose of A).

Step 3. Use the *rows* of B to form a maximization problem with $\leqslant$ problem constraints.

◆ EXAMPLE 14 Form the dual problem:

$$\begin{aligned} \text{Minimize} \quad & C = 40x_1 + 12x_2 + 40x_3 \\ \text{Subject to} \quad & 2x_1 + x_2 + 5x_3 \geq 20 \\ & 4x_1 + x_2 + x_3 \geq 30 \\ & x_1, x_2, x_3 \geq 0 \end{aligned}$$

Solution *Step 1.* Form the matrix A:

$$A = \left[\begin{array}{ccc|c} 2 & 1 & 5 & 20 \\ 4 & 1 & 1 & 30 \\ \hline 40 & 12 & 40 & 1 \end{array}\right]$$

Step 2. Form the matrix B (the transpose of A):

$$B = \left[\begin{array}{cc|c} 2 & 4 & 40 \\ 1 & 1 & 12 \\ 5 & 1 & 40 \\ \hline 20 & 30 & 1 \end{array}\right]$$

Step 3. State the dual problem:

$$\begin{aligned} \text{Maximize} \quad & P = 20y_1 + 30y_2 \\ \text{Subject to} \quad & 2y_1 + 4y_2 \leq 40 \\ & y_1 + y_2 \leq 12 \\ & 5y_1 + y_2 \leq 40 \\ & y_1, y_2 \geq 0 \end{aligned}$$

◆

PROBLEM 14 Find the dual problem:

$$\begin{aligned} \text{Minimize} \quad & C = 16x_1 + 9x_2 + 21x_3 \\ \text{Subject to} \quad & x_1 + x_2 + 3x_3 \geq 16 \\ & 2x_1 + x_2 + x_3 \geq 12 \\ & x_1, x_2, x_3 \geq 0 \end{aligned}$$

◆

◆ SOLUTION OF MINIMIZATION PROBLEMS

The following theorem establishes the relationship between the solution of a minimization problem and the solution of its dual:

THEOREM 6

> A minimization problem has a solution if and only if its dual problem has a solution. If a solution exists, then the optimal value of the minimization problem is the same as the optimal value of the dual problem.

In the preceding section, we saw that the simplex method can be used to solve standard maximization problems (problem constraints of the form $\le$ with nonnegative constants on the right and objective function coefficients any real numbers). When the dual of a minimization problem is formed, the coefficients of the objective function in the minimization problem (the last row in A) become the constants in the problem constraints in the dual problem (the last column in B).

Thus, a minimization problem with $\ge$ problem constraints whose objective function has nonnegative coefficients can be solved by applying the simplex method to the dual.

To illustrate this, let us return to Example 8, whose dual was found at the beginning of this section.

Original Problem	Dual Problem
Minimize $C = 3x_1 + x_2$	Maximize $P = 84y_1 + 120y_2$
Subject to $10x_1 + 2x_2 \ge 84$	Subject to $10y_1 + 8y_2 \le 3$
$8x_1 + 4x_2 \ge 120$	$2y_1 + 4y_2 \le 1$
$x_1, x_2 \ge 0$	$y_1, y_2 \ge 0$

Now we will use the simplex method to solve the dual problem. For reasons that will become clear later, we will use the variables x_1 and x_2 from the original problem as the slack variables in the dual:

$$\begin{array}{rrrrrl}
10y_1 + & 8y_2 + x_1 & & & = 3 \\
2y_1 + & 4y_2 & + x_2 & & = 1 \\
-84y_1 - & 120y_2 & & + P & = 0
\end{array}$$

Initial system for the dual.

$$\begin{array}{c}
\\ x_1 \\ x_2 \\ P
\end{array}
\begin{array}{c}
\begin{array}{ccccc} y_1 & y_2 & x_1 & x_2 & P \end{array}\\
\left[\begin{array}{rrrrr|r}
10 & 8 & 1 & 0 & 0 & 3 \\
2 & \boxed{4} & 0 & 1 & 0 & 1 \\ \hdashline
-84 & -120 & 0 & 0 & 1 & 0
\end{array}\right]
\end{array}
\quad \tfrac{1}{4}R_2 \to R_2$$

$$\sim \left[\begin{array}{rrrrr|r}
10 & 8 & 1 & 0 & 0 & 3 \\
\frac{1}{2} & \boxed{1} & 0 & \frac{1}{4} & 0 & \frac{1}{4} \\ \hdashline
-84 & -120 & 0 & 0 & 1 & 0
\end{array}\right]
\quad \begin{array}{l} R_1 + (-8)R_2 \to R_1 \\ \\ R_3 + 120R_2 \to R_3 \end{array}$$

$$\begin{array}{c} x_1 \\ \sim y_2 \\ P \end{array}
\left[\begin{array}{rrrrr|r}
\boxed{6} & 0 & 1 & -2 & 0 & 1 \\
\frac{1}{2} & 1 & 0 & \frac{1}{4} & 0 & \frac{1}{4} \\ \hdashline
-24 & 0 & 0 & 30 & 1 & 30
\end{array}\right]
\quad \tfrac{1}{6}R_1 \to R_1$$

$$\sim \left[\begin{array}{rrrrr|r}
\boxed{1} & 0 & \frac{1}{6} & -\frac{1}{3} & 0 & \frac{1}{6} \\
\frac{1}{2} & 1 & 0 & \frac{1}{4} & 0 & \frac{1}{4} \\ \hdashline
-24 & 0 & 0 & 30 & 1 & 30
\end{array}\right]
\quad \begin{array}{l} R_2 + (-\frac{1}{2})R_1 \to R_2 \\ R_3 + 24R_1 \to R_3 \end{array}$$

$$\begin{array}{c} y_1 \\ \sim y_2 \\ P \end{array}
\left[\begin{array}{rrrrr|r}
1 & 0 & \frac{1}{6} & -\frac{1}{3} & 0 & \frac{1}{6} \\
0 & 1 & -\frac{1}{12} & \frac{5}{12} & 0 & \frac{1}{6} \\ \hdashline
0 & 0 & 4 & 22 & 1 & 34
\end{array}\right]$$

Since all indicators in the bottom row are nonnegative, the solution to the dual problem is

$$y_1 = \frac{1}{6}, \quad y_2 = \frac{1}{6}, \quad x_1 = 0, \quad x_2 = 0, \quad P = 34$$

and the maximum value of P is 34. According to Theorem 6, the minimum value of C must also be 34. This agrees with the solution we found in Section 6-2, where we used a geometric approach.

If we examine the geometric solution, we see that the minimum value of C occurred at $x_1 = 4$ and $x_2 = 22$, which are the numbers in the bottom row of the final simplex tableau above. This is no accident.

The entire solution to the minimization problem always can be obtained from the bottom row of the final simplex tableau for the dual problem.

Thus, from the row

$$\begin{matrix} & & x_1 & x_2 & & \\ [0 & 0 & 4 & 22 & 1 & | \; 34] \end{matrix}$$

we can conclude that the solution to the minimization problem is

$$\text{Min } C = 34 \quad \text{at} \quad x_1 = 4, \quad x_2 = 22$$

Now we can see that using x_1 and x_2 as slack variables in the dual problem makes it easy to identify the solution of the original problem.

Solution of a Minimization Problem

Given a minimization problem with nonnegative coefficients in the objective function:

Step 1. Write all problem constraints as $\geq$ inequalities. (This may introduce negative numbers on the right side of some problem constraints.)

Step 2. Form the dual problem.

Step 3. Write the initial system of the dual problem, using the variables from the minimization problem as slack variables.

Step 4. Use the simplex method to solve the dual problem.

Step 5. Read the solution of the minimization problem from the bottom row of the final simplex tableau in step 4. [*Note:* If the dual problem has no solution, then the minimization problem has no solution.]

◆ EXAMPLE 15 Solve the following minimization problem by maximizing the dual:

$$\begin{aligned} \text{Minimize} \quad & C = 40x_1 + 12x_2 + 40x_3 \\ \text{Subject to} \quad & 2x_1 + x_2 + 5x_3 \geqslant 20 \\ & 4x_1 + x_2 + x_3 \geqslant 30 \\ & x_1, x_2, x_3 \geqslant 0 \end{aligned}$$

Solution From Example 14 the dual is

$$\begin{aligned} \text{Maximize} \quad & P = 20y_1 + 30y_2 \\ \text{Subject to} \quad & 2y_1 + 4y_2 \leqslant 40 \\ & y_1 + y_2 \leqslant 12 \\ & 5y_1 + y_2 \leqslant 40 \\ & y_1, y_2 \geqslant 0 \end{aligned}$$

Using x_1, x_2, and x_3 for slack variables, we obtain the initial system for the dual:

$$\begin{aligned} 2y_1 + 4y_2 + x_1 \qquad\qquad &= 40 \\ y_1 + y_2 \quad + x_2 \qquad &= 12 \\ 5y_1 + y_2 \qquad + x_3 \quad &= 40 \\ -20y_1 - 30y_2 \qquad\qquad + P &= 0 \end{aligned}$$

Now we form the simplex tableau and solve the dual problem:

$$\begin{array}{c} \\ x_1 \\ x_2 \\ x_3 \\ P \end{array} \begin{array}{c} \begin{array}{cccccc} y_1 & y_2 & x_1 & x_2 & x_3 & P \end{array} \\ \left[\begin{array}{cccccc|c} 2 & \textcircled{4} & 1 & 0 & 0 & 0 & 40 \\ 1 & 1 & 0 & 1 & 0 & 0 & 12 \\ 5 & 1 & 0 & 0 & 1 & 0 & 40 \\ \hdashline -20 & -30 & 0 & 0 & 0 & 1 & 0 \end{array}\right] \end{array} \quad \tfrac{1}{4}R_1 \to R_1$$

$$\sim \left[\begin{array}{cccccc|c} \tfrac{1}{2} & \textcircled{1} & \tfrac{1}{4} & 0 & 0 & 0 & 10 \\ 1 & 1 & 0 & 1 & 0 & 0 & 12 \\ 5 & 1 & 0 & 0 & 1 & 0 & 40 \\ \hdashline -20 & -30 & 0 & 0 & 0 & 1 & 0 \end{array}\right] \quad \begin{array}{l} R_2 + (-1)R_1 \to R_2 \\ R_3 + (-1)R_1 \to R_3 \\ R_4 + 30R_1 \to R_4 \end{array}$$

$$\sim \begin{array}{c} y_2 \\ x_2 \\ x_3 \\ P \end{array} \left[\begin{array}{cccccc|c} \tfrac{1}{2} & 1 & \tfrac{1}{4} & 0 & 0 & 0 & 10 \\ \textcircled{\tfrac{1}{2}} & 0 & -\tfrac{1}{4} & 1 & 0 & 0 & 2 \\ \tfrac{9}{2} & 0 & -\tfrac{1}{4} & 0 & 1 & 0 & 30 \\ \hdashline -5 & 0 & \tfrac{15}{2} & 0 & 0 & 1 & 300 \end{array}\right] \quad 2R_2 \to R_2$$

$$\sim \left[\begin{array}{cccccc|c} \tfrac{1}{2} & 1 & \tfrac{1}{4} & 0 & 0 & 0 & 10 \\ \textcircled{1} & 0 & -\tfrac{1}{2} & 2 & 0 & 0 & 4 \\ \tfrac{9}{2} & 0 & -\tfrac{1}{4} & 0 & 1 & 0 & 30 \\ \hdashline -5 & 0 & \tfrac{15}{2} & 0 & 0 & 1 & 300 \end{array}\right] \quad \begin{array}{l} R_1 + (-\tfrac{1}{2})R_2 \to R_1 \\ R_3 + (-\tfrac{9}{2})R_2 \to R_3 \\ R_4 + 5R_2 \to R_4 \end{array}$$

$$\sim \begin{array}{c} y_2 \\ y_1 \\ x_3 \\ P \end{array} \left[\begin{array}{cccccc|c} 0 & 1 & \tfrac{1}{2} & -1 & 0 & 0 & 8 \\ 1 & 0 & -\tfrac{1}{2} & 2 & 0 & 0 & 4 \\ 0 & 0 & 2 & -9 & 1 & 0 & 12 \\ \hdashline 0 & 0 & 5 & 10 & 0 & 1 & 320 \end{array}\right]$$

From the bottom row of this tableau, we see that

$$\text{Min } C = 320 \quad \text{at} \quad x_1 = 5, \quad x_2 = 10, \quad x_3 = 0$$ ◆

PROBLEM 15 Solve the following minimization problem by maximizing the dual (see Problem 14):

$$\begin{aligned} \text{Minimize} \quad & C = 16x_1 + 9x_2 + 21x_3 \\ \text{Subject to} \quad & x_1 + x_2 + 3x_3 \geq 16 \\ & 2x_1 + x_2 + x_3 \geq 12 \\ & x_1, x_2, x_3 \geq 0 \end{aligned}$$ ◆

In the preceding section, we noted that multiplying a problem constraint by a number (usually in order to simplify calculations) changes the units of the slack variable. This requires some special interpretation of the value of the slack variable in the optimal solution, but causes no serious problems. However, when using the dual method, multiplying a problem constraint in the dual problem by a number can have some very serious consequences—the bottom row of the final simplex tableau may no longer give the correct solution to the minimization problem. To see this, refer to the first problem constraint of the dual problem in Example 15:

$$2y_1 + 4y_2 \leq 40$$

If we multiply this constraint by $\frac{1}{2}$ and then solve, the final tableau is (verify this):

$$\begin{array}{cccccc|c} y_1 & y_2 & x_1 & x_2 & x_3 & P & \\ 0 & 1 & 1 & -1 & 0 & 0 & 8 \\ 1 & 0 & -1 & 2 & 0 & 0 & 4 \\ 0 & 0 & 4 & -9 & 1 & 0 & 12 \\ \hline 0 & 0 & 10 & 10 & 0 & 1 & 320 \end{array}$$

The bottom row of this tableau indicates that the optimal solution to the minimization problem is $C = 320$ at $x_1 = 10$ and $x_2 = 10$. This is not the correct answer ($x_1 = 5$ in the correct answer). Thus, **you should never multiply a problem constraint in a maximization problem by a number if that maximization problem is being used to solve a minimization problem.** You may still simplify problem constraints in a minimization problem before forming the dual problem.

◆ EXAMPLE 16 Solve the following minimization problem by maximizing the dual:

$$\begin{aligned} \text{Minimize} \quad & C = 5x_1 + 10x_2 \\ \text{Subject to} \quad & x_1 - x_2 \geq 1 \\ & -x_1 + x_2 \geq 2 \\ & x_1, x_2 \geq 0 \end{aligned}$$

Solution

$$A = \left[\begin{array}{rr|r} 1 & -1 & 1 \\ -1 & 1 & 2 \\ \hline 5 & 10 & 1 \end{array}\right] \qquad B = \left[\begin{array}{rr|r} 1 & -1 & 5 \\ -1 & 1 & 10 \\ \hline 1 & 2 & 1 \end{array}\right]$$

The dual problem is

$$\begin{aligned} &\text{Maximize} \quad P = y_1 + 2y_2 \\ &\text{Subject to} \quad y_1 - y_2 \leqslant 5 \\ &\qquad\qquad\quad -y_1 + y_2 \leqslant 10 \\ &\qquad\qquad\qquad\quad y_1, y_2 \geqslant 0 \end{aligned}$$

Introduce slack variables x_1 and x_2 and form the initial system for the dual:

$$\begin{aligned} y_1 - y_2 + x_1 \qquad\qquad &= 5 \\ -y_1 + y_2 \qquad + x_2 \qquad &= 10 \\ -y_1 - 2y_2 \qquad\qquad + P &= 0 \end{aligned}$$

Form the simplex tableau and solve:

$$\begin{array}{c} \\ x_1 \\ x_2 \\ P \end{array} \begin{array}{c} \begin{array}{ccccc} y_1 & y_2 & x_1 & x_2 & P \end{array} \\ \left[\begin{array}{rrrrr|r} 1 & -1 & 1 & 0 & 0 & 5 \\ -1 & \textcircled{1} & 0 & 1 & 0 & 10 \\ \hdashline -1 & -2 & 0 & 0 & 1 & 0 \end{array}\right] \end{array} \quad \begin{array}{l} R_1 + R_2 \to R_1 \\ \\ R_3 + 2R_2 \to R_3 \end{array}$$

$$\sim \left[\begin{array}{rrrrr|r} 0 & 0 & 1 & 1 & 0 & 15 \\ -1 & 1 & 0 & 1 & 0 & 10 \\ \hdashline -3 & 0 & 0 & 2 & 1 & 20 \end{array}\right] \quad \begin{array}{l} \text{No positive elements} \\ \text{above dashed line in} \\ \text{pivot column} \end{array}$$

↑ Pivot column

The -3 in the bottom row indicates that Column 1 is the pivot column. Since no positive elements are in the pivot column above the dashed line, we are unable to select a pivot row. We stop the pivot operation and conclude that this maximization problem has no solution (see Section 6-4, page 336). Theorem 6 now implies that the original minimization problem has no solution. The graph of the inequalities in the minimization problem (see Fig. 11, page 360) shows that the feasible region is empty; thus, it is not surprising that an optimal solution does not exist. ◆

PROBLEM 16 Solve the following minimization problem by maximizing the dual:

$$\begin{aligned} &\text{Minimize} \quad C = 2x_1 + 3x_2 \\ &\text{Subject to} \quad x_1 - 2x_2 \geqslant 2 \\ &\qquad\qquad\quad -x_1 + x_2 \geqslant 1 \\ &\qquad\qquad\qquad\quad x_1, x_2 \geqslant 0 \end{aligned}$$

◆

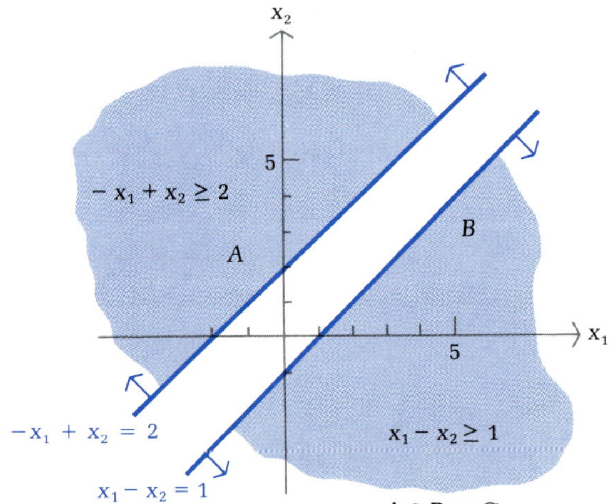

FIGURE 11

◆ APPLICATION: A TRANSPORTATION PROBLEM

One of the first applications of linear programming was to the problem of minimizing the cost of transporting materials. Problems of this type are referred to as **transportation problems.**

◆ EXAMPLE 17 *Transportation*

A computer manufacturing company has two assembly plants, plant *A* and plant *B*, and two distribution outlets, outlet I and outlet II. Plant *A* can assemble at most 700 computers a month, and plant *B* can assemble at most 900 computers a month. Outlet I must have at least 500 computers a month, and outlet II must have at least 1,000 computers a month. Transportation costs for shipping one computer from each plant to each outlet are as follows: $6 from plant *A* to outlet I; $5 from plant *A* to outlet II; $4 from plant *B* to outlet I; $8 from plant *B* to outlet II. Find a shipping schedule that will minimize the total cost of shipping the computers from the assembly plants to the distribution outlets. What is this minimum cost?

Solution

	DISTRIBUTION OUTLET I	DISTRIBUTION OUTLET II	ASSEMBLY CAPACITY
PLANT *A*	$6	$5	700
PLANT *B*	$4	$8	900
MINIMUM REQUIRED	500	1,000	

First, we summarize the relevant data in the table in the margin. In order to find a shipping schedule, we must determine the number of computers that should be shipped from each plant to each outlet. This will require the use of four decision variables:

x_1 = Number of computers shipped from plant *A* to outlet I
x_2 = Number of computers shipped from plant *A* to outlet II
x_3 = Number of computers shipped from plant *B* to outlet I
x_4 = Number of computers shipped from plant *B* to outlet II

The total number of computers shipped from plant A is $x_1 + x_2$. Since this cannot exceed the assembly capacity at A, we have

$x_1 + x_2 \leqslant 700$ Number shipped from plant A

Similarly, the total number shipped from plant B must satisfy

$x_3 + x_4 \leqslant 900$ Number shipped from plant B

The total number shipped to each outlet must satisfy

$x_1 + x_3 \geqslant 500$ Number shipped to outlet I

and

$x_2 + x_4 \geqslant 1{,}000$ Number shipped to outlet II

Using the shipping charges in the table, the total shipping charges are

$C = 6x_1 + 5x_2 + 4x_3 + 8x_4$

Thus, we must solve the following linear programming problem:

$$
\begin{array}{lll}
\text{Minimize} & C = 6x_1 + 5x_2 + 4x_3 + 8x_4 & \\
\text{Subject to} & x_1 + x_2 \leqslant 700 & \text{Available from } A \\
& x_3 + x_4 \leqslant 900 & \text{Available from } B \\
& x_1 + x_3 \geqslant 500 & \text{Required at I} \\
& x_2 + x_4 \geqslant 1{,}000 & \text{Required at II} \\
& x_1, x_2, x_3, x_4 \geqslant 0 &
\end{array}
$$

Before we can solve this problem, we must multiply the first two constraints by -1 so that all the problem constraints are of the $\geqslant$ type. This will introduce negative constants into the minimization problem but not into the dual. Since the coefficients of C are nonnegative, the constants in the dual problem will be nonnegative and the dual will be a standard maximization problem. The problem can now be stated as

$$
\begin{array}{ll}
\text{Minimize} & C = 6x_1 + 5x_2 + 4x_3 + 8x_4 \\
\text{Subject to} & -x_1 - x_2 \geqslant -700 \\
& -x_3 - x_4 \geqslant -900 \\
& x_1 + x_3 \geqslant 500 \\
& x_2 + x_4 \geqslant 1{,}000 \\
& x_1, x_2, x_3, x_4 \geqslant 0
\end{array}
$$

$$
A = \left[\begin{array}{cccc|r}
-1 & -1 & 0 & 0 & -700 \\
0 & 0 & -1 & -1 & -900 \\
1 & 0 & 1 & 0 & 500 \\
0 & 1 & 0 & 1 & 1{,}000 \\
\hline
6 & 5 & 4 & 8 & 1
\end{array}\right]
$$

$$B = \left[\begin{array}{rrrr|r} -1 & 0 & 1 & 0 & 6 \\ -1 & 0 & 0 & 1 & 5 \\ 0 & -1 & 1 & 0 & 4 \\ 0 & -1 & 0 & 1 & 8 \\ \hline -700 & -900 & 500 & 1{,}000 & 1 \end{array}\right] \qquad B \text{ is the transpose of } A$$

The dual problem is

$$\begin{array}{ll} \text{Maximize} & P = -700y_1 - 900y_2 + 500y_3 + 1{,}000y_4 \\ \text{Subject to} & -y_1 + y_3 \le 6 \\ & -y_1 + y_4 \le 5 \\ & - y_2 + y_3 \le 4 \\ & - y_2 + y_4 \le 8 \\ & y_1, y_2, y_3, y_4 \ge 0 \end{array}$$

Introduce slack variables x_1, x_2, x_3, and x_4 and form the initial system for the dual:

$$\begin{array}{rcl} -y_1 + y_3 + x_1 & = & 6 \\ -y_1 + y_4 + x_2 & = & 5 \\ -y_2 + y_3 + x_3 & = & 4 \\ -y_2 + y_4 + x_4 & = & 8 \\ 700y_1 + 900y_2 - 500y_3 - 1{,}000y_4 + P & = & 0 \end{array}$$

Form the simplex tableau and solve:

	y_1	y_2	y_3	y_4	x_1	x_2	x_3	x_4	P		
x_1	−1	0	1	0	1	0	0	0	0	6	
x_2	−1	0	0	(1)	0	1	0	0	0	5	
x_3	0	−1	1	0	0	0	1	0	0	4	
x_4	0	−1	0	1	0	0	0	1	0	8	$R_4 + (-1)R_2 \to R_4$
P	700	900	−500	−1,000	0	0	0	0	1	0	$R_5 + 1{,}000R_2 \to R_5$

	y_1	y_2	y_3	y_4	x_1	x_2	x_3	x_4	P		
x_1	−1	0	1	0	1	0	0	0	0	6	$R_1 + (-1)R_3 \to R_1$
y_4	−1	0	0	1	0	1	0	0	0	5	
~ x_3	0	−1	(1)	0	0	0	1	0	0	4	
x_4	1	−1	0	0	0	−1	0	1	0	3	
P	−300	900	−500	0	0	1,000	0	0	1	5,000	$R_5 + 500R_3 \to R_5$

	y_1	y_2	y_3	y_4	x_1	x_2	x_3	x_4	P		
x_1	−1	1	0	0	1	0	−1	0	0	2	$R_1 + R_4 \to R_1$
y_4	−1	0	0	1	0	1	0	0	0	5	$R_2 + R_4 \to R_2$
~ y_3	0	−1	1	0	0	0	1	0	0	4	
x_4	(1)	−1	0	0	0	−1	0	1	0	3	
P	−300	400	0	0	0	1,000	500	0	1	7,000	$R_5 + 300R_4 \to R_5$

$$\sim \begin{array}{c} \\ x_1 \\ y_4 \\ y_3 \\ y_1 \\ P \end{array} \left[\begin{array}{ccccccccc|c} y_1 & y_2 & y_3 & y_4 & x_1 & x_2 & x_3 & x_4 & P & \\ 0 & 0 & 0 & 0 & 1 & -1 & -1 & 1 & 0 & 5 \\ 0 & -1 & 0 & 1 & 0 & 0 & 0 & 1 & 0 & 8 \\ 0 & -1 & 1 & 0 & 0 & 0 & 1 & 0 & 0 & 4 \\ 1 & -1 & 0 & 0 & 0 & -1 & 0 & 1 & 0 & 3 \\ \hline 0 & 100 & 0 & 0 & 0 & 700 & 500 & 300 & 1 & 7{,}900 \end{array}\right]$$

From the bottom row of this tableau, we have

$$\text{Min } C = 7{,}900 \quad \text{at} \quad x_1 = 0, \quad x_2 = 700, \quad x_3 = 500, \quad x_4 = 300$$

The shipping schedule that minimizes the shipping charges is

700 from plant *A* to outlet II

500 from plant *B* to outlet I

300 from plant *B* to outlet II

The total shipping cost is $7,900. ◆

PROBLEM 17 Repeat Example 17 if the shipping charge from plant *A* to outlet I is increased to $7 and the shipping charge from plant *B* to outlet II is decreased to $3. ◆

◆ SUMMARY OF PROBLEM TYPES AND SOLUTION METHODS

In this and the preceding sections, we have solved both maximization and minimization problems, but with certain restrictions on the problem constraints, constants on the right, and/or objective function coefficients. Table 4 summarizes the types of problems and methods of solution we have considered so far.

TABLE 4

Summary of Problem Types and Solution Methods

PROBLEM TYPE	PROBLEM CONSTRAINTS	RIGHT-SIDE CONSTANTS	COEFFICIENTS OF OBJECTIVE FUNCTION	METHOD OF SOLUTION
1. Maximization	$\leqslant$	Nonnegative	Any real numbers	Simplex method with slack variables
2. Minimization	$\geqslant$	Any real numbers	Nonnegative	Form dual and solve by simplex method with slack variables

The next section develops a generalized version of the simplex method that can handle both maximization and minimization problems with any combination of $\leqslant$, $\geqslant$, and $=$ problem constraints.

Answers to Matched Problems

14. Maximize $P = 16y_1 + 12y_2$
Subject to
$$\begin{aligned} y_1 + 2y_2 &\le 16 \\ y_1 + \ y_2 &\le 9 \\ 3y_1 + \ y_2 &\le 21 \\ y_1, y_2 &\ge 0 \end{aligned}$$

15. Min $C = 132$ at $x_1 = 0$, $x_2 = 10$, $x_3 = 2$

16. Dual problem: Maximize $P = 2y_1 + y_2$
Subject to
$$\begin{aligned} y_1 - y_2 &\le 2 \\ -2y_1 + y_2 &\le 3 \\ y_1, y_2 &\ge 0 \end{aligned}$$
No solution

17. 600 from plant *A* to outlet II, 500 from plant *B* to outlet I, 400 from plant *B* to outlet II; total shipping cost \$6,200

EXERCISE 6-5

A

In Problems 1 and 2:

(A) Form the dual problem.
(B) Write the initial system for the dual problem.
(C) Write the initial simplex tableau for the dual problem and label the columns of the tableau.

1. Minimize $C = 8x_1 + 9x_2$
Subject to
$$\begin{aligned} x_1 + 3x_2 &\ge 4 \\ 2x_1 + \ x_2 &\ge 5 \\ x_1, x_2 &\ge 0 \end{aligned}$$

2. Minimize $C = 12x_1 + 5x_2$
Subject to
$$\begin{aligned} 2x_1 + x_2 &\ge 7 \\ 3x_1 + x_2 &\ge 9 \\ x_1, x_2 &\ge 0 \end{aligned}$$

In Problems 3 and 4, a minimization problem, the corresponding dual problem, and the final simplex tableau in the solution of the dual problem are given.

(A) Find the optimal solution of the dual problem.
(B) Find the optimal solution of the minimization problem.

3. Minimize $C = 21x_1 + 50x_2$
Subject to
$$\begin{aligned} 2x_1 + 5x_2 &\ge 12 \\ 3x_1 + 7x_2 &\ge 17 \\ x_1, x_2 &\ge 0 \end{aligned}$$

Maximize $P = 12y_1 + 17y_2$
Subject to
$$\begin{aligned} 2y_1 + 3y_2 &\le 21 \\ 5y_1 + 7y_2 &\le 50 \\ y_1, y_2 &\ge 0 \end{aligned}$$

$$\begin{array}{c} \begin{matrix} y_1 & y_2 & x_1 & x_2 & P & \end{matrix} \\ \left[\begin{array}{rrrrr|r} 0 & 1 & 5 & -2 & 0 & 5 \\ 1 & 0 & -7 & 3 & 0 & 3 \\ \hline 0 & 0 & 1 & 2 & 1 & 121 \end{array}\right] \end{array}$$

4. Minimize $C = 16x_1 + 25x_2$
Subject to $3x_1 + 5x_2 \geq 30$
$2x_1 + 3x_2 \geq 19$
$x_1, x_2 \geq 0$

Maximize $P = 30y_1 + 19y_2$
Subject to $3y_1 + 2y_2 \leq 16$
$5y_1 + 3y_2 \leq 25$
$y_1, y_2 \geq 0$

$$
\begin{array}{c} \\ \end{array}
\begin{array}{ccccc|c} y_1 & y_2 & x_1 & x_2 & P & \\ \hline 0 & 1 & 5 & -3 & 0 & 5 \\ 1 & 0 & -3 & 2 & 0 & 2 \\ \hline 0 & 0 & 5 & 3 & 1 & 155 \end{array}
$$

In Problems 5–12:

(A) Form the dual problem.
(B) Find the solution to the original problem by applying the simplex method to the dual problem.

5. Minimize $C = 9x_1 + 2x_2$
Subject to $4x_1 + x_2 \geq 13$
$3x_1 + x_2 \geq 12$
$x_1, x_2 \geq 0$

6. Minimize $C = x_1 + 4x_2$
Subject to $x_1 + 2x_2 \geq 5$
$x_1 + 3x_2 \geq 6$
$x_1, x_2 \geq 0$

7. Minimize $C = 7x_1 + 12x_2$
Subject to $2x_1 + 3x_2 \geq 15$
$x_1 + 2x_2 \geq 8$
$x_1, x_2 \geq 0$

8. Minimize $C = 3x_1 + 5x_2$
Subject to $2x_1 + 3x_2 \geq 7$
$x_1 + 2x_2 \geq 4$
$x_1, x_2 \geq 0$

9. Minimize $C = 11x_1 + 4x_2$
Subject to $2x_1 + x_2 \geq 8$
$-2x_1 + 3x_2 \geq 4$
$x_1, x_2 \geq 0$

10. Minimize $C = 40x_1 + 10x_2$
Subject to $2x_1 + x_2 \geq 12$
$3x_1 - x_2 \geq 3$
$x_1, x_2 \geq 0$

11. Minimize $C = 7x_1 + 9x_2$
Subject to $-3x_1 + x_2 \geq 6$
$x_1 - 2x_2 \geq 4$
$x_1, x_2 \geq 0$

12. Minimize $C = 10x_1 + 15x_2$
Subject to $-4x_1 + x_2 \geq 12$
$12x_1 - 3x_2 \geq 10$
$x_1, x_2 \geq 0$

B

Solve the following linear programming problems by applying the simplex method to the dual problem:

13. Minimize $C = 3x_1 + 9x_2$
Subject to $2x_1 + x_2 \geq 8$
$x_1 + 2x_2 \geq 8$
$x_1, x_2 \geq 0$

14. Minimize $C = 2x_1 + x_2$
Subject to $x_1 + x_2 \geq 8$
$x_1 + 2x_2 \geq 4$
$x_1, x_2 \geq 0$

15. Minimize $C = 7x_1 + 5x_2$
Subject to $x_1 + x_2 \geq 4$
$x_1 - 2x_2 \geq -8$
$-2x_1 + x_2 \geq -8$
$x_1, x_2 \geq 0$

16. Minimize $C = 10x_1 + 4x_2$
Subject to $2x_1 + x_2 \geq 6$
$x_1 - 4x_2 \geq -24$
$-8x_1 + 5x_2 \geq -24$
$x_1, x_2 \geq 0$

17. Minimize $C = 10x_1 + 30x_2$
Subject to
$$\begin{aligned} 2x_1 + x_2 &\geq 16 \\ x_1 + x_2 &\geq 12 \\ x_1 + 2x_2 &\geq 14 \\ x_1, x_2 &\geq 0 \end{aligned}$$

18. Minimize $C = 40x_1 + 10x_2$
Subject to
$$\begin{aligned} 3x_1 + x_2 &\geq 24 \\ x_1 + x_2 &\geq 16 \\ x_1 + 3x_2 &\geq 30 \\ x_1, x_2 &\geq 0 \end{aligned}$$

19. Minimize $C = 5x_1 + 7x_2$
Subject to
$$\begin{aligned} x_1 \quad\quad &\geq 4 \\ x_1 + x_2 &\geq 8 \\ x_1 + 2x_2 &\geq 10 \\ x_1, x_2 &\geq 0 \end{aligned}$$

20. Minimize $C = 4x_1 + 5x_2$
Subject to
$$\begin{aligned} 2x_1 + x_2 &\geq 12 \\ x_1 + 2x_2 &\geq 18 \\ x_2 &\geq 6 \\ x_1, x_2 &\geq 0 \end{aligned}$$

21. Minimize
$$C = 10x_1 + 7x_2 + 12x_3$$
Subject to
$$\begin{aligned} x_1 + x_2 + 2x_3 &\geq 7 \\ 2x_1 + x_2 + x_3 &\geq 4 \\ x_1, x_2, x_3 &\geq 0 \end{aligned}$$

22. Minimize
$$C = 18x_1 + 8x_2 + 20x_3$$
Subject to
$$\begin{aligned} x_1 + x_2 + 3x_3 &\geq 6 \\ 3x_1 + x_2 + x_3 &\geq 9 \\ x_1, x_2, x_3 &\geq 0 \end{aligned}$$

23. Minimize
$$C = 5x_1 + 2x_2 + 2x_3$$
Subject to
$$\begin{aligned} x_1 - 4x_2 + x_3 &\geq 6 \\ -x_1 + x_2 - 2x_3 &\geq 4 \\ x_1, x_2, x_3 &\geq 0 \end{aligned}$$

24. Minimize
$$C = 6x_1 + 8x_2 + 3x_3$$
Subject to
$$\begin{aligned} -3x_1 - 2x_2 + x_3 &\geq 4 \\ x_1 + x_2 - x_3 &\geq 2 \\ x_1, x_2, x_3 &\geq 0 \end{aligned}$$

C

25. Minimize
$$C = 16x_1 + 8x_2 + 4x_3$$
Subject to
$$\begin{aligned} 3x_1 + 2x_2 + 2x_3 &\geq 16 \\ 4x_1 + 3x_2 + x_3 &\geq 14 \\ 5x_1 + 3x_2 + x_3 &\geq 12 \\ x_1, x_2, x_3 &\geq 0 \end{aligned}$$

26. Minimize
$$C = 6x_1 + 8x_2 + 12x_3$$
Subject to
$$\begin{aligned} x_1 + 3x_2 + 3x_3 &\geq 6 \\ x_1 + 5x_2 + 5x_3 &\geq 4 \\ 2x_1 + 2x_2 + 3x_3 &\geq 8 \\ x_1, x_2, x_3 &\geq 0 \end{aligned}$$

27. Minimize
$$C = 5x_1 + 4x_2 + 5x_3 + 6x_4$$
Subject to
$$\begin{aligned} x_1 + x_2 \quad\quad\quad\quad &\leq 12 \\ x_3 + x_4 &\leq 25 \\ x_1 \quad\quad + x_3 \quad\quad &\geq 20 \\ x_2 \quad\quad + x_4 &\geq 15 \\ x_1, x_2, x_3, x_4 &\geq 0 \end{aligned}$$

28. Repeat Problem 27 with $C = 4x_1 + 7x_2 + 5x_3 + 6x_4$.

APPLICATIONS

Formulate each of the following as a linear programming problem. Then solve the problem by applying the simplex method to the dual problem.

Business & Economics

29. *Manufacturing—production scheduling.* A food processing company produces regular and deluxe ice cream at three plants. Per hour of operation,

the plant in Cedarburg produces 20 gallons of regular ice cream and 10 gallons of deluxe ice cream, the Grafton plant produces 10 gallons of regular and 20 gallons of deluxe, and the West Bend plant produces 20 gallons of regular and 20 gallons of deluxe. It costs $70 per hour to operate the Cedarburg plant, $75 per hour to operate the Grafton plant, and $90 per hour to operate the West Bend plant. The company needs at least 300 gallons of regular ice cream and at least 200 gallons of deluxe ice cream each day. How many hours per day should each plant operate in order to produce the required amounts of ice cream and minimize the cost of production? What is the minimum production cost?

30. *Mining—production scheduling.* A mining company operates two mines, which produce three grades of ore. The West Summit mine can produce 4 tons of low-grade ore, 3 tons of medium-grade ore, and 2 tons of high-grade ore per hour of operation. The North Ridge mine can produce 1 ton of low-grade ore, 1 ton of medium-grade ore, and 2 tons of high-grade ore per hour of operation. It costs $1,000 per hour to operate the mine at West Summit and $300 per hour to operate the North Ridge mine. To satisfy existing orders, the company needs at least 48 tons of low-grade ore, 45 tons of medium-grade ore, and 31 tons of high-grade ore. How many hours should each mine be operated to supply the required amounts of ore and minimize the cost of production? What is the minimum production cost?

31. *Purchasing.* Acme Micros markets computers with single-sided and double-sided disk drives. The disk drives are supplied by two other companies, Associated Electronics and Digital Drives. Associated Electronics charges $250 for a single-sided disk drive and $350 for a double-sided disk drive. Digital Drives charges $290 for a single-sided disk drive and $320 for a double-sided disk drive. Each month, Associated Electronics can supply at most 1,000 disk drives in any combination of single-sided and double-sided drives. The combined monthly total supplied by Digital Drives cannot exceed 2,000 disk drives. Acme Micros needs at least 1,200 single-sided drives and at least 1,600 double-sided drives each month. How many disk drives of each type should Acme Micros order from each supplier in order to meet its monthly demand and minimize the purchase cost? What is the minimum purchase cost?

32. *Transportation.* A feed company stores grain in elevators located in Ames, Iowa, and Bedford, Indiana. Each month the grain is shipped to processing plants in Columbia, Missouri, and Danville, Illinois. The monthly supply (in tons) of grain at each elevator, the monthly demand (in tons) at each processing plant, and the cost per ton for transporting the grain are given in the table at the top of the next page. Determine a shipping schedule that will minimize the cost of transporting the grain. What is the minimum cost?

	SHIPPING COST (PER TON) Columbia	Danville	SUPPLY (IN TONS)
AMES	$22	$38	700
BEDFORD	$46	$24	500
DEMAND (IN TONS)	400	600	

Life Sciences

33. *Nutrition—people.* A dietitian in a hospital is to arrange a special diet using three foods, L, M, and N. Each ounce of food L contains 20 units of calcium, 10 units of iron, 10 units of vitamin A, and 20 units of cholesterol. Each ounce of food M contains 10 units of calcium, 10 units of iron, 20 units of vitamin A, and 24 units of cholesterol. Each ounce of food N contains 10 units of calcium, 10 units of iron, 10 units of vitamin A, and 18 units of cholesterol. If the minimum daily requirements are 300 units of calcium, 200 units of iron, and 240 units of vitamin A, how many ounces of each food should be used to meet the minimum requirements and at the same time minimize the cholesterol intake? What is the minimum cholesterol intake?

34. *Nutrition—plants.* A farmer can buy three types of plant food, mix A, mix B, and mix C. Each cubic yard of mix A contains 20 pounds of phosphoric acid, 10 pounds of nitrogen, and 5 pounds of potash. Each cubic yard of mix B contains 10 pounds of phosphoric acid, 10 pounds of nitrogen, and 10 pounds of potash. Each cubic yard of mix C contains 20 pounds of phosphoric acid, 20 pounds of nitrogen, and 5 pounds of potash. The minimum monthly requirements are 480 pounds of phosphoric acid, 320 pounds of nitrogen, and 225 pounds of potash. If mix A costs $30 per cubic yard, mix B costs $36 per cubic yard, and mix C costs $39 per cubic yard, how many cubic yards of each mix should the farmer blend to meet the minimum monthly requirements at a minimal cost? What is the minimum cost?

Social Sciences

35. *Education—resource allocation.* A metropolitan school district has two high schools that are overcrowded and two that are underenrolled. In order to balance the enrollment, the school board has decided to bus students from the overcrowded schools to the underenrolled schools. North Division High School has 300 more students than it should have, and South Division High School has 500 more students than it should have. Central High School can accommodate 400 additional students and Washington High School can accommodate 500 additional students. The weekly cost of busing a student from North Division to Central is $5, from North Division to Washington is $2, from South Division to Central is $3, and from South Division to Washington is $4. Determine the number of students that should be bused from each of the overcrowded schools to each of the

underenrolled schools in order to balance the enrollment and minimize the cost of busing the students. What is the minimum cost?

36. *Education — resource allocation.* Repeat Problem 35 if the weekly cost of busing a student from North Division to Washington is \$7 and all the other information remains the same.

SECTION 6-6

Maximization and Minimization with Mixed Problem Constraints

- AN INTRODUCTION TO THE BIG M METHOD
- THE BIG M METHOD
- MINIMIZATION BY THE BIG M METHOD
- SUMMARY OF METHODS OF SOLUTION
- LARGER PROBLEMS — A REFINERY APPLICATION

In the preceding two sections, we have seen how to solve both maximization and minimization problems, but with rather severe restrictions on problem constraints, right-side constants, and/or objective function coefficients (see the summary in Table 4 of the last section). In this section we will present a generalized version of the simplex method that will solve both maximization and minimization problems with any combination of $\leq$, $\geq$, and $=$ problem constraints. The only requirement is that each problem constraint have a nonnegative constant on the right side. (This restriction is easily accommodated, as you will see.)

AN INTRODUCTION TO THE BIG M METHOD

We introduce the big M method through a simple maximization problem with mixed problem constraints. The key parts of the method will then be summarized and applied to more complex problems.

Consider the following problem:

$$\begin{aligned} \text{Maximize} \quad & P = 2x_1 + x_2 \\ \text{Subject to} \quad & x_1 + x_2 \leq 10 \\ & -x_1 + x_2 \geq 2 \\ & x_1, x_2 \geq 0 \end{aligned} \tag{1}$$

To form an equation out of the first inequality, we introduce a slack variable s_1, as before, and write

$$x_1 + x_2 + s_1 = 10$$

How can we form an equation out of the second inequality? We introduce a second variable s_2 and subtract it from the left side so that we can write

$$-x_1 + x_2 - s_2 = 2$$

The variable s_2 is called a **surplus variable,** because it is the amount (surplus) by which the left side of the inequality exceeds the right side.

We now express the linear programming problem (1) as a system of equations:

$$\begin{aligned} x_1 + x_2 + s_1 \qquad\qquad &= 10 \\ -x_1 + x_2 \qquad - s_2 \quad &= 2 \\ -2x_1 - x_2 \qquad\qquad + P &= 0 \\ x_1, x_2, s_1, s_2 &\geq 0 \end{aligned} \tag{2}$$

It can be shown that a basic solution of (2) is not feasible if any of the variables (excluding P) are negative. Thus, **a surplus variable is required to satisfy the nonnegative constraint.**

The basic solution found by setting the nonbasic variables x_1 and x_2 equal to 0 is

$$x_1 = 0, \quad x_2 = 0, \quad s_1 = 10, \quad s_2 = -2, \quad P = 0$$

But this basic solution is not feasible, since the surplus variable s_2 is negative (which is a violation of the nonnegative requirements of all variables except P). The simplex method works only when the basic solution for a tableau is feasible, so we cannot solve this problem simply by writing the tableau for (2) and starting pivot operations.

In order to use the simplex method on problems with mixed constraints, we turn to an ingenious device called an *artificial variable.* This variable has no physical meaning in the original problem (which explains the use of the word "artificial") and is introduced solely for the purpose of obtaining a basic feasible solution so that we can apply the simplex method. An **artificial variable** is a variable introduced into each equation that has a surplus variable. As before, to ensure that we consider only feasible basic solutions, **an artificial variable is required to satisfy the nonnegative constraint.** (As we shall see later, artificial variables are also used to augment equality problem constraints when they are present.)

Returning to the problem at hand, we introduce an artificial variable a_1 into the equation involving the surplus variable s_2:

$$-x_1 + x_2 - s_2 + a_1 = 2$$

To prevent an artificial variable from becoming part of an optimal solution to the original problem, a very large "penalty" is introduced into the objective function. This penalty is created by choosing a positive constant M so large that the artificial variable is forced to be 0 in any final optimal solution of the original problem. (Since the constant M can be made as large as we wish, in computer solutions, M is often selected as the largest number the computer can hold!) We then add the term $-Ma_1$ to the objective function:

$$P = 2x_1 + x_2 - Ma_1$$

We now have a new problem, which we call the **modified problem:**

$$\begin{aligned} \text{Maximize} \quad & P = 2x_1 + x_2 - Ma_1 \\ \text{Subject to} \quad & x_1 + x_2 + s_1 \qquad\qquad = 10 \\ & -x_1 + x_2 \qquad - s_2 + a_1 = 2 \\ & x_1, x_2, s_1, s_2, a_1 \geqslant 0 \end{aligned} \tag{3}$$

The initial system for the modified problem (3) is

$$\begin{aligned} x_1 + x_2 + s_1 \qquad\qquad\qquad & = 10 \\ -x_1 + x_2 \qquad - s_2 + \quad a_1 \qquad & = 2 \\ -2x_1 - x_2 \qquad\qquad + Ma_1 + P & = 0 \\ x_1, x_2, s_1, s_2, a_1 & \geqslant 0 \end{aligned} \tag{4}$$

We next write the augmented coefficient matrix for (4), which we call the **preliminary simplex tableau** for the modified problem. (The reason we call it the "preliminary" simplex tableau instead of the "initial" simplex tableau will be made clear shortly.)

$$\begin{array}{c} \begin{matrix} x_1 & x_2 & s_1 & s_2 & a_1 & P & \end{matrix} \\ \left[\begin{array}{rrrrrr|r} 1 & 1 & 1 & 0 & 0 & 0 & 10 \\ -1 & 1 & 0 & -1 & 1 & 0 & 2 \\ \hdashline -2 & -1 & 0 & 0 & M & 1 & 0 \end{array}\right] \end{array} \tag{5}$$

To start the simplex process, including any necessary pivot operations, the preliminary simplex tableau should either meet the two requirements given in the box below or be transformed by row operations into a tableau that meets these two requirements.

Initial Simplex Tableau Requirements

For a system tableau to be considered an **initial simplex tableau,** it must satisfy the following two requirements:

1. The requisite number of basic variables must be selectable by the process described in Section 6-4. That is, a variable can be selected as a basic variable only if it corresponds to a column in the tableau that has exactly one nonzero element and the nonzero element in the column is not in the same row as the nonzero element in the column of another basic variable. The remaining variables are then selected as nonbasic variables to be set equal to 0 in determining a basic solution.
2. The basic solution found by setting the nonbasic variables equal to 0 is feasible.

Tableau (5) satisfies the first initial simplex tableau requirement, since s_1, s_2, and P can be selected as basic variables according to the criterion stated. (Not all preliminary simplex tableaux satisfy the first requirement; see Example 19.) However, tableau (5) does not satisfy the second initial simplex tableau requirement, since the basic solution is not feasible ($s_2 = -2$). To use the simplex method, we must first use row operations to transform (5) into an equivalent matrix that satisfies both initial simplex tableau requirements. **Note that this transformation is not a pivot operation.**

To get an idea of how to proceed, notice in tableau (5) that -1 in the s_2 column is in the same row as 1 in the a_1 column. This is not an accident! The artificial variable a_1 was introduced so that this would happen. If we eliminate M from the bottom of the a_1 column, then the nonbasic variable a_1 will become a basic variable and the troublesome basic variable s_2 will become a nonbasic variable. Thus, we proceed to eliminate M from the a_1 column using row operations:

$$\begin{array}{c} \\ \\ \\ \end{array}\begin{matrix} x_1 & x_2 & s_1 & s_2 & a_1 & P & \end{matrix}$$

$$\left[\begin{array}{cccccc|c} 1 & 1 & 1 & 0 & 0 & 0 & 10 \\ -1 & 1 & 0 & -1 & 1 & 0 & 2 \\ \hline -2 & -1 & 0 & 0 & M & 1 & 0 \end{array}\right] \quad R_3 + (-M)R_2 \rightarrow R_3$$

$$\sim \left[\begin{array}{cccccc|c} 1 & 1 & 1 & 0 & 0 & 0 & 10 \\ -1 & 1 & 0 & -1 & 1 & 0 & 2 \\ \hline M-2 & -M-1 & 0 & M & 0 & 1 & -2M \end{array}\right]$$

From this last matrix we see that the basic variables are s_1, a_1, and P. The basic solution found by setting the nonbasic variables x_1, x_2, and s_2 equal to 0 is

$$x_1 = 0, \quad x_2 = 0, \quad s_1 = 10, \quad s_2 = 0, \quad a_1 = 2, \quad P = -2M$$

The basic solution is feasible (remember, P can be negative), and both requirements for an initial simplex tableau are met. We can now commence with the simplex process using pivot operations.

The pivot column is determined by the most negative indicator in the bottom row of the tableau. Since M is a positive number, $-M - 1$ is certainly a negative indicator. What about the indicator $M - 2$? Remember that M is a very large positive number. We will assume that M is so large that any expression of the form $M - k$ is positive. Thus, the only negative indicator in the bottom row is $-M - 1$.

$$\begin{matrix} & x_1 & x_2 & s_1 & s_2 & a_1 & P & \end{matrix}$$

$$\text{Pivot row} \rightarrow \left[\begin{array}{cccccc|c} 1 & 1 & 1 & 0 & 0 & 0 & 10 \\ -1 & \textcircled{1} & 0 & -1 & 1 & 0 & 2 \\ \hline M-2 & -M-1 & 0 & M & 0 & 1 & -2M \end{array}\right] \begin{array}{l} \frac{10}{1} = 10 \\ \frac{2}{1} = 2 \\ \\ \end{array}$$

$\uparrow$ Pivot column

Having identified the pivot element, we now begin pivotting:

$$\begin{array}{c} \\ s_1 \\ a_1 \\ P \end{array}\begin{array}{c} \begin{array}{cccccc} x_1 & x_2 & s_1 & s_2 & a_1 & P \end{array} \\ \left[\begin{array}{cccccc|c} 1 & 1 & 1 & 0 & 0 & 0 & 10 \\ -1 & \textcircled{1} & 0 & -1 & 1 & 0 & 2 \\ \hdashline M-2 & -M-1 & 0 & M & 0 & 1 & -2M \end{array}\right] \end{array} \begin{array}{l} \\ R_1 + (-1)R_2 \to R_1 \\ \\ R_3 + (M+1)R_2 \to R_3 \end{array}$$

$$\sim \begin{array}{c} s_1 \\ x_2 \\ P \end{array}\left[\begin{array}{cccccc|c} \textcircled{2} & 0 & 1 & 1 & -1 & 0 & 8 \\ -1 & 1 & 0 & -1 & 1 & 0 & 2 \\ \hdashline -3 & 0 & 0 & -1 & M+1 & 1 & 2 \end{array}\right] \begin{array}{l} \frac{1}{2}R_1 \to R_1 \\ \\ \\ \end{array}$$

$$\sim \left[\begin{array}{cccccc|c} \textcircled{1} & 0 & \frac{1}{2} & \frac{1}{2} & -\frac{1}{2} & 0 & 4 \\ -1 & 1 & 0 & -1 & 1 & 0 & 2 \\ \hdashline -3 & 0 & 0 & -1 & M+1 & 1 & 2 \end{array}\right] \begin{array}{l} \\ R_2 + R_1 \to R_2 \\ R_3 + 3R_1 \to R_3 \end{array}$$

$$\sim \begin{array}{c} x_1 \\ x_2 \\ P \end{array}\left[\begin{array}{cccccc|c} 1 & 0 & \frac{1}{2} & \frac{1}{2} & -\frac{1}{2} & 0 & 4 \\ 0 & 1 & \frac{1}{2} & -\frac{1}{2} & \frac{1}{2} & 0 & 6 \\ \hdashline 0 & 0 & \frac{3}{2} & \frac{1}{2} & M-\frac{1}{2} & 1 & 14 \end{array}\right]$$

Since all the indicators in the last row are nonnegative ($M - \frac{1}{2}$ is nonnegative because M is a very large positive number), we can stop and write the optimal solution:

$$\text{Max } P = 14 \quad \text{at} \quad x_1 = 4, \quad x_2 = 6, \quad s_1 = 0, \quad s_2 = 0, \quad a_1 = 0$$

This is an optimal solution to the modified problem (3). How is it related to the original problem (2)? Since $a_1 = 0$ in this solution,

$$x_1 = 4, \quad x_2 = 6, \quad s_1 = 0, \quad s_2 = 0, \quad P = 14 \tag{6}$$

is certainly a feasible solution for (2). [You can verify this by direct substitution into (2).] Surprisingly, it turns out that (6) is an optimal solution to the original problem. To see that this is true, suppose we were able to find feasible values of x_1, x_2, s_1, and s_2 that satisfy the original system (2) and produce a value of $P > 14$. Then by using these same values in (3) along with $a_1 = 0$, we have found a feasible solution of (3) with $P > 14$. This contradicts the fact that $P = 14$ is the maximum value of P for the modified problem. Thus, (6) is an optimal solution for the original problem.

As this example illustrates, if $a_1 = 0$ in an optimal solution for the modified problem, then deleting a_1 produces an optimal solution for the original problem. What happens if $a_1 \neq 0$ in the optimal solution for the modified problem? In this case, it can be shown that the original problem has no solution because its feasible set is empty.

In larger problems, each $\geq$ problem constraint will require the introduction of a surplus variable and an artificial variable. If one of the problem constraints is an equation rather than an inequality, there is no need to introduce a slack or surplus variable. However, each $=$ problem constraint will require the introduction of another artificial variable to prevent the initial basic solution from violating the equality constraint—the decision variables are often 0 in the initial basic solution (see Example 19). Finally, each artificial variable also must be included in the objective function for the modified problem. The same constant M can be used for each artificial variable. Because of the role that the constant M plays in this approach, this method is often called the **big *M* method.**

◆ THE BIG M METHOD

We now summarize the key steps of the big M method and use them to solve several problems.

The Big *M* Method—Introducing Slack, Surplus, and Artificial Variables to Form the Modified Problem

Step 1. If any problem constraints have negative constants on the right side, multiply both sides by -1 to obtain a constraint with a nonnegative constant. (If the constraint is an inequality, this will reverse the direction of the inequality.)

Step 2. Introduce a slack variable in each $\leqslant$ constraint.

Step 3. Introduce a surplus variable and an artificial variable in each $\geqslant$ constraint.

Step 4. Introduce an artificial variable in each $=$ constraint.

Step 5. For each artificial variable a_i, add $-Ma_i$ to the objective function. Use the same constant M for all artificial variables.

◆ EXAMPLE 18 Find the modified problem for the following linear programming problem. (Do not attempt to solve the problem.)

$$
\begin{aligned}
\text{Maximize} \quad & P = 2x_1 + 5x_2 + 3x_3 \\
\text{Subject to} \quad & x_1 + 2x_2 - x_3 \leqslant 7 \\
& -x_1 + x_2 - 2x_3 \leqslant -5 \\
& x_1 + 4x_2 + 3x_3 \geqslant 1 \\
& 2x_1 - x_2 + 4x_3 = 6 \\
& x_1, x_2, x_3 \geqslant 0
\end{aligned}
$$

Solution First, we multiply the second constraint by -1 to change -5 to 5:

$$(-1)(-x_1 + x_2 - 2x_3) \geqslant (-1)(-5)$$

$$x_1 - x_2 + 2x_3 \geqslant 5$$

Next, we introduce the slack, surplus, and artificial variables according to the rules stated in the box:

$$\begin{aligned}
x_1 + 2x_2 - x_3 + s_1 &= 7 \\
x_1 - x_2 + 2x_3 - s_2 + a_1 &= 5 \\
x_1 + 4x_2 + 3x_3 - s_3 + a_2 &= 1 \\
2x_1 - x_2 + 4x_3 + a_3 &= 6
\end{aligned}$$

Finally, we add $-Ma_1$, $-Ma_2$, and $-Ma_3$ to the objective function:

$$P = 2x_1 + 5x_2 + 3x_3 - Ma_1 - Ma_2 - Ma_3$$

The modified problem is

$$\begin{aligned}
\text{Maximize} \quad & P = 2x_1 + 5x_2 + 3x_3 - Ma_1 - Ma_2 - Ma_3 \\
\text{Subject to} \quad & x_1 + 2x_2 - x_3 + s_1 = 7 \\
& x_1 - x_2 + 2x_3 - s_2 + a_1 = 5 \\
& x_1 + 4x_2 + 3x_3 - s_3 + a_2 = 1 \\
& 2x_1 - x_2 + 4x_3 + a_3 = 6 \\
& x_1, x_2, x_3, s_1, s_2, s_3, a_1, a_2, a_3 \geq 0
\end{aligned}$$

◆

PROBLEM 18 Repeat Example 18 for:

$$\begin{aligned}
\text{Maximize} \quad & P = 3x_1 - 2x_2 + x_3 \\
\text{Subject to} \quad & x_1 - 2x_2 + x_3 \geq 5 \\
& -x_1 - 3x_2 + 4x_3 \leq -10 \\
& 2x_1 + 4x_2 + 5x_3 \leq 20 \\
& 3x_1 - x_2 - x_3 = -15 \\
& x_1, x_2, x_3 \geq 0
\end{aligned}$$

◆

We now list the key steps for solving a problem using the big M method. The various steps and remarks are based on a number of important theorems which we assume without proof. In particular, step 2 is based on the fact that (except for some degenerate cases not considered here) if the modified linear programming problem has a solution, then the preliminary simplex tableau will be transformed into an initial simplex tableau by eliminating the M's from the columns corresponding to the artificial variables in the preliminary simplex tableau. Having obtained an initial simplex tableau, we can then commence with pivot operations.

The Big M Method—Solving the Problem

Step 1. Form the preliminary simplex tableau for the modified problem.

Step 2. Use row operations to eliminate the M's in the bottom row of the preliminary simplex tableau in the columns corresponding to the artificial variables. The resulting tableau is the initial simplex tableau.

Step 3. Solve the modified problem by applying the simplex method to the initial simplex tableau found in step 2.

Step 4. Relate the solution of the modified problem to the original problem.

(A) If the modified problem has no solution, then the original problem has no solution.
(B) If all artificial variables are 0 in the solution to the modified problem, then delete the artificial variables to find a solution to the original problem.
(C) If any artificial variables are nonzero in the solution to the modified problem, then the original problem has no solution.

◆ EXAMPLE 19 Solve the following linear programming problem using the big M method:

$$\begin{aligned} \text{Maximize} \quad & P = x_1 - x_2 + 3x_3 \\ \text{Subject to} \quad & x_1 + x_2 \leq 20 \\ & x_1 + x_3 = 5 \\ & x_2 + x_3 \geq 10 \\ & x_1, x_2, x_3 \geq 0 \end{aligned}$$

Solution State the modified problem:

$$\begin{aligned} \text{Maximize} \quad & P = x_1 - x_2 + 3x_3 - Ma_1 - Ma_2 \\ \text{Subject to} \quad & x_1 + x_2 + s_1 = 20 \\ & x_1 + x_3 + a_1 = 5 \\ & x_2 + x_3 - s_2 + a_2 = 10 \\ & x_1, x_2, x_3, s_1, s_2, a_1, a_2 \geq 0 \end{aligned}$$

Write the preliminary simplex tableau for the modified problem, and find the initial simplex tableau by eliminating the M's from the artificial variable columns:

$$\begin{array}{c} \\ \\ \end{array}\begin{bmatrix} x_1 & x_2 & x_3 & s_1 & a_1 & s_2 & a_2 & P & \\ 1 & 1 & 0 & 1 & 0 & 0 & 0 & 0 & 20 \\ 1 & 0 & 1 & 0 & 1 & 0 & 0 & 0 & 5 \\ 0 & 1 & 1 & 0 & 0 & -1 & 1 & 0 & 10 \\ \hdashline -1 & 1 & -3 & 0 & M & 0 & M & 1 & 0 \end{bmatrix}$$

Eliminate M from the a_1 column.

$R_4 + (-M)R_2 \rightarrow R_4$

$$\begin{array}{c} \\ \sim \\ \\ \\ \sim \end{array}\begin{array}{cccccccc|c}
x_1 & x_2 & x_3 & s_1 & a_1 & s_2 & a_2 & P & \\
1 & 1 & 0 & 1 & 0 & 0 & 0 & 0 & 20 \\
1 & 0 & 1 & 0 & 1 & 0 & 0 & 0 & 5 \\
0 & 1 & 1 & 0 & 0 & -1 & 1 & 0 & 10 \\
\hdashline
-M-1 & 1 & -M-3 & 0 & 0 & 0 & M & 1 & -5M \\
\hline
1 & 1 & 0 & 1 & 0 & 0 & 0 & 0 & 20 \\
1 & 0 & 1 & 0 & 1 & 0 & 0 & 0 & 5 \\
0 & 1 & 1 & 0 & 0 & -1 & 1 & 0 & 10 \\
\hdashline
-M-1 & -M+1 & -2M-3 & 0 & 0 & M & 0 & 1 & -15M
\end{array}$$

Eliminate M from the a_2 column.

$R_4 + (-M)R_3 \rightarrow R_4$

From this last matrix we see that the basic variables are s_1, a_1, a_2, and P. The basic solution found by setting the nonbasic variables x_1, x_2, x_3, and s_2 equal to 0 is

$$x_1 = 0, \quad x_2 = 0, \quad x_3 = 0, \quad s_1 = 20, \quad a_1 = 5, \quad s_2 = 0, \quad a_2 = 10, \quad P = -15M$$

The basic solution is feasible, and both requirements for an initial simplex tableau are met. We can now commence with pivot operations to find the optimal solution.

$$\begin{array}{c}
\\ s_1 \\ a_1 \\ a_2 \\ P \\ s_1 \\ \sim x_3 \\ a_2 \\ P \\ s_1 \\ \sim x_3 \\ x_2 \\ P
\end{array}
\begin{array}{cccccccc|c}
x_1 & x_2 & x_3 & s_1 & a_1 & s_2 & a_2 & P & \\
1 & 1 & 0 & 1 & 0 & 0 & 0 & 0 & 20 \\
1 & 0 & \textcircled{1} & 0 & 1 & 0 & 0 & 0 & 5 \\
0 & 1 & 1 & 0 & 0 & -1 & 1 & 0 & 10 \\
\hdashline
-M-1 & -M+1 & -2M-3 & 0 & 0 & M & 0 & 1 & -15M \\
\hline
1 & 1 & 0 & 1 & 0 & 0 & 0 & 0 & 20 \\
1 & 0 & 1 & 0 & 1 & 0 & 0 & 0 & 5 \\
-1 & \textcircled{1} & 0 & 0 & -1 & -1 & 1 & 0 & 5 \\
\hdashline
M+2 & -M+1 & 0 & 0 & 2M+3 & M & 0 & 1 & -5M+15 \\
\hline
2 & 0 & 0 & 1 & 1 & 1 & -1 & 0 & 15 \\
1 & 0 & 1 & 0 & 1 & 0 & 0 & 0 & 5 \\
-1 & 1 & 0 & 0 & -1 & -1 & 1 & 0 & 5 \\
\hdashline
3 & 0 & 0 & 0 & M+4 & 1 & M-1 & 1 & 10
\end{array}$$

$R_3 + (-1)R_2 \rightarrow R_3$
$R_4 + (2M+3)R_2 \rightarrow R_4$
$R_1 + (-1)R_3 \rightarrow R_1$

$R_4 + (M-1)R_3 \rightarrow R_4$

Since the bottom row has no negative indicators, we can stop and write the optimal solution to the modified problem:

$$x_1 = 0, \quad x_2 = 5, \quad x_3 = 5, \quad s_1 = 15, \quad a_1 = 0, \quad s_2 = 0, \quad a_2 = 0, \quad P = 10$$

Since $a_1 = 0$ and $a_2 = 0$, the solution to the original problem is

$$\text{Max } P = 10 \quad \text{at} \quad x_1 = 0, \quad x_2 = 5, \quad x_3 = 5$$ ◆

PROBLEM 19 Solve the following linear programming problem using the big M method:

$$\begin{aligned}
\text{Maximize} \quad & P = x_1 + 4x_2 + 2x_3 \\
\text{Subject to} \quad & \phantom{x_1 - {}} x_2 + x_3 \le 4 \\
& x_1 \phantom{{}- x_2} - x_3 = 6 \\
& x_1 - x_2 - x_3 \ge 1 \\
& x_1, x_2, x_3 \ge 0
\end{aligned}$$ ◆

◆ EXAMPLE 20 Solve the following linear programming problem using the big M method:

$$\begin{aligned} \text{Maximize} \quad & P = 3x_1 + 5x_2 \\ \text{Subject to} \quad & 2x_1 + x_2 \le 4 \\ & x_1 + 2x_2 \ge 10 \\ & x_1, x_2 \ge 0 \end{aligned}$$

Solution Introducing slack, surplus, and artificial variables, we obtain the modified problem:

$$\begin{aligned} 2x_1 + x_2 + s_1 \qquad\qquad &= 4 \\ x_1 + 2x_2 \qquad - s_2 + a_1 \qquad &= 10 \\ -3x_1 - 5x_2 \qquad\qquad + Ma_1 + P &= 0 \end{aligned}$$

Modified problem

Preliminary simplex tableau

Eliminate M in the a_1 column.

$$\begin{array}{c} \\ \\ \\ \end{array}\begin{matrix} x_1 & x_2 & s_1 & s_2 & a_1 & P & \\ \end{matrix}$$

$$\left[\begin{array}{cccccc|c} 2 & 1 & 1 & 0 & 0 & 0 & 4 \\ 1 & 2 & 0 & -1 & 1 & 0 & 10 \\ \hline -3 & -5 & 0 & 0 & M & 1 & 0 \end{array}\right]$$

$R_3 + (-M)R_2 \to R_3$

Initial simplex tableau

$$\begin{matrix} s_1 \\ \sim a_1 \\ P \end{matrix}\left[\begin{array}{cccccc|c} 2 & \textcircled{1} & 1 & 0 & 0 & 0 & 4 \\ 1 & 2 & 0 & -1 & 1 & 0 & 10 \\ \hline -M-3 & -2M-5 & 0 & M & 0 & 1 & -10M \end{array}\right]$$

Begin pivot operations.

$R_2 + (-2)R_1 \to R_2$

$R_3 + (2M+5)R_1 \to R_3$

$$\begin{matrix} x_2 \\ \sim a_1 \\ P \end{matrix}\left[\begin{array}{cccccc|c} 2 & 1 & 1 & 0 & 0 & 0 & 4 \\ -3 & 0 & -2 & -1 & 1 & 0 & 2 \\ \hline 3M+7 & 0 & 2M+5 & M & 0 & 1 & -2M+20 \end{array}\right]$$

The optimal solution of the modified problem is

$$x_1 = 0, \quad x_2 = 4, \quad s_1 = 0, \quad s_2 = 0, \quad a_1 = 2, \quad P = -2M + 20$$

Since $a_1 \ne 0$, the original problem has no solution. Figure 12 shows that the feasible region for the original problem is empty.

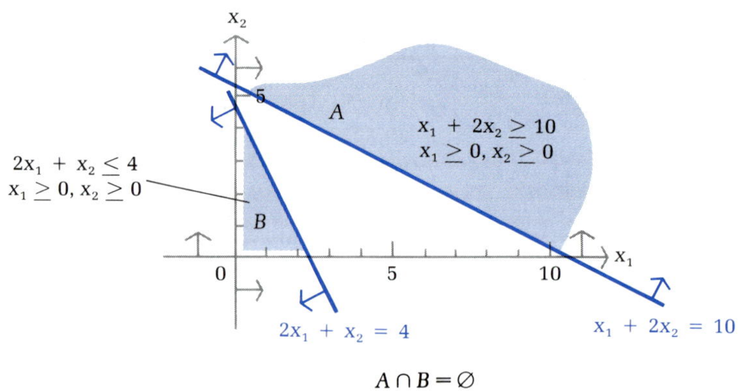

FIGURE 12 ◆

PROBLEM 20 Solve the following linear programming problem using the big M method:

$$\begin{aligned} \text{Maximize} \quad & P = 3x_1 + 2x_2 \\ \text{Subject to} \quad & x_1 + 5x_2 \leq 5 \\ & 2x_1 + x_2 \geq 12 \\ & x_1, x_2 \geq 0 \end{aligned}$$

◆

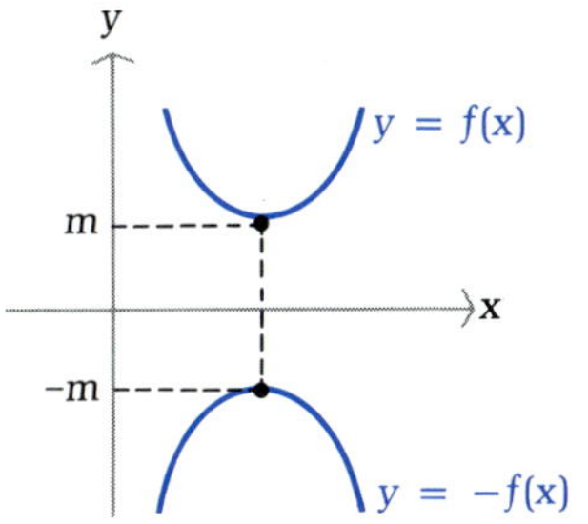

FIGURE 13

◆ MINIMIZATION BY THE BIG M METHOD

In addition to solving any maximization problem, the big M method can be used to solve minimization problems. To minimize an objective function, we have only to maximize its negative. Figure 13 illustrates the fact that the minimum value of a function f occurs at the same point as the maximum value of the function $-f$. Furthermore, if m is the minimum value of f, then $-m$ is the maximum value of $-f$, and conversely. Thus, we can find the minimum value of a function f by finding the maximum value of $-f$ and then changing the sign of the maximum value.

◆ EXAMPLE 21

Production Scheduling

A small jewelry manufacturing company employs a person who is a highly skilled gem cutter, and it wishes to use this person at least 6 hours per day for this purpose. On the other hand, the polishing facilities can be used in any amounts up to 8 hours per day. The company specializes in three kinds of semiprecious gemstones, J, K, and L. Relevant cutting, polishing, and cost requirements are listed in the table. How many gemstones of each type should be processed each day to minimize the cost of the finished stones? What is the minimum cost?

	J	K	L
CUTTING	2 hr	1 hr	1 hr
POLISHING	1 hr	1 hr	2 hr
COST PER STONE	\$30	\$30	\$10

Solution If we let x_1, x_2, and x_3 represent the number of type J, K, and L stones finished per day, respectively, then we have the following linear programming problem to solve, where C is the cost of the stones:

$$\begin{aligned} \text{Minimize} \quad & C = 30x_1 + 30x_2 + 10x_3 && \text{Objective function} \\ \text{Subject to} \quad & \left.\begin{aligned} 2x_1 + x_2 + x_3 &\geq 6 \\ x_1 + x_2 + 2x_3 &\leq 8 \end{aligned}\right\} && \text{Problem constraints} \\ & x_1, x_2, x_3 \geq 0 && \text{Nonnegative constraints} \end{aligned}$$

We convert this to a maximization problem by letting

$$P = -C = -30x_1 - 30x_2 - 10x_3$$

Thus, we get:

$$\begin{aligned} &\text{Maximize} \quad P = -30x_1 - 30x_2 - 10x_3 \\ &\text{Subject to} \quad 2x_1 + x_2 + x_3 \geq 6 \\ &\qquad\qquad\quad x_1 + x_2 + 2x_3 \leq 8 \\ &\qquad\qquad\qquad x_1, x_2, x_3 \geq 0 \end{aligned}$$

and Min $C = -$Max P. To solve, we first state the modified problem:

$$\begin{aligned} 2x_1 + x_2 + x_3 - s_1 + a_1 \qquad\qquad &= 6 \\ x_1 + x_2 + 2x_3 \qquad\qquad + s_2 \qquad &= 8 \\ 30x_1 + 30x_2 + 10x_3 \quad + Ma_1 \quad + P &= 0 \\ x_1, x_2, x_3, s_1, s_2, a_1 &\geq 0 \end{aligned}$$

$$\begin{array}{c} \\ \\ \\ \end{array}\begin{array}{ccccccc|c} x_1 & x_2 & x_3 & s_1 & a_1 & s_2 & P & \\ \hline 2 & 1 & 1 & -1 & 1 & 0 & 0 & 6 \\ 1 & 1 & 2 & 0 & 0 & 1 & 0 & 8 \\ \hline 30 & 30 & 10 & 0 & M & 0 & 1 & 0 \end{array}$$

Eliminate M in the a_1 column.
$R_3 + (-M)R_1 \to R_3$

Begin pivot operations. Assume M is so large that $-2M + 30$, $-M + 30$, and $-M + 10$ are all negative.

$$\begin{array}{c} a_1 \\ \sim s_2 \\ P \end{array}\left[\begin{array}{ccccccc|c} \textcircled{2} & 1 & 1 & -1 & 1 & 0 & 0 & 6 \\ 1 & 1 & 2 & 0 & 0 & 1 & 0 & 8 \\ \hline -2M+30 & -M+30 & -M+10 & M & 0 & 0 & 1 & -6M \end{array}\right] \quad \tfrac{1}{2}R_1 \to R_1$$

$$\sim\left[\begin{array}{ccccccc|c} \textcircled{1} & \frac{1}{2} & \frac{1}{2} & -\frac{1}{2} & \frac{1}{2} & 0 & 0 & 3 \\ 1 & 1 & 2 & 0 & 0 & 1 & 0 & 8 \\ \hline -2M+30 & -M+30 & -M+10 & M & 0 & 0 & 1 & -6M \end{array}\right] \quad \begin{array}{l} R_2 + (-1)R_1 \to R_2 \\ R_3 + (2M - 30)R_1 \to R_3 \end{array}$$

$$\begin{array}{c} x_1 \\ \sim s_2 \\ P \end{array}\left[\begin{array}{ccccccc|c} 1 & \frac{1}{2} & \frac{1}{2} & -\frac{1}{2} & \frac{1}{2} & 0 & 0 & 3 \\ 0 & \frac{1}{2} & \textcircled{\tfrac{3}{2}} & \frac{1}{2} & -\frac{1}{2} & 1 & 0 & 5 \\ \hline 0 & 15 & -5 & 15 & M-15 & 0 & 1 & -90 \end{array}\right] \quad \tfrac{2}{3}R_2 \to R_2$$

$$\sim\left[\begin{array}{ccccccc|c} 1 & \frac{1}{2} & \frac{1}{2} & -\frac{1}{2} & \frac{1}{2} & 0 & 0 & 3 \\ 0 & \frac{1}{3} & \textcircled{1} & \frac{1}{3} & -\frac{1}{3} & \frac{2}{3} & 0 & \frac{10}{3} \\ \hline 0 & 15 & -5 & 15 & M-15 & 0 & 1 & -90 \end{array}\right] \quad \begin{array}{l} R_1 + (-\frac{1}{2})R_2 \to R_1 \\ \\ R_3 + 5R_2 \to R_3 \end{array}$$

$$\begin{array}{c} x_1 \\ \sim x_3 \\ P \end{array}\left[\begin{array}{ccccccc|c} 1 & \frac{1}{3} & 0 & -\frac{2}{3} & \frac{2}{3} & -\frac{1}{3} & 0 & \frac{4}{3} \\ 0 & \frac{1}{3} & 1 & \frac{1}{3} & -\frac{1}{3} & \frac{2}{3} & 0 & \frac{10}{3} \\ \hline 0 & \frac{50}{3} & 0 & \frac{50}{3} & M-\frac{50}{3} & \frac{10}{3} & 1 & -\frac{220}{3} \end{array}\right]$$

Since the bottom row has no negative indicators, the optimal solution for the modified problem is

$$x_1 = \tfrac{4}{3}, \quad x_2 = 0, \quad x_3 = \tfrac{10}{3}, \quad s_1 = 0, \quad a_1 = 0, \quad s_2 = 0, \quad P = -\tfrac{220}{3}$$

Since $a_1 = 0$, deleting a_1 produces the solution to the original maximization problem and also to the minimization problem. Thus,

$$\text{Min } C = -\text{Max } P = -(-\tfrac{220}{3}) = 73\tfrac{1}{3} \quad \text{at} \quad x_1 = \tfrac{4}{3}, \quad x_2 = 0, \quad x_3 = \tfrac{10}{3}$$

That is, a minimum cost of \$$73\frac{1}{3}$ for gemstones will be realized if $1\frac{1}{3}$ type *J*, no type *K*, and $3\frac{1}{3}$ type *L* stones are processed each day. The fractional values for production make sense if we think of them as average daily production figures. ◆

PROBLEM 21 Repeat Example 21 with $C = 40x_1 + 40x_2 + 10x_3$. ◆

◆ SUMMARY OF METHODS OF SOLUTION

The big *M* method can be used to solve any minimization problem, including those that can be solved by the dual method. (Note that Example 21 could have been solved by the dual method.) Both methods of solving minimization problems are important. You will be instructed to solve most minimization problems in Exercise 6-6 by the big *M* method in order to gain more experience with that method. If the method of solution is not specified, then the dual method is usually easier.

Table 5 should help you select the proper method of solution for any linear programming problem.

TABLE 5

Summary of Problem Types and Solution Methods

PROBLEM TYPE	PROBLEM CONSTRAINTS	RIGHT-SIDE CONSTANTS	COEFFICIENTS OF OBJECTIVE FUNCTION	METHOD OF SOLUTION
1. Maximization	$\le$	Nonnegative	Any real numbers	Simplex method with slack variables
2. Minimization	$\ge$	Any real numbers	Nonnegative	Form dual and solve by method 1.
3. Maximization	Mixed ($\le, \ge, =$)	Nonnegative	Any real numbers	Form modified problem with slack, surplus, and artificial variables and solve by the big *M* method.
4. Minimization	Mixed ($\le, \ge, =$)	Nonnegative	Any real numbers	Maximize negative of objective function by method 3.

◆ LARGER PROBLEMS — A REFINERY APPLICATION

Up to this point, all the problems we have considered could be solved by hand. However, the real value of the simplex method lies in its ability to solve problems with a large number of variables and constraints, where a computer is generally used to perform the actual pivot operations. As a final application, we will consider a problem that would require the use of a computer to complete the solution.

◆ EXAMPLE 22

Petroleum Blending

A refinery produces two grades of gasoline, regular and premium, by blending together two components, *A* and *B*. Component *A* has an octane rating of 90 and costs $28 a barrel. Component *B* has an octane rating of 110 and costs $32 a barrel. The octane rating for regular gasoline must be at least 95, and the octane rating for premium must be at least 105. Regular gasoline sells for $34 a barrel and premium sells for $40 a barrel. Currently, the company has 30,000 barrels of component *A* and 20,000 barrels of component *B*. It also has orders for 20,000 barrels of regular and 10,000 barrels of premium that must be filled. Assuming that all the gasoline produced can be sold, determine the maximum possible profit.

Solution

First we organize the information given in the problem in Table 6.

TABLE 6

COMPONENT	OCTANE RATING	COST	AVAILABLE SUPPLY
A	90	$28	30,000 barrels
B	110	$32	20,000 barrels
GRADE	**MINIMUM OCTANE RATING**	**SELLING PRICE**	**EXISTING ORDERS**
Regular	95	$34	20,000 barrels
Premium	105	$40	10,000 barrels

Let

x_1 = Number of barrels of component *A* used in regular gasoline

x_2 = Number of barrels of component *A* used in premium gasoline

x_3 = Number of barrels of component *B* used in regular gasoline

x_4 = Number of barrels of component *B* used in premium gasoline

The total amount of component *A* used is $x_1 + x_2$. This cannot exceed the available supply. Thus, one constraint is

$$x_1 + x_2 \leq 30{,}000$$

The corresponding inequality for component *B* is

$$x_3 + x_4 \leq 20{,}000$$

The amounts of regular and premium gasoline produced must be sufficient to meet the existing orders:

$$x_1 + x_3 \geq 20{,}000 \quad \text{Regular}$$

$$x_2 + x_4 \geq 10{,}000 \quad \text{Premium}$$

Now consider the octane ratings. The octane rating of a blend is simply the proportional average of the octane ratings of the components. Thus, the octane rating for regular gasoline is

$$90\frac{x_1}{x_1 + x_3} + 110\frac{x_3}{x_1 + x_3}$$

where $x_1/(x_1 + x_3)$ is the percentage of component *A* used in regular gasoline and $x_3/(x_1 + x_3)$ is the percentage of component *B*. The final octane rating of regular gasoline must be at least 95; thus,

$$90\frac{x_1}{x_1 + x_3} + 110\frac{x_3}{x_1 + x_3} \geq 95 \qquad \text{Multiply by } x_1 + x_3.$$

$$90x_1 + 110x_3 \geq 95(x_1 + x_3) \qquad \text{Collect like terms on the left side.}$$

$$-5x_1 + 15x_3 \geq 0$$

The corresponding inequality for premium gasoline is

$$90\frac{x_2}{x_2 + x_4} + 110\frac{x_4}{x_2 + x_4} \geq 105$$

$$90x_2 + 110x_4 \geq 105(x_2 + x_4)$$

$$-15x_2 + 5x_4 \geq 0$$

The cost of the components used is

$$C = 28(x_1 + x_2) + 32(x_3 + x_4)$$

The revenue from selling all the gasoline is

$$R = 34(x_1 + x_3) + 40(x_2 + x_4)$$

and the profit is

$$\begin{aligned} P &= R - C \\ &= 34(x_1 + x_3) + 40(x_2 + x_4) - 28(x_1 + x_2) - 32(x_3 + x_4) \\ &= (34 - 28)x_1 + (40 - 28)x_2 + (34 - 32)x_3 + (40 - 32)x_4 \\ &= 6x_1 + 12x_2 + 2x_3 + 8x_4 \end{aligned}$$

To find the maximum profit, we must solve the following linear programming problem:

$$\begin{array}{llr} \text{Maximize} & P = 6x_1 + 12x_2 + 2x_3 + 8x_4 & \text{Profit} \\ \text{Subject to} & x_1 + x_2 \leq 30{,}000 & \text{Available } A \\ & x_3 + x_4 \leq 20{,}000 & \text{Available } B \\ & x_1 + x_3 \geq 20{,}000 & \text{Required regular} \\ & x_2 + x_4 \geq 10{,}000 & \text{Required premium} \\ & -5x_1 + 15x_3 \geq 0 & \text{Octane for regular} \\ & -15x_2 + 5x_4 \geq 0 & \text{Octane for premium} \\ & x_1, x_2, x_3, x_4 \geq 0 & \end{array}$$

The tableau for this problem would have seven rows and sixteen columns. Solving this problem by hand is possible but would require considerable effort. In actual practice, computers are used to solve most linear programming problems. Table 7 displays typical input and output for a computer solution of this problem.

TABLE 7

INPUT TO PROGRAM

```
        * - LINEAR PROGRAMMING - *
             PROBLEM DISPLAY

MAXIMIZE
P =  6X1 +  12X2 +  2X3 +  8X4
SUBJECT TO:
  X1 +    X2               <= 30000
               X3 +  X4 <= 20000
  X1        +   X3         >= 20000
          X2        +  X4 >= 10000
-5X1       + 15X3          >=     0
     -  15X2        + 5X4 >=     0
XI >= 0,  I = 1 TO 4
```

OUTPUT FROM PROGRAM

```
        * - LINEAR PROGRAMMING - *
             SOLUTION DISPLAY

DECISION VARIABLES   SLACK VARIABLES
X1 = 26250           S1 = 0
X2 = 3750            S2 = 0
X3 = 8750            SURPLUS VARIABLES
X4 = 11250           S3 = 15000
                     S4 = 5000
                     S5 = 0
                     S6 = 0
MAXIMUM VALUE OF OBJECTIVE FUNCTION
310000
```

According to the output in Table 7, the refinery should blend 26,250 barrels of component *A* and 8,750 barrels of component *B* to produce 35,000 barrels of regular. They should blend 3,750 barrels of component *A* and 11,250 barrels of component *B* to produce 15,000 barrels of premium. This will result in a maximum profit of $310,000. ◆

PROBLEM 22 Suppose the refinery in Example 22 has 35,000 barrels of component *A*, which costs $25 a barrel, and 15,000 barrels of component *B*, which costs $35 a barrel. If all the other information is unchanged, formulate a linear programming problem whose solution is the maximum profit. Do not attempt to solve the problem (unless you have access to a computer and a program that solves linear programming problems). ◆

Answers to Matched Problems

18. Maximize $P = 3x_1 - 2x_2 + x_3 - Ma_1 - Ma_2 - Ma_3$

Subject to
$$\begin{aligned} x_1 - 2x_2 + x_3 - s_1 + a_1 &= 5 \\ x_1 + 3x_2 - 4x_3 - s_2 + a_2 &= 10 \\ 2x_1 + 4x_2 + 5x_3 + s_3 &= 20 \\ -3x_1 + x_2 + x_3 + a_3 &= 15 \\ x_1, x_2, x_3, s_1, a_1, s_2, a_2, s_3, a_3 &\geq 0 \end{aligned}$$

19. Max $P = 22$ at $x_1 = 6$, $x_2 = 4$, $x_3 = 0$

20. No solution

21. Min $C = \$86\frac{2}{3}$ at $x_1 = \frac{4}{3}$, $x_2 = 0$, $x_3 = \frac{10}{3}$

22. Maximize $P = 9x_1 + 15x_2 - x_3 + 5x_4$

Subject to

$$\begin{aligned} x_1 + x_2 &\leq 35{,}000 \\ x_3 + x_4 &\leq 15{,}000 \\ x_1 + x_3 &\geq 20{,}000 \\ x_2 + x_4 &\geq 10{,}000 \\ -5x_1 + 15x_3 &\geq 0 \\ -15x_2 + 5x_4 &\geq 0 \\ x_1, x_2, x_3, x_4 &\geq 0 \end{aligned}$$

EXERCISE 6-6

A *In Problems 1–8:*

(A) Introduce slack, surplus, and artificial variables and form the modified problem.

(B) Write the preliminary simplex tableau for the modified problem and find the initial simplex tableau.

(C) Find the optimal solution of the modified problem by applying the simplex method to the initial simplex tableau.

(D) Find the optimal solution of the original problem, if it exists.

1. Maximize $P = 5x_1 + 2x_2$

Subject to $x_1 + 2x_2 \leq 12$

$x_1 + x_2 \geq 4$

$x_1, x_2 \geq 0$

2. Maximize $P = 3x_1 + 7x_2$

Subject to $2x_1 + x_2 \leq 16$

$x_1 + x_2 \geq 6$

$x_1, x_2 \geq 0$

3. Maximize $P = 3x_1 + 5x_2$

Subject to $2x_1 + x_2 \leq 8$

$x_1 + x_2 = 6$

$x_1, x_2 \geq 0$

4. Maximize $P = 4x_1 + 3x_2$

Subject to $x_1 + 3x_2 \leq 24$

$x_1 + x_2 = 12$

$x_1, x_2 \geq 0$

5. Maximize $P = 4x_1 + 3x_2$

Subject to $-x_1 + 2x_2 \leq 2$

$x_1 + x_2 \geq 4$

$x_1, x_2 \geq 0$

6. Maximize $P = 3x_1 + 4x_2$

Subject to $x_1 - 2x_2 \leq 2$

$x_1 + x_2 \geq 5$

$x_1, x_2 \geq 0$

7. Maximize $P = 5x_1 + 10x_2$

Subject to $x_1 + x_2 \leq 3$

$2x_1 + 3x_2 \geq 12$

$x_1, x_2 \geq 0$

8. Maximize $P = 4x_1 + 6x_2$

Subject to $x_1 + x_2 \leq 2$

$3x_1 + 5x_2 \geq 15$

$x_1, x_2 \geq 0$

B *Use the big M method to solve the following problems:*

9. Minimize and maximize

$P = 6x_1 - 2x_2$

Subject to $x_1 + x_2 \leq 10$

$3x_1 + 2x_2 \geq 24$

$x_1, x_2 \geq 0$

10. Minimize and maximize

$P = -4x_1 + 16x_2$

Subject to $3x_1 + x_2 \leq 28$

$x_1 + 2x_2 \geq 16$

$x_1, x_2 \geq 0$

11. Maximize $P = 2x_1 + 5x_2$
Subject to
$$\begin{aligned} x_1 + 2x_2 &\leq 18 \\ 2x_1 + x_2 &\leq 21 \\ x_1 + x_2 &\geq 10 \\ x_1, x_2 &\geq 0 \end{aligned}$$

12. Maximize $P = 6x_1 + 2x_2$
Subject to
$$\begin{aligned} x_1 + 2x_2 &\leq 20 \\ 2x_1 + x_2 &\leq 16 \\ x_1 + x_2 &\geq 9 \\ x_1, x_2 &\geq 0 \end{aligned}$$

13. Maximize
$$P = 10x_1 + 12x_2 + 20x_3$$
Subject to
$$\begin{aligned} 3x_1 + x_2 + 2x_3 &\geq 12 \\ x_1 - x_2 + 2x_3 &= 6 \\ x_1, x_2, x_3 &\geq 0 \end{aligned}$$

14. Maximize
$$P = 5x_1 + 7x_2 + 9x_3$$
Subject to
$$\begin{aligned} x_1 - x_2 + x_3 &\geq 20 \\ 2x_1 + x_2 + 3x_3 &= 36 \\ x_1, x_2, x_3 &\geq 0 \end{aligned}$$

15. Minimize
$$C = -5x_1 - 12x_2 + 16x_3$$
Subject to
$$\begin{aligned} x_1 + 2x_2 + x_3 &\leq 10 \\ 2x_1 + 3x_2 + x_3 &\geq 6 \\ 2x_1 + x_2 - x_3 &= 1 \\ x_1, x_2, x_3 &\geq 0 \end{aligned}$$

16. Minimize
$$C = -3x_1 + 15x_2 - 4x_3$$
Subject to
$$\begin{aligned} 2x_1 + x_2 + 3x_3 &\leq 24 \\ x_1 + 2x_2 + x_3 &\geq 6 \\ x_1 - 3x_2 + x_3 &= 2 \\ x_1, x_2, x_3 &\geq 0 \end{aligned}$$

17. Maximize
$$P = 3x_1 + 5x_2 + 6x_3$$
Subject to
$$\begin{aligned} 2x_1 + x_2 + 2x_3 &\leq 8 \\ 2x_1 + x_2 - 2x_3 &= 0 \\ x_1, x_2, x_3 &\geq 0 \end{aligned}$$

18. Maximize
$$P = 3x_1 + 6x_2 + 2x_3$$
Subject to
$$\begin{aligned} 2x_1 + x_2 + 3x_3 &\leq 12 \\ 2x_1 - 2x_2 + x_3 &= 0 \\ x_1, x_2, x_3 &\geq 0 \end{aligned}$$

19. Maximize
$$P = 2x_1 + 3x_2 + 4x_3$$
Subject to
$$\begin{aligned} x_1 + 2x_2 + x_3 &\leq 25 \\ 2x_1 + x_2 + 2x_3 &\leq 60 \\ x_1 + 2x_2 - x_3 &\geq 10 \\ x_1, x_2, x_3 &\geq 0 \end{aligned}$$

20. Maximize
$$P = 5x_1 + 2x_2 + 9x_3$$
Subject to
$$\begin{aligned} 2x_1 + 4x_2 + x_3 &\leq 150 \\ 3x_1 + 2x_2 + x_3 &\leq 90 \\ -x_1 + 5x_2 + x_3 &\geq 120 \\ x_1, x_2, x_3 &\geq 0 \end{aligned}$$

21. Maximize
$$P = x_1 + 2x_2 + 5x_3$$
Subject to
$$\begin{aligned} x_1 + 3x_2 + 2x_3 &\leq 60 \\ 2x_1 + 5x_2 + 2x_3 &\geq 50 \\ x_1 - 2x_2 + x_3 &\geq 40 \\ x_1, x_2, x_3 &\geq 0 \end{aligned}$$

22. Maximize
$$P = 2x_1 + 4x_2 + x_3$$
Subject to
$$\begin{aligned} 2x_1 + 3x_2 + 5x_3 &\leq 280 \\ 2x_1 + 2x_2 + x_3 &\geq 140 \\ 2x_1 + x_2 \quad &\geq 150 \\ x_1, x_2, x_3 &\geq 0 \end{aligned}$$

C *Problems 23–30 are mixed. Some can be solved by the methods presented in Sections 6-4 and 6-5, while others must be solved by the big M method.*

23. Minimize
$$C = 10x_1 - 40x_2 - 5x_3$$
Subject to
$$\begin{aligned} x_1 + 3x_2 \qquad &\le 6 \\ 4x_2 + x_3 &\le 3 \\ x_1, x_2, x_3 &\ge 0 \end{aligned}$$

24. Maximize
$$P = 7x_1 - 5x_2 + 2x_3$$
Subject to
$$\begin{aligned} x_1 - 2x_2 + x_3 &\ge -8 \\ x_1 - x_2 + x_3 &\le 10 \\ x_1, x_2, x_3 &\ge 0 \end{aligned}$$

25. Maximize
$$P = -5x_1 + 10x_2 + 15x_3$$
Subject to
$$\begin{aligned} 2x_1 + 3x_2 + x_3 &\le 24 \\ x_1 - 2x_2 - 2x_3 &\ge 1 \\ x_1, x_2, x_3 &\ge 0 \end{aligned}$$

26. Minimize
$$C = -5x_1 + 10x_2 + 15x_3$$
Subject to
$$\begin{aligned} 2x_1 + 3x_2 + x_3 &\le 24 \\ x_1 - 2x_2 - 2x_3 &\ge 1 \\ x_1, x_2, x_3 &\ge 0 \end{aligned}$$

27. Minimize
$$C = 10x_1 + 40x_2 + 5x_3$$
Subject to
$$\begin{aligned} x_1 + 3x_2 \qquad &\ge 6 \\ 4x_2 + x_3 &\ge 3 \\ x_1, x_2, x_3 &\ge 0 \end{aligned}$$

28. Maximize
$$P = 8x_1 + 2x_2 - 10x_3$$
Subject to
$$\begin{aligned} x_1 + x_2 - 3x_3 &\le 6 \\ 4x_1 - x_2 + 2x_3 &\le -7 \\ x_1, x_2, x_3 &\ge 0 \end{aligned}$$

29. Maximize
$$P = 12x_1 + 9x_2 + 5x_3$$
Subject to
$$\begin{aligned} x_1 + 2x_2 + x_3 &\le 40 \\ 2x_1 + x_2 + 3x_3 &\le 60 \\ x_1, x_2, x_3 &\ge 0 \end{aligned}$$

30. Minimize
$$C = 10x_1 + 12x_2 + 28x_3$$
Subject to
$$\begin{aligned} 4x_1 + 2x_2 + 3x_3 &\ge 20 \\ 5x_1 - x_2 - 4x_3 &\le 10 \\ x_1, x_2, x_3 &\ge 0 \end{aligned}$$

APPLICATIONS

In Problems 31–38, formulate each problem as a linear programming problem and solve by using the big M method.

Business & Economics

31. *Manufacturing—resource allocation.* An electronics company manufactures two types of add-on memory modules for microcomputers, a 16k module and a 64k module. Each 16k module requires 10 minutes for assembly and 2 minutes for testing. Each 64k module requires 15 minutes for assembly and 4 minutes for testing. The company makes a profit of $18 on each 16k module and $30 on each 64k module. The assembly department can work a maximum of 2,200 minutes per day, and the testing department can work a maximum of 500 minutes a day. In order to satisfy current orders, the company must produce at least 50 of the 16k modules per day. How many units of each module should the company manufacture each day in order to maximize the daily profit? What is the maximum profit?

32. *Manufacturing—resource allocation.* Repeat Problem 31 if the assembly department can work a maximum of 2,100 minutes daily.

33. *Advertising.* A company planning an advertising campaign to attract new customers wants to place a total of at most 10 ads in 3 newspapers. Each ad in the *Sentinel* costs $200 and will be read by 2,000 people. Each ad in the *Journal* costs $200 and will be read by 500 people. Each ad in the *Tribune* costs $100 and will be read by 1,500 people. The company wants at least 16,000 people to read its ads. How many ads should it place in each paper in order to minimize the advertising costs? What is the minimum cost?

34. *Advertising.* Repeat Problem 33 if the *Tribune* is unable to accept more than 4 ads from the company.

Life Sciences

	UNITS PER BOTTLE *A*	*B*	*C*
PROTEIN	10	10	20
CARBOHYDRATES	2	3	4
COST PER BOTTLE	$0.60	$0.40	$0.90

35. *Nutrition—people.* An individual on a high-protein, low-carbohydrate diet requires at least 100 units of protein and at most 24 units of carbohydrates daily. The diet will consist entirely of three special liquid diet foods, *A*, *B*, and *C*. The contents and costs of the diet foods are given in the table. How many bottles of each brand of diet food should be consumed daily in order to meet the protein and carbohydrate requirements at minimal cost? What is the minimum cost?

36. *Nutrition—people.* Repeat Problem 35 if brand *C* liquid diet food costs $1.50 a bottle.

37. *Nutrition—plants.* A farmer can use three types of plant food, mix *A*, mix *B*, and mix *C*. The amounts (in pounds) of nitrogen, phosphoric acid, and potash in a cubic yard of each mix are given in the table. Tests performed on the soil in a large field indicate that the field needs at least 800 pounds of potash. The tests also indicate that no more than 700 pounds of phosphoric acid should be added to the field. The farmer plans to plant a crop that requires a great deal of nitrogen. How many cubic yards of each mix should be added to the field in order to satisfy the potash and phosphoric acid requirements and maximize the amount of nitrogen added? What is the maximum amount of nitrogen?

	POUNDS PER CUBIC YARD *A*	*B*	*C*
NITROGEN	12	16	8
PHOSPHORIC ACID	12	8	16
POTASH	16	8	16

38. *Nutrition—plants.* Repeat Problem 37 if the field should have no more than 1,000 pounds of phosphoric acid.

In Problems 39–45, formulate each problem as a linear programming problem. Do not solve.

Business & Economics

39. *Manufacturing—production scheduling.* A company manufactures car and truck frames at plants in Milwaukee and Racine. The Milwaukee plant has a daily operating budget of $50,000 and can produce at most 300 frames daily in any combination. It costs $150 to manufacture a car frame and $200 to manufacture a truck frame at the Milwaukee plant. The Racine plant has a daily operating budget of $35,000, can produce a maximum combined total of 200 frames daily, and produces a car frame at a cost of $135 and a truck frame at a cost of $180. Based on past demand, the company wants to limit production to a maximum of 250 car frames and 350 truck frames per day. If the company realizes a profit of $50 on each car frame and $70 on each truck frame, how many frames of each type should be produced at each plant to maximize the daily profit?

TYPE OF LOAN	ANNUAL RETURN
Signature	18%
First mortgage	12%
Second mortgage	14%
Automobile	16%

40. *Finances—loan distributions.* A savings and loan company has $3 million to lend. The types of loans and annual returns offered by the company are given in the table. State laws require that at least 50% of the money loaned for mortgages must be for first mortgages and that at least 30% of the total amount loaned must be for either first or second mortgages. Company policy requires that the amount of signature and automobile loans cannot exceed 25% of the total amount loaned and that signature loans cannot exceed 15% of the total amount loaned. How much money should be allocated to each type of loan in order to maximize the company's return?

41. *Blending—petroleum.* A refinery produces two grades of gasoline, regular and premium, by blending together three components, A, B, and C. Component A has an octane rating of 90 and costs $28 a barrel, component B has an octane rating of 100 and costs $30 a barrel, and component C has an octane rating of 110 and costs $34 a barrel. The octane rating for regular must be at least 95 and the octane rating for premium must be at least 105. Regular gasoline sells for $38 a barrel and premium sells for $46 a barrel. The company has 40,000 barrels of component A, 25,000 barrels of component B, and 15,000 barrels of component C and must produce at least 30,000 barrels of regular and 25,000 barrels of premium. How should the components be blended in order to maximize profit?

TYPE OF MIX	INGREDIENTS
Regular	At least 20% nuts At most 40% cereal
Deluxe	At least 30% nuts At most 25% cereal

42. *Blending—food processing.* A company produces two brands of trail mix, regular and deluxe, by mixing dried fruits, nuts, and cereal. The recipes for the mixes are given in the table. The company has 1,200 pounds of dried fruits, 750 pounds of nuts, and 1,500 pounds of cereal to be used in producing the mixes. The company makes a profit of $0.40 on each pound of regular mix and $0.60 on each pound of deluxe mix. How many pounds of each ingredient should be used in each mix in order to maximize the company's profit?

Life Sciences

	UNITS PER OUNCE *L*	*M*	*N*
CALCIUM	30	10	30
IRON	10	10	10
VITAMIN A	10	30	20
CHOLESTEROL	8	4	6
CALORIES	60	40	50
COST PER OUNCE	$0.40	$0.60	$0.80

43. *Nutrition—people.* A dietitian in a hospital is to arrange a special diet using the foods L, M, and N. The table gives the nutritional contents and the cost of 1 ounce of each food. The daily requirements for the diet are at least 400 units of calcium, at least 200 units of iron, at least 300 units of vitamin A, at most 150 units of cholesterol, and at most 900 calories. How many ounces of each food should be used in order to meet the requirements of the diet at minimal cost?

44. *Nutrition—feed mixtures.* A farmer grows three crops, corn, oats, and soybeans, which he mixes together to feed his cows and pigs. At least 40% of the feed mix for the cows must be corn. The feed mix for the pigs must contain at least twice as much soybeans as corn. He has harvested 1,000 bushels of corn, 500 bushels of oats, and 1,000 bushels of soybeans. He needs 1,000 bushels of each feed mix for his livestock. The unused corn, oats, and soybeans can be sold for $4, $3.50, and $3.25 a bushel, respectively (thus, these amounts also represent the cost of the crops used to feed the livestock). How many bushels of each crop should be used in each feed mix in order to produce sufficient food for the livestock at minimal cost?

Social Sciences

	WEEKLY TRANSPORTATION COST PER STUDENT *School I*	*School II*
TOWN *A*	$4	$8
TOWN *B*	6	4
TOWN *C*	3	9

45. *Education—resource allocation.* Three towns are forming a consolidated school district with two high schools. Each high school has a maximum capacity of 2,000 students. Town *A* has 500 high school students, town *B* has 1,200, and town *C* has 1,800. The weekly costs of transporting a student from each town to each school are given in the table. In order to keep the enrollment balanced, the school board has decided that each high school must enroll at least 40% of the total student population. Furthermore, no more than 60% of the students in any town should be sent to the same high school. How many students from each town should be enrolled in each school in order to meet these requirements and minimize the cost of transporting the students?

SECTION 6-7 Chapter Review

Important Terms and Symbols

6-1 *Systems of Linear Inequalities in Two Variables.* Graph of a linear inequality in two variables; upper half-plane; lower half-plane; graphical solution of systems of linear inequalities; solution region; feasible region; corner point; bounded regions; unbounded regions

6-2 *Linear Programming in Two Dimensions—A Geometric Approach.* Linear programming problem; decision variables; objective function; problem constraints; nonnegative constraints; mathematical model; graphical solution; maximization problem; constant-profit line; optimal solution; feasible region; fundamental theorem of linear programming; multiple optimal solution; empty feasible region; unbounded objective function

6-3 *A Geometric Introduction to the Simplex Method.* Standard maximization problem in standard form; slack variables; basic variables; nonbasic variables; basic solution; basic feasible solution; simplex method (algorithm)

6-4 *The Simplex Method: Maximization with Problem Constraints of the Form* $\leqslant$. Initial system; basic solutions and basic feasible solutions for initial systems; initial basic feasible solution; simplex tableau; initial simplex tableau; selecting basic and nonbasic variables for the simplex method; pivot operation; entering variable; exiting variable; pivot column; indicators; pivot row; pivot element; pivoting

6-5 *The Dual; Minimization with Problem Constraints of the Form* $\geqslant$. Dual problem; transpose; solution of a minimization problem; transportation problem

6-6 *Maximization and Minimization with Mixed Problem Constraints.* Surplus variable; artificial variable; modified problem; preliminary simplex tableau; initial simplex tableau; big M method

Summary of Problem Types and Solution Methods

PROBLEM TYPE	PROBLEM CONSTRAINTS	RIGHT-SIDE CONSTANTS	COEFFICIENTS OF OBJECTIVE FUNCTION	METHOD OF SOLUTION
1. Maximization	$\leqslant$	Nonnegative	Any real numbers	Simplex method with slack variables
2. Minimization	$\geqslant$	Any real numbers	Nonnegative	Form dual and solve by method 1.
3. Maximization	Mixed ($\leqslant, \geqslant, =$)	Nonnegative	Any real numbers	Form modified problem with slack, surplus, and artificial variables and solve by the big M method.
4. Minimization	Mixed ($\leqslant, \geqslant, =$)	Nonnegative	Any real numbers	Maximize negative of objective function by method 3.

EXERCISE 6-7

Chapter Review

Work through all the problems in this chapter review and check your answers in the back of the book. (Answers to all review problems are there.) Where weaknesses show up, review appropriate sections in the text.

A *Solve the systems in Problems 1 and 2 graphically, and indicate whether each solution region is bounded or unbounded. Find the coordinates of each corner point.*

1. $$\begin{aligned} 2x_1 + x_2 &\leqslant 8 \\ 3x_1 + 9x_2 &\leqslant 27 \\ x_1, x_2 &\geqslant 0 \end{aligned}$$

2. $$\begin{aligned} 3x_1 + x_2 &\geqslant 9 \\ 2x_1 + 4x_2 &\geqslant 16 \\ x_1, x_2 &\geqslant 0 \end{aligned}$$

3. Solve the linear programming problem geometrically:

$$\begin{aligned} \text{Maximize} \quad & P = 6x_1 + 2x_2 \\ \text{Subject to} \quad & 2x_1 + x_2 \le 8 \\ & x_1 + 2x_2 \le 10 \\ & x_1, x_2 \ge 0 \end{aligned}$$

4. Convert the problem constraints in Problem 3 into a system of equations using slack variables.

5. How many basic variables and how many nonbasic variables are associated with the system in Problem 4?

6. Find all basic solutions for the system in Problem 4 and determine which basic solutions are feasible.

7. Write the simplex tableau for Problem 3 and circle the pivot element. Indicate the entering and exiting variables.

8. Solve Problem 3 using the simplex method.

9. For the simplex tableau below, identify the basic and nonbasic variables. Find the pivot element, the entering and exiting variables, and perform one pivot operation.

$$\begin{array}{c} \begin{matrix} x_1 & x_2 & x_3 & s_1 & s_2 & s_3 & P \end{matrix} \\ \left[\begin{array}{rrrrrrr|r} 2 & 1 & 3 & -1 & 0 & 0 & 0 & 20 \\ 3 & 0 & 4 & 1 & 1 & 0 & 0 & 30 \\ 2 & 0 & 5 & 2 & 0 & 1 & 0 & 10 \\ \hline -8 & 0 & -5 & 3 & 0 & 0 & 1 & 50 \end{array}\right] \end{array}$$

10. Find the basic solution for each tableau. Determine whether the optimal solution has been reached, additional pivoting is required, or the problem has no solution.

$$\begin{array}{c} \begin{matrix} x_1 & x_2 & s_1 & s_2 & P \end{matrix} \\ \text{(A)} \left[\begin{array}{rrrrr|r} 4 & 1 & 0 & 0 & 0 & 2 \\ 2 & 0 & 1 & 1 & 0 & 5 \\ \hline -2 & 0 & 3 & 0 & 1 & 12 \end{array}\right] \end{array}$$

$$\text{(B)} \left[\begin{array}{rrrrr|r} -1 & 3 & 0 & 1 & 0 & 7 \\ 0 & 2 & 1 & 0 & 0 & 0 \\ \hline -2 & 1 & 0 & 0 & 1 & 22 \end{array}\right]$$

$$\text{(C)} \left[\begin{array}{rrrrr|r} 1 & -2 & 0 & 4 & 0 & 6 \\ 0 & 2 & 1 & 6 & 0 & 15 \\ \hline 0 & 3 & 0 & 2 & 1 & 10 \end{array}\right]$$

11. Solve the linear programming problem geometrically:

$$\begin{aligned} \text{Minimize} \quad & C = 5x_1 + 2x_2 \\ \text{Subject to} \quad & x_1 + 3x_2 \geq 15 \\ & 2x_1 + x_2 \geq 20 \\ & x_1, x_2 \geq 0 \end{aligned}$$

12. Form the dual of Problem 11.
13. Write the initial system for the dual in Problem 12.
14. Write the first simplex tableau for the dual in Problem 12 and label the columns.
15. Use the simplex method to find the optimal solution of the dual in Problem 12.
16. Use the final simplex tableau from Problem 15 to find the optimal solution of Problem 11.

B

17. Solve the linear programming problem geometrically:

$$\begin{aligned} \text{Maximize} \quad & P = 3x_1 + 4x_2 \\ \text{Subject to} \quad & 2x_1 + 4x_2 \leq 24 \\ & 3x_1 + 3x_2 \leq 21 \\ & 4x_1 + 2x_2 \leq 20 \\ & x_1, x_2 \geq 0 \end{aligned}$$

18. Solve Problem 17 using the simplex method.
19. Solve the linear programming problem geometrically:

$$\begin{aligned} \text{Minimize} \quad & C = 3x_1 + 8x_2 \\ \text{Subject to} \quad & x_1 + x_2 \geq 10 \\ & x_1 + 2x_2 \geq 15 \\ & x_2 \geq 3 \\ & x_1, x_2 \geq 0 \end{aligned}$$

20. Form the dual of Problem 19.
21. Solve Problem 19 by applying the simplex method to the dual in Problem 20.

Solve the following linear programming problems:

22. $$\begin{aligned} \text{Maximize} \quad & P = 5x_1 + 3x_2 - 3x_3 \\ \text{Subject to} \quad & x_1 - x_2 - 2x_3 \leq 3 \\ & 2x_1 + 2x_2 - 5x_3 \leq 10 \\ & x_1, x_2, x_3 \geq 0 \end{aligned}$$

23. $$\begin{aligned} \text{Maximize} \quad & P = 5x_1 + 3x_2 - 3x_3 \\ \text{Subject to} \quad & x_1 - x_2 - 2x_3 \leq 3 \\ & x_1 + x_2 \leq 5 \\ & x_1, x_2, x_3 \geq 0 \end{aligned}$$

In Problems 24 and 25:

(A) Introduce slack, surplus, and artificial variables and form the modified problem.
(B) Write the preliminary simplex tableau for the modified problem and find the initial simplex tableau.
(C) Find the optimal solution of the modified problem by applying the simplex method to the initial simplex tableau.
(D) Find the optimal solution of the original problem, if it exists.

24. Maximize $P = x_1 + 3x_2$
Subject to
$$x_1 + x_2 \geq 6$$
$$x_1 + 2x_2 \leq 8$$
$$x_1, x_2 \geq 0$$

25. Maximize $P = x_1 + x_2$
Subject to
$$x_1 + x_2 \geq 5$$
$$x_1 + 2x_2 \leq 4$$
$$x_1, x_2 \geq 0$$

26. Find the modified problem for the following linear programming problem. (Do not solve.)

Maximize $P = 2x_1 + 3x_2 + x_3$
Subject to
$$x_1 - 3x_2 + x_3 \leq 7$$
$$-x_1 - x_2 + 2x_3 \leq -2$$
$$3x_1 + 2x_2 - x_3 = 4$$
$$x_1, x_2, x_3 \geq 0$$

C

27. Solve the following linear programming problem by the simplex method, keeping track of the obvious basic solution at each step. Then graph the feasible region and illustrate the path to the optimal solution determined by the simplex method.

Maximize $P = 2x_1 + 3x_2$
Subject to
$$x_1 + 2x_2 \leq 22$$
$$2x_1 + x_2 \leq 20$$
$$x_1 \leq 8$$
$$x_2 \leq 10$$
$$x_1, x_2 \geq 0$$

28. Solve by the dual method:

Minimize $C = 2x_1 + 3x_2$
Subject to
$$2x_1 + x_2 \leq 20$$
$$2x_1 + x_2 \geq 10$$
$$x_1 + 2x_2 \geq 8$$
$$x_1, x_2 \geq 0$$

29. Solve Problem 28 by the big M method.

30. Solve by the dual method:

$$\begin{array}{ll} \text{Minimize} & C = 15x_1 + 12x_2 + 15x_3 + 18x_4 \\ \text{Subject to} & x_1 + x_2 \leq 240 \\ & x_3 + x_4 \leq 500 \\ & x_1 + x_3 \geq 400 \\ & x_2 + x_4 \geq 300 \\ & x_1, x_2, x_3, x_4 \geq 0 \end{array}$$

APPLICATIONS

Formulate the following as linear programming problems. (Do not solve.)

Business & Economics

31. *Manufacturing—resource allocation.* South Shore Sail Loft manufactures regular and competition sails. Each regular sail takes 2 hours to cut and 4 hours to sew. Each competition sail takes 3 hours to cut and 9 hours to sew. The Loft makes a profit of \$100 on each regular sail and \$200 on each competition sail. If there are 150 hours available in the cutting department and 360 hours available in the sewing department, how many sails of each type should the company manufacture in order to maximize their profit?

32. *Investment.* An investor has \$150,000 to invest in oil stock, steel stock, and government bonds. The oil stock is expected to yield 12%, the steel stock is expected to yield 9%, and the government bonds are guaranteed to yield 5%. The investor's policy requires that the total amount of money invested in stock cannot exceed the amount invested in bonds and that at least twice as much money should be invested in steel stock as in oil stock. How much should be invested in each alternative in order to maximize the return?

33. *Transportation—shipping schedule.* A company produces motors for washing machines at factory A and factory B. Factory A can produce 1,500 motors a month and factory B can produce 1,000 motors a month. The motors are then shipped to one of three plants, where the washing machines are assembled. In order to meet anticipated demand, plant X must assemble at least 500 washing machines a month, plant Y must assemble at least 700 washing machines a month, and plant Z must assemble at least 800 washing machines a month. The shipping charges for one motor are given in the table. Determine a shipping schedule that will minimize the cost of transporting the motors from the factories to the assembly plants.

	SHIPPING CHARGES		
	Plant X	*Plant Y*	*Plant Z*
FACTORY *A*	\$5	\$8	\$12
FACTORY *B*	\$9	\$7	\$ 6

34. *Blending—food processing.* A company blends long grain rice and wild rice to produce two brands of rice mixes, brand A, which is marketed under the company's name, and brand B, which is marketed as a generic brand. Brand A must contain at least 10% wild rice, and brand B must contain at least 5% wild rice. Long grain rice costs the company \$0.70 per pound, and wild rice costs \$3.40 per pound. The company sells brand A for \$1.50 a pound and brand B for \$1.20 a pound. The company has 8,000 pounds of long grain rice and 500 pounds of wild rice on hand. How many pounds of each brand of rice mix should they produce in order to maximize profit?

Life Sciences

35. *Nutrition—animals.* A special diet for laboratory animals is to contain at least 300 units of vitamins, 200 units of minerals, and 900 calories. There are two feed mixes available, mix A and mix B. A gram of mix A contains 3 units of vitamins, 2 units of minerals, and 6 calories. A gram of mix B contains 4 units of vitamins, 5 units of minerals, and 10 calories. Mix A costs \$0.02 per gram and mix B costs \$0.04 per gram. How many grams of each mix should be used to satisfy the requirements of the diet at minimal cost?

APPENDIX C

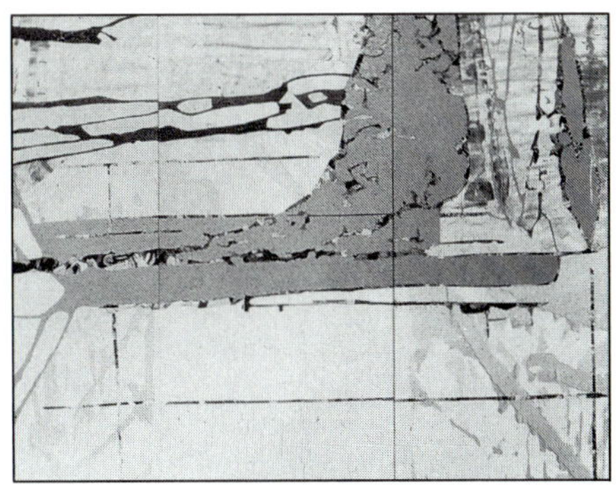

Tables

TABLE I

Basic Geometric Formulas

◆ 1. SIMILAR TRIANGLES

(A) Two triangles are similar if two angles of one triangle have the same measure as two angles of the other.
(B) If two triangles are similar, their corresponding sides are proportional:

$$\frac{a}{a'} = \frac{b}{b'} = \frac{c}{c'}$$

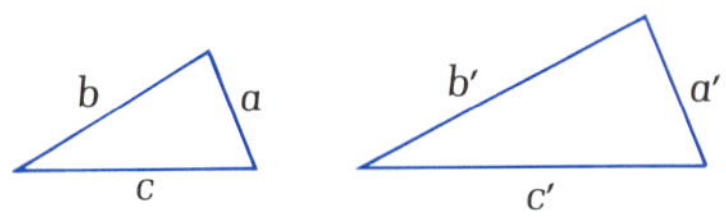

◆ 2. PYTHAGOREAN THEOREM

$c^2 = a^2 + b^2$

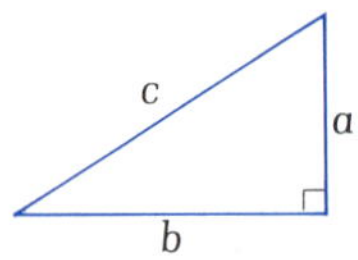

◆ 3. RECTANGLE

$A = ab$ Area

$P = 2a + 2b$ Perimeter

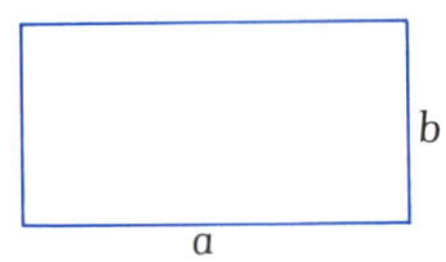

◆ 4. PARALLELOGRAM

$h =$ Height

$A = ah = ab \sin \theta$ Area

$P = 2a + 2b$ Perimeter

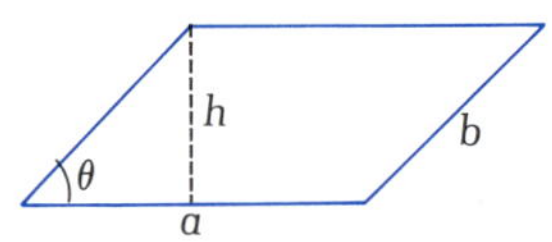

◆ 5. TRIANGLE

$h =$ Height

$A = \frac{1}{2}hc$ Area

$P = a + b + c$ Perimeter

$s = \frac{1}{2}(a + b + c)$ Semiperimeter

$A = \sqrt{s(s-a)(s-b)(s-c)}$ Area—Heron's formula

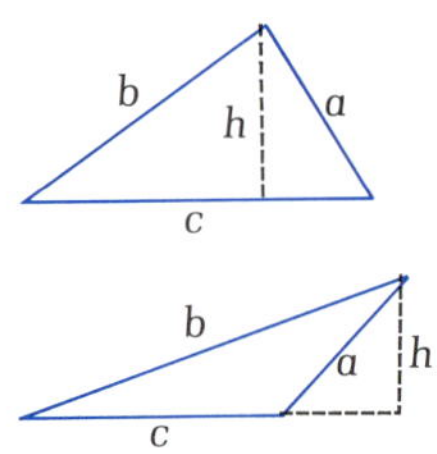

◆ 6. TRAPEZOID

Base a is parallel to base b.

$h =$ Height

$A = \frac{1}{2}(a + b)h$ Area

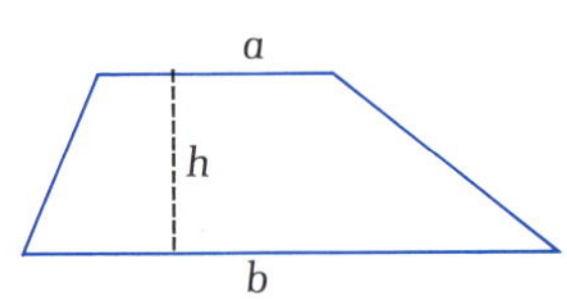

7. CIRCLE

R = Radius

D = Diameter

$D = 2R$

$A = \pi R^2 = \frac{1}{4}\pi D^2$ Area

$C = 2\pi R = \pi D$ Circumference

$\frac{C}{D} = \pi$ For all circles

$\pi \approx 3.141\ 59$

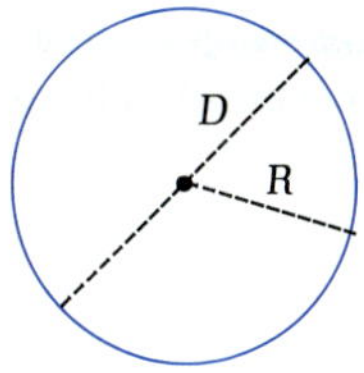

8. RECTANGULAR SOLID

$V = abc$ Volume

$T = 2ab + 2ac + 2bc$ Total surface area

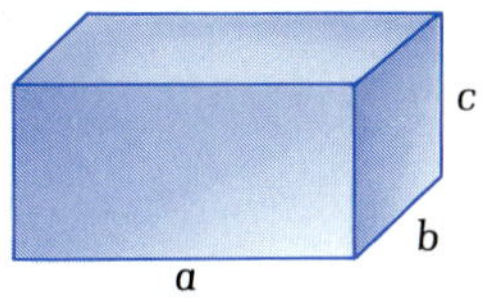

9. RIGHT CIRCULAR CYLINDER

R = Radius of base

h = Height

$V = \pi R^2 h$ Volume

$S = 2\pi Rh$ Lateral surface area

$T = 2\pi R(R + h)$ Total surface area

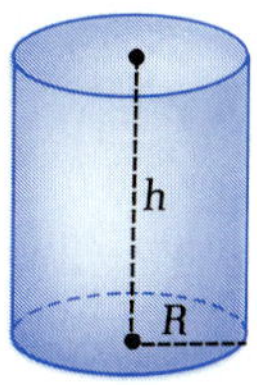

10. RIGHT CIRCULAR CONE

R = Radius of base

h = Height

s = Slant height

$V = \frac{1}{3}\pi R^2 h$ Volume

$S = \pi Rs = \pi R\sqrt{R^2 + h^2}$ Lateral surface area

$T = \pi R(R + s) = \pi R(R + \sqrt{R^2 + h^2})$ Total surface area

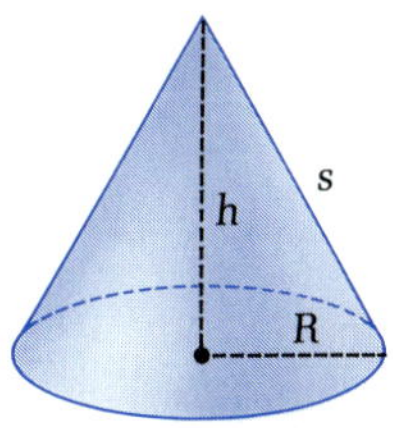

11. SPHERE

R = Radius

D = Diameter

$D = 2R$

$V = \frac{4}{3}\pi R^3 = \frac{1}{6}\pi D^3$ Volume

$S = 4\pi R^2 = \pi D^2$ Surface area

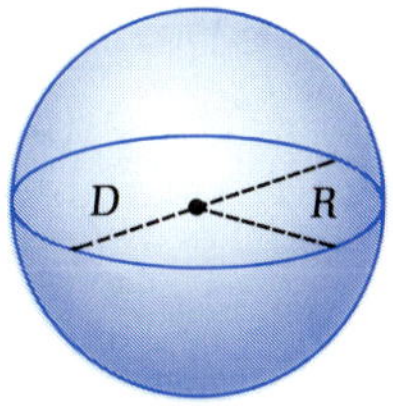

TABLE II

Integration Formulas

[*Note:* The constant of integration is omitted for each integral, but must be included in any particular application of a formula. The variable u is the variable of integration; all other symbols represent constants.]

◆ INTEGRALS INVOLVING u^n

1. $\int u^n\,du = \frac{u^{n+1}}{n+1}, \qquad n \neq -1$
2. $\int u^{-1}\,du = \int \frac{1}{u}\,du = \ln|u|$

◆ INTEGRALS INVOLVING $a + bu$; $a \neq 0$ AND $b \neq 0$

3. $\int \frac{1}{a+bu}\,du = \frac{1}{b}\ln|a+bu|$
4. $\int \frac{u}{a+bu}\,du = \frac{u}{b} - \frac{a}{b^2}\ln|a+bu|$
5. $\int \frac{u^2}{a+bu}\,du = \frac{(a+bu)^2}{2b^3} - \frac{2a(a+bu)}{b^3} + \frac{a^2}{b^3}\ln|a+bu|$
6. $\int \frac{u}{(a+bu)^2}\,du = \frac{1}{b^2}\left(\ln|a+bu| + \frac{a}{a+bu}\right)$
7. $\int \frac{u^2}{(a+bu)^2}\,du = \frac{(a+bu)}{b^3} - \frac{a^2}{b^3(a+bu)} - \frac{2a}{b^3}\ln|a+bu|$
8. $\int u(a+bu)^n\,du = \frac{(a+bu)^{n+2}}{(n+2)b^2} - \frac{a(a+bu)^{n+1}}{(n+1)b^2}, \qquad n \neq -1, -2$
9. $\int \frac{1}{u(a+bu)}\,du = \frac{1}{a}\ln\left|\frac{u}{a+bu}\right|$
10. $\int \frac{1}{u^2(a+bu)}\,du = -\frac{1}{au} + \frac{b}{a^2}\ln\left|\frac{a+bu}{u}\right|$
11. $\int \frac{1}{u(a+bu)^2}\,du = \frac{1}{a(a+bu)} + \frac{1}{a^2}\ln\left|\frac{u}{a+bu}\right|$
12. $\int \frac{1}{u^2(a+bu)^2}\,du = -\frac{a+2bu}{a^2u(a+bu)} + \frac{2b}{a^3}\ln\left|\frac{a+bu}{u}\right|$

◆ INTEGRALS INVOLVING $a^2 - u^2$, $a > 0$

13. $\int \frac{1}{u^2-a^2}\,du = \frac{1}{2a}\ln\left|\frac{u-a}{u+a}\right|$
14. $\int \frac{1}{a^2-u^2}\,du = \frac{1}{2a}\ln\left|\frac{u+a}{u-a}\right|$

◆ INTEGRALS INVOLVING $(a+bu)$ AND $(c+du)$; $b \neq 0$, $d \neq 0$, AND $ad-bc \neq 0$

15. $$\int \frac{1}{(a+bu)(c+du)}\,du = \frac{1}{ad-bc}\ln\left|\frac{c+du}{a+bu}\right|$$

16. $$\int \frac{u}{(a+bu)(c+du)}\,du = \frac{1}{ad-bc}\left(\frac{a}{b}\ln|a+bu| - \frac{c}{d}\ln|c+du|\right)$$

17. $$\int \frac{u^2}{(a+bu)(c+du)}\,du = \frac{1}{bd}\,u - \frac{1}{ad-bc}\left(\frac{a^2}{b^2}\ln|a+bu| - \frac{c^2}{d^2}\ln|c+du|\right)$$

18. $$\int \frac{1}{(a+bu)^2(c+du)}\,du = \frac{1}{ad-bc}\,\frac{1}{a+bu} + \frac{d}{(ad-bc)^2}\ln\left|\frac{c+du}{a+bu}\right|$$

19. $$\int \frac{u}{(a+bu)^2(c+du)}\,du = -\frac{a}{b(ad-bc)}\,\frac{1}{a+bu} - \frac{c}{(ad-bc)^2}\ln\left|\frac{c+du}{a+bu}\right|$$

20. $$\int \frac{a+bu}{c+du}\,du = \frac{bu}{d} + \frac{ad-bc}{d^2}\ln|c+du|$$

◆ INTEGRALS INVOLVING $\sqrt{a+bu}$, $a \neq 0$ AND $b \neq 0$

21. $$\int \sqrt{a+bu}\,du = \frac{2\sqrt{(a+bu)^3}}{3b}$$

22. $$\int u\sqrt{a+bu}\,du = \frac{2(3bu-2a)}{15b^2}\sqrt{(a+bu)^3}$$

23. $$\int u^2\sqrt{a+bu}\,du = \frac{2(15b^2u^2-12abu+8a^2)}{105b^3}\sqrt{(a+bu)^3}$$

24. $$\int \frac{1}{\sqrt{a+bu}}\,du = \frac{2\sqrt{a+bu}}{b}$$

25. $$\int \frac{u}{\sqrt{a+bu}}\,du = \frac{2(bu-2a)}{3b^2}\sqrt{a+bu}$$

26. $$\int \frac{u^2}{\sqrt{a+bu}}\,du = \frac{2(3b^2u^2-4abu+8a^2)}{15b^3}\sqrt{a+bu}$$

27. $$\int \frac{1}{u\sqrt{a+bu}}\,du = \frac{1}{\sqrt{a}}\ln\left|\frac{\sqrt{a+bu}-\sqrt{a}}{\sqrt{a+bu}+\sqrt{a}}\right|, \qquad a>0$$

28. $$\int \frac{1}{u^2\sqrt{a+bu}}\,du = -\frac{\sqrt{a+bu}}{au} - \frac{b}{2a\sqrt{a}}\ln\left|\frac{\sqrt{a+bu}-\sqrt{a}}{\sqrt{a+bu}+\sqrt{a}}\right|, \qquad a>0$$

◆ INTEGRALS INVOLVING $\sqrt{a^2-u^2}$, $a>0$

29. $$\int \frac{1}{u\sqrt{a^2-u^2}}\,du = -\frac{1}{a}\ln\left|\frac{a+\sqrt{a^2-u^2}}{u}\right|$$

30. $\displaystyle\int \frac{1}{u^2\sqrt{a^2-u^2}}\,du = -\frac{\sqrt{a^2-u^2}}{a^2u}$

31. $\displaystyle\int \frac{\sqrt{a^2-u^2}}{u}\,du = \sqrt{a^2-u^2} - a\ln\left|\frac{a+\sqrt{a^2-u^2}}{u}\right|$

◆ INTEGRALS INVOLVING $\sqrt{u^2+a^2}$, $a>0$

32. $\displaystyle\int \sqrt{u^2+a^2}\,du = \frac{1}{2}\left(u\sqrt{u^2+a^2} + a^2\ln|u+\sqrt{u^2+a^2}|\right)$

33. $\displaystyle\int u^2\sqrt{u^2+a^2}\,du = \frac{1}{8}\left[u(2u^2+a^2)\sqrt{u^2+a^2} - a^4\ln|u+\sqrt{u^2+a^2}|\right]$

34. $\displaystyle\int \frac{\sqrt{u^2+a^2}}{u}\,du = \sqrt{u^2+a^2} - a\ln\left|\frac{a+\sqrt{u^2+a^2}}{u}\right|$

35. $\displaystyle\int \frac{\sqrt{u^2+a^2}}{u^2}\,du = -\frac{\sqrt{u^2+a^2}}{u} + \ln|u+\sqrt{u^2+a^2}|$

36. $\displaystyle\int \frac{1}{\sqrt{u^2+a^2}}\,du = \ln|u+\sqrt{u^2+a^2}|$

37. $\displaystyle\int \frac{1}{u\sqrt{u^2+a^2}}\,du = \frac{1}{a}\ln\left|\frac{u}{a+\sqrt{u^2+a^2}}\right|$

38. $\displaystyle\int \frac{u^2}{\sqrt{u^2+a^2}}\,du = \frac{1}{2}\left(u\sqrt{u^2+a^2} - a^2\ln|u+\sqrt{u^2+a^2}|\right)$

39. $\displaystyle\int \frac{1}{u^2\sqrt{u^2+a^2}}\,du = -\frac{\sqrt{u^2+a^2}}{a^2u}$

◆ INTEGRALS INVOLVING $\sqrt{u^2-a^2}$, $a>0$

40. $\displaystyle\int \sqrt{u^2-a^2}\,du = \frac{1}{2}\left(u\sqrt{u^2-a^2} - a^2\ln|u+\sqrt{u^2-a^2}|\right)$

41. $\displaystyle\int u^2\sqrt{u^2-a^2}\,du = \frac{1}{8}\left[u(2u^2-a^2)\sqrt{u^2-a^2} - a^4\ln|u+\sqrt{u^2-a^2}|\right]$

42. $\displaystyle\int \frac{\sqrt{u^2-a^2}}{u^2}\,du = -\frac{\sqrt{u^2-a^2}}{u} + \ln|u+\sqrt{u^2-a^2}|$

43. $\displaystyle\int \frac{1}{\sqrt{u^2-a^2}}\,du = \ln|u+\sqrt{u^2-a^2}|$

44. $\displaystyle\int \frac{u^2}{\sqrt{u^2-a^2}}\,du = \frac{1}{2}\left(u\sqrt{u^2-a^2} + a^2\ln|u+\sqrt{u^2-a^2}|\right)$

45. $\displaystyle\int \frac{1}{u^2\sqrt{u^2-a^2}}\,du = \frac{\sqrt{u^2-a^2}}{a^2u}$

◆ INTEGRALS INVOLVING e^{au}, $a\neq 0$

46. $\displaystyle\int e^{au}\,du = \frac{e^{au}}{a}$

47. $\displaystyle\int u^n e^{au}\,du = \frac{u^n e^{au}}{a} - \frac{n}{a}\int u^{n-1}e^{au}\,du$

48. $\displaystyle\int \frac{1}{c + de^{au}}\,du = \frac{u}{c} - \frac{1}{ac}\ln|c + de^{au}|, \quad c \neq 0$

◆ INTEGRALS INVOLVING $\ln u$

49. $\displaystyle\int \ln u\,du = u\ln u - u$

50. $\displaystyle\int \frac{\ln u}{u}\,du = \frac{1}{2}(\ln u)^2$

51. $\displaystyle\int u^n \ln u\,du = \frac{u^{n+1}}{n+1}\ln u - \frac{u^{n+1}}{(n+1)^2}, \quad n \neq -1$

52. $\displaystyle\int (\ln u)^n\,du = u(\ln u)^n - n\int (\ln u)^{n-1}\,du$

◆ INTEGRALS INVOLVING TRIGONOMETRIC FUNCTIONS OF $au, \quad a \neq 0$

53. $\displaystyle\int \sin au\,du = -\frac{1}{a}\cos au$

54. $\displaystyle\int \cos au\,du = \frac{1}{a}\sin au$

55. $\displaystyle\int \tan au\,du = -\frac{1}{a}\ln|\cos au|$

56. $\displaystyle\int \cot au\,du = \frac{1}{a}\ln|\sin au|$

57. $\displaystyle\int \sec au\,du = \frac{1}{a}\ln|\sec au + \tan au|$

58. $\displaystyle\int \csc au\,du = \frac{1}{a}\ln|\csc au - \cot au|$

59. $\displaystyle\int (\sin au)^2\,du = \frac{u}{2} - \frac{1}{4a}\sin 2au$

60. $\displaystyle\int (\cos au)^2\,du = \frac{u}{2} + \frac{1}{4a}\sin 2au$

61. $\displaystyle\int (\sin au)^n\,du = -\frac{1}{an}(\sin au)^{n-1}\cos au + \frac{n-1}{n}\int (\sin au)^{n-2}\,du, \qquad n \neq 0$

62. $\displaystyle\int (\cos au)^n\,du = \frac{1}{an}\sin au(\cos au)^{n-1} + \frac{n-1}{n}\int (\cos au)^{n-2}\,du, \qquad n \neq 0$

TABLE III

Area Under the Standard Normal Curve

Table entries represent the area under the standard normal curve from 0 to z, $z \geq 0$.

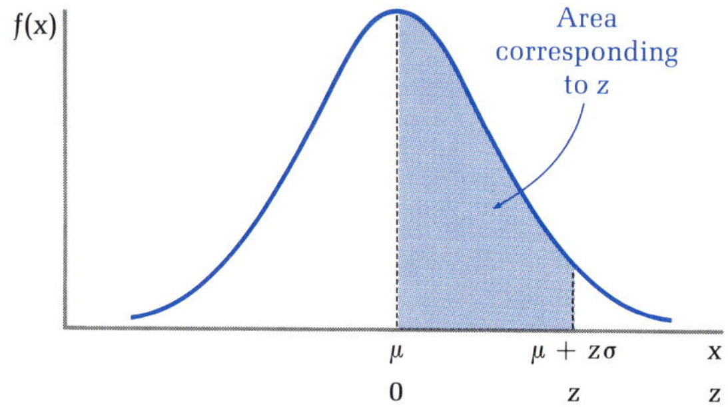

z	.00	.01	.02	.03	.04	.05	.06	.07	.08	.09
0.0	0.0000	0.0040	0.0080	0.0120	0.0160	0.0199	0.0239	0.0279	0.0319	0.0359
0.1	0.0398	0.0438	0.0478	0.0517	0.0557	0.0596	0.0636	0.0675	0.0714	0.0753
0.2	0.0793	0.0832	0.0871	0.0910	0.0948	0.0987	0.1026	0.1064	0.1103	0.1141
0.3	0.1179	0.1217	0.1255	0.1293	0.1331	0.1368	0.1406	0.1443	0.1480	0.1517
0.4	0.1554	0.1591	0.1628	0.1664	0.1700	0.1736	0.1772	0.1808	0.1844	0.1879
0.5	0.1915	0.1950	0.1985	0.2019	0.2054	0.2088	0.2123	0.2157	0.2190	0.2224
0.6	0.2257	0.2291	0.2324	0.2357	0.2389	0.2422	0.2454	0.2486	0.2517	0.2549
0.7	0.2580	0.2611	0.2642	0.2673	0.2704	0.2734	0.2764	0.2794	0.2823	0.2852
0.8	0.2881	0.2910	0.2939	0.2967	0.2995	0.3023	0.3051	0.3078	0.3106	0.3133
0.9	0.3159	0.3186	0.3212	0.3238	0.3264	0.3289	0.3315	0.3340	0.3365	0.3389
1.0	0.3413	0.3438	0.3461	0.3485	0.3508	0.3531	0.3554	0.3577	0.3599	0.3621
1.1	0.3643	0.3665	0.3686	0.3708	0.3729	0.3749	0.3770	0.3790	0.3810	0.3830
1.2	0.3849	0.3869	0.3888	0.3907	0.3925	0.3944	0.3962	0.3980	0.3997	0.4015
1.3	0.4032	0.4049	0.4066	0.4082	0.4099	0.4115	0.4131	0.4147	0.4162	0.4177
1.4	0.4192	0.4207	0.4222	0.4236	0.4251	0.4265	0.4279	0.4292	0.4306	0.4319
1.5	0.4332	0.4345	0.4357	0.4370	0.4382	0.4394	0.4406	0.4418	0.4429	0.4441
1.6	0.4452	0.4463	0.4474	0.4484	0.4495	0.4505	0.4515	0.4525	0.4535	0.4545
1.7	0.4554	0.4564	0.4573	0.4582	0.4591	0.4599	0.4608	0.4616	0.4625	0.4633
1.8	0.4641	0.4649	0.4656	0.4664	0.4671	0.4678	0.4686	0.4693	0.4699	0.4706
1.9	0.4713	0.4719	0.4726	0.4732	0.4738	0.4744	0.4750	0.4756	0.4761	0.4767
2.0	0.4772	0.4778	0.4783	0.4788	0.4793	0.4798	0.4803	0.4808	0.4812	0.4817
2.1	0.4821	0.4826	0.4830	0.4834	0.4838	0.4842	0.4846	0.4850	0.4854	0.4857
2.2	0.4861	0.4864	0.4868	0.4871	0.4875	0.4878	0.4881	0.4884	0.4887	0.4890
2.3	0.4893	0.4896	0.4898	0.4901	0.4904	0.4906	0.4909	0.4911	0.4913	0.4916
2.4	0.4918	0.4920	0.4922	0.4925	0.4927	0.4929	0.4931	0.4932	0.4934	0.4936
2.5	0.4938	0.4940	0.4941	0.4943	0.4945	0.4946	0.4948	0.4949	0.4951	0.4952
2.6	0.4953	0.4955	0.4956	0.4957	0.4959	0.4960	0.4961	0.4962	0.4963	0.4964
2.7	0.4965	0.4966	0.4967	0.4968	0.4969	0.4970	0.4971	0.4972	0.4973	0.4974
2.8	0.4974	0.4975	0.4976	0.4977	0.4977	0.4978	0.4979	0.4979	0.4980	0.4981
2.9	0.4981	0.4982	0.4982	0.4983	0.4984	0.4984	0.4985	0.4985	0.4986	0.4986
3.0	0.4987	0.4987	0.4987	0.4988	0.4988	0.4989	0.4989	0.4989	0.4990	0.4990
3.1	0.4990	0.4991	0.4991	0.4991	0.4992	0.4992	0.4992	0.4992	0.4993	0.4993
3.2	0.4993	0.4993	0.4994	0.4994	0.4994	0.4994	0.4994	0.4995	0.4995	0.4995
3.3	0.4995	0.4995	0.4995	0.4996	0.4996	0.4996	0.4996	0.4996	0.4996	0.4997
3.4	0.4997	0.4997	0.4997	0.4997	0.4997	0.4997	0.4997	0.4997	0.4997	0.4998
3.5	0.4998	0.4998	0.4998	0.4998	0.4998	0.4998	0.4998	0.4998	0.4998	0.4998
3.6	0.4998	0.4998	0.4999	0.4999	0.4999	0.4999	0.4999	0.4999	0.4999	0.4999
3.7	0.4999	0.4999	0.4999	0.4999	0.4999	0.4999	0.4999	0.4999	0.4999	0.4999
3.8	0.4999	0.4999	0.4999	0.4999	0.4999	0.4999	0.4999	0.4999	0.4999	0.4999
3.9	0.5000	0.5000	0.5000	0.5000	0.5000	0.5000	0.5000	0.5000	0.5000	0.5000

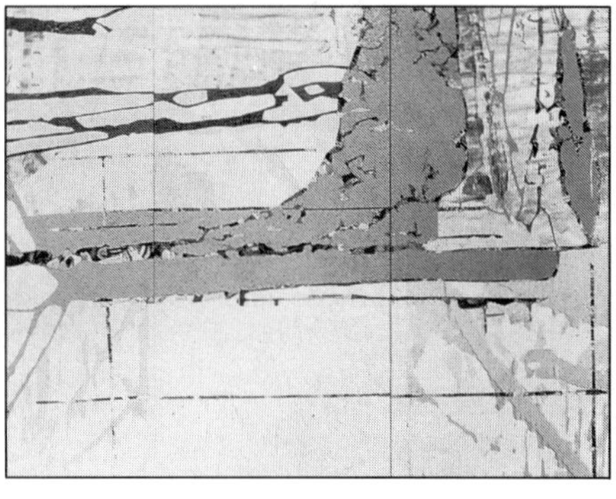

Answers

CHAPTER 1

◆ EXERCISE 1-1

1.

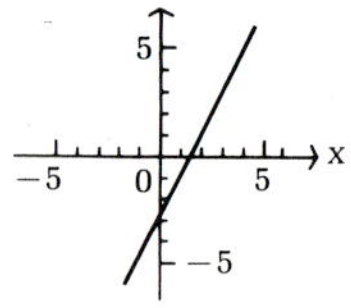

3.

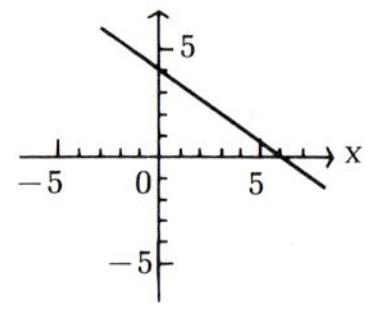

5. Slope = 2; y intercept = −3
7. Slope = $-\frac{2}{3}$; y intercept = 2
9. $y = -2x + 4$
11. $y = -\frac{3}{5}x + 3$

13.

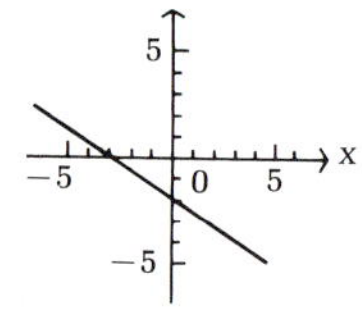

15.

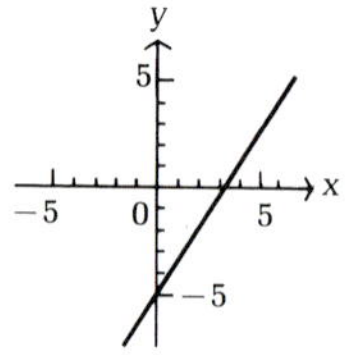

17.

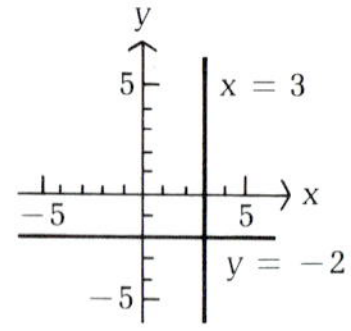

19. $y = -3x + 5$; $m = -3$
21. $y = -\frac{2}{3}x + 4$; $m = -\frac{2}{3}$
23. $y + 1 = -3(x - 4)$; $y = -3x + 11$
25. $y + 5 = \frac{2}{3}(x + 6)$; $y = \frac{2}{3}x - 1$
27. $\frac{1}{3}$
29. $-\frac{1}{5}$

31. $(y - 3) = \frac{1}{3}(x - 1)$; $x - 3y = -8$
33. $(y + 2) = -\frac{1}{5}(x + 5)$; $x + 5y = -15$
35. $x = 3$; $y = -5$
37. $x = -1$; $y = -3$
39. $y = -\frac{1}{2}x + 4$
41. (A) $y = -\frac{1}{2}x + 1$ (B) $y = 2x + 6$

43. (A) $y = \frac{1}{2}x$
(B) $y = -2x - 5$

45.

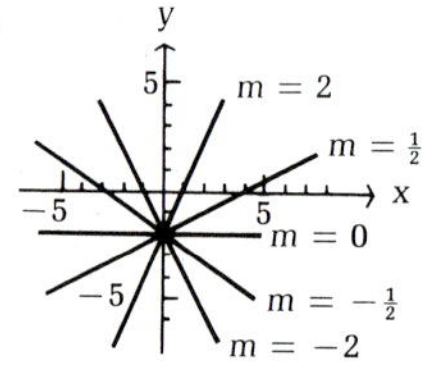

47. $x = 2$
49. $y = 3$

51. (A)

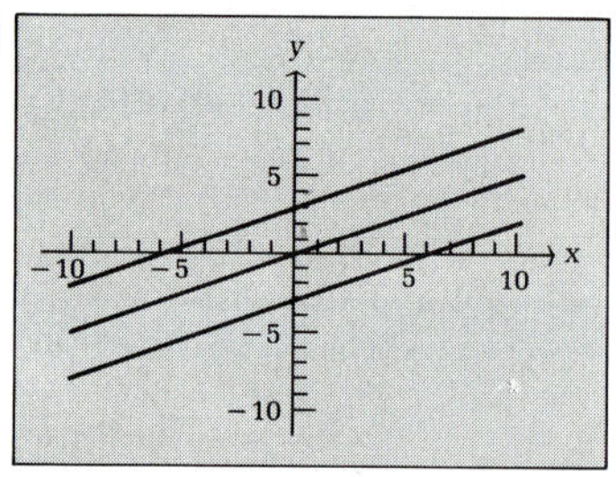

(B) $\frac{1}{2}$
(C) Yes, lines with the same slope are parallel.
(D) −3, 0, 3

53. (A)

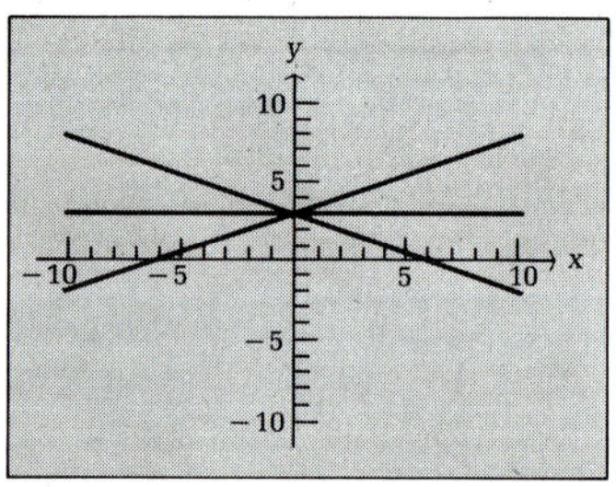

(B) $-\frac{1}{2}$, 0, $\frac{1}{2}$
(C) No, the lines do not have the same slope.
(D) 3

55. (A)

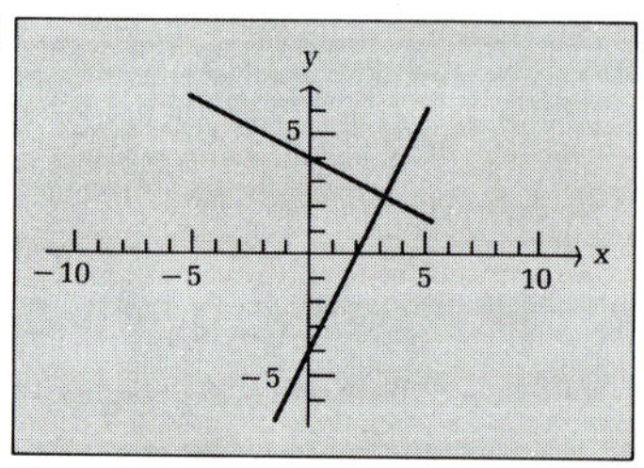

(B) Yes, since $m_1m_2 = (2)(-\frac{1}{2}) = -1$

57. (A) \$130; \$220
(B)

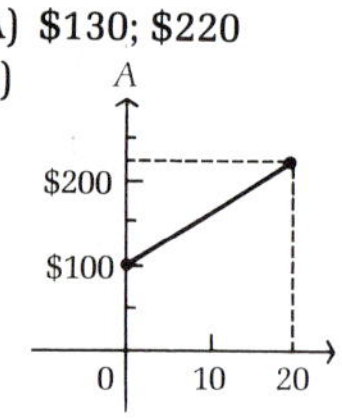

(C) 6

59. (A) $C = 180x + 200$
(B) \$2,360
(C)

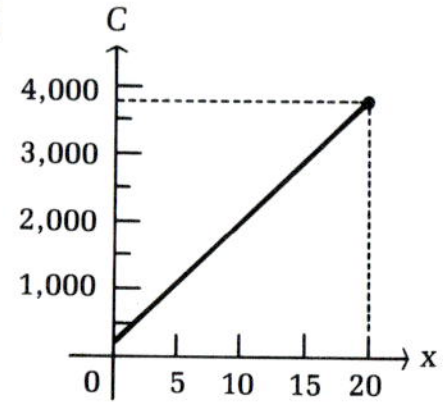

61. (A)

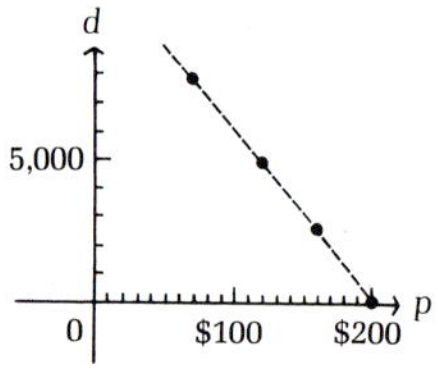

(B) $d = -60p + 12{,}000$

63. $0.2x + 0.1y = 20$

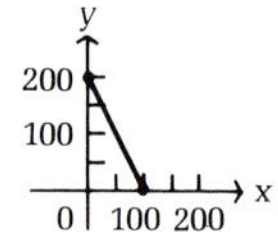

65. (A) 64 g; 35 g
(B)

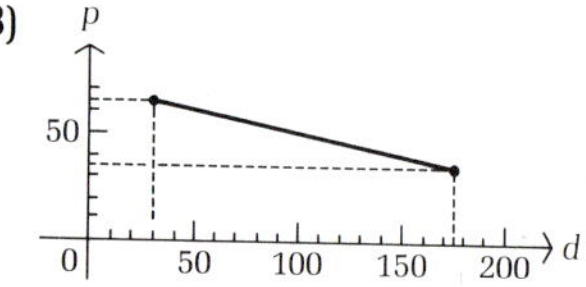

(C) $-\frac{1}{5}$

◆ EXERCISE 1-2

1. Function **3.** Not a function **5.** Function **7.** Function **9.** Not a function **11.** Function **13.** 4 **15.** -5 **17.** -6 **19.** -2 **21.** -12 **23.** -1 **25.** -6 **27.** 12 **29.** $\frac{3}{4}$ **31.** All nonnegative real numbers **33.** All real numbers except $x = -3, 5$ **35.** $x \geq -5$ or $[-5, \infty)$ **37.** All real numbers except $x = \pm 1$ **39.** All real numbers except $x = -4, 1$ **41.** All real numbers **43.** $x > -2$ or $(-2, \infty)$ **45.** A function with domain R **47.** A function with domain R **49.** Not a function; for example, when $x = 1$, $y = \pm 3$ **51.** A function with domain all real numbers except $x = 4$ **53.** Not a function; for example, when $x = 4$, $y = \pm 3$ **55.** 4 **57.** $h + 2$ **59.** $h - 1$ **61.** 4 **63.** $8a + 4h - 7$ **65.** $3a^2 + 3ah + h^2$

67. $\dfrac{1}{\sqrt{a+h} + \sqrt{a}}$ **69.** $P(w) = 2w + \dfrac{50}{w}$, $w > 0$

71. $A(\ell) = \ell(50 - \ell)$, $0 \leq \ell \leq 50$ **73.** $C(x) = 96{,}000 + 80x$; $136,000

75. $R(x) = x(200 - \frac{1}{40}x)$, $0 \leq x \leq 8{,}000$

77. (A) $R(x) = xp = 120x - \frac{1}{400}x^2$ (B) $P(x) = (120x - \frac{1}{400}x^2) - (100{,}000 + 20x) = -\frac{1}{400}x^2 + 100x - 100{,}000$
(C) $P(10{,}000) = \$650{,}000$; $P(20{,}000) = \$900{,}000$; $P(40{,}000) = -\$100{,}000$

79. (A) $V(x) = x(8 - 2x)(12 - 2x)$ (B) $0 \leq x \leq 4$ (C)

x	V(x)
1	60
2	64
3	36

81. $P(x) = 10x - 15 + \dfrac{270}{x}$;

x	P(x)
3	105
4	92.5
5	89
6	90

83. $v = \dfrac{75 - w}{15 + w}$; 1.9032 cm/sec

◆ EXERCISE 1-3

1. Slope: 2
y intercept: -4
x intercept: 2

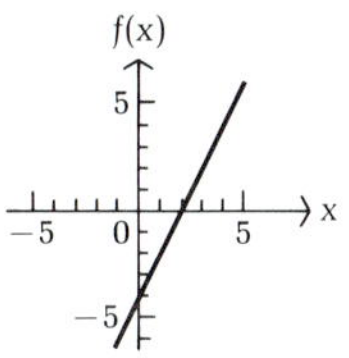

3. Slope: -2
y intercept: 4
x intercept: 2

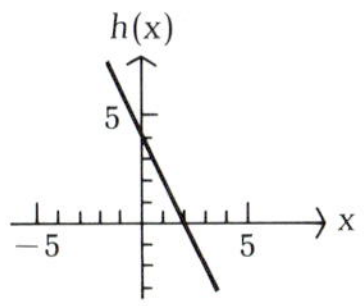

5. Slope: $-\frac{2}{3}$
y intercept: 4
x intercept: 6

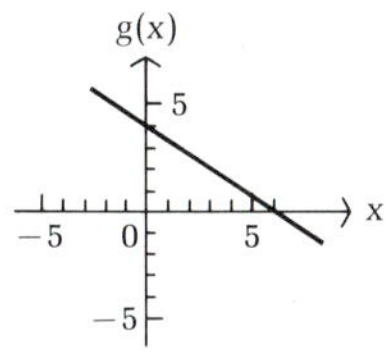

7. $f(x) = -2x + 6$
9. $f(x) = -\frac{1}{2}x + \frac{9}{2}$

11. Axis: $x = 3$
Vertex: $(3, -1)$
Min: $f(3) = -1$
y intercept: 8
x intercepts: 2, 4
Range: $[-1, \infty)$

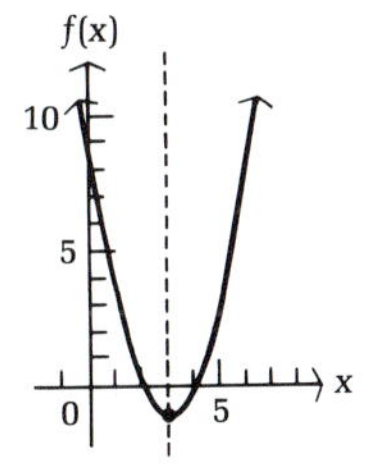

13. Axis: $x = -1$
Vertex: $(-1, 9)$
Max: $h(-1) = 9$
y intercept: 8
x intercepts: -4, 2
Range: $(-\infty, 9]$

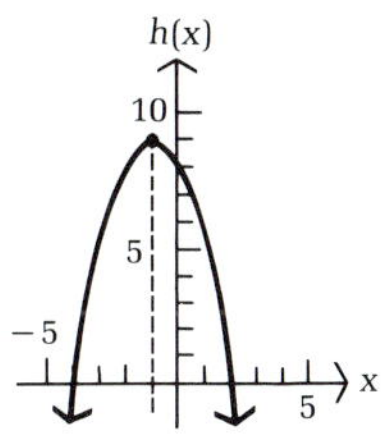

15. Axis: $x = -4$
Vertex: $(-4, 0)$
Min: $f(-4) = 0$
y intercept: 16
x intercept: -4
Range: $[0, \infty)$

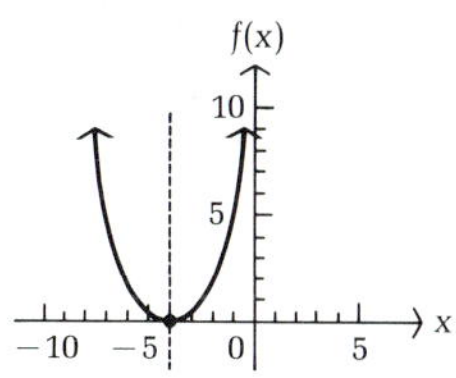

17. Axis: $u = 1$
Vertex: $(1, 3)$
Min: $f(1) = 3$
y intercept: 4
No u intercepts
Range: $[3, \infty)$

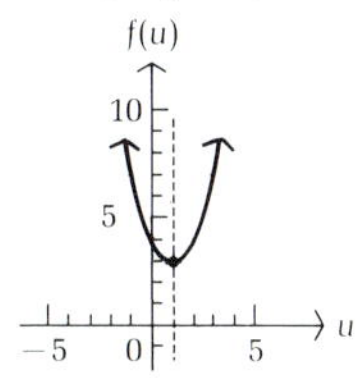

19. Axis: $x = 2$
Vertex: $(2, 6)$
Max: $h(2) = 6$
y intercept: 2
x intercepts: $2 \pm \sqrt{6}$
Range: $(-\infty, 6]$

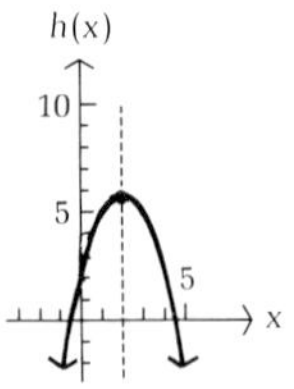

21. Axis: $x = 3$
Vertex: $(3, 9)$
Max: $f(3) = 9$
y intercept: 0
x intercepts: 0, 6
Range: $(-\infty, 9]$

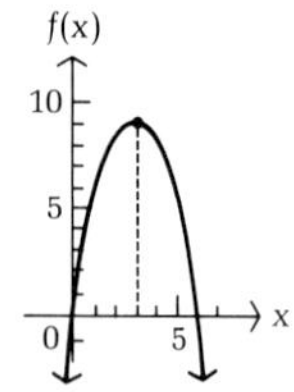

23. Axis: $s = 0$
Vertex: $(0, -4)$
Min: $F(0) = -4$
y intercept: -4
s intercepts: -2, 2
Range: $[-4, \infty)$

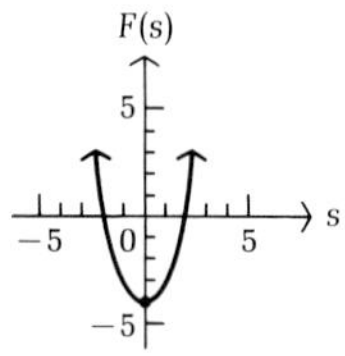

25. Axis: $x = 0$
Vertex: $(0, 4)$
Max: $F(0) = 4$
y intercept: 4
x intercepts: -2, 2
Range: $(-\infty, 4]$

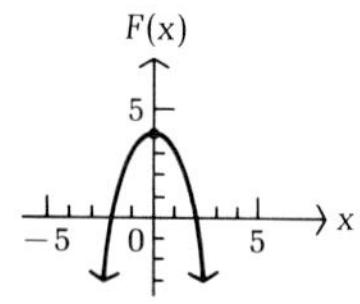

27. Domain: $[0,5]$
Range: $\{1, 3, 5\}$

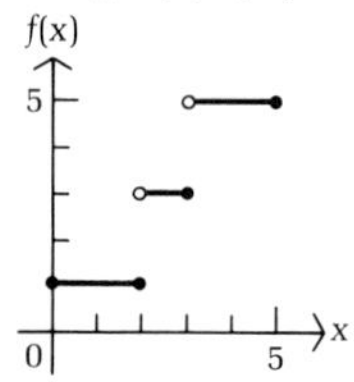

29. Domain: $[-2, 2]$
Range: $[-2, 1]$

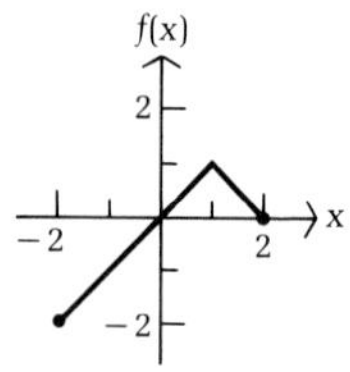

31. Domain: $(-\infty, 0) \cup (0, \infty)$
Range: $(-\infty, -2) \cup (2, \infty)$

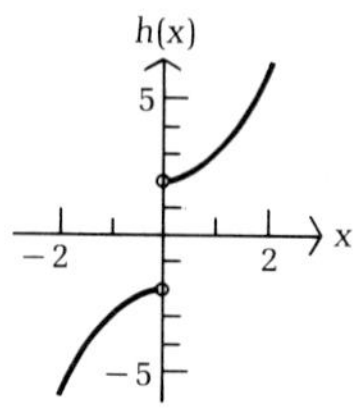

33. Domain: All real numbers
Range: $[-3, 3]$

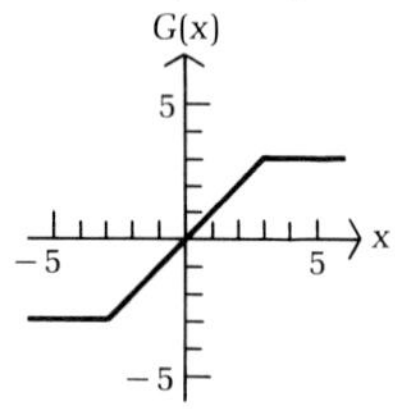

35. Axis: $x = 3.5$
Vertex: $(3.5, -2.25)$
Min: $f(3.5) = -2.25$
y intercept: 10
x intercepts: 2, 5
Range: $[-2.25, \infty)$

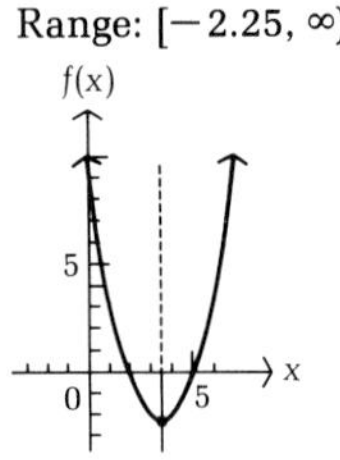

37. Axis: $x = -2.5$
Vertex: $(-2.5, 8.25)$
Max: $h(-2.5) = 8.25$
y intercept: 2
x intercepts: $-\frac{5}{2} \pm \frac{1}{2}\sqrt{33}$
Range: $(-\infty, 8.25]$

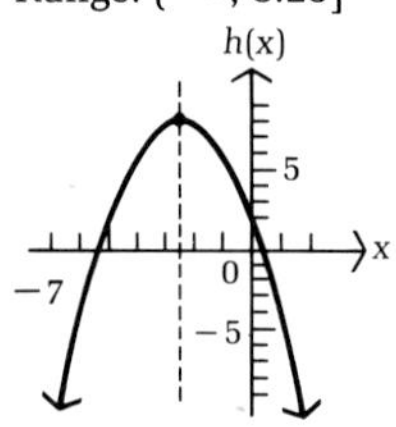

39.

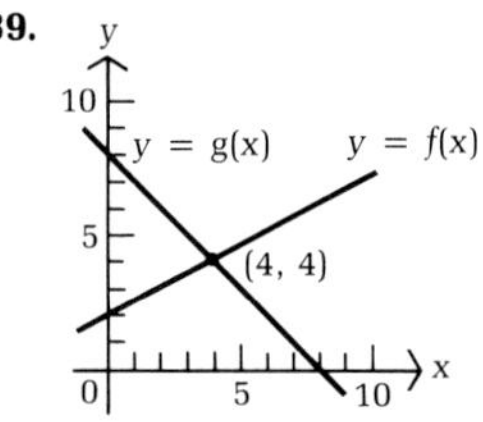

41.

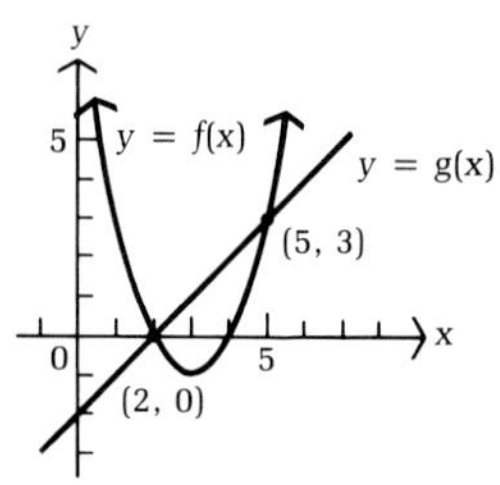

43.

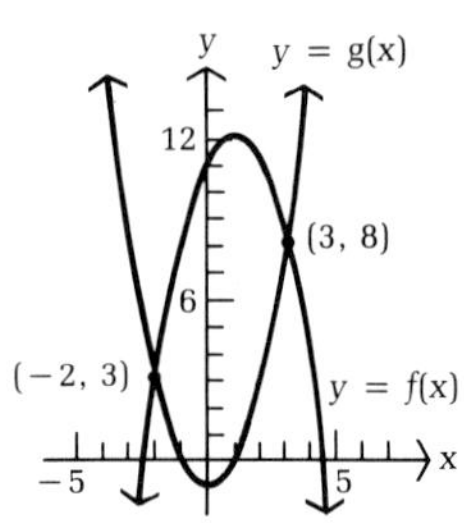

45.

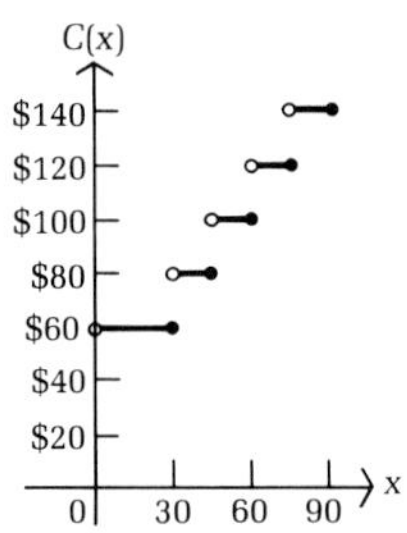

47. (A) $C = 360{,}000 - 900p$
(B) $R = xp = (9{,}000 - 30p)p = 9{,}000p - 30p^2$
(C)

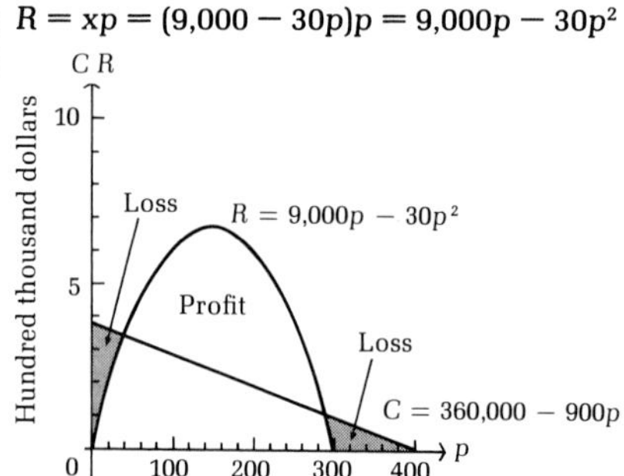

(D) \$42, \$288 (E) \$150 (F) \$165

49. (A) $C = 260{,}000 - 800p$
(B) $R = 8{,}000p - 40p^2$
(C)
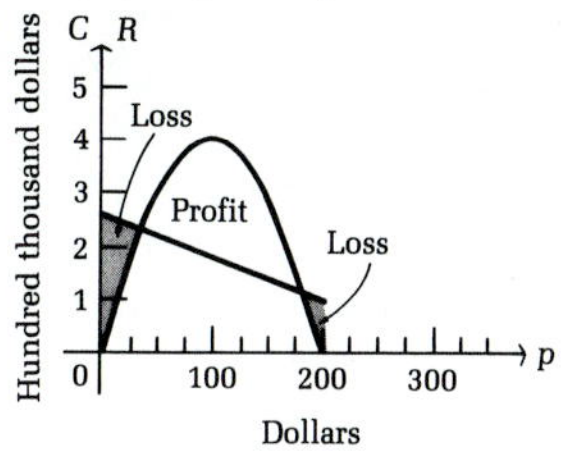

(D) \$35, \$185 (E) \$100 (F) \$110

51. \$42, \$288 **53.** \$35, \$185 **55.** (A) $C(x) = 240{,}000 + 20x$
(B) $R(x) = 35x$
(C)
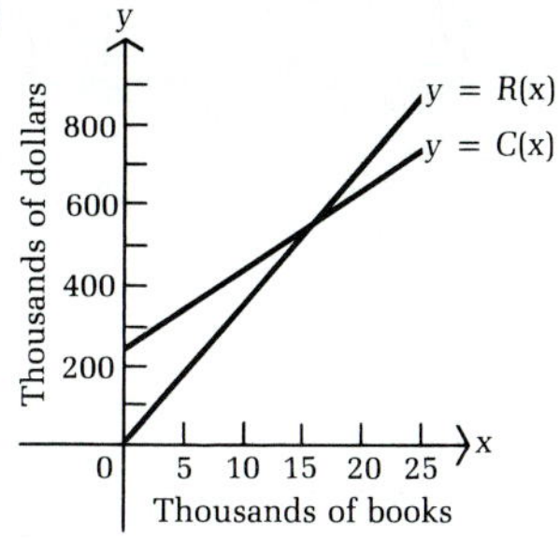

(D) 16,000 books

57.
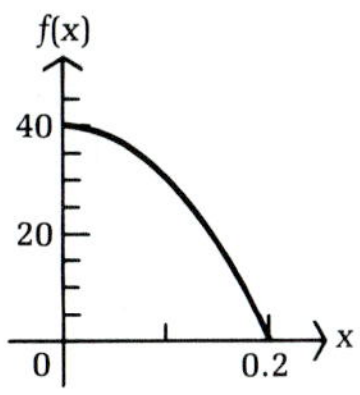

59. (A) 1 lb; 3 lb
(B)
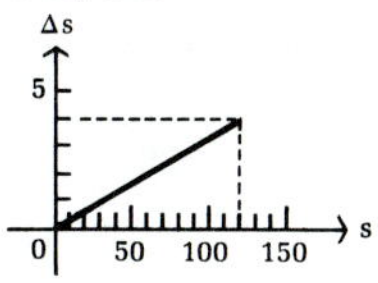

(C) $\frac{1}{30}$

◆ EXERCISE 1-4

1.
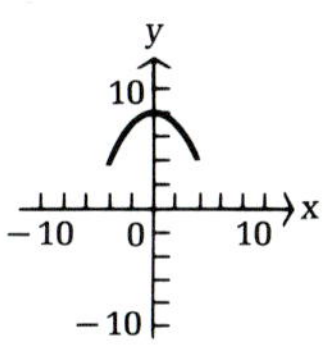

3.
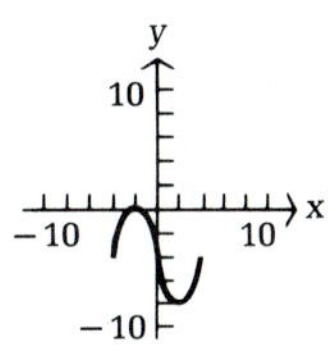

5.
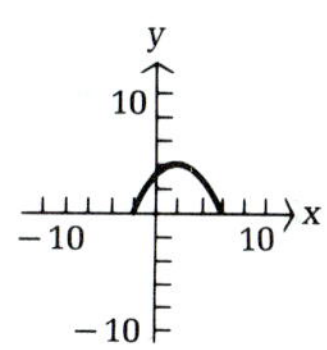

7.
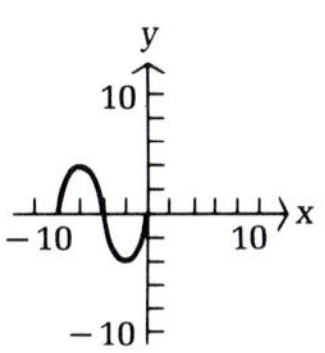

9.
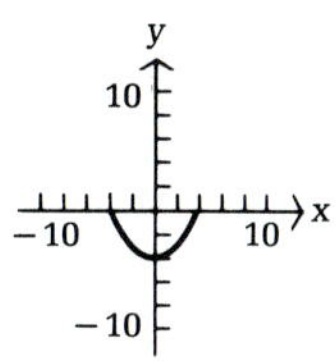

11.
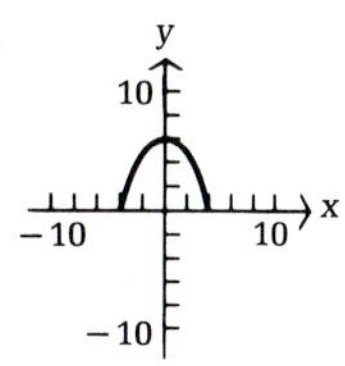

13.
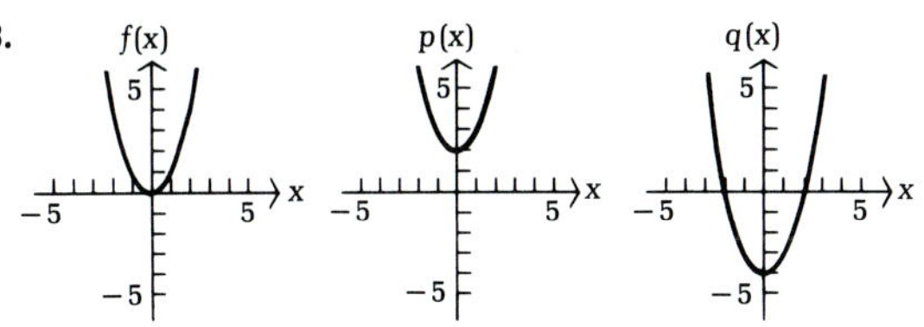

15.
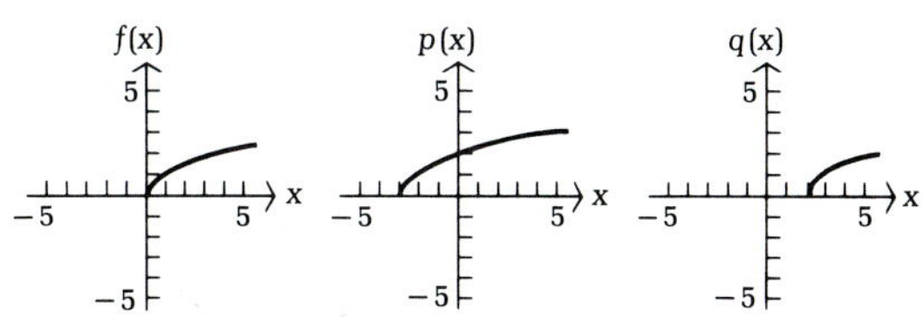

17. 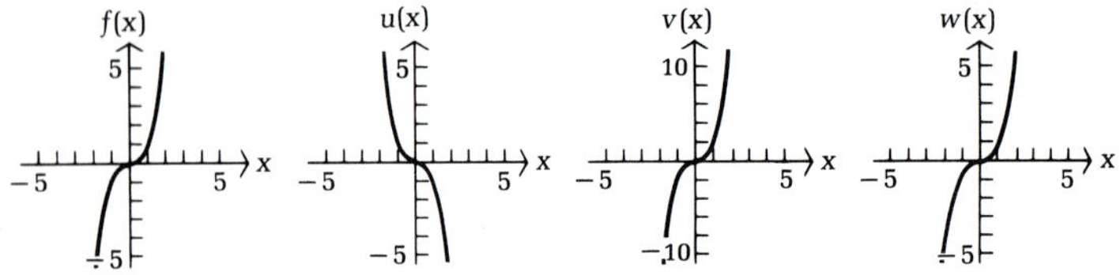

19. It is the same as the graph of $S(x) = x^2$ reflected in the x axis and shifted right 2 units.

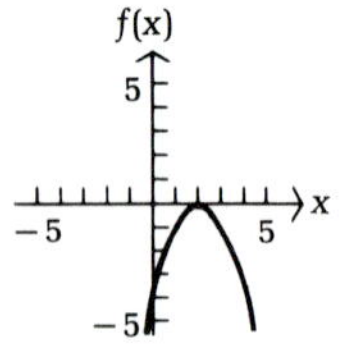

21. It is the same as the graph of $A(x) = |x|$ shifted up 2 units and to the left 1 unit.

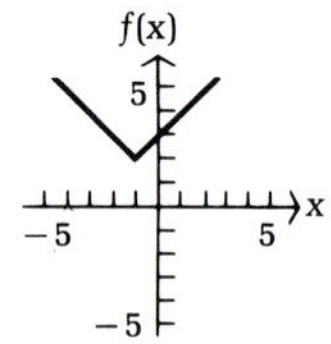

23. It is the same as the graph of $R(x) = \sqrt{x}$ expanded by a factor of 4 and then shifted down 4 units.

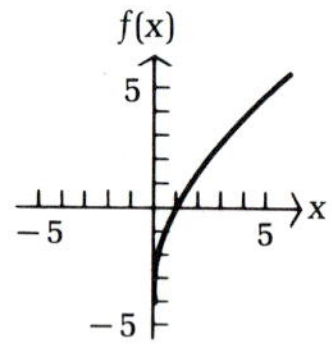

25. 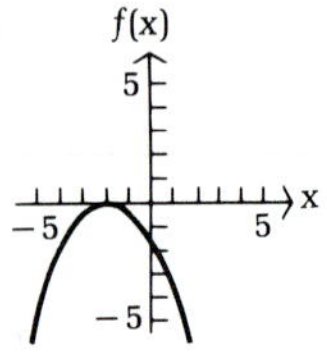

27. 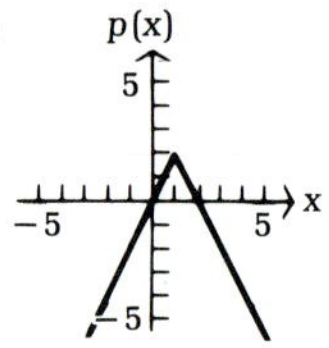

29. 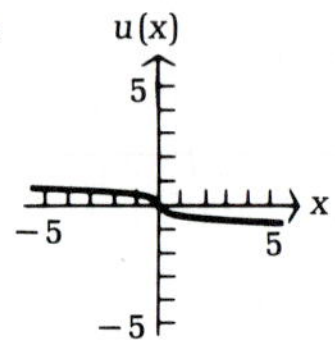

31. 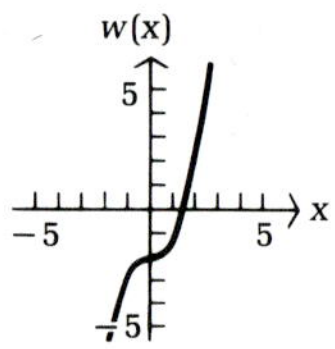

33. 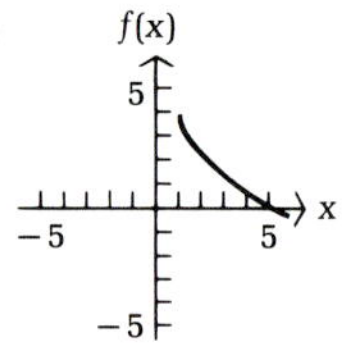

35. $f(x) = 2(x - 3)^2 - 9$; it is the same as the graph of $S(x) = x^2$ shifted 3 units to the right, expanded by a factor of 2, and then shifted down 9 units.

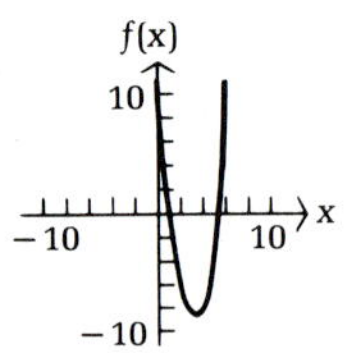

37. $f(x) = -\frac{1}{2}(x - 4)^2 + 5$; it is the same as the graph of $S(x) = x^2$ shifted right 4 units, contracted by a factor of $\frac{1}{2}$, reflected in the x axis, and then shifted up 5 units.

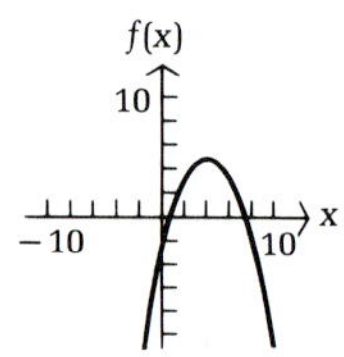

39.

x	f(x)
−2	16
−1.5	5.0625
−1	1
−0.5	0.0625
0	0
0.5	0.0625
1	1
1.5	5.0625
2	16

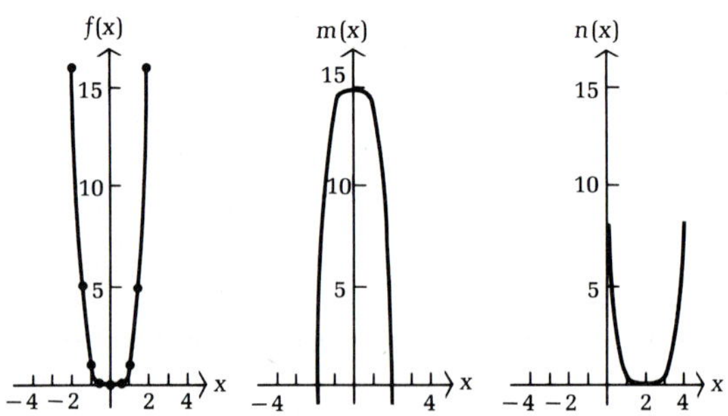

41.

x	f(x)
−3	−6.24
−2	−3.17
−1	−1
−0.5	−0.32
0	0
0.5	0.32
1	1
2	3.17
3	6.24

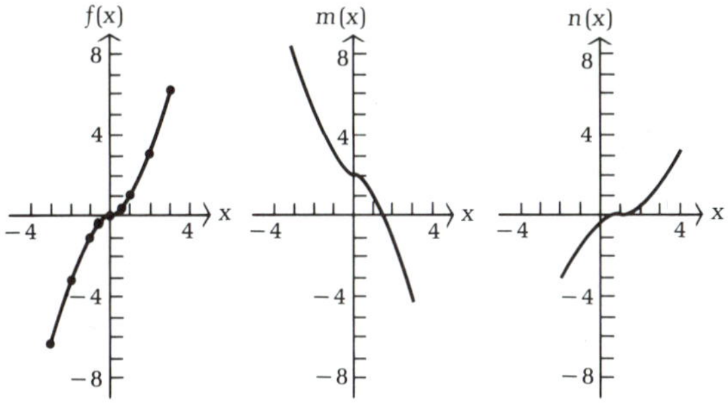

43.

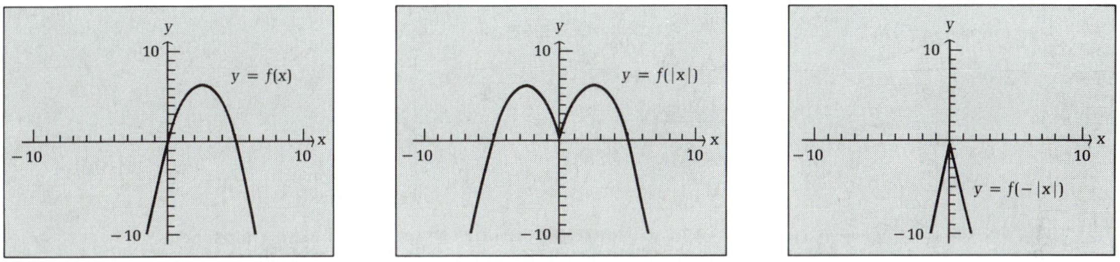

45.

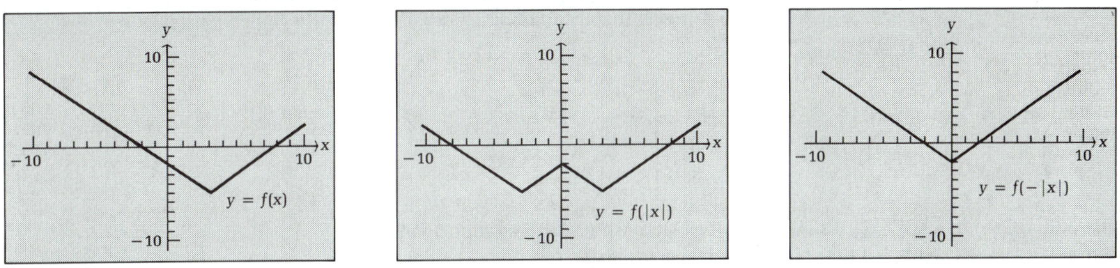

47. The right side of the graph of $y = f(|x|)$ is the same as the right side of the graph of $y = f(x)$, and the left side is the reflection of the right side of the graph of $y = f(x)$ with respect to the y axis.

49.

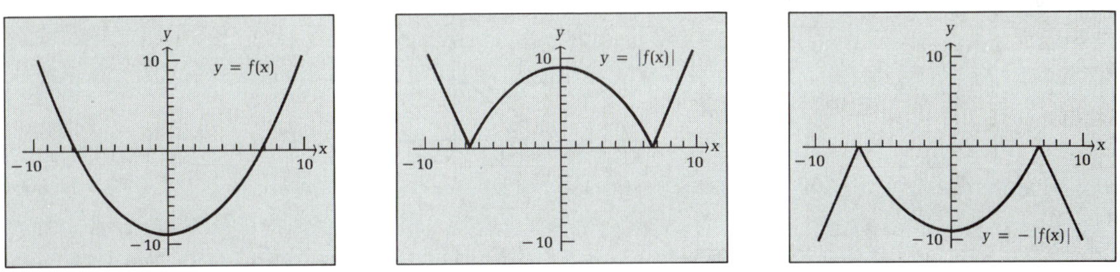

51.

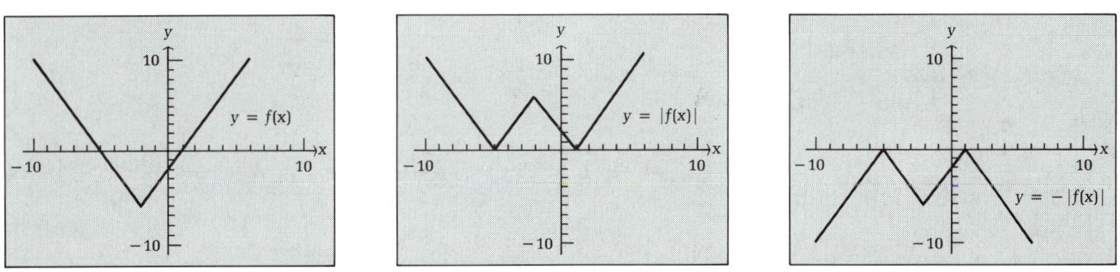

53. The graph of $y = |f(x)|$ is the same as the graph of $y = f(x)$ whenever $f(x) \geqslant 0$ and is the reflection of the graph of $y = f(x)$ with respect to the x axis whenever $f(x) < 0$.

55. 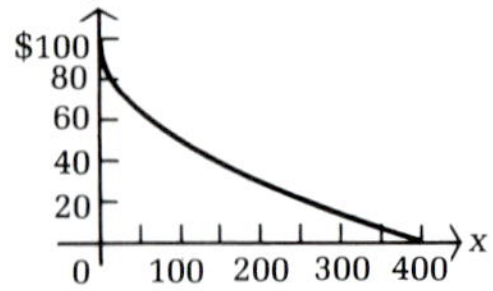

57. 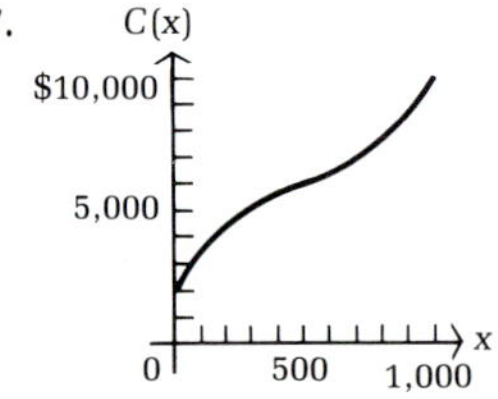

59. 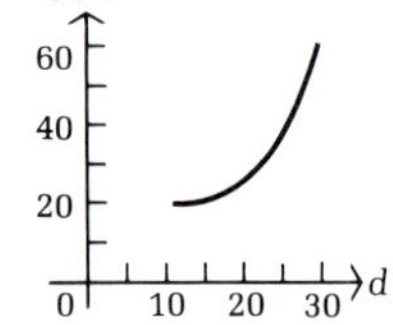

◆ EXERCISE 1-5 CHAPTER REVIEW

1. 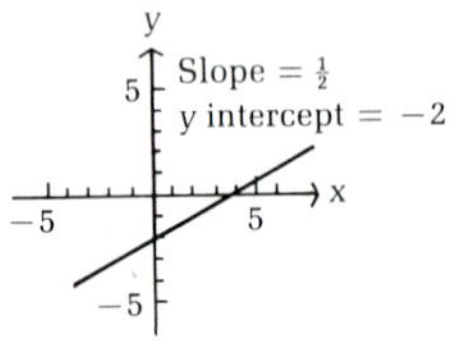

2. $y = \frac{1}{2}x + 1$

3. 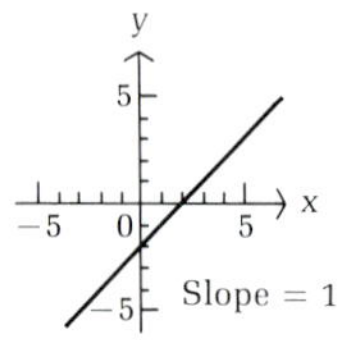

4. −2

5. 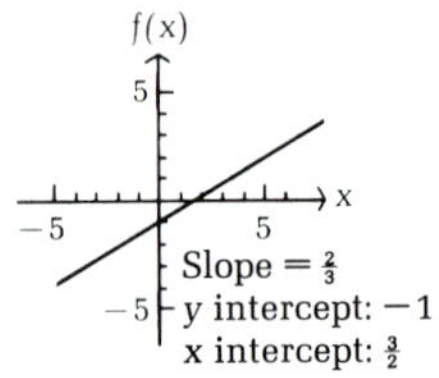

6. Min $f(x) = f(4) = -9$; vertex: $(4, -9)$

7. (A) It is the same as the graph of $S(x) = x^2$ reflected in the x axis.
(B) It is the same as the graph of $S(x) = x^2$ shifted up 4 units.
(C) It is the same as the graph of $S(x) = x^2$ shifted to the right 5 units.

8. 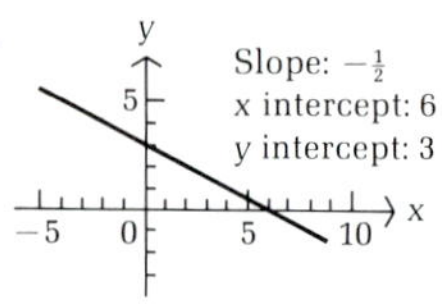

9. $x + 2y = 4$; slope $= -\frac{1}{2}$

10. 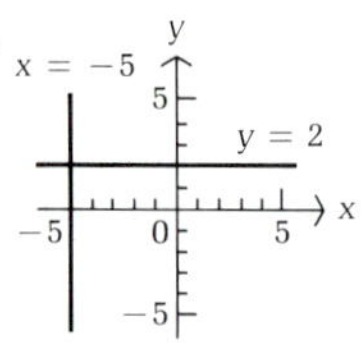

11. $y = -\frac{3}{2}x + 2$

12. (A) 22 (B) −36 (C) −91 (D) $-\frac{3}{4}$ 13. Domain *f*: *R*; Domain g: All *R* except 2 14. 2

15. (A) A function with domain all real numbers (B) Not a function; for example, when $x = 2$, $y = \pm 3$
(C) A function with domain all real numbers except $x = -3$

16. Axis: $x = 2$
Vertex: (2, 8)
Max: $g(2) = 8$
y intercept: 0
x intercepts: 0, 4
Range: $(-\infty, 8]$

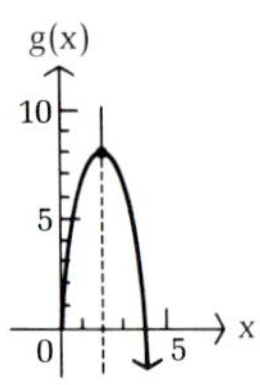

17. Domain: $[-1, 1]$; Range: $[0, 2]$

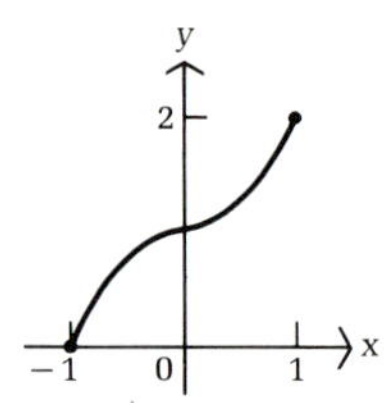

18. 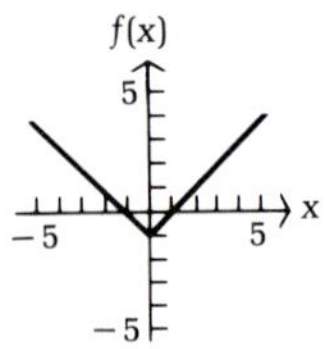

19. 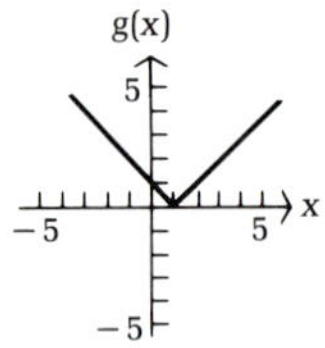

20. 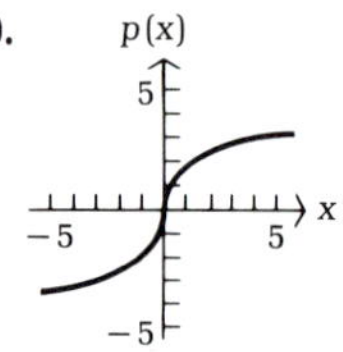

21. 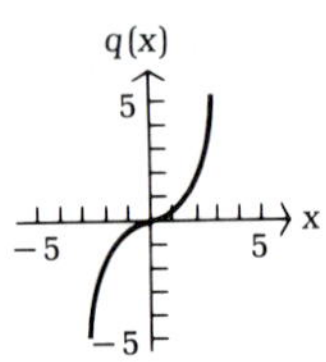

22. 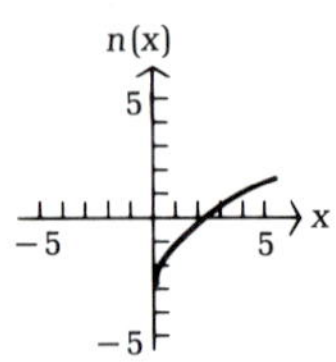

23. 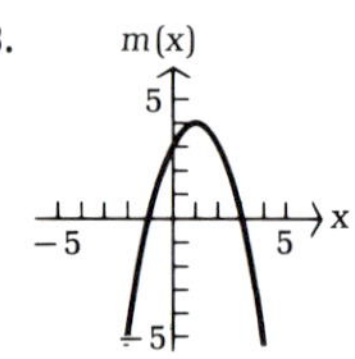

24. $x = 4$ **25.** (A) $x - 2y = 8$ (B) $2x + y = 1$ **26.** (A) All real numbers except $x = 3$ (B) $[1, \infty)$ **27.** $2a + 7 + h$

28.

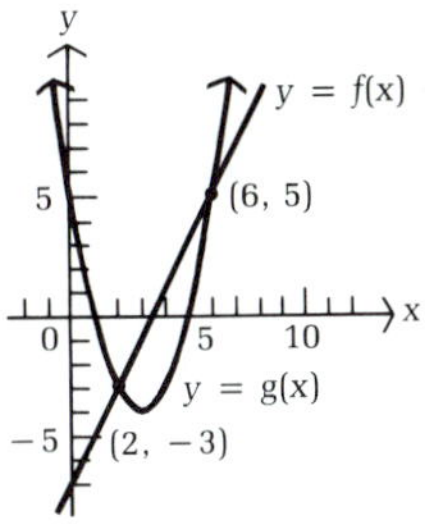

29. It is the same as the graph of $C(x) = x^3$ shifted left 2 units, expanded by a factor of 3, reflected in the x axis, and then shifted up 4 units.

30.

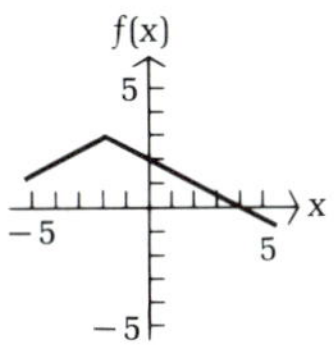

31. (A) $V(t) = -1{,}250t + 12{,}000$ (B) \$5,750 **32.** (A) $R = \frac{8}{5}C$ (B) \$168

33. (A) $C(x) = 30 + 0.4x$
(B) $R(x) = 1.4x$
(C)

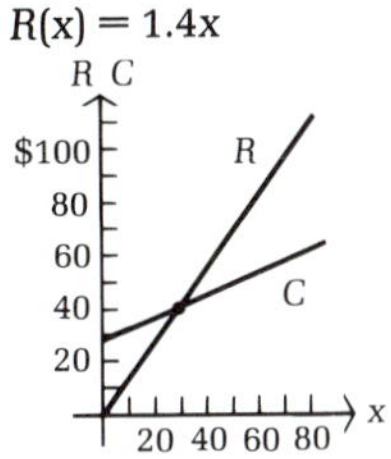

(D) 30

34.

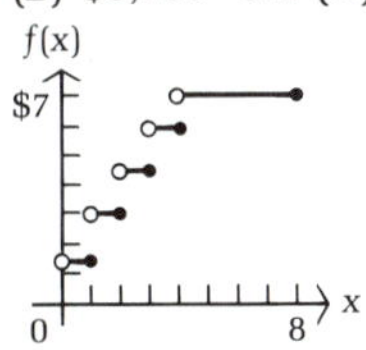

35. (A) $R(p) = 500p - 10p^2$
(B) $C(p) = 8{,}000 - 100p$
(C)

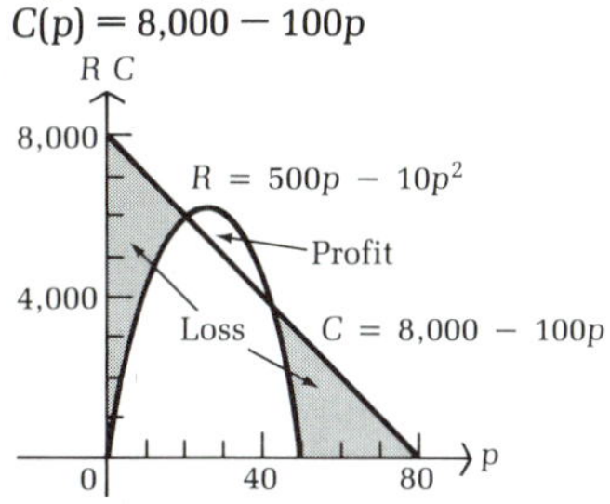

(D) \$20; \$40 (E) \$30

36. (A) $A(x) = 90x - \frac{3}{2}x^2$ (B) $0 < x < 60$ (C) $x = 30$, $y = 22.5$

CHAPTER 2

◆ EXERCISE 2-1

1.

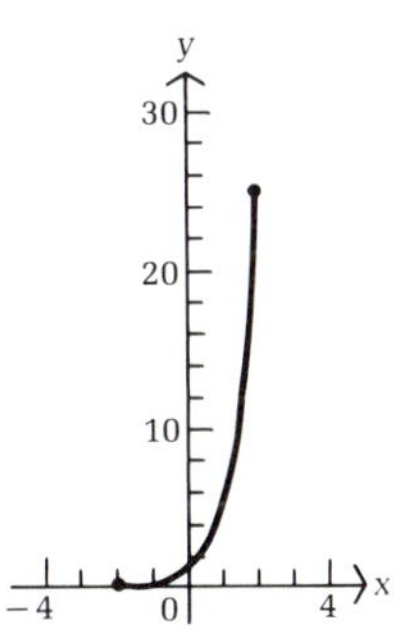

3.

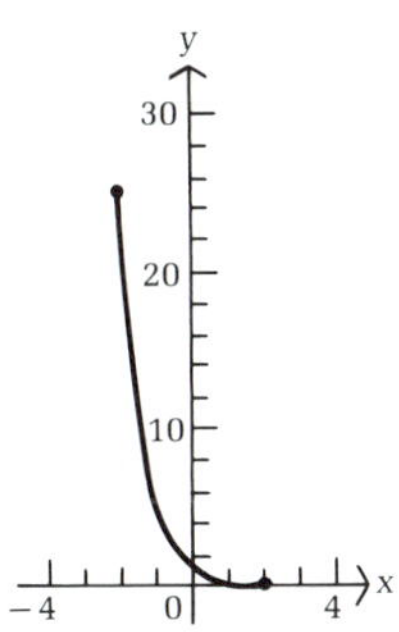

5.

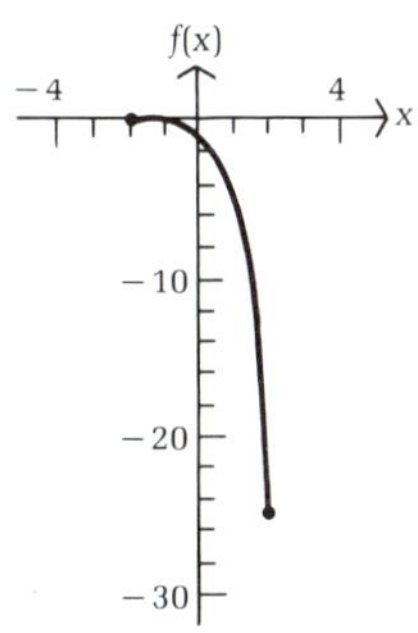

7.

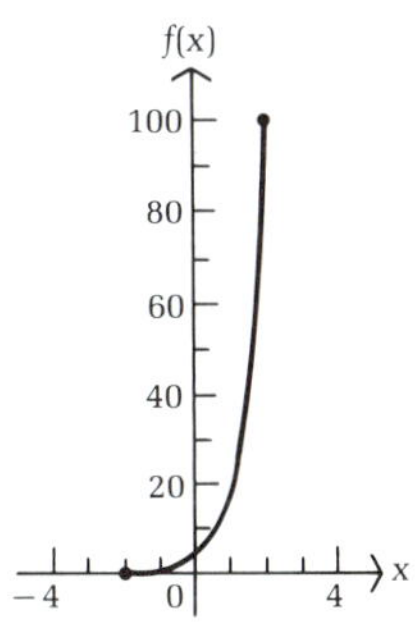

9.

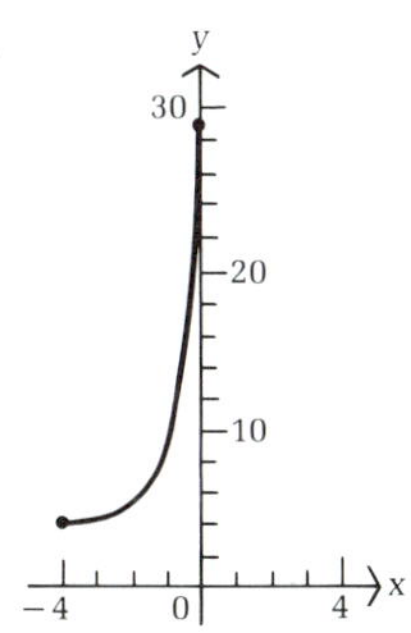

11. 4^{6xy} **13.** 5 **15.** $2^{xz}3^{yz}$ **17.** $x = 1$ **19.** $x = -1, 6$ **21.** $x = 3$ **23.** $x = 2$

25.
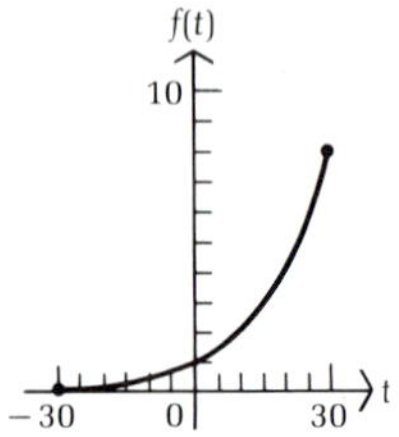

27.
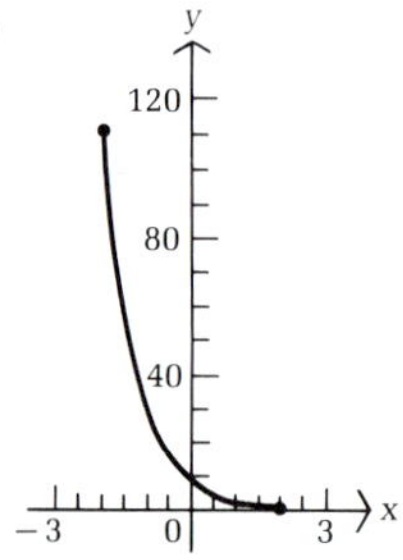

29.
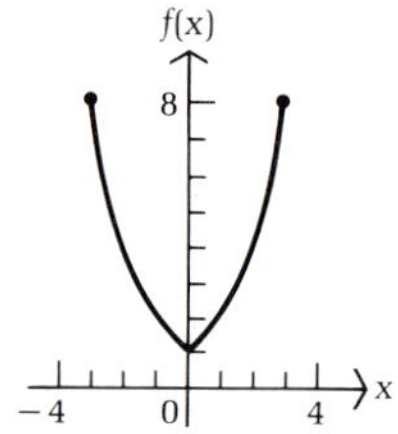

31.
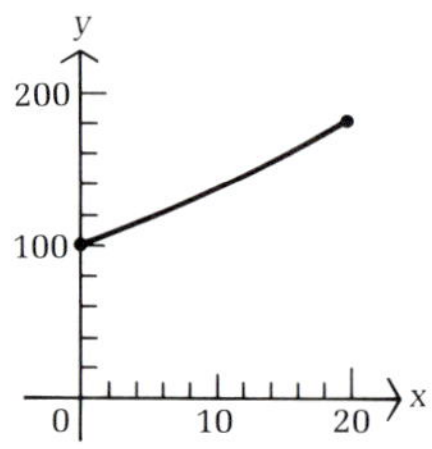

33.
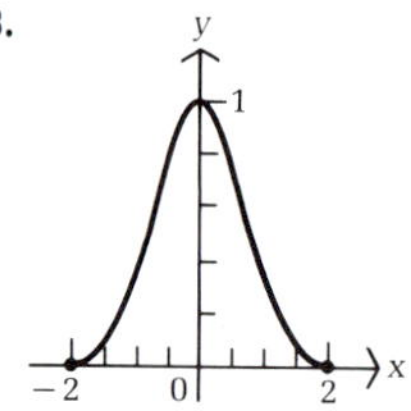

35. $3^{2x} - 3^{-2x}$ **37.** $2(3^{2x}) + 2(3^{-2x})$ **39.**
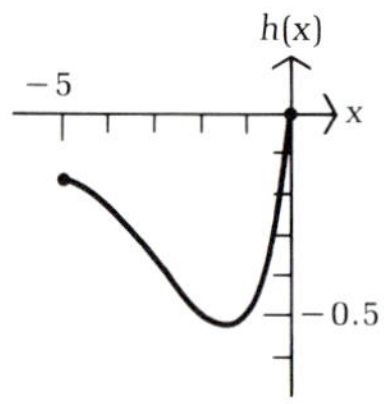

41.
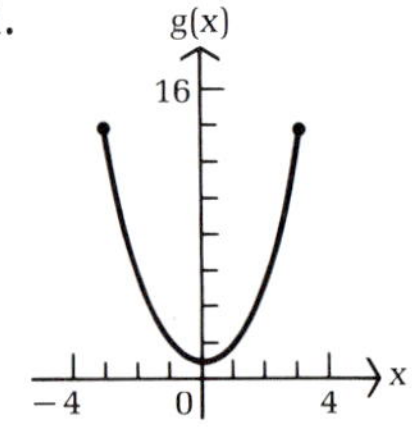

43. 0.2115

45. 0.1377 **47.** 15.7563 **49.** (A) $2,633.56 (B) $7,079.54 **51.** $9,217 **53.** (A) 33,000,000 (B) 69,000,000
55. (A) 8.49 mg (B) 0.75 mg

57.
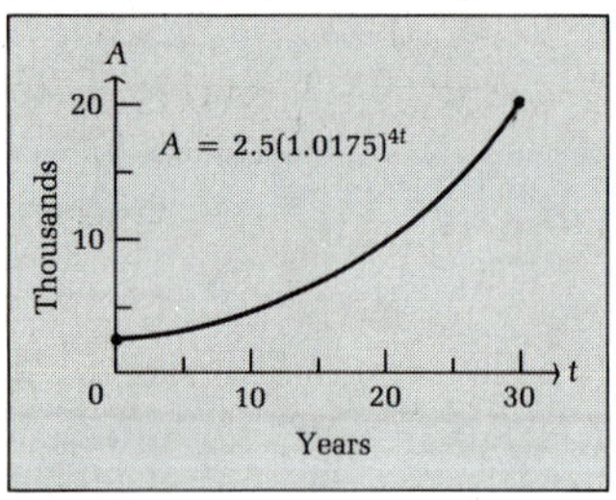

59.
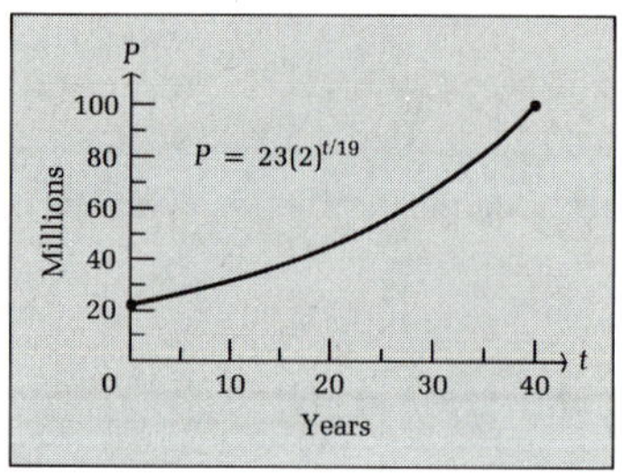

61.
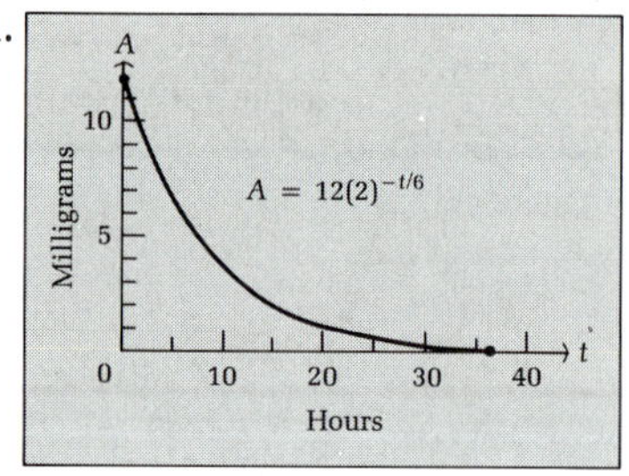

◆ EXERCISE 2-2

1.
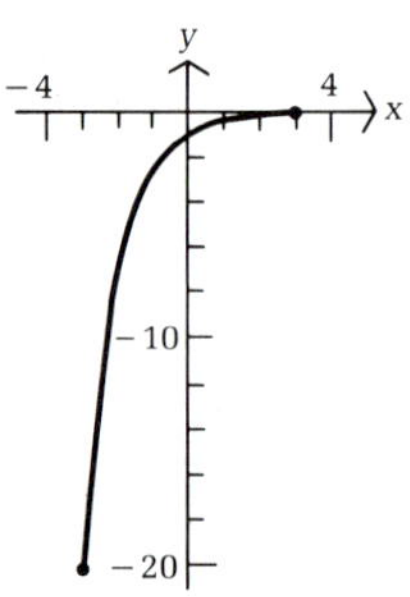

3.
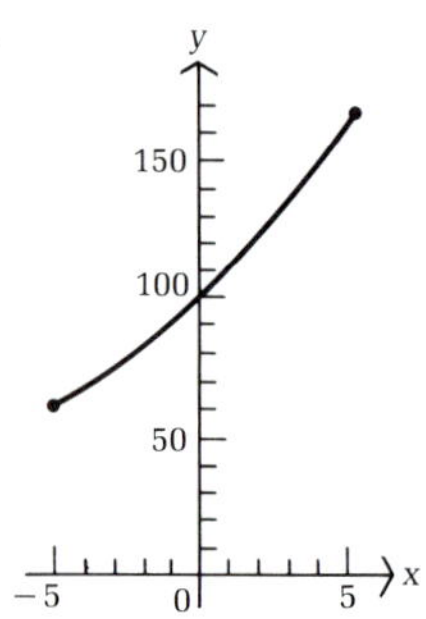

5.
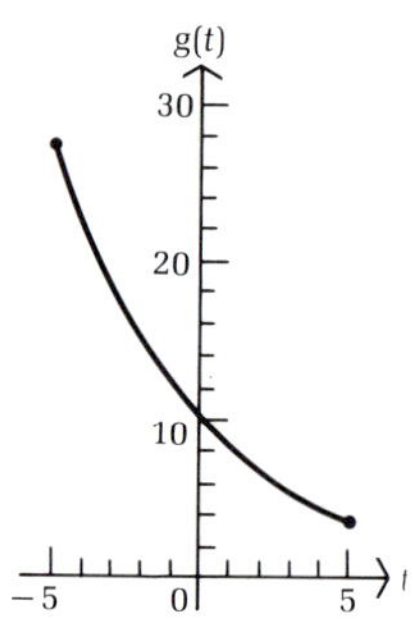

7.
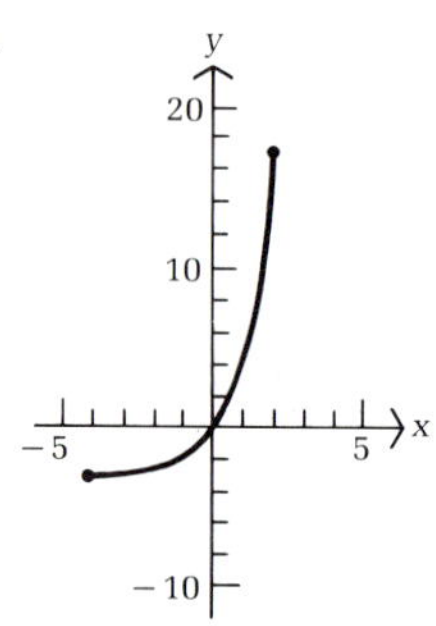

9. 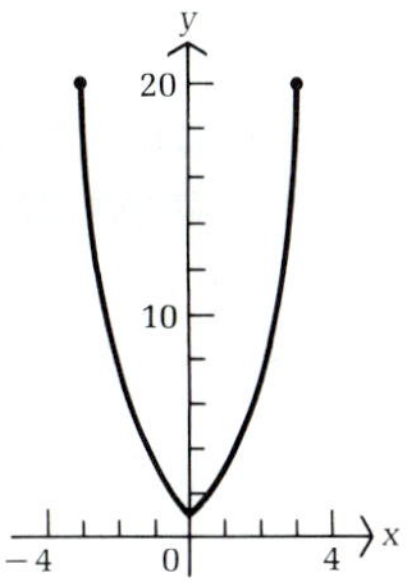

11. 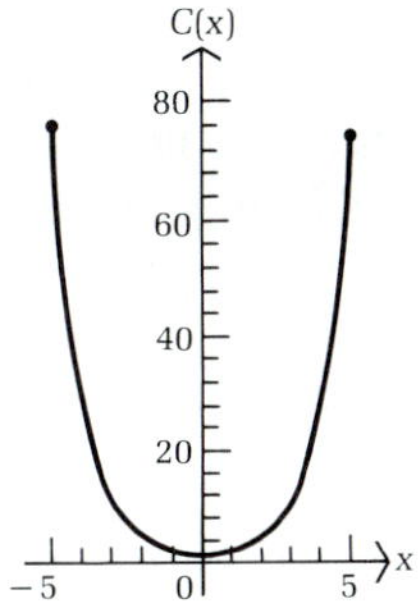

13. $e^x - e^{-x}$
15. $2/e^{2x}$
17. $x = 3$
19. $x = -3, 0$

21. 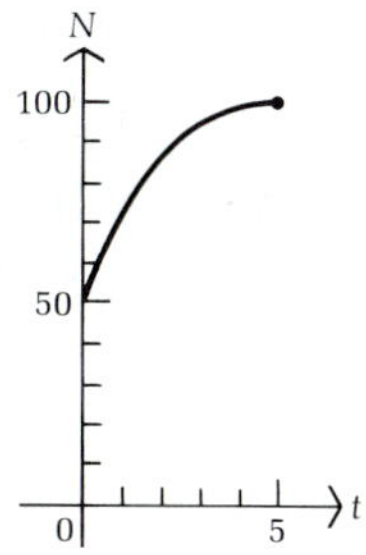

23. (A) \$11,871.65 (B) \$20,427.93
25. (A) \$10,850.88 (B) \$10,838.29
27. \$28,847.49

29. N approaches 2 as t increases without bound.

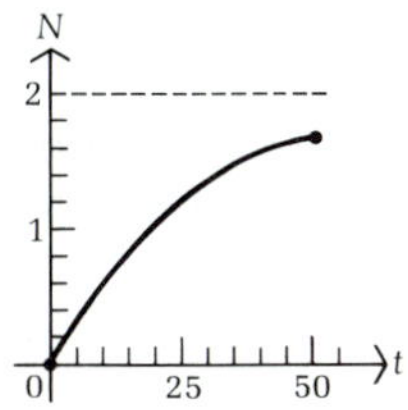

31. (A) 10% (B) 1%
33. (A) 104,382 (B) 241,786
35. 120 million

◆ EXERCISE 2-3

1. $27 = 3^3$ 3. $10^0 = 1$ 5. $8 = 4^{3/2}$ 7. $\log_7 49 = 2$ 9. $\log_4 8 = \frac{3}{2}$ 11. $\log_b A = u$ 13. 3 15. -3 17. 3 19. $\log_b P - \log_b Q$ 21. $5 \log_b L$ 23. $\log_b p - \log_b q - \log_b r - \log_b s$ 25. $x = 9$ 27. $y = 2$ 29. $b = 10$ 31. $x = 2$ 33. $y = -2$ 35. $b = 100$ 37. $5 \log_b x - 3 \log_b y$ 39. $\frac{1}{3} \log_b N$ 41. $2 \log_b x + \frac{1}{3} \log_b y$ 43. $\log_b 50 - 0.2t \log_b 2$ 45. $\log_b P + t \log_b(1 + r)$ 47. $\log_e 100 - 0.01t$ 49. $x = 2$ 51. $x = 8$ 53. $x = 7$ 55. No solution

57. 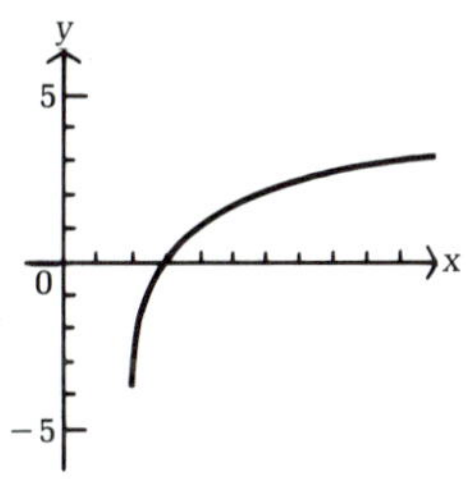

59. (A) 3.547 43 (B) −2.160 32 (C) 5.626 29 (D) −3.197 04
61. (A) 13.4431 (B) 0.0089 (C) 16.0595 (D) 0.1514
63. 4.959 65. 7.861 67. 3.301 69. 1.0792 71. 1.4595 73. 30.6589
75. 1.548 77. 0.401 79. 1.407

81. 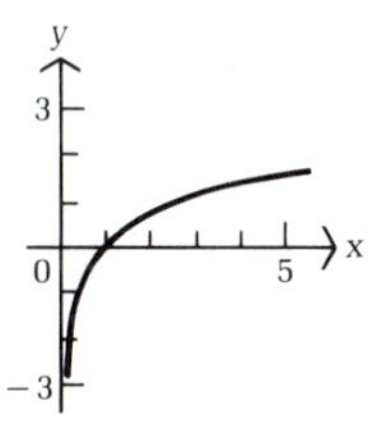

83. 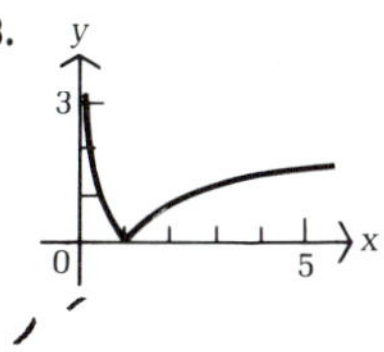

85. 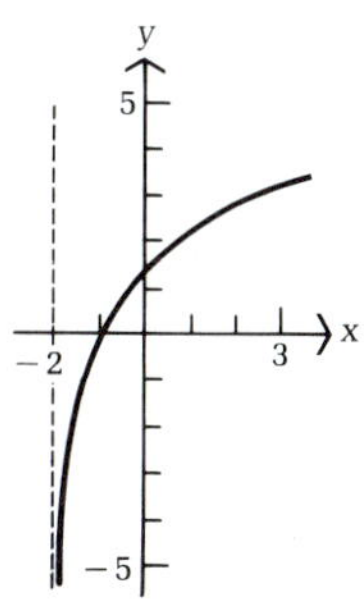

87. 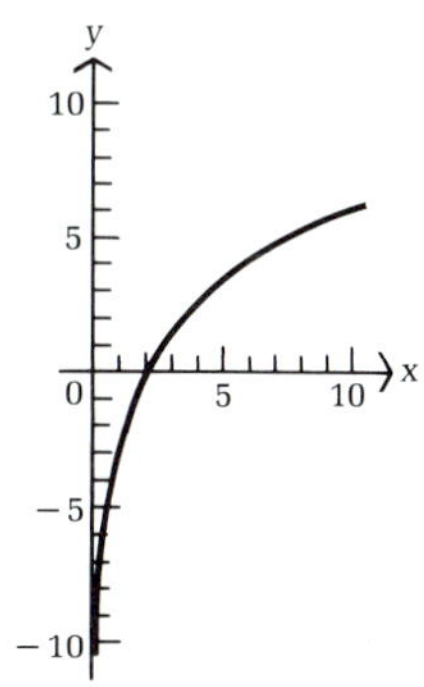

89. $\log_b 1 = 0, b > 0, b \neq 1$ 91. $y = c \cdot 10^{0.8x}$ 93. 12 yr 95. $t = \dfrac{\ln 3}{\ln(1 + r)}$ 97. 9.87 yr; 9.80 yr 101. Approx. 538 yr

◆ EXERCISE 2-4 CHAPTER REVIEW

1. $v = \ln u$ **2.** $y = \log x$ **3.** $M = e^N$ **4.** $u = 10^v$ **5.** 5^{2x} **6.** e^{2u^2} **7.** $x = 9$ **8.** $x = 6$ **9.** $x = 4$ **10.** $x = 2.157$ **11.** $x = 13.128$
12. $x = 1{,}273.503$ **13.** $x = 0.318$ **14.** $x = 8$ **15.** $x = 3$ **16.** $x = 3$ **17.** $x = -1, 3$ **18.** $x = 0, \frac{3}{2}$ **19.** $x = -2$ **20.** $x = \frac{1}{2}$ **21.** $x = 27$
22. $x = 13.3113$ **23.** $x = 158.7552$ **24.** $x = 0.0097$ **25.** $x = 1.4359$ **26.** $x = 1.4650$ **27.** $x = 92.1034$ **28.** $x = 9.0065$
29. $x = 2.1081$ **30.** 2.8074 **31.** -1.0387 **32.** $1 + 2e^x - e^{-x}$ **33.** $2e^{-2x} - 2$

34.

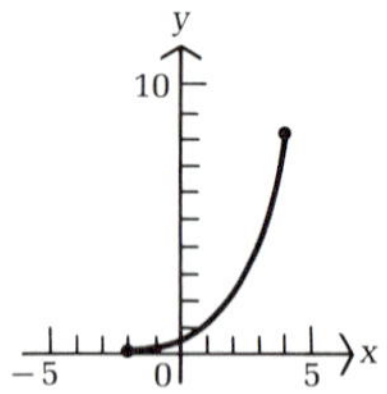

35.

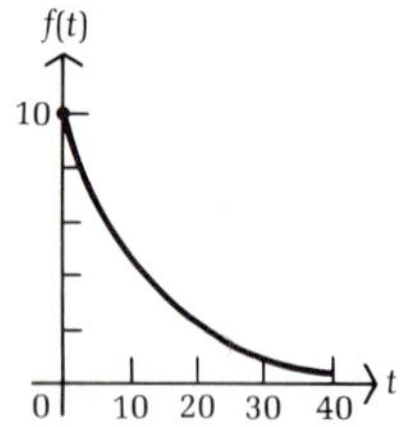

36.

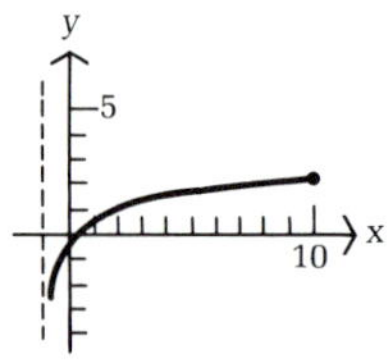

37. $x = 2$
38. $x = 2$
39. $x = 1$
40. $x = 300$
41. $y = ce^{-5t}$

42. If $\log_1 x = y$, then $1^y = x$; that is, $1 = x$ for all positive real numbers x, which is not possible. **43.** \$10,263.65 **44.** \$10,272.17
45. 8 yr **46.** 6.93 yr **47.** (A) $N = 2^{2t}$ or $N = 4^t$ (B) 15 days **48.** $k = 0.009\,42$; 489 ft **49.** 23.4 yr **50.** 23.1 yr

CHAPTER 3

◆ EXERCISE 3-1

1. (A) 2 (B) 2 (C) Does not exist (D) 4 **3.** (A) 2 (B) 2 (C) 2 (D) Not defined **5.** $c = 0, 1$ **7.** 2, 4 **9.** All x
11. All x, except $x = 5$ **13.** All x, except $x = -2$ and $x = 3$
15. (A) $\lim_{x\to 0^-} f(x) = 1$, $\lim_{x\to 0^+} f(x) = 1$, $\lim_{x\to 0} f(x) = 1$, $f(0) = 1$ (B) Yes, all three conditions are satisfied.
17. (A) $\lim_{x\to 1^-} f(x) = 2$, $\lim_{x\to 1^+} f(x) = 1$, $\lim_{x\to 1} f(x)$ does not exist, $f(1) = 1$ (B) No, because $\lim_{x\to 1} f(x)$ does not exist.
19. (A) $\lim_{x\to -2^-} f(x) = 1$, $\lim_{x\to -2^+} f(x) = 1$, $\lim_{x\to -2} f(x) = 1$, $f(-2) = 3$ (B) No, because $\lim_{x\to -2} f(x) \neq f(-2)$.

21. (A)

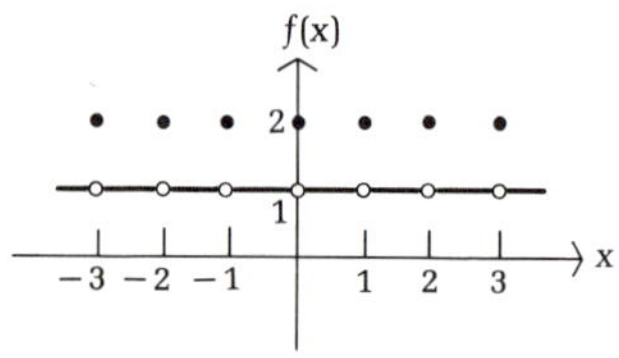

(B) 1
(C) 2
(D) No
(E) All integers

23. 5
25. 7
27. -6
29. 1
31. $(-\infty, \infty)$
33. $[5, \infty)$
35. $(-\infty, \infty)$
37. $(-\infty, 1), (1, 2), (2, \infty)$

39.

x	0.9	0.99	0.999 → 1 ← 1.001	1.01	1.1
f(x)	−1	−1	−1 → ? ← 1	1	1

(A) -1 (B) 1 (C) Does not exist

41.

x	0.9	0.99	0.999 → 1 ← 1.001	1.01	1.1
f(x)	2.71	2.97	2.997 → ? ← 3.003	3.03	3.31

(A) 3 (B) 3 (C) 3

43. Discontinuous at $x = 1$

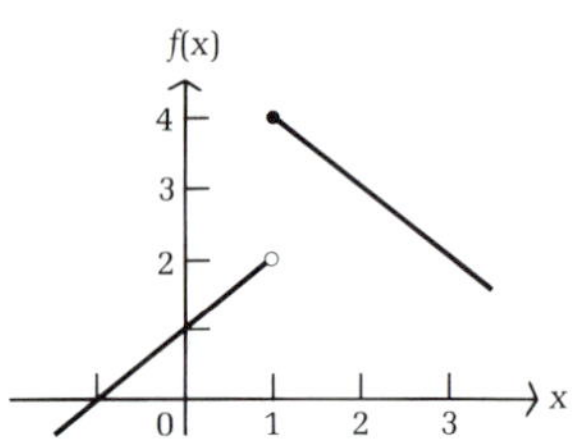

45. Continuous for all x

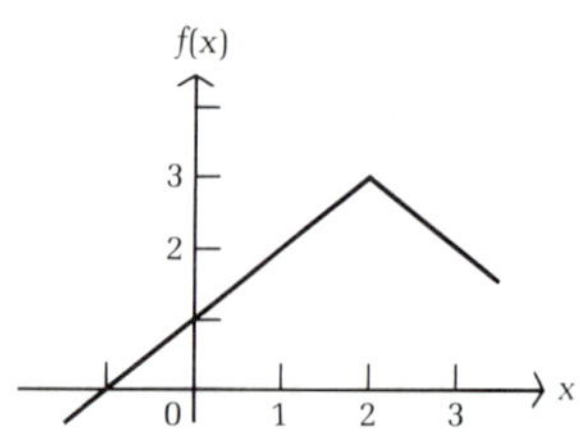

47. Discontinuous at $x = 0$

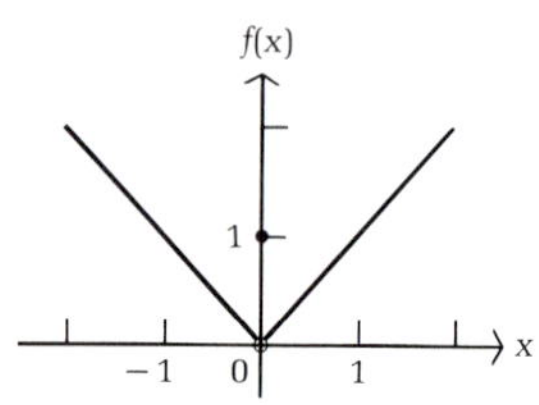

49. (A) Yes (B) Yes (C) Yes (D) Yes **51.** (A) Yes (B) No (C) Yes (D) No (E) Yes **53.** No; no

55. $\lim_{x\to 0^-} f(x) = -3$, $\lim_{x\to 0^+} f(x) = 3$

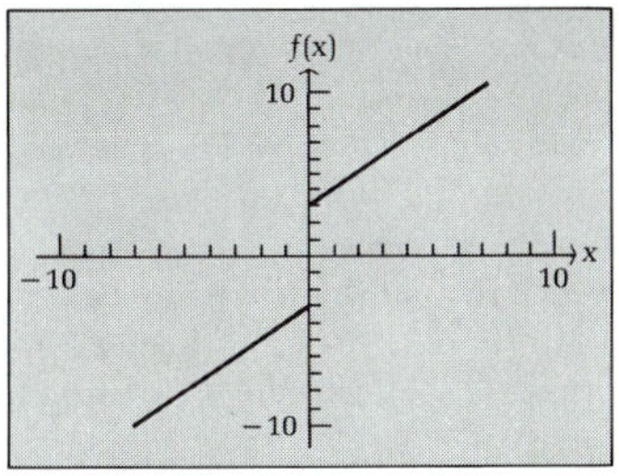

57. $\lim_{x\to 0^-} f(x) = -4$, $\lim_{x\to 0^+} f(x) = 4$

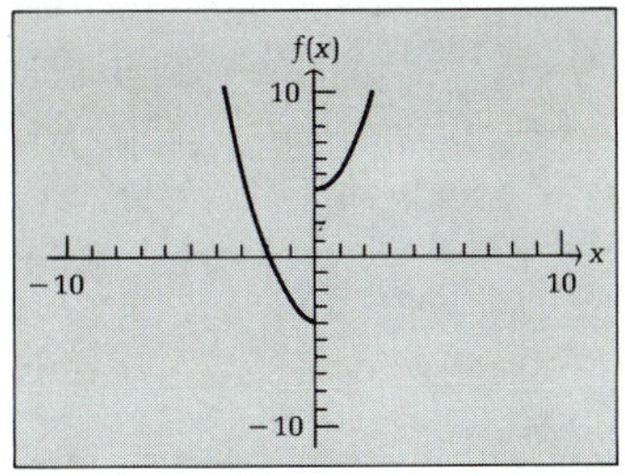

59. $\lim_{x\to 2^-} f(x) = -4$, $\lim_{x\to 2^+} f(x) = 4$

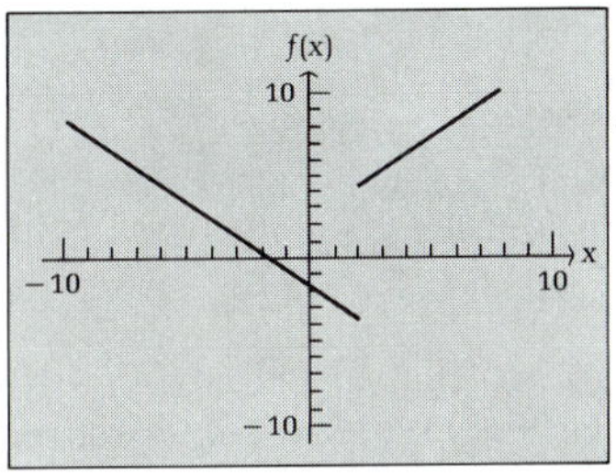

61. $\lim_{x\to -3^-} f(x) = -3$, $\lim_{x\to -3^+} f(x) = 3$, $\lim_{x\to 3^-} f(x) = -3$, $\lim_{x\to 3^+} f(x) = 3$

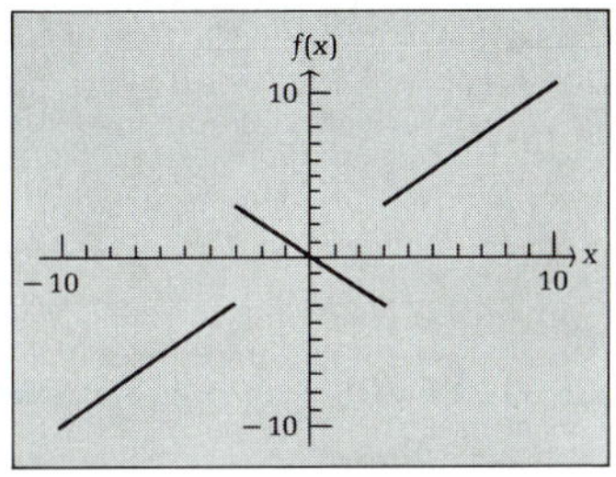

63. (A)

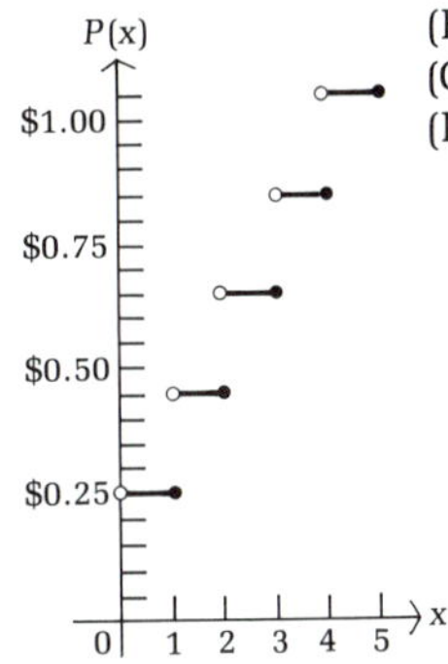

(B) $\lim_{x\to 4.5} P(x) = \$1.05$; $P(4.5) = \$1.05$
(C) $\lim_{x\to 4} P(x)$ does not exist; $P(4) = \$0.85$
(D) Continuous at $x = 4.5$; not continuous at $x = 4$

65. (A)

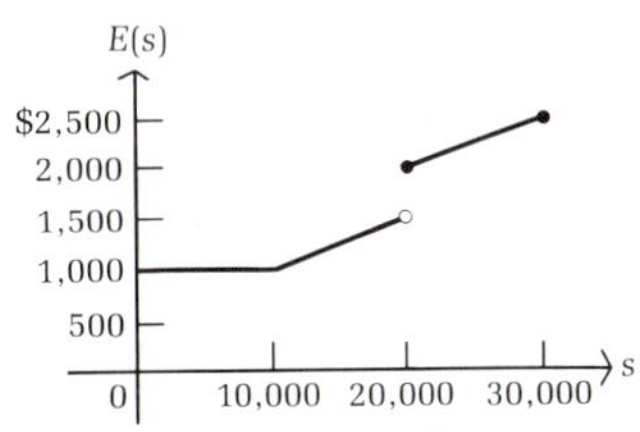

(B) $\lim_{s\to 10{,}000} E(s) = \$1{,}000$; $E(10{,}000) = \$1{,}000$
(C) $\lim_{s\to 20{,}000} E(s)$ does not exist; $E(20{,}000) = \$2{,}000$
(D) Yes; no

67. (A) t_2, t_3, t_4, t_6, t_7 (B) $\lim_{t\to t_5} N(t) = 7$; $N(t_5) = 7$ (C) $\lim_{t\to t_3} N(t)$ does not exist; $N(t_3) = 4$

◆ EXERCISE 3-2

1. -4 **3.** 36 **5.** $\frac{5}{9}$ **7.** $\sqrt{5}$ **9.** $\frac{7}{5}$ **11.** 47 **13.** -4 **15.** $\frac{5}{3}$ **17.** 243 **19.** -3 **21.** 0 **23.** 5 **25.** 2 **27.** 0 **29.** 1 **31.** $\frac{1}{2}$ **33.** 0 **35.** -1 **37.** -5 **39.** ∞ **41.** 2 **43.** $\frac{3}{5}$ **45.** 4 **47.** $\frac{2}{3}$ **49.** $-\infty$ **51.** ∞ **53.** Does not exist **55.** (A) 0 (B) 0 (C) $y = 0$ **57.** (A) ∞ (B) ∞ (C) ∞ (D) $x = -1$ **59.** 3 **61.** 4 **63.** 0 **65.** $1/(2\sqrt{2})$ **67.** Does not exist **69.** $\sqrt{3}$ **71.** $\sqrt[3]{4}$ **73.** ∞ **75.** 0 **77.** 0 **79.** $\frac{1}{4}$ **81.** 12 **83.** (A) $-\infty$ (B) ∞ (C) Does not exist (D) Yes **85.** $2a$ **87.** $1/(2\sqrt{a})$ **89.** (A) \$23 (B) \$3.20 (C) \$5 (D) \$3

91. (A)

COMPOUNDED	n	$A(n)$
Annually	1	\$108.00
Semiannually	2	\$108.16
Quarterly	4	\$108.24
Monthly	12	\$108.30
Weekly	52	\$108.32
Daily	365	\$108.33
Hourly	8,760	\$108.33

(B) \$108.33 **93.** (A) 0.056 (B) 0.07 (C) 0.07 (D) 0
95. (A) 30 (B) 44 (C) 44 (D) 60

◆ EXERCISE 3-3

1. (A) 3 (B) 3 **3.** (A) $8 + 2h$ (B) 8 **5.** (A) 1 (B) $2 + h$ (C) 2 **7.** $f'(x) = 2; f'(1) = 2, f'(2) = 2, f'(3) = 2$
9. $f'(x) = -2x; f'(1) = -2, f'(2) = -4, f'(3) = -6$ **11.** (A) 5 (B) $3 + h$ (C) 3 (D) $y = 3x - 1$
13. (A) 5 m/sec (B) $3 + h$ m/sec (C) 3 m/sec **15.** $f'(x) = 6 - 2x; f'(1) = 4, f'(2) = 2, f'(3) = 0$
17. $f'(x) = 1/(2\sqrt{x}); f'(1) = \frac{1}{2}, f'(2) = 1/(2\sqrt{2}), f'(3) = 1/(2\sqrt{3})$ **19.** $f'(x) = 1/x^2; f'(1) = 1, f'(2) = \frac{1}{4}, f'(3) = \frac{1}{9}$ **21.** Yes **23.** No **25.** No
27. Yes **29.** (A) $f'(x) = 2x - 4$ (B) $-4, 0, 4$ (C)

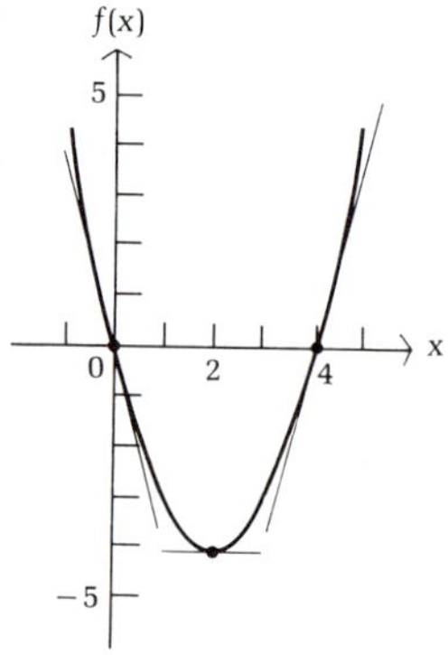

31. $v = f'(x) = 8x - 2$; 6 ft/sec, 22 ft/sec, 38 ft/sec
33. f is nondifferentiable at $x = 1$

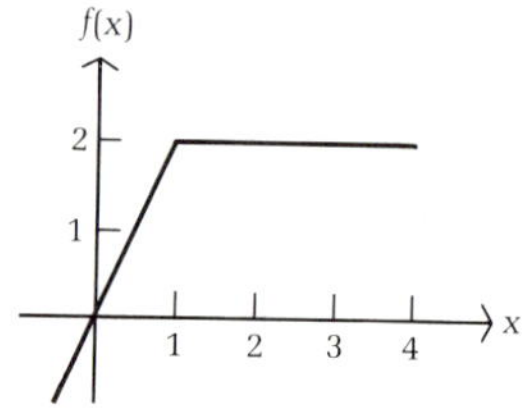

35. No
37. No
39. (A) $2 - 2x$ (B) 2 (C) -2

41. (A) 3 (\$300/board) (B) $4 - h$ (C) $C'(3) = 4$ (\$400/board) (D) $C'(x) = 10 - 2x$
(E) $C'(1) = 8$ (the rate of total cost increase at the level of production of 1 board per day is \$800/board); $C'(2) = 6$ (the rate of total cost increase at the level of production of 2 boards per day is \$600/board); $C'(3) = 4$ (the rate of total cost increase at the level of production of 3 boards per day is \$400/board); $C'(4) = 2$ (the rate of total cost increase at the level of production of 4 boards per day is \$200/board)

◆ EXERCISE 3-4

1. 0 **3.** 0 **5.** $12x^{11}$ **7.** 1 **9.** $-7x^{-8}$ **11.** $\frac{5}{2}x^{3/2}$ **13.** $-5x^{-6}$ **15.** $8x^3$ **17.** $2x^5$ **19.** $x^4/3$ **21.** $-10x^{-6}$ **23.** $-16x^{-5}$ **25.** x^{-3}
27. $-x^{-2/3}$ **29.** $4x - 3$ **31.** $15x^4 - 6x^2$ **33.** $-12x^{-5} - 4x^{-3}$ **35.** $-\frac{1}{2}x^{-2} + 2x^{-4}$ **37.** $2x^{-1/3} - \frac{5}{3}x^{-2/3}$ **39.** $-\frac{9}{5}x^{-8/5} + 3x^{-3/2}$
41. $-\frac{1}{3}x^{-4/3}$ **43.** $-6x^{-3/2} + 6x^{-3} + 1$
45. (A) $f'(x) = 6 - 2x$ (B) $f'(2) = 2; f'(4) = -2$ (C) $y = 2x + 4; y = -2x + 16$ (D) $x = 3$
47. (A) $f'(x) = 12x^3 - 12x$ (B) $f'(2) = 72; f'(4) = 720$ (C) $y = 72x - 127; y = 720x - 2{,}215$ (D) $x = -1, 0, 1$
49. (A) $v = f'(x) = 176 - 32x$ (B) $f'(0) = 176$ ft/sec; $f'(3) = 80$ ft/sec (C) $x = 5.5$ sec
51. (A) $v = f'(x) = 3x^2 - 18x + 15$ (B) $f'(0) = 15$ ft/sec; $f'(3) = -12$ ft/sec (C) $x = 1, 5$ sec **53.** $-20x^{-2}$ **55.** $2x - 3 - 10x^{-3}$
59. $f'(x) = 2x - 3 - 2x^{-1/2}; x = 2.18$ **61.** $f'(x) = 4\sqrt[3]{x} - 3x - 3; x = -2.90$
63. $f'(x) = 0.2x^3 + 0.3x^2 - 3x - 1.6; x = -4.46, -0.52, 3.48$ **65.** $f'(x) = 0.8x^3 - 9.36x^2 + 32.5x - 28.25; x = 1.30$
67. (A) $C'(x) = 60 - (x/2)$
(B) $C'(60) = \$30$/racket (at a production level of 60 rackets, the rate of change of total cost relative to production is \$30 per racket; thus, the cost of producing 1 more racket at this level of production is approx. \$30).
(C) \$29.75 (the marginal cost of \$30 per racket found in part B is a close approximation to this value)
(D) $C'(80) = \$20$ per racket (at a production level of 80 rackets, the rate of change of total cost relative to production is \$20 per racket; thus, the cost of producing 1 more racket at this level of production is approx. \$20)

69. (A) $N'(x) = 60 - 2x$

(B) $N'(10) = 40$ (at the \$10,000 level of advertising, there would be an approximate increase of 40 units of sales per \$1,000 increase in advertising); $N'(20) = 20$ (at the \$20,000 level of advertising, there would be an approximate increase of only 20 units of sales per \$1,000 increase in advertising); the effect of advertising levels off as the amount spent increases

71. (A) -1.37 beats/min (B) -0.58 beat/min 73. (A) 25 items/hr (B) 8.33 items/hr

◆ EXERCISE 3-5

1. $2x^3(2x) + (x^2 - 2)(6x^2) = 10x^4 - 12x^2$ 3. $(x - 3)(2) + (2x - 1)(1) = 4x - 7$ 5. $\frac{(x-3)(1) - x(1)}{(x-3)^2} = \frac{-3}{(x-3)^2}$

7. $\frac{(x-2)(2) - (2x+3)(1)}{(x-2)^2} = \frac{-7}{(x-2)^2}$ 9. $(x^2 + 1)(2) + (2x - 3)(2x) = 6x^2 - 6x + 2$

11. $\frac{(2x-3)(2x) - (x^2+1)(2)}{(2x-3)^2} = \frac{2x^2 - 6x - 2}{(2x-3)^2}$ 13. $(x^2 + 2)2x + (x^2 - 3)2x = 4x^3 - 2x$

15. $\frac{(x^2-3)2x - (x^2+2)2x}{(x^2-3)^2} = \frac{-10x}{(x^2-3)^2}$ 17. $(2x + 1)(2x - 3) + (x^2 - 3x)(2) = 6x^2 - 10x - 3$

19. $(2x - x^2)(5) + (5x + 2)(2 - 2x) = -15x^2 + 16x + 4$ 21. $\frac{(x^2+2x)(5) - (5x-3)(2x+2)}{(x^2+2x)^2} = \frac{-5x^2 + 6x + 6}{(x^2+2x)^2}$

23. $\frac{(x^2-1)(2x-3) - (x^2-3x+1)(2x)}{(x^2-1)^2} = \frac{3x^2 - 4x + 3}{(x^2-1)^2}$ 25. $f'(x) = (1 + 3x)(-2) + (5 - 2x)(3)$; $y = -11x + 29$

27. $f'(x) = \frac{(3x-4)(1) - (x-8)(3)}{(3x-4)^2}$; $y = 5x - 13$ 29. $f'(x) = (2x - 15)(2x) + (x^2 + 18)(2) = 6(x - 2)(x - 3)$; $x = 2$, $x = 3$

31. $f'(x) = \frac{(x^2+1)(1) - x(2x)}{(x^2+1)^2} = \frac{1 - x^2}{(x^2+1)^2}$; $x = -1$, $x = 1$ 33. $7x^6 - 3x^2$ 35. $-27x^{-4}$

37. $(2x^4 - 3x^3 + x)(2x - 1) + (x^2 - x + 5)(8x^3 - 9x^2 + 1)$ 39. $\frac{(4x^2+5x-1)(6x-2) - (3x^2-2x+3)(8x+5)}{(4x^2+5x-1)^2}$

41. $9x^{1/3}(3x^2) + (x^3 + 5)(3x^{-2/3})$ 43. $\frac{(x^2-3)(2x^{-2/3}) - 6x^{1/3}(2x)}{(x^2-3)^2}$ 45. $x^{-2/3}(3x^2 - 4x) + (x^3 - 2x^2)(-\frac{2}{3}x^{-5/3})$

47. $\frac{(x^2+1)[(2x^2-1)(2x) + (x^2+3)(4x)] - (2x^2-1)(x^2+3)(2x)}{(x^2+1)^2}$

49. (A) $S'(t) = \frac{7{,}200 - 200t^2}{(t^2+36)^2}$

(B) $S(2) = 10$; $S'(2) = 4$; at $t = 2$ months, monthly sales are 10,000 and increasing at 4,000 albums per month

(C) $S(8) = 16$; $S'(8) = -0.56$; at $t = 8$ months, monthly sales are 16,000 and decreasing at 560 albums per month

51. (A) $d'(x) = \frac{-50{,}000(2x+10)}{(x^2+10x+25)^2} = \frac{-100{,}000}{(x+5)^3}$

(B) $d'(5) = -100$ radios per \$1 increase in price; $d'(15) = -12.5$ radios per \$1 increase in price

53. $y = 4t + 2$; $y = -0.56t + 20.48$

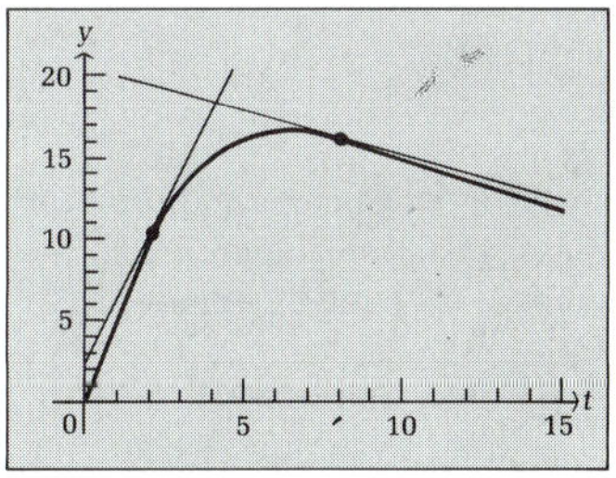

55. $y = -100x + 1{,}000$; $y = -12.5x + 312.5$

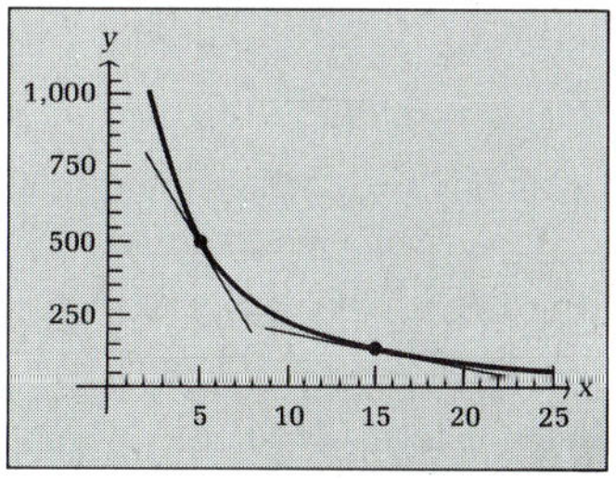

57. (A) $C'(t) = \dfrac{0.14 - 0.14t^2}{(t^2+1)^2}$

(B) $C'(0.5) = 0.0672$ (concentration is increasing at 0.0672 unit/hr); $C'(3) = -0.0112$ (concentration is decreasing at 0.0112 unit/hr)

59. (A) $N'(x) = \dfrac{(x+32)(100) - (100x+200)}{(x+32)^2} = \dfrac{3{,}000}{(x+32)^2}$ (B) $N'(4) = 2.31$; $N'(68) = 0.30$

◆ EXERCISE 3-6

1. $6(2x+5)^2$ **3.** $-8(5-2x)^3$ **5.** $30x(3x^2+5)^4$ **7.** $8(x^3-2x^2+2)^7(3x^2-4x)$ **9.** $(2x-5)^{-1/2}$ **11.** $-8x^3(x^4+1)^{-3}$
13. $f'(x) = 6(2x-1)^2$; $y = 6x - 5$; $x = \frac{1}{2}$ **15.** $f'(x) = 2(4x-3)^{-1/2}$; $y = \frac{2}{3}x + 1$; none **17.** $24x(x^2-2)^3$ **19.** $-6(x^2+3x)^{-4}(2x+3)$
21. $x(x^2+8)^{-1/2}$ **23.** $(3x+4)^{-2/3}$ **25.** $\frac{1}{2}(x^2-4x+2)^{-1/2}(2x-4) = (x-2)/(x^2-4x+2)^{1/2}$ **27.** $(-1)(2x+4)^{-2}(2) = -2/(2x+4)^2$

29. $-15x^2(x^3+4)^{-6}$ **31.** $(-1)(4x^2-4x+1)^{-2}(8x-4) = -4/(2x-1)^3$ **33.** $-2(x^2-3x)^{-3/2}(2x-3) = \dfrac{-2(2x-3)}{(x^2-3x)^{3/2}}$

35. $f'(x) = (4-x)^3 - 3x(4-x)^2$; $y = -16x + 48$ **37.** $f'(x) = \dfrac{(2x-5)^3 - 6x(2x-5)^2}{(2x-5)^6}$; $y = -17x + 54$

39. $f'(x) = (2x+2)^{1/2} + x(2x+2)^{-1/2}$; $y = \frac{5}{2}x - \frac{1}{2}$
41. $f'(x) = 2x(x-5)^3 + 3x^2(x-5)^2 = 5x(x-5)^2(x-2)$; $x = 0$, $x = 2$, $x = 5$

43. $f'(x) = \dfrac{(2x+5)^2 - 4x(2x+5)}{(2x+5)^4} = \dfrac{5-2x}{(2x+5)^3}$; $x = \frac{5}{2}$ **45.** $f'(x) = \dfrac{x-4}{\sqrt{x^2-8x+20}}$; $x = 4$

47. $18x^2(x^2+1)^2 + 3(x^2+1)^3 = 3(x^2+1)^2(7x^2+1)$ **49.** $\dfrac{2x^3 4(x^3-7)^3 3x^2 - (x^3-7)^4 6x^2}{4x^6} = \dfrac{3(x^3-7)^3(3x^3+7)}{2x^4}$

51. $(2x-3)^2[3(2x^2+1)^2(4x)] + (2x^2+1)^3[2(2x-3)(2)] = 4(2x^2+1)^2(2x-3)(8x^2-9x+1)$

53. $4x^2[\frac{1}{2}(x^2-1)^{-1/2}(2x)] + (x^2-1)^{1/2}(8x) = \dfrac{12x^3-8x}{\sqrt{x^2-1}}$ **55.** $\dfrac{(x-3)^{1/2}(2) - 2x[\frac{1}{2}(x-3)^{-1/2}]}{x-3} = \dfrac{x-6}{(x-3)^{3/2}}$

57. $(2x-1)^{1/2}(x^2+3)(11x^2-4x+9)$
59. (A) $C'(x) = (2x+16)^{-1/2} = 1/\sqrt{2x+16}$
(B) $C'(24) = \frac{1}{8}$ or \$12.50 per calculator (the rate of change of total cost relative to production at a production level of 24 calculators is \$12.50 per calculator; the cost of producing 1 more calculator at this level of production is approx. \$12.50); $C'(42) = \frac{1}{10}$ or \$10.00 per calculator (the rate of change of total cost relative to production at a production level of 42 calculators is \$10.00 per calculator; the cost of producing 1 more calculator at this level of production is approx. \$10.00)

61. $4{,}000(1 + \frac{1}{12}r)^{47}$ **63.** $\dfrac{(4 \times 10^6)x}{(x^2-1)^{5/3}}$

65. (A) $f'(n) = n(n-2)^{-1/2} + 2(n-2)^{1/2} = \dfrac{3n-4}{(n-2)^{1/2}}$

(B) $f'(11) = \frac{29}{3}$ (rate of learning is $\frac{29}{3}$ units/min at the $n = 11$ level); $f'(27) = \frac{77}{5}$ (rate of learning is $\frac{77}{5}$ units/min at the $n = 27$ level)

◆ EXERCISE 3-7

1. (A) \$29.50 (B) \$30
3. (A) \$420
(B) $\overline{C}'(500) = -0.24$; at a production level of 500 units, a unit increase in production will decrease average cost by approx. 24¢
5. (A) $R'(1{,}600) = 20$; at a production level of 1,600 units, a unit increase in production will increase revenue by approx. \$20
(B) $R'(2{,}500) = -25$; at a production level of 2,500 units, a unit increase in production will decrease revenue by approx. \$25
7. (A) \$4.50 (B) \$5
9. (A) $P'(450) = 0.5$; at a production level of 450 units, a unit increase in production will increase profit by approx. 50¢
(B) $P'(750) = -2.5$; at a production level of 750 units, a unit increase in production will decrease profit by approx. \$2.50

11. (A) \$1.25
(B) $\overline{P}'(150) = 0.015$; at a production level of 150 units, a unit increase in production will increase average profit by approx. 1.5¢

13. (A) $C'(x) = 60$ (B) $R(x) = 200x - (x^2/30)$ (C) $R'(x) = 200 - (x/15)$
(D) $R'(1{,}500) = 100$ (at a production level of 1,500 units, a unit increase in production will increase revenue by approx. \$100); $R'(4{,}500) = -100$ (at a production level of 4,500 units, a unit increase in production will decrease revenue by approx. \$100)
(E) Break-even points: (600, 108,000) and (3,600, 288,000)

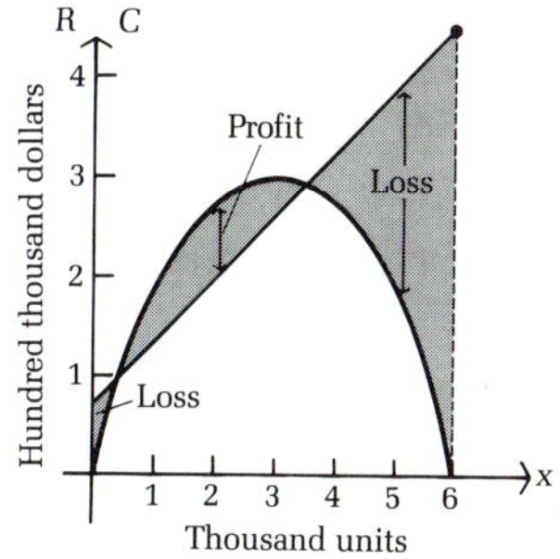

(F) $P(x) = -(x^2/30) + 140x - 72{,}000$
(G) $P'(x) = -(x/15) + 140$
(H) $P'(1{,}500) = 40$ (at a production level of 1,500 units, a unit increase in production will increase profit by approx. \$40); $P'(3{,}000) = -60$ (at a production level of 3,000 units, a unit increase in production will decrease profit by approx. \$60)

15. (A) $p = 20 - (x/50)$ (B) $R(x) = 20x - (x^2/50)$ (C) $C(x) = 4x + 1{,}400$
(D) Break-even points: (100, 1,800) and (700, 4,200)

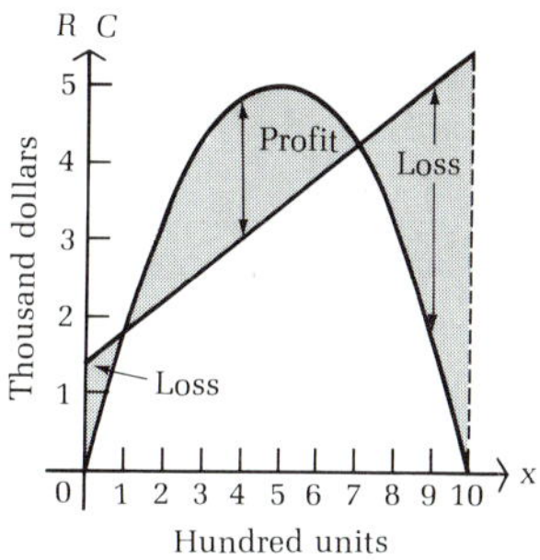

(E) $P(x) = 16x - (x^2/50) - 1{,}400$
(F) $P'(250) = 6$ (at a production level of 250 units, a unit increase in production will increase profit by approx. \$6); $P'(475) = -3$ (at a production level of 475 units, a unit increase in production will decrease profit by approx. \$3

17. (A) $x = 500$ (B) $P(x) = 176x - 0.2x^2 - 21{,}900$
(C) $x = 440$
(D) Break-even points: (150, 25,500) and (730, 39,420); x intercepts for $P(x)$: $x = 150$ and $x = 730$

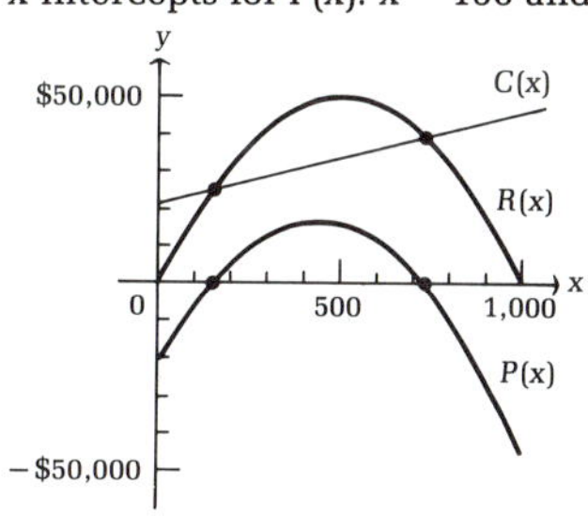

19. (A) $R(x) = 20x - x^{3/2}$
(B) Break-even points: (44, 588), (258, 1,016)

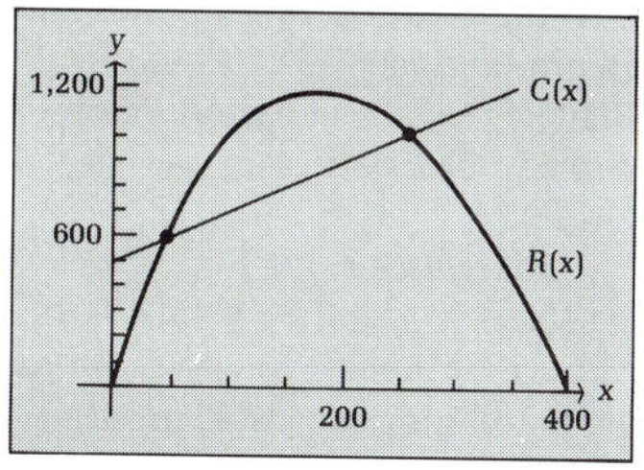

◆ EXERCISE 3-8 CHAPTER REVIEW

1. $12x^3 - 4x$ 2. $x^{-1/2} - 3 = \dfrac{1}{x^{1/2}} - 3$ 3. 0 4. $-x^{-3} + x$ 5. $(2x - 1)(3) + (3x + 2)(2) = 12x + 1$

6. $(x^2 - 1)(3x^2) + (x^3 - 3)(2x) = 5x^4 - 3x^2 - 6x$ 7. $\dfrac{(x^2 + 2)2 - 2x(2x)}{(x^2 + 2)^2} = \dfrac{4 - 2x^2}{(x^2 + 2)^2}$ 8. $(-1)(3x + 2)^{-2}(3) = \dfrac{-3}{(3x + 2)^2}$

9. $3(2x-3)^2(2)=6(2x-3)^2$ **10.** $-2(x^2+2)^{-3}(2x)=\dfrac{-4x}{(x^2+2)^3}$ **11.** (A) Does not exist (B) 3 (C) No

12. (A) 2 (B) Not defined (C) No **13.** (A) 1 (B) 1 (C) Yes **14.** $12x^3+6x^{-4}$

15. $(2x^2-3x+2)(2x+2)+(x^2+2x-1)(4x-3)=8x^3+3x^2-12x+7$ **16.** $\dfrac{(x-1)^2(2)-(2x-3)(2)(x-1)}{(x-1)^4}=\dfrac{4-2x}{(x-1)^3}$

17. $x^{-1/2}-2x^{-3/2}=\dfrac{1}{\sqrt{x}}-\dfrac{2}{\sqrt{x^3}}$ **18.** $(x^2-1)[2(2x+1)(2)]+(2x+1)^2(2x)=2(2x+1)(4x^2+x-2)$

19. $\dfrac{1}{3}(x^3-5)^{-2/3}(3x^2)=\dfrac{x^2}{\sqrt[3]{(x^3-5)^2}}$ **20.** $-8x^{-3}$ **21.** $\dfrac{(2x-3)(4)(x^2+2)^3(2x)-(x^2+2)^4(2)}{(2x-3)^2}=\dfrac{2(x^2+2)^3(7x^2-12x-2)}{(2x-3)^2}$

22. (A) $m=f'(1)=2$ (B) $y=2x+3$ **23.** (A) $m=f'(1)=16$ (B) $y=16x-12$ **24.** $x=5$ **25.** $x=-5, x=3$
26. $x=-2, x=2$ **27.** $x=0, x=3, x=\frac{15}{2}$ **28.** (A) $v=f'(x)=32x-4$ (B) $f'(3)=92$ ft/sec
29. (A) $v=f'(x)=96-32x$ (B) $x=3$ sec **30.** (A) 4 (B) 6 (C) Does not exist (D) 6 (E) No
31. (A) 3 (B) 3 (C) 3 (D) 3 (E) Yes **32.** (A) ∞ (B) ∞ (C) ∞ **33.** (A) $-\infty$ (B) ∞ (C) Does not exist **34.** $(-\infty, \infty)$

35. $(-\infty,-5), (-5,\infty)$ **36.** $(-\infty,-2), (-2,3), (3,\infty)$ **37.** $[3,\infty)$ **38.** $(-\infty,\infty)$ **39.** $\dfrac{2(3)-3}{3+5}=\dfrac{3}{8}$ **40.** $2(3^2)-3+1=16$ **41.** -1 **42.** 4

43. 4 **44.** $\frac{1}{6}$ **45.** Does not exist **46.** $1/(2\sqrt{7})$ **47.** $\sqrt{2}$ **48.** 3 **49.** ∞ **50.** $\frac{2}{3}$ **51.** 0 **52.** ∞ **53.** $2x-1$ **54.** $1/(2\sqrt{x})$ **55.** No **56.** No
57. No **58.** Yes

59. Discontinuous at $x=0$

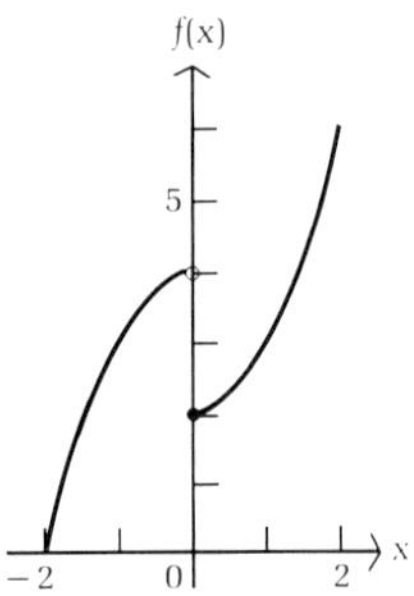

60. Continuous for all x

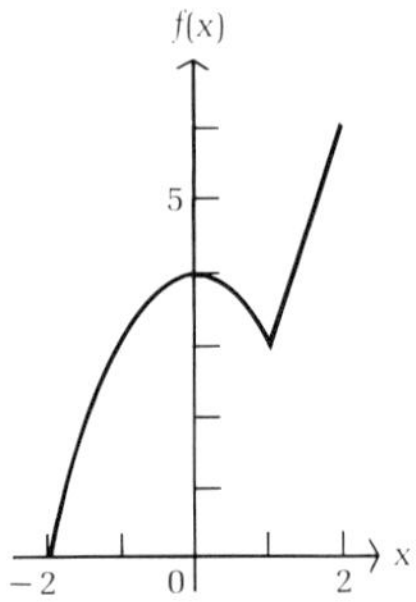

61. $7x(x-4)^3(x+3)^2$

62. $\dfrac{x^4(2x+5)}{(2x+1)^5}$

63. $\dfrac{1}{x^2\sqrt{x^2-1}}$

64. $\dfrac{4}{(x^2+4)^{3/2}}$

65. (A) 1 (B) 1 (C) 1 (D) Yes
66. (A) Yes (B) Yes (C) Yes (D) Yes
67. (A) 1 (B) -1 (C) Does not exist (D) No
68. (A) \$179.90 (B) \$180

69. (A) $C(100)=9{,}500$ and $C'(100)=50$; at a production level of 100 bicycles, the total cost is \$9,500 and is increasing at a rate of \$50 per unit increase in production.
(B) $\overline{C}(100)=95$ and $\overline{C}'(100)=-0.45$; at a production level of 100 bicycles, the average cost is \$95 and is decreasing at a rate of \$0.45 per unit increase in production.

70. (A) $C'(x)=2$; $\overline{C}(x)=2+56x^{-1}$; $\overline{C}'(x)=-56x^{-2}$
(B) $R(x)=xp=20x-x^2$; $R'(x)=20-2x$; $\overline{R}(x)=20-x$; $\overline{R}'(x)=-1$
(C) $P(x)=R(x)-C(x)=18x-x^2-56$; $P'(x)=18-2x$; $\overline{P}(x)=18-x-56x^{-1}$; $\overline{P}'(x)=-1+56x^{-2}$
(D) Solving $R(x)=C(x)$, we find break-even points at $x=4, 14$.
(E) $P'(7)=4$ (increasing production increases profit); $\overline{P}'(9)=0$ (stable); $P'(11)=-4$ (increasing production decreases profit)
(F)

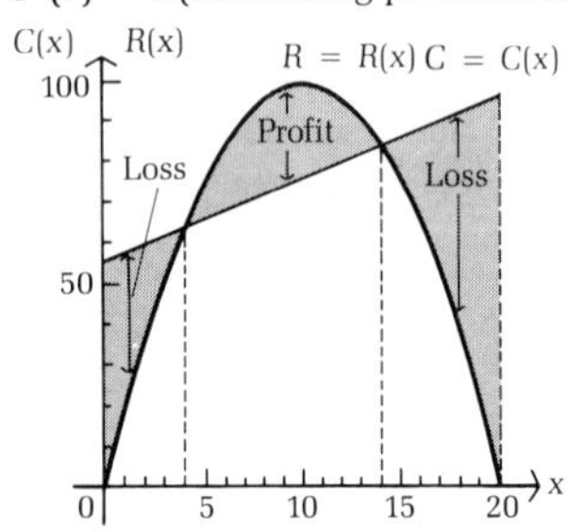

71. (A) 2 components/day (B) 3.2 components/day (C) 40 components/day

72. $N(3) = 14$ and $N'(3) = 2.5$; monthly sales during the third month are 14,000 and are increasing at a rate of 2,500 pools per month.
73. $C'(9) = -1$ ppm/m; $C'(99) = -0.001$ ppm/m **74.** (A) 10 items/hr (B) 5 items/hr

CHAPTER 4

◆ EXERCISE 4-1

1. (a, b); (d, f); (g, h); **3.** c, d, f **5.** b, f **7.** Local maximum at $x = a$; local minimum at $x = c$; no local extrema at $x = b$ and $x = d$

9.

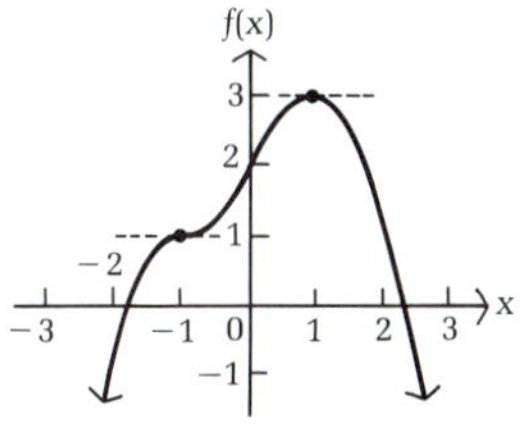

11.

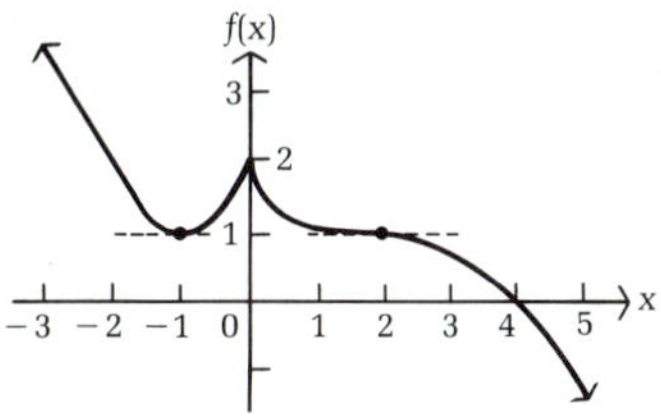

13. $-3 < x < 4$; $(-3, 4)$ **15.** $x < 3$ or $x > 7$; $(-\infty, 3) \cup (7, \infty)$ **17.** $-5 < x < 0$ or $x > 3$; $(-5, 0) \cup (3, \infty)$
19. Decreasing on $(-\infty, 8)$; increasing on $(8, \infty)$; local minimum at $x = 8$
21. Increasing on $(-\infty, 5)$; decreasing on $(5, \infty)$; local maximum at $x = 5$
23. Increasing for all x; no local extrema **25.** Decreasing for all x; no local extrema
27. Increasing on $(-\infty, -2)$ and $(2, \infty)$; decreasing on $(-2, 2)$; local maximum at $x = -2$; local minimum at $x = 2$
29. Increasing on $(-\infty, -2)$ and $(4, \infty)$; decreasing on $(-2, 4)$; local maximum at $x = -2$; local minimum at $x = 4$
31. Increasing on $(-\infty, -1)$ and $(0, 1)$; decreasing on $(-1, 0)$ and $(1, \infty)$; local maxima at $x = -1$ and $x = 1$; local minimum at $x = 0$

33. Increasing on $(-\infty, 4)$
Decreasing on $(4, \infty)$
Horizontal tangent at $x = 4$

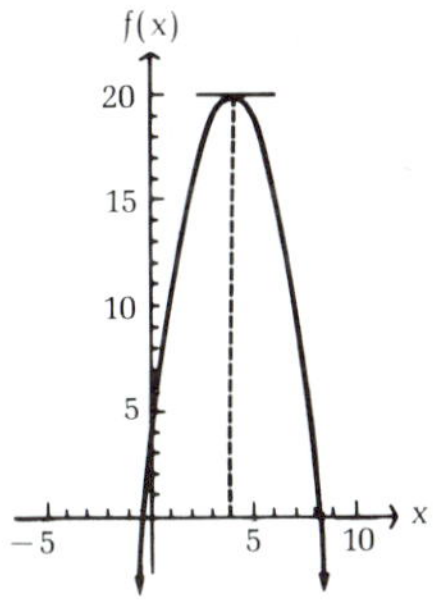

35. Increasing on $(-\infty, -1)$, $(1, \infty)$
Decreasing on $(-1, 1)$
Horizontal tangents at $x = -1, 1$

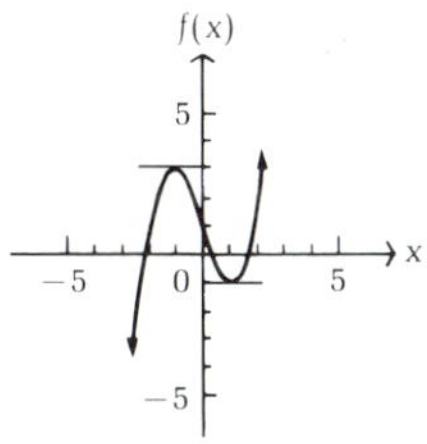

37. Decreasing for all x
Horizontal tangent at $x = 2$

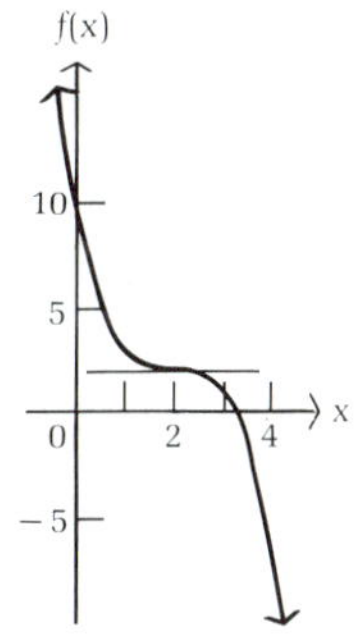

39. No critical values; increasing on $(-\infty, -2)$ and $(-2, \infty)$ no local extrema
41. Critical values: $x = -2$, $x = 2$; increasing on $(-\infty, -2)$ and $(2, \infty)$; decreasing on $(-2, 0)$ and $(0, 2)$; local maximum at $x = -2$; local minimum at $x = 2$
43. Critical value: $x = -2$; increasing on $(-2, 0)$; decreasing on $(-\infty, -2)$ and $(0, \infty)$; local minimum at $x = -2$
45. Critical values: $x = 0$, $x = 4$; increasing on $(-\infty, 0)$ and $(4, \infty)$; decreasing on $(0, 2)$ and $(2, 4)$; local maximum at $x = 0$; local minimum at $x = 4$
47. Critical values: $x = 0$, $x = 4$, $x = 6$; increasing on $(0, 4)$ and $(6, \infty)$; decreasing on $(-\infty, 0)$ and $(4, 6)$; local maximum at $x = 4$; local minima at $x = 0$ and $x = 6$
49. Critical value: $x = 2$; increasing on $(2, \infty)$; decreasing on $(-\infty, 2)$; local minimum at $x = 2$
51. Critical value: $x = 1$; increasing on $(0, 1)$; decreasing on $(1, \infty)$; local maximum at $x = 1$

53. Critical values: $x = -1.26$; increasing on $(-1.26, \infty)$; decreasing on $(-\infty, -1.26)$; local minimum at $x = -1.26$

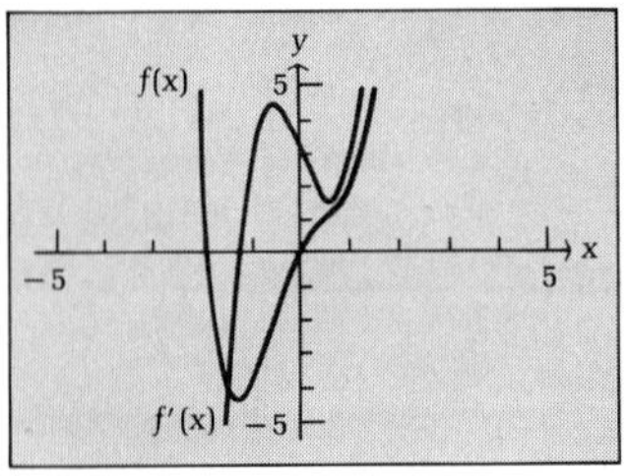

55. Critical values: $x = -0.43$, $x = 0.54$, $x = 2.14$; increasing on $(-0.43, 0.54)$ and $(2.14, \infty)$; decreasing on $(-\infty, -0.43)$ and $(0.54, 2.14)$;
local maximum at $x = 0.54$, local minima at $x = -0.43$ and $x = 2.14$

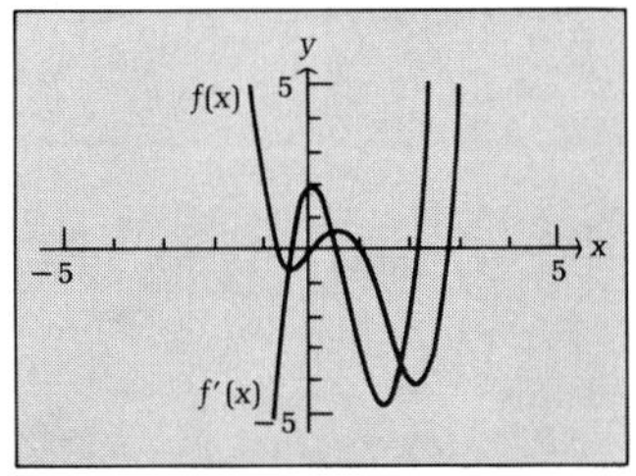

57. Critical value: $x = 80$; decreasing for $0 < x < 80$; increasing for $80 < x < 150$; local minimum at $x = 80$
59. $P(x)$ is increasing over (a, b) if $P'(x) = R'(x) - C'(x) > 0$ over (a, b); that is, if $R'(x) > C'(x)$ over (a, b).
61. Critical value: $t = 1$; increasing for $0 < t < 1$; decreasing for $1 < t < 24$; local maximum at $t = 1$
63. Critical value: $t = 7$; increasing for $0 < t < 7$; decreasing for $7 < t < 24$; local maximum at $t = 7$

◆ EXERCISE 4-2

1. (a, c), (c, d), (e, g) 3. d, e, g 5. Local minimum 7. Unable to determine 9. Neither

11.

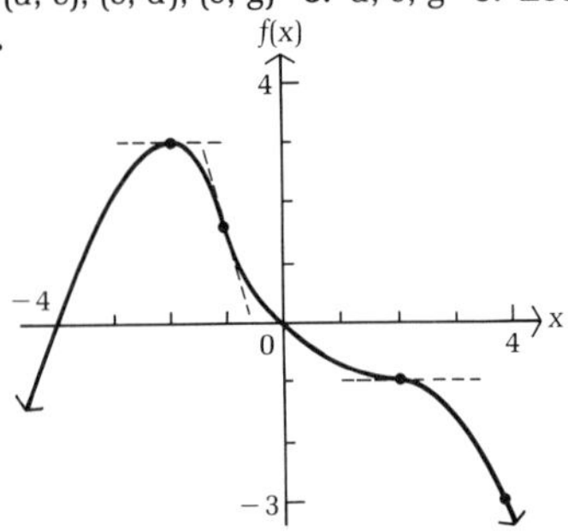

13.

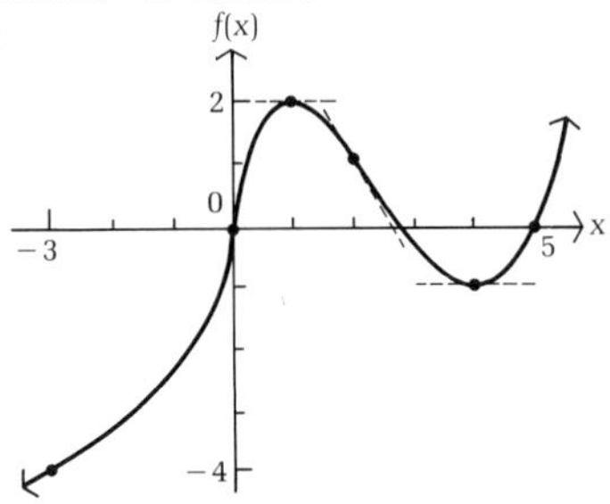

15. $6x - 4$ 17. $40x^3$ 19. $6x$ 21. $24x^2(x^2 - 1) + 6(x^2 - 1)^2 = 6(x^2 - 1)(5x^2 - 1)$ 23. $6x^{-3} + 12x^{-4}$
25. $f(2) = -2$ is a local minimum 27. $f(-1) = 2$ is a local maximum; $f(2) = -25$ is a local minimum 29. No local extrema
31. $f(-2) = -6$ is a local minimum; $f(0) = 10$ is a local maximum; $f(2) = -6$ is a local minimum 33. $f(0) = 2$ is a local minimum
35. $f(-4) = -8$ is a local maximum; $f(4) = 8$ is a local minimum 37. Concave upward for all x; no inflection points
39. Concave upward on $(6, \infty)$; concave downward on $(-\infty, 6)$; inflection point at $x = 6$
41. Concave upward on $(-\infty, -2)$ and $(2, \infty)$; concave downward on $(-2, 2)$; inflection points at $x = -2$ and $x = 2$
43. Concave upward on $(0, 2)$; concave downward on $(-\infty, 0)$ and $(2, \infty)$; inflection points at $x = 0$ and $x = 2$

45. Local maximum at $x = 0$
Local minimum at $x = 4$
Inflection point at $x = 2$

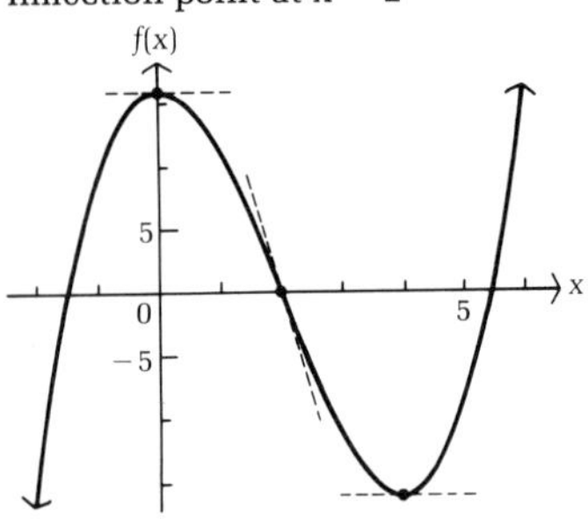

47. Inflection point at $x = 0$

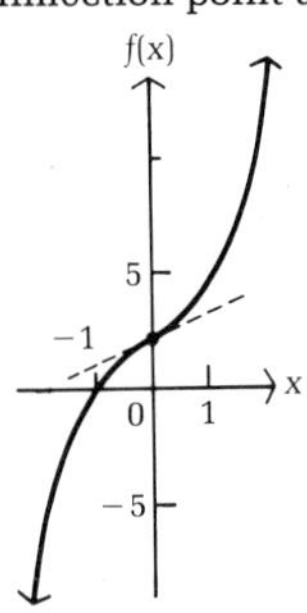

49. Inflection point at $x = 2$

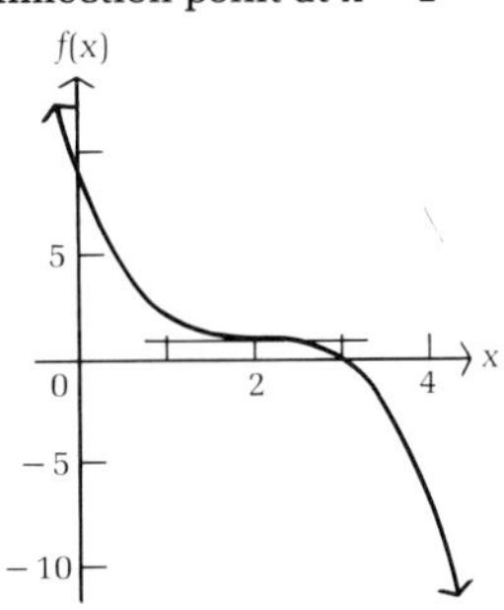

51. Local maximum at $x = -2$
Local minimum at $x = 2$
Inflection point at $x = 0$

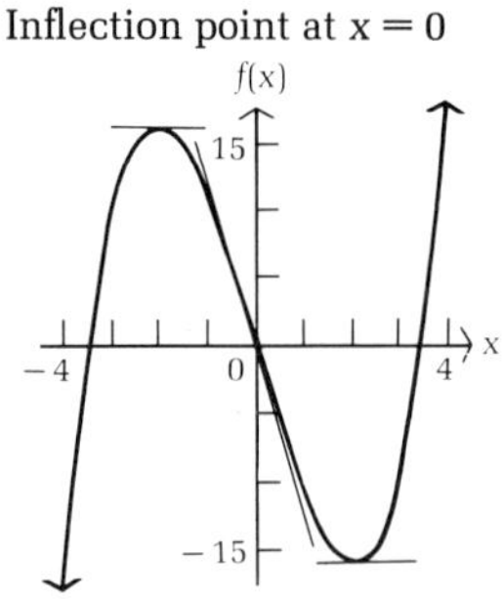

53. Inflection points at $x = -2$ and $x = 2$ **55.** Inflection points at $x = -6$, $x = 0$, and $x = 6$

57. Inflection point at $x = -1.40$
Concave upward on $(-1.40, \infty)$
Concave downward on $(-\infty, -1.40)$

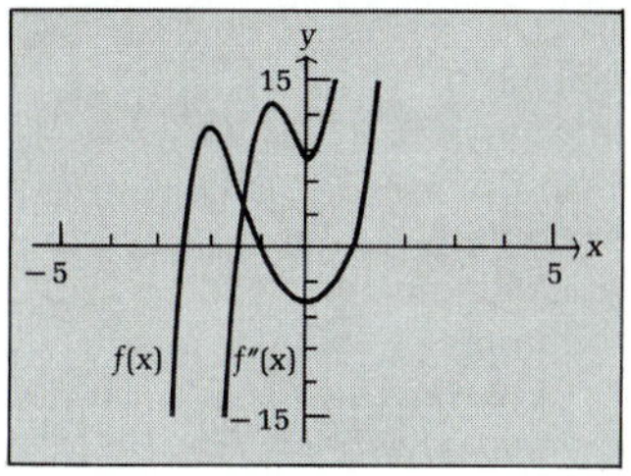

59. Inflection points at $x = -0.61$, $x = 0.66$, and $x = 1.74$
Concave upward on $(-0.61, 0.66)$ and $(1.74, \infty)$
Concave downward on $(-\infty, -0.61)$ and $(0.66, 1.74)$

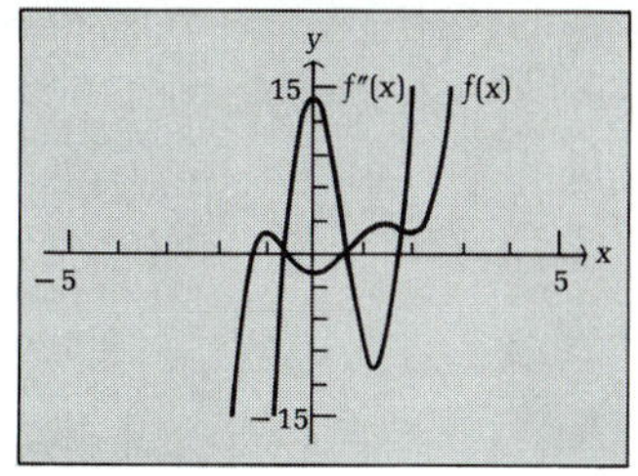

61. (A) Local maximum at $x = 60$ (B) Concave downward on the whole interval $(0, 80)$

63. (A) Increasing on $(10, 25)$; decreasing on $(25, 40)$
(B) Inflection point at $x = 25$
(C)

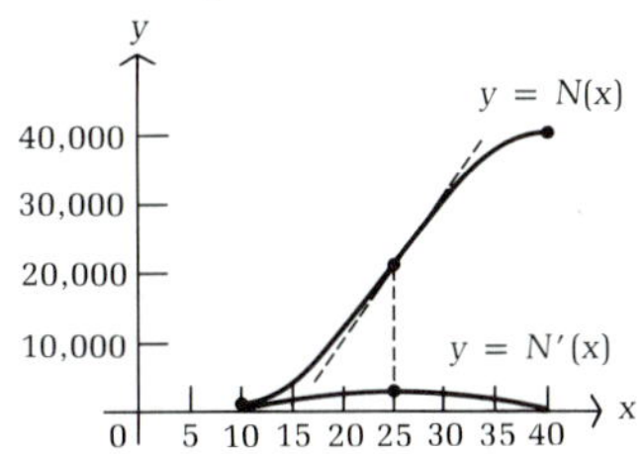

(D) Max $N'(x) = N'(25) = 2{,}025$

65. (A) Increasing on $(0, 10)$ decreasing on $(10, 20)$
(B) Inflection point at $t = 10$
(C)

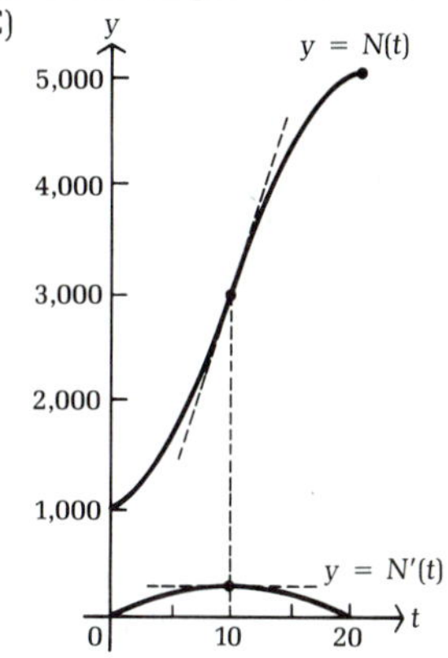

(D) $N'(10) = 300$

67. (A) Increasing on $(5, \infty)$; decreasing on $(0, 5)$
(B) Inflection point at $n = 5$

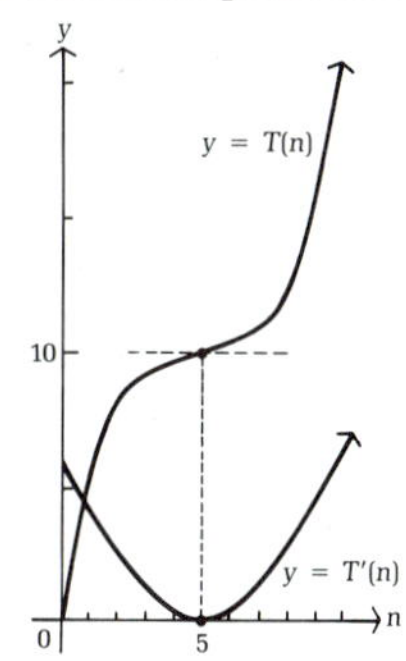

(C) $T'(5) = 0$

◆ EXERCISE 4-3

1. (b, d), $(d, 0)$, (g, ∞) **3.** $x = 0$ **5.** (a, d), (e, h) **7.** $x = a$, $x = h$ **9.** $x = d$, $x = e$

11.

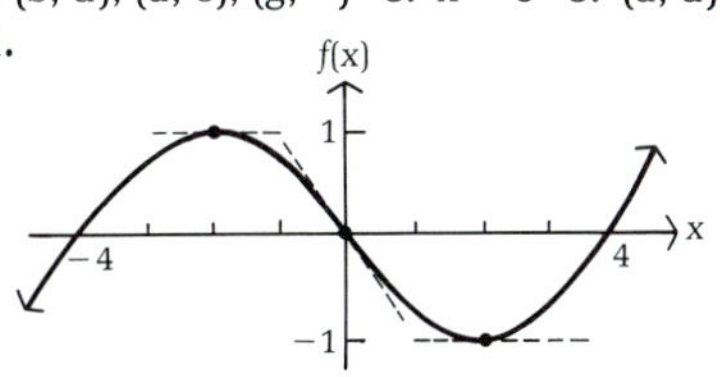

13.

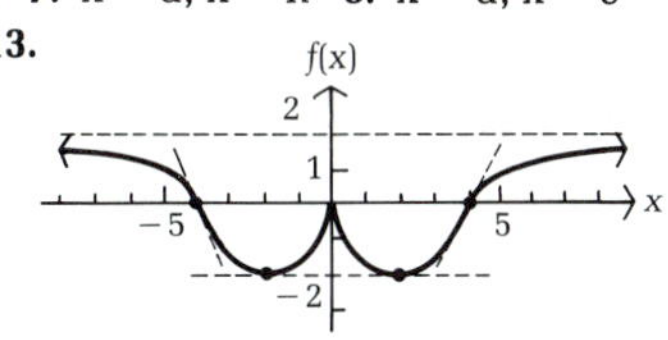

15.

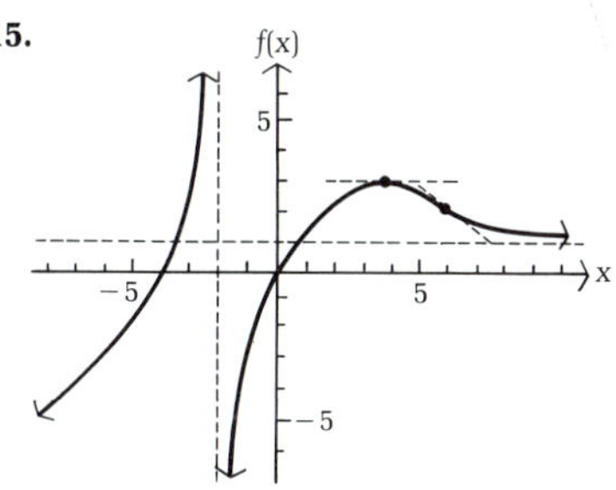

17. Horizontal asymptote: $y = 2$; vertical asymptote: $x = -2$
19. Horizontal asymptote: $y = 1$; vertical asymptotes: $x = -1$ and $x = 1$ **21.** No horizontal or vertical asymptotes
23. Horizontal asymptote: $y = 0$; no vertical asymptotes **25.** No horizontal asymptote; vertical asymptote: $x = 3$

27. Domain: All real numbers
y intercept: 5
x intercepts: 1, 5
Decreasing on $(-\infty, 3)$
Increasing on $(3, \infty)$
Local minimum at $x = 3$
Concave upward on $(-\infty, \infty)$

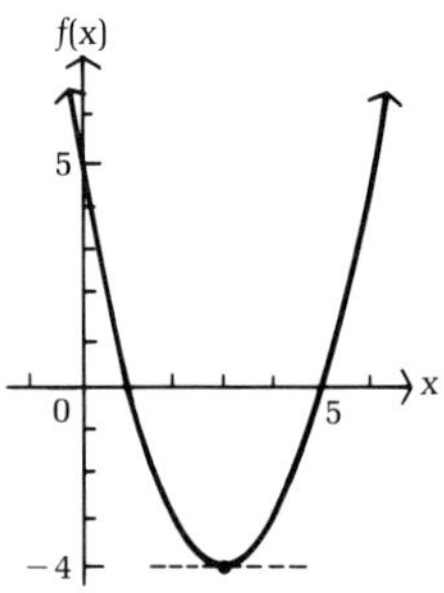

29. Domain: All real numbers
y intercept: 0
x intercepts: 0, 6
Increasing on $(-\infty, 0)$ and $(4, \infty)$
Decreasing on $(0, 4)$
Local maximum at $x = 0$
Local minimum at $x = 4$
Concave upward on $(2, \infty)$
Concave downward on $(-\infty, 2)$
Inflection point at $x = 2$

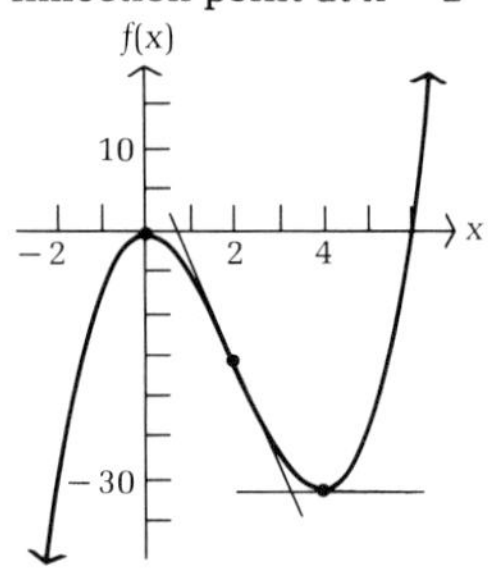

31. Domain: All real numbers
y intercept: 16
x intercepts: -4, 2
Increasing on $(-\infty, -2)$ and $(2, \infty)$
Decreasing on $(-2, 2)$
Local maximum at $x = -2$
Local minimum at $x = 2$
Concave upward on $(0, \infty)$
Concave downward on $(-\infty, 0)$
Inflection point at $x = 0$

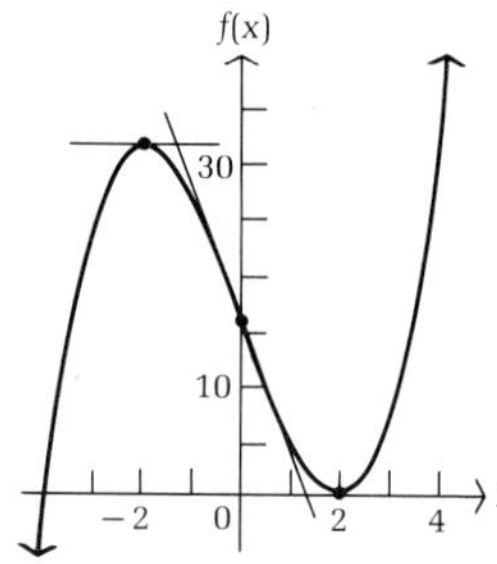

33. Domain: All real numbers
y intercept: 0
x intercepts: 0, 4
Increasing on $(-\infty, 3)$
Decreasing on $(3, \infty)$
Local maximum at $x = 3$
Concave upward on $(0, 2)$
Concave downward on $(-\infty, 0)$ and $(2, \infty)$
Inflection points at $x = 0$ and $x = 2$

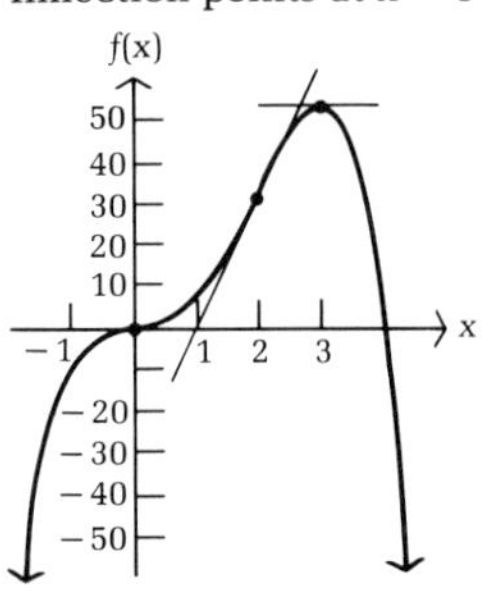

35. Domain: All real numbers except 3
y intercept: -1
x intercept: -3
Horizontal asymptote: $y = 1$
Vertical asymptote: $x = 3$
Decreasing on $(-\infty, 3)$ and $(3, \infty)$
Concave upward on $(3, \infty)$
Concave downward on $(-\infty, 3)$

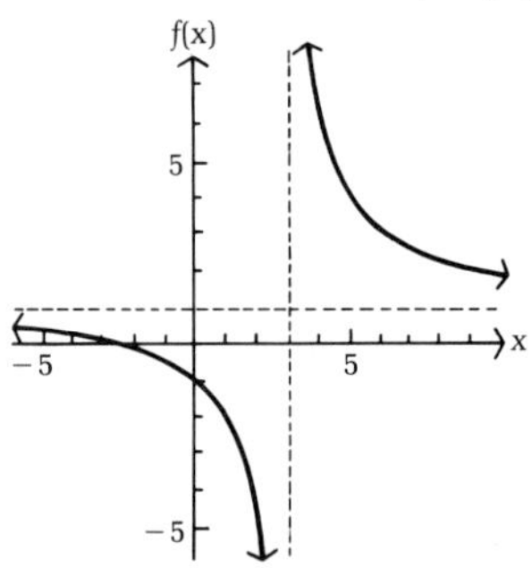

37. Domain: All real numbers except 2
y intercept: 0
x intercept: 0
Horizontal asymptote: $y = 1$
Vertical asymptote: $x = 2$
Decreasing on $(-\infty, 2)$ and $(2, \infty)$
Concave downward on $(-\infty, 2)$
Concave upward on $(2, \infty)$

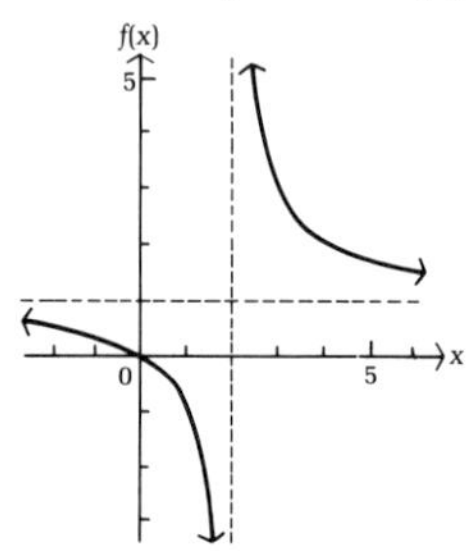

39. Domain: All real numbers except 0
Vertical asymptote: $x = 0$
Increasing on $(-\infty, -1)$ and $(1, \infty)$
Decreasing on $(-1, 0)$ and $(0, 1)$
Local maximum at $x = -1$
Local minimum at $x = 1$
Concave upward on $(0, \infty)$
Concave downward on $(-\infty, 0)$

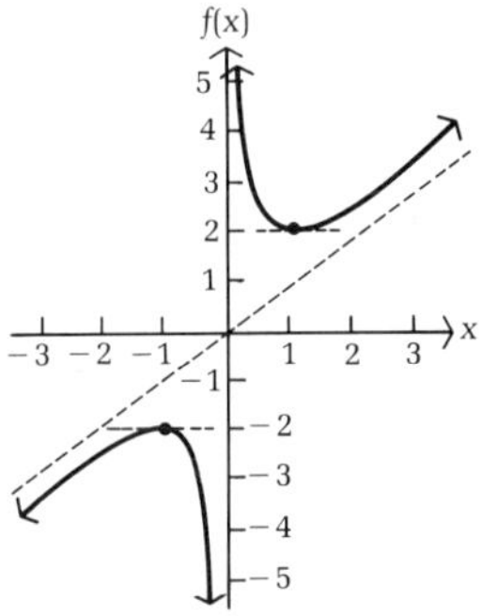

41. Domain: All real numbers
y intercept: 0
x intercepts: $-1, 0, 1$
Increasing on $(-\infty, -\sqrt{3}/3)$ and $(\sqrt{3}/3, \infty)$
Decreasing on $(-\sqrt{3}/3, \sqrt{3}/3)$
Local maximum at $x = -\sqrt{3}/3$
Local minimum at $x = \sqrt{3}/3$
Concave downward on $(-\infty, 0)$
Concave upward on $(0, \infty)$
Inflection point at $x = 0$

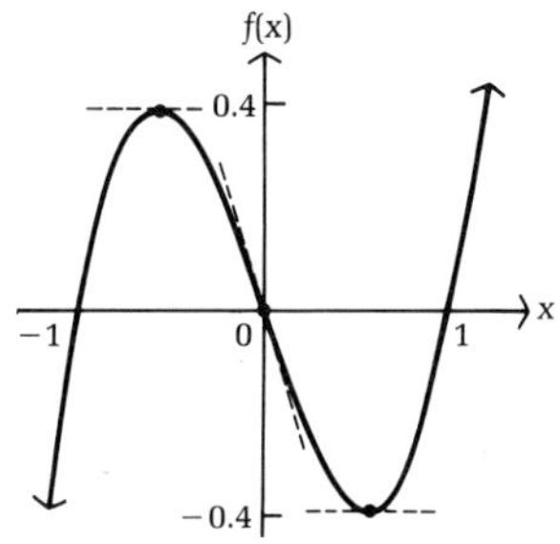

43. Domain: All real numbers
y intercept: 27
x intercepts: $-3, 3$
Increasing on $(-\infty, -\sqrt{3})$ and $(0, \sqrt{3})$
Decreasing on $(-\sqrt{3}, 0)$ and $(\sqrt{3}, \infty)$
Local maxima at $x = -\sqrt{3}$ and $x = \sqrt{3}$
Local minimum at $x = 0$
Concave upward on $(-1, 1)$
Concave downward on $(-\infty, -1)$ and $(1, \infty)$
Inflection points at $x = -1$ and $x = 1$

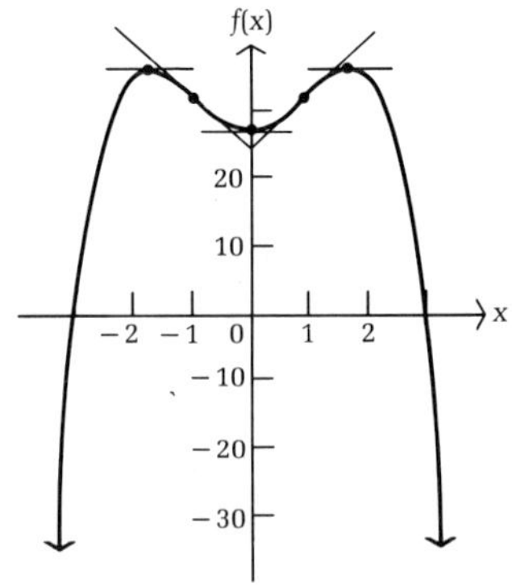

45. Domain: All real numbers
y intercept: 16
x intercepts: $-2, 2$
Decreasing on $(-\infty, -2)$ and $(0, 2)$
Increasing on $(-2, 0)$ and $(2, \infty)$
Local minima at $x = -2$ and $x = 2$
Local maximum at $x = 0$
Concave upward on $(-\infty, -2\sqrt{3}/3)$ and $(2\sqrt{3}/3, \infty)$
Concave downward on $(-2\sqrt{3}/3, 2\sqrt{3}/3)$
Inflection points at $x = -2\sqrt{3}/3$ and $x = 2\sqrt{3}/3$

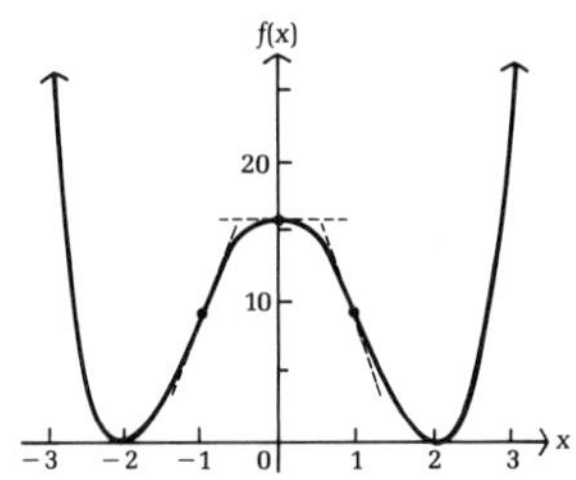

47. Domain: All real numbers
y intercept: 0
x intercepts: 0, 1.5
Decreasing on $(-\infty, 0)$ and $(0, 1.25)$
Increasing on $(1.25, \infty)$
Local minimum at $x = 1.25$
Concave upward on $(-\infty, 0)$ and $(1, \infty)$
Concave downward on $(0, 1)$
Inflection points at $x = 0$ and $x = 1$

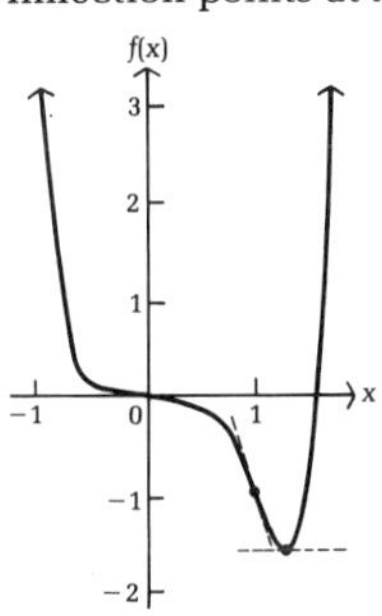

49. Domain: All real numbers except ± 2
y intercept: 0
x intercept: 0
Horizontal asymptote: $y = 0$
Vertical asymptotes: $x = -2$, $x = 2$
Decreasing on $(-\infty, -2)$, $(-2, 2)$, and $(2, \infty)$
Concave upward on $(-2, 0)$ and $(2, \infty)$
Concave downward on $(-\infty, -2)$ and $(0, 2)$
Inflection point at $x = 0$

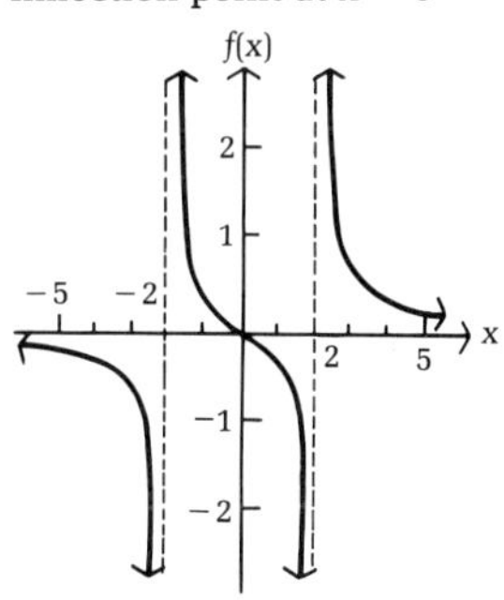

51. Domain: All real numbers
y intercept: 1
Horizontal asymptote: $y = 0$
Increasing on $(-\infty, 0)$
Decreasing on $(0, \infty)$
Local maximum at $x = 0$
Concave upward on $(-\infty, -\sqrt{3}/3)$ and $(\sqrt{3}/3, \infty)$
Concave downward on $(-\sqrt{3}/3, \sqrt{3}/3)$
Inflection points at $x = -\sqrt{3}/3$ and $x = \sqrt{3}/3$

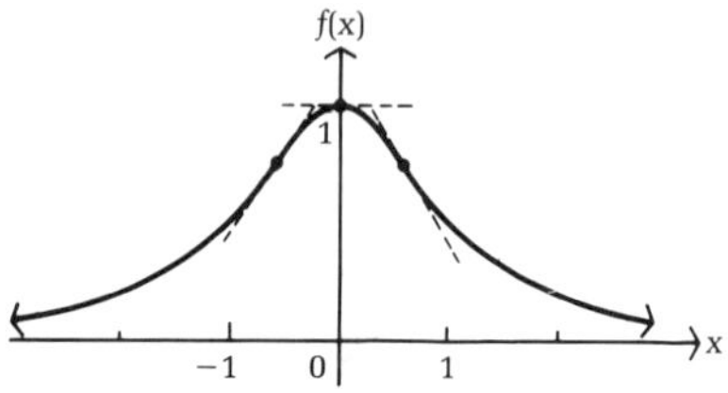

53. Horizontal asymptotes: $y = -1$ and $y = 1$
Vertical asymptote: $x = 0$

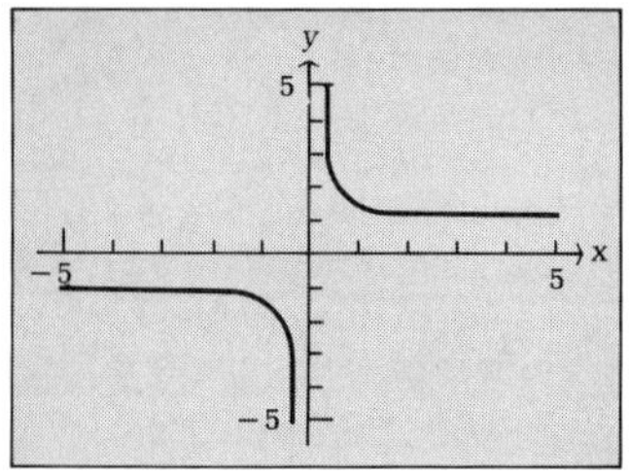

55. Horizontal asymptote: $y = -1$
Vertical asymptote: $x = 1$

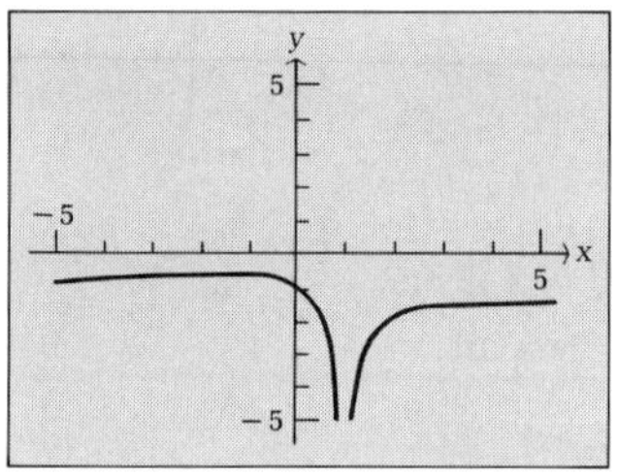

57. Horizontal asymptotes: $y = -1$ and $y = 1$
Vertical asymptotes: $x = -1$ and $x = 1$

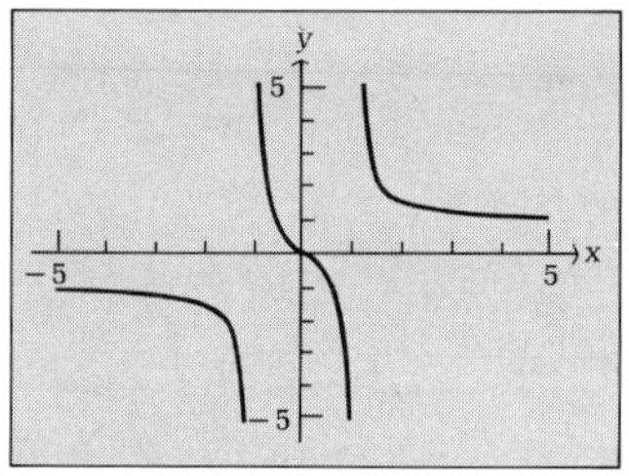

59.

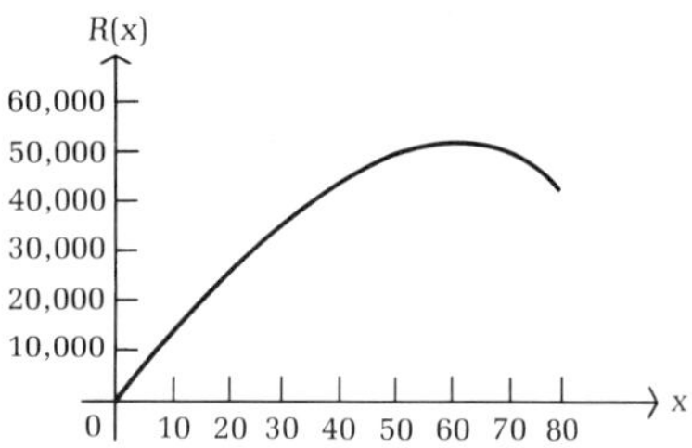

61. (A) Increasing on (0, 1)
(B) Concave upward on (0, 1)
(C) $x = 1$ is a vertical asymptote
(D) The origin is both an x and a y intercept
(E)

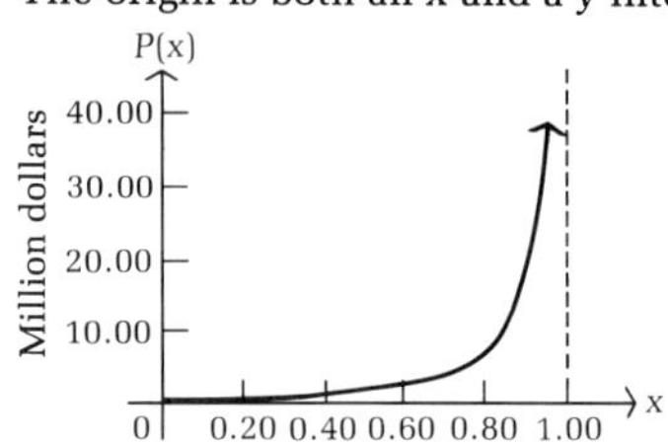

63. (A) $\overline{C}(n) = \dfrac{3{,}200}{n} + 250 + 50n$
(B)

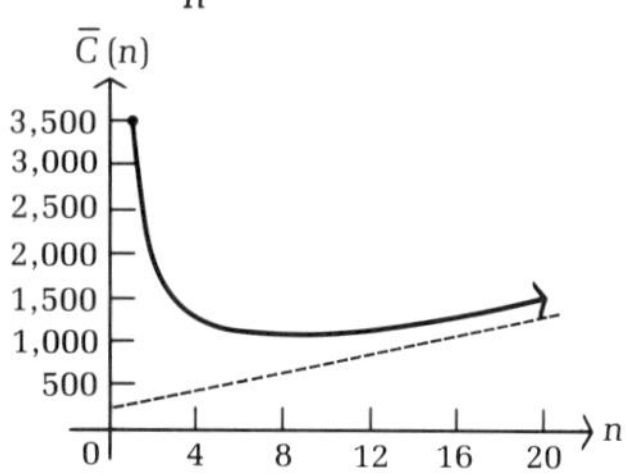

(C) 8 yr

65. (A)

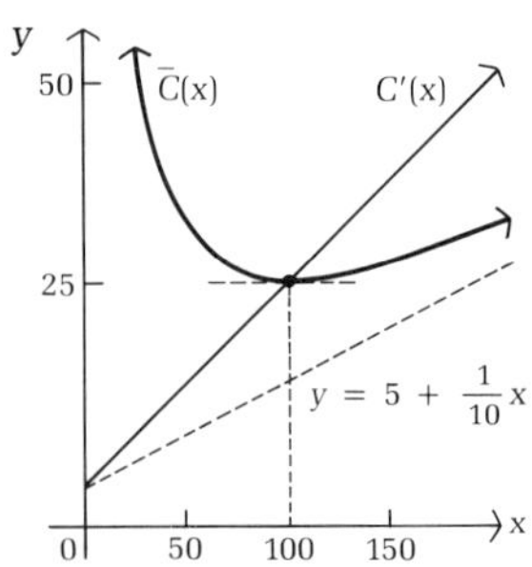

(B) 25 at $x = 100$

67.

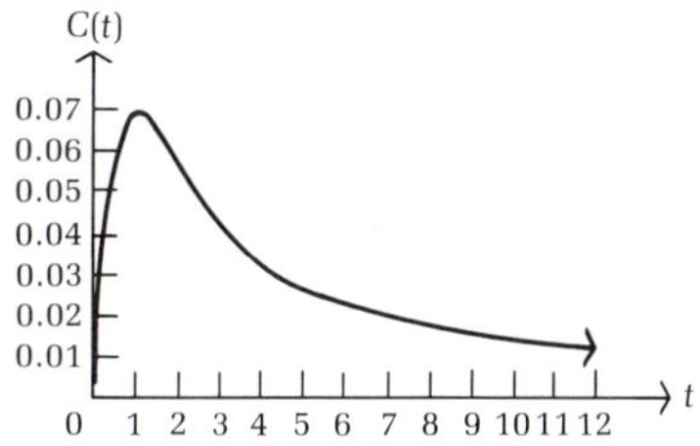

69.

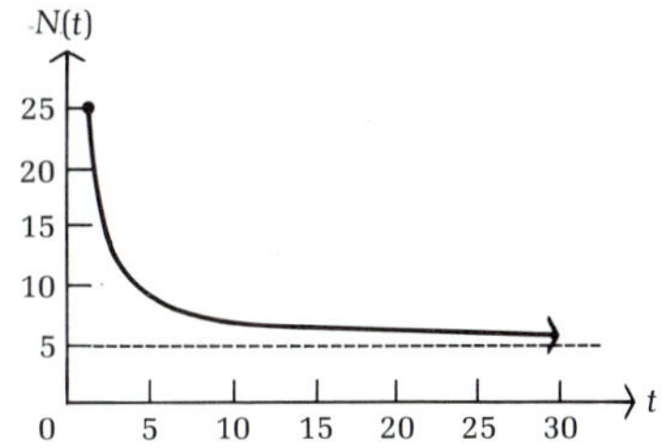

◆ EXERCISE 4-4

1. Min $f(x) = f(2) = 1$; no maximum **3.** Max $f(x) = f(4) = 26$; no minimum **5.** No absolute extrema exist. **7.** Max $f(x) = f(2) = 16$
9. Min $f(x) = f(2) = 14$
11. (A) Max $f(x) = f(5) = 14$; Min $f(x) = f(-1) = -22$ (B) Max $f(x) = f(1) = -2$; Min $f(x) = f(-1) = -22$
(C) Max $f(x) = f(5) = 14$; Min $f(x) = f(3) = -6$
13. (A) Max $f(x) = f(0) = 126$; Min $f(x) = f(2) = -26$ (B) Max $f(x) = f(7) = 49$; Min $f(x) = f(2) = -26$
(C) Max $f(x) = f(6) = 6$; Min $f(x) = f(3) = -15$
15. Exactly in half **17.** 15 and -15 **19.** A square of side 25 cm; maximum area = 625 cm^2
21. (A) Max $R(x) = R(3{,}000) = \$300{,}000$ (B) Maximum profit is \$75,000 when 2,100 sets are manufactured and sold for \$130 each.
(C) Maximum profit is \$64,687.50 when 2,025 sets are manufactured and sold for \$132.50 each.
23. \$35; \$6,125 **25.** 40 trees; 1,600 lb **27.** $(10 - 2\sqrt{7})/3 = 1.57$ in. squares
29. 20 ft by 40 ft (with the expensive side being one of the short sides) **31.** 10,000 books in 5 printings
33. (A) $x = 5.1$ mi (B) $x = 10$ mi **35.** 4 days; 20 bacteria/cm^3 **37.** 50 mice per order **39.** 1 month; 2 ft **41.** 4 yr from now

◆ EXERCISE 4-5

1. $\Delta x = 3$; $\Delta y = 45$; $\Delta y/\Delta x = 15$ **3.** 12 **5.** 12 **7.** $dy = (24x - 3x^2)\,dx$ **9.** $dy = \left(2x - \dfrac{x^2}{3}\right) dx$ **11.** $dy = -\dfrac{295}{x^{3/2}}\,dx$

13. (A) $12 + 3\,\Delta x$ (B) 12 **15.** $dy = 6(2x+1)^2\,dx$ **17.** $dy = \dfrac{9 - x^2}{(x^2+9)^2}\,dx$ **19.** $dy = 1.4$; $\Delta y = 1.44$ **21.** $dy = -3$; $\Delta y = -3\frac{1}{3}$

23. 120 in.^{3} **25.** $dy = \dfrac{6x - 2}{3(3x^2 - 2x + 1)^{2/3}}\,dx$ **27.** $dy = 3.9$; $\Delta y = 3.83$ **29.** 40 unit increase; 20 unit increase **31.** $-\$2.50$; \$1.25

33. -1.37/min; -0.58/min **35.** 1.26 mm^2 **37.** 3 wpm **39.** (A) 2,100 increase (B) 4,800 increase (C) 2,100 increase

◆ EXERCISE 4-6 CHAPTER REVIEW

1. (a, c_1), (c_3, c_6) **2.** (c_1, c_3), (c_6, b) **3.** (a, c_2), (c_4, c_5), (c_7, b) **4.** c_3 **5.** c_6 **6.** c_1, c_3, c_5 **7.** c_6 **8.** c_2, c_4, c_5, c_7

9.

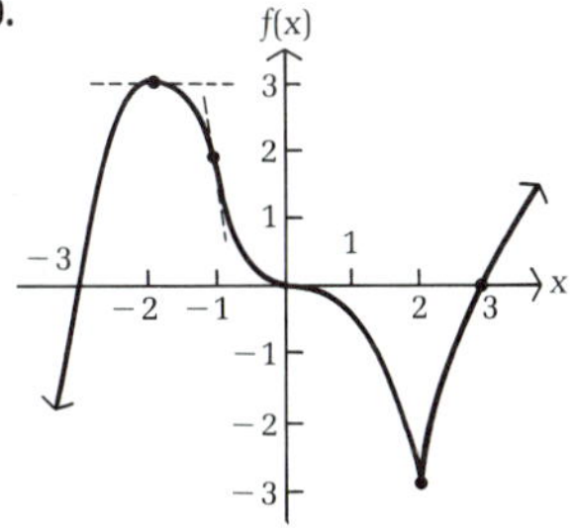

10. $f''(x) = 12x^2 + 30x$
11. $y'' = 8/x^3$
12. $dy = (3x^2 + 4)\,dx$
13. $dy = 18x(3x^2 - 7)^2\,dx$
14. $-3 < x < 4$; $(-3, 4)$

15. $-3 < x < 0$ or $x > 5$; $(-3, 0) \cup (5, \infty)$
16. (A) All real numbers (B) y intercept: 0; x intercepts: 0, 9 (C) No horizontal or vertical asymptotes
17. (A) 3, 9 (B) 3, 9 (C) Increasing on $(-\infty, 3)$ and $(9, \infty)$; decreasing on $(3, 9)$
(D) Local maximum at $x = 3$; local minimum at $x = 9$
18. (A) Concave downward on $(-\infty, 6)$; concave upward on $(6, \infty)$ (B) Inflection point at $x = 6$
19.

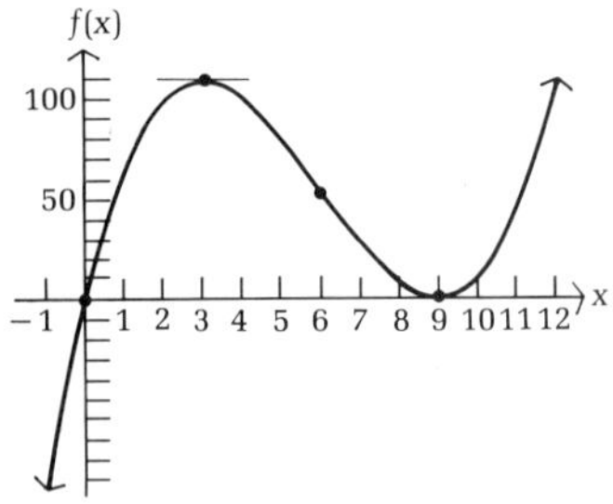

20. (A) All real numbers except -2 (B) y intercept: 0; x intercept: 0 (C) Horizontal asymptote: $y = 3$; vertical asymptote: $x = -2$
21. (A) None (B) -2 (C) Increasing on $(-\infty, -2)$ and $(-2, \infty)$ (D) None
22. (A) Concave upward on $(-\infty, -2)$; concave downward on $(-2, \infty)$ (B) No inflection points
23.

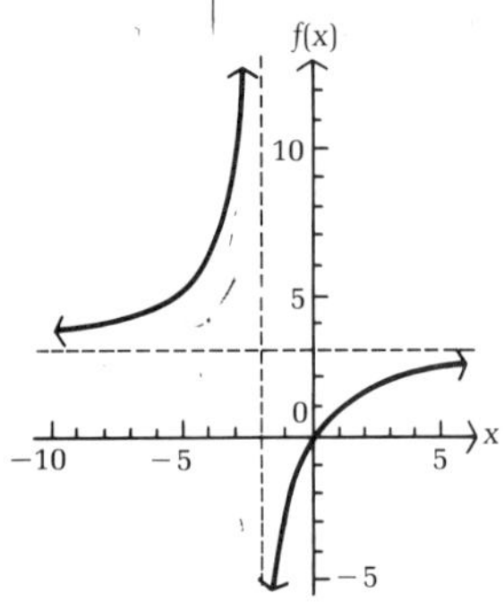

24. Local maximum at $x = -1$; local minimum at $x = 5$
25. Min $f(x) = f(2) = -4$; Max $f(x) = f(5) = 77$
26. Min $f(x) = f(2) = 8$
27. Horizontal asymptote: $y = 0$; no vertical asymptotes
28. No horizontal asymptotes; vertical asymptotes: $x = -3$ and $x = 3$
29. $dy = 7.3$; $\Delta y = 7.45$

30. Max $f'(x) = f'(2) = 12$

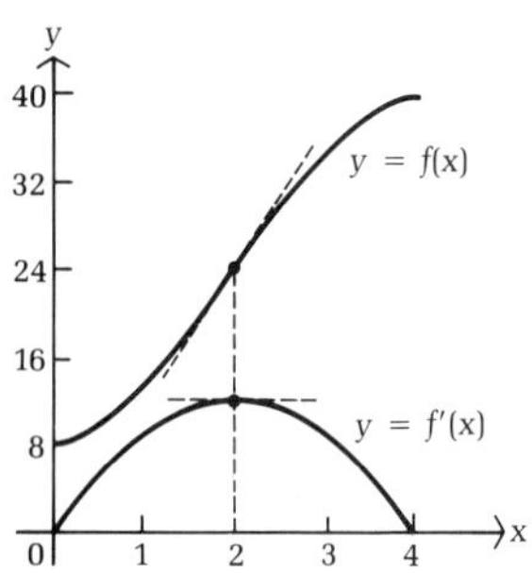

31. Each number is 20; minimum sum is 40
32. Domain: All real numbers
y intercept: -3
x intercepts: -3, 1
No vertical or horizontal asymptotes
Increasing on $(-2, \infty)$
Decreasing on $(-\infty, -2)$
Local minimum at $x = -2$
Concave upward on $(-\infty, -1)$ and $(1, \infty)$
Concave downward on $(-1, 1)$
Inflection points at $x = -1$ and $x = 1$

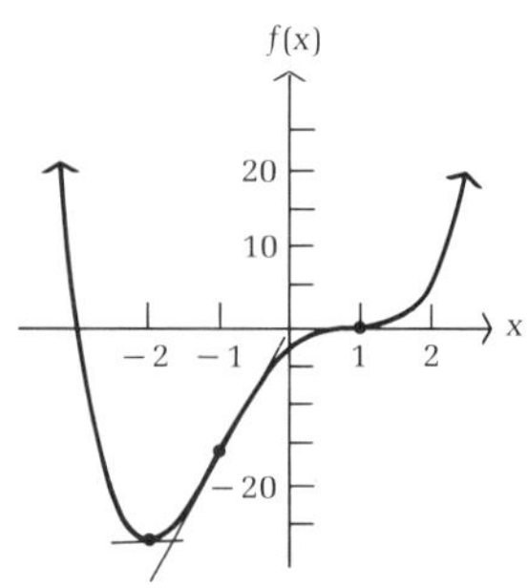

33. $dy = -0.0031$; $\Delta y = -0.0031$
34. Max $P(x) = P(3{,}000) = \$175{,}000$
35. (A) The expensive side is 50 ft; the other side is 100 ft. (B) The expensive side is 75 ft; the other side is 150 ft.

36. Min $\overline{C}(x) = \overline{C}(200) = 50$

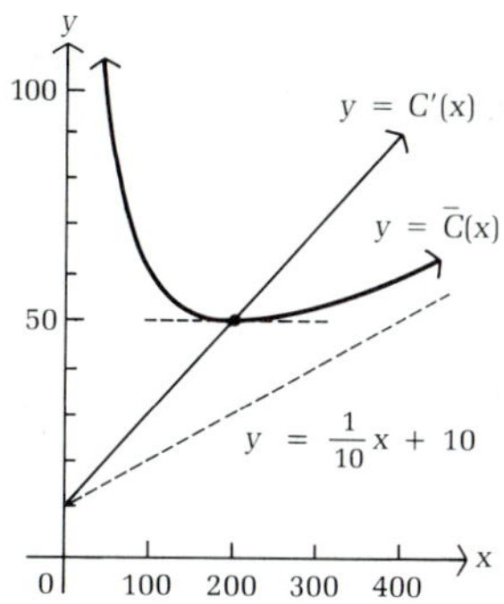

37. \$49; \$6,724
38. 12 orders/yr
39. \$110
40. 3 days
41. 2 yr from now

CHAPTER 5

◆ EXERCISE 5-1

1. \$1,221.40; \$1,648.72; \$2,225.54 **3.** 11.55 **5.** 10.99 **7.** 0.14

9.

n	$[1 + (1/n)]^n$
10	2.593 74
100	2.704 81
1,000	2.716 92
10,000	2.718 15
100,000	2.718 27
1,000,000	2.718 28
10,000,000	2.718 28
↓	↓
∞	e = 2.718 281 828 459 . . .

11.

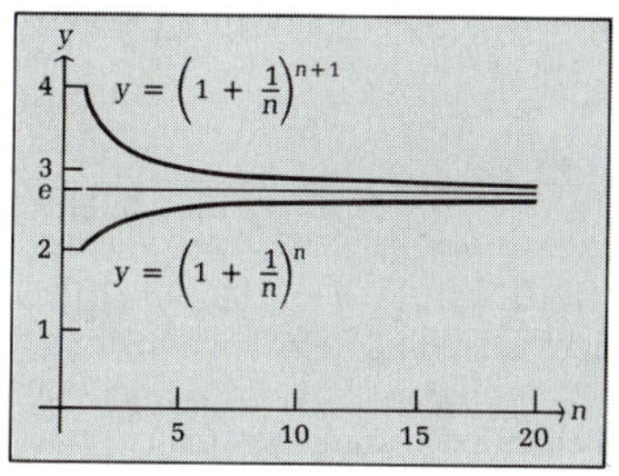

13. \$55,463.90 **15.** \$9,931.71 **17.** $r = \frac{1}{4} \ln 1.5 \approx 0.1014$ or 10.14%

19. (A)

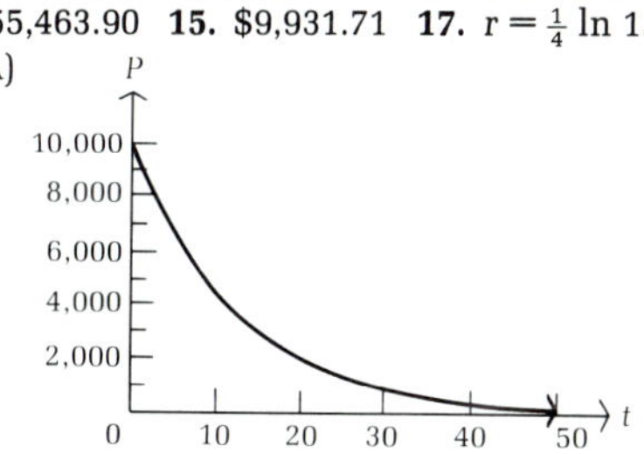

(B) $\lim_{t\to\infty} 10{,}000e^{-0.08t} = 0$

21. 2.77 yr **23.** 13.86% **25.** $A = Pe^{rt}$; $2P = Pe^{rt}$; $e^{rt} = 2$; $\ln e^{rt} = \ln 2$; $rt = \ln 2$; $t = (\ln 2)/r$ **27.** 34.66 yr **29.** 3.47%
31. $t = -(\ln 0.5)/0.000\,433\,2 \approx 1{,}600$ yr **33.** $r = (\ln 0.5)/30 \approx -0.0231$ **35.** Approx. 521 yr

◆ EXERCISE 5-2

1. $6e^x - \frac{7}{x}$ **3.** $2ex^{e-1} + 3e^x$ **5.** $\frac{5}{x}$ **7.** $\frac{2 \ln x}{x}$ **9.** $x^3 + 4x^3 \ln x = x^3(1 + 4 \ln x)$ **11.** $x^3e^x + 3x^2e^x = x^2e^x(x + 3)$

13. $\frac{(x^2 + 9)e^x - 2xe^x}{(x^2 + 9)^2} = \frac{e^x(x^2 - 2x + 9)}{(x^2 + 9)^2}$ **15.** $\frac{x^3 - 4x^3 \ln x}{x^8} = \frac{1 - 4 \ln x}{x^5}$ **17.** $3(x + 2)^2 \ln x + \frac{(x + 2)^3}{x} = (x + 2)^2\left(3 \ln x + \frac{x + 2}{x}\right)$

19. $(x + 1)^3e^x + 3(x + 1)^2e^x = (x + 1)^2e^x(x + 4)$ **21.** $\frac{2xe^x - (x^2 + 1)e^x}{(e^x)^2} = \frac{2x - x^2 - 1}{e^x}$ **23.** $(\ln x)^3 + 3(\ln x)^2 = (\ln x)^2(\ln x + 3)$

25. $-15e^x(4 - 5e^x)^2$ **27.** $\frac{1}{2x\sqrt{1 + \ln x}}$ **29.** xe^x **31.** $4x \ln x$ **33.** $y = ex$ **35.** $y = \frac{1}{e}x$

37. Max $f(x) = f(e^3) = e^3 \approx 20.086$ **39.** Min $f(x) = f(1) = e \approx 2.718$ **41.** Max $f(x) = f(e^{1/2}) = 2e^{-1/2} \approx 1.213$

43. Domain: All real numbers
y intercept: 0
x intercept: 0
Horizontal asymptote: $y = 1$
Decreasing on $(-\infty, \infty)$
Concave downward on $(-\infty, \infty)$

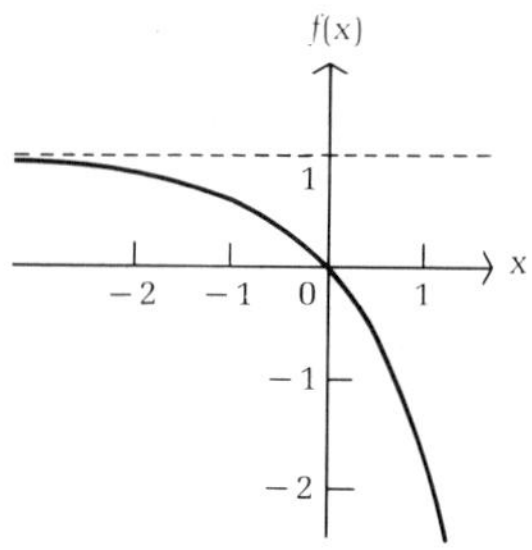

45. Domain: $(0, \infty)$
Vertical asymptote: $x = 0$
Increasing on $(1, \infty)$
Decreasing on $(0, 1)$
Local minimum at $x = 1$
Concave upward on $(0, \infty)$

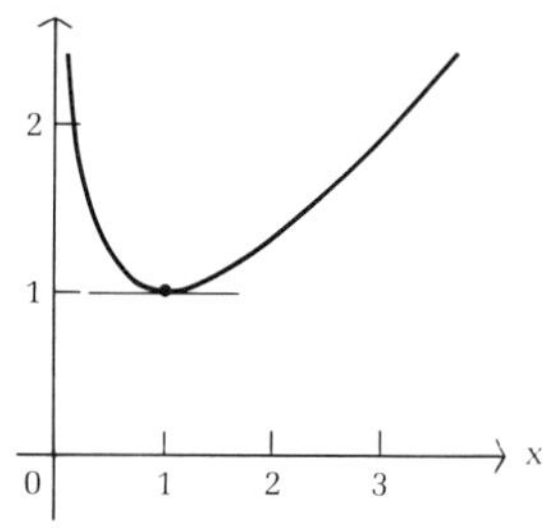

47. Domain: All real numbers
y intercept: 3
x intercept: 3
Horizontal asymptote: $y = 0$
Increasing on $(-\infty, 2)$
Decreasing on $(2, \infty)$
Local maximum at $x = 2$
Concave upward on $(-\infty, 1)$
Concave downward on $(1, \infty)$
Inflection point at $x = 1$

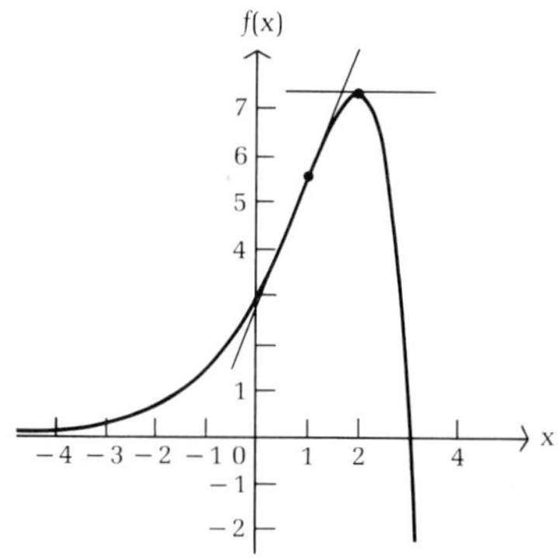

49. Domain: $(0, \infty)$
x intercept: 1
Increasing on $(e^{-1/2}, \infty)$
Decreasing on $(0, e^{-1/2})$
Local minimum at $x = e^{-1/2}$
Concave upward on $(e^{-3/2}, \infty)$
Concave downward on $(0, e^{-3/2})$
Inflection point at $x = e^{-3/2}$

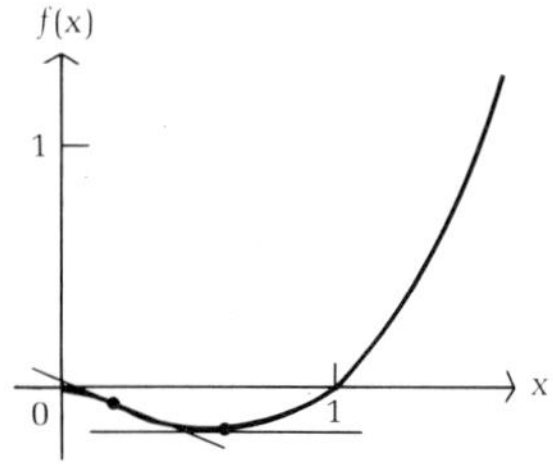

51. Critical values: $x = 0.36$, $x = 2.15$
Increasing on $(-\infty, 0.36)$ and $(2.15, \infty)$
Decreasing on $(0.36, 2.15)$
Local maximum at $x = 0.36$
Local minimum at $x = 2.15$

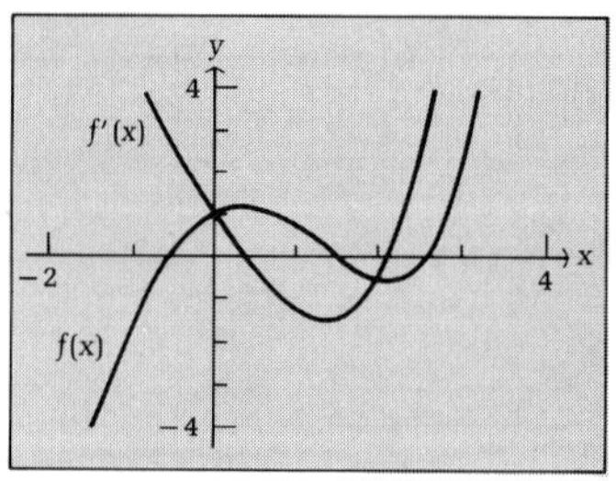

53. Critical value: $x = 2.21$
Increasing on $(0, 2.21)$
Decreasing on $(2.21, \infty)$
Local maximum at $x = 2.21$

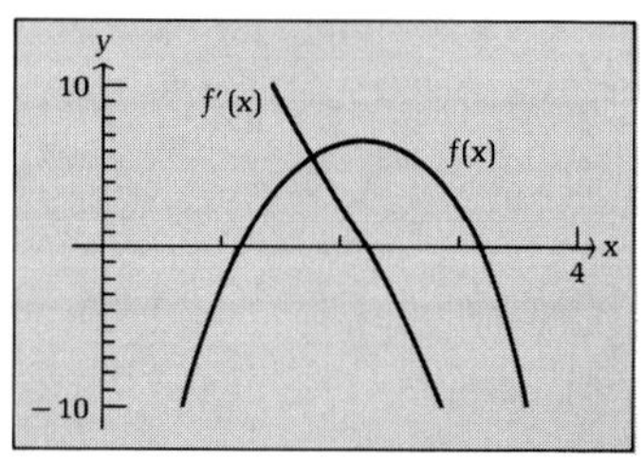

55. $p = \$2$ **57.** Min $\overline{C}(x) = \overline{C}(e^7) \approx \$99.91$ **59.** (A) At \$3.68 each, the maximum revenue will be \$3,680/wk (in the test city).
(B)

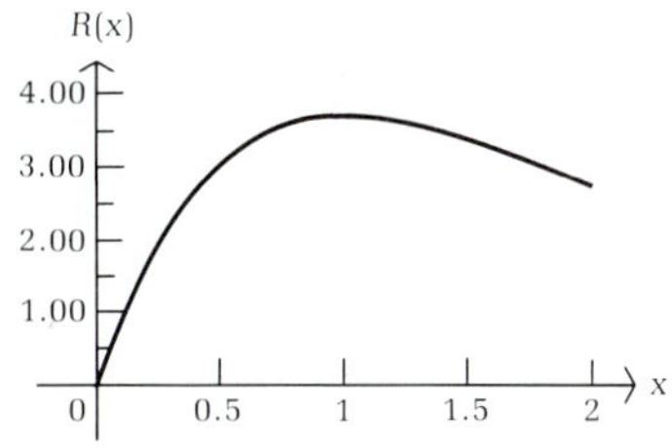

61. At the 40 lb weight level, blood pressure would increase at the rate of 0.44 mm of mercury/lb of weight gain.
At the 90 lb weight level, blood pressure would increase at the rate of 0.19 mm of mercury/lb of weight gain.

63. (A) After 1 hr, the concentration is decreasing at the rate of 1.60 mg/ml/hr; after 4 hr, the concentration is decreasing at the rate of 0.08 mg/ml/hr.
(B)

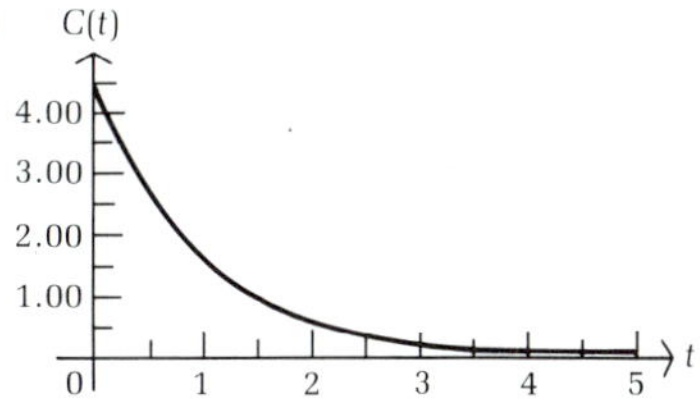

65. $dR/dS = k/S$

◆ EXERCISE 5-3

1. $y = u^3;\ u = 2x + 5$ **3.** $y = \ln u;\ u = 2x^2 + 7$ **5.** $y = e^u;\ u = x^2 - 2$ **7.** $y = (2 + e^x)^2;\ dy/dx = 2e^x(2 + e^x)$

9. $y = e^{2-x^4};\ dy/dx = -4x^3e^{2-x^4}$ **11.** $y = \ln(4x^5 - 7);\ \dfrac{dy}{dx} = \dfrac{20x^4}{4x^5 - 7}$ **13.** $\dfrac{1}{x - 3}$ **15.** $\dfrac{-2}{3 - 2t}$ **17.** $6e^{2x}$ **19.** $-8e^{-4t}$ **21.** $-3e^{-0.03x}$

23. $\dfrac{4}{x + 1}$ **25.** $4e^{2x} - 3e^x$ **27.** $(6x - 2)e^{3x^2-2x}$ **29.** $\dfrac{2t + 3}{t^2 + 3t}$ **31.** $\dfrac{x}{x^2 + 1}$

33. $\dfrac{4[\ln(t^2 + 1)]^3(2t)}{t^2 + 1} = \dfrac{8t[\ln(t^2 + 1)]^3}{t^2 + 1}$ **35.** $4(e^{2x} - 1)^3(2e^{2x}) = 8e^{2x}(e^{2x} - 1)^3$ **37.** $\dfrac{(x^2 + 1)(2e^{2x}) - e^{2x}(2x)}{(x^2 + 1)^2} = \dfrac{2e^{2x}(x^2 - x + 1)}{(x^2 + 1)^2}$

39. $(x^2 + 1)(-e^{-x}) + e^{-x}(2x) = e^{-x}(2x - x^2 - 1)$ **41.** $\dfrac{e^{-x}}{x} - e^{-x}\ln x$ **43.** $\dfrac{-2x}{(1 + x^2)[\ln(1 + x^2)]^2}$ **45.** $\dfrac{-2x}{3(1 - x^2)[\ln(1 - x^2)]^{2/3}}$

47. Domain: $(-\infty, \infty)$
y intercept: 0
x intercept: 0
Horizontal asymptote: $y = 1$
Increasing on $(-\infty, \infty)$
Concave downward on $(-\infty, \infty)$

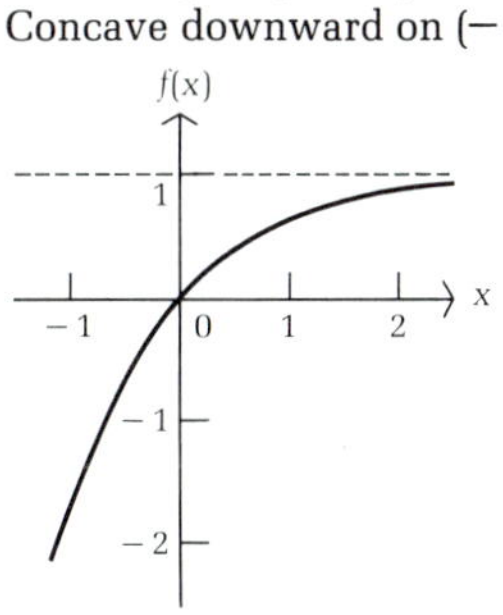

49. Domain: $(-\infty, 1)$
y intercept: 0
x intercept: 0
Vertical asymptote: $x = 1$
Decreasing on $(-\infty, 1)$
Concave downward on $(-\infty, 1)$

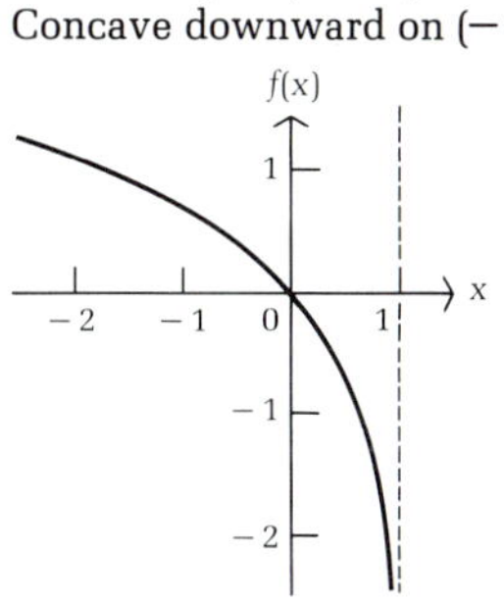

51. Domain: $(-\infty, \infty)$
y intercept: 1
Horizontal asymptote: $y = 0$
Increasing on $(-\infty, 0)$
Decreasing on $(0, \infty)$
Local maximum at $x = 0$
Concave upward on $(-\infty, -1)$ and $(1, \infty)$
Concave downward on $(-1, 1)$
Inflection points at $x = -1$ and $x = 1$

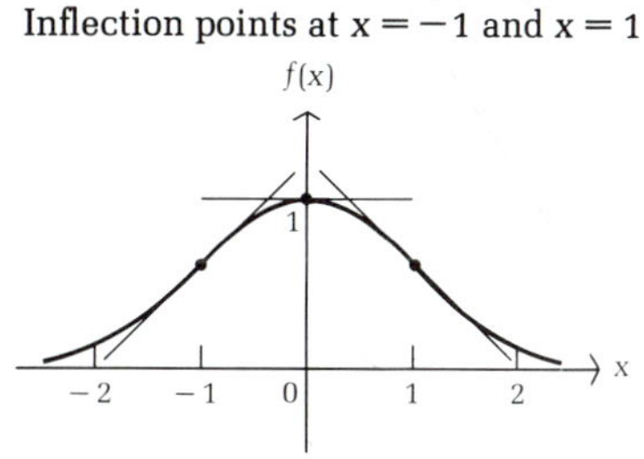

53. $y = 1 + [\ln(2 + e^x)]^2$; $\dfrac{dy}{dx} = \dfrac{2e^x \ln(2 + e^x)}{2 + e^x}$ **55.** $\dfrac{1}{\ln 2}\left(\dfrac{6x}{3x^2 - 1}\right)$ **57.** $(2x + 1)(10^{x^2+x})(\ln 10)$

61. A maximum revenue of $735.80 is realized at a production level of 20 units at $36.79 each.
63. A maximum profit of $224.61 is realized at a production level of 17 units at $42.74 each.
65. −$27,145/yr; −$18,196/yr; −$11,036/yr **67.** (A) 23 days; $26,685; about 50%
(B)

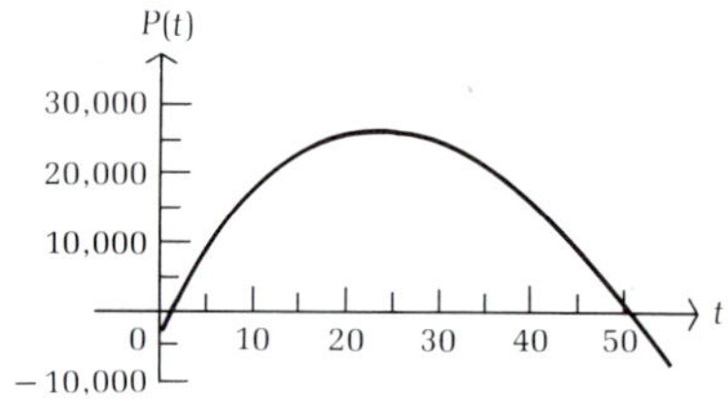

69. 2.27 mm of mercury/yr; 0.81 mm of mercury/yr; 0.41 mm of mercury/yr
71. $A'(t) = 2(\ln 2)5{,}000e^{2t \ln 2} = 10{,}000(\ln 2)2^{2t}$; $A'(1) = 27{,}726$ bacteria/hr (rate of change at the end of the first hour); $A'(5) = 7{,}097{,}827$ bacteria/hr (rate of change at the end of the fifth hour)
73.

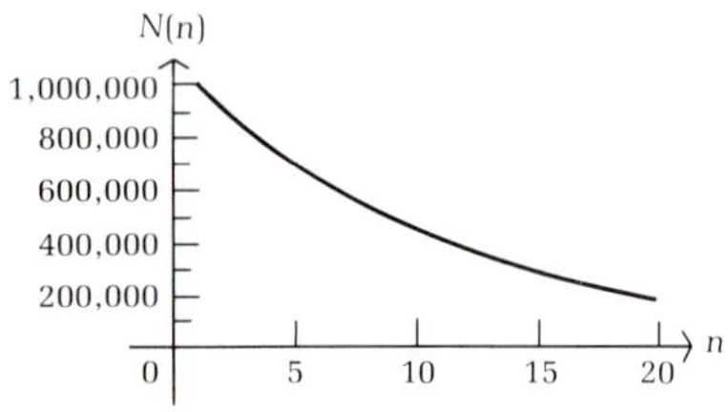

◆ EXERCISE 5-4

1. $y' = 6x$; 6 **3.** $y' = \dfrac{3x}{y}$; 3 **5.** $y' = \dfrac{1}{2y + 1}$; $\frac{1}{3}$ **7.** $y' = -\dfrac{y}{x}$; $-\frac{3}{2}$ **9.** $y' = -\dfrac{2y}{2x + 1}$; 4 **11.** $y' = \dfrac{6 - 2y}{x}$; −1 **13.** $y' = \dfrac{2x}{e^y - 2y}$; 2
15. $y' = \dfrac{3x^2y}{y + 1}$; $\frac{3}{2}$ **17.** $y' = \dfrac{6x^2y - y \ln y}{x + 2y}$; 2 **19.** $x' = \dfrac{2tx - 3t^2}{2x - t^2}$; 8 **21.** $y = -x + 5$ **23.** $y = \frac{2}{5}x - \frac{12}{5}$; $y = \frac{3}{5}x + \frac{12}{5}$
25. $y' = \dfrac{1}{3(1 + y)^2 + 1}$; $\frac{1}{13}$ **27.** $y' = \dfrac{3(x - 2y)^2}{6(x - 2y)^2 + 4y}$; $\frac{3}{10}$ **29.** $y' = \dfrac{3x^2(7 + y^2)^{1/2}}{y}$; 16 **31.** $y' = \dfrac{y}{2xy^2 - x}$; 1 **33.** $p' = \dfrac{1}{2p - 2}$
35. $p' = -\dfrac{\sqrt{10{,}000 - p^2}}{p}$ **37.** $\dfrac{dL}{dV} = \dfrac{-(L + m)}{V + n}$

◆ EXERCISE 5-5

1. 240 **3.** $\frac{9}{4}$ **5.** $\frac{1}{2}$ **7.** Decreasing at 9 units/sec **9.** Approx. −3.03 ft/sec **11.** $dA/dt \approx 126$ ft²/sec **13.** 3,768 cc/min
15. 6 lb/in.²/hr **17.** $-\frac{9}{4}$ ft/sec **19.** $\frac{20}{3}$ ft/sec **21.** (A) $dC/dt = \$15{,}000$/wk (B) $dR/dt = -\$50{,}000$/wk (C) $dP/dt = -\$65{,}000$/wk
23. $ds/dt = \$2{,}207$/wk **25.** (A) $dx/dt = -12.73$ units/month (B) $dp/dt = \$1.53$/month **27.** Approx. 100 ft³/min

◆ EXERCISE 5-6

1. $\frac{8}{3}$ **3.** $\frac{1}{2}$ **5.** 1 **7.** 4 **9.** 0 **11.** ∞ **13.** ∞ **15.** 8 **17.** 0 **19.** ∞ **21.** ∞ **23.** $\frac{1}{3}$ **25.** −2 **27.** −∞ **29.** 0 **31.** 0 **33.** 0 **35.** $\frac{1}{4}$ **37.** $\frac{1}{3}$

39. 0 **41.** 0 **43.** ∞ **45.** $\lim\limits_{x\to\infty} \frac{\sqrt{1+x^2}}{x} = 1$

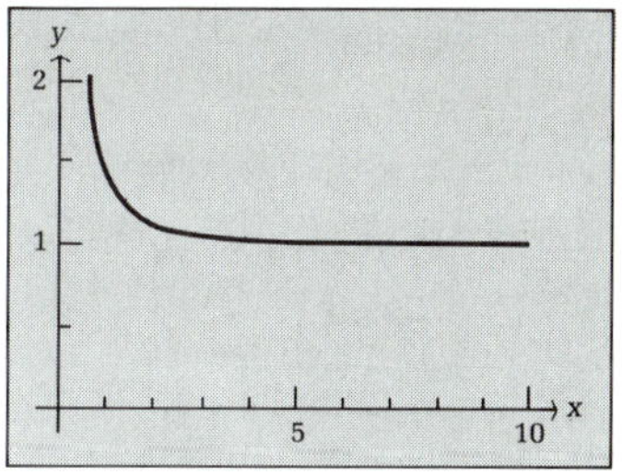

47. $\lim\limits_{x\to-\infty} \frac{\sqrt[3]{x^3+1}}{x} = 1$

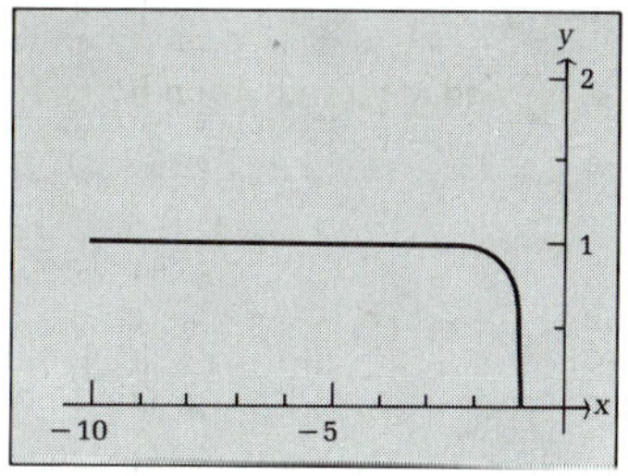

◆ EXERCISE 5-7

1. (A) $x = f(p) = 6{,}000 - 200p$ (B) $E(p) = -p/(30 - p)$ (C) $E(10) = -0.5$; 5% decrease in demand (D) $E(25) = -5$; 50% decrease in demand (E) $E(15) = -1$; 10% decrease in demand
3. (A) $x = f(p) = 3{,}000 - 50p$ (B) $R(p) = 3{,}000p - 50p^2$ (C) $E(p) = -p/(60 - p)$ (D) Elastic on (30, 60); inelastic on (0, 30) (E) Increasing on (0, 30); decreasing on (30, 60) (F) Decrease (G) Increase
5. $E(p) = -2p^2/(1{,}200 - p^2)$; (A) $E(10) = -\frac{2}{11}$ (inelastic) (B) $E(20) = -1$ (unit elasticity) (C) $E(30) = -6$ (elastic)
7. $E(p) = -(20p + 2p^2)/(9{,}500 - 20p - p^2)$; (A) $E(30) = -\frac{3}{10}$ (inelastic) (B) $E(50) = -1$ (unit elasticity) (C) $E(70) = -\frac{7}{2}$ (elastic)
9. Elastic on (10, 30); inelastic on (0, 10) **11.** Elastic on (48, 72); inelastic on (0, 48) **13.** Elastic on $(25, 25\sqrt{2})$; inelastic on (0, 25)

15.

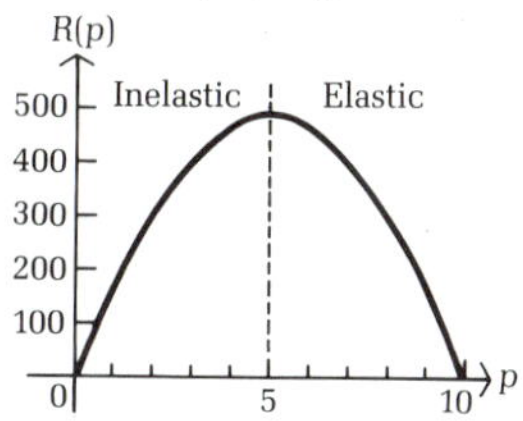

17.

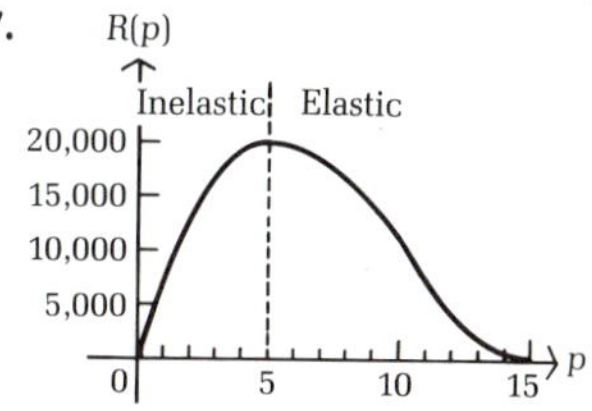

19.

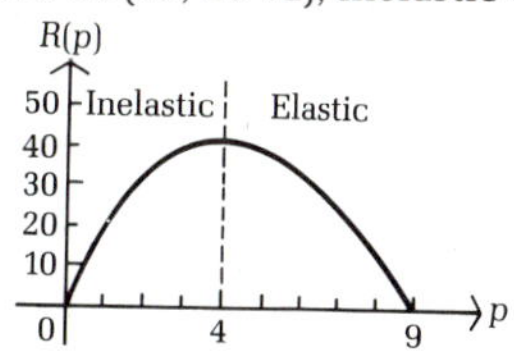

21. $-\frac{3}{2}$ **23.** $-\frac{1}{2}$ **25.** $-k$ **27.** Elastic on (10, 30); inelastic on (0, 10)

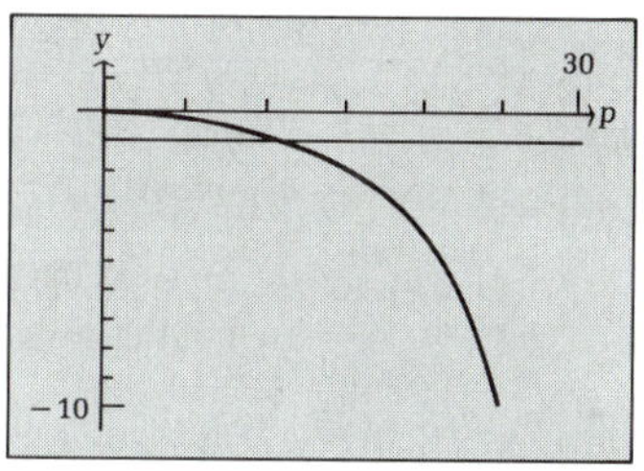

29. Elastic on (48, 72); inelastic on (0, 48)

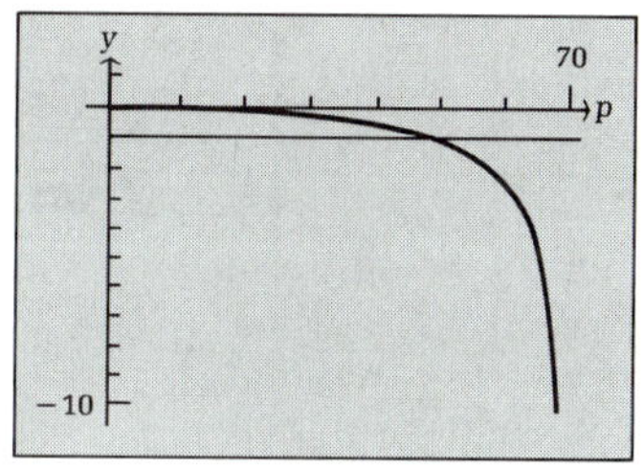

31. Elastic on (36.2, 49); inelastic on (0, 36.2)

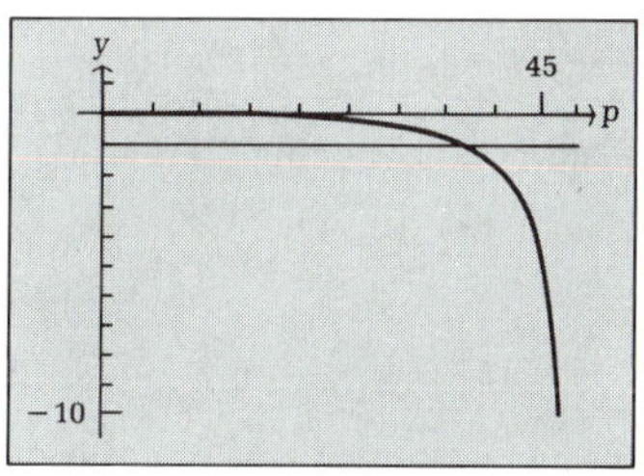

33. (A) $E(2.2) = -0.8$ (B) Demand decreases by approx. 8% to 25,760 (C) Revenue increases
35. (A) $E(90) = -\frac{3}{7}$ (B) Demand increases by approx. 4% to 2,184 (C) Revenue decreases

1. \$3,136.62; \$4,919.21; \$12,099.29 **2.** $\frac{2}{x} + 3e^x$ **3.** $2e^{2x-3}$ **4.** $\frac{2}{2x+7}$ **5.** (A) $y = \ln(3 + e^x)$ (B) $\frac{dy}{dx} = \frac{e^x}{3 + e^x}$ **6.** $y' = \frac{9x^2}{4y}; \frac{9}{8}$

7. $dy/dt = 216$ **8.** (A) $x = f(p) = 1{,}000 - 25p$ (B) $E(p) = -p/(40 - p)$
(C) $E(15) = -0.6$; 6% decrease in demand (D) $R(p) = 1{,}000p - 25p^2$ (E) Increase

9. Domain: All real numbers
y intercept: 100
Horizontal asymptote: $y = 0$
Decreasing on $(-\infty, \infty)$
Concave upward on $(-\infty, \infty)$

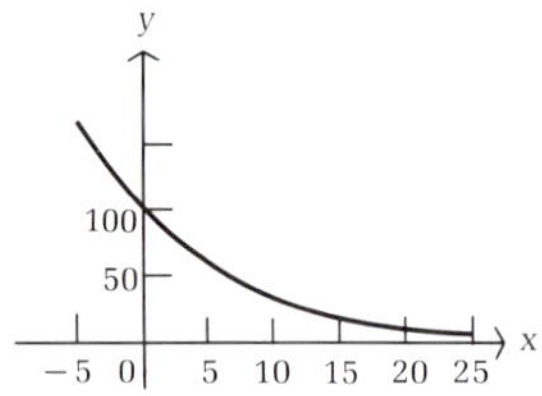

10. $\frac{7[(\ln z)^6 + 1]}{z}$ **11.** $x^5(1 + 6 \ln x)$

12. $\frac{e^x(x - 6)}{x^7}$ **13.** $\frac{6x^2 - 3}{2x^3 - 3x}$

14. $(3x^2 - 2x)e^{x^3 - x^2}$ **15.** $\frac{1 - 2x \ln 5x}{xe^{2x}}$

16. $y = -x + 2$; $y = -ex + 1$

17. $y' = \frac{3y - 2x}{8y - 3x}; \frac{8}{19}$ **18.** $x' = \frac{4tx}{3x^2 - 2t^2}; -4$ **19.** $y' = \frac{1}{e^y + 2y}; 1$ **20.** $y' = \frac{2xy}{1 + 2y^2}; \frac{2}{3}$ **21.** $dy/dt = -2$ units/sec **22.** 0.27 ft/sec

23. $dR/dt = 1/\pi \approx 0.318$ in./min **24.** Elastic on (5, 15); inelastic on (0, 5) **25.**

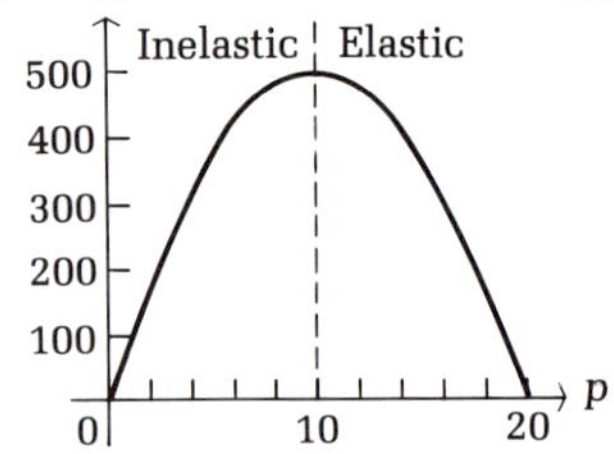

26. 3 **27.** $-\frac{1}{5}$

28. $-\infty$ **29.** 0 **30.** ∞ **31.** 1 **32.** 0 **33.** 0 **34.** 1 **35.** 2

36. Max $f(x) = f(e^{4.5}) = 2e^{4.5} \approx 180.03$ **37.** Max $f(x) = f(0.5) = 5e^{-1} \approx 1.84$

38. Domain: All real numbers
y intercept: 0
x intercept: 0
Horizontal asymptote: $y = 5$
Increasing on $(-\infty, \infty)$
Concave downward on $(-\infty, \infty)$

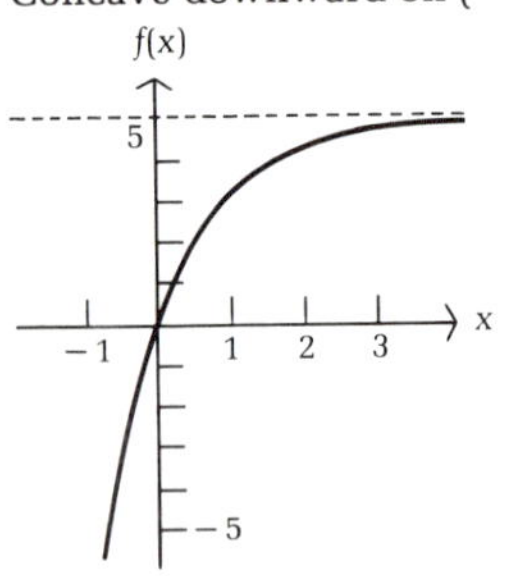

39. Domain: $(0, \infty)$
x intercept: 1
Increasing on $(e^{-1/3}, \infty)$
Decreasing on $(0, e^{-1/3})$
Local minimum at $x = e^{-1/3}$
Concave upward on $(e^{-5/6}, \infty)$
Concave downward on $(0, e^{-5/6})$
Inflection point at $x = e^{-5/6}$

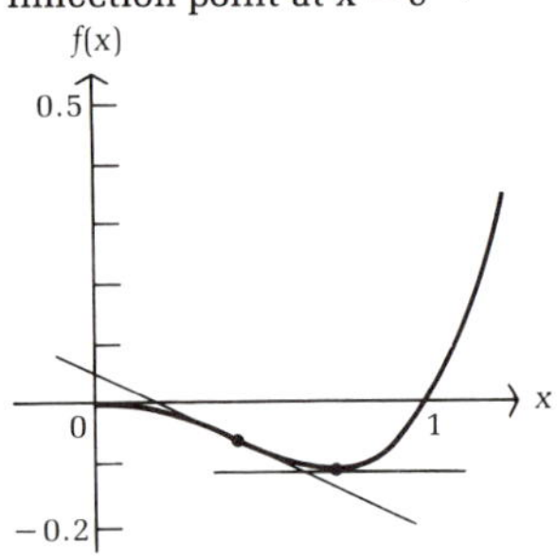

40. (A) $y = [\ln(4 - e^x)]^3$ (B) $\frac{dy}{dx} = \frac{-3e^x[\ln(4 - e^x)]^2}{4 - e^x}$

41. $2x(5^{x^2-1})(\ln 5)$

42. $\left(\frac{1}{\ln 5}\right)\frac{2x - 1}{x^2 - x}$

43. $\frac{2x + 1}{2(x^2 + x)\sqrt{\ln(x^2 + x)}}$

44. $y' = \frac{2x - e^{xy}y}{xe^{xy} - 1}; 0$

45. (A) 15 yr (B) 13.9 yr

46. $A'(t) = 10e^{0.1t}$; $A'(1) = \$11.05$/yr;
$A'(10) = \$27.18$/yr

47. $R'(x) = (1{,}000 - 20x)e^{-0.02x}$

48. A maximum revenue of \$18,394 is realized at a production level of 50 units at \$367.88 each.

49.

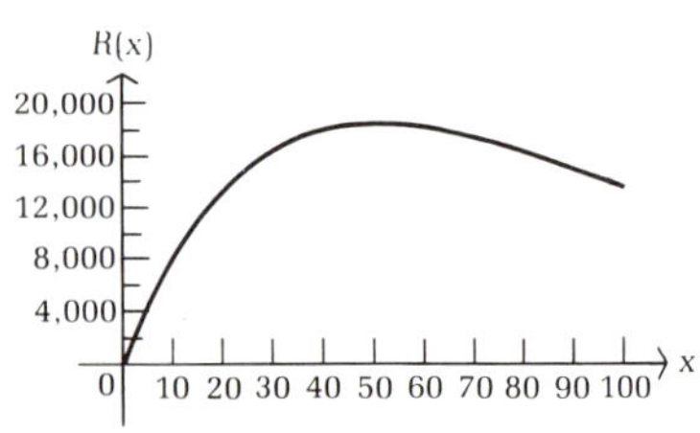

50. Min $\overline{C}(x) = \overline{C}(e^5) \approx \$49.66$ **51.** $p' = \frac{-(5{,}000 - 2p^3)^{1/2}}{3p^2}$ **52.** $dR/dt = \$110$/day

53. (A) -1.1 (B) Demand decreases by approx. 11% to 3,560 (C) Decrease

54. -1.111 mg/ml/hr; -0.335 mg/ml/hr

55. $dR/dt = -3/(2\pi)$; approx. 0.477 mm/day

56. (A) Increasing at the rate of 2.68 units/day at the end of 1 day of training; increasing at the rate of 0.54 unit/day after 5 days of training

(B)

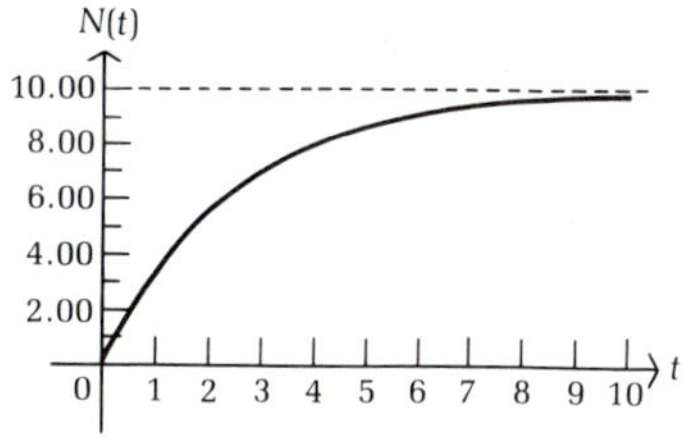

57. $dT/dt = -1/27 \approx -0.037$ min/operation hour

CHAPTER 6

◆ EXERCISE 6-1

1. $7x + C$ **3.** $(x^7/7) + C$ **5.** $2t^4 + C$ **7.** $u^2 + u + C$ **9.** $x^3 + x^2 - 5x + C$ **11.** $(s^5/5) - \frac{4}{3}s^6 + C$ **13.** $3e^t + C$ **15.** $2 \ln|z| + C$
17. $y = 40x^5 + C$ **19.** $P = 24x - 3x^2 + C$ **21.** $y = \frac{1}{3}u^6 - u^3 - u + C$ **23.** $y = e^x + 3x + C$ **25.** $x = 5 \ln|t| + t + C$ **27.** $4x^{3/2} + C$
29. $-4x^{-2} + C$ **31.** $2\sqrt{u} + C$ **33.** $-(x^{-2}/8) + C$ **35.** $-(u^{-4}/8) + C$ **37.** $x^3 + 2x^{-1} + C$ **39.** $2x^5 + 2x^{-4} - 2x + C$
41. $2x^{3/2} + 4x^{1/2} + C$ **43.** $\frac{3}{5}x^{5/3} + 2x^{-2} + C$ **45.** $(e^x/4) - (3x^2/8) + C$ **47.** $-z^{-2} - z^{-1} + \ln|z| + C$ **49.** $y = x^2 - 3x + 5$
51. $C(x) = 2x^3 - 2x^2 + 3{,}000$ **53.** $x = 40\sqrt{t}$ **55.** $y = -2x^{-1} + 3 \ln|x| - x + 3$ **57.** $x = 4e^t - 2t - 3$ **59.** $y = 2x^2 - 3x + 1$
61. $x^2 + x^{-1} + C$ **63.** $\frac{1}{2}x^2 + x^{-2} + C$ **65.** $e^x - 2 \ln|x| + C$ **67.** $M = t + t^{-1} + \frac{3}{4}$ **69.** $y = 3x^{5/3} + 3x^{2/3} - 6$ **71.** $p(x) = 10x^{-1} + 10$
73. $P(x) = 50x - 0.02x^2$; $P(100) = \$4{,}800$ **75.** $R(x) = 100x - (x^2/10)$; $p = 100 - (x/10)$; $p = \$30$
77. $S(t) = 2{,}000 - 15t^{5/3}$; $80^{3/5} \approx 14$ months **79.** $L(x) = 4{,}800x^{1/2}$; $L(25) = 24{,}000$ labor-hours **81.** $W(h) = 0.0005h^3$; $W(70) = 171.5$ lb
83. 19,400

◆ EXERCISE 6-2

1. $\frac{1}{6}(x^2 - 4)^6 + C$ **3.** $e^{4x} + C$ **5.** $\ln|2t + 3| + C$ **7.** $\frac{1}{24}(3x - 2)^8 + C$ **9.** $\frac{1}{16}(x^2 + 3)^8 + C$ **11.** $-20e^{-0.5t} + C$ **13.** $\frac{1}{10} \ln|10x + 7| + C$
15. $\frac{1}{4}e^{2x^2} + C$ **17.** $\frac{1}{3} \ln|x^3 + 4| + C$ **19.** $-\frac{1}{18}(3t^2 + 1)^{-3} + C$ **21.** $\frac{1}{3}(4 - x^3)^{-1} + C$ **23.** $\frac{2}{5}(x + 4)^{5/2} - \frac{8}{3}(x + 4)^{3/2} + C$
25. $\frac{2}{3}(x - 3)^{3/2} + 6(x - 3)^{1/2} + C$ **27.** $\frac{1}{11}(x - 4)^{11} + \frac{2}{5}(x - 4)^{10} + C$ **29.** $\frac{1}{8}(1 + e^{2x})^4 + C$ **31.** $\frac{1}{2} \ln|4 + 2x + x^2| + C$ **33.** $e^{x^2+x+1} + C$
35. $\frac{1}{4}(e^x - 2x)^4 + C$ **37.** $-\frac{1}{12}(x^4 + 2x^2 + 1)^{-3} + C$ **39.** $\frac{1}{9}(3x^2 + 7)^{3/2} + C$ **41.** $\frac{1}{8}x^8 + \frac{4}{5}x^5 + 2x^2 + C$ **43.** $\frac{1}{9}(x^3 + 2)^3 + C$
45. $\frac{1}{4}(2x^4 + 3)^{1/2} + C$ **47.** $\frac{1}{4}(\ln x)^4 + C$ **49.** $e^{-1/x} + C$ **51.** $x = \frac{1}{3}(t^3 + 5)^7 + C$ **53.** $y = 3(t^2 - 4)^{1/2} + C$ **55.** $p = -(e^x - e^{-x})^{-1} + C$
59. $p(x) = 2{,}000/(3x + 50)$; 250 bottles **61.** $C(x) = 12x + 500 \ln(x + 1) + 2{,}000$; $\overline{C}(1{,}000) = \$17.45$
63. $S(t) = 10t + 100e^{-0.1t} - 100$, $0 \le t \le 24$; $S(12) \approx \$50$ million **65.** $Q(t) = 100 \ln(t + 1) + 5t$, $0 \le t \le 20$; $Q(9) \approx 275$ thousand barrels
67. $W(t) = 2e^{0.1t}$; $W(8) \approx 4.45$ g **69.** $N(t) = 5{,}000 - 1{,}000 \ln(1 + t^2)$; $N(10) \approx 385$ bacteria/ml
71. $N(t) = 100 - 60e^{-0.1t}$, $0 \le t \le 15$; $N(15) \approx 87$ words/min **73.** $E(t) = 12{,}000 - 10{,}000(t + 1)^{-1/2}$; $E(15) = 9{,}500$ students

◆ EXERCISE 6-3

1. $A = 1{,}000e^{0.08t}$ **3.** $A = 8{,}000e^{0.06t}$ **5.** $p(x) = 100e^{-0.05x}$ **7.** $N = L(1 - e^{-0.051t})$ **9.** $I = I_0e^{-0.00942x}$; $x \approx 74$ ft
11. $Q = 3e^{-0.04t}$; $Q(10) = 2.01$ ml **13.** $-0.023\ 117$ **15.** Approx. 24,200 yr **17.** 104 times; 67 times
19. (A) 7 people; 353 people (B) 400

21.

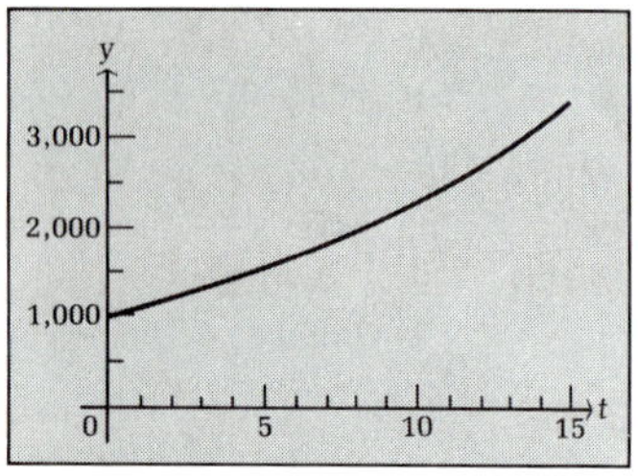

23.

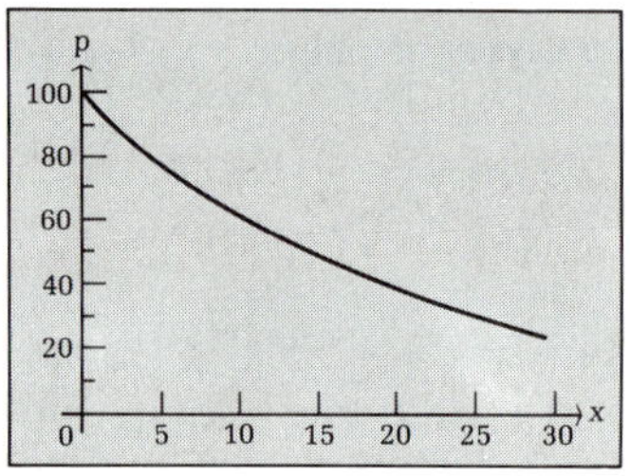

25.

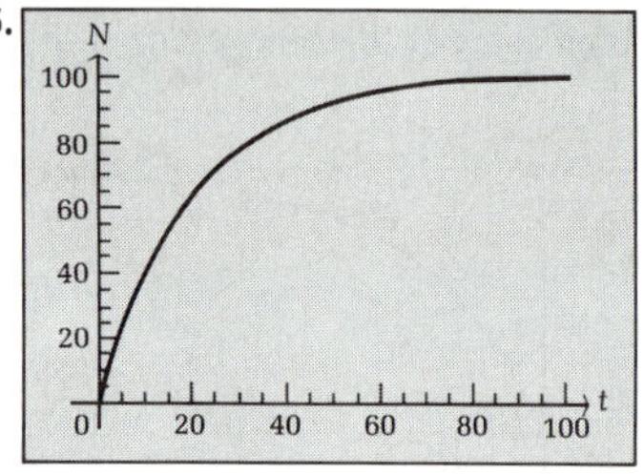

27.

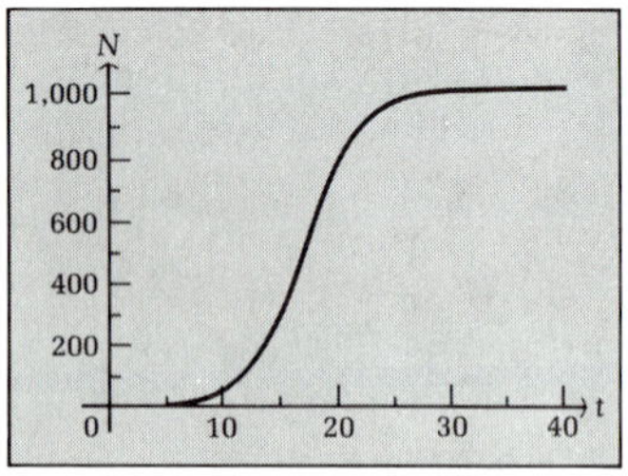

◆ EXERCISE 6-4

1. $(2 \cdot 2 - 1) + (2 \cdot 3 - 1) + (2 \cdot 4 - 1) = 15$ **3.** $2^1 + 2^2 + 2^3 + 2^4 = 30$ **5.** $f(x_1)\Delta x + f(x_2)\Delta x + f(x_3)\Delta x$
7. $f(2 + 1 \cdot \Delta x)\Delta x + f(2 + 2 \cdot \Delta x)\Delta x + f(2 + 3 \cdot \Delta x)\Delta x$
9. (A) $\Delta x = 2$ (B) $2 + 2k$ (C) $x_1 = 4, x_2 = 6, x_3 = 8, x_4 = 10$ (D) $f(x_1) = 6, f(x_2) = 8, f(x_3) = 10, f(x_4) = 12$ (E) 72
11. (A) $\Delta x = 1.5$ (B) $1.5k$ (C) $x_1 = 1.5, x_2 = 3, x_3 = 4.5, x_4 = 6$ (D) $f(x_1) = 3.5, f(x_2) = 5, f(x_3) = 6.5, f(x_4) = 8$ (E) 34.5
13. 990 **15.** 9,900 **17.** 10,890 **19.** 8,610 **21.** 330 **23.** 8,290 **25.** $\frac{1}{2} + \frac{1}{2n}; \frac{1}{2}$ **27.** $4 + \frac{2}{n}; 4$
29. (A) $\frac{4}{n}$ (B) $3 + k\left(\frac{4}{n}\right)$ (C) $\left[3 + k\left(\frac{4}{n}\right)\right] + 2 = 5 + k\left(\frac{4}{n}\right)$ (D) $\left[5 + k\left(\frac{4}{n}\right)\right]\frac{4}{n} = \frac{20}{n} + \left(\frac{16}{n^2}\right)k$
33. 28 **35.** 16 **37.** 12 **39.** $9 + \frac{27}{2n} + \frac{9}{2n^2}; 9$ **41.** $\frac{2}{3} + \frac{4}{3n^2}; \frac{2}{3}$
43. (A) $\frac{4}{n}$ (B) $2 + k\left(\frac{4}{n}\right)$ (C) $\left[2 + k\left(\frac{4}{n}\right)\right]^2 + 1 = 5 + \left(\frac{16}{n}\right)k + \left(\frac{16}{n^2}\right)k^2$ (D) $\frac{20}{n} + \left(\frac{64}{n^2}\right)k + \left(\frac{54}{n^3}\right)k^2$
47. $\frac{220}{3}$ **49.** $\frac{50}{3}$ **51.** $\frac{7}{3}$

◆ EXERCISE 6-5

1. 8 **3.** 10 **5.** -18 **7.** 24 **9.** 14 **11.** 23.4 **13.** 55.3 **15.** $\int_2^5 (1 - x^2)\,dx$ **17.** $\int_2^{12} (3x^2 - 2x + 3)\,dx$
19. $\lim_{n\to\infty} \sum_{k=1}^{n} (2c_k - 3)\frac{3 - 1}{n}; c_k = 1 + k\frac{2}{n}$ **21.** $\lim_{n\to\infty} \sum_{k=1}^{n} (3c_k^2 - 4)\frac{4 - 0}{n}; c_k = 0 + k\frac{4}{n}$ **23.** 2 **25.** 48 **27.** $-\frac{7}{3}$
29. 1 **31.** 12 **33.** $-\frac{1}{3}$ **35.** 4 **39.** $\frac{1}{3}\int_0^3 (1 + 12x - x^2)\,dx$ **41.** 3,120,000 ft^2 **43.** $2\int_0^{0.5} (1 + \frac{1}{2}t + \frac{1}{4}t^2)\,dt$
45. 1.1 liters **47.** $\frac{1}{6}\int_0^6 (-\frac{1}{4}t^2 + t + 4)\,dt$

◆ EXERCISE 6-6

1. 5 **3.** 5 **5.** 2 **7.** 48 **9.** $-\frac{7}{3}$ **11.** 2 **13.** $\frac{1}{2}(e^2 - 1)$ **15.** $2 \ln 3.5$ **17.** 9 **19.** $e^2 - e^{-1}$ **21.** $-\ln 0.5$ **23.** -2 **25.** 14 **27.** $5^6 = 15{,}625$

29. $\ln 4$ **31.** $20(e^{0.25} - e^{-0.5}) \approx 13.55$ **33.** $\frac{56}{3} \approx 18.667$ **35.** $\frac{28}{3} \approx 9.333$ **37.** $\frac{1}{6}[(e^2 - 2)^3 - 1]$ **39.** $-3 - \ln 2$ **41.** 250 **43.** 2
45. $\frac{45}{28} \approx 1.61$ **47.** $2(1 - e^{-2})$ **49.** $\int_2^5 (1 - x^2)\,dx = -36$ **51.** $\int_2^{12} (3x^2 - 2x + 3)\,dx = 1{,}610$ **53.** $\frac{1}{6}(15^{3/2} - 5^{3/2})$ **55.** $\frac{1}{2}(\ln 2 - \ln 3)$
57. 0 **59.** $[f(b) - f(a)]/(b - a)$ **61.** (A) $I = -200t + 600$ (B) $\frac{1}{3}\int_0^3 (-200t + 600)\,dt = 300$
63. $\int_0^5 500(t - 12)\,dt = -\$23{,}750$; $\int_5^{10} 500(t - 12)\,dt = -\$11{,}250$ **65.** (A) \$420 (B) \$135,000
67. $20 + 100e^{-1.2} \approx \$50$ million; $120 + 100e^{-2.4} - 100e^{-1.2} \approx \$99$ million **69.** \$149.18; \$122.96
71. $(40 + 100e^{-0.7})/7 \approx 12.8$ or 12,800 hamburgers; $(140 + 100e^{-1.4} - 100e^{-0.7})/7 \approx 16.4$ or 16,400 hamburgers
73. $100 \ln 11 + 50 \approx 290$ thousand barrels; $100 \ln 21 - 100 \ln 11 + 50 \approx 115$ thousand barrels **75.** 194 billion ft^3 **77.** 10°C
79. An increase of $60 - 60e^{-0.5} \approx 24$ words per minute; an increase of $60e^{-0.5} - 60e^{-1} \approx 14$ words per minute; an increase of $60e^{-1} - 60e^{-1.5} \approx 9$ words per minute
81. $0.6 \ln 2 + 0.1 \approx 0.516$; $(4.2 \ln 625 + 2.4 - 4.2 \ln 49)/24 \approx 0.546$ **83.**

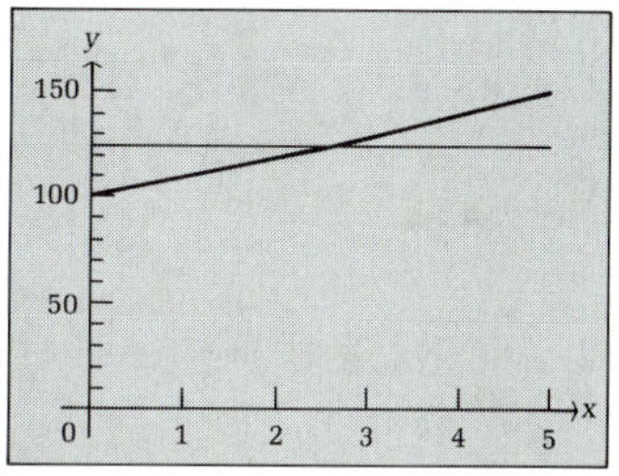

85.

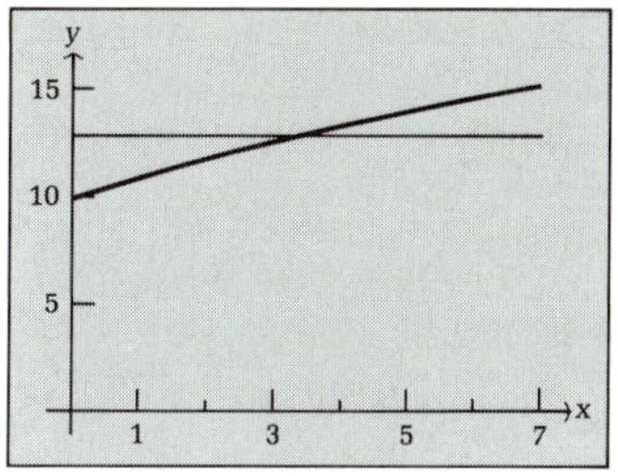

87.

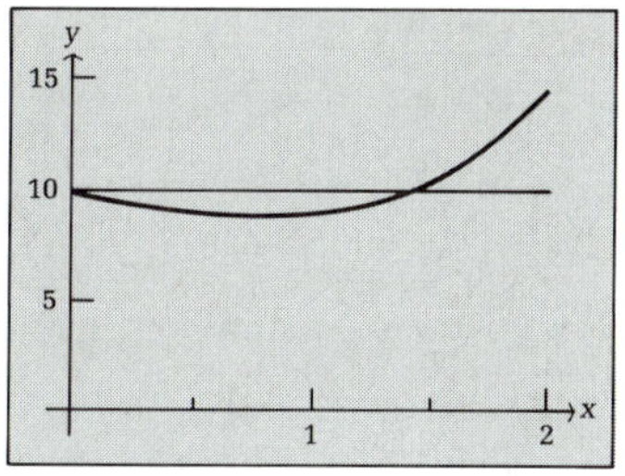

89.

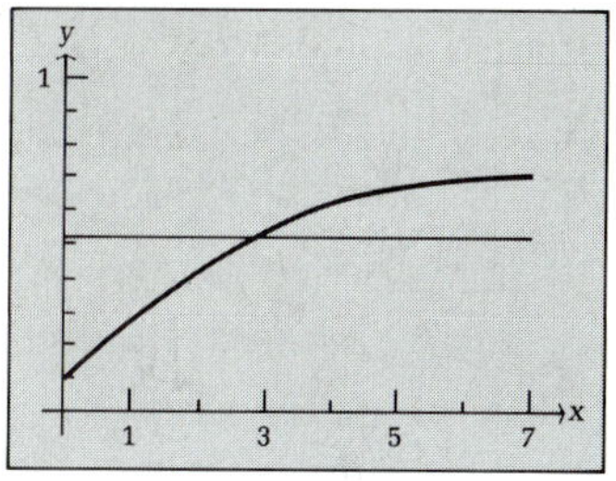

◆ EXERCISE 6-7 CHAPTER REVIEW

1. $t^3 - t^2 + C$ **2.** 12 **3.** $-3t^{-1} - 3t + C$ **4.** $\frac{15}{2}$ **5.** $-2e^{-0.5x} + C$ **6.** $2 \ln 5$ **7.** $y = f(x) = x^3 - 2x + 4$ **8.** $7 + 11 + 15 + 19 + 23 = 75$
9. 24.75 **10.** 30.8 **11.** 12 **12.** 7 **13.** $\frac{1}{8}(6x - 5)^{4/3} + C$ **14.** 2 **15.** $-2x^{-1} - e^{x^2} + C$ **16.** $(20^{3/2} - 8)/3$ **17.** $-\frac{1}{2}e^{-2x} + \ln|x| + C$
18. $-500(e^{-0.2} - 1) \approx 90.63$ **19.** $y = f(x) = 3 \ln|x| + x^{-1} + 4$ **20.** $y = 3x^2 + x - 4$ **21.** 343,500 **22.** $12 + 8/n$; 12

23. (A) $\lim_{n\to\infty} (32 + 16/n) = 32$ (B) $\int_0^4 (2x + 4)\,dx = 32$ **24.** (A) 36 (B) -9 (C) 3 **25.** 24 **26.** $\frac{13}{2}$ **27.** $\frac{1}{2} \ln 10$ **28.** 0.45

29. $\frac{1}{48}(2x^4 + 5)^6 + C$ **30.** $-\ln(e^{-x} + 3) + C$ **31.** $-(e^x + 2)^{-1} + C$ **32.** $\frac{1}{3}(\ln x)^3 + C$ **33.** $\frac{1}{8}x^8 - \frac{2}{5}x^5 + \frac{1}{2}x^2 + C$
34. $\frac{2}{3}(6 - x)^{3/2} - 12(6 - x)^{1/2} + C$ **35.** $\frac{1{,}234}{15} \approx 82.267$ **36.** $\frac{64}{15}$ **37.** $y = 3e^{x^3} - 1$ **38.** $N = 800e^{0.06t}$

39. (A) 150 (B) $\int_0^5 (3x^2 + 5)\,dx = 150$ **40.** $P(x) = 100x - 0.01x^2$; $P(10) = \$999$ **41.** $\int_0^{15} (60 - 4t)\,dt = 450$ thousand barrels

42. $\int_{10}^{40} \left(150 - \frac{x}{10}\right) dx = \$4{,}425$ **43.** 109 items **44.** $(65 + 10e^{-3})/3 \approx 21.833$ or 21,833 customers
45. $S(t) = 50 - 50e^{-0.08t}$; $50 - 50e^{-0.96} \approx \$31$ million; $-(\ln 0.2)/0.8 \approx 20$ months **46.** 1 cm^2 **47.** 800 gal
48. (A) $226e^{0.11} \approx 252$ million (B) $(\ln 2)/0.011 \approx 63$ yr **49.** $(-\ln 0.04)/0.000\,123\,8 \approx 26{,}000$ yr
50. $N(t) = 95 - 70e^{-0.1t}$; $N(15) \approx 79$ words per minute

CHAPTER 7

◆ EXERCISE 7-1

1. $\frac{3}{2}$ **3.** 24 **5.** 15 **7.** 9 **9.** $2e + \ln 2 - 2e^{0.5}$ **11.** $\frac{5}{2}$ **13.** $\frac{23}{3}$ **15.** 32 **17.** 36 **19.** $\frac{4}{3}$ **21.** $\frac{343}{3}$ **23.** $\frac{9}{2}$ **25.** $\frac{863}{6}$ **27.** 8 **29.** $\frac{407}{4}$

31. Total production from the end of the fifth year to the end of the tenth year is $50 + 100 \ln 20 - 100 \ln 15 \approx 79$ thousand barrels.

33. Useful life $= 2 \ln \frac{8}{3} \approx 2$ yr; Total profit $= 10 - 16e^{-1} \approx 4.114$ or \$4,114

35. 1935: 0.412; 1947: 0.231; income is more equally distributed in 1947.

37. 1963: 0.818; 1983: 0.846; total assets are less equally distributed in 1983.

39. Total weight gain during the first 10 hr is $3e - 3 \approx 5.15$ g.

41. $20e^{-1} \approx 7.36$ million acres of farmland could be saved over the next 20 years by providing financial aid for farmers.

43. Average number of words learned during the second 2 hr is $15 \ln 4 - 15 \ln 2 \approx 10$.

45. $A = 4.99$

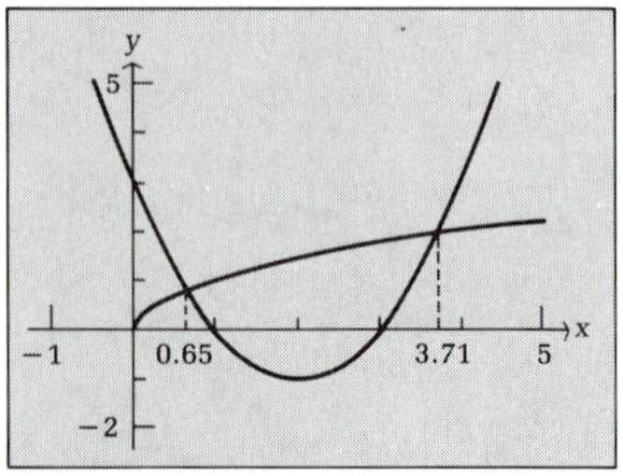

47. $A = 2.85$

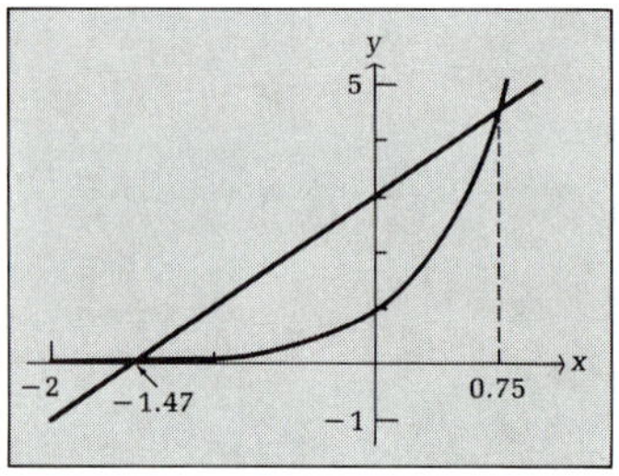

49. $A = 19.01$

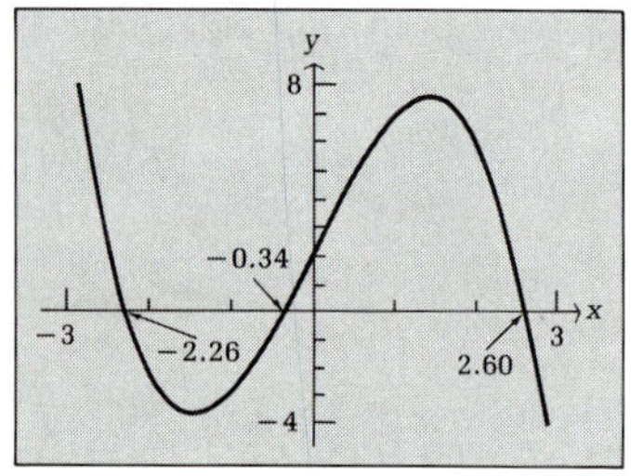

51. $A = 6.56$

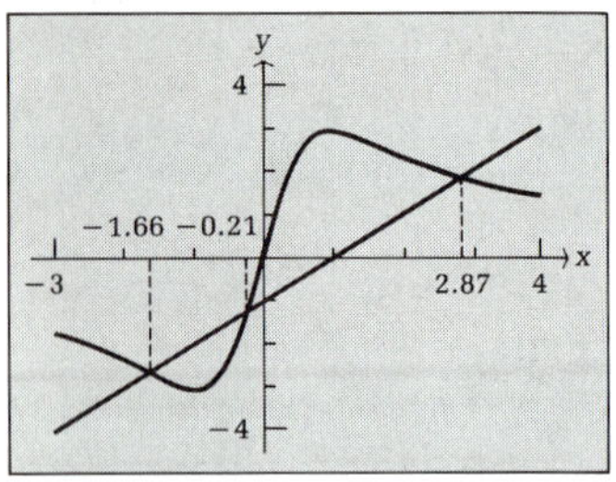

◆ EXERCISE 7-2

1. \$12,500 **3.** $8{,}000(e^{0.15} - 1) \approx \$1{,}295$ **5.** $10{,}000(e^{0.8} - 1) \approx \$12{,}255$ **7.** $12{,}500(e^{0.4} - e^{-0.08}) \approx \$7{,}109$

9. $25{,}000(e^{0.36} - 1) - 9{,}000 \approx \$1{,}833$

11. Clothing store: $FV = 120{,}000(e^{0.5} - 1) \approx \$77{,}847$; computer store: $FV = 200{,}000(e^{0.5} - e^{0.25}) \approx \$72{,}939$; the clothing store is the better investment.

13. Bond: $FV = 10{,}000e^{0.4} \approx \$14{,}918$; business: $FV = 25{,}000(e^{0.4} - 1) \approx \$12{,}296$; the bond is the better investment.

15. \$46,283 **17.** $\frac{k}{r}(e^{rT} - 1)$ **19.** \$625,000 **21.** \$9,500 **23.** $\bar{p} = 24$; $CS = \$3{,}380$; $PS = \$1{,}690$

25. $\bar{p} \approx 49$; $CS = 55{,}990 - 80{,}000e^{-0.49} \approx \$6{,}980$; $PS = 54{,}010 - 30{,}000e^{0.49} \approx \$5{,}041$ **27.** \$2,000

29. $\overline{p} = 24$

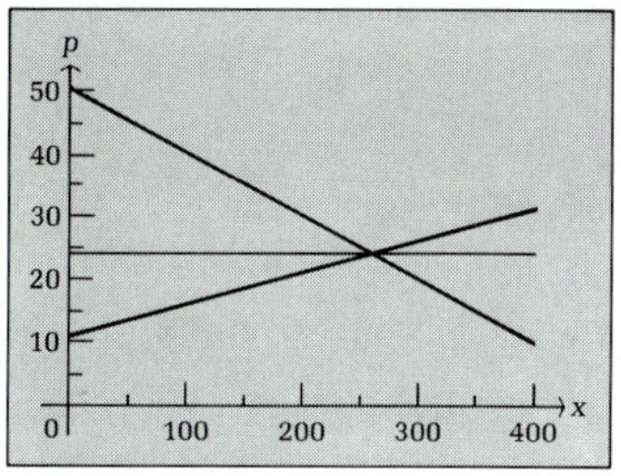

31. $\overline{p} = 49$

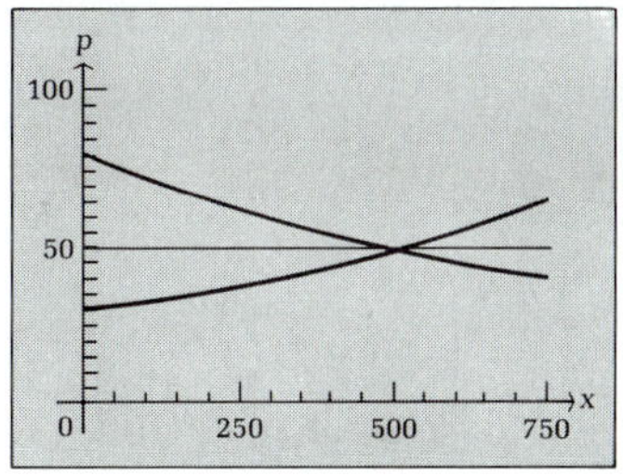

◆ EXERCISE 7-3

1. $\frac{1}{3}xe^{3x} - \frac{1}{9}e^{3x} + C$ **3.** $\frac{x^3}{3}\ln x - \frac{x^3}{9} + C$ **5.** $-xe^{-x} - e^{-x} + C$ **7.** $\frac{1}{2}e^{x^2} + C$
9. $(xe^x - 4e^x)|_0^1 = -3e + 4 \approx -4.1548$ **11.** $(x \ln 2x - x)|_1^3 = (3 \ln 6 - 3) - (\ln 2 - 1) \approx 2.6821$ **13.** $\ln(x^2 + 1) + C$ **15.** $(\ln x)^2/2 + C$
17. $\frac{2}{3}x^{3/2} \ln x - \frac{4}{9}x^{3/2} + C$ **19.** $(x^2 - 2x + 2)e^x + C$ **21.** $\frac{xe^{ax}}{a} - \frac{e^{ax}}{a^2} + C$ **23.** $\left(-\frac{\ln x}{x} - \frac{1}{x}\right)\Big|_1^e = -\frac{2}{e} + 1 \approx 0.2642$
25. $6 \ln 6 - 4 \ln 4 - 2 \approx 3.205$ **27.** $xe^{x-2} - e^{x-2} + C$ **29.** $\frac{1}{2}(1 + x^2) \ln(1 + x^2) - \frac{1}{2}(1 + x^2) + C$ **31.** $(1 + e^x) \ln(1 + e^x) - (1 + e^x) + C$
33. $x(\ln x)^2 - 2x \ln x + 2x + C$ **35.** $x(\ln x)^3 - 3x(\ln x)^2 + 6x \ln x - 6x + C$ **37.** $(e^2 - 3)/2$ **39.** $2e^2 - 3$

41.

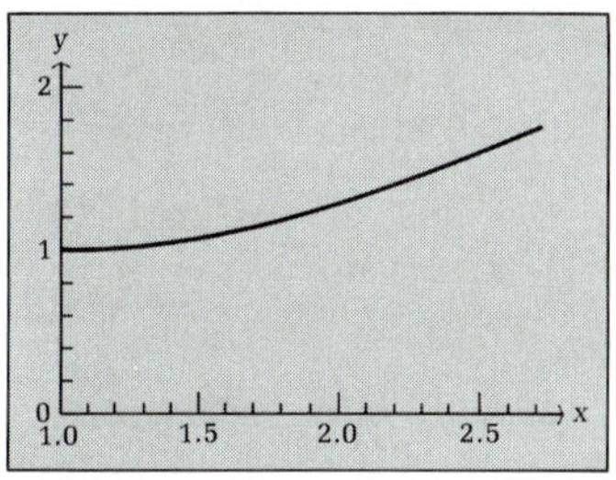

43.

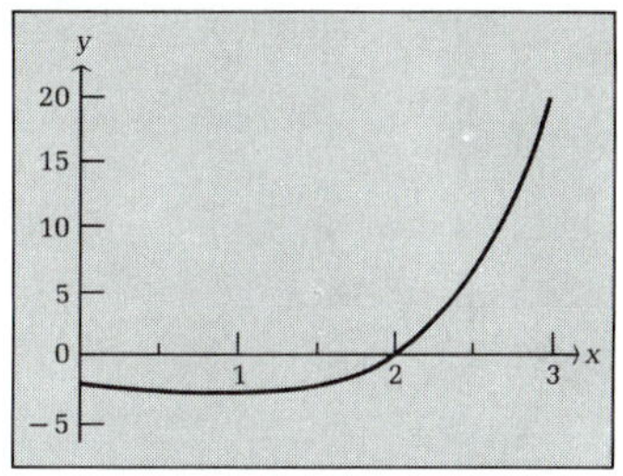

45. $P(t) = t^2 + te^{-t} + e^{-t} - 1$ **47.** $31{,}250 - 18{,}750e^{0.4} \approx \$3{,}278$ **49.** $1 - 2e^{-1} \approx 0.264$ **51.** $(10 - 2 \ln 6)/3 \approx 2.1388$ ppm **53.** 20,980

◆ EXERCISE 7-4

1. $\ln\left|\frac{x}{1 + x}\right| + C$ **3.** $\frac{1}{3 + x} + 2 \ln\left|\frac{5 + 2x}{3 + x}\right| + C$ **5.** $\frac{2(x - 32)}{3}\sqrt{16 + x} + C$ **7.** $\frac{1}{2} \ln\left|\frac{2}{2 + \sqrt{x^2 + 4}}\right| + C$
9. $\frac{1}{3}x^3 \ln x - \frac{1}{9}x^3 + C$ **11.** $9 \ln \frac{3}{2} - 2 \approx 1.6492$ **13.** $\frac{1}{2} \ln \frac{12}{5} \approx 0.4377$ **15.** $\ln 3 \approx 1.0986$
17. $-\frac{\sqrt{4x^2 + 1}}{x} + 2 \ln|2x + \sqrt{4x^2 + 1}| + C$ **19.** $\frac{1}{2} \ln|x^2 + \sqrt{x^4 - 16}| + C$ **21.** $\frac{1}{6}(x^3\sqrt{x^6 + 4} + 4 \ln|x^3 + \sqrt{x^6 + 4}|) + C$
23. $-\frac{\sqrt{4 - x^4}}{8x^2} + C$ **25.** $\frac{1}{5} \ln\left|\frac{3 + 4e^x}{2 + e^x}\right| + C$ **27.** $\frac{2}{3}(\ln x - 8)\sqrt{4 + \ln x} + C$ **29.** $\frac{1}{5}x^2e^{5x} - \frac{2}{25}xe^{5x} + \frac{2}{125}e^{5x} + C$
31. $-x^3e^{-x} - 3x^2e^{-x} - 6xe^{-x} - 6e^{-x} + C$ **33.** $x(\ln x)^3 - 3x(\ln x)^2 + 6x \ln x - 6x + C$ **35.** $\frac{64}{3}$ **37.** $\frac{1}{2} \ln \frac{9}{5} \approx 0.2939$
39. $\frac{1}{2} \ln|x^2 + 2x| + C$ **41.** $\frac{2}{3} \ln|3 + x| + \frac{1}{3} \ln|x| + C$ **43.** $3{,}000 + 1{,}500 \ln \frac{1}{3} \approx \$1{,}352$ **45.** $100{,}000e - 250{,}000 \approx \$21{,}828$

47. $\frac{19}{135} \approx 0.1407$ **49.** $S(t) = 1 + t - \frac{1}{1+t} - 2 \ln|1 + t|$; $24.96 - 2 \ln 25 \approx \$18.5$ million **51.** $100 \ln 3 \approx 110$ ft **53.** $60 \ln 5 \approx 97$ items

55. $\bar{x} = 200$

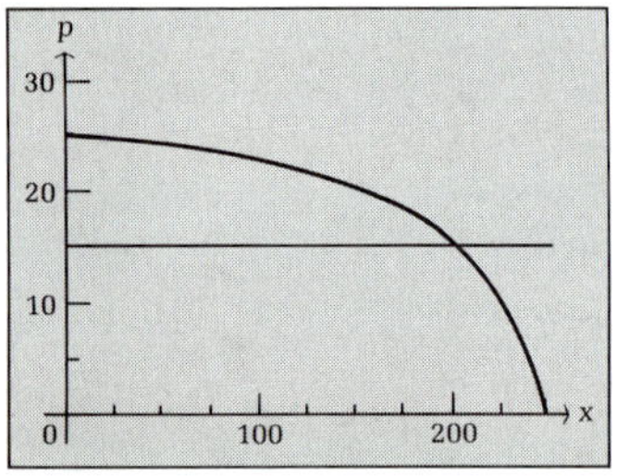

◆ EXERCISE 7-5

1. $\frac{1}{3}$ **3.** 2 **5.** Diverges **7.** 1 **9.** Diverges **11.** Diverges **13.** 1 **15.** $\int_2^{3.5} \left(-\frac{x}{2} + 2\right) dx \approx .94$ **17.** $\frac{1}{4} \int_1^{\infty} e^{-t/4}\, dt \approx .78$

19. 1 **21.** 1 **23.** Diverges **25.** $\frac{1}{2}$ **27.** Diverges **29.** Diverges

31. $F(b) = \frac{1}{3} - \frac{1}{3b^3}$, $\lim_{b\to\infty} F(b) = \frac{1}{3}$

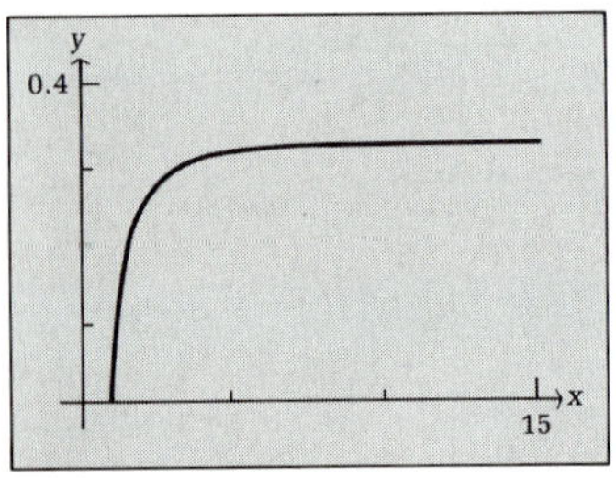

33. $F(b) = 2 - 2e^{-b/2}$, $\lim_{b\to\infty} F(b) = 2$

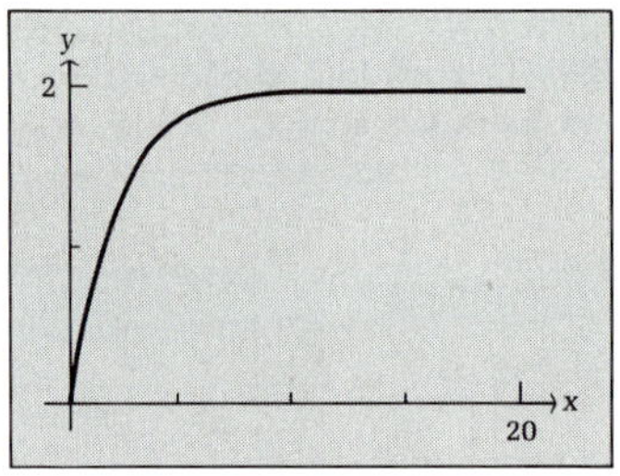

35. $F(b) = 2\sqrt{b} - 2$, $\lim_{b\to\infty} F(b) = \infty$

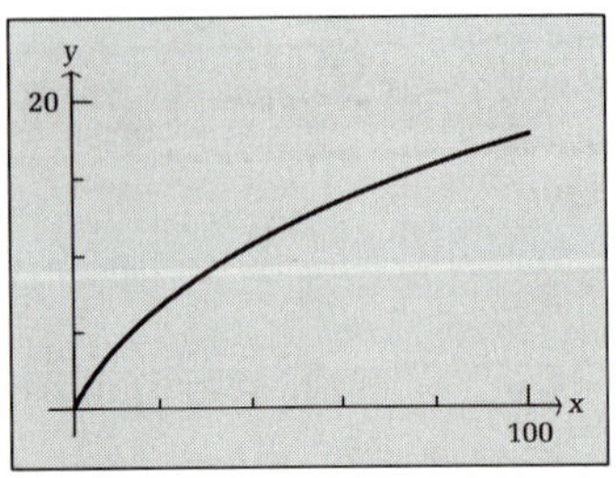

37. $F(b) = 1 - \frac{1}{b+1}$, $\lim_{b\to\infty} F(b) = 1$

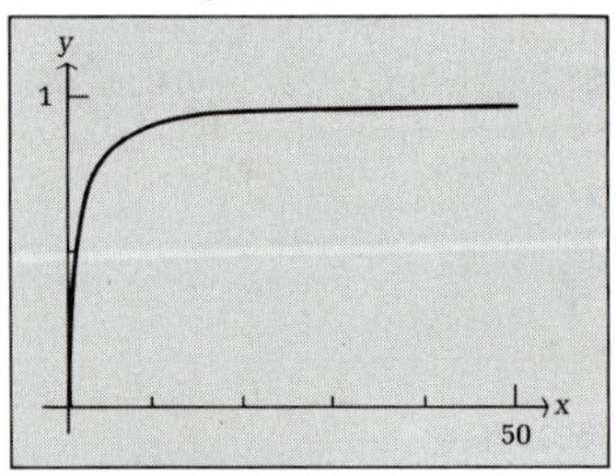

39. \$50,000 **41.** \$30,000 **43.** 6.25 million ft^3 **45.** (A) .75 (B) .11 **47.** (A) .11 (B) .10 **49.** .89 **51.** 500 gal

53. $.2 \int_0^5 e^{-.2t}\, dt \approx .63$ **55.** $\int_9^{\infty} \frac{dx}{(x+1)^2} = .1$

◆ EXERCISE 7-6 CHAPTER REVIEW

1. $\frac{1}{4}xe^{4x} - \frac{1}{16}e^{4x} + C$ **2.** $\frac{1}{2}x^2 \ln x - \frac{1}{4}x^2 + C$ **3.** $\frac{1}{1+x} + \ln\left|\frac{x}{1+x}\right| + C$

4. $-\dfrac{\sqrt{1+x}}{x} - \dfrac{1}{2}\ln\left|\dfrac{\sqrt{1+x}-1}{\sqrt{1+x}+1}\right| + C$ **5.** $\frac{1}{2}$ **6.** Diverges **7.** 8 **8.** 72 **9.** $5 + e^{-2}$ **10.** 1 **11.** 1
12. $\frac{15}{2} - 8 \ln 8 + 8 \ln 4 \approx 1.955$ **13.** $\frac{1}{6}(3x\sqrt{9x^2 - 49} - 49 \ln|3x + \sqrt{9x^2 - 49}|) + C$ **14.** $-2te^{-0.5t} - 4e^{-0.5t} + C$
15. $\frac{1}{3}x^3 \ln x - \frac{1}{9}x^3 + C$ **16.** $x - \ln|1 + 2e^x| + C$ **17.** $\frac{1}{3}$ **18.** $1.5 + 0.5 \ln 0.5 \approx 1.153$ **19.** $\frac{125}{6}$ **20.** $\frac{1}{3}(\ln x)^3 + C$
21. $\frac{1}{2}x^2 (\ln x)^2 - \frac{1}{2}x^2 \ln x + \frac{1}{4}x^2 + C$ **22.** $-\frac{1}{4}e^{-2x^2} + C$ **23.** $-\frac{1}{2}x^2e^{-2x} - \frac{1}{2}xe^{-2x} - \frac{1}{4}e^{-2x} + C$ **24.** $\sqrt{x^2 - 36} + C$
25. $\frac{1}{2} \ln|x^2 + \sqrt{x^4 - 36}| + C$ **26.** $50 \ln 10 - 42 \ln 6 - 24 \approx 15.875$ **27.** $x(\ln x)^2 - 2x \ln x + 2x + C$ **28.** 0 **29.** 2
30. $\frac{1}{4}$ **31.** 8 **32.** Useful life $= 10 \ln \frac{20}{3} \approx 19$ yr; Total profit $= 143 - 200e^{-1.9} \approx 113.086$ or \$113,086
33. Current: 0.3; projected: 0.2; income will be more equally distributed 10 yr from now.
34. (A) $50{,}000(e^{0.25} - 1) \approx \$14{,}201$ (B) $25{,}000(e^{0.75} - e^{0.25}) \approx \$20{,}824$ (C) \$6,623
35. (A) \$1,000 (B) \$800 (C) $\overline{p} = 40$; CS = \$2,250; PS = \$2,700
36. 3,374 thousand barrels; 10,000 thousand barrels **37.** \$20,000
38. $S(t) = 2t + 10 \ln|1 + e^{-0.2t}| - 10 \ln 2$; $48 + 10 \ln|1 + e^{-4.8}| - 10 \ln 2 \approx \$41.15$ million **39.** $.02 \int_0^1 e^{-.02t}\, dt \approx .02$
40. $120 \ln 2 - 120 \ln 1.5 - 30 \approx 4.552$ ml; $120 \ln 2 - 60 \approx 23.178$ ml **41.** $\int_1^\infty f(t)\, dt = \int_1^3 f(t)\, dt = \frac{1}{3}$
42. 45 thousand; 50 thousand **43.** $.5 \int_2^\infty e^{-.5t}\, dt \approx .37$

CHAPTER 8

◆ EXERCISE 8-1

1. 10 **3.** 1 **5.** 0 **7.** 1 **9.** 6 **11.** 150 **13.** 16π **15.** 791 **17.** 0.192 **19.** 118 **21.** $100e^{0.8} \approx 222.55$ **23.** $2x + h$ **25.** $2y^2$
27. $E(0, 0, 3)$; $F(2, 0, 3)$ **29.** \$4,400; \$6,000; \$7,100 **31.** $R(p, q) = -5p^2 + 6pq - 4q^2 + 200p + 300q$; $R(2, 3) = \$1{,}280$; $R(3, 2) = \$1{,}175$
33. 30,065 units **35.** \$272,615.08 **37.** $T(70, 47) \approx 29$ min; $T(60,27) = 33$ min **39.** $C(6, 8) = 75$; $C(8.1, 9) = 90$
41. $Q(12, 10) = 120$; $Q(10, 12) \approx 83$

◆ EXERCISE 8-2

1. 3 **3.** 2 **5.** $-4xy$ **7.** -6 **9.** $10xy^3$ **11.** 60 **13.** $2x - 2y + 6$ **15.** 6 **17.** -2 **19.** 2 **21.** $2e^{2x+3y}$ **23.** $6e^{2x+3y}$ **25.** $6e^2$ **27.** $4e^3$
29. $f_x(x, y) = 6x(x^2 - y^3)^2$; $f_y(x, y) = -9y^2(x^2 - y^3)^2$ **31.** $f_x(x, y) = 24xy(3x^2y - 1)^3$; $f_y(x, y) = 12x^2(3x^2y - 1)^3$
33. $f_x(x, y) = 2x/(x^2 + y^2)$; $f_y(x, y) = 2y/(x^2 + y^2)$ **35.** $f_x(x, y) = y^4e^{xy^2}$; $f_y(x, y) = 2xy^3e^{xy^2} + 2ye^{xy^2}$
37. $f_x(x, y) = 4xy^2/(x^2 + y^2)^2$; $f_y(x, y) = -4x^2y/(x^2 + y^2)^2$ **39.** $f_{xx}(x, y) = 2y^2 + 6x$; $f_{xy}(x, y) = 4xy = f_{yx}(x, y)$; $f_{yy}(x, y) = 2x^2$
41. $f_{xx}(x, y) = -2y/x^3$; $f_{xy}(x, y) = (-1/y^2) + (1/x^2) = f_{yx}(x, y)$; $f_{yy}(x, y) = 2x/y^3$
43. $f_{xx}(x, y) = (2y + xy^2)e^{xy}$; $f_{xy}(x, y) = (2x + x^2y)e^{xy} = f_{yx}(x, y)$; $f_{yy}(x, y) = x^3e^{xy}$ **45.** $x = 2$ and $y = 4$
47. $f_{xx}(x, y) + f_{yy}(x, y) = (2y^2 - 2x^2)/(x^2 + y^2)^2 + (2x^2 - 2y^2)/(x^2 + y^2)^2 = 0$ **49.** (A) $2x$ (B) $4y$
51. $P_x(1{,}200, 1{,}800) = 24$: profit will increase approx. \$24 per unit increase in production of type *A* calculators at the (1,200, 1,800) output level; $P_y(1{,}200, 1{,}800) = -48$: profit will decrease approx. \$48 per unit increase in production of type *B* calculators at the (1,200, 1,800) output level
53. $\partial x/\partial p = -5$: a \$1 increase in the price of brand *A* will decrease the demand for brand *A* by 5 lb at any price level (p, q); $\partial y/\partial p = 2$: a \$1 increase in the price of brand *A* will increase the demand for brand *B* by 2 lb at any price level (p, q)
55. (A) $f_x(x, y) = 7.5x^{-0.25}y^{0.25}$; $f_y(x, y) = 2.5x^{0.75}y^{-0.75}$
(B) Marginal productivity of labor $= f_x(600, 100) \approx 4.79$; Marginal productivity of capital $= f_y(600, 100) \approx 9.58$ (C) Capital
57. Competitive **59.** Complementary
61. (A) $f_w(w, h) = 6.65w^{-0.575}h^{0.725}$; $f_h(w, h) = 11.34w^{0.425}h^{-0.275}$
(B) $f_w(65, 57) = 11.31$: for a 65 lb child 57 in. tall, the rate of change in surface area is 11.31 in.^{2} for each pound gained in weight (height is held fixed); $f_h(65, 57) = 21.99$: for a 65 lb child 57 in. tall, the rate of change in surface area is 21.99 in.^{2} for each inch gained in height (weight is held fixed)
63. $C_W(6, 8) = 12.5$: index increases approx. 12.5 units for 1 in. increase in width of head (length held fixed) when $W = 6$ and $L = 8$; $C_L(6, 8) = -9.38$: index decreases approx. 9.38 units for 1 in. increase in length (width held fixed) when $W = 6$ and $L = 8$

◆ EXERCISE 8-3

1. $f(-2, 0) = 10$ is a local maximum **3.** $f(-1, 3) = 4$ is a local minimum **5.** f has a saddle point at $(3, -2)$
7. $f(3, 2) = 33$ is a local maximum **9.** $f(2, 2) = 8$ is a local minimum **11.** f has a saddle point at $(0, 0)$
13. f has a saddle point at $(0, 0)$; $f(1, 1) = -1$ is a local minimum
15. f has a saddle point at $(0, 0)$; $f(3, 18) = -162$ and $f(-3, -18) = -162$ are local minima
17. The test fails at $(0, 0)$; f has saddle points at $(2, 2)$ and $(2, -2)$ **19.** 2,000 type A and 4,000 type B; Max $P = P(2, 4) = \$15$ million

21. (A)

p	q	x	y
\$10	\$12	56	16
\$11	\$11	6	56

(B) A maximum weekly profit of \$288 is realized for $p = \$10$ and $q = \$12$. **23.** $P(x, y) = P(4, 2)$ **25.** 8 by 4 by 2 in. **27.** 20 by 20 by 40 in.

◆ EXERCISE 8-4

1. Max $f(x, y) = f(3, 3) = 18$ **3.** Min $f(x, y) = f(3, 4) = 25$
5. Max $f(x, y) = f(3, 3) = f(-3, -3) = 18$; Min $f(x, y) = f(3, -3) = f(-3, 3) = -18$
7. Maximum product is 25 when each number is 5 **9.** Min $f(x, y, z) = f(-4, 2, -6) = 56$
11. Max $f(x, y, z) = f(2, 2, 2) = 6$; Min $f(x, y, z) = f(-2, -2, -2) = -6$
13. 60 of model A and 30 of model B will yield a minimum cost of \$32,400 per week
15. (A) 8,000 units of labor and 1,000 units of capital; Max $N(x, y) = N(8{,}000, 1{,}000) \approx 263{,}902$ units
(B) Marginal productivity of money ≈ 0.6598; increase in production $\approx 32{,}990$ units
17. 8 by 8 by $\frac{8}{3}$ in. **19.** $x = 50$ ft and $y = 200$ ft; maximum area is 10,000 ft^2

◆ EXERCISE 8-5

1.
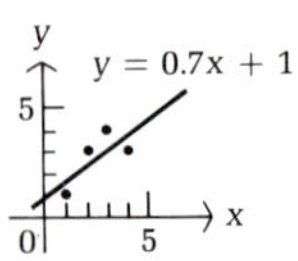

3.
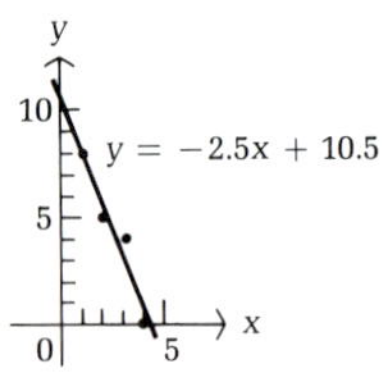

5.
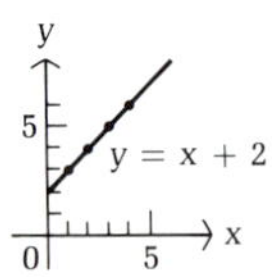

7. $y = 2.12x + 10.8$; $y = 63.8$ when $x = 25$
9. $y = -1.2x + 12.6$; $y = 10.2$ when $x = 2$
11. $y = -1.53x + 26.67$; $y = 14.4$ when $x = 8$

13. $y = 0.75x^2 - 3.45x + 4.75$
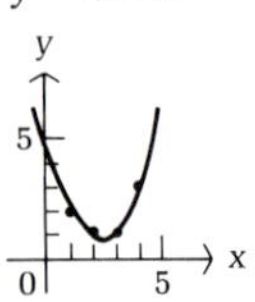

17. (A) $y = 743x + 10{,}012$ (B) \$17,442
19. (A) $y = -0.48x + 4.38$ (B) \$6.56/bottle
21. (A) $P = -0.66T + 48.8$ (B) 11.18 beats/min
23. (A) $D = -3.1A + 54.6$ (B) 45%

◆ EXERCISE 8-6

1. (A) $3x^2y^4 + C(x)$ (B) $3x^2$ **3.** (A) $2x^2 + 6xy + 5x + E(y)$ (B) $35 + 30y$ **5.** (A) $\sqrt{y + x^2} + E(y)$ (B) $\sqrt{y + 4} - \sqrt{y}$ **7.** 9 **9.** 330
11. $(56 - 20\sqrt{5})/3$ **13.** 16 **15.** 49 **17.** $\frac{1}{8}\int_1^5 \int_{-1}^1 (x + y)^2\, dy\, dx = \frac{32}{3}$ **19.** $\frac{1}{15}\int_1^4 \int_2^7 (x/y)\, dy\, dx = \frac{1}{2} \ln \frac{7}{2} \approx 0.6264$

21. $\frac{4}{3}$ cubic units **23.** $\frac{32}{3}$ cubic units **25.** $\int_0^1 \int_1^2 xe^{xy}\,dy\,dx = \frac{1}{2} + \frac{1}{2}e^2 - e$ **27.** $\int_0^1 \int_{-1}^1 \frac{2y + 3xy^2}{1 + x^2}\,dy\,dx = \ln 2$

29. $\frac{1}{0.4} \int_{0.6}^{0.8} \int_5^7 \frac{y}{1 - x}\,dy\,dx = 30 \ln 2 \approx \$20.8$ billion

31. $\frac{1}{10} \int_{10}^{20} \int_1^2 x^{0.75}y^{0.25}\,dy\,dx = \frac{8}{175}(2^{1.25} - 1)(20^{1.75} - 10^{1.75}) \approx 8.375$ or 8,375 units

33. $\frac{1}{192} \int_{-8}^{8} \int_{-6}^{6} [10 - \frac{1}{10}(x^2 + y^2)]\,dy\,dx = \frac{20}{3}$ insects/ft² **35.** $\frac{1}{8} \int_{-2}^{2} \int_{-1}^{1} [100 - 15(x^2 + y^2)]\,dy\,dx = 75$ ppm

37. $\frac{1}{10{,}000} \int_{2{,}000}^{3{,}000} \int_{50}^{60} 0.000\,013\,3xy^2\,dy\,dx \approx 100.86$ ft **39.** $\frac{1}{16} \int_8^{16} \int_{10}^{12} 100\frac{x}{y}\,dy\,dx = 600 \ln 1.2 \approx 109.4$

◆ EXERCISE 8-7

1. $R = \{(x, y) | 0 \le y \le 4 - x^2,\ \ 0 \le x \le 2\}$
$= \{(x, y) | 0 \le x \le \sqrt{4 - y},\ \ 0 \le y \le 4\}$

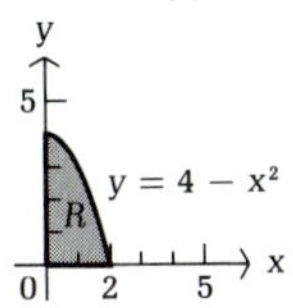

R is both a regular x-region and a regular y-region

3. $R = \{(x, y) | x^3 \le y \le 12 - 2x,\ \ 0 \le x \le 2\}$

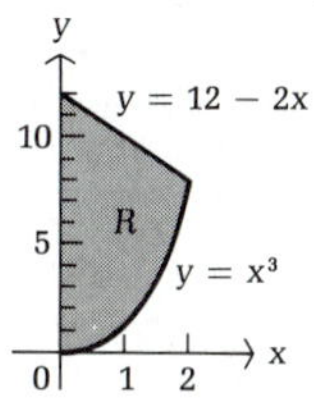

R is a regular x-region

5. $R = \{(x, y) | \frac{1}{2}y^2 \le x \le y + 4,\ \ -2 \le y \le 4\}$

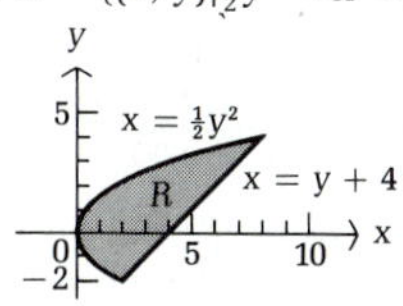

R is a regular y-region

7. $\frac{1}{2}$ **9.** $\frac{39}{70}$ **11.** $\frac{56}{3}$ **13.** $-\frac{3}{4}$ **15.** $\frac{1}{2}e^4 - \frac{5}{2}$ **17.** $R = \{(x, y) | 0 \le y \le x + 1,\ \ 0 \le x \le 1\}$
$\int_0^1 \int_0^{x+1} \sqrt{1 + x + y}\,dy\,dx = (68 - 24\sqrt{2})/15$

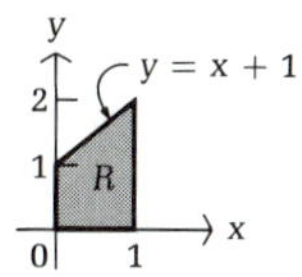

19. $R = \{(x, y) | 0 \le y \le 4x - x^2,\ \ 0 \le x \le 4\}$
$\int_0^4 \int_0^{4x - x^2} \sqrt{y + x^2}\,dy\,dx = \frac{128}{5}$

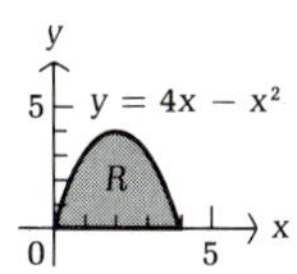

21. $R = \{(x, y) | 1 - \sqrt{x} \le y \le 1 + \sqrt{x},\ \ 0 \le x \le 4\}$
$\int_0^4 \int_{1-\sqrt{x}}^{1+\sqrt{x}} x(y - 1)^2\,dy\,dx = \frac{512}{21}$

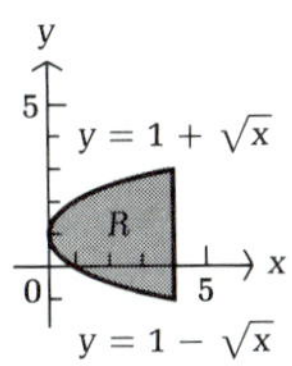

23. $\int_0^3 \int_0^{3-y} (x + 2y)\,dx\,dy = \frac{27}{2}$

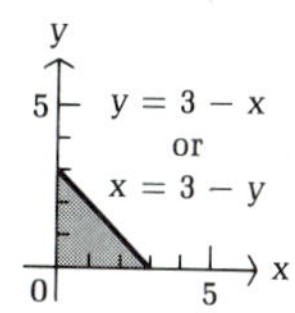

25. $\int_0^1 \int_0^{\sqrt{1-y}} x\sqrt{y}\,dx\,dy = \frac{2}{15}$

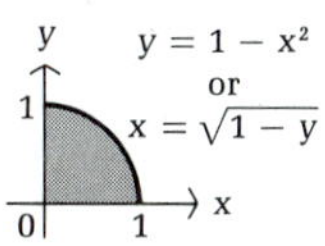

27. $\int_0^1 \int_{4y^2}^{4y} x\,dx\,dy = \frac{16}{15}$

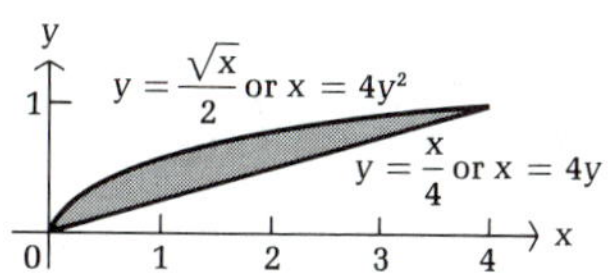

29. $\int_0^4 \int_0^{4-x} (4 - x - y)\,dy\,dx = \frac{32}{3}$

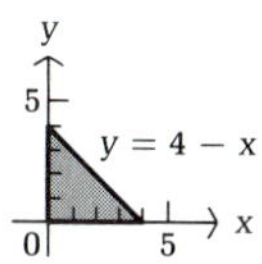

31. $\int_0^1 \int_0^{1-x^2} 4\,dy\,dx = \frac{8}{3}$

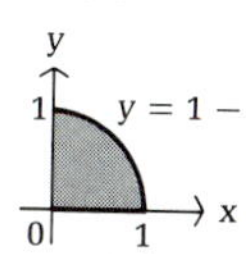

33. $\int_0^4 \int_0^{\sqrt{y}} \frac{4x}{1 + y^2}\,dx\,dy = \ln 17$ **35.** $\int_0^1 \int_0^{\sqrt{x}} 4ye^{x^2}\,dy\,dx = e - 1$

37. $R = \{(x, y)|x^2 \le y \le 1 + \sqrt{x},\quad 0 \le x \le 1.49\}$
$\int_0^{1.49}\int_{x^2}^{1+\sqrt{x}} x\,dy\,dx \approx 0.96$

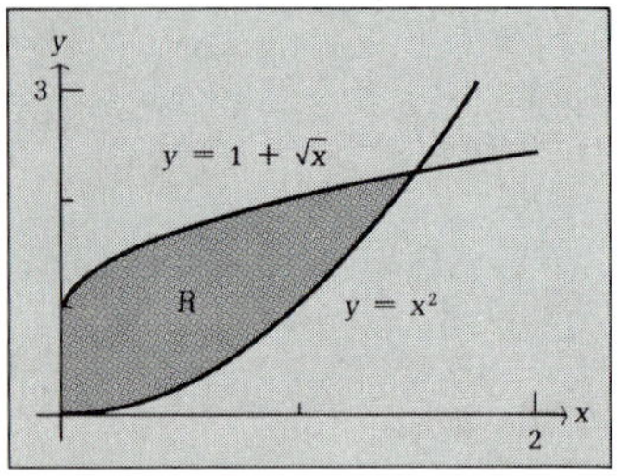

39. $R = \{(x, y)|y^3 \le x \le 1 - y,\quad 0 \le y \le 0.68\}$
$\int_0^{0.68}\int_{y^3}^{1-y} 24xy\,dx\,dy \approx 0.83$

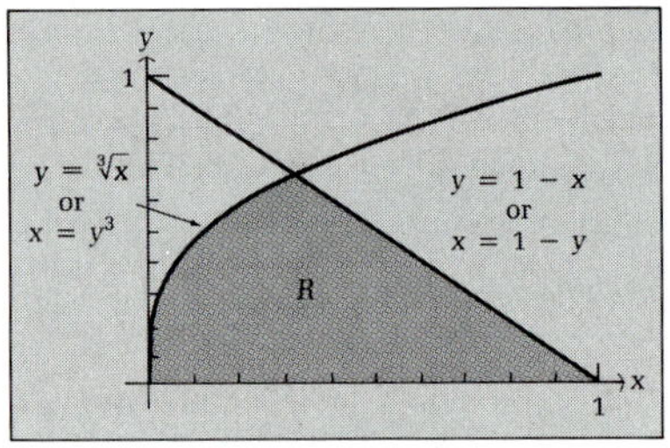

41. $R = \{(x, y)|e^{-x} \le y \le 3 - x,\quad -1.51 \le x \le 2.95\}$ Regular x region
$= \{(x, y)|-\ln y \le x \le 3 - y,\quad 0.05 \le y \le 4.51\}$ Regular y region
$\int_{-1.51}^{2.95}\int_{e^{-x}}^{3-x} 4y\,dy\,dx = \int_{0.05}^{4.51}\int_{-\ln y}^{3-y} 4y\,dx\,dy \approx 40.67$

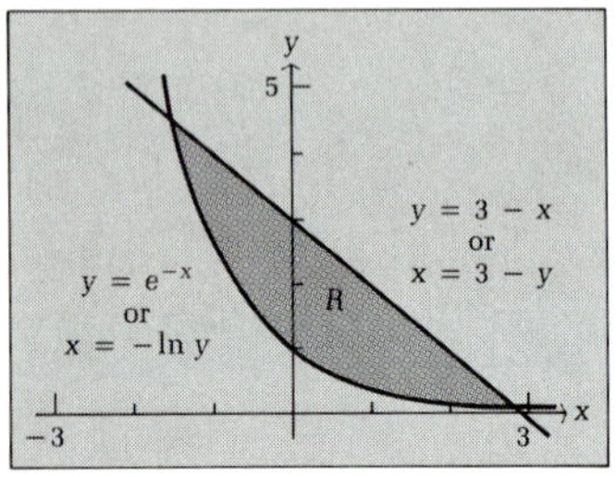

◆ EXERCISE 8-8 CHAPTER REVIEW

1. $f(5, 10) = 2{,}900$; $f_x(x, y) = 40$; $f_y(x, y) = 70$ **2.** $\partial^2z/\partial x^2 = 6xy^2$; $\partial^2z/\partial x\,\partial y = 6x^2y$ **3.** $2xy^3 + 2y^2 + C(x)$ **4.** $3x^2y^2 + 4xy + E(y)$ **5.** 1
6. $\frac{1}{2}$ **7.** $f(2, 3) = 7$; $f_y(x, y) = -2x + 2y + 3$; $f_y(2, 3) = 5$ **8.** $(-8)(-6) - (4)^2 = 32$ **9.** $(1, 3, -\frac{1}{2})$, $(-1, -3, \frac{1}{2})$
10. $y = -1.5x + 15.5$; $y = 0.5$ when $x = 10$ **11.** 18
12. $R = \{(x, y)|-x \le y \le x^2,\quad 0 \le x \le 2\}$
$\int_0^2\int_{-x}^{x^2}(4x + 5y)\,dy\,dx = 36$

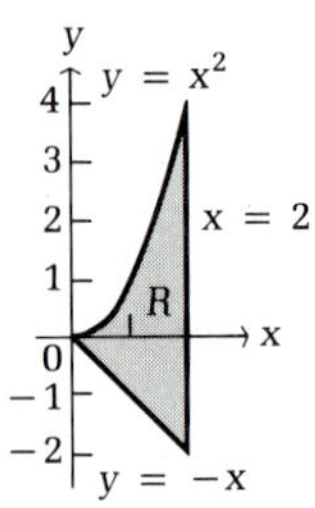

13. $\int_1^3\int_1^x 30x^2y\,dy\,dx = \int_1^3\int_y^3 30x^2y\,dx\,dy = 596$

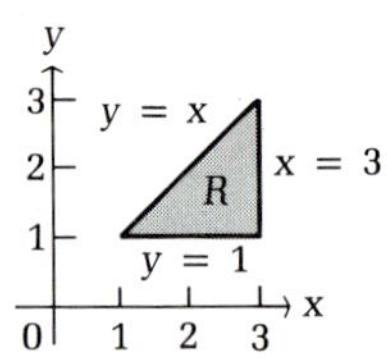

14. $f_x(x, y) = 2xe^{x^2+2y}$; $f_y(x, y) = 2e^{x^2+2y}$; $f_{xy}(x, y) = 4xe^{x^2+2y}$ **15.** $f_x(x, y) = 10x(x^2 + y^2)^4$; $f_{xy}(x, y) = 80xy(x^2 + y^2)^3$
16. $f(2, 3) = -25$ is a local minimum; f has a saddle point at $(-2, 3)$ **17.** Max $f(x, y) = f(6, 4) = 24$ **18.** Min $f(x, y, z) = f(2, 1, 2) = 9$

19. $y = \frac{116}{165}x + \frac{100}{3}$ **20.** $\frac{27}{5}$ **21.** 4 cubic units **22.** $\int_0^6 \int_0^{6-x} (6 - x - y)\, dy\, dx = 36$

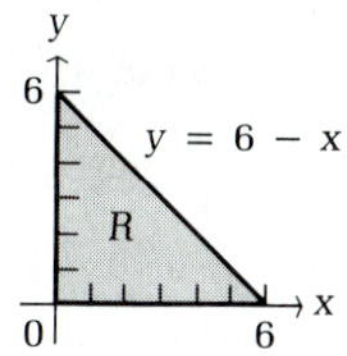

23. (A) $P_x(1, 3) = 8$; profit will increase \$8,000 for 100 unit increase in product *A* if production of product *B* is held fixed at an output level of (1, 3)

(B) For 200 units of *A* and 300 units of *B*, $P(2, 3) = \$100$ thousand is a local maximum.

24. 8 by 6 by 2 in. **25.** $y = 0.63x + 1.33$; profit in sixth year is \$5.11 million

26. (A) Marginal productivity of labor ≈ 8.37; marginal productivity of capital ≈ 1.67; management should encourage increased use of labor.

(B) 80 units of labor and 40 units of capital; Max $N(x, y) = N(80, 40) \approx 696$ units; marginal productivity of money ≈ 0.0696; increase in production ≈ 139 units

(C) $\frac{1}{1{,}000} \int_{50}^{100} \int_{20}^{40} 10x^{0.8}y^{0.2}\, dy\, dx = \frac{(40^{1.2} - 20^{1.2})(100^{1.8} - 50^{1.8})}{216} = 621$ items

27. $T_x(70, 17) = -0.924$ min/ft increase in depth when $V = 70$ ft³ and $x = 17$ ft **28.** $\frac{1}{16} \int_{-2}^{2} \int_{-2}^{2} [100 - 24(x^2 + y^2)]\, dy\, dx = 36$ ppm

29. 50,000 **30.** $y = \frac{1}{2}x + 48$; $y = 68$ when $x = 40$

CHAPTER 9

◆ EXERCISE 9-1

1.

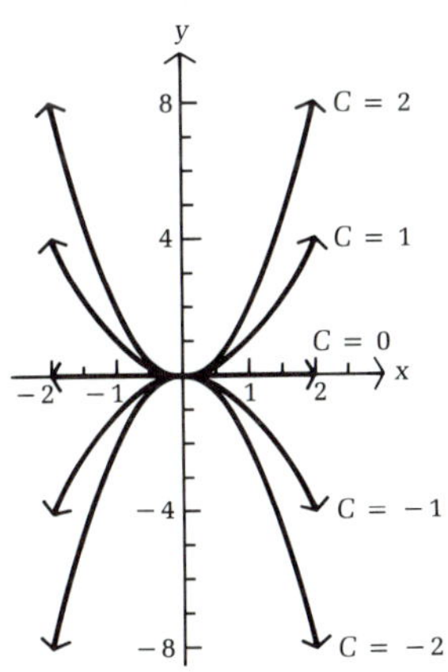

3.

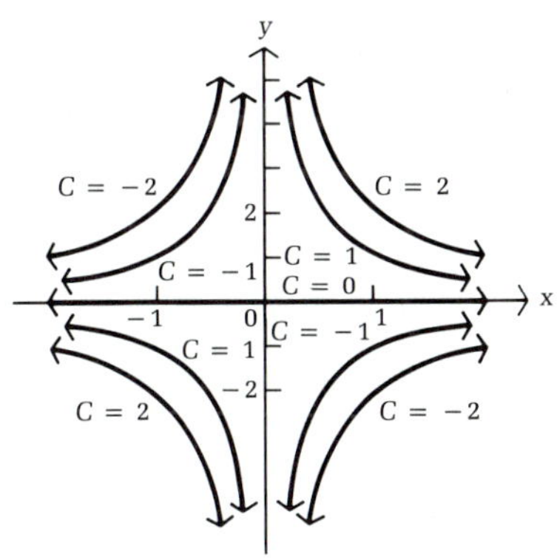

5. $y = 7e^x - 5x - 5$ **7.** $y = e^x - 2e^{2x}$ **9.** $y = x + \frac{2}{x}$ **15.** $y = \sqrt{9 - x^2}$ **17.** $y = 2 - e^x$

19. (A) $y = 2 - e^{-x}$
(B) $y = 2$
(C) $y = 2 + e^{-x}$

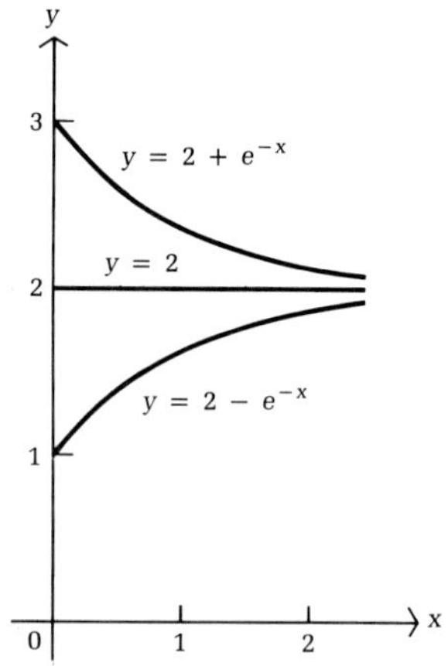

21. (A) $y = 2 - e^{x}$
(B) $y = 2$
(C) $y = 2 + e^{x}$

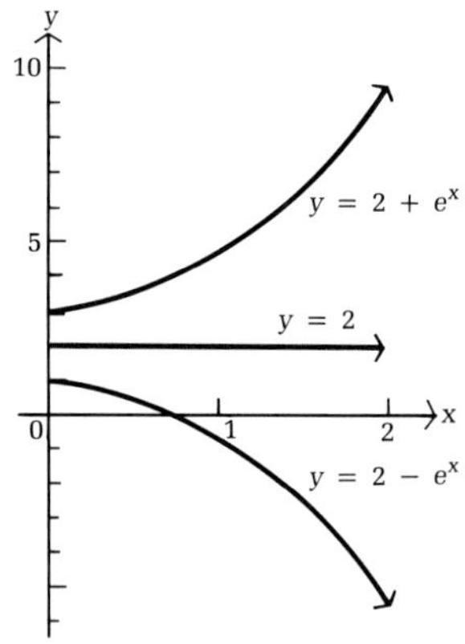

23. (A) $y = \dfrac{10}{1 + 9e^{-x}}$
(B) $y = 10$
(C) $y = \dfrac{10}{1 - 0.5e^{-x}}$

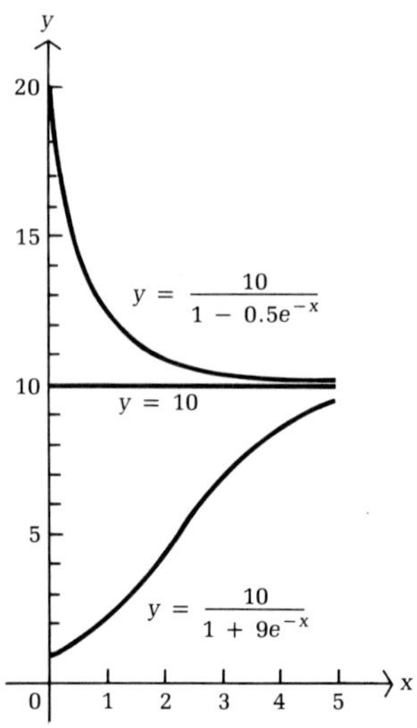

25. (A) $y = Cx^3 + 2$ for any C (B) No particular solution exists (C) $y = 2 - x^3$

27. (A)

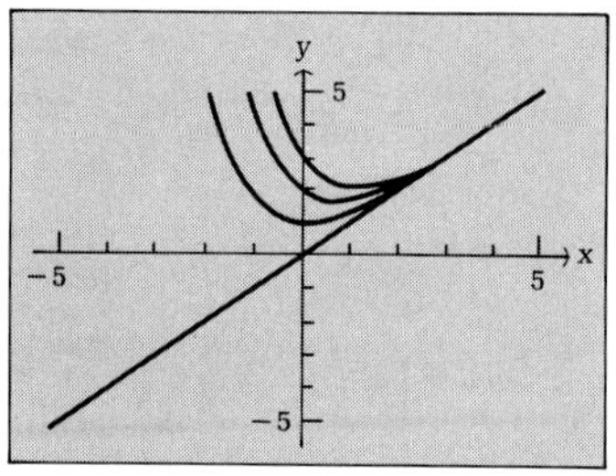

(B) Each graph decreases to a local minimum and then increases, approaching the line $y = x$ as x approaches ∞.

(C)

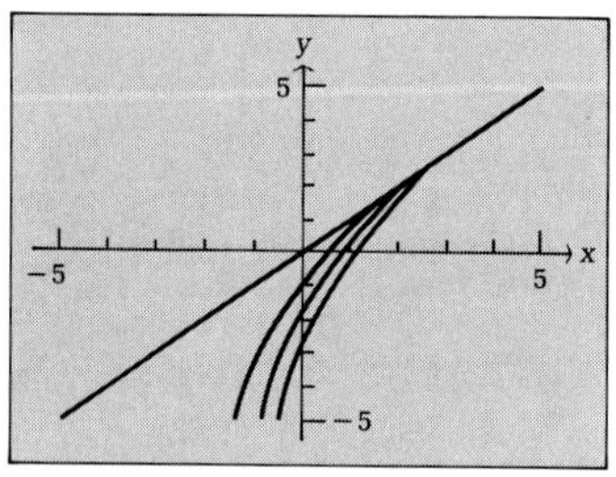

(D) Each graph is increasing for all x and approaches the line $y = x$ as x approaches ∞.

29. $\overline{p} = 5$
$p(t) = 5 - 4e^{-0.1t}$
$p(t) = 5 + 5e^{-0.1t}$

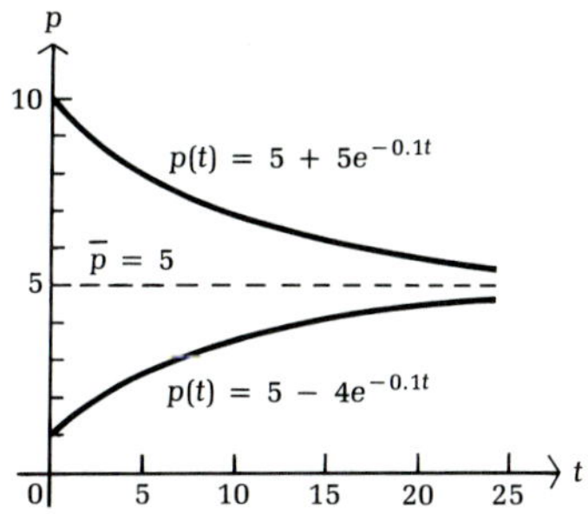

31. $A(t) = 2{,}500e^{0.08t} - 2{,}500$
$A(t) = 3{,}500e^{0.08t} - 2{,}500$

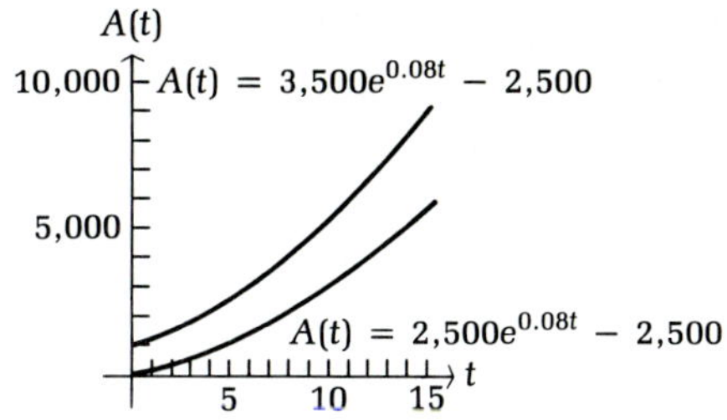

33. $\overline{N} = 200$
$N(t) = 200 - 150e^{-0.5t}$
$N(t) = 200 + 100e^{-0.5t}$

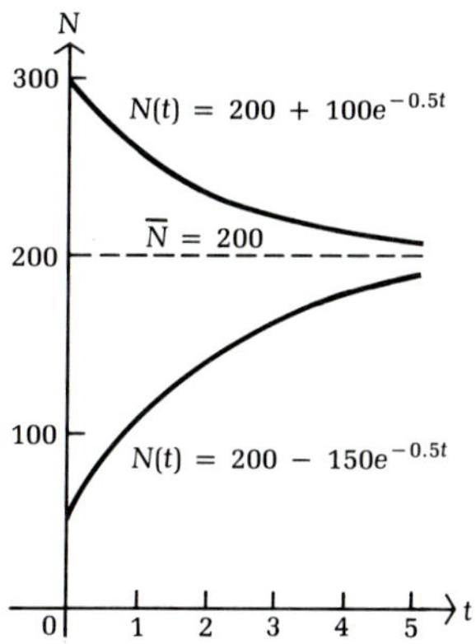

35. $N(t) = 200e^{2-2e^{-0.5t}}$
$\overline{N} = 200e^2 \approx 1{,}478$

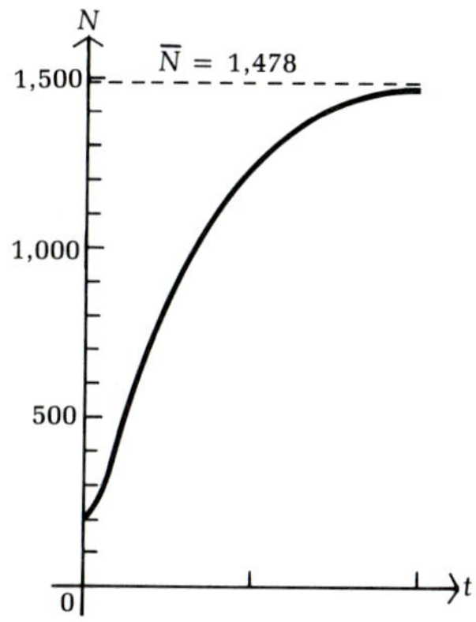

◆ EXERCISE 9-2

1. General solution: $y = x + C$; particular solution: $y = x + 2$ **3.** General solution: $y = 2x^{1/2} + C$; particular solution: $y = 2x^{1/2} - 4$
5. General solution: $y = Ce^x$; particular solution: $y = 10e^x$ **7.** General solution: $y = 25 - Ce^{-x}$; particular solution: $y = 25 - 20e^{-x}$
9. General solution: $y = Cx$; particular solution: $y = 5x$ **11.** General solution: $y = (3x + C)^{1/3}$; particular solution: $y = (3x + 24)^{1/3}$
13. General solution: $y = Ce^{e^x}$; particular solution: $y = 3e^{e^x}$
15. General solution: $y = \ln(e^x + C)$; particular solution: $y = \ln(e^x + 1)$
17. General solution: $y = Ce^{x^2/2} - 1$; particular solution: $y = 3e^{x^2/2} - 1$
19. General solution: $y = 2 - 1/(e^x + C)$; particular solution: $y = 2 - e^{-x}$ **21.** $y + \frac{1}{3}y^3 = x + \frac{1}{3}x^3 + C$
23. $\frac{1}{2}\ln|1 + y^2| = \ln|x| + (x^2/2) + C$ or $\ln(1 + y^2) = \ln(x^2) + x^2 + C$ **25.** $\ln(1 + e^y) = \frac{1}{2}x^2 + C$ **27.** $y = \sqrt{1 + (\ln x)^2}$
29. $y = \frac{1}{4}[x + \ln(x^2) + 3]^2$ **31.** $y = \sqrt{\ln(x^2 + e)}$ **33.** $5{,}000e^{1.2} \approx \$16{,}600$ **35.** $\frac{7\ln 0.5}{\ln 0.8} \approx 22$ days **37.** $\frac{30\ln 0.1}{\ln 0.5} \approx 100$ days
39. $\frac{4\ln 0.2}{\ln 0.8} \approx 29$ yr **41.** $\frac{2\ln\frac{5}{12}}{\ln\frac{5}{6}} \approx 9.6$ min **43.** $25 + 300e^{4\ln(2/3)} \approx 84.26°\text{F}$ **45.** (A) $100e^{5\ln 1.4} \approx 538$ bacteria (B) $\frac{\ln 10}{\ln 1.4} \approx 6.8$ hr

47. (A) $\dfrac{50{,}000}{1 + 499e^{2\ln(99/499)}} \approx 2{,}422$ people (B) $\dfrac{10\ln(1/499)}{\ln(99/499)} \approx 38.4$ days **49.** $I = As^k$

51. (A) $\dfrac{1{,}000}{1 + 199e^{7\ln(99/199)}} \approx 400$ people (B) $\dfrac{\ln(3/3{,}383)}{\ln(99/199)} \approx 10$ days

53. 38 days **55.** $k = -0.2$; $M = \$11$ million **57.** 47 days

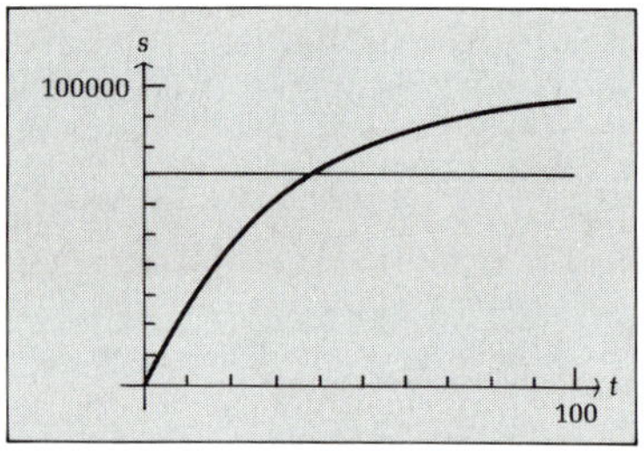

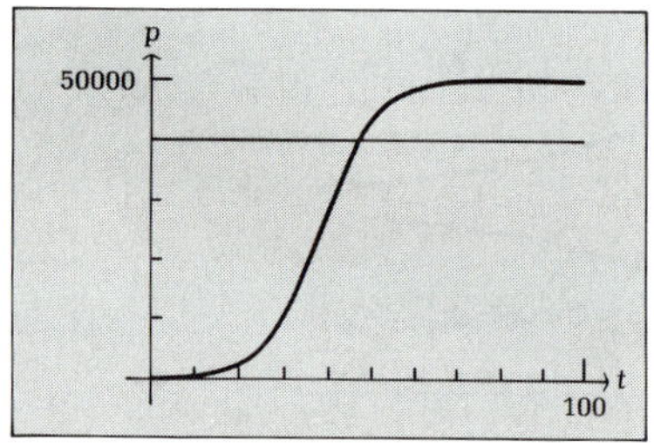

◆ EXERCISE 9-3

1. $I(x) = e^{2x}$; $y = 2 + Ce^{-2x}$; $y = 2 - e^{-2x}$ **3.** $I(x) = e^x$; $y = -e^{-2x} + Ce^{-x}$; $y = -e^{-2x} + 4e^{-x}$
5. $I(x) = e^{-x}$; $y = 2xe^x + Ce^x$; $y = 2xe^x - 4e^x$ **7.** $I(x) = e^x$; $y = 3x^3e^{-x} + Ce^{-x}$; $y = 3x^3e^{-x} + 2e^{-x}$
9. $I(x) = x$; $y = x + (C/x)$; $y = x$ **11.** $I(x) = x^2$; $y = 2x^3 + (C/x^2)$; $y = 2x^3 - (32/x^2)$
13. $I(x) = e^{x^2/2}$; $y = 5 + Ce^{-x^2/2}$ **15.** $I(x) = e^{-2x}$; $y = -2x - 1 + Ce^{2x}$ **17.** $I(x) = x$; $y = e^x - (e^x/x) + (C/x)$
19. $I(x) = x$; $y = \frac{1}{2}x\ln x - \frac{1}{4}x + (C/x)$ **21.** $y = 1 + (C/x)$ **23.** $y = C(1 + x^2) - 1$ **25.** $y = Ce^{x^2} - 1$ **27.** $y = Ce^{kt}$
29. $A = 100{,}000 - 80{,}000e^{0.04t}$; $(\ln 1.25)/0.04 \approx 5.579$ yr; \$22,316 **31.** $30{,}000(1 - e^{-0.5}) \approx \$11{,}804$
33. $A = 32{,}000e^{0.08t} - 25{,}000$; \$22,738.39 **35.** $p(t) = 20 + 10e^{-3t}$; $\bar{p} = 20$ **37.** 622 lb; $\frac{622}{250} \approx 2.5$ lb/gal
39. $600 - 200e^{-0.5} \approx 479$ lb; $3 - e^{-0.5} \approx 2.4$ lb/gal
41. $120 + 40e^{-0.15} \approx 154$ lb; $-(\ln \frac{3}{4})/0.005 \approx 58$ days; weight will approach 120 lb if the diet is maintained for a long period of time
43. $17.5\left(\dfrac{125 - 130e^{-0.15}}{1 - e^{-0.15}}\right) \approx 1{,}647$ cal **45.** Student A: $0.9 - 0.8e^{-4.8} \approx 0.8934$ or 89.34%; student B: $0.7 - 0.3e^{-4.8} \approx 0.6975$ or 69.75%

47. 10.6 yr **49.** 6.2 hr

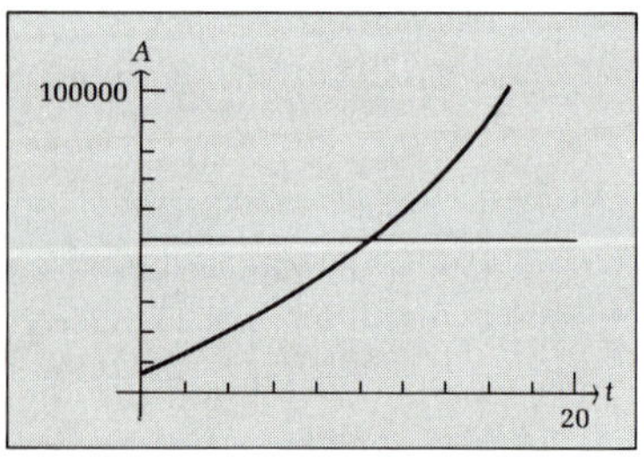

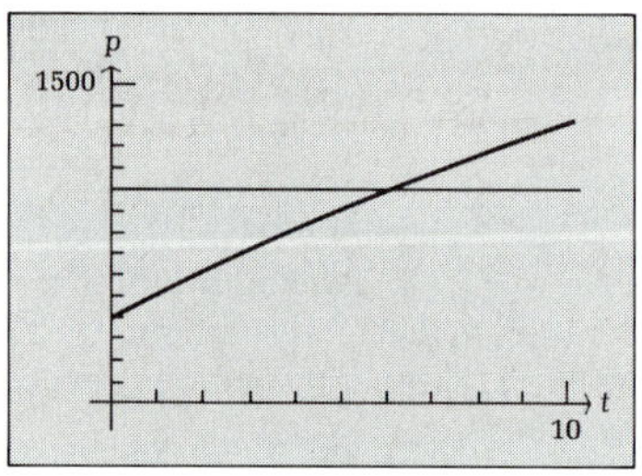

◆ EXERCISE 9-4

1. $y = C_1e^{-x} + C_2e^{-2x}$ **3.** $y = C_1e^{-5x} + C_2e^{3x}$ **5.** $y = C_1 + C_2e^{-6x}$ **7.** $y = C_1e^{2x} + C_2xe^{2x}$ **9.** $y = 2e^x + e^{-x}$ **11.** $y = \frac{3}{2}e^{x/3} - \frac{1}{2}e^{3x}$
13. $y = 2e^{-x} + 6xe^{-x}$ **15.** $y = C_1e^{4x} + C_2e^{-x} - 3$ **17.** $y = 2e^x + e^{-2x} - 3$ **19.** $y = x + C_1e^{-0.5x} + C_2$ **21.** $y = -\frac{3}{2}x + C_1e^{2x} + C_2$
23. $y = e^{4x}$ **25.** $y = e^{2x} - e^{0.5x}$ **27.** $y = C_1\cos x + C_2\sin x$ **29.** $y = e^{2x}(C_1\cos 3x + C_2\sin 3x)$
31. $p(t) = -50e^{-0.1t} + 100e^{-0.2t} + 25$; $\bar{p} = 25$ **33.** $y = 2 - e^{-t}$; 2

◆ EXERCISE 9-5

1. $x = C_1e^t + C_2e^{-2t}$, $y = 2C_1e^t - C_2e^{-2t}$ **3.** $x = C_1e^t + C_2e^{-t}$, $y = C_1e^t + 3C_2e^{-t}$; $x = 2e^t - e^{-t}$, $y = 2e^t - 3e^{-t}$
5. $x = C_1e^{3t} + C_2$, $y = C_1e^{3t} - 2C_2$; $x = e^{3t} + 1$, $y = e^{3t} - 2$ **7.** $x = C_1e^t + C_2e^{-t}$, $y = 3C_1e^t + C_2e^{-t}$
9. $x = C_1e^{2t} + C_2e^{-t} + 2$, $y = C_1e^{2t} + 2C_2e^{-t} + 4$
11. $p = 5e^{-2t} + 10e^{-3t} + 85$, $q = 10e^{-2t} + 10e^{-3t} + 80$; p decreases to the limiting value of 85; q decreases to the limiting value of 80
13. $x = 25e^{-4t} + 25e^{-6t}$; $y = 25e^{-4t} - 25e^{-6t}$; 30.5 units in compartment 1 and 3.0 units in compartment 2 after 6 min; 4.6 units in compartment 1 and 2.1 units in compartment 2 after 30 min; 0.5 unit in compartment 1 and 0.4 unit in compartment 2 after 1 hr

15.

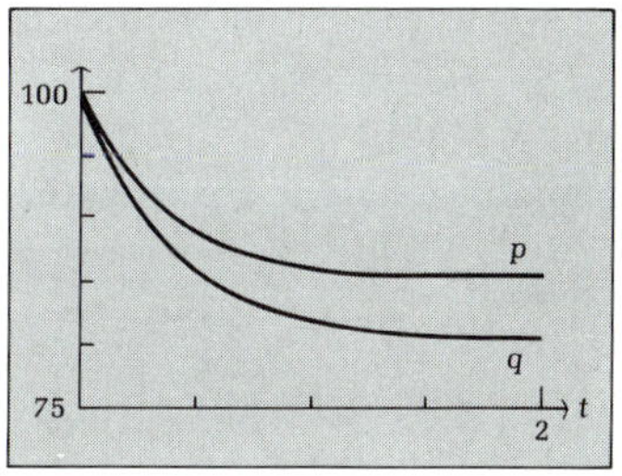

17.

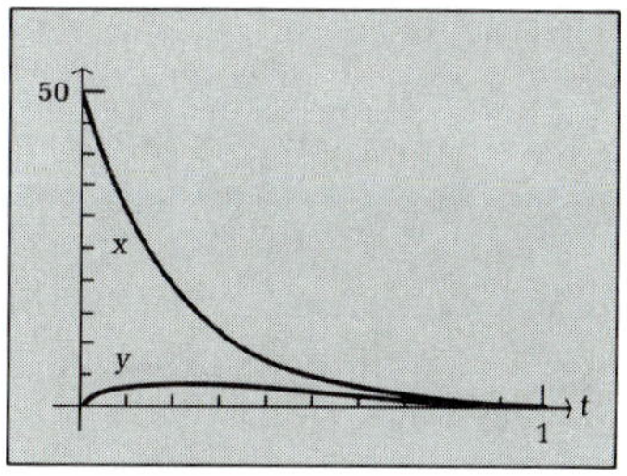

◆ EXERCISE 9-6 CHAPTER REVIEW

1.

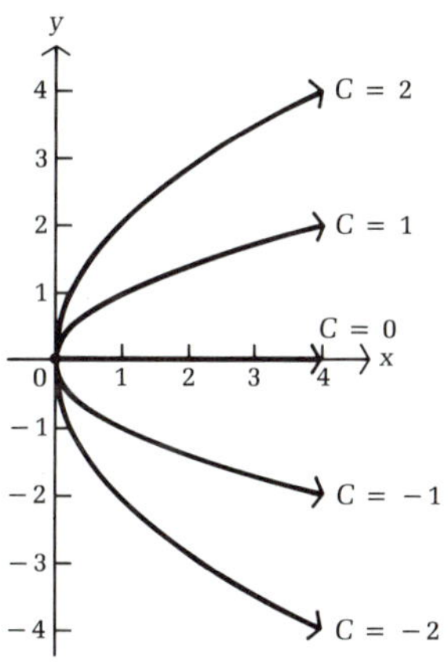

2.

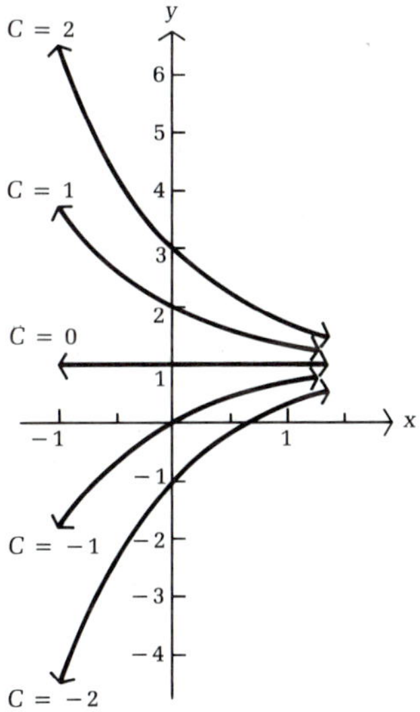

3. $y = C/x^4$
4. $y = \frac{1}{6}x^2 + (C/x^4)$
5. $y = -1/(x^3 + C)$
6. $y = e^x + Ce^{2x}$
7. $y = \frac{1}{2}x^7 + Cx^5$
8. $y = C(2 + x) - 3$

9. $y = C_1e^{7x} + C_2e^{-3x}$ **10.** $y = C_1e^{-6x} + C_2xe^{-6x}$ **11.** $y = 10 - 10e^{-x}$ **12.** $y = x - 1 + e^{-x}$ **13.** $y = e^2e^{-2e^{-x}}$ **14.** $y = (x^2 + 4)/(x + 4)$
15. $y = \sqrt{x^2 + 16} - 4$ **16.** $y = \frac{1}{3}x \ln x - \frac{1}{9}x + \frac{19}{9}(1/x^2)$ **17.** $y = \sqrt{1 + 2x^2}$ **18.** $y = (2x + 1)e^{-x^2}$ **19.** $y = \frac{1}{2}e^{-4x} + \frac{1}{2}$
20. $y = e^{4x} + e^{-4x} - 1$ **21.** $x = 1 - t$; $y = 1 - 2t$ **22.** $x = e^{2t} - e^{-4t} + 2$; $y = 2e^{2t} + e^{-4t} - 1$ **23.** $500e^{(-0.25)\ln 20} \approx \$236.44$
24. $(\ln \frac{1}{4})/(\ln \frac{3}{4}) \approx 5$ yr **25.** $p = 50 + 25e^{-t}$; $\overline{p} = 50$ **26.** $A = 100{,}000 - 40{,}000e^{0.05t}$; $(\ln 2.5)/0.05 \approx 18.326$ yr; \$91,630
27. $p = -25e^{-3t} - 25e^{-4t} + 100$, $q = 25e^{-3t} + 50e^{-4t} + 75$, p increases to a limiting value of 100; q decreases to a limiting value of 75
28. $y = 100 + te^{-t} - 100e^{-t}$ **29.** 300 lb

30. $x = 50e^{0.01t} + 25e^{0.04t}$, $y = 100e^{0.01t} - 25e^{0.04t}$; the first species increases without bound; the second species dies out after approx. 46.2 yr

31. (A) $200 - 199e^{(5/2)\ln(190/199)} \approx 23$ people (B) $\dfrac{2\ln(100/199)}{\ln(190/199)} \approx 30$ days

CHAPTER 10

◆ EXERCISE 10-1

1. $\dfrac{24}{(2-x)^5}$ **3.** $16e^{-2x}$ **5.** $1 - x + \frac{1}{2}x^2 - \frac{1}{6}x^3 + \frac{1}{24}x^4$ **7.** $2x - 2x^2 + \frac{8}{3}x^3$ **9.** $1 + \frac{1}{3}x - \frac{1}{9}x^2 + \frac{5}{81}x^3$
11. $1 + (x-1) + \frac{1}{2}(x-1)^2 + \frac{1}{6}(x-1)^3 + \frac{1}{24}(x-1)^4$ **13.** $3(x-\frac{1}{3}) - \frac{9}{2}(x-\frac{1}{3})^2 + 9(x-\frac{1}{3})^3$ **15.** $1 - 2x + 2x^2 - \frac{4}{3}x^3$; 0.604 166 67
17. $4 + \frac{1}{8}x - \frac{1}{512}x^2$; 4.123 046 9 **19.** $1 + \frac{1}{2}(x-1) - \frac{1}{8}(x-1)^2 + \frac{1}{16}(x-1)^3 - \frac{5}{128}(x-1)^4$; 1.095 437 5 **21.** $5^n e^{5x}$
23. $(-1)^{n-1}(n-1)!(5+x)^{-n}$ **25.** $x - \frac{1}{2}x^2 + \frac{1}{3}x^3 - \cdots + \dfrac{(-1)^{n+1}}{n}x^n$ **27.** $1 + x + x^2 + \cdots + x^n$

29. $\ln 3 + \dfrac{4}{3}x - \dfrac{1}{2}\left(\dfrac{4}{3}\right)^2 x^2 + \dfrac{1}{3}\left(\dfrac{4}{3}\right)^3 x^3 - \cdots + \dfrac{(-1)^{n+1}}{n}\left(\dfrac{4}{3}\right)^n x^n$

31. $e^3 + e^3(x-3) + \dfrac{e^3}{2!}(x-3)^2 + \dfrac{e^3}{3!}(x-3)^3 + \cdots + \dfrac{e^3}{n!}(x-3)^n$

33. $\ln\dfrac{1}{2} + 2\left(x - \dfrac{1}{2}\right) - \dfrac{2^2}{2}\left(x - \dfrac{1}{2}\right)^2 + \dfrac{2^3}{3}\left(x - \dfrac{1}{2}\right)^3 - \cdots + \dfrac{(-1)^{n+1}}{n}2^n\left(x - \dfrac{1}{2}\right)^n$

35. $e^{-10} - e^{-10}(x-10) + \dfrac{e^{-10}}{2!}(x-10)^2 - \cdots + \dfrac{e^{-10}(-1)^n}{n!}(x-10)^n$

37. $p_1(x) = x$, $p_2(x) = x - \frac{1}{2}x^2$, $p_3(x) = x - \frac{1}{2}x^2 + \frac{1}{3}x^3$

x	$p_1(x)$	$p_2(x)$	$p_3(x)$	$f(x)$
−0.2	−0.2	−0.22	−0.222 667	−0.223 144
−0.1	−0.1	−0.105	−0.105 333	−0.105 361
0	0	0	0	0
0.1	0.1	0.095	0.095 333	0.095 31
0.2	0.2	0.18	0.182 667	0.182 322

x	$\lvert p_1(x) - f(x)\rvert$	$\lvert p_2(x) - f(x)\rvert$	$\lvert p_3(x) - f(x)\rvert$
−0.2	0.023 144	0.003 144	0.000 477
−0.1	0.005 361	0.000 361	0.000 028
0	0	0	0
0.1	0.004 69	0.000 31	0.000 023
0.2	0.017 678	0.002 322	0.000 345

39.

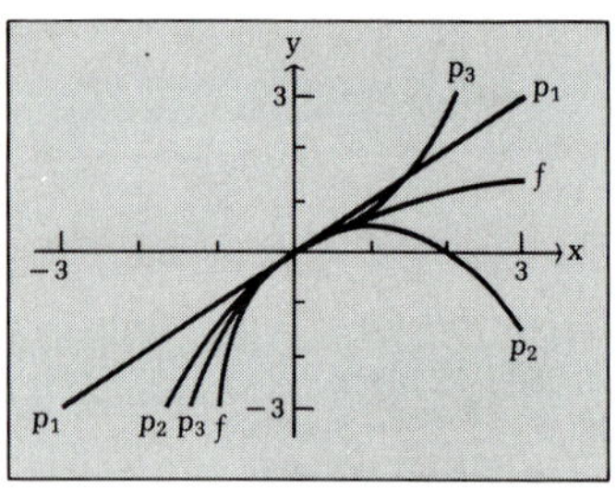

41. $e^a\left[1 + (x-a) + \dfrac{1}{2!}(x-a)^2 + \dfrac{1}{3!}(x-a)^3 + \cdots + \dfrac{1}{n!}(x-a)^n\right] = e^a \displaystyle\sum_{k=0}^{n} \dfrac{1}{k!}(x-a)^k$

43. $\dfrac{1}{a} - \dfrac{1}{a^2}(x-a) + \dfrac{1}{a^3}(x-a)^2 - \dfrac{1}{a^4}(x-a)^3 + \cdots + \dfrac{(-1)^n}{a^{n+1}}(x-a)^n = \displaystyle\sum_{k=0}^{n} \dfrac{(-1)^k}{a^{k+1}}(x-a)^k$

45. $p_2(x) = 10 - 0.0005x^2$; \$9.85 **47.** $p_2(x) = 8 - \frac{3}{40}(x-60) - \frac{1}{1024}(x-60)^2$; \$7.12 **49.** $p_2(t) = 10 - 0.8t^2$; approx. \$18 million
51. $p_2(t) = -3 + \frac{3}{25}t^2$; 6.32 cm² **53.** $p_2(x) = 120 + 30(x-5) + \frac{3}{4}(x-5)^2$; 126.25 ppm **55.** $p_2(t) = 6 - 0.06t^2$; 27.5 wpm

57.

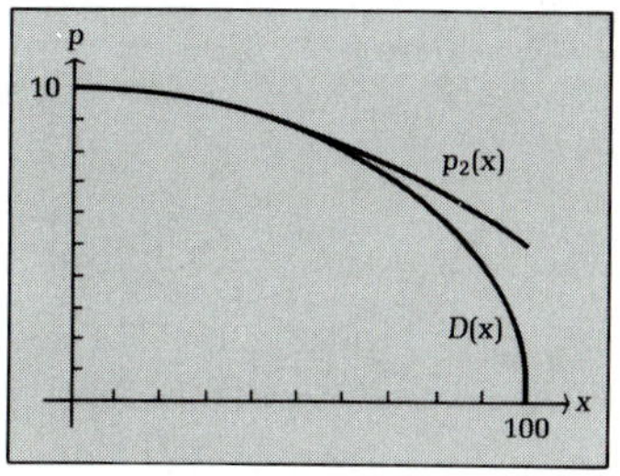

59.

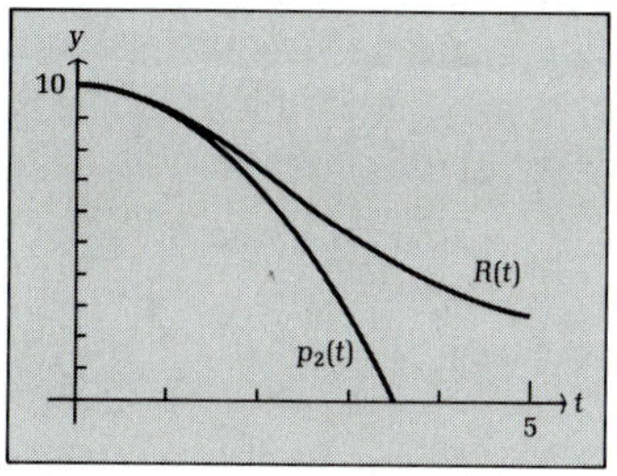

61.

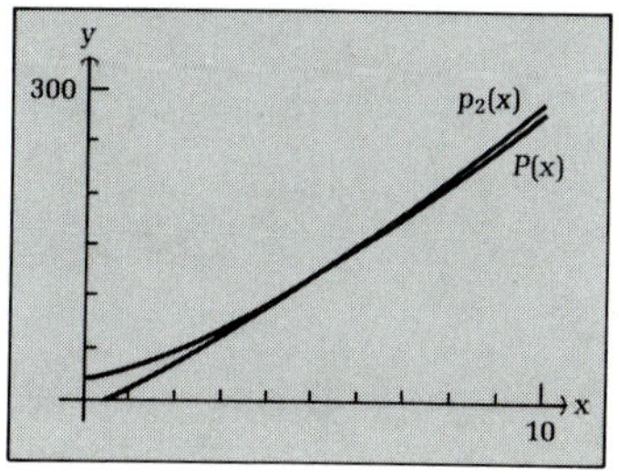

63.

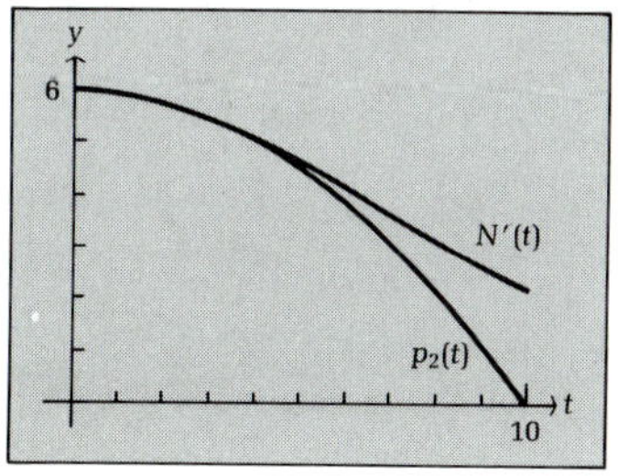

◆ EXERCISE 10-2

1. $|x| < 1$ or $-1 < x < 1$ **3.** $|x| < \frac{1}{2}$ or $-\frac{1}{2} < x < \frac{1}{2}$ **5.** $|x| < 1$ or $-1 < x < 1$ **7.** $|x| < 1$ or $-1 < x < 1$ **9.** $-\infty < x < \infty$
11. $|x-4| < 1$ or $3 < x < 5$ **13.** $|x+1| < 3$ or $-4 < x < 2$ **15.** $-\infty < x < \infty$ **17.** $|x-1| < \frac{1}{4}$ or $\frac{3}{4} < x < \frac{5}{4}$

19. $p_n(x) = 1 + 4x + \frac{4^2}{2!}x^2 + \cdots + \frac{4^n}{n!}x^n$; $1 + 4x + \frac{4^2}{2!}x^2 + \cdots + \frac{4^n}{n!}x^n + \cdots$; $-\infty < x < \infty$

21. $p_n(x) = 2x - \frac{2^2}{2}x^2 + \frac{2^3}{3}x^3 - \cdots + \frac{(-1)^{n-1}2^n}{n}x^n$; $2x - \frac{2^2}{2}x^2 + \frac{2^3}{3}x^3 - \cdots + \frac{(-1)^{n-1}2^n}{n}x^n + \cdots$; $|x| < \frac{1}{2}$ or $-\frac{1}{2} < x < \frac{1}{2}$

23. $p_n(x) = 1 + \frac{1}{2}x + \frac{1}{2^2}x^2 + \cdots + \frac{1}{2^n}x^n$; $1 + \frac{1}{2}x + \frac{1}{2^2}x^2 + \cdots + \frac{1}{2^n}x^n + \cdots$; $|x| < 2$ or $-2 < x < 2$

25. $p_n(x) = -(x-1) - \frac{1}{2}(x-1)^2 - \frac{1}{3}(x-1)^3 - \cdots - \frac{1}{n}(x-1)^n$; $-(x-1) - \frac{1}{2}(x-1)^2 - \frac{1}{3}(x-1)^3 - \cdots - \frac{1}{n}(x-1)^n - \cdots$; $|x-1| < 1$ or $0 < x < 2$

27. $p_n(x) = -1 + (x-2) - (x-2)^2 + \cdots + (-1)^{n-1}(x-2)^n$; $-1 + (x-2) - (x-2)^2 + \cdots + (-1)^{n-1}(x-2)^n + \cdots$; $|x-2| < 1$ or $1 < x < 3$

◆ EXERCISE 10-3

1. $1 + 2x + \frac{1}{2}x^2 + \cdots + \left[1 + \frac{(-1)^{n-1}}{n}\right]x^n + \cdots, -1 < x < 1$ **3.** $x^3 + x^4 + x^5 + \cdots + x^{n+3} + \cdots, -1 < x < 1$

5. $x + x^3 + \frac{1}{2!}x^5 + \cdots + \frac{1}{n!}x^{2n+1} + \cdots, -\infty < x < \infty$ **7.** $1 + x^3 + x^6 + \cdots + x^{3n} + \cdots, -1 < x < 1$

9. $\frac{1}{2} + \frac{1}{2^2}x + \frac{1}{2^3}x^2 + \cdots + \frac{1}{2^{n+1}}x^n + \cdots, -2 < x < 2$ **11.** $1 + 8x^3 + 8^2x^6 + \cdots + 8^nx^{3n} + \cdots, -\frac{1}{2} < x < \frac{1}{2}$

13. $\frac{1}{4} - \frac{1}{4^2}x^2 + \frac{1}{4^3}x^4 - \cdots + \frac{(-1)^n}{4^{n+1}}x^{2n} + \cdots, -2 < x < 2$

15. (A) $1 + x^2 + x^4 + x^6 + \cdots + x^{2n} + \cdots, -1 < x < 1$ (B) $2x + 4x^3 + 6x^5 + \cdots + 2nx^{2n-1} + \cdots, -1 < x < 1$

17. $x - \frac{1}{3}x^3 + \frac{1}{5}x^5 - \cdots + \frac{(-1)^n}{2n+1}x^{2n+1} + \cdots, -1 < x < 1$ **19.** $1 + (x-3) + (x-3)^2 + \cdots + (x-3)^n + \cdots, 2 < x < 4$

21. $(x-1) - \frac{1}{2}(x-1)^2 + \frac{1}{3}(x-1)^3 - \cdots + \frac{(-1)^{n-1}}{n}(x-1)^n + \cdots, 0 < x < 2$

23. $1 + 3(x-1) + 3^2(x-1)^2 + \cdots + 3^n(x-1)^n + \cdots, \frac{2}{3} < x < \frac{4}{3}$ **25.** $1 + 3x + 6x^2 + \cdots + \frac{n(n-1)}{2}x^{n-2} + \cdots, -1 < x < 1$

27. $1 + 2x + 2x^2 + \cdots + 2x^n + \cdots, -1 < x < 1$ **29.** $x + \frac{1}{3}x^3 + \frac{1}{5}x^5 + \cdots + \frac{1}{2n+1}x^{2n+1} + \cdots, -1 < x < 1$

31. $1 + \frac{1}{2!}x^2 + \frac{1}{4!}x^4 + \cdots + \frac{1}{(2n)!}x^{2n} + \cdots, -\infty < x < \infty$

33. $\ln a + \frac{1}{a}(x-a) - \frac{1}{2a^2}(x-a)^2 + \cdots + \frac{(-1)^{n-1}}{na^n}(x-a)^n + \cdots, 0 < x < 2a$

◆ EXERCISE 10-4

1. 0.4; $|R_1(0.6)| \le 0.18$ **3.** 0.544; $|R_3(0.6)| \le 0.0054$ **5.** 0.255; $|R_2(0.3)| \le 0.009$ **7.** 0.261 975; $|R_4(0.3)| \le 0.000\ 486$
9. 0.818 667; $|R_3(0.2)| \le 0.000\ 067$ **11.** 0.970 446; $|R_3(0.03)| \le 0.000\ 000\ 033$ **13.** 0.492; $|R_3(0.6)| \le 0.0324$
15. 0.058 272; $|R_3(0.06)| \le 0.000\ 003\ 24$ **17.** 0.6704; $n = 4$ **19.** 0.9608; $n = 2$ **21.** 0.095; $n = 2$ **23.** 0.01; $n = 1$
25. 0.1973 **27.** 0.0656 **29.** 0.0193 **31.** 1.345; $|R_2(0.3)| \le 0.0135$ **33.** 1.051 25; $|R_2(0.05)| \le 0.000\ 063$
35. 4.123 047; $|R_2(1)| \le 0.000\ 061$ **37.** $\int_0^1 \frac{10x^6}{4+x^2}\,dx \approx 0.302$; coefficient of inequality ≈ 0.698
39. $S(4) \approx 2.031$ within ± 0.004 or \$2,031,000 within $\pm$\$4,000 **41.** 2 yr; \$1,270 **43.** 30.539°C **45.** $A(2) \approx 6.32$ cm² within ± 0.031
47. $N(5) \approx 27.5$ wpm within ± 0.1875

◆ EXERCISE 10-5 CHAPTER REVIEW

1. $-6/(x+5)^4$ **2.** $1 + \frac{1}{3}x - \frac{1}{9}x^2 + \frac{5}{81}x^3$; 1.003 322 **3.** $2 + \frac{1}{4}(x-3) - \frac{1}{64}(x-3)^2 + \frac{1}{512}(x-3)^3$; 1.974 842 **4.** $3 + \frac{1}{6}x^2$; 3.001 667
5. $-\frac{1}{4} < x < \frac{1}{4}$ **6.** $1 < x < 11$ **7.** $-1 < x < 1$ **8.** $-\infty < x < \infty$ **9.** $(-1)^n 9^n e^{-9x}$

10. $\frac{1}{7} + \frac{1}{7^2}x + \frac{1}{7^3}x^2 + \cdots + \frac{1}{7^{n+1}}x^n + \cdots, -7 < x < 7$

11. $\ln 2 + \frac{1}{2}(x-2) - \frac{1}{8}(x-2)^2 + \cdots + \frac{(-1)^{n-1}}{n2^n}(x-2)^n + \cdots, 0 < x < 4$

12. $\frac{1}{10} - \frac{1}{10^2}x + \frac{1}{10^3}x^2 - \cdots + \frac{(-1)^n}{10^{n+1}}x^n + \cdots, -10 < x < 10$

13. $\frac{1}{4}x^2 + \frac{1}{4^2}x^4 + \frac{1}{4^3}x^6 + \cdots + \frac{1}{4^{n+1}}x^{2n+2} + \cdots, -2 < x < 2$

14. $x^2 + 3x^3 + \frac{3^2}{2!}x^4 + \cdots + \frac{3^n}{n!}x^{n+2} + \cdots, -\infty < x < \infty$ **15.** $x + \frac{1}{e}x^2 - \frac{1}{2e^2}x^3 + \cdots + \frac{(-1)^{n-1}}{ne^n}x^{n+1} + \cdots, -e < x < e$

16. $\frac{1}{2} + \frac{1}{2^2}(x-2) + \frac{1}{2^3}(x-2)^2 + \cdots + \frac{1}{2^{n+1}}(x-2)^n + \cdots, 0 < x < 4$

17. $f(x) = \frac{1}{2} + \frac{1}{2^2}x + \frac{1}{2^3}x^2 + \frac{1}{2^4}x^3 + \cdots + \frac{1}{2^{n+1}}x^n + \cdots, -2 < x < 2;$

$g(x) = \frac{1}{2^2} + \frac{1}{2^2}x + \frac{3}{2^4}x^2 + \cdots + \frac{n}{2^{n+1}}x^{n-1} + \cdots, -2 < x < 2$

18. $f(x) = x^2 - x^4 + x^6 - \cdots + (-1)^n x^{2n+2} + \cdots, -1 < x < 1;$
$g(x) = 2x - 4x^3 + 6x^5 - \cdots + (2n+2)(-1)^n x^{2n+1} + \cdots, -1 < x < 1$

19. $\frac{1}{3 \cdot 9}x^3 - \frac{1}{5 \cdot 9^2}x^5 + \frac{1}{7 \cdot 9^3}x^7 - \cdots + \frac{(-1)^n}{(2n+3)9^{n+1}}x^{2n+3} + \cdots, -3 < x < 3$

20. $\frac{1}{5 \cdot 16}x^5 + \frac{1}{7 \cdot 16^2}x^7 + \frac{1}{9 \cdot 16^3}x^9 + \cdots + \frac{1}{(2n+5)16^{n+1}}x^{2n+5} + \cdots, -4 < x < 4$

21. 1.78; $|R_2(0.6)| \le 0.108$ **22.** 1.0618; $|R_2(0.06)| \le 0.000\ 108$ **23.** 0.261 975; $n = 4$ **24.** 0.03; $n = 1$ **25.** 0.0612 **26.** 0.314

27. $p_2(x) = 5 - \frac{1}{1{,}000}x^2$; \$4.93 **28.** $p_2(t) = 9 - 0.03t^2$; \$80,000 **29.** $\int_0^1 \frac{18x^3}{8 + x^2}\,dx \approx 0.516$; coefficient of inequality ≈ 0.484

30. $S(8) \approx 3.348$ within ± 0.001 or \$3,348,000 within $\pm$\$1,000 **31.** $A(2) \approx 10.17$ cm² within ± 0.01 **32.** $\frac{1}{5}\int_0^5 \frac{5{,}000t^2}{10{,}000 + t^4}\,dt \approx 4.06$

33. $\frac{1}{5}\int_0^5 (10 + 2t - 5e^{-0.01t^2})\,dt \approx 10.4$

CHAPTER 11

◆ EXERCISE 11-1

1. $x_n = \frac{x_{n-1}^2 + 4}{2x_{n-1}}$; $x_5 = 2.000\ 000\ 1$ **3.** $x_n = \frac{2x_{n-1}^3 + 8}{3x_{n-1}^2}$; $x_5 = 2.000\ 000\ 2$ **5.** $x_n = \frac{e^{x_{n-1}}(x_{n-1} - 1)}{e^{x_{n-1}} + 1}$; $x_5 = -0.567\ 143\ 29$

7. $x_n = \frac{x_{n-1}(1 - \ln x_{n-1})}{1 + x_{n-1}}$; $x_5 = 0.567\ 143\ 29$ **9.** $x_n = \frac{x_{n-1}(1 + x_{n-1}^2 - \ln x_{n-1})}{1 + 2x_{n-1}^2}$; $x_5 = 0.652\ 918\ 64$

11. 0.298 437 88, 6.701 562 1

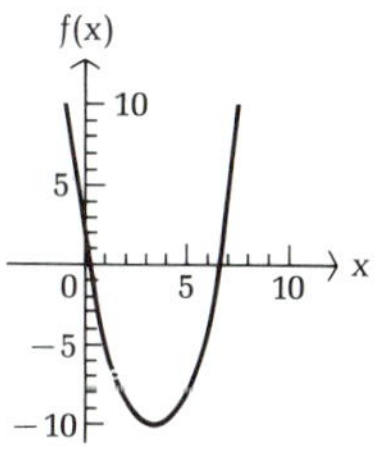

13. −1.556 773 3

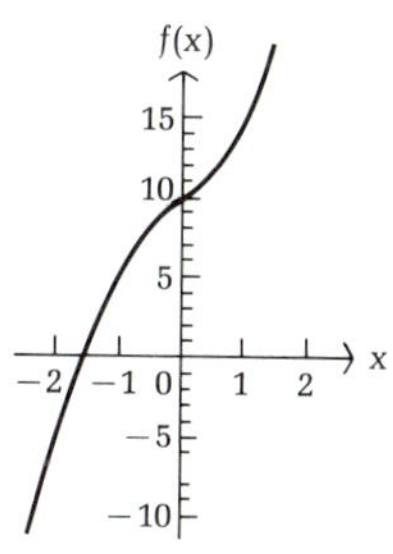

15. −1.286 772, 1.443 623 4, 11.843 149

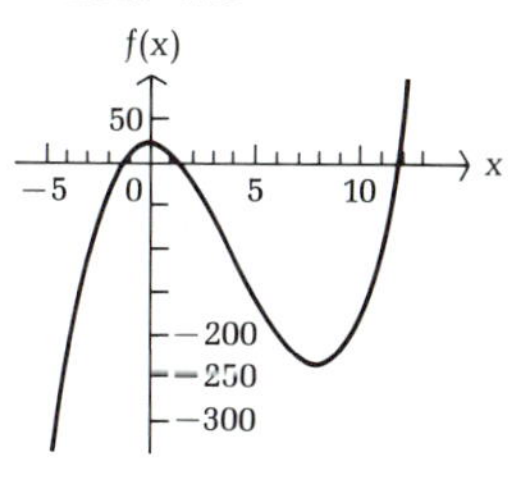

17. 0.628 708 31, 5.444 569 9

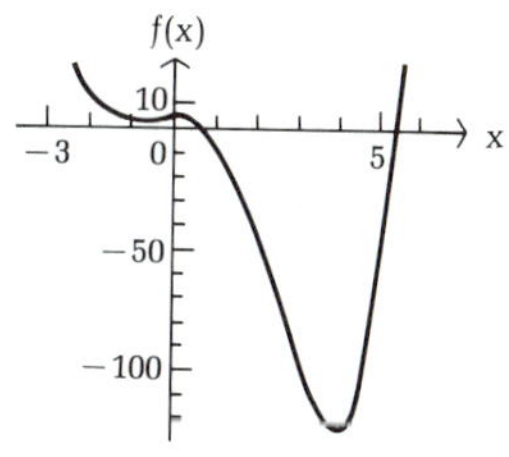

19. 1.796 321 9

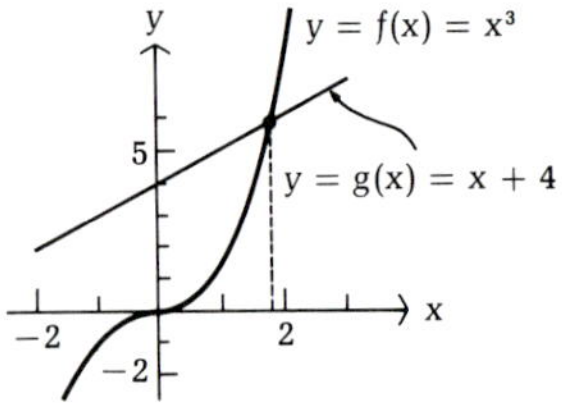

21. 9.855 959 5

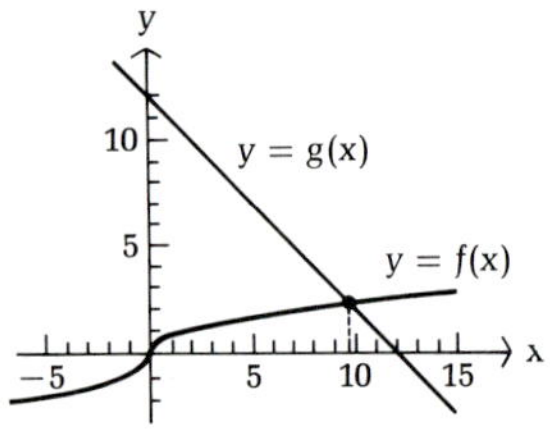

23. −1.841 405 7, 1.146 193 2

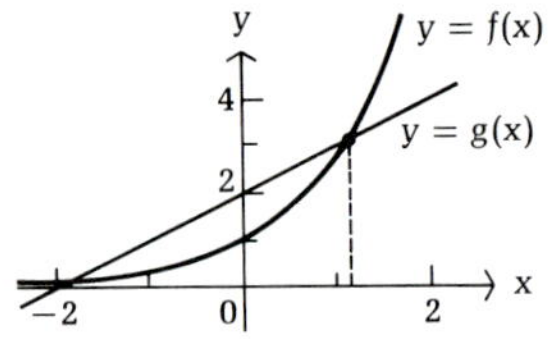

27. $x_n = \dfrac{p-1}{p} x_{n-1} + \dfrac{A}{p x_{n-1}^{p-1}}$ **29.** x_n oscillates between −1 and +1 **31.** At $x = 1$: $y = \dfrac{\sqrt{2}}{4}(x+1)$; at $x = -1$: $y = \dfrac{\sqrt{2}}{4}(x-1)$

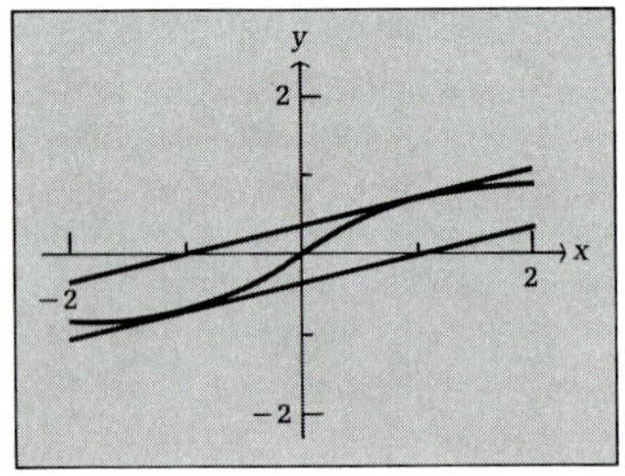

33. 9,788 units or 35,627 units **35.** \$2.06 **37.** 23.74% **39.** 11.52% **41.** $2.3 < t < 9.4$ **43.** 3.3 hr after the first injection **45.** 9.2 hr

◆ EXERCISE 11-2

1. $p(x) = 2 + 2(x-1) + (x-1)(x-3)$ **3.** $p(x) = 6 - (x+1) + 2(x+1)x - 2(x+1)x(x-2)$

5.

x_k	y_k	1st D.D.	2nd D.D.
1	4		
		4	
2	8		1
		6	
3	14		

$p(x) = 4 + 4(x-1) + (x-1)(x-2)$

7.

x_k	y_k	1st D.D.	2nd D.D.	3rd D.D.
−1	−3			
		4		
0	1		−1	
		2		1
1	3		2	
		6		
2	9			

$p(x) = -3 + 4(x+1) - (x+1)x + (x+1)x(x-1)$

9.

x_k	y_k	1st D.D.	2nd D.D.	3rd D.D.
−2	25			
		−5		
1	10		3	
		7		−1
2	17		−3	
		−2		
4	13			

$p(x) = 25 - 5(x+2) + 3(x+2)(x-1) - (x+2)(x-1)(x-2)$

11. $p(x) = -64 + 24(x+4) - 4(x+4)x + (x+4)x(x-4)$; (A) 8 (B) 56 **13.** $p(x) = \frac{1}{4}(x+1)x(x-1)$; (A) 1.5 (B) 6
15. $p(x) = 24 - 11(x+4) + \frac{5}{2}(x+4)(x+2) - \frac{1}{2}(x+4)(x+2)x + \frac{1}{8}(x+4)(x+2)x(x-2)$; (A) $\frac{57}{8} = 7.125$ (B) $-\frac{23}{8} = -2.875$

17. $p(x) = -24 + 18(x+3) - 6(x+3)(x+2) + (x+3)(x+2)(x+1)$; (A) $\frac{3}{8} = 0.375$ (B) $\frac{105}{8} = 13.125$

19. $p(x) = 2 - (x+2) + \frac{1}{2}(x+2)x = \frac{1}{2}x^2$

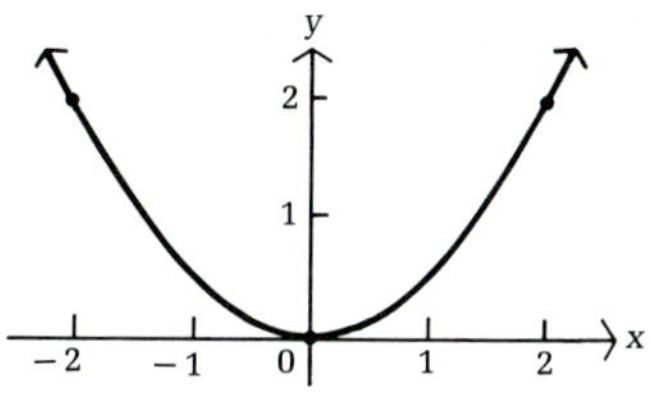

21. $p(x) = -4 + 2x$

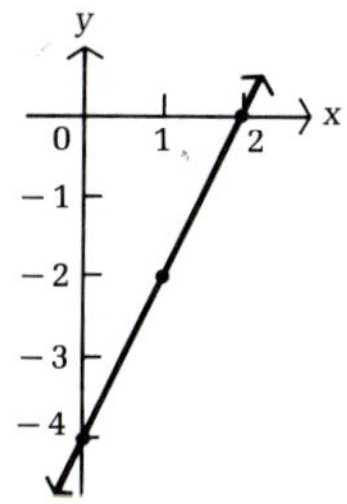

23. $p(x) = 2(x+1) - (x+1)x = 2 + x - x^2$

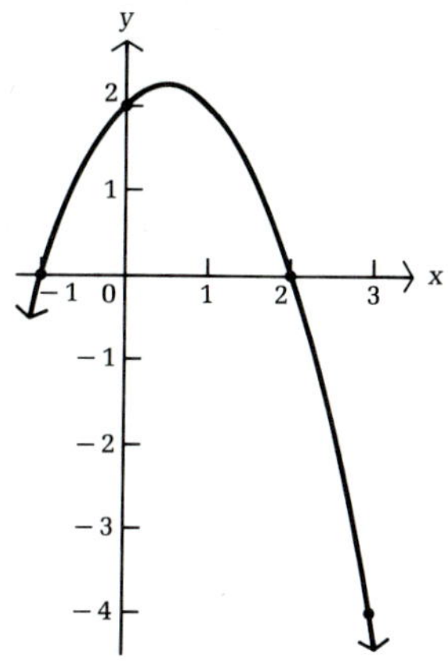

25. $p(x) = 1 + 4(x+2) - 3(x+2)(x+1) + (x+2)(x+1)x$
$= x^3 - 3x + 3$

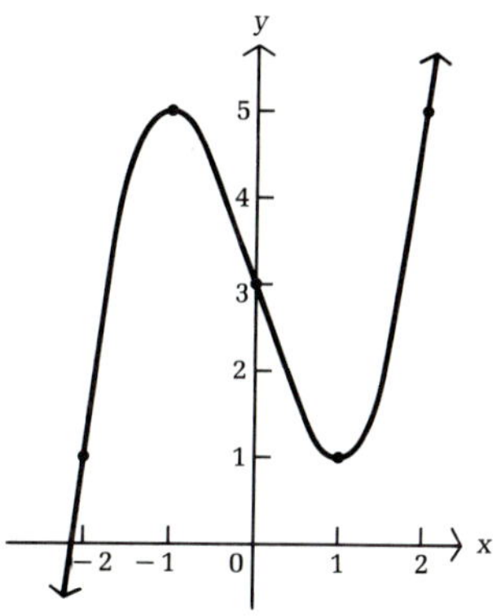

27. $p(x) = -3 + 3(x+2) + (x+2)(x+1) - 2(x+2)(x+1)x + (x+2)(x+1)x(x-1)$
$= 5 - 6x^2 + x^4$

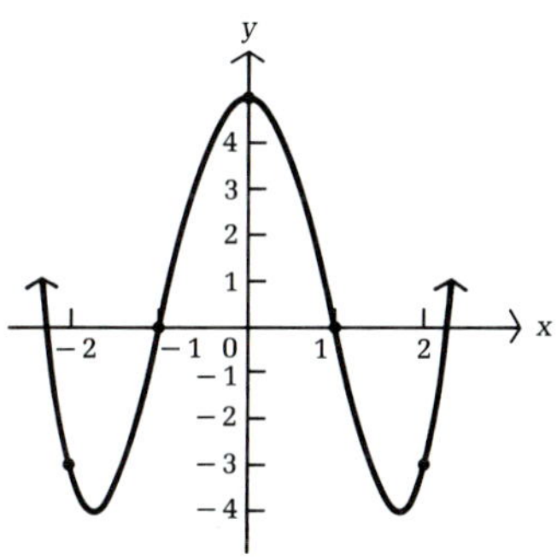

29. $p(x) = 1 + \frac{1}{3}(x-1) - \frac{1}{60}(x-1)(x-4)$

x	1	2	3	4	5	6	7	8	9
p(x)	1	1.4	1.7	2	2.3	2.5	2.7	2.9	3
$\sqrt{x}$	1	1.4	1.7	2	2.2	2.4	2.6	2.8	3

31. $p(x) = -4 - (x + 2) + 3(x + 2)(x + 1) - (x + 2)(x + 1)x$
$= 6x - x^3$

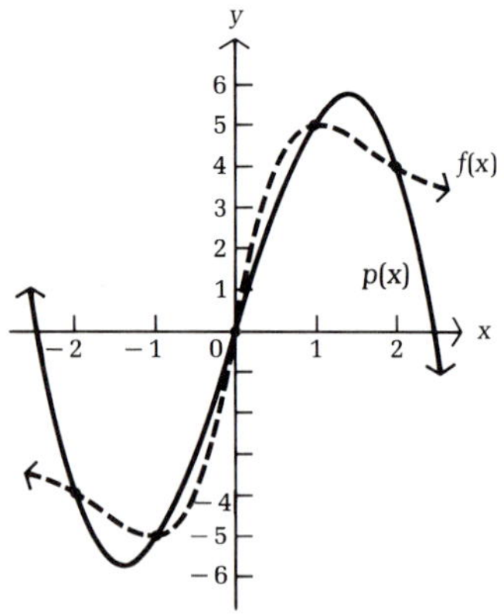

33. $y = y_2 + (y_1 - y_2)x^2/x_1^2$ **35.** $p(t) = 2 + 7.5t - 0.75t(t - 4) = 2 + 10.5t - 0.75t^2$; $C(6) \approx 38$ or \$38,000; approx. \$15,500
37. $p(x) = 0.5x + 3.125x(x - 0.2)(x - 0.8) = x - 3.125x^2 + 3.125x^3$; 0.5208
39. $p(x) = 24.4 + 5.8(x - 2) - 1.6(x - 2)(x - 4) = 15.4x - 1.6x^2$; $R(5) \approx 37$ or \$37,000; approx. 4,813 units
41. $p(t) = 14 - t + 2t(t - 1) - t(t - 1)(t - 2) = 14 - 5t + 5t^2 - t^3$; approx. 14.7°C
43. $p(t) = 450 - 130t + 20t(t - 2) = 450 - 170t + 20t^2$; approx. 4.25 days
45. $p(t) = 10{,}000 + 350t + 15t(t - 10) - t(t - 10)(t - 20) = 10{,}000 + 45t^2 - t^3$; approx. 14,000 registered voters

◆ EXERCISE 11-3

1. $\frac{\Delta x}{2}[f(x_0) + 2f(x_1) + 2f(x_2) + 2f(x_3) + 2f(x_4) + f(x_5)]$ **3.** $\frac{\Delta x}{3}[f(x_0) + 4f(x_1) + 2f(x_2) + 4f(x_3) + 2f(x_4) + 4f(x_5) + f(x_6)]$

5.

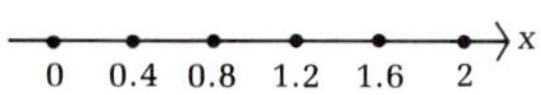

$0.2[f(0) + 2f(0.4) + 2f(0.8) + 2f(1.2) + 2f(1.6) + f(2)]$

7.

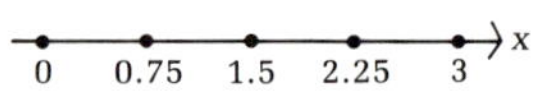

$0.25[f(0) + 4f(0.75) + 2f(1.5) + 4f(2.25) + f(3)]$

9. 1.1150 **11.** 4.5469 **13.** 1.4637 **15.** 1.0043 **17.** 10.25 **19.** 74.5 **21.** 5.535 70 **23.** 0.387 52
25. Exact value = 1.718 282

n	T_n	$Err(T_n)$	S_n	$Err(S_n)$
4	1.727 222	−0.008 940	1.718 319	−0.000 037
10	1.719 713	−0.001 431	1.718 283	−0.000 001

27. $|Err(T_n)| \le 2/(3n^2)$; $|Err(S_n)| \le 16/(45n^4)$

n	$2/(3n^2)$	$16/(45n^4)$
10	0.01	0.000 04
20	0.002	0.000 002
30	0.0007	0.000 000 4
40	0.0004	0.000 000 1
50	0.0003	0.000 000 06

29. $f''(x) = 0$ implies $Err(T_n) = 0$ implies $\int_a^b f(x)\,dx = T_n$ **31.** 1.1114 **33.** 4.5513 **35.** 1.4627 **37.** 1.0034
39. $\bar{p} = \$12$; $CS \approx 24.77$; $PS \approx 22.08$ **41.** \$9,333 **43.** 140,667 ft² **45.** 15 m **47.** 10.76 **49.** 21.25°C **51.** 25 items

◆ EXERCISE 11-4

1.

k	x_k	y_k
0	0	2
1	0.2	2.4
2	0.4	2.92
3	0.6	3.58
4	0.8	4.42
5	1	5.46

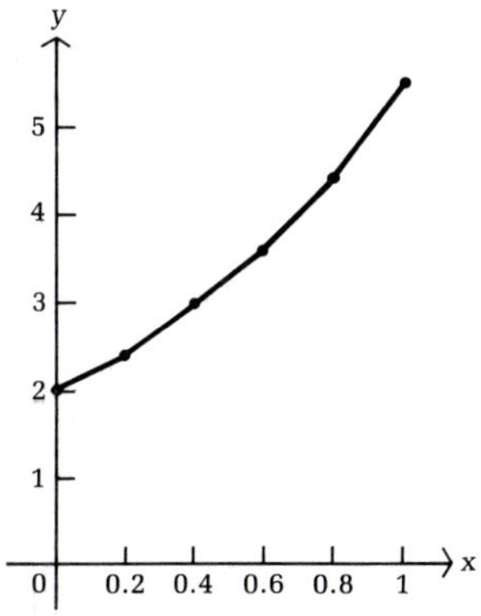

3.

k	x_k	y_k
0	0	1
1	0.2	1.2
2	0.4	1.44
3	0.6	1.73
4	0.8	2.07
5	1	2.49

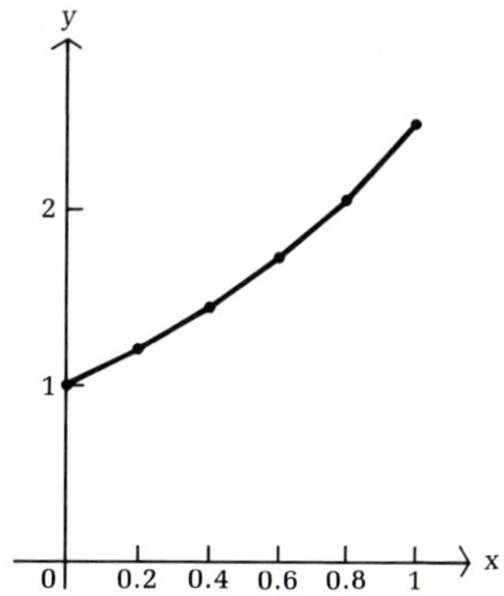

5.

k	x_k	y_k
0	0	1
1	0.2	1.4
2	0.4	1.55
3	0.6	1.41
4	0.8	0.93
5	1	0.03

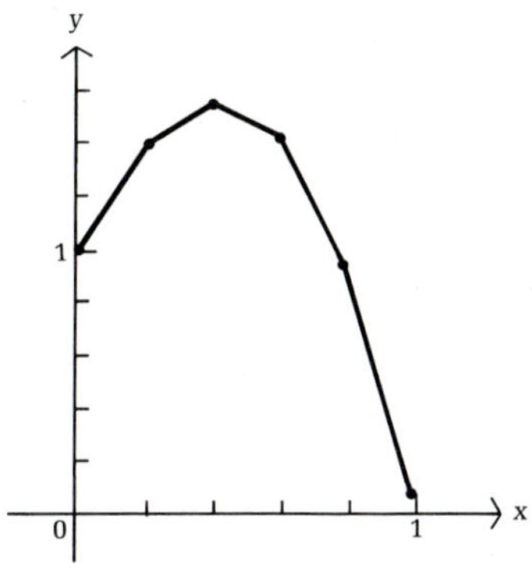

7.

k	x_k	y_k
0	0	1
1	0.2	0.8
2	0.4	0.64
3	0.6	0.51
4	0.8	0.41
5	1	0.33

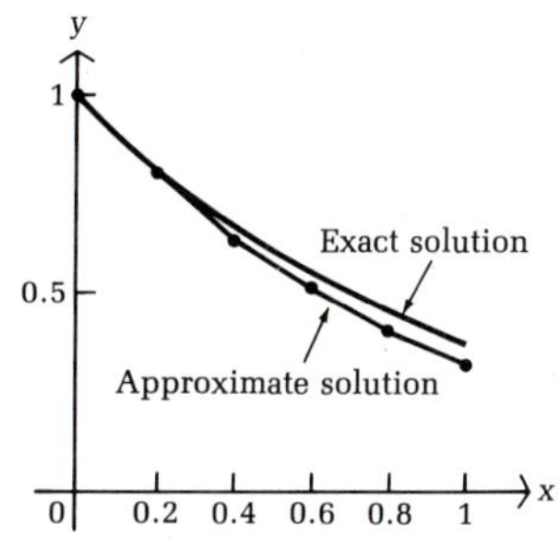

9.

k	x_k	y_k
0	0	0
1	0.2	0.2
2	0.4	0.44
3	0.6	0.73
4	0.8	1.07
5	1	1.49

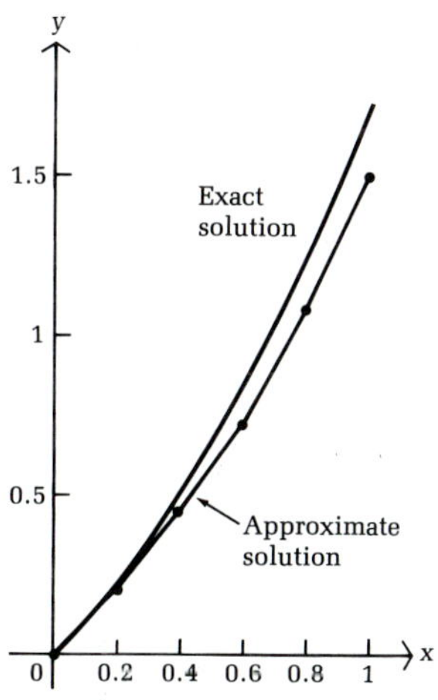

11. $\Delta x = 0.2$:

k	x_k	y_k
0	0	1
1	0.2	1.2
2	0.4	1.2
3	0.6	0.96
4	0.8	0.43
5	1	−0.44

$\Delta x = 0.1$:

k	x_k	y_k
0	0	1
1	0.1	1.1
2	0.2	1.15
3	0.3	1.15
4	0.4	1.08
5	0.5	0.95
6	0.6	0.74
7	0.7	0.46
8	0.8	0.08
9	0.9	−0.39
10	1	−0.97

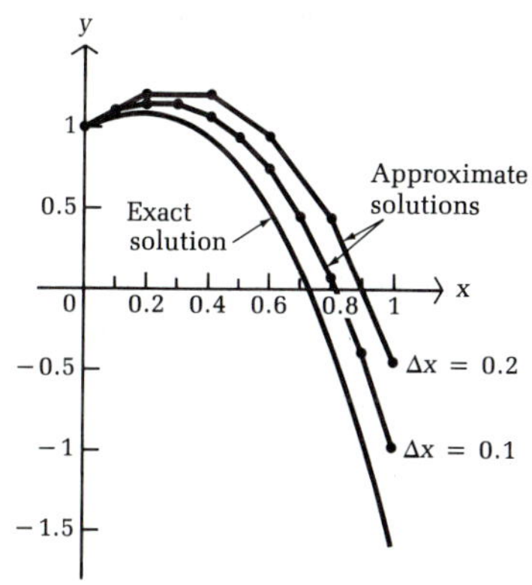

13. 1.29 **15.** −2.81 **17.** 1.10 **19.** 1.49 **21.**

x_k	1	1.6	2.24	2.94	3.73	4.64
y_k	−1	0	0.96	1.92	2.92	3.99

23.

x_k	2	2.6	3.56	5.10	7.56	11.50
y_k	−1	−0.4	0.56	2.10	4.56	8.50

25. (A)

k	x_k	y_k
0	0	1
1	0.2	0.6
2	0.4	0.49
3	0.6	0.61
4	0.8	0.90
5	1	1.32

(B)

k	x_k	y_k
0	0	2
1	0.2	1.43
2	0.4	1.16
3	0.6	1.13
4	0.8	1.30
5	1	1.64

(C)

k	x_k	y_k
0	0	3
1	0.2	2.31
2	0.4	1.90
3	0.6	1.75
4	0.8	1.82
5	1	2.08

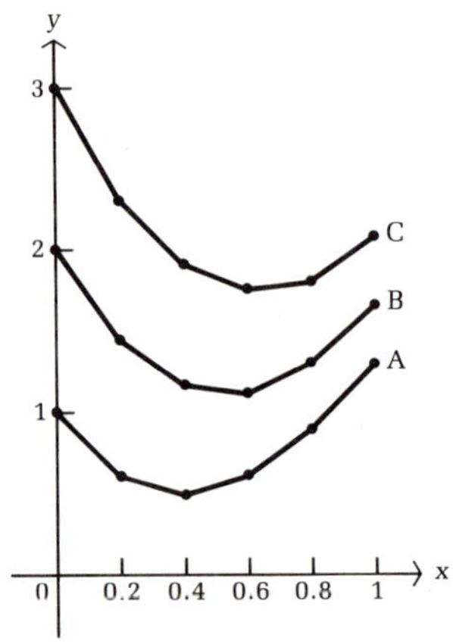

27.

Δx	1	0.5	0.25
Approximation	3.4	3.3	3.2

29. 18,649 **31.** $p(2) \approx 222.5$; $q(2) \approx 308.8$ **33.** 6,213 **35.** 87,900 rabbits; 24,200 foxes **37.** 7,567

◆ EXERCISE 11-5 CHAPTER REVIEW

1. $x_n = \dfrac{x_{n-1}^2 + 10}{2x_{n-1}}$; $x_4 \approx 3.162\ 277\ 7$ **2.** $x_n = \dfrac{-x_{n-1} \ln x_{n-1}}{1 + x_{n-1}}$; $x_5 = 0.278\ 464\ 54$ **3.** $p(x) = 11 + (x-1) - (x-1)(x-2)$

4.

x_k	y_k	1st D.D.	2nd D.D.	3rd D.D.
−1	1			
		5		
0	6		−3	
		−1		2
1	5		3	
		5		
2	10			

$p(x) = 1 + 5(x+1) - 3(x+1)x + 2(x+1)x(x-1)$

5. 0.881 **6.** 9.276

7.

k	x_k	y_k
0	0	3
1	0.2	1.8
2	0.4	1.08
3	0.6	0.65
4	0.8	0.39
5	1	0.23

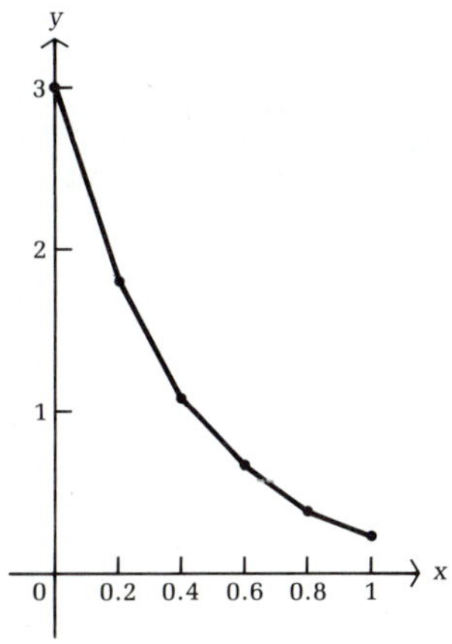

8.

k	x_k	y_k
0	0	1
1	0.2	1.2
2	0.4	1.24
3	0.6	1.09
4	0.8	0.71
5	1	0.05

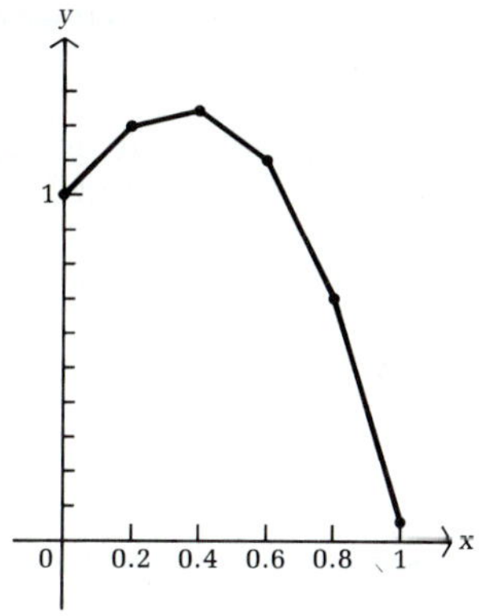

9. $-3.582\ 918\ 7, 0.251\ 322\ 86, 3.331\ 595\ 8$ **10.** $-1.132\ 933\ 2$ **11.** $0.772\ 882\ 96$

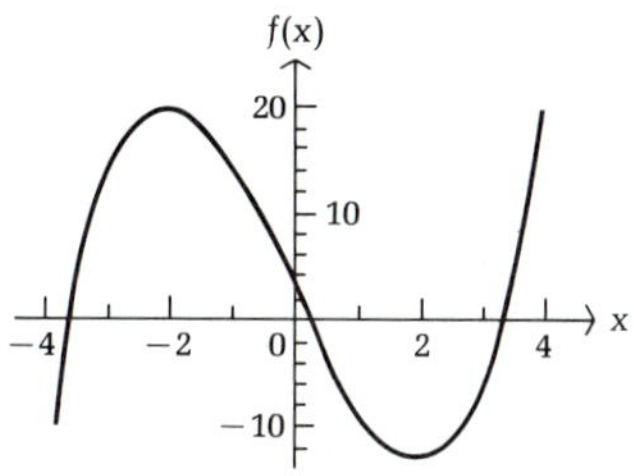

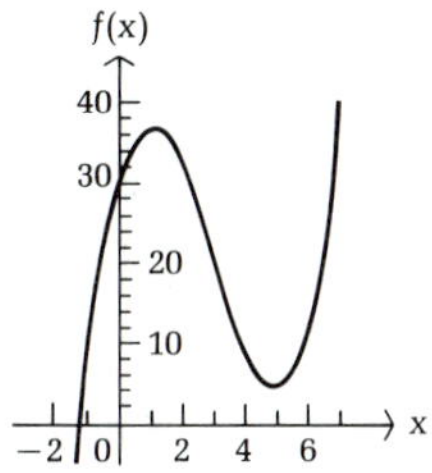

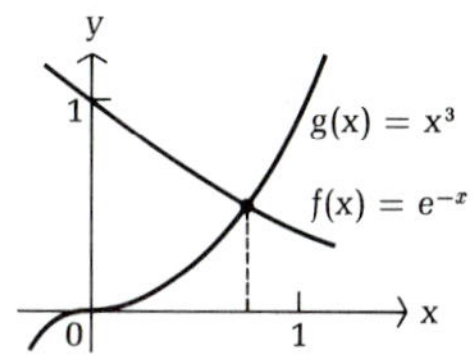

12. $-1.380\ 277\ 6, 0.819\ 172\ 51$ **13.** $p(x) = 16 - 6(x-1) + 2(x-1)(x-2) + (x-1)(x-2)(x-3)$; $f(4) \approx 16$

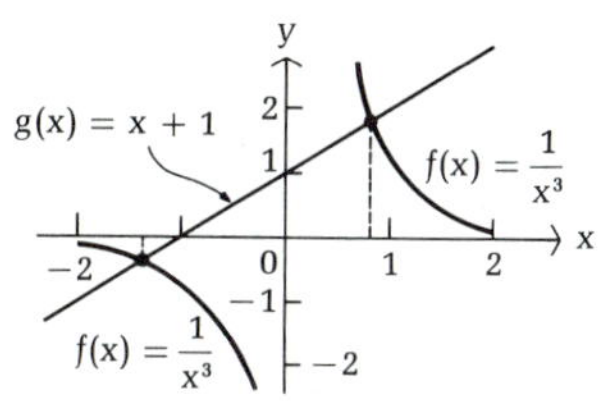

14. $p(x) = 20 + 10(x+4) + 8(x+4)(x+2) - 4(x+4)(x+2)x + (x+4)(x+2)x(x-2)$; $f(1) \approx 115$

15. $p(x) = 10 - 4x + x(x-2)$ **16.** $p(x) = 16 - 2(x+3) + 3(x+3)(x+1) - 2(x+3)(x+1)(x-1)$

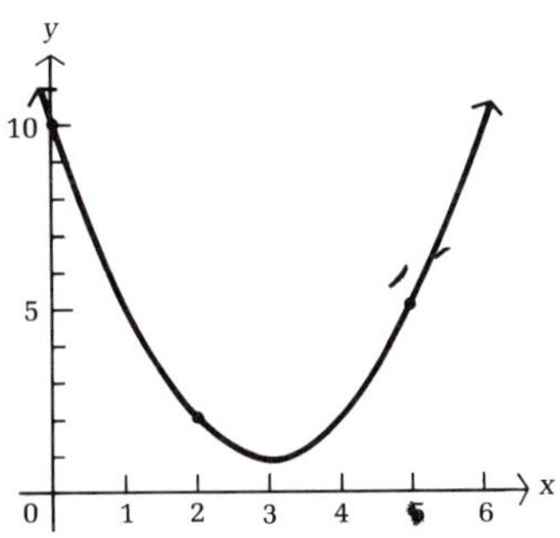

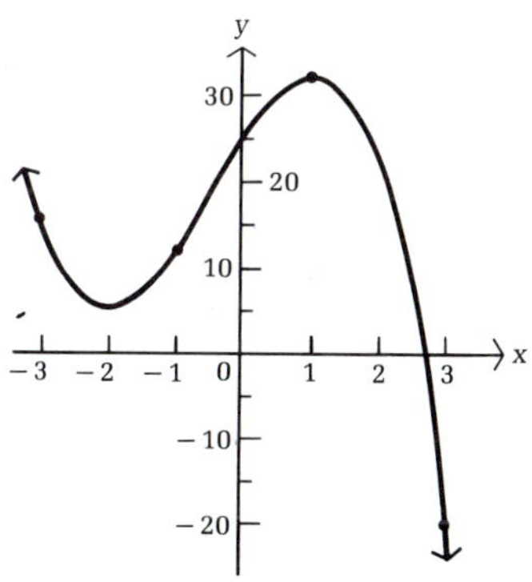

17. 0.650 45 **18.** 5.368 99 **19.** 48.5 **20.** 1.118 43 **21.**

k	x_k	y_k
0	0	1
1	0.2	1
2	0.4	0.96
3	0.6	0.88
4	0.8	0.74
5	1	0.52

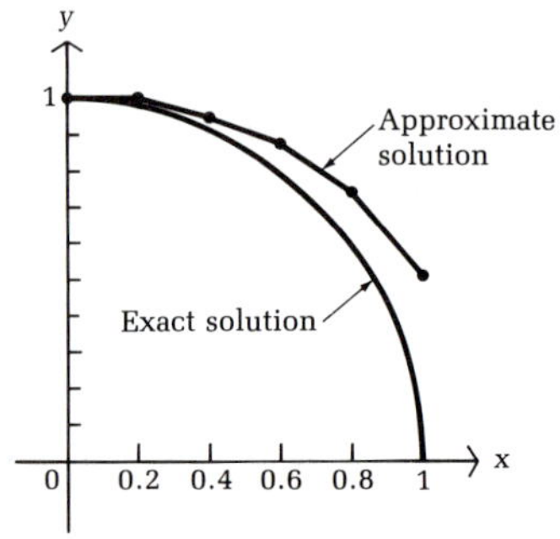

22. 0.64 **23.** 3.318 967 6 **24.**

x_k	1	1.2	1.8	2.9	5.0	8.7
y_k	0	0.8	1.9	3.7	6.8	12.2

25. $x_n = \dfrac{ax_{n-1}^2 - c}{2ax_{n-1} + b}$

26. $p(x) = 8 - 3(x+2) - (x+2)(x+1) + 2(x+2)(x+1)x - (x+2)(x+1)x(x-1) = 6x^2 - x^4$

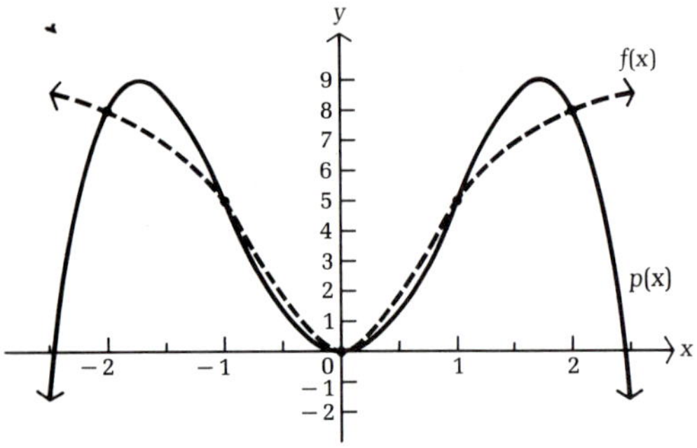

27. Exact value = 6.4

n	T_n	$\text{Err}(T_n)$	S_n	$\text{Err}(S_n)$
4	7.063	−0.663	6.417	−0.017
8	6.566	−0.166	6.401	−0.001

28. $\Delta x = 0.2$:

k	x_k	y_k
0	0	1
1	0.2	1.6
2	0.4	2.04
3	0.6	2.26
4	0.8	2.16
5	1	1.62

$\Delta x = 0.1$:

k	x_k	y_k
0	0	1
1	0.1	1.3
2	0.2	1.56
3	0.3	1.77
4	0.4	1.93
5	0.5	2.01
6	0.6	2.01
7	0.7	1.92
8	0.8	1.70
9	0.9	1.34
10	1	0.81

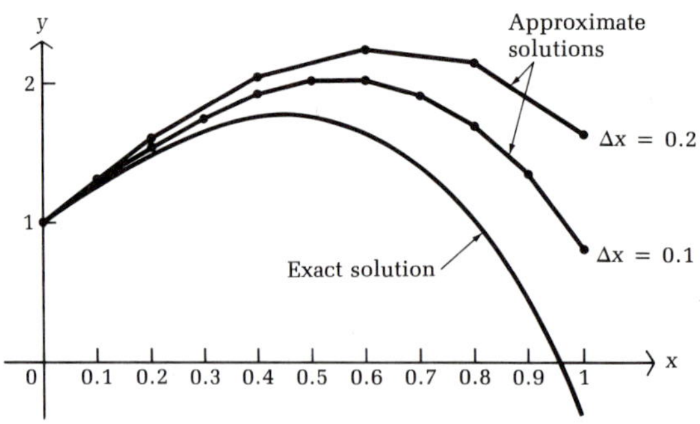

29. $x \approx 5.491\ 667\ 2$ **30.** 16.28%
31. $p(x) = 22x + 3x(x-2) - 2x(x-2)(x-4) = 15x^2 - 2x^3$; $R(3) \approx 81$ or \$81,000; approx. 5,000 units
32. $\overline{p} = \$6$; $CS \approx 14.06$; $PS \approx 6.41$ **33.** 2 yr; \$2,300 **34.** \$658,000 **35.** $0.6 < t < 6.4$ **36.** 95 torrs **37.** 24,100 foxes; 50,900 rabbits **38.** 6

CHAPTER 12

◆ EXERCISE 12-1

1. $\mu = 0$
$\sigma = 1.095\ 445\ 1$

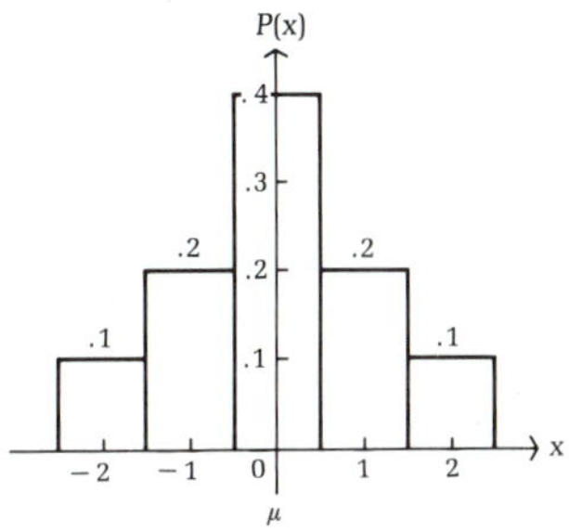

3. $\mu = -.9$
$\sigma = 1.374\ 772\ 7$

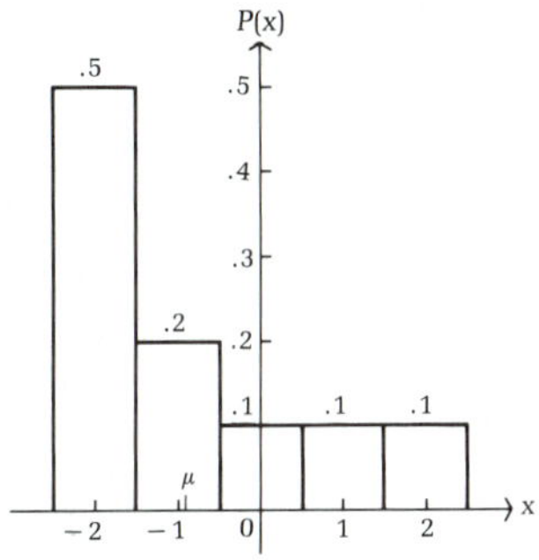

5. $S = \{1, 2, 3, 4, 5, 6, 7, 8, 9, 10\}$
7. .5 **9.** 5.5 **11.** 1

13.

x_i	0	1	2	3	4
p_i	$\frac{1}{16}$	$\frac{4}{16}$	$\frac{6}{16}$	$\frac{4}{16}$	$\frac{1}{16}$

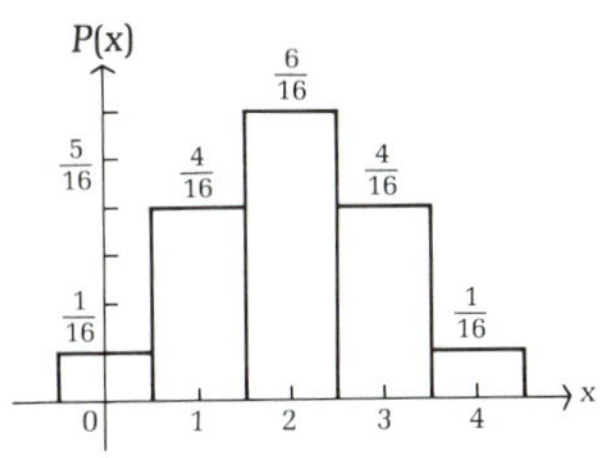

15. $\frac{1}{2}$ **17.** 2

19.

x_i	2	3	4	5	6	7	8	9	10	11	12
p_i	$\frac{1}{36}$	$\frac{2}{36}$	$\frac{3}{36}$	$\frac{4}{36}$	$\frac{5}{36}$	$\frac{6}{36}$	$\frac{5}{36}$	$\frac{4}{36}$	$\frac{3}{36}$	$\frac{2}{36}$	$\frac{1}{36}$

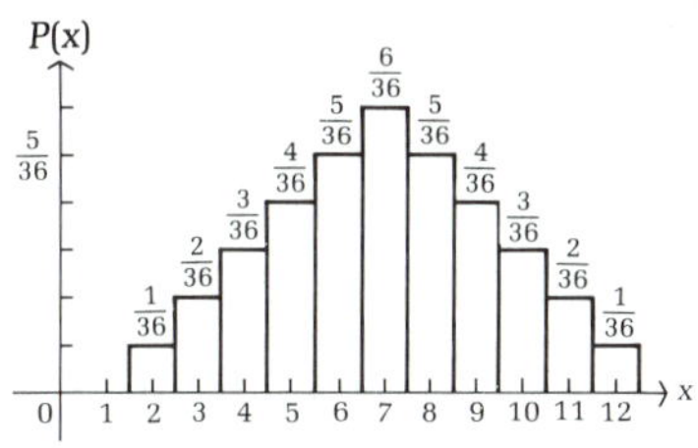

21. $\frac{5}{18}$ **23.** 7 **25.** −\$0.50 **27.** −\$0.25 **29.** −\$0.052 631 58

31. (A)

x_i	\$4,950	−\$50
p_i	.005	.995

(B) $E(X) = -\$25$

33. Payoff table:

x_i	\$4,850	−\$150
p_i	.01	.99

$E(X) = -\$100$

35. Site A, with $E(X) = \$3.6$ million **37.** 1.54 **39.** For A_1, $E(X) = \$4$, and for A_2, $E(X) = \$4.80$; A_2 is better

◆ EXERCISE 12-2

1. $\frac{3}{8} = .375$
3. $\frac{1}{8} = .125$
5. .230
7. $C_{3,2}(.5)^2(.5) = .375$
9. $C_{3,0}(.5)^0(.5)^3 = .125$
11. $C_{3,2}(.5)^2(.5) + C_{3,3}(.5)^3(.5)^0 = .500$

13. $\mu = .6$; $\sigma = .65$

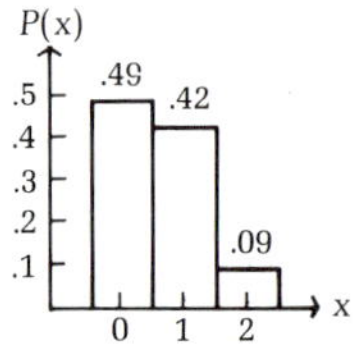

15. $\mu = 2$; $\sigma = 1$

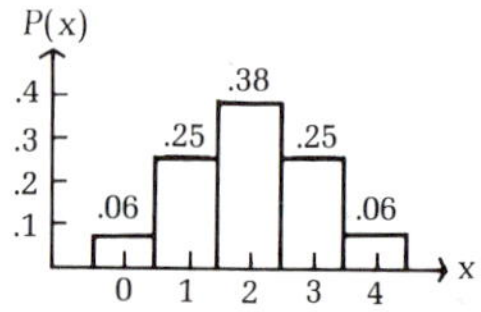

17. $C_{4,3}(\frac{1}{6})^3(\frac{5}{6}) \approx .0154$
19. $C_{4,0}(\frac{1}{6})^0(\frac{5}{6})^4 \approx .482$
21. $1 - [C_{4,0}(\frac{1}{6})^0(\frac{5}{6})^4] \approx .518$
23. (A) .311 (B) .437

25. $\mu = 2.4$; $\sigma = 1.2$

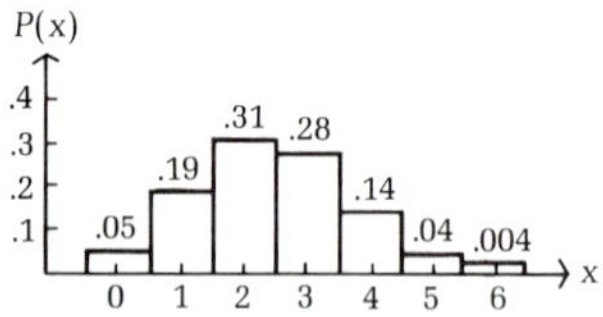

27. $\mu = 2.4$; $\sigma = 1.3$

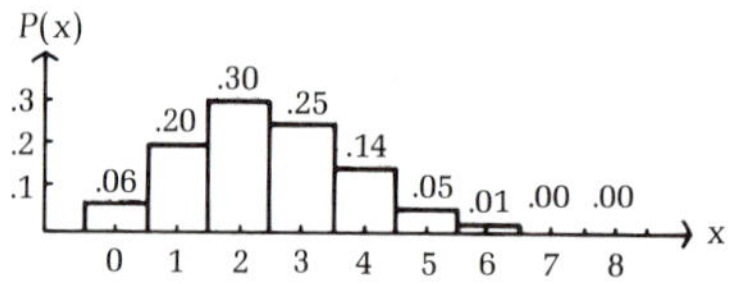

29. .238

31. (A) .318 (B) .647 33. .0188
37. .998 39. (A) .001 (B) .264 (C) .897

35. (A) $P(x) = C_{6,x}(.5)^x(.95)^{6-x}$
(D) $\mu = .30$; $\sigma = .53$

(B)

x	P(x)
0	.735
1	.232
2	.031
3	.002
4	.000
5	.000
6	.000

(C)

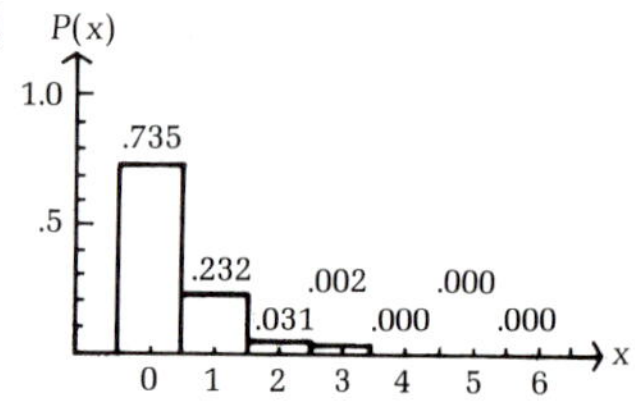

41. (A) $P(x) = C_{6,x}(.6)^x(.4)^{6-x}$

(B)

x	P(x)
0	.004
1	.037
2	.138
3	.276
4	.311
5	.187
6	.047

(C)

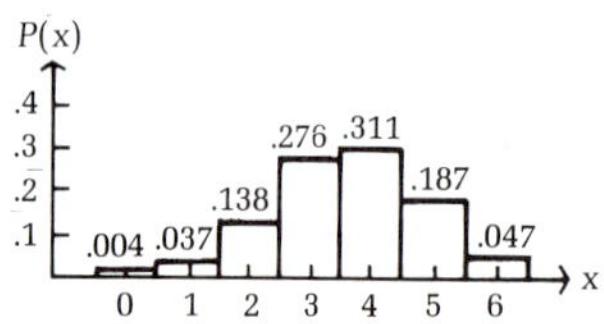

(D) $\mu = 3.6$; $\sigma = 1.2$

43. .000 864 45. (A) $P(x) = C_{5,x}(.2)^x(.8)^{5-x}$

(B)

x	P(x)
0	.328
1	.410
2	.205
3	.051
4	.006
5	.000

(C)

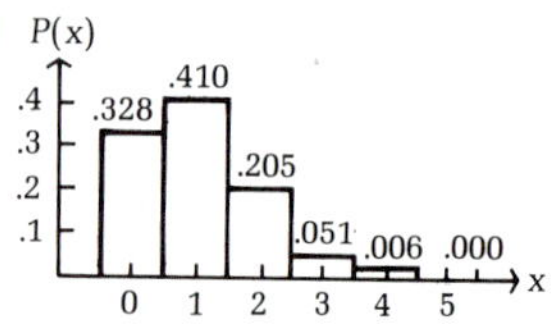

(D) $\mu = 1$; $\sigma = .89$

47. (A) .0041 (B) .0467 (C) .138 (D) .959

◆ EXERCISE 12-3

1. .0183 3. .0243 5. .6767 7. .3233 9. Binomial: .1814; Poisson: .1804 11. Binomial: .0126; Poisson: .0126
13. (A) .0498 (B) .1008 15. $N \approx 11{,}513$ 17. (A) .6065 (B) .9856 (C) .0144 19. (A) .3679 (B) .0153 (C) .0190
21. (A) .0369 (B) .7408 23. (A) .6703 (B) .0536 (C) .0616 25. (A) .1353 (B) .9473 (C) .0527 27. (A) .0107 (B) .0138
29. (A) .0498 (B) 461,000 per liter 31. (A) .0067 (B) .4972 (C) .1247 33. (A) .9161 (B) .0839 (C) .0620

◆ EXERCISE 12-4

1. $f(x) \geq 0$ from graph
$\int_0^4 f(x)\,dx = 1$

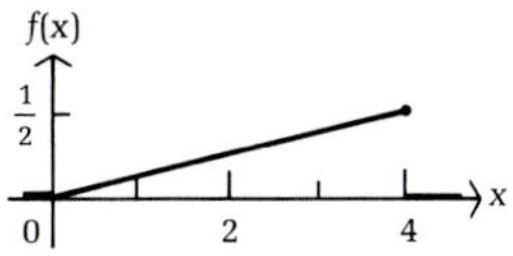

3. (A) $\int_1^3 \frac{1}{8}x\,dx = \frac{1}{2}$

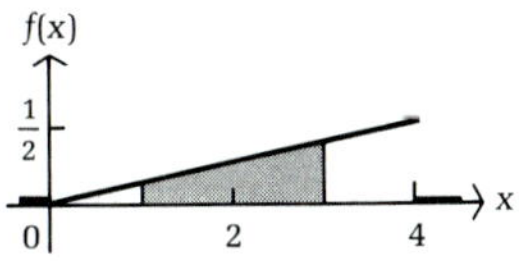

(B) $\int_0^2 \frac{1}{8}x\,dx = \frac{1}{4}$

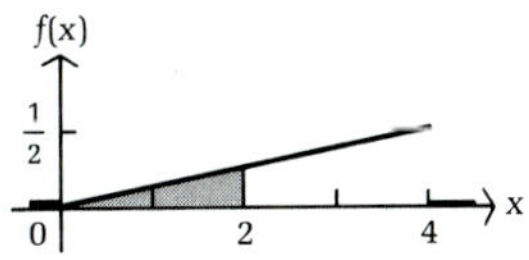

(C) $\int_3^4 \frac{1}{8}x\,dx = \frac{7}{16}$

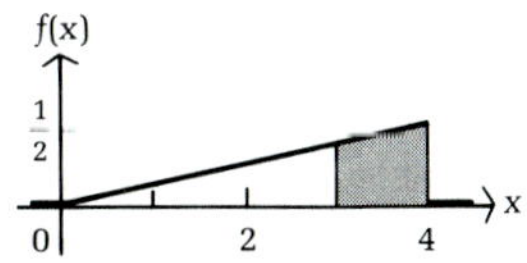

5. (A) $\int_1^1 f(x)\,dx = 0$ (B) $\int_5^\infty f(x)\,dx = 0$ (C) $\int_{-\infty}^5 f(x)\,dx = 1$ **7.** $F(x) = \begin{cases} 0 & \text{if } x < 0 \\ \frac{1}{16}x^2 & \text{if } 0 \leq x \leq 4 \\ 1 & \text{if } x > 4 \end{cases}$

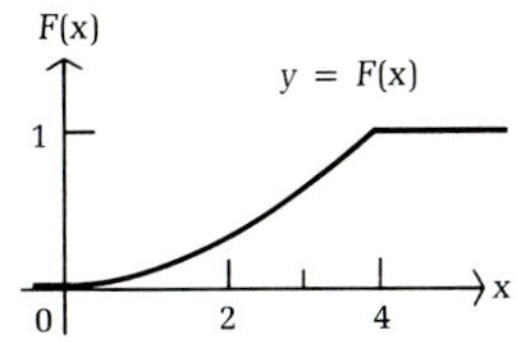

9. (A) $F(4) - F(2) = \frac{3}{4}$ (B) $F(2) - F(0) = \frac{1}{4}$ **11.** (A) 2 (B) $\frac{4}{3}$

13. $f(x) \geq 0$ from graph **15.** (A) $\int_1^4 \frac{2}{(1+x)^3}\,dx = .21$ (B) $\int_3^\infty \frac{2}{(1+x)^3}\,dx = \frac{1}{16}$ (C) $\int_0^2 \frac{2}{(1+x)^3}\,dx = \frac{8}{9}$

$\int_0^\infty \frac{2}{(1+x)^3}\,dx = 1$

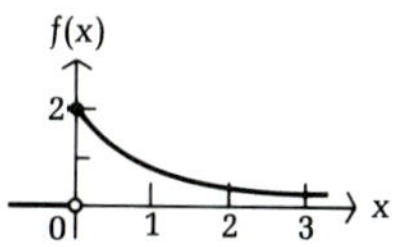

17. $F(x) = \begin{cases} 0 & \text{if } x < 0 \\ 1 - [1/(1+x^2)] & \text{if } x \geq 0 \end{cases}$

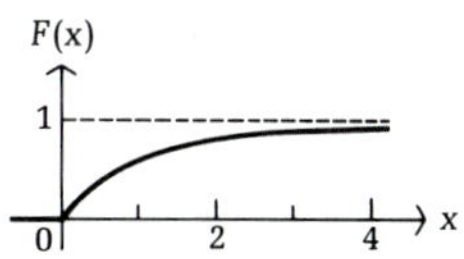

19. (A) 1 (B) 3 **21.** $F(x) = \begin{cases} 0 & \text{if } x < 0 \\ \frac{3}{4}x^2 - \frac{1}{4}x^3 & \text{if } 0 \leq x \leq 2 \\ 1 & \text{if } x > 2 \end{cases}$

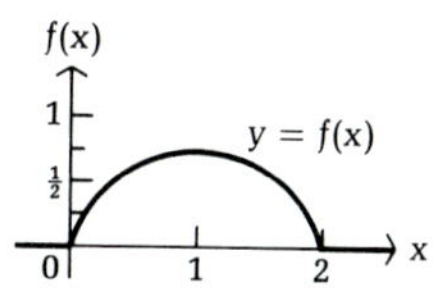

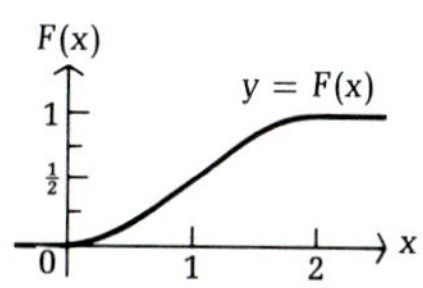

23. $F(x) = \begin{cases} 0 & \text{if } x < -1 \\ \frac{3}{8} + \frac{1}{2}x + \frac{1}{8}x^4 & \text{if } -1 \leq x \leq 1 \\ 1 & \text{if } x > 1 \end{cases}$

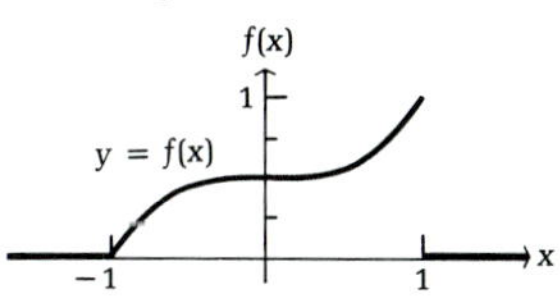

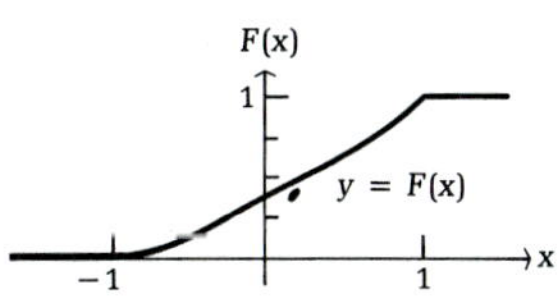

25. $x \approx 0.57$ **27.** $x \approx 0.44$

29. $F(x) = \begin{cases} 0 & \text{if } x < 1 \\ x \ln x - x + 1 & \text{if } 1 \leq x \leq e \\ 1 & \text{if } x > e \end{cases}$ **31.** $F(x) = \begin{cases} 1 - xe^{-x} - e^{-x} & \text{if } x \geq 0 \\ 0 & \text{otherwise} \end{cases}$

$F(2) - F(1) = 2 \ln 2 - 1 \approx .3863$ $\qquad 1 - F(1) = 2e^{-1} \approx .7358$

33. $f(x) = \begin{cases} 2x & \text{if } 0 \leq x \leq 1 \\ 0 & \text{otherwise} \end{cases}$ **35.** $f(x) = \begin{cases} 12x - 24x^2 + 12x^3 & \text{if } 0 \leq x \leq 1 \\ 0 & \text{otherwise} \end{cases}$

37. $F(x) = \begin{cases} 0 & \text{if } x < 0 \\ \frac{1}{2}x^2 & \text{if } 0 \leq x < 1 \\ 2x - \frac{1}{2}x^2 - 1 & \text{if } 1 \leq x \leq 2 \\ 1 & \text{if } x > 2 \end{cases}$ **39.** (A) $\int_0^8 (.2 - .02x)\, dx = .96$ (B) $\int_5^{10} (.2 - .02x)\, dx = .25$

41. (A) $\int_0^1 \frac{1}{10} e^{-x/10}\, dx = 1 - e^{-1/10} \approx .0952$ (B) $\int_4^\infty \frac{1}{10} e^{-x/10}\, dx = e^{-2/5} \approx .6703$

43. (A) $\int_4^{10} .003x\sqrt{100 - x^2}\, dx = \frac{(84)^{3/2}}{1,000} \approx .7699$ (B) $\int_0^8 .003x\sqrt{100 - x^2}\, dx = .784$ (C) $\sqrt{100 - (100)^{2/3}} \approx 8,858$ lb

45. (A) $\int_7^{10} \frac{1}{5,000}(10x^3 - x^4)\, dx = .47178$ (B) $\int_0^5 \frac{1}{5,000}(10x^3 - x^4)\, dx = \frac{3}{16} = .1875$

47. (A) $\int_0^{20} \frac{800x}{(400 + x^2)^2}\, dx = .5$ (B) $\int_{15}^\infty \frac{800x}{(400 + x^2)^2}\, dx = .64$ (C) 10 days

49. (A) $\int_{30}^\infty \frac{1}{20} e^{-x/20}\, dx = e^{-1.5} \approx .223$ (B) $\int_{80}^\infty \frac{1}{20} e^{-x/20}\, dx = e^{-4} \approx .018$

◆ EXERCISE 12-5

1. $\frac{4}{3}$; $\frac{2}{9}$; .471 **3.** 3.5; .75; .866 **5.** $\frac{4}{3}$; $\frac{1}{18}$; .236 **7.** $1/\sqrt{2} \approx .707$ **9.** $\sqrt{10} \approx 3.162$ **11.** $4 - 2\sqrt{2} \approx 1.172$ **13.** $\frac{4}{3}$; $\frac{2}{9}$; .471 **15.** $\frac{8}{3}$; $\frac{8}{9}$; .943 **17.** $e^{.5} \approx 1.649$ **19.** $\frac{2}{3}$ **21.** 1 **23.** $(\ln 2)/2 \approx .347$ **25.** $a\mu + b$ **27.** $x_1 = 1$; $x_2 = \sqrt{2} \approx 1.414$; $x_3 = \sqrt{3} \approx 1.732$ **29.** $x_1 = 1$; $x_2 = 3$; $x_3 = 9$ **31.** .54 **33.** 2.16 **35.** (A) $\frac{22}{3} \approx \$7.333$ or \$7,333 (B) $10 - 2\sqrt{2} \approx \$7.172$ or \$7,172 **37.** $3 \ln 2 \approx 2.079$ min **39.** 1 million gal **41.** $\frac{20}{3} \approx 6.7$ min **43.** 20 days **45.** 1.8 hr

◆ EXERCISE 12-6

1. $f(x) = \begin{cases} \frac{1}{2} & \text{if } 0 \leq x \leq 2 \\ 0 & \text{otherwise} \end{cases}$ $F(x) = \begin{cases} 0 & \text{if } x < 0 \\ x/2 & \text{if } 0 \leq x \leq 2 \\ 1 & \text{if } x > 2 \end{cases}$

3. $f(x) = \begin{cases} 20x^3(1 - x) & \text{if } 0 \leq x \leq 1 \\ 0 & \text{otherwise} \end{cases}$ $F(x) = \begin{cases} 0 & \text{if } x < 0 \\ 5x^4 - 4x^5 & \text{if } 0 \leq x \leq 1 \\ 1 & \text{if } x > 1 \end{cases}$

5. $f(x) = \begin{cases} 2e^{-2x} & \text{if } x \geq 0 \\ 0 & \text{otherwise} \end{cases}$ $F(x) = \begin{cases} 1 - e^{-2x} & \text{if } x \geq 0 \\ 0 & \text{otherwise} \end{cases}$

7. $\mu = 3$; $x_m = 3$; $\sigma = 2/\sqrt{3} \approx 1.155$ **9.** $\mu = 5$; $x_m = 5 \ln 2 \approx 3.466$; $\sigma = 5$ **11.** $\mu = \frac{3}{7}$; $\sigma = \sqrt{\frac{8}{147}} \approx .233$ **13.** $\mu = 2$; $\frac{1}{2}$ **15.** $\mu = \frac{1}{3}$; $\frac{5}{9}$ **17.** $\mu = 1$; $1 - e^{-1} \approx .632$ **19.** $\mu = 0$; $\sigma = 5/\sqrt{3} \approx 2.887$; $1/\sqrt{3} \approx .577$ **21.** $\mu = 6$; $\sigma = 6$; $1 - e^{-2} \approx .865$ **23.** $\beta = 2$; $\frac{297}{625} = .4752$ **29.** .61 **31.** .74 **33.** $\frac{3}{8} = .375$ **35.** (A) .6 or 60% (B) $1 - 4(.8)^3 + 3(.8)^4 \approx .1808$ **37.** (A) 1 (B) $\frac{27}{32} \approx .844$ **39.** $1 - e^{-2/3} \approx .487$ **41.** $1 - (1/\sqrt{2}) \approx .293$ **43.** (A) $\frac{3}{8}$ or 37.5% (B) $1 - 2.2(.5)^{1.2} + 1.2(.5)^{2.2} \approx .304$ **45.** (A) 37 (B) $1 - 39(.9)^{38} + 38(.9)^{39} \approx .912$ **47.** (A) $-1/(\ln .7) \approx 2.8$ yr (B) $e^{-1} \approx .368$ **49.** (A) .9 or 90% (B) $1 - 19(.95)^{18} + 18(.95)^{19} \approx .245$ **51.** $e^{-2.5} \approx .082$

◆ EXERCISE 12-7

1. .3413 **3.** .4987 **5.** .3159 **7.** .4932 **9.** .4332 **11.** .4995 **13.** .1915 **15.** .2881 **17.** .6811 **19.** .3142 **21.** .2266 **23.** .9699 **25.** .7888 **27.** .5328 **29.** .0122 **31.** .1056 **33.** No **35.** Yes **37.** No **39.** Yes **41.** .89 **43.** .16 **45.** .01 **47.** .01

49.

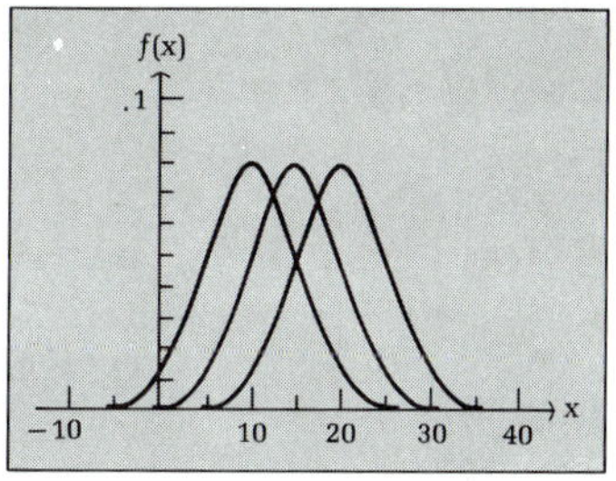

51.

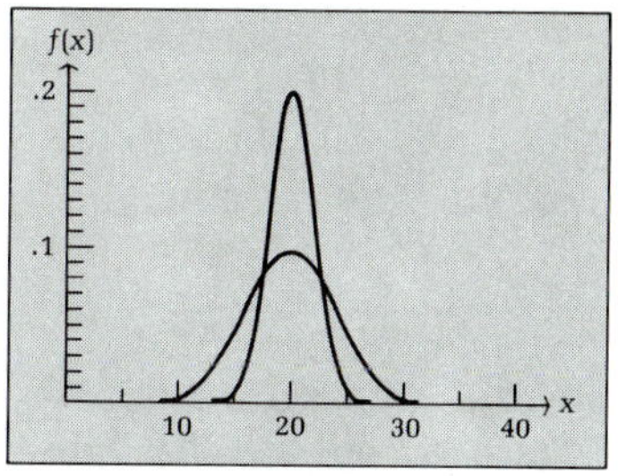

53. 2.28% **55.** 1.247 **57.** .0031; either a rare event has happened or the company's claim is false **59.** 0.82% **61.** .0158 **63.** 2.28% **65.** A's, 80.2 or greater; B's, 74.2–80.2; C's, 65.8–74.2; D's 59.8–65.8; F's, 59.8 or lower

◆ EXERCISE 12-8 CHAPTER REVIEW

1. $S = \{1, 2, 3, 4\}$

x_i	1	2	3	4
p_i	$\frac{1}{8}$	$\frac{2}{8}$	$\frac{3}{8}$	$\frac{2}{8}$

2. $\frac{1}{2} = .5$ **3.** $E(X) = \frac{11}{4} = 2.75$; $V(X) = \frac{15}{16} = .9375$; $\sigma = \sqrt{15}/4 \approx .9682$ **4.** $\frac{7}{64}$

5. (A)

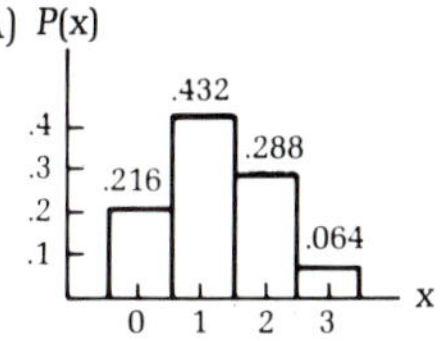

(B) $\mu = 1.2$; $\sigma = .85$

6. Binomial: .1808; Poisson: .1804

7. .0916; .7619 **8.** $\int_0^1 (1 - \frac{1}{2}x)\,dx = \frac{3}{4} = .75$

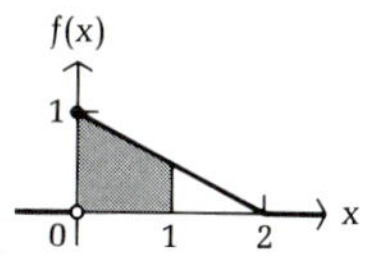

9. $\mu = \int_0^2 (x - \frac{1}{2}x^2)\,dx = \frac{2}{3} \approx .6667$; $V(X) = \int_0^2 (x^2 - \frac{1}{2}x^3)\,dx - (\frac{2}{3})^2 = \frac{2}{9} \approx .2222$; $\sigma = \sqrt{2}/3 \approx .4714$

10. $F(x) = \begin{cases} 0 & \text{if } x < 0 \\ x - \frac{1}{4}x^2 & \text{if } 0 \le x \le 2 \\ 1 & \text{if } x > 2 \end{cases}$ **11.** $2 - \sqrt{2} \approx .5858$ **12.** .4938 **13.** .4641

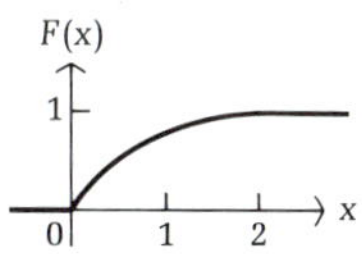

14. (A)

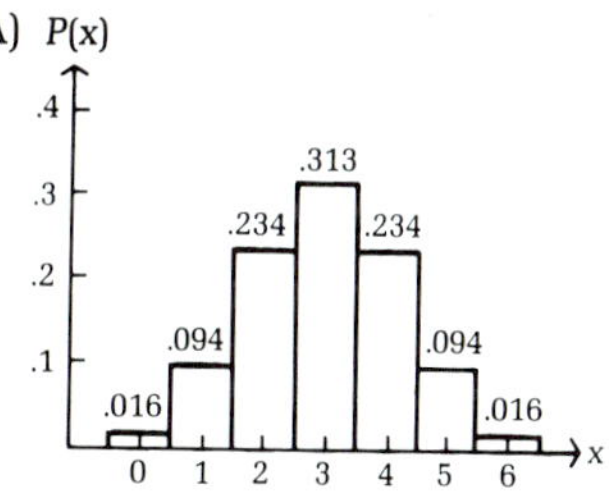

(B) $\mu = 3$; $\sigma = 1.22$

15. .0607 **16.** $\int_1^4 \frac{5}{2}x^{-7/2}\,dx = \frac{31}{32} \approx .9688$

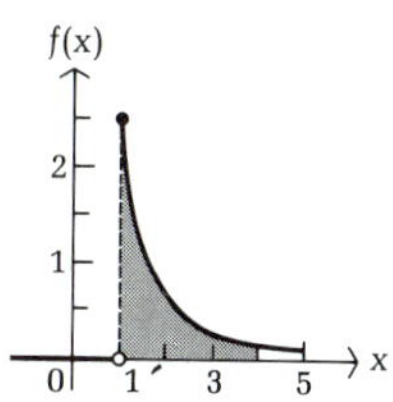

17. $\mu = \int_1^\infty \frac{5}{2}x^{-5/2}\,dx = \frac{5}{3} \approx 1.667$; $V(X) = \int_1^\infty \frac{5}{2}x^{-3/2}\,dx - (\frac{5}{3})^2 = \frac{20}{9} \approx 2.222$; $\sigma = \frac{2}{3}\sqrt{5} \approx 1.491$

18. $F(x) = \begin{cases} 1 - x^{-5/2} & \text{if } x \geqslant 1 \\ 0 & \text{otherwise} \end{cases}$ **19.** $2^{2/5} \approx 1.32$ **20.** $f(x) = \begin{cases} 42x^5(1-x) & \text{if } 0 \leqslant x \leqslant 1 \\ 0 & \text{otherwise} \end{cases}$

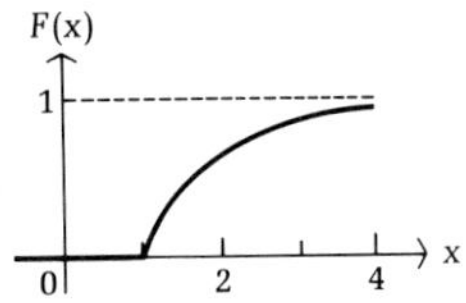

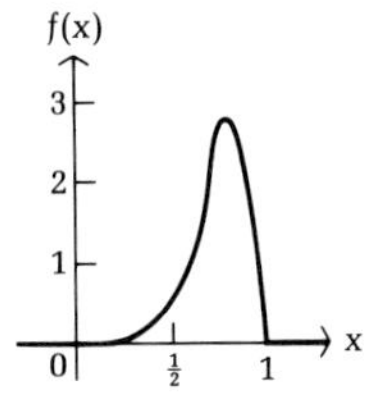

21. $F(.75) - F(.25) = 7(.75)^6 - 6(.75)^7 - 7(.25)^6 + 6(.25)^7 \approx .4436$ **22.** $F(x) = \begin{cases} 0 & \text{if } x < 0 \\ 7x^6 - 6x^7 & \text{if } 0 \leqslant x \leqslant 1 \\ 1 & \text{if } x > 1 \end{cases}$

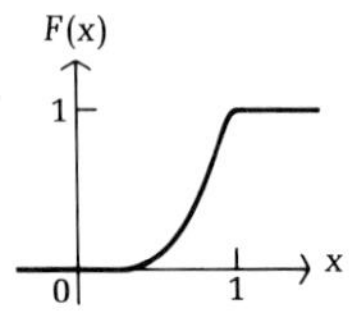

23. $\mu = \frac{3}{4} = .75$; $\sigma = \sqrt{3}/12 \approx .1443$ **24.** $f(x) = \begin{cases} \frac{1}{2}e^{-x/2} & \text{if } x \geqslant 0 \\ 0 & \text{otherwise} \end{cases}$ **25.** $\int_0^2 \frac{1}{2}e^{-x/2}\,dx = 1 - e^{-1} \approx .6321$

26. $F(x) = \begin{cases} 1 - e^{-x/2} & \text{if } x \geqslant 0 \\ 0 & \text{otherwise} \end{cases}$

27. $\mu = 2$; $\sigma = 2$; $x_m = 2 \ln 2 \approx 1.386$ **28.** $\mu = 600$; $\sigma = 15.49$ **29.** .999 **30.** (A) .9104 (B) .0668 **31.** 7

32. $\mu = \int_0^\infty \frac{50x}{(x+5)^3}\,dx = 5$; $x_m = 5\sqrt{2} - 5 \approx 2.071$ **33.** $a\sigma^2 + a\mu^2 + b\mu + c$ **34.** .0188 **35.** (A) .6065 (B) .3033 (C) .0144

36. (A) $\frac{1}{50}\int_0^5 (1 - .01x)\,dx = \frac{3}{4} = .75$ (B) 80 lb **37.** (A) $\int_{.2}^1 6x(1-x)\,dx = .896$ (B) 50%
38. (A) $1 - F(4) = e^{-1} \approx .3679$ (B) $F(1) - F(0) = 1 - e^{-.25} \approx .2212$ **39.** (A) .0902 (B) .2231 **40.** .0228 **41.** (A) 60.46% (B) 7.35%
42. (A) $\mu = 140$; $\sigma = 6.48$ (B) Yes (C) .939 (D) .0125 **43.** (A) $1 - F(5) = \frac{2}{3} \approx .6667$ (B) 10 months
44. (A) $1 - F(2) = e^{-4} \approx .0183$ (B) $\frac{1}{2}$ month **45.** .1492 **46.** (A) .0498 (B) .1494 (C) .3528 **47.** $F(1) - F(.5) = \frac{15}{16} \approx .9375$
48. .0136 **49.** (A) .4493 (B) .3595 (C) .0014

◆ EXERCISE 5-1

1. $x = 3$, $y = 2$ **3.** $x = 2$, $y = 4$ **5.** No solution (parallel lines) **7.** $x = 4$, $y = 5$ **9.** $x = 1$, $y = 4$ **11.** $u = 2$, $v = -3$ **13.** $m = 8$, $n = 6$ **15.** $x = 1$, $y = -5$ **17.** No solution (inconsistent) **19.** $x = -\frac{4}{3}$, $y = 1$ **21.** Infinitely many solutions (dependent) **23.** $x = 4{,}000$, $y = 280$ **25.** $x = \frac{4}{7}$, $y = \frac{3}{7}$ **27.** $x = 1.1$, $y = 0.3$ **29.** $x = -\frac{5}{4}$, $y = \frac{5}{3}$

31.

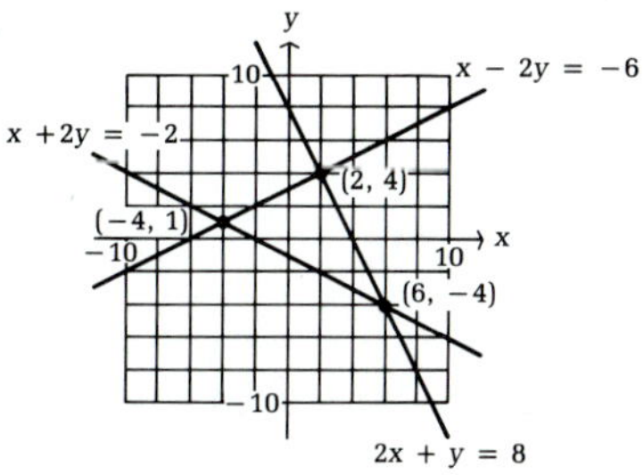

33.

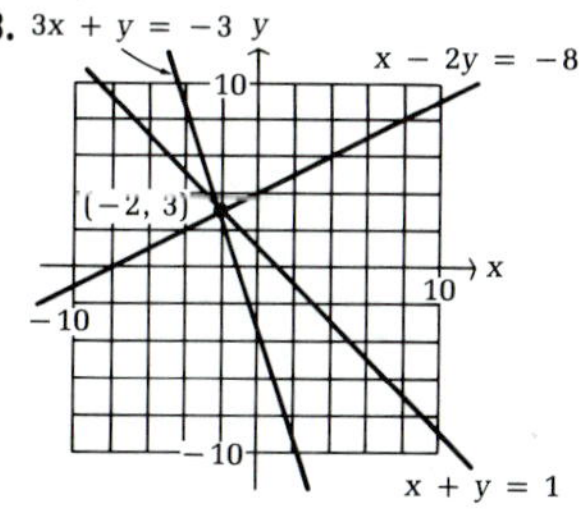

35.

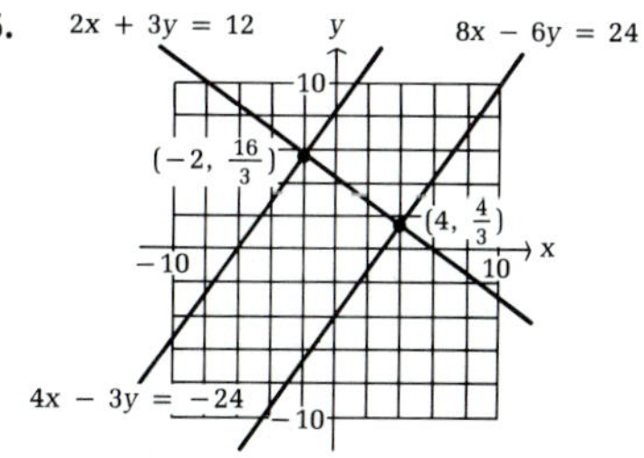

37. (A) Equilibrium price = \$6.50; Equilibrium quantity = 500

(B)

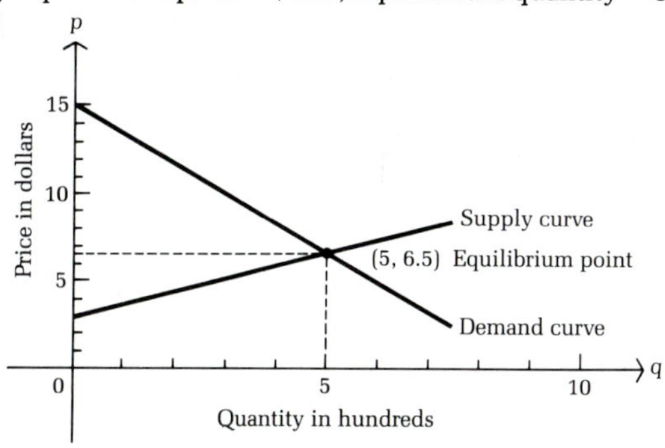

39. (A) $p = 0.001q + 0.15$

(B) $p = -0.002q + 1.74$

(C) Equilibrium price = \$0.68; Equilibrium quantity = 530

(D)

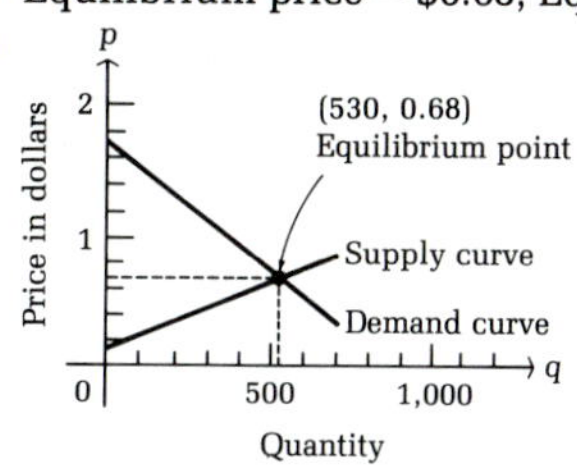

41. (A) For $x = 120$ units, Cost = \$216,000 = Revenue

(B)

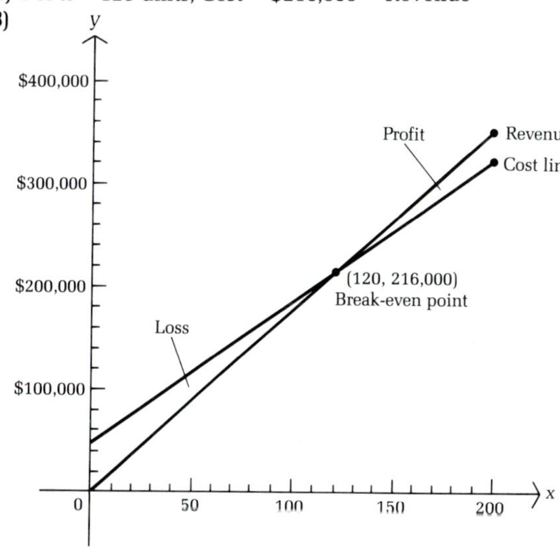

43. (A) For $x = 1{,}920$ tapes, Cost = \$38,304 = Revenue

(B)

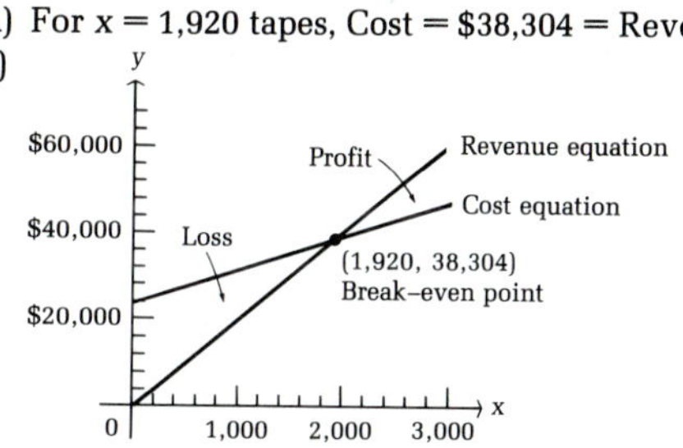

45. Mix A: 80 g; mix B: 60 g **47.** (A)

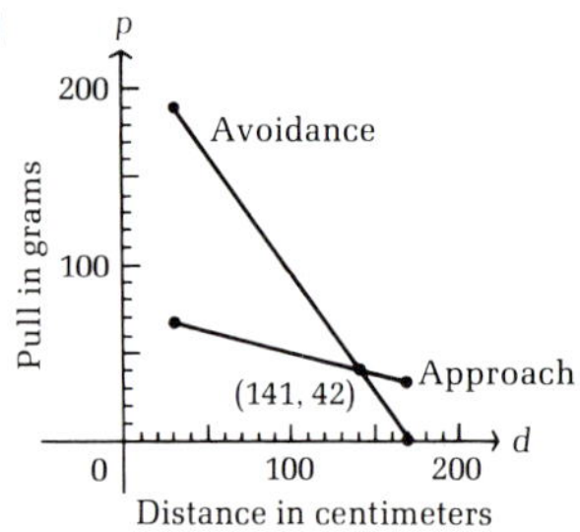

(B) $d = 141$ cm (approx.) (C) Vacillate

◆ EXERCISE 5-2

1. $\left[\begin{array}{cc|c} 4 & -6 & -8 \\ 1 & -3 & 2 \end{array}\right]$ **3.** $\left[\begin{array}{cc|c} -4 & 12 & -8 \\ 4 & -6 & -8 \end{array}\right]$ **5.** $\left[\begin{array}{cc|c} 1 & -3 & 2 \\ 8 & -12 & -16 \end{array}\right]$ **7.** $\left[\begin{array}{cc|c} 1 & -3 & 2 \\ 0 & 6 & -16 \end{array}\right]$ **9.** $\left[\begin{array}{cc|c} 1 & -3 & 2 \\ 2 & 0 & -12 \end{array}\right]$
11. $\left[\begin{array}{cc|c} 1 & -3 & 2 \\ 3 & -3 & -10 \end{array}\right]$ **13.** $x_1 = 3, x_2 = 2$ **15.** $x_1 = 3, x_2 = 1$ **17.** $x_1 = 2, x_2 = 1$ **19.** $x_1 = 2, x_2 = 4$ **21.** No solution
23. $x_1 = 1, x_2 = 4$ **25.** Infinitely many solutions: $x_2 = s, x_1 = 2s - 3$ for any real number s
27. Infinitely many solutions: $x_2 = s, x_1 = \frac{1}{2}s + \frac{1}{2}$ for any real number s **29.** $x_1 = -1, x_2 = 3$ **31.** No solution
33. Infinitely many solutions: $x_2 = t, x_1 = \frac{3}{2}t + 2$ for any real number t **35.** $x_1 = 2, x_2 = -1$ **37.** $x_1 = 2, x_2 = -1$
39. $x_1 = 1.1, x_2 = 0.3$

◆ EXERCISE 5-3

1. Yes **3.** No **5.** No **7.** Yes **9.** $x_1 = -2, x_2 = 3, x_3 = 0$ **11.** $x_1 = 2t + 3, x_2 = -t - 5, x_3 = t$ for t any real number **13.** No solution
15. $x_1 = 2s + 3t - 5, x_2 = s, x_3 = -3t + 2, x_4 = t$ for s and t any real numbers
17. $\left[\begin{array}{cc|c} 1 & 0 & -7 \\ 0 & 1 & 3 \end{array}\right]$ **19.** $\left[\begin{array}{ccc|c} 1 & 0 & 0 & -5 \\ 0 & 1 & 0 & 4 \\ 0 & 0 & 1 & -2 \end{array}\right]$ **21.** $\left[\begin{array}{ccc|c} 1 & 0 & 2 & -\frac{5}{3} \\ 0 & 1 & -2 & \frac{1}{3} \\ 0 & 0 & 0 & 0 \end{array}\right]$
23. $x_1 = -2, x_2 = 3, x_3 = 1$ **25.** $x_1 = 0, x_2 = -2, x_3 = 2$ **27.** $x_1 = 2t + 3, x_2 = t - 2, x_3 = t$ for t any real number
29. $x_1 = (-4t - 4)/7, x_2 = (5t + 5)/7, x_3 = t$ for t any real number **31.** $x_1 = -1, x_2 = 2$ **33.** No solution **35.** No solution
37. $x_1 = t - 1, x_2 = 2t + 2, x_3 = t$ for t any real number **39.** $x_1 = -2s + t + 1, x_2 = s, x_3 = t$ for s and t any real numbers
41. $x_1 = 0, x_2 = 2, x_3 = -3$ **43.** $x_1 = 1, x_2 = -2, x_3 = 1$
45. $x_1 = 2s - 3t + 3, x_2 = s + 2t + 2, x_3 = s, x_4 = t$ for s and t any real numbers
47. 20 one-person boats, 220 two-person boats, 100 four-person boats
49. $(t - 80)$ one-person boats, $(-2t + 420)$ two-person boats, t four-person boats, $80 \le t \le 210$, t an integer
51. No solution; no production schedule will use all the labor-hours in all departments
53. $(5t - 29)$ 6,000 gal cars, $(-6t + 53)$ 8,000 gal cars, and t 18,000 gal cars, where $t = 6, 7,$ or 8
55. Federal: \$370,500; state: \$123,500; local; \$58,500 **57.** 8 oz food A, 2 oz food B, 4 oz food C **59.** No solution
61. 8 oz food A, $(-2t + 10)$ oz food B, t oz food C, $0 \le t \le 5$
63. $(10 - t)$ barrels of mix A, $(t - 5)$ barrels of mix B, $(25 - 2t)$ barrels of mix C, and t barrels of mix D, where t is an integer, $5 \le t \le 10$
65. Company A: 10 hr; company B: 15 hr
67. (A) 6th St. and Washington Ave.: $x_1 + x_2 = 1{,}200$; 6th St. and Lincoln Ave.: $x_2 + x_3 = 1{,}000$; 5th St. and Lincoln Ave.: $x_3 + x_4 = 1{,}300$
(B) $x_1 = 1{,}500 - t, x_2 = t - 300, x_3 = 1{,}300 - t$, and $x_4 = t$, where $300 \le t \le 1{,}300$ (C) 1,300; 300
(D) Washington Ave.: 500; 6th St.: 700; Lincoln Ave.: 300

◆ EXERCISE 5-4

1. 2×2; 1×4 **3.** $-3, 0, 5$ **5.** $\begin{bmatrix} 0 & 0 \\ 0 & 0 \end{bmatrix}$ **7.** C, D **9.** A, B **11.** $\begin{bmatrix} -1 & 0 \\ 5 & -3 \end{bmatrix}$ **13.** Not defined

15. $\begin{bmatrix} -2 \\ 3 \\ 0 \end{bmatrix}$ **17.** $\begin{bmatrix} -1 \\ 6 \\ 5 \end{bmatrix}$ **19.** $\begin{bmatrix} -15 & 5 \\ 10 & -15 \end{bmatrix}$ **21.** $\begin{bmatrix} 1 & 3 & -1 & 1 \\ -1 & -5 & 7 & 2 \\ 4 & 8 & 0 & -2 \end{bmatrix}$ **23.** $\begin{bmatrix} 5.4 & 0.7 & -1.8 \\ 7.6 & -4.0 & 7.9 \end{bmatrix}$ **25.** $\begin{bmatrix} 250 & 360 \\ 40 & 350 \end{bmatrix}$

27. $\begin{bmatrix} 2{,}280 & 3{,}460 \\ 1{,}380 & 2{,}310 \end{bmatrix}$ **29.** $a = -1, b = 1, c = 3, d = -5$ **31.** $x = 2, y = -3$ **33.** No **35.** No

37.

Guitar	Banjo	
\$33	\$26	Materials
\$57	\$77	Labor

39.

Markup

	Basic car	Air	AM/FM radio	Cruise control
Model *A*	\$1,180	\$ 82	\$ 54	\$30
Model *B*	\$1,075	\$ 98	\$ 81	\$30
Model *C*	\$1,715	\$106	\$108	\$36

41.

	Round	Wrinkled
Yellow	689	225
Green	218	68

$= M + N$;

	Round	Wrinkled
Yellow	57%	19%
Green	18%	6%

◆ EXERCISE 5-5

1. [10] **3.** [−1] **5.** [12 13] **7.** $\begin{bmatrix} 5 \\ -3 \end{bmatrix}$ **9.** $\begin{bmatrix} 2 & 4 \\ 1 & -5 \end{bmatrix}$ **11.** $\begin{bmatrix} 1 & -5 \\ -2 & -4 \end{bmatrix}$ **13.** [−7] **15.** $\begin{bmatrix} -15 & 6 \\ -20 & 8 \end{bmatrix}$ **17.** [6] **19.** [15]

21. $\begin{bmatrix} 0 & 9 \\ 5 & -4 \end{bmatrix}$ **23.** $\begin{bmatrix} 5 & 8 & -5 \\ -1 & -3 & 2 \\ -2 & 8 & -6 \end{bmatrix}$ **25.** [11] **27.** $\begin{bmatrix} 3 & -2 & -4 \\ 6 & -4 & -8 \\ -9 & 6 & 12 \end{bmatrix}$ **29.** Not defined

31. $\begin{bmatrix} 7 \\ 35 \\ 23 \end{bmatrix}$ **33.** $\begin{bmatrix} 0 & 0 & -5 \\ -6 & 15 & 13 \\ 5 & -6 & -14 \end{bmatrix}$ **35.** 0 **37.** $\begin{bmatrix} \frac{1}{3} & \frac{1}{3} \\ \frac{2}{3} & \frac{2}{3} \end{bmatrix}$

39. $AB = \begin{bmatrix} 5 & 7 \\ 2 & 3 \end{bmatrix}$; $BA = \begin{bmatrix} 1 & 3 \\ 2 & 7 \end{bmatrix}$ **41.** Both sides equal $\begin{bmatrix} 0 & 12 \\ 1 & 5 \end{bmatrix}$ **43.** $A^2 - B^2 = \begin{bmatrix} -2 & 0 \\ -8 & -10 \end{bmatrix}$; $(A - B)(A + B) = \begin{bmatrix} 2 & 4 \\ -8 & -14 \end{bmatrix}$

45. (A) \$11.80 per boat (B) $[1.5 \quad 1.2 \quad 0.4] \begin{bmatrix} 9 \\ 12 \\ 6 \end{bmatrix} = \$30.30$ (C) 3×2 (D)

	Plant I	Plant II
One-person boat	\$11.80	\$13.80
Two-person boat	\$18.50	\$21.60
Four-person boat	\$26.00	\$30.30

Labor costs per boat at each plant

47. (A) $A^2 = \begin{bmatrix} 0 & 0 & 2 & 0 & 0 \\ 1 & 0 & 0 & 0 & 1 \\ 0 & 1 & 0 & 2 & 0 \\ 1 & 0 & 0 & 0 & 1 \\ 0 & 0 & 1 & 0 & 0 \end{bmatrix}$ There is one way to travel from Baltimore to Atlanta with one intermediate connection; there are two ways to travel from Atlanta to Chicago with one intermediate connection. In general, the elements in A^2 indicate the number of different ways to travel from the ith city to the jth city with one intermediate connection.

(B) $A^3 = \begin{bmatrix} 2 & 0 & 0 & 0 & 2 \\ 0 & 1 & 0 & 2 & 0 \\ 0 & 0 & 3 & 0 & 0 \\ 0 & 1 & 0 & 2 & 0 \\ 1 & 0 & 0 & 0 & 1 \end{bmatrix}$ There is one way to travel from Denver to Baltimore with two intermediate connections; there are two ways to travel from Atlanta to El Paso with two intermediate connections. In general, the elements in A^3 indicate the number of different ways to travel from the ith city to the jth city with two intermediate connections.

49. $A = \begin{bmatrix} 0 & 0 & 0 & 1 & 1 \\ 1 & 0 & 0 & 1 & 0 \\ 0 & 1 & 0 & 0 & 0 \\ 1 & 0 & 0 & 0 & 1 \\ 0 & 1 & 1 & 0 & 0 \end{bmatrix}$; $A + A^2 + A^3 = \begin{bmatrix} 2 & 3 & 2 & 3 & 3 \\ 3 & 2 & 2 & 3 & 4 \\ 2 & 1 & 0 & 2 & 2 \\ 3 & 3 & 2 & 2 & 3 \\ 3 & 2 & 1 & 3 & 2 \end{bmatrix}$ It is possible to travel from any origin to any destination with at most two intermediate connections.

51. (A) \$2,025 (B) $[2{,}000 \quad 800 \quad 8{,}000] \begin{bmatrix} \$0.40 \\ \$0.75 \\ \$0.25 \end{bmatrix} = \$3{,}400$ (C) $\begin{bmatrix} \$2{,}025 \\ \$3{,}400 \end{bmatrix}$ $\begin{matrix} \text{Berkeley} \\ \text{Oakland} \end{matrix}$
Cost per town
(D) Telephone call, House call, Letter: [3,000 1,300 13,000]
Number of each type of contact made

◆ EXERCISE 5-6

1. $\begin{bmatrix} 2 & -3 \\ 4 & 5 \end{bmatrix}$ **3.** $\begin{bmatrix} 2 & -3 \\ 4 & 5 \end{bmatrix}$ **5.** $\begin{bmatrix} -2 & 1 & 3 \\ 2 & 4 & -2 \\ 5 & 1 & 0 \end{bmatrix}$ **7.** $\begin{bmatrix} -2 & 1 & 3 \\ 2 & 4 & -2 \\ 5 & 1 & 0 \end{bmatrix}$ **15.** $\begin{bmatrix} -1 & 0 \\ -3 & 1 \end{bmatrix}$ **17.** $\begin{bmatrix} 3 & -2 \\ -1 & 1 \end{bmatrix}$ **19.** $\begin{bmatrix} 7 & -3 \\ -2 & 1 \end{bmatrix}$

21. $\begin{bmatrix} 7 & 6 & -3 \\ 2 & 2 & -1 \\ -6 & -5 & 3 \end{bmatrix}$ **23.** $\begin{bmatrix} \frac{3}{2} & -\frac{1}{2} & -\frac{1}{2} \\ -\frac{1}{2} & \frac{1}{2} & \frac{1}{2} \\ -\frac{3}{2} & \frac{1}{2} & \frac{3}{2} \end{bmatrix}$ **25.** $\begin{bmatrix} 3 & 2 \\ -4 & -3 \end{bmatrix}$ **27.** The inverse does not exist. **29.** $\begin{bmatrix} 1 & -\frac{1}{2} \\ -2 & \frac{3}{2} \end{bmatrix}$

31. $\begin{bmatrix} 1 & 2 & 2 \\ -2 & -3 & -4 \\ -1 & -2 & -1 \end{bmatrix}$ **33.** The inverse does not exist. **35.** $\begin{bmatrix} 1 & 1 & -\frac{1}{2} \\ -2 & -\frac{3}{2} & 1 \\ -1 & -1 & \frac{3}{4} \end{bmatrix}$ **39.** $\begin{bmatrix} 0.5 & -0.3 & 0.85 & -0.25 \\ 0 & 0.1 & 0.05 & -0.25 \\ -1 & 0.9 & -1.55 & 0.75 \\ -0.5 & 0.4 & -1.3 & 0.5 \end{bmatrix}$

41. $\begin{bmatrix} 1.75 & 5.25 & 8.75 & -1 & -18.75 \\ 1.25 & 3.75 & 6.25 & -1 & -13.25 \\ -4.75 & -13.25 & -22.75 & 3 & 48.75 \\ -1.375 & -4.625 & -7.875 & 1 & 16.375 \\ 3.25 & 8.75 & 15.25 & -2 & -32.25 \end{bmatrix}$ **43.** 36 44 59 86 61 82 68 95 25 37 49 64 63 81 47 66 43 62

45. GONE WITH THE WIND **47.** 27 19 82 9 89 50 114 101 70 123 50 100 75 72 89 63 96 99 69 99 52 105 106 56 97 59 127 110 81 132
49. THE GREATEST SHOW ON EARTH

◆ EXERCISE 5-7

1. $3x_1 + x_2 = 5$, $2x_1 - x_2 = -4$ **3.** $-3x_1 + x_2 = 3$, $2x_1 + x_3 = -4$, $-x_1 + 3x_2 - 2x_3 = 2$ **5.** $\begin{bmatrix} 3 & -4 \\ 2 & 1 \end{bmatrix}\begin{bmatrix} x_1 \\ x_2 \end{bmatrix} = \begin{bmatrix} 1 \\ 5 \end{bmatrix}$ **7.** $\begin{bmatrix} 1 & -3 & 2 \\ -2 & 3 & 0 \\ 1 & 1 & 4 \end{bmatrix}\begin{bmatrix} x_1 \\ x_2 \\ x_3 \end{bmatrix} = \begin{bmatrix} -3 \\ 1 \\ -2 \end{bmatrix}$

9. $x_1 = -8, x_2 = 2$ **11.** $x_1 = 0, x_2 = 4$ **13.** (A) $x_1 = -3, x_2 = 2$ (B) $x_1 = -1, x_2 = 2$ (C) $x_1 = -8, x_2 = 3$
15. (A) $x_1 = 17, x_2 = -5$ (B) $x_1 = 7, x_2 = -2$ (C) $x_1 = 24, x_2 = -7$
17. (A) $x_1 = 1, x_2 = 0, x_3 = 0$ (B) $x_1 = -1, x_2 = 0, x_3 = 1$ (C) $x_1 = -1, x_2 = -1, x_3 = 1$
19. (A) $x_1 = 1, x_2 = 1, x_3 = 3$ (B) $x_1 = -1, x_2 = 1, x_3 = -1$ (C) $x_1 = 5, x_2 = -1, x_3 = -5$
21. $X = (A - B)^{-1}C$ $[X \neq C(A - B)^{-1}]$ **23.** $X = (A + I)^{-1}C$ **25.** $X = (A + B)^{-1}(C + D)$ **27.** $x_1 = 18.2, x_2 = 27.9, x_3 = -15.2$
29. $x_1 = 24, x_2 = 5, x_3 = -2, x_4 = 15$
31. Concert 1: 6,000 \$4 tickets and 4,000 \$8 tickets; Concert 2: 5,000 \$4 tickets and 5,000 \$8 tickets; Concert 3: 3,000 \$4 tickets and 7,000 \$8 tickets
33. Order 1: 280 hr at plant *A* and 25 hr at plant *B*; Order 2: 160 hr at plant *A* and 150 hr at plant *B*; Order 3: 80 hr at plant *A* and 225 hr at plant *B*
35. Diet 1: 60 oz mix *A* and 80 oz mix *B*; Diet 2: 20 oz mix *A* and 60 oz mix *B*; Diet 3: 0 oz mix *A* and 100 oz mix *B*
37. President: \$56,600; executive vice president: \$47,000; associative vice president: \$37,400; assistant vice president: \$27,900
39. *A*: \$3,300,000; *B*: \$3,000,000; *C*: \$3,700,000; *D*: \$4,200,000; Difference = \$200,000
41. (A) Freezers: \$346,000; Refrigerators: \$254,000; Accounting: \$85,000; Maintenance: \$127,000
(B) Sum of direct costs = \$600,000; Sum of total costs = \$346,000 + \$254,000 = \$600,000; the direct costs of the service departments are distributed among the production departments.

◆ EXERCISE 5-8

1. 40¢ from *A*; 20¢ from *E* **3.** $\begin{bmatrix} 0.6 & -0.2 \\ -0.2 & 0.9 \end{bmatrix}$; $\begin{bmatrix} 1.8 & 0.4 \\ 0.4 & 1.2 \end{bmatrix}$ **5.** $X = \begin{bmatrix} x_1 \\ x_2 \end{bmatrix} = \begin{bmatrix} 16.4 \\ 9.2 \end{bmatrix}$ **7.** 20¢ from *A*; 10¢ from *B*; 10¢ from *E*
11. Agriculture: \$18 billion; Building: \$15.6 billion; Energy: \$22.4 billion **13.** $\begin{bmatrix} 1.4 & 0.4 \\ 0.6 & 1.6 \end{bmatrix}$; $\begin{bmatrix} 24 \\ 46 \end{bmatrix}$
15. $\begin{bmatrix} 1.58 & 0.24 & 0.58 \\ 0.4 & 1.2 & 0.4 \\ 0.22 & 0.16 & 1.22 \end{bmatrix}$; $\begin{bmatrix} 38.6 \\ 18 \\ 17.4 \end{bmatrix}$ **17.** Coal: \$28 billion; Steel: \$26 billion
19. Agriculture: \$148 million; Tourism: \$146 million
21. Agriculture: \$40.1 billion; Manufacturing: \$29.4 billion; Energy: \$34.4 billion
23. Year 1: Agriculture: \$65 billion; Energy: \$83 billion; Labor: \$71 billion; Manufacturing: \$88 billion.
Year 2: Agriculture: \$81 billion; Energy: \$97 billion; Labor: \$83 billion; Manufacturing: \$99 billion.
Year 3: Agriculture: \$117 billion; Energy: \$124 billion; Labor: \$106 billion; Manufacturing: \$120 billion.

◆ EXERCISE 5-9 CHAPTER REVIEW

1. $x = 4, y = 4$ **2.** $x = 4, y = 4$ **3.** $\begin{bmatrix} 3 & 3 \\ 4 & 2 \end{bmatrix}$ **4.** Not defined **5.** $\begin{bmatrix} -3 & 0 \\ 1 & -1 \end{bmatrix}$ **6.** $\begin{bmatrix} 4 & 3 \\ 7 & 4 \end{bmatrix}$ **7.** Not defined **8.** $\begin{bmatrix} 5 \\ 5 \end{bmatrix}$
9. $\begin{bmatrix} 2 & 3 \\ 4 & 6 \end{bmatrix}$ **10.** [8] **11.** Not defined **12.** $\begin{bmatrix} 3 & -2 \\ -4 & 3 \end{bmatrix}$ **13.** $x_1 = -1, x_2 = 3$ **14.** $x_1 = -1, x_2 = 3$
15. $x_1 = -1, x_2 = 3$; $x_1 = 1, x_2 = 2$; $x_1 = 8, x_2 = -10$ **16.** Not defined **17.** $\begin{bmatrix} 10 & -8 \\ 4 & 6 \end{bmatrix}$ **18.** $\begin{bmatrix} -2 & 8 \\ 8 & 6 \end{bmatrix}$
19. $\begin{bmatrix} -2 & -1 & -3 \\ 4 & 2 & 6 \\ 6 & 3 & 9 \end{bmatrix}$ **20.** [9] **21.** $\begin{bmatrix} 10 & -5 & 1 \\ -1 & -4 & -5 \\ 1 & -7 & -2 \end{bmatrix}$ **22.** $\begin{bmatrix} -\frac{5}{2} & 2 & -\frac{1}{2} \\ 1 & -1 & 1 \\ \frac{1}{2} & 0 & -\frac{1}{2} \end{bmatrix}$

23. (A) $x_1 = 2, x_2 = 1, x_3 = -1$ (B) $x_1 = -5t - 12, x_2 = 3t + 7, x_3 = t$ for t any real number
24. $x_1 = 2, x_2 = 1, x_3 = -1$; $x_1 = 1, x_2 = -2, x_3 = 1$; $x_1 = -1, x_2 = 2, x_3 = -2$
25. $(I - M)^{-1} = \begin{bmatrix} 1.4 & 0.3 \\ 0.8 & 1.6 \end{bmatrix}, X = \begin{bmatrix} 48 \\ 56 \end{bmatrix}$ **26.** $\begin{bmatrix} -\frac{11}{12} & -\frac{1}{12} & 5 \\ \frac{10}{12} & \frac{2}{12} & -4 \\ \frac{1}{12} & -\frac{1}{12} & 0 \end{bmatrix}$ **27.** $x_1 = 1{,}000, x_2 = 4{,}000, x_3 = 2{,}000$
28. $x_1 = 1{,}000, x_2 = 4{,}000, x_3 = 2{,}000$
29. $(I - M)^{-1} = \begin{bmatrix} 1.3 & 0.4 & 0.7 \\ 0.2 & 1.6 & 0.3 \\ 0.1 & 0.8 & 1.4 \end{bmatrix}, X = \begin{bmatrix} 81 \\ 49 \\ 62 \end{bmatrix}$ **30.** $0.01x_1 + 0.02x_2 = 4.5$
$0.02x_1 + 0.05x_2 = 10$
$x_1 = 250$ tons of ore A, $x_2 = 100$ tons of ore B

31. (A) $\underset{X}{\begin{bmatrix} x_1 \\ x_2 \end{bmatrix}} = \underset{A^{-1}}{\begin{bmatrix} 500 & -200 \\ -200 & 100 \end{bmatrix}} \begin{bmatrix} 4.5 \\ 10 \end{bmatrix} = \begin{bmatrix} 250 \\ 100 \end{bmatrix}$
$x_1 = 250$ tons of ore A, $x_2 = 100$ tons of ore B
(B) $\underset{X}{\begin{bmatrix} x_1 \\ x_2 \end{bmatrix}} = \underset{A^{-1}}{\begin{bmatrix} 500 & -200 \\ -200 & 100 \end{bmatrix}} \begin{bmatrix} 2.3 \\ 5 \end{bmatrix} = \begin{bmatrix} 150 \\ 40 \end{bmatrix}$
$x_1 = 150$ tons of ore A, $x_2 = 40$ tons of ore B
32. 1 model A truck, 6 model B trucks, and 5 model C trucks; or 3 model A trucks, 3 model B trucks, and 6 model C trucks; or 5 model A trucks and 7 model C trucks.
33. (A)

	Supplier A	Supplier B	
$MN =$	\$7,620	\$7,530	Alloy 1
	\$13,880	\$13,930	Alloy 2

Cost of each alloy from each supplier

(B)

Supplier A	Supplier B
[\$21,500	\$21,460]

Total cost for both alloys from each supplier

34. (A) $[0.25 \quad 0.20 \quad 0.05] \begin{bmatrix} 15 \\ 12 \\ 4 \end{bmatrix} = \$6.35$ (B)

Calif.	Texas	
\$3.65	\$3.00	Model A
\$6.35	\$5.20	Model B

Total labor costs for each model at each plant
35. \$2,000 at 5%; \$3,000 at 10% **36.** \$2,000 at 5%; \$3,000 at 10%
37. Concert 1: 5,000 \$8 tickets, 15,000 \$12 tickets, and 5,000 \$20 tickets; Concert 2: 7,500 \$8 tickets, 10,000 \$12 tickets, and 7,500 \$20 tickets; Concert 3: 10,000 \$8 tickets, 5,000 \$12 tickets, and 10,000 \$20 tickets
38. (A) Agriculture: \$80 billion; Fabrication: \$60 billion (B) Agriculture: \$135 billion; Fabrication: \$145 billion

CHAPTER 6

◆ EXERCISE 6-1

1.
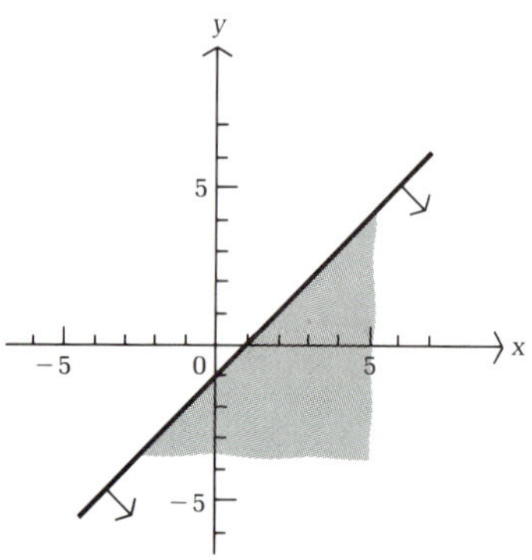
3.
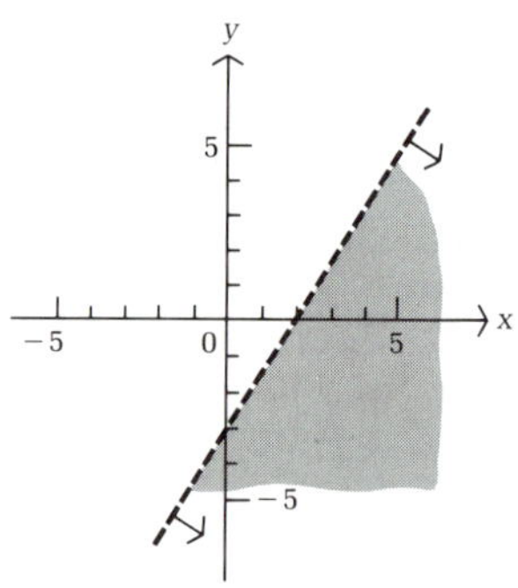
5.
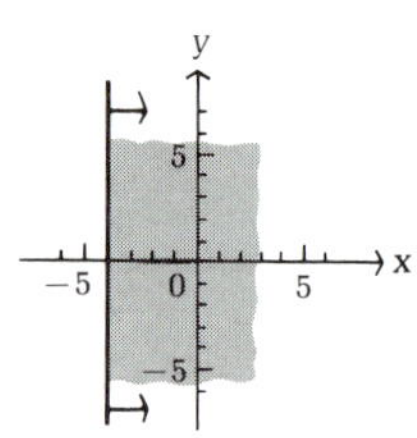
7.
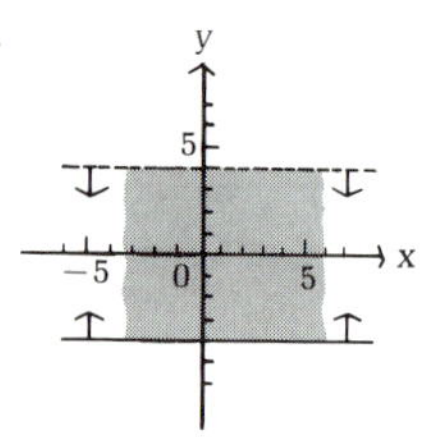

9.

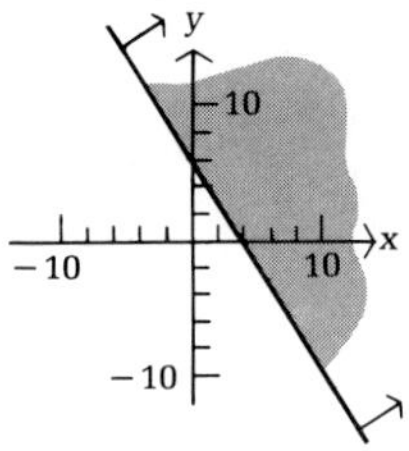

11.

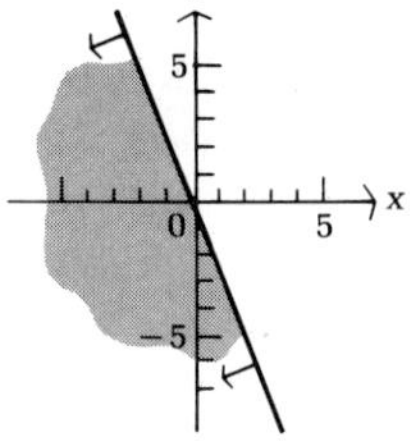

13. IV **15.** I **17.**

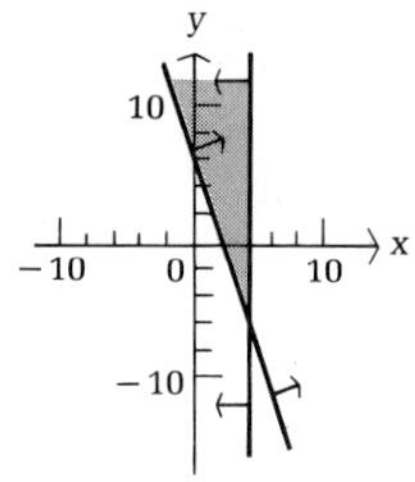

19.

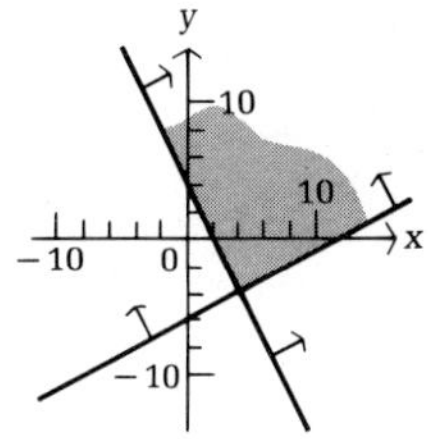

21. IV; (8, 0), (18, 0), (6, 4) **23.** I; (0, 16), (6, 4), (18, 0) **25.** Bounded

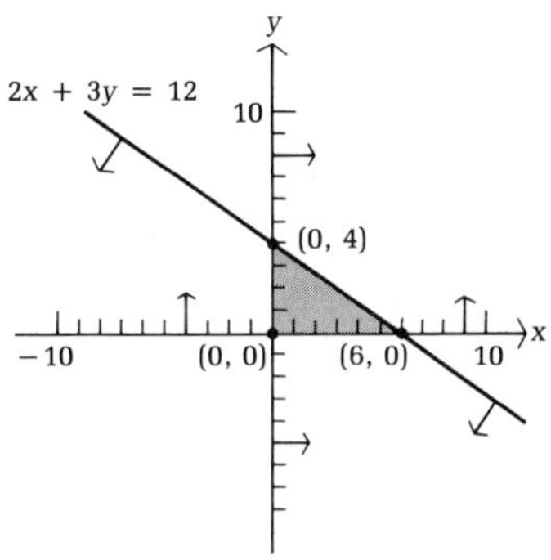

27. Bounded

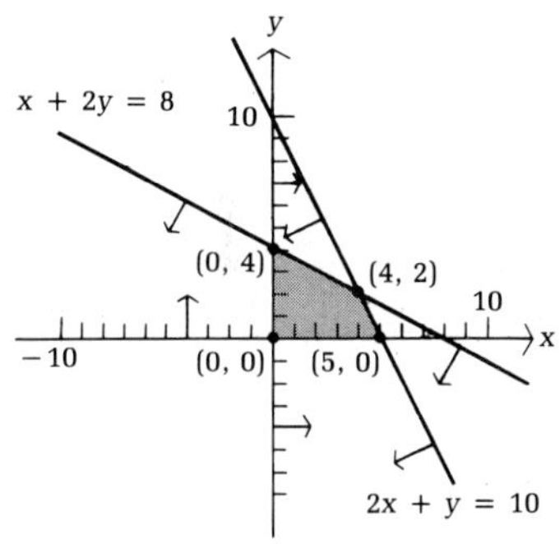

29. Unbounded

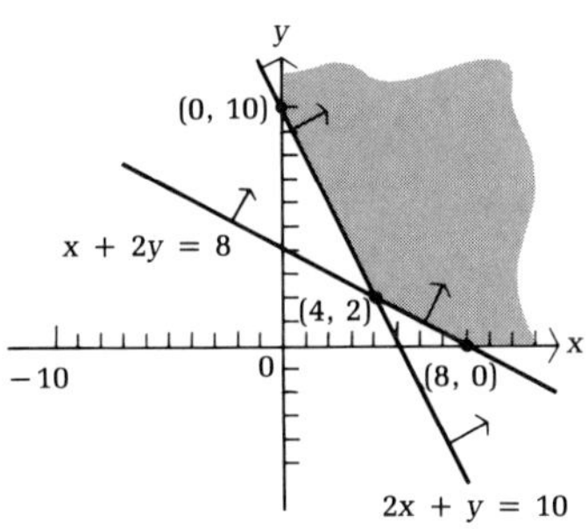

31. Bounded

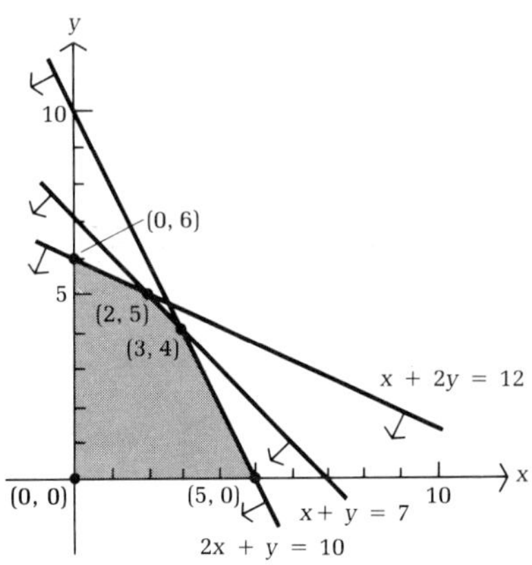

33. Unbounded

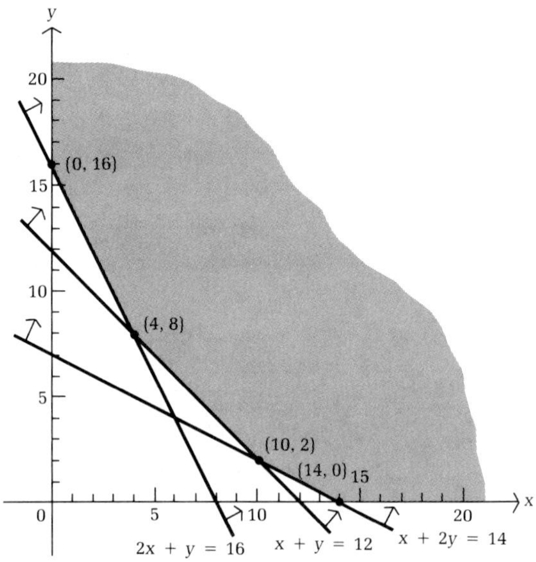

35. Bounded

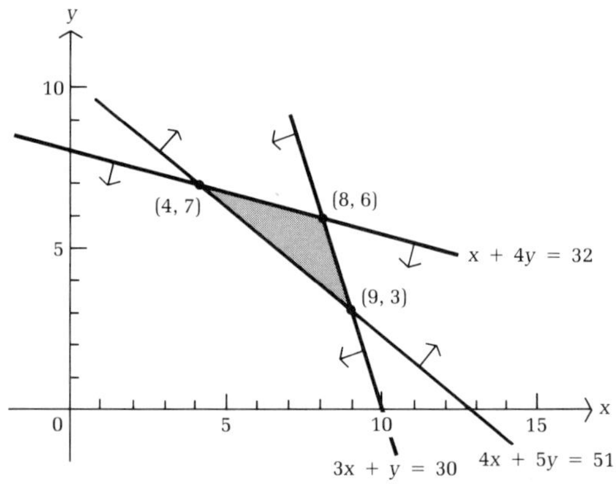

37. Empty

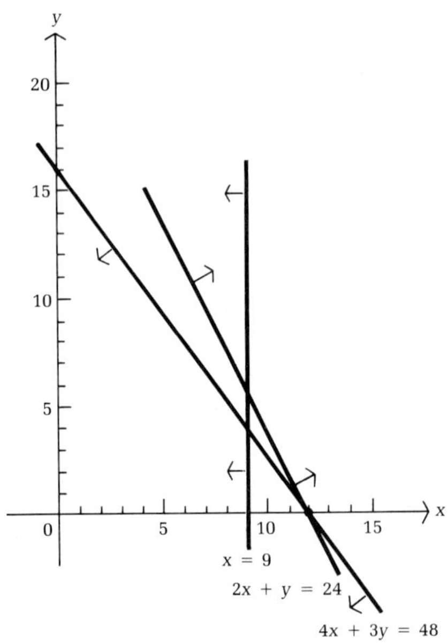

39. Unbounded

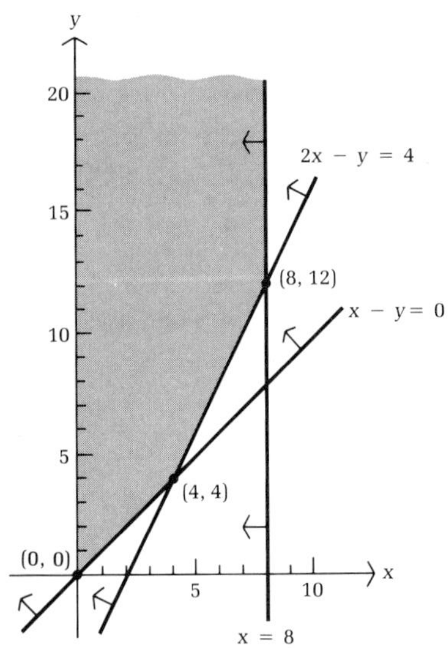

41. Bounded

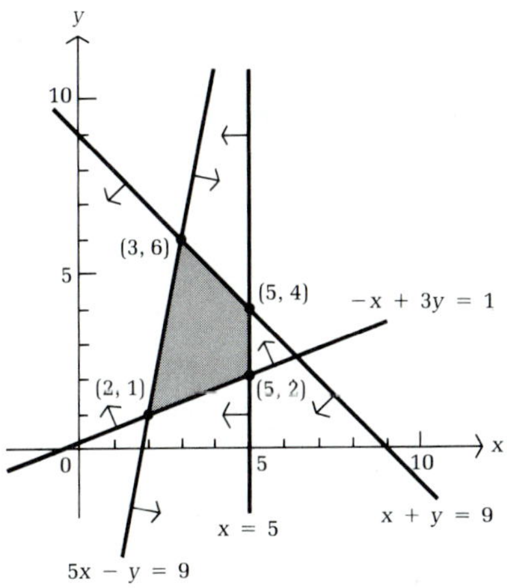

43. $6x + 4y \leq 108$
$x + y \leq 24$
$x \geq 0$
$y \geq 0$

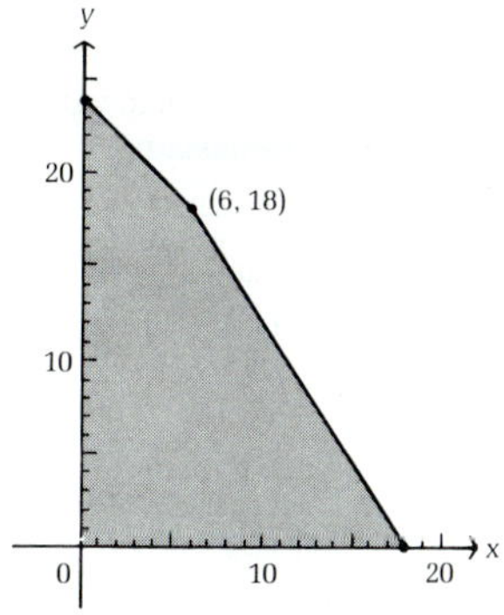

45. $20x + 10y \geq 460$
$30x + 30y \geq 960$
$5x + 10y \geq 220$
$x \geq 0$
$y \geq 0$

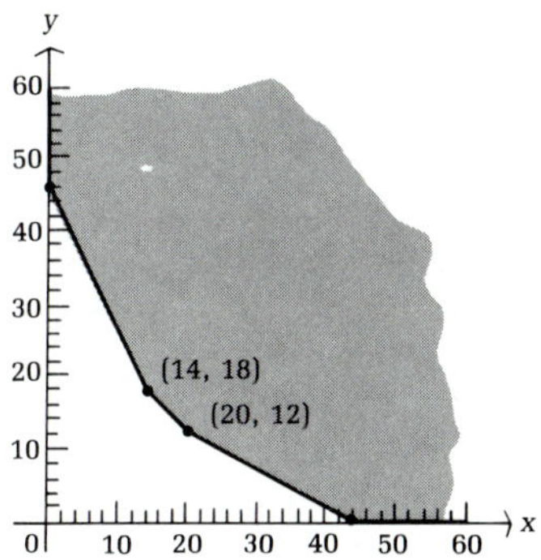

47. $10x + 20y \leq 800$
$20x + 10y \leq 640$
$x \geq 0$
$y \geq 0$

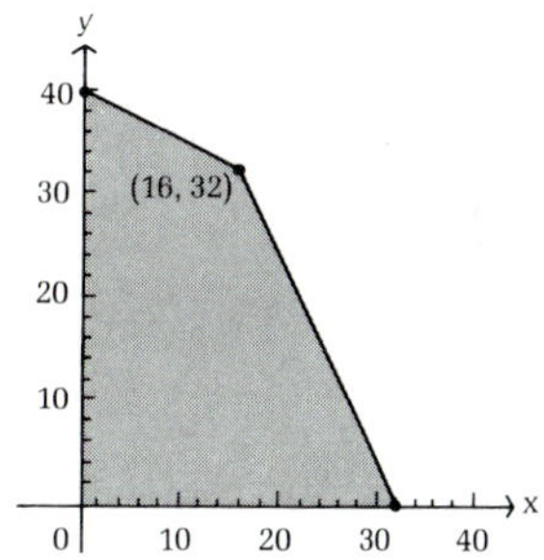

◆ EXERCISE 6-2

1. 16 **3.** 84 **5.** 32 **7.** 36 **9.** Max $P = 30$ at $x_1 = 4$ and $x_2 = 2$ **11.** Min $z = 14$ at $x_1 = 4$ and $x_2 = 2$; no max

13. Max $P = 260$ at $x_1 = 2$ and $x_2 = 5$ **15.** Min $z = 140$ at $x_1 = 14$ and $x_2 = 0$; no max

17. Min $P = 20$ at $x_1 = 0$ and $x_2 = 2$; Max $P = 150$ at $x_1 = 5$ and $x_2 = 0$ **19.** Feasible region empty; no optimal solutions

21. Min $P = 140$ at $x_1 = 3$ and $x_2 = 8$; Max $P = 260$ at $x_1 = 8$ and $x_2 = 10$, at $x_1 = 12$ and $x_2 = 2$, or at any point on the line segment from (8, 10) to (12, 2)

23. Max $P = 26{,}000$ at $x_1 = 400$ and $x_2 = 600$ **25.** (A) $2a < b$ (B) $\frac{1}{3}a < b < 2a$ (C) $b < \frac{1}{3}a$ (D) $b = 2a$ (E) $b = \frac{1}{3}a$

27. 6 trick, 18 slalom; \$780

29. (A) Plant *A*: 5 days; Plant *B*: 4 days; min cost \$8,600 (B) Plant *A*: 10 days; Plant *B*: 0 days; min cost \$6,000
(C) Plant *A*: 0 days; Plant *B*: 10 days; min cost \$8,000

31. 7 buses, 15 vans; min cost \$9,900 **33.** 1,500 gal by new process, none by old process; max profit \$300

35. (A) 150 bags brand *A*, 100 bags brand *B*; max nitrogen 1,500 lb (B) 0 bags brand *A*, 250 bags brand *B*; min nitrogen 750 lb

37. 20 cubic yards of *A*, 12 cubic yards of *B*; \$1,020 **39.** 48; 16 mice, 32 rats

◆ EXERCISE 6-3

1. (A) 5 (B) 4 (C) 5 basic and 4 nonbasic variables (D) 5 linear equations with 5 variables

3.

	NONBASIC	BASIC	FEASIBLE?
(A)	x_1, x_2	s_1, s_2	Yes
(B)	x_1, s_1	x_2, s_2	Yes
(C)	x_1, s_2	x_2, s_1	No
(D)	x_2, s_1	x_1, s_2	No
(E)	x_2, s_2	x_1, s_1	Yes
(F)	s_1, s_2	x_1, x_2	Yes

5.

	x_1	x_2	s_1	s_2	FEASIBLE?
(A)	0	0	50	40	Yes
(B)	0	50	0	−60	No
(C)	0	20	30	0	Yes
(D)	25	0	0	15	Yes
(E)	40	0	−30	0	No
(F)	20	10	0	0	Yes

7.

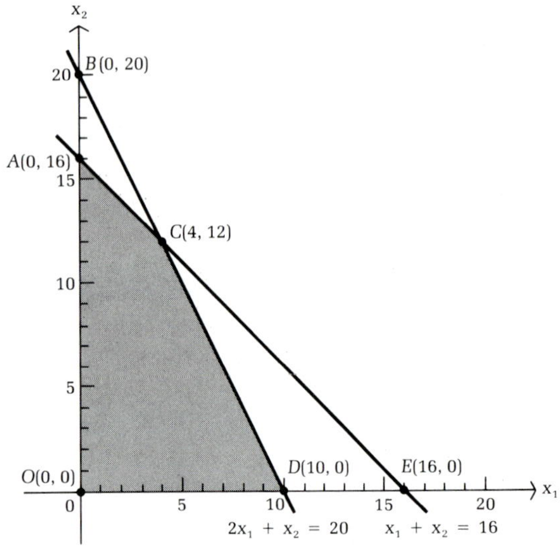

$$x_1 + x_2 + s_1 \quad = 16$$
$$2x_1 + x_2 \quad + s_2 = 20$$

x_1	x_2	s_1	s_2	INTERSECTION POINT	FEASIBLE?
0	0	16	20	O	Yes
0	16	0	4	A	Yes
0	20	−4	0	B	No
16	0	0	−12	E	No
10	0	6	0	D	Yes
4	12	0	0	C	Yes

9.

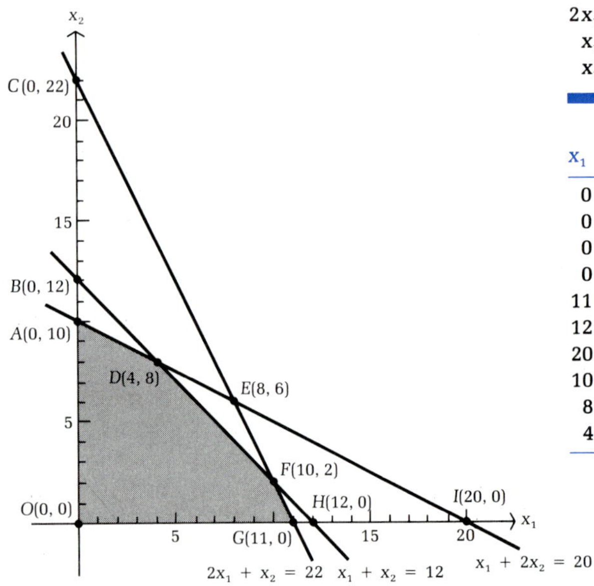

$$2x_1 + x_2 + s_1 \quad = 22$$
$$x_1 + x_2 \quad + s_2 \quad = 12$$
$$x_1 + 2x_2 \quad + s_3 = 20$$

x_1	x_2	s_1	s_2	s_3	INTERSECTION POINT	FEASIBLE?
0	0	22	12	20	O	Yes
0	22	0	−10	−24	C	No
0	12	10	0	−4	B	No
0	10	12	2	0	A	Yes
11	0	0	1	9	G	Yes
12	0	−2	0	8	H	No
20	0	−18	−8	0	I	No
10	2	0	0	6	F	Yes
8	6	0	−2	0	E	No
4	8	6	0	0	D	Yes

◆ EXERCISE 6-4

1. (A) Basic: x_2, s_1, P; nonbasic: x_1, s_2 (B) $x_1 = 0$, $x_2 = 12$, $s_1 = 15$, $s_2 = 0$, and $P = 20$ (C) Additional pivot required

3. (A) Basic: x_2, x_3, s_3, P; nonbasic: x_1, s_1, s_2 (B) $x_1 = 0$, $x_2 = 15$, $x_3 = 5$, $s_1 = 0$, $s_2 = 0$, $s_3 = 12$, and $P = 45$ (C) No solution

5. Enter ↓ x_1

$$\begin{array}{r} \\ \text{Exit}\rightarrow s_1 \\ s_2 \\ P \end{array} \begin{array}{c} \begin{array}{cccccc} x_1 & x_2 & s_1 & s_2 & P & \end{array} \\ \left[\begin{array}{ccccc|c} \textcircled{1} & 4 & 1 & 0 & 0 & 4 \\ 3 & 5 & 0 & 1 & 0 & 24 \\ \hdashline -8 & -5 & 0 & 0 & 1 & 0 \end{array}\right] \end{array}$$

$$\sim \begin{array}{r} x_1 \\ s_2 \\ P \end{array} \left[\begin{array}{ccccc|c} 1 & 4 & 1 & 0 & 0 & 4 \\ 0 & -7 & -3 & 1 & 0 & 12 \\ \hdashline 0 & 27 & 8 & 0 & 1 & 32 \end{array}\right]$$

7. Enter ↓ x_1

$$\begin{array}{r} \\ \text{Exit}\rightarrow x_2 \\ s_2 \\ s_3 \\ P \end{array} \begin{array}{c} \begin{array}{ccccccc} x_1 & x_2 & s_1 & s_2 & s_3 & P & \end{array} \\ \left[\begin{array}{cccccc|c} \textcircled{2} & 1 & 1 & 0 & 0 & 0 & 4 \\ 3 & 0 & 1 & 1 & 0 & 0 & 8 \\ 0 & 0 & 2 & 0 & 1 & 0 & 2 \\ \hdashline -4 & 0 & -3 & 0 & 0 & 1 & 5 \end{array}\right] \end{array}$$

$$\sim \begin{array}{r} x_1 \\ s_2 \\ s_3 \\ P \end{array} \left[\begin{array}{cccccc|c} 1 & \frac{1}{2} & \frac{1}{2} & 0 & 0 & 0 & 2 \\ 0 & -\frac{3}{2} & -\frac{1}{2} & 1 & 0 & 0 & 2 \\ 0 & 0 & 2 & 0 & 1 & 0 & 2 \\ \hdashline 0 & 2 & -1 & 0 & 0 & 1 & 13 \end{array}\right]$$

9. (A)

$$\begin{aligned} 2x_1 + x_2 + s_1 &= 10 \\ x_1 + 2x_2 + s_2 &= 8 \\ -15x_1 - 10x_2 + P &= 0 \end{aligned}$$

(B) Enter ↓ x_1

$$\begin{array}{r} \\ \text{Exit}\rightarrow s_1 \\ s_2 \\ P \end{array} \begin{array}{c} \begin{array}{cccccc} x_1 & x_2 & s_1 & s_2 & P & \end{array} \\ \left[\begin{array}{ccccc|c} \textcircled{2} & 1 & 1 & 0 & 0 & 10 \\ 1 & 2 & 0 & 1 & 0 & 8 \\ \hdashline -15 & -10 & 0 & 0 & 1 & 0 \end{array}\right] \end{array}$$

(C) Max $P = 80$ at $x_1 = 4$ and $x_2 = 2$

11. (A)

$$\begin{aligned} 2x_1 + x_2 + s_1 &= 10 \\ x_1 + 2x_2 + s_2 &= 8 \\ -30x_1 - x_2 + P &= 0 \end{aligned}$$

(B) Enter ↓ x_1

$$\begin{array}{r} \\ \text{Exit}\rightarrow s_1 \\ s_2 \\ P \end{array} \begin{array}{c} \begin{array}{cccccc} x_1 & x_2 & s_1 & s_2 & P & \end{array} \\ \left[\begin{array}{ccccc|c} \textcircled{2} & 1 & 1 & 0 & 0 & 10 \\ 1 & 2 & 0 & 1 & 0 & 8 \\ \hdashline -30 & -1 & 0 & 0 & 1 & 0 \end{array}\right] \end{array}$$

(C) Max $P = 150$ at $x_1 = 5$ and $x_2 = 0$

13. Max $P = 260$ at $x_1 = 2$ and $x_2 = 5$

15. No optimal solution exists.

17. Max $P = 7$ at $x_1 = 3$ and $x_2 = 5$

19. Max $P = \frac{190}{3}$ at $x_1 = \frac{40}{3}$, $x_2 = 0$, and $x_3 = \frac{10}{3}$

21. Max $P = 17$ at $x_1 = 4$, $x_2 = 3$, and $x_3 = 0$

23. Max $P = 22$ at $x_1 = 1$, $x_2 = 6$, and $x_3 = 0$

25. Max $P = 26{,}000$ at $x_1 = 400$ and $x_2 = 600$

27. Max $P = 450$ at $x_1 = 0$, $x_2 = 180$, and $x_3 = 30$

29. Max $P = 88$ at $x_1 = 24$ and $x_2 = 8$

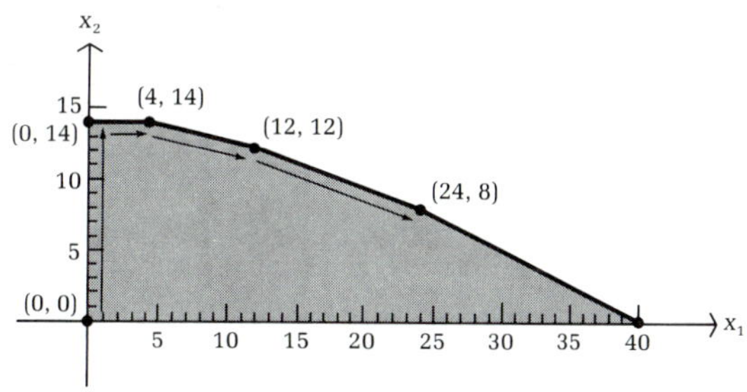

31. Let x_1 = Number of *A* components
x_2 = Number of *B* components
x_3 = Number of *C* components

Maximize $P = 7x_1 + 9x_2 + 10x_3$
Subject to

$$\begin{aligned} 2x_1 + x_2 + 2x_3 &\le 1{,}000 \\ x_1 + 2x_2 + 2x_3 &\le 800 \\ x_1, x_2, x_3 &\ge 0 \end{aligned}$$

400 *A* components, 200 *B* components, and 0 *C* components; max profit \$4,600

33. Let x_1 = Amount invested in government bonds
x_2 = Amount invested in mutual funds
x_3 = Amount invested in money market funds

$$\begin{aligned} \text{Maximize} \quad & P = 0.08x_1 + 0.13x_2 + 0.15x_3 \\ \text{Subject to} \quad & x_1 + x_2 + x_3 \le 100{,}000 \\ & -x_1 + x_2 + x_3 \le 0 \\ & x_1, x_2, x_3 \ge 0 \end{aligned}$$

$50,000 in government bonds, $0 in mutual funds, and $50,000 in money market funds; max return $11,500

35. Let x_1 = Number of ads placed in daytime shows
x_2 = Number of ads placed in prime-time shows
x_3 = Number of ads placed in late-night shows

$$\begin{aligned} \text{Maximize} \quad & P = 14{,}000x_1 + 24{,}000x_2 + 18{,}000x_3 \\ \text{Subject to} \quad & x_1 + x_2 + x_3 \le 15 \\ & 1{,}000x_1 + 2{,}000x_2 + 1{,}500x_3 \le 20{,}000 \\ & x_1, x_2, x_3 \ge 0 \end{aligned}$$

10 daytime ads, 5 prime-time ads, and 0 late-night ads; max number of potential customers 260,000

37. Let x_1 = Number of colonial houses
x_2 = Number of split-level houses
x_3 = Number of ranch houses

$$\begin{aligned} \text{Maximize} \quad & P = 20{,}000x_1 + 18{,}000x_2 + 24{,}000x_3 \\ \text{Subject to} \quad & \tfrac{1}{2}x_1 + \tfrac{1}{2}x_2 + x_3 \le 30 \\ & 60{,}000x_1 + 60{,}000x_2 + 80{,}000x_3 \le 3{,}200{,}000 \\ & 4{,}000x_1 + 3{,}000x_2 + 4{,}000x_3 \le 180{,}000 \\ & x_1, x_2, x_3 \ge 0 \end{aligned}$$

20 colonial, 20 split-level, and 10 ranch houses; max profit $1,000,000

39. Let x_1 = Number of boxes of assortment I
x_2 = Number of boxes of assortment II
x_3 = Number of boxes of assortment III

$$\begin{aligned} \text{Maximize} \quad & P = 4x_1 + 3x_2 + 5x_3 \\ \text{Subject to} \quad & 4x_1 + 12x_2 + 8x_3 \le 4{,}800 \\ & 4x_1 + 4x_2 + 8x_3 \le 4{,}000 \\ & 12x_1 + 4x_2 + 8x_3 \le 5{,}600 \\ & x_1, x_2, x_3 \ge 0 \end{aligned}$$

200 boxes of assortment I, 100 boxes of assortment II, and 350 boxes of assortment III; max profit $2,850

41. Let x_1 = Number of grams of food *A*
x_2 = Number of grams of food *B*
x_3 = Number of grams of food *C*

$$\begin{aligned} \text{Maximize} \quad & P = 3x_1 + 3x_2 + 5x_3 \\ \text{Subject to} \quad & x_1 + 3x_2 + 2x_3 \le 30 \\ & 2x_1 + x_2 + x_3 \le 24 \\ & x_1, x_2, x_3 \ge 0 \end{aligned}$$

6 g food *A*, 0 g food *B*, and 12 g food *C*; max protein 78 units

43. Let x_1 = Number of undergraduate students
x_2 = Number of graduate students
x_3 = Number of faculty members

$$\begin{aligned} \text{Maximize} \quad & P = 18x_1 + 25x_2 + 30x_3 \\ \text{Subject to} \quad & x_1 + x_2 + x_3 \le 20 \\ & 60x_1 + 90x_2 + 120x_3 \le 1{,}620 \\ & x_1, x_2, x_3 \ge 0 \end{aligned}$$

6 undergraduate and 14 graduate students, 0 faculty members; max number of interviews 458

◆ EXERCISE 6-5

1. (A) $$\begin{aligned} \text{Maximize} \quad & P = 4y_1 + 5y_2 \\ \text{Subject to} \quad & y_1 + 2y_2 \le 8 \\ & 3y_1 + y_2 \le 9 \\ & y_1, y_2 \ge 0 \end{aligned}$$

(B) $$\begin{aligned} y_1 + 2y_2 + x_1 \qquad\qquad &= 8 \\ 3y_1 + y_2 \qquad + x_2 \qquad &= 9 \\ -4y_1 - 5y_2 \qquad\qquad + P &= 0 \end{aligned}$$

(C) $$\begin{array}{c} \begin{matrix} y_1 & y_2 & x_1 & x_2 & P \end{matrix} \\ \left[\begin{array}{rrrrr|r} 1 & 2 & 1 & 0 & 0 & 8 \\ 3 & 1 & 0 & 1 & 0 & 9 \\ \hline -4 & -5 & 0 & 0 & 1 & 0 \end{array}\right] \end{array}$$

3. (A) Max $P = 121$ at $y_1 = 3$ and $y_2 = 5$ (B) Min $C = 121$ at $x_1 = 1$ and $x_2 = 2$

5. (A) $$\begin{aligned} \text{Maximize} \quad & P = 13y_1 + 12y_2 \\ \text{Subject to} \quad & 4y_1 + 3y_2 \le 9 \\ & y_1 + y_2 \le 2 \\ & y_1, y_2 \ge 0 \end{aligned}$$
(B) Min $C = 26$ at $x_1 = 0$ and $x_2 = 13$

7. (A) $$\begin{aligned} \text{Maximize} \quad & P = 15y_1 + 8y_2 \\ \text{Subject to} \quad & 2y_1 + y_2 \le 7 \\ & 3y_1 + 2y_2 \le 12 \\ & y_1, y_2 \ge 0 \end{aligned}$$
(B) Min $C = 54$ at $x_1 = 6$ and $x_2 = 1$

9. (A) $$\begin{aligned} \text{Maximize} \quad & P = 8y_1 + 4y_2 \\ \text{Subject to} \quad & 2y_1 - 2y_2 \le 11 \\ & y_1 + 3y_2 \le 4 \\ & y_1, y_2 \ge 0 \end{aligned}$$
(B) Min $C = 32$ at $x_1 = 0$ and $x_2 = 8$

11. (A) $$\begin{aligned} \text{Maximize} \quad & P = 6y_1 + 4y_2 \\ \text{Subject to} \quad & -3y_1 + y_2 \le 7 \\ & y_1 - 2y_2 \le 9 \\ & y_1, y_2 \ge 0 \end{aligned}$$
(B) No optimal solution exists.

13. Min $C = 24$ at $x_1 = 8$ and $x_2 = 0$ **15.** Min $C = 20$ at $x_1 = 0$ and $x_2 = 4$ **17.** Min $C = 140$ at $x_1 = 14$ and $x_2 = 0$
19. Min $C = 44$ at $x_1 = 6$ and $x_2 = 2$ **21.** Min $C = 43$ at $x_1 = 0$, $x_2 = 1$, and $x_3 = 3$ **23.** No optimal solution exists.
25. Min $C = 44$ at $x_1 = 0$, $x_2 = 3$, and $x_3 = 5$ **27.** Min $C = 166$ at $x_1 = 0$, $x_2 = 12$, $x_3 = 20$, and $x_4 = 3$

29. Let x_1 = Number of hours the Cedarburg plant is operated
x_2 = Number of hours the Grafton plant is operated
x_3 = Number of hours the West Bend plant is operated

$$\begin{array}{ll} \text{Minimize} & C = 70x_1 + 75x_2 + 90x_3 \\ \text{Subject to} & 20x_1 + 10x_2 + 20x_3 \geq 300 \\ & 10x_1 + 20x_2 + 20x_3 \geq 200 \\ & x_1, x_2, x_3 \geq 0 \end{array}$$

Cedarburg plant 10 hr per day, West Bend plant 5 hr per day, Grafton plant not used; \$1,150

31. Let x_1 = Number of single-sided drives ordered from Associated Electronics
x_2 = Number of double-sided drives ordered from Associated Electronics
x_3 = Number of single-sided drives ordered from Digital Drives
x_4 = Number of double-sided drives ordered from Digital Drives

$$\begin{array}{ll} \text{Minimize} & C = 250x_1 + 350x_2 + 290x_3 + 320x_4 \\ \text{Subject to} & x_1 + x_2 \leq 1{,}000 \\ & x_3 + x_4 \leq 2{,}000 \\ & x_1 + x_3 \geq 1{,}200 \\ & x_2 + x_4 \geq 1{,}600 \\ & x_1, x_2, x_3, x_4 \geq 0 \end{array}$$

1,000 single-sided drives from Associated Electronics, 200 single-sided and 1,600 double-sided drives from Digital Drives; \$820,000

33. Let x_1 = Number of ounces of food *L*
x_2 = Number of ounces of food *M*
x_3 = Number of ounces of food *N*

$$\begin{array}{ll} \text{Minimize} & C = 20x_1 + 24x_2 + 18x_3 \\ \text{Subject to} & 20x_1 + 10x_2 + 10x_3 \geq 300 \\ & 10x_1 + 10x_2 + 10x_3 \geq 200 \\ & 10x_1 + 20x_2 + 10x_3 \geq 240 \\ & x_1, x_2, x_3 \geq 0 \end{array}$$

10 oz *L*, 4 oz *M*, 6 oz *N*; 404 units

35. Let x_1 = Number of students bused from North Division to Central
x_2 = Number of students bused from North Division to Washington
x_3 = Number of students bused from South Division to Central
x_4 = Number of students bused from South Division to Washington

$$\begin{array}{ll} \text{Minimize} & C = 5x_1 + 2x_2 + 3x_3 + 4x_4 \\ \text{Subject to} & x_1 + x_2 \geq 300 \\ & x_3 + x_4 \geq 500 \\ & x_1 + x_3 \leq 400 \\ & x_2 + x_4 \leq 500 \\ & x_1, x_2, x_3, x_4 \geq 0 \end{array}$$

300 students bused from North Division to Washington, 400 from South Division to Central, and 100 from South Division to Washington; \$2,200

◆ EXERCISE 6-6

1. (A)
$$\begin{array}{ll} \text{Maximize} & P = 5x_1 + 2x_2 - Ma_1 \\ \text{Subject to} & x_1 + 2x_2 + s_1 = 12 \\ & x_1 + x_2 - s_2 + a_1 = 4 \\ & x_1, x_2, s_1, s_2, a_1 \geq 0 \end{array}$$

(B)
$$\begin{array}{c} \begin{matrix} x_1 & x_2 & s_1 & s_2 & a_1 & P \end{matrix} \\ \left[\begin{array}{cccccc|c} 1 & 2 & 1 & 0 & 0 & 0 & 12 \\ 1 & 1 & 0 & -1 & 1 & 0 & 4 \\ \hline -M-5 & -M-2 & 0 & M & 0 & 1 & -4M \end{array}\right] \end{array}$$

(C) $x_1 = 12$, $x_2 = 0$, $s_1 = 0$, $s_2 = 8$, $a_1 = 0$, $P = 60$ (D) Max $P = 60$ at $x_1 = 12$ and $x_2 = 0$

3. (A)
$$\begin{array}{ll} \text{Maximize} & P = 3x_1 + 5x_2 - Ma_1 \\ \text{Subject to} & 2x_1 + x_2 + s_1 = 8 \\ & x_1 + x_2 + a_1 = 6 \\ & x_1, x_2, s_1, a_1 \geq 0 \end{array}$$

(B)
$$\begin{array}{c} \begin{matrix} x_1 & x_2 & s_1 & a_1 & P \end{matrix} \\ \left[\begin{array}{ccccc|c} 2 & 1 & 1 & 0 & 0 & 8 \\ 1 & 1 & 0 & 1 & 0 & 6 \\ \hline -M-3 & -M-5 & 0 & 0 & 1 & -6M \end{array}\right] \end{array}$$

(C) $x_1 = 0$, $x_2 = 6$, $s_1 = 2$, $a_1 = 0$, $P = 30$ (D) Max $P = 30$ at $x_1 = 0$ and $x_2 = 6$

5. (A)
$$\begin{array}{ll} \text{Maximize} & P = 4x_1 + 3x_2 - Ma_1 \\ \text{Subject to} & -x_1 + 2x_2 + s_1 = 2 \\ & x_1 + x_2 - s_2 + a_1 = 4 \\ & x_1, x_2, s_1, s_2, a_1 \geq 0 \end{array}$$

(B)
$$\begin{array}{c} \begin{matrix} x_1 & x_2 & s_1 & s_2 & a_1 & P \end{matrix} \\ \left[\begin{array}{cccccc|c} -1 & 2 & 1 & 0 & 0 & 0 & 2 \\ 1 & 1 & 0 & -1 & 1 & 0 & 4 \\ \hline -M-4 & -M-3 & 0 & M & 0 & 1 & -4M \end{array}\right] \end{array}$$

(C) No optimal solution exists. (D) No optimal solution exists.

7. (A) Maximize $P = 5x_1 + 10x_2 - Ma_1$

Subject to
$$\begin{aligned} x_1 + x_2 + s_1 \qquad\qquad &= 3 \\ 2x_1 + 3x_2 \qquad - s_2 + a_1 &= 12 \\ x_1, x_2, s_1, s_2, a_1 &\geq 0 \end{aligned}$$

(B)
$$\begin{array}{c} \\ \\ \\ \\ \end{array}\left[\begin{array}{cccccc|c} x_1 & x_2 & s_1 & s_2 & a_1 & P & \\ 1 & 1 & 1 & 0 & 0 & 0 & 3 \\ 2 & 3 & 0 & -1 & 1 & 0 & 12 \\ \hline -2M-5 & -3M-10 & 0 & M & 0 & 1 & -12M \end{array}\right]$$

(C) $x_1 = 0, x_2 = 3, s_1 = 0, s_2 = 0, a_1 = 3, P = -3M + 30$ (D) No optimal solution exists.

9. Min $P = 12$ at $x_1 = 4$ and $x_2 = 6$; Max $P = 60$ at $x_1 = 10$ and $x_2 = 0$ **11.** Max $P = 44$ at $x_1 = 2$ and $x_2 = 8$

13. No optimal solution exists. **15.** Min $C = -9$ at $x_1 = 0$, $x_2 = \frac{7}{4}$, and $x_3 = \frac{3}{4}$ **17.** Max $P = 32$ at $x_1 = 0$, $x_2 = 4$, and $x_3 = 2$

19. Max $P = 65$ at $x_1 = \frac{35}{2}$, $x_2 = 0$, and $x_3 = \frac{15}{2}$ **21.** Max $P = 120$ at $x_1 = 20$, $x_2 = 0$, and $x_3 = 20$

23. Min $C = -30$ at $x_1 = 0$, $x_2 = \frac{3}{4}$, and $x_3 = 0$ **25.** Max $P = 17$ at $x_1 = \frac{49}{5}$, $x_2 = 0$, and $x_3 = \frac{22}{5}$

27. Min $C = \frac{135}{2}$ at $x_1 = \frac{15}{4}$, $x_2 = \frac{3}{4}$, and $x_3 = 0$ **29.** Max $P = 380$ at $x_1 = \frac{80}{3}$, $x_2 = \frac{20}{3}$, and $x_3 = 0$

31. Let x_1 = Number of 16k modules manufactured daily
x_2 = Number of 64k modules manufactured daily

Maximize $P = 18x_1 + 30x_2$

Subject to
$$\begin{aligned} 10x_1 + 15x_2 &\leq 2{,}200 \\ 2x_1 + 4x_2 &\leq 500 \\ x_1 \qquad &\geq 50 \\ x_1, x_2 &\geq 0 \end{aligned}$$

130 of the 16k modules and 60 of the 64k modules; \$4,140

33. Let x_1 = Number of ads placed in the *Sentinel*
x_2 = Number of ads placed in the *Journal*
x_3 = Number of ads placed in the *Tribune*

Minimize $C = 200x_1 + 200x_2 + 100x_3$

Subject to
$$\begin{aligned} x_1 + x_2 + x_3 &\leq 10 \\ 2{,}000x_1 + 500x_2 + 1{,}500x_3 &\geq 16{,}000 \\ x_1, x_2, x_3 &\geq 0 \end{aligned}$$

2 ads in the *Sentinel*, 0 ads in the *Journal*, 8 ads in the *Tribune*; \$1,200

35. Let x_1 = Number of bottles of brand *A*
x_2 = Number of bottles of brand *B*
x_3 = Number of bottles of brand *C*

Minimize $C = 0.6x_1 + 0.4x_2 + 0.9x_3$

Subject to
$$\begin{aligned} 10x_1 + 10x_2 + 20x_3 &\geq 100 \\ 2x_1 + 3x_2 + 4x_3 &\leq 24 \\ x_1, x_2, x_3 &\geq 0 \end{aligned}$$

0 bottles of *A*, 4 bottles of *B*, 3 bottles of *C*; \$4.30

37. Let x_1 = Number of cubic yards of mix *A*
x_2 = Number of cubic yards of mix *B*
x_3 = Number of cubic yards of mix *C*

Maximize $P = 12x_1 + 16x_2 + 8x_3$

Subject to
$$\begin{aligned} 12x_1 + 8x_2 + 16x_3 &\leq 700 \\ 16x_1 + 8x_2 + 16x_3 &\geq 800 \\ x_1, x_2, x_3 &\geq 0 \end{aligned}$$

25 cubic yards of *A*, 50 cubic yards of *B*, 0 cubic yards of *C*; 1,100 lb

39. Let x_1 = Number of car frames produced at the Milwaukee plant
x_2 = Number of truck frames produced at the Milwaukee plant
x_3 = Number of car frames produced at the Racine plant
x_4 = Number of truck frames produced at the Racine plant

Maximize $P = 50x_1 + 70x_2 + 50x_3 + 70x_4$

Subject to
$$\begin{aligned} x_1 + x_3 &\leq 250 \\ x_2 + x_4 &\leq 350 \\ x_1 + x_2 &\leq 300 \\ x_3 + x_4 &\leq 200 \\ 150x_1 + 200x_2 &\leq 50{,}000 \\ 135x_3 + 180x_4 &\leq 35{,}000 \\ x_1, x_2, x_3, x_4 &\geq 0 \end{aligned}$$

41. Let x_1 = Number of barrels of *A* used in regular gasoline
x_2 = Number of barrels of *A* used in premium gasoline
x_3 = Number of barrels of *B* used in regular gasoline
x_4 = Number of barrels of *B* used in premium gasoline
x_5 = Number of barrels of *C* used in regular gasoline
x_6 = Number of barrels of *C* used in premium gasoline

Maximize $P = 10x_1 + 18x_2 + 8x_3 + 16x_4 + 4x_5 + 12x_6$

Subject to
$$\begin{aligned} x_1 + x_2 &\leq 40{,}000 \\ x_3 + x_4 &\leq 25{,}000 \\ x_5 + x_6 &\leq 15{,}000 \\ x_1 + x_3 + x_5 &\geq 30{,}000 \\ x_2 + x_4 + x_6 &\geq 25{,}000 \\ -5x_1 + 5x_3 + 15x_5 &\geq 0 \\ -15x_2 - 5x_4 + 5x_6 &\geq 0 \\ x_1, x_2, x_3, x_4, x_5, x_6 &\geq 0 \end{aligned}$$

43. Let x_1 = Number of ounces of food *L*
x_2 = Number of ounces of food *M*
x_3 = Number of ounces of food *N*

Minimize $C = 0.4x_1 + 0.6x_2 + 0.8x_3$

Subject to
$$\begin{aligned} 30x_1 + 10x_2 + 30x_3 &\geq 400 \\ 10x_1 + 10x_2 + 10x_3 &\geq 200 \\ 10x_1 + 30x_2 + 20x_3 &\geq 300 \\ 8x_1 + 4x_2 + 6x_3 &\leq 150 \\ 60x_1 + 40x_2 + 50x_3 &\leq 900 \\ x_1, x_2, x_3 &\geq 0 \end{aligned}$$

45. Let x_1 = Number of students from town *A* enrolled in school I
x_2 = Number of students from town *A* enrolled in school II
x_3 = Number of students from town *B* enrolled in school I
x_4 = Number of students from town *B* enrolled in school II
x_5 = Number of students from town *C* enrolled in school I
x_6 = Number of students from town *C* enrolled in school II

Minimize $C = 4x_1 + 8x_2 + 6x_3 + 4x_4 + 3x_5 + 9x_6$

Subject to

$$\begin{aligned}
x_1 + x_2 &= 500\\
x_3 + x_4 &= 1{,}200\\
x_5 + x_6 &= 1{,}800\\
x_1 + x_3 + x_5 &\le 2{,}000\\
x_2 + x_4 + x_6 &\le 2{,}000\\
x_1 + x_3 + x_5 &\ge 1{,}400\\
x_2 + x_4 + x_6 &\ge 1{,}400\\
x_1 &\le 300\\
x_2 &\le 300\\
x_3 &\le 720\\
x_4 &\le 720\\
x_5 &\le 1{,}080\\
x_6 &\le 1{,}080\\
x_1, x_2, x_3, x_4, x_5, x_6 &\ge 0
\end{aligned}$$

◆ EXERCISE 6-7 CHAPTER REVIEW

1. Bounded

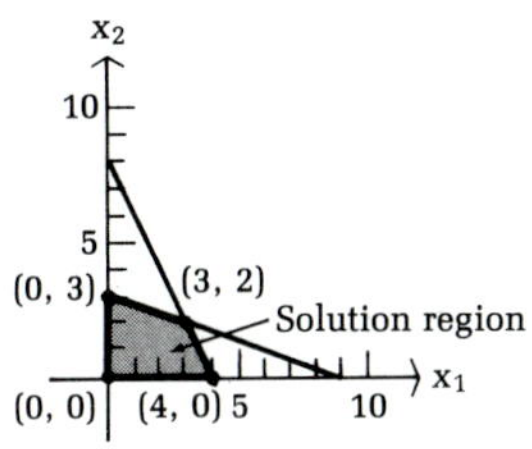

2. Unbounded

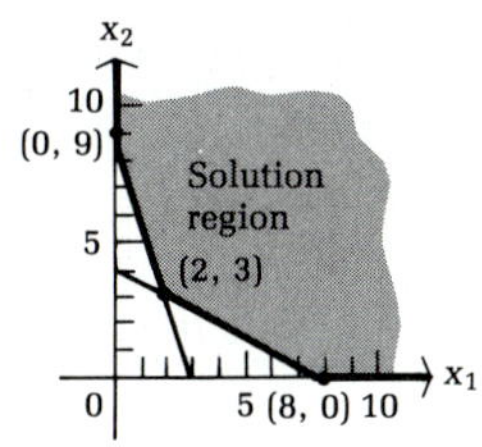

3. Max $P = 24$ at $x_1 = 4$ and $x_2 = 0$

4. $2x_1 + x_2 + s_1 = 8$
$x_1 + 2x_2 + s_2 = 10$

5. 2 basic and 2 nonbasic variables

6.

x_1	x_2	s_1	s_2	FEASIBLE?
0	0	8	10	Yes
0	8	0	−6	No
0	5	3	0	Yes
4	0	0	6	Yes
10	0	−12	0	No
2	4	0	0	Yes

7. Enter ↓ x_1; Exit → s_1

$$\begin{array}{c} \\ s_1 \\ s_2 \\ P \end{array}
\begin{array}{c} \begin{array}{cccccc} x_1 & x_2 & s_1 & s_2 & P & \end{array} \\
\left[\begin{array}{ccccc|c}
\textcircled{2} & 1 & 1 & 0 & 0 & 8\\
1 & 2 & 0 & 1 & 0 & 10\\
\hdashline
-6 & -2 & 0 & 0 & 1 & 0
\end{array}\right]\end{array}$$

8. Max $P = 24$ at $x_1 = 4$ and $x_2 = 0$

9. Basic variables: x_2, s_2, s_3, P; nonbasic variables: x_1, x_3, s_1

Enter ↓ x_1; Exit → s_3

$$\begin{array}{c|ccccccc|c}
 & x_1 & x_2 & x_3 & s_1 & s_2 & s_3 & P & \\
\hline
x_2 & 2 & 1 & 3 & -1 & 0 & 0 & 0 & 20\\
s_2 & 3 & 0 & 4 & 1 & 1 & 0 & 0 & 30\\
s_3 & \textcircled{2} & 0 & 5 & 2 & 0 & 1 & 0 & 10\\
\hdashline
P & -8 & 0 & -5 & 3 & 0 & 0 & 1 & 50
\end{array}
\sim
\begin{array}{c|ccccccc|c}
 & x_1 & x_2 & x_3 & s_1 & s_2 & s_3 & P & \\
\hline
x_2 & 0 & 1 & -2 & -3 & 0 & -1 & 0 & 10\\
s_2 & 0 & 0 & -\frac{7}{2} & -2 & 1 & -\frac{3}{2} & 0 & 15\\
x_1 & 1 & 0 & \frac{5}{2} & 1 & 0 & \frac{1}{2} & 0 & 5\\
\hdashline
P & 0 & 0 & 15 & 11 & 0 & 4 & 1 & 90
\end{array}$$

10. (A) $x_1 = 0$, $x_2 = 2$, $s_1 = 0$, $s_2 = 5$, $P = 12$; additional pivoting required
(B) $x_1 = 0$, $x_2 = 0$, $s_1 = 0$, $s_2 = 7$, $P = 22$; no optimal solution exists
(C) $x_1 = 6$, $x_2 = 0$, $s_1 = 15$, $s_2 = 0$, $P = 10$; optimal solution

11. Min $C = 40$ at $x_1 = 0$ and $x_2 = 20$

12. Maximize $P = 15y_1 + 20y_2$
Subject to $y_1 + 2y_2 \leq 5$
$3y_1 + y_2 \leq 2$
$y_1, y_2 \geq 0$

13.
$$\begin{aligned} y_1 + 2y_2 + x_1 \qquad\quad &= 5 \\ 3y_1 + y_2 \qquad + x_2 \quad &= 2 \\ -15y_1 - 20y_2 \qquad\quad + P &= 0 \end{aligned}$$

14.
$$\begin{array}{c} \\ \\ \\ \end{array}\begin{matrix} y_1 & y_2 & x_1 & x_2 & P & \\ \end{matrix}$$
$$\left[\begin{array}{ccccc|c} 1 & 2 & 1 & 0 & 0 & 5 \\ 3 & 1 & 0 & 1 & 0 & 2 \\ \hline -15 & -20 & 0 & 0 & 1 & 0 \end{array}\right]$$

15. Max $P = 40$ at $y_1 = 0$ and $y_2 = 2$ 16. Min $C = 40$ at $x_1 = 0$ and $x_2 = 20$ 17. Max $P = 26$ at $x_1 = 2$ and $x_2 = 5$

18. Max $P = 26$ at $x_1 = 2$ and $x_2 = 5$ 19. Min $C = 51$ at $x_1 = 9$ and $x_2 = 3$

20. Maximize $P = 10y_1 + 15y_2 + 3y_3$
Subject to $y_1 + y_2 \leq 3$
$y_1 + 2y_2 + y_3 \leq 8$
$y_1, y_2, y_3 \geq 0$

21. Min $C = 51$ at $x_1 = 9$ and $x_2 = 3$

22. No optimal solution exists.

23. Max $P = 23$ at $x_1 = 4$, $x_2 = 1$, and $x_3 = 0$

24. (A) Modified problem:
Maximize $P = x_1 + 3x_2 - Ma_1$
Subject to
$$\begin{aligned} x_1 + x_2 - s_1 + a_1 \qquad &= 6 \\ x_1 + 2x_2 \qquad\qquad + s_2 &= 8 \\ x_1, x_2, s_1, s_2, a_1 &\geq 0 \end{aligned}$$

(B) Preliminary simplex tableau:
$$\begin{array}{cccccc|c} x_1 & x_2 & s_1 & a_1 & s_2 & P & \\ 1 & 1 & -1 & 1 & 0 & 0 & 6 \\ 1 & 2 & 0 & 0 & 1 & 0 & 8 \\ \hline -1 & -3 & 0 & M & 0 & 1 & 0 \end{array}$$

Initial simplex tableau:
$$\begin{array}{cccccc|c} x_1 & x_2 & s_1 & a_1 & s_2 & P & \\ 1 & 1 & -1 & 1 & 0 & 0 & 6 \\ 1 & 2 & 0 & 0 & 1 & 0 & 8 \\ \hline -M-1 & -M-3 & M & 0 & 0 & 1 & -6M \end{array}$$

(C) $x_1 = 4$, $x_2 = 2$, $s_1 = 0$, $a_1 = 0$, $s_2 = 0$, $P = 10$
(D) Since $a_1 = 0$, the solution to the original problem is Max $P = 10$ at $x_1 = 4$ and $x_2 = 2$.

25. (A) Modified problem:
Maximize $P = x_1 + x_2 - Ma_1$
Subject to
$$\begin{aligned} x_1 + x_2 - s_1 + a_1 \qquad &= 5 \\ x_1 + 2x_2 \qquad\qquad + s_2 &= 4 \\ x_1, x_2, s_1, s_2, a_1 &\geq 0 \end{aligned}$$

(B) Preliminary simplex tableau:
$$\begin{array}{cccccc|c} x_1 & x_2 & s_1 & a_1 & s_2 & P & \\ 1 & 1 & -1 & 1 & 0 & 0 & 5 \\ 1 & 2 & 0 & 0 & 1 & 0 & 4 \\ \hline -1 & -1 & 0 & M & 0 & 1 & 0 \end{array}$$

Initial simplex tableau:
$$\begin{array}{cccccc|c} x_1 & x_2 & s_1 & a_1 & s_2 & P & \\ 1 & 1 & -1 & 1 & 0 & 0 & 5 \\ 1 & 2 & 0 & 0 & 1 & 0 & 4 \\ \hline -M-1 & -M-1 & M & 0 & 0 & 1 & -5M \end{array}$$

(C) $x_1 = 4$, $x_2 = 0$, $s_1 = 0$, $s_2 = 0$, $a_1 = 1$, $P = -M + 4$ (D) Since $a_1 \neq 0$, the original problem has no solution.

26. Maximize $P = 2x_1 + 3x_2 + x_3 - Ma_1 - Ma_2$
Subject to
$$\begin{aligned} x_1 - 3x_2 + x_3 + s_1 \qquad\qquad &= 7 \\ x_1 + x_2 - 2x_3 \qquad - s_2 + a_1 \quad &= 2 \\ 3x_1 + 2x_2 - x_3 \qquad\qquad + a_2 &= 4 \\ x_1, x_2, x_3, s_1, s_2, a_1, a_2 &\geq 0 \end{aligned}$$

27. Max $P = 36$ at $x_1 = 6$ and $x_2 = 8$

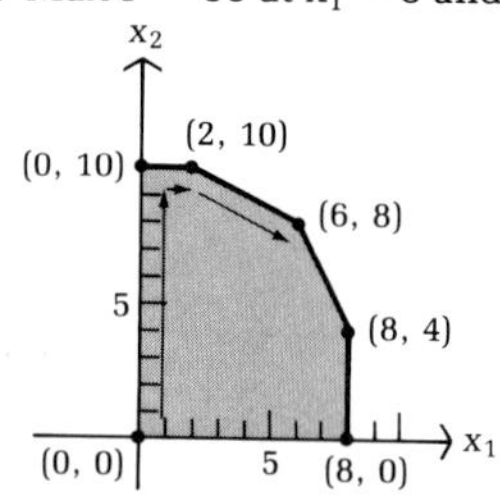

28. Min $C = 14$ at $x_1 = 4$ and $x_2 = 2$ 29. Min $C = 14$ at $x_1 = 4$ and $x_2 = 2$ 30. Min $C = 9{,}960$ at $x_1 = 0$, $x_2 = 240$, $x_3 = 400$, and $x_4 = 60$

31. Let x_1 = Number of regular sails
x_2 = Number of competition sails

Maximize $P = 100x_1 + 200x_2$
Subject to $2x_1 + 3x_2 \leq 150$
$4x_1 + 9x_2 \leq 360$
$x_1, x_2 \geq 0$

32. Let x_1 = Amount invested in oil stock
x_2 = Amount invested in steel stock
x_3 = Amount invested in government bonds

$$\begin{aligned} \text{Maximize} \quad & P = 0.12x_1 + 0.09x_2 + 0.05x_3 \\ \text{Subject to} \quad & x_1 + x_2 + x_3 \leqslant 150{,}000 \\ & x_1 + x_2 - x_3 \leqslant 0 \\ & 2x_1 - x_2 \leqslant 0 \\ & x_1, x_2, x_3 \geqslant 0 \end{aligned}$$

33. Let x_1 = Number of motors shipped from factory *A* to plant *X*
x_2 = Number of motors shipped from factory *A* to plant *Y*
x_3 = Number of motors shipped from factory *A* to plant *Z*
x_4 = Number of motors shipped from factory *B* to plant *X*
x_5 = Number of motors shipped from factory *B* to plant *Y*
x_6 = Number of motors shipped from factory *B* to plant *Z*

$$\begin{aligned} \text{Minimize} \quad & C = 5x_1 + 8x_2 + 12x_3 + 9x_4 + 7x_5 + 6x_6 \\ \text{Subject to} \quad & x_1 + x_2 + x_3 \leqslant 1{,}500 \\ & x_4 + x_5 + x_6 \leqslant 1{,}000 \\ & x_1 + x_4 \geqslant 500 \\ & x_2 + x_5 \geqslant 700 \\ & x_3 + x_6 \geqslant 800 \\ & x_1, x_2, x_3, x_4, x_5, x_6 \geqslant 0 \end{aligned}$$

34. Let x_1 = Number of pounds of long grain rice used in brand *A*
x_2 = Number of pounds of long grain rice used in brand *B*
x_3 = Number of pounds of wild rice used in brand *A*
x_4 = Number of pounds of wild rice used in brand *B*

$$\begin{aligned} \text{Maximize} \quad & P = 0.8x_1 + 0.5x_2 - 1.9x_3 - 2.2x_4 \\ \text{Subject to} \quad & 0.1x_1 - 0.9x_3 \leqslant 0 \\ & 0.05x_2 - 0.95x_4 \leqslant 0 \\ & x_1 + x_2 \leqslant 8{,}000 \\ & x_3 + x_4 \leqslant 500 \\ & x_1, x_2, x_3, x_4 \geqslant 0 \end{aligned}$$

35. Let x_1 = Number of grams of mix *A*
x_2 = Number of grams of mix *B*

$$\begin{aligned} \text{Minimize} \quad & C = 0.02x_1 + 0.04x_2 \\ \text{Subject to} \quad & 3x_1 + 4x_2 \geqslant 300 \\ & 2x_1 + 5x_2 \geqslant 200 \\ & 6x_1 + 10x_2 \geqslant 900 \\ & x_1, x_2 \geqslant 0 \end{aligned}$$

Index for Chapters 1–12

Index for Appendices A and B